HANDBOOK OF
NANOPHYSICS

Handbook of Nanophysics

HANDBOOK OF NANOPHYSICS

Nanoelectronics and Nanophotonics

Edited by

Klaus D. Sattler

CRC Press
Taylor & Francis Group
Boca Raton London New York

CRC Press is an imprint of the
Taylor & Francis Group, an **informa** business

CRC Press
Taylor & Francis Group
6000 Broken Sound Parkway NW, Suite 300
Boca Raton, FL 33487-2742

First issued in paperback 2020

© 2011 by Taylor and Francis Group, LLC
CRC Press is an imprint of Taylor & Francis Group, an Informa business

No claim to original U.S. Government works

ISBN 13: 978-1-138-11343-5 (pbk)
ISBN 13: 978-1-4200-7550-2 (hbk)

Library of Congress Cataloging-in-Publication Data

Handbook of nanophysics. Nanoelectronics and nanophotonics / editor, Klaus D. Sattler.
        p. cm.
    "A CRC title."
    Includes bibliographical references and index.
    ISBN 978-1-4200-7550-2 (alk. paper)
    1. Nanoelectronics--Handbooks, manuals, etc. 2. Nanophotonics--Handbooks, manuals, etc. I. Sattler, Klaus D. II. Title.

TK7874.84.H36 2010
621.381--dc22
                                                                                      2010001108

Visit the Taylor & Francis Web site at
http://www.taylorandfrancis.com

and the CRC Press Web site at
http://www.crcpress.com

Contents

PART I Computing and Nanoelectronic Devices

PART II Nanoscale Transistors

PART III Nanolithography

PART IV Optics of Nanomaterials

Preface

The *Handbook of Nanophysics* is the first comprehensive reference to consider both fundamental and applied aspects of nanophysics. As a unique feature of this work, we requested contributions to be submitted in a tutorial style, which means that state-of-the-art scientific content is enriched with fundamental equations and illustrations in order to facilitate wider access to the material. In this way, the handbook should be of value to a broad readership, from scientifically interested general readers to students and professionals in materials science, solid-state physics, electrical engineering, mechanical engineering, computer science, chemistry, pharmaceutical science, biotechnology, molecular biology, biomedicine, metallurgy, and environmental engineering.

What Is Nanophysics?

Modern physical methods whose fundamentals are developed in physics laboratories have become critically important in nanoscience. Nanophysics brings together multiple disciplines, using theoretical and experimental methods to determine the physical properties of materials in the nanoscale size range (measured by millionths of a millimeter). Interesting properties include the structural, electronic, optical, and thermal behavior of nanomaterials; electrical and thermal conductivity; the forces between nanoscale objects; and the transition between classical and quantum behavior. Nanophysics has now become an independent branch of physics, simultaneously expanding into many new areas and playing a vital role in fields that were once the domain of engineering, chemical, or life sciences.

This handbook was initiated based on the idea that breakthroughs in nanotechnology require a firm grounding in the principles of nanophysics. It is intended to fulfill a dual purpose. On the one hand, it is designed to give an introduction to established fundamentals in the field of nanophysics. On the other hand, it leads the reader to the most significant recent developments in research. It provides a broad and in-depth coverage of the physics of nanoscale materials and applications. In each chapter, the aim is to offer a didactic treatment of the physics underlying the applications alongside detailed experimental results, rather than focusing on particular applications themselves.

The handbook also encourages communication across borders, aiming to connect scientists with disparate interests to begin interdisciplinary projects and incorporate the theory and methodology of other fields into their work. It is intended for readers from diverse backgrounds, from math and physics to chemistry, biology, and engineering.

The introduction to each chapter should be comprehensible to general readers. However, further reading may require familiarity with basic classical, atomic, and quantum physics. For students, there is no getting around the mathematical background necessary to learn nanophysics. You should know calculus, how to solve ordinary and partial differential equations, and have some exposure to matrices/linear algebra, complex variables, and vectors.

External Review

All chapters were extensively peer reviewed by senior scientists working in nanophysics and related areas of nanoscience. Specialists reviewed the scientific content and nonspecialists ensured that the contributions were at an appropriate technical level. For example, a physicist may have been asked to review a chapter on a biological application and a biochemist to review one on nanoelectronics.

Organization

The *Handbook of Nanophysics* consists of seven books. Chapters in the first four books (*Principles and Methods, Clusters and Fullerenes, Nanoparticles and Quantum Dots,* and *Nanotubes and Nanowires*) describe theory and methods as well as the fundamental physics of nanoscale materials and structures. Although some topics may appear somewhat specialized, they have been included given their potential to lead to better technologies. The last three books (*Functional Nanomaterials, Nanoelectronics and Nanophotonics,* and *Nanomedicine and Nanorobotics*) deal with the technological applications of nanophysics. The chapters are written by authors from various fields of nanoscience in order to encourage new ideas for future fundamental research.

After the first book, which covers the general principles of theory and measurements of nanoscale systems, the organization roughly follows the historical development of nanoscience. *Cluster* scientists pioneered the field in the 1980s, followed by extensive work on *fullerenes*, *nanoparticles*, and *quantum dots* in the 1990s. Research on *nanotubes* and *nanowires* intensified in subsequent years. After much basic research, the interest in applications such as the *functions of nanomaterials* has grown. Many bottom-up and top-down techniques for nanomaterial and nanostructure generation were developed and made possible the development of *nanoelectronics* and *nanophotonics*. In recent years, real applications for *nanomedicine* and *nanorobotics* have been discovered.

MATLAB® is a registered trademark of The MathWorks, Inc. For product information, Please contact:

The MathWorks, Inc.
3 Apple Hill Drive
Natick, MA, 01760-2098 USA
Tel: 508-647-7000
Fax: 508-647-7001
E-mail: info@mathworks.com
Web: www.mathworks.com

Acknowledgments

Many people have contributed to this book. I would like to thank the authors whose research results and ideas are presented here. I am indebted to them for many fruitful and stimulating discussions. I would also like to thank individuals and publishers who have allowed the reproduction of their figures. For their critical reading, suggestions, and constructive criticism, I thank the referees. Many people have shared their expertise and have commented on the manuscript at various stages. I consider myself very fortunate to have been supported by Luna Han, senior editor of the Taylor & Francis Group, in the setup and progress of this work. I am also grateful to Jessica Vakili, Jill Jurgensen, Joette Lynch, and Glenon Butler for their patience and skill with handling technical issues related to publication. Finally, I would like to thank the many unnamed editorial and production staff members of Taylor & Francis for their expert work.

Klaus D. Sattler
Honolulu, Hawaii

Editor

Klaus D. Sattler pursued his undergraduate and master's courses at the University of Karlsruhe in Germany. He received his PhD under the guidance of Professors G. Busch and H.C. Siegmann at the Swiss Federal Institute of Technology (ETH) in Zurich, where he was among the first to study spin-polarized photoelectron emission. In 1976, he began a group for atomic cluster research at the University of Konstanz in Germany, where he built the first source for atomic clusters and led his team to pioneering discoveries such as "magic numbers" and "Coulomb explosion." He was at the University of California, Berkeley, for three years as a Heisenberg Fellow, where he initiated the first studies of atomic clusters on surfaces with a scanning tunneling microscope.

Dr. Sattler accepted a position as professor of physics at the University of Hawaii, Honolulu, in 1988. There, he initiated a research group for nanophysics, which, using scanning probe microscopy, obtained the first atomic-scale images of carbon nanotubes directly confirming the graphene network. In 1994, his group produced the first carbon nanocones. He has also studied the formation of polycyclic aromatic hydrocarbons (PAHs) and nanoparticles in hydrocarbon flames in collaboration with ETH Zurich. Other research has involved the nanopatterning of nanoparticle films, charge density waves on rotated graphene sheets, band gap studies of quantum dots, and graphene foldings. His current work focuses on novel nanomaterials and solar photocatalysis with nanoparticles for the purification of water.

Among his many accomplishments, Dr. Sattler was awarded the prestigious Walter Schottky Prize from the German Physical Society in 1983. At the University of Hawaii, he teaches courses in general physics, solid-state physics, and quantum mechanics.

In his private time, he has worked as a musical director at an avant-garde theater in Zurich, composed music for theatrical plays, and conducted several critically acclaimed musicals. He has also studied the philosophy of Vedanta. He loves to play the piano (classical, rock, and jazz) and enjoys spending time at the ocean, and with his family.

Contributors

Grigory E. Adamov
Open Joint-Stock Company
Central Scientific Research Institute
    of Technology "Technomash"
Moscow, Russia

Paolo Amato
Numonyx

and

Department of Materials Science
University of Milano-Bicocca
Milano, Italy

José Aumentado
National Institute of Standards
    and Technology
Boulder, Colorado

Bruno Azzerboni
Department of Matter Physics
    and Electronic Engineering
Faculty of Engineering
University of Messina
Messina, Italy

Dieter Bimberg
Institut für Festkörperphysik
Technische Universität Berlin
Berlin, Germany

Vladimir G. Bordo
A.M. Prokhorov General Physics
    Institute
Russian Academy of Sciences
Moscow, Russia

Vincent Bouchiat
Nanosciences Department
Centre National de la Recherche
    Scientifique
Néel-Institut
Grenoble, France

Michel Broyer
Laboratoire de Spectrométrie Ionique et
    Moléculaire
Centre National de la Recherche
    Scientifique
Université de Lyon
Villeurbanne, France

John H. Burnett
Atomic Physics Division
National Institute of Standards
    and Technology
Gaithersburg, Maryland

Maria A. Cataluna
Division of Electronic Engineering
    and Physics
School of Engineering, Physics
    and Mathematics
University of Dundee
Dundee, United Kingdom

Gianfranco Cerofolini
Department of Materials Science
University of Milano-Bicocca
Milano, Italy

Giulio Cerullo
Dipartimento di Fisica
Politecnico di Milano
Milano, Italy

Swapan Chakrabarti
Department of Chemistry
University of Calcutta
Kolkata, West Bengal, India

Chia-Ching Chang
Department of Biological Science
    and Technology
National Chiao Tung University
Hsinchu, Taiwan

and

Institute of Physics
Academia Sinica
Taipei, Taiwan

Jean-Pierre Colinge
Tyndall National Institute
University College Cork
Cork, Ireland

Giancarlo Consolo
Department of Matter Physics
    and Electronic Engineering
Faculty of Engineering
University of Messina
Messina, Italy

Emmanuel Cottancin
Laboratoire de Spectrométrie Ionique et
    Moléculaire
Centre National de la Recherche
    Scientifique
Université de Lyon
Villeurbanne, France

Damien Deleruyelle
Centre National de la Recherche
    Scientifique
Institut Matériaux Microélectronique
    Nanosciences de Provence
Universités d'Aix Marseille
Marseille, France

Alain C. Diebold
College of Nanoscale Science
    and Engineering
University at Albany
Albany, New York

Lamberto Duò
Dipartimento di Fisica
Politecnico di Milano
Milano, Italy

Mikhail Erementchouk
NanoScience Technology Center

and

Department of Physics
University of Central Florida
Orlando, Florida

Marco Finazzi
Dipartimento di Fisica
Politecnico di Milano
Milano, Italy

Giovanni Finocchio
Department of Matter Physics
    and Electronic Engineering
Faculty of Engineering
University of Messina
Messina, Italy

Martin Geller
Experimental Physics

and

Center for Nanointegration
    Duisburg-Essen
University of Duisburg-Essen
Duisburg, Germany

Gabriel González
NanoScience Technology Center

and

Department of Physics
University of Central Florida
Orlando, Florida

Evgeny P. Grebennikov
Open Joint-Stock Company
Central Scientific Research Institute
    of Technology "Technomash"
Moscow, Russia

Jim Greer
Tyndall National Institute
University College Cork
Cork, Ireland

Akihiro Hashimoto
Department of Electrical and Electronics
    Engineering
Graduate School of Engineering
University of Fukui
Fukui, Japan

Jun He
Department of Chemistry
Institute for Optical Sciences

and

Centre for Quantum Information
    and Quantum Control
University of Toronto
Toronto, Ontario, Canada

Woong-Ki Hong
Department of Materials Science
    and Engineering
Gwangju Institute of Science
    and Technology
Gwangju, Korea

Zhijun Hu
Center for Soft Matter Physics
    and Interdisciplinary Research
Soochow University
Suzhou, China

Vanessa M. Huxter
Department of Chemistry
Institute for Optical Sciences

and

Centre for Quantum Information
    and Quantum Control
University of Toronto
Toronto, Ontario, Canada

Frank Jahnke
Institute for Theoretical Physics
University of Bremen
Bremen, Germany

Gunho Jo
Department of Materials Science
    and Engineering
Gwangju Institute of Science
    and Technology
Gwangju, Korea

Alain M. Jonas
Institute of Condensed Matter
    and Nanosciences
Division of Bio- and Soft Matter
Catholic University of Louvain
Louvain-la-Neuve, Belgium

Yoshihiko Kanemitsu
Institute for Chemical Research
Kyoto University
Uji, Kyoto, Japan

Tadashi Kawazoe
Department of Electrical Engineering
    and Information Systems
School of Engineering
The University of Tokyo
Tokyo, Japan

Stephen Knight
Office of Microelectronics Programs
National Institute of Standards
    and Technology
Gaithersburg, Maryland

Kiyoshi Kobayashi
Department of Electrical Engineering
    and Information Systems
The University of Tokyo
Tokyo, Japan

and

Core Research of Evolutional Science
    and Technology
Japan Science and Technology

and

Department of Electrical and Electronic
    Engineering
University of Yamanashi
Kofu, Japan

Hermann Kohlstedt
Christian-Albrechts-Universität Zu Kiel
Faculty of Engineering Nanoelectronics
Kiel, Germany

Takhee Lee
Department of Materials Science
    and Engineering
Gwangju Institute of Science
    and Technology
Gwangju, Korea

Jean Lermé
Laboratoire de Spectrométrie Ionique et
  Moléculaire
Centre National de la Recherche
  Scientifique
Université de Lyon
Villeurbanne, France

Michael N. Leuenberger
NanoScience Technology Center

and

Department of Physics
University of Central Florida
Orlando, Florida

James Alexander Liddle
Center for Nanoscale Science
  and Technology
National Institute of Standards
  and Technology
Gaithersburg, Maryland

Jongsun Maeng
Department of Materials Science
  and Engineering
Gwangju Institute of Science
  and Technology
Gwangju, Korea

Andreas Marent
Institut für Festkörperphysik
Technische Universität Berlin
Berlin, Germany

Vincent Meunier
Oak Ridge National Laboratory
Oak Ridge, Tennessee

Gilles Micolau
Institut Matériaux Microélectronique
  Nanosciences de Provence
Centre National de la Recherche
  Scientifique
Universités d'Aix Marseille
Marseille, France

Makoto Naruse
National Institute of Information
  and Communications Technology
Koganei, Japan

and

Department of Electrical Engineering
  and Information Systems
School of Engineering
The University of Tokyo
Tokyo, Japan

Wataru Nomura
School of Engineering
The University of Tokyo
Tokyo, Japan

Motoichi Ohtsu
Department of Electrical Engineering
  and Information Systems
School of Engineering
The University of Tokyo
Tokyo, Japan

Tomáš Ostatnický
Faculty of Mathematics and Physics
Department of Chemical Physics and
  Optics
Charles University
Prague, Czech Republic

Andrew R. Parker
Department of Zoology
The Natural History Museum
London, United Kingdom

and

School of Biological Science
University of Sydney
Sydney, New South Wales, Australia

André Avelino Pasa
Laboratório de Filmes Finos e
  Superfícies
Departamento de Física
Universidade Federal de Santa Catarina
Santa Catarina, Brazil

Ivan Pelant
Institute of Physics
Academy of Sciences of the Czech
  Republic
Prague, Czech Republic

Michel Pellarin
Laboratoire de Spectrométrie Ionique et
  Moléculaire
Centre National de la Recherche
  Scientifique
Université de Lyon
Villeurbanne, France

Nikolay A. Pertsev
A.F. Ioffe Physico-Technical Institute
Russian Academy of Sciences
St. Petersburg, Russia

Adrian Petraru
Christian-Albrechts-Universität Zu Kiel
Faculty of Engineering Nanoelectronics
Kiel, Germany

Vivek M. Prabhu
Polymers Division
National Institute of Standards
  and Technology
Gaithersburg, Maryland

Edik U. Rafailov
Division of Electronic Engineering
  and Physics
School of Engineering, Physics
  and Mathematics
University of Dundee
Dundee, United Kingdom

Arndt Remhof
Division of Hydrogen and Energy
Department of Environment, Energy
  and Mobility
Swiss Federal Laboratories for Materials
  Testing and Research
Dübendorf, Switzerland

Elisabetta Romano
Department of Materials Science
University of Milano-Bicocca
Milano, Italy

Harry E. Ruda
Centre for Advanced Nanotechnology
University of Toronto
Toronto, Ontario, Canada

Amir Sa'ar
Racah Institute of Physics

and

The Harvey M. Kruger Family Center
    for Nanoscience and Nanotechnology
The Hebrew University of Jerusalem
Jerusalem, Israel

Suguru Sangu
Device and Module Technology
    Development Center
Ricoh Company, Ltd.
Yokohama, Japan

Gregory D. Scholes
Department of Chemistry
Institute for Optical Sciences

and

Centre for Quantum Information
    and Quantum Control
University of Toronto
Toronto, Ontario, Canada

Sabyasachi Sen
Department of Chemistry
JIS College of Engineering
Kolkata, West Bengal, India

Alexander Shik
Centre for Advanced Nanotechnology
University of Toronto
Toronto, Ontario, Canada

Christopher L. Soles
Polymers Division
National Institute of Standards
    and Technology
Gaithersburg, Maryland

Sunghoon Song
Department of Materials Science
    and Engineering
Gwangju Institute of Science
    and Technology
Gwangju, Korea

Bobby G. Sumpter
Oak Ridge National Laboratory
Oak Ridge, Tennessee

Kien Wen Sun
Department of Applied Chemistry
National Chiao Tung University
Hsinchu, Taiwan

Jan Valenta
Faculty of Mathematics and Physics
Department of Chemical Physics
    and Optics
Charles University
Prague, Czech Republic

Marek S. Wartak
Department of Physics and Computer
    Science
Wilfrid Laurier University
Waterloo, Ontario, Canada

Andreas Westphalen
Department of Physics and Astronomy
Institute for Condensed Matter Physics
Ruhr-Universität Bochum
Bochum, Germany

Obert R. Wood II
GLOBALFOUNDRIES
Albany, New York

Naoki Yamamoto
Department of Physics
Tokyo Institute of Technology
Tokyo, Japan

Takashi Yatsui
School of Engineering
The University of Tokyo
Tokyo, Japan

Hartmut Zabel
Department of Physics and Astronomy
Institute for Condensed Matter Physics
Ruhr-Universität Bochum
Bochum, Germany

Natalya A. Zimbovskaya
Department of Physics and Electronics
University of Puerto Rico
Humacao, Puerto Rico

and

Institute for Functional Nanomaterials
University of Puerto Rico
San Juan, Puerto Rico

Computing and Nanoelectronic Devices

I

1

Quantum Computing in Spin Nanosystems

Gabriel González
University of Central Florida

Michael N. Leuenberger
University of Central Florida

1.1 Introduction

The history of quantum computers begins with the articles of Richard Feynman who, in 1982, speculated that quantum systems might be able to perform certain tasks more efficiently than would be possible in classical systems (Feynman, 1982). Feynman was the first to propose a direct application of the laws of quantum mechanics to a realization of quantum algorithms. The fundamentals of quantum computing were introduced and developed by several authors after Feynman's idea. A model and a description of a quantum computer as a quantum Turing machine was developed by Deutsch (1985). In 1994, Shor introduced the quantum algorithm for the integer-number factorization and in 1997, Grover proposed the fast quantum search algorithm (Grover, 1997). Later on, Wooters and Zurek proved the noncloning theorem, which puts definite limits on the quantum computations, but Shor's work challenges all that with the quantum error correction code (Shor, 1995). In the last years, the development of quantum computing has grown to enormous practical importance as an interdisciplinary field, which links the elements of physics, mathematics, and computer science. Currently, various physical models of quantum computers are under intensive study. Several types of elementary quantum computing devices have been developed based on atomic, molecular, optical, and semiconductor physics and technologies.

Before contemplating the physical realization of a quantum computer, it is necessary to decide how information is going to be stored within the system and how the system will process that information during a desired computation. In classical computers, the information is typically carried in microelectronic circuits that store information using the charge properties of electrons. Information processing is carried out by manipulating electrical fields within semiconductor materials in such a way as to perform useful computational tasks. Presently it seems that the most promising physical model for quantum computation is based on the electron's spin. A strong research effort toward the implementation of the electron spin as a new information carrier has been the subject of a new form of electronics based on spin called *spintronics*. Experiments that have been conducted on quantum spin dynamics in semiconductor materials demonstrate that electron spins have several characteristics that are promising for quantum computing applications. Electron spin states possess the following advantages: very long relaxation time in the absence of external fields, fairly long decoherence time $\tau_d \approx 1\,\mu s$, and the possibility of easy spin manipulation by an external magnetic field. These characteristics are very promising since longer decoherence times relax constraints on the switching speeds of quantum gates necessary for reliable error correction. Typically quantum gates are required to switch 10^4 times faster than the loss of qubit coherence. Spin coherent transport over lengths as large as $100\,\mu m$ have been reported in semiconductors. This makes electron spin a perfect candidate as an information carrier in semiconductors (Adamowski et al., 2005).

The spin of particles exhibiting quantum behavior is specially suitable for the construction of quantum computers. The electron spin states can be used to construct qubits and logic operations in different ways. They can be constructed either directly with the application of a magnetic field or indirectly with the application of symmetry properties of the many-electron wave function (namely by the resulting singlet or triplet spin states). Traditionally, in nanoelectronic devices, the charge of the electrons has been used to carry and transform information, which makes the interaction of spintronic devices with charge-based devices important for compatibility with existing classical computing schemes. Overall, we can say that spin nanosystems are good candidates for the physical implementation of quantum computation.

1.2 Qubits and Quantum Logic Gates

We know that the information stored in a classical computer can take one of the two values, i.e., 0 or 1, with probability 0 or 1 each. Quantum bits or qubits are the quantum mechanical analogue of classical bits. In contrast, the qubit can be defined as a quantum state vector in a two-dimensional Hilbert space. Suppose $|0\rangle$, $|1\rangle$ forms a basis for the Hilbert space, then the qubit can be expressed as the superposition of the two states as

$$|\psi\rangle = c_0\,|0\rangle + c_1\,|1\rangle, \tag{1.1}$$

where c_0 and c_1 are complex numbers and the modulus squared of each complex number represents the probability to obtain the qubit $|0\rangle$ or $|1\rangle$, respectively. Additionally, they must satisfy the normalization condition $|c_0|^2 + |c_1|^2 = 1$. Contrary to the classical bit, the quantum bit takes on a continuum of values, which are determined by the probability amplitudes given by c_0 and c_1. If we perform a measurement on qubit $|\psi\rangle$, we obtain either outcome $|0\rangle$ with probability $|c_0|^2$ or outcome $|1\rangle$ with probability $|c_1|^2$. However, if the qubit is prepared to be exactly equal to one of the states of the computational basis, i.e., $|\psi\rangle = |0\rangle$ or $|\psi\rangle = |1\rangle$, then we can predict the exact result of the measurement with probability 1. This nondeterministic characteristic between the general state of the qubit and the precise result of the measurement in the basis state plays an essential role in quantum computations. To carry out a quantum computation, we require at least a two-qubit state, i.e., the states of a two-particle quantum system. The two-qubit states can be constructed as tensor products of the basis states $|0\rangle$, $|1\rangle$. The two-qubit basis consists of the states $|00\rangle$, $|01\rangle$, $|10\rangle$, $|11\rangle$, where in the shorthanded notation $|0\rangle \otimes |1\rangle \equiv |01\rangle$, etc. implied. An arbitrary two-qubit state has the form

$$|\Psi\rangle = c_0\,|00\rangle + c_1\,|00\rangle + c_2\,|10\rangle + c_3\,|11\rangle, \tag{1.2}$$

where the normalization condition takes the form $|c_0|^2 + |c_1|^2 + |c_2|^2 + |c_3|^2 = 1$. Another basic characteristic for quantum computing is the so-called *entanglement*. The two quantum two-level systems can become *entangled* by interacting with each other. This means that we cannot fully describe one system independently of the other. For example, suppose that the state $(|01\rangle - |10\rangle)/\sqrt{2}$ gives a complete description of the whole system. Then a measurement over the first subsystem forces the second subsystem into one of the two states $|0\rangle$ or $|1\rangle$. This means that one measurement over one subsystem influences the other, even though it may be arbitrarily far away.

Qubits can be transformed by unitary transformations (observables) $\mathcal{U}$ that play the role of quantum logic gates and which transforms the initial qubit into a final qubit according to

$$|\Psi_f\rangle = \mathcal{U}\,|\Psi_i\rangle. \tag{1.3}$$

Depending on the type of qubit on which they operate, we deal with either a 2×2 or a 4×4 matrix. For example, the quantum NOT gate is defined as

$$\mathcal{U}_{\mathrm{NOT}} = \begin{pmatrix} 0 & 1 \\ 1 & 0 \end{pmatrix}. \tag{1.4}$$

The effect of the quantum logic gate (4) in the one-qubit state given in Equation 1.1 is to exchange the probability amplitudes i.e.,

$$\mathcal{U}_{\mathrm{NOT}} \begin{pmatrix} c_0 \\ c_1 \end{pmatrix} = \begin{pmatrix} 0 & 1 \\ 1 & 0 \end{pmatrix} \begin{pmatrix} c_0 \\ c_1 \end{pmatrix} = \begin{pmatrix} c_1 \\ c_0 \end{pmatrix}. \tag{1.5}$$

An example of an important quantum logic gate that operates on a two-qubit state is the so-called controlled-NOT gate $\mathcal{U}_{\mathrm{CNOT}}$, for which the first qubit is the control qubit and the second qubit is the target qubit. The controlled-NOT gate transforms the two-qubit basis states as follows:

$$\mathcal{U}_{\mathrm{CNOT}}\,|00\rangle = |00\rangle, \quad \mathcal{U}_{\mathrm{CNOT}}\,|01\rangle = |01\rangle,$$

$$\mathcal{U}_{\mathrm{CNOT}}\,|10\rangle = |11\rangle, \quad \text{and} \quad \mathcal{U}_{\mathrm{CNOT}}\,|11\rangle = |10\rangle, \tag{1.6}$$

which means that the CNOT (conditionalNOT) gate changes the second qubit if and only if the first qubit is in state $|1\rangle$. The matrix representation of the CNOT gate is

$$\mathcal{U}_{\mathrm{CNOT}} = \begin{pmatrix} 1 & 0 & 0 & 0 \\ 0 & 1 & 0 & 0 \\ 0 & 0 & 0 & 1 \\ 0 & 0 & 1 & 0 \end{pmatrix}. \tag{1.7}$$

It has been shown that the set of logic operations, which consists of all the one-qubit gates and the single two-qubit gate $\mathcal{U}_{\mathrm{CNOT}}$ is universal in the sense that all unitary transformations on

N-qubit states can be expressed by different compositions of the set of universal gates (DiVincenzo, 1995).

Using the concepts of superposition and entanglement, we can describe another important characteristic of quantum computation, which is known as *quantum parallelism*. Quantum parallelism is based on the fact that a single unitary transformation can simultaneously operate on all the qubits in the system. In a sense, it can perform several calculations in a single step. In fact, it can be proved that the computing power of a quantum computer scales exponentially with the number of qubits, whereas a classical computer the scale is only linear.

1.3 Conditions for the Physical Implementation of Quantum Computing

The successful implementation of a quantum computer has to satisfy some basic requirements. These are known as the DiVincenzo criteria and can be summarized in the following way (DiVincenzo, 2001).

1. *Physical realizability of the qubits.* We need to find some quantum property of a scalable physical system in which to encode our information so that it lives long enough to enable us to perform computations.
2. *Initial state preparation.* It should be possible to precisely prepare the initial qubit state.
3. *Isolation.* We need a controlled evolution of the qubit; this will require enough isolation of the qubit from the environment to reduce the effects of decoherence.
4. *Gate implementation.* We need to be able to manipulate the states of individual qubits with reasonable precision, as well as to induce interaction between them in a controlled way, so that the implementation of gates is possible. Also, the gate operation time τ_s has to be much shorter than the decoherence time τ_d, so that $\tau/\tau_d \ll r$, where r is the maximum tolerable error rate for quantum error correction schemes to be effective.
5. *Readout.* We must be able to accurately measure the final qubit state.

The conditions listed above put certain limitations on the quantum computing technology. For example, the complete isolation of the qubit with respect to the environment disables the read/write operations. Therefore, some slight interaction of the quantum system and the environment is necessary. On the other hand, this interaction leads to decay and decoherence processes, which reduce the performance of the quantum computer.

In the decay process, the quantum system jumps in a very short time to a new state, releasing part of its energy to the environment. The decay is characterized by the relaxation time, which for the spin states can be very long. Decoherence is a more subtle process because the energy is conserved but the relative phase of the computational basis is changed. As a result of the decoherence the qubit changes as follows:

$$|\psi\rangle = c_0 |0\rangle + e^{i\theta} c_1 |1\rangle, \tag{1.8}$$

where the real number θ denotes the relative phase. The appearance of the nonzero relative phase results due to the coupling between the quantum system and the environment and can lead to significant changes in the measurement process. The ratio of the decoherence time to the elementary operation time τ_s, i.e., $R = \tau_d/\tau_s$, is an approximate measure of the number of computation steps performed before the coupling with the environment destroys the qubit. This ratio changes abruptly for different quantum computing schemes. For example, $R = 10^3$ for the electron states in quantum dots, $R = 10^7$ for nuclear spin sates, and $R = 10^{13}$ for trapped ions.

An important factor that should always be kept in mind when constructing quantum computers is the scalability of the device. We should be able to enlarge the physical device to contain many qubits and still fulfill the DiVincenzo requirements described above.

1.4 Zeeman Effects

An external magnetic field superimposed on an atom perturbs its state in a definite way. The Hamiltonian for such an atom may be divided into the operator $\mathcal{H}_0$ for the unperturbed atom and the perturbation operator $\mathcal{H}'$ due to the magnetic field. The external magnetic field $\vec{H}$ causes the vectors $\vec{L}$ and $\vec{S}$ to precess about its direction. We shall examine the two cases where the magnetic field is weak and is strong.

Let us first write down the perturbation operator explicitly. Consider an electron in the simplest possible atom (hydrogen) rotating around the nucleus. The electron orbit can be regarded as a current loop. The current is the charge per unit time past any fixed point on the orbit, therefore

$$j = \frac{e}{T} = \frac{ev}{2\pi r} = \frac{e(p/m_e)}{2\pi r}, \tag{1.9}$$

where p and m_e are the momentum and mass of the electron, respectively. The magnitude of the magnetic moment of the loop is the current times the enclosed area, i.e., $\mu_{orb} = \pi r^2 j$, therefore

$$\vec{\mu}_{orb} = \frac{e}{2m_e} \vec{r} \times \vec{p} = \frac{\mu_B}{\hbar} \vec{L}, \tag{1.10}$$

where we have introduced the Bohr magneton, $\mu_B = e\hbar/2m_e$. We do not have a good semiclassical picture for spin and therefore we can only conclude that the spin magnetic moment is

$$\vec{\mu}_{sp} = g_s \frac{\mu_B}{\hbar} \vec{S}, \tag{1.11}$$

where g_s is called the gyromagnetic factor and its value is approximately equal to 2. Therefore, the total magnetic moment is

$$\vec{\mu} = \frac{\mu_B}{\hbar}(\vec{L} + 2\vec{S}). \qquad (1.12)$$

The perturbation energy caused by the magnetic field is given then by

$$\mathcal{H}' = -\vec{\mu} \cdot \vec{H} = \frac{\mu_B}{\hbar}(\vec{J} + \vec{S}), \qquad (1.13)$$

where the plus sign resulted because the charge of the electron is $-e$.

1.4.1 Weak External Field

In a hydrogen atom, the electron circles around the proton; but in a system fixed to an electron, the proton circles around the electron and generates an inner magnetic field at the position of the electron given by

$$\vec{H}_{in} = \frac{\mu_0 e}{4\pi m_p r^3}\vec{L}. \qquad (1.14)$$

Equation 1.14 gives the following energy for the electron's spin

$$\mathcal{H}_{so} = \vec{\mu}_{sp} \cdot \vec{H}_{in} = \frac{e^2}{4\pi\epsilon_0 r^3 m^2 c^2}\vec{S} \cdot \vec{L}, \qquad (1.15)$$

which is called the spin–orbit interaction. We interpret the spin–orbit interaction as an internal Zeeman effect because it splits the energy levels without an external magnetic field. Let the external field be weak compared with the effective internal field. Since the Larmor frequency is proportional to the magnetic field, the triangle $\vec{L}\,\vec{S}\,\vec{J}$ in Figure 1.1 rotates about $\vec{J}$ considerably faster than the precession around $\vec{H}$. Therefore, the coupling of the vectors $\vec{L}$, $\vec{S}$, and $\vec{J}$ in the triangle is not disrupted.

Let us now find the correction to the energy due to the external magnetic field. Equation 1.13 involves the vectors $\vec{J}$ and $\vec{S}$, if we consider that the z-axis coincides with the direction of the external magnetic field, then, the only projections of these two vectors that contribute to the energy are the ones along the z-axis. Thus, the mean value of $\mathcal{H}'$ is equal to

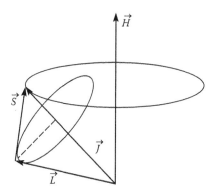

FIGURE 1.1   Schematic of the spin–orbit coupling.

$$\langle\mathcal{H}'\rangle = \frac{\mu_B H J_z}{\hbar}\left(1 + \frac{\vec{S} \cdot \vec{J}}{J^2}\right), \qquad (1.16)$$

and using

$$\frac{\vec{S} \cdot \vec{J}}{J^2} = \frac{j(j+1) + s(s+1) - l(l+1)}{2j(j+1)} = g \qquad (1.17)$$

we get

$$\langle\mathcal{H}'\rangle = \mu_B H g m_j \quad \text{for } m_j = -j, -j+1, \dots, j. \qquad (1.18)$$

Thus, the energy levels with a given J are split into as many levels as there are different projections of $\vec{J}$ on the magnetic field, i.e., $2j + 1$. The factor g is called the Landé factor and takes a given value for a given L and S and each corresponding value of J. The Zeeman effect in a weak magnetic field is called anomalous.

1.4.2 Strong External Field

We shall now consider the opposite extreme case, when the external field is strong compared with the internal field, so that the coupling between the vectors $\vec{L}, \vec{S}, \vec{J}$ is disrupted. This can be explained by the fact that $\vec{S}$ precesses twice as fast $\vec{L}$. Then, from the classical analogy, each of the vectors $\vec{S}$ and $\vec{L}$ precesses independently about the magnetic field. For this case, the correction to the energy is given by

$$\langle\mathcal{H}'\rangle = \frac{\mu_B e H}{\hbar}(L_z + 2S_z). \qquad (1.19)$$

In Equation 1.19, L_z and S_z are the projections of the orbital angular momentum and spin along the z-axis, respectively, and are given by

$$L_z = \hbar m_l \quad \text{for } m_l = -l, \dots, l, \quad S_z = \hbar m_s, \quad \text{for } m_s = \pm\frac{1}{2}. \qquad (1.20)$$

The projections S_z and L_z are changed by unity, therefore all the levels in Equation 1.19 are equidistant. Of course, certain values of $\langle\mathcal{H}'\rangle$ may be repeated several times if the sum $L_z + 2S_z$ assumes the same value. This Zeeman effect in a strong magnetic field is called the normal Zeeman effect.

1.5 Atom–Light Interaction

Though an atom has infinitely many energy levels, we can have under certain assumptions a two-level atom. These assumptions are (1) the difference of the energy levels approximately matches the energy of the incident photon, (2) the selection rules allow transition of the electrons between the two levels, and (3) all other energy levels are sufficiently detuned in frequency separation with respect to the incoming frequency such that there is no transition to these levels.

Consider that we apply the two-level approximation to a simple atom where the interaction with visible light involves a single electron. The corresponding wave function of the system is

$$|\psi\rangle = c_0|0\rangle + c_1|1\rangle, \tag{1.21}$$

where $|0\rangle$ ($|1\rangle$) corresponds to the nonexcited (excited) state and the coefficients $c_0(c_1)$ represent the probability amplitude to find the system in either state, respectively. The probability amplitudes satisfy the normalization condition $|c_0|^2 + |c_1|^2 = 1$. The wavelength of visible light, typical for atomic transitions, is about a few thousand times the diameter of an atom. Therefore, there is no significant spatial variation of the electric field across an atom, and $\vec{E}(\vec{r},t) \approx \vec{E}(t)$ can be taken as independent of the position (dipole approximation). Consistent with this long-wavelength approximation is that the magnetic field $\vec{H}$ is approximately zero, i.e., $\vec{H} \approx 0$, so that the interaction energy between the field and the electron of the atom is given by $\mathcal{H}' = -e\vec{r} \cdot \vec{E}$, hence the total Hamiltonian is

$$\mathcal{H} = \mathcal{H}_0 + \mathcal{H}', \tag{1.22}$$

where $\mathcal{H}_0$ represents the unperturbed atom system, i.e., $\mathcal{H}_0|0\rangle = \varepsilon_0|0\rangle$ and $\mathcal{H}_0|1\rangle = \varepsilon_1|1\rangle$, where ε_0 and ε_1 are the energy values for the nonexcited and excited states, respectively.

Seeking a solution to the Schrödinger equation

$$\mathcal{H}|\psi(t)\rangle = i\hbar\frac{\partial|\psi(t)\rangle}{\partial t}, \tag{1.23}$$

in the following form

$$|\psi(t)\rangle = c_0(t)e^{-i\varepsilon_0 t/\hbar}|0\rangle + c_1(t)e^{-i\varepsilon_1 t/\hbar}|1\rangle, \tag{1.24}$$

we get

$$i\hbar\frac{c_l(t)}{dt} = \sum_{k=0,1} c_k(t)e^{-i(\varepsilon_k - \varepsilon_l)t/\hbar}\langle l|\mathcal{H}'|k\rangle, \quad \text{for } l = 0,1. \tag{1.25}$$

Representing the light with frequency ν by means of the complex electric field vector E_0 in the form

$$\vec{E}(t) = \frac{1}{2}(\vec{E}_0 e^{-i\nu t} + \vec{E}_0^* e^{i\nu t}), \tag{1.26}$$

and using the fact that the diagonal elements of the interacting Hamiltonian are zero due to the parity of the eigenfunction, i.e., $\langle k|\mathcal{H}'|k\rangle = 0$, we end up with the following coupled differential equations

$$\frac{dc_0(t)}{dt} = \frac{i}{2\hbar}\left[\vec{E}_0 \cdot \langle 0|\vec{d}|1\rangle e^{-i(\omega_{10}+\nu)t} + \vec{E}_0^* \cdot \langle 0|\vec{d}|1\rangle e^{-i(\omega_{10}-\nu)t}\right]c_1(t), \tag{1.27}$$

$$\frac{dc_1(t)}{dt} = \frac{i}{2\hbar}\left[\vec{E}_0^* \cdot \langle 1|\vec{d}|0\rangle e^{i(\omega_{10}+\nu)t} + \vec{E}_0 \cdot \langle 1|\vec{d}|0\rangle e^{i(\omega_{10}-\nu)t}\right]c_0(t), \tag{1.28}$$

where $\omega_{10} = (\varepsilon_1 - \varepsilon_0)/\hbar$ and $\langle 0|\vec{d}|1\rangle = \langle 1|\vec{d}|0\rangle^* = \langle 0|e\vec{r}|1\rangle$.

For the case when $\nu \approx \omega_{10}$, we can drop the terms containing $\exp[\pm(\omega_{10}+\nu)t]$ in Equation 1.28 since they oscillate rapidly in time and can be neglected with respect to the near resonant terms, i.e., the terms of the form $\exp[\pm(\omega_{10}-\nu)t]$. This approximation is called the rotating wave approximation (RWA), and the coupled differential equations become

$$\frac{dc_0(t)}{dt} = \frac{i\Omega_R^*}{2}e^{-i\Delta t}c_1(t), \tag{1.29}$$

$$\frac{dc_1(t)}{dt} = \frac{i\Omega_R}{2}e^{i\Delta t}c_0(t),$$

where

$\Omega_R = \vec{E}_0 \cdot \langle 1|\vec{d}|0\rangle/\hbar$ is known as the Rabi frequency
$\Delta = (\omega_{10} - \nu)$ is the so-called detuning

The general solution for Equation 1.29 for strictly monochromatic fields, i.e., $\Omega_R = |\Omega_R|e^{i\phi} = \text{constant}$, is given by

$$c_0(t) = \frac{1}{\Omega}(\mu_1 e^{i\mu_2 t} - \mu_2 e^{i\mu_1 t}),$$
$$c_1(t) = \frac{i\Omega_R}{\Omega}e^{i\Delta t/2}\sin\left(\frac{\Omega t}{2}\right), \tag{1.30}$$

where

$$\mu_{1,2} = -\frac{\Delta}{2} \pm \sqrt{\Delta^2 + \Omega_R^2},$$
$$\Omega = \mu_1 - \mu_2 = \sqrt{\Delta^2 + \Omega_R^2}. \tag{1.31}$$

Equation 1.31 gives the transition probabilities from the nonexcited to the excited state

$$|c_0(t)|^2 = \left(\frac{\Omega_R}{\Omega}\right)^2 \sin^2\left(\frac{\Omega t}{2}\right), \quad |c_1(t)|^2 = 1 - |c_0(t)|^2, \tag{1.32}$$

which oscillate with frequency Ω (Rabi flopping frequency) between levels ε_0 and ε_1. Any two-level system can be represented by a 2×2 matrix Hamiltonian and hence can be expressed in terms of Pauli matrices. For example, choosing the energy zero to be half way between the excited and nonexcited state $|0\rangle$ and $|1\rangle$, we have the following matrix Hamiltonian for this system

$$\mathcal{H}_0 = \frac{\hbar\omega}{2}\begin{pmatrix} 1 & 0 \\ 0 & -1 \end{pmatrix} = \frac{\hbar\omega}{2}\sigma_z, \tag{1.33}$$

where
$$\hbar\omega = \varepsilon_1 - \varepsilon_0$$
σ_z is a Pauli matrix

Therefore, we can write the total Hamiltonian of the atom–light interaction in the RWA as

$$\mathcal{H} = \frac{\hbar}{2}\begin{pmatrix} \omega & -\Omega_R e^{i\Delta t} \\ -\Omega_R^* e^{-i\Delta t} & -\omega \end{pmatrix} = \frac{\hbar\omega}{2}\sigma_z - \frac{\hbar}{2}(\Omega_R e^{i\Delta t}\sigma_+ + \Omega_R^* e^{-i\Delta t}\sigma_-),$$

(1.34)

where

$$\sigma_+ = \begin{pmatrix} 0 & 1 \\ 0 & 0 \end{pmatrix} \quad \text{and} \quad \sigma_- = \begin{pmatrix} 0 & 0 \\ 1 & 0 \end{pmatrix}.$$

(1.35)

1.5.1 Jaynes–Cummings Model

So far, we have used a semiclassical description of the interaction between a two-level atom and an electric field. A quantum description of the interaction would require a quantization of the radiation field. This description is known as the *Jaynes–Cummings model*. Consider the sourceless electromagnetic field in a cubic cavity of volume $V = L \times L \times L$. Working with the Coulomb gauge for which the vector potential satisfies the requirement

$$\nabla \cdot \vec{A} = 0,$$

(1.36)

and, since there are no sources, it satisfies the homogeneous wave equation

$$\nabla^2 \vec{A} - \frac{1}{c^2}\frac{\partial^2 \vec{A}}{\partial t^2} = 0.$$

(1.37)

Then the fields are fully specified by the vector potential in the form

$$\vec{E} = -\frac{\partial \vec{A}}{\partial t} \quad \text{and} \quad \vec{B} = \nabla \times \vec{A}.$$

(1.38)

We can expand the vector potential in the form

$$\vec{A}(\vec{r},t) = \frac{1}{\sqrt{\epsilon_0 V}}\sum_{\vec{k}}\left[\vec{A}_{\vec{k}}(t)e^{i\vec{k}\cdot\vec{r}} + \vec{A}_{\vec{k}}^*(t)e^{-i\vec{k}\cdot\vec{r}} \right].$$

(1.39)

Equation 1.36 implies that $\vec{A}_{\vec{k}}$ and $\vec{A}_{\vec{k}}^*$ are perpendicular to the vector of propagation $\vec{k}$, therefore we can rewrite Equation 1.39 in the following form:

$$\vec{A}(\vec{r},t) = \frac{1}{\sqrt{\epsilon_0 V}}\sum_{\vec{k}}\left[\vec{\epsilon}_{\vec{k}} A_{\vec{k}}(t)e^{i\vec{k}\cdot\vec{r}} + \vec{\epsilon}_{\vec{k}}^* A_{\vec{k}}^*(t)e^{-i\vec{k}\cdot\vec{r}} \right],$$

(1.40)

where $\vec{\epsilon}_{\vec{k}} = (e_1, e_2)$ is called the polarization vector, which is perpendicular to the direction of propagation and $A_{\vec{k}}(t)$ is a time

dependent amplitude that varies harmonically with time, i.e., $\dot{A}_{\vec{k}} = -i\omega_{\vec{k}}A_{\vec{k}}$, where $\omega_{\vec{k}} = \sqrt{k_x^2 + k_y^2 + k_z^2} = |\vec{k}|c$. Assuming that the field is periodic in space, and the lengths of the periods in three perpendicular directions are equal to the dimensions of the cubic box, then $k_j = 2n_j\pi/L$ for $j = x, y, z$ and n_j are integers of any sign. Using Equation 1.38 we get

$$\vec{E}(\vec{r},t) = \frac{i}{\sqrt{\epsilon_0 V}}\sum_{\vec{k}}v_{\vec{k}}\left[\vec{\epsilon}_{\vec{k}} A_{\vec{k}}e^{i\vec{k}\cdot\vec{r}} + \vec{\epsilon}_{\vec{k}}^* A_{\vec{k}}^* e^{-i\vec{k}\cdot\vec{r}} \right],$$

(1.41)

and

$$\vec{H}(\vec{r},t) = \frac{i}{\sqrt{\epsilon_0 V}}\sum_{\vec{k}}\left[(\vec{k}\times\vec{\epsilon}_{\vec{k}})A_{\vec{k}}e^{i\vec{r}\cdot\vec{r}} - (\vec{k}\times\vec{\epsilon}_{\vec{k}}^*)A_{\vec{k}}^* e^{-i\vec{r}\cdot\vec{r}} \right].$$

(1.42)

The classical Hamiltonian for the field is given by

$$\mathcal{H}_{E-M} = \frac{\epsilon_0}{2}\int_V \left(|\vec{E}|^2 + c^2|\vec{H}|^2 \right) dx\,dy\,dz.$$

(1.43)

Substituting Equations 1.41 and 1.42 into Equation 1.43 one gets

$$\mathcal{H}_{E-M} = \sum_{\vec{k}}\omega_k^2 A_{\vec{k}}A_{\vec{k}}^*.$$

(1.44)

If we introduce the variable

$$A_{\vec{k}} = \frac{1}{\sqrt{2}}\left(Q_{\vec{k}} + i\frac{P_{\vec{k}}}{\omega_{\vec{k}}} \right),$$

(1.45)

then Equation 1.44 becomes

$$\mathcal{H}_{E-M} = \frac{1}{2}\sum_{\vec{k}}(Q_{\vec{k}}^2\omega_{\vec{k}}^2 + P_{\vec{k}}^2).$$

(1.46)

Equation 1.46 corresponds to the sum of independent harmonic oscillators. This suggests that each mode of the field is dynamically equivalent to a mechanical harmonic oscillator. The canonical quantization of the field consists of the substitution of the variables $Q_{\vec{k}}$ and $P_{\vec{k}}$ for operators, which fulfill the commutation relation

$$[Q_{\vec{k}}, P_{\vec{k}'}] = i\hbar\delta_{\vec{k}\vec{k}'},$$

(1.47)

this means that

$$[A_{\vec{k}}, A_{\vec{k}'}^\dagger] = \frac{\hbar}{\omega_{\vec{k}}}\delta_{\vec{k}\vec{k}'}.$$

(1.48)

Introducing the creation and annihilation operators

$$a_{\vec{k}} = \sqrt{\frac{\omega_{\vec{k}}}{\hbar}}A_{\vec{k}} \quad \text{and} \quad a_{\vec{k}}^\dagger = \sqrt{\frac{\omega_{\vec{k}}}{\hbar}}A_{\vec{k}}^\dagger,$$

(1.49)

we can write the electromagnetic field Hamiltonian in the form

$$\mathcal{H}_{\text{E-M}} = \sum_{\vec{k}} \left(\hbar\omega_{\vec{k}} a_{\vec{k}} a_{\vec{k}}^{\dagger} + \frac{1}{2} \hbar\omega_{\vec{k}} \right). \qquad (1.50)$$

The total Hamiltonian that describes the interaction of an atom with the quantized electromagnetic field can be written in the form

$$\mathcal{H} = \sum_{i} \epsilon_{i} |i\rangle\langle i| + \sum_{\vec{k}} \hbar\omega_{\vec{k}} a_{\vec{k}}^{\dagger} a_{\vec{k}} - \sum_{i,j} \langle i|\vec{d}|j\rangle |i\rangle\langle j|\cdot\vec{E}, \qquad (1.51)$$

where ϵ_{i} is the energy corresponding to the state $|i\rangle$ of the atom and we have dropped the constant terms corresponding to the zero point energy.

The electric field operator is given by

$$\vec{E} = i \sum_{\vec{k}} \vec{\epsilon}_{\vec{k}} \sqrt{\frac{\hbar\omega_{\vec{k}}}{\epsilon_{0} V}} \left[a_{\vec{k}} e^{i\vec{k}\cdot\vec{r}} - a_{\vec{k}}^{\dagger} e^{-i\vec{k}\cdot\vec{r}} \right]. \qquad (1.52)$$

Making the dipole approximation, i.e., $e^{i\vec{k}\cdot\vec{r}} \approx 1$, and assuming that we are dealing with a two-level atom, i.e., $i, j = 0, 1$, then we can write Equation 1.51 as

$$\mathcal{H} = \frac{\hbar\omega_{10}}{2} \sigma_{z} + \sum_{\vec{k}} \hbar\omega_{\vec{k}} a_{\vec{k}}^{\dagger} a_{\vec{k}} + \sum_{\vec{k}} (g_{\vec{k}}^{*} \sigma_{+} - g_{\vec{k}} \sigma_{-})(a_{\vec{k}} - a_{\vec{k}}^{\dagger}), \qquad (1.53)$$

where $g_{\vec{k}} = i\sqrt{\hbar\omega_{k}/\epsilon_{0}V} \vec{d}_{01} \cdot \vec{\epsilon}_{\vec{k}}$. The scalar product between the dipole moment and the polarization vector yields a complex number that can be written as $\vec{d}_{01} \cdot \vec{\epsilon}_{\vec{k}} = |\vec{d}_{01} \cdot \vec{\epsilon}_{\vec{k}}| e^{i\phi}$. This allows one to choose the phase ϕ in such a way that $g_{\vec{k}} = g_{\vec{k}}^{*}$. If the atom interacts only with one mode of the electromagnetic field, then we can drop the sum over the vector of propagation to get

$$\mathcal{H} = \frac{\hbar\omega_{10}}{2} \sigma_{z} + \hbar\omega a_{k}^{\dagger} a_{k} + g_{k}(\sigma_{+} - \sigma_{-})(a_{k} - a_{k}^{\dagger}) = \mathcal{H}_{0} + \mathcal{H}'. \qquad (1.54)$$

In the interaction picture, i.e., $\mathcal{H}_{I} = e^{-i\mathcal{H}_{0}t/\hbar} \mathcal{H} e^{i\mathcal{H}_{0}t/\hbar}$, the Hamiltonian of interaction will have the form

$$\mathcal{H}_{I}' = g_{k} \left(\sigma_{+} a_{k} e^{-i(\omega_{10}-\omega)t} + \sigma_{-} a_{k}^{\dagger} e^{i(\omega_{10}-\omega)t} - \sigma_{+} a_{k}^{\dagger} e^{i(\omega_{10}+\omega)t} \right.$$
$$\left. -\sigma_{-} a_{k} e^{-i(\omega_{10}+\omega)t} \right), \qquad (1.55)$$

using the RWA, i.e., dropping the rapidly oscillating terms containing $e^{\pm i(\omega_{10}+\omega)t}$, we end up with

$$\mathcal{H}_{I} = \frac{\hbar\omega_{10}}{2} \sigma_{z} + \hbar\omega a_{k}^{\dagger} a_{k} + g_{k}(\sigma_{+} a_{k} e^{-i\Delta t} + \sigma_{-} a_{k}^{\dagger} e^{i\Delta t}), \qquad (1.56)$$

where $\Delta = \omega_{10} - \omega$ is the detuning. The Hamiltonian of Equation 1.56 corresponds in the Schrödinger picture to

$$\mathcal{H} = \frac{\hbar\omega_{10}}{2} \sigma_{z} + \hbar\omega a_{k}^{\dagger} a_{k} + g_{k}(\sigma_{+} a_{k} + \sigma_{-} a_{k}^{\dagger}). \qquad (1.57)$$

Equation 1.57 is the Jaynes–Cummings model.

1.6 Loss–DiVincenzo Proposal

The first proposals for quantum computing made use of cavity quantum electrodynamics (QED), trapped ions, and nuclear magnetic resonance (NMR). All of these proposals benefit from long decoherence times due to a very weak coupling of the qubits to their environment. The long decoherence times have led to big successes in achieving experimental realizations. A conditional phase (CPHASE) gate was demonstrated early on in cavity QED systems. The two-qubit controlled-NOT gate has been realized in single-ion and two-ion versions. The most remarkable realization of the power of quantum computing to date is the implementation of Shor's algorithm to factor the number 15 in a liquid-state NMR quantum computer to yield the known result 5 and 3. However, these proposals may not be scalable and therefore do not meet the DiVincenzo criteria. The requirement for scalability motivated the Loss–DiVincenzo proposal for a solid state quantum computer based on electron spin qubits (Loss and DiVincenzo, 1998). The spin of an electron in a quantum dot can point up or down with respect to an external magnetic field; these eigenstates, $|\uparrow\rangle$ and $|\downarrow\rangle$, correspond to the two basis states of the qubit. The electron trapped in a quantum dot, which is basically a small electrically defined box that can be filled with electrons, can be defined by metal gate electrodes on top of a semiconductor (GaAs/AlGaAs) heterostructure. At the interface between GaAs and AlGaAs conduction band, electrons accumulate and can only move in the lateral direction. Applying negative voltages to the gates locally depletes this two-dimensional electron gas underneath. The resulting gated quantum dots are very controllable and versatile systems, which can be manipulated and probed electrically. When the size of the dot is comparable to the wavelength of the electrons that occupy it, the system exhibits a discrete energy spectrum, resembling that of an atom.

Initialization of the quantum computer can be achieved by allowing all spins to reach their equilibrium thermodynamic ground state at a low temperature T in an applied magnetic field, so that all the spins will be aligned if the condition $|g\mu_{B}H| \gg k_{B}T$ is satisfied (where k_{B} is Boltzmann constant).

To perform single-qubit operations, we can apply a microwave magnetic field on resonance with the Zeeman splitting, i.e., with a frequency $f = \Delta E_{Z}/h$ (h is Planck's constant). The oscillating magnetic component perpendicular to the static magnetic field H results in a spin nutation. By applying the oscillating field for a fixed duration, a superposition of $|\uparrow\rangle$ and $|\downarrow\rangle$ can be created. This magnetic technique is known as electron spin resonance (ESR). In the Loss–Divincenzo proposal, two-qubit operations can be carried out purely electrically by varying the gate voltages between neighboring dots. When the barrier is high, the spins are decoupled. When the interdot barrier is pulsed low,

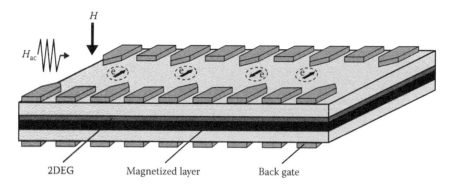

FIGURE 1.2 Theoretical proposal by Loss–DiVincenzo for quantum computing using quantum dots and electric gates.

an appreciable overlap develops between the two electron wave functions, resulting in a nonzero Heisenberg exchange coupling J (see Figure 1.2). The Hamiltonian describing this time dependent process is given by

$$\mathcal{H}(t) = J(t)\vec{S}_n \cdot \vec{S}_{n+1}. \qquad (1.58)$$

Equation 1.58 is sometimes referred to as the direct interaction.

The evolution of the quantum state is described by the propagator given by $U(t) = \mathcal{T}\exp[-i\int \mathcal{H}(t)\mathrm{d}t/\hbar]$, where $\mathcal{T}$ is the time ordering operator. If the exchange is pulsed on for a time τ_s such that $\int J(t)\mathrm{d}t/\hbar = J_0\tau_s/\hbar = \pi$, the states of the two spins will be exchanged. This is the SWAP operation. Pulsing the exchange for a shorter time $\tau_s/2$ generates the square root of SWAP operation, which can be used in conjunction with single-qubit operator to generate the controlled-NOT gate.

A last crucial ingredient requires a method to read out the state of the spin qubit. This implies measuring the spin orientation of a single electron. Therefore, an indirect spin measurement is proposed. First, the spin orientation of the electron is correlated with its position, via spin to charge conversion. Then an electrometer is used to measure the position of the charge, thereby revealing its spin. In this way, the problem of measuring the spin orientation has been replaced by the much easier measurement of charge.

The Loss–DiVincenzo ideas have influenced an enormous research effort aimed at implementing the different parts of the proposal and has been quickly followed by a series of alternative solid state realizations for trapped atoms in optical lattices that may also be scalable. It should also be stressed that the efforts to create a spin qubit are not purely application driven. If we have the ability to control and read out a single electron spin, we are in a unique position to study the interaction of the spin with its environment. This may lead to a better understanding of decoherence and will also allow us to study the semiconductor environment using the spin as a probe.

1.6.1 RKKY Interaction

The RKKY interaction is a long-range magnetic interaction that involves nearest-neighbor ions as well as magnetic atoms that are further apart; this interaction is sometimes

called indirect interaction. The basis of this interaction lies in an exchange interaction, which was proposed by Ruderman and Kittel (1954), and extended by Kasuya (1956) and Yosida (1957), and now known as the RKKY interaction. This interaction refers to an exchange energy written as

$$\mathcal{H}_{\mathrm{exch}} = -J(r)\sum_i \vec{s}_i \cdot \vec{S}, \qquad (1.59)$$

where

The exchange parameter $J(r)$ falls off rapidly with distance r between the center of a localized magnetic ion and electron spin
$\vec{s}_i$ is the spin state of a conduction electron
$\vec{S}$ is the spin of a localized magnetic ion

The minus sign in Equation 1.59 is related to the Pauli exclusion principle (lowest energy of electron occupancy). Below some critical magnetic ordering temperature, the itinerant or localized spins may condense into an ordered array, i.e., ferromagnetic or antiferromagnetic. As in the case of atomic scattering, any disorder within this array will cause additional electron scattering. This scattering may be elastic, in which case there is no change in energy or spin-flip, or it may be inelastic, in which case the spin state of the conduction electrons changes. Usually RKKY interaction is of significance only in compounds with a high concentration of the magnetic atom. Thus, localized ions may start to interact indirectly via the conduction electrons.

Now, consider the case in which there exist two localized spins at lattice points $\vec{R}_n$ and $\vec{R}_m$. By the interaction between spin S_{mz} localized at $\vec{R}_m$ and the spin density of conduction electrons polarized by spin S_{nz} localized at $\vec{R}_n$, the following interaction between the spins $\vec{S}_n$ and $\vec{S}_m$ is found:

$$\mathcal{H}_{\mathrm{exch}} = -9\pi\frac{J^2}{\varepsilon_F}\left(\frac{N_e}{N}\right)^2 F\left(2k_F \mid \vec{R}_n - \vec{R}_m \mid\right)S_{mz}S_{nz}, \qquad (1.60)$$

where

$$F(x) = \frac{-x\cos(x) + \sin(x)}{x^4}. \qquad (1.61)$$

Recently, the optical RKKY interaction between two spins was introduced as a means to produce an effective exchange interaction (Piermarocchi et al., 2002). Similar to the Loss–DiVincenzo scheme, in this scheme, the two qubits are defined by the excess electrons of semiconductor quantum dots. Instead of using the direct exchange interaction between the two electrons, the exchange interaction is indirectly mediated by the itinerant electrons of virtual excitons that are optically excited in the host material, which can be made of bulk, quantum well, or quantum wire structures. This scheme has the advantage that two-qubit gates can be performed on the femtosecond timescale due to the possibility of using ultrafast laser optics. The Coulomb interaction between the photoexcited itinerant electrons and the localized electrons in the two quantum dots contains direct and indirect terms. While the direct terms give rise to state renormalization of the localized electrons, the exchange terms lead to an effective Heisenberg interaction of the form

$$\mathcal{H}_{\text{ORKKY}} = -J_{12}\mathbf{S}_1 \cdot \mathbf{S}_2 \propto P_{12} \frac{H_C^2 H_X^2}{\delta^3} P_{12}, \qquad (1.62)$$

which is calculated in fourth-order perturbation theory using the diagram depicted in Figure 1.3.

P_{12} is the projection operator on the two-spin Hilbert space of the two localized spins. The control Hamiltonian

$$\mathcal{H}_C = \sum_{\mathbf{k},\sigma} \frac{\Omega_{\mathbf{k},\sigma}(t)}{2} e^{-i\omega P\sigma} c^\dagger_{\mathbf{k},-\sigma} h^\dagger_{-\mathbf{k},\sigma} + h.c. \qquad (1.63)$$

describes the creation of the virtual excitons by means of an external laser field that is detuned by the energy δ from the continuum states of the host material, or more precisely from the exciton 1s level. $c^\dagger_{\mathbf{k},\sigma}$ and $h^\dagger_{\mathbf{k},\sigma}$ are electron and hole creation operators, respectively. The RKKY interaction between a localized electron in the quantum dot and the itinerant electrons in the host material is given in second quantization by

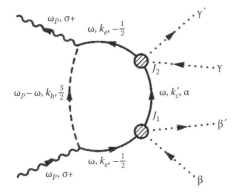

FIGURE 1.3 Effective spin–spin interaction for the localized electrons in the dots 1 and 2 (indicated by dotted lines) induced by a photoexcited electron–hole pair (the solid and dashed lines, respectively). The indices β and γ denote the spin states of the electrons localized in the dots. The photon propagator is depicted by a wavy line. (From Piermarocchi, C. et al., *Phys. Rev. Lett.*, 89, 167402, 2002.)

$$\mathcal{H}_X = -\frac{1}{V} \sum_{\alpha,\alpha',\mathbf{k},\mathbf{k}'} J(\mathbf{k},\mathbf{k}')\mathbf{S}_i \cdot \mathbf{s}_{\alpha,\alpha'} c^\dagger_{\mathbf{k},\alpha} c_{\mathbf{k}',\alpha'}. \qquad (1.64)$$

The predicted exchange interaction $J_{12}(\mathbf{R})$ (see Figure 1.2) can be of the order of 1 meV, which is of the same order as the Heisenberg interaction in the Loss–DiVincenzo scheme.

1.7 Quantum Computing with Molecular Magnets

Shor and Grover demonstrated that a quantum computer can outperform any classical computer in factoring numbers (Shor, 1997) and in searching a database (Grover, 1997) by exploiting the parallelism of quantum mechanics. Recently, the latter has been successfully implemented (Ahn et al., 2000) using Rydberg atoms. Leuenberger and Loss (2001) proposed an implementation of Grover's algorithm using molecular magnets (Friedman et al., 1996; Thomas et al., 1996; Sangregorio et al., 1997; Thiaville and Miltat, 1999; Wernsdorfer et al., 2000); their spin eigenstates make them natural candidates for single-particle systems. It was shown theoretically that molecular magnets can be used to build dense and efficient memory devices based on the Grover algorithm. In particular, one single crystal can serve as a storage unit of a dynamic random access memory device. Fast ESR pulses can be used to decode and read out stored numbers of up to 10^5, with access times as short as 10^{-10} s. This proposal should be feasible using the molecular magnets Fe_8 and Mn_{12}.

Suppose we want to find a phone number in a phone book consisting of $N = 2^n$ entries. Usually it takes $N/2$ queries on average to be successful. Even if the N entries were encoded binary, a classical computer would need approximately $\log_2 N$ queries to find the desired phone number (Grover, 1997). But the computational parallelism provided by the superposition and interference of quantum states enables the Grover algorithm to reduce the search to one single query (Grover, 1997). This query can be implemented in terms of a unitary transformation applied to the single spin of a molecular magnet. Such molecular magnets, forming identical and largely independent units, are embedded in a single crystal so that the ensemble nature of such a crystal provides a natural amplification of the magnetic moment of a single spin. However, for the Grover algorithm to succeed, it is necessary to find ways to generate arbitrary superpositions of spin eigenstates. For spins larger than ½, this turns out to be a highly nontrivial task as spin excitations induced by magnetic dipole transitions in conventional ESR can change the magnetic quantum number m by only ±1. To circumvent such physical limitations, it was proposed to use multifrequency coherent magnetic radiation that allows the controlled generation of arbitrary spin superpositions. In particular, it was shown that by means of advanced ESR techniques, it is possible to coherently populate and manipulate many spin states simultaneously by applying one single pulse of a magnetic a.c. field containing an appropriate number of matched frequencies. This a.c. field creates a nonlinear response of the magnet via multiphoton absorption processes involving particular sequences of σ and π photons, which allows the encoding and, similarly, the decoding of states. Finally, the subsequent read-out of the decoded quantum state can be achieved by means of pulsed ESR techniques. These exploit the nonequidistance of energy levels, which is typical of molecular magnets.

Molecular magnets have the important advantage that they can be grown naturally as single crystals of up to 10–100 μm length containing about 10^{12}–10^{15} (largely) independent units so that only minimal sample preparation is required. The molecular magnets are described by a single-spin Hamiltonian of the form $\mathcal{H}_{\text{spin}} = \mathcal{H}_a + V + \mathcal{H}_{\text{sp}} + \mathcal{H}_T$ (Leuenberger and Loss, 1999, 2000a,b), where $\mathcal{H}_a = -AS_z^2 - BS_z^4$ represents the magnetic anisotropy ($A \gg B > 0$). The Zeeman term $V = g\mu_B \mathbf{H} \cdot \mathbf{S}$ describes the coupling between the external magnetic field $\mathbf{H}$ and the spin $\mathbf{S}$ of length s. The calculational states are given by the $2s + 1$ eigenstates of $\mathcal{H}_a + g\mu_B H_z S_z$ with eigenenergies $\varepsilon_m = -Am^2 - Bm^4 + g\mu_B H_z m$, $-s \leq m \leq s$. The corresponding classical anisotropy potential energy $E(\theta) = -As \cos^2 \theta - Bs \cos^4 \theta + g\mu_B H_z s \cos \theta$ is obtained by the substitution $S_z = s \cos \theta$, where θ is the polar spherical angle. We have introduced the notation $m, m' = m - m'$. By applying a bias field H_z such that $g\mu_B H_z > E_{mm'}$, tunneling can be completely suppressed and thus $\mathcal{H}_T$ can be neglected (Leuenberger and Loss, 1999, 2000a,b). For temperatures below 1 K, transitions due to spin–phonon interactions ($\mathcal{H}_{\text{sp}}$) can also be neglected. In this regime, the level lifetime in Fe_8 and Mn_{12} is estimated to be about $\tau_d = 10^{-7}$ s, limited mainly by hyperfine and/or dipolar interactions (Leuenberger and Loss, 2001).

Since the Grover algorithm requires that all the transition probabilites are almost the same, Leuenberger and Loss (2001) and Leuenberger et al. (2003) propose that all the transition amplitudes between the states $|s\rangle$ and $|m\rangle$, $m = 1, 2, ..., s-1$, are of the same order in perturbation V. This allows us to use perturbation theory. A different approach uses the magnetic field amplitudes to adjust the appropriate transition amplitudes (Leuenberger et al., 2002). Both methods work only if the energy levels are not equidistant, which is typically the case in molecular magnets owing to anisotropies. In general, if we choose to work with the states $m = m_0, m_0+1, ..., s-1$, where $m_0 = 1, 2, ..., s-1$, we have to go up to nth order in perturbation, where $n = s - m_0$ is the number of computational states used for the Grover search algorithm (see below) to obtain the first nonvanishing contribution. Figure 1.4 shows the transitions for $s = 10$ and $m_0 = 5$. The nth-order transitions correspond to the nonlinear response of the spin system to strong magnetic fields. Thus, a coherent magnetic pulse of duration T is needed with a discrete frequency spectrum $\{\omega_m\}$, say, for Mn_{12} between 20 and 300 GHz and a single low-frequency 0 around 100 MHz.

The low-frequency field $\mathbf{H}_z(t) = H_0(t) \cos(\omega_0 t)\mathbf{e}_z$, applied along the easy-axis, couples to the spin of the molecular magnet through the Hamiltonian

$$V_{\text{low}} = g\mu_B H_0(t)\cos(\omega_0 t)S_z, \qquad (1.65)$$

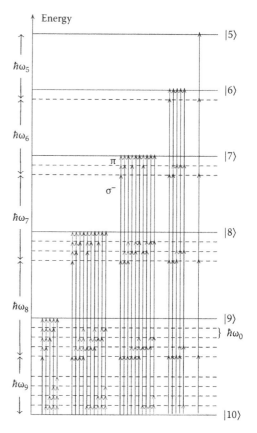

FIGURE 1.4 Feynman diagrams $\mathcal{F}$ that contribute to $S_{m,s}^{(5)}$ for $s = 10$ and $m_0 = 5$ describing transitions (of fifth order in V) in the left well of the spin system (see Figure 1.5). The solid and dotted arrows indicate σ⁻ and π transitions governed by Equations 1.66 and 1.65, respectively. We note that $S_{m,s}^{(j)} = 0$ for $j < n$, and $S_{m,s}^{(j)} \ll S_{m,s}^{(n)}$ for $j > n$.

where $\hbar\omega_0 \ll \varepsilon_{m_0} - \varepsilon_{m_0+1}$ and $\mathbf{e}_z$ is the unit vector pointing along the z-axis. The π photons of V_{low} supply the necessary energy for the resonance condition (see below). They give rise to virtual transitions with $\Delta m = 0$, that is, they do not transfer any angular momentum (see Figure 1.4). The perturbation Hamiltonian for the high-frequency transitions from $|s\rangle$ to virtual states that are just below $|m\rangle$, $m = m_0, ..., s-1$, given by the transverse fields $\mathbf{H}_\perp(t) = \sum_{m=m_0}^{s-1} H_m(t)[\cos(\omega_m t + \Phi_m)\mathbf{e}_x - \sin(\omega_m t + \Phi_m)\mathbf{e}_y]$, reads

$$V_{\text{high}}(t) = \sum_{m=m_0}^{s-1} g\mu_B H_m(t)[\cos(\omega_m t + \Phi_m)S_x - \sin(\omega_m t + \Phi_m)S_y]$$

$$= \sum_{m=m_0}^{s-1} \frac{g\mu_B H_m(t)}{2}\left[e^{i(\omega_m t + \Phi_m)}S_+ + e^{-i(\omega_m t + \Phi_m)}S_-\right], \qquad (1.66)$$

with phases Φ_m (see below), where we have introduced the unit vectors $\mathbf{e}_x$ and $\mathbf{e}_y$ pointing along the x- and y-axis, respectively. These transverse fields rotate clockwise and thus produce left circularly polarized σ⁻ photons, which induce only transitions in the left well (see Figure 1.5). In general, absorption (emission) of σ⁻ photons gives rise to $\Delta m = -1$ ($\Delta m = +1$) transitions, and vice versa in the

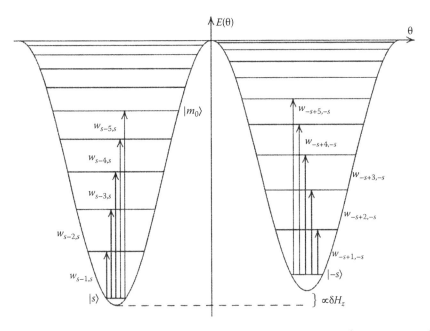

FIGURE 1.5 Double-well potential seen by the spin due to magnetic anisotropies in Mn$_{12}$. Arrows depict transitions between spin eigenstates driven by the external magnetic field **H**.

case of σ^+ photons. Anticlockwise rotating magnetic fields of the form $\mathbf{H}_\perp^+(t) = \sum_{m=m_0}^{s-1} H_m(t)[\cos(\omega_m t + \Phi_m)\mathbf{e}_x + \sin(\omega_m t + \Phi_m)\mathbf{e}_y]$ can be used to induce spin transitions only in the right well (see Figure 1.5). In this way, both wells can be accessed independently.

Next we calculate the quantum amplitudes for the transitions induced by the magnetic a.c. fields (see Figure 1.4) by evaluating the S-matrix perturbatively. The jth-order term of the perturbation series of the S-matrix in powers of the total perturbation Hamiltonian $V(t) = V_{\text{low}}(t) + V_{\text{high}}(t)$ is expressed by

$$S_{m,s}^{(j)} = \left(\frac{1}{i\hbar}\right)^j \prod_{k=1}^{j-1} \int_{-\infty}^{\infty} dt_k \int_{-\infty}^{\infty} dt_j \Theta(t_k - t_{k+1})$$

$$\times U(\infty, t_1)V(t_1)U(t_1, t_2)V(t_2)\ldots V(t_j)U(t_j, -\infty), \quad (1.67)$$

which corresponds to the sum over all Feynman diagrams $\mathcal{F}$ of order j, and where $U(t, t_0) = e^{-i(\mathcal{H}_a + g\mu_B \delta H_z)(t-t_0)/\hbar}$ is the free propagator, $\Theta(t)$ is the Heavyside function. The total S-matrix is then given by $S = \sum_{j=0}^{\infty} S^{(j)}$. The high-frequency virtual transition changing m from s to $s - 1$ is induced by the frequency $\omega_{s-1} = \omega_{s-1,s} - (n - 1)\omega_0$. The other high frequencies ω_m, $m = m_0$, …, $s - 2$, of the high-frequency fields H_m mismatch the level separations by ω_0, that is, $\hbar\omega_m = \varepsilon_m - \varepsilon_{m+1} + \hbar\omega_0$ (see Figure 1.4). As the levels are not equidistant, it is possible to choose the low and high frequencies in such a way that $S_{m,s}^{(j)} = 0$ for $j < n$, in which case the resonance condition is not satisfied, that is, energy is not conserved. In addition, the higher-order amplitudes $|S_{m,s}^{(j)}|$ are negligible compared to $|S_{m,s}^{(j)}|$ for $j > n$. Using rectangular pulse shapes, $H_k(t) = H_k$, if $-T/2 < t < T/2$, and 0 otherwise, for $k = 0$ and $k \geq m_0$, one obtains ($m \geq m_0$)

$$S_{m,s}^{(n)} = \sum_{\mathcal{F}} \Omega_m \frac{2\pi}{i} \left(\frac{g\mu_B}{2\hbar}\right)^n \prod_{k=m}^{s-1} \frac{H_k e^{-i\Phi_k} H_0^{m-m_0} p_{m,s}(\mathcal{F})}{(-1)^{q\mathcal{F}} q\mathcal{F}! r_s(\mathcal{F})! \omega_0^{n-1}}$$

$$\times \delta^{(T)}\left(\omega_{m,s} - \sum_{k=m}^{s-1} \omega_k - (m - m_0)\omega_0\right), \quad (1.68)$$

where $\Omega_m = (m - m_0)!$, is the symmetry factor of the Feynman diagrams $\mathcal{F}$ (see Figure 1.4), $q\mathcal{F} = m - m_0 - r_s(\mathcal{F})$, $p_{m,s}(\mathcal{F}) = \prod_{k=m}^{s} \langle k | S_z | k \rangle^{r_k(\mathcal{F})}$, $r_k(\mathcal{F}) = 0, 1, 2, \ldots, \leq m - m_0$ is the number of π transitions directly above or below the state $|k\rangle$, depending on the particular Feynman diagram $\mathcal{F}$, and $\delta^{(T)}(\omega) = \frac{1}{2\pi} \int_{-T/2}^{+T/2} e^{i\omega t} dt = \sin(\omega T/2)/\pi\omega$ is the delta-function of width $1/T$, ensuring overall energy conservation for $\omega T \gg 1$. The duration T of the magnetic pulses must be shorter than the lifetimes τ_d of the states $|m\rangle$ (see Figure 1.5). In general, the magnetic field amplitudes H_k must be chosen in such a way that perturbation theory is still valid and the transition probabilities are almost equal, which is required by the Grover algorithm. According to Leuenberger and Loss (2001), the amplitudes H_k do not differ too much between each other due to the partial destructive interference of the different transition diagrams shown in Figure 1.5. Leuenberger et al. (2002) show that the transition probabilities can be increased by increasing both the magnetic field amplitudes and the detuning energies under the condition that the magnetic field amplitudes remain smaller than the detuning energies. In this way, both high-multiphoton Rabi oscillation frequencies and small quantum computation times can be attained. This makes both methods (Leuenberger and Loss, 2001; Leuenberger et al., 2002) very robust against decoherence sources.

In order to perform the Grover algorithm, one needs the relative phases φ_m between the transition amplitudes $S_{m,s}^{(n)}$, which is determined by $\Phi_m = \Sigma_{k=s-1}^{m+1}\Phi_k + \varphi_m$, where Φ_m are the relative phases between the magnetic fields $H_m(t)$. In this way, it is possible to read-in and decode the desired phases Φ_m for each state $|m\rangle$. The read-out is performed by standard spectroscopy with pulsed ESR, where the circularly polarized radiation can now be incoherent because only the absorption intensity of only one pulse is needed. We emphasize that the entire Grover algorithm (read-in, decoding, read-out) requires three subsequent pulses each of duration T with $\tau_d > T > \omega_0^{-1} > \omega_m^{-1} > \omega_{m,m\pm1}^{-1}$. This gives a "clock-speed" of about 10 GHz for Mn_{12}, that is, the entire process of read-in, decoding, and read-out can be performed within about 10^{-10} s.

The proposal for implementing Grover's algorithm works not only for molecular magnets but for any electron or nuclear spin system with nonequidistant energy levels, as is shown by Leuenberger et al. (2002) for nuclear spins in GaAs semiconductors. Instead of storing information in the phases of the eigenstates $|m\rangle$ (Leuenberger and Loss, 2001), Leuenberger et al. (2002) use the eigenenergies of $|m\rangle$ in the generalized rotating frame for encoding information. The decoding is performed by bringing the delocalized state $(1/\sqrt{n})\Sigma_m |m\rangle$ into resonance with $|m\rangle$ in the generalized rotating frame. Although such spin systems cannot be scaled arbitrarily, large spin s (the larger a spin becomes, the faster it decoheres and the more classical its behavior will be) systems of given s can be used to great advantage in building dense and highly efficient memory devices.

For a first test of the nonlinear response, one can irradiate the molecular magnet with an a.c. field of frequency $\omega_{s-2,s}/2$, which gives rise to a two-photon absorption and thus to a Rabi oscillation between the states $|s\rangle$ and $|s-2\rangle$. For stronger magnetic fields, it is in principle possible to generate

superpositions of Rabi oscillations between the states $|s\rangle$ and $|s-1\rangle$, $|s\rangle$ and $|s-2\rangle$, $|s\rangle$ and $|s-3\rangle$, and so on (see also Leuenberger et al., 2002).

1.8 Semiconductor Quantum Dots

In the following, we mainly focus on quantum dots made of III–V semiconductor compounds with zincblende structure, like GaAs or InAs. The electronic bandstructure of a three-dimensional semiconductor with zincblende structure is illustrated in Figure 1.6. The bands are parabolic close to their extrema, which are all located at the Γ point. The conduction (c) states have orbital s symmetry and are spin degenerate. The valence (v) band consists of three subbands: the heavy-hole (hh), the light-hole (lh), and the split-off (so) band. The v-band states have orbital p symmetry. The bottom of the c band and the top of the v band are split by the band-gap energy E_{gap}. The v-band states with different j ($j = \frac{1}{2}$ for the so-band, $j = \frac{3}{2}$ for the hh and lh band) are split by Δ_{so} in energy due to spin–orbit interaction.

The hh states have the angular momentum projections $J_z = \pm\frac{3}{2}$ and the lh states $J_z = \pm\frac{1}{2}$. For finite electron wavevectors $k \neq 0$, and the hh and lh subbands split into two branches according to the different curvatures of the energy dispersion, which implies different effective masses of heavy and light holes. The v-band states with spin can be written in terms of the orbital angular momentum basis by using the Clebsch–Gordon coefficients which gives us

$$\text{Heavy hole} \left\{ \begin{array}{l} |\frac{3}{2},\frac{3}{2}\rangle = |1,1\rangle|\uparrow\rangle \\ |\frac{3}{2},-\frac{3}{2}\rangle = |1,-1\rangle|\downarrow\rangle \end{array} \right\}, \quad (1.69)$$

$$\text{Light hole} \left\{ \begin{array}{l} |\frac{3}{2},\frac{1}{2}\rangle = \frac{1}{\sqrt{3}}|1,1\rangle|\downarrow\rangle + \sqrt{\frac{2}{3}}|1,0\rangle|\uparrow\rangle \\ |\frac{3}{2},-\frac{1}{2}\rangle = \sqrt{\frac{2}{3}}|1,0\rangle|\downarrow\rangle + \frac{1}{\sqrt{3}}|1,-1\rangle|\uparrow\rangle \end{array} \right\}, \quad (1.70)$$

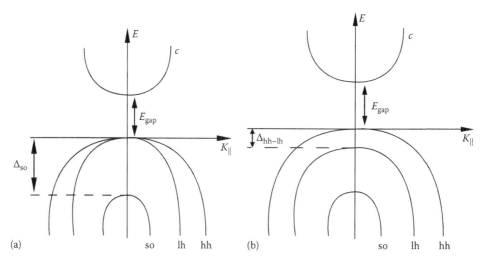

FIGURE 1.6 Electronic band structure in the vicinity of the Γ point for (a) a three-dimensional crystal and (b) a quantum well. The conduction and valence bands are shown as a function of the wavevector.

$$\text{Split off} \left\{ \begin{array}{l} |\tfrac{1}{2},\tfrac{1}{2}\rangle = \tfrac{1}{\sqrt{3}}|1,0\rangle|\Uparrow\rangle + \sqrt{\tfrac{2}{3}}|1,1\rangle|\Downarrow\rangle \\ |\tfrac{1}{2},-\tfrac{1}{2}\rangle = \sqrt{\tfrac{2}{3}}|1,-1\rangle|\Uparrow\rangle + \tfrac{1}{\sqrt{3}}|1,0\rangle|\Downarrow\rangle \end{array} \right\}. \qquad (1.71)$$

Quantum confinement along the crystal axis quantizes the wavevector component, consequently the hh and lh states of the lowest subband are split by an energy Δ_{hh-lh} at the Γ point. Uniaxial strain in the semiconductor crystal can also lift the degeneracy of the heavy and light holes, and thus define the spin quantization axis. If we have a spherically symmetric quantum dot known as *colloidal quantum dots*, then we can have degeneracy between the heavy and light hole band, i.e., $\Delta_{hh-lh} = 0$.

Via photon absorption, an electron in a v-band state can be excited to a c-band state. Such interband transitions are determined by optical selection rules. The source or the optical transition rules are due to spin–orbit interaction. The electron–hole pair created with an interband transition is called an exciton. The electron and hole of an exciton form a bound state due to the Coulomb interaction, similar to that of a hydrogen atom. We refer to the system of two bound excitons as a biexciton.

1.8.1 Classical Faraday Effect

Michael Faraday first observed the effect in 1845 when studying the influence of a magnetic field on plane-polarized light waves. Light waves vibrate in two planes at right angles to one another, and passing ordinary light through certain substances eliminates the vibration in one plane. He discovered that the plane of vibration is rotated when the light path and the direction of the applied magnetic field are parallel. In particular, a linearly polarized wave can be decomposed into right and left circularly polarized waves where each wave propagates with different speeds. The waves can be considered to recombine upon emergence from the medium; however, owing to the difference in propagation speed, they do so with a net phase offset, resulting in a rotation of the angle of linear polarization.

The Faraday effect occurs in many solids, liquids, and gases. The magnitude of the rotation depends upon the strength of the magnetic field, the nature of the transmitting substance, and Verdets constant, which is a property of the transmitting substance, its temperature, and the frequency of light. The relation between the angle of rotation of the polarization and the magnetic field in a diamagnetic material is

$$\beta = \mathcal{V}Hd, \qquad (1.72)$$

where

β is the angle of rotation
$\mathcal{V}$ is the Verdet constant for the material
H is the magnitude of the applied magnetic field
d is the length of the path where the light and magnetic field interact (see Figure 1.7)

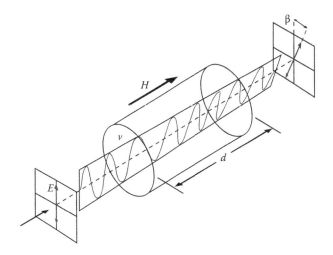

FIGURE 1.7 The figure shows how the polarization of a linearly polarized beam of light rotates when it goes through a material exposed to an external magnetic field.

A positive Verdet constant corresponds to an anticlockwise rotation when the direction of propagation is parallel to the magnetic field and to a clockwise rotation when the direction of propagation is antiparallel. The Faraday effect is used in spintronics research to study the polarization of electron spins in semiconductors.

1.8.2 Quantum Faraday Effect

The Faraday effect is expected to emerge in low dimensional systems such as semiconductor quantum dots in which the spin states of the electron in the conduction band and the light and heavy hole in the valence band provide a system where different circular polarizations of light couple differently during the process of virtual absorption. The quantum analogue of the Faraday effect does not require an external magnetic field because it is created by selection rules (one circular polarization interacts with the heavy hole band while the other circular polarization interacts with the light hole band) and by the Pauli exclusion principle (the absorption of a right polarized wave is excluded because the allowed transition state between bands have the same spin) (Leuenberger et al., 2005b). In particular, we will be interested in the quantum Faraday effect in a semiconductor colloidal two-level quantum dot system. We can have a two-level system in a colloidal quantum dot where the heavy hole and the light hole bands are degenerate at the Γ point. This two-level system is achieved by the valence and the conduction band under certain assumptions: (1) the split-off band can be ignored since typical split-off energies are around 10^2 meV, thus bringing the energy level out of resonance with the single photon; (2) under the appropriate doping and thermal conditions, it can be assumed that the top of the valence band is filled with four electrons, while there is one excess electron in the conduction band; and (3) the energy of the electromagnetic wave is taken to be slightly below the effective band-gap energy, so that the transition from the top of the valence band to the bottom of the conduction band is the strongest transition by far (Leuenberger, 2006).

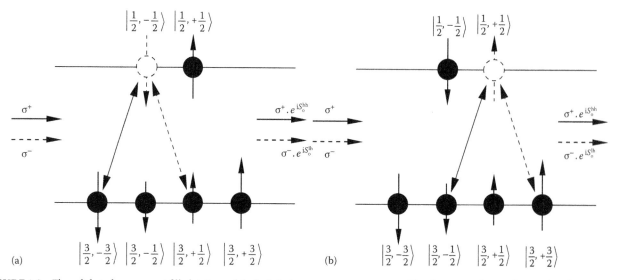

FIGURE 1.8 The solid circles represent filled states and dashed circles represent empty states. The electron in the conduction band interacts with the right or left circularly polarized electromagnetic wave allowing virtual transitions between the conduction and valence band states. After the interaction, the RCP and LCP electromagnetic wave acquire different phases, which produce the rotation of the incident wave.

Interestingly, the idea of the conditional single-photon Faraday rotation first developed by Leuenberger et al. (2005b) and already patented by the authors has been copied (Hu et al., 2008), which demonstrates the importance of this scheme.

We now turn our focus to the optical selection rules. In Figure 1.8 we show how different circular polarizations of the photon interact with the two-level system. The conduction band state has one electron with spin either up ($|\frac{1}{2},\frac{1}{2}\rangle$) or down ($|\frac{1}{2},-\frac{1}{2}\rangle$). The presence of a spin up (down) electron in the conduction band forces the virtual transitions of the incident electromagnetic wave to couple $|\frac{3}{2},-\frac{3}{2}\rangle$ ($|\frac{3}{2},-\frac{1}{2}\rangle$) or $|\frac{3}{2},\frac{1}{2}\rangle$ ($|\frac{3}{2},\frac{3}{2}\rangle$) heavy hole and light hole states depending if it is a right or left circularly polarized photon. Examination of the coefficients given in Equations 1.69 and 1.70 shows that the matrix elements of the heavy hole band and the matrix element involving the light hole band are different, leading to different phases and causing a change in the refractive index, which induces the Faraday rotation effect. The Faraday rotation is clockwise if the spin is up, and counterclockwise if the spin is down. The complex refractive index ($\tilde{\eta}$) can be calculated as follows:

$$\tilde{\eta}^2 = 1 - \frac{e^2 N}{\epsilon_0 m_e} \sum_{i,j} \frac{f_{ji}}{(\omega^2 - \omega_{ji}^2) + i\gamma\omega}, \quad (1.73)$$

where

|e| is the electron's charge
N is the electron number density
ϵ_0 is the permittivity of free space
m_e is the mass of the electron
ω_{ji} is the transition frequency from the ith state to the jth state
γ is the full-width at half maximum (assumed to be much smaller than the detuning energy)
ω is the frequency of the incident electromagnetic wave
f_{ji} is the oscillator strength given by

$$f_{ji} = \frac{2}{m\hbar\omega_{ji}} |\langle j|(-e)\vec{r}\cdot\vec{E}|i\rangle|^2. \quad (1.74)$$

This quantum Faraday rotation can be used to measure the spin polarization of charge carriers in quantum wells (Kikkawa et al., 1997; Kikkawa and Awschalom, 1998, 1999). Recently, an experiment based on the quantum Faraday rotation was able to measure the spin state of a single excess electron inside a quantum dot (Berezovsky et al., 2006), which is crucial for the read-out of a qubit in a quantum computing scheme (see below).

1.9 Single-Photon Faraday Rotation

The single-photon Faraday effect (SPFE) is similar to the quantum Faraday effect but involves only the nonresonant interaction of a single photon with the quantum dot two-level system and can result in an entanglement of the photon with the spin of the extra electron (Leuenberger et al., 2005b; Leuenberger, 2006). The SPFE can be described by the Jaynes–Cummings Hamiltonian

$$\mathcal{H} = \hbar\omega\left(a_{\sigma^+}^\dagger a_{\sigma^+} + a_{\sigma^-}^\dagger a_{\sigma^-}\right) + \hbar\omega_{hh}\sigma_{\frac{3}{2}v} + \hbar\omega_{hh}\sigma_{-\frac{3}{2}v}$$

$$+ \hbar\omega_{lh}\sigma_{\frac{1}{2}v} + \hbar\omega_{lh}\sigma_{-\frac{1}{2}v} + \hbar\omega_e\sigma_{\frac{1}{2}c} + \hbar\omega_e\sigma_{-\frac{1}{2}c} + \hbar g_{\frac{3}{2}v,\frac{1}{2}c}$$

$$\times\left(a_{\sigma^-}^\dagger\sigma_{\frac{3}{2}v,\frac{1}{2}c} + a_{\sigma^-}\sigma_{\frac{1}{2}c,\frac{3}{2}v} + a_{\sigma^+}^\dagger\sigma_{-\frac{3}{2}v,-\frac{1}{2}c} + a_{\sigma^+}\sigma_{-\frac{1}{2}c,-\frac{3}{2}v}\right)$$

$$+ \hbar g_{\frac{1}{2}v,\frac{1}{2}c}\left(a_{\sigma^-}^\dagger\sigma_{\frac{1}{2}v,-\frac{1}{2}c} + a_{\sigma^-}\sigma_{-\frac{1}{2}c,\frac{1}{2}v} + a_{\sigma^+}^\dagger\sigma_{-\frac{1}{2}v,-\frac{1}{2}c} + a_{\sigma^+}\sigma_{\frac{1}{2}c,-\frac{1}{2}v}\right).$$

$$(1.75)$$

Equation 1.76 describes how the right circularly polarized (σ^+) and left circularly polarized (σ^-) components of a linearly polarized photon field interact with the degenerate quantum dot levels in a cavity. It is assumed that the spin of the excess electron in the conduction band is up ($\uparrow$). The evolution of the system will be given by the state vector

$$|\Psi(t)\rangle = \frac{1}{\sqrt{2}}\left[C_{\uparrow hh}(t)|\uparrow, hh\rangle + C_{\uparrow\sigma^+}(t)|\uparrow,\sigma_z^+\rangle \right.$$
$$\left. + C_{\uparrow lh}(t)|\uparrow, lh\rangle + C_{\uparrow\sigma^-}(t)|\uparrow,\sigma_z^-\rangle \right], \qquad (1.76)$$

where $|\uparrow, hh\rangle$ and $|\uparrow, lh\rangle$ are the states in which the quantum dot is in an excited state with a heavy-hole exciton or a light-hole exciton, and the spin of the excess electron in the conduction band is up. $|\uparrow,\sigma_z^+\rangle$ and $|\uparrow,\sigma_z^-\rangle$ are the states in which the quantum dot is in the ground state with the photon present in the cavity, and the spin of the extra electron in the conduction band is up.

In order to solve for the phase accumulated for RCP and LCP components of the linearly polarized photon during the interaction with the quantum dot in the nanocavity, we must find the time evolution of the probability amplitudes $C_{\uparrow\sigma^+}(t)$ and $C_{\uparrow\sigma^-}(t)$, respectively. Assuming the following initial conditions, $C_{\uparrow hh}(0) = 0$, $C_{\uparrow\sigma^+}(0) = 1$, $C_{\uparrow lh}(0) = 0$, and $C_{\uparrow\sigma^-}(0) = 1$, then the probability amplitudes of interest are given by (Seigneur et al., 2008)

$$C_{\uparrow\sigma^+}(t) = e^{-i\Delta t/2}\left[\cos\left(\frac{\Omega_{\frac{3}{2}}t}{2}\right) + \frac{i\Delta}{\Omega_{\frac{3}{2}}}\sin\left(\frac{\Omega_{\frac{3}{2}}t}{2}\right)\right] \qquad (1.77)$$

$$C_{\uparrow\sigma^-}(t) = e^{-i\Delta t/2}\left[\cos\left(\frac{\Omega_{\frac{1}{2}}t}{2}\right) + \frac{i\Delta}{\Omega_{\frac{1}{2}}}\sin\left(\frac{\Omega_{\frac{1}{2}}t}{2}\right)\right], \qquad (1.78)$$

where $\Omega_{\frac{3}{2}}^2 = \Delta^2 + 4g_{\frac{3}{2}v,\frac{1}{2}c}^2$ and $\Omega_{\frac{1}{2}}^2 = \Delta^2 + 4g_{\frac{1}{2}v,\frac{1}{2}c}^2$, with $g_{\frac{3}{2}v,\frac{1}{2}c}$ and $g_{\frac{1}{2}v,\frac{1}{2}c}$ being the coupling strength involving the heavy-hole electron and the light-hole electron, respectively, with $\Delta = \omega - \omega_{ph}$ being the detuning frequency and $g_{\frac{3}{2}v,\frac{1}{2}c} = \sqrt{3}g_{\frac{1}{2}v,\frac{1}{2}c}$.

Rewriting the complex coefficients given in Equations 1.77 and 1.78 in the form $C_{\uparrow\sigma^+}(t) = |C_{\uparrow\sigma^+}(t)|\exp[iS_0^{hh}]$ and $C_{\uparrow\sigma^-}(t) = |C_{\uparrow\sigma^-}(t)|\exp[iS_0^{lh}]$, we obtain an expression for the phase accumulated during the interaction of the right and left circularly polarized component with the heavy-hole and the light-hole band, i.e.,

$$S_0^{hh} = \tan^{-1}\left[\frac{\Delta}{\Omega_{\frac{3}{2}}}\tan\left(\frac{\Omega_{\frac{3}{2}}t}{2}\right)\right] \quad \text{and} \quad S_0^{lh} = \tan^{-1}\left[\frac{\Delta}{\Omega_{\frac{1}{2}}}\tan\left(\frac{\Omega_{\frac{1}{2}}t}{2}\right)\right].$$
$$(1.79)$$

To determine the entanglement of the single photon with the electron, we know that the electron–photon state after the interaction reads

$$|\psi_{ep}^{(1)}\rangle = e^{iS_0^{hh}}|\psi_{hh}^{(1)}\rangle + e^{iS_0^{lh}}|\psi_{lh}^{(1)}\rangle, \qquad (1.80)$$

where

$$|\psi_{hh}^{(1)}\rangle = \frac{1}{\sqrt{2}}\left(\alpha|\uparrow\rangle|\sigma_{(z)}^+\rangle + \beta|\downarrow\rangle|\sigma_{(z)}^-\rangle\right) \qquad (1.81)$$

represents the photon scattering off a heavy hole, and

$$|\psi_{lh}^{(1)}\rangle = \frac{1}{\sqrt{2}}\left(\alpha|\uparrow\rangle|\sigma_{(z)}^-\rangle + \beta|\downarrow\rangle|\sigma_{(z)}^+\rangle\right) \qquad (1.82)$$

represents the photon scattering off a light hole. Let

$$|\varphi\rangle = \cos\varphi|\leftrightarrow\rangle + \sin\varphi|\updownarrow\rangle \qquad (1.83)$$

be the photon state with linear polarization that is rotated by φ around the z-axis with respect to the state $|\leftrightarrow\rangle$ of linear polarization in x direction. Changing to circular polarization states, Equation 1.83 can be rewritten as

$$|\varphi\rangle = \frac{\cos\varphi}{\sqrt{2}}\left(|\sigma_{(z)}^+\rangle + |\sigma_{(z)}^-\rangle\right) + \frac{\sin\varphi}{i\sqrt{2}}\left(|\sigma_{(z)}^+\rangle - |\sigma_{(z)}^-\rangle\right)$$
$$= \frac{1}{\sqrt{2}}\left(e^{-i\varphi}|\sigma_{(z)}^+\rangle + e^{i\varphi}|\sigma_{(z)}^-\rangle\right). \qquad (1.84)$$

Using the representation of this linear polarization with $\varphi = \pm(S_0^{hh} - S_0^{lh})/2 = \pm S_0/2$, we find that Equation 1.10 becomes

$$|\psi_{ep}^{(1)}(T)\rangle = \alpha|\uparrow\rangle\left(e^{iS_0^{hh}}|\sigma_{(z)}^+\rangle + e^{iS_0^{lh}}|\sigma_{(z)}^-\rangle\right)\Big/\sqrt{2}$$
$$+ \beta|\downarrow\rangle\left(e^{iS_0^{lh}}|\sigma_{(z)}^+\rangle + e^{iS_0^{hh}}|\sigma_{(z)}^-\rangle\right)\Big/\sqrt{2}$$
$$= e^{i\left(S_0^{hh}+S_0^{lh}\right)/2}\left(\alpha|\uparrow\rangle|-S_0/2\rangle + \beta|\downarrow\rangle|+S_0/2\rangle\right). \qquad (1.85)$$

From this equation, we draw the conclusion that the accumulated CPHASE shift S_0 corresponds to a single-photon Faraday rotation around the z-axis by the angle $S_0/2$ due to a single spin. Let j be an integer. For $S_0/2 = (2j+1)\pi/4$, the electron–photon state $|\psi_{ep}^{(1)}\rangle$ is maximally entangled. If j is even,

$$|\psi_{ep}^{(1)}\rangle = \alpha|\uparrow\rangle|\searrow\rangle + \beta|\downarrow\rangle|\nearrow\rangle. \qquad (1.86)$$

If j is odd,

$$|\psi_{ep}^{(1)}\rangle = \alpha|\uparrow\rangle|\nearrow\rangle + \beta|\downarrow\rangle|\searrow\rangle. \qquad (1.87)$$

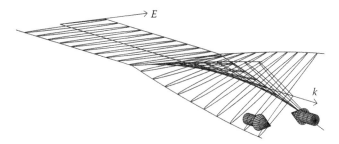

FIGURE 1.9 The conditional Faraday rotation of the linear polarization of the photon depends on the spin state $|\psi_e\rangle = \alpha |{\uparrow}\rangle + \beta |{\downarrow}\rangle$. If the spin state is up (down), the linear polarization turns (counter) clockwise. The photon and the spin get maximally entangled if the Faraday rotation angle is $S_0/2 = \pi/4$. This entanglement between photon polarization and electron spin can be used as an optospintronic link to transfer quantum information from electron spins to photons and back.

For $S_0/2 = 2j\pi/4$, the electron–photon state $|\psi_{ep}^{(1)}\rangle$ is not entangled. If j is even,

$$|\psi_{ep}^{(1)}(T)\rangle = \alpha |{\uparrow}\rangle |{\leftrightarrow}\rangle + \beta |{\downarrow}\rangle |{\leftrightarrow}\rangle. \tag{1.88}$$

If j is odd,

$$|\psi_{ep}^{(1)}(T)\rangle = \alpha |{\uparrow}\rangle |{\updownarrow}\rangle - \beta |{\downarrow}\rangle |{\updownarrow}\rangle. \tag{1.89}$$

This means the entanglement oscillates continually between 0 and 1 if no decoherence is present. The entanglement process between the spin of the electron and the polarization of the photon during the conditional Faraday rotation is visualized in Figure 1.9.

1.9.1 Quantifying the EPR Entanglement

Rewriting Equation 1.85 in the basis states of the linear polarization in x and y directions yields

$$\left|\psi_{ep}^{(1)}\right\rangle = \alpha |{\uparrow}\rangle \left(\cos\frac{S_0}{2} |{\leftrightarrow}\rangle - \sin\frac{S_0}{2} |{\updownarrow}\rangle \right)$$
$$+ \beta |{\downarrow}\rangle \left(\cos\frac{S_0}{2} |{\leftrightarrow}\rangle + \sin\frac{S_0}{2} |{\updownarrow}\rangle \right). \tag{1.90}$$

The entanglement between the photon and the spin of the excess electron can be calculated by means of the von Neumann entropy (Bennett et al., 1996):

$$E[\psi_{ep}^{(1)}] = -\mathrm{Tr}_e[\mathrm{Tr}_p\, \rho_{ep} \log_2(\mathrm{Tr}_p\, \rho_{ep})] \tag{1.91}$$

or the normalized linear entropy (Silberfarb and Deutsch, 2004)

$$E_L[\psi_{ep}^{(1)}] = 2[1 - \mathrm{Tr}_e(\mathrm{Tr}_p\, \rho_{ep})^2]. \tag{1.92}$$

We will use the linear entropy in our calculation. The density matrix is given by

$$\rho_{ep} = \begin{pmatrix} \rho_{\uparrow\uparrow} & \rho_{\uparrow\downarrow} \\ \rho_{\downarrow\uparrow} & \rho_{\downarrow\downarrow} \end{pmatrix}, \tag{1.93}$$

where

$$\rho_{\uparrow\uparrow} = \begin{pmatrix} |\alpha|^2 \cos^2\frac{S_0}{2} & -2|\alpha|^2 \sin S_0 \\ -2|\alpha|^2 \sin S_0 & |\alpha|^2 \sin^2\frac{S_0}{2} \end{pmatrix}, \tag{1.94}$$

$$\rho_{\uparrow\downarrow} = \begin{pmatrix} \alpha\beta^* \cos^2\frac{S_0}{2} & 2\alpha\beta^* \sin S_0 \\ -2\alpha\beta^* \sin S_0 & -\alpha\beta^* \sin^2\frac{S_0}{2} \end{pmatrix}, \tag{1.95}$$

$$\rho_{\downarrow\uparrow} = \begin{pmatrix} \alpha^*\beta \cos^2\frac{S_0}{2} & -2\alpha^*\beta \sin S_0 \\ 2\alpha^*\beta \sin S_0 & -\alpha^*\beta \sin^2\frac{S_0}{2} \end{pmatrix}, \tag{1.96}$$

$$\rho_{\downarrow\downarrow} = \begin{pmatrix} |\beta|^2 \cos^2\frac{S_0}{2} & 2|\beta|^2 \sin S_0 \\ 2|\beta|^2 \sin S_0 & |\beta|^2 \sin^2\frac{S_0}{2} \end{pmatrix}, \tag{1.97}$$

and thus by $\rho_{ep} =$

$$\begin{pmatrix} |\alpha|^2 \cos^2\frac{S_0}{2} & -2|\alpha|^2 \sin S_0 & \alpha\beta^* \cos^2\frac{S_0}{2} & 2\alpha\beta^* \sin S_0 \\ -2|\alpha|^2 \sin S_0 & |\alpha|^2 \sin^2\frac{S_0}{2} & -2\alpha\beta^* \sin S_0 & -\alpha\beta^* \sin^2\frac{S_0}{2} \\ \alpha^*\beta \cos^2\frac{S_0}{2} & -2\alpha^*\beta \sin S_0 & |\beta|^2 \cos^2\frac{S_0}{2} & 2|\beta|^2 \sin S_0 \\ 2\alpha^*\beta \sin S_0 & -\alpha^*\beta \sin^2\frac{S_0}{2} & 2|\beta|^2 \sin S_0 & |\beta|^2 \sin^2\frac{S_0}{2} \end{pmatrix} \tag{1.98}$$

in the basis $|{\uparrow}\rangle |{\leftrightarrow}\rangle$; $|{\uparrow}\rangle |{\updownarrow}\rangle$; $|{\downarrow}\rangle |{\leftrightarrow}\rangle$; $|{\downarrow}\rangle |{\updownarrow}\rangle$. Since the measures for entanglement must be independent of the chosen basis, all of the measures involve the trace of ρ_{ep}. Taking the trace over the photon states yields

$$\mathrm{Tr}_p\, \rho_{ep} = \begin{pmatrix} |\alpha|^2 & \alpha\beta^* \cos S_0 \\ \alpha^*\beta \cos S_0 & |\beta|^2 \end{pmatrix}. \tag{1.99}$$

So we obtain the entanglement

$$E_L[\psi_{ep}^{(1)}] = 2\left(1 - |\alpha|^4 - |\beta|^4 - 2|\alpha|^2|\beta|^2 \cos^2 S_0\right). \tag{1.100}$$

With the parametrization $\alpha = \cos\frac{\eta}{2} e^{-i\chi/2}$, $\beta = \sin\frac{\eta}{2} e^{-i\chi/2}$ we obtain

$$E_L[\psi_{ep}^{(1)}] = 2\left(1 - \cos^4\frac{\eta}{2} - \sin^4\frac{\eta}{2} - \frac{1}{2}\sin^2\eta\cos^2 S_0\right). \tag{1.101}$$

This result is plotted in Figure 1.10.

1.9.2 Single Photon Faraday Effect and GHZ Quantum Teleportation

The theory of quantum teleportation was first developed by Bennett et al. (1993) and later experimentally verified by Bouwmeester et al. (1997). This scheme relies on EPR pairs,

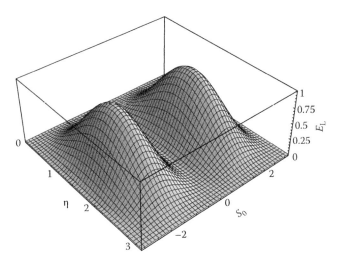

FIGURE 1.10 Entanglement of the photon with the spin of the excess electron.

i.e., pairs of entangled particles, which can be entangled in, for example, the polarization degree of freedom in the case of photons or the spin degree of freedom in the case of electrons or holes. Leuenberger et al. (2005b) show that all the three particles that take part in the teleportation process are entangled in a so-called Greenberger–Horne–Zeilinger (GHZ) state (Greenberger et al., 1989, 1990).

We will perform now a GHZ teleportation. In this GHZ quantum teleportation, the photon (qubit 2) interacts first with the electron spin of the destination (qubit 3) and then with the electron spin of the origin (qubit 1). We identify qubit 1, 2, and 3 with the qubits used by Bennett et al. (1993). Qubit 1 is with Alice, qubit 2 with Charlie, and qubit 3 with Bob. We start from the state $|\Psi^{(1)}_{pe'}(t_A)\rangle = |\leftrightarrow\rangle|\leftarrow{}'\rangle$. After interaction in Bob's microcavity, we obtain

$$|\Psi^{(1)}_{pe'}(t_C)\rangle = \frac{1}{\sqrt{2}}\left(|\sigma^+_z\rangle|\uparrow{}'_y\rangle + |\sigma^-_{(z)}\rangle|\downarrow{}'_y\rangle\right)$$
$$= \frac{1}{\sqrt{2}}\left(|\nwarrow\rangle|\nearrow\rangle|\downarrow{}'\rangle\right). \qquad (1.102)$$

This hybrid photon–spin entangled state corresponds to the shared EPR state of qubit 2 and 3 in the original version of the teleportation (EPR teleportation). Note that there is no way to distinguish between a single-spin or a single-photon Faraday rotation because we do not have any preferred basis defined by α and β. We have not yet used qubit 1 at all. Alice's photon can be stored for as long as it maintains its entanglement with Bob's spin. This step is like the distribution of the EPR pair in EPR teleportation. Instead of performing now a Bell measurement on qubit 1 and 2 to complete the EPR teleportation, we let the photon (qubit 2) interact with Alice's spin (qubit 1), giving rise to a GHZ state in the hybrid spin–photon–spin system. After interaction of the photon in Alice's microcavity, we obtain

$$|\psi^{(1)}_{epe'}(t_C + T)\rangle = \frac{1}{\sqrt{2}}\left[\alpha|\uparrow\rangle\left(e^{iS^{hh}_0}|\sigma^+_{(z)}\rangle|\uparrow{}'_y\rangle\right.\right.$$
$$\left.+ e^{iS^{lh}_0}|\sigma^-_{(z)}\rangle|\downarrow{}'_y\rangle\right)$$
$$+ \beta|\downarrow\rangle\left(e^{iS^{lh}_0}|\sigma^+_{(z)}\rangle\left|\uparrow{}'_y\right\rangle\right.$$
$$\left.\left.+ e^{iS^{hh}_0}|\sigma^-_{(z)}\rangle|\downarrow{}'_y\rangle\right)\right]. \qquad (1.103)$$

Now we change to the S_z representation of Bob's spin and to the linear polarization representation of Charlie's photon, which yields ($t_C = 0$)

$$|\psi^{(1)}_{epe'}(T)\rangle = \frac{e^{i\left(S^{hh}_0 + S^{lh}_0\right)/2}}{\sqrt{2}}\left(\alpha|\uparrow\rangle|-S_0/2 - \pi/4\rangle|\uparrow{}'\rangle\right.$$
$$+ \alpha|\uparrow\rangle|-S_0/2 + \pi/4\rangle|\downarrow{}'\rangle$$
$$+ \beta|\downarrow\rangle|+S_0/2 - \pi/4\rangle|\uparrow{}'\rangle$$
$$+ \beta|\downarrow\rangle|+S_0/2 + \pi/4\rangle|\downarrow{}'\rangle. \qquad (1.104)$$

Choosing $S_0 = \pi/2$, we obtain

$$|\psi^{(1)}_{epe'}(T)\rangle = \frac{1}{\sqrt{2}}\left[|\updownarrow\rangle\left(-\alpha|\uparrow\rangle|\uparrow{}'\rangle + \beta|\downarrow\rangle|\downarrow{}'\rangle\right)\right.$$
$$\left.+ |\leftrightarrow\rangle\left(\alpha|\uparrow\rangle|\downarrow{}'\rangle + \beta|\downarrow\rangle|\uparrow{}'\rangle\right)\right]. \qquad (1.105)$$

[From this equation, it becomes obvious that we can produce all the four Bell states between qubit 1 (Alice's electron spin) and qubit 3 (Bob's electron spin).] Now we change to the S_x representation of Alice's spin:

$$|\psi^{(1)}_{epe'}(T)\rangle = \frac{1}{\sqrt{2}}|\updownarrow\rangle\left[|\leftarrow\rangle\left(-\alpha|\uparrow{}'\rangle + \beta|\downarrow{}'\rangle\right)\right.$$
$$\left.+ |\rightarrow\rangle\left(-\alpha|\uparrow{}'\rangle - \beta|\downarrow{}'\rangle\right)\right]$$
$$+ \frac{1}{\sqrt{2}}|\leftrightarrow\rangle\left[|\leftarrow\rangle\left(\beta|\uparrow{}'\rangle + \alpha|\downarrow{}'\rangle\right)\right.$$
$$\left.+ |\rightarrow\rangle\left(\beta|\uparrow{}'\rangle - \alpha|\downarrow{}'\rangle\right)\right]. \qquad (1.106)$$

The difference between our method and the original version of the teleportation (Bennett et al., 1993) is that we have entangled qubit 2 (the photon) with qubit 1 (Alice's electron spin) by means of the electron–photon interaction, which is not done in the original version. That is why we do not need Bell measurements. So after measuring the photon polarization state (qubit 2) and the spin state of Alice's electron (qubit 1), the spin state of the electron of the destination gets projected onto $-\alpha|\uparrow{}'\rangle + \beta|\downarrow{}'\rangle$, $-\alpha|\uparrow{}'\rangle - \beta|\downarrow{}'\rangle$, $\beta|\uparrow{}'\rangle + \alpha|\downarrow{}'\rangle$, or $\beta|\uparrow{}'\rangle - \alpha|\downarrow{}'\rangle$ with equal probability. This corresponds exactly to the outcome of the original version of the teleportation. However, we do not use any Bell measurements.

Let us check what happens if there is no interaction between Charlie's photon and Alice's spin. Then we get

$$|\psi_{epe'}^{(1)}(T)\rangle = \frac{1}{\sqrt{2}}\left[|\diagdown\rangle\left(\alpha|{\uparrow}\rangle|{\uparrow}'\rangle + \beta|{\downarrow}\rangle|{\uparrow}'\rangle\right)\right.$$
$$\left. + |\diagup\rangle\left(\alpha|{\uparrow}\rangle|{\downarrow}'\rangle + \beta|{\downarrow}\rangle|{\downarrow}'\rangle\right)\right].$$

$$(1.107)$$

In this case, we would have to apply a Bell measurement in order to perform the teleportation.

1.9.3 Single-Photon Faraday Effect and Quantum Computing

Recently, the *destructive probabilistic* CNOT gate for all-optical quantum computing proposed by Knill et al. (2001), which relies on postselection by measurements, has been implemented experimentally (O'Brien et al., 2003; Nemoto and Munro, 2004). Based on a proposal by Pittman et al. (2001), experiments have demonstrated that this destructive CNOT gate and a quantum parity check can be combined with a pair of entangled photons to produce a *nondestructive probabilistic* CNOT gate (Pittman et al., 2002a,b, 2003; Gasparoni et al., 2004; Zhao et al., 2005). It has been shown theoretically that deterministic quantum computing with postselection is feasible for quantum systems where the qubits are represented by several degrees of freedom of a single photon (Cerf et al., 1998). An interferometric approach to linear-optical deterministic CNOT gate between the polarization and momentum of a single photonic qubit has recently been demonstrated experimentally (Fiorentino and Wong, 2004). A general scheme to teleport a quantum state through a universal gate has been given by Gottesman and Chuang (1999). This idea gave Raussendorf and Briegel the inspiration to develop the one-way quantum computer (Raussendorf and Briegel, 2001; Raussendorf et al., 2003), which performs quantum computations by measurement-induced teleportation processes on an entangled network of qubits. The name one-way quantum computing refers to the irreversibility of the computations due to the measurement processes.

Leuenberger developed a scheme for fault-tolerant quantum computing with the excess electron spins in quantum dots based on the SPFE (Leuenberger, 2006). This scheme shares a similarity with the one-way quantum computing scheme in the sense that it makes use of measurement processes to perform quantum computations. Both the one-way quantum computing scheme and Leuenberger's scheme are fully deterministic and scalable to an arbitrary number of qubits. However, while the one-way quantum computing scheme is irreversible, Leuenberger's scheme is fully reversible. In addition, the one-way quantum computing scheme needs to maintain the coherence of the whole cluster of qubits and is therefore more sensitive to the environment than single isolated qubits, since in a cluster the decoherence of one qubit leads also to decoherence of neighboring qubits

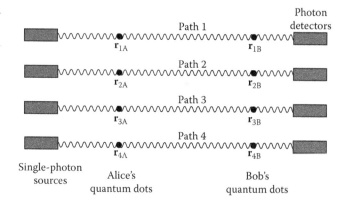

FIGURE 1.11 The nonresonant interaction of a photon with Alices and Bobs quantum dot produces the CPHASE gate required for universal quantum computing. This CPHASE gate applies to the case of the noncoded spins, where each microcavity contains only a single quantum dot with a single excess electron.

that are entangled to it. Therefore, Leuenberger's scheme is more robust in this respect.

We describe now Leuenberger's quantum computing scheme (see Figure 1.11). DiVincenzo showed that the CNOT gate and a single qubit rotation are sufficient for universal quantum computing (DiVincenzo, 1995). In the framework of the conditional Faraday rotation, it is easier to work with the implementation of the CPHASE gate than with the implementation of the CNOT gate. The CPHASE gate

$$u_{\text{CPHASE}} = \begin{pmatrix} 1 & 0 & 0 & 0 \\ 0 & 1 & 0 & 0 \\ 0 & 0 & 1 & 0 \\ 0 & 0 & 0 & -1 \end{pmatrix}$$

$$(1.108)$$

is equivalent (similar) to the CNOT gate (see Equation 1.7), i.e., they can be transformed into each other by means of a basis transformation. The CPHASE gate shifts the phase only if both spins point down. Let us define two persons Alice and Bob. Both of them have one photonic crystal, in which n noninteracting quantum dots are embedded. The single excess electrons of Alice's quantum dots are in a general single-spin state $|\psi\rangle_A = \alpha|{\uparrow}\rangle_A + \beta|{\downarrow}\rangle_A$, where the quantization axis is the z-axis. Bob's spins are in a general single-spin state $|\psi\rangle_B = \gamma|{\uparrow}\rangle_B + \delta|{\downarrow}\rangle_B$. The photons that interact with both Alice's and Bob's quantum dots are initially in a horizontal linear polarization state $|{\leftrightarrow}\rangle$.

Since there is no need to detect optically the spin states in transverse direction (Leuenberger et al., 2005a), our quantum dots can be nonspherical. So each photon can virtually create only a heavy-hole exciton on each quantum dot (see Figure 1.6). The strong selection rules imply that only a $\sigma^+_{(z)}(\sigma^-_{(z)})$ photon can interact with the quantum dot if the excess electron's spin is up (down). This leads to a conditional Faraday rotation of the linear polarization of the photon depending on the spin state of the quantum dot. The photonic crystal ensures that the photon's propagation direction is always in z-direction, perpendicular to the quantum dot plane.

In Leuenberger's implementation of the CPHASE gate, a single photon interacts sequentially with Alice's and then with Bob's quantum dot. The spin on Alice's quantum dot is prepared in the state $|\psi_A(0)\rangle = \alpha|\uparrow\rangle_A + \beta|\downarrow\rangle_A$. So we start with the electron–photon state $|\psi_{Ap}(0)\rangle = (\alpha|\uparrow\rangle_A + \beta|\downarrow\rangle_A)|\leftrightarrow\rangle$. After the interaction with Alice's quantum dot, the resulting electron–photon state is

$$|\psi_{Ap}(T)\rangle = (\alpha|\uparrow\rangle_A |\diagdown\rangle + \beta|\downarrow\rangle_A |\diagup\rangle)/\sqrt{2}, \qquad (1.109)$$

which is maximally entangled.

We let the photon interact also with Bob's quantum dot, which yields

$$\begin{aligned}|\psi_{ApB}(2T)\rangle &= |\updownarrow\rangle\big(-\alpha\gamma\,|\uparrow\rangle_A |\uparrow\rangle_B + \beta\delta\,|\downarrow\rangle_A |\downarrow\rangle_B\big) \\ &\quad + |\leftrightarrow\rangle\big(\alpha\delta\,|\uparrow\rangle_A |\downarrow\rangle_B + \beta\gamma\,|\downarrow\rangle_A |\uparrow\rangle_B\big) \\ &= \frac{1}{\sqrt{2}}\Big[|\diagdown\rangle\big(-\alpha\gamma\,|\uparrow\rangle_A |\uparrow\rangle_B + \beta\delta\,|\downarrow\rangle_A |\downarrow\rangle_B \\ &\quad + \alpha\delta\,|\uparrow\rangle_A |\downarrow\rangle_B + \beta\gamma\,|\downarrow\rangle_A |\uparrow\rangle_B\big) \\ &\quad + |\diagup\rangle\big(\alpha\gamma\,|\uparrow\rangle_A |\uparrow\rangle_B - \beta\delta\,|\downarrow\rangle_A |\downarrow\rangle_B \\ &\quad + \alpha\delta\,|\uparrow\rangle_A |\downarrow\rangle_B + \beta\gamma\,|\downarrow\rangle_A |\uparrow\rangle_B\big)\Big]. \end{aligned} \qquad (1.110)$$

If the linear polarization of the photon is measured in the $\diagdown$ axis, we obtain the two results

$$\begin{aligned}|\psi_{AB}(2T)\rangle &= -\alpha\gamma\,|\uparrow\rangle_A |\uparrow\rangle_B + \beta\delta\,|\downarrow\rangle_A |\downarrow\rangle_B \\ &\quad + \alpha\delta\,|\uparrow\rangle_A |\downarrow\rangle_B + \beta\gamma\,|\downarrow\rangle_A |\uparrow\rangle_B, \end{aligned} \qquad (1.111)$$

$$\begin{aligned}|\psi_{AB}(2T)\rangle &= \alpha\gamma\,|\uparrow\rangle_A |\uparrow\rangle_B - \beta\delta\,|\downarrow\rangle_A |\downarrow\rangle_B \\ &\quad + \alpha\delta\,|\uparrow\rangle_A |\downarrow\rangle_B + \beta\gamma\,|\downarrow\rangle_A |\uparrow\rangle_B \end{aligned} \qquad (1.112)$$

with equal probability, where the CPHASE shift is π if both spins are up or both spins are down, respectively. For deterministic quantum computing, we need to choose either Equation 1.111 or 1.112 to be the correct phase gate. Let us choose Equation 1.112 to be the correct implementation. The measurement of the photon's polarization state must be shared through a classical channel between Alice and Bob, in order for the CPHASE gate to be deterministic. Then Equation 1.111 can be transformed into Equation 1.112 by two local single-qubit phase shifts $\sigma_{A,z}\sigma_{B,z}$ applied to Alice's and Bob's qubit, where σ_z is a Pauli matrix. This scheme for implementing the CPHASE gate is similar to the scheme of quantum teleportation in the sense that the measurement outcome of the photon's polarization needs to be shared between Alice and Bob through a classical channel. This scheme for implementing the CPHASE gate is a generalization of the scheme of quantum teleportation in the sense that it takes not only one local unitary operation but two unitary operations

to reconstruct the desired two-qubit state, which makes sense, because we perform a two-qubit manipulation, as opposed to a single qubit manipulation in the case of quantum teleportation.

Thus, we can produce deterministically the CPHASE gate, which is equivalent to the CNOT gate in the basis $\{|\uparrow\rangle_A |\leftarrow\rangle_B, |\uparrow\rangle_A |\rightarrow\rangle_B, |\downarrow\rangle_A |\leftarrow\rangle_B, |\downarrow\rangle_A |\rightarrow\rangle_B\}$, where $|\leftarrow\rangle = (|\uparrow\rangle + |\downarrow\rangle)/\sqrt{2}$ and $|\rightarrow\rangle = (|\uparrow\rangle - |\downarrow\rangle)/\sqrt{2}$ in the S_x representation. This can be proven by writing down the mappings

$$\begin{aligned} |\uparrow\rangle_A |\leftarrow\rangle_B &= \frac{1}{\sqrt{2}}|\uparrow\rangle_A\big(|\uparrow\rangle_B + |\downarrow\rangle_B\big) \\ &\rightarrow \frac{1}{\sqrt{2}}|\uparrow\rangle_A\big(-|\uparrow\rangle_B + |\downarrow\rangle_B\big) = -|\uparrow\rangle_A |\rightarrow\rangle_B, \\ |\uparrow\rangle_A |\rightarrow\rangle_B &= \frac{1}{\sqrt{2}}|\uparrow\rangle_A\big(|\uparrow\rangle_B - |\downarrow\rangle_B\big) \\ &\rightarrow \frac{1}{\sqrt{2}}|\uparrow\rangle_A\big(-|\uparrow\rangle_B - |\downarrow\rangle_B\big) = -|\uparrow\rangle_A |\leftarrow\rangle_B, \\ |\downarrow\rangle_A |\leftarrow\rangle_B &= \frac{1}{\sqrt{2}}|\downarrow\rangle_A\big(|\uparrow\rangle_B + |\downarrow\rangle_B\big) \\ &\rightarrow \frac{1}{\sqrt{2}}|\downarrow\rangle_A\big(|\uparrow\rangle_B + |\downarrow\rangle_B\big) = |\downarrow\rangle_A |\leftarrow\rangle_B, \\ |\downarrow\rangle_A |\rightarrow\rangle_B &= \frac{1}{\sqrt{2}}|\downarrow\rangle_A\big(|\uparrow\rangle_B - |\downarrow\rangle_B\big) \\ &\rightarrow \frac{1}{\sqrt{2}}|\downarrow\rangle_A\big(|\uparrow\rangle_B - |\downarrow\rangle_B\big) = |\downarrow\rangle_A |\rightarrow\rangle_B, \end{aligned} \qquad (1.113)$$

corresponding to the CPHASE gate in Equation 1.111, and

$$\begin{aligned} |\uparrow\rangle_A |\leftarrow\rangle_B &= \frac{1}{\sqrt{2}}|\uparrow\rangle_A\big(|\uparrow\rangle_B + |\downarrow\rangle_B\big) \\ &\rightarrow \frac{1}{\sqrt{2}}|\uparrow\rangle_A\big(|\uparrow\rangle_B + |\downarrow\rangle_B\big) = -|\uparrow\rangle_A |\leftarrow\rangle_B, \\ |\uparrow\rangle_A |\rightarrow\rangle_B &= \frac{1}{\sqrt{2}}|\uparrow\rangle_A\big(|\uparrow\rangle_B - |\downarrow\rangle_B\big) \\ &\rightarrow \frac{1}{\sqrt{2}}|\uparrow\rangle_A\big(|\uparrow\rangle_B - |\downarrow\rangle_B\big) = |\uparrow\rangle_A |\rightarrow\rangle_B, \\ |\downarrow\rangle_A |\leftarrow\rangle_B &= \frac{1}{\sqrt{2}}|\downarrow\rangle_A\big(|\uparrow\rangle_B + |\downarrow\rangle_B\big) \\ &\rightarrow \frac{1}{\sqrt{2}}|\downarrow\rangle_A\big(|\uparrow\rangle_B - |\downarrow\rangle_B\big) = |\downarrow\rangle_A |\rightarrow\rangle_B, \\ |\downarrow\rangle_A |\rightarrow\rangle_B &= \frac{1}{\sqrt{2}}|\downarrow\rangle_A\big(|\uparrow\rangle_B - |\downarrow\rangle_B\big) \\ &\rightarrow \frac{1}{\sqrt{2}}|\downarrow\rangle_A\big(|\uparrow\rangle_B + |\downarrow\rangle_B\big) = |\downarrow\rangle_A |\leftarrow\rangle_B, \end{aligned} \qquad (1.114)$$

corresponding to the CPHASE gate in Equation 1.112.

For universal quantum computing, the CPHASE gate described above and single-qubit operations are sufficient (Barenco et al., 1995; DiVincenzo, 1995). The single-qubit operations on Alice's and Bob's spins can be implemented by means of the optical Stark effect (Gupta et al., 2001). In contrast to Imamoglu et al.'s two-qubit gate (Imamoglu et al., 1999), which relies on a shared mode volume for the exchange of a single virtual photon between the two qubits, this scheme requires only local mode volumes for each qubit separately and is therefore fully scalable to an arbitrary number of qubits. Since the strong interaction between a quantum dot and a cavity inside a photonic crystal has already been experimentally observed (Yoshie et al., 2004), it would be possible to build a fully scalable quantum network inside a photonic crystal hosting qubits, which can be for example spins of excess electrons in semiconductor quantum dots or N-V center states in diamond.

1.10 Concluding Remarks

Quantum computers have excited many researchers because they would perform a kind of parallel processing that would be extremely effective for certain tasks, such as searching databases and factoring large numbers. Also, quantum computers can be used to simulate or model quantum systems and could bring huge advances in physics, chemistry, and nanotechnology. Recently, diamond has become very attractive for solid state electronics. Pure diamond is an electrical insulator, but on doping, it can become a semiconductor. The particular impurity that researchers are interested in is the nitrogen-vacancy (N-V) center. The N-V center has a number of properties and characteristics that make it very promising to build a quantum computer (Awschalom et al., 2007). The N-V center electrons have a spin state that is extremely stable against environmental disturbances. One of the most exciting aspects of the N-V center is that it exhibits quantum behavior even at room temperature; this characteristic is very important because it makes this kind of systems easy to study and easy to turn into a practical technology. Other important aspects of N-V centers come from the fact that they have weak spin–orbit interaction and dipole–dipole interaction, which make the spin state very stable and it can be used to encode quantum information even at room temperatures. The decoherence time of the N-V center spin is about 1 ns and the operation time is about 10 ns, therefore $R = \tau_d/\tau_s = 100,000$ operations that can be performed in the millisecond lifetime of the spin quantum system. This rate of decay is well below the threshold and better than any other system of solid-state qubits to date.

Acknowledgments

We acknowledge support from NSF-ECCS 0725514, the DARPA/MTO Young Faculty Award HR0011-08-1-0059, NSF-ECCS 0901784, and AFOSR FA9550-09-1-0450. We would also like to thank Sergio Tafur for his comments on improving the manuscript.

References

Adamowski, J., S. Bednarek, and B. Szafram. Quantum computing with quantum dots. *Schedae Inform.*, 14:95–111, 2005.

Ahn, J., Weinacht, T.C., and P.H. Bucksbaum. Information storage and retrieval through quantum phase. *Science*, 287: 463–465, 2000.

Awschalom, D.D., R. Epstein, and R. Hanson. The diamond age of spintronics. *Sci. Am.*, 297:84–91, 2007.

Barenco, A., D. Deutsch, A. Ekert, and R. Josza. Conditional quantum dynamics and logic gates. *Phys. Rev. Lett.*, 74:4083, 1995.

Bennett, C.H., G. Brassard, C. Crépeau, R. Jozsa, A. Peres, and W.K. Wootters. Teleporting an unknown quantum state via dual classical and Einstein–Podolsky–Rosen channels. *Phys. Rev. Lett.*, 70:1895–1899, 1993.

Bennett, C.H., D.P. DiVincenzo, J.A. Smolin, and W.K. Wootters. Mixed-state entanglement and quantum error correction. *Phys. Rev. A*, 54:3824–3851, 1996.

Berezovsky, J., M.H. Mikkelsen, O. Gywat, N.G. Stoltz, L.A. Coldren, and D.D. Awschalom. Nondestructive optical measurements of a single electron spin in a quantum dot. *Science*, 314:1916, 2006.

Bouwmeester, D., J.-W. Pan, K. Mattle, M. Eibl, H. Weinfurter, and A. Zeilinger. Experimental quantum teleportation. *Nature*, 390:575, 1997.

Cerf, N.J., C. Adami, and P.G. Kwiat. Optical simulation of quantum logic. *Phys. Rev. A*, 57:R1477, 1998.

Deutsch, D. Quantum theory, the Church–Turing principle, and the universal quantum computer. *Proc. R. Soc. Lond. A*, 400:97–117, 1985.

DiVincenzo, D.P. Two-bit quantum gates are universal for quantum computation. *Phys. Rev. A*, 51:1015–1022, 1995.

DiVincenzo, D.P. The physical implementation of quantum computation, in *Scalable Quantum Computers*, S.L. Braunstein and H.K. Lo (eds.), Wiley, Berlin, Germany, 2001.

Feynman, R.P. Simulating physics with computers. *Int. J. Theor. Phys.*, 21:467–488, 1982.

Fiorentino, M. and F.N.C. Wong. Deterministic controlled-not gate for single-photon two-qubit quantum logic. *Phys. Rev. Lett.*, 93:070502, 2004.

Friedman, J.R., M.P. Sarachik, J. Tejada, and R. Ziolo. Macroscopic measurement of resonant magnetization tunneling in high-spin molecules. *Phys. Rev. Lett.*, 76:3830–3833, 1996.

Gasparoni, S., J.-W. Pan, P. Walther, T. Rudolph, and A. Zeilinger. Realization of a photonic controlled-not gate sufficient for quantum computation. *Phys. Rev. Lett.*, 93:020504, 2004.

Gottesman, D. and I.L. Chuang. Demonstrating the viability of universal quantum computation using teleportation and single-qubit operations. *Nature*, 402:390, 1999.

Greenberger, D.M., M.A. Horne, and A. Zeilinger. in *Bell's Theorem, Quantum Theory, and Conceptions of the Universe*. Kluwer, Dordrecht, the Netherlands, p. 73, 1989.

Greenberger, D.M., M.A. Horne, A. Shimony, and A. Zeilinger. Bell theorem without inequalities. *Am. J. Phys.*, 58:1131, 1990.

Grover, L.K. Quantum mechanics helps in searching for a needle in a haystack. *Phys. Rev. Lett.*, 79:325–328, 1997.

Gupta, J.A., R. Knobel, N. Samarth, and D.D. Awschalom. Ultrafast manipulation of electron spin coherence. *Science*, 292:2458, 2001.

Hu, C.Y., A. Young, J.L. O'Brien, W.J. Munro, and J.G. Rarity. Giant optical Faraday rotation induced by a single-electron spin in a quantum dot: Applications to entangling remote spins via a single photon. *Phys. Rev. B*, 78:085307, 2008.

Imamoglu, A., D.D. Awschalom, G. Burkard, D.P. DiVincenzo, D. Loss, M. Sherwin, and A. Small. Quantum information processing using quantum dot spins and cavity QED. *Phys. Rev. Lett.*, 83:4204, 1999.

Kasuya, T. A theory of metallic ferro- and antiferromagnetism on Zener's model, *Prog. Theor. Phys.* 16:45, 1956.

Kikkawa, J.M. and D.D. Awschalom. Resonant spin amplification in n-type gas. *Phys. Rev. Lett.*, 80:4313, 1998.

Kikkawa, J.M. and D.D. Awschalom. Lateral drag of spin coherence in gallium arsenide. *Nature*, 397:139, 1999.

Kikkawa, J.M., I.P. Smorchkova, N. Samarth, and D.D. Awschalom. Room-temperature spin memory in two-dimensional electron gases. *Science*, 277:1284, 1997.

Knill, E., R. Laflamme, and G.J. Milburn. A scheme for efficient quantum computation with linear optics. *Nature*, 409:46, 2001.

Leuenberger, M.N. Fault-tolerant quantum computing with coded spins using the conditional Faraday rotation in quantum dots. *Phys. Rev. B*, 73:075312–8, 2006.

Leuenberger, M.N. and D. Loss. Spin relaxation in Mn12-acetate. *EuroPhys. Lett.*, 46:692–698, 1999.

Leuenberger, M.N. and D. Loss. Spin tunneling and phonon-assisted relaxation in Mn12-acetate. *Phys. Rev. B*, 61:1286–1302, 2000a.

Leuenberger, M.N. and D. Loss. Incoherent Zener tunneling and its application to molecular magnets. *Phys. Rev. B*, 61:12200–12203, 2000b.

Leuenberger, M.N. and D. Loss. Quantum computing in molecular magnets. *Nature*, 410:789–793, 2001.

Leuenberger, M.N., D. Loss, M. Poggio, and D.D. Awschalom. Quantum information processing with large nuclear spins in gas semiconductors. *Phys. Rev. Lett.*, 89:207601–4, 2002.

Leuenberger, M.N., F. Meier, and D. Loss. Quantum spin dynamics in molecular magnets. *Monatshefte Chem.*, 134:217–233, 2003.

Leuenberger, M.N., M.E. Flatté, and D.D. Awschalom. Teleportation of electronic many-qubit states encoded in the electron spin of quantum dots via single photons. *Phys. Rev. Lett.*, 94:107401–4, 2005a.

Leuenberger, M.N., M.E. Flatté, and D.D. Awschalom. Teleportation of electronic many-qubit states encoded in the electron spin of quantum dots via single photons. *Phys. Rev. Lett.*, 94:107401–4, 2005b.

Loss, D. and D.P. DiVincenzo. Quantum computation with quantum dots. *Phys. Rev. A*, 57:120–126, 1998.

Nemoto, K. and W.J. Munro. Nearly deterministic linear optical controlled-not gate. *Phys. Rev. Lett.*, 93:250502, 2004.

O'Brien, J.L., G.J. Pryde, A.G. White, T.C. Ralph, and D. Branning. Demonstration of an all-optical quantum controlled-not gate. *Nature*, 426:264, 2003.

Piermarocchi, C., P. Chen, and L.J. Sham. Optical RKKY interaction between charged semiconductor quantum dots. *Phys. Rev. Lett.*, 89:167402, 2002.

Pittman, T.B., B.C. Jacobs, and J.D. Franson. Probabilistic quantum logic operations using polarizing beam splitters. *Phys. Rev. A*, 64:062311, 2001.

Pittman, T.B., B.C. Jacobs, and J.D. Franson. Demonstration of feed-forward control for linear optics quantum computation. *Phys. Rev. A*, 66:052305, 2002a.

Pittman, T.B., B.C. Jacobs, and J.D. Franson. Demonstration of nondeterministic quantum logic operations using linear optical elements. *Phys. Rev. Lett.*, 88:257902, 2002b.

Pittman, T.B., M.J. Fitch, B.C. Jacobs, and J.D. Franson. Experimental controlled-not logic gate for single photons in the coincidence basis. *Phys. Rev. A*, 68:032316, 2003.

Raussendorf, R. and H.J. Briegel. A one-way quantum computer. *Phys. Rev. Lett.*, 86:5188–5191, 2001.

Raussendorf, R., D.E. Browne, and H.J. Briegel. Measurement-based quantum computation on cluster states. *Phys. Rev. A*, 68:022312, 2003.

Ruderman, M.A. and C. Kittel. Indirect exchange coupling of nuclear magnetic moments by conduction electrons, *Phys. Rev.* 96:99, 1954.

Sangregorio, C., T. Ohm, C. Paulsen, R. Sessoli, and D. Gatteschi. Quantum tunneling of the magnetization in an iron cluster nanomagnet. *Phys. Rev. Lett.*, 78:4645–4648, 1997.

Seigneur, H.P., M.N. Leuenberger, and W.V. Schoenfeld. Single-photon Mach–Zehnder interferometer for quantum networks based on the single-photon Faraday effect. *J. Appl. Phys.*, 104:014307–13, 2008.

Shor, P. Scheme for reducing decoherence in quantum computer memory. *Phys. Rev. A*, 52:2493–2496, 1995.

Shor, P. Polynomial-time algorithms for prime factorization and discrete logarithms on a quantum computer. *SIAM J. Comput.*, 26:1484–1509, 1997.

Silberfarb, A. and I.H. Deutsch. Entanglement generated between a single atom and a laser pulse. *Phys. Rev. A*, 69:042308–8, 2004.

Thiaville, A. and J. Miltat. Small is beautiful. *Science*, 284:1939–1940, 1999.

Thomas, L., F. Lionti, R. Ballou, D. Gatteschi, R. Sessoli, and B. Barbara. Macroscopic quantum tunnelling of magnetization in a single crystal of nanomagnets. *Nature*, 383: 145–147, 1996.

Wernsdorfer, W., R. Sessoli, A. Caneshi, D. Gatteschi, and A. Cornia. Nonadiabatic Landau–Zener tunneling in Fe8 molecular nanomagnets. *EuroPhys. Lett.*, 50:552–558, 2000.

Yoshie, T., A. Scherer, J. Hendrickson, G. Khitrova, H.M. Gibbs, G. Rupper, C. Ell, O.B. Shchekin, and D.G. Deppe. Vacuum Rabi splitting with a single quantum dot in a photonic crystal nanocavity. *Nature*, 432:200, 2004.

Yosida, K. Magnetic properties of Cu-Mn alloys, *Phys. Rev.* 106:893, 1957.

Zhao, Z., A.-N. Zhang, Y.-A. Chen, H. Zhang, J.-F. Du, T. Yang, and J.-W. Pan. Experimental demonstration of a nondestructive controlled-not quantum gate for two independent photon qubits. *Phys. Rev. Lett.*, 94:030501, 2005.

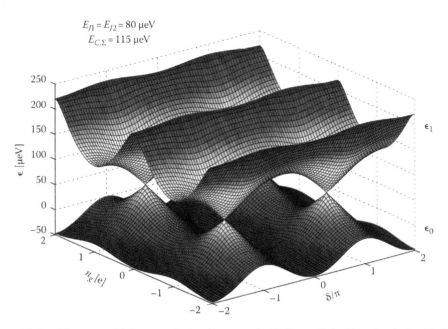

FIGURE 16.4 The ground (ϵ_0) and first excited (ϵ_1) states calculated by numerically diagonalizing Equation 16.11 with the energies indicated in the figure.

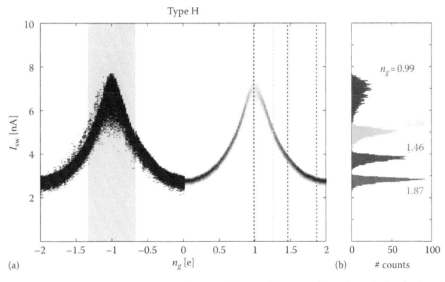

FIGURE 16.11 (a) Type H I_{sw} histograms versus n_g. Histogram height is displayed in grayscale on the right-hand side, whereas all counts are displayed equally on the left-hand side. As in Figure 16.7, the gray box in (a) denotes regions where the island potential is trap like for quasiparticles. (b) Selected histograms corresponding to several gate voltages. Device parameters: $\Delta_i = 246\,\mu eV$, $\Delta_\ell = 205\,\mu eV$, $E_C \simeq 115\,\mu eV$, and $E_{J1} = E_{J2} = 82\,\mu eV$.

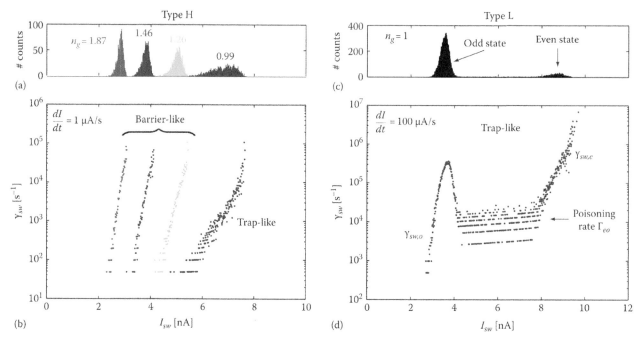

FIGURE 16.13 (a,c) Switching current (I_{sw}) histograms and (b,d) derived switching/escape rates for a type H (barrier-like) and type L (trap-like) CPTs. For the type L device, the quasiparticle trapping behavior is evident in the bimodal I_{sw} distribution. In this case, the poisoning rate Γ_{eo} can be read directly from the derived escape rate in (d) as shown. Although the type H device is barrier-like for most n_g, it still looks like a trap near $n_g = 1$ (see Figure 16.7). This is apparent in the "curvy" structure of the escape curve for $n_g = 0.99$ as compared with the escape rates at other n_g in (b).

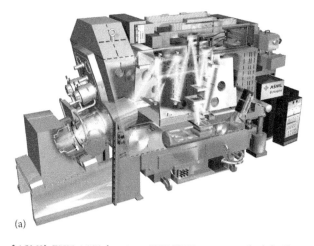

FIGURE 20.6 (a) Artist illustration of ASML's EUV ADT showing a DPP EUV source on the left, illumination optics in the middle, and mask, projection optics, and wafer on the right. (Courtesy of ASML, Veldhoven, the Netherlands.)

FIGURE 20.18 Artist illustration of ASML NXE: 3100 preproduction EUV exposure tool showing the LPP EUV source on the right, the illuminator optics in the center, and the mask and the projection optics on the right. (Courtesy of ASML, Veldhoven, the Netherlands.)

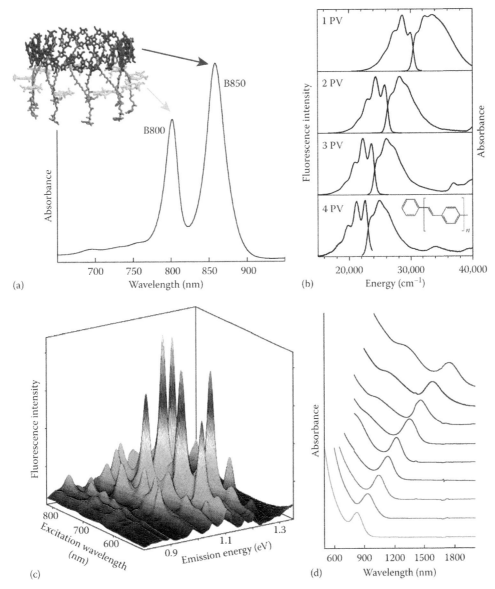

FIGURE 23.1 Excitons and structural size variations on the nanometer length scale. (a) The photosynthetic antenna of purple bacteria, LH2, is an example of a molecular exciton. The absorption spectrum clearly shows the dramatic distinction between the B800 absorption band, arising from essentially "monomeric" bacteriochlorophyll-a (Bchl) molecules, and the redshifted B850 band that is attributed to the optically bright lower exciton states of the 18 electronically coupled Bchl molecules. (b) The size-scaling of polyene properties, for example, oligophenylenevinylene oligomers, derives from the size-limited delocalization of the molecular orbitals. However, as the length of the chains increases, disorder in the chain conformation impacts the picture for exciton dynamics. Absorption and fluorescence spectra are shown as a function of the number of repeat units. (c) SWCNT size and "wrapping" determine the exciton energies. Samples contain many different kinds of tubes, therefore optical spectra are markedly inhomogeneously broadened. By scanning excitation wavelengths and recording a map of fluorescence spectra, the emission bands from various different CNTs can be discerned, as shown. (Courtesy of Dr. M. Jones). (d) Rather than thinking in terms of delocalizing the wavefunction of a semiconductor through interactions between the unit cells, the small size of the nanocrystal confines the exciton relative to the bulk. Size-dependent absorption spectra of PbS quantum dots are shown. (Adapted from Scholes, G. D. and Rumbles, G., *Nat. Mater.*, 5, 683, 2006.)

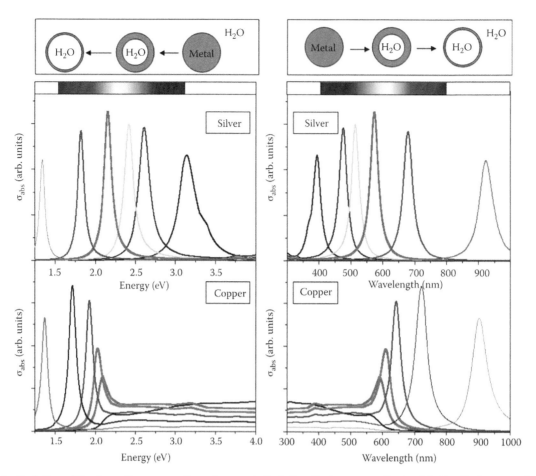

FIGURE 24.9 Evolution of the absorption cross section of a nanoshell of silver (top) or copper (bottom) in the dipolar approximation versus energy (left) or versus wavelength (right) for various thicknesses of the shell. The core is filled with water and the external medium is also water ($n = 1.33$). The dielectric functions of copper and silver have been extracted from Palik (1985–1991). The correspondence between the energy in eV and the wavelength in nm is E (eV) = $1239.85/\lambda$ (nm). The total radius of the cluster is always 15 nm and the thickness takes the following values: $e = R - R_c$ = 5; 4; 3; 2; 1 nm corresponding to ratios between the shell thickness and the total cluster radius: (e/R) = 0.33; 0.27; 0.2; 0.13; 0.07. The spectra in black correspond to the fully homogeneous cluster.

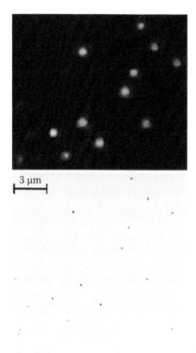

FIGURE 24.15 (Top) Color image of a typical sample of silver nanoparticles as viewed under the dark-field microscope. The brightness of the particles increases from blue to red due to both the intrinsic optical scattering cross section and the spectral output of the light source (the red particle is overexposed). This image is correlated to its electron microscopy image (Bottom). (Reprinted from Mock, J.J. et al., *J. Chem. Phys.*, 116, 6755, 2002. With permission.)

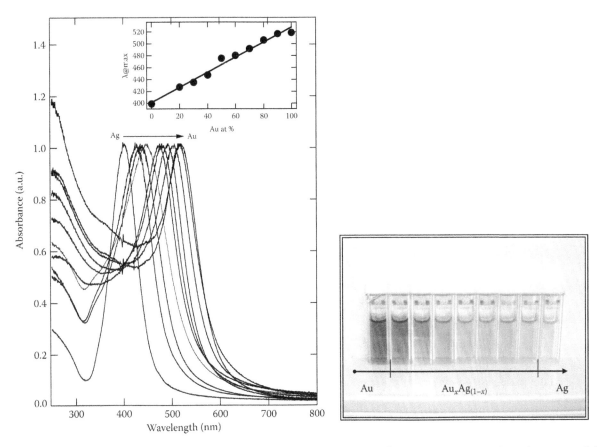

FIGURE 24.16 (Left) Normalized UV–visible spectra of Au–Ag alloy nanoparticles with varying composition. (Insert) Location of the SPR maximum as a function of the gold content. (Right) Corresponding solutions whose colors vary from the red (pure gold nanoparticles) to the yellow (pure silver nanoparticle). (From Russier-Antoine, I. et al., *Phys. Rev. B*, 78, 35436, 2008. With permission.)

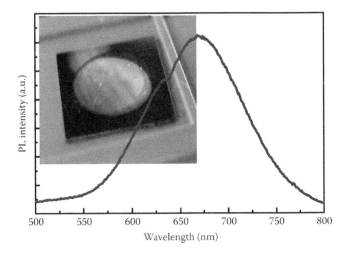

FIGURE 25.2 The PL spectrum from porous silicon with a maximum PL at a wavelength of about 675 nm. The photograph at the inset demonstrates the red color of the emitted PL from a circular layer of porous silicon (the sample has been illuminated with a UV lamp).

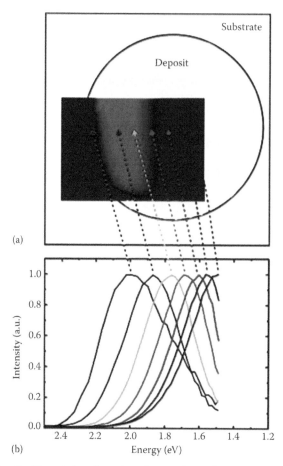

(a)

(b)

FIGURE 25.6 (a) A photograph showing the PL variation along the substrate from silicon nanocrystals deposited by the laser pyrolysis technique. (b) The normalized PL spectra from different positions along the substrate. (Reprinted from Ledoux, G. et al., *Appl. Phys. Lett.*, 80, 4834, 2002. With permission.)

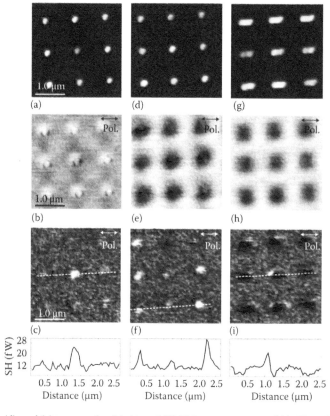

FIGURE 28.10 Nanoparticles: (a), (d), and (g) topography; (b), (e), and (h) FW transmission; and (c), (f), and (i) SH emission SNOM images with corresponding cross sections along the dashed lines, from the raw data. Incident light is polarized parallel to the major axis. The particle major axis lengths are 100 nm (a)–(c), 150 nm (d)–(f), and 400 nm (g)–(i). Image size: 3 × 3 μm². (Reprinted from Zavelani-Rossi, M. et al., *Appl. Phys. Lett.*, 92, 093119, 2008. With permission.)

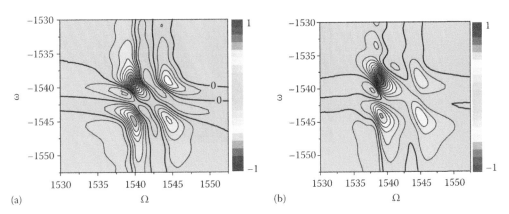

FIGURE 29.3 The imaginary part of the two-dimensional Fourier spectrum, $P(\Omega, \omega)$. (a) The spectrum calculated using Equation 29.92. (b) The experimental results of Zhang et al. (2005).

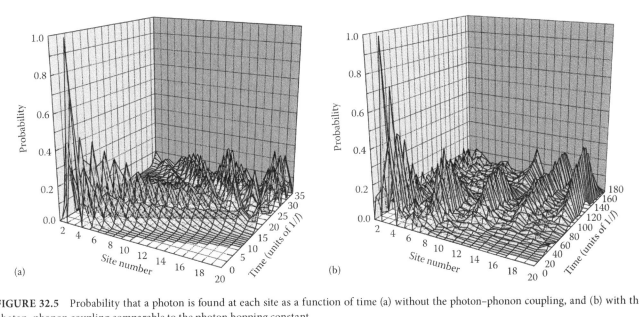

FIGURE 32.5 Probability that a photon is found at each site as a function of time (a) without the photon–phonon coupling, and (b) with the photon–phonon coupling comparable to the photon hopping constant.

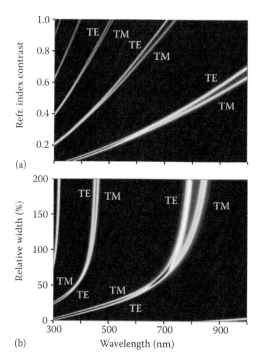

(a)

(b)

FIGURE 37.14 Calculated spectral positions of the substrate modes as a function of (a) the refractive index contrast $\Delta n = n_2 - n_3$ and (b) the relative thickness of the waveguide core compared to the sample 5×10^{17} Si cm^{-2}. Gray scale indicates intensity increasing from black up to white for the highest intensity. Several orders of modes are seen starting from the first one in infrared region.

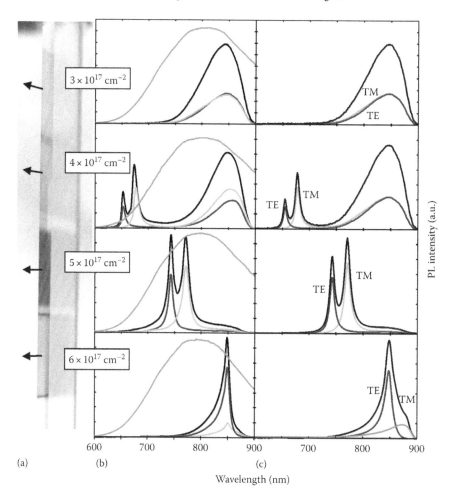

(a)　　　(b)　　　(c)

FIGURE 37.16 (a) Photograph of the edge of a set of Si$^+$ ion implanted layers with direction of PL indicated by arrows, the edge is on the left. (b) Measured PL from samples implanted to different Si ion fluences in standard (the broadest curves) and waveguiding geometry (black lines, the slightly lighter gray lines stand for TE and TM resolved polarizations). (c) Theoretically calculated PL spectra. We note that these results were obtained on different set of samples than in Figure 37.10. The mode positions are not exactly the same for samples with identical implantation dose because the annealing conditions were slightly different. Therefore, the refractive index profiles are not identical. (Adapted from Pelant, I. et al., *Appl. Phys. B*, 83, 87, 2006.)

2

Nanomemories Using Self-Organized Quantum Dots

Martin Geller
University of Duisburg-Essen

Andreas Marent
Technische Universität Berlin

Dieter Bimberg
Technische Universität Berlin

2.1 Introduction

One of the fundamental achievements of today's information society is providing storage and processing of ever-increasing amounts of data that is based on basic research in physics in combination with fundamental technology developments over the last four decades. As no low-cost universal memory that is suited for all kinds of applications exists, digital data storage is separated mainly into three different lines:

1. The digital versatile disk (DVD) and the compact disk (CD) are nonvolatile storage media based on an *optical* write and read-out process using laser light at wavelengths of 650 and 780 nm, respectively. The newest representative of optical data storage is the Blu-ray disk, using a laser wavelength at 405 nm with a maximum capacity of up to 50 GB disk^{-1}.
2. The hard-disk drive in a personal computer is a nonvolatile mass storage device with a capacity of up to 1000 GB in production year 2008 that uses the *magnetization* of rapidly rotating platters.

3. *Semiconductor* memories are the third line of digital storage media and will be the focus of this chapter with a perspective of future nanomemories.

Semiconductor memories can be divided into two groups: volatile memories, like the dynamic random access memory (DRAM), and nonvolatile memories, the so-called Flash. The DRAM presents the main working memory in a personal computer. Flash memories are found, for example, in mp3 players, cell phones, and memory sticks; in automobiles and microwave ovens; and have started to replace the hard-disk drive in notebooks as well as the DVD and CD as easily portable high-capacity nonvolatile memories. Driven by the increasing demand for such portable electronic applications, where nonvolatile data storage with low power consumption is needed, the market for Flash memories is growing rapidly and is replacing the DRAM as the market driver.

Up to now, the semiconductor memory industry improved the storage density and performance while simultaneously reducing the cost per information bit just by scaling down the feature size. This leads to an exponentially growing number of components on a memory chip as predicted by Moore in 1965 (Moore 1965) known nowadays as "Moore's Law." Contrary to all predictions, Moore's Law has held remarkably well over the last decades. The feature size for Flash memories has reached 45 nm in 2008. The International Roadmap for Semiconductors (ITRS 2008) predicts a further shrinkage down to 14 nm in 2020. The problems encountered during this shrinking process were mainly solved by developing new materials, for example, for isolation and interconnects, and for new types of cell structures. Upon further size reduction, quantum mechanics will dominate at least some of the physical properties. In addition, the amount of technological difficulties to realize such structures increases enormously. Therefore, a considerable effort is devoted to the search for alternative memory technologies that are even based on completely new non-semiconducting materials. Phase change access memories (PCRAMs), magnetic random access memories (MRAMs), and ferroelectric RAM (FeRAMs) are just three alternative memory concepts that are explored for having the potential to replace today's DRAM and/or Flash memory.

One completely different technology is the use of self-organized nanomaterials in future nanoelectronic devices. Especially self-organized quantum dots (QDs) based on III–V materials (e.g., GaAs, InAs, InP, etc.) provide a number of important advantages for new generations of nanomemories. Billions of self-organized QDs can be formed simultaneously and fast in a single technological step, allowing for massive parallel production in a bottom-up approach, and offering an elegant method to create huge ensembles of electronic traps without lithography. They can store just a few or even single charge carriers with a retention time depending on the material combination—potentially up to many years at room temperature. With an area density of up to 10^{11} cm^{-2}, an enormous storage density in the order of 1 TBit per square inch could be possible, if each single QD would represent one information bit and could be addressed individually. The carrier capture process into the QDs is of the order of pico to sub-picoseconds, an important prerequisite for a very fast write time in such memories. The use of self-organized QDs could thus lead to novel nonvolatile memories with high storage density combined with a fast read/write access time.

Using self-organized QDs for future semiconductor memory applications is discussed in this chapter. Section 2.2 gives a brief overview of the main semiconductor memories, DRAM and Flash, while the following Section 2.3 concentrates on three nonconventional memories that may replace Flash or DRAM in the future. Section 2.5 presents an alternative nanomemory concept that is based on III–V materials, especially self-organized QDs. By using capacitance spectroscopy, introduced in Section 2.6, the carrier storage time in different QD systems is studied in Section 2.7 and a storage time of seconds at room temperature is demonstrated. Finally, the chapter will show results on fast write times in QD-based nanomemories in Section 2.8 and will close with a summary and outlook in Section 2.9.

2.2 Conventional Semiconductor Memories

Presently the semiconductor memory industry focuses essentially on two memory types: the DRAM (Mandelman et al. 2002, Waser 2003) and the Flash (Geppert 2003). Both memory concepts have their advantages and disadvantages in speed, endurance, storage time, and cost. A memory concept that adds the advantages of a DRAM to a Flash would combine high storage density, fast read/write access, long data storage time, and good endurance with low production cost. In addition, for portable applications like mobile phones and mp3 players, low power consumption is demanded. This section describes the two conventional semiconductor memories, while Section 2.3 will focus on nonconventional alternatives that have the potential to replace DRAM and/or Flash.

2.2.1 Dynamic Random Access Memory

Since the invention of the DRAM in the late 1960s (Dennard 1968), its cell structure mainly stayed the same, consisting of a transistor and a capacitor (cf. Figure 2.1a). The capacitor stores the information by means of electric charge. The charge state is defined by the voltage level on the capacitor. The stored charge

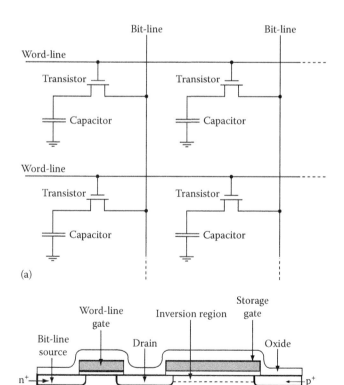

FIGURE 2.1 (a) Schematic picture of an array of DRAM cells, where the capacitors act as the storage units. Each cell can be addressed individually by a matrix of word- and bit-lines; it is the so-called random-access architecture. (b) Cell layout for a planar DRAM cell.

disappears typically in a few milliseconds mainly due to leakage and recombination currents. A DRAM is volatile and requires periodic "refreshing" of the stored charge. After any read process, the information has to be rewritten into the DRAM cell. A chip on the integrated circuit controls the refresh rewrite process automatically and this continuous read and write operation has led to the name "Dynamic."

Figure 2.1b shows schematically the layout of a simple planar DRAM cell (Sze 2002). The storage capacitor uses the inversion channel region as one plate, the storage gate as the other plate, and the gate oxide as the dielectric. The access transistor is a metal oxide semiconductor field effect transistor (MOSFET) with a source, drain, and a gate contact. The drain contact is connected to the storage capacitor. The source contact is connected to the bit-line and the gate contact to the word-line. Via word- and bit-line, a single cell can be addressed in an organized matrix of DRAM cells (Figure 2.1a). This so-called "random access" provides short access times independent of the location of the data (Waser 2003). For a write/read operation, a voltage pulse is passed through the selected bit- and word-line. Only at the crossing point of a given DRAM cell, the access transistor switches to "open" and the capacitor is charged or discharged for a write or read event, respectively.

DRAM cells provide fast read, write, and erase access times below 20 ns in combination with a very good endurance of more than 10^{15} write/erase cycles. The endurance is defined as the minimum number of possible write/erase operations until the memory cell is destroyed. DRAM memory cells draw power continuously due to the refresh process and the information is lost after switching off the computer. Another disadvantage is the relatively large number of electrons, presently in the order of 10^5, needed to store one information bit, leading additionally to permanent high power consumption. Both, the volatility and the power consumption make DRAMs unsuitable for mobile applications.

The goal to shrink the feature size, based on the assumption of scalability, of a DRAM (Kim et al. 1998) down to 14 nm by 2020 is a major challenge. It is speculated that the leakage currents of the capacitor might inhibit the scalability very soon. A fixed capacitance of the order of 50 pF is needed to maintain a sufficiently high voltage signal during the read-out process. Shrinking the area of the capacitor will decrease its capacitance,

the number of stored electrons per DRAM cell, and the amplitude of the read-out signal. Downscaling increases leakage currents due to quantum-mechanical tunneling through thinner dielectrics (Frank et al. 2001). Therefore, there is a search for novel dielectrics.

2.2.2 Nonvolatile Semiconductor Memories (Flash)

Nonvolatile memories (NVMs) can retain their data for typically more than 10 years without power consumption. The most important nonvolatile memory is the Flash-EEPROM (electrically erasable and programmable read-only memory); in short just Flash (Geppert 2003, Lai 2008). A Flash offers the possibility of repeated electrical read, write, and erase processes. The Flash memory market is the fastest growing memory market today, since it is the ideal memory device for portable applications. In a mobile phone, it holds the instructions and data needed to send and receive calls, or stores phone numbers. But not only portable applications are ideal for Flash. In each computer, a Flash chip holds the data on how to boot up. Other electronic products of all types, from microwaves ovens to industrial machines, store their operating instructions in Flash memories.

Flash is based on a floating-gate structure (Pavan et al. 1997, Sze 1999), where the charge carriers are trapped inside a polysilicon floating gate embedded between two SiO_2 barriers (Figure 2.2a). The SiO_2 barriers having a height of ~3.2 eV and an average thickness of 10 nm guarantee a storage time of more than 10 years at room temperature. However, these barriers are also the origin of the two main disadvantages of a Flash cell: a slow write time (in the order of microseconds) and a poor endurance (in the order of 10^6 write/erase cycles).

The write process is realized by means of "hot-electron injection." Here a small voltage is applied between the source contact and the bit-line (drain contact) (cf. Figure 2.2b). The word-line (control gate) is set to a high positive bias of 10–20 V, the MOSFET is set to "open" and the electrons flow though the inversion channel from the source to drain and are, in addition, accelerated in the direction of the floating-gate due to the high electric field in the order of 10^7 V cm^{-1}. These "hot-electrons" have sufficient kinetic energy to reach the floating gate over one of the SiO_2 barriers, but destroy this barrier step by step by

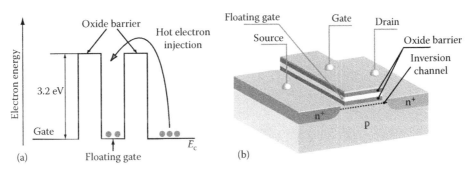

FIGURE 2.2 (a) Schematic band structure of a Flash memory based on Si as matrix material and SiO_2 for the oxide barriers. (b) Schematic cell layout of a floating-gate Flash memory.

Memory	Write	Read	Storage	Electrons	Endurance
DRAM	~10 ns	~10 ns	~ms	>10,000	>10^{15}
Flash	μs–ms	~20 ns	>10 years	~1,000	~10^6

FIGURE 2.3 Comparison of DRAM and Flash.

creating defects leading eventually to leakage. This is the reason for the poor endurance of a Flash memory cell. The slow write time is due to the low probability for energy relaxation of these hot-electrons into the floating gate.

The read-out of the stored information is normally done by measuring the resistance of the two-dimensional electron gas (2DEG) of the MOSFET—the inversion channel. A bias is applied at the gate contact, forming inversion between the source and the drain. Stored electrons in the floating gate reduce the conductivity of the 2DEG and a higher resistance between the source and the drain is measured.

Figure 2.3 compares the main properties of the DRAM and Flash. The Flash memory has already replaced the DRAM as a technology driver for the semiconductor memory industry (Mikolajick et al. 2007) in the year 2003. It is leading in storage density and the main innovations are coming from the Flash industry. The feature size of the Flash has decreased from 1.5 μm in the year 1990 down to 45 nm in the year 2008. Accordingly, the number of electrons decreased from about 10^5 to less than 1000 per information bit (Atwood 2004) (cf. Figure 2.4). A further shrinkage to 14 nm in the year 2020, as predicted, would lead to only 100 electrons per information bit.

Flash memory scaling beyond 32 nm feature size will become more and more difficult, mainly due to the physical limitations (Atwood 2004). For instance, during the hot-electron injection a voltage of about 4.5 V is applied between the source and the drain. Shrinking the gate length while keeping the write voltage

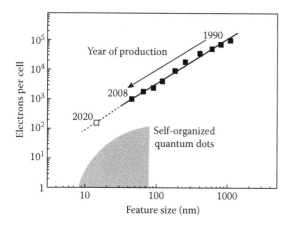

FIGURE 2.4 Electron number per Flash memory cell versus the minimum feature size. The feature size decreases from 1.5 μm in the production year 1990 down to 45 nm in the year 2008. Accordingly, the number of electrons decreased from about 10^5 to less than 1000 per information bit.

fixed at 4.5 V, leads to an increased electric field in the source–drain channel and in the end to a punch-through. Another limitation for future scaling is the capacitive coupling between different floating gates (cross talk) and the scalability of the floating gate. While the number of electrons decreases to about 200 per cell in year 2020, the main charge leakage mechanisms will remain. Dangling bonds and other defects at the Si/SiO_2 interface will dramatically decrease the storage time while scaling to smaller cell sizes.

2.3 Nonconventional Semiconductor Memories

The increasing number of difficulties and challenges for further scaling of the DRAM and Flash has led to a continuous search for alternative memory concepts. A large variety of proposed concepts using different physical phenomena to store an information bit have been reported (Burr et al. 2008). This section concentrates on three nonconventional memories, which are the most advanced ones (Geppert 2003, Burr et al. 2008): the ferroelectric RAM (FeRAM), the magnetic RAM (MRAM), and the phase-change RAM (PCRAM).

2.3.1 Ferroelectric RAM

The ferroelectric RAM (FeRAM) (Jones et al. 1995, Sheikholeslami and Gulak 2000) has almost the same structure as a DRAM cell, except for the capacitor. The dielectric inside the capacitor of a DRAM is a non-ferroelectric material like silicon dioxide. When the charge is stored on the metal plates of the capacitor, it leaks away into the silicon substrate within a few milliseconds, causing the nonvolatility of a DRAM cell. In a FeRAM, a ferroelectric film such as zirconate titanate, also known as PZT, replays the dielectric of the capacitor. This ferroelectric material has a remnant polarization that occurs when an electric field has been applied. In PZT the center atom is zirconium or titanium which can be moved by an external electric field into two different stable states, representing a "0" or "1" in a ferroelectric memory. One state is near the top face of the PZT cube, the other one is near the bottom face. Even after removal of the external electric field, these stable states store the information for more than 10 years. When an opposite electric field is applied, the dipoles flip to the opposite direction, that is, the zirconium or titanium atoms are switching into the other stable state.

The write operations of a "0" and "1" state in a FeRAM are in principle the same as in a DRAM cell. An electric pulse via a word and a bit-line switches the ferroelectric state of the PZT. To read an information bit, an electrical pulse is applied via the access transistor to the ferroelectric capacitor and a sense amplifier can measure a current pulse. The amplitude depends on the position of the zirconium or titanium atoms in the PZT cube. For instance, if the external electric field from the pulse is pointing in the same direction of the PZT, that is, the atoms are already at the top of the cube and a smaller current pulse is detected than for opposite direction.

Reading the information in a FeRAM cell destroys the data stored in its ferroelectric capacitor. After a read operation, the information has to be rewritten into the FeRAM cell, which is a major disadvantage in comparison to the Flash memory based on a floating-gate structure. However, the FeRAM is nonvolatile having a random read/write access time below 50 ns and an almost unlimited endurance (>10^{12} write/erase cycles).

2.3.2 Magnetoresistive RAM

The magnetoresistive RAM (Gallagher and Parkin 2006, Wolf et al. 2006, Burr et al. 2008) uses the giant magnetoresistance (GMR) effect that was discovered in the late 1980s by the groups of Albert Fert (Baibich et al. 1988) and Peter Grunberg (Binasch et al. 1989). Grunberg and Fert received the Nobel Prize in Physics 2007 for their discovery.

The MRAM is based on a magnetic tunnel junction (MTJ) (see Figure 2.5a), which consists of two magnetic layers separated

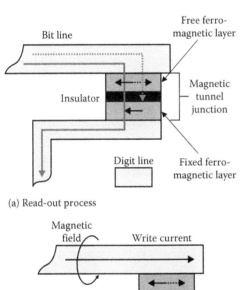

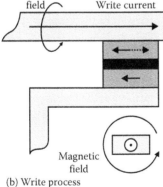

(a) Read-out process

(b) Write process

FIGURE 2.5 Schematic picture of the read (a) and write (b) process of an MRAM cell. During read out a current passes through the MTJ. If the magnetization of the two magnetic layers is parallel to each other a low resistance is measured; if the magnetization is antiparallel the resistance is high. The write process is based on current flow through both the bit and digit line. The sum of both currents is strong enough to flip the magnetic domains inside the free magnetic layer to a "1" or "0" state, depending on the current direction inside the lines.

by a thin insulating layer (like AlOx) with a thickness in the order of 1 nm. One of the ferromagnetic layers has a fixed magnetization, while the other layers can flip its magnetization by an external writing event. The read-out of the information in a MRAM cell works as follows: if the access transistor for a certain cell is turned on, a current is driven through the MTJ and the resistance of the MRAM cell is measured. If the two magnetic moments are parallel, the resistance is low, representing a "0" state. If the moments are antiparallel, the resistance is high, representing a "1" state. The difference between these two values can be up to 70%, therefore, GMR.

The physical effect that leads to this GMR effect is called tunnel magnetoresistance (TMR). The TMR effect can be understood in terms of spin polarized tunneling of the electrons. The electron spins are polarized inside one magnetic layer and if the spin is conserved during tunneling through the thin insulator layer, an initially spin up electron can only tunnel to a spin up final state. If the magnetic layers have parallel magnetic moments, a higher current is detected than for antiparallel directions.

The write mechanism can be understood by looking at Figure 2.5b. During the write mode, a current is passing through two wires: a bit-line that runs over the MTJ and a digit line that is below. The sum of both currents is strong enough to flip the magnetic domains inside the free magnetic layer to a "1" or "0" state, depending on the current direction. Both magnetic states are stable for more than 10 years; hence, an MRAM is a nonvolatile memory device with a random-access architecture and unlimited endurance. In addition, the write/read time has been demonstrated to be below 50 ns. However, scaling MRAM to smaller feature sizes is a big challenge as the write current has to remain very high (>1 mA), even if the feature size goes below 40 nm. Such large currents in very small devices might damage the structure.

2.3.3 Phase-Change RAM

The architecture of the phase-change RAM (PCRAM)—also called ovonic unified memory (OUM)—is again an array of access transistors that are connected to a word- and a bit-line. The capacitor is now replaced by a phase-change material (Geppert 2003, Burr et al. 2008). The PCRAM uses the large resistance change between a (poly) crystalline and an amorphous state in a chalcogenide glass, which is also used in rewritable CDs and DVDs. The chalcogenides used for this type of memory are alloys containing elements like selenium, tellurium, or antimony (GeSbTe, GeTe, Sb_2Te_3, etc.). The crystalline and amorphous states show a large difference of up to five orders of magnitude in the electrical resistance, representing the binary state "1" for low resistance and the state "0" for high resistance. The read-out can easily be done by a measurement of the resistance of the PCRAM cell.

Besides the chalcogenide, the PCRAM cell consists of a top and a bottom electrode that are connected to the word- and bit-line and a resistive heater below the phase-change material (see Figure 2.6). To switch the state of the cell to the crystalline

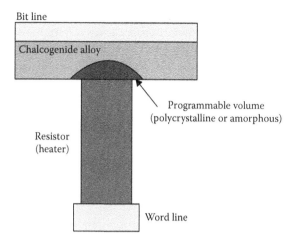

FIGURE 2.6 The PCRAM is based on a chalcogenide that can exist in two different states: in the (poly) crystalline state the resistance is low (representing a "1"), while the resistance is high in the amorphous state (representing a "0"). Switching to the amorphous state is done by heating the programmable volume up to its melting point with the resistive heater and cooling down rapidly. To make it crystalline, the heater is switched on for a short time period (~50 ns) to heat the volume above its crystallization temperature.

one, a short write pulse is applied that heats the programmable volume just below its melting point and holds it there for a certain time. This SET operation limits the write speed of the PCRAM because of the required duration to crystallize the phase-change material. The write speed is in the order of tens of nanoseconds (~50 ns), depending on the used material. In the RESET operation, the memory cell is switched to the amorphous "0" state by applying a larger electrical current (~mA) in the order of 100 ns that heats up the phase-change material just above the melting point.

The PCRAM access time is presently longer than for the MRAM. However, the write/erase time improves with scaling, as the active programmable volume is getting smaller and shorter write/erase pulses are sufficient to switch the state of the phase-change material. The read operation shows a very good scaling that gives PCRAM the highest market potential among all nonconventional memory concepts that could replace the Flash memory in the future. A big disadvantage is the endurance of a PCRAM cell (10^7–10^{12}), being above the conventional Flash cell but several orders of magnitude away from the DRAM ($>10^{15}$).

2.4 Semiconductor Nanomemories

The semiconductor memory industry has already entered the "nano-world" years ago and has introduced a feature size of 45 nm in the year 2008. To improve the device performance of a conventional Flash cell, nanomaterials have been used as replacements for the floating gate. For instance, replacing the Si floating gate with Si nanocrystals—Si particles with diameters in the order of 1–10 nm—leads to the advantage that charge leakage from any particular nanocrystal does not discharge the complete floating gate. The ultimate goal would be the usage of only one nanocrystal as one information bit to build a single electron memory.

This section will briefly review semiconductor memories based on charge traps and silicon nanocrystals in Section 2.4.1 and a single electron memory based on silicon QD in Section 2.4.2.

2.4.1 Charge Trap Memories

One of the major drawbacks is the limited number of write/erase cycles of a conventional Flash cell, the poor endurance. This will be a major problem during scaling down the feature size and simultaneously the thickness of the SiO_2 tunneling barrier. A single defect in the tunneling oxide will always destroy the entire Flash cell. The simplest way to improve the endurance of a conventional Flash memory and, hence, to extend the scalability is to replace the floating gate by a charge trapping material, schematically illustrated in Figure 2.7.

A charge trapping memory cell can be realized in different approaches. One possibility is to use a dielectric material that stores the charges in deep traps. Silicon nitride is the most established among such materials and the memory cell is often referred to as oxide–nitride–oxide, short just ONO (Lai 2008). A variation of this type of charge trapping device is the SONOS (silicon–oxide–nitride–oxide–silicon) (White et al. 2000). Figure 2.7a illustrates the floating-gate structure and the SONOS device principle. The thin silicon-nitrite (Si_3N_4) film is the storage unit where a single defect in the oxide layer will not discharge the complete memory cell. In addition, SONOS cells show a reduced cross talk and lower read/write voltages.

A second alternative to create a charge trapping memory is the usage of nanoparticles as the floating gate. The first attempts of such memory devices were made by Lambe and Jaklevic

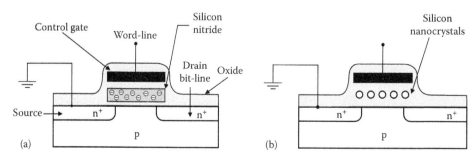

FIGURE 2.7 (a) Schematic picture of a SONOS nonvolatile charge trapping memory that consists instead of a Si floating gate of an oxide–nitride–oxide (ONO) sandwich. The nitride layer can store charges (electrons or holes) in deep traps. (b) A nanocrystal Flash based on silicon nanoparticles.

(1969) 40 years ago. They used an array of Al droplets—with a diameter in the order of 10 nm—embedded in thin oxide and observed a memory effect in capacitance measurements. Tiwari et al. (1996) embedded Si nanocrystals into a floating-gate structure to improve the Flash concept with nanomaterials. The principle of such memory devices is depicted in Figure 2.7b. The Si nanocrystals are fabricated through spontaneous decomposition during chemical vapor deposition onto the tunneling oxide barrier. These particles have a typical size in the order of about 1–10 nm. Replacing the floating gate with these Si nanocrystals has the advantage that charge leakage through defect inside the oxide layer will not completely discharge the whole memory cell. Thinner tunneling barriers can be used and an improved endurance is observed.

2.4.2 Single Electron Memories

One charge carrier inside a single nanocrystal or QD for one information bit is the ultimate goal for a memory cell. Such devices were realized with a single polysilicon QD embedded into an oxide matrix (Guo et al. 1997). This device showed a storage time of 5 s at room temperature and is schematically depicted in Figure 2.8. It was fabricated by electron beam lithography (EBL) and reactive ion etching (RIE). A source–drain channel with a width between 25 and 120 nm was created onto an oxide layer, followed by a second oxide layer and the polysilicon dot. The dot had a width of about 20 nm and is covered again by silicon oxide and polysilicon as a gate electrode. By an appropriate bias on the control gate, the single dot is charged and discharged at room temperature and the charge state—the read-out of the information—was possible via a resistance measurement of the source–drain channel.

This measurement demonstrates perfectly the possibility of single charge storage and read-out even at room temperature without conductance quantization inside the source–drain channel. However, these structures were fabricated via EBL, which is not suitable for mass production and the storage time is limited to 5 s at 300 K.

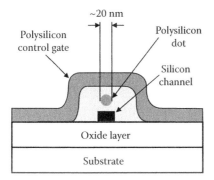

FIGURE 2.8 Schematic picture of single electron memory device based on a polysilicon QD above a silicon channel. The silicon channel is connected to a source and drain contact to measure the resistance of the channel. The charge state can be written and erased by the polysilicon control gate. (After Guo, L. et al., *Appl. Phys. Lett.*, 70(7), 850, 1997.)

2.5 A Nanomemory Based on III–V Semiconductor Quantum Dots

In this section, a memory concept based on III–V nanomaterials—self-organized QDs—that has the potential to overcome the drawbacks of the current conventional Flash and DRAM is presented (Geller et al. 2006a). Such a nanomemory cell should provide long storage times (>10 years) and good endurance (>10^{15} write/erase cycles) in combination with even better read/write access time than the DRAM (<10 ns). In addition, an ultimate memory cell should store one information bit by means of as few carriers as possible, let it be electrons or holes, to compete with the ongoing miniaturization in the semiconductor memory industry. With a lateral feature QD size of about 7–20 nm—depending on the material combination and growth conditions—to have a single QD as one information bit is a must for the future. Comparing this QD size with the predicted feature size of conventional Flash presented in Figure 2.4, one expects that the QD memory could be competitive for the next 10 years. A major challenge for such QD-based memory will be to address a single self-organized QD. However, a lot of work has been focused on QD alignment in two-dimensional arrays that would allow addressing a single self-organized QD in the future. An overview on QD alignment can be found in Kiravittaya et al. (2009).

This section concentrates on a brief introduction to III–V semiconductor hetereostructures and self-organized QDs in Sections 2.5.1 and 2.5.2, respectively. The last subsection will focus on a nanomemory concept based on these QDs.

2.5.1 III–V Semiconductor Materials

Metal organic chemical vapor deposition (MOCVD) or molecular beam epitaxy (MBE) are epitaxial growth techniques to deposit different defect-free semiconductor materials with different electronic properties (e.g., band offsets) on each other. Such a semiconductor crystal, made of more than one material, is referred to as a "heterostructure," and the interface between the two materials as "heterojunction." The electronic properties of the entire heterostructure are mainly determined by this heterojunction. Heterostructures pave the way to design devices with tailored optical or electronic properties.

The use of heterojunctions has led to the development of many device families like the double heterostructure (DHS) laser or the hetero-bipolar transistor, both invented in 1963 by Alferov and Kroemer (Alferov and Kazarinov 1963, Kroemer 1963) and honored with the Nobel prize in physics in the year 2000. Alferov built the first DHS semiconductor laser using gallium arsenide (GaAs) and aluminum arsenide (AlAs) in 1969.

By using epitaxial growth techniques, it is also easily possible to create ternary alloys, for instance $Al_xGa_{1-x}As$, which exhibit a bandgap smoothly varying between the binary alloys. Figure 2.9 shows the energy bandgap as a function of lattice constant for different III–V compound semiconductors (points) and their ternary alloys (lines) at zero temperature. The III–V compound

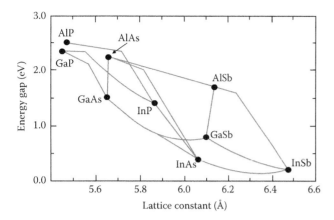

FIGURE 2.9 Energy bandgap as a function of lattice constant for different III–V compound semiconductors (points) and their ternary alloys (lines) at zero temperature. (After Vurgaftman, I. et al. *J. Appl. Phys.*, 89, 5815, 2001.)

semiconductors and their ternary alloys offer a large variety of material combinations in heterostructure devices with the possibility of a controlled band-structure engineering.

For a semiconductor heterostructure, besides the energy bandgap, the relative energetic position of the valence and conduction band at the interface is of central importance: the so-called band alignment. The band alignment and the bandgap control the band offsets in the conduction and valence band. The band offset is simply defined as the discontinuity in the band edges at the interface between the two semiconductors. Figure 2.10 shows the various types of band alignments that can arise for semiconductor heterojunctions: type I, type II staggered, and type II broken-gap. For a type I alignment, the bandgap of the narrower gap semiconductor E_g^2 lies completely within the bandgap of the wider gap E_g^1, shown in Figure 2.10a. This band alignment occurs in a large number of technically important heterojunctions, for example, InAs/GaAs, GaAs/AlGaAs, and GaSb/AlSb. For a type II staggered band alignment, the bandgaps of the two materials

show only partial overlap (Figure 2.10b). An example for a type II staggered band alignment is GaSb/GaAs, which plays an important role for memories based on self-organized QDs. A type II broken-gap alignment occurs, when the bandgaps of the two materials do not overlap at all (Figure 2.10c).

2.5.2 Self-Organized Quantum Dots

Self-organized QDs are low-dimensional heterostructures (Bimberg et al. 1998) that confine electrons and/or holes in all three spatial directions. Such a zero-dimensional system exhibits the electronic and optical properties like "artificial" atoms and offers lots of possibilities for new applications, for example, in electronic devices such as field effect transistors (Kim et al. 2000, Koike et al. 2000) as well as in optoelectronic devices such as lasers (Kirstaedter et al. 1994, Heinrichsdorff et al. 2000, Kuntz et al. 2004, Hopfer et al. 2006), detectors (Campbell et al. 1997, Chu et al. 2001), amplifiers (Lämmlin et al. 2006), and single-photon sources (Santori et al. 2002, Lochmann et al. 2006) or memory cells as described in this chapter.

The fabrication of QDs is based on self-organization effects on crystal surfaces first reported by Stranski and Krastanow (1938). On a flat substrate surface, a layer of a material with a different lattice constant is deposited, leading to a strained wetting layer. The strain energy can be reduced by forming small islands from the material of the wetting layer. These small structures with dimensions in the nanometer range can be very regular in size and shape and can confine carriers in all three dimensions (Shchukin et al. 1995, 2003). The term Stranski–Krastanow growth mode (SK-growth) was reintroduced in heteroepitaxy for the formation of islands on an initially two-dimensional layer. The formation of coherent, that is, defect free islands as a result of SK-growth is today exploited to fabricate QDs (Eaglesham and Cerullo 1990, Mo et al. 1990). Self-organized QDs have often a pyramidal or truncated pyramidal shape (Heinrichsdorff et al. 1996, Eisele et al. 1999) with typical base widths and heights in the order of 7–30 nm and 2–5 nm, respectively, depending on the growth conditions and material combinations.

The reduction of the extension of the QDs to values below the de Broglie wavelength leads to charge carrier localization in all three directions and the disappearance of the continuous $E(k)$ dispersion, called band structure. As in a real atom, the resulting energy level structure is discrete. The QD provides a confining potential of finite depth for electrons and/or holes. Figure 2.11 shows schematically two types of QDs: type I QDs (like InAs/GaAs) confine electrons and holes, while a type II QD confines either electrons or holes. The depicted type II GaSb/GaAs QD system has a localization potential for holes only and provides a barrier for the electrons in the conduction band. The localization energy represents the energy difference of the confined electron or hole states with respect to the conduction or valence band edge, equivalent to the binding energy.

The single particle states in self-organized QDs depend on two structural properties: size and shape, and the chemical composition. The influence of the QD geometry on the electron and

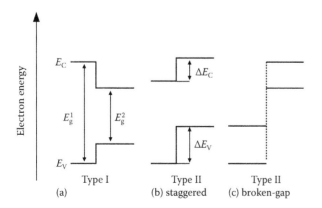

FIGURE 2.10 Types of band alignment at a heterojunction consisting of two semiconductors with different bandgap E_g^1 and E_g^2. The conduction and valence band edges have been labeled with E_C and E_V, while the conduction and valence band offsets have the label ΔE_C and ΔE_V, respectively.

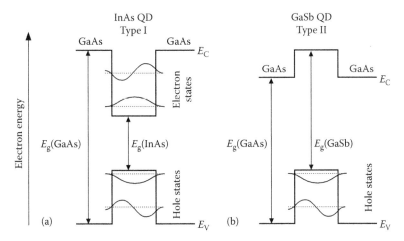

FIGURE 2.11 Schematic picture of the electron and hole states in a type I InAs/GaAs (a) and in a type II GaSb/GaAs (b) QD system. A type I QD confines electrons as well as holes, while a type II QD confines either electrons or holes. The depicted type II GaSb/GaAs QD confines only holes and has a potential barrier for electrons in the conduction band.

hole states can be understood in principle from a simple particle-in-a-box model (Grundmann et al. 1995, Stier et al. 1999). For an infinite potential well, the quantum-mechanical solution predicts energy separation and nonzero energy for the lowest level, depending on the size of the box. In contrast, the confining barrier in buried QDs is of finite height, given by the band offsets. Only a finite number of bound states exist, depending on the QD size and shape. As the QD becomes smaller, the separations between the electron and hole levels increase—quantum size effect—and all states are lifted in energy toward the valence/conduction band edge. Below a certain size, no bound electron and/or hole state exists.

Self-organized QDs can trap electrons or holes, storing these charge carriers for a certain time depending on the emission processes involved. Therefore, QDs may store information like a memory cell. Carrier emission limits the storage time in a QD-based memory. Confined charge carriers can escape by the following four mechanisms: thermal activation, tunneling, phonon-assisted tunneling, and optical activation. Phonon-assisted tunneling is a combination process of tunneling and thermal activation. The charge carrier is thermally activated by phonons to a higher energy state E_1 (a real or virtual state) and tunnels subsequent to the conduction or valence band through a barrier. The mechanisms are schematically depicted in Figure 2.12 for electron escape into the conduction band. Whether a charge state and, hence, the information bit is destroyed by one of the emission processes depends on the electric field and the temperature in comparison to the localization energy of the charge carriers E_{Loc}. If the electric field is weak and the thermal energy $k_B T$ is large, the charge carriers will be emitted due to thermal activation. If the electric field is strong and the thermal energy is small in comparison to the localization energy, the charge carriers will be released due to tunneling. Phonon-assisted tunneling will occur for a situation, were the tunneling rate from the state E_1 is equal to the thermal activation rate from the ground state E_0 to the state E_1.

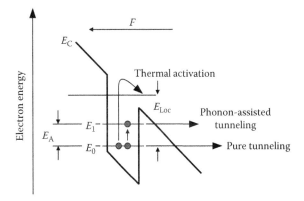

FIGURE 2.12 The possible carrier emission processes in an applied electric field: thermal activation, phonon-assisted tunneling and tunneling are shown for electron escape from a QD ground state into the conduction band. The term E_{Loc} represents the localization energy of the charge carriers in the ground state while E_A is the thermal activation energy for phonon-assisted tunneling.

To understand the limitations in storage time in a nanomemory based on self-organized QDs all above mentioned emission processes have to be taken into account. However, in the storage situation of a memory, the electric field should be sufficiently small such that tunneling and phonon-assisted tunneling can be ruled out and only pure thermal activation is of interest. The average time for a thermally activated emission process from the ground state of the QDs to the band edge (which presents an upper limit for the average storage time of the charge carriers) depends in this situation exponentially on the localization energy of the charge carriers; cf. Equation 2.27 in Section 2.6.2.1. Therefore, III–V material combinations with large band offsets that provide large localization energies are needed to build a nonvolatile nanomemory based on self-organized QDs. After a presentation of the QD-based memory concept, the next sections will concentrate on storage times and localization energies in different III–V self-organized QDs and their suitability for a nonvolatile memory.

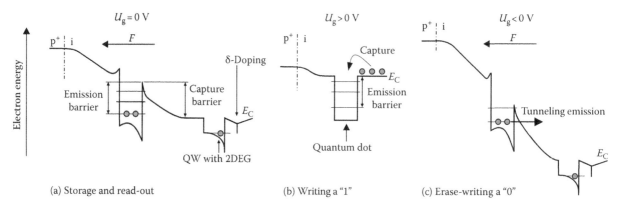

(a) Storage and read-out (b) Writing a "1" (c) Erase-writing a "0"

FIGURE 2.13 Schematic illustration of the (a) storage and read-out, (b) write, and (c) erase process in a possible future QD-based nanomemory.

2.5.3 A Memory Cell Based on Self-Organized QDs

Here, a memory concept is presented where the self-organized QDs are embedded in a p–n or p–i–n diode structure (Geller et al. 2006a). The QDs act as storage units, since they can be charged with electrons or holes representing the "0" (uncharged QDs) and "1" (charged QDs) of an information bit. An emission barrier is needed to store a "1" in such a memory concept. The barrier height—which is related to the localization energy of the charge carrier (Figure 2.13a)—can be varied by varying the material and size of the QDs and the material of the surrounding matrix (see Section 2.5.1). If the localization energy is increased, a longer storage time of the charge carriers is expected. Furthermore, a capture barrier is needed to store a "0" (Figure 2.13a). This barrier protects an empty QD cell from unwanted capture of charge carriers. In this concept, the capture barrier is realized by using the band bending of a p–i–n or p–n diode.

The major advantage in using a diode structure is the possibility to tune the height of the barriers by an external bias. Therefore, during the write process (Figure 2.13b), the capture barrier can be eliminated by a positive external bias between the gate and source contact ($V_g > 0\,V$) and the charge carriers can directly relax into the QD states. This allows to benefit from another advantage of self-organized QDs: the charge carrier relaxation time into the QD states is in the range of picoseconds at room temperature (Müller et al. 2003, Geller et al. 2006b) (see Section 2.7), enabling very fast write times in a QD-based memory. In addition, a very good endurance of 10^{15} write operations should be feasible. To erase the information, the electric field is increased within the QD layer by a negative external bias between the gate and source ($V_g < 0\,V$), such that tunneling of the charge carriers occurs (cf. Figure 2.13c).

Figure 2.14 shows schematically the device structure of such a QD-based memory, where the QDs are charged with holes to represent an information bit. The doping sequence of the p- and n-regions would be inverted for electron storage. The distance to the junction is adjusted, such that the QDs are inside the depletion region for zero bias ($V_g = 0\,V$ in Figure 2.13a). The read-out of the stored information is done by a two-dimensional electron gas (2DEG) for an electron and a two-dimensional hole gas

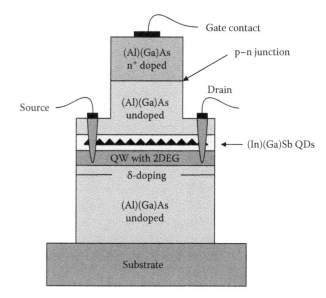

FIGURE 2.14 Schematic picture of the layer sequence for a possible QD-based memory. The QDs are located below the p–i–n junction and a 2DEG is placed below the QD layer to detect the charge state.

(2DHG) for a hole storage device. The 2DEG or 2DHG is situated 10–50 nm below the QD layer and is filled with charge carriers, provided by the additional n- or p-δ-doping, respectively. Stored charge carriers inside the QDs reduce the conductivity of the 2DEG/2DHG; hence, a higher resistance is measured between the source/drain contacts, in analogy to the read-out in a conventional Flash cell.

2.6 Capacitance Spectroscopy

To elucidate the potential of nanomaterials like self-organized QDs for future memory applications the localization energy and the carrier storage time at room temperature has to be determined for different III–V heterostructures. The first goal is to find a material combination that yields a minimum storage time of milliseconds at 300 K, the benchmark for present DRAM.

The capacitance spectroscopy is a powerful tool to study the energy levels and charge carrier emission times in QDs. It has been widely used for different device geometries, for instance in p–i–n diodes to investigate the tunneling dynamics and localization energies in self-organized InAs/GaAs QDs (Fricke et al. 1996, Luyken et al. 1999) or to probe single-electron levels in QDs that are laterally confined inside a 2DEG (Ashoori et al. 1992, Ashoori 1996).

In this section, capacitance spectroscopy is used to determine storage and emission times as well as localization energies in self-organized QDs. The QDs are embedded in a GaAs matrix nearby an abrupt p–n junction that forms a depletion region. This different capacitance spectroscopy method consists of measurements of the depletion capacitance of a p–n or Schottky diode, it is the "capacitance spectroscopy of a depletion capacitance" (Lang 1974).

2.6.1 Depletion Region

The work function or Fermi energy with respect to the vacuum level of a metal or doped semiconductor differs for different metals or semiconductors having a different doping concentration. If a metal and a semiconductor (or two differently doped semiconductors—a p–n diode) are in electric contact, the free charge carriers are exchanged between the different materials until a thermodynamic equilibrium is reached and the Fermi energy is equal throughout the entire structure. Sometimes only ionized donors and acceptors are present in the vicinity of the interface while all free charge carriers are moved away in the semiconductor (Figure 2.15). The layer depleted from free charge carriers is usually referred to as the "depletion region" (Sze 1985, Blood and Orton 1992). The width of the depletion region depends on the doping concentration and the potential difference between the materials, of which the latter can easily be modified by an externally applied bias.

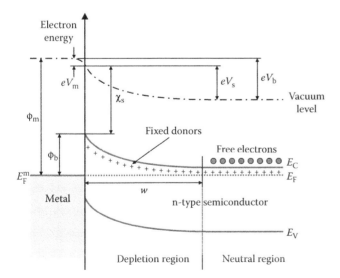

FIGURE 2.15 Energy band diagram of a depleted metal-semiconductor Schottky contact. The width of the depletion region *w* depends on the doping concentration in the semiconductor. A build-in voltage V_b is present without external bias due to the band bending of the depletion region.

A metal-semiconductor contact can be described by the Schottky model and is referred to as the "Schottky contact." A junction of a p-doped and n-doped semiconductor is a "p–n junction." Both types of contacts provide a depletion region and they are briefly described in the following.

2.6.1.1 Schottky Contact

A metal-semiconductor contact is usually described in the framework of the Schottky model, which is an acceptable approach to construct the band diagram of the contact (Figure 2.15). Within this model, the barrier height ϕ_b inside the metal is independent or just weakly dependent on the applied external bias.

According to the Schottky model, the energy band diagram is constructed by reference to the vacuum level, defined as the energy of a free electron to rest outside the material. The work function of the metal ϕ_m and the electron affinity of the semiconductor χ_s are defined as the energies required to remove an electron from the Fermi level of the metal or the semiconductor conduction band edge, respectively, to the vacuum level. These properties are assumed constant in a given material and it is further assumed that the vacuum level is continuous across the interface. The Fermi levels in the metal and semiconductor must be equal in thermal equilibrium. These conditions result in a band diagram for the interface as shown in Figure 2.15.

Since the vacuum level is the same at the interface of the metal and the semiconductor, a step between the Fermi level in the metal and the conduction band edge E_C of the semiconductor occurs. This is the barrier height ϕ_b given by*

$$\phi_b = \phi_m - \chi_s - eV_m. \qquad (2.1)$$

The band bending in the metal is very small due to the large density of electron states; hence, eV_m can be neglected. Therefore, the Schottky barrier height is usually written as

$$\phi_b = \phi_m - \chi_s. \qquad (2.2)$$

For increasing distance from the interface, the conduction band energy decreases. At the end of the depletion region, it has the same value as in the neutral semiconductor with respect to the Fermi level. The resulting band bending is an effect of the removed free charge carriers, leaving behind a distribution of fixed positive charges from ionized donors. The depletion region ends at that position, where the bands become flat and the associated electric field is zero. The width of the depletion region w is determined by the net ionized charge density according to Poisson's equation (see Section 2.6.1.3).

Since the density of states in the metal is much greater than the doping density in the semiconductor, the depletion width in the metal is much smaller. Therefore, it can be assumed that the potential difference across the metal near the contact (V_m) is negligibly small compared to that in the semiconductor (V_s).

---

* For a Schottky barrier, ϕ_m must exceed χ_s otherwise the bands bend in the opposite direction.

The total zero bias band bending of the Schottky contact, also referred to as "build-in potential" or "build-in voltage" V_b can be written as

$$eV_b \approx eV_s = \phi_m - \chi_s - (E_C - E_F) = \phi_b - (E_C - E_F). \quad (2.3)$$

Experimental values for various metal Schottky contacts on GaAs can be found in Myburg et al. (1998).

2.6.1.2 p–n Junction

The band diagram of an abrupt p–n junction, in Figure 2.16, is considered in a similar manner. Again, two rules are used to construct the diagram: (1) the vacuum level is continuous and (2) the Fermi energy is constant across the junction in thermal equilibrium. For the same p- and n-doped semiconductor the electron affinity χ_s is the same on both sides of the junction. Therefore, the band bending is caused entirely by the difference in the Fermi level with respect to the conduction band of the two differently doped materials. From Figure 2.16, using subscripts to denote the n and p side, one obtains for zero external bias

$$(E_C^p - E_F) + \chi_s = eV_b + \chi_s + (E_C^n - E_F), \quad (2.4)$$

hence, the build-in voltage is

$$eV_b = E_g - (E_F - E_V^p) - (E_C^n - E_F), \quad (2.5)$$

where E_g is the energy gap of the semiconductor. The build-in voltage as the total electrostatic potential difference between the p-side and the n-side is temperature-dependent and a function of the fixed donor and acceptor charges of density N_d and N_a (Sze 1985):

$$V_b = \frac{k_B T}{e} \ln\left(\frac{N_a N_d}{n_i^2}\right), \quad (2.6)$$

where n_i is the intrinsic carrier density

$$n_i = \sqrt{N_C N_V} \exp\left(-\frac{E_g}{2k_B T}\right), \quad (2.7)$$

with the effective density of states in the valence N_V and conduction band N_C, respectively. A typical doping density of 10^{17} cm^{-3} yields for GaAs a build-in voltage of 1.3 V at 300 K.

In a p–n junction, depletion regions on each side of the contact exist, where the fixed donor and acceptor charges lead to the band bending. Since the total charge in the p–n diode is zero (charge neutrality), in the depletion approximation (Blood and Orton 1992) of an abrupt depletion layer edge

$$N_a w_p = N_d w_n \quad (2.8)$$

must hold. For similar doping concentrations, the depletion width on the p- and n-side of the semiconductor will be comparable. However, for the purpose of material characterization (such as capacitance experiments) doping concentrations are often chosen such that the depletion region is situated almost entirely on one side of the junction. The depletion region of such an asymmetrical p–n junction resembles the depletion region of a Schottky contact. Such asymmetrical junctions are briefly denoted as p$^+$–n or n$^+$–p junction with $N_a \gg N_d$ or $N_d \gg N_a$, respectively.

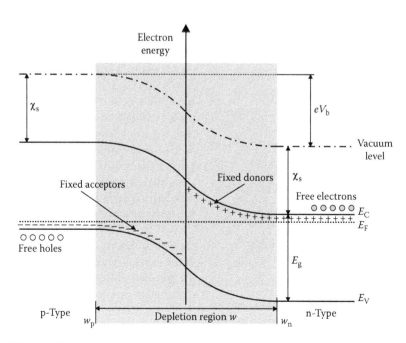

FIGURE 2.16 Energy band diagram of a p–n junction.

2.6.1.3 Width of the Depletion Region

The entire band bending across the depletion region is defined by the sum of the build-in voltage of the contact V_b (Equation 2.5 or 2.6), and the applied external bias V_a in reverse direction: $V = V_b + V_a$. The width of the depletion region and the electric field can be calculated using Poisson's equation. The electrostatic potential ψ at any point is given by

$$-\frac{\partial^2 \psi}{\partial x^2} = \frac{\partial F}{\partial x} = \frac{\rho(x)}{\varepsilon \varepsilon_0}, \qquad (2.9)$$

where
  F is the electric field
  ε_0 is the vacuum permittivity
  ε is the dielectric constant of the semiconductor material

If the donors or acceptors are entirely ionized, the charge density ρ is eN_d or eN_a, respectively:

$$-\frac{\partial^2 \psi}{\partial x^2} = \frac{eN_d}{\varepsilon \varepsilon_0} \quad \text{for } 0 \leq x \leq w_n. \qquad (2.10)$$

For a Schottky contact or an abrupt asymmetric p^+–n junction, integration of Equation 2.9 yields

$$F(x) = F_0 + \frac{eN_d x}{\varepsilon \varepsilon_0} \quad \text{for } 0 \leq x \leq w_n, \qquad (2.11)$$

while $F(x) = 0$ for $x < 0$ and $x > w_n$. Hence, using the approximation $F(x) = 0$ at the edge of the depletion region, gives the boundary condition for the integration constant:

$$F_0 = -\frac{eN_d w_n}{\varepsilon \varepsilon_0}, \qquad (2.12)$$

and represents the electric field at the interface $F(0)$, where it has its maximum. Therefore, the electric field F across the depletion region in an n-doped semiconductor with the donor concentration N_d is

$$F(x) = \frac{eN_d}{\varepsilon \varepsilon_0}(x - w_n) \quad \text{for } 0 \leq x \leq w_n. \qquad (2.13)$$

Analogously, the electric field distribution across the depletion region on the p-doped side is

$$F(x) = -\frac{eN_a}{\varepsilon \varepsilon_0}(x + w_p) \quad \text{for } -w_p \leq x \leq 0. \qquad (2.14)$$

The potential distribution across the depletion region for a p^+–n junction inside the n-doped semiconductor is now obtained by integration of Equation 2.13:

$$\psi(x) = -\int_0^x \frac{eN_d}{\varepsilon \varepsilon_0}(x - w_n)\, dx = \frac{eN_d w_n}{\varepsilon \varepsilon_0}\left(x - \frac{x^2}{2w_n}\right) + \psi(0), \qquad (2.15)$$

with $\psi(0) = 0$ as a reference for the potential distribution. For an asymmetric p^+–n junction, one notices that the band bending in the present approximation only occurs in the n-doped region of the junction. The contact potential is equal to the total band bending, that means, equal to the build-in voltage plus the external bias $\psi(0) = -V = -(V_b + V_a)$. Since the potential on the edge of the depletion region is zero, $\psi(w_n) = 0$, one obtains

$$V = V_b + V_a = \frac{eN_d}{2\varepsilon \varepsilon_0} w_n^2, \qquad (2.16)$$

or the expression for the depletion width

$$w_n = \sqrt{\frac{2\varepsilon \varepsilon_0}{eN_d} V}. \qquad (2.17)$$

If the p–n junction is symmetric and the depletion region extends into the p- and n-doped semiconductor, Equation 2.16 is modified to

$$V = V_b + V_a = \frac{eN_d}{2\varepsilon \varepsilon_0} w_n^2 + \frac{eN_a}{2\varepsilon \varepsilon_0} w_p^2, \qquad (2.18)$$

with the net acceptor doping concentration N_a and the width of the depletion region w_p in the p-doped side. The total depletion width w is given by

$$w = (w_n + w_p) = \sqrt{\frac{2\varepsilon \varepsilon_0}{e}\left(\frac{N_a + N_d}{N_a N_d}\right) V}. \qquad (2.19)$$

The assumption of an abrupt depletion layer edge, is the so-called depletion approximation. A more general description of the depletion region can be found in reference Blood and Orton (1992).

2.6.1.4 Depletion Layer Capacitance

The capacitance that arises from the depletion region is the so-called "depletion capacitance." In comparison to a plate capacitor, here, the charge is accumulated inside the depletion region of the p–n diode. The number of ionized donors depends linearly on the width w_n, while again the width is a function of the square root of the voltage (Equations 2.17 and 2.19). Hence, the capacitance does not depend linearly on the voltage and has to be defined differentially for a small bias signal ΔV:

$$C = \lim_{\Delta V \to 0} \frac{\Delta Q}{\Delta V} = \frac{dQ}{dV}. \qquad (2.20)$$

Using Equation 2.17, the entire stored charge inside the depletion region of a Schottky diode or an abrupt asymmetric p^+–n junction with area A is

$$Q = eN_d A \cdot w_n(V) = A\sqrt{2e\varepsilon \varepsilon_0 N_d V}, \qquad (2.21)$$

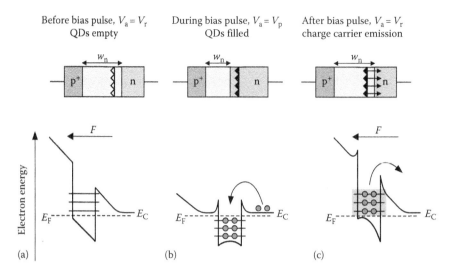

Before bias pulse, $V_a = V_r$
QDs empty

FIGURE 2.17   DLTS work cycle of a p⁺–n diode with embedded QDs. The upper row schematically shows sketches of the devices for three different bias situations: (a) before, (b) during, and (c) after the bias pulse, respectively. The lower row shows the corresponding band diagrams.

and the capacitance is

$$C(V) = \left| \frac{dQ}{dV} \right|_V = A\sqrt{\frac{\varepsilon\varepsilon_0 \, e N_d}{2V}} = \frac{\varepsilon\varepsilon_0 \, A}{w_n}. \qquad (2.22)$$

The depletion capacitance resembles the capacitance of a plate capacitor with a distance of w_n between the plates and a dielectric with relative permittivity ε, although the charge is actually stored in the volume rather than on the edges of the space charge region.

2.6.2 Capacitance Transient Spectroscopy

The depletion width depends on the applied voltage and the doping concentration, that is, on the charge stored inside the space charge region. As a consequence, the measured capacitance is sensitive to the carrier population inside deep levels or QDs situated inside the depletion region (Kimerling 1974). Semiconductor QDs can be considered as deep levels (or traps), storing few or single charge carriers for a certain time. Time-resolved capacitance spectroscopy allows investigating the electronic structure of QDs, since the carrier dynamics can be studied in detail.

Historically, time-resolved capacitance spectroscopy was initially used to study and characterize deep levels caused by impurities or defects. In this context it is usually referred to as "deep level transient spectroscopy" (DLTS) or "capacitance transient spectroscopy" (Sah et al. 1970, Lang 1974, Miller et al. 1977, Rhoderick and Williams 1988, Grimmeiss and Ovrén 1981, Blood and Orton 1992). Besides the determination of activation energies and capture cross sections, DLTS also permits to obtain a depth profile of the trap concentration and to investigate the influence of the electric field on the emission process. As QDs act more or less as deep levels, the DLTS has been used successfully

to study thermal emission processes, activation energies, and capture cross sections of various QD systems (Anand et al. 1995, 1998, Kapteyn et al. 1999, 2000a,b, Schulz et al. 2004).

2.6.2.1 Measurement Principle

In the following, the DLTS measurement principle will be described for QDs in a p–n or Schottky diode structure. A more detailed description can be found in Blood and Orton (1992).

First, a single layer of QDs with density N_{QD} per area in an n-doped material (with doping concentration N_d) shall be considered. The work cycle of a DLTS experiment of a p⁺–n structure with QDs embedded is depicted in Figure 2.17. The upper row displays schematically the p–n diode while the lower row depicts the corresponding potential distribution of the conduction band.

In an initial step (Figure 2.17a), a reverse bias V_r is chosen such that the depletion region extends well over the QDs. The QDs are completely depleted from charge carriers and the Fermi level is below the QD levels. During the pulse V_p, the depletion region is shorter than the distance of the QD layer from the p–n interface and the QDs are consequently filled with carriers—electrons in Figure 2.17b. The Fermi level is now above the QD states. After switching back to the reverse bias situation (Figure 2.17c), the QDs are again situated inside the depletion region but still filled with electrons. The depletion width is larger for QDs filled with charge carriers, hence, the capacitance after the pulse V_p is smaller and increases again when electrons/holes are emitted due to thermal activation or tunneling emission.[*] By recording the depletion capacitance as a function of time, transients are observed, which represent the carrier emission

---

[*] The emission of carriers from QD states is usually slow, whereas the free carriers in the matrix material are considered to follow the change in the external bias instantaneously at the timescale of the experiment.

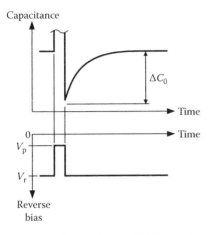

FIGURE 2.18 DLTS work cycle during a DLTS experiment. The lower part displays the external bias on the device as function of time, the upper part the corresponding capacitance.

processes from the QD states. The pulse sequence during the experiment and the measured capacitance transient is schematically depicted in Figure 2.18.

2.6.2.2 Rate Window and Double-Boxcar Method

The capacitance transient, recorded from a single emission process is usually mono-exponential having the time-constant τ. Hence, the capacitance transient is given by (Blood and Orton 1992)

$$C(t) = C(\infty) - \Delta C_0 \exp\left(-\frac{t}{\tau}\right), \qquad (2.23)$$

where

$C(\infty)$ is the steady state capacitance at V_r
ΔC_0 is the entire change in the capacitance $C(t)$ for $t = \infty$ (see Figure 2.18)

The emission time constant can be determined by using Equation 2.23 and plotting the data on a semilogarithmic scale. However,

in general, the capacitance transient for deep levels and QDs are multi-exponentials due to the ensemble broadening (Omling et al. 1983) and a linear fit is consequently impossible in most cases.

In order to obtain the emission time constant, the activation energies and capture cross section from multi-exponential transients, the rate window concept is commonly applied. One investigates the contribution to the observed emission process at a chosen reference time constant τ_{ref}. By plotting the boxcar amplitude $C(t_2) - C(t_2)$ for that reference time constant as function of temperature, the relation between temperature and emission rate can be evaluated for a thermal activated process.

In the rate window concept, the selection of the contribution for a chosen reference time constant is done by a simple technique: the DLTS signal at a certain temperature $S(T, t_1, t_2)$ is given by the difference of the capacitance at two times t_1 and t_2 by

$$S(T, t_1, t_2) = C(T, t_2) - C(T, t_1), \qquad (2.24)$$

or with Equation 2.23

$$S(T, t_1, t_2) = \Delta C_0 \left[\exp\left(-\frac{t_2}{\tau(T)}\right) - \exp\left(-\frac{t_1}{\tau(T)}\right) \right]. \qquad (2.25)$$

The two times t_1 and t_2 define the rate window, which has the reference time constant

$$\tau_{ref} = \frac{t_2 - t_1}{\ln(t_2/t_1)}. \qquad (2.26)$$

Plotting $S(T, t_1, t_2)$ as function of temperature yields the DLTS spectrum, schematically depicted in Figure 2.19. A maximum appears at that temperature, where the emission time constant of the thermally activated process equals almost the applied reference time constant: $\tau(T) = \tau_{ref}$. A maximum appears only for a thermally activated process, a temperature independent tunneling process leads to a constant DLTS signal (Kapteyn

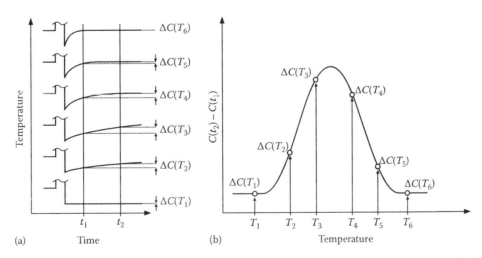

FIGURE 2.19 The evaluation of capacitance transients (a) for increasing temperature by a rate window, defined by t_1 and t_2, leads to a DLTS plot $[C(t_1) - C(t_2)]$ of a thermally activated emission process (b).

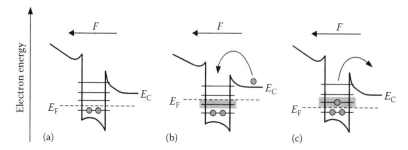

FIGURE 2.20 Work cycle during a charge-selective DLTS experiment (a) before, (b) during, and (c) after the bias pulse, respectively. The filling pulse V_p is chosen, that the emission or capture of approximately one charge carrier per QD is probed.

2001). This relation is also valid for inhomogeneous broadened DLTS spectra of self-organized QDs. The measured emission time constant is the average time constant at the maximum of the Gaussian ensemble distribution (Omling et al. 1983).

The thermal emission rate of the charge carriers, e_{th} is now given by (Lang 1974, Blood and Orton 1992)

$$e_{th} = \gamma T^2 \sigma_\infty \exp\left(-\frac{E_A}{kT}\right), \qquad (2.27)$$

where

E_A is the thermal activation energy
σ_∞ is the apparent capture cross section for $T = \infty$
γ is a temperature-independent constant

Knowing the emission time constant $\tau = 1/e_{th}$ for different temperatures enables to derive the thermal activation energy and the capture cross section. For varying τ_{ref} different peak positions T_{max} are obtained. The plot of $\ln(T_{max}^2 \tau_{ref})$ as a function of T_{max}^{-1}, cf. Figure 2.22b, is called an Arrhenius plot and is a linearization of Equation 2.27. The slope yields the activation energy E_A and from the y-axis intersection, the apparent capture cross section σ_∞ at infinite temperature can be obtained.

In order to improve the signal-to-noise ratio (SNR), the capacitance transients near t_1 and t_2 can be averaged over an interval t_{av} (Day et al. 1979). The SNR in this case is found to scale as $\sim\sqrt{t_{av}}$ (Miller et al. 1977), and the reference time constant is in good approximation

$$\tau_{ref} = \frac{t_2 - t_1}{\ln\left((t_2 + (1/2)t_{av})/(t_1 + (1/2)t_{av})\right)}. \qquad (2.28)$$

This method to improve the SNR is referred to as "Double-Boxcar" approach.

2.6.2.3 Charge-Selective Deep Level Transient Spectroscopy

As described in the previous section, usually the QDs are completely filled with charge carriers during the pulse bias V_p (Figure 2.17b). Consequently, after the pulse, the emission of many charge carriers from multiply charged QDs is probed.

The activation energy of each emitted charge carrier depends on the actual charge state in such conventional DLTS experiments. The DLTS spectrum is broadened due to different emission processes from different QD states having different emission time constants, for instance, nicely observed in DLTS experiments on Ge/Si QDs (Kapteyn et al. 2000b).

In order to study the charge states in more detail, charge-selective DLTS probes the emission of approximately one charge carrier per QD. The principle of this method is schematically illustrated in Figure 2.20. Before the filling pulse, the QDs might be already filled with charge carriers up to the Fermi level. During the filling pulse (Figure 2.20b), the Fermi level is adjusted by the applied pulse bias, such that approximately one carrier per QD will be captured. To probe all QD states from the ground state up to the excited state the pulse bias with respect to the reverse bias is set to a constant height and the reverse bias is decreased step by step. After the bias pulse, the reverse bias is set to the initial condition. Now, the previously captured carrier is emitted again and the emission process of approximately one charge carrier per QD is observed. Narrow peaks appear now in the DLTS spectra, which are due to differently charged QDs (cf. Figure 2.23).

Moreover, even for an energy-broadened ensemble of QDs (normally measured in DLTS experiments) the charge-selective DLTS offers a simple method to probe the emission from many charge carriers in different QDs, all having the same activation energy. Decreasing the reverse bias and keeping the pulse bias height fixed (to ensure the emission of only one charge carrier per QDs) gives the activation energy starting at the ground states of the ensemble to the excited states and finishing at completely charged QDs.

2.7 Charge Carrier Storage in Quantum Dots

In this section, experimental results from capacitance spectroscopy measurements on different QD material systems are presented. This time-resolved method allows to derive thermal activation energies and capture cross sections of electron and hole states in QDs. In addition, the important storage time at room temperature can be quantified and connected to the localization energy of the charge carriers.

2.7.1 Carrier Storage in InGaAs/GaAs Quantum Dots

Emission of electrons and holes from InGaAs/GaAs QDs was observed using conventional DLTS measurements by Kapteyn et al. (1999, 2000a,b). The electron/hole emission from InGaAs/GaAs QDs was studied in more detail by using charge-selective DLTS (Geller et al. 2006d) and time-resolved tunneling capacitance measurements (TRTCM) (Geller et al. 2006c). Two samples H1 and E1 were investigated to study the hole and electron emission, respectively, where the QD layer was incorporated in the slightly p-doped/n-doped ($\sim 3 \times 10^{16}$ cm^{-3}) region of a n$^+$–p or p$^+$–n diode structure. The QD layer was situated 500 nm/415 nm below the p–n junction for the hole/electron sample. Mesa structures with a diameter of 800 μm and ohmic contacts were formed by employing standard optical lithography. The results of these measurements are presented in Figure 2.21.

It turns out that phonon-assisted tunneling controls the charge-carrier emission process from self-organized QDs in an electric field. The influence of the tunneling part, however, depends strongly on the effective mass and the strength of the electric field. The observed hole ground state activation energy $E_A^{H1} = 120 \pm 10$ meV, hence, underestimates the localization energy. It is the thermal activation part in a phonon-assisted tunneling process: thermal activation into an excited state and subsequent tunneling through the remaining triangular barrier, cf. schematic pictures in Figure 2.21. The true hole localization energy was determined by using the TRTCM method to $E_{loc}^{H1} = 210 \pm 20$ meV.

For the electrons, two contributions to the DLTS signal were observed: a DLTS signal with an activation energy of $E_{A1}^{E1} = 82 \pm 10$ meV that is in good agreement with the theoretically predicted value for the ground/excited state energy splitting (70 meV). In addition, a DLTS signal with a smaller activation energy of $E_{A2}^{E1} = 44 \pm 10$ meV is observed. This value is attributed to the first/second-excited state energy splitting and in satisfactory agreement with the theoretically predicted value of 50 meV. The ground state localization energy was determined by the TRTCM method to $E_{loc}^{E1} = 290 \pm 30$ meV (Geller et al. 2006c).

The electron/hole storage time at room temperature can be estimated by using Equation 2.27 and the capture cross section and the localization energy for sample E1/H1, respectively. An average storage time for electrons of about 200 ns and for holes of about 0.5 ns is obtained. This means, InAs QDs embedded in a GaAs matrix do not have a sufficiently long storage time to act as storage units in future nanomemories.

2.7.2 Hole Storage in GaSb/GaAs Quantum Dots

The storage time can be further increased by changing the material of the QDs and/or the surrounding matrix. A larger difference in the energy bandgap than in InGaAs/GaAs is more promising, for example, InAs/AlAs QDs. Moreover, large band discontinuities and, hence, strong hole localization is expected in type II QD heterostructures (Hatami et al. 1998), for example, GaSb/GaAs or InSb/GaAs. Only holes are confined in GaSb/GaAs QDs, while a repulsive potential barrier exists for the electrons in the conduction band. Type II material combinations are therefore very attractive for future memory applications.

Hole storage in and emission from GaSb$_{0.6}$As$_{0.4}$/GaAs QDs was investigated by DLTS and charge-selective DLTS (Geller et al. 2003). The sample was an n$^+$–p diode structure containing a single layer of GaSb QDs. An area density of about 3×10^{10} cm^{-2}, an average QD height of about 3.5 nm, and an average base width of about 26 nm were determined by structural characterization of uncapped samples grown under identical conditions (Müller-Kirsch et al. 2001). The QD layer was placed 500 nm below the n$^+$–p junction in a slightly p-doped ($p = 3 \times 10^{16}$ cm^{-3}) GaAs region. Mesa structures with a diameter of 800 μm and ohmic contacts were formed by employing standard optical lithography.

2.7.2.1 Multiple Hole Emission

In Figure 2.22, conventional DLTS measurements of the QD sample for a reverse bias $V_r = 10$ V and a pulse bias $V_p = 4$ V are displayed. The pulse length was 10 ms. For these conditions, the QDs are completely filled during the bias pulse and release all trapped holes after the pulse.

The DLTS spectrum of the QD sample (Figure 2.22a) shows three maxima, in the following denoted by "A," "B," and "C." From an Arrhenius plot, an activation energy $E_A^C = 530 \pm 20$ meV and a capture cross section $\sigma_\infty^C \approx 6 \times 10^{-16}$ cm^2 is obtained for peak C. This signature can be identified as the bulk hole trap H3 in p-type GaAs (Stievenard et al. 1986).

Peaks A and B, observed at temperatures $T = 140$ and 230 K, are obviously related to QD formation. An activation energy of about 600 meV is found for peak B. This relatively high activation energy is comparable to the unstrained GaSb/GaAs valence band offset (Hatami et al. 1995, North et al. 1998) and could be explained by the existence of relaxed GaSb islands in the QD sample.

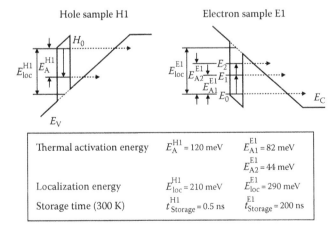

	Hole sample H1	Electron sample E1
Thermal activation energy	$E_A^{H1} = 120$ meV	$E_{A1}^{E1} = 82$ meV
		$E_{A2}^{E1} = 44$ meV
Localization energy	$E_{loc}^{H1} = 210$ meV	$E_{loc}^{E1} = 290$ meV
Storage time (300 K)	$t_{Storage}^{H1} = 0.5$ ns	$t_{Storage}^{E1} = 200$ ns

FIGURE 2.21 Summary of the results from the charge-selective DLTS and tunneling emission experiments on self-organized InGaAs/GaAs QDs.

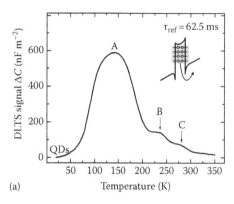 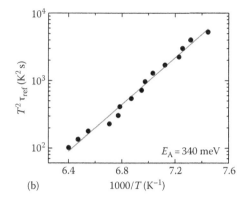

(a)   (b)

FIGURE 2.22 (a) DLTS signal of the GaSb$_{0.6}$As$_{0.4}$/GaAs QD sample, measured at a reverse bias $V_r = 10$ V and a pulse bias $V_p = 4$ V with a filling pulse width of 10 ms and a reference time constant $\tau_{ref} = 62.5$ ms. (b) The corresponding Arrhenius plot for peak A, which is related to multiple hole emission from the GaSb QD states. An average activation energy of $E_A = 340$ meV is obtained.

Peak A in the DLTS spectrum of the QD sample in Figure 2.22a is attributed to the hole emission from the GaSb QD states, where the activation energy of each successively emitted hole depends on the actual charge state. The DLTS peak is broadened as previously observed for multiply charged Ge/Si QDs (Kapteyn et al. 2000b). From a standard Arrhenius plot in Figure 2.22b, an activation energy of $E_A = 340$ meV is obtained. Consequently, this activation energy of peak A represents only an average value for hole emission from completely charged QDs, where 15 holes from different states with different confinement are involved.

2.7.2.2 Charge-Selective DLTS

In order to study the charge states in more detail, the QD sample was studied using the charge-selective DLTS method. The pulse bias was always set to $V_p = V_r - 0.5$ V, while the reverse bias was increased from 4.5 to 9.5 V. Narrow peaks appear in the DLTS spectra in Figure 2.23. At $V_r = 4.5$ V the DLTS signal shows a maximum at about 80 K and for increasing reverse bias the DLTS peak shifts to higher temperature. The activation energy increases accordingly from 150 meV at 4.5 V to 450 meV for 9.5 V in Figure 2.24 obtained by standard Arrhenius plots.

The DLTS spectra for a reverse bias between 4.5 and 9.5 V in Figure 2.23 are attributed to the hole emission from differently charged QDs. All these spectra exhibit a maximum in the temperature range between 80 and 180 K, the range covered by peak A in Figure 2.22a. The maximum activation energy of 450 meV represents the average hole ground state energy of the QD ensemble, that is, the localization energy. The decrease in the activation energy from 450 meV down to 150 meV corresponds to an increase in the average occupation of the QDs, see the schematic insets in Figure 2.23. With increasing amount of charge in the QDs, state filling lowers the thermal activation barrier. The completely charged QDs are filled with 15 holes up to the Fermi level at the valence band edge, where Coulomb charging generates the barrier height.

In order to compare the storage time for GaSb$_{0.6}$As$_{0.4}$/GaAs with In(Ga)As/GaAs QDs, the observed emission rates were extrapolated to room temperature, using Equation 2.27. A

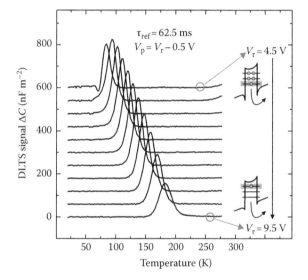

FIGURE 2.23 Charge-selective DLTS spectra of hole emission from GaSb$_{0.6}$As$_{0.4}$/GaAs QDs for a reference time constant of $\tau_{ref} = 62.5$ ms. The reverse bias V_r is increased from 4.5 V up to 9.5 V while the pulse height is fixed at 0.5 V for all spectra. The pulse width is 10 ms and the data is displayed vertically shifted for clarity.

storage time of about 1 μs for localized holes with the ground state energy of 450 meV is estimated, three orders of magnitude longer than the hole storage time in InGaAs/GaAs QDs.

2.7.3 InGaAs/GaAs Quantum Dots with Additional AlGaAs Barrier

Hole storage in InGaAs/GaAs QDs with an additional AlGaAs barrier is presented in this section. The additional AlGaAs barrier increases the activation energies and a longer storage time at room temperature is observed. Two different samples having an AlGaAs barrier with different aluminum content were studied. The first contains an Al$_{0.6}$Ga$_{0.4}$As, the second an Al$_{0.9}$Ga$_{0.1}$As barrier below the QD layer. The activation energy in the latter is increased sufficiently to reach a retention time of seconds at room temperature.

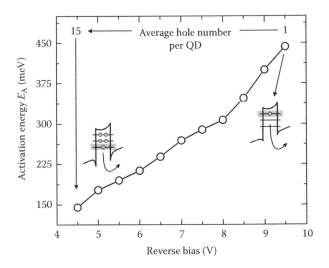

FIGURE 2.24 Dependence of the activation energy E_A on the reverse bias GaSb$_{0.6}$As$_{0.4}$/GaAs QDs. For a reverse bias of $V_r = 9.5\,V$ hole emission from the QD ground states is probed.

2.7.3.1 Storage Time: Milliseconds at Room Temperature

The first sample is an n$^+$–p diode structure, grown by MBE. It contains a single layer of InGaAs QDs embedded in slightly p-doped GaAs ($p = 2 \times 10^{15}$ cm^{-3}). The QDs are placed 1500 nm below the p–n junction and an additional undoped Al$_{0.6}$Ga$_{0.4}$As barrier of 20 nm thickness is situated 7 nm below the QD layer to increase the hole storage time. Again, mesa structures with a diameter of 800 µm and ohmic contacts were formed.

Figure 2.25a shows the charge-selective DLTS spectra with a reference time constant of 5 ms (Marent et al. 2006). The pulse bias height was fixed to 0.2 V for all spectra ($V_p = V_r - 0.2\,V$). For a reverse bias above $V_r = 3.2\,V$ no DLTS signal is visible, as the Fermi level is energetically above the QD states and no QD states are occupied. By decreasing the reverse bias the Fermi level reaches the QD ground state at $V_r = 3.2\,V$ and a peak in the DLTS spectrum appears at 300 K. This peak is related to thermally activated hole emission from the ground states of the QD ensemble

across the AlGaAs barrier, see inset in the upper left corner of Figure 2.25b. A further decrease of the reverse bias leads to QD state-filling and emission from higher QD states is observed. A peak in the DLTS appears at that temperature, where the averaged time constant of the thermally activated emission equals the applied reference time constant, cf. Equation 2.26. Therefore, the peak at 300 K for $\tau_{ref} = 5\,ms$ ($V_r = 3.2\,V$ in Figure 2.25a) represents an average emission time constant (storage time) of 5 ms for hole emission from the QD ground states across the Al$_{0.6}$Ga$_{0.4}$As barrier.

The thermal activation energies are shown in Figure 2.25b. The highest value of 560 ± 60 meV at $V_r = 3.2\,V$ is related to thermal activation from the hole ground states of the QD ensemble across the AlGaAs barrier. The decrease in the activation energy corresponds to an increase in the average occupation of the QDs. At $V_r = 1.4\,V$ the energetic position of the Fermi level is at the valence band edge and no further QD state-filling is possible. As a consequence, between $V_r = 1.4$ and 0.2 V carrier emission from the valence band edge is probed. The activation energy remains roughly constant with a mean value of 340 meV. This energy represents the energetic height of the Al$_{0.6}$Ga$_{0.4}$As barrier.

2.7.3.2 Storage Time: Seconds at Room Temperature

The second sample is also an n$^+$–p diode structure, grown by MOCVD. It contains a single layer of InGaAs QDs embedded in p-doped GaAs ($p = 3 \times 10^{16}$ cm^{-3}). The QDs are placed 400 nm below the p–n junction and an additional undoped Al$_{0.9}$Ga$_{0.1}$As barrier of 20 nm thickness is situated 7 nm below the QD layer.

Figure 2.26a shows the charge-selective DLTS spectra (Marent et al. 2007). The pulse bias height was fixed here to 0.3 V for all spectra ($V_p = V_r - 0.3\,V$). At $V_r = 5.7\,V$ a peak in the DLTS spectrum appears at 380 K, related to thermally activated hole emission from the ground states of the QD ensemble across the Al$_{0.9}$Ga$_{0.1}$As barrier. As mentioned before, the peak at 380 K for $\tau_{ref} = 5\,ms$ represents an average emission time constant (storage time) of 5 ms for hole emission from the ground states of the QD ensemble across the Al$_{0.9}$Ga$_{0.1}$As barrier. To obtain the storage time at room temperature, the capacitance transient recorded at

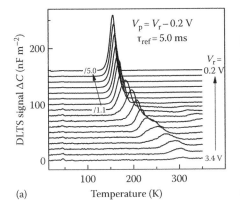

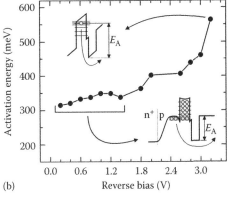

FIGURE 2.25 Charge-selective DLTS spectra of thermally activated hole emission from InGaAs/GaAs QDs with an additional Al$_{0.6}$Ga$_{0.4}$As barrier below the QD layer (a). Spectra below $V_r = 1.6\,V$ are divided by a factor from 1.1 up to 5. (b) Dependence of the thermal activation energy on the reverse bias V_r. (Reprinted from Marent, A. et al., *Appl. Phys. Lett.*, 89, 072103, 2006. With permission.)

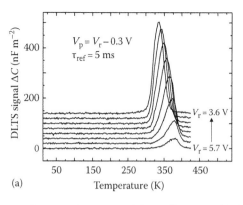
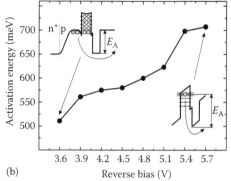

FIGURE 2.26 Charge-selective DLTS spectra of thermally activated hole emission from InGaAs/GaAs QDs with an additional $Al_{0.9}Ga_{0.1}As$ barrier below the QD layer (a). Dependence of the thermal activation energy on the reverse bias V_r.

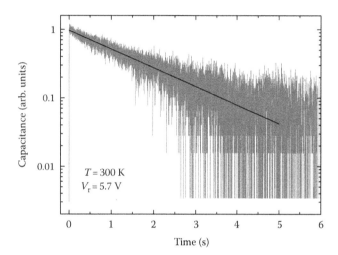

FIGURE 2.27 Capacitance transient of hole emission from InGaAs/GaAs QDs with an additional $Al_{0.9}Ga_{0.1}As$ barrier (for a reverse bias of $V_r = 5.7\,V$). An average hole storage time of 1.6 s at $T = 300\,K$ is obtained from a linear fit of the transient on a semilogarithmic scale.

300 K is plotted on a semilogarithmic scale in Figure 2.27. From a linear fit an average hole storage time of 1.6 s at room temperature is determined.

Furthermore, the thermal activation energies are shown in Figure 2.26b. The highest value of 710 ± 40 meV at $V_r = 5.7\,V$ is related to thermal activation from the hole ground states of the QD ensemble across the $Al_{0.9}Ga_{0.1}As$ barrier. The decrease

in the activation energy corresponds again to an increase in the average occupation of the QDs. At $V_r = 3.6\,V$, carrier emission from the valence band edge is probed and the activation energy has a value of 520 meV. This energy now represents the energetic height of the $Al_{0.9}Ga_{0.1}As$ barrier.

2.7.4 Storage Time in Quantum Dots

Carrier emission from different QD systems has been studied in order to determine the carrier storage time at room temperature. In addition, the localization energy was obtained by using time-resolved capacitance spectroscopy (DLTS) and related to the hole/electron retention time. The results of the experiments are summarized in Figure 2.28.

An electron/hole localization energy of 290/210 meV, respectively, was obtained for InGaAs QDs embedded in a GaAs matrix. Based on these values the storage time at $T = 300\,K$ was estimated to ~200 ns for electrons and ~0.5 ns for holes. Furthermore, the more promising type II $GaSb_{0.6}As_{0.4}/GaAs$ has been studied in detail. Ground state activation energy of 450 meV was determined, which accounts for a room temperature emission time in the order of 1 μs. The ground state localization is about twice as large and the retention time at room temperature is about three orders of magnitude longer than in InGaAs/GaAs QDs. Finally, the hole emission from InGaAs/GaAs QDs with an additional AlGaAs barrier was investigated. The activation energy for the QD ground states increases from 210 to 560 meV

Material System	Charge Carrier Type	Localization Energy	Storage Time at 300 K
InAs/GaAs	Hole	210 meV	~0.5 ns
	Electron	290 meV	~200 ns
Ge/Si	Hole	350 meV	~0.1 μs
$GaSb_{0.6}As_{0.4}/GaAs$	Hole	450 meV	~1 μs
InAs/GaAs with $Al_{0.6}Ga_{0.4}As$ barrier	Hole	560 meV	5 ms
InAs/GaAs with $Al_{0.9}Ga_{0.1}As$ barrier	Hole	710 meV	1.6 s

FIGURE 2.28 Summary of the measured electron/hole storage time (at 300 K) and localization energy for different QD systems.

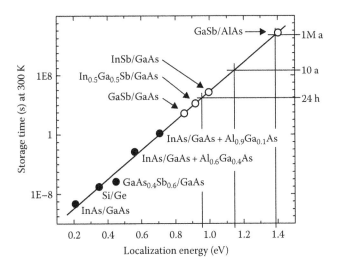

FIGURE 2.29 Dependence of the hole storage time on the localization energy for a variety of QD systems. The solid line is a fit to the experimentally obtained data (full circles). The open circles are estimated storage times for the labeled material systems according to the calculated localization energies and the fit. (Reprinted from Marent, A. et al., *Appl. Phys. Lett.*, 91, 42109, 2007. With permission.)

for an additional $Al_{0.6}Ga_{0.4}As$ barrier and the hole storage time is in the order of milliseconds. Using an $Al_{0.9}Ga_{0.1}As$ barrier increases the activation energy accordingly from 210 to 710 meV and the hole storage time at room temperature increases by nine orders of magnitude to 1.6 s. This value is already three orders of magnitude longer than today's DRAM refresh time, which is in the millisecond range.

From the experimental results, material combinations can be predicted to obtain a storage time of more than 10 years at room temperature. Figure 2.29 displays the hole storage time in dependence of the localization energy (full circles). The solid line is a fit to the data and corresponds to Equation 2.27. The storage time shows an exponential dependence on the localization energy as predicted by the common rate equation of thermally activated emission. The storage time increases by one order of magnitude for an increase of the localization energy of about 50 meV. From these results, hole localization energies can be estimated, which provide storage times of 24 h or 10 years. These storage times are reached for localization energies of 0.96 and 1.14 eV, respectively.

To find a material system with such large localization energies, the hole localization energies for $GaAs_xSb_{1-x}/GaAs$ and

$In_xGa_{1-x}Sb/GaAs$ QDs have been calculated using eight-band k · p theory (Stier et al. 1999, Schliwa et al. 2007). Based on structural characterization of GaSb/GaAs QDs (Müller-Kirsch et al. 2001) the QDs are modeled as truncated pyramids with a base width of 21 nm and a height of 3.9 nm. The results are summarized in Figure 2.30 and plotted in Figure 2.29 as open circles. The localization energy in $GaAs_xSb_{1-x}/GaAs$ QDs increases from 350 meV up to 853 meV for an antimony content of 50% and 100%, respectively. Since the bandgap in III–V materials decreases with increasing lattice constant (Vurgaftman et al. 2001), InSb/GaAs QDs should offer a larger localization energy than GaSb/GaAs. Eight-band k·p calculations for $In_xGa_{1-x}Sb/GaAs$ QDs provide a localization energy of 919 and 996 meV for an indium content of 50% and 100%, respectively. A hole storage time in InSb/GaAs QDs of more than 24 h is predicted (see Figure 2.29). Using an AlAs matrix instead of a GaAs matrix a storage time of more than 10 years can be reached. Since the valence band offset between GaAs and AlAs is about 550 meV (Batey and Wright 1986) the entire hole localization energy in GaSb/AlAs QDs is about 1.4 eV. This value leads to the prediction of an average hole storage time of more than 1 million years at room temperature (see Figure 2.29), orders of magnitude longer than needed for a nonvolatile memory.

2.8 Write Times in Quantum Dot Memories

The carrier capture of electrons/holes from the valence/conduction band into the QD states limits physically the possible write time in a QD-based nanomemory. Normally, carrier capture into QDs is studied by interband absorption optical techniques, like time-resolved photoluminescence (PL) spectroscopy (Heitz et al. 1997, Giorgi et al. 2001). However, such experiments probe the exciton dynamics, that means, the electron–hole capture and relaxation into the QD states is measured simultaneously. A detailed knowledge of either electron or hole capture is of great importance, for future QD-based memory applications as only one sort of charge carriers is stored. Electron or hole capture has been investigated separately, for example, in interband pump and intraband probe experiments of Müller et al. (2003). In addition, the hole capture into $GaSb_{0.6}As_{0.4}/GaAs$ self-organized QDs has been studied (Geller et al. 2008) using DLTS experiments. The investigations demonstrate a fast

Material System	Charge Carrier Type	Localization Energy	Storage Time at 300 K
$GaSb_{0.5}As_{0.5}/GaAs$	Hole	350 meV	0.1 μs
GaSb/GaAs	Hole	853 meV	13 min
$In_{0.5}Ga_{0.5}Sb/GaAs$	Hole	919 meV	4 h
InSb/GaAs	Hole	996 meV	6 days
GaSb/AlAs	Hole	~1.4 eV	$>10^6$ years

FIGURE 2.30 Hole localization energies in different Sb-based QDs calculated by eight-band k.p theory. The storage time is estimated according to the fit of the experimental data in Figure 2.29.

capture and relaxation process in self-organized QDs in the range of picoseconds at room temperature, more than three orders of magnitude faster than the write time in a DRAM cell. This fast carrier capture should enable very fast write times in a QD-based nanomemory (<ns) which are independent of the storage time, possibly up to more than 10 years at room temperature. The QD-based memory concept should therefore allow a nonvolatile semiconductor memory with orders of magnitude faster write times than the present Flash concept.

To confirm the presumption of a fast write time that is independent of the storage time, this section shows the write time measurements for two different QD memory structures containing type I InAs/GaAs and type II $GaSb_{0.6}As_{0.4}$/GaAs QDs—both already described in Section 2.7—where the average storage time at 300 K ranges from nanoseconds (InAs/GaAs QDs) to microseconds ($GaSb_{0.6}As_{0.4}$/GaAs QDs).

2.8.1 Hysteresis Measurements

As already discussed in Section 2.7, where the charge carrier storage time was determined, the read-out of the charge state (which represents the stored information) inside the QDs is done by the measurement of the capacitance of the p–n diode. The capacitance of a diode structure with embedded QDs depends on the number of holes stored inside the depletion region. A larger capacitance corresponds to unoccupied QDs ("0") while a smaller capacitance represents a "1," where the QDs are filled with holes.

A fundamental property of storage devices is the appearance of hysteresis when switching between the two information states. Accordingly, a nanomemory device containing self-organized QDs shows a hysteresis in the capacitance measurement after writing (yielding occupied QDs, a "1" state) and erasing (yielding unoccupied QDs, a "0" state) process. Such a hysteresis measurement is shown in Figure 2.31 for InAs/GaAs QDs (a) and $GaSb_{0.6}As_{0.4}$/GaAs QDs (b). Both memory structures are n-p diodes where holes are stored inside the QDs. The sample description can be found in Section 2.7.1 for the InAs/GaAs QDs and in Section 2.7.2 for the $GaSb_{0.6}As_{0.4}$/GaAs QDs. One hysteresis sweep takes a few seconds; hence, the temperature was reduced down to 15 K for the InAs QDs where the hole storage time is in the order of minutes. Analogously, for the $GaSb_{0.6}As_{0.4}$/ GaAs QDs—having hole localization energy twice as high as the InAs/GaAs QDs—a higher temperature of 100 K is already sufficient to obtain hole storage times of several minutes.

Figure 2.31 shows the switching between the two information states by a hysteresis curve of the capacitance for both samples. At the reverse bias of 14 V/16 V (point 1) the charge carriers tunnel out of the QDs (erasing the information). If the reverse bias is now swept from 14 V/16 V to the storage situation at 8.2 V (InAs QDs) or 7.2 V (GaSb QDs), respectively, the QDs are empty and a larger capacitance is observed (point 2). If the bias is swept to 0 V (point 3), the QDs are charged with holes and a smaller capacitance is observed upon sweeping back to the storage situation (point 4). The maximum hysteresis opening is now defined at

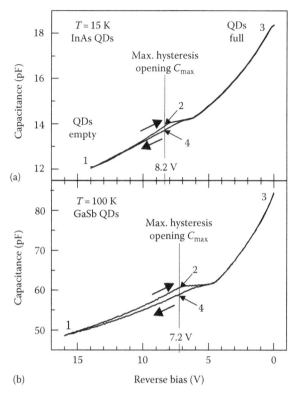

FIGURE 2.31 Capacitance sweeps of a p–n-diode containing InAs/ GaAs (a) and $GaSb_{0.6}As_{0.4}$/GaAs (b) QDs. (Reprinted from Geller, M. et al., *Appl. Phys. Lett.*, 92, 092108, 2008. With permission.)

the storage position as the difference between both capacitance values for a "0" and "1" state. The structures are not yet optimized since the carriers are not stored at zero bias. However, the memory structure can be easily adjusted in the future to yield a storage situation at zero bias, realized by changing the width between the QD layer and the p–n junction.

2.8.2 Write Time Measurements

To study the limit of the write-time, series of write pulses are applied with decreasing pulse widths down to 300 ps. The operation principle is schematically depicted in Figure 2.32. The cycle

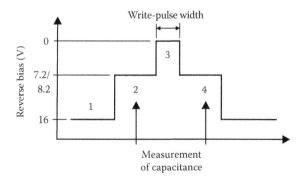

FIGURE 2.32 Operation principle for measuring the write time, while the labeled numbers refer to the numbers in Figure 2.31. (Reprinted from Geller, M. et al., *Appl. Phys. Lett.*, 92, 092108, 2008. With permission.)

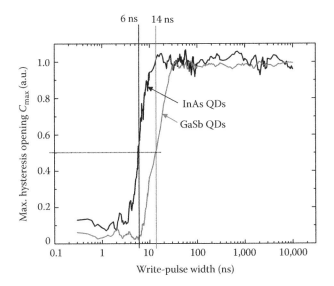

FIGURE 2.33 Maximum hysteresis opening C_{max} for varying write-pulse width. The InAs QD memory is successfully charged for write-pulses down to 6 ns. Analogously, the information in the GaSb QD memory structure can be written down to 14 ns.

started with a 10 s long erase pulse at $V_r = 14\,V/16\,V$ (point 1). Then the capacitance of the device structure is measured at the storage position (point 2) at 7.2 V (GaSb QDs) and 8.2 V (InAs QDs), respectively. A write pulse of $V_r = 0\,V$ was applied to the device (point 3) and the capacitance is again measured at the storage situation point 4.

The maximum hysteresis opening C_{max} was plotted versus the write-pulse width in Figure 2.33. The limit of the write time is reached when the QDs are not sufficiently charged anymore. The hysteresis opening C_{max} vanishes and the write time is defined as a drop in C_{max} to 50%. A minimum write time of 6 ns at $T = 15\,K$ for the InAs QD sample and 14 ns at $T = 100\,K$ for the GaSb QD sample is obtained, respectively. These values are in the order of the write time in a DRAM cell and are now only limited by the experimental set-up and the parasitic cut-off frequency of the RC low pass of the devices. Additional measurements for different temperatures on the same device showed a temperature-independent write time, for example, the difference between the InAs and GaSb structure is due to these different cut-off frequencies. This assumption is also confirmed by previous measurements (Geller et al. 2006b) of the same structure, where average hole capture and relaxation times were found in the order of picoseconds. Therefore, much faster write times below 1 ns are feasible for improved nanomemory structures having higher external cut-off frequencies.

2.9 Summary and Outlook

The semiconductor memory market is evenly split following two different layout approaches: the DRAM and the Flash. Both memories have their limitations in storage time, write/read/erase time, and endurance and will be facing fundamental physical problems in the near future when the feature size has to go down

to 14 nm by the year 2020. Therefore, a large variety of novel concepts using different physical phenomena to store an information bit has been reported. One of the promising options is the use of self-organized nanostructures in future nanoelectronic devices. Especially self-organized QDs based on III–V materials (e.g., GaAs, InAs, InP, etc.) could realize a nonvolatile memory (like Flash) with fast read/write/erase access time (like DRAM). In addition, with an area density of up to 10^{11} cm^{-2}, a storage density in the order of 1 TBit in.$^{-2}$ could be possible, if each single QD—with a lateral feature size of about 10 nm—would represent one information bit and it could be addressed individually.

This chapter showed the potential and advantages of using self-organized QDs in a III–V semiconductor nanomemory concept. The charge carrier storage and emission from different QD systems has been studied in order to determine the carrier storage time at room temperature. Localization energies were determined by using time-resolved capacitance spectroscopy (DLTS) and were related to the hole/electron retention time. A maximum storage time of 1.6 s at room temperature has already been demonstrated in InAs/GaAs QDs with $Al_{0.9}Ga_{0.1}As$ barriers. A storage time of more than 10 years is predicted for hole storage in GaSb/AlAs QDs.

The use of III–V heterostructures enables tuning the band structure to a desired storage time at room temperature while simultaneously getting rid of one of the largest drawback of Si-based Flash memories: the fixed SiO_2 barrier that leads to the poor endurance and a bad write time. A very fast write time below 1 ns independent of the storage time (possibly beyond 10 years) is possible. The physical limitation of the write time in such a QD-based nanomemory is only given by the relaxation time of the charge carriers, known to be in the picosecond range. A write time of 14 ns in $GaSb_{0.6}As_{0.4}/GaAs$ QDs and 6 ns for InAs/GaAs QDs has already been achieved. These numbers are now limited by the experimental set-up and the parasitic cut-off frequency of the RC low pass of the devices. Faster write times below 1 ns in combination with a storage time of more than 10 years at room temperature are expected in the future. A fast semiconductor Flash nanomemory—even faster than the present day DRAM—based on III–V self-organized QDs seems to be possible.

Acknowledgments

This work was partly funded by the SANDiE Network of Excellence of the European Commission, Contract No. NMP4-CT-2004-500101; the Nanomat project of the European Commission Growth Program, Contract No. G5RD-CT-00545; and SFB 296 of DFG. We also acknowledge financial support by the QD-Flash project in the framework of ProFIT of the Investitionsbank Berlin (IBB) and the DFG, BI 284/29-1. We want to acknowledge the support of and helpful discussions with Dr. C. Kapteyn, T. Nowozin, and E. Stock. Dr. L. Müller-Kirsch, K. Pötschke, D. Feise, A. P. Vasi'ev, E. S. Semenova, Prof. A. E. Zhukov, and Prof. V. M. Ustinov are acknowledged for providing us with the samples.

References

Alferov, Z. I. and Kazarinov, R. F. 1963. Semiconductor laser with electric pumping. Inventor's certificate no. 181737 [in Russian]. Application no. 950840.

Anand, S., Carlsson, N., Pistol, M.-E. et al. 1995. Deep level transient spectroscopy of InP quantum dots. *Appl. Phys. Lett.* **67**: 3016.

Anand, S., Carlsson, N., Pistol, M.-E. et al. 1998. Electrical characterization of InP/GaInP quantum dots by space charge spectroscopy. *J. Appl. Phys.* **84**: 3747.

Ashoori, R. C. 1996. Electrons in artificial atoms. *Nature* **379**: 413.

Ashoori, R. C., Störmer, H. L., Weiner, J. S. et al. 1992. Single-electron capacitance spectroscopy of discrete quantum levels. *Phys. Rev. Lett.* **68**: 3088.

Atwood, G. 2004. Future directions and challenges for Etox flash memory scaling. *IEEE Trans. Device Mater. Reliab.* **4**: 301.

Baibich, M. N., Broto, J. M., Fert, A. et al. 1988. Giant magnetoresistance of (001)Fe/(001)Cr magnetic superlattices. *Phys. Rev. Lett.* **61**: 2472.

Batey, J. and Wright, S. L. 1986. Energy band alignment in GaAs:(Al,Ga)As heterostructures: The dependence on alloy composition. *J. Appl. Phys.* **59**: 200.

Bimberg, D., Grundmann, M., and Ledentsov, N. N. 1998. *Quantum Dot Heterostructures*. Chichester, U.K.: John Wiley & Sons.

Binasch, G., Grunberg, P., Saurenbach, F. et al. 1989. Enhanced magnetoresistance in layered magnetic-structures with antiferromagnetic interlayer exchange. *Phys. Rev. B* **39**: 4828.

Blood, P. and Orton, J. W. 1992. *The Electrical Characterization of Semiconductors: Majority Carriers and Electron States*. London, U.K.: Academic Press.

Burr, G. W., Kurdi, B. N., Scott, J. C. et al. 2008. Overview of candidate device technologies for storage-class memory. *IBM J. Res. Dev.* **52**: 449.

Campbell, J. C., Huffaker, D. L., Deng, H. et al. 1997. Quantum dot resonant cavity photodiode with operation near 1.3 μm wavelength. *Electron. Lett.* **33**: 1337.

Chu, L., Zrenner, A., Bichler, M. et al. 2001. Quantum-dot infrared photodetector with lateral carrier transport. *Appl. Phys. Lett.* **79**: 2249.

Day, D. S., Tsai, M. Y., Streetman, B. G. et al. 1979. Deep-level-transient spectroscopy: System effects and data analysis. *J. Appl. Phys.* **50**: 5093.

Dennard, R. H. 1968. Field-effect transistor memory. U.S. patent no. 3,387,286.

Eaglesham, D. J. and Cerullo, M. 1990. Dislocation-free Stranski–Krastanow growth of Ge on Si (100). *Phys. Rev. Lett.* **64**: 1943.

Eisele, H., Flebbe, O., Kalka, T. et al. 1999. Cross-sectional scanning-tunneling microscopy of stacked InAs quantum dots. *Appl. Phys. Lett.* **75**: 106.

Frank, D. J., Dennard, R. H., Nowak, E. et al. 2001. Device scaling limits of Si MOSFETs and their application dependencies. *Proc. IEEE* **89**: 259.

Fricke, M., Lorke, A., Kotthaus, J. P. et al. 1996. Shell structure and electron–electron interaction in self-assembled InAs quantum dots. *Europhys. Lett.* **36**: 197.

Gallagher, W. J. and Parkin, S. S. P. 2006. Development of the magnetic tunnel junction MRAM at IBM: From first junctions to a 16-MB MRAM demonstrator chip. *IBM J. Res. Dev.* **50**: 333.

Geller, M., Kapteyn, C., Müller-Kirsch, L. et al. 2003. 450 meV hole localization energy in GaSb/GaAs quantum dots. *Appl. Phys. Lett.* **82**: 2706.

Geller, M., Marent, A., and Bimberg, D. 2006a. Speicherzelle und Verfahren zum Speichern von Daten. German patent application no. 10 2006 059 110.0.

Geller, M., Marent, A., Stock, E. et al. 2006b. Hole capture into self-organized InGaAs quantum dots. *Appl. Phys. Lett.* **89**: 232105.

Geller, M., Stock, E., Kapteyn, C. et al. 2006c. Tunneling emission from self-organized In(Ga)As/GaAs quantum dots observed via time-resolved capacitance measurements. *Phys. Rev. B* **73**: 205331.

Geller, M., Stock, E., Sellin, R. L. et al. 2006d. Direct observation of tunneling emission to determine localization energies in self-organized In(Ga)As quantum dots. *Physica E* **32**: 171.

Geller, M., Marent, A., Nowozin, T. et al. 2008. A write time of 6 ns for quantum dot-based memory structures. *Appl. Phys. Lett.* **92**: 092108.

Geppert, L. 2003. The new indelible memories—It's a three-way race in the multibillion-dollar memory sweepstakes. *IEEE Spectr.* **40**: 48.

Giorgi, M. D., Lingk, C., Plessen, G. V. et al. 2001. Capture and thermal re-emission of carriers in long-wavelength InGaAs/GaAs quantum dots. *Appl. Phys. Lett.* **79**: 3968.

Grimmeiss, H. G. and Ovrén, C. 1981. Fundamentals of junction measurements in the study of deep energy levels in semiconductors. *J. Phys. E: Sci. Instrum.* **14**: 1032.

Grundmann, M., Stier, O., and Bimberg, D. 1995. InAs/GaAs quantum pyramids: Strain distribution, optical phonons and electronic structure. *Phys. Rev. B* **52**: 11969.

Guo, L., Leobandung, E., and Chou, S. Y. 1997. A room-temperature silicon single-electron metal-oxide-semiconductor memory with nanoscale floating-gate and ultranarrow channel. *Appl. Phys. Lett.* **70**(7): 850.

Hatami, F., Grundmann, M., Ledentsov, N. N. et al. 1998. Carrier dynamics in type-II GaSb/GaAs quantum dots. *Phys. Rev. B* **57**: 4635.

Hatami, F., Ledentsov, N. N., Grundmann, M. et al. 1995. Radiative recombination in type-II GaSb/GaAs quantum dots. *Appl. Phys. Lett.* **67**: 656.

Heinrichsdorff, F., Krost, A., Grundmann, M. et al. 1996. Self-organization processes of InGaAs/GaAs quantum dots grown by metalorganic chemical vapor deposition. *Appl. Phys. Lett.* **68**: 3284.

Heinrichsdorff, F., Ribbat, C., Grundmann, M. et al. 2000. High-power quantum-dot lasers at 1100 nm. *Appl. Phys. Lett.* **76**: 556.

Heitz, R., Veit, M., Ledentsov, N. N. et al. 1997. Energy relaxation by multiphonon processes in InAs/GaAs quantum dots. *Phys. Rev. B* **56**: 10435.

Hopfer, F., Mutig, A., Kuntz, M. et al. 2006. Single-mode sub-monolayer quantum-dot vertical-cavity surface-emitting lasers with high modulation bandwidth. *Appl. Phys. Lett.* **89**: 141106.

ITRS. 2008. International Technology Roadmap for Semi-conductors (ITRS) 2007 Edition (2008). Technical report.

Jones, R. E., Maniar, P. D., Moazzami, R. et al. 1995. Ferroelectric non-volatile memories for low-voltage, low-power applications. *Thin Solid Films* **270**: 584.

Kapteyn, C. 2001. *Carrier Emission and Electronic Properties of Self-Organized Semiconductor Quantum Dots.* Berlin, Germany: Mensch & Buch Verlag.

Kapteyn, C. M. A., Heinrichsdorff, F., Stier, O. et al. 1999. Electron escape from InAs quantum dots. *Phys. Rev. B* **60**: 14265.

Kapteyn, C. M. A., Lion, M., Heitz, R. et al. 2000a. Hole and electron emission from InAs quantum dots. *Appl. Phys. Lett.* **76**: 1573.

Kapteyn, C. M. A., Lion, M., Heitz, R. et al. 2000b. Many-particle effects in Ge quantum dots investigated by time-resolved capacitance spectroscopy. *Appl. Phys. Lett.* **77**: 4169.

Kim, K., Hwang, C. G., and Lee, J. G. 1998. DRAM technology perspective for gigabit era. *IEEE Trans. Electron Devices* **45**: 598.

Kim, H., Noda, T., Kawazu, T. et al. 2000. Control of current hysteresis effects in a GaAs/n-AlGaAs quantum trap field effect transistor with embedded InAs quantum dots. *Jpn. J. Appl. Phys.* **39**: 7100.

Kimerling, L. C. 1974. Influence of deep traps on the measurement of free-carrier distributions in semiconductors by junction capacitance techniques. *J. Appl. Phys.* **45**: 1839.

Kiravittaya, S., Rastelli, A., and Schmidt, O. G. 2009. Advanced quantum dot configurations. *Rep. Prog. Phys.* **72**: 046502.

Kirstaedter, N., Ledentsov, N. N., Grundmann, M. et al. 1994. Low threshold, large T_0 injection laser emission from (InGa)As quantum dots. *Electron. Lett.* **30**: 1416.

Koike, K., Saitoh, K., Li, S. et al. 2000. Room-temperature operation of a memory-effect AlGaAs/GaAs heterojunction field-effect transistor with self-assembled InAs nanodots. *Appl. Phys. Lett.* **76**: 1464.

Kroemer, H. 1963. Semiconductor laser with optical pumping. U.S. patent no. 3309553.

Kuntz, M., Fiol, G., Lämmlin, M. et al. 2004. 35 GHz mode-locking of 1.3 μm quantum dot lasers. *Appl. Phys. Lett.* **85**: 843.

Lai, S. K. 2008. Flash memories: Successes and challenges. *IBM J. Res. Dev.* **52**: 529.

Lambe, J. and Jaklevic, R. C. 1969. Charge-quantization studies using a tunnel capacitor. *Phys. Rev. Lett.* **22**: 1371.

Lämmlin, M., Fiol, G., Meuer, C. et al. 2006. Distortion-free optical amplification of 20–80 GHz modelocked laser pulses at 1.3 μm using quantum dots. *Electron. Lett.* **42**: 697.

Lang, D. V. 1974. Deep-level transient spectroscopy: A new method to characterize traps in semiconductors. *J. Appl. Phys.* **45**: 3023.

Lochmann, A., Stock, E., Schulz, O. et al. 2006. Electrically driven single quantum dot polarised single photon emitter. *Electron. Lett.* **42**: 774.

Luyken, R. J., Lorke, A., Govorov, A. O. et al. 1999. The dynamics of tunneling into self-assembled InAs dots. *Appl. Phys. Lett.* **74**: 2486.

Mandelman, J. A., Dennard, R. H., Bronner, G. B. et al. 2002. Challenges and future directions for the scaling of dynamic random-access memory (DRAM). *IBM J. Res. Dev.* **46**: 187.

Marent, A., Geller, M., Bimberg, D. et al. 2006. Carrier storage time of milliseconds at room temperature in self-organized quantum dots. *Appl. Phys. Lett.* **89**: 072103.

Marent, A., Geller, M., Schliwa, A. et al. 2007. 10^6 years extrapolated hole storage time in GaSb/AlAs quantum dots. *Appl. Phys. Lett.* **91**: 42109.

Mikolajick, T., Nagel, N., Riedel, S. et al. 2007. Scaling of non-volatile memories to nanoscale feature sizes. *Mater. Sci.-Pol.* **25**: 33.

Miller, G. L., Lang, D. V., and Kimerling, L. C. 1977. Capacitance transient spectroscopy. *Ann. Rev. Mater. Sci.* **7**: 377.

Mo, Y.-W., Savage, D. E., Swartzentruber, B. S. et al. 1990. Kinetic pathway in Stranski–Krastanov growth of Ge on Si(001). *Phys. Rev. Lett.* **65**: 1020.

Moore, G. E. 1965. Cramming more components onto integrated circuits. *Electronics* **38**: 82.

Müller, T., Schrey, F. F., Strasser, G. et al. 2003. Ultrafast intraband spectroscopy of electron capture and relaxation in InAs/GaAs quantum dots. *Appl. Phys. Lett.* **83**: 3572.

Müller-Kirsch, L., Heitz, R., Pohl, U. W. et al. 2001. Temporal evolution of GaSb/GaAs quantum dot formation. *Appl. Phys. Lett.* **79**: 1027.

Myburg, G., Auret, F. D., Meyer, W. E. et al. 1998. Summary of Schottky barrier height data on epitaxially grown n- and p-GaAs. *Thin Solid Films* **325**: 181.

North, S. M., Briddon, P. R., Cusack, M. A. et al. 1998. Electronic structure of GaSb/GaAs quantum dots. *Phys. Rev. B* **58**: 12601.

Omling, P., Samuelson, L., and Grimmeis, H. G. 1983. Deep level transient spectroscopy evaluation of nonexponential transients in semiconductor alloys. *J. Appl. Phys.* **54**: 5117.

Pavan, P., Bez, R., Olivo, P. et al. 1997. Flash memory cells—An overview. *Proc. IEEE* **85**: 1248.

Rhoderick, E. H. and Williams, R. H. 1988. *Metal Semiconductor Contacts.* Oxford, U.K.: Oxford Science Publications.

Sah, C. T., Forbes, L., Rosier, L. L. et al. 1970. Thermal and optical emission and capture rates and cross sections of electrons and holes at imperfection centers in semiconductors from photo and dark junction current and capacitance experiments. *Solid-State Electron.* **13**: 759.

Santori, C., Fattal, D., Vučković, J. et al. 2002. Indistinguishable photons from a single-photon device. *Nature* **419**: 594.

Schliwa, A., Winkelnkemper, M., and Bimberg, D. 2007. Impact of size, shape, and composition on piezoelectric effects and electronic properties of In(Ga)As/AaAs quantum dots. *Phys. Rev. B* **76**: 205324.

Schulz, S., Schnüll, S., Heyn, C. et al. 2004. Charge-state dependence of InAs quantum-dot emission energies. *Phys. Rev. B* **69**: 195317.

Shchukin, V. A., Ledentsov, N. N., Kopev, P. S. et al. 1995. Spontaneous ordering of arrays of coherent strained islands. *Phys. Rev. Lett.* **75**: 2968.

Shchukin, V. A., Ledentsov, N. N., and Bimberg, D. 2003. *Epitaxy of Nanostructures*. Berlin, Germany: Springer.

Sheikholeslami, A. and Gulak, P. G. 2000. A survey of circuit innovations in ferroelectric random-access memories. *Proc. IEEE* **88**: 667.

Stier, O., Grundmann, M., and Bimberg, D. 1999. Electronic and optical properties of strained quantum dots modeled by 8-band-k.p theory. *Phys. Rev. B* **59**: 5688.

Stievenard, D., Boddaert, X., and Bourgoin, J. 1986. Irradiation-induced defects in p-type GaAs. *Phys. Rev. B* **34**: 4048.

Stranski, I. N. and Krastanow, L. 1938. Zur Theorie der orientierten Ausscheidung von Ionenkristallen aufeinander. *Sitzungsber. Akad. Wiss. Wien, Math.-Naturwiss. K1, Abt. 2B* **146**: 797.

Sze, S. M. 1985. *Semiconductor Devices—Physics and Technology*. New York: John Wiley & Sons.

Sze, S. M. 1999. Evolution of nonvolatile semiconductor memory: From floating-gate concept to single-electron memory cell. In *Future Trends in Microelectronics*, S. Luryi, J. Xu, and A. Zaslavsky (Eds.), p. 291. New York: John Wiley & Sons.

Sze, S. M. 2002. *Semiconductor Devices*. New York: John Wiley & Sons.

Tiwari, S., Rana, F., Hanafi, H. et al. 1996. A silicon nanocrystal based memory. *Appl. Phys. Lett.* **68**: 1377.

Vurgaftman, I., Meyer, J. R., and Ram-Mohan, L. R. 2001. Band parameters for III–V compound semiconductors and their alloys. *J. Appl. Phys.* **89**: 5815.

Waser, R. 2003. *Microelectronics and Information Technology*. Berlin, Germany: Wiley-VCH.

White, M. H., Adams, D. A., and Bu, J. K. 2000. On the go with SONOS. *IEEE Circuits Devices* **16**: 22.

Wolf, S. A., Chtchelkanova, A. Y., and Treger, D. M. 2006. Spintronics—A retrospective and perspective. *IBM J. Res. Dev.* **50**: 101.

3

Carbon Nanotube Memory Elements

Vincent Meunier
Oak Ridge National Laboratory

Bobby G. Sumpter
Oak Ridge National Laboratory

3.1 Introduction

An incredibly rapid expansion in the demand of portable consumer electronics ranging from personal computers, cellular phones, MP3 players, digital cameras, PDAs, USB memory sticks to applications in the networking arena continues to drive the pursuit for the development of higher density, faster, more efficient, and economical electronic memory devices (Scott 2004). Currently, most commercially available memory elements consist of dynamic random access memory (DRAM), static random access memory (SRAM), or flash memory (non-volatile memory). DRAM is considered to be very fast and cheap but its contents are lost when power is switched off (volatility). SRAM is much faster and needs less power but is far more expensive and also suffers from volatility. Flash memory is nonvolatile but currently operates at low write speed and thus has a slow random access. In addition, it is power-hungry and very expensive. In the past, incremental improvements in memory capacity and capability were primarily achieved by the simple scaling of the physical dimensions of the devices. However, the semiconductor industry is quickly reaching the fundamental limits on the miniaturization, encountering, for example, extreme difficulties due to short channel effects in the scaling of metal-oxide-semiconductor field-effect transistor

(MOSFET). This difficulty has stimulated an intense growth in research and development in the area of new technologies and materials that can overcome these limitations and deliver unprecedented capabilities for memory storage and access. Some promising new types of memory-storage devices include magnetoresistive RAM (MRAM), ferroelectric RAM (FRAM), phase-change memory (PRAM), and novel high speed–density nonvolatile memory (these are primarily based on carbon nanotubes [CNTs]) (Waser 2003).

CNT-based data-storage devices or memory elements offer high potential for considerable improvements compared to current memory paradigms (Bichoutskaia et al. 2008, Zhang 2006). First, CNTs are considerably smaller than conventional cells, e.g., each nanotube is less than 1/10,000th the width of a human hair, thus allowing an impressive density to be achieved in a modest space. The intrinsic structural, thermal, and mechanical stability of CNTs are expected to confer to these memory devices a very long lifetime while the power requirements for their operation should be considerably less than DRAM. In addition, CNTs are known to be radiation-resistant and nanotube-based memory elements are therefore expected to be robust in numerous types of environments. As such, these novel types of memory elements have recently been touted as a "universal memory" (Bichoutskaia et al. 2008).

In this chapter, we review the developments in memory elements that exploit the properties of CNTs. These nanoscale systems fit into the realm of nanoelectronics and offer exceptional properties and well-characterized structures that have rapidly emerged as a viable route to satisfy the future requirements and needs for data-storage and memory devices. Given the difficulty in the integration of nanoscale systems, such as CNTs, into adequate arrangements for useful structures and devices, and the complexity in the assessment of the properties of the devices, a tremendous effort has been devoted in the past years toward the development of experimental and theoretical methods making it possible to creatively imagine, design, and test new systems for desired and tailored characteristics. In this chapter, we review the different types of CNT-based memory devices, which include memory elements based on CNT field-effect transistors (CNFET) (Section 3.2), nanoelectromechanical systems (NEMS) (Section 3.3), and electromigration (Section 3.4). The fundamental operational principles and characteristics of the different devices and concepts are examined in detail along with a discussion of the current status of experimental fabrication and practical realization.

3.2 CNFET-Based Memory Elements

A number of CNT memory elements have been developed from the functioning principle of CNFETs. The following section will be devoted to a description of a few of them. First, we will expose

the basic working principle of the CNFET before showing how CNFETs can be implemented for memory applications.

3.2.1 General Description of a CNFET

Nanotube transistors are among the possible replacements for silicon field-effect transistors (FETs) as bulk-device scaling approaches quantum limits. A CNFET is a device consisting of a single nanotube and three contacts (Figure 3.1a). It has been demonstrated repeatedly in the laboratory to possess characteristics similar to those of conventional silicon FETs. Because nanotubes can be used as either n- or p-type CNFETs, existing complementary metal-oxide-semiconductor (CMOS) design techniques can be utilized for a nanotube-based FET technology. A CNT-based CMOS industry is very appealing, in particular because with their electrical and thermal properties, CNTs can sustain very large current in the billion amperes per square centimeter regime (about three orders of magnitude larger than conventional metallic wires).

The main hurdle facing nanotube-FET technology remains the difficulty of developing scalable manufacturing methods to compete with the well-developed silicon large-scale production.

CNTs can behave as either conductors or as semiconductors, depending on the arrangement of the atoms along the nanotube (Bernholc et al. 2002, Charlier et al. 2007). While about one-third of nanotubes are metals, the bandgap of the semiconducting tubes varies inversely with the tube diameter, making it possible to design a large range of devices and heterostructures. Conversely,

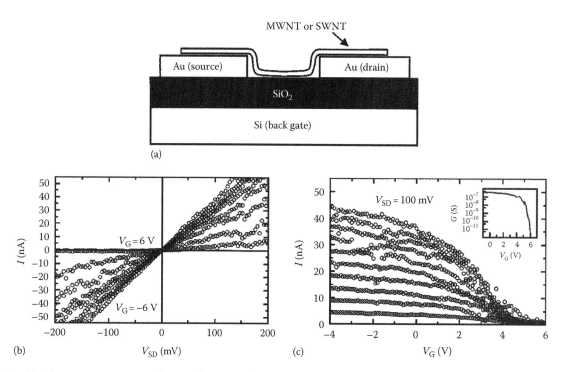

FIGURE 3.1 (a) Schematic cross section of a typical CNFET device. A single CNT of either multiwall- or single-wall-type bridges the gap between two gold electrodes. The silicon substrate is used as back gate. Output and transfer characteristics are shown in (b) and (c), respectively. The current vs. source–drain voltage curves in (b) were measured for gate voltage values of −6, 0, 1, 2, 3, 4, 5, and 6 V. In (c) the current vs. gate voltage are given for 10 different source–drain voltages between 10 and 100. The inset shows that the gate modulates the conductance by five orders of magnitude (for a source–drain voltage of 10 mV). (Reprinted from Martel, R. et al., *Appl. Phys. Lett.*, 73(17), 2447, 1998. With permission.)

conducting nanotubes possess a very low resistance for electron flow (ballistic conductors) because the decoherence length of the nanotube wavefunctions is typically larger than practical nanotube lengths. As long as the nanotube length is shorter than the decoherence length, the resistance will not depend upon the cylinder length. In that case, the only electrical resistance is at the contact points where the electrons enter and leave the tube. As shown below, it is precisely at the interface between the nanotube and the metallic electrodes that the physical processes responsible for the functioning of a nanotube transistor take place.

CNFETs can be fabricated with a structure very similar to silicon FETs. As shown in Figure 3.1a, such a device consists of a single-walled semiconducting nanotube whose ends are attached to two metallic contacts. In the terminology of transistors, the nanotube acts as the channel and the two contacts are the drain and the source, respectively. The device is completed by a gate electrode that is positioned somewhere between the source and the drain. There are a number of possible designs: the nanotube can be positioned above or under the metallic contacts and the gate can be placed as a back or front gate. For a transistor to work properly the semiconductor CNT must be doped. In air, at room temperature, the device behaves as a depletion mode p-FET (the *I–V* is typical of hole conduction, Figure 3.1b). The transistor action is due to the electric field around the gate modulating the Schottky barrier at the points where the source and drain terminals contact the nanotube. For reasonably short tubes (a few hundred nanometers or less), the effect of the gate voltage on channel resistance is minimal. Experimentally, the first-reported CNFETs were all p-type devices. This is because when exposed to air, nanotubes are p-doped due to the presence of oxygen that shifts the Fermi level at the metal–nanotube interface toward the valence band, leading to a hole-conduction mechanism. When the same device is heated in a vacuum and allowed to outgas, the Fermi level shifts toward the conduction band, which leads to an electron-conduction mechanism, i.e., acts as an n-type FET. Different types of FETs can be obtained

depending on chemical doping, leading to p-type, n-type, or even p–n-type (ambipolar) systems.

Here we describe the general functioning of a p-type transistor. An understanding for the n-type systems can be easily obtained by considering electrons as majority carriers. As shown in Figure 3.2, the application of a negative gate voltage increases the number of majority carriers in the channel, thereby leading to an increased current density across the transistor. For positive gate voltage, the current is negligibly small, due to the depletion in the number of holes. At the same time there is no evidence of electron conduction, even at very large positive gate voltages.

While nanotubes show great promise in the mechanical and microelectronic realms, major problems still exist (Bernholc et al. 2002). Assuming the ability to produce large quantities of nanotubes with specific geometries, a separate difficulty lies in actually assembling large quantities of transistors out of them and placing those transistors, along with an interconnect, on a substrate to form a complete circuit. Solutions to these problems are only beginning to surface. With respect to electronics, the biggest difficulty is devising a production method that can provide the same facility for large-scale fabrication as silicon photolithography. This goal appears distant but it continues to be a topic of intense research. With the demonstrated potential payoff for a range of applications, it seems likely that a solution to this problem will eventually be found. Further details on the general aspects of CNFETs can be found in the vast literature published on the topic (e.g., Derycke et al. 2001, 2002, Martel et al. 2001, and Tans et al. 1998). Below we review how CNTs have been implemented for FET-based memory applications.

3.2.2 FET-Based Memory Elements: Pioneering Works

Nonvolatile and volatile memory elements can be designed from the functioning of a CNFET (Wunnicke 2006). The storage media is typically a charge that is stored in the vicinity of the nanotube

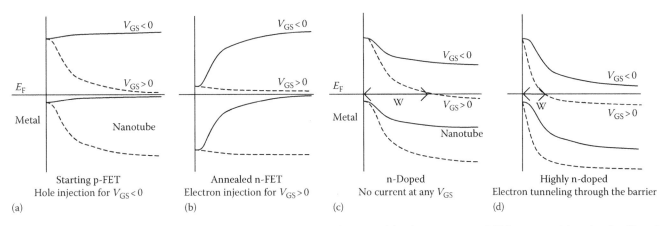

FIGURE 3.2 Schematic of the bands in the vicinity of the contacts as a function of the electrostatic gate field for a p-type (a), ambipolar (b), and n-type (c) CNFETs. Annealing the device in vacuum results in a transition from the situation of (a) to that of (c) in case of low doping and (d) for high doping. The transition can be reversed by the introduction of air, clearly showing that the presence of oxygen p-doping of the nanotube. It is clear that the Schottky barrier at the nanotube–metal interface plays a key role. Note also the importance of Fermi-level pinning for the functioning of the device. (Reprinted from Derycke, V. et al., *Appl. Phys. Lett.*, 80(15), 2773, 2002. With permission.)

channel. The charge causes a threshold potential shift (the threshold potential is defined as the potential at which the majority carriers start flowing) of the nanotube FET. Because the nanotube has very high carrier mobility, the information can be stored in a few (as few as one) electrons configuration. The states can be reversibly written and erased. The reading mechanism is performed via a measurement of the source–drain current, while the writing mechanism is completed using a large bias voltage applied to the gate.

A number of groups have obtained convincing results on CNFET-based memory elements. How the information is stored and consequently under which conditions the memory is usable depends on the details of the manufacturing, particularly in the gating material. The first two examples of FET-memory elements were proposed independently by Fuhrer et al. (2002) and Radosavljevic et al. (2002). The functioning of the devices hinges on the operation of a single-electron memory. In this case, the capacitance of the storage node must be small enough so that its Coulomb-charging energy is significantly larger than the thermal energy at the operating temperature, and the readout device must be sensitive enough to detect a single nearby electronic charge. In

Fuhrer's device, the charge is reversibly injected and removed from the dielectric (placed between the tube and the gate electrode) by applying a moderate (10 V) bias between the nanotube and the substrate. The nanotube is ideal as a charge-detecting device due to its high carrier mobility, large geometrical capacitance, and its one-dimensional (1D) nature ensuring that local changes in charge affects the global conductance (due to slow screening). In Fuhrer's pioneering work, for instance, discrete charge states corresponding to differences of a single, or at most a few, stored electrons are observed. The device is based on the characteristics of a p-type FET (i.e., it conducts at negative gate voltage and becomes insulating at positive gate voltage) and can be operated at temperatures up to 100 K. Properties relevant to memory operation are demonstrated by the large hysteresis I–V_g curve that is obtained by sweeping the gate voltage V_g between −10 and +10 V. As is shown in Figure 3.3, the threshold voltage is shifted by more than 6 V. The mechanism of charge storage is related to the rearrangement of charges in the dielectric or by the injection or removal of charges from the dielectric through the electrodes or the nanotube. Due to the geometry of the device, the electric field at the surface of the

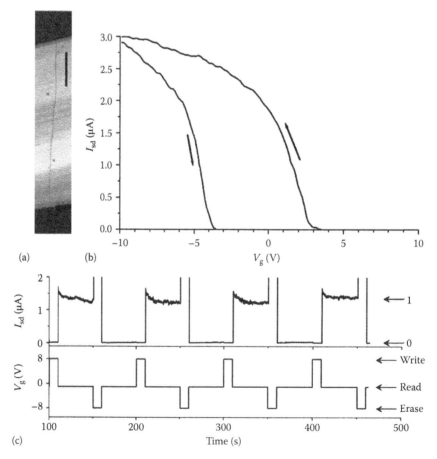

FIGURE 3.3 (a) Atomic force microscope topographic image of the nanotube device used in Fuhrer et al. (2002). The nanotube extends between the two dark blocks at the top and bottom of the image (i.e., the electrodes). The scale bar represents 1 μm. (b) Drain current as a function of gate voltage at room temperature and a source–drain bias of 500 mV. As the gate voltage is swept from positive to negative and back, a strong hysteresis is observed, as indicated by the arrows denoting the sweep direction. (c) Series of four read–write cycles of the nanotube memory at room temperature. The upper panel shows the drain current at a source–drain bias of 500 mV, while the lower panel shows the gate voltage. The memory state was read at −1 V, and written with pulses of ±8 V. (Reprinted from Fuhrer, M.S. et al., *Nano Lett.*, 2(7), 755, 2002. With permission.)

nanotube is very large, at least large enough to cause the movement of charge in the dielectric (Fuhrer et al. evaluated the field to be comparable to the breakdown field in SiO$_2$). At these high fields the electrons are easily injected into the dielectric from the nanotube and remain trapped in metastable states until the polarity is reversed.

In Radosavljevic's device, the FET used is based on an n-type transistor (obtained from annealing the nanotube in hydrogen gas). Again, it was shown that CNFETs are extremely sensitive to the presence of individual charges around the channel, largely because of the nanoscale capacitance of the CNTs. A technique known as scanning gate microscopy (SGM) (Freitag et al. 2002, Meunier et al. 2004) was used to study the local transport properties across the channel, revealing the key role of nanotube–metal contacts (Schottky barriers) and also the positions of the possible sites where the electrons are trapped (i.e., where the information is physically stored, the storage nodes). Figure 3.4 shows the reproducible hysteresis loop in the I–V_g curve that becomes larger as the range of V_g is increased, indicating that it originates from avalanche injection into bulk oxide traps. The hysteresis can become so large that the device varies from depletion mode (normally on at $V_g = 0$) and enhancement mode (normally off) behavior. The location and sign of the trapped charge can be determined by examining the directions of the hysteresis loop. After a sweep to positive V_g, the CNFET threshold voltage moves toward more positive gate values, indicating injection of

negative charges into oxide traps. To read the memory, a 1 MΩ load resistor is added to create a voltage divider. Read ($V_{in} = 0$) and write ($V_{in} = +20$ or -20 V) are applied to the input terminal (back gate). A logical gate (1 or 0) is defined as $V_{out} = 1$ or 0. To write a "1" to the memory cell, V_{in} is switched rapidly to -20 or $+20$ V and back to "0," so the CNFET is in the on or off at the read voltage. This memory device was found to be nonvolatile at room temperature and with a bit-storage retention time of at least 16 h. The trap charging time limits the speed of the device and the bit was evaluated to be stored in no more than 2, 70, and 200 e.

In the pioneering works discussed above, it is noteworthy that the storage nodes are not precisely the nanotube itself, but consist of trapping sites in the dielectric layer of the gating material. The functional temperatures, storage density, operating speeds, etc., are very sensitive to the details of the trapping mechanisms. As summarized below, a number of studies have been reported following the seminal studies of Fuhrer and Radosavljevic, where the trapping mechanism was further analyzed and modified.

3.2.3 FET-Based Memory Elements: Further Improvements

Charge-storage stability of up to 12 days at room temperature was reported in a 150 nm long nanotube channel (Cui et al. 2002). The observed conductance decreases for increasing gate

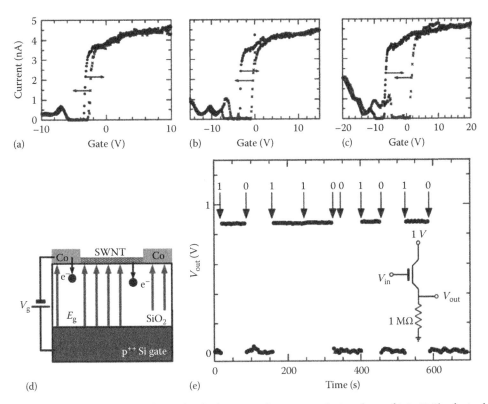

FIGURE 3.4 (a–c) Current vs. gate voltage data obtained in high vacuum for a source drain voltage of 0.5 mV. The device hysteresis increases steadily with increasing gate voltage due to avalanche charge injection into bulk oxide traps, schematically shown in (d). (e) Demonstration of the CNFET-based nonvolatile molecular memory cell. A series of bits is written into the cell and the cell contents are continuously monitored as a voltage signal (V_{out}) in the circuit shown in the inset. (Reprinted from Radosavljevic, M. et al., *Nano Lett.*, 2(7), 761, 2002. With permission.)

potential, a normal feature of p-type, air-exposed single-wall carbon nanotubes (SWCNTs). The hysteresis loop is obtained by sweeping continuously the gate voltage from −3 to +3 V. This shows two conductance states at $V_g = 0$ differing by more than two orders of magnitude, associated with a threshold voltage of 1.25 V. The method used by Cui et al. is more technologically straightforward compared to the previously reported techniques, as it does not require separating SWCNT bundles into individual SWCNTs during sample preparation. The sample treatment is likely to be responsible for the modified type of trapping centers, compared to earlier similar devices. The heat treatment and exposure to oxygen plasma cause the metallic nanotubes present in the bundle to be preferentially oxidized, leading to increased gate dependence due to the remaining intact semiconducting tubes. In addition, oxidation-related defects are likely to be formed in the remaining amorphous carbon particles on the bundle surface or at the SiO_2 interface. These defects act as charge-storage traps and their close proximity to the surface of the channel accounts for the large threshold voltage shifts.

Another type of charge trapping was presented by Kim et al. (2003). In that case, water molecules were shown to be responsible for the hysteresis properties (i.e., responsible for the threshold potential shift) of the I–V_g curves. The water molecules contributing to the trapping could be located on the nanotube surface or on the SiO_2 close to the tube. Heating under dry conditions significantly reduces the hysteresis. A completely hysteresis-free CNFET was possible by passivation of the device using a polymer coating, clearly indicating that the storage nodes were removed by the treatment. This work also confirms the central role played by surface chemistry on the properties of the device and that truly robust passivation is needed in order to use CNT-based devices in practical electronic devices, unless it is simply used as a detection device (e.g., of humidity).

More recent work of Yang et al. confirmed the role of adsorbed (including water and alcohol) molecules for charge trapping, also showing charge retention of up to 7 days under ambient conditions (Yang et al. 2004).

3.2.4 FET-Based Memory Elements: Controlling Storage Nodes

Soon after the publication of the pioneering works of 2002, a number of groups confirmed that the as-prepared nanotube FETs possessed intrinsic charge-trapping centers that were responsible for the shift in the threshold voltage measured in the hysteresis loop of the I–V_g characteristics. As discussed briefly above, the trapping centers can be defects in the SiO_2, water, or alcohol molecules adsorbed on the tube or the dielectric, or oxidized amorphous carbon. It was soon understood that a better control of the storage medium was required in order to harness the full potential of CNFET-based memory elements.

In 2003, Choi et al. proposed a SWCNT nonvolatile memory device using SiO_2-Si_3N_4-SiO_2 (ONO) layers as the storage node (Choi et al. 2003). In that device, the top gate structure is placed above the ONO layer, which is positioned directly above

a few-nanometers-long channel. Charges can tunnel from the CNT surface into the traps present in the ONO layers. The stored charges impose a threshold voltage shift of 60 mV, which is independent of charging time, suggesting that the ONO traps present a quasi-quantized state. The choice of SiO_2-Si_3N_4-SiO_2 is motivated by the fact that it presents a high breakdown voltage, low defect density, and high charge-retention capability (Bachhofer et al. 2001).

Another type of device was demonstrated experimentally in 2005 (Ganguly et al. 2005). In this case, the charge-storage nodes consist of gold nanocrystals placed on the top gate above the nanotube channel. The device was found to have a large memory window with low voltage operations and single-electron-controlled drain currents. The device is based on a Coulomb blockade behavior, in other words on the difficulty of adding supplementary electrons due to the large charging effect of the system with large capacitance and Coulomb repulsion. The Coulomb blockade in the nanocrystals, along with the single-charge sensitivity of the nanotube FET, is suggested to allow for multilevel operations. In this device, the reported retention is rather modest: 6200 s at 10 K and only 800 s at room temperature. Better dielectric and preprocessing should improve these low retention times.

More recently, Sakurai et al. proposed a modified design where the dielectric insulator was made up of a ferroelectric thin film (Sakurai et al. 2006). The experimental results show that the carriers in the CNTs are controlled by the spontaneous polarization of the ferroelectric films. Ferroelectric-gate FETs offer potential advantages as nonvolatile memory elements, such as low power consumption, the capability of high-density integration, and nondestructive readable operation. So far, the use of ferroelectric thin films in conventional Si-based FETs for memory applications has been complicated because of the extreme difficulty of fabricating the ferroelectric/Si structure with good interface properties due to the chemical reaction and interdiffusion of Si and ferroelectrics. It was suggested that this problem could be resolved by inserting an insulating layer between the ferroelectric and the semiconductor layers, but at the cost of increased voltage operation and depolarization field. Another method was to use an oxide semiconductor, such as indium tin oxide. However, even though this system also shows good memory operation, the carrier mobility is much lower than in Si, causing problems that are unacceptable for high-performance nonvolatile memory elements. The use of a semiconducting tube as the conducting channel and a ferroelectric as a gate insulator alleviates all of these problems: the new design allows the carriers in the channel to be controlled by the spontaneous polarization of the ferroelectric film, while keeping the high-conduction characteristics of CNTs. In the demonstrated prototypical device, a 400 nm thick PZT ($PbZr_{0.5}TiO_5O_3$) film was used. The threshold voltage was found to be shifted in the positive direction when V_g was swept from negative to positive bias and was shifted in the negative direction when V_g was swept from positive to negative. This gave a clear hysteresis loop and bistable current values at $V_g = 0$, due to the spontaneous polarization of the electric field (the effects of charge trapping in the dielectric was found to be negligible compared to the effects due to the ferroelectric, at least for small sweeping range of V_g).

3.2.5 FET-Based Memory Elements: Optoelectronic Memory

In all of the CNFET-based memory devices mentioned thus far in this chapter, the information is stored by applying a large enough gate voltage to store or remove the charge in, or from, the traps (for the ferroelectric-based FET, the gate is used to flip the polarization field). In Star's CNFET-memory (p-type) device, the information is still stored as electric charges (Star et al. 2004), however, it differs in that the writing mechanism relies on the use of optical illumination of a photosensitive polymer (which is part of the gate electrode) that converts photons into electric charge stored in the vicinity of the channel. The information is read by measuring the source–drain current, while it is erased (the charge is removed) using a large gate voltage. In other words, this optoelectronic memory is written optically and read and erased electrically. The threshold voltage change compared to the non-illuminated system was measured to be as large as +2 V after optical excitation, suggesting a charge transfer of about 300 electrons per micron in the tube.

3.2.6 Redox Active Molecules as Storage Nodes

In this experimentally built system, the overall configuration and operating principle consists of a nanotube or nanowire FET functionalized with redox active molecules, where an applied gate voltage or source–drain voltage pulse is used to inject net positive or negative charge into the molecular layer (Duan et al. 2002). The oxide layer, on the surface of the channel, serves as a barrier to reduce charge leakage between the molecules and the channel, and thus maintains the charge state of the redox molecules, thereby guaranteeing nonvolatility. The charged redox molecules gate the FET to a logic on state with higher channel conductance or the off state with lower channel conductance. Positive charges in the redox material act just like a positive gate, leading to an accumulation of electrons and an on state in n-type semiconducting nanowire-based FETs. For p-type FETs, the same effect is obtained with negative charges in the redox layer, which cause an increase in majority carrier density, in this case holes, in the channel (Figure 3.5). Mannick et al. also described a similar device, using SWCNTs as channels, where the conductance is switched on or off, guided by chemical reactivity of a CNT in H_2SO_4 (Goldsmith et al. 2007, Mannik et al. 2006).

3.2.7 Two-Terminal Memory Devices

Transistor devices consist of three electrodes: a source, a drain, and a gate terminal. The source and the drain are good metals (Au, Pt, Co, etc.) and the gate terminal is typically made up of a dielectric layer (e.g., SiO_2) sandwiched between the channel and a heavily doped Si-wafer, which is hooked up to the gate battery. The present authors have theoretically proposed a modified version of the FET memory device where the physical gate is replaced by a virtual electrode. The role of the

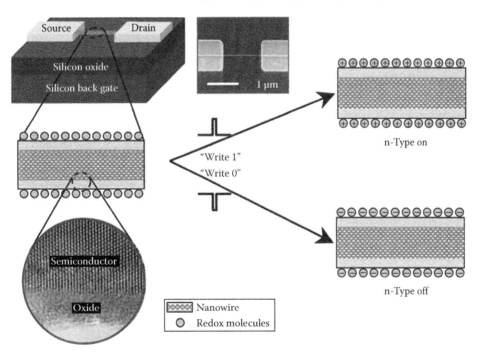

FIGURE 3.5 Nanowire-based nonvolatile devices. The devices consist of a semiconductor nanowire (NW) configured as a FET with the oxide surface functionalized with redox active molecules. The top-middle inset shows a scanning electron microscope (SEM) image of a device, and the lower circular inset shows a TEM image of an InP NW highlighting the crystalline core and surface oxide. Positive or negative charges are injected into, and stored in, the redox molecules with an applied gate or bias voltage pulse. In an n-type NW, positive charges create an on or logic "1" state, while negative charges produce an off or logic "0" state. (Reprinted from Duan, X.F. et al., *Nano Lett.*, 2(5), 487, 2002. With permission.)

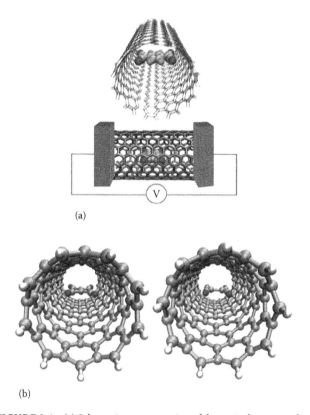

(a)

(b)

FIGURE 3.6 (a) Schematic representation of the typical setup used to probe the information stored in the nonvolatile memory element based on the encapsulation of a donor–acceptor molecule inside a metallic nanotube. In (a) the system is made up a tetrafluorotetracyano-p-quinodimethane molecule encapsulated inside a (10, 10) armchair nanotube. (b) Examples of bistable orientations of the C=C bond of tetracyanoethylene (TCNE) inside a (9, 0) zigzag CNT. (Reprinted from Meunier, V. and Sumpter, B.G., *Nanotechnology*, 18(42), 424032, 2007. With permission.)

gate is filled by a single donor or acceptor molecule embedded inside a metallic nanotube channel (Meunier and Sumpter 2007, Meunier et al. 2007) (Figure 3.6). The resulting device is a two-terminal system where the conduction properties can be turned on and off by modifying the position of the donor (or acceptor) molecule relative to the nanotube inner core. The intrinsic gating effect works for any nanotube-molecule couple, as long as the molecule possess two stable positions corresponding to a different interacting scheme with the nanotube host (bistability). The gating mechanism works as follows: In one orientation, charge transfer between the molecule and the nanotube imposes a suppression of the current. In another orientation, when the interaction is not only purely electrostatic (charge transfer), but also includes significant directional binding, the gating effect is suppressed and current flow is allowed. It was proposed that the molecule's orientation can be modified by imposing a lateral stress to the nanotube, changing the local atomic arrangement around the molecule, which in turn results in its flipping. Other optional writing mechanisms include the use of a magnetic field, optoelectronics, etc. This example underlines the importance of developing a memory device with a well-defined and well-characterized storage node (here the storage node is a single molecule). It is also a memory device that lies at the boundary of FET-type system and NEMS device, as it involves both transistor-like technology and "moving parts," on which NEMS are based, as shown in Section 3.3.

Tour and coworkers proposed a memory element, not based on FET operation, but relying on the conformation of a molecule as the storage media (He et al. 2006). The idea is based on the development of a metal-free silicon-molecule-nanotube test bed for exploring the electrical properties of single molecules that demonstrated a useful hysteresis *I*–*V* loop for memory storage.

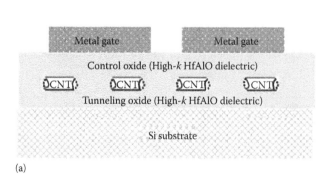

(a)

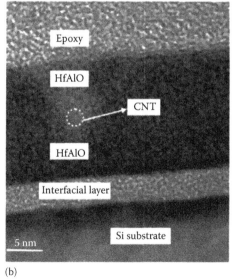

(b)

FIGURE 3.7 (a) Schematic structure of the HfAlO/CNTs/HfAlO/Si MOS memory structure of Lu and Dai where a nanotube is used as the storage node. (b) High-resolution transmission electron microscopy (HRTEM) image of CNTs embedded in HfAlO control and tunneling layers. (Reprinted from Lu, X.B. and Dai, J.Y., *Appl. Phys. Lett.*, 88(11), 113104-3, 2006. With permission.)

Nonvolatile memory for more than 3 days with cyclability of greater than 1000 read and write–erase operations was obtained for a molecular interface between the Si and SWCNTs consisting of π-conjugated organic molecules. In this case, the hysteresis in the *I–V* curve resulted from a conductivity change at high-voltage bias. The conduction state could be easily switched from high to low conductivity, by using a +5 V pulse, and makes it possible for the Si-molecule–SWCNT junction to be used as a nonvolatile memory element. The underlying mechanism for the switching was examined and it was concluded that it was not due to a thermoionic process but likely from a tunneling process across the molecule. Akdim and Pachter suggested that the switching in the device may be driven by conformational changes in the molecule upon the application of an electric field and that the nature of the contact at the interface of the SWCNT mat plays an important role in the switching (Akdim and Pachter 2008).

3.2.8 Using Nanotubes as Storage Nodes

Some authors have reported CNFETs, where nanotubes were used as the storage node. Yoneya et al. proposed a 40 K device with a crossed nanotube junction where one of the tubes is used as channel and the other one is used as floating gate (memory-storage node) (Yoneya et al. 2002). In this case it is the large capacitance property of the nanotube that is exploited. The floating gate is charged and discharged using a back gate. The main disadvantage of this approach is the requirement of a low operating temperature. Other reports have focused on room-temperature use of CNTs as storage nodes. Most notably, Lu et al. developed FET-based memory elements (Chakraborty et al. 2008, Lu and Dai 2006) in which the nanotubes are embedded inside the gate oxide in the metal oxide semiconductor structures. The memory structure is made up of an HfAlO/CNTs/HfAlO/Si structure. HfAlO was chosen as the tunneling and control oxides in the memory structures because of its promising performance for high-*k* gate dielectric applications and floating device applications. The schematic structure of a floating-gate memory device using CNTS as the floating gate is shown in Figure 3.7. The p-type substrate is designed as the current channel and the CNTs are designed to be embedded in the HfAlO film and act as the charge-storage nodes. Excellent long-term charge retention characteristics are expected for the memory structure using CNTs as a floating gate due to their hole-trapping characteristics, as is demonstrated in Lu and Dai's papers. While short-term charge retention was not found to be excellent in their prototype device, the memory window was found to remain at a reasonably large value over the long term.

3.2.9 Conclusions

CNFET-based devices for memory applications have received tremendous interest as is witnessed by the numerous published works outlined above. In all the examples presented, the working principle is that the channel of the transistor consists of a doped semiconducting CNT. For all the examples presented above,

with the exception of those shown in Section 3.2.8, the storage node is usually located close to the channel but is not the channel itself, and depends dramatically on the surface chemistry of the dielectric, nanotube, and dielectric–nanotube interface. In those cases, it is the charge injection into the defects or charge traps in the dielectrics or interface responsible for the bistability and the memory properties.

3.3 NEMS-Based Memory

3.3.1 NEMS: Generalities

NEMS are made of electromechanical devices that have critical dimensions from hundreds to few nanometers (Ke and Espinosa 2006a,b). By exploiting nanoscale effects, NEMS offer a number of unique properties, which in some cases can differ significantly from those of the conventional microelectromechanical systems (MEMS). Those properties pave the way to applications such as force sensors, chemical sensors, and ultrahigh frequency resonators. For instance, NEMS operate in the microwave range and have a mechanical quality that allows low-energy dissipation, active mass in the femtogram range, unprecedented sensitivity (forces in the attonewton range, mass up to attograms, heat capacities below yoctocalories), power consumption on the order of 10 attowatts, and high integration levels (Ke and Espinosa 2006a,b). The most interesting properties of NEMS arise from the behavior of the active parts, which is typically in the form of cantilevers or doubly clamped beams with nanoscale dimensions. In NEMS, the charge controls the mechanical motion, and vice versa. The presence of mechanical motion demands that the moving element to possess high-quality mechanical properties, including high strengths, high Young's moduli, and low density. While limitations in strength and flexibility compromise the performance of Si-based NEMS actuators, CNTs are better candidates to realize the full potential of NEMS, in part due to their one-dimensional structure, high aspect ratio, perfect terminated surfaces, and exceptional electronic and mechanical properties. These properties, now complemented by significant advances in growth and manipulation techniques, make CNTs the most promising building blocks for next-generation NEMS (Bernholc et al. 2002, Chakraborty et al. 2007, and Yousif et al. 2008).

A majority of CNT-based NEMS devices exploit the specific propensity of CNTs to respond efficiently to capacitive forces. Capacitive forces can develop between a nanotube and a gate electrode when the presence of a gate potential induces the rearrangement of a net charge on the nanotube's surface, which in turn causes the appearance of a capacitance force between the nanotube and the gate. Because the nanotube is very flexible, the capacitance force bends the nanotube toward the gate electrode. This phenomenon hinges on two key properties of the nanotube: a large capacitance due to its shape and size and its exceptional mechanical flexibility. In addition to capacitance forces, other forces play a role in nanotube-based NEMS: elastic, van der Waals (vdW), and short-range forces. It is the delicate balance between those forces that makes the functioning of nanotube-based NEMS possible. Here we review a few examples of NEMS implementation for memory applications. We will discuss crossbar nonvolatile RAM, nano-relays, feedback-controlled nano-cantilevers, electro-actuated multi-walled nanotubes, linear-bearing nano-switches, and telescoping nanotube devices.

3.3.2 Carbon Nanotube Crossbars for Nonvolatile Random Access Memory Applications

Rueckes et al. proposed one of the most promising applications of CNTs for memory applications (Rueckes et al. 2000). The device exploits a suspended SWCNT crossbar array for both I/O and switchable bistable elements with well-defined on and off states. The crossbar consists of a set of parallel SWCNTs on a substrate and a set of parallel SWCNTs that are suspended on a periodic array of supports (Figure 3.8). The storage node is found at each place where two nanotubes cross. The bistability is obtained from a balance between the elastic energy (corresponding to the bending energy of the tube, having a minimum for a non-bent tube), and the attractive vdW energy (which creates a minimum corresponding to a situation where the upper nanotube is deflected downward into contact with the lower nanotube), as shown qualitatively in Figure 3.9. Since one minimum corresponds to a large vertical distance between the two mutually perpendicular arrays (defined roughly by the height of the supports), it corresponds to a situation where no current can flow between the two wires (this is the off state). At the second minimum (roughly 0.34 nm, the distance between two tubes in a bundle), tunneling current is possible as the tube–tube distance is small enough to allow wave-function overlap between the two tubes. The "bent" geometry therefore corresponds to the on state.

The device can be switched between the on and off situation by transiently charging the nanotubes to produce attractive or repulsive electrostatic forces. In the integrated system, electrical contacts are made only at one end of each of the lower and upper sets of the nanoscale wires in the crossbar array, which makes it possible to address many device elements from a limited number of contacts. It was found that there is a wide range of parameters that yield a bistable potential for the proposed device configuration. The robustness of the two states suggests: this architecture is tolerant to variations in the structure. The other important feature of this device is the large difference in resistance between the two states (this is the key property of electron tunneling: the

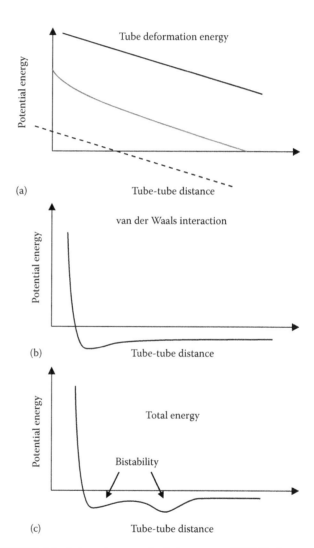

(a)

(b)

(c)

FIGURE 3.9 Interaction energies between two SWCNTs (one top and one bottom) in the device shown in Figure 3.8. (a) Mechanical strain in the top nanotube and (b) attractive van der Waals interaction, showing a minimum at the nanotube–nanotube distance of about 0.34 nm. Plot in (c) shows the resulting bi-stable potential well used for storing information. The first minimum is in the on state (contact) and the second minimum is the off state (physical separation). The existence of the energy barrier between the two minima ensures nonvolatility.

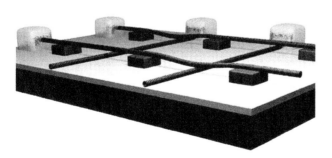

FIGURE 3.8 NRAM device made from a suspended nanotube device architecture with a three-dimensional view of a suspended crossbar array showing four junctions with two elements in the on state and two elements in the off state. The on state corresponds to nanotubes in contact and the off state to nanotubes separated. The substrate consists of a conducting layer (dark gray) that terminates in a thin dielectric layer (light gray). The lower nanotubes are supported directly on the dielectric film, whereas the upper nanotubes are suspended by periodic inorganic or organic supports (gray blocks). A metal electrode, represented by yellow blocks, contacts each nanotube. (Reprinted from Rueckes, T. et al., *Science*, 289(5476), 94, 2000. With permission.)

tunneling resistance decreases exponentially with the separation), which makes the device operation very reliable. The main drawback of the proposed geometry remains however, that at such small dimensions, the junction gap size imposes significant challenges in the nanofabrication of parallel device arrays.

The concept proposed by Rueckes et al. was further developed by the Nantero company, which has devoted a particular important effort toward integrating nanotube array-based memory devices for practical applications. In an IEEE communication in 2004, Ward and coworkers improved the initial device by proposing a novel technique to overcome the hurdle of manipulating individual nanotube structures at the molecular level (Ward et al. 2004). This technique allows CNT-based NEMS devices to be fabricated directly on existing production CMOS fabrication lines. The approach relies on the deposition and lithographic patterning of a 1–2 nm thick fabric of nanotubes, which retain their molecular-scale electromechanical characteristics, even when patterned with 80 nm feature sizes. Because the nonvolatile memory elements are created in an all-thin-film process, it can be monolithically integrated directly within existing CMOS circuitry to facilitate addressing and readout. The transfer of the NRAM fabrication process to a commercial CMOS foundry is ongoing, and, when commercialized, will be the first actual application of CNTs for their unique electronic properties.

A modified NEMS switch using a suspended CNT was studied by Cha et al. (2005). In that memory element, the device has a triode structure and is designed so that a suspended CNT is mechanically switched to one of two self-aligned electrodes by repulsive electrostatic forces between the tube and the other self-aligned nanotube electrodes. One of the self-aligned electrodes is set as the source electrode and the third is the gate electrode. The nanotube is suspended between the two other electrodes (acting as a drain electrode). As the gate bias increases, the force between the gate and the drain electrode deflects the suspended CNT toward the source electrode and establishes electric contact, which results in current flow between the source and the drain electrodes. The electrical measurements show well defined on and off states that can be changed with the application of the gate voltage.

Dujardin et al. exposed another type of memory device based on suspended nanotubes. In their NEMS, multiwall carbon nanotubes (MWCNTs) are suspended across metallic trenches at an adjustable height above the bottom electrode (Dujardin et al. 2005). When a voltage is applied between the nanotube and the bottom electrode, the nanotube switches between conducting and nonconducting states by physically getting closer or farther from the bottom electrode. The device acts as a very efficient electrical switch and can be improved by surface functionalizing the bottom electrode with a self-assembled monolayer. The nanotube bridge was also experimentally studied by Kang et al. who highlighted the importance of the interatomic interactions between the CNT bridge and the substrate and the damping rate on the operation of the NEM memory device as a nonvolatile memory (Kang et al. 2005).

3.3.3 Nanorelays

CNT nanorelays are three terminal devices made up of a conducting CNT positioned on a terrace on a silicon substrate (Figure 3.10). It is connected to a fixed-source electrode (single clamping) and a gate electrode is placed beneath the nanotube. When a bias is applied to it, charge is induced in the nanotube and the resulting capacitance force established between the gate and the tube triggers the bending of the tube, whose end is brought in closer contact to the drain electrode, thereby closing an electric circuit. This device was first proposed by Kinaret et al. (2003), and later demonstrated experimentally by Lee et al. (2004) and by Axelsson et al. (2005). As in other memory devices, the mechanism hinges on the existence of two stable positions (bistability), corresponding to the on and off states, respectively. The advantage of the device is that it is characterized by a sharp transition between the conducting and the nonconducting states. The large variation in the resistance is due to the exponential dependence of the tunneling resistance on the tube-end–drain distance. The transition occurs at fixed source–drain potential, when the gate voltage is varied. In this case, the tube is bent toward the drain and a large current is established, which subsequently disappears when the tube is moved far from it. Aside from possessing the property of a memory element, this device also allows for the amplification of weak signals superimposed on the gate voltage.

A number of investigations, theoretical and experimental, have been performed to further characterize and optimize the practical applications of the nanorelay for memory purposes. The fundamental property of the nanorelay is reached for a balance of the vdW, adhesive (close range), electrostatic (capacitance), and elastic forces. The so-called *pull-in* voltage (voltage required to bring the tube in contact with the drain) was first studied by Dequesnes et al. for a two-terminal device where the entire substrate acted as the gate electrode (Dequesnes et al. 2002). It was found that vdW forces reduce the pull-in voltage, but do not qualitatively modify the on–off transition. The importance of vdW forces decreases with decreasing tube length and increasing terrace height and tube diameter. The transition is, however, very sensitive to short-range attractive forces, which enhance the tendency of the CNT to remain in contact (*stiction* effect). Stiction makes

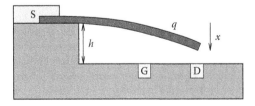

FIGURE 3.10 Nanorelay: schematic picture of the model system consisting of a conducting CNT placed on a terraced Si substrate. The terrace height is labeled h, and q denotes the excess charge on the tube. The CNT is connected to a source electrode (S), and the gate (G) and drain (D) electrodes are placed on the substrate beneath the CNT. The displacement x of the nanotube tip is measured toward the substrate. Typical practical lengths are $L \sim 50$–100 nm, $h \sim 5$ nm. (Reprinted from Kinaret, J.M. et al., *Appl. Phys. Lett.*, 82(8), 1287, 2003. With permission.)

the device unusable as it remains stuck in the on position (Jonsson et al. 2004a,b). The effect of the surface forces can be visualized by means of a stability diagram, which reveals the existence and positions of zero net force (local minima) on the cantilever as a function of gate voltage and tube deflection. A typical curve was computed by Jonsson et al. and is reproduced in Figure 3.11. A stability diagram with more than one local equilibrium for a specific voltage results in a hysteretic behavior in the I–V_g characteristics (more than one position of zero net force means that there actually are three such positions, one of which is not stable). The net force is positive to the right of the curve and negative to the left of the curve, so that, when lowering the voltage for a tube in contact with the drain electrode, it will not release until there is no stable position at the surface. The surface forces exacerbate this effect. The stiction problem corresponds to a stable nanotube position at the surface that can no longer be modified by the application of a gate voltage. The stiction problem is a general issue encountered in many nanotube-based devices. The problem of stiction hinders the use of SWNTs (because they are too flexible) as a bistable NEMS switch. MWNT and SWNT bundles are better candidates for bistable switching (Yousif et al. 2008). This problem can be alleviated, for example, by using stiffer (by avoiding total collapse of the tube, e.g., by using MWCNTs) or shorter tubes. It can also be avoided by applying a layer of adsorbates.

Fujita et al. suggested a very different approach with a modified device design. They proposed using a floating gate that is built between a bundle of CNTs and electrically isolated from the input and output electrodes. The bundle of CNTs is, therefore, bent by the electrostatic force between the floating and control gate, thereby allowing the on–off threshold voltage to be controlled by changing the back-gate voltage, similar to silicon MOSFET (Fujita et al. 2007). A consequence of reducing the possibility of stiction is that a higher pull-in voltage is required, and field emission from the end of the tip can be important. Field emission will become important when the effective potential at the tube end is equal or larger to the work function of the nanotube (i.e., about 4.5 V). For a set of design parameters, field emission can be deliberately sought after (noncontact mode). In the contact mode described above, tunneling current passes from the tube to the drain electrode. In the noncontact mode, the device is designed (short tube) in such a way that the tube is never in physical contact with the drain electrodes. In that case, an electron flow is established via a field emission mechanism. The field emission current onsets with a sufficiently large source–drain voltage and then increases nonlinearly as the source–drain voltage (i.e., the applied field) is further increased. The nanotube can be switched very quickly between the on and off states, and the nanorelay operation was evaluated to work in the gigahertz regime (Jonsson et al. 2004a).

3.3.4 Feedback-Controlled Nanocantilevers

A variation of the nanorelay device presented in the previous section was proposed (Ke and Espinosa 2004), as shown in Figure 3.12a. It is made of an MWCNT placed as a cantilever over a micro-fabricated step. A bottom electrode, a resistor, and a power supply complete the device circuit. Compared to the nanorelay, this is a two-terminal device, providing more flexibility in terms of device realization and control. When the applied potential difference between the tube and the bottom electrode exceeds a certain potential, the tube becomes unstable and collapses onto the electrode. The potential that causes the tube to collapse is defined as the pull-in voltage, already encountered above (Dequesnes et al. 2002). Above the pull-in voltage, the electrostatic force becomes larger than the elastic forces and the CNT accelerates toward the electrode. At small nanotube-tip electrode distance (of the order of 0.3–1.0 nm), substantial tunneling current passes between the tip of the tube and the bottom electrode. The main difference with the nanorelay design is the presence of a resistance R placed in the circuit; with the increase in the current, the voltage drop at R increases, which causes a decrease in the effective tube-electrode potential and the tube moves away from the electrode. It follows that the current decreases and a new equilibrium is reached. Without damping in the system, the cantilever would keep on oscillating at high frequency. However, damping is usually present, and the kinetic energy of the CNT dissipates. The tip ends up at a position where the electrostatic force is equal to the

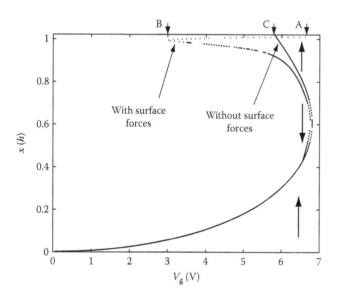

FIGURE 3.11 Nanorelay stability diagram with and without surface forces computed with parameters given in Jonsson et al. (2004b). The curve shows the positions of zero net force on the tube (or local minima) as functions of gate voltage (at constant source voltage = 0.01 V) and deflection x (in units of h, see Figure 3.10). The large arrows show the direction of the force on each side of the curves, indicating that one local equilibrium state is unstable. The required voltage for pulling the tube to the surface (*pull-in* voltage) is given by A (~6.73 V). A tube at the surface will not leave the surface until the voltage is lower than the "release voltage," B and C in the figure. Note that A > B, C, which indicates a hysteretic behavior in the current-gate voltage characteristics, a feature significantly enhanced by surface forces. (Reprinted from Jonsson, L.M. et al., *Nanotechnology*, 15(11), 1497, 2004a. With permission.)

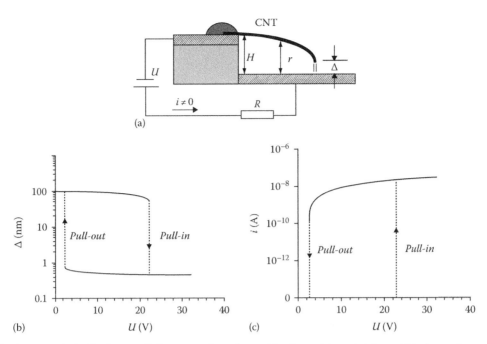

FIGURE 3.12 (a) Schematic of a feedback controlled nanocantilever device. *H* is the initial step height and Δ is the gap between the deflected tip and bottom conductive substrate. *R* is the feedback resistor, as explained in the text. (b,c) Characteristic of *pull-in* and *pull-out* processes for a device with *L* = 500 nm, *H* = 100 nm, and *R* = 1 GΩ. (b) The relation between the gap Δ and the applied voltage *U*. (c) The relation between the current *I* in the circuit and the applied voltage *U*. (Reprinted from Ke, C.H. and Espinosa, H.D., *Appl. Phys. Lett.*, 85(4), 681, 2004. With permission.)

elastic one and a stable tunneling current is established. This is the "lower" equilibrium position for the CNT cantilever. If the applied voltage decreases, the cantilever starts retracting. When the applied potential is lower than the so-called *pull-out* voltage, the CNT leaves the "lower" position and returns back to the upper equilibrium position where the tunneling current significantly decreases and becomes negligibly small. The pull-in and pull-out processes follow a hysteretic loop for the applied voltage and current in the device. The lower and upper positions correspond to the on and off states of the switch, respectively. The lower equilibrium state is very robust, thanks to the existence of the tunneling current and the feedback resistor. (The concept of robustness is related to the value of the ratio of current in on and off states. A ratio of 104 is usually needed for such a device to be considered robust.) Representative characteristics are reproduced in Figure 3.12b. This device has been demonstrated experimentally by Ke and Espinosa (2006a,b) and Ke et al. (2005), showing striking agreement between theoretical prediction and experimental realization. The presence of the resistor allows adjusting the electrostatic field to achieve the second stable equilibrium position. This is an advantage compared to the NRAM system developed by Rueckes et al., since it reduces the constraints in fabricating devices with nanometer gaps between the freestanding CNTs and the substrate, thereby increasing reliability and tolerance to variability in fabrication parameters. The main drawback of this device for memory applications is that the memory becomes volatile; when the applied potential is turned off, the tube retracts back to the upper position and the information stored in the position of the tube is erased.

3.3.5 Data Storage Based on Vertically Aligned Carbon Nanotubes

Another type of memory cell using a nanotube cantilever is based on the work of Kim et al. on nanotweezers (Kim and Lieber 1999). The nanotweezer consists of two vertically aligned MWCNTs that are brought together or separated by the application of a bias potential between them. Motivated by that realization, Jang et al. proposed a memory system that improves the integration density and provides a simple fabrication technique applicable to other types of NEMS (Jang et al. 2005). The device consists of three MWCNTs grown vertically from predefined positions on the electrodes, as shown in Figure 3.13. The first nanotube is electrically connected to the ground and acts as a negative electrode. Positive electrostatic charges build up in the second and the third tubes when they are connected to a positive voltage. The third tube pushes the second tube toward the first tube, because of the forces induced when the positive bias of the third tube increases, while maintaining a constant positive bias on the second one. Above a threshold bias, the second tube makes electrical contact to the first tube, establishing the on state. The balance of the electrostatic, elastic, and vdW forces involved in the device operation determines the threshold bias. If the attractive force between the first and second tube is larger than the repulsive electrostatic forces between them, they remain held together even after the driving bias is removed (otherwise they would return to the original position). Therefore, the system can act as either a volatile or nonvolatile memory element, depending on the design parameters. Note that a two-terminal memory device can be devised using the same principles. In that

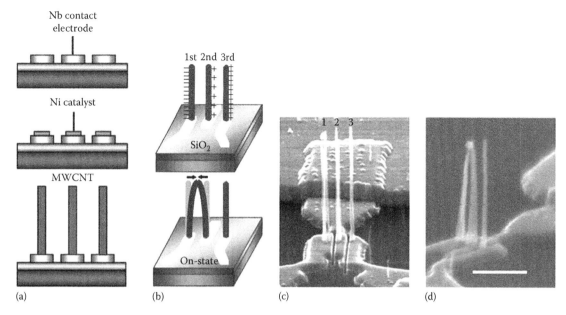

FIGURE 3.13 A schematic illustration of the CNT-based electromechanical switch device proposed by Jang et al. (a) Schematic of fabrication process: Three Nb electrodes are patterned by electron-beam lithography, followed by sputtering and lift off. Similarly, Ni catalyst dots were also formed on the predefined locations for the growth of MWCNTs. The MWCNTs were then vertically grown from the Ni catalyst dots. (b) Illustration of CNT-based electromechanical switch action. (c)–(d) SEM image of the actual device: The length and diameter of the MWCNTs are about 2 μm and 70 nm, respectively. The scale bar corresponds to 1 μm. (Reprinted from Jang, J.E. et al., *Appl. Phys. Lett.*, 87(16), 163114-3, 2005. With permission.)

case the "gate" electrode is no longer the third tube but a conventional gate that is used to separate the two tubes from their contact position.

Jang et al. also reported a nanoelectromechanical-switched capacitor structure based on vertically aligned MWCNTs in which the mechanical movement of a nanotube relative to a CNT-based capacitor defines on and off states. The CNTs are grown with controlled dimensions at predefined locations on a silicon substrate in a process that could be made compatible with existing silicon technology, and the vertical orientation allows for a significant decrease in cell area over conventional devices. It was predicted that it should be possible to read data with standard dynamic random-access memory-sensing circuitry. Jang et al.'s simulations suggest that the use of high-*k* dielectrics in the capacitors will increase the capacitance to the level needed for dynamic random-access memory applications (Jang et al. 2008).

3.3.6 Linear Bearing Nanoswitch

This type of memory device exploits the friction properties of MWCNTs. In MWCNTs, the intershell resistance force, or friction for the sliding of one shell inside another, is known to be negligibly small, as small as 10^{-14} N/atom (Cumings and Zettl 2000, Yu et al. 2000). The inner core of an open MWCNT can therefore be expected to move quasi-freely forward and backward along the tube axis, provided that at least one end of the tube is open. Deshpande et al. presented

a remarkable application of the low-friction-bearing capabilities of MWCNTs to realize a NEM switch as shown in Figure 3.14 (Deshpande et al. 2006). The switch consists of two open-ended MWCNT segments separated by a nanometer-scale gap. The switching occurs through electrostatically actuated sliding of the inner nanotube shells to close the gap, producing

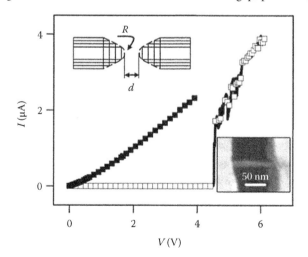

FIGURE 3.14 CNT linear bearing nanoswitch. Main panel: abrupt rise in conductance upon sweeping of source–drain voltage (open squares) and subsequent latching in the on state (filled squares). Lower-right inset: SEM image of the device after latching showing that the gap has closed. Upper-right inset: schematic cup and cone model of the system. (Reprinted from Deshpande, V.V. et al., *Nano Lett.*, 6(6), 1092, 2006. With permission.)

a conducting on state. For double-walled nanotubes, in particular, a gate voltage was found to restore the insulating off state. The device was shown to act as a nonvolatile memory element capable of several switching cycles. The authors indicate that the nanotubes are straightforward to implement, are self-aligned, and do not require complex fabrication, potentially allowing for scalability. The nanotube-bearing device is fabricated in high yield by using electric breakdown to create gaps in a single freestanding MWCNT device, producing an insulating off state. The device is actuated using electrostatic forces between the two ends of the tube that are connected to external electrodes. The force triggers a linear bearing motion that telescope the inner shells in the two multiwall or double-wall segments, so that they bridge the gap (Forro 2000). This restores electrical contact and produces the on state. Adhesion forces between the tube ends maintain the conductive state. The insulating state is controllably restored using a gate voltage, thereby completing the three-terminal nonvolatile memory devices. The gap is closed for a source–drain voltage around 9 V, leading to a conducting on state. At a large gate voltage (110 V) and small source–drain voltage (10 mV), the device snaps back to the zero conductance state. The explanation for the transition from the on state back to the off state is that the gate voltage imposes a bending stress on the nanotube and acts to break the connection. The use of elementary beam mechanics confirms that the force acting upon the two portions of the nanotube is significantly larger than the adhesion forces, allowing for the gap to be reopened.

The use of the exceptional low friction between individual walls in MWNTs is also the basic property exploited in a number of theoretically proposed structures for nanotube-based memory elements. In the following paragraphs, we will briefly review the work of Maslov (2006) and Kang and Jiang (2007).

Maslov proposed a concept that uses vertically aligned MWCNTs, that is first opened (or "peeled") layer by layer, so that the inner core is able to move along the vertical tube axis. By mounting another dielectric above the nanotube at a specific distance from the tube caps, two stable vdW states for the inner core are created and provide for nonvolatile data storage. The two stable states are (1) when the inner tube is away from the top electrode, stabilized by the interaction with the outer shell (off state, where the circuit is open) and (2) when the inner tube is in contact with the top electrode, stabilized by vdW interaction with the top electrode (on state, where the circuit is closed). The switching between the two states is realized by exploiting the electrostatic force resulting from the positive charging of the top (or bottom) electrode and negative charging of the tube inner core, pulling the core toward (away from) the top electrode. For nonvolatility purposes, it is important that the adhesion force is large enough to maintain the tube position, even when the battery potential is turned off.

Another interesting theoretical proposal using the friction properties of multiwalled systems was made by Kan and Jiang. The conceptual design and operation principle of the

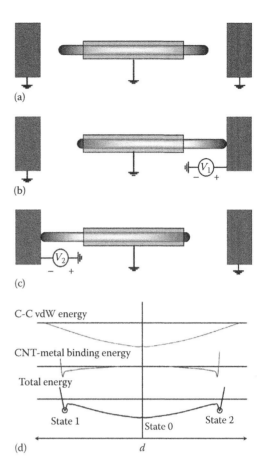

FIGURE 3.15 Electrostatically telescoping nanotube nonvolatile memory device. (a) The initial equilibrium position, (b) the core CNT contacts with the right electrode with V_1 and (c) the core CNT contacts with the left electrode with V_2. (d) Energetic schematics of the telescoping MWCNT-based memory, showing the presence of two local minima (bistability) and an energy barrier between them (nonvolatility). (Reprinted from Kang, J.W. and Jiang, Q., *Nanotechnology*, 18(9), 095705, 2007. With permission.)

MWCNT-based storage are illustrated in Figure 3.15 (Kang and Jiang 2007). The main structural element of the devices is composed of a MWCNT deposited on a metallic electrode. The metallic electrode clamps the outer shell of the multiwall tube, while the inner core of the tube is allowed to freely slide back and forth (both ends of the tube outer-shell are opened). This telescoping device possesses three stable positions, resulting from the balance between the wall–wall interactions and the tube–metal electrode interactions. Two minima correspond to inner tube positions close to the electrode (allowing current to flow between that electrode and the bottom one, attached to the outer core of the tube). A third minimum corresponds to the tube positioned completely inside the outer shell. The movement of the inner core is realized using electrostatic forces induced by the voltage differences between the various electrodes. This leads to reversibility and nonvolatility, provided that the electrode material is carefully chosen (large enough vdW force to ensure contact when bias is turned off, while avoiding stiction where the nanotube cannot be separated from the electrode anymore due to small range forces, as illustrated in Figure 3.15b).

3.3.7 Conclusions

In this section, we have presented the state of the art of fundamental research of the use of CNTs NEMS for memory applications. The CNT-NEMS are particularly interesting because they exploit the two most remarkable properties of CNTs, namely, their electronic and mechanical properties. The viability of CNT-based NEMS switches and their comparison with their CMOS equivalents was recently presented by Yousif et al. (2008). A detailed analysis of performance metrics regarding threshold voltage control, static and dynamic power dissipation, and speed and integration density revealed that apart from packaging and reliability issues (which are the subject of intense current research), nanotube-based switches are competitive in low power, particularly low-standby power, logic, and memory applications.

3.4 Electromigration CNT-Based Data Storage

During the past decade there have been numerous promising concepts developed, based on utilizing ionized endohedral fullerenes and nanotubes as high-density nonvolatile memory elements. In this section the fundamental concepts underlying the operation and development of these types of systems are described and two specific examples are discussed in detail.

3.4.1 A "Bucky Shuttle" Memory Element

One of the earliest concepts for a CNT nonvolatile memory element was based on an ionized fullerene that could be electrically shuttled from one energy state to another inside a larger fullerene or CNT, as shown in Figure 3.16 (Kwon et al. 1999).

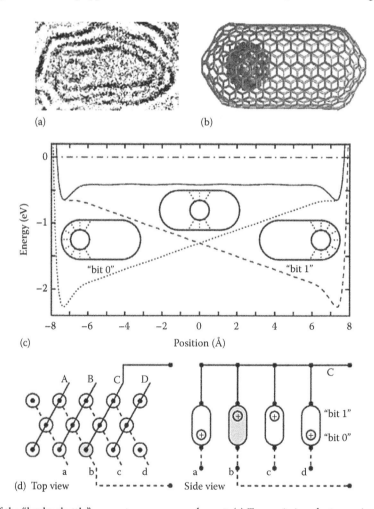

FIGURE 3.16 Illustration of the "bucky shuttle" concept as a memory element: (a) Transmission electron micrograph of a synthesized carbon structure showing a fullerene encapsulated inside a short nanotubule. (b) Molecular model depicting the ideal structure. (c) Position-dependent energy of an endohedral $K@C_{60}^{+}$ within the structure shown in (b). The results (total energy) were obtained from molecular dynamics simulations without an electric field (solid line) and following the application of a small electric field on the system (dashed line). A schematic of the corresponding C_{60} position and intrinsic state ("bit 0" or "bit 1") is shown on the figure. (d) A schematic of the overall "bucky shuttle" memory element concept that corresponds to a high-density memory board. (Reprinted from Kwon, Y.K. et al., *Phys. Rev. Lett.*, 82(7), 1470, 1999. With permission.)

One advantage of this concept was an appropriate material could be self-assembled from elemental carbon via thermal treatment of diamond power. A unique geometry of a nanotube encapsulating a fullerene, where a small fullerene structure could reside at one of the two ends of a capped nanotube or larger fullerene was experimentally identified (Figure 3.16a), and this structural arrangement became the basis for the original concept of a "bucky shuttle" memory element. In this system the on (bit 1) and off (bit 0) states correspond to the two ends of the nanotube that are local minima in the energy landscape of the endohedral fullerene (Figure 3.16c). The position of the ionized endohedral fullerene (obtained by doping with potassium, e.g., K^+@C60) can be manipulated by applying a bias voltage across the ends of the nanotube. The energetics of a switching field of 0.1 V/Å generated by applying a voltage of 1.5 V is shown in Figure 3.16c. With such a small electric field, integrity problems are not expected since graphitic structures are known not to degrade under fields of <3 V/Å. The overall thermal stability and nonvolatility of the system directly depends on the relative depth of the energy minima for the two states. Molecular dynamics simulations were used to explore the functional operation regime for the "bucky shuttle" and also to perform detailed performance analyses. Those calculations successfully demonstrated that information stored would be thermally stable well beyond room temperature, requiring temperatures above 3000 K to be destroyed. In addition, the results indicated that the "bucky shuttle" can be switched between the two states on a very fast timescale, ~10 ps, by applying a small electric field. This operation regime would allow for memory-switching access of close to 0.1 THz, giving a data throughput rate of 10 Gbyte/s in its simplest serial mode. Parallel arrays of the "bucky shuttles" could be envisioned from ordered close-packed nanotube arrays and such an assembly could potentially provide unprecedented memory storage (Figure 3.16d). This ideal system would offer a nonvolatile memory device with a combination of high switching speed and high density.

3.4.2 Memory Elements Based on the "Bucky Shuttle" Concept

While the "bucky shuttle" concept provides a relatively simple mechanism for storing information at the nanoscale, and it is strongly supported by molecular dynamics simulations that have solidified its theoretical operation, a practical measurement of the intrinsic on and off states is not clear. Also, the experimental production, although shown possible using self-assembly, is not able to exclusively produce the required "bucky shuttle" structures and would require novel separation and sorting techniques to facilitate their practical use. These shortcomings stimulated the development of a number of related concepts that have continued to evolve, such as a "nanotube shuttle" employing a transition from single-walled to double-walled nanotube for detection of the "shuttle" position (Kang

and Hwang 2004c, 2005a), multi-endo-fullerenes- or peapod-based shuttles (Kang and Hwang 2004c, 2005b), and variations of these including boron nitride nanotubes (Choi et al. 2004, Hwang et al. 2005b), bipolar endo-fullerenes (Hwang et al. 2005a), metallofullerenes (Byun et al. 2004), and ionic fluidic media (Kang and Hwang 2004b). Again, the basic underlying concept relies on the electromigration of an encapsulated architecture(s) from one local energy minimum to another. The "nanotube shuttle," for example (see Figure 3.17), is based on the use of a capped peapod (nanotube with multiple endohedral C_{60}s, see Figure 3.17a), which is subjected to thermal and e-beam treatment (Figure 3.17b) that subsequently causes coalescence of the encapsulated C_{60}s to form an inner tubule. This is a well-known experimental process and one of the primary methods used for producing high-quality double-walled CNTs. The capped double-walled carbon nanotube (DWCNT) is then cut using lithography and etching technology in order to produce two aligned open ends with the inner tubule initially residing within one end (Figure 3.17c through e). The shuttle mechanism allows the inner tubule to move from one tube end across the gap into the other tube (Figure 3.17f). The separation of the two open ends is required in order to have two independent nanotube structures, which is critical for detecting the state that the overall system is in. The separation must be large enough to preempt strong interactions between the two nanotube structures (the two ends need to be independent in order for the electronic structure to allow unique detection) but small enough so as to prevent the shuttling inner tubule from completely escaping. For such an ideal system, the position of the inner tubule could in principle be measured by detecting the electronic structure of one of the ends since it will change from that corresponding to a single-walled to a double-walled nanotube (Lee et al. 2002).

3.4.3 Conclusions

The general "bucky shuttle" concept relies on utilizing the basic principle of electromigration in a two-level system defined by an ionized encapsulated fullerene-based structure. Its operation has been demonstrated to be very efficient (very short switching times) and is in general highly robust in regard to thermal and electrical stability. In addition, the possibility of assembling very high density architectures promises to allow construction of memory elements with high switching speed, high density, and nonvolatile storage of data. Unfortunately, despite the tremendous interest in the fundamentally simple and attractive systems underlying the use of CNT-based electromigration for memory elements, experimental realization has been difficult. This is because of the complexities in practical fabrication of the required structures and for making appropriate electrical contacts with them. As such, these types of systems have remained as conceptual ideas, to date, relying on the proof-of-principles results primarily obtained from classical molecular dynamics simulations.

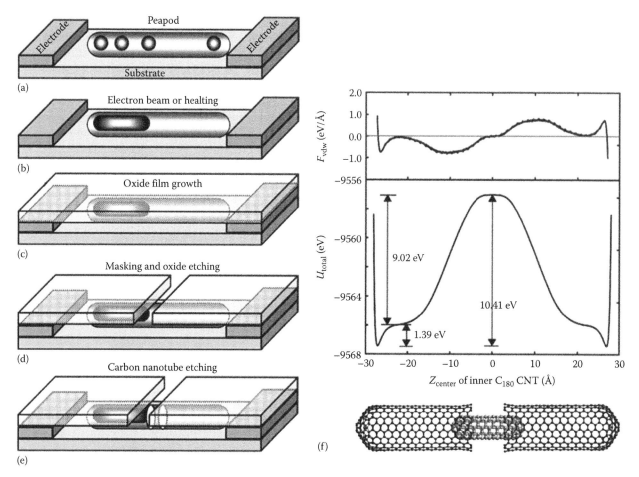

FIGURE 3.17 A schematic illustration of the design (a–e) and operation of a "nanotube shuttle" memory element (f): (a) a peapod CNT that can be coalesced by e-beams and thermal treatment. (b) the coalesced DWCNT; (c) oxide film growth; (d) Masking and oxide etching; (e) etching of the DWCNT; (f) The van der Waals force (F_{vdW}) and total energy (U_{total}) as a function of the center position of the inter-nanotubule. (Reprinted from Kang, J.W. and Hwang, H.J., *Carbon*, 42, 3018, 2004a. With permission.)

3.5 General Conclusions

At the nanoscale, where electronics is integrated on a lateral scale associated with a minimum feature size significantly smaller than tens of nanometers, a number of new phenomena stemming from quantum effects and confinement dominate the processes governing the electron flow across an active device (Waser 2003). It is when those quantum effects are understood, that new device concepts can be developed and novel mechanisms tuned to desired functionality. For practical applications, the role of fundamental research is to address the question of how a new material can be employed in useful devices. This is particularly important for nanoelectronics and information storage since any technology that has ambitions to compete with conventional approaches must display significant benefits and distinct advantages. As discussed in this chapter, CNTs are among the most cited materials that have been used for prototype nanoelectronics and information-storage devices, a dominant position that originates from their intrinsic structural and electronic properties. It is clear that numerous memory devices based on CNTs have been proposed and successfully demonstrated during the

past decade. Many of these devices show tremendous promise for providing enhanced densities, lower power requirements, more efficient read–write processes, and nonvolatility of data. These advantages indicate that CNT-based memory could, in principle, compete with current memory devices such as DRAM. However, cycling stability and data-retention times that are competitive with flash memory has proven to be difficult to achieve. Other critical issues revolve around practical fabrication and the scalability for CNT-based memory devices that can provide the same facility for large-scale fabrication as silicon photolithography. Presently, this goal appears to be in the distant future but it continues to be a topic of intense research. With the demonstrated potential payoff for a range of useful applications, it seems likely that a solution to these problems will eventually be found.

Acknowledgments

This research was sponsored in part by the Laboratory Directed Research and Development Program of Oak Ridge National Laboratory (ORNL), by the Division of Materials Sciences and

Engineering, and by the Center of Nanophase Materials Science, U.S. Department of Energy under Contract No. DEAC05-00OR22725 with UT-Battelle, LLC at ORNL.

References

Akdim, B. and Pachter, R., Switching behavior in pi-conjugated molecules bridging nonmetallic electrodes: A density functional theory study, *Journal of Physical Chemistry C* 112(9), 3170–3174, 2008.

Axelsson, S., Campbell, E. E. B., Jonsson, L. M. et al., Theoretical and experimental investigations of three-terminal carbon nanotube relays, *New Journal of Physics* 7, 245, 2005.

Bachhofer, H., Reisinger, H., Bertagnolli, E., and von Philipsborn, H., Transient conduction in multidielectric silicon-oxide-nitride-oxide semiconductor structures, *Journal of Applied Physics* 89(5), 2791–2800, 2001.

Bernholc, J., Brenner, D., Nardelli, M. B., Meunier, V., and Roland, C., Mechanical and electrical properties of nanotubes, *Annual Review of Materials Research* 32, 347–375, 2002.

Bichoutskaia, E., Popov, A. M., and Lozovik, Y. E., Nanotube-based data storage devices, *Materials Today* 11(6), 38–43, 2008.

Byun, K. R., Kang, J. W., and Hwang, H. J., Nanoscale data storage device of metallofullerene peapods, *Journal of the Korean Physical Society* 45(2), 416–422, 2004.

Cha, S. N., Jang, J. E., Choi, Y. et al., Fabrication of a nanoelectromechanical switch using a suspended carbon nanotube, *Applied Physics Letters* 86(8), 083105-3, 2005.

Chakraborty, R. S., Narasimhan, S., and Bhunia, S., Hybridization of CMOS with CNT-based nano-electromechanical switch for low leakage and robust circuit design, *IEEE Transactions on Circuits and Systems I—Regular Papers* 54(11), 2480–2488, 2007.

Chakraborty, G., Sarkar, C. K., Lu, X. B., and Dai, J. Y., Study of the tunnelling initiated leakage current through the carbon nanotube embedded gate oxide in metal oxide semiconductor structures, *Nanotechnology* 19(25), 255401, 2008.

Charlier, J. C., Blase, X., and Roche, S., Electronic and transport properties of nanotubes, *Reviews of Modern Physics* 79(2), 677–732, 2007.

Choi, W. B., Chae, S., Bae, E. et al., Carbon-nanotube-based nonvolatile memory with oxide-nitride-oxide film and nanoscale channel, *Applied Physics Letters* 82(2), 275–277, 2003.

Choi, W. Y., Kang, J. W., and Hwang, H. J., Bucky shuttle memory system based on boron-nitride nanopeapod, *Physica E—Low-Dimensional Systems & Nanostructures* 23(1–2), 135–140, 2004.

Cui, J. B., Sordan, R., Burghard, M., and Kern, K., Carbon nanotube memory devices of high charge storage stability, *Applied Physics Letters* 81(17), 3260–3262, 2002.

Cumings, J. and Zettl, A., Low-friction nanoscale linear bearing realized from multiwall carbon nanotubes, *Science* 289(5479), 602–604, 2000.

Dequesnes, M., Rotkin, S. V., and Aluru, N. R., Calculation of pull-in voltages for carbon-nanotube-based nanoelectromechanical switches, *Nanotechnology* 13(1), 120–131, 2002.

Derycke, V., Martel, R., Appenzeller, J., and Avouris, P., Carbon nanotube inter- and intramolecular logic gates, *Nano Letters* 1(9), 453–456, 2001.

Derycke, V., Martel, R., Appenzeller, J., and Avouris, P., Controlling doping and carrier injection in carbon nanotube transistors, *Applied Physics Letters* 80(15), 2773–2775, 2002.

Deshpande, V. V., Chiu, H. Y., Postma, H. W. C. et al., Carbon nanotube linear bearing nanoswitches, *Nano Letters* 6(6), 1092–1095, 2006.

Duan, X. F., Huang, Y., and Lieber, C. M., Nonvolatile memory and programmable logic from molecule-gated nanowires, *Nano Letters* 2(5), 487–490, 2002.

Dujardin, E., Derycke, V., Goffman, M. F., Lefevre, R., and Bourgoin, J. P., Self-assembled switches based on electro-actuated multiwalled nanotubes, *Applied Physics Letters* 87(19), 193107-3, 2005.

Forro, L., Nanotechnology: Beyond Gedanken experiments, *Science* 289(5479), 560–561, 2000.

Freitag, M., Johnson, A. T., Kalinin, S. V., and Bonnell, D. A., Role of single defects in electronic transport through carbon nanotube field-effect transistors, *Physical Review Letters* 89(21), 216801, 2002.

Fuhrer, M. S., Kim, B. M., Durkop, T., and Brintlinger, T., High-mobility nanotube transistor memory, *Nano Letters* 2(7), 755–759, 2002.

Fujita, S., Nomura, K., Abe, K., and Lee, T. H., 3-d nanoarchitectures with carbon nanotube mechanical switches for future on-chip network beyond CMOS architecture, *IEEE Transactions on Circuits and Systems I—Regular Papers* 54(11), 2472–2479, 2007.

Ganguly, U., Kan, E. C., and Zhang, Y., Carbon nanotube-based nonvolatile memory with charge storage in metal nanocrystals, *Applied Physics Letters* 87(4), 043108-3, 2005.

Goldsmith, B. R., Coroneus, J. G., Khalap, V. R. et al., Conductance-controlled point functionalization of single-walled carbon nanotubes, *Science* 315(5808), 77–81, 2007.

He, J. L., Chen, B., Flatt, A. K. et al., Metal-free silicon-molecule-nanotube testbed and memory device, *Nature Materials* 5(1), 63–68, 2006.

Hwang, H. J., Byun, K. R., Lee, J. Y., and Kang, J. W., A nanoscale field effect data storage of bipolar endo-fullerenes shuttle device, *Current Applied Physics* 5(6), 609–614, 2005a.

Hwang, H. J., Choi, W. Y., and Kang, J. W., Molecular dynamics simulations of nanomemory element based on boron-nitride nanotube-to-peapod transition, *Computational Materials Science* 33(1–3), 317–324, 2005b.

Jang, J. E., Cha, S. N., Choi, Y. et al., Nanoelectromechanical switches with vertically aligned carbon nanotubes, *Applied Physics Letters* 87(16), 163114-3, 2005.

Jang, J. E., Cha, S. N., Choi, Y. J. et al., Nanoscale memory cell based on a nanoelectromechanical switched capacitor, *Nature Nanotechnology* 3(1), 26–30, 2008.

Jonsson, L. M., Axelsson, S., Nord, T., Viefers, S., and Kinaret, J. M., High frequency properties of a CNT-based nanorelay, *Nanotechnology* 15(11), 1497–1502, 2004a.

Jonsson, L. M., Nord, T., Kinaret, J. M., and Viefers, S., Effects of surface forces and phonon dissipation in a three-terminal nanorelay, *Journal of Applied Physics* 96(1), 629–635, 2004b.

Kang, J. W. and Hwang, H. J., Carbon nanotube shuttle' memory device, *Carbon* 42, 3018–3021, 2004a.

Kang, J. W. and Hwang, H. J., Nano-memory-element applications of carbon nanocapsule encapsulating potassium ions: Molecular dynamics study, *Journal of the Korean Physical Society* 44(4), 879–883, 2004b.

Kang, J. W. and Hwang, H. J., Schematics and atomistic simulations of nanomemory element based on carbon tube-to-peapod transition, *Japanese Journal of Applied Physics Part 1—Regular Papers Short Notes & Review Papers* 43(7A), 4447–4452, 2004c.

Kang, J. W. and Hwang, H. J., Carbon nanotube shuttle memory device based on singlewall-to-doublewall carbon nanotube transition, *Computational Materials Science* 33(1–3), 338–345, 2005a.

Kang, J. W. and Hwang, H. J., Schematics and simulations of nanomemory device based on nanopeapods, *Materials Science & Engineering C—Biomimetic and Supramolecular Systems* 25(5–8), 843–847, 2005b.

Kang, J. W. and Jiang, Q., Electrostatically telescoping nanotube nonvolatile memory device, *Nanotechnology* 18(9), 095705, 2007.

Kang, J. W., Lee, J. H., Lee, H. J., and Hwang, H. J., A study on carbon nanotube bridge as a electromechanical memory device, *Physica E: Low-Dimensional Systems and Nanostructures* 27(3), 332–340, 2005.

Ke, C. H. and Espinosa, H. D., Feedback controlled nanocantilever device, *Applied Physics Letters* 85(4), 681–683, 2004.

Ke, C.-H. and Espinosa, H. D., Nanoelectromechanical systems and modeling, in *Handbook of Theoretical and Computational Nanotechnology* American Scientific Publishers, Valencia, CA, 2006a.

Ke, C. H. and Espinosa, H. D., In situ electron microscopy electromechanical characterization of a bistable NEMS device, *Small* 2(12), 1484–1489, 2006b.

Ke, C. H., Pugno, N., Peng, B., and Espinosa, H. D., Experiments and modeling of carbon nanotube-based NEMS devices, *Journal of the Mechanics and Physics of Solids* 53(6), 1314–1333, 2005.

Kim, P. and Lieber, C. M., Nanotube nanotweezers, *Science* 286(5447), 2148–2150, 1999.

Kim, W., Javey, A., Vermesh, O. et al., Hysteresis caused by water molecules in carbon nanotube field-effect transistors, *Nano Letters* 3(2), 193–198, 2003.

Kinaret, J. M., Nord, T., and Viefers, S., A carbon-nanotube-based nanorelay, *Applied Physics Letters* 82(8), 1287–1289, 2003.

Kwon, Y. K., Tomanek, D., and Iijima, S., "Bucky shuttle" memory device: Synthetic approach and molecular dynamics simulations, *Physical Review Letters* 82(7), 1470–1473, 1999.

Lee, J., Kim, H., Kahng, S.-J. et al., Bandgap modulation of carbon nanotubes by encapsulated metallofullerenes, *Nature* 415, 1005–1008, 2002.

Lee, S. W., Lee, D. S., Morjan, R. E. et al., A three-terminal carbon nanorelay, *Nano Letters* 4(10), 2027–2030, 2004.

Lu, X. B. and Dai, J. Y., Memory effects of carbon nanotubes as charge storage nodes for floating gate memory applications, *Applied Physics Letters* 88(11), 113104-3, 2006.

Mannik, J., Goldsmith, B. R., Kane, A., and Collins, P. G., Chemically induced conductance switching in carbon nanotube circuits, *Physical Review Letters* 97(1), 016601, 2006.

Martel, R., Schmidt, T., Shea, H. R., Hertel, T., and Avouris, P., Single- and multi-wall carbon nanotube field-effect transistors, *Applied Physics Letters* 73(17), 2447–2449, 1998.

Martel, R., Derycke, V., Lavoie, C. et al., Ambipolar electrical transport in semiconducting single-wall carbon nanotubes, *Physical Review Letters* 87(25), 256805, 2001.

Maslov, L., Concept of nonvolatile memory based on multiwall carbon nanotubes, *Nanotechnology* 17(10), 2475–2482, 2006.

Meunier, V. and Sumpter, B. G., Tuning the conductance of carbon nanotubes with encapsulated molecules, *Nanotechnology* 18(42), 424032, 2007.

Meunier, V., Kalinin, S. V., Shin, J., Baddorf, A. P., and Harrison, R. J., Quantitative analysis of electronic properties of carbon nanotubes by scanning probe microscopy: From atomic to mesoscopic length scales, *Physical Review Letters* 93(24), 246801, 2004.

Meunier, V., Kalinin, S. V., and Sumpter, B. G., Nonvolatile memory elements based on the intercalation of organic molecules inside carbon nanotubes, *Physical Review Letters* 98(5), 056401, 2007.

Radosavljevic, M., Freitag, M., Thadani, K. V., and Johnson, A. T., Nonvolatile molecular memory elements based on ambipolar nanotube field effect transistors, *Nano Letters* 2(7), 761–764, 2002.

Rueckes, T., Kim, K., Joselevich, E. et al., Carbon nanotube-based nonvolatile random access memory for molecular computing, *Science* 289(5476), 94–97, 2000.

Sakurai, T., Yoshimura, T., Akita, S., Fujimura, N., and Nakayama, Y., Single-wall carbon nanotube field effect transistors with non-volatile memory operation, *Japanese Journal of Applied Physics Part 2—Letters & Express Letters* 45(37–41), L1036–L1038, 2006.

Scott, J. C., Materials science: Is there an immortal memory? *Science* 304(5667), 62–63, 2004.

Star, A., Lu, Y., Bradley, K., and Gruner, G., Nanotube optoelectronic memory devices, *Nano Letters* 4(9), 1587–1591, 2004.

Tans, S. J., Verschueren, A. R. M., and Dekker, C., Room-temperature transistor based on a single carbon nanotube, *Nature* 393(6680), 49–52, 1998.

Ward, J. W., Meinhold, M., Segal, B. M. et al., A nonvolatile nanoelectromechanical memory element utilizing a fabric of carbon nanotubes, in *Non-Volatile Memory Technology Symposium*, Orlando, FL, 2004, pp. 34–38.

Waser, R., *Nanoelectronics and Information Technology: Advanced Electronic Materials and Novel Devices*, Wiley-VCH, Weinheim, Germany, 2003.

Wunnicke, O., Gate capacitance of back-gated nanowire field-effect transistors, *Applied Physics Letters* 89(8), 083102, 2006.

Yang, D. J., Zhang, Q., Wang, S. G., and Zhong, G. F., Memory effects of carbon nanotube-based field effect transistors, *Diamond and Related Materials* 13(11–12), 1967–1970, 2004.

Yoneya, N., Tsukagoshi, K., and Aoyagi, Y., Charge transfer control by gate voltage in crossed nanotube junction, *Applied Physics Letters* 81(12), 2250–2252, 2002.

Yousif, M. Y. A., Lundgren, P., Ghavanini, F., Enoksson, P., and Bengtsson, S., CMOS considerations in nanoelectrome-chanical carbon nanotube-based switches, *Nanotechnology* 19(28), 285204, 2008.

Yu, M. F., Yakobson, B. I., and Ruoff, R. S., Controlled sliding and pullout of nested shells in individual multiwalled carbon nanotubes, *Journal of Physical Chemistry B* 104(37), 8764–8767, 2000.

Zhang, Y., Carbon nanotube based nonvolatile memory devices, *International Journal of High Speed Electronics and Systems* 16(4), 959–975, 2006.

4

Ferromagnetic Islands

Arndt Remhof
*Swiss Federal Laboratories for
Materials Testing and Research*

Andreas Westphalen
Ruhr-Universität Bochum

Hartmut Zabel
Ruhr-Universität Bochum

4.1 Introduction

A ferromagnetic island is a detached piece of ferromagnetic material, supported by a nonmagnetic* substrate. The term *ferromagnetic island* is usually used to describe a prolate object with a thickness of a few nanometers and lateral dimensions of up to a few micrometers. If the geometrical confinement becomes comparable or smaller than the intrinsic magnetic length scales, such as the exchange length, the domain size, or the domain wall thickness, novel magnetic properties and magnetization reversal mechanisms emerge, different from the properties of the respective bulk material. The domain configuration, the reversal mechanism, and the demagnetizing field of ferromagnetic islands can be tailored by their size, their shape, and their chemical composition.

Nanomagnetism is a rapidly growing area in solid-state research and development. There exist several excellent reviews covering the whole field, including preparation and characterization techniques (Dennis et al. 2002, Martín et al. 2003, Adeyeye and Singh 2008). We do not aim to give another comprehensive overview. This chapter is intended to give an introduction to ferromagnetic islands and to illustrate their properties and interactions on some of our own selected recent experimental results. First, we will present an overview of the fundamentals of ferromagnetism, particularly with regard to its impacts on nanostructured materials. Subsequent, examples of selected systems starting from single dots and rectangular-shaped islands to more complex geometries will be discussed. Preparation and characterization techniques are only covered if they are necessary for the understanding of the physical properties. This chapter concludes with an outlook on the future perspectives.

---

* Throughout the text the term "nonmagnetic" will be used to address non-ferromagnetic materials.

4.2 Background

4.2.1 Ferromagnetism

The discovery of magnetism is one of the oldest cultural achievements in human history. The attraction of iron to loadstone was observed before recorded history began. Magnetism arises from the magnetic dipole moments of atoms and molecules influenced by their symmetry and by the relative orientation of their electron orbits. Depending on the response of substances in an external magnetic field, they are classified as diamagnetic, paramagnetic, ferromagnetic, antiferromagnetic, or ferrimagnetic. The main characteristic of ferromagnetic materials is their permanent magnetization, caused by a natural tendency of the magnetic moments of its atoms or molecules to align under their mutual interactions. The ferromagnetism of the 3d transition metals Fe, Co, and Ni is essentially due to a quantum mechanical phenomenon which has its origin in electrostatics (Heisenberg 1928).

The permanent magnetic moment of a free atom of the 3d transition metals is caused by the uncompensated spin moments of partly filled 3d shells and the related orbital moments. The wave function of the 3d electrons can be separated into an orbital wave function and a spin wave function. Paulis exclusion principle forces the total wave function to be antisymmetric, so that no two electrons may have the same set of quantum numbers in an atom. Parallel alignment of the spins forces the occupation of higher energy levels, allowing for a larger separation of the electrons and a reduction of their Coulomb interaction. The win in Coulomb energy is confronted with an increasing kinetic energy, as some electrons have to occupy higher energy levels for the symmetric spin configuration. Thus, the developing of ferromagnetism depends on which energy term dominates. In the most simple case of a two-electron system, the energy difference between the parallel and the antiparallel spin alignment is expressed by the exchange integral J:

$$J = 2 \int \left(\Psi_1^*(r_1) \Psi_2^*(r_2) V(r_1, r_2) \Psi_2(r_1) \Psi_2(r_2) \right) dr_1 dr_2. \qquad (4.1)$$

where

$V(r_1, r_2)$ denotes the interaction potential
$\Psi_1(r_1)$ and $\Psi_2(r_2)$ are the respective single-particle wave functions

A strong exchange interaction implies a large spatial overlap of the electron wave functions, which cannot be realized if these electrons are strictly localized. At least in the metals Fe, Co, and Ni, conduction electrons of the outer shell are not bound to their atoms, they are free to move and must interact in some way with the electrons at the lattice sites. In these metals the 3d band is overlapped in energy by a much wider 4s band. The bands are filled up to the Fermi level, so that the electrons which each atom contributes to the conduction band are not all from the 4s band, but partly from the 3d band. Therefore, the number of d electrons per atom contributing to the bulk magnetism is not an integral number (Aharoni 2000). In the bulk, the direction of the magnetization vector is defined by the crystalline symmetry. In materials with a large magnetic anisotropy there is a strong coupling between the spin and orbital angular moments within an atom. In addition, the atomic orbitals are generally nonspherical. Because of their shape, the orbits prefer to lie in certain crystallographic directions. The spin–orbit coupling then assures a preferred direction for the magnetization, called the easy direction. To rotate the magnetization away from the easy direction costs energy, the so-called anisotropy energy. The anisotropy energy depends on the lattice structure. In a perfect single crystal, the anisotropy axis is well defined macroscopically, while in a polycrystalline sample the anisotropy axis is still defined locally in each crystallite, but macroscopically it is averaged out due to the random orientation of the crystallites. In ferromagnetic islands the induced anisotropy of the geometrical shape plays a mayor role and will be discussed in more detail in the following sections. Anisotropies may also be induced by the interface to the substrate, the surface, or strains. A more detailed discussion of the relevant energy terms will be given in Section 4.2.4.

4.2.2 Ferromagnetic Domains

Most ferromagnetic materials show no resulting magnetic moment outside the material. The reason for this observation are domain structures (Hubert and Schäfer 1998). The equilibrium magnetization configuration is a result of the competition between the local effects of exchange and anisotropy with the nonlocal effect of the magnetostatic field (in the case without applied magnetic field). Exchange and anisotropy alone would favor a homogeneously magnetized sample with the magnetization along one of the system's easy axes. But this configuration creates a large, long-range dipolar stray field. The energy

contained in the field can be reduced thoroughly by separation of the sample into so-called magnetic domains or Weiss regions, in which the magnetization is still orientated along one of the easy axes, and where the demagnetizing fields of the different domains compensate each other. Only at the boundary between the domains, a domain wall is created, where locally some exchange and anisotropy energy is invested. A domain structure is stable if the interplay between magnetic field energy and domain wall energy has reached an optimal value.

Inside a domain the homogeneous magnetization is parallel to an easy axis direction of the magnetic anisotropy. Thus, domain formation leads to a minimization of exchange and anisotropy energy. If an external magnetic field is applied to the sample, it will also bring energy into the system, the so-called Zeeman energy. The Zeeman energy favors domains parallelly aligned to the external field and leads to domain wall moving or rotation of the magnetization. In a soft magnetic material where no anisotropy exists, a continuous magnetic vector field is expected. But also in the absence of anisotropy, regular domain pattern develops due to the shape of the element. The reason for this is the requirement of a minimum demagnetizing field. The following conditions have to be fulfilled: the magnetic vector field has to be free of divergence, the field has to be aligned parallel to the edges and to the surface, and the vector field has to have a constant length. The conditions will be only achieved simultaneously, if line discontinuities are admitted (van den Berg 1986). These discontinuities are replaced by domain walls. Thus, the shape of the sample enforces a regular domain pattern in a magnetic film of no anisotropy. Van den Berg invented a geometrical procedure to construct this equilibrium domain pattern: the basic domain structure is the locus of centers of all circles inside the sample that touch the sample edge at least at two points. The most important result of van den Berg is that in most cases domains will form along the edges (closure domains) in order to minimize the demagnetizing field. Figure 4.1 schematically shows the reduction of the stray field by domain formation with the help of van den Berg's construction. Examples of the domain structure of a three-dimensional Fe island, discussed on the basis of van den Berg's construction, are provided by Hertel et al. (2005).

Inside a domain wall the magnetization changes its direction in numerous small steps (Bloch 1932). The cost of exchange energy for building up a domain wall decreases as the domain wall thickness increases. On the other hand a minimum value for the anisotropy energy is observed in thin domain walls as the magnetization direction inside the wall is different from the easy direction. In the equilibrium state, the domain wall thickness δ_W follows from the balance of the exchange energy J, and the anisotropy energy K_1:

$$\delta_W = \sqrt{\frac{JS^2 \pi^2}{K_1 a}} = \pi \sqrt{\frac{A}{K_1}} \qquad (4.2)$$

A is a material constant, the so-called exchange constant, and equals JS^2/a for a simple cubic lattice with lattice constant a.

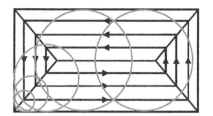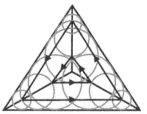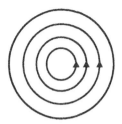

FIGURE 4.1   Van den Berg's construction on the example of ferromagnetic islands with simple geometric shapes.

The domain wall thickness can vary from a few nanometers in Co, to approximately 30 nm in Fe up to a few micrometers in extreme soft permalloy. If the domain wall thickness exceeds the lateral extension of the island, it will stay as a single domain. Larger islands will develop a domain structure if the gain in field energy overcompensates the domain wall energy. There is also a lower limit for the existence of a stable ferromagnetic single-domain island. As the anisotropy energy of an island scales with the volume of the island, there is a critical volume below which the anisotropy energy becomes comparable to the thermal energy $k_{\mathrm{B}}T$. Below this *superparamagnetic limit*, the magnetization of the island is subject to thermal fluctuations. For magnetic data storage devices, the superparamagnetic limit marks the smallest possible bit size. A more detailed discussion of the relevant length scales in nanostructures is given by Martín et al. (2003) and Dennis et al. (2002).

4.2.3 Hysteresis Loops

The magnetic response of a ferromagnetic material to an external field is represented by a hysteresis loop in which the magnetization is plotted as a function of the magnetic field strength. The magnetization is a vector. Usually, the component parallel to the applied field (the *x*-component) is recorded. A hysteresis loop consists of two branches. They are called the ascending and descending branch, as it depicts the magnetization during field increase or field decrease. Generally, both branches do not overlap, i.e., the magnetization is not unambiguously defined on the field strength alone, it also depends on the magnetic history of the sample. The area enclosed by the two branches represents the magnetic energy stored in the system. A hysteresis loop shows

three characteristic points: (1) The point where a further increase in magnetic field strength does not result in a further increase of the magnetization is called the saturation field. At this point all domains are aligned with the external field. (2) The persisting magnetization in zero field is called the remanent magnetization or the remanence. (3) The field needed to reduce the magnetization to zero is the coercive field or the coercivity. Coercivities may range from the mOe range for soft magnetic FeNi alloys to several 10 kOe for hard magnetic $SmCo_5$. Figure 4.2 shows two model hysteresis loops for a ferromagnetic island magnetized along the easy axis (left) and along the hard axis (right). The remagnetization along the easy axis usually is dominated by the rapid growth of oppositely oriented domains as soon as the applied field exceeds the nucleation field. In the case of elongated islands often domains nucleate at the two ends of the islands that grow by domain wall movement. The hysteresis loop is therefore characterized by a sudden change of the magnetization close to the coercive field and a high remanence, resulting in a square-shaped hysteresis loop. If on the other hand the field is applied along the hard axis, the remagnetization is realized by a coherent rotation of all the individual spins. The resulting hystereses loop therefore shows a continuous variation of the magnetization and a low remanence.

4.2.4 Micromagnetic Simulations

For the prediction of the expected domain configuration and the time-dependent development of the magnetization, micromagnetic computer simulations are used. A realistic approach for the description of the magnetic behavior of a magnetic material is to ignore the atomic nature of matter, to neglect quantum

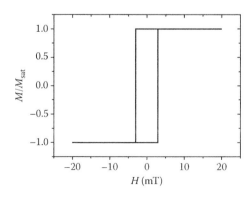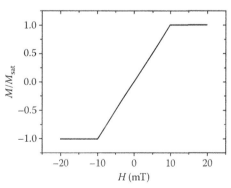

FIGURE 4.2   Model hysteresis loops for a single-domain island magnetized along the easy axis (left) and along the hard axis (right).

effects, and to use classical physics in a continuum description. Micromagnetism is the link between the atomistic, quantum mechanical description of the single atom on the one hand and the macroscopic view of a material characterized by the shape of its hysteresis curve on the other hand. Essentially, the magnetization is to be assumed as a continuous vector field $\vec{M}(\vec{r})$; thus, only the local average over the magnetic moments in a certain neighborhood instead of individual spins is taken into account. The history of micromagnetics starts from 1935 with the paper of L. Landau and E. Lifshitz on the structure of a wall between two antiparallel domains (Landau and Lifschitz 1935), and the work of W. F. Brown Jr. around 1940. A detailed treatment of micromagnetism is given by Brown (1963) and Prohl (2001). The energy will be formulated in terms of the continuous magnetization vector field, and this energy will be minimized in order to determine static magnetization structures. Thermodynamically, the system is described by the free energy $F = E - TS$ with the inner energy E of the system. There are several contributions to the inner energy:

- The exchange energy, responsible for the ferromagnetic coupling, is given by

$$E_{\mathrm{ex}} = -2 \sum_{i<j} J_{ij} \vec{S}_i \cdot \vec{S}_j, \qquad (4.3)$$

  where
  the coefficients J_{ij} are the exchange integrals
  the coefficients $\vec{S}_i$ are the atomic spins

  Positive J_{ij} lead to ferromagnetic order while negative J_{ij} cause antiferromagnetic order. The exchange energy is short-ranged and involves a summation over the nearest neighbors.

- The magnetostatic or demagnetization energy, which is the potential energy that the magnetic moments experience in their own magnetic field, can be written as

$$E_{\mathrm{demag}} = -\frac{\mu_0}{2} \int_V \vec{H}_{\mathrm{d}} \cdot \vec{M} dV, \qquad (4.4)$$

  where μ_0 is the magnetic constant. The magnetostatic or demagnetizing field, $\vec{H}_{\mathrm{d}}$, is related to the magnetization via $\nabla \cdot \vec{H}_{\mathrm{d}} = -\nabla \cdot \vec{M}$ as Maxwell's equation require that $\nabla \cdot \vec{B} = \mu_0 \nabla \cdot (\vec{H} + \vec{M}) = 0$. In the case of a homogenous magnetized ferromagnetic sample the demagnetizing energy can be expressed as

$$E_{\mathrm{demag}} = \frac{V}{2} \left(N_{11} M_x^2 + N_{22} M_y^2 + N_{33} M_z^2 \right) \qquad (4.5)$$

  where N_{ii} are the components of the so-called magnetometric demagnetizing tensor **N**. The components of the tensor are purely determined by the geometry.

- The magnetocrystalline anisotropy originates at the atomic level as discussed in Section 4.2.1. In a system with uniaxial anisotropy along the z direction, the leading energy terms are

$$E_{\mathrm{mca}} = V K_{u1} \sin^2 \theta (1 + \sin^2 \theta) \qquad (4.6)$$

  where
  θ is the angle between the z direction and the magnetization
  K_{u1} is the anisotropy constant
  V is the volume of the sample

  A large positive anisotropy constant K_{u1} describes an easy axis, and a large negative constant K_{u1} an easy plane, perpendicular to the anisotropy axis. Uniaxial anisotropy occurs in hexagonal crystals such as Co, for example. Cubic anisotropy occurs for example in Fe and Ni, the leading energy terms are

$$E_{\mathrm{mca}} = V K_1 \left(m_x^2 m_y^2 + m_y^2 m_z^2 + m_x^2 m_z^2 \right) + V K_2 m_x^2 m_y^2 m_z^2 \qquad (4.7)$$

  The anisotropy constants K_1 and K_2 and their ratio determine the easy axis of the system. Typical values for K_1 are in the order of 10^3–10^5 J/m³. Most work on ferromagnetic islands has been either performed on polycrystalline samples, where the magnetocrystalline anisotropy is averaged out, or on materials with a negligible anisotopy. Therefore, contributions of magnetocrystalline anisotropy are not considered throughout this contribution.

- The surface anisotropy arises from the broken symmetry at the interface between the ferromagnetic and a nonmagnetic material. It can phenomenologically be described by the anisotropy constant K_{S}:

$$E_{\mathrm{S}} = \frac{1}{d} V K_{\mathrm{S}} \sin^2 \theta, \qquad (4.8)$$

  where
  d denotes the thickness of the island
  θ denotes the angle between the magnetization and the surface normal

  K_{S} lies typically in the order of 10^{-4} to 10^{-3} J/m². The sign of K_{S} defines whether the easy axis is in-plane or out-of-plane. For the ferromagnetic islands discussed in this chapter, the surface anisotropy is negligible compared to the shape anisotropy, forcing in-plane magnetization.

- Finally the Zeeman energy is the energy of the magnetic moments in an external magnetic field H_{a}. It is given by

$$E_{\mathrm{Zeeman}} = -\mu_0 M_{\mathrm{sat}} \int_V \vec{H}_{\mathrm{a}} \cdot \vec{m} dV, \qquad (4.9)$$

As experimentally the applied magnetic field $\vec{H}_a$ is the control variable, it is more adapted to use Gibbs free energy or thermodynamic potential

$$G(T,\vec{H}_a,\vec{M}) = F - \mu_0 \int \vec{H}_a \cdot \vec{M} \, dV. \qquad (4.10)$$

We assume that all experiments are performed at constant temperature, so that the entropy term can be neglected. Of course, it is necessary to take into account that the material constants are temperature-dependent. The system has the thermodynamic tendency to evolve toward the minimum of the potential G when approaching equilibrium. The metastable states of the ferromagnetic system are local minima of the potential G.

4.2.5 Experimental Techniques

To investigate the magnetic domain structure as well as the remagnetization process of micro- and nanostructured materials, a number of experimental techniques are available to the experimenter. On the one hand, magnetic domains can be imaged in real space by various techniques, an overview is given in Ref. Hopster and Oepen (2005): Kerr microscopy (Feldtkeller and Stein 1967), Lorentz microscopy (Chapman et al. 1994), scanning electron microscopy with polarization analysis (SEMPA) (Oepen and Kirschner 1991), scanning transmission x-ray microscopy (STXM) (Fischer et al. 1999), photo emission electron microscopy (PEEM) (Mundschau et al. 1996), spin-polarized low-energy electron microscopy (SPLEEM) (Bauer 1994), or magnetic force microscopy (MFM) (Martin and Wickramasinghe 1987). If real space images are obtained at different external magnetic field values, the hysteresis loop is derived by evaluating the total size of the magnetic domains having a particular direction of the magnetization vector with respect to the applied field. On the other hand, magnetic hysteresis loops can be measured with magneto-optical Kerr effect (MOKE) (Kerr 1877), micro MOKE (Allwood et al. 2003), superconducting quantum interference device (SQUID) magnetometry (Philo 1977), or vibrating sample magnetometry (VSM) (Foner 1959). Micro MOKE is an advanced magnetometry for the analysis of single magnetic nanostructures. In the case of the other techniques, a large number of identical elements are investigated in parallel in order to increase the resolution and accuracy. Therefore, the measured hysteresis loops yield information on the average remagnetization process of all elements and not on a single element. Thus, correlation effects between elements also become visible. The periodicity of an artificially structured array of ferromagnetic islands enables the use of diffraction techniques, such as Bragg-MOKE, where the Kerr effect measurements are performed at the diffraction spots from a lateral structure. Diffraction techniques suppress magnetic contributions of the substrate and contain information about the domain configuration inside

the magnetic structures via the magnetic form factor (Remhof et al. 2007). Most experimental techniques can be extended to a vector-magnetometer: by applying a field perpendicular to the direction of measurement, the transverse (the y) component of the magnetization in the ferromagnetic structure can be investigated. From the measurements with the longitudinal and the transverse field orientation, the complete in-plane magnetization vector for the remagnetization can be derived (Westphalen et al. 2007). Using a photoelastic modulator and a digital lock-in amplifier for θ and 2θ response, all three components can be analyzed by measuring different polarizations of the incident beam (Vavassori 2000).

4.3 State of the Art

4.3.1 Sample Preparation

The preparation of sub-micrometer-sized ferromagnetic islands has benefited from the advancements in miniaturization techniques over the last decades. Usually optical lithography is used to define small patterns such as integrated circuits. Lithography, originally meaning "writing with stone," refers to a printing technique in which the imprint of an original pattern is reproduced on paper. In microtechnology it is used do describe a process, in which the desired pattern is first defined as a template on a mask, that can be transferred to the work piece. In optical lithography this is usually an opaque pattern on a transparent glass plate. An optical system is that used to downsize the pattern by projection onto a substrate coated with a photo-sensitive resist. The great advantage of this method is the high throughput. It is limited by the wavelength of the light and by the refractive power of the optical components. E-beam lithography (EBL) is a direct writing technique, in which the resist is exposed by a focused electron beam. In this case the mask only exists as a design file in the computer. The advantage is the high flexibility and the high resolution as compared to the optical methods. There are two kinds of resist: *positive* and *negative*. In the positive technique, first a resist template is produced by one of the aforementioned lithographic methods. Then the magnetic film is deposited onto the template. In a final chemical lift-off step only those magnetic islands remain, which have direct contact to the substrate or another metal layer. In the negative process a pattern is etched out of a continuous ferromagnetic film. Therefore, the homogeneous film is covered with an etch-resistant resist. After exposure, developing, and dry etching, only those magnetic islands that were exposed to the e-beam and covered with resist remain. Both methods are schematically illustrated in Figure 4.3. Apart from visible and UV light and electrons, other radiations such as x-rays and focused ion beams are well established in micro- and nanolithography. Also self-assembly and nanoimprint technologies are in use to prepare ferromagnetic islands. For a more detailed description of the various lithographic techniques we refer to Waser (2003), Bucknall (2005), and Suzuki and Smith (2005).

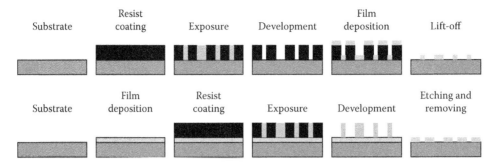

FIGURE 4.3 Patterning steps using positive resist (top panel) and negative resist (bottom panel).

4.3.2 Rectangular and Elliptical Islands

The simple geometry of rectangular and elliptical islands, featuring a well-defined and pronounced uniaxial shape anisotropy, is predestined as model systems. Especially in islands composed of materials with little or no crystalline anisotropy or in polycrystalline systems, the shape anisotropy becomes dominant. In elongated islands the shape anisotropy usually forces the easy axis along the major axis, while the hard magnetic axis lies along the minor axis. The strength of the anisotropy is thereby governed by the aspect ratio (Martín et al. 2003).

At remanence small rectangular or elliptic small islands are in a single-domain state with pronounced magnetic poles at the opposite ends of the island. MFM images these magnetic poles as areas of maximum contrast as depicted in the right panel of Figure 4.4. The left panel shows a secondary electron micrograph, recorded by scanning electron microscopy (SEM) of an array of 50 nm thick, 3 μm × 300 nm-sized permalloy islands placed on a square grid of 3 μm × 3 μm on a silicon wafer. While the SEM image shows the topological contrast, the MFM images the strength of the stay field (i.e., the density of the field lines) normal to the sample. Therefore, the poles at the end, where the stray field enters and leaves the island can, respectively, be identified as black and white spots.

Larger islands will decay into domains at remanence. The terms *small* and *larger* depend on the material's properties and the aspect ratio, determining the anisotropy. Last et al. (2004) have investigated polycrystalline permalloy (Py) micro- and nanostructures (thickness: 38 and 50 nm, length: 0.2–60 μm, width: 0.2–30 μm) in order to map the multi-domain to single-domain transition with MFM. They found a single-domain state, if the aspect ratio (length:width) is greater than 20:1; the multi-domain state is visible inside the rectangles with a aspect ratio between 1:1 and 3:1. Between these two cases they observed a transition state with a featureless inner region and fan-type closure domains at both ends of the rectangle. Figure 4.5 also shows that for very small particles, where the geometric extensions approach the domain wall width, smaller aspect ratios are sufficient to obtain a single-domain state. Extreme aspect ratios lead to *ferromagnetic nanowires*. In the case of thin wires the magnetization reversal process is initiated by the nucleation of oppositely oriented domains at the two ends of the wire and the formation of 180° domain walls. During the reversal process these domains grow by domain wall motion, until the walls meet in the middle of the film where they annihilate. Thicker wires show a more complex reversal mechanism, involving perpendicular in-plane domains that fill up the inner part of the wire and 180° as well as 90° domain walls are formed (Hausmanns et al. 2002). The domain walls can be manipulated (Tsoi et al. 2003) and used to perform logic operations (Allwood et al. 2002) and they can be employed for data storage (Parkin et al. 2008).

Returning to the original discussion of rectangular and elliptic islands and their dipolar behavior in remanence, the pronounced stray field of elongated micrometer-sized islands as discussed before are ideal to study the mutual magnetostatic interactions as a function of their separation.

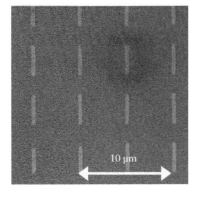 

FIGURE 4.4 Micrographs of an array of 50 nm thick, 3 μm × 300 nm-sized permalloy islands placed on a square grid of 3 μm × 3 μm on a silicon wafer. The left panel shows a secondary electron micrograph, the right panel an magnetic force microscopy image.

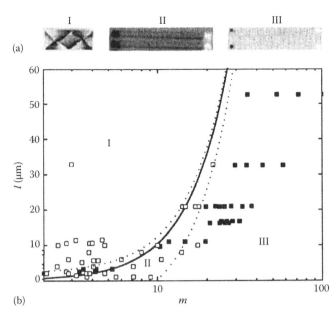

(a)

(b)

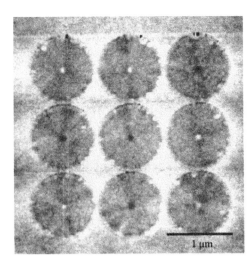

FIGURE 4.6 MFM image of an array of permalloy dots, 1 μm in diameter and 50 nm thick. (From Shinjo, T. et al., *Science*, 289, 930, 2000. With permission.)

FIGURE 4.5 (a) Magnetic force microscopy images of Py structures in remanence with (I) low-aspect ratio ($m = 2.5$, with $l = 10$ μm, $t = 38$ nm), (II) medium-aspect ratio ($m \approx 14$, $l = 20.9$ μm, $t = 38$ nm), and (III) high-aspect ratio ($m = 57$ (top wire), $m = 43$ (bottom wire), $t = 50$ nm). (b) Experimentally observed domain states of Py microstructures and nanostructures for varied length l vs. aspect ratios m. Open squares indicate multidomain states (I), gray squares denote high remanent states (II), and closed squares indicate a single-domain state (III). The black solid line depicts the transition line between multidomain and single-domain configuration from a phenomenological model. The dotted lines are guides to the eye indicating transitions between regions I, II, and III. (Reused from Last, T. et al., *J. Appl. Phys.*, 96, 6706, 2004. With permission.)

4.3.3 Ferromagnetic Dots

For technical applications such as the magnetic data recording, each island has to be addressed individually. Therefore, interactions between the islands have to be avoided. This can be done either by increasing the distances between the islands, or by avoiding strong and extended stray fields. Due to their geometric shape, circular dots and rings of certain aspect ratio exhibit a flux-closed state known as vortex state (Cowburn et al. 1999, Raabe et al. 2000, Shinjo et al. 2000, Fraerman et al. 2002). In this case a curling magnetic structure is established, in which the spin directions only change gradually, minimizing the loss in exchange energy. This configuration suppresses the occurrence of a net in-plane magnetization (zero remanence) without the need to form magnetic domains and to build domain walls. Close to the dot center, the increasing angle between adjacent spins forces them out-of-plane, so that they can align parallel to each other along the plane normal. Therefore, in the center of the vortex a spot of perpendicular magnetization exists, first observed by Shinjo et al. (2000) in circular permalloy dots with diameters ranging from 0.3 to 1 μm and a thickness of 50 nm. MFM images obtained by Shinjo et al. are represented in Figure 4.6.

Let us discuss the vortex state in a single permalloy dot of diameter 0.8 μm as an example for micromagnetic simulation, using the OOMMF 1.2a3 software (Donahue and Porter 1999) with a saturation magnetization $M_s = 860 \times 10^3$ A/m, exchange stiffness $A = 13 \times 10^{-12}$ J/m, anisotropy constant $K_1 = 0$, and a damping coefficient $\alpha = 0.5$. Figure 4.7 shows the result of the micromagnetic simulation.

As it is typical for an ideal vortex configuration, the y component of the magnetization is zero. Starting from negative saturation field (1) to point (2) the core of the vortex develops abruptly at the so-called *nucleation field*. In the remanent state (3) the core has moved into the center of the circular dot. This state lowers the system energy by reducing stray fields and hence lowering magnetostatic energy. Increasing the field then deforms the vortex, and its core moves away from the center of the dot to the upper edge (4). Finally it becomes unstable and again abruptly the core vanishes at the *annihilation field* (step in the hysteresis loop between (4) and (5) at approximately 90 Oe). Note that the nucleation and annihilation fields are different. The resulting minor loops are the main characteristics of the vortex state. Also note that the system is degenerate: the vortex may nucleate at the bottom and move to the top, or vice versa nucleate at the top and propagate to the bottom.

The geometrical conditions which favor the occurrence of the vortex state were studied by Cowburn at al. (1999) on the example of Supermalloy ($Ni_{80}Fe_{15}Mo_5$) dots with diameters ranging between 55 and 500 nm and a thickness range between 6 and 15 nm. The larger dots exhibit the remagnetization behavior of a vortex as described above, while the smaller dots show a more square-like hysteresis loop, a high remanence ($M_{rem}/M_{sat} \approx 0.8$) and a small coercive field, as depicted in Figure 4.8. This is characteristic of a uniform single-domain behavior. In agreement with micromagnetic simulations, the critical diameter drops with increasing thickness from 200 nm for 6 nm thick dots to 100 nm for 15 nm thick ones.

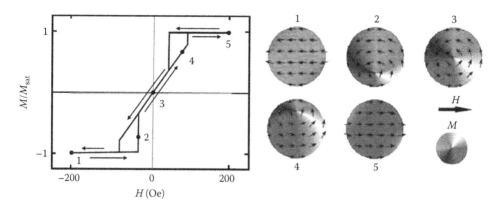

FIGURE 4.7 Hysteresis loop for the *x* (black line) and *y* component (gray line) of the magnetization, and some selected calculated magnetization profiles.

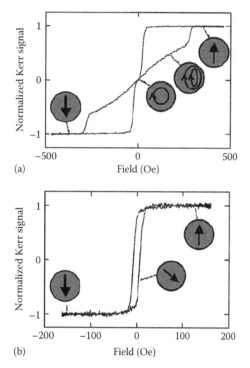

FIGURE 4.8 Hysteresis loops measured from nanomagnets of diameter *d* and thickness *t*: (a) $d = 300\,$nm, $t = 10\,$nm; (b) $d = 100\,$nm, $t = 10\,$nm. The schematic annotation shows the magnetization within a circular nanomagnet, assuming a field oriented up the page. (Reprinted from Cowburn, R.P. et al., *Phys. Rev. Lett.*, 83, 1042, 1999. With permission.)

The magnetic vortex is characterized by its rotational sense (chirality) which can be either clockwise or anticlockwise and the out-of-plane magnetization of the vortex core (polarity) pointing either upwards or downwards. Combining the chirality and the polarity, four different ground states can be realized. Often MOKE and Hall bar measurements are employed to determine the hysteresis of the magnetization reversal of circular magnetic disks (Cowburn et al. 1999, Rahm et al. 2003). The measured hysteresis has the features characteristic for the reversal process occurring through a vortex state, i.e., a vanishing remanence and closed lobes on either side of the remanence with the nucleation field and the annihilation field. However, these measurements are neither sensitive to the chirality nor to the polarity of magnetic vortices. Using MFM (Raabe et al. 2000, Shinjo et al. 2000), the polarity can be determined but not the chirality, whereas Lorentz microscopy operated in the Fresnel mode is sensitive to the chirality but not to the polarity (Raabe et al. 2000, Schneider et al. 2000).

Longitudinal Bragg MOKE on the other hand is sensitive to the average vortex chirality in a periodic array of magnetic islands. In case of longitudinal MOKE (L-MOKE) the measured variable, i.e., the Kerr rotation θ_x is proportional to the component of the magnetization vector $\vec{m} = (m_x, m_y)$ projected onto the intersection between the plane of incidence and the plane of the sample surface: $\theta_x \propto m_x$ (longitudinal component). For patterned samples such as an array of islands, the Kerr angle can be measured not only for the specularly reflected beam but also at the diffraction spots whose pattern represents the Fourier transform of the two-dimensional lattice structure of the real pattern. The Kerr rotation θ_x^{hk} measured for the diffraction orders *h* and *k*, corresponds to the magnetic form factor f_x^{hk}, i.e., the Fourier transform of the magnetization distribution within a single element of the array

$$f_x^{hk} \propto \int dS\, \vec{m}_x(\vec{r}) \exp(i\vec{G}\,\vec{r}) \tag{4.11}$$

where
  $\vec{G} = 2\pi(h\vec{e}_x + k\vec{e}_y)/a$ is the reciprocal lattice vector of the dot array
  a is the periodicity of the dot array
  the integral is taken over the area of a single dot

The relation between the magnetic form factor f_x^{hk} and the corresponding Kerr rotation θ_x^{hk} can be expressed as follows:

$$\theta_x^{hk} = c_{hk}(\phi_i)\big(a_{hk}\Re(f_m) + b_{hk}\Im(f_m)\big) \tag{4.12}$$

where ϕ_i is the angle of incidence,

$$c_{hk}(\phi_i) = \left(h\frac{\lambda}{a} - \sin(\phi_i) \right) \bigg/ \left(1 - \left(k\frac{\lambda}{a} \right)^2 \right). \quad (4.13)$$

and a_{hk} and b_{hk} are coefficients describing the extent of the real and the imaginary contribution, respectively, of the magnetic form factor to the Bragg MOKE signal. They depend on a, ϕ_i, λ, the (magneto-) optical constants of the dots and the substrate, the shape, and the thickness of the dots. With the complete knowledge of all these parameters it is possible to determine the coefficients a_{hk} and b_{hk} (Lee et al. 2008). However, since the computation is rather complicated and cumbersome, they are usually treated here as adjustable parameters.

Grimsditch et al. concluded from Bragg MOKE measurements in the transverse geometry that the vortex chirality manifests itself in the sign of the imaginary part $\Im(f_m^{hk})$ (Grimsditch et al. 2002). This effect can also be applied in the longitudinal geometry. It can be shown that for a magnetization reversal process occurring through a vortex state, $\Im(f_m^{hk})$ always possesses the global extremum at the zero point of the external magnetic field and vanishes as the sample magnetization approaches the saturation level (Grimsditch et al. 2002, Grimsditch and Vavassori 2004). Hence, this imaginary contribution is expected to give rise to a splitting of the ascending branch and the descending branch of the hysteresis loop. For an array of perfectly circular disks a random distribution of left- and right-handed chiralities is expected. In this case the splitting of the hysteresis at remanence will vanish. Since the extent of the splitting corresponds to the degree of the imbalance between clockwise and anticlockwise vortices in magnetic islands, the Bragg MOKE technique provides a statistical means of measuring the chirality distribution in an array of magnetic islands. For a perfectly circular dot there are equal chances for the occurrence of either rotational senses. The vortex chirality can be controlled by breaking the circular symmetry (Lee et al. 2008). This can be most easily realized by shaping the dot with a flat edge at the top as depicted in Figure 4.9.

When a magnetic field is applied in the direction parallel to the flat edge (along the x-axis), the vortex nucleation always starts at the top close to the flat edge with a negative chirality (counterclockwise). With decreasing field strength the core moves down to the bottom perpendicularly to the field direction where it is annihilated. Micromagnetic simulation confirm that along the ascending branch the rotational sense is clockwise, whereas in the descending branch it is counterclockwise in agreement with the experimental observation (Lee et al. 2008).

4.3.4 Ferromagnetic Rings

Within a disk-shaped island, the vortex is only stable for diameters above approximately 100 nm, depending on the thickness of the disk. The vortex state can be stabilized by removing the vortex core in a ring-shaped element. Apart from the low remanent vortex state, which is the ground state of the system, ferromagnetic rings exhibit another, metastable magnetic configuration, the so-called *onion state* (Rothman et al. 2001). The onion state is characterized by the formation of two equally sized domains as shown in Figure 4.10. Due to the shape anisotropy, the magnetization in both domains follows the perimeter of the ring. The onion state allows the majority of the spins to be aligned with the applied magnetic field during the decrease of the magnetization from saturation to zero. Magnetization reversal is achieved by domain wall movement. The domain wall movement starts as soon as the applied reverse field exceeds the pinning strength of the walls. For perfectly symmetric rings, the pinning strength for both walls will be identical and both walls will move synchronous. Consequently, the ring will switch from the onion state to the reverse onion state. In a real system, however, shape irregularities and defects will give rise to differences in the pinning strength, which allows the wall that moves first to annihilate with the second wall, forming the vortex state. Therefore, specific shape irregularities in ferromagnetic rings allow to tailor their reversal behavior, to stabilize the vortex state, and to determine its chirality (Kläui et al. 2001, Vaz et al. 2007).

4.3.5 Noncircular Rings

Closed flux lines in the vortex state or the dipolar onion state are not restricted to circular-shaped objects. They can also be

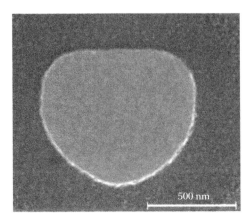

FIGURE 4.9 Scanning electron micrograph of a flat dot. (Reused from Lee, M.-S. et al., *J. Appl. Phys.*, 103, 093913, 2008. With permission.)

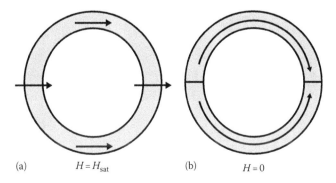

(a) $H = H_{sat}$ (b) $H = 0$

FIGURE 4.10 Schematic representation of the magnetization of a ring-shaped island. (a) In saturation by an applied field and (b) in remanence.

 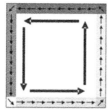 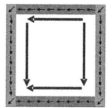 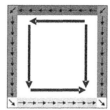

FIGURE 4.11 From left to right: Scanning electron micrograph of a square ring and sketches of the three different symmetrical ground states: vortex, onion, and horseshoe.

observed in square rings (or closed frames), as displayed in Figure 4.11. The shape anisotropy always forces the magnetization parallel to the sides of the frame. In the vortex state the magnetic moments form a closed ring; no stray field is expected to be observed outside the frame. The onion state consists of magnetically parallel-aligned sides of the frame, resulting in two magnetic poles in the opposite corners of the frames. The last panel of Figure 4.11 depicts a remanent state that is not observed in rings, the so-called *horseshoe state*. This state is formed by two magnetically parallel-aligned sides and by two antiparallel-aligned sites. Consequently two magnetic poles occur on one branch of the frame.

The remanent state is not only determined by the size and shape of the island, it can also be influenced by the direction of the external field during the magnetization reversal process. As long as the external field is applied in-plane, only the onion state is observed in remanence regardless of the azimuthal angle (Vavassori et al. 2003). However, if the field is applied perpendicular to the sample surface, the vortex state is the favorite remanent state. At intermediate angles (around 70°) the horseshoe state coexists with the onion state. MFM images of the different remanent states, together with an indication of the field direction, are presented in Figure 4.12.

Noncircular rings also allow the study of different symmetries. Triangular structures (Westphalen et al. 2006) have a sixfold magnetic shape anisotropy. In contrast to a square ring, where at the same time two sides are orientated in the hard axis direction and the other two sides point in the easy axis direction if the external magnetic field is orientated parallel to these

sides, in a triangular ring structure two sides are always in an intermediate position if the third side is orientated parallel to the magnetic field, i.e., in the easy direction. This will lead to a more complex domain structure and to more possibilities for the remagnetization process according to the orientation of the structures to the applied magnetic field. By choosing different sizes for the same structure the shape anisotropy also affects the remagnetization process because it is more pronounced in the small triangular rings. Rings and islands have received most attention, because these elements show a vortex state in remanence which leads to a reduced stray field energy. Such elements can be stacked close together without disturbing each other, unlike elements with dipole character.

4.3.6 Ferromagnetic Spirals

Spirals are hybrids between circular islands and magnetic dipoles. In contrast to ring structures a closed magnetic flux configuration cannot be realized in the spirals during the remagnetization process. In fact they act as magnetic dipoles at the rim while the center is screened from the environment. On the other hand, a spiral is from a geometrical point of view a vortex. Therefore, spirals allow to investigate the interplay between a geometric and a magnetic vortex and to answer the question whether the geometry influences the magnetization: does the shape of a spiral lead to a vortex state in the remanent magnetization? In such a case the outer end piece of a long spiral would act as a magnetic monopole, because in the near region of the outer end piece the interaction with the second magnetic pole at the inner end piece could be neglected. The second question addressed by magnetic spirals is whether the core of the spiral is magnetically harder or softer than the outer parts and how the coercivity is affected by stray field interaction between the spirals. By choosing Fe as the material from which the spirals are fabricated a strong domain formation during the magnetization reversal is expected.

In order to address the aforementioned questions various quadratic arrays of polycrystalline Fe spirals (maximum radius = $2.8\,\mu m$, linewidth = $100\,nm$) were realized (Westphalen et al. 2008a). The polycrystallinity of the Fe suppresses the intrinsic magnetocrystalline anisotropy. In the first case (S1) only one spiral is set on the square grid. In a second sample (S2), it consists of two spirals facing each other without any connection while they are connected in the third sample (S3). Micrographs of the resulting pattern are shown in Figure 4.13.

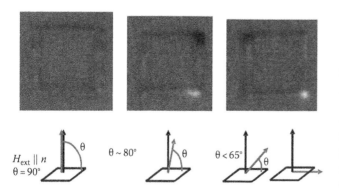

FIGURE 4.12 MFM images of square rings recorded in remanence after saturation in an external field. The field direction is indicated below each micrograph.

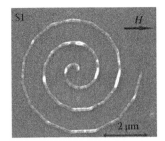
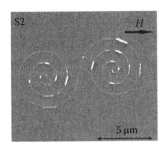
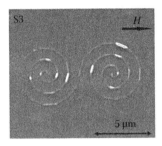

FIGURE 4.13 Scanning electron micrograph images of the four Fe spiral structures used in the experiments. The different spiral structures are named S1, S2, and S3. (Reused from Westphalen, A. et al., *J. Appl. Phys.*, 104, 013906, 2008. With permission.)

The hysteresis loops are similar for all four patterns; however, strong variations occur in the coercive field values. S1 exhibits a coercive field of 240 Oe, while S2 and S3 exhibit coercive fields of 305 and 190 Oe, respectively. It is therefore possible to tailor the coercivity without disturbing the overall magnetization process.

The variations in coercivity are not only due to differences and inhomogeneities in the shape of the spiral structures. Micromagnetic simulations of ideal spiral structures, i.e., spirals without kinks and with constant linewidth, show the same behavior.

One important aspect which is obtained from the micromagnetic simulation is the observation that the remanent state of all spiral patterns is characterized by an onion-like magnetization configuration. The spirals are divided by a horizontal axis parallel to the applied field into an upper and lower part. In each part the magnetization follows the curvature of the spiral forming poles at the horizontal axis. Additionally, micromagnetic simulations of spiral structures with a diameter one order of magnitude smaller than for the ones considered so far still show a preference for the onion state. This behavior is different from closed ring structures, where a vortex state forms spontaneously during the magnetization reversal (Rothman et al. 2001, Castaño et al. 2003, Kläui et al. 2003). In order to generate a magnetic vortex in a spiral structure (geometric vortex), a staggered magnetic field is required that changes orientation after each half turn.

In general the inner parts of the spiral are more stable and switch later than the outer ones. In pattern S2 where the end pieces of two spirals face each other through a small gap, dipolar stray fields stabilize the magnetic configuration and the outer parts switch later than the inner parts. In contrast, joining two spirals by a magnetic bridge in pattern S3 results in a decrease of the coercive field, as this bridge allows domain walls to propagate across. Thus, the boundary condition for the outer pieces of the spiral controls the coercivity. Vice versa, using the same spiral parameters but arranging them in different ways the coercivity can be tailored. In fact the spiral structures have some similarity to the exchange spring heterostructures consisting of magnetically hard and soft materials, one providing the anisotropy and the other the magnetic moment (Fullerton et al. 1999). In comparison, the core of the spiral represents the hard part, whereas the outer tail of the spiral is either soft or can be hardened by interaction.

4.3.7 Magnetostatic Interactions

A key issue in fundamental physics and data storage technology is to understand the remagnetization behavior of submicron-sized magnetic elements and to functionalize their mutual interactions. Recently, the possibility to build logic devices such as AND and NOR gatters on the basis of elongated, dipolar-coupled permalloy (Py) has been demonstrated (Imre et al. 2006). Due to the demagnetizing field of elongated magnetic islands the remanent magnetic state and the reversal mechanism become sensitive to the shape of the elements. In well-ordered artificially structured nanomagnetic dipoles, the interactions arising from the stray field lead to highly correlated systems. In particular, geometric frustration leads to the spin ice state, in which the individual dipoles mimic the frustration of hydrogen ions in frozen water. In the ice structure, two hydrogen atoms are close to the oxygen atoms and two further away. This rule is usually quoted as the "two in–two out" ice rule. Figuratively speaking, the magnetic dipoles on the square lattice obey the spin ice rule, if two spins or magnetic dipoles point into a vertex and two point out. To realize this state, a two-dimensional square lattice of 25 nm thick Py islands (80 nm × 220 nm) was carefully demagnetized and visualized by means of MFM (Wang et al. 2006). The spin ice state is one of many possible magnetic superstructures which may appear in nanomagnetic arrays. Apart from the shape of the elements the symmetry of the superstructure depends on the relative distances between the particles. If the interparticle distance is comparable with or smaller than the length of an element, higher-order magnetostatic contributions become important (Vedmedenko et al. 2005). These contributions might induce additional anisotropy and/or select specific vertex configurations from the manifold of the spin ice symmetries. This, in turn, might influence the magnetization reversal in magnetic arrays.

The open frame system consisting of two perpendicularly crossed stripe arrays is a suitable prototype of interacting islands (Remhof et al. 2008). The open frame system is inherently frustrated. Each vertex where the two sublattices intersect has to accommodate four magnetic poles in close proximity. In the most unfavorable case, four equal magnetic poles meet in a vertex, i.e., the four magnetic moments point all inward or all outward. Energetically more favorable is a configuration in which two moments point inward and two point outward. A situation in which each magnetic pole is surrounded only by opposite poles is impossible. Analogous to the steric frustration of hydrogen ions in frozen water, the geometric frustration of spins in a magnetic material has been denoted as spin ice. The spin ice state is a demagnetized state. It is characterized by a lack of long-range order and by the so-called ice rules which require a minimization of the spin–spin interaction energy when two spins point inward and two spins point outward on each vertex. Model calculations based on the magnetostatic interactions including multipolar contributions clearly show that the spin ice state is indeed the demagnetized ground state of the system. At remanence, however, the samples exhibit a persistent net magnetization, indicative of the existence of a metastable, ordered state. The possible symmetric remanent states are shown schematically in Figure 4.14. The onion state, in which each of the two sublattices is parallel-aligned, has the highest symmetry and the highest magnetization. The resulting net magnetization of the whole array is tilted 45° against the long axes of the individual islands. At each vertex, the magnetization of two adjacent elements from different sublattices point inward, while the two others point outward. In the so-called horseshoe state only one of the two sublattices is parallel-aligned, while the other is antiparallel-ordered. Only the parallel-aligned sublattice contributes to the net magnetization, which consequently aligns with the anisotropy axis of the respective islands. There are two possible configurations in the horseshoe state. In the first one, in the following referred to as horseshoe 1, at each vertex the magnetization of two adjacent elements from different sublattices points inward, while the two others point outward. In the second one, in the following denoted as horseshoe 2, at each vertex the magnetization of three elements points inward (or outward), while only one points outward (inward). Experimentally, mainly the horseshoe 2 state is observed.

The micro-vortex state (Vedmedenko 2007) exhibits no net magnetic moment. In order to realize the micro-vortex state, the chirality of two adjacent vortices has to be antiparallel. Also in this state, the magnetization of two elements point inward and two outward. However, opposite to the aforementioned states, this time the magnetization of the two next nearest elements, i.e., from the same sublattice, points inward, while the two perpendicular elements from the other sublattice point outward, or vice versa. Consequently, the distance between two equal poles is larger than in the onion or the horseshoe state and the distance between two opposite poles is smaller. Thus, this configuration is energetically more favorable than the two arrangements discussed before. The micro-vortex state can be seen as an "ordered" spin ice state. It fulfills the condition of demagnetization and the ice rules.

The onion state can easily be realized experimentally. Applying an external field along the diagonal axis, i.e., under an angle of 45° with respect to the individual stripes, naturally leads to the onion state in remanence. In this case, the external field exhibits a strong component along the easy axis of each Py element, resulting in two magnetic poles in the opposite corners of the frames. For an applied magnetic field pointing along the horizontal stripes (0°), the situation is different. In this case the horizontal stripes are magnetized along the easy axis, while the vertical stripes are magnetized along their hard axis. In remanence, the shape anisotropy of the elongated islands supports the magnetization along the islands. While it is energetically favorable for the magnetization of the horizontal stripes to stay in the field direction, there is no preference for the vertical ones. If the effect of the stray field of the neighboring islands is negligible (i.e., at sufficiently large distances), 50% of the stripes are magnetically oriented north–south, while the other 50% are oriented south–north. For large distances ($2\,\mu$m), this leads to the coexistence of the onion state together with the horseshoe 2 state (see Figure 4.15 right). For small distances ($0.5\,\mu$m), the situation is different. Here, both sublattices are magnetically aligned parallel. Obviously, at small distances the stray field stabilizes the onion state irrespective of the direction of the field. At zero distance, i.e., in closed square rings, the onion state has also been observed in remanence. In this case the sample had to be carefully demagnetized to reach the vortex state.

The variation of distances between elongated Py elements without any variation of the particle shape changes the remanent configuration of the whole array. This means that with increasing interparticle distance the relative depth of energy minima corresponding to different configurations changes. For a deeper understanding of the observed phenomena, theoretical calculations are necessary. In the geometrically frustrated crossed bar configuration small distances lead to a quenching of the impressed fourfold anisotropy by the formation of the highly symmetric, long-range ordered onion state. At larger distances one of the sublattice starts to randomize and the onion state coexists with the horseshoe 2 state. The experimental results can be understood on the basis of the magnetostatic energy of the whole system, including higher-order terms. The one-dimensional energy curve displays three minima that can be identified as the micro-vortex state, the onion state, and the

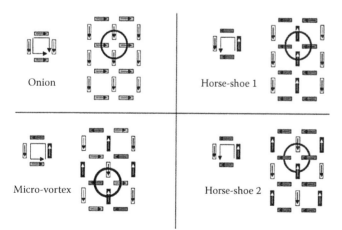

FIGURE 4.14 Schematic representation of the possible symmetrical (meta-) stable remanent states. On the right a representative square formed of four individual islands is depicted, on the left the long-range order of the respective structure is shown. The circles within the extended structure indicate the vertices.

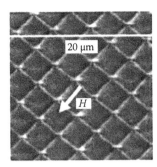
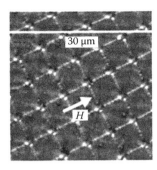
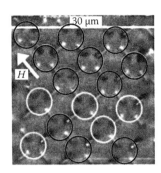

FIGURE 4.15 MFM images of Py dipole arrays on a square lattice, recorded in the remanent state ($H = 0$). The individual elements have separations of 0.42, 0.84, and 2.34 μm (from left to right). The field direction is indicated by the white arrow. In the panel to the right vertices displaying an onion state are highlighted with a black circle, vertices with a horseshoe state are highlighted with a white circle. (Reprinted from Remhof, A. et al., *Phys. Rev. B*, 77, 134409, 2008. With permission.)

horseshoe 2 state. Thereby, the formation of the micro-vortex state as the ground state is separated from the observed onion state by a substantial energy barrier.

Indeed, one of the initially longitudinally magnetized elements have to be completely reversed. Depending on the mechanism (coherent or incoherent magnetization rotation) the magnetization reversal requires an additional energy comparable with that of the magnetostatic coupling. The reversed element should be antiparallel to the applied magnetic field. This makes the magnetization reversal even more energetically expensive. The probability for climbing an energy barrier in the presence of available local minima is close to zero. Thus, it follows from the calculations that for initial 0° saturation of a Py array in the case of small separations, the most probable state is the onion magnetization configuration. However, for large interparticle distances a relevant number (≈30%) of horseshoe 2 states should appear. In remanence the energetically most favorable vortex state is inaccessible because of the large energy barrier of the magnetization reversal. A similar analysis has been performed for initially 45° saturated arrays. The main difference to the previous discussion is that the total energy of the 45° state is lower than that of the 0° configuration. Therefore, the energy minimum for the horseshoe 2 ordering disappears already for the more widely spaced arrays with dominating dipolar interactions. Hence, if an array has been initially magnetized under 45° to its axes the onion state should prevail in wider- as well as in denser-packed arrays. Energetically preferred micro-vortex configuration can be obtained only after a thorough demagnetization of an array. A detailed discussion of the theoretical calculations is given in Vedmedenko and Mikuszeit (2008).

Recently, magnetostatic interactions have also been studied in other frustrated systems. Especially the kagome lattice, a two-dimensional structure composed of corner-sharing triangles and the related honeycomb structure, gained considerable attraction. The kagome lattice and the honeycomb structure intrinsically exhibit a sixfold symmetry; both patterns are schematically represented in Figure 4.16. In the interconnected honeycomb net, three lines meet in each vertex. The ice rules change consequently from two-in–two-out to two-in–one-out or to one-in–two-out. In the magnetic ground state, prepared by

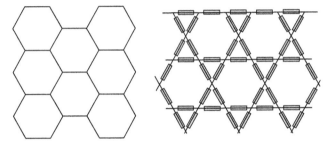

FIGURE 4.16 Schematic representation of the honeycomb structure (left) and of a periodic array of islands, placed between the vertices of a kagome lattice (right).

careful demagnetization, Qi et al. observed that the magnetization along each connecting element of adopts a single domain, and that the domain walls are constrained within the vertices. Thereby, the ice rules are strictly obeyed. Compared to the square lattices discussed above, where the realization of the artificial spin ice is hindered by large energy barriers, the magnetic honeycomb structure turned out to be an ideal artificial spin ice system (Qi et al. 2008).

Rectangular islands, placed between the vertices of a kagome lattice, lead to a periodic structure in which the number of nearest neighbors of a lattice point is four as on a square lattice. However, the angles between the connecting lines to the nearest neighbors on the triangular lattice are 60° instead of 90°. Varying the size of the rectangular bars and the interparticle distance the effect of stray field interaction on the magnetization reversal was investigated by (Westphalen et al. 2008b). The islands on the kagome pattern can be subdivided into three sublattices with triangular symmetry. The magnetization reversal depends on the direction of the applied field. Two simple geometric cases may be distinguished: The field may be applied either parallel to one of the sublattices 0° orientation or perpendicular to one sublattice 30° orientation. For the 0° orientation, two reversal processes were observed. In case that the interaction between the net magnetic moments is weak, the particles inclined to the field direction by 60° flip first, followed by the sublattice parallel to the applied field. For stronger interaction at reduced particle distance, the parallel sublattice switches first, followed by the domain process in the inclined bars. Applying a magnetic field in the 30° orientation leads to a strong frustration of the particles with the hard axis orientation. This frustration causes domain formation preferentially in the perpendicular bars.

4.4 Summary and Discussion

The technological progress in the field of magnetoelectronics and the advancement in miniaturization techniques brought topological-ordered nano- and microstructured magnetic thin films in the focus of many research activities (Martín et al. 2003).

Understanding, modeling, and controlling the magnetization reversal of magnetic micro- and nanostructures as a function of shape, aspect ratio, and interparticle separation are of high current interest. Various shapes have been investigated in the past, including stripes, closed and open squares, and circular and elliptic dots.

Even though there exists a large variety of shapes, sizes, and materials, leading to a huge variety of properties, ferromagnetic islands have a few characteristics in common:

- In small ferromagnetic islands, the shape anisotropy dominates the remanent state and the reversal mechanism. Therefore, the geometrical extensions of magnetic structures determine their magnetic properties. Elongated structures tend to form magnetic dipoles, while small aspect ratios favor closed flux states.
- By lateral structuring of originally homogeneous, magnetically isotropic ferromagnetic thin films new anisotropy axes can be impressed in arbitrary directions.
- Micrometer and nanometer-sized ferromagnetic islands exhibit much larger coercive fields than bulk magnets composed of the same material.
- The magnetization reversal is dominated by coherent rotation in hard axis orientation and by domain wall motion in easy axis orientation.
- At small distances, magnetostatic interactions occur, leading to the collective behavior of an array of islands.

In future, the technological advancements of sample preparation will lead to even smaller individual structures and to larger arrays. Three major routes are currently followed. The first one tries to optimize the lithographic techniques by using electromagnetic radiation with smaller wavelength (Bakshi 2008) or by nanoimprint technologies (Chou et al. 1996). The second route involves self-assembly of clusters and surfaces to complex, periodic structures (Boncheva et al. 2005). The third route is the bottom-up technique. In this approach, the ferromagnetic island is built from its atoms. A tip from a scanning tunneling microscope can be used to move individual atoms on a surface, thereby reaching island sizes that will never be accessible with classical lithography (Stroscio and Eigler 1991). Further miniaturization will not only allow to study smaller objects, and bridge the gap between individual atoms, clusters and nano- to micrometer-sized islands, it also leads to a higher definition of the circumference of the island, reducing the often undesired influence of rough edges.

The research on ferromagnetic islands is nowadays embedded in the field of spintronics (Wolf et al. 2001). In the classical electronics the information is carried by the charge of the electrons, while its spin is ignored. In the future, the standard microelectronics will be combined with spin-dependent effects that arise from the interaction between spin of the carrier and the magnetic properties of the material. Therefore, the magnetic properties of ferromagnetic (FM) islands will be joined to other materials such as antiferromagnets (AF), semiconductors (SC), and superconductors (S). For extended films, the mutual interaction within F/AF and F/S heterostructures leading to the exchange bias effect and the superconducting proximity effect have been studied in great detail in the past (Zabel and Bader 2008). Patterning the ferromagnetic layer allows to control its coercivity and its anisotropy. It also allows to tune the size of the islands to the typical length scales of the AF, i.e., its domain size, and to the S, i.e., its coherence length. The exchange bias, for example, stabilizes the magnetization of ferromagnetic islands close to the superparamagnetic limit, fueling the hope of higher storage densities in magnetic recording devices (Skumryev et al. 1991). Ferromagnetic islands play an important role for the preparation of spin-polarized electric currents as the electrical conductivity of the majority spin electrons differs substantially from minority spin, resulting in a spin-polarized electric current. The spin-polarized current may then be transferred into a SC device ballistically via an ohmic contact. However, F/SC contacts involve doping of the semiconductor surface, leading to spin-flip scattering and loss of the spin polarization. Understanding and optimizing spin injection and spin transport are some of the key challenges for the development of truly spintronic devices.

Another degree of freedom are vertical structures, in which the islands are stacked on top of each other, separated by nonmagnetic material. In such a way, one-dimensional spin chains may be realized, in which the islands interact via their magnetostatic fields (Fraerman et al. 2008). It has already been shown, that interacting islands may be used as logic devices (Wang et al. 2006). Research in this area is still at a very early stage and far away from a technical application; the perspective, however, to functionalize the magnetostatic interactions will have stimulating effects on the research worldwide.

References

Adeyeye, A. O. and Singh, N. 2008. Large area patterned magnetic nanostructures. *J. Phys. D: Appl. Phys.* 41: 153001.

Aharoni, A. 2000. *Introduction to the Theory of Ferromagnetism.* Oxford, U.K.: Oxford Science Publications.

Allwood, D. A., Xiong, G., Cooke, M. D. et al. 2002. Submicrometer ferromagnetic NOT gate and shift register. *Science* 296: 2003–2006.

Allwood, D. A., Xiong, G., Cooke, M. D., and Cowburn, R. P. 2003. Magneto-optical Kerr effect analysis of magnetic nanostructures. *J. Phys. D: Appl. Phys.* 36(18): 2175–2182.

Bakshi, V. 2008. *EUV Lithography,* Bellingham, WA: SPIE Publications.

Bauer, E. 1994. Low energy electron microscopy. *Rep. Prog. Phys.* 57: 895–938.

Bloch, F. 1932. Zur Theorie des Austauschproblems und der Remanenzerscheinung der Ferromagnetika. *Z. Phys.* 74 (5–6): 295–335.

Boncheva, M., Andreev, S. A., Mahadevan, L. et al. 2005. Magnetic self-assembly of three-dimensional surfaces from planar sheets. *Proc. Natl. Acad. Sci. USA* 102(11): 3924–3929.

Brown, W. F. 1963. *Micromagnetics.* New York: John Wiley & Sons.

Bucknall, D. G. 2005. *Nanolithography and Patterning Techniques in Microelectronics.* Cambridge, U.K.: Woodhead Publishing in Materials.

Castaño, F. J., Ross, C. A., Frandsen, C. et al. 2003. Metastable states in magnetic nanorings. *Phys. Rev. B* 67(18): 184425.

Chapman, J. N., Johnston, A. B., and Heyderman, L. J. 1994. Coherent Foucault imaging: A method for imaging magnetic domain structures in thin Films. *J. Appl. Phys.* 76(9): 5349–5355.

Chou, S. Y., Krauss, P. R., and Renstrom, P. J. 1996. Imprint lithography with 25-nanometer resolution. *Science* 272: 85–87.

Cowburn, R. P., Koltsov, D. K., Adeyeye, A. O. et al. 1999. Single-domain circular nanomagnets. *Phys. Rev. Lett.* 83(5): 1042–1045.

Dennis, C. L., Borges, R. P., Buda, L. D. et al. 2002. The defining length scales of mesomagnetism: A review. *J. Phys. Condens. Matter* 14: R1175–R1262.

Donahue, M. J. and Porter, D. G. 1999. *OOMMF User's Guide, Version 1.0, Interagency Report NISTIR 6376.* Gaithersburg, MD: National Institute of Standards and Technology.

Feldtkeller, E. and Stein, K. U. 1967. Verbesserte Kerr-Technik zur Beobachtung magnetischer Domänen. *Z. Angew. Phys.* 23: 100–102.

Fischer, P., Eimüller, T., Schütz, G. et al. 1999. Magnetic domain imaging with a transmission X-ray microscope. *J. Magn. Magn. Mater.* 198–199: 624–627.

Foner, S. 1959. Versatile and sensitive vibrating-sample magnetometer. *Rev. Sci. Instrum.* 30(7): 548–557.

Fraerman, A. A., Gusev, S. A., Mazo, L. A. et al. 2002. Rectangular lattices of permalloy nanoparticles: Interplay of single-particle magnetization distribution and interparticle interaction. *Phys. Rev. B* 65(6): 064424.

Fraerman, A. A., Gribkov, B. A., Gusev, S. A. et al. 2008. Magnetic force microscopy of helical states in multilayer nanomagnets. *J. Appl. Phys.* 103(7): 073916.

Fullerton, E. E., Jiang, J. S., and Bader, S. D. 1999. Hard/soft magnetic heterostructures: Model exchange-spring magnets. *J. Magn. Magn. Mater.* 200(1–3): 392–404.

Grimsditch, M. and Vavassori, P. 2004. The diffracted magneto-optic Kerr effect: What does it tell you? *J. Phys. Condens. Matter* 16(9): R275–R294.

Grimsditch, M., Vavassori, P., Novosad, V. et al. 2002. Vortex chirality in an array of ferromagnetic dots. *Phys. Rev. B* 65(17): 172419.

Hausmanns, B., Krome, T. P., Dumpich, G. et al. 2002. Magnetization reversal process in thin Co nanowires. *J. Magn. Magn. Mater.* 240(1–3): 297–300.

Heisenberg, W. 1928. Zur Theorie des Ferromagnetismus. *Z. Physik* 49: 619–636.

Hertel, R., Fruchart, O., Cherifi, S. et al. 2005. Three-dimensional magnetic-flux-closure patterns in mesoscopic Fe islands. *Phys. Rev. B* 72: 214409.

Hopster, H and Oepen, H. P. 2005. *Magnetic Microscopy of Nanostructures.* Berlin Heidelberg: Springer Verlag.

Hubert, A. and Schäfer, R. 1998. *Magnetic Domains, The Analysis of Magnetic Microstructures.* Heidelberg, Germany: Springer Verlag.

Imre, A., Csaba, G., Ji, L. et al. 2006. Majority logic gate for magnetic quantum-dot cellular automata. *Science* 311(5758): 205–208.

Kerr, J. 1877. On the rotation of the plane of polarisation by reflection from the pole of a magnet. *Philos. Mag.* 3: 321–343.

Kläui, M., Rothman, J., Lopez-Diaz, L. et al. 2001. Vortex circulation control in mesoscopic ring magnets. *Appl. Phys. Lett.* 78(21): 3268–3270.

Kläui, M., Vaz, C. A. F., Bland, J. A. C. et al. 2003. Direct observation of spin configurations and classification of switching processes in mesoscopic ferromagnetic rings. *Phys. Rev. B* 68(13): 134426.

Landau, L. and Lifschitz, L. 1935. On the theory of dispersion of magnetic permeability in ferromagnetic bodies. *Physik. Z. Sowjetunion* 8: 153–169.

Last, T., Hacia, S., Wahle, M. et al. 2004. Optimization of nanopatterned permalloy electrodes for a lateral hybrid spin-valve structure. *J. Appl. Phys.* 96(11): 6706–6711.

Lee, M.-S., Westphalen, A., Remhof, A. et al. 2008. Extended longitudinal vector and Bragg magneto-optic Kerr effect for the determination of the chirality distribution in magnetic vortices. *J. Appl. Phys.* 103(9): 093913.

Martin, Y. and Wickramasinghe, H. K. 1987. Magnetic imaging by "force microscopy" with 1000 Å resolution. *Appl. Phys. Lett.* 50(20): 1455–1457.

Martín, J. I., Nogués, J., Liu, K. et al. 2003. Ordered magnetic nanostructures: Fabrication and properties. *J. Magn. Magn. Mater.* 256: 449–501.

Mundschau, M., Romanowicz, J., Wang, J. Y. et al. 1996. Imaging of ferromagnetic domains using photoelectrons: Photoelectron emission microscopy of neodymium-iron-boron ($Nd_2Fe_{14}B$). *J. Vac. Sci. Technol. B* 14(4): 3126–3130.

Oepen, H. and Kirschner, J. 1991. Imaging of magnetic microstructures at surfaces—The scanning electron microscope with spin polarization analysis. *Scanning Microsc.* 5(1): 1–16.

Parkin, S. S. P., Hayashi, M., and Thomas, L. 2008. Magnetic domain-wall racetrack memory. *Science* 320(5873): 190–194.

Philo, J. S. and Fairbank, W. M. 1977. High-sensitivity magnetic susceptometer employing superconducting technology *Rev. Sci. Instrum.* 48: 1529–1536.

Prohl, A. 2001. *Computational Micromagnetism.* Stuttgart, Germany: B. G. Teubner.

Qi, Y., Brintlinger, T., and Cumings, J. 2008. Direct observation of the ice rule in an artificial kagome spin ice. *Phys. Rev. B* 77: 094418.

Raabe, J., Pulwey, R., Sattler, R. et al. 2000. Magnetization pattern of ferromagnetic nanodisks. *J. Appl. Phys.* 88(7): 4437–4439.

Rahm, M., Biberger, J., Umansky, V. et al. 2003. Vortex pinning at individual defects in magnetic nanodisks. *J. Appl. Phys.* 93(10): 7429–7431.

Remhof, A., Westphalen, A., Theis-Bröhl, K. et al. 2007. Magnetization reversal studies of periodic magnetic arrays via scattering methods. In *Springer Series in Materials Science: Magnetic Nanostructures*, B. Aktas, L. Tagirov, and F. Mikailov (eds.), pp. 65–96. Heidelberg, Germany: Springer Verlag.

Remhof, A., Schumann, A., Westphalen, A. et al. 2008. Magnetostatic interactions on a square lattice. *Phys. Rev. B* 77: 134409.

Rothman, J., Kläui, M., Lopez-Diaz, L. et al. 2001. Observation of a Bi-domain state and nucleation free switching in mesoscopic ring magnets. *Phys. Rev. Lett.* 86(6): 1098–1101.

Schneider, M., Hoffmann, H., and Zweck, J. 2000. Lorentz microscopy of circular ferromagnetic permalloy nanodisks. *Appl. Phys. Lett.* 77(18): 2909–2911.

Shinjo, T., Okuno, T., Hassdorf, R. et al. 2000. Magnetic vortex core observation in circular dots of permalloy. *Science* 289(5481): 930–932.

Skumryev, V., Stoyanov, S., Zhang, Y. et al. 1991. Beating the superparamagnetic limit with exchange bias. *Nature (London)* 423: 850–853.

Stroscio, J. A. and Eigler, D. M. 1991. Atomic and molecular manipulation with the scanning tunneling microscope. *Science* 254(5036): 1319–1326.

Suzuki, K. and Smith, B. W. 2005. *Microlithography: Science and Technology*. Boca Raton, FL: CRC Press.

Tsoi, M., Fontana, R. E., and Parkin, S. S. P. 2003. Magnetic domain wall motion triggered by an electric current. *Appl. Phys. Lett.* 83(13): 2617–2619.

van den Berg, H. A. M. 1986. Self-consistent domain theory in soft-ferromagnetic media. II. Basic domain structures in thin-film objects. *J. Appl. Phys.* 60: 1104–1113.

Vavassori, P. 2000. Polarization modulation technique for magneto-optical quantitative vector magnetometry. *Appl. Phys. Lett.* 77(11): 1605–1607.

Vavassori, P., Grimsditch, M., Novosad, V., Metlushko, V., and Ilic, B. 2003. Metastable states during magnetization reversal in square permalloy rings. *Phys. Rev. B* 67(13): 134429.

Vaz, C. A. F., Hayward, T. J., Llandro, J. et al. 2007. Ferromagnetic nanorings. *J. Phys. Condens. Matter* 19(14): 255207.

Vedmedenko, E. Y. 2007. *Competing Interactions and Pattern Formation in Nanoworld*. Weinheim, Germany: Wiley-VHC.

Vedmedenko, E. Y. and Mikuszeit, N. 2008. Multipolar ordering in electro- and magnetostatic coupled nanosystems. *Chem. Phys. Chem.* 9(9): 1222–1240.

Vedmedenko, E. Y., Mikuszeit, N., Oepen, H. P., and Wiesendanger, R. 2005. Multipolar ordering and magnetization reversal in two-dimensional nanomagnet arrays. *Phys. Rev. Lett.* 95: 207202.

Wang, R. F., Nisoli, C., Freitas, R. S. et al. 2006. Artificial spin ice in a geometrically frustrated lattice of nanoscale ferromagnetic islands. *Nature (London)* 439: 303–306.

Waser, R. 2003. *Nanoelectronics and Information Technology*. Weinheim, Germany: Wiley-VCH.

Westphalen, A., Lee, M.-S., Remhof, A., and Zabel, H. 2007. Vector and Bragg magneto-optical Kerr effect for the analysis of nanostructured magnetic arrays. *Rev. Sci. Instrum.* 78(12): 121301.

Westphalen, A., Remhof, A., and Zabel, H. 2008a. Magnetization reversal in nanowires with a spiral shape. *J. Appl. Phys.* 104(1): 013906.

Westphalen, A., Schumann, A., Remhof, A., Zabel, H., Last, T., and Kunze, U. 2006. Magnetization reversal of equilateral Fe triangles. *Phys. Rev. B* 74(10): 104417.

Westphalen, A., Schumann, A., Remhof, A. et al. 2008b. Magnetization reversal of microstructured kagome lattices. *Phys. Rev. B* 77: 174407.

Wolf, S. A., Awschalom, D. D., Buhrman, R. A. et al. 2001. Spintronics: A spin-based electronics vision for the future. *Science* 294(5546): 1488–1495.

Zabel, H. and Bader, S. D. 2008. *Magnetic Heterostructures, Advances and perspectives in Spinstructures and Spintransport*. Heidelberg, Germany: Springer Verlag.

5

A Single Nano-Dot Embedded in a Plate Capacitor

Gilles Micolau
Universités d'Aix Marseille

Damien Deleruyelle
Universités d'Aix Marseille

5.1 Introduction

Nowadays, nonvolatile memories (NVM) are used on a massive scale in different domains related to personal and industrial applications such as cellular phones, massive computer data storage on mobile support, digital photography, aeronautics, and embedded electronic in automotive applications.

The current NVM technology is based on electric charge storage in silicon-based devices. The scaling of these devices has been predicted by Gordon Moore. In his famous paper published in 1965, he mentions: "...The complexity for minimum component costs as increased at a rate of roughly a factor two per year [...] Over the longer term [...] there is no reason to believe it [the rate] will not remain nearly constant for at least 10 years" [1].

As a matter of fact, even after 40 years, this prediction is still valid [2]. However, the typical size of memory cells has been so much reduced that a further miniaturization becomes critical (recently, NVM cells have been proposed with 43 nm typical size [technology node]) [3]. Indeed, new phenomena such as quantum or finite size effects and single-electron effects [4–6] challenge the possibility of further reduction.

In the field of industrial and academic research, two major parallel axes are proposed to solve these issues. The first one is based on the use of new materials using singular physical properties (phase changed memory [7], resistive random access memory [8], etc.). These promising approaches are not yet completely mature to replace the well-known traditional devices, studied for at least a half century. The other approach consists in modifying, step by step, the current devices in order to push the scaling limits of silicon-based devices. This chapter is devoted to this second approach.

The essential element of the current NVM is the metal oxide semiconductor transistor (TMOS), which can be seen as a nonlinear valve of electronic current controlled by an external voltage. To simplify, the schematic behavior of the MOS transistor is presented in Figure 5.1. It can be described as follows: the current flowing between the two polarized pads—source and drain—is null when the voltage applied on the control gate is lower than a threshold voltage (V_{T0}). It becomes arbitrarily high when the voltage is equal or greater than V_{T0}. The value of V_{T0} is an intrinsic quantity of the TMOS. The main idea of the NVM is to modulate the value of V_{T0} by storing electric charges, spatially confined in a floating gate located between the control gate and the channel of electronic transport. Injecting negative (resp. positive) charges in the floating gate shifts the value of the threshold voltage to values greater (resp. lower) than V_{T0} as shown in Figure 5.2. Consequently, two different electrical states are possible: the "up" one defined by $V_{Tu} > V_{T0}$ and the "down"

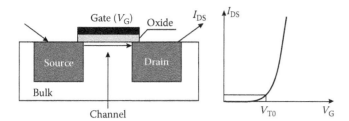

FIGURE 5.1　Schematic behavior of a MOS transistor.

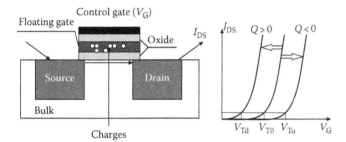

FIGURE 5.2　Schematic behavior of a TMOS with a floating gate.

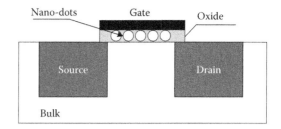

FIGURE 5.3　Floating nano-dots.

one with $V_{Td} < V_{T0}$. These two states define a level of programmation for the cell. Depending on the technology, the "up" (resp. "down") state is called programmed or erased (resp. erased or programmed) and the cell can store one fundamental memory unit, the well-known *bit*.

Classically, the floating gate is designed with degenerated polysilicon whose electrical behavior is quasi-metallic. It is electrically insulated by oxide layers (silicon oxide for the bottom and a stack of different oxides on its top) designed to maintain the electric charges as long as possible in the floating gate. One major issue in scaling is the reduction of the thickness of the oxides leading to an alteration of the retention ability of the device. Another critical issue is the size reduction of the floating gate. To work properly, the memory cell needs a certain quantity of electric charges—injected in its floating gate. Scaling the floating gate and the surrounding oxides implies using less charges for the programmation of the cell. The repartition of less charges in the volume of a small floating gate makes the discrimination between the two possible states of the cell more difficult (the classical term is "closure of the programming window"). Moreover, another issue is linked to the apparition of traps in the bottom oxide when cycling the memory device. This phenomenon is known as *stress-induced leakage current* (SILC) [9,10]. Because the charges are delocalized in the floating gate, the presence of the unique trap in the bottom dielectric can discharge the entire memory device and lead to the data loss. To tackle these issues, one solution could be to constrain charges in the vicinity of the channel. For accomplishing this goal, the use of nanoscale dots instead of a continuous floating gate has been experimentally and theoretically studied for many years [11–14]. The principle of the configuration is shown in Figure 5.3.

For modeling the electrical characteristics of these devices, such as programming (erasing) or retention, knowledge of the electrostatics and, more generally, the physics of this type of

devices is needed. The physical properties of such nanoobjects have been studied for many years, with the help of different innovative techniques such as scanning tunneling microscopy (STM), atomic force microscopy (AFM), transmission electron microscopy (TEM), and so on. A detailed review about nanoclusters can be found in Ref. [15].

The modeling of dynamics phenomenon such as tunneling currents needs an accurate modeling of the electric field in the device. For metallic or quasi-metallic dots, a classical (non-quantum) approach based on equivalent capacitance is efficient and has been even used in the case of large semiconductor islands [16–19]. Theoretically, for small semiconductor dots, solving the electrostatics of the configuration must be achieved by a self-consistent resolution of Poisson and Schrödinger equations as shown in Refs. [20–22]. However, this huge complete resolution can be avoided. First of all, the number of dots involved in the problem has a crucial influence. The electrostatics of a device with numerous dots can be governed by the classical interaction between each dot, even if the typical size of the dot is lower than the typical quantum range. Second, the current typical size of dots does not necessarily imply that the quantum approach is justified. So, many models assume that the dots are metallic and interact as plane capacitors. The validity of this approach has been proved in Ref. [23]. However, it was shown in Ref. [24], in a 2D case and for a single dot arbitrarily shaped, that this approach is no longer valid in the range of aspect ratios encountered in nanodot arrays. To enclose this state of art, in the context of ultrascaled devices, the individual electrical behavior of a dot can become significant in front of the collective behavior of a matrix of dots [6,25].

We aim to model, academically and as precisely as possible, the electric behavior of a single dot embedded in a dielectric oxide. So, we choose a very canonical and academic physical configuration, particularly with the twin aims of

- Determining the critical size of the dot leading to quantum effects
- Comparing the classical approximations with models giving more physical insights

In this chapter, we focused on a single spherical island, electrically charged, embedded in a plane capacitor. In the first part, we propose a rigorous—by this word, we mean that there are no physical approximations—non-quantum model for the electrostatics of a metallic dot and give a semi-analytical expression for its capacitance. In the second part, we propose a quantum approach for a dielectric dot, based on solving the Poisson and Schrödinger equations, with a discrete number of charges in the dot. Finally, results obtained from both approaches are quantitatively confronted to determine in which extent the quantum effects have to be considered.

5.2 Studied Configuration: Geometry and Notations

A spherical dot, of radius a, is embedded in a dielectric medium of relative dielectric permittivity ε_r, located between two infinite conductor planes geometrically supported by two planes Π_1 and Π_2. Each is held at a constant electric potential (V_1 for Π_1, V_2 for Π_2). The spherical dot is initially charged with an electrical charge Q_0, with a process not studied here. We assume that $\Delta V = V_1 - V_2$ and Q_0 are independent. We denote by σ the surface charge density of the dot when metallic and ρ the volumic charge density when dielectric. h is the distance between the two plate electrodes. All distances and notations are defined in Figure 5.4. Any point M in the dielectric medium is characterized by its Cartesian coordinates (x, y, z) and also by the spherical ones (r, θ, ϕ). Because the physics of the configuration is invariant by any rotation along the longitudinal angle ϕ, only the two spherical coordinates $\left(r = OM, \theta = \widehat{(\mathbf{e}_z, \mathbf{OM})} \right)$ are required to locate M.

5.3 Metallic Dot

In this section, we assume that the dot is a perfect conductor medium. As a consequence, the potential is constant in the volume of the dot but also floating because it is electrically insulated. We solve the classic electrostatics (non-quantum approach) of the configuration by a rigorous semi-analytical expansion of the potential. By rigorous, we mean that there are no physical approximations in our model. There are only numerical truncations and numerical approximations arising from the implementation. The model is based on the work exposed in Ref. [26] where the author rigorously solves the electrostatics of this problem, with the help of the method of images and the Legendre polynomial expansions. The author gives the solution for the potential in the medium (and discuss its mathematical validity) and the electrostatic force acting on the charged dot. Nevertheless,

the evaluation of floating potential is not explicit at all. Starting from this model, we derive a rigorous semi-analytical expression for the capacitance and a very simple expression for the polarization coupling coefficient. The floating potential is therefore directly determined without a recurring iterative process. Then, we validate the numerical implementation and compare results with classical capacitance approximations.

In this section, we first define the physical parameters required for the resolution of the electrostatics. Secondly, we find a semi-analytical expression for the capacitance and the coupling coefficient. Finally, we conclude with the numerical investigations of the model.

5.3.1 Definitions of Capacitance and Coupling Coefficient

Due to the linearity of the electrostatic equations and the unicity of the Poisson equation—with given boundary conditions—the potential, at any point M in the dielectric medium, can be described as the sum of two terms. The first one, denoted by $V^Q(M)$, is the potential created by the dot when charged (with electric charge Q_0) without polarization ($\Delta V = V_1 - V_2 = 0$). The second term, denoted by $V^{po}(M)$, is the potential created by the polarization between the two electrodes ($\Delta V = V_1 - V_2$) when $Q_0 = 0$. Thus, the electric potential anywhere in the dielectric medium can be written as

$$V(M) = V^Q(M) + V^{po}(M) \tag{5.1}$$

The computation of the function $V(M)$ involves the resolution of the Poisson equation with three Dirichlet boundary conditions: the potential is constant and imposed on the electrodes (Π_1 and Π_2), is constant, and floating on the dot. The floating potential, denoted by V_f, can be split into two parts: v_f^Q, due to the charge, and v_f^{po}, due to the polarization. It can now be written as

$$V_f - V_2 = v_f^Q + v_f^{po} - V_2 \tag{5.2}$$

Determining V_f requires another boundary condition. The charge density over the dot, σ, is linked to the normal derivative of the floating potential (normal component of the electric field):

$$\left(\frac{\partial V}{\partial r} \right)_{r=a} = -\frac{\sigma(\theta)}{\varepsilon_0 \varepsilon_r} \quad \text{and} \quad Q_0 = \int_{\theta=0}^{\pi} \sigma(\theta) 2\pi a^2 \sin\theta \, d\theta \tag{5.3}$$

Classically the capacitance is defined as follows: "…the capacitance of a conductor is the charge of the conductor when it is maintained at unit potential, all others conductors being held at zero potential" [27,28]. We define the capacitance of the dot, C_S, as

$$C_S = \left(\frac{\partial Q_0}{\partial v_f^Q} \right)_{\Delta V} = \left(\frac{\partial Q_0}{\partial V_f} \right)_{\Delta V} = \frac{Q_0}{v_f^Q} \tag{5.4}$$

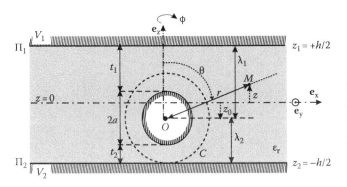

FIGURE 5.4 Configuration of study.

The floating potential linearly depends on the total difference of potential ΔV. We define the polarization coupling coefficient α as

$$\alpha = \left(\frac{\partial v_f^{po}}{\partial \Delta V} \right)_{Q_0} = \left(\frac{\partial V_f}{\partial \Delta V} \right)_{Q_0} = \frac{\Delta V}{v_f^{po}} \qquad (5.5)$$

We thus obtain the classical equation for the floating potential

$$V_f - V_2 = \frac{Q_0}{C_S} + \alpha \Delta V \qquad (5.6)$$

This equation still remains valid for any shape of the dot. Nevertheless, a rigorous semi-analytical expression of C_S can just be extracted for the canonical spherical shape. In other cases, computations can be carried out with the help of rigorous models that need numerical methods of discretization of the shape (method of moments, boundaries models, variational computations [24,29]) or by approximated analytical expansions (Rayleigh hypothesis on boundary conditions for instance [30]).

Let us denote by Q_1 (resp. Q_2) the charge supported by the top (resp. bottom) half-part of the dot. We write these quantities with an unique notation $Q_{(1,2)}$. Introducing here the angles $\theta_1 = 0$ (resp. $\theta_2 = \pi/2$), denoted by the unique notation $\theta_{(1,2)}$, we express the partial charges, as we did with Q_0, integrating σ on the top (resp. bottom) part of the dot:

$$Q_{(1,2)} = 2\pi a^2 \int\limits_{\theta_{(1,2)}}^{\theta_{(1,2)}+\pi/2} \sigma(\theta)\sin\theta\,d\theta \qquad (5.7)$$

Deriving the relation (5.7) along v_f^Q defines the partial capacitances C_1 and C_2, formalized with the following unique notation $C_{(1,2)}$:

$$C_{(1,2)} = \left(\frac{\partial Q_{(1,2)}}{\partial v_f^Q} \right)_{\Delta V} \qquad (5.8)$$

The Chasles relation on the integral over θ leads obviously to the equality $C_1 + C_2 = C_S$.

We can define the ratio $\kappa_{(1,2)}$ between the partial capacitances and the total one:

$$\kappa_{(1,2)} = \frac{C_{(1,2)}}{C_S} = \left(\frac{\partial Q_{(1,2)}}{\partial Q_0} \right)_{\Delta V} \quad \text{and so} \quad \kappa_1 = 1 - \kappa_2 \qquad (5.9)$$

As ΔV and Q_0 are independent, the coefficients $\kappa_{(1,2)}$ and α are not equal. Mathematically this property arises from the fact that the partial derivatives do not involve the same constant physical parameters (Q_0 for α and ΔV for κ_2).

The two following sections are dedicated respectively to the semi-analytical expansions of C_S and α.

5.3.2 Influence of the Charge

In this section, we assume that there is no polarization between the electrodes ($V_1 - V_2 = 0$). We focus on the term V^Q in the relation (5.1). The boundary conditions for the potential are such as

(Qa) The potential is constant on the dot. This floating potential is denoted by $v_f^Q = V^Q(r = a) = \dfrac{Q_0}{C_S}$.

(Qb) The potential is null everywhere on each plate electrode.

(Qc) The relations (5.3) are still valid.

Because the geometry simultaneously involves plane and spherical structures (electrodes and dot), it is interesting to include the plate boundary conditions in the expansion of the potential. For this, we use the classical method of images. The expansion of the potential is based on the work in Ref. [26] where the author treats the electrostatics of a charged dot rigorously in a nonpolarized plate capacitor. He uses the method of images and the Legendre polynomials expansion. The author arrives at the solution for the potential in the medium (and discuss its mathematical validity) and the electrostatic force acting on the charged dot. Nevertheless, the evaluation of the floating potential is not explicit so that the capacitance cannot be known.

5.3.2.1 Method of Images

The main idea of the method is to find a virtual distribution of charges involving the same boundaries conditions than the initial distribution [28]. Because of the unicity of the solution of the Poisson equation for given boundary conditions, the virtual distribution and the real one generate the same potential in the domain investigated. Here, we build a distribution of charges ensuring that the two planes Π_1 and Π_2 are planes of antisymmetry of the charge distribution. This is done by taking recursively the images of the sphere by the symmetries along the planes Π_1 and Π_2, as done in Refs. [26,31], and described in Figure 5.5. The z-coordinate z_p of the center O_p of the pth image is: $z_p = ph + (-1)^p z_0$, p describing the relative natural integers.

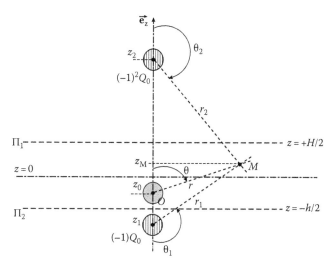

FIGURE 5.5 Coordinates of the images.

We denote by r_p the distance between the point of observation M and the center O_p, and θ_p the oriented angle between the line (O_pM) and vertical axis. To obtain a correctly oriented angle, it is necessary to alternate the sense of the axis $\mathbf{e}_z$ attached to the considered image. We get so $\theta_p = ((-1)^p \widehat{\mathbf{e}_z, \mathbf{O_PM}})$. These variables are shown in Figure 5.5. The classical Pythagore theorem and the definition of the geometric scalar product give the two coupled relations:

$$\begin{cases} r_p^2 = (z_0 - z_p)^2 + r^2 + 2r(z_0 - z_p)\cos\theta \\ \cos\theta_p = (-1)^p \dfrac{r\cos\theta + (z_0 - z_p)}{r_p} \end{cases} \quad (5.10)$$

To simplify the notations, we note for every point located on the surface of the dot $(r = a) : r_p^a(\theta) = (r_p)_{r=a}$ and $c_p^a(\theta) = (\cos\theta_p)_{r=a}$. Derivating the relations (5.10) along the variable r leads to

$$\begin{cases} dr_p^a(\theta) = \left(\dfrac{\partial r_p}{\partial r}\right)_{(r=a)} = \dfrac{a + (z_0 - z_p)\cos\theta}{r_p^a(\theta)} \\ dc_p^a(\theta) = \left(\dfrac{\partial \cos(\theta_p)}{\partial r}\right)_{(r=a)} = (-1)^p \left(\dfrac{\cos\theta}{r_p^a(\theta)} - \dfrac{a\cos\theta + (z_0 - z_p)}{r_p^a(\theta)^2} dr_p^a(\theta)\right) \end{cases}$$

$$(5.11)$$

5.3.2.2 Semi-Analytical Expression for the Capacitance

Each sphere centered in O_p generates a potential $v_p(M)$ expandable as an infinite set of Legendre polynomials [28]. Because the spheres get the same distribution of charges (or opposite), each sphere generates the same potential v_p (or opposite), vanishing far from its centre. The potential becomes [26,28]

$$v_p(M) = (-1)^p \sum_{l=0}^{+\infty} B_l \left(\frac{a}{r_p}\right)^{l+1} P_l(\cos\theta_p) \quad (5.12)$$

where P_l is the Legendre polynomial of order l and B_l are constant coefficients. The resolution of the electrostatics of the configuration is mathematically equivalent to determining B_l using boundary conditions.

The potential $V^Q(M)$, at every $M(r, \theta)$ in the dielectric medium, is the sum of the contributions of all spheres:

$$V^Q(r, \theta) = \sum_{l=0}^{+\infty} B_l \left(\frac{a}{r}\right)^{l+1} P_l(\cos\theta)$$

$$+ \sum_{p\star} (-1)^p \sum_{l=0}^{+\infty} B_l \left(\frac{a}{r_p(\theta)}\right)^{l+1} P_l(\cos\theta_p) \quad (5.13)$$

where $p\star$ denotes the summation over all the non-null relative integers.

The relation (5.13), combined with Equation 5.3, gives the charge of the sphere:

$$\frac{Q_0}{2\pi\varepsilon a} = \sqrt{2}B_0 + a \sum_{p\star} (-1)^p \sum_{l=0}^{+\infty} B_l \int_{\theta=0}^{\pi} \left[(l+1)\frac{dr_p^a(\theta)}{r_p^a(\theta)} P_l(c_p^a(\theta)) \right.$$

$$\left. - \cdots dc_p^a(\theta) P_l^{(1)}(c_p^a(\theta)) \right] \left(\frac{a}{r_p^a(\theta)}\right)^{l+1} \sin\theta\, d\theta \quad (5.14)$$

Because of the linearity between v_f^Q and Q_0, derivating Equation 5.14 along the floating potential is equivalent to dividing it by v_f^Q. So noting $\dot{B}_l = (B_l/v_f^Q) = (\partial B_l/\partial v_f^Q)$, it becomes

$$\frac{C_S}{2\pi\varepsilon a} = \sqrt{2}\dot{B}_0 + a \sum_{p\star} (-1)^p \sum_{l=0}^{+\infty} \dot{B}_l \int_{\theta=0}^{\pi} \left[(l+1)\frac{dr_p^a(\theta)}{r_p^a(\theta)} P_l(c_p^a(\theta)) \right.$$

$$\left. - \cdots dc_p^a(\theta) P_l^{(1)}(c_p^a(\theta)) \right] \left(\frac{a}{r_p^a(\theta)}\right)^{l+1} \sin\theta\, d\theta \quad (5.15)$$

Defining the row vector Γ, whom generic kth term Γ_k is

$$\Gamma_k = \frac{\delta_{0,k-1}}{\sqrt{2}} + \frac{a}{2\sqrt{2}} \sum_{p\star} (-1)^p \int_{\theta=0}^{\pi} \left[k\frac{dr_p^a(\theta)}{r_p^a(\theta)} P_{k-1}(c_p^a(\theta)) \right.$$

$$\left. - \cdots dc_p^a(\theta) P_{k-1}^{(1)}(c_p^a(\theta)) \right] \left(\frac{a}{r_p^a(\theta)}\right)^k \sin\theta\, d\theta$$

and defining the column vector $\dot{\mathbf{B}}$, the pth term of which is $\dot{B}_p$, we can write the relation (5.15) as a vector-product:

$$C_S = 4\pi a \Gamma \dot{\mathbf{B}} \quad (5.16)$$

5.3.2.2.1 Expression of B_l Coefficients

The relation (5.13), written at $r = a$ and divided by v_f^Q, gives

$$\sum_{l=0}^{+\infty} \dot{B}_l P_l(\cos\theta) + \sum_{p\star} (-1)^p \sum_{l=0}^{+\infty} \dot{B}_l \left(\frac{a}{r_p^a(\theta)}\right)^{l+1} P_l(c_p^a(\theta)) = 1, \quad \forall\theta$$

$$(5.17)$$

Projecting this relation on all the orders of the Legendre's polynomials (using their orthonormality and the fact that $P_0 = 1/\sqrt{2}$), we obtain the linear system:

$$\mathbb{M}\dot{\mathbf{B}} = \sqrt{2}\mathbf{1}_1$$

where

$\mathbf{1}_1$ is a column vector whose first coefficient is 1 and the others are zeros

$\mathbb{M}$ is a square matrix whose generic term M_{km} is

$$M_{km} = \delta_{k-1,m-1} + \sum_{p*} (-1)^p \int_{\theta=0}^{\pi} \left(\frac{a}{r_p^a(\theta)}\right)^m P_{m-1}(c_p^a(\theta)) P_{k-1}(\cos\theta) \sin\theta \, d\theta$$

$$(5.18)$$

with $\delta_{k,m}$ the Kronecker symbol.

The coefficients $\dot{B}_l$ are so the unknown variables of a linear well-posed problem. The matrix $\mathbb{M}$ is numerically well-conditioned (its size and its numerical properties are explained in the numerical section of this part). The coefficients $\dot{B}_l$ are given by

$$\dot{\mathbf{B}} = \sqrt{2}\mathbb{M}^{-1}\mathbf{1}_1$$

$$(5.19)$$

5.3.2.2.2 *Matricial Expression for the Capacitance*

Hence, the relation (5.16) becomes a simple matrix relation, numerically easy to manipulate:

$$C_S = 4\pi\varepsilon a \, \mathbf{\Gamma}\mathbb{M}^{-1}\mathbf{1}_1$$

$$(5.20)$$

5.3.2.2.3 *Corollary: Expression of Partial Capacitances*

Using the relation (5.19) and the row vector $\Gamma^{(1,2)}$ (the superscript (1,2) indicates the part of the dot where we integrate σ) whose generic kth term $\Gamma_k^{(1,2)}$ is

$$\Gamma_k^{(1,2)} = \frac{1}{2}\delta_{0,k-1} + a\frac{1}{\sqrt{2}}\sum_{p*}(-1)^p \int_{\theta=\theta^{(1,2)}}^{\theta^{(1,2)}+\pi/2} \left[k\frac{dr_p^a(\theta)}{r_p^a(\theta)} P_{k-1}(c_p^a(\theta)) \right.$$
$$\left. - \cdots dc_p^a(\theta)P_{k-1}^{(1)}(c_p^a(\theta)) \right]\left(\frac{a}{r_p^a(\theta)}\right)^k \sin\theta \, d\theta$$

the relation (5.8) becomes matricially

$$C_{(1,2)} = 4\pi\varepsilon a \, \mathbf{\Gamma}^{(1,2)}\mathbb{M}^{-1}\mathbf{1}_1$$

$$(5.21)$$

5.3.2.3 Semi-Analytical Expression for the Potential

To finish this section, we rewrite the expression of the potential due to the charge as a matricial product. Using the relations (5.19) and (5.4) and the row vector $\mathbf{G}(r, \theta)$ whose generic term is

$$G_k(r,\theta) = \sqrt{2}\left(\frac{a}{r}\right)^k P_{k-1}(\cos\theta) + \sqrt{2}\sum_{p*}(-1)^p\left(\frac{a}{r_p}\right)^k P_{k-1}(\cos\theta_p)$$

the expression (5.13) for the potential $V^Q(M)$ in the dielectric medium becomes

$$V^Q(M) = \frac{Q_0}{C_S}\mathbf{G}(r,\theta)\mathbb{M}^{-1}\mathbf{1}_1$$

$$(5.22)$$

5.3.3 Influence of Polarization

5.3.3.1 Analytical Expression for the Coupling Coefficient

In this section, we consider that $Q_0 = 0$ and that a polarization, $\Delta V = V_1 - V_2$, is applied between the two plate electrodes. As it is usually done in classical electrodynamics, the potential $V^{po}(M)$—defined in relation (5.1)—is split into two parts. The first, denoted by v^i, is the potential due to the polarization ΔV in the dielectric medium considered as empty. It is often called "incident" or "primary" potential (classically used in scattering models [30,32]). It is the potential that should exist if the dot was not there. The second part of the potential, denoted by v^d is due to the presence of the dot in the polarized medium. Classically this term is called "scattered" potential or "secondary" potential. Then

$$V^{po}(r,\theta) - V_2 = v^i(r,\theta) + v^d(r,\theta)$$

$$(5.23)$$

v^i is the potential in a plane capacitor.

$$v^i(r,\theta) = (\lambda_2 + z - z_0)\frac{\Delta V}{h} = (\lambda_2 + r\cos\theta)\frac{\Delta V}{h}$$

$$(5.24)$$

The boundary conditions satisfied by $V^{po}(M)$ are the following:

(Pa) $V^{po}(M) = V_1$ for $z = z_1$ and $V^{po}(M) = V_2$ for $z = z_2$
(Pb) The potential is floating and constant on the dot: $V^{po}(a,\theta) - v_f^{po} \; \forall\theta$
(Pc) Relations (5.3) with $Q_0 = 0$ are still valid

The scattered potential is expandable as a series of Legendre polynomials, for every M located in the same medium [28,30]. That is to say this expansion is valid only in the greatest sphere centered on O that does not cut the electrodes. This particular sphere, denoted by C in Figure 5.4, is tangent to the nearest electrode from the point O. Assuming that the nearest electrode is the bottom one (V_2), the radius of C is $\lambda_2 = a + t_2$. In this sphere ($r \leq \lambda_2$), we have

$$v^d(r,\theta) = \sum_{m=0}^{+\infty}\left[U_m\left(\frac{r}{a}\right)^m + W_m\left(\frac{a}{r}\right)^{m+1} \right]P_m(\cos\theta)$$

$$(5.25)$$

Using the boundary conditions (Pb) and (Pc) on the dot and the orthonormal properties of Legendre's polynomials, it follows that

$$\begin{cases} W_0 = 0 \\[2mm] U_0 = \left(v_f^{po} - V_2 - \lambda_2\frac{\Delta V}{h}\right)\sqrt{2} \\[2mm] U_m = -W_m - a\frac{\Delta V}{h}\sqrt{\frac{2}{3}}\delta_{m,1} \quad \forall m \geq 1 \end{cases}$$

$$(5.26)$$

The condition (Pa), for $\theta = \pi$, gives $U_m = -W_m$, $\forall m$. At zero-order it comes (with relations (5.26)):

$$v_f^{po} - V_2 = \frac{\lambda_2}{h} \Delta V \Rightarrow \alpha = \frac{\lambda_2}{h} \qquad (5.27)$$

Remark: If the nearest electrode is the electrode 1, the condition (Pa) gives $v_f^{po} - V_2 = (1 - (\lambda_1/h))\Delta V$ which implies the same result for α.

Finally, inside the sphere C, v^d can be written as

$$v^d(r, \theta) = \frac{a}{h} \Delta V \frac{\lambda_2^3}{\lambda_2^3 - a^3} \left(\left(\frac{a}{\lambda_2} \right)^3 \frac{r}{a} - \left(\frac{a}{r} \right)^2 \right) \cos\theta \qquad (5.28)$$

5.3.3.2 Semi-Analytical Expression for the Potential in the Dielectric Medium

As a last corollary, the semi-analytical expansion for C_S and the simple expression of α allows us to express the total potential everywhere in the dielectric medium, when ΔV and Q_0 are non null.

To express the potential V^{po} everywhere in the dielectric, we use again the method of images for determining v^d. Indeed, v^i already contains the boundary conditions on the planes Π_1 and Π_2. The boundary conditions on v^d are $v^d|_{\Pi_1} = v^d|_{\Pi_2} = 0$. So we can expand the potential v^d as we did for V^Q in the previous section, with the same notations. Using the conditions (Pa), (Pb), (Pc), and the relation (5.27), we can finally write V^{po} as

$$V^{po}(M) - V_2 = \frac{\lambda_2}{V} \Delta V \times \mathbf{G}\mathbb{M}^{-1}\mathbf{A}^h \qquad (5.29)$$

where $\mathbf{A}^h$ is the column vector of generic term of order k, $A_k^h = -\sqrt{(1/3)}(a/h)\delta_{2,k}$.

Finally using the two relations (5.22) and (5.29), the total potential in the dielectric medium is

$$V(M) - V_2 = \mathbf{G}\mathbb{M}^{-1}\left(\frac{Q_0}{C_S}\mathbf{1}_1 + \frac{\lambda_2}{h}\mathbf{A}^h \right) \qquad (5.30)$$

This expression allows one to estimate the distribution of charges on the electrodes, derivating this relation along z, at $z = 0$ and h. This is not the subject of this chapter but a parametric study of this distribution could be interesting to understand the classical approximations of C_S. Moreover, this should allow to obtain immediately the electric field everywhere in the medium. This physical quantity is necessary to model charge transport phenomenon through the dielectric layers.

To conclude this part, let us remind that the semi-analytical relations (5.20) and (5.27) solve the physical parameters of the configuration. As a final corollary, the matricial relation (5.30) gives the electrostatics of the dot.

We notice that this classical approach is invariant by scale change. So, these results can be directly used for larger configurations with the same aspect ratio. We can cite for instance chemical sciences where models of metallic bubbles embedded in dielectric solutions are often used.

5.3.4 Numerical Implementation

The numerical computation of the capacitance is numerically conditioned by the obtaining an invertible matrix $\mathbb{M}$. Numerically, the matrix $\mathbb{M}$ is a square matrix of the typical size 10×10. This is not a critical size for classical numeric inversion. Moreover, in practice, the conditioning (ratio between greatest and smallest values) of this matrix is about unity. Hence, the inversion of the matrix $\mathbb{M}$ is not the key point of this numerical implementation. Its quality is conditioned by the double truncation of the expansion of the potential (on the maximum order of polynomials in one hand and on the number of images in the other one). A discussion about the influence of this truncation is presented in Appendix A. Let us say here that the two truncations are independent one of each other, and numerically they are not critical at all for computations. Moreover, the semi-analytical expression was numerically validated by comparing results provided by finite element method (FEM) computations. This validation is also presented in Appendix A. To briefly sum up the conclusions, let us say that FEM may have some accuracy problems and it is clearly slower.

5.3.5 Numerical Results and Comparison with a Classical Approximation

Classically C_S and α are estimated by different physically based approaches using very simple expressions. Nevertheless, the validity of these latest approaches is not always well known. A very classical phenomenological approach is the equivalent capacitances approach (for details, see [24]) that we will briefly expose here.

It is assumed that a coupling capacitance can be defined between the dot and each electrode. Let us denote C_C^b (resp. C_C^t) the capacitance between the dot and the bottom (resp. top) electrode. The sum of these two capacitances is equal to C_C. We further assume that C_C^b and C_C^t are classical capacitances, as defined for metallic bodies under mutual influence and that the polarization coupling coefficient α is given by the ratio $\beta = (C_C^t/C_C)$. The parallel plate approximation (PPA) consists in approximating the capacitances C_C^b and C_C^t by plate capacitances in which surface S is the projected surface of the dot on each electrode and whose thickness t_C^b (resp. t_C^u) is the typical distance between the dot and each electrode. This rough approximation is even used to model single-electron effects such as Coulomb blockade [16]. Note that it is possible to use several surfaces and typical distances to calculate this plate capacitances. Here, we choose two different thicknesses for the same configuration ($t_C^t = t_1$ or $t_C^t = t_1 + a = \lambda_1$) to illustrate the capacitances.

1. The thicknesses of the capacitances are the distance between the border of the dot and the electrodes: $t_C^b = t_2$ and $t_C^t = t_1$. This approximation is named PPA(1).

$$C_C^b(1) = \varepsilon \frac{S}{t_2}, \quad C_C^t(1) = \varepsilon \frac{S}{t_1}, \quad C_C(1) = \varepsilon \left(\frac{S}{t_1} + \frac{S}{t_2} \right)$$

$$\text{and} \quad \beta(1) = \frac{t_2}{t_1 + t_2} \tag{5.31}$$

2. The thicknesses of the capacitances are the distance between the center of the dot and the electrodes: $t_C^b = \lambda_2$ and $t_C^t = \lambda_1$. This approximation is named PPA(2).

$$C_C^b(2) = \varepsilon \frac{S}{\lambda_2}, \quad C_C^t(2) = \varepsilon \frac{S}{\lambda_1}, \quad C_C(2) = \varepsilon \left(\frac{S}{\lambda_1} + \frac{S}{\lambda_2} \right)$$

$$\text{and} \quad \beta(2) = \frac{\lambda_2}{\lambda_1 + \lambda_2} \tag{5.32}$$

To compare the validity of these approximations, we compare some results given by PPA and results given by the semi-analytical (SA) method. For a given radius, we compare the evolution of the capacitances and different ratios versus the value of t_2. The numerical values are the following: $a = 2.8\,\text{nm}$, $t_1 = 12\,\text{nm}$, $\varepsilon_r = 3.9$, and t_2 is varying from 0.5 to 10.5 nm. For the surface S, we have chosen the surface of the disk of radius a as $S = \pi a^2$.

In Figure 5.6, we plot the evolution of the three capacitances C_S (denoted by SA), $C_C(1)$ and $C_C(2)$, divided by the factor $4\pi\varepsilon a$, versus the ratio t_2/a. We also plot the ratios $C_C(1)/C_S$ and $C_C(2)/C_S$. The behavior given by the approximation (1) is qualitatively correct. Quantitatively, this is not very convincing. Indeed for $t_2/a < 1$, the ratio $C_C(1)/C_S$ is not quite constant. It varies from 0.7 to 0.2. There is no way to fit the values of C_S by $C_C(1)$ just by adjusting the numerical value of S. PPA(2) seems

qualitatively and quantitatively more interesting to estimate C_S. For $t_2/a < 1$, the ratio $C_C(2)/C_S$ is varying from 0.12 to 0.14 (a variation about 20%).

For large values of t_2/a, the behavior of PPA(1) and PPA(2) are asymptotically equivalent. The asymptotic limit for $t_2 \gg a$ is $a/(4t_1) \sim 5.83 \times 10^{-2}$ for $C_C(1)/(4\pi\varepsilon a)$ and $a/(4(t_1 + a)) \sim 4.73 \times 10^{-2}$ for $C_C(2)/(4\pi\varepsilon a)$. In the figure, we see the beginning of this asymptotic behavior. Because the limiting value of t_2, in our simulation, is of the same order of t_1, the smallest values seen on the figure are approximatively twice the limiting value. For $t_2/a > 2$, the ratio $C_C(2)/C_S$ remains quite constant, varying from 0.11 to 0.09. The ratio $C_C(1)/C_S$ varies from 0.15 to 0.1.

Finally, in all the range of values of t_2/a presented here, only the PPA(2) allows to approximate C_S with finite relative error. The difficulty consists in finding a numerical value for the surface S that makes PPA match the capacitance.

To end this comparison, the dimensionless coefficients κ_1, $\beta(1)$, and $\beta(2) = \alpha$ are compared. For great values of t_2/a, $\beta(1)$ tends to 1 while $\beta(2)$ tends to 0.5. The limiting value of κ_1 is also 0.5 (because C_2 tends to the capacitance of a single half-sphere). On the other hand, for small t_2/a, the limiting value for $\beta(2)$ is $a/(t_1 + 2a) \sim 0.15$, while $\beta(1)$ tends to 0 and C_1 also to 0 (for small t_2, only C_2 contributes to the value of C_S).

In Figure 5.7, we plot the coefficients κ_1, $\beta(1)$, and $\beta(2)$ versus t_2/a for the same previous configuration. On this figure, it appears that the coefficient $\beta(2)$ underevaluates κ_1 for $t_2/a > 0.25$. The greatest relative difference between κ_1 and $\beta(2)$ is observed for $t_2/a \sim 1$ and is about 30%. For $t_2/a > 3$, the relative difference is lower than 10%. The coefficient $\beta(1)$ is far from κ_1, even in the range $t_2/a \ll 1$.

For values of t_2/a in the range of the unity, neither PPA(1) nor PPA(2) gives an approximation with an accuracy better than 10%. Only PPA(2) could give a good approximation, if we are able to evaluate an optimized value for the surface S.

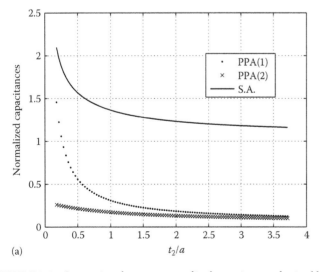

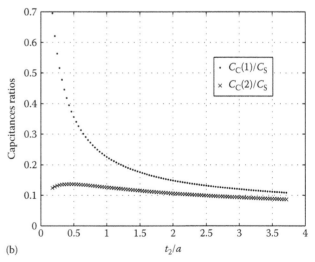

(a) t_2/a (b) t_2/a

FIGURE 5.6 Comparison between normalized capacitances obtained by the approximations PPA(1) and PPA (2) and by SA. In the first figure, the capacitances are normalized by the factor $4\pi\varepsilon a$. The second figure represents the ratios C_C/C_S for the two approximations. The capacitances and the ratios are plotted versus the ratio t_2/a, for a fixed radius a and variable t_2.

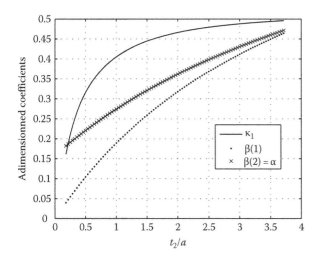

FIGURE 5.7 Undimensioned coefficients κ_1, $\beta(1)$, $\beta(2)$, and α versus the ratio t_2/a.

5.3.6 Conclusion

We devised a semi-analytical model for the potential and the capacitance of a spherical metallic dot located in a dielectric medium between two metallic planes at fixed potential. From this model, we establish a simple analytical expression for the polarization coupling coefficient. We have derived a semi-analytical formulation for the potential everywhere in the dielectric medium. The validity of our approach has been established by comparison with a variational numerical approach. Finally, we have discussed the validity of classical approximations for the capacitance, particularly when the distance between the border of the dot and the nearest electrode has the same order of magnitude as the radius of the dot.

Three major developments are under study. The first one, developed in the next part of this chapter, consists in using this approach in the quantum model of a dielectric dot. First, it allows to generate an initial evaluation of the floating potential used in iterative resolution of Schrödinger and Poisson systems. Second, the comparison between the classical and quantum approaches determines the critical dimensions leading to a quantum behavior. The second development is to adapt this model for dielectric dots. The method of images remains interesting but the spatial distribution of charges in the dot—even in a non-quantum approach—is a serious node of the resolution. The third point is to use the evaluation of the potential around the dot to model the electric field, first step to a modelization of an electronic transport through the dielectric.

5.4 Semiconductor Dot: Theoretical Approach

5.4.1 General Considerations on Quantization Effects

In the metallic case, the electric charge is distributed on the surface of the dot. On the contrary, in the case of a semiconductor dot, the electric charge is distributed in the volume of the

nanocrystal and in the surrounding dielectrics. The resolution of such a configuration in a classical approach needs conditions of continuity of the electric field involving model of macroscopic polarization of the dielectric medium. This approach is classically exposed in different works (for instance, Refs. [28,30]). However, when the dot size falls into the nanometer range, classical electrostatics and semiclassical approaches of charge in semiconductor devices are no longer sufficient to describe the charge density and thus the electrostatic potential inside the nanocrystal. Two arguments favor a quantum approach to correctly describe the problem. First of all, let us consider the De Broglie wavelength, λ_B, that describes the wave–particle duality of any particle of mass m; it is equal to the ratio of Planck constant, h, on the kinetic momentum p. In the case of a free thermal electron of mass m_e, at temperature T, De Broglie wavelength is

$$\lambda_B = \frac{h}{p} = \frac{h}{\sqrt{2kTm_e}}$$

where k is the Boltzmann constant.

At ambient temperature ($T \sim 300\,K$), λ_B is around 8 nm; this values gives an order of magnitude of the extension of the electronic wavefunction.

In the frame of this work, we will consider semiconductor islands with typical radius around 10 nm, so it becomes a priori necessary to take into account quantum effects to solve the electrostatics of the dot.

Another way to convince ourselves of this latest point is to remind us of some results obtained in the case of an infinite quantum well. Indeed, electrons confined in quantum dots can be roughly compared to the problem of a single particle in an infinite quantum well.

Let us consider an infinite one-dimensional quantum well of width $2a$. Due to the strong confinement of the wavefunction into the well, the only possible solution of Schrödinger equation inside the dot are standing waves with wave number, k_n, equal to

$$k_n = \frac{n\pi}{2a} \tag{5.33}$$

The non-normalized density of probability, $\tilde{\psi}_n(x)$ is given by

$$\tilde{\psi}_n(x) = \sin\left[k_n x - n\frac{\pi}{2}\right] \tag{5.34}$$

From now, the normalized density of probability is denoted by ψ. The normalization is such as the density is unitary.

The energy of the nth bound state, with respect to the bottom of the well, is proportional to the square of the wave number:

$$E_n = \frac{\hbar^2 k_n^2}{2m_e} = \frac{n^2 h^2}{32 m_e a^2} \tag{5.35}$$

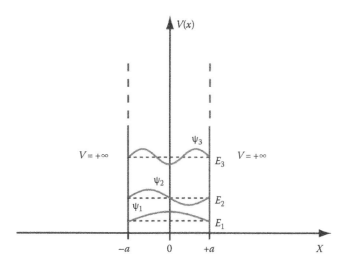

FIGURE 5.8 Schematic representation of an infinite quantum well of width $2a$ and of the three first bound states and their energy.

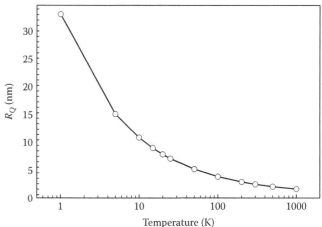

FIGURE 5.10 Evolution of quantum dot radius at which quantization energy becomes greater than thermal energy, R_Q, with temperature.

As an explicative pattern, the eigen wavefunction ψ_n and the level of energy E_n for the three first levels of quantification are plotted simultaneously in Figure 5.8. The evolution of the energy of the first bound state, E_1, is plotted as a function of the dimension of the quantum well in Figure 5.9. If we compare the energy level of the first bound state with the thermal energy, $k_B T$, we can see that when the quantum well's typical size, a, falls under 2–3 nm, E_1 becomes greater than the thermal energy at ambient temperature (~25.6 meV). From this consideration, we define R_Q the value of the quantum well radius at which the energy shift due to quantum confinement in the dot becomes equal to the thermal energy:

$$R_Q = \frac{nh}{4}\sqrt{\frac{1}{2m_e k_B T}} \tag{5.36}$$

As seen in Figure 5.10, for temperatures ranging from 5 to 300 K, R_Q values are in the 10 nm range, which is the typical size of the quantum dots that we will consider here.

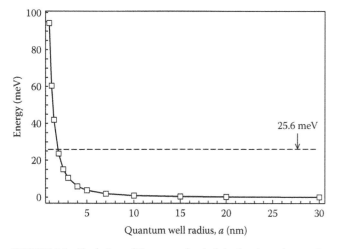

FIGURE 5.9 Evolution of the energy level of the first bound state, E_1, obtained in an infinite well of width $2a$ versus a.

5.4.2 Quantum Dot Containing a Single Electron: A Simplified Approach

To start with, in this study of the electronic properties of semiconductor dots, we present the main equations describing the behavior of a single electron in the conduction band of a silicon nanocrystal. As explained in the introduction of this section, due to the fact that the De Broglie wavelength is of the same order as the dimension of the nanocrystal, a quantum approach has to be used. Due to the difference in electronic affinity between silicon and silicon oxide, a single silicon dot embedded in a SiO_2 matrix acts as a quantum well for electrons in the conduction band, as shown in Figure 5.11. This system is close to the well-known case of a finite potential well described in many graduate textbooks [33,34]. If analytical expression can be easily found for E_n and $\psi_n(\vec{r})$ in the case of an empty finite or infinite quantum well, it is not the case for a finite quantum well containing several electrons; as a consequence, solving Schrödinger equation is done numerically.

In this chapter, we consider an undoped semiconductor quantum dot embedded in a silicon dioxide matrix. Due to the extremely low concentration of free carriers in intrinsic semiconductor ($n_i \approx 10^{10}$ cm^{-3} for silicon, at 300 K), we assume that there is no free electron in the conduction band of the nanocrystal. Let us make a quick calculation to illustrate this assertion.

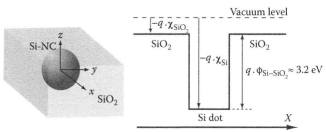

FIGURE 5.11 Energy band diagram representing a silicon nanocrystal embedded in a SiO_2 matrix along one direction.

Let us consider a silicon nanocrystal of radius $a = 10\,\text{nm}$, the intrinsic number of electrons N is

$$N = n_i \times \frac{4}{3}\pi a^3 \sim 4 \times 10^{-8} \ll 1$$

For the sake of simplicity, we only consider the bound states existing in the conduction band. We neglect any perturbation arising from the valence band and spin degeneracy. Let us note $E_C(\vec{r})$ the spatial density of the potential energy in the conduction band. Considering the above finite quantum well with a single additional free electron in the conduction band, the spatial evolution of probability density, $\psi(\vec{r})$ is described by the Schrödinger equation in stationary regime. If we look at the first electronic states—at low energy—it is possible to consider that the electron behaves as a quasi-free particle in the dot with an isotropic effective mass m_{eff}^* [35] ($m_{\text{eff}}^* \approx 0.26 m_e$ in silicon):

$$\mathbb{H}\psi(\vec{r}) = E\psi(\vec{r}) \tag{5.37}$$

with

$$\mathbb{H} = \frac{\vec{p}^2}{2m_{\text{eff}}^*} + V_{\text{conf}}(\vec{r}) \tag{5.38}$$

where

$V_{\text{conf}}(\vec{r})$ is the potential confining the electron evaluated at the point $\vec{r}$ (see Figure 5.11)

$\vec{p}$ is the kinetic momentum operator (classically equivalent to a Laplacian [33,34])

$V_{\text{conf}}(\vec{r})$ must be evaluated with the help of hypothesis, described in the following, for each configuration.

In the case of a single electron in a quantum well, the potential energy $V_{\text{conf}}(\vec{r})$ is only equal to the conduction band energy value at position $(\vec{r})$, $E_C(\vec{r})$. For convenience, we take the bottom of the conduction band energy in the quantum dot as the energy reference.

As there is no other energy contribution in the Hamiltonian $\mathbb{H}$ due to the presence of other electrons in the conduction band, this problem is the same as the finite quantum well. The only available energy states for the electron, are the same (E_n; ψ_n) than those of the finite well. For the following, we suppose that the wavefunction ψ_n has been properly normalized.

The presence probability $\mathcal{P}(\vec{r})$, of the electron at position $\vec{r}$, is thus equal to

$$\mathcal{P}(\vec{r}) = |\psi_n(\vec{r})|^2 \tag{5.39}$$

Under these conditions, we can built the charge density at any position, $\rho(\vec{r})$, as the product of the electronic charge by the presence probability at $\vec{r}$ (which is also known as the charge density *observable*):

$$\rho(r) = -e \times \mathcal{P}(\vec{r}) = -e \times |\psi_n(\vec{r})|^2 \tag{5.40}$$

where e is the fundamental charge. As described above, charge density depends on the bound state occupied by the electron. Once the wavefunction of the bound state is known, a numerical resolution of the Poisson equation using the above charge density gives the electrostatic potential of the structure and the evolution of the conduction band energy. For our problem, we make the assumption that the electron, without any external perturbation, is on the first available quantum state, i.e., the fundamental bound state at energy E_1 with the normalized presence probability density $\psi_1(\vec{r})$ and solve the Poisson equation to derive the electrostatic potential, $V(\vec{r})$, in the whole structure:

$$\Delta V(\vec{r}) = -\frac{\rho(\vec{r})}{\varepsilon_0 \varepsilon_r} \tag{5.41}$$

where

ε_r is the relative dielectric permittivity of the material

Δ is the Laplacian operator

5.4.3 Addition of a Second Electron in the Dot

In this section, we consider the previous system made of a silicon nanocrystal embedded in a SiO_2 matrix containing a single electron on the first bound state. When a second electron is added in the conduction band, the probability density, ψ, rigorously becomes a function of two positions:

$$\psi_n = \psi_n(\vec{r_1}, \vec{r_2}) \tag{5.42}$$

Under these conditions, the Hamiltonian of the system is written as

$$\mathbb{H} = \frac{\vec{p_1}^2}{2m_{\text{eff}}^*} + \frac{\vec{p_2}^2}{2m_{\text{eff}}^*} + V_{\text{conf}}(\vec{r_1}) + V_{\text{conf}}(\vec{r_2}) + \frac{q^2}{4\pi\varepsilon_0\,\varepsilon_r\,\|\vec{r_1} - \vec{r_2}\|^2} \tag{5.43}$$

where $\vec{p_1}$ and $\vec{p_2}$ are the kinetic momentum operators of each electron. At this stage, it is very important to notice that an energy term has been added in the Hamiltonian. This term is due to the *Coulombian interaction* of electrons with each other and corresponds to the *electrostatic energy*. Including the interaction term in the Hamiltonian and considering two position variables, solving the problem becomes very heavy, even numerically. Moreover, when we consider further more than two electrons in the nanocrystal, the problem quickly become a N-body problem. For this reason, some approximations are required to be developed further. In the following, we use an approximation that allows to write the wavefunction in a simpler way.

In the Hartree approximation [36], we make the assumption that each electron can be described by an individual wavefunction, respectively $\psi_1(\vec{r})$ and $\psi_2(\vec{r})$, as if they were *independent*. Therefore, the global wavefunction can be written as the product of these individual wavefunctions:

$$\psi(\vec{r}_1, \vec{r}_2) = \psi_1(\vec{r}_1) \times \psi_2(\vec{r}_2) \qquad (5.44)$$

In this approximation, we calculate ψ_1 and ψ_2 in such a way that the *total energy* of the system is *minimal*. Moreover the separation of the wavefunction allows to use a *mean field* approximation in which we consider that the electrostatic interaction felt by one electron can be described as if it felt the electrostatic potential induced by the other one(s).

The problem now consists in calculating ψ_1 and ψ_2 by solving individual Schrödinger equations with Hamiltonians $\mathbb{H}_1$ and $\mathbb{H}_2$ with

$$\mathbb{H}_{1,2} \psi_{1,2}(\vec{r}_{1,2}) = E_{1,2} \psi_{1,2}(\vec{r}_{1,2}) \qquad (5.45)$$

with

$$\mathbb{H}_{1,2} = \frac{\vec{p}_{1,2}^2}{2m_e} + V_{\text{conf}}(\vec{r}_{1,2}) + V_{\text{inter}_{1,2}}(\vec{r}_{1,2}) \qquad (5.46)$$

$V_{\text{inter}_{1,2}}(\vec{r}_{1,2})$ is the term of Coulomb interaction defined in relation (5.43).

If we compare this latest equation with the one used for the single-electron case (5.37), V_{inter} has been added to take account of the Coulomb interaction between both electrons. As V_{inter} is the mean electrostatic potential induced by the antagonist electron, it is a solution of the Poisson equation:

$$\Delta V_{\text{inter}_{1,2}}(\vec{r}_{1,2}) = -\frac{\rho_{2,1}(\vec{r}_{1,2})}{\varepsilon_0 \varepsilon_r} \qquad (5.47)$$

where $\rho_{2,1}(\vec{r})$ is given by (5.40)

$$\rho_{2,1}(\vec{r}) = -e \times \left| \psi_{2,1}(\vec{r}) \right|^2 \qquad (5.48)$$

The total energy of the system is equal to the sum of the energy corresponding to the bound states of both electrons *plus* the electrostatic energy arising form the Coulomb interaction between both electrons, E_{inter}:

$$\mathbb{E}_{\text{tot}} = \mathbb{E}_1 + \mathbb{E}_2 + \mathbb{E}_{\text{inter}} \qquad (5.49)$$

with

$$\mathbb{E}_{\text{inter}} = \iiint_{\text{space}} -e \left| \psi_1(\vec{r}) \right|^2 \times V_{\text{inter}_1}(\vec{r}) \cdot \mathrm{d}^3 \vec{r}$$

$$= \iiint_{\text{space}} -e \left| \psi_2(\vec{r}) \right|^2 \times V_{\text{inter}_2}(\vec{r}) \cdot \mathrm{d}^3 \vec{r} \qquad (5.50)$$

From Equations 5.46 to 5.48 it is obvious that this problem can be solved only *self-consistently*; ψ_1 is computed by solving (5.46) using an initial guess for ψ_2. Then, Poisson equation is

solved using ρ_1 as charge density and the corresponding potential V_{inter_2} is used in (5.46) to compute ψ_2. This algorithm is iteratively repeated until the total energy $\mathbb{E}_{\text{tot}}$ calculated at the end of each loop converges.

5.4.4 Multiple Electrons in the Dot

In this section, we extend the results presented in the previous section to the case of multiple electrons in the dot with the same methodology. When the quantum dot contains N electrons, Hamiltonian can be derived in a similar way than in the two-electron problem (relation 5.43):

$$\mathbb{H} = \sum_{j=1}^{N} \frac{\vec{p}_j^2}{2m_{\text{eff}}^*} + \sum_{j=1}^{N} V_{\text{conf}}(\vec{r}_j) + \frac{1}{2} \times \sum_{j=1}^{N} \sum_{\substack{i=1 \\ i \neq j}}^{N} \frac{q^2}{4\pi \varepsilon_0 \varepsilon_r \left\| \vec{r}_j - \vec{r}_i \right\|^2} \qquad (5.51)$$

where

$\vec{p}_j$ is the kinetic momentum operator of electron j
V_{conf} is the confining potential

the last term of Equation 5.51 is the interaction potential coming from the electrostatic interaction of the N electrons with each other.

Note that the prefactor 1/2 has been added to avoid to count twice the same interaction.

Rigorously, the solution of the problem is given by solving Schrödinger equation with a N-variable wavefunction:

$$\psi = \psi(\vec{r}_1, \vec{r}_2, \dots, \vec{r}_N) \qquad (5.52)$$

As mentioned in the previous section, solving this N-body problem needs an approximations on the wavefunction. In the Hartree approximation, the wavefunction can be written as the product of N individual *normalized* wavefunction:

$$\psi(\vec{r}_1, \vec{r}_2, \dots, \vec{r}_N) = \prod_{j=1}^{N} \psi_j(\vec{r}_j) \qquad (5.53)$$

each wavefunction can now be associated to an Hamiltonian operator describing the behavior of the jth electron:

$$\mathbb{H}_j \psi_j(\vec{r}_j) = \mathbb{E}_j \psi_j(\vec{r}_j) \qquad (5.54)$$

with

$$\mathbb{H}_j = \frac{\vec{p}_j^2}{2m_{\text{eff}}^*} + V_{\text{conf}}(\vec{r}_j) + V_{\text{inter}_j}(\vec{r}_j) \qquad (5.55)$$

V_{inter_j} refers to the electrostatic interaction between electron j and the mean potential induced by the $(N-1)$ other electrons.

Assuming that an equivalent charge density, $\rho_j(\vec{r})$ can be defined for each electron:

$$\rho_j(\vec{r}) = -e \times \left| \psi_j(\vec{r}) \right|^2 \tag{5.56}$$

the electrostatic interaction potential V_{inter} is solution of the following Poisson equation:

$$\Delta V_{inter_j}(\vec{r}) = -\sum_{\substack{i=1 \\ i \neq j}}^{N} \frac{\rho_i(\vec{r})}{\varepsilon_0 \varepsilon_r} \tag{5.57}$$

The total energy of the system is the sum of the individual eigen energies of the electrons *plus* the electrostatic interaction energy:

$$\mathbb{E}_{tot} = \sum_{i=1}^{N} \mathbb{E}_i + \mathbb{E}_{inter} \tag{5.58}$$

where

$$\mathbb{E}_{inter} = -\frac{1}{2} \sum_{i=1}^{N} \sum_{\substack{j=1 \\ j \neq i}}^{N} \iiint_{space} q \left| \psi_i(\vec{r}) \right|^2 \times V_{inter_j}(\vec{r}) \cdot d^3\vec{r} \tag{5.59}$$

To solve the entire problem in the Hartree approximation, it is now necessary to self-consistently solve the set of coupled Equations 5.54 and 5.57 in order to minimize the total energy of the system (5.58).

In this study, electrical results presented in the next section are obtained by using the algorithm presented in Figure 5.12. Starting from an initial assumption for wavefunctions (assuming, for example, an homogenous dilution of the electrons in the quantum dot) and for the potential of the dot (metallic approach or empty dot with coupling polarization), the Poisson equation is solved to determine the interaction potential, V_{inter}. This potential is used in the system Hamiltonian (5.55) as well as the confining potential, V_{conf}, to determine the bound states, i.e., their energies, $\mathbb{E}_j$ and the normalized wavefunctions, ψ_j. This process is repeated for the N electrons and then, the total energy of the system, $\mathbb{E}_{tot}$ is calculated using Equations 5.58 and 5.59. If the difference of the total energy found at the nth and $(n-1)$th iteration is smaller than an user-defined value (ε), then the Poisson equation is solved for all electrons in the nanocrystal, otherwise the resolution restarts until this condition is satisfied.

We notice here that the influence of the initial guess on the final solution is very weak when the number of electrons is low. However, it can become critical in the case where the number of electrons becomes of the order of magnitude of 10.

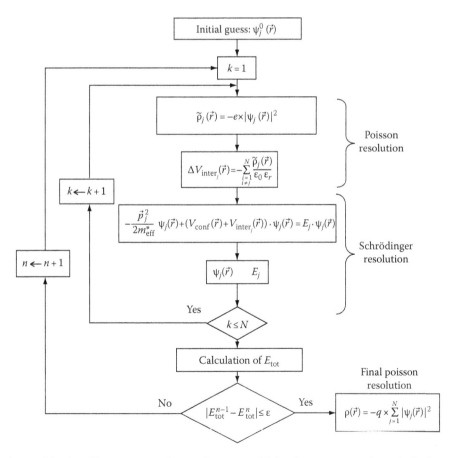

FIGURE 5.12 Algorithm used for the self-consistent resolution of Poisson and Schrödinger equations for multiple electrons in a quantum dot.

5.5 Finite Element Modeling of a Silicon Quantum Dot

In this section, we will present some simulation results obtained on a silicon dot by finite element modeling (FEM). For this study, we used COMSOL Multiphysics© [37] to generate the geometry as well as to solve Poisson and Schrödinger equations [38,39]. Firstly, we will present the studied structure and focus on the methodology used to optimize meshing, which is a crucial point in FEM. Then, the simulation results will be presented, in particular, we will see the influence of the dot radius and the number of electrons on the evolution of the electrostatic potential into the nanocrystal. Finally, we will compare the electrical results obtained in the previous chapter on metallic dots with those obtained in a full quantum treatment of the charge in a semiconductor dot.

5.5.1 Studied Structure and Meshing Strategy

The system studied consists in a spherical silicon dot embedded into a dielectric matrix of SiO_2. The notations are recalled in Figure 5.13. In this work, the thickness of the top and bottom oxides (t_1 and t_2) are respectively fixed to 10 and 5 nm, which are typical values used in the semiconductor industry [17].

As shown in Figure 5.14, the device can be reduced to a quarter of structure for plane-symmetry reasons along the (YZ) and (XZ) planes. This optimization of simulated geometry considerably reduces the computation time that can be advantageously converted into a finer mesh of the structure.

The boundary conditions used for this work are the following:

- Dirichlet conditions on both top and bottom electrodes: electrostatic potential fixed, respectively, to V_1 and V_2

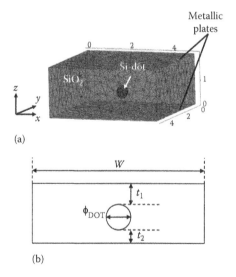

(a)

(b)

FIGURE 5.13 (a) Studied structure for the FEM simulation of the quantum dot containing multiple electrons. (b) 2D schematic view showing the dimensions of the structure.

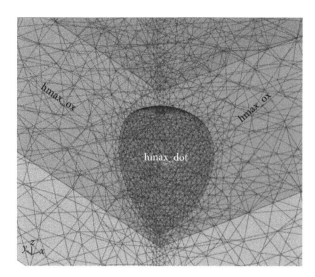

FIGURE 5.14 Detailed view of the quarter-structure studied in FEM showing some relevant meshing parameters considered during the meshing operation: maximum mesh size in the dot (hmax_dot) and maximum mesh size in the oxide (hmax_ox).

- Symmetry conditions on the symmetry planes of the structure (YZ) (XZ)
- Neumann conditions (i.e., null electric field) on the external planes of the structures parallel to (XZ) and (YZ)

For these simulations, we paid a particular attention to meshing which is a crucial point in FEM simulations. From a practical point of view, each simulation result was validated only after having verified that its result remains unchanged after successive mesh refinements [40]. Meshing the structure described in Figure 5.13 is sensitive to many parameters related to the dot size, oxide thickness and the total width of the structure. In each of these domains (dot or silicon dioxide), many parameters such as maximum (resp. minimum) mesh size as well as maximum (resp. minimum) mesh growth rate can be defined. For the sake of simplicity, we will only focus on two main parameters to see the way to optimize them as well as their impact on the numerical results. The parameters studied here are (see Figure 5.14)

- The maximum mesh size in the oxide: hmax_ox
- The maximum mesh size in the dot: hmax_dot

From a general point of view, meshing optimization consists in finding the best compromise between numerical accuracy and computation time by tuning the relevant meshing parameters. Indeed, for a three-dimensional problem the number of mesh points is N^3 where N is the average number of mesh per dimension. So, reducing the number of mesh in regions that do not need much refinement can drastically increase computation time. The two structures depicted in Figure 5.15 show typical results obtained for a set of value for (hmax_ox; hmax_dot) giving a coarse mesh (Figure 5.15a) and a finer one (Figure 5.15b). The natural trend is to refine mesh as much as possible, like in Figure 5.15b, but sometimes it considerably increases computation time or dramatically leads to numerical instabilities.

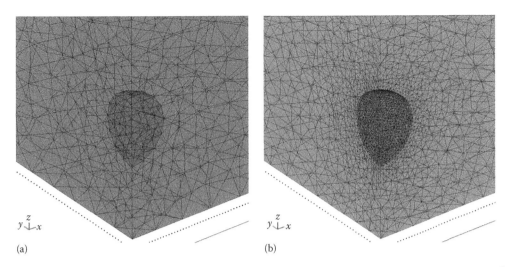

FIGURE 5.15 Comparison of the same structure meshed with different parameters: (a) coarse mesh giving fast computations (hmax_ox = 1 nm, hmax_ox = 3 nm) and (b) fine mesh giving a better numerical accuracy but more time consuming (hmax_dot = 0.25 nm, hmax_ox = 2 nm).

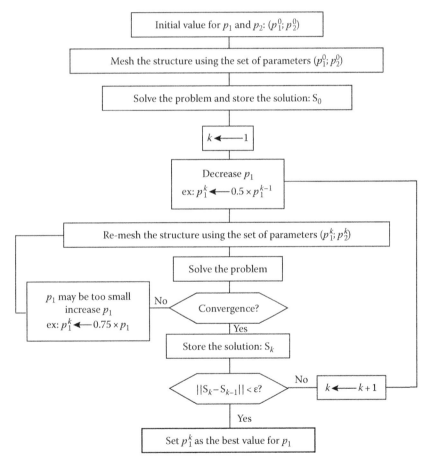

FIGURE 5.16 Algorithm used to mesh the structure.

As a consequence, for each of the two meshing parameters chosen in our study, a systematic study needs to be carried out to determine the best set of values for (hmax_ox; hmax_dot). To this aim, each parameter is separately adjusted, while the value of the other one is fixed, by an iterative study depicted in the algorithm of Figure 5.16. For each value of hmax_ox (resp. hmax_dot), the convergence is tested and the result is compared to the one obtained during the previous iteration. The value of the studied parameter is reduced until the difference between the simulation results obtained at iteration (k) and

$(k - 1)$ is lower than a user-fixed value. At this point, it means than reducing again the value of the studied parameter will not increase numerical accuracy anymore. The optimal set of meshing parameters values can be found by repeating this method as many times as the number of meshing parameters. To conclude with this method, let us mention that the factor used to reduce the studied parameter (i.e., 2) as well as the numerical accuracy criterion (i.e., ε) are very structure-dependent and need to be adjusted to the specificity of each problem.

In the typical case studied in this work, the meshing optimization has been performed on structure shown in Figure 5.14 with an applied bias of 5 V between the top and the bottom electrodes. To quantitatively evaluate the impact of the meshing parameters on the simulation results, the energy value of the first bound state in the dot without any electron inside (E_0) is followed when changing the values of hmax_ox and hmax_dot. Obviously, E_0 is obtained by solving Schrödinger equation in the dot after the meshing step. Note that the zero-energy reference has been taken at the bottom electrode. In Figure 5.17a and b, E_0 values are plotted for different values of hmax_ox and hmax_dot. As can be seen in Figure 5.17a and b, the maximum mesh size in the oxide does not seem to have

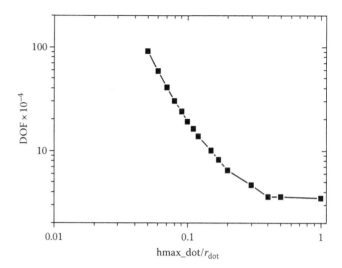

FIGURE 5.18 Number of degree of freedom (DOF) versus normalized maximum mesh size in the dot.

any quantitative influence on E_0 (the total energy scale represented in the graph is below 1 meV). On the contrary, hmax_ox plays a crucial role in the numerical results; reducing its value from $1 \times a$ (i.e., 2.5 nm) to $0.1 \times a$ (i.e., 0.25 nm) changes the value found for E_0 from ≈5 meV. It is very interesting to notice that beyond $0.1 \times a$, reducing hmax_dot does not significantly change the numerical result obtained for E_0 meaning that an optimum value can be found around $0.1 \times a$. At this stage, it is useful to have a look at Figure 5.18 showing the exponential increase of the number of degree of freedom in the structure (i.e., number of mesh points) due to the reduction of the maximum mesh size in the nanocrystal; reducing hmax_dot from $0.1 \times a$ to $0.05 \times a$ makes the number of mesh points increases from ~20,000 to ~1 million units.

For this work, the criterion chosen to determine the best value for each meshing parameter is that the variation of E_0 should not exceed 0.1 meV between two successive meshing. This criterion corresponds to an arbitrary choice allowing a good accuracy in the energy levels obtained in our problem; in this way, the maximum allowed energy uncertainty remains smaller than the thermal energy at 300 K (~25.6 meV). Using this criterion the value retained for hmax_ox and hmax_dot are, respectively, 3 nm and $0.07 \times a$.

5.5.2 Numerical Results

Using the algorithm detailed in Figure 5.12 and the structure meshed with the methodology exposed in previous section, Poisson and Schrödinger equations have been solved for an increasing number of electrons in the dot. Figure 5.19a shows the evolution of the charge density in the nanocrystal along the high-symmetry axis. When no external bias is applied, the shape of the charge density is very similar to the wavefunction of the first state in an infinite quantum well. However, when increasing the number of charges in the dot, Coulomb interaction between

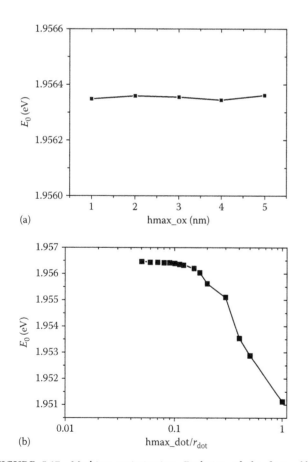

FIGURE 5.17 Meshing optimization. Evolution of the first self-energy found in the dot versus (a) hmax_ox and (b) hmax_dot. Note that values of are normalized with respect to the dot radius a.

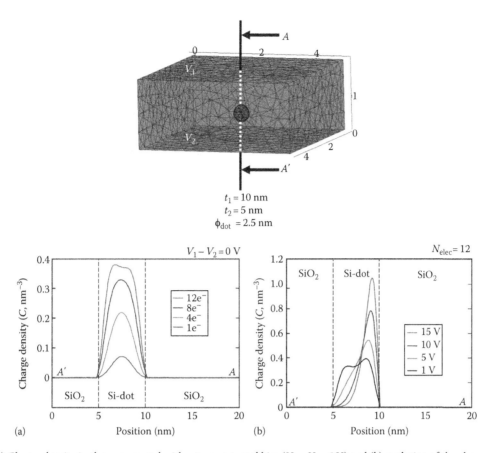

FIGURE 5.19 (a) Charge density in the nanocrystal without any external bias ($V_1 - V_2 = 0\,V$) and (b) evolution of the charge density in the dot containing 12 electrons for various external biases.

electrons splits the charge density centroid into two peaks situated on each side of the symmetry plan of the dot (see Figure 5.19a for 12 electrons or Figure 5.19b for 1 V). When an external bias is applied on the electrodes, the charge density is moved in a classical way in the direction of the increasing potential, corresponding to expected electrostatic result. As shown in Figure 5.19b, the more the potential difference increases, the more the charge centroid moves toward the edge of the dot.

Then, using the relationship given in (5.58), the total energy of the system has been computed and reported in Figure 5.20. These results are very close to those obtained by other groups on similar structures [20,21]. As a reminder, this energy takes into account the confinement energy of each electron and the Coulomb interaction between all electrons. Let us remark that the increase in the total energy due to the addition of a single electron significantly increases as the dot radius decreases. As an example, adding a sixth electron in a dot already containing five electrons increases the total energy of 500 meV ($\sim 20 k_B T$ at 300 K) if the dot radius is 6 nm. This value goes up to 3 eV when the radius is reduced to 1 nm ($\sim 120 k_B T$ at 300 K). As explained in Figure 5.10, this result clearly underlines the role played by the quantum effects in nanostructures and the necessity to use a full quantum approach when the dimension of the system studied is in the nanometer range.

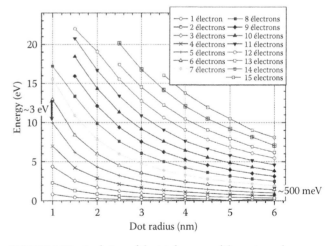

FIGURE 5.20 Evolution of the total energy of the system when varying the dot radius and the number of electrons stored in the nanocrystal.

5.5.3 Comparison with the Metallic Approximation

To conclude this study, let us compare the electrical results obtained in the case of a metallic dot with those obtained in a full quantum treatment of the charge density. Here, we will compare the electrostatic potential obtained in both methods. Let us note that, in the metallic case, the electrostatic potential is constant over the entire dot whereas it varies along the dot volume in the semiconductor cases. For ease of comparison, let us define the mean electrostatic potential, V_{mean}, as

$$V_{mean} = \frac{1}{Q_0} \iiint_{space} \rho(\vec{r}) \times V(\vec{r}) \, d^3\vec{r} \qquad (5.60)$$

where Q_0 is the total electronic charge carried by the dot

$$Q_0 = -e \times N_{elec} \qquad (5.61)$$

As defined in Equation 5.60, V_{mean} is nothing else than the mean potential over the whole structure weighted by the charge density.

Figure 5.21 compares the potential of the metallic dot with the mean potential in the semiconductor dot, V_{mean}, for different values of the dot radius, a, and various number of electrons stored in the dot. As it can be seen, there is a small difference between both electrostatic potential values. In each case, the difference observed in the potential is smaller than 200 mV and this difference decreases as the dot radius increase. Even for small dot size, a good agreement can be found when the number of electrons is low; a difference of less than 50 mV is observed when the dot is charged by one or two electrons, whatever the radius is.

This comparison leads to an interesting conclusion regarding the different levels of approximations developed in each approaches. As described in the previous section, the metallic approach gives immediate results thanks to the potential equation once the coupling capacitances between the dot and the electrodes are extracted. On the contrary, the Poisson–Schrödinger solver is an iterative process much more time-consuming but giving a more realistic description of the physical processes occurring inside the nanocrystal. However, both approaches can successfully coexist; we should keep in mind that a Poisson–Schrödinger

resolution always start with an initial guess for the Poisson or the Schrödinger problem. Giving the fact that the metallic approach gives electrostatic results quite close to the final solution (see Figure 5.21), the metallic solution can be used as an initial guess for the iterative Poisson–Schrödinger solver to improve the speed of convergence.

5.6 Conclusion

The purpose of this chapter was to study the electrostatic properties of individual nanodots embedded in a dielectric medium in a 3D configuration. To this purpose, various approaches have been developed and compared. The first one, based on a metallic floating nanoconductor, consist in solving the classical electrostatics equation with a rigorous semi-analytical model. As shown in Section 5.1, this method is able to predict the electrostatic potential of the floating conductor from a simple equation involving the coupling capacitance and the total charge carried by the dot. These results were successfully compared with FEM results obtained on the same structure with an excellent agreement. As a corollary, it should also be stated that the results derived from this approach are not sensitive to the length scale of the system and can be applied to structures of greater (or smaller) dimensions unless they have the same aspect ratio.

In the second part of this chapter, the focus is put on semiconductor nanocrystals. In this case, the quantum effects cannot be neglected any more like in the metallic case. Indeed, it has been shown that when the quantum well dimension falls in the nanometer scale, quantum confinement of the carriers in the nanocrystal is expected, even at ambient temperature. In this case, the electrostatic problem has been solved by a self-consistent resolution of Poisson and Schrödinger equations using the Hartree approximation. The approach developed in this work on a 3D structure by FEM gives the evolution of the charge density (and thus the electrostatic potential) in the volume of the structure with more physical insight than the rough metallic approximation. Moreover, the energy distribution of the carrier can be derived from the energy levels found for each bound state. However, as shown in the last part of this chapter, if we only need the mean electrostatic potential of the nanocrystal, the first metallic approach surprisingly gives very accurate results, even for nanocrystal with dimensions as small as few nanometers, whatever the quantity of charges in the dot.

We should keep in mind these last results as a conclusion. Indeed, they demonstrate that even macroscopic approaches can be used efficiently to predict some physical properties in nanostructures. For the typical case of the electrostatics of a semiconductor dot, there is, a priori no need to handle a complex approach based on a Poisson–Schrödinger solver if the only unknown is the electrostatic potential of the nanodot; the simple metallic approach can give accurate and immediate results if the precision required does not exceed few hundreds of millivolts.

However, if quantum transport between the electrodes and the dot, or even inter-dot transport, has to be studied, the

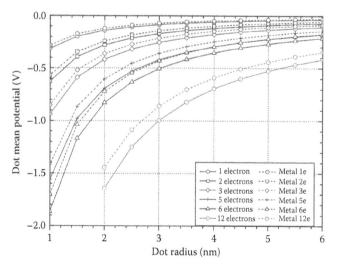

FIGURE 5.21 Comparison between the metallic and the quantum approach regarding the calculation of the electrostatic potential of the nanocrystal. No external bias is applied to the structure ($V_1 = V_2 = 0\,V$).

knowledge of the energy distribution of the electrons over their bound-states becomes primordial and will need the use of a quantum treatment of the charge. This is the first point of perspective of this work.

The second evolution of this work would consist in taking into account an irregular shape of the dot. Indeed, in the range of sizes of the current nanodot devices, it is expected that the shape of the dot could have a great influence on the electrostatics behavior. Such a study has been yet proposed in Ref. [24] in a 2D case, combining semi-analytical approach and FEM computations. The results suggest that, extended to a 3D case, a spherical approximation could be sufficient for some configurations.

Appendix A: A Numerical Validation of the Semi-Analytical Approach

5.A.1 Convergence

The tests of convergence of the numerical code are done by repeating a numerical simulation and refining precision until the convergence is satisfactory. For the numerical computation of C_S, there are two numerical truncations in relation (5.15). The first one concerns the required value of the maximum order, L_{max}, ensuring the convergence of the series. The value of L_{max} is physically constrained by the ratio a/λ_2 (the relative size of the dot seen from the plane). In practice, a value of around a few units is enough to ensure convergence (to fix the idea let us say that $L_{max} \sim 10 \times a/\lambda_2$ gives good results). L_{max} leads the discretization over θ that appears in the integration. This integration is assured here by a simple Riemann series with a finite step $\delta\theta$. In practice, $\delta\theta \sim \pi(10L_{max})^{-1}$ is sufficient. The second truncation is about the number of images taken in the summation over p. We denote by P_0 the greatest natural integer taken for the images. The difficulty of evaluating correctly this summation is the fact that the generic term is alternating because of the factor $(-1)^p$. Doing the summation described in the relations (5.15) and (5.18) for $p = \{-P_0, ..., -1, 1, ..., +P_0\}$ shows that the convergence is very slow. However, a little numerical arrangement allows a reasonable convergence of the series. Doing the same computation with an odd number of images $p = \{-P_0 -1, -P_0, ... -1, 1, ... P_0\}$ allows a convergence almost twice faster. An instance of this behavior is shown in Figure 5.22. We compare the results for the capacitance C_S versus the greatest order P_0 in the case of a summation over an even number of images and the case of a summation over an odd one. From $P_0 \sim 30$, the "odd" summation gives result that varies on the fourth digit. The "even" summation gives result still varying on the second digit around $P_0 \sim 40$, and still oscillating on the third digit for $P_0 \sim 70$. This difference between the two types of summation can be clearly explained expressing the potential on the electrodes, highlighting the property of symmetry of each term. The required value of P_0 is constrained by the ratio λ_2/h (the relative distance between the border of the dot and the electrodes), but we have to verify

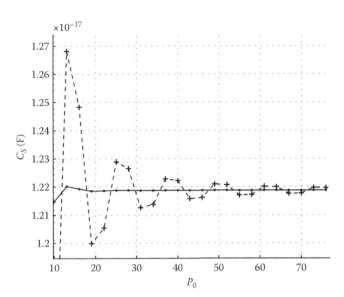

FIGURE 5.22 Behavior of the result for the capacitance versus the greatest order P_0. Slashed and stars, the summation is done for $p^\star = \{-57...57\}$. Full and dot, for $p^\star = \{-58...57\}$. The configuration is the following: $t_1 = 6\,\text{nm}$, $t_2 = 10\,\text{nm}$, $a = 15\,\text{nm}$. $L_{max} = 7$.

a posteriori the validity of this truncation, and did not found heuristic law to fix it.

In Figure 5.23, we plot the result for C_S versus L_{max} for three different values of P_0. The summation over $p^\star$ is made for $p^\star = -P_0 -1, ... -1, 1, ..., P_0$. In the left figure, we see that the influence of P_0 is weak. That is to say that each truncation can be treated separately. In the zoom, in the right side of the figure, we can notice that from $L_{max} \geq 8$, the result varies only on the fourth significant digit for the both three curves. The curve corresponding to $P_0 = 15$ differs from ~0.3% from the other one corresponding to $P_0 = 75$. The difference between the curve $P_0 = 35$ and $P_0 = 75$ is on the fifth significant digit. In this instance, the dot is particularly big and near from the electrode ($t_1 \sim a/15$ and $2a/h \sim 0.65$). This kind of configuration is the worst case for convergence.

5.A.2 Numerical Validation with FEM Approach

The numerical implementation has been finally validated comparing the results provided by a FEM computation over the same configuration. The capacitance C_S and the coefficient α have been computed for various radius a, keeping t_1 and t_2 constant. The methodology involved for the extraction of C_S and α by FEM calculation is exhaustively described in Ref. [24]. Let us explore this methodology. For each radius, V_f is computed for a given Q_0 and for a given ΔV, by an iterative process stopped on the condition of charge (relation (5.3)). This operation is repeated for each value of Q_0 and ΔV. The linear regressions between Q_0 and V_f (with ΔV constant) and between V_f and ΔV (with Q_0 constant) give, respectively C_S and α. The entire process is repeated for a new radius, giving so the results exposed in Figure 5.24.

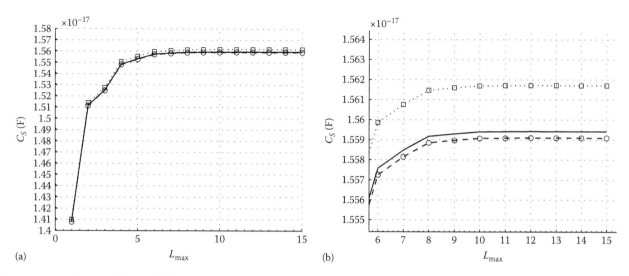

FIGURE 5.23 Behavior of the result for the capacitance versus the greatest order L_{max} for three values of P_0. Slashed and circle, the summation is done for $p^* = \{-76...75\}$. Full, for $p^* = \{-36...35\}$, dotted and square for $p^* = \{-16...15\}$. The configuration is the following: $t_1 = 1$ nm, $t_2 = 10$ nm, $a = 15$ nm. Left-hand side: global representation; right-hand side: zoom for values of $L_{max} \geq 6$.

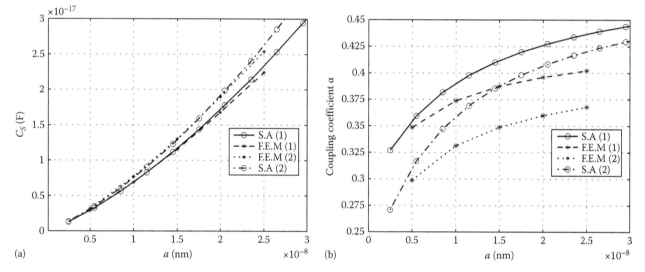

FIGURE 5.24 Comparison of FEM results with semi-analytical expression (denoted by SA). Left: Result for total capacitance versus the radius a for two sets of values for t_1 and t_2, right: Comparison of the coefficient α obtained by FEM and the value λ_2/h versus the radius a. For SA, $L_{max} = 9$ and $P_0 = 80$.

The two configurations are the following:

1. $t_2 = 6$ nm, $t_1 = 15$ nm, $h = t_1 + t_2 + 2a$. We plot C_S versus a, a varying from 0.5 to 30 nm.
2. $t_2 = 4$ nm, $t_1 = 15$ nm, $h = t_1 + t_2 + 2a$. We plot C_S versus a, a varying from 0.5 to 30 nm.

In Figure 5.24 are presented the results obtained by the two methods. Concerning C_S, the agreement between the two methods is better than 1% for values of a up to 20 nm and 2% for higher values. This difference for great values of a may be explained by the methodology involved for the extraction of C_S by FEM calculation. For large values of radius the meshing of the structure becomes a critical parameter. Indeed the space has to be meshed

far from the dot and meshed very finely near the dot. The FEM computation is based on the conservation of Q_0. For big sphere, the computed floating potential may be not sufficiently constant, because of the meshing problem. As a conclusion for this analysis, let us insist on the point that the two methods are completely different, and such an agreement is not frequent in numerical simulations.

For the coupling coefficient α, the agreement is not as satisfying as the previous comparison. We observe that the greatest difference between λ_2/h and α obtained by FEM is around 10% for configuration (2) and around 5% for configuration (1). The FEM calculation pains to extract α especially for great values of a and small value of t_2. This fact could be enlightened by the asymptotic behavior of the coupling coefficient for great values of a. Whatever the configuration the coefficient $\alpha = \lambda_2/(t_1 + t_2 + 2a)$ tends to the value 0.5, for large radius. This behavior appears earlier more quickly with SA than with FEM.

This instances seems to show that for great values of a ($a > t_2$), FEM may have some problems of accuracy for computing those two parameters.

To finish this appendix, let us say that a typical calculation of C_S by SA takes few seconds, with a non-optimized.

MATLAB®-like code, while the same extraction with FEM could takes several minutes.

References

1. Moore G.E., Cramming more components onto integrated circuits, *Electronics*, 38 (8), 114–117, April 1965.
2. *2007 ITRS ORTC, Public Conference*, Makuhari, Japan, http://www.itrs.net.
3. Noguchi M., Yaegashi T., Koyama H., Morikado M., Ishibashi Y., Ishibashi S., Ino, K., et al., A high-performance multi-level NAND Flash Memory with 43 nm-node floating-gate technology, in: *Proceedings of the International Electron Device Meeting (IEDM)*, pp. 445–448, Washington, DC, December 10–12, 2007, doi: 10.1109/iedm.2007.4418969.
4. Ishii T., Osabe T., Mine T., Murai F., Yano K., Engineering variations: Towards practical single-electron (few-electron) memory, in: *Technical Digest of IEDM'00*, San Francisco, CA, December 11–13, 2000.
5. Wang H., Takahashi N., Najima H., Inukai T., Saitoh M., Hiramoto T., Effects of dot size and its distribution on electron number control in metal-oxide-semiconductor-field-effect-transistor memories based on silicon nano-crystal floating dots, *Jpn. J. Appl. Phys.*, 40, 2038–2040, 2001.
6. Molas G., Deleruyelle D., DeSalvo B., Ghibaudo G., Gely M., Perniola L., et al., Degradation of floating gate reliability by few electron phenomena, *IEEE. Trans. Electron Devices*, 53 (10), 2610–2619, 2006.
7. Redaelli A., Pirovano A., Benvenuti A., Lacaita, A., Threshold switching and phase transition numerical models for change memory simulations, *J. Appl. Phys.*, 103, 111101-1–111101-18, 2008.
8. Sawa A., Resistive switching in transition metal oxides, *Mater. Today*, 11 (6), 28–36, June 2008.
9. Naruke K., et al., Stress induced leakage current limiting to scale down EEPROM tunnel oxide thickness, in: *Proceedings of International Electron Device Meeting (IEDM)*, pp. 424–427, San Francisco, CA, December 11–14, 1988.
10. Bez R., et al., Introduction to flash memory, *Proc. IEEE*, 91 (4), 489–502, 2003.
11. Tiwari S., et al., Volatile and non-volatile memories in silicon with nano-crystal storage, in: *Proceedings of the International Electron Device Meeting (IEDM)*, pp. 521–524, IEDM 1995, Washington, DC, December 10–13, 1995.
12. De Salvo B., Gerardi C., Lombardo S., Baron T., Perniola L., Mariolle D., Mur P., et al., How far will silicon nanocrystals push the scaling limits of NVMs technologies? in: *Technical Digest of IEDM'03*, Washington, DC, December 8–10, 2003.
13. Muralidhar R., Steimle R.F., Sadd M., Rao R., Swift C.T., Prinz E.J., Yater J., et al., A 6 V embedded 90 nm silicon nanocrystal nonvolatile memory, in: *Technical Digest of IEDM'03*, Washington, DC, December 8–10, 2003.
14. Choi S., Choi H., Kim T.-W., Yang H., Lee T., Jeon S., Kim C., Hwang H., High density silicon nanocrystal embedded in SiN prepared by low energy (<500 eV) SiH$_4$ plasma immersion ion implantation for non-volatile memory applications, in: *Technical Digest of IEDM'05*, Washington, DC, December 5–7, 2005.
15. Sattler K., *Handbook of Thin Films Materials*, Volume 5: *Nanomaterials and Magnetic Thin Film*, H.S. Nalwa (ed.), Academic Press, New York, Chapter 2, pp. 61–97, 2002, ISBN: 0-12-512913-0.
16. Averin D.V., Korotkov A.N., Likharev K.K., Theory of single-electron charging of quantum wells and dots, *Phys. Rev. B*, 44 (12), 6199–6211, 1991.
17. De Salvo B., Ghibaudo G., Luthereau P., Baron T., Guillaumot B., Reimbold G., Transport mechanisms and charge trapping in thin dielectric/Si nano-crystals structures, *Solid-State Electron.*, 45 (8), 1513–1519, 2001.
18. Han K., Kim I., Shin H., Characteristics of P-channel Si nano-crystal memory, *IEEE Trans. Electron Devices*, 48 (5), 874–879, 2001.
19. Busseret C., Ferraton S., Montès L., Zimmermann J., Granular description of charging kinetics in silicon nano-crystals memories, *Solid-State Electron.*, 50 (2), 134–141, 2006.
20. Sée J., Dollfus P., Galdin S., Hesto P., Electronic properties of semiconductor quantum dots for coulomb blockade applications, *Physica E*, 21 (2–4), 496–500, 2004.
21. Sée J., Dollfus P., Galdin S., Theoretical investigation of negative differential conductance regime of silicon nanocrystal single-electron devices, *IEEE Trans. Electron Devices*, 53 (5), 1268–1273, 2006.
22. Cordan A.S., Leroy Y., Leriche B., Electrostatic coupling between nanocrystals in a quantum flash memory, *Solid-State Electron.*, 50 (2), 205–208, 2006.
23. Ferry D.K., Goodnick S.M., *Transport in Nanostructures*, Section 4.2, p. 226, Cambridge University Press, Cambridge, U.K., 1999, ISBN: 0-521-66365-2.
24. Deleruyelle D., Micolau G., On the electrostatic behavior of floating nanoconductors, *Solid State Electron.*, 52 (1), 17–24, 2008.
25. Ganguly U., Narayanan V., Lee C., Hou T., Kan E., Three-dimensional analytical modeling of nanocrystal memory electrostatics, *J. Appl. Phys.*, 99, 114516-1–114516-6, 2006.
26. Godin Yu.A., Electrostatic problem of a conducting sphere in the field of a plane capacitor, *Sov. Phys. Tech. Phys.*, 33 (6), June 1988 [*Zh. Tech. Fiz.*, 58, 1216–1219, June 1988].

27. Pan Y., Chew W.C., Wan, L.X., A fast multipole-method-based calculation of the capacitance matrix for multiple conductors above stratified dielectric media, *IEEE Trans. Microw. Theory Tech.*, 49 (3), March 2001.

28. Jackson J.D., *Classical Electrodynamics*, 3rd edn., pp. 95–144, Wiley, New York, 1998 (Chapter 3), ISBN: 047130932X.

29. Ngakosso E., Saillard M., Vincent P., Electromagnetic diffraction by homogeneous cylinders of the field radiated by a dipole: A rigorous computation, *J. Electromagn. Waves Appl.*, 9, 1189–1205, 1995.

30. Petit R., *Ondes Electromagnétiques*, Masson Publications, 1988.

31. Aqua J.N., *Physique statistique des fluides coulombiens classiques et quantiques au voisinnage d'une paroi*, Thèse Université Paris XI Orsay, décembre 20, 2000, pp. 23–100, partie B.

32. Chew W.C., *Waves and Fields in Inhomogeneous Media*, Van Nostrand Reinhold, New York, 1990.

33. Cohen-Tannoudji C., Diu B., Laloé F., *Mécanique Quantique*, I Editions, Hermann, Paris, 1998, ISBN: 2-7056-6074-7.

34. Messiah A., *Quantum Mechanics*, Dover Publications, Mineola, NY, 1999, ISBN: 978-0486409245.

35. Sze S.M., *Physics of Semiconductor Devices*, John Wiley & Sons, New York, NJ, 1981, ISBN: 0-471-87424-8.

36. Kittel C., *Quantum Theory of Solids*, John Wiley & Sons, New York, 1987, ISBN: 0-471-62412-8.

37. www.comsol.com.

38. Deleruyelle D., Guiraud A., Micolau G., Study of electronic properties of nanocrystals from different point of view: From a simple 2D electrostatic approach to 3D Poisson Schrödinger simulations, in: *Proceedings of 2007 European COMSOL Conference*, Volume 2, pp. 636–641, Grenoble, France, October 23–24, 2007.

39. Deleruyelle D., Guiraud A., Bassani F., Theoretical and experimental investigation of electronic properties of semiconductor quantum-dots for non-volatile memory applications, in: *Proceedings of Nanoscale VI*, p. 146, Berlin, Germany, July 9–11, 2008.

40. Nougier J.P., *Méthodes de calcul numérique, vol. 2: Fonctions, équations aux dérivées*, pp. 206–212, Paris Hermès Science Publications, Paris, 2001 (Chapter 7), ISBN: 2-7462-0279-4.

6

Nanometer-Sized Ferroelectric Capacitors

Nikolay A. Pertsev
A.F. Ioffe Physico-Technical Institute

Adrian Petraru
Christian-Albrechts-Universität Zu Kiel

Hermann Kohlstedt
Christian-Albrechts-Universität Zu Kiel

6.1 Introduction

Ferroelectric materials possess unique dielectric, piezoelectric, pyroelectric, and electro-optic properties, which make them suitable for applications in various microelectronic and micromechanical devices (Lines and Glass 1977). Owing to the sustained trend toward the miniaturization of electronic devices, ferroelectrics are mostly used nowadays in thin-film form (Dawber et al. 2005, Setter et al. 2006). In particular, the switching of spontaneous polarization in a ferroelectric film is employed in low-power and fast nonvolatile random access memories (FeRAMs) produced for smart cards (Scott 2000, Ishiwara and Okuyama 2004, Kohlstedt et al. 2005a). Owing to their high piezoelectric and pyroelectric responses, ferroelectric thin films are promising for applications in various sensors and actuators and as the heart of microelectromechanical systems (MEMS) that combine Si-technology with the latest achievements in micromechanics (Scott 2007).

Numerous experimental and theoretical studies performed during the past decade have demonstrated that the physical properties of ferroelectric thin films may be very different from those of bulk ferroelectrics. When the film thickness or the lateral size of a ferroelectric capacitor (Figure 6.1) is reduced down to the nanoscale range, these properties generally become size-dependent. Therefore, the physics of nanoscale ferroelectrics represents an exciting field of research in solid-state physics with a close link to device applications. Impressive progress made

in this field is mainly the result of the following three recent achievements: (1) new theoretical investigations based on the phenomenological Landau theory of phase transitions (Landau et al. 1984) and the abinitio calculations of ferroelectric crystals and thin films; (2) tremendous improvement in the thin-film technology of complex oxides by using sophisticated deposition techniques and pattering; and (3) the advent of advanced analytical tools. The corresponding "research triangle" is sketched in Figure 6.1.

The phenomenological theory of bulk ferroelectrics was founded in the middle of the twentieth century and proved to be very successful in describing the physical properties of these materials (Devonshire 1954, Lines and Glass 1977). However, this mean-field theory, which is usually termed the Landau–Ginzburg–Devonshire theory, cannot be directly applied to thin-film and nanoscale ferroelectrics. The presence of surfaces (interfaces) and considerable lattice strains in these material systems require significant modifications of this thermodynamic theory (Kretschmer and Binder 1979, Pertsev et al. 1998). In particular, Pertsev et al. (1998) have shown that the substrate-induced lattice strains may give rise to new polarization states and phase transitions in epitaxial ferroelectric thin films. Subsequent experimental studies of single-crystalline films grown on dissimilar substrates confirmed many of the theoretical predictions and demonstrated the importance of strain effects in ferroelectric films (Schlom et al. 2007).

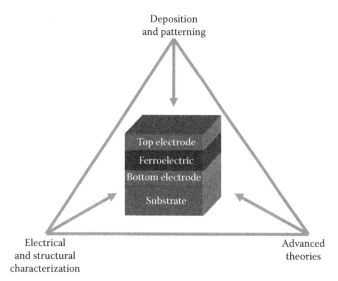

FIGURE 6.1 The "research triangle" summarizing the most important developments in the field of ferroelectric thin films and devices. Inside the triangle, a three-dimensional sketch of a ferroelectric capacitor is shown.

In addition to the advanced thermodynamic calculations, the past decade has been distinguished by the breakthrough in the first-principles density-functional theory investigations of ferroelectrics. Such theoretical studies, which do not involve any empirical input, were first performed for the two classical ferroelectrics, $BaTiO_3$ and $PbTiO_3$, and then were extended to other ferroelectric perovskites (Cohen 1992, King-Smith and Vanderbilt 1994, Waghmare and Rabe 1997). The treatment of ferroelectrics was greatly facilitated by the formulation of the correct definition of the electric polarization as a bulk quantity through the Berry-phase formalism of King-Smith and Vanderbilt (1993) and Resta (1994). This result and the related Wannier-function expression (King-Smith and Vanderbilt 1994) make it possible to calculate the spontaneous polarization and its derivatives after finishing the usual total-energy calculations (see the review paper by Dawber et al. (2005)).

Since the current computational limitations strongly restrict the size of material systems that can be directly calculated from first principles, the modeling based on first-principles results plays an important role in the theoretical studies of ferroelectrics as well. In particular, Monte Carlo simulations using an effective Hamiltonian approach make it possible to extend theoretical predictions to finite temperatures (Diéguez et al. 2004, Lai et al. 2005). It should be noted that the accuracy of first-principles calculations is still limited, which is clearly evidenced by the typical underestimation of the unit-cell size by about 1% and a substantial underestimation of the band gap of insulators. Despite these deficiencies and a considerable disparity between the size of systems studied theoretically and that of real devices, the ab initio calculations represent a very powerful tool for investigating bulk and thin-film ferroelectrics. Furthermore, by combining the first-principles results with the phenomenological Landau–Ginzburg–Devonshire theory, it is possible to

develop a fruitful approach making use of the strength of both theories (Gerra et al. 2007, Tagantsev et al. 2008).

In parallel to important developments in the theoretical physics of ferroelectrics, we can notice essential improvements in thin-film growth and characterization techniques. Today, these improvements allow an atomic level of control of ferroelectric inorganic materials such as $BaTiO_3$ and $PbZr_xTi_{1-x}O_3$ (PZT) at the interfaces with lattice-matched substrates ($SrTiO_3$, $NdGaO_3$, $DyScO_3$, etc.) and electrodes made of metal oxides ($SrRuO_3$, $LaSr_xMn_{1-x}O_3$, etc.). Molecular beam epitaxy (MBE), a sophisticated method borrowed from semiconductor technology, pulsed laser deposition, and sputtering, is the work horse in the growth of thin-film ferroelectrics. Essential progress in the structural analysis by transmission electron microscopy and the advent of nanoscale scanning probe techniques of ferroelectric thin films led to a considerable progress in the thin-film deposition, as well as in the understanding of ferroelectrics at the nanometer and angstrom scale (Gruverman and Kholkin 2004, Schlom et al. 2007, Jia et al. 2008).

In this chapter, we first briefly discuss the basic physics of ferroelectric materials and then focus on nanoscale ferroelectric capacitors and future prospects in this research area. Because of the length limit, we restrict our discussion to the class of inorganic ferroelectrics and related devices. Readers interested in ferroelectric polymers, ferroelectric liquid crystals, and ferroelectric II–VI mixed compounds are referred to the relevant publications (Weil et al. 1989, Xu 1991).

This chapter is organized in the following way. In Section 6.2, the fundamentals of ferroelectricity are presented. Section 6.3 is devoted to the state-of-the-art deposition and pattering techniques, whereas Section 6.4 describes the various aspects of the structural and electrical characterization of ferroelectric thin films and capacitors. In Section 6.5, the most important physical phenomena occurring in ferroelectric capacitors are discussed. New developments and emerging trends in the field of nanoscale ferroelectric devices are briefly described in Section 6.6.

6.2 Fundamentals of Ferroelectricity

Ferroelectricity was discovered by J. Valasek in Rochelle salt in 1920 (Valasek 1921). The phenomenological definition of ferroelectricity in the textbook by Lines and Class reads: "A crystal is said to be ferroelectric material when it has two or more orientational states in the absence of an electric field. Any two of the orientation states are identical (or enantiomorphous) in crystal structure and differ only in electric polarization vector at null electric field" (Lines and Glass 1977). The crystal symmetry imposes restrictions on the existence of the spontaneous polarization P_s. First, this property is not allowed in the presence of a center of symmetry, which excludes 11 centrosymmetric groups from the total set of 32 crystallographic point groups. Second, only 10 point groups out of the remaining 21 noncentrosymmetric ones posses a polar axis along which the spontaneous polarization may develop. This subset corresponds to *pyroelectric* crystals, where the build-in polarization manifests itself in temperature-induced changes of the total dipole moment of the unit cell. Ferroelectrics represents a subclass of pyroelectric crystals: they are distinguished from other pyroelectric materials by their ability to switch between two or more stable states with different spontaneous polarization under the action of sufficiently strong external electric field. Indeed, this switching field must be lower than the breakdown field of the material.

Currently, about 700 ferroelectric materials are known (Scott 2007), which can be divided into several groups in accordance with the microscopic origin of ferroelectricity and their atomic structure. The most important examples are ferroelectric oxides such as $BaTiO_3$ and $SrBi_2Ta_2O_9$, hydrogen-bonded ferroelectrics (e.g., KH_2PO_4), ferroelectric polymers like poly(vinylidenefluoride), and ferroelectric liquid crystals (Lines and Glass 1977, Xu 1991). There are many industrial applications of ferroelectric materials, ranging from $BaTiO_3$ capacitors to ferroelectric-liquid-crystal displays for flat TV screens.

In this chapter, we focus on ferroelectric oxides with the perovskite structure, i.e., with the crystal structure characteristic of the mineral $CaTiO_3$ found in the Ural Mountains and named "perovskite." The general chemical formula for perovskite oxides is ABO_3, where A and B represent two cations of very different sizes. In the cubic unit cell of a perovskite oxide, the A-type atoms are situated at the cube corners, the B-type atom sits at the cube center, and oxygen atoms are located at the face centers. In Figure 6.2, the unit cell of the perovskite oxide $BaTiO_3$ is shown schematically for the cubic and tetragonal phases. In the high-temperature cubic state, $BaTiO_3$ is indeed paraelectric with a zero dipole moment of the unit cell. Below about 120°C, stress-free bulk crystals of $BaTiO_3$ become ferroelectric owing to a small shift of the Ti^{4+} ion relative to the center of the surrounding oxygen cage, which leads to the appearance of dipole moment and lattice spontaneous polarization. Since the direction of this shift depends on temperature T, $BaTiO_3$ has three different ferroelectric phases: tetragonal (10°C ≤ T ≤ 120°C), orthorhombic (−71°C ≤ T ≤ 10°C), and rhombohedral (T ≤ −71°C) (Jona and Shirane 1962).

The build-in switchable polarization **P** represents the most important physical characteristic of a ferroelectric material.

In the thermodynamic theory of ferroelectrics, the polarization vector **P** is used as an *order parameter* and the additional Gibbs free energy density $\Delta G(\mathbf{P})$ of the ferroelectric phase is expanded in terms of the polarization components P_i (i = 1, 2, 3). For a stress-free tetragonal ferroelectric crystal subjected to an electric field **E** parallel to the polar x_3 axis ($P_1 = P_2 = 0$, $P_3 \neq 0$), the Gibbs free energy G_{ferro} can be written as

$$G_{ferro} = G_{para} + a_1 P_3^2 + a_{11} P_3^4 + a_{111} P_3^6 + \cdots - P_3 E_3,$$

where a_1, a_{11}, and a_{111} are the dielectric stiffness and higher-order stiffness coefficients of the centrosymmetric paraelectric phase at constant stress. The dielectric stiffness a_1 is given the linear temperature dependence $a_1 = (T - \theta)/(2\varepsilon_0 C)$, where θ and C are the Curie–Weiss temperature and constant and ε_0 is the permittivity of the vacuum. The minimization of $\Delta G(P_3)$ makes it possible to calculate the polarization P_3 of a homogeneously polarized crystal as a function of temperature T and applied field E_3. When the coefficient a_{11} of the fourth-order polarization term is positive, the spontaneous polarization $P_s = P_3$ ($E_3 = 0$) remains zero down to the transition temperature $T_c = \theta$, below which it increases gradually as $P_s \sim \sqrt{\theta - T}$. The energetics of this *second-order* ferroelectric phase transition is described schematically in Figure 6.3b. In contrast, at $a_{11} < 0$ and $a_{111} > 0$, the spontaneous polarization displays a step-like increase up to a value of $P_c = \sqrt{|a_{11}|/2a_{111}}$ at $T_c = \theta + \varepsilon_0 C a_{11}^2/(2a_{111})$. Therefore, a *first-order* ferroelectric phase transition takes place in this situation, as illustrated in Figure 6.3a.

Under an external field directed against the spontaneous polarization, the magnitude of $P_3(E_3)$ first gradually decreases with increasing field intensity E_3. When it reduces down to a certain minimum value P_{min}, the "antiparallel" polarization state becomes unstable. As a result, the polarization switches by 180° into the direction parallel to the applied field. In the case of ferroelectrics with a second-order transition, $P_{min} = P_s/\sqrt{3}$, and the critical switching field equals $E_{th} = -(4/3)a_1 P_{min} \sim (\theta - T)^{3/2}$. The field E_{th} represents the *thermodynamic coercive field* that corresponds to a homogeneous polarization reversal in the whole crystal. The polarization-field curve resulting from the thermodynamic calculations is hysteretic, as shown in Figure 6.4a. The theoretical hysteresis loop is qualitatively similar to the experimental ones, which differ mainly by a gradual polarization

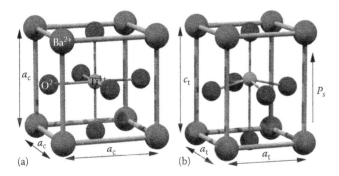

(a) (b)

FIGURE 6.2 Schematic representation of the unit cell of $BaTiO_3$ in the paraelectric cubic (a) and ferroelectric tetragonal (b) phases.

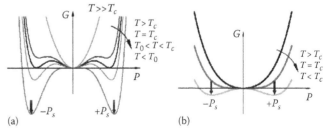

(a) (b)

FIGURE 6.3 Temperature evolution of the free energy density as a function of polarization shown schematically for ferroelectrics with the first-order (a) and second-order (b) phase transition.

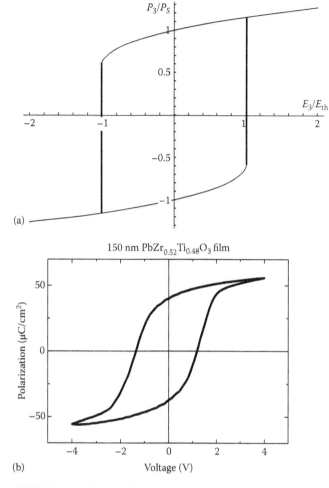

(a)

150 nm PbZr$_{0.52}$Ti$_{0.48}$O$_3$ film

(b)

FIGURE 6.4 Theoretical (a) and measured (b) ferroelectric hysteresis loops. The theoretical dependence of the normalized polarization on the normalized applied field corresponds to a homogeneous polarization reversal in the whole crystal. The experimental polarization-voltage loop was measured for the 150 nm thick PZT film.

reversal seen in Figure 6.4b. The measured coercive fields E_c of bulk crystals, however, are typically several orders of magnitude lower than E_{th}, because the polarization switching develops in reality via the nucleation and growth of ferroelectric domains (Lines and Glass 1977).

When the size of a ferroelectric crystal is reduced down to a length scale comparable to the so-called *ferroelectric correlation length*, its physical properties generally become size-dependent. This feature is due to the fact that ferroelectricity is a collective phenomenon resulting from a delicate balance between long-range Coulomb forces (dipole–dipole interactions), which are responsible for the ferroelectric state, and a short-range repulsion favoring the paraelectric state (Lines and Glass 1977). The scaling of physical characteristics such as the remanent mean polarization $P_r = \langle P_3(E_3 = 0)\rangle$ and coercive field E_c is currently in the focus of experimental and theoretical studies in the field of nanometer-sized ferroelectric capacitors. The most important results of these studies will be discussed in Sections 6.4 and 6.5.

6.3 Deposition and Patterning

6.3.1 General Aspects

The deposition of thin films belongs to the heart of today's micro- and nano-electronics. In order to grow thin films with desired properties, several important issues have to be considered. Besides the choice of the deposition method, a number of process parameters are essential, such as the background pressure of the vacuum system, the deposition rate (measured in nm/s), the substrate material and temperature, and the composition of the material source. Furthermore, the process pressure is important, irrespective of the conditions in the chamber, i.e., the state of an ultra-high vacuum or the presence of a noble gas (typically argon) or reactive gases (oxygen or nitrogen). The basic principles of thin-film deposition are described in several textbooks (Chopra 1969, Maissel and Glang 1979, Bunshah 1994). Here we focus on the deposition of complex oxides by means of physical methods.

As already mentioned in the introduction, thin-film deposition techniques for the growth of heteroepitaxial oxides have made tremendous progress recently. The purpose of this section is to give a short overview of the current status of the MBE, pulsed laser deposition (PLD), and sputter deposition (SD). For chemical-based methods, such as metal-organic chemical vapor deposition (MOCVD), atomic layer deposition (ALD), and chemical solution deposition (CSD), we refer the reader to the relevant papers (Oikawa et al. 2004, Kato et al. 2007, Schneller and Waser 2007). We also note that ferroelectric polymers are deposited by either a spin-on technique or the Langmuir-Blodgett method (Ducharme et al. 2002).

Consider a planar ferroelectric capacitor sketched in Figure 6.1. The choice of materials for the substrate, electrodes, and ferroelectric layer strongly depends on the application or the research task. For the epitaxial growth of perovskite ferroelectrics, single-crystalline substrates having small lattice mismatches with these complex oxides are preferable. At present, the commercially available single crystals of SrTiO$_3$ are most popular, although other substrates such as MgO, KTaO$_3$, GdScO$_3$, and DyScO$_3$ have also been successfully employed. When the ferroelectric overlayer is commensurate with a dissimilar thick substrate, it appears to be strained to a certain extent defined by the mismatch in their in-plane lattice parameters. Above some critical thickness, however, these lattice strains start to relax due to the generation of misfit dislocations (see Section 6.4).

As for the electrodes, noble metals such as Pt, Ir (also IrO$_2$), and Ru are used in most device applications, e.g., in FeRAMs and MEMS (Kohlstedt et al. 2005a). On the contrary, conducting complex oxides were favored so far as electrode materials in the basic research studies of scaling effects. Prominent examples of such electrode materials are SrRuO$_3$, LaSr$_x$Mn$_{1-x}$O$_3$, and LaCa$_x$Mn$_{1-x}$O$_3$, which are routinely used in the complex-oxide heterostructures (Eom et al. 1992, Sun 1998). A comparison of the advantages and disadvantages of metal and oxide electrodes shows a delicate trade-off. Let us compare, for instance, Pt and SrRuO$_3$.

On the one hand, the resistivity of a sputtered thin-film Pt at room temperature (~10 μΩ cm) is much smaller than that of even a high-quality $SrRuO_3$ (≈300 μΩ cm). On the other hand, the screening-space-charge capacitance density, which is important for the stabilization of ferroelectricity in ultrathin films (Pertsev and Kohlstedt 2007), is equal to 0.9 F/m² for the $SrRuO_3$ electrode and only to 0.4 F/m² for the Pt electrode (Pertsev et al. 2007). This feature seems to make $SrRuO_3$ electrodes preferable for nanoscale ferroelectric capacitors. In addition, the electrode surface roughness, crystallographic orientation of the ferroelectric layer grown on a particular electrode, and the quality of the electrode–ferroelectric interface must be taken into account. Currently, conducting complex oxides are preferred for the fabrication of the bottom electrode, whereas the top electrode can be made of Pt or other noble metal as well. In the rest of this section, we focus on entirely complex-oxide heterostructures for ferroelectric capacitors.

6.3.2 Deposition Techniques

6.3.2.1 Molecular Beam Epitaxy

MBE has developed from a simple evaporation technique via the use of ultra-high vacuum (UHV) to avoid disturbances by residual gases and additional incorporation of various effusion (Knudson) cells as material sources. Figure 6.5 schematically shows an MBE system involving several material sources that allow controlled deposition of multi-element compounds.

In contrast to the deposition of most semiconductor materials such as GaN, GaAs, and InP, the growth of oxides by MBE requires relatively high partial pressure of oxygen (~10⁻⁷ mbar) during the deposition. This is necessary to avoid the oxygen deficiency in the final film, which could seriously deteriorate the quality of a ferroelectric capacitor. Partial pumping, the use of reactive oxygen (e.g., ozone), and post-annealing of the films in a high-pressure oxygen atmosphere (several mbar) are used to supply the films with a sufficient amount of oxygen.

Owing to the UHV conditions in the MBE chamber, all UHV surface techniques can be employed. This feature, indeed, constitutes the strength of MBE (Haeni et al. 2000). This tool offers the highest degree of freedom to apply sophisticated in situ analytical techniques to study films during the growth and just after the deposition without breaking the vacuum. Complex MBE systems using low-energy electron microscopy (LEEM) and Auger electron spectroscopy (AES), for example, were developed (Habermeier 2007 and Clayhold et al. 2008). The standard technique currently is the reflection high-energy electron diffraction (RHEED), which allows control of the surface chemistry of the last layer during the deposition. This technique provides an opportunity to fabricate oxide films with a definite termination at the surface (e.g., the BaO or TiO_2 termination in $BaTiO_3$ films). Many groups successfully demonstrated this approach (Logvenov and Bozovic 2008) with similar oxide materials. An obvious research goal now is to find correlations between atomic terminations at ferroelectric–metal interfaces and the electrical properties of capacitors.

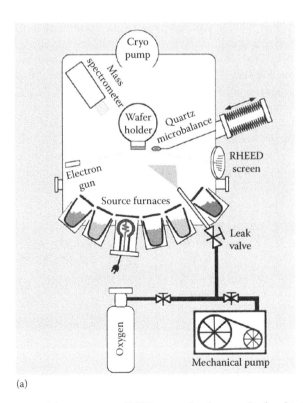

(a)

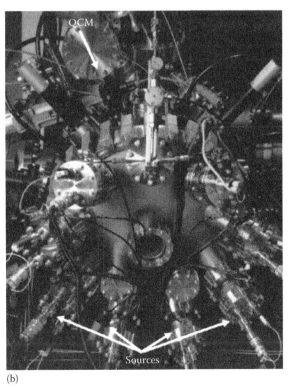

(b)

FIGURE 6.5 Schematic view of MBE system for the growth of multi-element-compound thin films (a) and a photograph of MBE chamber (b). (From Lettieri, J. et al., *J. Vac. Sci. Technol. A*, 20, 1332, 2002. With permission.)

An additional important feature of MBE is the low energy of deposited species. Indeed, the temperature of the material source in effusion cells or electron-beam evaporators does not exceed 3500 K. Hence, the corresponding thermal energy of the deposited species is about 300 meV. This value is an order of magnitude lower than the energies characterizing pulsed laser deposition and typical sputter deposition. Among available deposition techniques, MBE is the most flexible one with respect to the incorporation of analytical tools and offers the highest degree of atomic layer control. On the other hand, MBE systems are difficult to handle, considerable time is necessary for their maintenance, and, last but not least, special methods are needed to supply complex oxides with a sufficient amount of oxygen. Due to the highly complex machinery, experienced researchers working with MBE systems for years translated the acronym MBE as "many boring evenings."

6.3.2.2 Pulsed Laser Deposition

PLD is a very useful and flexible tool for growing oxide materials (Hubler and Chrisey 1994). A sketch of a PLD system is shown in Figure 6.6. A pulsed laser beam, from a KrF (248 nm) or ArF (193 nm) excimer laser, for example, is focused on a rotating target made of an oxide material (e.g., $BaTiO_3$, PZT, or $SrRuO_3$). The oxygen gas pressure during ablation can be varied from 10^{-7} to 0.5 mbar. Owing to the intense laser beam, a plasma containing energetic ions, electrons, neutral atoms, and molecules is formed. The energy density is in the range of 2–5 J/cm^2 at the target surface. As a result, the energy of the ablated material may reach values well above 10 eV at the substrate surface. The wavelength of the used laser beam may be 248 or 193 nm (at this UV wavelength, the absorption in the oxide target materials is sufficiently large). A repetition rate of several Hz and a pulse length of 25 ns represent typical parameters. Initially, a serious problem of the method was the formation of droplets on the substrate, which can easily deteriorate the device properties. Currently, various methods to reduce this effect are known, e.g., the time-of-flight selection of ablated

material. PLD systems are widely used for basic research studies of thin films. The main advantage of this method is that a certain film stoichiometry can be easily achieved by PLD. Many targets (six or more) can be placed on the target carrousel holder. In this way, numerous materials can be deposited without time-consuming rearrangements of the deposition chamber or complicated source exchange procedures, as in the case of MBE or MOCVD.

6.3.2.3 Sputter Deposition

Plasma sputtering is a physical vapor deposition technique that has been known for 150 years since the time when W.R. Grove first observed the sputtering of surface atoms. Different sputtering techniques, such as dc- and rf-sputtering with or without a magnetron arrangement, have been used to grow a variety of materials. Figure 6.7a illustrates the principle of dc-sputtering. A potential of several hundred volts is applied between the target (cathode) and the heater (anode), accelerating positively charged ions toward the target. These accelerated particles sputter off the target material, which finally arrives at the substrate. The discharge is maintained because the accelerated electrons continuously collide with the gas circulating in the chamber and ionize new atoms.

For insulating targets such as ferroelectric ones, the dc-sputtering is not suitable. Insulating targets have to be sputtered using alternating electric fields to generate the plasma. Typically, an rf-frequency of 13.56 MHz is employed. This frequency is not a magic number, rather a frequency that is approved by the government for industrial purposes. A symmetrical arrangement of cathode and anode and the use of a low-frequency alternating field would result in similar sputtering and re-sputtering rates so that the film will not grow. In the case of a high-frequency alternating field, however, light electrons can respond to the field at this frequency, whereas heavy Ar^+ ions see only an average electric field (Kawamura et al. 1999). Moreover, the geometrical asymmetry between small cathodes (target side) and large anodes (heater and chamber) leads to a higher electron concentration at the former, resulting in a self-generated dc bias that accelerates Ar^+ ions toward the target.

The high-pressure sputtering technique of oxide materials was developed by Poppe et al. (1988) and served initially for the

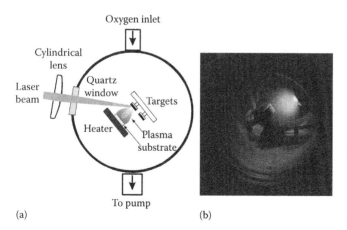

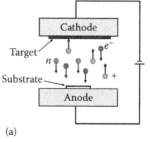

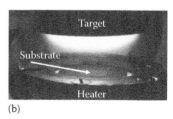

(a)   (b)

FIGURE 6.6 Pulsed laser deposition system for the growth of oxide films: PLD setup scheme (a) and a photograph showing the ablation process and the plasma plume (b). (From Schlom, D., Cornell University, Ithaca, NY.)

FIGURE 6.7 Schematic representation of the dc-sputtering process showing the electrons, positive plasma ions (light gray), and neutral atoms (dark gray) moving toward cathode or anode (a) and a photograph of a high-pressure sputtering system (b). (From Rodríguez Contreras, J. et al., PhD thesis)

growth of oxide superconductors. A planar on-axis arrangement of the target and substrate is used, as shown in Figure 6.7b. A high sputtering pressure of 2.5–3.5 mbar, corresponding to a mean-free-path $\lambda_{mean\,free} = 6 \times 10^{-3}$ cm at 600°C and exceeding largely the pressure of 10^{-2} mbar used for conventional sputtering ($\lambda_{mean\,free} = 2$ cm at 600°C), leads to multiple scattering of the negatively charged oxygen ions accelerated toward the substrate. As a result of the thermalization of ions, the re-sputtering of the deposited films, which is caused by negatively charged ions, is negligible. This technique yields excellent thin films due to the low kinetic energy (as in the case of MBE) of sputtered particles.

A disadvantage of the high-pressure sputtering technique could be a low deposition rate of several nanometers per hour, which may lead to interdiffusion at heterogeneous interfaces. To enhance the deposition rate, a low ionization degree of less than 1% of the atoms in the plasma is increased by the use of magnetic fields forcing electrons onto helical paths close to the cathode, which leads to much higher ionization probability. This so-called magnetron sputtering can be employed for high-pressure sputtering (Poppe et al. 1988), as well as for conventional, low-pressure sputtering (Fisher et al. 1994). Sputtering is routinely used as a vapor deposition method for the growth of complex-oxide films.

6.3.3 Patterning

The patterning of oxide heterostructures represents an important step in the fabrication of ferroelectric capacitors with small lateral dimensions ranging from a few micrometers to tens of nanometers. As in many other areas of nanoelectronics, two different approaches exist for the device fabrication. A conventional approach relies on the well-established processes used in the modern semiconductor industry: deposition, lithography, and etching.

Using these techniques sequentially, ferroelectric capacitors can be fabricated. It should be emphasized that this fabrication procedure employs the so-called *top-down* approach, where external tools are used to create a nanoscale device out of a larger structure. In contrast, the *bottom-up* approach is based on the self-organization of constituents or their positional assembly necessary for a desired nanodevice. Such techniques recently became very fashionable (Spatz et al. 2000) because they do not require advanced and expensive patterning tools. The bottom-up approach has had considerable success, but improvements are needed to achieve registered arrays of devices, such as those produced by the state-of-the-art complementary metal-oxide-semiconductor (CMOS) technology. In some works, mixed top-down and bottom-up methods were used to produce nanoscale ferroelectric dots and crystals (Kronholz et al. 2006, Szafraniak et al. 2008).

One of the simplest ways to fabricate ferroelectric capacitors is the lift-off technique. The main steps of this technique are shown in Figure 6.8a. First, the bottom electrode and the ferroelectric layer are deposited on a substrate. A subsequent photo-lithographic step defines the area of the capacitor. Next, the top electrode is deposited, for example, by the sputtering of Pt. After a lift-off in acetone, the metal with photoresist underneath is removed and the capacitor is ready for electrical characterization. Because the top interface is subjected to photoresist and chemical developer during this procedure, relatively poor electrical properties (e.g., large leakage) are observed here (Rodríguez et al. 2003a and Rodríguez Contreras 2004). The post-annealing of capacitors at high temperatures and in an oxygen atmosphere was successfully used to improve the electrical properties considerably (Schneller and Waser 2007).

The aforementioned drawback, however, can be avoided using another method, which involves the fabrication steps shown schematically in Figure 6.8b. Here the whole sandwich (bottom

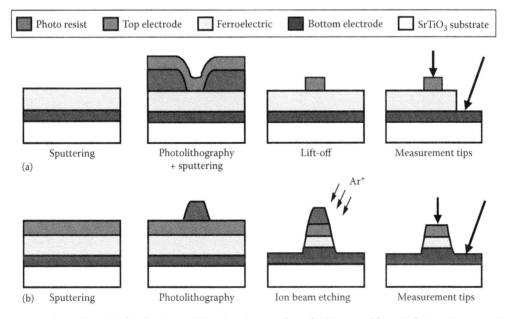

FIGURE 6.8 Patterning of capacitors by lift-off technique (a) and ion beam etching (b). (Reprinted from Rodríguez Contreras, J. et al., *Appl. Phys. Lett.*, 83, 126, 2003a. With permission. American Institute of Physics.)

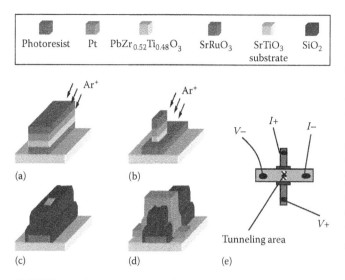

Photoresist   Pt $PbZr_{0.52}Ti_{0.48}O_3$ $SrRuO_3$ $SrTiO_3$ substrate SiO_2

FIGURE 6.9 Patterning of tunnel junctions with the aid of photolithography and ion beam etching. (From Rodriquez Contreras, J. et al., *Mater. Res. Soc. Symp. Proc.*, 688, C8.10, 2002. With permission.)

electrode/ferroelectric/top electrode) is deposited without breaking the vacuum. Then the capacitor area is defined by a photolithographic process and dry-etching, typically using Ar ion beam milling. Again acetone is employed to remove the photoresist from the top electrode. Although this method delivers better electrical interface properties than the first one, etch residuals at the sidewalls of the mesa can cause short-circuiting along the sidewall from the top electrode to the bottom one. An etch stop right after reaching the top surface of the ferroelectric can avoid this problem in a simple way. Ion-beam etching is the preferred method for the studies of scaling effects, because it results in the smallest degradation at the top electrode–ferroelectric interface.

A more complex procedure should be used to pattern oxide tunnel junctions. As shown in Figure 6.9, three photo-mask steps in total allow the fabrication of tunnel junctions by conventional photolithography and ion-beam etching (Sun 1998). First, the whole layer sequence such as the $SrRuO_3(20\,nm)/BaTiO_3(2\,nm)/SrRuO_3(20\,nm)$ trilayer is deposited in situ. Using a photolithographic step and ion-beam milling, the shape of the bottom electrode is defined. Here the whole trilayer has to be etched down to the substrate. Next, the tunnel junction is defined by the second photo-mask step and an etching down to the bottom electrode. The subsequent deposition of SiO_2 is needed to isolate the surroundings of the mesa. A third photo-mask step and the deposition of the wiring layer complete the fabrication of a tunnel junction. Figure 6.9e shows the top view of the junction layout. The four-point probe arrangement, which avoids electrical artifacts during current-voltage measurements, is essential (Rodríguez et al. 2003b and Rodríguez Contreras 2004).

6.3.4 Current Trends in Deposition and Patterning

The physical vapor deposition methods described above make it possible to grow complex oxides with a high structural quality and a low surface roughness. Further developments in this field

involve combinatorial PLD and MBE systems for the fabrication of epitaxial layers with composition gradients and wedge-like films with a thickness varying across the substrate (Nicolaou et al. 2002, Ohtani et al. 2005, Habermeier 2007). This interesting approach helps to reduce run-to-run uncertainties and allows the variation of certain film parameters in a single deposition run while keeping other conditions fixed.

New patterning methods can be developed as well, in particular, a combination of conventional top-down methods and bottom-up techniques. It should be noted that electron beam lithography has been used to produce capacitors with nanoscale dimensions (Szafraniak-Wiza et al. 2008). Focused ion-beam direct-writing techniques already showed their strength, but they suffer from sidewall contaminations at the mesa structure created by etching ions or atoms, e.g., Ga (Hambe et al. 2008).

6.4 Characterization of Ferroelectric Films and Capacitors

6.4.1 Rutherford Backscattering Spectrometry

Rutherford backscattering spectrometry (RBS) is an accurate nondestructive technique for measuring the stoichiometry, layer thickness, quality of interfaces, and crystalline perfection of thin films. RBS offers a quantitative determination of the absolute concentrations of different elements in multi-elemental thin films. A collimated mono-energetic beam of low-mass ions hits the specimen to be analyzed. Typically, He^+ ions with energy of 1.4 MeV are used in RBS experiments. A small fraction of the ions that impinge on the sample is scattered back elastically by the atomic nuclei and are then collected by a detector. The detector determines the energy of the backscattered ions, which provides an RBS energy spectrum. The RBS spectra describe the yield of backscattered particles as a function of their energy. The analysis of RBS spectra is done using modern software. A more detailed description of the technique is given by Chu et al. (1978).

In the so-called random experiments, the ion beam is not aligned with respect to the crystallographic directions of the specimen. The energy distribution of the collected ions provides information on the masses of atoms constituting the sample and on the thicknesses of deposited layers. Information on the sharpness of interfaces between these layers is given by the abruptness of the low-energy edge in the "random" spectrum.

Epitaxial films usually have the same major channeling axis as the substrate. The degree of epitaxy is determined from ion channeling experiments by a ratio of the elemental signals from the film for the channeled and random sample orientations. This ratio is called the "minimum yield," χ_{min}, and its value provides information on the crystalline perfection of a film. Defects inside the film lead to higher values of χ_{min}.

6.4.2 X-Ray Diffraction for Thin-Film Analysis

X-ray diffraction (XRD) represents a powerful tool for the characterization of thin films. It can be used to determine whether the film grown on a crystalline substrate is amorphous, polycrystalline, or single-crystalline (epitaxial growth). Moreover, this technique makes it possible to determine the film thickness, lattice parameters, and the amount of strain in an epitaxially grown film with a high precision.

In particular, the 2θ scans performed at a fixed glancing incident angle of the incoming x-ray beam (in the range of 0.5°–2°) are suited for the investigations of polycrystalline films, since the spectrum contains only the peaks coming from the XRD of randomly oriented crystallites. (The single-crystal substrate does not contribute to the XRD because the Bragg condition is not satisfied for this angle of incidence.) For (001)-oriented epitaxial films, the normal θ–2θ scans reveal only the (00l) reflections. Thus, we can distinguish between an epitaxial film on a crystalline substrate and a polycrystalline one. Moreover, from the 2θ position of these reflections, one can precisely determine the out-of-plane lattice parameter. Once this parameter is measured, the in-plane lattice constants could be determined as well by finding the peak positions of the (h0l) reflections, for example. For ultrathin films, however, this becomes difficult because of the overlap with substrate peaks. Therefore, the grazing incidence diffraction, which is characterized by a low penetration depth of the incoming x-ray beam, has to be used to measure the in-plane lattice constants of ultrathin films.

The amount of strain in an epitaxial film can be nicely visualized by the reciprocal space maps measured, for example, around an asymmetric (103) reflection. These maps can indicate whether the film is fully strained by the substrate or is partially relaxed owing to the generation of misfit dislocations. Representative reciprocal space maps of strained and relaxed epitaxial $BaTiO_3$ films grown on $SrRuO_3$-covered $SrTiO_3$ substrates are given in Figure 6.10a and b. The out-of-plane and in-plane lattice parameters of $BaTiO_3$ films extracted from such maps are plotted in Figure 6.10c as a function of the film thickness t (Petraru et al. 2007). It can be seen that ultrathin films with $t < 30\,nm$ are commensurate with the substrate, which results in a compressive biaxial in-plane strain and an out-of-plane elongation of the unit cell.

The synchrotron x-ray scattering measurements give additional possibilities for the characterization of ultrathin films (Fong et al. 2005). In particular, it was demonstrated that electrode-free $PbTiO_3$ films grown on $SrTiO_3$ remain ferroelectric for thicknesses down to only 3 unit cells (Fong et al. 2004).

Finally, we note that the film thickness itself can be measured precisely using the x-ray specular reflectivity method based on interference fringes whose spacing is characteristic for this thickness (Fewster 1996). This method can be applied to films with any structure, crystalline or amorphous, but requires a flat surface over the region studied. It was demonstrated to work even

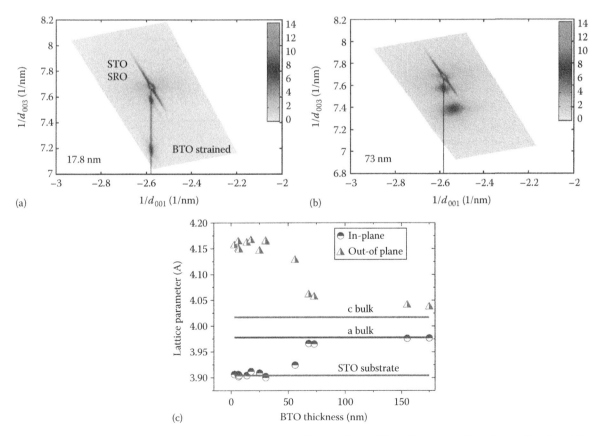

FIGURE 6.10 X-ray reciprocal space maps around the (103) Bragg reflection obtained for fully strained (a) and partially relaxed (b) $BaTiO_3$ films epitaxially grown on $SrRuO_3$-covered $SrTiO_3$. The film lattice parameters are plotted as a function of the film thickness in panel (c). (From Petraru, A. et al., *J. Appl. Phys.*, 101, 114106, 2007. With permission.)

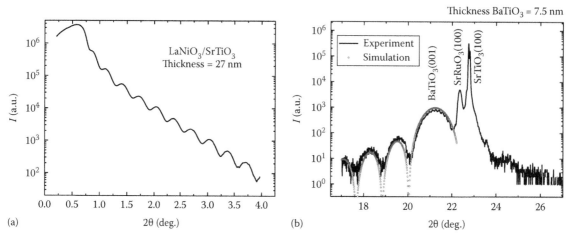

FIGURE 6.11 (a) Interference fringes appearing in an x-ray specular reflectivity scan for the 27 nm thick $LaNiO_3$ film deposited on $SrTiO_3$. The film thickness was calculated from the spacing of these fringes. (b) High-angle finite-size oscillations occurring in the θ–2θ scan around the (001) peak of the $BaTiO_3$ film (7.5 nm thick) grown on $SrRuO_3$-covered $SrTiO_3$. The solid line shows the measured signal, whereas the dots denote the results of simulations.

in the case of ultrathin films with thicknesses down to 24 Å. The amplitude of oscillations depends mainly on the density contrast between the layers, and the number of oscillations correlates with the roughness of the surface and interfaces involved. In the case of rough surfaces, the average intensity of reflectivity decreases rapidly with an increasing 2θ angle (Nevot and Croce 1980). For epitaxial films, high-angle finite-size oscillations occurring in the θ–2θ scans around the (001) peak allow determination of the number of planes involved in the diffraction, and, therefore, of the film thickness (Schuller 1980). Examples of the low- and high-angle finite-size oscillations are given in Figure 6.11.

6.4.3 Ferroelectric Capacitors: *P-E* Hysteresis Loop Measurements

A ferroelectric capacitor usually displays a polarization-field (*P-E*) hysteresis loop similar to that shown in Figure 6.4b. There are several techniques that are used to measure the *P-E* loops of ferroelectric capacitors. The simplest method employs a circuit proposed by Sawyer and Tower, which is shown schematically in Figure 6.12a. The circuit consists of a fixed capacitor with known capacitance, the test ferroelectric capacitor, an oscilloscope, and a function generator. The method relies on the fact that two capacitors in a series have the same charge. The ac voltage created by the generator and the potential across the standard capacitor are shown on the *x*- and *y*-axes of the oscilloscope. The capacitance of the standard capacitor is chosen to be large enough so that the voltage drop across this capacitor is much smaller than the potential difference across the tested ferroelectric capacitor.

Another method uses a fast current-to-voltage converter connected in series with the ferroelectric capacitor (see the circuit shown in Figure 6.12b). In this case, the current-voltage curve is measured as a response of the ferroelectric capacitor to a triangular signal excitation. It is very useful to look at the switching current

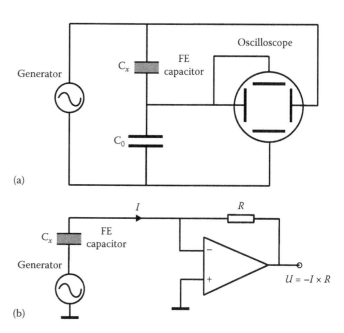

FIGURE 6.12 Experimental techniques for the measurement of polarization-voltage loops of ferroelectric capacitors: (a) Sawyer-Tower circuit allowing the visualization of ferroelectric hysteresis loops, and (b) an alternative method using a fast current-to-voltage converter, where the loop is obtained via the integration of measured electric current.

peaks that appear in this curve in order to distinguish the ferroelectric switching from artifacts, especially in the case of leaky ferroelectric samples. By numerical integration of the current over the time, the classical *P-E* hysteresis loop is obtained, from which the remanent polarization and the coercive field can be determined. The remanent polarization of nanoscale $SrRuO_3$/$BaTiO_3$/$SrRuO_3$ capacitors fabricated on the $SrTiO_3$ substrate is shown in Figure 6.13 as a function of the $BaTiO_3$ thickness *t* (Petraru et al. 2008). Remarkably, even at *t* = 3.5 nm, the strained $BaTiO_3$ film

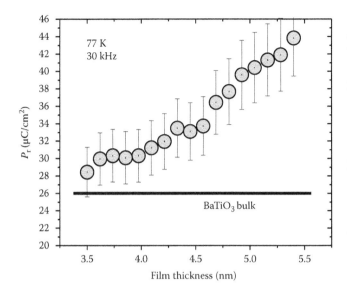

FIGURE 6.13 Thickness dependence of remanent polarization in the SrRuO$_3$/BaTiO$_3$/SrRuO$_3$ ferroelectric capacitors measured at 77 K. (From Petraru, A. et al., *Appl. Phys. Lett.*, 93, 072902, 2008. With permission.)

remains ferroelectric and has a remanent polarization larger than the spontaneous polarization $P_s = 26\,\mu C/cm^2$ of bulk BaTiO$_3$. The coercive field E_c of BaTiO$_3$ capacitors relatively weakly depends on the thickness t (Jo et al. 2006b), which contrasts with a strong increase of E_c (Figure 6.14) in ultrathin PZT capacitors with Pt top electrodes (Pertsev et al. 2003b).

Ferroelectric capacitors are often rather leaky, because thin films, especially at small thicknesses, are not perfect insulators. The conduction here results from the Schottky injection or Fowler–Nordheim tunneling through the interfacial barrier followed by the charge transport across the film via the Poole–Frenkel conduction mechanism, space-charge-limited conduction, or variable range hopping (Dawber et al. 2005). The leakage contribution to the total current can be singled out with the aid of the positive-up negative-down (PUND) pulsed method (Smolenskii et al. 1984). It involves the application of a series of voltage pulses from a function generator and the measurement

of the transient current response of a ferroelectric device, which allows the separation of different contributions. As a representative example, we consider the sequence of five train pulses shown in Figure 6.15. The first one (0) is the pre-polarization pulse—it puts the sample into a definite polarization state. The pulse (1) switches the polarization of the sample, and its current response is the sum of the ferroelectric displacement current caused by the switching of spontaneous polarization, the dielectric displacement current, and the leakage current. The pulse (2) has the same polarity but comes after a certain delay time. Therefore, in case of a stable polarization, the current response contains only the components arising from the dielectric response and leakage current. In order to find the switchable polarization (the quantity of primary interest), the current response due to pulse (2) is subtracted from the current created by pulse (1), and the result is numerically integrated over the measuring time. Moreover, this method makes it possible to study the stability of ferroelectric polarization against back-switching. To that end, we can vary the delay time between pulses (1) and (2) and determine the relaxation time of the polarization. A similar analysis can be done for currents resulting from pulses (3) and (4) applied to the capacitor with opposite polarization.

6.4.4 Scanning Probe Techniques: Atomic Force Microscopy and Piezoresponse Force Microscopy

Atomic force microscopy (AFM) is one of the most widely used scanning probe microscopy (SPM) techniques (Garcia and Perez 2002). The primary purpose of an AFM instrument is to quantitatively measure the roughness of various surfaces. The lateral and vertical resolutions are typically about 5 and 0.01 nm, respectively. An atomically sharp tip is scanned over a surface with feedback mechanisms that enable the piezoelectric scanners to maintain the tip at a constant force (to obtain height information) or height (to obtain force information) above the sample surface.

Tips are typically made of Si$_3$N$_4$ or Si and extend down from the end of a cantilever. The AFM head employs an optical

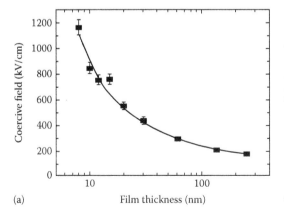

(a)

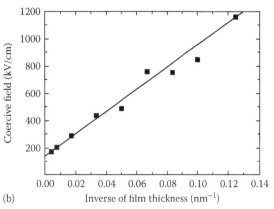

(b)

FIGURE 6.14 Coercive field of PZT 52/48 epitaxial films measured at 20 kHz and plotted versus the film thickness t (a) and the inverse of film thickness $1/t$ (b). The straight line in (b) shows a linear fit to the experimental data, whereas the curve in (a) is a guide to the eyes.

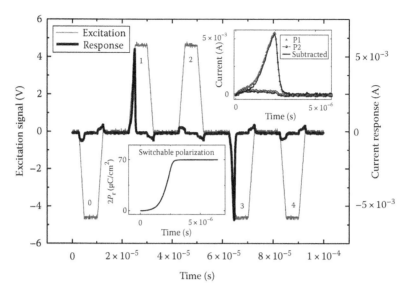

FIGURE 6.15 Measurements of switchable ferroelectric polarization by the PUND pulsed method. The excitation signal consists of five pulses denoted by thin lines, and the current response is shown by a thick line. The upper inset demonstrates the switching (1) and nonswitching (2) current responses. The integration of their difference gives the switchable polarization plotted in the lower inset.

detection system, in which the tip is attached to the bottom of a reflective cantilever. A laser diode is focused onto the back of this cantilever. As the tip scans the surface of a sample, the laser beam is deflected by the cantilever into a four-quadrant photodiode. In contact mode, feedback from the photodiode difference signal, through the software control from a computer, enables the tip to maintain either a constant force or a constant height above the sample. In the constant force mode, the piezoelectric transducer monitors real-time height variations. In the constant height mode, the deflection force acting on the tip is recorded. The instrument gives a topographical map of the sample surface by plotting the local sample height versus the horizontal probe tip position. For many soft materials like polymers and biological samples, the operation in contact mode often modifies or destroys the surface. These complications can be avoided using the tapping-mode AFM. In tapping mode, the AFM tip–cantilever assembly oscillates at the sample surface during the scanning. As a result, the tip lightly taps the surface while scanning and only touches the sample at the bottom of each oscillation. This prevents damage of soft specimens and avoids the "pushing" of specimens along the substrate. By using a constant oscillation amplitude, a constant tip–sample distance is maintained until the scan is complete. Tapping-mode AFM can be performed on both wet and dry surfaces.

Scanning probe microscopy techniques also offer several different possibilities for the investigation of domain patterns in crystals. For imaging domain structures in ferroelectrics, piezoresponse force microscopy (PFM) is most widely used nowadays (Figure 6.16). Introduced in 1992 by Güthner and Dransfeld (1992), the PFM method has been developed by several groups to visualize domain structures in ferroelectric thin films. It became a popular tool in the science and technology of ferroelectrics and is considered to be a main instrument for getting information on ferroelectric properties at the nanoscale. Several reviews on

the SPM-based methods for the characterization of ferroelectric domains are available in the literature (Gruverman and Kholkin 2004, Kholkin et al. 2007).

Ideally, when a modulation voltage V is applied to a piezoelectric material, the vertical displacement of the probing tip, which is in mechanical contact with the sample, accurately follows the motion of the sample surface resulting from the converse piezoelectric effect. The applied voltage V generates an electric field $\mathbf{E}(\mathbf{r})$ in the ferroelectric film, which creates the lattice strain $\delta u_3 = d_{13}E_1 + d_{23}E_2 + d_{33}E_3$ in the film thickness direction. Here, d_{ij} are the local piezoelectric coefficients of the ferroelectric material, which depend on the polarization orientation. The strain field $\delta u_3(\mathbf{r})$ changes the film thickness at the tip position by an amount δt so that the local piezoresponse signal proportional to $\delta t(V)$ can be recorded. The amplitude of the tip vibration measured by the lock-in technique provides information on the effective piezoelectric coefficient $d_{33}^{\mathrm{eff}} = \delta t / V$. The phase yields information on the polarization direction in a studied ferroelectric domain (Rodriguez et al. 2002). It should be noted that not only the surface electromechanical response but also the electrostatic forces could contribute to the measured signal in the PFM setup, which complicates the analysis of the PFM results. In particular, there exists a nonlocal contribution caused by the capacitive cantilever–sample interaction (Kalinin and Bonnell 2002).

Most PFM measurements are performed in a local-excitation configuration where the modulation voltage is applied between the bottom electrode and conductive SPM tip, which scans the bare surface of the film having no top electrode. In this case, the PFM image has a lateral resolution of about 10 nm (Gruverman et al. 1998). It should be noted that the electric field generated by the SPM tip in such film is highly inhomogeneous, which makes the *quantitative* analysis of the field-induced signal extremely difficult. In other words, PFM measurements on a sample

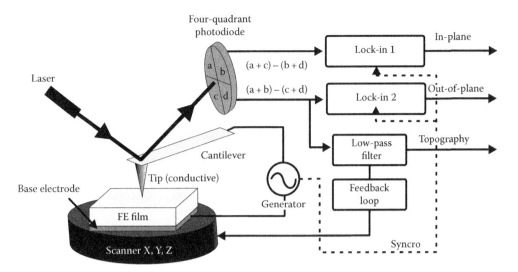

FIGURE 6.16 PFM setup for simultaneous measurements of the surface topography and the out-of-plane and in-plane piezoelectric responses of a ferroelectric sample. The cantilever deflection caused by the voltage-induced surface displacements is detected by a laser beam reflecting into a four-quadrant photodiode. The vertical and horizontal piezoresponses are determined with the aid of two lock-in amplifiers by demodulating the corresponding signals $(a + c) - (b + d)$ and $(a + b) - (c + d)$, respectively. The instrument is operated in a constant force mode.

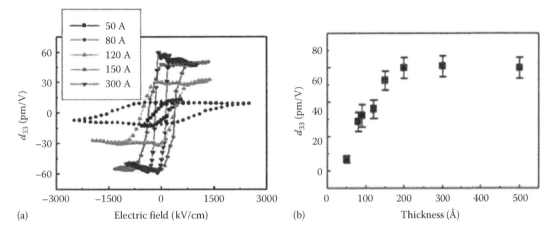

FIGURE 6.17 Out-of-plane piezoelectric response of $SrRuO_3/PbZr_{0.2}Ti_{0.8}O_3/SrRuO_3$ capacitors measured by the PFM technique: (a) piezoelectric hysteresis loops obtained for five different film thicknesses, and (b) the piezoelectric coefficient d_{33} as a function of the film thickness. (From Nagarajan, V. et al., *J. Appl. Phys.*, 100, 051609, 2006, With permission.)

without extended top electrodes collect signals from a subsurface layer of unknown thickness that is a function of dielectric permittivity and contact conditions (Gruverman et al. 1998).

Alternatively, a ferroelectric film with a deposited top electrode may be studied at the expense of a lower lateral resolution. By applying a voltage to the SPM tip contacting the top electrode, a submicron variation of piezoelectric properties in PZT capacitors was demonstrated (Christman et al. 2000, Setter et al. 2006). Under these conditions, a homogeneous electric field is generated in a ferroelectric layer, and the electrostatic tip–sample interaction is suppressed. This approach allows the investigations of domain-wall dynamics and polarization reversal mechanisms in ferroelectric capacitors and quantitative studies of the scaling of piezoelectric properties in ultrathin ferroelectric films. In particular, it was found (Nagarajan et al. 2006) that

the piezoelectric coefficient d_{33} of epitaxial $PbZr_{0.2}Ti_{0.8}O_3$ films sandwiched between $SrRuO_3$ electrodes decreases rapidly as the thickness is reduced from 20 to 5 nm (see Figure 6.17).

6.5 Physical Phenomena in Ferroelectric Capacitors

There are several physical effects that make the phase states and electric properties of thin-film ferroelectric capacitors different from those of bulk ferroelectrics. First, the ferroelectric film is generally subjected to an in-plane straining and clamping due to the presence of a dissimilar thick substrate. Second, an internal electric field exists in the capacitor, which depends on the electrode material and the film thickness. Third, the scaling of physical properties may result from the short-range interatomic interactions at the film–electrode interfaces. The current status of the theoretical description of these effects in thin films of perovskite ferroelectrics is given below.

6.5.1 Strain Effect

Owing to the electrostrictive coupling between lattice strains and polarization, the mechanical film–substrate interaction may strongly affect the physical properties of ferroelectric thin films (Pertsev et al. 1998). In a film deposited on a dissimilar thick substrate, the in-plane strains u_1, u_2, and u_6 are totally governed by the substrate, whereas the stresses σ_3, σ_4, and σ_5 are usually equal to zero. (We use the Voigt matrix notation and the reference frame with the x_3 axis orthogonal to the film surfaces.) Under such "mixed" mechanical boundary conditions, the equilibrium polarization state corresponds to a minimum of the modified thermodynamic potential $\tilde{G}$ (Pertsev et al. 1998), but not of the standard elastic Gibbs function G (Haun et al. 1987). In the most important case of a film grown in the (001)-oriented cubic paraelectric phase on a (001)-oriented cubic substrate ($u_1 = u_2 = u_m$, $u_6 = 0$), the stability ranges of different polarization states can be conveniently described with the aid of two-dimensional phase diagrams, where the misfit strain $u_m = (b - a_0)/a_0$ and temperature T are used as two independent parameters (a_0 is the equivalent cubic cell constant of the free standing film and b is the substrate lattice parameter). Such "misfit strain-temperature" diagrams were developed with the aid of thermodynamic calculations for single-domain BaTiO$_3$, PbTiO$_3$, and Pb(Zr$_{1-x}$Ti$_x$)O$_3$ (PZT) films (Pertsev et al. 1998, 2003a). Since the substrate-induced strains lower the symmetry of the paraelectric phase from cubic to tetragonal, the film polarization state may be very different from the ferroelectric phases observed in the corresponding bulk material (see Figure 6.18).

At large negative misfit strains, films of perovskite ferroelectrics stabilize in the tetragonal c phase with the spontaneous polarization $\mathbf{P}_s$ orthogonal to the film–substrate interface, whereas at large positive strains the orthorhombic aa phase forms, where $\mathbf{P}_s$ is directed along the in-plane face diagonal of the prototypic cubic cell. At low temperatures, the stability ranges of the c and aa phases are separated by a "monoclinic gap," where the monoclinic r phase with three nonzero polarization components P_i becomes the energetically most favorable state. These predictions of the thermodynamic theory were confirmed by the first-principles calculations (Bungaro and Rabe 2004, Diéguez et al. 2004).

It should be emphasized that the orthorhombic and monoclinic phases do not exist in the bulk PbTiO$_3$ crystals, where only the tetragonal ferroelectric state is stable (Haun et al. 1987). In the case of BaTiO$_3$, the aa phase may be compared with the orthorhombic phase forming in the bulk crystal in the low-temperature range between 10°C and –71°C, whereas the r phase can be regarded as a distorted modification of the rhombohedral phase that exists in a free crystal below –71°C (Jona and Shirane 1962).

The most remarkable manifestation of the strain effect appears in thin films of strontium titanate. In a mechanically free state, bulk SrTiO$_3$ crystals remain paraelectric down to zero absolute temperature despite a strong softening of the transverse optic polar mode near $T = 0\,K$ (Müller and Burkard 1979). The thermodynamic calculations show that this "incipient ferroelectricity" exists in epitaxial SrTiO$_3$ films grown on dissimilar cubic substrates only at small misfit strains ranging from -2×10^{-3} to -2×10^{-4} (Pertsev et al. 2000). Outside this "paraelectric gap," the ferroelectric phase transition takes place in the SrTiO$_3$ film at a finite temperature, which rises rapidly with the increase of the strain magnitude. The predicted phenomenon of strain-induced ferroelectricity was observed experimentally in SrTiO$_3$ films grown on (110)-oriented DyScO$_3$, which were found to display ferroelectric properties at room temperature (Haeni et al. 2004).

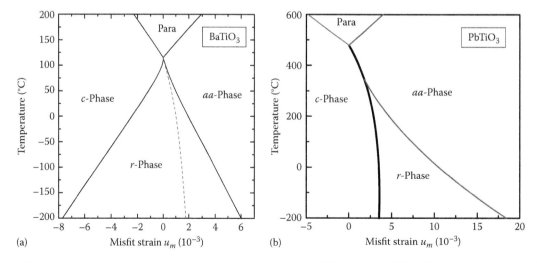

FIGURE 6.18 Misfit strain-temperature phase diagrams of single-domain BaTiO$_3$ (a) and PbTiO$_3$ (b) thin films epitaxially grown on (001)-oriented cubic substrates. The second- and first-order phase transitions are shown by thin and thick lines, respectively. (From Pertsev, N.A. et al., *Phys. Rev. Lett.*, 80, 1988, 1998. With permission.)

The strain-induced increase of the temperature T_c, at which the paraelectric to ferroelectric phase transition takes place, is characteristic of all studied perovskite ferroelectrics (Pertsev et al. 1998, 2003a). In addition, the two-dimensional clamping of the film by a thick substrate may change the order of this transition (Pertsev et al. 1998). Strong dependence of T_c on the misfit strain u_m explains the very high transition temperatures observed in epitaxial films grown on dissimilar substrates (Choi et al. 2004, He and Wells 2006).

The magnitude of the spontaneous polarization is also sensitive to the lattice strains. This effect is especially pronounced in ferroelectric films grown on "compressive" substrates ($u_m < 0$), where the polarization $\mathbf{P}_s$ is orthogonal to the film surfaces. For fully strained BaTiO$_3$ films grown on SrTiO$_3$ ($u_m = -2.6\%$), the thermodynamic theory predicts $P_s = 35\,\mu\text{C/cm}^2$, which is close to the experimental values of 43–44 $\mu\text{C/cm}^2$ (Kim et al. 2005, Petraru et al. 2007). Remarkably, the film polarization exceeds the polarization $P_b = 26\,\mu\text{C/cm}^2$ of bulk BaTiO$_3$ significantly. At the same time, the strain sensitivity of polarization in highly polar Pb-based perovskites, where the ferroelectric ionic displacements are already large in the bulk, is relatively low (Lee et al. 2007).

The enhancement of polarization P_s and the decrease of the in-plane permittivity ε_{11} in the strained c phase (Koukhar et al. 2001) should lead to a considerable increase of the coercive field E_c in thin films (Pertsev et al. 2003b). From the Landauer model of domain nucleation (Landauer 1957) it follows that $E_c \sim \gamma^{6/5}/\left(\varepsilon_{11}^{1/5} P_s^{3/5}\right)$, where $\gamma \sim P_s^3$ is the domain-wall energy. Hence, the coercive field $E_c \sim P_s^3/\varepsilon_{11}^{1/5}$ of BaTiO$_3$ films grown on SrTiO$_3$ ($\varepsilon_{11} \approx 170$) may be about eight times larger than that of the bulk crystal ($\varepsilon_{11} \approx 3600$). Although it is certainly a very strong increase, the strain effect alone cannot explain the observed drastic difference between the measured coercive fields $E_c \sim$ 150–300 kV/cm of epitaxial BaTiO$_3$ films (Jo et al. 2006b, Petraru et al. 2007) and the bulk $E_c \sim 1$ kV/cm.

Finally, it should be noted that ferroelectric properties of epitaxial thin films may strongly depend on the orientation of the crystal lattice with respect to the substrate surface. In particular, the phase states and dielectric properties of single-domain PbTiO$_3$ films with the (111)-orientation of the paraelectric phase were found to be very different from those of the (001)-oriented films (Tagantsev et al. 2002).

6.5.2 Depolarizing-Field Effect

When the polarization charges $\rho = -\text{div}\,\mathbf{P}$ existing at the film surfaces are not perfectly compensated for by other charges, an internal electric field appears in the ferroelectric layer (Figure 6.19). This "depolarizing" field $\mathbf{E}_{\text{dep}}$ may be significant even in short-circuited ferroelectric capacitors with perfect interfaces because the electronic screening length in metals is finite (Guro et al. 1970, Mehta et al. 1973).

For a homogeneously polarized film, the electrostatic calculation gives $E_{\text{dep}} = -P_3/(\varepsilon_0\varepsilon_b + c_i t)$, where P_3 is the equilibrium out-of-plane polarization in the film of thickness t, ε_0 is the permittivity of the vacuum, $\varepsilon_b \sim 10$ is the background

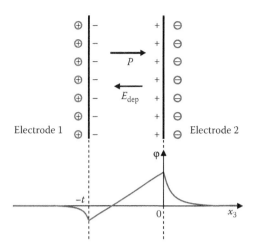

FIGURE 6.19 Imperfect screening of polarization charges in a ferroelectric capacitor. Distribution of the electrostatic potential φ is shown schematically for the case of dissimilar electrodes kept at a bias voltage compensating for the difference of their work functions. (From Pertsev, N.A. and Kohlstedt, H., *Phys. Rev. Lett.*, 98, 257603, 2007. With permission.)

dielectric constant of a ferroelectric material (Tagantsev and Gerra 2006), and c_i is the total capacitance of the screening space charge in the electrodes per unit area (Ku and Ullman 1964). When $P_1 = P_2 = 0$ (the c phase), the polarization $P_3(t)$ can be calculated from the nonlinear equation of state $\partial\tilde{G}/\partial P_3 = 0$ written for a strained film with an internal field E_{dep}. Since the capacitance c_i affects the polarization only via the product $c_i t$, the dependencies $P_3(t)$ corresponding to different electrode materials can be described by one universal curve $P_3(t_{\text{eff}})$. Here the effective film thickness t_{eff} may be defined as $t_{\text{eff}} = (c_i/c_1)t$, where $c_1 = 1$ F/m^2.

Figure 6.20 shows the dependencies $P_3(t_{\text{eff}})$ calculated for fully strained PZT 50/50 and BaTiO$_3$ films grown on SrTiO$_3$ (Pertsev and Kohlstedt 2007). It can be seen that the spontaneous polarization decreases in thinner films and vanishes at a critical film thickness t_0. In the case of capacitors with SrRuO$_3$ electrodes ($c_i = 0.444$ F/m^2), the thickness t_0 is about 2 nm for PZT 50/50 films and about 2.6 nm for BaTiO$_3$ films. However, the size-induced phase transition at t_0, also predicted by the first-principles calculations (Junquera and Ghosez 2003), cannot be observed in reality since at a slightly larger film thickness $t_c < 3$ nm the single-domain ferroelectric state becomes unstable and transforms into the 180° polydomain state (Pertsev and Kohlstedt 2007).

The experimental studies of ultrathin BaTiO$_3$ films (Kim et al. 2005, Petraru et al. 2008) showed that the remanent polarization decreases monotonically with decreasing thickness at $t < 30$ nm, where the misfit strain becomes constant. This behavior is similar to the dependence shown in Figure 6.20, but the thermodynamic theory does not provide a precise quantitative description of the experimental data (Kim et al. 2005, Petraru et al. 2008). The first-principles calculations performed by Junquera and Ghosez (2003) also cannot explain the measured dependence quantitatively (Kim et al. 2005).

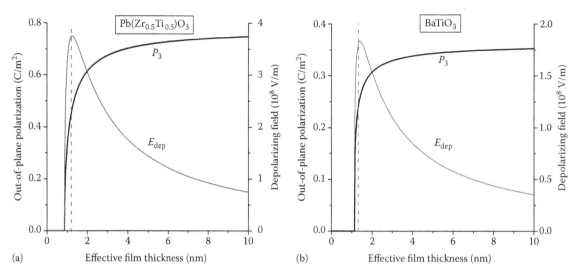

FIGURE 6.20 Thickness dependence of the out-of-plane polarization and depolarizing field in ultrathin $Pb(Zr_{0.5}Ti_{0.5})O_3$ (a) and $BaTiO_3$ (b) films epitaxially grown on $SrTiO_3$. The misfit strain is taken to be -39×10^{-3} for $Pb(Zr_{0.5}Ti_{0.5})O_3$ films and -26×10^{-3} for $BaTiO_3$ films; $T = 25°C$. The dashed line shows the film thickness below which the single-domain state becomes unstable.

For $BaTiO_3$ capacitors with $SrRuO_3$ electrodes, the depolarizing field E_{dep} has been evaluated experimentally as well and was found to increase dramatically with decreasing film thickness (Kim et al. 2005). Such a strong rise is consistent with the theoretical thickness dependence of E_{dep} shown in Figure 6.20, which was calculated with the account of the polarization variation $P_3(t_{eff})$. The predicted saturation of the depolarizing field and its decrease in ultrathin films, however, were not observed. In this thickness range, the single-domain state becomes unstable (Pertsev and Kohlstedt 2007), and the mean value of the depolarizing field vanishes due to the formation of 180° domain structure.

The presence of a depolarizing field in a capacitor is expected to affect the shape of polarization-voltage (P-V) hysteresis loops and the magnitude of coercive fields (Tagantsev and Gerra 2006, Jo et al. 2006a,b). Indeed, a simple calculation shows that imperfect compensation of polarization charges at the interfaces not only reduces the remanent polarization, but also leads to a tilt of the hysteresis loop (Tagantsev and Gerra 2006). The influence of E_{dep} on the magnitude of the coercive field seems to be less pronounced. An accurate electrostatic modeling of the polarization switching on the macroscopic level indicates that E_c reduces with decreasing film thickness, but only when the P-V loop is not saturated (Tagantsev and Gerra 2006). In the limit of high amplitudes of the driving field applied to a capacitor, the influence of the depolarizing field on E_c disappears. This feature is explained by the absence of voltage drop across the metal–ferroelectric interface at $E = E_c$, where the macroscopic polarization $P(E)$ goes to zero by definition.

At the same time, Jo et al. (2006a,b) proposed that the depolarizing field may strongly reduce the potential barriers hindering polarization switching in ultrathin films. Such reduction may happen at the early stage of switching, when the macroscopic polarization P is large, but the influence of E_{dep} becomes

negligible in a film with $P = 0$. Therefore, it is not clear whether the coercive field changes. It would decrease if the switching develops as an "avalanche" process after the nucleation of a few reversed domains. On the contrary, the change of E_c should be small when the applied field must increase significantly in the course of switching to facilitate further domain nucleation or domain-wall motion.

It should be noted that the nucleation of reversed domains in a perfect bulk crystal is impossible since the activation energy is many orders of magnitude larger than the thermal energy (Landauer's paradox). Jo et al. (2006a) suggested that such "homogeneous" domain nucleation becomes possible in ultrathin ferroelectric films, where the field-dependent activation energy decreases dramatically due to strong depolarizing fields. This conclusion, however, is based on a rough estimate of the activation energy obtained in the approximation of a homogeneous depolarizing field, which is valid for a single-domain film only.

Domain nucleation is evidently easier near crystal surfaces, ferroelectric–electrode interfaces, and lattice defects. Gerra et al. (2005) proposed that the interface coupling between a ferroelectric layer and electrodes changes its sign after the polarization reversal and, therefore, stimulates the switching. According to their model, the surface-stimulated nucleation may reduce the coercive field of $BaTiO_3$ down to a value two orders of magnitude smaller than the thermodynamic coercive field. High local electric fields $E_{loc} \gg E$ created by spikes on the electrodes may also permit the domain formation at small applied fields E. Besides, residual domains probably exist in ferroelectric films, especially in polycrystalline ones. The growth of these domains may play an important role in the polarization switching, as it happens in ferroelectric polymers (Pertsev and Zembilgotov 1991).

6.5.3 Intrinsic Size Effect in Ultrathin Films

Since the unit cells adjacent to the film–electrode interfaces have an atomic environment different from that of the inner cells, the ferroelectric polarization may depend on the film thickness even in the absence of a depolarizing field. This "intrinsic" size effect can be described with the aid of a modified thermodynamic theory based on the concept of extrapolation length (Kretschmer and Binder 1979). In this theory, the total energy of the ferroelectric layer involves an additional surface contribution, and the polarization distribution across the film, in general, is taken to be inhomogeneous.

In the most important case of the (001)-oriented film grown on a compressive substrate ($P_1 = P_2 = 0$, $P_3 \neq 0$), the polarization profile $P_3(x_3)$ can be calculated from the Euler–Lagrange equation (Zembilgotov et. al. 2002). For a film having the same atomic terminations at both surfaces and sandwiched between identical electrodes, the boundary conditions can be written as $dP_3/dx_3 = P_b/\delta$ at $x_3 = 0$ and $dP_3/dx_3 = -P_b/\delta$ at $x_3 = t$, where P_b is the polarization value at the film boundaries and δ is the extrapolation length. The polarization suppression (enhancement) near the film surfaces is described by positive (negative) values of the extrapolation length. In a weakly conducting ferroelectric, such as $BaTiO_3$ or $PbTiO_3$, the inner polarization charges $\rho = -dP_3/dx_3$ are largely compensated by charge carriers so that the associated depolarizing field should be negligible. In this case, the spatial scale of polarization variations is determined by the ferroelectric correlation length $\xi^* = \sqrt{g_{11}/|a_3^*|}$ of a strained film, and the strength of the intrinsic size effect is governed by the ratio $\xi^*/|\delta|$. (Here g_{11} and $a_3^*(u_m)$ are the coefficients of the gradient term and the renormalized second-order polarization term in the free energy expansion, respectively.)

The numerical calculations demonstrate that the polarization suppression in the surface layers reduces the temperature T_c of ferroelectric transition at a given misfit strain u_m. This reduction, however, is significant only in ultrathin films with thicknesses about a few $\xi^*(T = 0)$. Below the transition temperature $T_c(u_m)$, the mean polarization reduces with decreasing film thickness and vanishes at a critical thickness $t_c \approx \xi^*$ (Zembilgotov et al. 2002). Thus, the intrinsic surface effect may lead to a size-induced ferroelectric to paraelectric transformation.

Since the discussed thermodynamic theory is a continuum theory, it is valid only when the characteristic length ξ^* of the polarization variations is larger than the interatomic distances. In the case of a negligible depolarizing field, this condition is satisfied at least near the transition temperature T_c because the coefficient a_3^* goes to zero at this temperature (Pertsev et al. 1998). The situation, however, changes dramatically in a perfectly insulating ferroelectric, where the uncompensated polarization charges $\rho = -\mathrm{div}\,\mathbf{P}$ inside the film create a nonzero depolarizing field (Kretschmer and Binder 1979, Tagantsev et al. 2008). The characteristic length of polarization variations becomes $\xi_d^* = \sqrt{g_{11}/\left|a_3^* + (\varepsilon_0\varepsilon_b)^{-1}\right|}$, which, in contrast to ξ^*, does not

increase significantly near T_c. Since $\xi_d^* \sim 0.1\,\mathrm{nm}$ only, the continuum approach based on the concept of extrapolation length cannot be used to describe the surface effect in perfectly insulating perovskite ferroelectrics (Tagantsev et al. 2008).

In the latter case, however, the ferroelectric film may be assumed to be homogeneously polarized in the thickness direction, and the intrinsic size effect can be described with the aid of a phenomenological approach as well (Tagantsev et al. 2008). To that end, the film free energy is written as the sum of the "bulk" and "surface" contributions, each represented by a polynomial in terms of ferroelectric polarization. In general, the surface contribution should involve not only the even-power terms, but also the odd-power terms, because the surface breaks the inversion symmetry of the ferroelectric (Levanyuk and Sigov 1988, Bratkovsky and Levanyuk 2005). However, when the film–electrode interfaces are identical, the linear term vanishes and the surface energy can be approximated by a quadratic polarization term. The coefficient of this term may be evaluated from the comparison of the phenomenological theory with the results of first-principles calculations performed for ultrathin ferroelectric films. For $BaTiO_3$ capacitors with $SrRuO_3$ electrodes, this procedure reveals that the surface energy is positive (Tagantsev et al. 2008), which implies the polarization suppression at the interfaces.

In other metal/ferroelectric/metal heterostructures, however, polarization could be enhanced near the interfaces. Such enhancement was demonstrated by the first-principles-based calculations performed for ultrathin $PbTiO_3$ and $BaTiO_3$ films (Ghosez and Rabe 2000, Lai et al. 2005). The short-range interactions between the film and electrodes must be also taken into account to prove that this effect exists in ferroelectric capacitors as well. The importance of ionic displacements in the boundary layers of $SrRuO_3$ electrodes for the stabilization of ferroelectricity in ultrathin films was revealed by the first-principles investigations (Sai et al. 2005, Gerra et al. 2006).

6.6 Future Perspective

When the thickness of the ferroelectric layer in a biased capacitor becomes as small as a few nanometers, the quantum mechanical electron tunneling across the insulating barrier should become significant (Kohlstedt et al. 2005b, Zhuravlev et al. 2005a). Since the tunnel current exponentially depends on the barrier thickness, the crossover from a capacitor to a *tunnel junction* takes place near some threshold thickness. The presence of spontaneous polarization in a ferroelectric barrier and its piezoelectric properties are expected to make the current-voltage (*I-V*) characteristic of a ferroelectric tunnel junction (FTJ) very different from those of conventional tunnel junctions involving nonpolar dielectrics (Kohlstedt et al. 2005b and Rodríguez Contreras 2004). Theoretically, the junction conductance can change strongly after the polarization reversal in the barrier so that the FTJs are promising for the memory storage with nondestructive readout. For the memory applications, asymmetric FTJs with dissimilar electrodes seem to be preferable since such junctions should exhibit much larger conductance on/off ratios (Kohlstedt et al. 2005b, Zhuravlev et al. 2005a). This feature is due to the fact that here the mean barrier height changes after the polarization reversal by the amount $\Delta\phi \cong eP_3(t)(c_{m2}^{-1} - c_{m1}^{-1})$, where c_{m1} and c_{m2} are the capacitances of two electrodes and e is the electron charge.

The ferroelectric tunnel barrier may also be combined with ferromagnetic electrodes. For such a *multiferroic* tunnel junction (MFTJ), new functionalities may be expected since the tunneling probability becomes different for the spin-up and spin-down electrons owing to the exchange splitting of electronic bands in ferromagnetic electrodes. In particular, Zhuravlev et al. proposed a new spintronic device, where an electric current is injected from a diluted magnetic semiconductor through the ferroelectric barrier to a normal (nonmagnetic) semiconductor (Zhuravlev et al. 2005b). Their theoretical calculations indicated that the switching of ferroelectric polarization in the barrier may change the spin polarization of the injected current markedly, which provides a two-state electrical control of the device performance.

When both electrodes are ferromagnetic, the tunnel current becomes dependent on the mutual orientation of the electrode magnetizations. This phenomenon, which is termed *tunneling magnetoresistance* (TMR), is important for the applications in spin-electronic devices such as magnetic sensors and magnetic random-access memories. Since the TMR ratio depends not only on the properties of ferromagnetic electrodes, but also on the barrier characteristics (Slonczewski 1989), it may be sensitive to the orientation of the ferroelectric polarization in the MFTJ. This supposition was confirmed by the theoretical calculations performed for junctions involving two magnetic semiconductor electrodes (Zhuravlev et al. 2005b). It was shown that, under certain conditions, the MFTJ works as a device that allows the switching of TMR between positive and negative values.

The experimental realization of ferroelectric and multiferroic tunnel junctions, however, is a task with many obstacles because it requires the fabrication of ultrathin films retaining pronounced ferroelectric properties at a thickness of only a few unit cells. Moreover, the ferroelectric state with a nonzero net polarization must be stable at such a small thickness and switchable by a moderate external voltage. In our opinion, reliable FTJs showing resistive switching in the tunneling regime have not been fabricated yet, despite several attempts made in this direction (Rodríguez et al. 2003b, Gajek et al. 2007). The observation of a hysteretic *I-V* curve or resistance jumps after short-voltage pulses alone is not sufficient to prove the existence of an FTJ (Kohlstedt et al. 2008). Hysteretic *I-V* curves were also measured for *non*ferroelectric LaSr$_x$Mn$_{1-x}$O$_3$/SrTiO$_3$/LaSr$_x$Mn$_{1-x}$O$_3$ tunnel junctions and explained by other effects (Sun 2001). Further experiments are necessary to demonstrate the functioning of FTJs unambiguously.

In case the above challenges will be overcome in the future, a number of new exciting opportunities for technological applications and interesting physical phenomena are anticipated. Figure 6.21 summarizes the variety of novel functional oxide tunnel junctions. Here, the Josephson tunnel junctions with a ferroelectric or multiferroic barrier are included as well. The

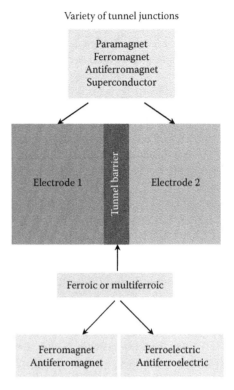

Variety of tunnel junctions

FIGURE 6.21 A "zoo" of novel tunnel junctions involving multifunctional tunnel barriers and electrodes of various types.

influence of a ferroic tunnel barrier on the Cooper pair and quasiparticle tunneling might lead to interesting new physics. Besides oxide materials, ferroelectric polymers such as PVDF and P(VDF-TrFE) (Xu 1991, Bune et al. 1998) can be incorporated in low-temperature superconducting Josephson junctions [e.g., Nb/Al-AlO$_x$-P(VDF-TrFE)/Nb] (Huggins and Gurvitch 1985) and in magnetic tunnel junctions [e.g., Co$_x$Fe$_{1-x}$/AlO$_x$-P(VDF-TrFE)/Ni$_{20}$Fe$_{80}$] (Moodera et al. 1995). In principle, one would expect the appearance of physical effects similar to those

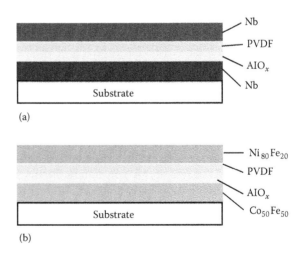

FIGURE 6.22 Low-temperature superconducting Josephson junctions (a) and magnetic tunnel junctions (b) with the AlO$_x$-ferroelectric polymer composite barriers.

described above for entirely oxide tunnel junctions. On the other hand, metallic (Josephson and magnetic) tunnel junctions are more reliable than oxide ones, and, in addition, ultrathin PVDF films are compatible with AlO_x so that composite ferroelectric-oxide barriers can be fabricated. The possible structures of superconducting and magnetic metallic tunnel junctions involving PVDF are shown schematically in Figure 6.22.

In conclusion of this section, it should be noted that in addition to the planar metal/ferroelectric/metal multilayers discussed above, heterostructures of other geometries may be useful for certain applications in nanoelectronics and may even display specific physical properties. Remarkably, one-dimensional structures in the form of ferroelectric nanowires (Urban et al. 2003) and nanotubes with inner and outer electrodes (Alexe et al. 2006) have been successfully fabricated. Ferroelectric quantum dots are of great interest as well, in particular, for electro-optical devices (Ye et al. 2000).

6.7 Summary and Outlook

The overview presented in this chapter demonstrates impressive achievements in the deposition, characterization, and theoretical description of nanoscale ferroelectric films and heterostructures. Remarkably, capacitors involving only a few nanometers of perovskite ferroelectrics were successfully fabricated, displaying a high remanent polarization. Several advanced analytical tools are now available, showing that high structural quality and sharp interfaces can be retained even in nanoscale capacitors. The basic physical effects in ferroelectric thin films, such as the strain and depolarizing-field effects, are already well understood theoretically, and the influence of short-range interactions at the ferroelectric–metal interfaces is intensively studied by first-principles calculations. Thus, all three constituents of the research triangle shown in Figure 6.1 are functioning effectively, which promises new advances in the physics of nanoscale ferroelectrics and their device applications in the near future.

References

Alexe, M., Hesse, D., Schmidt, V. et al. 2006. Ferroelectric nanotubes fabricated using nanowires as positive templates. *Applied Physics Letters* 89: 172907.

Bratkovsky, A. M. and Levanyuk, A. P. 2005. Smearing of phase transition due to a surface effect or a bulk inhomogeneity in ferroelectric nanostructures. *Physical Review Letters* 94: 107601.

Bune, A. V., Fridkin, V. M., Ducharme, S. et al. 1998. Two-dimensional ferroelectric films. *Nature* 391: 874–877.

Bungaro, C. and Rabe, K. M. 2004. Epitaxially strained [001]-$(PbTiO_3)_1(PbZrO_3)_1$ superlattice and $PbTiO_3$ from first principles. *Physical Review B* 69: 184101.

Bunshah, R. F. 1994. *Handbook of Deposition Technologies for Films and Coatings*, Park Ridge, NY: Noyes Publications.

Choi, K. J., Biegalski, M., Li, Y. L. et al. 2004. Enhancement of ferroelectricity in strained $BaTiO_3$ thin films. *Science* 306: 1005–1009.

Chopra, K. L. 1969. *Thin Film Phenomena*, New York: McGraw-Hill.

Christman, J. A., Kim, S. H., Maiwa, H. et al. 2000. Spatial variation of ferroelectric properties in $Pb(Zr_{0.3},Ti_{0.7})O_3$ thin films studied by atomic force microscopy. *Journal of Applied Physics* 87: 8031–8034.

Chu, W., Mayer, J., and Nicolet, M. 1978. *Backscattering Spectrometry*, New York: Academic Press.

Clayhold, J. A., Kerns, B. M., Schroer, M. D. et al. 2008. Combinatorial measurements of Hall effect and resistivity in oxide films. *Review of Scientific Instruments* 73: 033908.

Cohen, R. E. 1992. Origin of ferroelectricity in perovskite oxides. *Nature* 358: 136–138.

Dawber, M., Rabe, K. M., and Scott, J. F. 2005. Physics of thin-film ferroelectric oxides. *Reviews of Modern Physics* 77: 1083–1130.

Devonshire, A. F. 1954. Theory of ferroelectrics. *Advances in Physics* 3: 85–130.

Diéguez, O., Tinte, S., Antons, A. et al. 2004. Ab initio study of the phase diagram of epitaxial $BaTiO_3$. *Physical Review B* 69: 212101.

Ducharme, S., Palto, S. P., Fridkin, V. M., and Blinov, L. M. 2002. Ferroelectric Polymer Langmuir-Blodgett Films. In *Ferroelectric and Dielectric Thin Films*, Vol. 3 of *Handbook of Thin Films Materials*, ed. H. S. Nalwa, San Diego, CA: Academic Press.

Eom, C. B., Cava, R. J., Fleming, R. M. et al. 1992. Single-crystal epitaxial thin films of the isotropic metallic oxides $Sr_{1-x}Ca_xRuO_3$ ($0 < x < 1$). *Science* 258: 1766–1769.

Fewster, P. F. 1996. X-ray analysis of thin films and multilayers. *Reports on Progress in Physics* 59: 1339–1407.

Fisher, O., Triscone, J. M., Fivat, P. et al. 1994. Investigations of coupled $DyBe_2Cu_3O_7$-$(Y_{1-x}Pr_x)$-$Be_2Cu_3O_7$ multilayer structures. *Superconducting Superlattices and Multilayers* 2157: 134–141.

Fong, D. D., Stephenson, G. B., Streiffer, S. K. et al. 2004. Ferroelectricity in ultrathin perovskite films. *Science* 304: 1650–1653.

Fong, D. D., Cionca, C., Yacoby, Y. et al. 2005. Direct structural determination in ultrathin ferroelectric films by analysis of synchrotron x-ray scattering measurements. *Physical Review B* 71: 144112.

Gajek, M., Bibes, M., Fusil, S. et al. 2007. Tunnel junctions with multiferroic barriers. *Nature Materials* 6: 296–302.

Garcia, R. and Perez, R. 2002. Dynamic atomic force microscopy methods. *Surface Science Reports* 47: 197–301.

Gerra, G., Tagantsev, A. K., and Setter, N. 2005. Surface-stimulated nucleation of reverse domains in ferroelectrics. *Physical Review Letters* 94: 107602.

Gerra, G., Tagantsev, A. K., Setter, N. et al. 2006. Ionic polarizability of conductive metal oxides and critical thickness for ferroelectricity in $BaTiO_3$. *Physical Review Letters* 96: 107603.

Gerra, G., Tagantsev, A. K., and Setter, N. 2007. Ferroelectricity in asymmetric metal-ferroelectric-metal heterostructures: A combined first-principles-phenomenological approach. *Physical Review Letters* 98: 207601.

Ghosez, P. and Rabe, K. M. 2000. Microscopic model of ferroelectricity in stress-free PbTiO$_3$ ultrathin films. *Applied Physics Letters* 76: 2767–2769.

Gruverman, A. and Kholkin, A. 2004. Ferroelectric nanodomains. In *Encyclopedia of Nanoscience and Nanotechnology*, ed. H. S. Nalwa, pp. 359–375. Los Angeles, CA: American Scientific Publishers.

Gruverman, A., Auciello, O., and Tokumoto, H. 1998. Imaging and control of domain structures in ferroelectric thin films via scanning force microscopy. *Annual Review of Materials Science* 28: 101–123.

Guro, G. M., Ivanchik, I. I., and Kovtonyuk, N. F. 1970. C-Domain barium titanate crystal in a short-circuited capacitor. *Soviet Physics Solid State* 11: 1574.

Güthner, P. and Dransfeld, K. 1992. Local poling of ferroelectric polymers by scanning force microscopy. *Applied Physics Letters* 61: 1137.

Habermeier, H. U. 2007. Thin films of perovskite-type complex oxides. *Materials Today* 10: 34–43.

Haeni, J. H., Theis, C. D., and Schlom, D. G. 2000. RHEED intensity oscillations for the stoichiometric growth of SrTiO$_3$ thin films by reactive molecular beam epitaxy. *Journal of Electroceramics* 4: 385–391.

Haeni, J. H., Irvin, P., Chang, W. et al. 2004. Room-temperature ferroelectricity in strained SrTiO$_3$. *Nature* 430: 758–761.

Hambe, M., Wicks, S., Gregg, J. M. et al. 2008. Creation of damage-free ferroelectric nanostructures via focused ion beam milling. *Nanotechnology* 19: 175302.

Haun, M. J., Furman, E., Jang, S. J. et al. 1987. Thermodynamic theory of PbTiO$_3$. *Journal of Applied Physics* 62: 3331–3338.

He, F. Z. and Wells, B. O. 2006. Lattice strain in epitaxial BaTiO$_3$ thin films. *Applied Physics Letters* 88: 152908.

Hubler, G. K. and Chrisey, D. B. 1994. *Pulsed Laser Deposition of Thin Films*, New York: John Wiley & Sons, Inc.

Huggins, H. A. and Gurvitch, M. 1985. Preparation and characterization of Nb/Al-oxide/Nb tunnel junctions. *Journal of Applied Physics* 57: 2103–2109.

Ishiwara, H. and Okuyama, M. 2004. *Ferroelectric Random Access Memories: Fundamentals and Applications*, Berlin, Germany: Springer.

Jia, C. L., Mi, S. B., Urban, K. et al. 2008. Atomic-scale study of electric dipoles near charged and uncharged domain walls in ferroelectric films. *Nature Materials* 7: 57–61.

Jo, J. Y., Kim, D. J., Kim, Y. S. et al. 2006a. Polarization switching dynamics governed by the thermodynamic nucleation process in ultrathin ferroelectric films. *Physical Review Letters* 97: 247602.

Jo, J. Y., Kim, Y. S., Noh, T. W. et al. 2006b. Coercive fields in ultrathin BaTiO$_3$ capacitors. *Applied Physics Letters* 89: 232909.

Jona, F. and Shirane, G. 1962. *Ferroelectric Crystals*, New York: MacMillan.

Junquera, J. and Ghosez, P. 2003. Critical thickness for ferroelectricity in perovskite ultrathin films. *Nature* 422: 506–509.

Kalinin, S. V. and Bonnell, D. A. 2002. Imaging mechanism of piezoresponse force microscopy of ferroelectric surfaces. *Physical Review B* 65: 125408.

Kato, Y., Kaneko, Y., Tanaka, H. et al. 2007. Overview and future challenge of ferroelectric random access memory technologies. *Japanese Journal of Applied Physics Part 1—Regular Papers Brief Communications & Review Papers* 46: 2157–2163.

Kawamura, E., Vahedi, V., Lieberman, M. A. et al. 1999. Ion energy distributions in rf sheaths; review, analysis and simulation. *Plasma Sources Science & Technology* 8: R45–R64.

Kholkin, A., Kalinin, S., Roelofs, A. et al. 2007. Review of ferroelectric domain imaging by piezoresponse force microscopy. In *Scanning Probe Microscopy: Electrical and Electromechanical Phenomena at the Nanoscale*, eds. S. Kalinin, and A. Gruverman, pp. 173–214. New York: Springer.

Kim, D. J., Jo, J. Y., Kim, Y. S. et al. 2005a. Polarization relaxation induced by a depolarization field in ultrathin ferroelectric BaTiO$_3$ capacitors. *Physical Review Letters* 95: 7602.

Kim, Y. S., Kim, D. H., Kim, J. D. et al. 2005b. Critical thickness of ultrathin ferroelectric BaTiO$_3$ films. *Applied Physics Letters* 86: 102907.

King-Smith, R. D. and Vanderbilt, D. 1993. Theory of polarization of crystalline solids. *Physical Review B* 47: 1651–1654.

King-Smith, R. D. and Vanderbilt, D. 1994. First-principles investigation of ferroelectricity in perovskite compounds. *Physical Review B* 49: 5828–5844.

Kohlstedt, H., Mustafa, Y., Gerber, A. et al. 2005a. Current status and challenges of ferroelectric memory devices. *Microelectronic Engineering* 80: 296–304.

Kohlstedt, H., Pertsev, N. A., Rodríguez Contreras, J. R. et al. 2005b. Theoretical current-voltage characteristics of ferroelectric tunnel junctions. *Physical Review B* 72: 125341.

Kohlstedt, H., Petraru, A., Szot, K. et al. 2008. Method to distinguish ferroelectric from nonferroelectric origin in case of resistive switching in ferroelectric capacitors. *Applied Physics Letters* 92: 062907.

Koukhar, V. G., Pertsev, N. A., and Waser, R. 2001. Thermodynamic theory of epitaxial ferroelectric thin films with dense domain structures. *Physical Review B* 64: 214103.

Kretschmer, R. and Binder, K. 1979. Surface effects on phase transitions in ferroelectrics and dipolar magnets. *Physical Review B* 20: 1065.

Kronholz, S., Rathgeber, S., Karthauser, S. et al. 2006. Self-assembly of diblock-copolymer micelles for template-based preparation of PbTiO$_3$ nanograins. *Advanced Functional Materials* 16: 2346–2354.

Ku, H. Y. and Ullman, F. G. 1964. Capacitance of thin dielectric structures. *Journal of Applied Physics* 35: 265–267.

Lai, B. K., Kornev, I. A., Bellaiche, L. et al. 2005. Phase diagrams of epitaxial BaTiO$_3$ ultrathin films from first principles. *Applied Physics Letters* 86: 132904.

Landau, L. D., Lifshitz, E. M., and Pitaevskii, L. P. 1984. *Electrodynamics of Continuous Media*, Oxford, NY: Pergamon.

Landauer, R. 1957. Electrostatic considerations in $BaTiO_3$ domain formation during polarization reversal. *Journal of Applied Physics* 28: 227–234.

Lee, H. N., Nakhmanson, S. M., Chisholm, M. F. et al. 2007. Suppressed dependence of polarization on epitaxial strain in highly polar ferroelectrics. *Physical Review Letters* 98: 217602.

Lettieri, J., Haeni, J. H., and Schlom, D. G. 2002. Critical issues in the heteroepitaxial growth of alkaline-earth oxides on silicon. *Journal of Vacuum Science and Technology A* 20: 1332.

Levanyuk, A. P. and Sigov, A. S. 1988. *Defects and Structural Phase Transitions*, Amsterdam, the Netherlands: Gordon and Breach.

Lines, M. E. and Glass, A. M. 1977. *Principles and Applications of Ferroelectrics and Related Materials*, Oxford, NY: Oxford University Press.

Logvenov, G. and Bozovic, I. 2008. Artificial superlattices grown by MBE: Could we design novel superconductors? *Physica C—Superconductivity and Its Applications* 468: 100–104.

Maissel, L. I. and Glang, R. 1979. *Handbook of Thin Film Technology*, New York: McGraw-Hill.

Mehta, R. R., Silverman, B. D., and Jacobs, J. T. 1973. Depolarization fields in thin ferroelectric films. *Journal of Applied Physics* 44: 3379–3385.

Moodera, J. S., Lisa R. K., Wong, T. M., and Meservey, R. 1995. Large magnetoresistance at room temperature in ferromagnetic thin film tunnel junctions. *Physical Review Letters* 74: 3273–3276.

Muller, K. A. and Burkard, H. 1979. $SrTiO_3$—Intrinsic quantum paraelectric below 4 K. *Physical Review B* 19: 3593–3602.

Nagarajan, V., Junquera, J., He, J. Q. et al. 2006. Scaling of structure and electrical properties in ultrathin epitaxial ferroelectric heterostructures. *Journal of Applied Physics* 100: 051609.

Nevot, L. and Croce, P. 1980. Characterization of surfaces by grazing x-ray reflection—Application to study of polishing of some silicate-glasses. *Revue De Physique Appliquee* 15: 761–779.

Nicolaou, K. C., Hanko, R., and Hartwig, W. 2002. *Handbook of Combinatorial Chemistry: Drugs, Catalysts, Materials*, Weinheim, Germany: Wiley-VCH Verlag GmbH.

Ohtani, M., Lippmaa, M., Ohnishi, T. et al. 2005. High throughput oxide lattice engineering by parallel laser molecular-beam epitaxy and concurrent x-ray diffraction. *Review of Scientific Instruments* 76: 062218.

Oikawa, T., Morioka, H., Nagai, A. et al. 2004. Thickness scaling of polycrystalline Pb(Zr,Ti)O_3 films down to 35 nm prepared by metalorganic chemical vapor deposition having good ferroelectric properties. *Applied Physics Letters* 85: 1754–1756.

Pertsev, N. A. and Kohlstedt, H. 2007. Elastic stabilization of a single-domain ferroelectric state in nanoscale capacitors and tunnel junctions. *Physical Review Letters* 98: 257603.

Pertsev, N. A. and Zembilgotov, A. G. 1991. Microscopic mechanism of polarization switching in polymer ferroelectrics. *Soviet Physics—Solid State* 33: 165–175.

Pertsev, N. A., Zembilgotov, A. G., and Tagantsev, A. K. 1998. Effect of mechanical boundary conditions on phase diagrams of epitaxial ferroelectric thin films. *Physical Review Letters* 80: 1988–1991.

Pertsev, N. A., Tagantsev, A. K., and Setter, N. 2000. Phase transitions and strain-induced ferroelectricity in $SrTiO_3$ epitaxial thin films. *Physical Review B* 61: R825–828; 2002. Erratum. *Physical Review B* 65: 219901.

Pertsev, N. A., Kukhar, V. G., Kohlstedt, H. et al. 2003a. Phase diagrams and physical properties of single-domain epitaxial Pb(Zr$_{1-x}$Ti$_x$)O_3 thin films. *Physical Review B* 67: 054107.

Pertsev, N. A., Rodríguez Contreras, J., Kukhar, V. G. et al. 2003b. Coercive field of ultrathin Pb(Zr$_{0.52}$Ti$_{0.48}$)O_3 epitaxial films. *Applied Physics Letters* 83: 3356–3358.

Pertsev, N. A., Dittmann, R., Plonka, R. et al. 2007. Thickness dependence of intrinsic dielectric response and apparent interfacial capacitance in ferroelectric thin films. *Journal of Applied Physics* 101: 074102.

Petraru, A., Pertsev, N. A., Kohlstedt, H. et al. 2007. Polarization and lattice strains in epitaxial $BaTiO_3$ films grown by high-pressure sputtering. *Journal of Applied Physics* 101: 114106.

Petraru, A., Kohlstedt, H., Poppe, U. et al. 2008. Wedgelike ultrathin epitaxial $BaTiO_3$ films for studies of scaling effects in ferroelectrics. *Applied Physics Letters* 93: 072902.

Poppe, U., Schubert, J., Arons, R. R. et al. 1988. Direct production of crystalline superconducting thin-films of YBa$_2$Cu$_3$O$_7$ by high-pressure oxygen sputtering. *Solid State Communications* 66: 661–665.

Resta, R. 1994. Macroscopic polarization in crystalline dielectrics—The geometric phase approach. *Reviews of Modern Physics* 66: 899–915.

Rodríguez, B. J., Gruverman, A., Kingon, A. I. et al. 2002. Piezoresponse force microscopy for polarity imaging of GaN. *Applied Physics Letters* 80: 4166–4168.

Rodríquez Contreras, J., Schubert, J., Poppe, U. et al. 2002. Structural and ferroelectric properties of epitaxial PbZr$_{0.52}$Ti$_{0.48}$O$_3$ and $BaTiO_3$ thin films prepared on SrRuO$_3$/SrTiO$_3$(100) substrates. *Materials Research Society Symposium Proceedings* 688: C8.10.

Rodríguez Contreras, J. 2004. *Ferroelectric Tunnel Junctions*, Jülich, Germany: Forschungszentrum Jülich GmbH Verlag.

Rodríguez Contreras, J., Kohlstedt, H., Poppe, U. et al. 2003a. Surface treatment effects on the thickness dependence of the remanent polarization of PbZr$_{0.52}$Ti$_{0.48}$O$_3$ capacitors. *Applied Physics Letters* 83: 126–128.

Rodríguez Contreras, J., Kohlstedt, H., Poppe, U. et al. 2003b. Resistive switching in metal-ferroelectric-metal junctions. *Applied Physics Letters* 83: 4595–4597.

Sai, N., Kolpak, A. M., and Rappe, A. M. 2005. Ferroelectricity in ultrathin perovskite films. *Physical Review B* 72: 020101(R).

Schlom, D. G., Chen, L. Q., Eom, C. B. et al. 2007. Strain tuning of ferroelectric thin films. *Annual Review of Materials Research* 37: 589–626.

Schneller, T. and Waser, R. 2007. Chemical modifications of $Pb(Zr_{0.3},Ti_{0.7})O_3$ precursor solutions and their influence on the morphological and electrical properties of the resulting thin films. *Journal of Sol-Gel Science and Technology* 42: 337–352.

Schuller, I. K. 1980. New class of layered materials. *Physical Review Letters* 44: 1597–1600.

Scott, J. F. 2000. *Ferroelectric Memories*, Berlin, Germany: Springer.

Scott, J. F. 2007. Applications of modern ferroelectrics. *Science* 315: 954–959.

Setter, N., Damjanovic, D., Eng, L. et al. 2006. Ferroelectric thin films: Review of materials, properties, and applications. *Journal of Applied Physics* 100: 051606.

Slonczewski, J. C. 1989. Conductance and exchange coupling of two ferromagnets separated by a tunneling barrier. *Physical Review B* 39: 6995–7002.

Smolenskii, G. A., Bokov, V. A., Isupov, V. A. et al. 1984. *Ferroelectrics and Related Materials*, New York: Gordon and Breach.

Spatz, J. P., Mossmer, S., Hartmann, C. et al. 2000. Ordered deposition of inorganic clusters from micellar block copolymer films. *Langmuir* 16: 407–415.

Sun, J. Z. 1998. Thin-film trilayer manganate junctions. *Philosophical Transactions of the Royal Society of London Series A—Mathematical Physical and Engineering Sciences* 356: 1693–1711.

Sun, J. Z. 2001. Spin-dependent transport in trilayer junctions of doped manganites. *Physica C—Superconductivity and Its Applications* 350: 215–226.

Szafraniak-Wiza, I., Hesse, D., and Alexe, M. 2008. *Nanosized Ferroelectric Crystals*, Cambridge, U.K.: Woodhead Publishing, Limited.

Tagantsev, A. K. and Gerra, G. 2006. Interface-induced phenomena in polarization response of ferroelectric thin films. *Journal of Applied Physics* 100: 051607.

Tagantsev, A. K., Pertsev, N. A., Muralt, P. et al. 2002. Strain-induced diffuse dielectric anomaly and critical point in perovskite ferroelectric thin films. *Physical Review B* 65: 012104.

Tagantsev, A. K., Gerra, G., and Setter, N. 2008. Short-range and long-range contributions to the size effect in metal-ferroelectric-metal heterostructures. *Physical Review B* 77: 174111.

Urban, J. J., Spanier, J. E., Lian, O. Y. et al. 2003. Single-crystalline barium titanate nanowires. *Advanced Materials* 15: 423–426.

Valasek, J. 1921. Piezo-electric and allied phenomena in Rochelle salt. *Physical Review* 17: 475–481.

Waghmare, U. V. and Rabe, K. M. 1997. Ab initio statistical mechanics of the ferroelectric phase transition in $PbTiO_3$. *Physical Review B* 55: 6161–6173.

Weil, R., Nkum, R., Muranevich, E. et al. 1989. Ferroelectricity in zinc-cadmium telluride. *Physical Review Letters* 62: 2744–2746.

Xu, Y. 1991. *Ferroelectric Materials and Their Applications*, Amsterdam, the Netherlands: North-Holland.

Ye, H., Xu, Y. H., and Mackenzie, J. D. 2000. Semiconducting ferroelectric SbSI quantum dots in organically modified TiO_2 matrix. *Proceedings of SPIE* 3943: 95–101.

Zembilgotov, A. G., Pertsev, N. A., Kohlstedt, H. et al. 2002. Ultrathin epitaxial ferroelectric films grown on compressive substrates: Competition between the surface and strain effects. *Journal of Applied Physics* 91: 2247–2254.

Zhuravlev, M. Y., Jaswal, S. S., Tsymbal, E. Y. et al. 2005b. Ferroelectric switch for spin injection. *Applied Physics Letters* 87: 222114.

Zhuravlev, M. Y., Sabirianov, R. F., Jaswal, S. S. et al. 2005a. Giant electroresistance in ferroelectric tunnel junctions. *Physical Review Letters* 94: 246802.

7

Superconducting Weak Links Made of Carbon Nanostructures

Vincent Bouchiat
*Centre National de la
Recherche Scientifique*

7.1 Introduction

It has been known since the 1960s (De Gennes and Guyon 1963) that charge carriers emitted from a superconductor can propagate in a non-superconducting medium, provided that the electron wave functions of the two electrons originating from the superconductor (the so-called Cooper pairs) keep their correlation during their diffusion outside the superconductor. This process requires the absence of defects in the propagating medium that break time-reversal symmetry, such as magnetic impurities. These correlated electrons form an evanescent state that "bleeds" from the superconductors into the non-superconducting electrode on mesoscopic distances. The physics of these evanescent states, known as the "superconducting proximity effect" depends on many parameters (the coherence length of the superconductor, the contact barrier at the interface, the temperature, the electron diffusion length, etc.).

It appeared then possible to sever a superconducting electrode with non-superconducting elements (a narrow constriction, a normal metal, a thin insulating barrier, known as "weak links") while keeping a supercurrent flow. As tunneling processes are known to depend exponentially on the thickness of oxide barriers, controlling the structure and the geometry of weak links down to the atomic scale appears to be of capital importance if one wants to get reproducible devices. Therefore, improving the miniaturization and the crystalline quality of these weak links was clearly identified as a critical issue. Among the possible candidates are sp^2-hybridized carbon nanostructures (i.e., graphenes, carbon nanotubes [CNTs], or fullerenes), which, among other advantages, offer good control of the crystallinity and provide ideal media to control a supercurrent flow. Quantum confinement of electronic states within the carbon nanostructure generates sharp variations of the electronic density of states. Tuning the weak link chemical potential with an electrostatic gate allows to adjust the weak link transparency.

The purpose of this chapter is to propose a basic overview of the principles that govern the physics of these carbon based devices and illustrate them by presenting experimental demonstrations and some applications.

7.2 Superconducting Transport through a Weak Link

7.2.1 Density of Electronic States in a Superconductor

At energies around the chemical potential μ and below the energy gap Δ, a superconductor (noted in the following by the symbol S) only has a single ground state, which depends on a macroscopic phase. In a bulk superconductor, all electrons are in the same quantum state characterized by a global phase φ. The charge carriers are condensed in a bound state of two electrons, the so-called Cooper pair, which has a charge of $-2e$ (Figure 7.1a). The ground state wave function depends on the phase φ and can be written as

$$\Delta \approx \sqrt{n_s}\exp(i\varphi), \qquad (7.1)$$

where n_s is the total number of Cooper pairs.

Δ is the "superconductor order parameter" and specifies the choice for the quantum phase φ. There are no other states present in the interval $[\mu - \Delta, \mu + \Delta]$ because it takes an energy 2Δ to break a Cooper pair. An electron in this energy window cannot be injected in a superconductor alone, generating a "gap" in the transmission of charge carriers at a normal/superconductor metal interface.

7.2.2 Coherent Transport at a Normal/ Superconducting Interface

At the interface between a normal metal (N) and a superconductor (S), an electron at the Fermi level can nevertheless be converted into a Cooper pair with zero momentum if a hole (charge $+e$) is reflected with energy $-\varepsilon$ and opposite momentum, conserving the charge, energy, and momentum. The reflected hole acquires an additional phase factor φ, the order parameter phase. This

conversion of an electron–hole pair into a Cooper pair is a process known as the Andreev reflection (Figure 7.1b). It dominates the charge transport at low bias through the interface. This Andreev reflection can lead to a supercurrent transport if correlated charge carriers are collected by a second superconductor placed in a series. The process is still effective at the interface between superconducting electrodes and nanoscale constriction, which we will refer to as a "weak link."

7.2.3 A Brief Introduction to the Josephson Effect

The order parameter characterizing the superconducting state varies on a spatial scale equal to $\xi = h v_F/\Delta$, where v_F is the Fermi velocity. The quantity ξ, which is a measure of the size of Cooper pairs, is called the coherence length. At the interface with an insulator, a metal, or a nano-object, the superconducting order parameter $\Delta(\mathbf{r})$ (its wave function) weakens gradually close to the interface on a characteristic scale, which is on the order of ξ.

When two superconductors with phase φ_1 and φ_2 are separated by a weak link (see Figure 7.2), they interact through direct tunneling or by the Andreev reflection process presented in the previous paragraph. This sandwich structure is called a Josephson junction. It is dominated by the interaction energy between the two superconductors. In 1962, Brian Josephson predicted that due to the interaction between the two superconducting condensates in close proximity, a non-superconducting nanostructure can let flow a superconducting current I, the value of which depends on the phase twist between the two condensates (Figure 7.2) (Josephson 1962). For a thin insulating tunnel barrier, he has shown that the current has the following form:

$$I = I_c \sin(\varphi_2 - \varphi_1). \qquad (7.2)$$

This surprising prediction (Josephson 1962) was rapidly experimentally confirmed, leading to a new field in superconducting electronics. Until the mid-1990s, such superconducting "weak links" were either made of insulating barriers (the so-called Josephson tunnel junctions, at the origin of all the recent advances in superconducting Qubits), narrow diffusive metallic

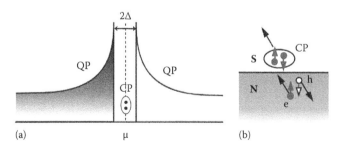

(a)   (b)

FIGURE 7.1 (a) Density of state of a superconducting electrode: all the electrons are in the ground state wave function at the chemical potential μ. (b) Schematics depicting the Andreev reflection. Describing coherent transport at a normal/superconducting interface below the gap. An electron (e) impinging the interface between a normal conductor (N) and a superconductor (S) produces a Cooper pair (CP, oval) in the top superconducting electrode and a retro-reflected hole (h) in the normal conductor. Vertical arrows indicate the spin band occupied by each particle.

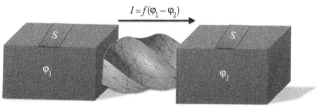

FIGURE 7.2 Schematics of a Josephson weak link. In this device, two superconducting electrodes, characterized by their phase φ_1 and φ_2 are separated by a weak link (symbolized by a twisted solid). The superconducting current I flowing through the weak link is a given function f of the "phase twist" i.e., the phase difference $\varphi_1 - \varphi_2$. This current–phase function f exhibits a 2π-periodicity and is highly non linear. It is the main characteristics of a Josephson weak link.

or semiconducting wires, or point contacts inserted between bulk electrodes (Likharev 1979). The first realization of a ballistic weak link with a limited number of conduction channels and controlled at the atomic scale came in 1997, with the measurement transport through gold atomic contacts in close vicinity of superconducting aluminum contacts (Scheer et al. 1997). In the last 5 years, it has been shown that carbon nanostructures can provide another class of weak links of good structural quality, for which strong quantum confinement offers gate tunability.

7.2.4 Normal Electron Transport through an Atomic or Molecular Conductor

At the atomic level, electron transport occurs by hopping through the atomic or molecular orbitals that are quantized in the case of a laterally confined nanostructure. They provide discrete conducting channels for electron transport (Imry and Landauer 1999). At the interface with a nanostructure, these contributions are related to the overlaps between molecular orbitals and the electronic wave functions in the metal. The physics of this quantum transport follows the description of Rolf Landauer from the 1970s, which refers to treating incoming charge carriers as plane waves traveling through "conduction channels" that are the eigen-modes for electronic propagation within the nanostructure. The electronic wave is partially transmitted and reflected at the interface. The theoretical work by Landauer and subsequently confirmed by numerous experiments have shown that the conductance of an electronic mesoscopic conductor is indeed divided into a discrete sum over all conduction channels i placed in parallel:

$$G = \sum_i G_i = \sum_i G_0 T_i = \frac{2e^2}{h} \sum_i T_i. \qquad (7.3)$$

Each channel bears a quantized conductance G_0 called the quantum of conductance given by a ratio of two fundamental constants, $G_0 = 2e^2/h = (12.8\,k\Omega)^{-1}$. In practice, each conduction channel is partially coupled to the modes of the reservoirs. This induces a reduction of the conductance of a dimension-less prefactor T_i with $0 < T_i < 1$. The origin of these reduction factors is essentially of technologic origin and can be optimized by choosing the right interface materials as shown below. Nevertheless, even for a perfectly transmitting interface ($T = 1$), the normal transport through a nanoscale conductor is limited to a resistive transport of $12\,k\Omega$ per channel, the resistance being localized at the interface between the reservoir and the leads (see Figure 7.8a for an example).

In the case of coupling to superconducting electrodes, the Landauer formula 7.3 is no more valid: it is possible to cancel this contact resistance and obtain a zero-resistance electron flow through an atomic-sized contacts. This simple observation might have important technical consequences in the future of nanoelectronics, for example, to limit heat dissipation in the case of the ultra-high density integration of nanodevices involving few conduction channels.

7.2.5 Superconducting Transport through Andreev Bound States

As shown in Section 7.2.2, the microscopic transfer of charges at a superconducting interface arises by an electron–hole process known as the Andreev reflection. In a weak link, there are two normal/superconducting interfaces back-to-back (S/weak link and weak link/S). The two proximity effects on each side percolate and give rise to a superconducting current flow through the weak link. Furthermore, one has to consider the confinement in the weak link of the electron and holes created by the successive Andreev reflections at the interfaces: the electron and the holes remain spatially and energetically confined (Figure 7.3), thus defining a quantum well. The eigenstates of this well, the so-called Andreev bound states, require constructive interferences to propagate. The electron state, once converted into its hole conjugate, is converted back into the same electron state with the same quantum phase. This defines a quantization condition that depends on the phase difference between the superconductors: they shift the electron–hole Andreev wave function by their phase $\exp(i\varphi)$.

When no scattering occurs between the superconductors, there are two bound states per conductance channel with the energies $E_\pm = \pm\Delta\cos(\phi/2)$. More realistically, in the presence of a finite transmission T_i, their energies become $E_\pm = \pm\Delta\sqrt{1-\sin^2(\phi/2)}$ (Goffman et al. 2000). At low temperatures, only the Andreev-bound state lowest in energy is occupied. The Josephson superconducting currents carried by the different channels add their contribution to each other and for the superconducting current give

$$I = -\frac{\partial \sum_i E_-(\phi, T_i)}{\partial \phi} = \frac{e\Delta}{2\hbar}\sum_i T_i \frac{\sin\phi}{\sqrt{1 - T_i\sin^2(\phi/2)}}. \qquad (7.4)$$

This "current-phase" relation, linking the supercurrent I to the phase shift ϕ across the ballistic weak link is a characteristic of Andreev processes and more generally will be the fingerprint of any superconducting weak link (Golubov et al. 2004). From that last formula, one can see that the "critical" current I_c, defined as the maximum supercurrent that can flow through the weak link with a length shorter than the superconducting coherence length, is quantized. Its maximum value is equal to

$$I_c = \frac{Ne\Delta}{\hbar}. \qquad (7.5)$$

In the case of a nanodevice that can be modeled by N ballistic conduction channels placed in parallel, the quantum interference of the Andreev electrons bouncing back and forth between the two S/N and N/S serial interfaces at the junction boundaries is also at the origin of resonance currents called "Multiples Andreev reflections" (Klapwijk et al. 1982), which occur at voltages below the superconducting gap Δ (see examples in Figure 7.18). Due to resonating energy conditions they occur at bias

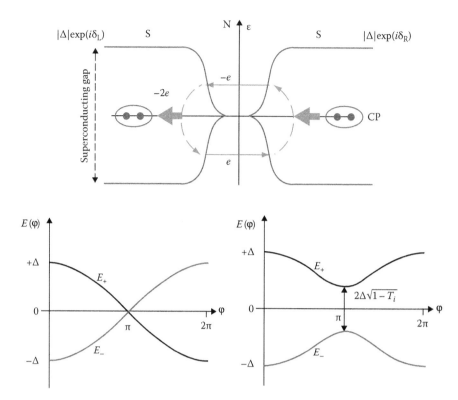

FIGURE 7.3 Top: Schematics depicting superconducting transport through a normal weak link N sandwiched between two superconducting electrodes. An electron is reflected as a hole at the interface with the left superconductor and the hole is reflected back as an electron on the right interface. The net result is a transfer of one Cooper pair through the S/N/S junction. Bottom: Phase dependence of the Andreev bound states energy levels ($\pi = \delta_r - \delta_l$). Left: For a perfectly transmitted channel without scattering potential ($T_i = 1$). Right: For a partially transmitted channel $T_i < 1$. The lower energy state is the only occupied state at low temperature.

voltages corresponding to the sub-harmonics of the superconducting gap ($1/n^*(2\Delta/e)$), where n is an integer. These multiple Andreev Reflections, which position are independent of the weak link density of state but are strongly affected by scattering, will then provide the most convincing experimental signatures of a ballistic weak link.

7.2.6 Introduction to Low-Dimensional sp² Carbon Structures

Low-dimensional allotropes based on carbon nanostructures with sp^2-hybridized bonds (Figure 7.4a), such as CNTs, graphenes, and fullerenes, have been extensively studied in the last decade (see appropriate chapters in the Handbook for immediate reference). Their exceptional electronic properties could provide alternative solutions in many fields of electronics including industrial applications for logic. They also behave as ideal media for transmitting superconducting charge carriers for several reasons.

First, they experimentally provide a quasi-perfect crystalline semiconducting lattice on a rather large scale length (typically a fraction of a micron, which corresponds to several thousands of C–C sp^2-hybridized bonds in every direction). Second, the inertness and absence of dangling bonds on the carbon surface solves the problem of surface termination (passivation), which is usually found in almost all other systems including

all semiconductors candidates. This allow to make atomically clean edges and boundaries. This important feature enables the fabrication of reliable nanoscale metal/sp^2 carbon contacts with reproducible transparency.

7.2.7 Gate Control of the Weak Link Transparency

As we will detail in the following paragraph, the quantum confinement created by the low dimensionality of carbon nano-objects is at the origin of a sharp dependence of the electron density of states with the chemical potential of the weak link (Figure 7.4a). The resulting Andreev bound states will then be directly affected by this confinement and will create gate-tunable "filters" for the coupling between the two superconducting electrodes (Figure 7.4b). In the superconducting state, a supercurrent will flow if an Andreev bound state is aligned with respect to the Fermi level of the leads as depicted in paragraph 7.2.5. A voltage V_g applied on an electrostatic gate allows for a shift in the chemical potential of the weak link, which is equivalent to translating the confined energy levels (Figure 7.4b) with respect to the Fermi level of the leads. This will in turn tune the maximum superconducting current through the junction, thus implementing a "gate tunable Cooper pair filter." In the following section, we provide details of the typical cases for the different types of nanocarbon structures considered.

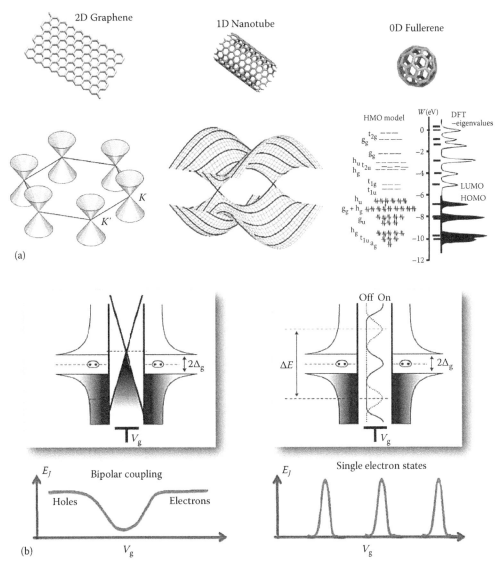

FIGURE 7.4 (a) Crystalline structure of the three low-dimensional phases of sp^2 carbon allotropes (top), and its associated densities of states (bottom). Left: in Graphene, the two dimensional density of states is described by a set of two "Dirac" cones that intersect the Brillouin zone at points K and K'. These cones provide a linear gap-less dispersion curve. Middle: Metallic single-walled carbon nanotube. The periodic boundary conditions generated by the rolled-up graphene sheet generate solutions symbolized by thick black lines. For selected chiral angles and diameter of the nanotube, these lines intersect the K and K' points of the Brillouin zone, leading to metallic nanotubes which behave as perfect 1D channels. Right: C60 and derivated fullerenes. The energy spectrum shows discrete states, whose positions could be derived through molecular orbitals (HMO, left side) or density functional theory (DFT, right side of axis). The molecule shows a gap between the occupied and empty states of about 1 eV. (b) "Semiconducting" representations of the energy levels of a Superconductor/Nano-carbon/Superconductor junctions for two dimensional (left) and zero dimensional (right) weak links. The chemical potential of the weak link can be translated by adjustment of a gate voltage V_g which allows to tune the transmission of the Cooper pairs through the weak link. Left: For a 2D junction (i.e., graphene-based weak links), the transport of Cooper pairs can either be transmitted by the valence or the conduction band, leading to an ambipolar semiconductor-like transmission function. At the charge neutrality point (position where the valence and conduction cones intersect, superconducting transport is minimal and is mediated through evanescent waves. Right: for a quantum dot junction (short nanotube junction or fullerene), superconducting transport occurs when a single electron or single hole energy levels is aligned with the Fermi level of the electrode. Such discrete energy spectrum of the weak link acts as a tunable filter for superconducting transport and leads to a series of peaks in the gate dependence of the maximum superconducting current.

7.3 Superconducting Transport in a Carbon Nanotube Weak Link

7.3.1 Introduction to Carbon Nanotube Devices

CNTs are cylinders of carbon of about one nanometer in diameter and up to several micron in length, which can be seen as a rolled-up sheet of sp^2 carbon "graphene." Their cylindrical shape and original 1D electronic structure (for a review, see Charlier et al. 2007) constitute the unique experimental realizations of a quasi-perfect 1D electronic system.

Depending on the chiral angle of the crystalline direction with the cylinder axis, CNTs can exhibit either metallic or semiconducting properties. For the sake of simplicity, one assumes in the following that the nanotube is single-walled and is of a metallic variety (Figure 7.4).

In practice, semiconducting nanotubes are so heavily doped by the metal contacting electrodes that they can also behave in a similar fashion.

A metallic single-walled CNT has two channels of conduction electrons available for electron conduction. According to the Landauer Formula (see Equation 7.3), when perfectly connected to a normal state electrode, it has a resistance that equals the fundamental constant called resistance quantum $h/4e^2 = 6.5\,k\Omega$. Due to the mismatch of the electron orbitals in the nanotube and in the metal electrodes, the transmission is generally not perfect. The result is equivalent to a tunnel barrier at the contact ($T < 1$). This induces a probability of transmission between the electrode and the nanotube that can be less than unity. These two semi-transparent electronic junctions placed in a series are in all means reminiscent to what is occurring for photon beams in a "Fabry–Perot" cavity (Liang et al. 2001) in which a photon is trapped between two semi-reflective mirrors. Such a cavity has internal resonating modes that correspond to stationary waves given by integer multiples of the half-wavelength and the resonating mode width (invert of the so-called finesse of the cavity) increases linearly with the mirror

transparency. Similarly, the CNT junction behaves as a "quantum dot" that the gap between levels (normal modes) increases proportionally to $1/L$ where L is the nanotube length. The width of the levels is proportional to the transmission of contacts. For a short nanotube junction, we obtain discrete levels when the nanotube is completely decoupled from the electrodes like the energy level of an isolated molecule (see the case of C60 Figure 7.4a). These levels widen a bit when the quantum dot interacts with contacts. The position of the nanotube levels with respect to the energy level of the contacts can be adjusted by varying the electrostatic voltage applied to a gate electrode. The transmission of electrons is maximal when the energy level of the quantum dot coincides with the contacts. It thus realizes a molecular transistor (Figure 7.9, left) in which junction transparency can be periodically tuned with a voltage applied to the electrostatic gate (Figure 7.6).

7.3.2 Experimental Realizations of Carbon Nanotube Weak Links

Progress in the late 1990s in improving the contact of CNTs to superconducting electrodes showed that such junctions can exhibit superconducting fluctuations that can be gate controlled (Morpurgo et al. 1999) and that they can accommodate a superconducting current (Kasumov et al. 1999). In 2006, it was then shown that such junctions behave as gate-controlled Josephson junctions (Jarillo-Herrero et al. 2006, Jørgensen et al. 2006). During this period, fabrication techniques have been optimized to generate low-resistive contacts at the metal/sp^2 carbon interfaces. They required the use of specific metals for connection such as titanium (Jarillo-Herrero et al. 2006) or palladium (Cleuziou et al. 2006, Javey et al. 2003) that both provide good wetting properties onto the sp^2 carbon layer and minimize the Schottky barrier that could arise from the differences of the electronic work functions at the metal/carbon interface. More subtle effects linked to the hybridization of the metal orbitals to sp^2 carbon (Nemec et al. 2006) could explain the differences

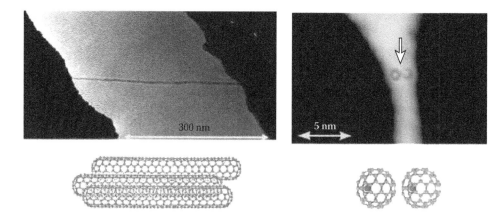

FIGURE 7.5 Micrographs made using a transmission electron microscope (TEM) of superconducting junctions, involving a carbon nanotube bundle, left and for a fullerene dimer formed by two C82 molecules in series, right. The weak links are free-standing over a slit and bridge superconducting electrodes deposited on a silicon membrane. This allows TEM imaging and electrical measurements on the same samples. (After Kasumov, A.Y. et al., *Science*, 284, 1508, 1999. With permission.)

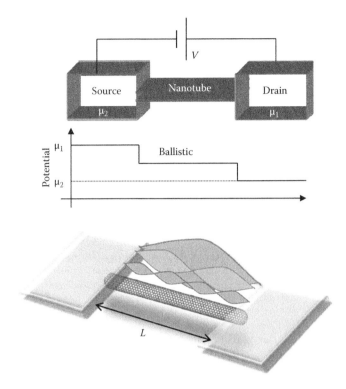

FIGURE 7.6 Top schematic of electron transport through a "defect-less" carbon nanotube contacted with two metallic electrodes. Electron transport is ballistic along the nanotube: the voltage drop $V = \mu_1 - \mu_2$ is occurring at the metal/nanotube contacts which act as tunnel barriers partially transmitting the electron. Bottom: For low resistance barrier contacts, the device act as a Fabry–Perot interferometer in which standing electron waves are separated by the energy difference $\Delta E = hv_F/2L$.

of interface resistances that have been reproducibly observed between these different metals (see Figure 7.7).

While not superconducting by itself, the Pd or Ti interfacing layer is made thin enough (a few nanometers) so that it lets the superconductivity induced by a covering top layer made of aluminum (Jarillo-Herrero et al. 2006) or niobium (Pallecchi et al. 2008)

establish through the whole sandwich by the proximity effect. Using those materials, high transparency contacts with almost transparency T close to unity (Javey et al. 2003) can be reproducibly achieved by metal evaporation thus placing connected nanotube junctions in the high coupling limit (Liang et al. 2001). Advances in synthesis methods of nanotubes (in situ grown by CVD), which allow nanotubes to be connected during their growth onto platinum contacts (Cao et al. 2005) are another promising fabrication route. This last method makes clean suspended nanotube junctions, which in the near future could make possible the coupling of nanomechanical vibration modes to superconducting charge transport, as has been recently observed with normal charge carriers, by using non-superconducting metal electrodes as contacts (Lassagne et al. 2009, Steele et al. 2009).

7.3.3 Nanotube Quantum Dot Connected to Superconducting Electrodes

In a CNT weak link, the device considered is a portion of a nanotube with a length L connected by the two superconducting contacts (see Figure 7.8).

At one dimension, a defect along the channel cannot be overcome by the diffusion of the incoming electrons like it occurs in a 2D or 3D system. Therefore, the slightest defect along the nanotube generates back scattering which cancel the transmission of a supercurrent by Andreev bound states. It is then extremely important to realize the injection of Cooper pairs in a perfectly clean system. The nanotube portion length L is usually chosen not to exceed 300 nm in order to limit the occurrence of these structural defects. In such a case, a ballistic transport is reached (i.e., the voltage drop is localized at the contacts, see the top of Figure 7.6). The sub-micron length and nanometer diameter of the nanotube junction makes the intrinsic capacitance C of the junction very small, creating a Coulomb blockade effect (Beenakker and van Houten 1992). A charging energy $e^2/2C$ contributes to their energy and adds to the kinetic energy which leads to the Andreev bound states becoming

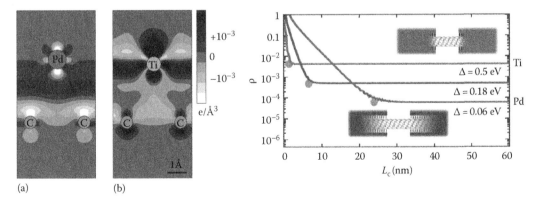

FIGURE 7.7 Left: Cross section showing the charge density redistribution at a metal/carbon nanotube interface for Palladium (a) and Titanium contacts (b). These results are obtained using an ab-initio density functional theory. The titanium contact show a much stronger hybridization with the carbon surface compared to Palladium. Right: Contact reflection coefficient $r = 1 - t$, (where t is contact transmission) as a function of the contact length L_c, the transparency first increases linearly with the contact length then saturates at a constant value independent on the contact length. Their results shows that the optimal metal/carbon interface transparency is obtained for a contact combining low hybridization with a large contact length. (Adapted from Nemec, N. et al., *Phys. Rev. Lett.*, 96, 076802, 2006. With permission.)

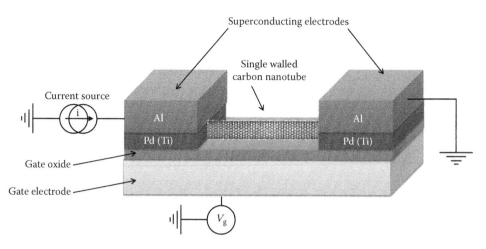

FIGURE 7.8 Schematic diagram of a carbon nanotube superconducting transistor. The contacting electrodes are made of a bilayer of aluminum (for the source of superconductivity) and underlayer of few nanometer thick film either made of palladium or titanium to increase contact transparency. These latter materials are known for their good contact abilities on sp^2 carbon layers (see Figure 7.7). The highly doped silicon substrate is connected to a voltage source V_g and is used as a "backgate" electrode.

$$E_{e,h} = \frac{e^2}{2C}(\pm1 - n_g)^2 \pm \hbar v_F k, \qquad (7.6)$$

where n_g is the gate-induced charge on the nanotube [$n_g = C_g V_g/e$]; electrostatic degeneracy between charge states occurs at zero gate bias.

The wavevector k is quantized by the Andreev boundary condition at the superconducting interfaces in multiples of $1/L$. In the presence of the Coulomb repulsion in the Quantum dot, the Andreev bound states described in Section 7.2.5 split their energy in a sum of the charging energy term and the confinement energy term, which for the electron states give

$$E_e = \frac{e^2}{2C}(1 - n_g)^2 + \frac{\hbar v_F T}{2L}(\phi + \psi_e(n_g, \phi)), \qquad (7.7)$$

and for hole states

$$E_h = \frac{e^2}{2C}(1 + n_g)^2 + \frac{\hbar v_F T}{2L}(\phi + \psi_h(n_g, \phi)), \qquad (7.8)$$

where $\psi_h(n_g, \phi)$ is a gate-dependent phase. In most experiments, the charging energy e^2/C and quantization energies are of the same order of magnitude (typically 1–10 meV). The eigenstates are then mixed states of charge and Andreev bound states. Close to a degeneracy point ($n_g = 1/2$), the CNT Andreev bound states become resonant with the Fermi level of the leads, thus maximizing the supercurrent flow. As this energy separation can be tuned by applying a gate voltage V_g to the nanotube, we are able to tune the maximum supercurrent (referred to as "switching current") with the gate voltage.

The gate dependence of this last quantity is then periodic (see Figure 7.9, left) and directly reflects the conductance variations observed in the normal state (linked to the Fabry–Perot-like oscillations described in the previous paragraph). By varying the

electrostatic voltage V_g applied to the nanotube, it is possible to shift the position of these energy levels relative to the energy levels of the electrons in superconducting contacts (Figure 7.4b, right). When levels coincide, the maximum supercurrent that can pass before obtaining a resistive state (called "current jump") reaches its maximum. On the opposite, if Andreev bound states spectral weight decrease due to levels misalignment, the current will then be minimal. It has been shown (Cleuziou 2006) that even in the case of abscence of Andreev bound states, a tiny supercurrent (of the order of tens of pico-amperes) can be maintained through the CNT probably due to higher order cotunneling processes. It has been experimentally shown by Jarillo-Herrero et al. (2006) that the maximum superconducting current can indeed be controlled by the gate voltage (Figure 7.9).

Therefore, such CNT junctions implement superconducting transistors for which the switching current can be tuned with the gate voltage over a large dynamic range (typically up to 2–3 orders of magnitude, depending on the coupling; see Figure 7.9 right). The current–voltage curves show some marked hysteresis especially in the off-state. The origin of this hysteresis (Gang et al. 2009), which is usually attributed to the existence of a capacitance in parallel with the junction for tunnel weak links (Barone and Paterno 1982), is still debated in the case of these carbon weak links in which the capacitance is extremely small. Hysteresis could rather be associated with the Joule dissipation that occurs in the weak link after switching to the normal state as it has been recently observed in metallic junctions (Courtois et al. 2008). Heat dissipation within the tiny volume of the junction could delay the re-trapping of the weak link in the superconducting state once the current bias is decreased. This occurs especially when the nanotube is in the high resistance state (Off State, see Figure 7.9).

The maximum superconducting current that can be carried by a single-walled CNT is given by Equation 7.5 and is equal to $2\Delta/\hbar \sim 30$ nA for the two conducting channels of a metallic

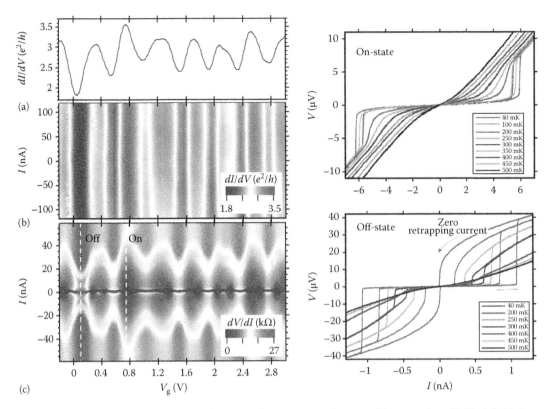

FIGURE 7.9 Electron transport through a transparent carbon nanotube junction as a function of the gate-voltage V_g. Top left: (a) Gate modulation of the nanotube conductivity dI/dV at zero bias in the normal state. (b) Map of the conductivity for current bias I. At resonance, the nanotube conductance is close to the theoretical maximum $4e^2/h$ thus corresponds to the high transparency limit ($t \sim 1$). Gate oscillations are broadly distributed in gate due to the strong hybridization of the nanotube levels with the contacting leads. (c) Differential resistivity dV/dI versus current I and V_g in the superconducting state. Darkest zones around zero bias corresponds to the supercurrent flow which are occurring to the same gate voltages as the high conducting state for non-superconducting electrodes. Right column: Voltage–Current characteristics of a carbon nanotube superconducting transistor measured at low temperature, showing a modulation of the superconducting current for gate voltage corresponding respectively to the on and off state (depicted by the white dotted lines on the resistance map shown in left). The off-state show a strong hysteresis which completely vanished in the on-state.

single-walled CNT connected to Aluminum electrode. In early experiments, measured switching currents were of a few nano-amperes, typically 10%–20% of the expected value. The discrepancy between the predicted and the observed values have been attributed to the sensitivity of the superconducting current to external perturbations, such as micro-wave photons emitted by thermal fluctuations and traveling along the lines. A better control of the electromagnetic environment by using LC and RC filtering on the measurement lines (Jarillo-Herrero et al. 2006) and by implementing locally dissipative elements, such as on-chip resistors close to the nanotube junctions (Jorgensen et al. 2007), has proven to be helpful to stabilize the superconducting current and increase the measured value of the switching current.

7.3.4 High-Frequency Irradiation of Nanotube Weak Links

According to what is known as the AC Josephson Effect, the phase twist across the weak links varies in time at the frequency given by the characteristic frequency of the coupling. Upon irradiation with radio frequency (RF) electromagnetic waves (Barone and Paterno 1982), superconducting weak links are known to

develop a series of voltages steps that are so precisely positioned that they are used to define the voltage standards (Hamilton 2000). In the case of the irradiation of a single quantum channel, a peculiar behavior is expected for the RF-irradiated Andreev states (Averin and Bardas 1995).

Experimental measurements (Cleuziou et al. 2007) showed that the voltage steps across the nanotube weak link follow a completely different behavior upon the RF irradiation power depending on if the nanotube is placed either in the "on" or "off" state (see Figure 7.10). This behavior appears to be linked to the existence of the hysteresis found in the direct current (DC) measurements (Figure 7.9) and is discussed in the previous paragraph. They have been associated with the change of dissipation (Liu et al. 2009) that occurs when the nanotube levels are aligned or misaligned with the level of the leads. The RF properties of the superconducting carbon-based junction is expected to be very interesting principally because of their tiny size, which induces an extremely low intrinsic capacitance C, thus leading to extremely high cut-off frequencies $(RC)^{-1}$ that could exceed tens of GHz (Burke 2004). The coupling of these weak links to superconducting resonators will allow for the assessment of their additional RF properties in the near future.

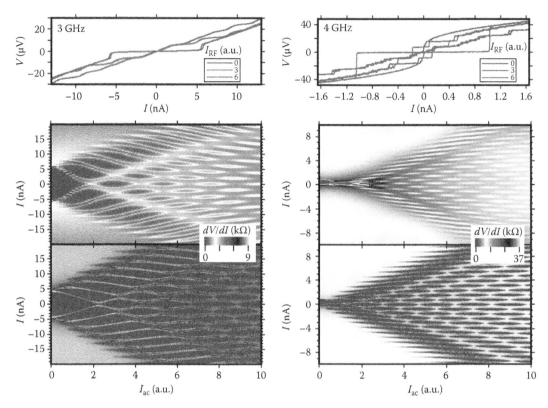

FIGURE 7.10 Carbon nanotube superconducting transistor behavior upon radio-frequency (RF) irradiation in the on and off state (respectively shown in the left and right columns). Top curves: Voltage current characteristics at several RF amplitudes showing the typical constant voltage steps, known as "Shapiro steps." Hysteresis in these steps is seen on the "on" state and vanishes in the "off" state. Middle panels: differential resistance dV/dI maps for increasing RF power showing a different behavior in the off and on state. Bottom panels: Numerical simulation of the data of the middle panels using the RSJ (resp. RCSJ) non linear model for the on (resp. off) states.

7.4 Nanotube-Based Superconducting Quantum Interferometers

7.4.1 Quantum Interference with Single Nanotube Weak Links

An interferometer is an apparatus that measures the phase differences between two wave components. Interferometers work by splitting a wave into two components, sending them off along different paths, and then recombining them to record an interference pattern. The pattern is dependent on the phase difference arising from changes in the conditions along their paths. A superconducting interferometer (referred to by its acronym "DC-SQUID" [Clarke and Braginski 2004], which stands for direct current superconducting quantum interferometer device) achieves a kind of "Young slits" experiment with Cooper pairs, the slit being implemented by 2 weak links placed in parallel.

The coherence of the photon beam is provided here by the coherence of the Cooper pair supercurrent. Such a device was first implemented in the 1960s by using oxide-based tunnel weak links (Jaklevic et al. 1964), which rapidly became ubiquitous in the field of superconducting electronics: SQUIDs have been successfully used as ultra-sensitive magnetometers as well as amplifiers of low-level signals and low-impedance sources (Clarke and Braginski 2004).

The main feature of this device is that the maximum superconducting current can be periodically modulated by a tiny magnetic flux threaded by the loop, with a period equal to the flux quantum ϕ_0 $h/2e$ (Figure 7.11, right). We briefly recall in the next paragraph the principle of operation of a DC-SQUID.

The two currents circulating in each weak link interfere with each other and each weak link induces a phase shift $\Delta\phi_i$, according to Equation 7.2.

The total supercurrent is therefore written as the sum of two currents flowing in each branch.

If one assumes a sinusoidal current-phase relation, one obtains

$$I = I_{c1} \sin\left(\Delta\phi_1\right) + I_{c2} \sin\left(\Delta\phi_2\right). \tag{7.9}$$

Furthermore, an additional controlled phase (analogous to a geometric phase) can be introduced by applying a magnetic flux Φ_{ext} through the loop formed by the two weak link branches. Because of the uniqueness of the superconducting wave function, Φ_{ext} couples the phase shifts induced by each of these junctions in the two branches

$$2\pi\left(\frac{\Phi_{ext}}{\Phi_0}\right) = \Delta\phi_1 - \Delta\phi_2 \,(\mathrm{mod}.2\pi). \tag{7.10}$$

In practice, the SQUID is current-biased and one measures the maximum superconducting current before the onset of a finite voltage

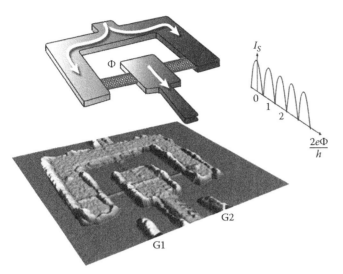

FIGURE 7.11 Schematic diagram of a superconducting quantum interferometer (SQUID) based on carbon nanotube weak links (top) aligned with an atomic force micrograph of a real device. (Adapted from Cleuziou, J.-P. et al., *Nat. Nanotechnol.*, 1, 53, 2006.) Thanks to a peculiar fork geometry design, the weak links in each branch can be made of two portions of the same carbon nanotube generating two independent quantum dots. The phase shift between the two currents flowing in each arm (white arrows) is adjusted by the magnetic flux *F* threaded by the loop. The two electrodes, G1 and G2 allow the application of locally addressed gate voltages which tune independently the transparency of each portion of the nanotube according to the principles of the carbon nanotube junctions shown above.

across the device. This switching current is directly related to the critical current of the SQUID, which can be obtained by computing the maximum value of the solutions of Equations 7.9 and 7.10.

$$I_c = \sqrt{\left(I_{c1} - I_{c2}\right)^2 + 4 I_{c1} I_{c2} \, \cos^2\left(\frac{\pi \Phi_{\text{ext}}}{\Phi_o}\right)}. \qquad (7.11)$$

Due to thermal or quantum fluctuations of the phase across the weak links, the measured switching current does not reach that critical value I_c but its magnetic flux dependence still follows the relation in Equation 7.11. The quantum interference, therefore, produces periodic oscillations of the current as a function of external magnetic flux, analogous to the bright fringes observed in an experiment of Young's slit and whose period is equal to the flux quantum $h/2e$.

At the nanometer scale, a single CNT can be used to implement the two weak links that are placed on both arms of the interferometer by using a forked geometry (Figure 7.11).

The two branches implement two nanotube quantum dots; their behavior has been detailed in Section 7.3.3. It is therefore possible to modulate the quantum interference by independently adjusting the intensity of coupling in each quantum dot (see the three limit cases in Figure 7.12). Such electrostatic control of a DC-SQUID is extremely useful, both to optimize its operation and its coupling to local magnetic fields to study the detailed

physics of the molecular Josephson junction. More generally, it is an experimental model system for the study of the variation of the phase of charge carriers across a quantum dot. Many physical phenomena such as an equal number of electrons in the nanotube (including those related to interactions between the spins of electrons) have been predicted and may now be studied.

The integration of these nanotubes between superconducting electrodes opens the way for the realization of superconducting circuits operating at radio frequencies, particularly useful to follow the time evolution of magnetization of individual nano-objects. They could afford to explore how quantum states of magnetization of a molecule magnet can couple to those generated in a superconducting circuit (which are the basis of superconducting quantum bits).

7.4.2 Tunability of the Phase Shift across Nanotube Weak Links

For an odd electron number of charge sitting on the nanotube, a CNT weak link can behave as a quantum dot that has a nonzero spin state. It thus behaves as a "spin impurity" seen from the superconducting electrodes. Quantum dots populated with a odd number of electrons are prone to an additional electronic resonance called the Kondo Effect (Kouwenhoven and Glazman 2001). This resonance originates from the tendency of conduction electrons to screen the nonzero spin states of the quantum dot in a similar fashion as it occurs in a 3D metal, which contains a low level of magnetic impurities. This screening current adds to the usual conduction channel of the quantum dot and generates a conduction channel at zero bias, which bypasses the Coulomb blockade regime (Nygard et al. 2000). In the superconducting state, the nonzero spin tunneling induces higher order electron transfers events (Buitelaar et al. 2002) that allow the flipping of both spins of the Cooper pair during the tunneling through the nanotube. This amounts to changing the sign of the Cooper pair wave function. As can be seen in Equation 7.1, a change of sign in this wave function is equivalent to a phase shift of π in the superconducting phase.

Such an effect, in which the minimum energy state of one of the Josephson junctions (the so-called π-junction) is obtained for a phase difference of π instead of 0, has been the focus of intense studies. Reminiscent of ferromagnetic impurities in a Josephson junction, it was predicted some 20 years ago (Glazman and Matveev 1989) that a reverse Josephson current would take place in a junction involving tunneling through a quantum dot populated with an odd number of electrons. For a strong Kondo effect (Figure 7.2), the Josephson coupling is expected to be positive (0-junction) since the localized spin is screened due to the Kondo effect. On the other hand, for a weak Kondo effect, the large on-site interaction only allows the electrons in a Cooper pair to tunnel one-by-one via virtual processes in which the spin ordering of the Cooper pair is reversed, leading to a negative Josephson coupling (π-junction) and hence a reversed supercurrent. Controlling the phase-shift across a nanotube weak link can be achieved by tuning the charge induced on the weak link (see Figure 7.13).

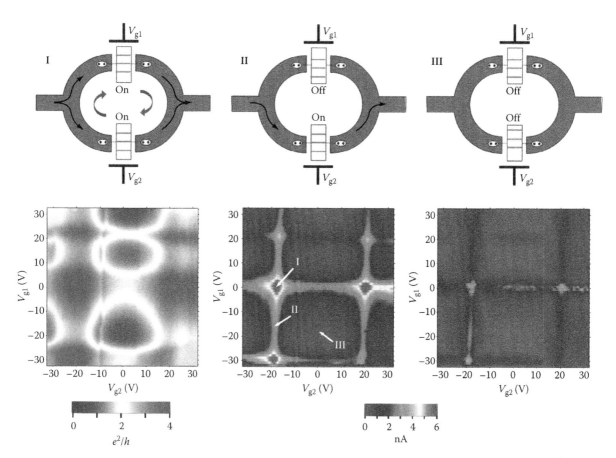

FIGURE 7.12 Top: Schematics of three typical working points of a carbon nanotube SQUID, in case **I**, both junctions have a quantum level adjusted to the Fermi energy of the leads (on-resonance) and supercurrent can flow through the device. In **II** and **III**, one and two junctions are tuned off-resonance, respectively. Bottom left: Modulation of the conductance of the device in the normal state. Center: Modulation of the maximum superconducting current of the same device. The number refers to the three cases described above. (Adapted from Cleuziou, J.-P. et al., *Nat. Nanotechnol.*, 1, 53, 2006. With permission.)

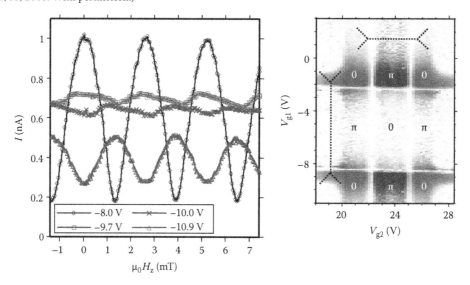

FIGURE 7.13 Gate control of the phase in a carbon nanotube SQUID. Left: magnetic field dependence of the maximum supercurrent in a nanotube SQUID as a function for different lateral gate voltages V_g, the modulation can be tuned from an even (0-junction SQUID) to an odd (π-junction SQUID) curve. The modulation is highly non-sinusoidal between the two states (Right), Color-scale representation of differential resistivity dV/dI as a function of the lateral gate voltages V_{g1} and V_{g2} showing the periodic switching between zero and π modulation. The dotted lines corresponds to voltage gate range for which an odd number of electrons is in the quantum dot. Note that when both SQUID junctions has π modulation, the SQUID recovers normal (0) modulation. (Adapted from Cleuziou, J.-P. et al., *Nat. Nanotechnol.*, 1, 53, 2006. With permission.)

7.4.3 Application of Nanotube Weak Links to Magnetometry

SQUID-based magnetometry in the near field regime (i.e., for the detection of fields emitted by source at the distance smaller than the SQUID loop radius) has been exploited for nearly 15 years in the field of molecular magnetism (Wernsdorfer 2001). It is based on the inductive coupling of a nano-magnetic object placed near a constriction of the loop of a SQUID. A local change in flux, for example, induced by the magnetization reversal of a magnetic dipole in the vicinity of the loop (Wernsdorfer et al. 1995) will be detected by a variation in the SQUID switching current. By recording these changes while sweeping the magnetic field in direction and intensity, one can accurately study the physical conditions of magnetization switching of a nanoparticle or a assembly of molecules. Therefore, it is possible to study the details of the dynamics of magnetization reversal on nano-magnets and highlight, in the case of molecular magnets, the existence of a tunnel effect of this reversal (Wernsdorfer et al. 1997).

The current system is limited in sensitivity and the threshold corresponds to the detection of a few thousands of elementary spins. The miniaturization of the most sensitive SQUID (superconducting weak seal, represented here by the CNT) seems to be a promising way of gaining sensitivity.

The strong 1D geometrical form factor of CNT junctions combined with their tiny cross-section open some interesting perspectives concerning the magnetometry at the nanometer scale. Indeed one can show that shrinking the cross-section of a weak link allows for the optimization of the inductive coupling between the loop of the SQUID magnetometer and a magnetic object sitting in close vicinity of the weak link (Bouchiat 2009). This inductive coupling factor is known to be proportional to the inverse of the weak link cross-section radius. Therefore, miniaturization of the weak link footprint is indeed necessary to match the size of a single molecule magnet (see Figure 7.14).

Noise measurements on practical devices (Cleuziou et al. 2006) have shown that the detection threshold in CNT magnetometers should allow measurements in the near field regime of magnetized nano-objects of typically 10 Bohr magnetons, for example, strays emitted by a single molecular magnet (see Figure 7.14). The control and measurement of the magnetization of a single molecular magnet is indeed a promising way to implement spin Qubits, which are readily interfaced to coherent superconducting electronic currents. The fact that nanotube SQUIDs can see their current gated within a large dynamic from 10 nA down to picoampere critical currents (Cleuziou et al. 2006) (see Figure 7.9) is also a promising feature. This could allow the limiting of the back-action of the measuring instrument to the evolving quantum spin, thus helping to preserve long coherence times. More generally, magnetometers based on these ultra-miniaturized weak links could provide new instrumentation set-ups, useful for the building of a spintronics at the molecular scale (Bogani and Wernsdorfer 2008).

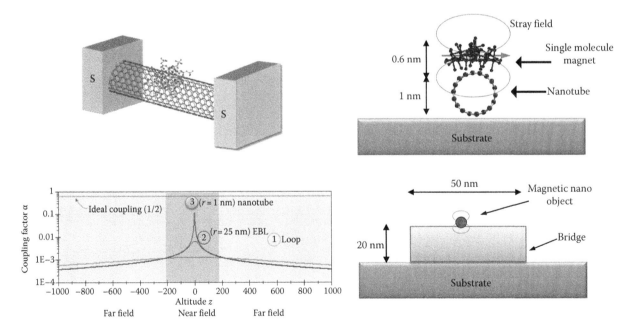

FIGURE 7.14 Optimization of the inductive coupling of single molecule magnet to the nanotube superconducting junction. Top left: artist view of the grafting of a single molecule magnet (here a "manganese-12" molecule) in close contact onto a carbon nanotube junction. Top right, cross section of the same assembly showing the stray field emitted from the magnetic molecule. Bottom right: cross section of a similar situation for the previous generation of weak links made of a nanowire with state-of-the art electron beam lithography. Bottom left: Simulation of the inductive coupling factor α for the two cases shown in right as a function of the distance z between the centers of the two coupled objects: the black curve stand for a lithography-made nanobridge ($r = 25$ nm) while the curve (labelled 3) depicts the situation for a single-walled carbon nanotube ($r = 1$ nm). The nanotube based weak link which diameter matches the molecule size, better grab the stray field and thus increase the inductive coupling by two order of magnitude with respect to the weak link made by top down lithography. The top straight line depicts the theoretical coupling limit ($\alpha = 1/2$) which is almost reached with a carbon nanotube junction.

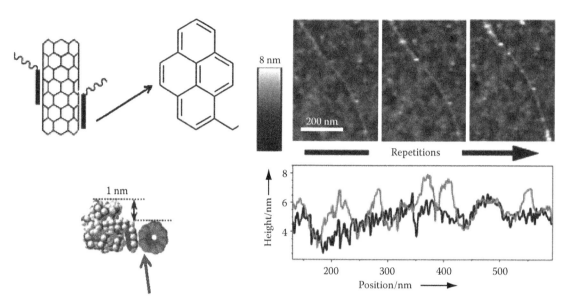

FIGURE 7.15 Chemical functionalization of carbon nanotube weak links for single molecule magnetism applications. Top left: Schematics showing the grafting of carbon nanotubes with organic molecules using pyrene group (inset) that specifically bind onto nanotube side walls by p-p interaction. (After Chen, R. et al., *J. Am. Chem. Soc.*, 123, 3838, 2001.) Right: AFM topographic analysis of molecular magnet/nanotube hybrids. Top right: Height images of the same CNT acquired on repeating the grafting process: left one time, center four times, right ten times. Bottom right: Section profile along the same CNT before (black line) and after (gray line) multiple graftings of 1. Bottom left: Heights of the grafted objects. (Adapted from Bogani, L. et al., *Angew. Chem.*, 48, 746, 2009.)

The full integration of magnetic molecules coupled with the device is still to be made. This requires control of the binding of single magnetic molecules to the nanotube while preserving its good conduction properties. Progress has been made in that direction by using chemically functionalized organic entities (Chen et al. 2001) that can mimic on a few cells, the graphitic honeycomb lattice (pyrene groups, see Figure 7.15), thus through π–π interaction, they provide specific grafting along the

nanotube side walls. Pyrene functionalized molecular magnets have been shown (Bogani et al. 2009) to specifically bind in solution in a rather controlled way onto nanotube junctions.

7.4.4 Application of Nanotube Weak Links to Quantum Information

As shown above, CNT-based weak links offer the possibility of realizing the controlled injection of Cooper pairs that travel as

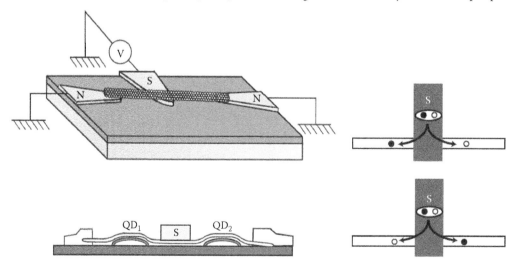

FIGURE 7.16 Top left: Schematics of a carbon nanotube based Cooper pair splitter: the nanotube is coupled in its middle part to a voltage biased superconducting electrode (noted S) which injects Cooper pairs in the nanotube. The nanotube splits the current into two components which are collected by symmetrically positioned normal metal electrodes. Bottom left: Cross section of the same device. The two branches act as nanotube quantum dots noted QD₁ and QD₂ in a similarly geometry as previously presented nanotube SQUID. Right: schematics showing the two possible paths for the two correlated electrons that forms the Cooper pair which splits in the nanotube and travel in opposite directions. (Adapted from Bouchiat, V. et al., *Nanotechnology*, 14, 77, 2003.)

correlated quasi-particles in a 1D or 0D conductor. Many experiments taking advantage of splitting the quantum correlated charge carriers emitted from a superconducting electrode can be devised. They are mostly inspired from those already realized in quantum optics and propose to realize their fully-integrated solid-state counterparts involving Cooper pairs instead of correlated photon pairs. In Cooper pairs splitting experiment, an additional quantum degree of freedom is also provided by the spin of the electron.

Among interesting basic devices is the nanotube-based Cooper pair beam splitter (Figure 7.16). It consists of a single-walled CNT connected at both ends to normal state electrodes and coupled in its middle part to a superconducting nanowire. Such a device acts as an electronic beam splitter for correlated electrons originating from the superconductor. Note that this geometry is very similar to the nanotube SQUID described in Section 7.4, but with the notable difference of having spatially separated normal-state "collecting electrodes" instead of a single superconducting one found in the SQUID geometry (Figure 7.11).

The Cooper pair splitter device was first discussed on a theoretical level in 2002–2003 (Bena et al. 2002, Recher and Loss 2002, Bouchiat et al. 2003) and its first preliminary experimental realizations just appeared recently using either a CNT (Herrmann et al. 2010) or a III–V semiconducting nanowire (Hofstetter et al. 2009).

In this device, the splitting of Cooper pairs in the superconductor provides two electrons with opposite spins. In the case of a full splitting, they can travel into opposite directions in the nanotube and are individually collected at both ends of the nanotube (Figure 7.16). Thanks to the Andreev process, the Cooper pair splitter generates electrons that are simultaneously entangled in energy and in spin. Correlation and noise measurements of the current emerging from the nanotube should allow the testing of Bell inequalities (Bouchiat et al. 2003) with electrons instead of photons.

7.5 Graphene-Based Superconducting Weak Links

7.5.1 Proximity Effect in Graphene Weak Links

Graphene, a single atomic layer of graphite, exhibits unique electrical and mechanical properties on account of its reduced dimensionality and "relativistic" band structure. The transport properties of the graphene sheets have recently received a lot of attention on both experimental (Geim and Novoselov 2007) and theoretical points of view. On one hand, this is due to its 2D character and on the other hand, to its remarkable band structure combining a linear dispersion relation of electronic wave functions similar to mass-less particles and perfect electron hole symmetry. The Fermi surface consists of two cones touching at one singular point, the so-called Dirac point, where the density of states is zero (see Figure 7.4a). Over the last 5 years, graphene has emerged as a unique platform for testing new electronic properties of condensed matter (Figure 7.17).

By tuning a gate voltage positively or negatively, one can respectively induce a finite density of electrons and holes with a

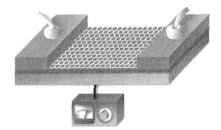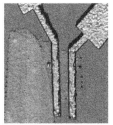

FIGURE 7.17 Left: Schematic diagram of a graphene-based superconducting weak link. The graphene carrier density is controlled by adjusting the backgate voltage on the silicon substrate, which is isolated from the device by a thermal oxide layer. Right: Atomic force micrograph of a first realized graphene superconducting junction. (Adapted from Heersche, H.B. et al., *Nature*, 446, 56, 2007.)

linear energy dispersion curve. Furthermore, this 2D electron gas can be exposed to external physicochemical functionalization and be easily connected with superconducting electrodes. Soon after first electron transport measurements, graphene has been shown to allow Cooper pair transport (Heersche et al. 2007) with a well-defined Josephson effect exhibiting microamperes of critical current (Figure 7.18) and clearly defined (Figure 7.18) multiple Andreev reflections (Du et al. 2008a, Heersche et al. 2007).

The supercurrent of such a junction can be modulated in intensity with the gate, which tunes the carrier type and density of the graphene sheet (see Figure 7.4b). Depending on the gate voltage applied to the graphene, a supercurrent can be mediated by either electrons or holes (Heersche et al. 2007), which makes graphene the first 2D ambipolar superconducting material. Unlike nanotube weak links, the modulation no more shows any gate periodicy in the modulation of the switching supercurrent because of the absence of lateral quantum confinement. The superconducting current is monotonically increased by increasing the gate and, as it is predicted by the number of conduction channels contributing in parallel to the superconducting current (Equations 7.4 and 7.5), the intensity of the maximum superconducting current is usually thousands of times the one measured for a CNT junction.

New features associated with the existence of Dirac Fermions have been predicted (Beenakker 2008). For example, due to the multiple conical valley band structure, the Andreev reflection, which usually is a retro-reflection (as seen in Figure 7.1), can be induced specularly with a high probability in undoped samples when the Fermi energy lies in the vicinity of the Dirac point (Beenakker 2006). Such effects, as well as the observation of peculiarities induced by the Andreev bound states (Titov and Beenakker 2006), require the realization of the ballistic graphene weak links, a challenging fabrication task since the mean free path of graphene samples usually made are below 0.1 μm. Promising routes are offered by making suspended graphene obtained by under-etching the supporting substrate. Such a process has been shown to greatly enhance the normal state electronic properties in the non-superconducting state by boosting mean free paths (Du et al. 2008b) and electronic mobilities (Bolotin et al. 2008).

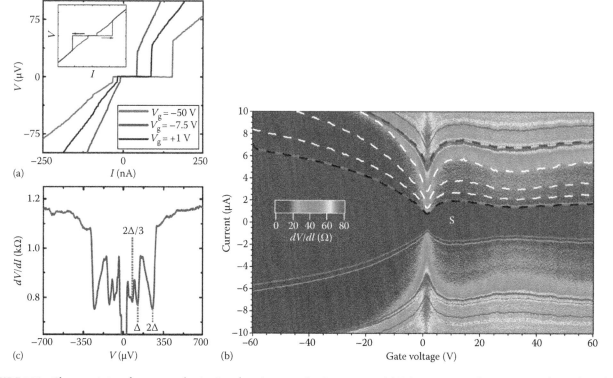

FIGURE 7.18 Characteristics of a superconducting/graphene/superconducting junction. (a) Voltage–current characteristics of a graphene-based superconducting weak link. (From Heersche, H.B. et al., *Nature*, 446, 56, 2007. With permission.) (b) Resistance map of a two-junction graphene device showing the ambipolar gate dependence of the superconducting state (central zone denoted by the letter S). Differential resistance of a device showing dependence of critical current on back gate voltage (dashed line) at 20 mK. Traces of multiple Andreev reflections at constant voltages $2\Delta g/ne$ for $n = 1, 2, 3$ are indicated by the white dashed lines (*n*) 1 at top, $\Delta g \sim 75\,\mu eV$. (Adapted from Girit, Ç. et al., *Nano Lett.*, 9, 198, 2009.) (c) Differential resistance as a function of the voltage bias showing the multiple Andreev reflection peaks below the Gap 2D. (From Heersche, H.B. et al., *Nature*, 446, 56, 2007.)

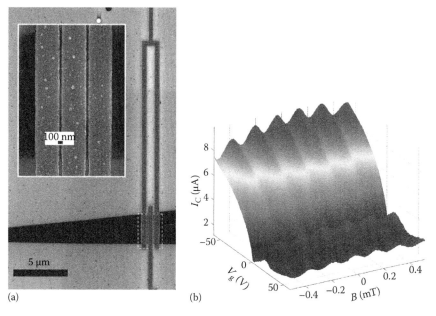

FIGURE 7.19 (a) Scanning electron microscope image of a superconducting quantum interference device with weak links made of portion of the same single layer graphene flake (black, graphene; light gray, silicon oxide substrate; dark gray, Pd/Al electrodes Inset: close-up of sub-100 nm wide graphene Josephson junctions. (b) Maximum switching current as a function of magnetic field *B* and gate voltage V_g. The sinusoidal oscillations indicate quantum interference between the two graphene Josephson junctions. For a fixed magnetic field, the magnitude of the modulation can be controlled by the back gate. (After Girit, Ç. et al., *Nano Lett.*, 9, 198, 2009.)

In the case of graphene, this is predicted to lead to an unusual bias dependence of the differential conductance of the weak link. In most experiments, superconducting charge carriers are injected by the measuring electrodes. However, it has been predicted (Feigel'man et al. 2010) and experimentally shown (Kessler et al. 2010) that the superconductivity could be induced globally by decorating the graphene with an array of superconducting electrodes. In this latter case, the proximity effect generates a supercurrent, which percolates through the entire graphene sheet.

7.5.2 Graphene Nanotube Superconducting Quantum Interferometers

Graphene-based SQUIDs devices can be obtained in a similar way as has been done for CNT weak links. Indeed, the fabrication and operation of a two junction SQUID formed by a single graphene sheet contacted with aluminum/palladium electrodes in the geometry of a loop (see Figure 7.19) has been demonstrated in 2009 (Girit et al. 2009).

As for the nanotube SQUID, graphene SQUIDs allow the tuning of the quantum interference with an electrostatic gate with a dependence (Figure 7.19b) that follows the transparency given by the graphene weak link (Figure 7.18). It also suggests a new modality for the ultrasensitive magnetometry of nanomagnets chemically or physically attached to the carbon surface.

7.6 Fullerene-Based Superconducting Weak Links

This last category considers the smallest sp^2 carbon weak links that involve fullerene molecules (Smalley 1997). We consider a device made of a fullerene molecule (see Figure 7.4a, right) placed in a transistor geometry (depicted in Figure 7.4b). Interaction of the discrete levels of a molecular entity with the superconducting electrode has triggered a lot a theoretical studies (Choi et al. 2004, Novotny et al. 2005). Unlike all the previous carbon-based weak links discussed above, for which the properties depend not only upon the material properties but also on the electrode geometries

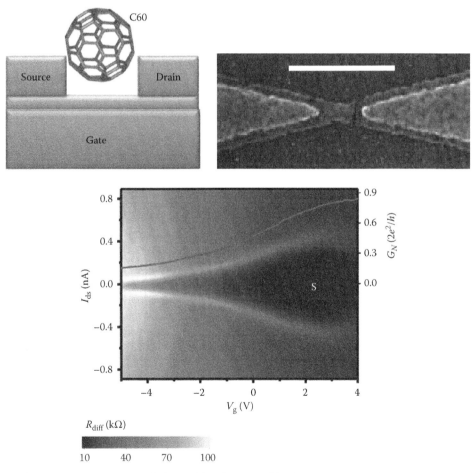

FIGURE 7.20 Fullerene based superconducting weak link. Top left: Schematics of the device C60 connected to superconducting source and drain. The device sits over an aluminum gate electrode whose thin native oxide forms the gate dielectrics. Top right: scanning electron micrograph of the device, showing the nanoscale gap created by electromigration. (scale bar = 300 nm). Bottom: Map of the differential resistance of the device as a function of the gate voltage (*x* axis) and the current bias (*y* axis). The black region corresponds to the superconducting state. The gate dependence of the superconducting current is in agreement with the expected dependence for a proximity coupled quantum dot. (After Winkelmann, C.B. et al., *Nat. Phys.*, 5, 876, 2009.)

(for example, in a nanotube weak link, we have seen that the weak link gate modulation is inversely proportional to the length of the weak link, which is defined by the gap between the electrode), fullerene-based weak links involve the entire molecule. Therefore, electron transport through the junction is supposed to exhibit fingerprints (Zazunov et al. 2006) linked to the intrinsic molecule states, such as vibrational modes (Park et al. 2000) or molecular magnetism (Lee et al. 2008). Indeed a set of experiments (see Figure 7.5, right) involving magnetically active molecules such as endofullerenes (Grose et al. 2008, Kasumov et al. 2005), which consist of additional atoms inserted in the carbon cage formed by the fullerene shell, are a sensitive probe of molecular magnetism.

7.6.1 Molecular Transistors Made by Electromigration

Recent progress in the field of molecular electronics now allows for the direct integration of molecular objects into electrical on-chip circuits. It has been shown that by inserting molecules into a nanometer gap by sectioning a metallic wire, it is possible to create a molecular transistor which conducting states are linked to molecular orbitals available for conduction (Park et al. 2000). This possibility was provided by the electromigration process (Park et al. 1999), which allows the generation of nanogaps in metallic wires with good yield and an unprecedented reproducibility.

7.6.2 Implementation of a C60 Superconducting Transistor

Such electromigration processes have been recently adapted to the fabrication of superconducting transistors involving a single C60 molecule inserted between gold/aluminum electrodes that can be measured at milliKelvin temperatures (Winkelmann et al. 2009). Figure 7.20 shows the gate control of the superconducting state achieved in this molecular transistor, in the case of low dissipation in the environment of the linking conductor. The observed switching current is significantly lower than the critical current and has been shown to follow the law

$$\frac{I_S}{I_0} = \left(1 - \sqrt{1 - \frac{G_N h}{2e^2}}\right)^{3/2}.$$

The results of a C60 molecular transistor (Winkelmann et al. 2009) show good agreement between the measured critical current as a function of gate voltage and the prediction for $I_c(V_g)$ when taking into account the phase oscillations across the weak link.

7.7 Concluding Remarks

A new class of molecular weak links has now become available for research, in which superconductors can be coupled with reliable transparent interfaces to nanostructures composed of a crystalline arrangement of sp^2-hybridized carbon. The main original feature offered by this new type of device is the tunability of the superconducting transport being possible through gate control of the weak link chemical potential offered by quantum confinement. The integration control over these nanostructures, which has been made possible by advances in nanofabrication, allows for the fabrication of hybrid devices that mix bottom-up and top-down approaches. They provide building blocks that pave the way for the realization of more complex devices that should be useful for quantum information and nanomagnetism.

Acknowledgment

The author warmly acknowledges Wolfgang Wernsdorfer for stimulating discussions. The author would like to thank Sathyanarayanamoorthy Sridharan and Klaus Sattler for their kind help.

References

Averin, D. and Bardas, A. 1995. ac Josephson effect in a single quantum channel. *Physical Review Letters*, 75: 1831.

Barone, A. and Paterno, G. 1982. *Physics and Applications of the Josephson Effect*. New York: John Wiley & Sons.

Beenakker, C. W. J. 2006. Specular Andreev reflection in graphene. *Physical Review Letters*, 97: 067007.

Beenakker, C. W. J. 2008. Colloquium: Andreev reflection and Klein tunneling in graphene. *Reviews of Modern Physics*, 80: 1337.

Beenakker, C. W. J. and van Houten, H. 1992. *Single-Electron Tunneling and Mesoscopic Devices* (eds. Koch, H. and Lübbig, H.) http://xxx.lanl.gov/abs/condmat/0111505l. Berlin, Germany: Springer.

Bena, C., Vishveshwara, S., Balents, L., and Fisher, M. P. A. 2002. Quantum entanglement in carbon nanotubes. *Physical Review Letters*, 89: 037901.

Bogani, L. and Wernsdorfer, W. 2008. Molecular spintronics using single-molecule magnets. *Nature Materials*, 7: 179.

Bogani, L., Danieli, C., Biavardi, E. et al. 2009. Single-molecule-magnet carbon-nanotube hybrids. *Angewandte Chemie*, 48: 746–750.

Bolotin, K. I., Sikes, K. J., Jiang, Z. et al. 2008. Ultrahigh electron mobility in suspended graphene. *Solid State Communications*, 146: 351.

Bouchiat, V. 2009. Detection of magnetic moments using a nano-SQUID: Limits of resolution and sensitivity in near-field SQUID magnetometry. *Superconductor Science and Technology*, 22: 064002.

Bouchiat, V., Chtchelkatchev, N., Feinberg, D. et al. 2003. Single-walled carbon nanotube–superconductor entangler: Noise correlations and Einstein–Podolsky–Rosen states. *Nanotechnology*, 14: 77–85.

Buitelaar, M. R., Nussbaumer, T., and Schonenberger, C. 2002. Quantum dot in the Kondo regime coupled to superconductors. *Physical Reviews Letters*, 89: 256801.

Burke, P. J. 2004. Ac performance of nanoelectronics: Towards a ballistic THz nanotube transistor. *Solid State Electronics*, 40: 1981.

Cao, J., Wang, Q., and Dai, H. 2005. Electron transport in very clean, as-grown suspended carbon nanotubes. *Nature Materials*, 4: 745–749.

Charlier, J.-C., Blase, X., and Roche, S. 2007. Electronic and transport properties of nanotubes. *Reviews of Modern Physics*, 79: 677.

Chen, R., Zhang, Y., Wang, D., and Dai, H. 2001. Non-covalent sidewall functionalization of single-walled carbon nanotubes for protein immobilization. *Journal of the American Chemical Society*, 123: 3838–3839.

Choi, M. S., Lee, M., Kang, K., and Belzig, W. 2004. Kondo effect and Josephson current through a quantum dot between two superconductors. *Physical Review B*, 70: 020502.

Clarke, J. and Braginski, A. I. 2004. *The SQUID Handbook*. Weinheim, Germany: Wiley-VCH.

Cleuziou, J.-P., Wernsdorfer, W., Bouchiat, V., Ondarcuhu, T., and Monthioux, M. 2006. Carbon nanotube superconducting quantum interference device. *Nature Nanotechnology*, 1: 53.

Cleuziou, J. P., Wernsdorfer, W., Andergassen, S. et al. 2007. Gate-tuned high frequency response of carbon nanotube Josephson junctions. *Physical Review Letters*, 99: 117001.

Courtois, H., Meschke, M., Peltonen, J. T., and Pekola, J. P. 2008. Origin of hysteresis in a proximity Josephson junction. *Physical Review Letters*, 101: 067002.

De Gennes, P. G. and Guyon, E. 1963. Superconductivity in "normal" metals. *Physics Letters*, 3: 168–169.

Du, X., Skachko, I., and Andrei, E. Y. 2008a. Josephson current and multiple Andreev reflections in graphene SNS junctions. *Physical Review B*, 77: 184507.

Du, X., Skachko, I., Barker, A., and Andrei, E. Y. 2008b. Approaching ballistic transport in suspended graphene. *Nature Nanotechnology*, 3: 491–495.

Feigel'man, M. V., Skvortsov, M. A., and Tikhonov, K. S. 2008. Proximity-induced superconductivity in graphene. *JETP Letters*, 88: 862.

Gang, L., Yong, Z., and Chun Ning, L. 2009. Gate-tunable dissipation and "superconductor-insulator" transition in carbon nanotube Josephson junctions. *Physical Review Letters*, 102: 016803.

Geim, A. K. and Novoselov, K. S. 2007. The rise of graphene. *Nature Materials*, 6: 183.

Girit, Ç., Bouchiat, V., Naaman, O. et al. 2009. Tunable graphene superconducting quantum interference device. *Nano Letters*, 9: 198–199.

Glazman, L. I. and Matveev, K. A. 1989. Resonant Josephson current through Kondo impurities in a tunnel barrier. *JETP Letters*, 49: 659.

Goffman, M. F., Cron, R., Levy Yeyati, A. et al. 2000. Supercurrent in atomic point contacts and Andreev states. *Physical Review Letters*, 85: 170.

Golubov, A. A., Kupriyanov, M. Y., and Il'ichev, E. 2004. The current-phase relation in Josephson junctions. *Reviews of Modern Physics*, 76: 411–469.

Grose, J. E., Tam, E. S., Timm, C. et al. 2008. Tunnelling spectra of individual magnetic endofullerene molecules. *Nature Materials*, 7: 884.

Hamilton, C. A. 2000. Josephson voltage standards. *Review of Scientific Instruments*, 71: 3611.

Heersche, H. B., Jarillo-Herrero, P., Oostinga, J. B., Vandersypen, L. M., and Morpurgo, A. F. 2007. Graphene JJ. *Nature*, 446: 56.

Herrmann, L. G., Portier, F., Roche, P. et al. 2010. Carbon nanotubes as cooper pair beam splitters. *Physical Review Letters*, 104: 026801.

Hofstetter, L., Csonka, S., Nygard, J., and Schonenberger, C. 2009. Cooper pair splitter realized in a two-quantum-dot Y-junction. *Nature*, 461: 960–963.

Imry, Y. and Landauer, R. 1999. Conductance viewed as transmission. *Reviews of Modern Physics*, 71: S306.

Jaklevic, R. C., Lambe, J., Silver, A. H., and Mercereau, J. E. 1964. Quantum interference effects in Josephson tunneling. *Physical Review Letters*, 12: 159–160.

Jarillo-Herrero, P., van Dam, J. A., and Kouwenhoven, L. P. 2006. Quantum supercurrent transistors in carbon nanotubes. *Nature*, 439: 953–956.

Javey, A., Guo, J., Wang, Q. et al. 2003. Ballistic carbon nanotube field-effect transistors. *Nature*, 424: 654–657.

Jørgensen, H. I., Grove-Rasmussen, K., Novotny, T., Flensberg, K., and Lindelof, P. E. 2006. Electron transport in single-wall carbon nanotube weak links in the Fabry-Perot regime. *Physical Review Letters*, 96: 207003.

Jorgensen, H. I., Novotny, T., Grove-Rasmussen, K., Flensberg, K., and Lindelof, P. E. 2007. Critical current 0-pi transition in designed Josephson quantum dot junctions. *Nano Letters*, 7: 2441–2445.

Josephson, B. D. 1962. Possible new effects in superconductive tunnelling. *Physics Letters*, 1: 251–253.

Kasumov, A. Y., Deblock, R., Kociak, M. et al. 1999. Supercurrents through single-walled carbon nanotubes. *Science*, 284: 1508.

Kasumov, A. Y., Tsukagoshi, K., Kawamura, M. et al. 2005. Proximity effect in a superconductor-metallofullerene-superconductor molecular junction. *Physical Review B*, 72: 033414.

Kessler, B. M., Girit, C. O., Zettl, A., and Bouchiat, V. 2010. Tunable superconducting phase transition in metal-decorated graphene sheets. *Physical Review Letters*, 104: 047001.

Klapwijk, T. M., Blonder, G. E., and Tinkham, M. 1982. Explanation of subharmonic energy gap structure in superconducting contacts. *Physica B*, 109: 1657.

Kouwenhoven, L. and Glazman, L. 2001. Revival of the Kondo Effect. *Physics World*, 14: 33.

Lassagne, B., Tarakanov, Y., Kinaret, J., Garcia-Sanchez, D., and Bachtold, A. 2009. Coupling mechanics to charge transport in carbon nanotube mechanical resonators. *Science*, 325: 1107–1110.

Lee, M., Jonckheere, T., and Martin, T. 2008. Josephson effect through a magnetic metallofullerene molecule. *Physical Review Letters*, 101: 146804.

Liang, W., Bockrath, M., Bozovic, D. et al. 2001. Fabry-Perot interference in a nanotube electron waveguide. *Nature*, 411: 665.

Likharev, K. K. 1979. Superconducting weak links. *Reviews of Modern Physics*, 51: 101.

Liu, G., Zhang, Y., and Lau, C. N. 2009. Gate-tunable dissipation and "Superconductor-insulator" transition in carbon nanotube Josephson junctions. *Physical Review Letters*, 102: 016803.

Morpurgo, A. F., Kong, J., Marcus, C. M., and Dai, H. 1999. Gate-controlled superconducting proximity effect in carbon nanotubes. *Science*, 286: 263–265.

Nemec, N., Tomanek, D., and Cuniberti, G. 2006. Contact dependence of carrier injection in carbon nanotubes: An Ab initio study. *Physical Review Letters*, 96: 076802.

Novotny, T., Rossini, A., and Flensberg, K. 2005. Josephson current through a molecular transistor in a dissipative environment. *Physical Review B*, 72: 224502.

Nygard, J., Cobden, D. H., and Lindelof, P. E. 2000. Kondo physics in carbon nanotubes. *Nature*, 408: 342–346.

Pallecchi, E., Gaaß, M., Ryndyk, D., and Strunk, C. 2008. Carbon nanotube Josephson junctions with Nb contacts. *Applied Physics Letters*, 93: 072501.

Park, H., Lim, A. K. L., Alivisatos, A. P., Park, J., and McEuen, P. L. 1999. Fabrication of metallic electrodes with nanometer separation by electromigration. *Applied Physics Letters*, 75: 301.

Park, H., Park, J., Lim, A. K. L. et al. 2000. Nanomechanical oscillations in a single-C60 transistor. *Nature*, 407: 57–60.

Recher, P. and Loss, D. 2002. Superconductor coupled to two Luttinger liquids as an entangler for electron spins. *Physical Review B*, 65: 165327.

Scheer, E., Joyez, P., Esteve, D., Urbina, C., and Devoret, M. H. 1997. Conduction channel transmissions of atomic-size aluminum contacts. *Physical Review Letters*, 78: 3535.

Smalley, R. E. 1997. Discovering the fullerenes. *Reviews of Modern Physics*, 69: 723.

Steele, G. A., Huttel, A. K., Witkamp, B. et al. 2009. Strong coupling between single-electron tunneling and nanomechanical motion. *Science*, 325: 1103–1107.

Titov, M. and Beenakker, C. W. J. 2006. Josephson effect in ballistic graphene. *Physical Review B*, 74: 041401.

Wernsdorfer, W. 2001. Classical and quantum magnetization reversal studies in nanometer-sized particles and clusters. *Advances in Chemical Physics*, 188: 99–190.

Wernsdorfer, W., Hasselbach, K., Mailly, D. et al. 1995. Seminal microSQUID particle WW. *Journal of Magnetism and Magnetic Materials*, 145: 33.

Wernsdorfer, W., Bonet Orozco, E., Hasselbach, K. et al. 1997. Macroscopic quantum tunneling of magnetization of single ferrimagnetic nanoparticles of Barium Ferrite. *Physical Review Letters*, 79: 4014–4017.

Winkelmann, C. B., Roch, N., Wernsdorfer, W., Bouchiat, V., and Balestro, F. 2009. Superconductivity in a single-C60 transistor. *Nature Physics*, 5: 876–879.

Zazunov, A., Fierberg, D., and Martin, T. 2006. Phonon squeezing in a superconducting molecular transistor. *Physical Review Letters*, 97: 196801.

Micromagnetic Modeling of Nanoscale Spin Valves

Bruno Azzerboni
University of Messina

Giancarlo Consolo
University of Messina

Giovanni Finocchio
University of Messina

8.1 Introduction

Magnetism is an open field in which physicists, electrical engineers, material scientists, mathematicians, chemists, metallurgists, and others practice together.

Today information technologies ranging from personal computers to mainframes use magnetic materials to store information on tapes, floppy diskettes, and hard disks. Our seemingly insatiable appetite for more computer memory will probably be met by a variety of magnetic recording technologies based on nanocrystalline thin-film media and magneto-optic materials. Personal computers and many of our consumer and industrial electronics components are now powered largely by lightweight switch-mode power supplies using new magnetic materials technology that was unavailable 20 years ago. Magnetic materials touch many other aspects of our lives. Each automobile contains dozens of motors, actuators, sensors, inductors, and other electromagnetic and magneto-mechanical components using hard (permanent) as well as soft magnetic materials. Electric power generation, transformation, and distribution systems rely on hundreds of millions of transformers and generators that use the principles of electromagnetic fields as well. Finally, magnetism is present in telecommunication systems, such as mobile phones, electronic article surveillance, asset protection, and access control because of the presence of oscillating circuits and resonators working at microwave frequencies.

Magnetism, along with electricity, belongs to a larger phenomenon, electromagnetism, by which we describe the force generated by the passage of an electric current through matter. When two electric charges are at rest, it appears to the observer that the force between them is merely electric. If the charges are in motion, however, and in this instance motion or rest is referred to in relation to the observer, then it appears as though a different force, known as magnetism, exists between them. In fact, the difference between magnetism and electricity is purely artificial. Both are manifestations of a single fundamental force, with magnetism simply being an abstraction that people use for the changes in electromagnetic force created by the motion of electric charges.

To gain a deeper understanding of magnetism and the materials classification, it is necessary to first introduce some basic concepts. First, within the area of research of magnetism, one can identify at least four different scales that give correspondingly four different levels of investigation, as shown in Figure 8.1.

The top level, magnetic hysteresis models, treats the phenomenology of technical magnetization curves. Such an approach is encountered whenever connections between hysteresis and domain phenomena can be established. There are several models to treat the magnetization phenomena (curves) at this level. The first scalar model of hysteresis was proposed by Preisach (1929). From 1929, lots of efforts have been focused on the physical interpretation of its parameters as well as the inclusion of aftereffects, accommodation, and magnetostriction phenomena (Bertotti 1998).

At the bottom level, we find the atomic description, which deals with the spin structure of magnetically ordered material and the arrangements of spins on the crystal lattice sites. It also describes the origin, interactions, mutual arrangement, and statistical thermodynamics of elementary magnetic moments by means of quantum mechanic theories.

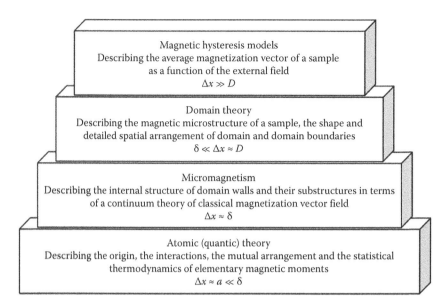

FIGURE 8.1 The hierarchy of descriptive levels of magnetically ordered materials. The values in parenthesis indicate the sample dimensions for which the different theories are applicable. Δx is the characteristic scale for each theory, D, δ, and a are the domain size, the wall width, and the interatomic distance, respectively.

There exist two models working at the intermediate mesoscopic scale: domain theory and micromagnetism. The former combines discrete, uniformly magnetized domains, domain walls, and their microstructure. The latter is the continuum theory of magnetic moments, underlying the description of the magnetic microstructure. The domain theory is located at a scale larger than the domain wall width, whereas the micromagnetic scale is comparable with that scale (see Figure 8.2).

The starting point of the domain theory is to suppose a given domain's distribution in a specimen according to both experimental observations and energy balance. The main assumption is that the wall that divides the adjacent domains has a zero width. In other words, this is a model where the magnetization vector presents discontinuities when passing from a domain to the adjacent one. As cited briefly above, this theory is valid under the assumption that the domain wall width is negligible with respect to the other dimensions involved in the problem. Without going into detail, it has to be noted that the domain theory offers the possibility of describing a lot of real systems. In spite of this, its limitations lead to the need for building a more accurate and general theory.

The micromagnetic theory was developed first by Landau and Lifshitz (1935) on a variational principle: it searches for magnetization distributions with the total smallest energy. This variational principle leads to a set of differential equations, the micromagnetic equations, which were initially developed by the same authors for one-dimensional problems. Stimulated again by experimental work and its analysis, W.F. Brown extended the equations to three dimensions (Brown 1940, 1963).

The micromagnetic equations are complicated nonlinear and nonlocal equations; they are therefore difficult to solve analytically, except in cases in which a linearization is possible. However, a number of problems in research need micromagnetic methods for their adequate treatment, such as the following:

1. The analysis of the behavior of small magnetic particles that are, at the same time, too small to accommodate a regular domain structure and too large to be described as uniformly magnetized
2. The determination of the spatial distribution of magnetization inside a domain wall
3. The calculation of the magnetic stability limits
4. The analysis of rapid magnetization dynamics occurring in nanoscale devices

To treat such problems, works on numerical solutions of the micromagnetic equations are increasingly pursued. It appears utopian, however, to apply micromagnetic methods to large-scale domain structures because the gap between the size of samples and the size of computational grids is simply too large and unaffordable computational times and memory allocation would be consequently required. According to these discussions, micromagnetic and domain theories should not be considered as mutually exclusive at all, but as complementary tools.

In this chapter, we will address our discussion on the usage of micromagnetic frameworks to describe the behavior of a class of novel devices that are currently investigated for their attracting

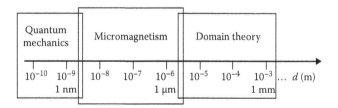

FIGURE 8.2 Characteristic length scales for mathematical models of magnetism.

potential applications (mainly as magnetic memories and microwave oscillators).

In 1996, in fact, almost contemporarily, Slonczewski (1996) and Berger (1996) predicted that a direct electric current passing through a magnetic material, adequately thick, becomes spin-polarized along the direction of its local magnetization vector. The current moving outward such a polarizer is thus referred to as "spin-polarized current." To some extent, it is possible to consider this phenomenon in analogy to what occurs in optical systems where a unpolarized light traverses a slab of polarizing material and causes the outgoing electromagnetic waves to oscillate along the direction of the transmission axis of the polarizer. If a spin-polarized current presents a sufficiently large amplitude and now traverses a second magnetic layer, it can transfer its spin-angular momentum and destabilize the static equilibrium orientation of the magnetization of this latter layer. Such phenomenon, which is correspondingly called "spin-transfer torque effect," has been indeed applied, from the very beginning, in multilayer systems called "spin-valve" like the one shown in Figure 8.3.

A classical spin-valve system consists of: a thicker magnetic layer, called fixed or pinned layer (PL), which acts as the spin-polarizer; a thinner magnetic layer, called free layer (FL),

which is the one generally subjected to the action of the spin-transfer torque; and a nonmagnetic spacer, used to decouple the exchange interactions between PL and FL. Depending on the electrical conductivity of the intermediate layer, it is possible to build-up either "classical" spin-valve (CSV) devices, if the spacer is a metal, or nonclassical magnetic tunnel-junctions (MTJ), if the spacer is an insulator. The mechanism of the torque and the resulting magnetic dynamics have also been studied experimentally in magnetic wires based on the capability of the spin-transfer torque to move domain walls.

Moreover, depending on the geometry, properties of the layered magnetic structure, and magnitude and orientation of the external bias field, such a "spin-transfer torque effect" can lead either to the switching of the magnetization direction or to a stable magnetization precession of the magnetization of the thinner FL.

The discovery that spin-polarized current can alter a magnetic state opened several perspectives for a new class of nano-scale magnetic devices in which the additional degree of freedom constituted by the "spin" plays an important role. The corresponding new research area was consequently called *spintronics*, a short notation for *"spin-based electronics"*. In general, spintronics refers to the study of the role played by electron (and more generally nuclear) spin in solid-state physics and in those devices that specifically exploit spin properties instead of, or in addition to, the degree of freedom given by the charge. The spin-transfer torque, by which spin-polarized currents can replace the role of a bias magnetic field in managing the spatial configuration of magnetization, offers the possibility of designing and building up a new class of devices and applications that includes, for example, magnetoresistive random access memory, logic gates, diodes and transistors, nano-oscillators, radiofrequency modulators, and detectors. To control device performance, it is critical to understand the nature of the spin-torque-driven magnetic excitations (Berkov and Miltat 2008, Ralph and Stiles 2008, Slavin and Tiberkevich 2008).

8.2 Background

We need, thus, a mathematical model to describe the magnetization dynamics driven by spin-polarized currents in nanoscale spin-valve systems.

To do that, as briefly introduced in Section 8.1, at a mesoscopic scale we need to introduce a continuum model, in terms of magnetic polarization per unit volume, and to characterize the state of a generic ferromagnetic body by means of its free energy.

To this aim, let us consider a region occupied by a magnetic body. Let us now focus on a "small" region dV_r within the body denoted by the position vector $r \in \Omega$. We refer to a "small" region to indicate that the volume dV_r is large enough to contain a huge number N of elementary magnetic moments μ_j, $j = 1, \dots, N$, but small enough so that the average magnetic moment varies smoothly. In this respect, we define the magnetization vector field $\mathbf{M}(\mathbf{r})$, so that the product $\mathbf{M}(\mathbf{r})dV_r$ represents the net magnetic moment of the elementary volume dV_r:

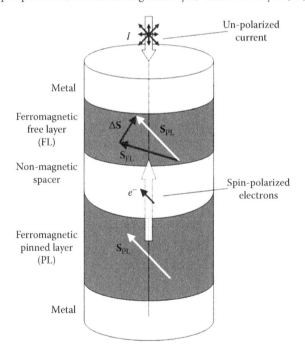

FIGURE 8.3 Schematic representation of the active region of a spin-valve when a perpendicular-to-plane electric current is injected through the structure. Please note that, because electrons bear negative charge, a downward current flow corresponds to an upward electron flow. A thicker ferromagnetic layer ("fixed" layer), having a magnetization vector $\mathbf{S}_{PL}$ fixed in space, and a thinner ferromagnetic layer ("free" layer), having a magnetization vector $\mathbf{S}_{FL}$ "free" to move, are separated by a nonmagnetic layer (spacer). The electrons of the current I exert a torque on the magnetization of the free layer. The current I is taken as positive when directed from the fixed to the free layer (electrons flow from the free into the fixed layer).

$$\mathbf{M}(\mathbf{r}) = \frac{\sum_{j=1}^{N} \boldsymbol{\mu}_j}{dV_r} \qquad (8.1)$$

Moreover, we assume that the magnetization is also a function of time t:

$$\mathbf{M} = \mathbf{M}(\mathbf{r}, t) \qquad (8.2)$$

The main hypotheses of *micromagnetism* consist of considering the magnetization vector $\mathbf{M}$ to be a continuous function of the position $\mathbf{r}$ within the body and whose modulus $|\mathbf{M}| = M_S$ is constant in time.

It is well known that the magnetization configuration of a ferromagnetic body should be such as to minimize its free energy $G(X, H, T) = F(X, T) - HX$. In this equation, $F(X, T)$ is the Helmholtz free energy of the system, H is the external field, and X is the state variable. The local minima obey the condition $\partial G / \partial X = 0$, by which all the metastable states are found. The great complication is that, in a space-dependent approach, X represents the full magnetization vector field $\mathbf{M}(\mathbf{r})$ defined over the entire body volume. Thus, the energy minimization has to be carried out in the infinite-dimensional functional space of all possible magnetization configurations, which, from the mathematical point of view, constitutes a nontrivial issue.

To characterize the energy landscape, it is necessary to first introduce the principal energy terms that control the behavior of a ferromagnetic body and show how they compete with each other.

It is well known that the most significant contributions to the energy balance arise from Maxwellian fields (magnetostatic and Zeeman) and non-Maxwellian ones (exchange and anisotropy). Let us briefly discuss each of them.

8.2.1 Exchange Interactions

The exchange interactions, which are the *source of the ferromagnetism,* should be vigorously analyzed by means of quantum theory, since this latter strongly concerns with spin–spin interactions. More specifically, on a scale in the order of the atomic scale, the exchange interaction tends to align neighboring magnetic moments. In other words, whenever $\mathbf{M}(\mathbf{r})$ changes orientation from point to point, we have some misalignment of the neighboring magnetic moments and this costs extra exchange energy. In view of a continuum average analysis in terms of a magnetization vector field, we expect that the exchange interactions tend to produce small uniformly magnetized regions, indeed observed experimentally and called magnetic domains.

The contribution of exchange interactions to the free energy of the whole magnetic body can be expressed as (Brown 1962)

$$F_{ex} = \int_{\Omega} A \left[(\nabla m_x)^2 + (\nabla m_y)^2 + (\nabla m_z)^2 \right] dV \qquad (8.3)$$

where
    the parameter A is called "exchange constant"
    the integration is extended over the whole volume region Ω

8.2.2 Anisotropy

Anisotropic effects in ferromagnetic bodies arise from the structure of the lattice and from particular symmetries that are produced in certain crystals. The experimental evidence of the existence of such energy-favored directions can be observed in certain ferromagnetic materials when, in the absence of external fields, the magnetization vector tends to be magnetized along precise directions, which in literature are referred to as *easy directions*. The fact that there is a "force" that tends to align magnetization along easy directions can be taken into account, in the micromagnetic framework, by means of an additional phenomenological term in the free energy functional

$$F_{an}(\mathbf{m}) = \int_{\Omega} f_{an}(\mathbf{m}) dV \qquad (8.4)$$

where $f_{an}(\mathbf{m})$ is the anisotropy free energy density. In this phenomenological analysis, the easy directions correspond to the minima of the anisotropy energy density, whereas the saddle-points and maxima of $f_{an}(\mathbf{m})$ determine the medium-hard axes and the hard axes, respectively.

8.2.3 Magnetostatic Interactions

Magnetostatic interactions represent the way the elementary magnetic moments interact over long distances within the body. In fact, the magnetostatic field at a given location within the body depends on the contributions from the whole magnetization vector field. We refer to this property as "nonlocal" character. Magnetostatic interactions can be taken into account by introducing the appropriate magnetostatic field $\mathbf{H}_m$ according to the Maxwell equations for magnetized media

$$\begin{cases} \nabla \cdot \mathbf{H}_m = -\nabla \cdot \mathbf{M} & \text{in } \Omega \\ \nabla \cdot \mathbf{H}_m = 0 & \text{in } \Omega^c \\ \nabla \times \mathbf{H}_m = 0 \end{cases} \qquad (8.5)$$

where the quantity $-\nabla \cdot \mathbf{M} = \rho_M$ plays the role of "magnetic charge" density.

The system (8.5) has to be completed with the boundary conditions at the body discontinuity surface $\partial \Omega$

$$\begin{cases} \mathbf{n} \cdot [\mathbf{H}_m]_{\partial \Omega} = \mathbf{n} \cdot \mathbf{M} \\ \mathbf{n} \times [\mathbf{H}_m]_{\partial \Omega} = 0 \end{cases} \qquad (8.6)$$

where
    the quantity $\mathbf{n} \cdot \mathbf{M} = \sigma_M$ plays the role of "surface magnetic charge" density
    $\mathbf{n}$ is the outward normal to the boundary $\partial \Omega$ of the magnetic body
    $[\mathbf{H}_m]_{\partial \Omega}$ is the jump of the vector field $\mathbf{H}_m$ across $\partial \Omega$

The expression for the contribution of magnetostatic interactions to the free energy of the system is given by

$$F_m = -\int_{\Omega_\infty} \frac{1}{2}\mu_0 \mathbf{M} \cdot \mathbf{H}_m \, dV \qquad (8.7)$$

Equation (8.7) denotes the *nonlocal* character of the magnetostatic field, as it functionally depends, through the boundary value problem (8.6), on the spatial distribution of the magnetization vector field through the entire body volume.

8.2.4 External Field

When a magnetic body is subjected to an external bias field $\mathbf{H}_{\text{ext}}$, the contribution to the energy is simply

$$F_{\text{ext}} = -\int_{\Omega} \mu_0 \mathbf{M} \cdot \mathbf{H}_{\text{ext}} \, dV \qquad (8.8)$$

This energy term is referred to in literature as Zeeman energy.

8.2.5 Energy Balance and Free Energy Functional Formulation

It is now possible to summarize the role played by each of the previous energy terms described above. In fact, because each energy contribution favors different energy minima configurations, the metastable configuration arises from a sort of competition among them. For example, a magnetic material uniformly magnetized along its easy axis presents a low energy contribution arising from both exchange and magnetocrystalline anisotropy fields, but, at the same time, a higher contribution coming from both the magnetostatic and Zeeman fields (if the applied field direction is different from the easy axis). On the contrary, a configuration having a minimum magnetostatic energy is a closed-loop structure, but such a configuration presents higher contributions coming from all the other energy terms.

The expression for the free energy of the ferromagnetic body can hence be formulated by collecting the previous equations

$$G(\mathbf{M}, \mathbf{H}_{\text{ext}}) = F_{\text{ex}} + F_{\text{an}} + F_m + F_{\text{ext}}$$

$$= \int_{\Omega} \left\{ A \left[\left(\nabla m_x \right)^2 + \left(\nabla m_y \right)^2 + \left(\nabla m_z \right)^2 \right] \right.$$

$$\left. + f_{\text{an}} - \frac{1}{2}\mu_0 \mathbf{M} \cdot \mathbf{H}_m - \mu_0 \mathbf{M} \cdot \mathbf{H}_{\text{ext}} \right\} dV \qquad (8.9)$$

which can be put in the compact form by expressing the exchange interaction energy density as $A(\nabla \mathbf{m})^2$

$$G(\mathbf{M}, \mathbf{H}_{\text{ext}}) = \int_{\Omega} \left\{ A(\nabla \mathbf{m})^2 + f_{\text{an}} - \frac{1}{2}\mu_0 \mathbf{M} \cdot \mathbf{H}_m - \mu_0 \mathbf{M} \cdot \mathbf{H}_{\text{ext}} \right\} dV \qquad (8.10)$$

By using a variational calculus, we are able to introduce now the so-called *effective field* as the variational derivative of the energy density of the system

$$\mathbf{H}_{\text{eff}} = -\frac{\delta G}{\delta \mathbf{M}} = \frac{2}{\mu_0 M_S} \nabla \cdot (A\nabla \mathbf{m}) - \frac{1}{\mu_0 M_S} \frac{\partial f_{\text{an}}}{\partial \mathbf{m}} + \mathbf{H}_m + \mathbf{H}_{\text{ext}} \qquad (8.11)$$

where M_S expresses the saturation magnetization of the body.

8.2.6 Equation of Motion

Once the effective field is defined, it is possible to present the so-called Landau–Lifshitz (LL) equation (Landau and Lifshitz 1935)

$$\frac{\partial \mathbf{M}}{\partial t} = -\gamma \mathbf{M} \times \mathbf{H}_{\text{eff}} \qquad (8.12)$$

which expresses the undamped precessional dynamics of the magnetization vector along the direction of the effective field (γ is the gyromagnetic ratio).

We observe that the Landau–Lifshitz equation (8.12) is a *conservative* (Hamiltonian) equation. Nevertheless, dissipative processes take place within the dynamic magnetization processes. The microscopic nature of this dissipation is still not clear and is currently the focus of considerable research. The approach generally followed to take into account dissipation precesses consists of introducing damping in a *phenomenological* way. Landau and Lifshitz introduced an additional torque term that pushes magnetization in the direction of the effective field. Then, the damped *Landau–Lifshitz equation* becomes, in the Gilbert formulation, (Gilbert 2004)

$$\frac{\partial \mathbf{M}}{\partial t} = -\gamma \mathbf{M} \times \mathbf{H}_{\text{eff}} + \frac{\alpha}{M_S} \left(\mathbf{M} \times \frac{\partial \mathbf{M}}{\partial t} \right) \qquad (8.13)$$

where α is the Gilbert damping parameter. To better understand the meaning of undamped and damped precessional dynamics, in Figure 8.4 we sketched the trajectories described by Equations 8.12 and 8.13, respectively.

By means of simple algebraic steps, it could be easily demonstrated that Equation 8.13 can be rewritten in the equivalent form

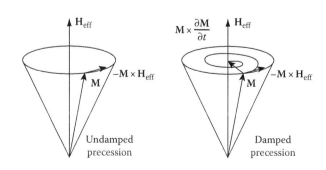

FIGURE 8.4 Comparison between undamped and damped precession.

$$\frac{\partial \mathbf{M}}{\partial t} = -\frac{\gamma}{1+\alpha^2}\mathbf{M}\times\mathbf{H}_{\text{eff}} - \frac{\gamma\alpha}{1+\alpha^2}\frac{1}{M_S}\mathbf{M}\times(\mathbf{M}\times\mathbf{H}_{\text{eff}}) \qquad (8.14)$$

where the damping term is now expressed as a double vector product. Such formulation brings the advantage of being an explicit expression for the time derivative of the magnetization so that, as it will be more clear in the next paragraph, it is more suitable to be implemented in a numerical framework.

8.2.7 Equation of Motion Including Spin-Transfer Torque Effects

At this stage, spin-polarized current has not yet been considered.

It has been demonstrated that current-induced spin-transfer effects in a spin-valve layered structure, such as the one sketched in Figure 8.3, are taken into account by including another torque term (Slonczewski 1996) into the equation of motion. The dynamic equation for magnetization dynamics driven by spin-polarized currents is called the Landau–Lifshitz–Gilbert–Slonczewski (LLGS) equation, and it is in the form

$$\frac{\partial \mathbf{M}}{\partial t} = -\gamma(\mathbf{M}\times\mathbf{H}_{\text{eff}}) + \frac{\alpha}{M_S}\left(\mathbf{M}\times\frac{\partial \mathbf{M}}{\partial t}\right) + \frac{\sigma I}{M_S}f(r)\Big[\mathbf{M}\times(\mathbf{M}\times\mathbf{p})\Big]$$

$$(8.15)$$

In this case, $\mathbf{M}$ is the magnetization vector of the FL while $\mathbf{p}$ is the unit vector in the direction of spin-polarization (magnetization of the PL), I is the electric current traversing the spin-valve structure, the dimensionless function $f(r)$ describes the spatial distribution of the current across the area of the FL, and the parameter σ is given by

$$\sigma = \frac{g\mu_B}{2eM_SLS}\varepsilon(\mathbf{M},\mathbf{p}) \qquad (8.16)$$

where
  g is the Landè factor
  μ_B is the Bohr magneton
  e is the modulus of the electron charge
  L is the FL thickness
  S is the area of the region involved by the current flow
  $\varepsilon(\mathbf{M},\mathbf{p})$ is the dimensionless spin-polarization efficiency (Slonczewski 1996, 1999)

The latter parameter is, in general, a function of the relative alignment between PL and FL and of the material under investigation. In its original formulation, Slonczewski derived the expression

$$\varepsilon(\mathbf{M},\mathbf{p}) = \left[\frac{-4+\left(1+\eta^3\right)\left(3+\left(\mathbf{M}\cdot\mathbf{p}/M_{S_{\text{FL}}}M_{S_{\text{PL}}}\right)\right)}{4\eta^{3/2}}\right]^{-1} \qquad (8.17)$$

with η being the polarization factor characteristic of the polarizing material. It has to be noticed, however, that in most cases,

this dependence does not lead to any qualitative or significant quantitative effects and, especially when only a qualitative understanding of the problem is required, can it be approximated by a constant value (generally treated as a free parameter to fit experimental data).

By comparing the formulation of the dissipation torque in Equation 8.14 and the formulation of the spin-transfer torque in Equation 8.15, one observes that, for a proper direction of the applied current (and exactly for *positive* current, which conventionally means electrons flowing from the FL to the PL), these expressions present a similar vector structure but with opposite signs (Slonczewski 1999, Slavin and Kabos 2005). A way to understand how these effects work is to consider that, while dissipation pushes the magnetization vector *toward* the direction of $\mathbf{H}_{\text{eff}}$, the spin-transfer pushes the magnetization vector *away* from the direction of $\mathbf{p}$. For such a reason, we might associate with the spin-transfer torque term the character of a "negative" damping, which opposes the natural "positive" Gilbert dissipation term (Slavin and Kabos 2005). The equality between these two terms represents the necessary condition under which stable persistent magnetization oscillations take place in a dissipative medium.

To give a complete picture on the mechanism underlying the process of spin-polarization and the subsequent spin-transfer torque effect, let us summarize in Figure 8.5 the FL magnetization dynamics occurring for both positive and negative currents.

The spin-polarization occurs via the absorption of the incident transverse spin current. The mechanism responsible for the transfer of the spin-angular momentum is the exchange interaction felt by electrons in the ferromagnet, which exerts a torque on the electron spins and, in turn, induces a reaction torque on the magnetization. In fact, let us consider the situation in which electrons incident from a nonmagnet are arranged

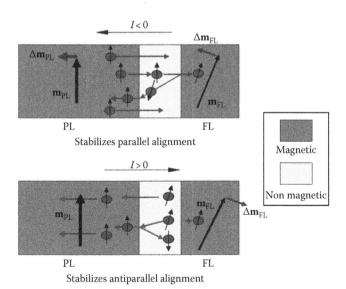

FIGURE 8.5 Spin-transfer torque effects for both directions of the current.

over a distribution of states represented by three different incident directions. All of these electrons are in the same spin state, which is transverse to the ferromagnetic spin density (e.g., the magnetization of either FL or PL). The reflected electron spins have a predominantly minority character and undergo a random spin rotation whose precession axis is on average directed along the direction antiparallel to the one of the ferromagnet. The transmitted majority electron spins, on the other hand, precess along the direction parallel to the ferromagnetic spin density as they go into the ferromagnet because the wave vectors for the majority and minority components are different. It has to be pointed out that, although the proposed picture is qualitatively correct for most of the commonly used metal/ferromagnet interfaces (e.g., Cu/Co), additional phenomena (such as spin-dephasing and spin-accumulation) should be taken into account to get a quantitative understanding of the investigated dynamics (Stiles and Miltat 2006).

By looking at Figure 8.5 it turns out that, for negative current (top figure), the electrons crossing the interface metal/FL are on average polarized along the direction of the PL and the applied torque ($\Delta\mathbf{m}_{FL}$) favors the parallel alignment between the two magnetic layers. The back-torque operated by the reflected electrons on the PL ($\Delta\mathbf{m}_{PL}$) is generally negligible as the PL presents usually large values for saturation magnetization and thickness, which keep it fixed in its equilibrium state. On the contrary, for positive current (bottom figure), the electrons reflected from the PL hit the interface metal/FL with an antiparallel polarization with respect to that of the PL. In this case, the consequent torque causes a destabilization of the FL magnetization, which tries to favor an antiparallel configuration between the magnetization of the two layers.

These phenomena hold under two main assumptions: (1) the length of the nonmagnetic path that electrons experience between the two magnetic layers has to be smaller than a given characteristic length scale, named "spin-diffusion length" (according to the classical drift diffusion model of spin transport, the ensemble averaged spin of electrons drifting and diffusing in a solid decays exponentially with distance due to spin dephasing interactions) and (2) the thickness of the magnetic layers is assumed to be too great for appreciable tunneling of minority-spin electrons, therefore, the component antiparallel to the magnetization is totally reflected back into the nonmagnetic spacer, while the component parallel to the magnetization is totally transmitted into the magnetic layer.

Of course, it has to be recalled that the current-driven destabilization mechanism, apart from occurring only for a proper direction of the spin-polarized current, also exhibits a *threshold* character, i.e., there exists a given critical current above which the phenomena observable in such devices (reversal or persistent oscillations) take place. The simplest explanation lies in the energy that the applied current has to supply to the system in order to modify the energy landscape created by all the contributions of the effective field and to balance the natural positive dissipation (Slonczewski 1999, Slavin and Kabos 2005).

8.3 Computational Micromagnetics of Nanoscale Spin-Valves: State of the Art

Let us come back to the mathematical viewpoint by recalling that the influence of current-induced spin-transfer effects on the magnetization dynamics of nanoscale devices is expressed by the LLGS equation (Equation 8.15). This differential equation presents a complex nonlinear and nonlocal structure, which, in general, prevents the formulation of an exact analytical solution unless strong simplifying assumptions are imposed to the analyzed problem.

To gain a better physical understanding of the complicated magnetization dynamics, alternative strategies make use of the numerical integration of the equation of motion by means of different techniques, such as boundary element (BE), finite elements (FE), finite differences (FD), or hybrid procedures. Such approaches, to some extent, offer better-accuracy solutions, with respect to analytical approaches, simply because they generally use a smaller set of restrictions and simplifying hypotheses. Numerical tools can also be used to explore the range of validity of analytical theories, an analysis which sometimes could not be carried out otherwise.

On a practical level, when numerically solving partial differential equations, the primary challenge is to create an equation that approximates the equation to be studied. The main requirement is the numerical stability of the integration scheme, meaning that errors in the input data and intermediate calculations do not accumulate and cause the resulting output to deviate substantially from the exact solution or, in the worst case, to be meaningless.

Let us give a brief overview of the numerical methods cited above.

The BE method is a numerical computational method used to solve linear partial differential equations that have been formulated as integral equations, and is applicable to problems for which Green's functions can be calculated. The boundary element method attempts to use the given boundary conditions to fit boundary values into the integral equation, rather than values throughout the space defined by a partial differential equation. Once this is done, in the post-processing stage, the integral equation can then be used again to numerically calculate the solution directly at any desired point in the interior of the solution domain. The boundary element method is often more efficient than other methods, including finite elements, in terms of computational resources for problems where there is a small surface/volume ratio even though it generally gives rise to fully populated matrices, which in turn yield a substantial increase of the storage and computational time requirements. In these cases, compression techniques might be applied, even though additional restrictions to the problem have to be added.

The FE method is a numerical technique for finding the approximate solutions of partial differential equations as well as of integral equations. The solution approach is based either on eliminating the differential equation completely (steady-state problems) or rendering the partial differential equation into an approximating system of ordinary differential equations, which are then numerically integrated using standard techniques (e.g., Euler's method, Runge-Kutta, etc.). In other words, the basic idea is to replace the infinite dimensional linear problem with a finite dimensional version. The algorithm can be subdivided into two steps: (1) rephrase the original problem in its weak or variational form and (2) discretize the weak form in a finite dimensional space. In step (2), one has to choose both finite-dimensional subspace of the original space and the shape functions that constitute the basis of the searched solution in that subspace. Because the original space is subdivided into a mesh of polygonal cells having whatever geometrical shape (triangular, tetragonal, etc.) the FE method is a good choice for solving partial differential equations when the computational domain presents a complex geometrical structure (e.g., for nonflat boundaries problems), but also when the desired precision varies over the entire domain or when the solution lacks smoothness (see Figure 8.6).

FD methods are numerical methods for approximating the solutions to differential equations using finite difference equations to approximate derivatives. The FD method implies a discretization of the computational region by using prismatic cells and then each derivative in the equation of motion has to be approximated (by using a given integration scheme) by a certain discretization.

A comparison between the FE and FD methods brings out the following main considerations:

- The FD method is an approximation of the differential equation, while the FE method is an approximation of the solution.
- The most attractive feature of the FE method lies in its ability to handle complex geometries (and boundaries) by means of a proper choice of the discretization cells. The FD method, on the contrary, in its basic formulation, is restricted to handle rectangular shapes.
- The most attractive feature of the FD method lies in the fact that it might be very easy to be implemented.

Let us now focus on the usage of FD methods with the aim of showing how the numerical solution of the LLGS equation can be carried out and, at the same time, offer a brief overview of the state-of-the-art micromagnetic studies applied to nanoscale devices.

8.3.1 Numerical Solution of LLGS Equation Based on FD Methods

The first step of a numerical approach consists of subdividing the entire volume of the specimen by using a regular mesh of prismatic cells ($\Delta_x \times \Delta_y \times \Delta_z$), as in Figure 8.7.

To establish a proper dimension for the grid size, it is necessary to define before a fundamental quantity, named "exchange length," as follows:

$$l_{ex} = \sqrt{\frac{2A}{\mu_0 M_S^2}} \qquad (8.18)$$

The exchange length gives an estimation of the characteristic dimension on which exchange interactions are dominant. For typical magnetic materials (e.g., silicon-iron alloys, called "Permalloy") l_{ex} is in the order of $5 \div 10\,nm$. Therefore, one expects that on a spatial scale in the order of l_{ex}, the magnetization is spatially uniform. For such a reason, within a micromagnetic framework, the cell size has to be smaller than the exchange length value, i.e., typically a cell-size as large as a $2 \div 5\,nm$-side is used. At the same time, the cell size should not be considered too small for at least two reasons: (1) the computational time should be kept in an affordable range and (2) the cell dimensions should be consistent with the mesoscopic scale.

Once the spatial discretization is addressed, the next step is to choose a proper time-step $\Delta\tau$. This value changes according to the integration scheme adopted, and the main constrain it has to satisfy is the convergence of the solution. To check that convergence, the criterion generally used is to analyze the time-evolution of the total system energy. Because of dissipation, the energy can only diminish, so that a gradual increase of the total energy reflects a divergent algorithm.

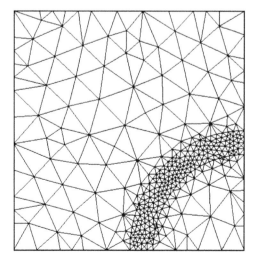

FIGURE 8.6 Example of mesh for FE methods with variable grid size to get a better resolution in correspondence of regions of interest.

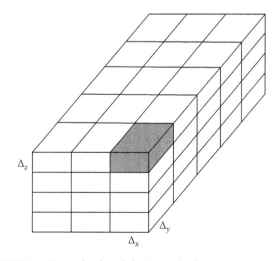

FIGURE 8.7 Example of mesh for FD methods.

Then, starting from an initial configuration of magnetization $\mathbf{m}(\mathbf{r}, \tau) \; \forall \mathbf{r} \in V$, we compute the effective field $\mathbf{h}_{\text{eff}}(\mathbf{r}, \tau) = \mathbf{h}_{\text{eff}}(\mathbf{m}(\mathbf{r}, \tau), \tau)$ (see below for details) in each computational cell.

After that, it is necessary to apply a numerical integration scheme to solve the LLGS equation and to obtain the updated value of the magnetization distribution in the next time step $\tau + \Delta\tau$: $\mathbf{m}(\mathbf{r}, \tau + \Delta\tau) \; \forall \mathbf{r} \in V$. Many integration schemes can be applied, either explicit or implicit, either with a fixed time-step or with an adaptive time-step (e.g., Euler, predictor-corrector, Runge-Kutta, mid-point, Adams-Bashforth-Moulton, Milne-Simpson, etc.). In most cases, one has to compute a finite differences representation for the derivate: that is the main reason why an explicit formulation for $\partial\mathbf{M}/\partial t$, like the one in Equation 8.14, is preferable with respect to the one shown in Equation 8.13.

Because our goal is to model the dynamic behavior of spintronic devices having a spin-valve structure as the one presented in Section 8.2, it is now necessary to focus on some aspects related to the implementation of the effective field for such devices. In the cases under investigation, the effective field has to be updated to take into account the additional contributions:

$$\mathbf{H}_{\text{eff}} = \mathbf{H}_{\text{ex}} + \mathbf{H}_{\text{an}} + \mathbf{H}_{\text{m}} + \mathbf{H}_{\text{ext}} + \mathbf{H}_{\text{mc}} + \mathbf{H}_{\text{Amp}} + \mathbf{H}_{\text{eddy}} + \mathbf{H}_{\text{th}} \quad (8.19)$$

Here $\mathbf{H}_{\text{ex}}$, $\mathbf{H}_{\text{an}}$, $\mathbf{H}_{\text{m}}$, and $\mathbf{H}_{\text{ext}}$ represent the exchange, anisotropy, magnetostatic, and external field contributions already introduced previously. The new contributions arise from $\mathbf{H}_{\text{mc}}$, which represents the magnetostatic coupling between PL and FL (it has to be distinguished from $\mathbf{H}_{\text{m}}$, which is the self-magnetostatic, or demagnetizing, field induced by the magnetization of each layer within its volume); $\mathbf{H}_{\text{Amp}}$, which represents the Ampère (or Oersted) magnetic field created by the current flow; $\mathbf{H}_{\text{eddy}}$, the field contribution arising from eddy currents; and $\mathbf{H}_{\text{th}}$, which represents the stochastic contribution arising from thermal noise.

Let us now give some details about the numerical implementation of each contribution within a FD micromagnetic framework. To do that, we will use a normalized notation, where each term is divided by the saturation magnetization value and written in lowercase letter, e.g., $\mathbf{m} = \mathbf{M}/M_S$.

The exchange field, $\mathbf{h}_{\text{exc}}$, is computed by considering the discrete expression for the exchange energy

$$f_{\text{ex}}(i,j,k) = A\left[\left(\nabla m_x(i,j,k)\right)^2 + \left(\nabla m_y(i,j,k)\right)^2 + \left(\nabla m_z(i,j,k)\right)^2 \right] \quad (8.20)$$

where we used the notation for each Cartesian component $\alpha = x, y, z$

$$\left(\nabla m_\alpha\right)^2 = \left(\frac{\partial m_\alpha}{\partial x}\right)^2 + \left(\frac{\partial m_\alpha}{\partial y}\right)^2 + \left(\frac{\partial m_\alpha}{\partial z}\right)^2 \quad (8.21)$$

Under the approximation of finite differences, the derivative is substituted by the incremental differences (computed at the center of each cell), that is

$$\left(\nabla m_\alpha\right)^2 \approx \left(\frac{\Delta_x m_\alpha}{\Delta x}\right)^2 + \left(\frac{\Delta_y m_\alpha}{\Delta y}\right)^2 + \left(\frac{\Delta_z m_\alpha}{\Delta z}\right)^2 \quad (8.22)$$

where $\Delta x, \Delta y, \Delta z$ are the sizes of the computational cells, whereas $\Delta_x, \Delta_y, \Delta_z$ represent the finite difference operators along the Cartesian axes x, y, z. Let us consider the magnetization vector $\mathbf{M}(i,j,k) = M_S \mathbf{m}(i,j,k)$ at the point (i,j,k) and $\mathbf{M}(i+1,j,k) = M_S \mathbf{m}(i+1,j,k)$ the corresponding vector at the neighboring cell $(i+1,j,k)$ in the x direction. We can rewrite the expression (8.22) as follows:

$$\left(\frac{\Delta_x m_\alpha}{\Delta x}\right)^2 = \left(\frac{m_x(i+1,j,k) - m_x(i,j,k)}{\Delta x}\right)^2$$

$$= \frac{m_x^2(i+1,j,k) - 2m_x(i+1,j,k)m_x(i,j,k) + m_x^2(i,j,k)}{\Delta x^2} \quad (8.23)$$

By taking into account that $|\mathbf{m}(i,j,k)| = 1 \; \forall \, (i,j,k)$, the expression (8.23) can be simplified as

$$\left(\frac{\Delta_x m_\alpha}{\Delta x}\right)^2 = \frac{1}{\Delta x^2}\left[2 - 2m_x(i+1,j,k)m_x(i,j,k)\right] \quad (8.24)$$

Also, by taking into account the neighboring cell in the negative directions, one ends up with

$$\left(\nabla m_x\right)^2 \approx \frac{1}{\Delta x^2}\Big[\left(2 - 2m_x(i+1,j,k)m_x(i,j,k)\right) + \left(2 - 2m_x(i,j,k)m_x(i-1,j,k)\right)\Big]$$
$$+ \frac{1}{\Delta y^2}\Big[\left(2 - 2m_x(i,j+1,k)m_x(i,j,k)\right) + \left(2 - 2m_x(i,j,k)m_x(i,j-1,k)\right)\Big]$$
$$+ \frac{1}{\Delta z^2}\Big[\left(2 - 2m_x(i,j,k+1)m_x(i,j,k)\right) + \left(2 - 2m_x(i,j,k)m_x(i,j,k-1)\right)\Big] \quad (8.25)$$

Under the hypothesis of cubic cells ($\Delta x = \Delta y = \Delta z$), we can generalize Equation 8.25 to the case of $N_x \times N_y \times N_z$ cells (along the x, y, and z axis, respectively) as follows:

$$f_{\text{ex}}(i,j,k) = \frac{2A}{\Delta x^2} \sum_{i'=1}^{N_x} \sum_{j'=1}^{N_y} \sum_{z'=1}^{N_z} \left[1 - \mathbf{m}(i,j,k)\mathbf{m}(i',j'k')\right]$$

$$= \frac{2A}{\Delta x^2} N_x N_y N_z - \frac{2A}{\Delta x^2} \mathbf{m}(i,j,k) \sum_{i'=1}^{N_x} \sum_{j'=1}^{N_y} \sum_{z'=1}^{N_z} \mathbf{m}(i',j'k') \quad (8.26)$$

The exchange field is computed by using the functional derivative of the energy ($\delta/\delta\mathbf{m}$), which becomes the ordinary derivative ($\partial/\partial\mathbf{m}$) in the discrete domain:

$$h_{\text{exc}}(i,j,k) = -\frac{1}{\mu_0 M_S^2} \frac{\partial f_{\text{ex}}(i,j,k)}{\partial \mathbf{m}(i,j,k)} \quad (8.27)$$

By substituting Equation 8.26 into Equation 8.27, one obtains the final implementation of the six-neighbors exchange field:

$$h_{exc}(i,j,k) = \frac{2A}{\mu_0 M_S^2} \left[\frac{\mathbf{m}(i+1,j,k) + \mathbf{m}(i-1,j,k)}{\Delta x^2} \right.$$
$$+ \frac{\mathbf{m}(i,j+1,k) + \mathbf{m}(i,j-1,k)}{\Delta y^2}$$
$$\left. + \frac{\mathbf{m}(i,j,k+1) + \mathbf{m}(i,j,k-1)}{\Delta z^2} \right] \qquad (8.28)$$

It should be pointed out that the partial derivative approximation for finite differences holds for small arguments, that is, when the angle formed between the magnetization vectors in two adjacent cells is small.

The anisotropy field, $\mathbf{h}_{an}$, in the case of uniaxial anisotropy, is simply computed by

$$h_{ani}(i,j,k) = \frac{2K_C}{\mu_0 M_S^2} \left(\mathbf{m}(i,j,k) \cdot \mathbf{u}_K \right) \mathbf{u}_K \qquad (8.29)$$

where

K_C is the anisotropy constant
$\mathbf{u}_K$ is the direction of uniaxial anisotropy

Both magnetostatic fields ($\mathbf{H}_m$ and $\mathbf{H}_{mc}$) are computed by using the generalization of the demagnetizing tensor given in (Newell et al. 1993)

$$h_{m,\alpha}(i,j,k) = \sum_\beta^{(x,y,z)} \sum_{i'=1}^{N_x} \sum_{j'=1}^{N_y} \sum_{z'=1}^{N_z} N_{\alpha\beta}(i-i',j-j',k-k') m_\beta(i',j',k') \qquad (8.30)$$

where

$h_{m,\alpha}$ represents the component of the magnetostatic field
$N_{\alpha\beta}(i-i',j-j',k-k')$ are the components of the Newell magnetostatic tensor, which only depend on both the distance between the source and the cell position, and the specimen geometry

The expression (8.30) presents the form of a three-dimensional convolution in the "*coordinates space*" (x,y,z). By using the *Convolution Theorem*, it is possible to express the previous convolution as a product scalar in the "*Fourier space*," or phases space (k_x, k_y, k_z):

$$\tilde{h}_{m,\alpha}(k_x, k_y, k_z) = \Im\left[h_{m,\alpha}(i,j,k) \right]$$
$$= \sum_\beta^{(x,y,z)} \Im\left[N_{\alpha\beta}(i,j,k) \right] \Im\left[m_\beta(i,j,k) \right]$$
$$= \sum_\beta^{(x,y,z)} \tilde{N}_{\alpha\beta}(k_x, k_y, k_z) \tilde{m}_\beta(k_x, k_y, k_z) \qquad (8.31)$$

To calculate the magnetostatic field, $\mathbf{h}_m$, it is necessary to follow these steps: (1) calculate the Newell tensor coefficients $N_{\alpha\beta}$; (2) perform the FFT of the previous tensor $\tilde{N}_{\alpha\beta}$; (3) perform the

FFT of the magnetization distribution in each cell $\tilde{m}_\beta$; and (4) perform the inverse FFT of the inner product between $\tilde{N}_{\alpha\beta}$ and $\tilde{m}_\beta$:

$$h_{m,\alpha}(i,j,k) = \Im^{-1}\left[\tilde{h}_{m,\alpha}(k_x, k_y, k_z) \right]$$
$$= \Im^{-1}\left[\sum_\beta^{(x,y,z)} \Im\left[N_{\alpha\beta}(i,j,k) \right] \Im\left[m_\beta(i,j,k) \right] \right] \qquad (8.32)$$

By denoting with N the total number of grid cells, such approach brings a computational complexity that grows as $N \log N$.

The external field $\mathbf{h}_{ext}$ is simply computed, in the limit of uniform fields, by applying the corresponding components of the external field to each computational cell:

$$h_x(i,j,k) = \mathbf{h}_{ext} \cdot \hat{\mathbf{e}}_x, \quad h_y(i,j,k) = \mathbf{h}_{ext} \cdot \hat{\mathbf{e}}_y, \quad h_z(i,j,k) = \mathbf{h}_{ext} \cdot \hat{\mathbf{e}}_z \qquad (8.33)$$

Moreover, once we inject an electric current density $\mathbf{J} = J_0 \mathbf{e}_z$ into a spin-valve stack composed by ferromagnetic and metallic materials, a new magnetic field is generated according to the Ampère (or Oersted) law. The values of the *Ampère field*, $\mathbf{H}_{Amp}$, in each point of the volume can be calculated by using the Ampère law:

$$\nabla \times \mathbf{H}_{Amp}(\mathbf{r},t) = \mathbf{J}(\mathbf{r},t) \qquad (8.34)$$

The Ampère field is a solenoidal field and thus exhibits the following properties:

$$\begin{cases} \nabla \cdot \mathbf{H}_{Amp}(\mathbf{r},t) = 0 \\ \nabla \times \mathbf{H}_{Amp}(\mathbf{r},t) = \mathbf{J}(\mathbf{r},t) \end{cases} \qquad (8.35)$$

An additional contribution arises from *eddy currents*. An eddy current (also known as a Foucault current) is an electrical phenomenon discovered by Léon Foucault in 1851. It is caused when a moving (or changing) magnetic field intersects a conductor or vice-versa. The relative motion causes a circulating flow of electrons, or current, within the conductor. These circulating eddies of current create electromagnets with magnetic fields that oppose the effect of the applied magnetic field (see Lenz's law). The stronger the applied magnetic field, or the greater the electrical conductivity of the conductor, or the greater the relative velocity of motion, the greater the currents developed and the greater the opposing field. The Maxwell equations for eddy currents are given by

$$\begin{cases} \nabla \times \mathbf{H}_{eddy}(\mathbf{r},t) = \sigma' \mathbf{E}(\mathbf{r},t) + \varepsilon' \dfrac{\partial \mathbf{E}(\mathbf{r},t)}{\partial t} \\[2mm] \nabla \cdot \mathbf{H}_{eddy}(\mathbf{r},t) = 0 \\[2mm] \nabla \times \mathbf{E}(\mathbf{r},t) = -\dfrac{\partial \mathbf{B}(\mathbf{r},t)}{\partial t} \\[2mm] \nabla \cdot \mathbf{E}(\mathbf{r},t) = 0 \end{cases} \qquad (8.36)$$

where

σ′ is the electrical conductivity
ε′ is the electric permeability
E is the electric field

To perform an analogy between $\mathbf{H}_{Amp}(\mathbf{r}, t)$ and $\mathbf{H}_{eddy}(\mathbf{r}, t)$, in the former the source of the field is given by the applied current density $\mathbf{J}(\mathbf{r}, t)$, whereas in the latter the sources of the field are given by both the time variation of the magnetic field $-\partial \mathbf{B}(\mathbf{r}, t)/\partial t$ and the conduction current $\mathbf{J}_C = \sigma′\mathbf{E}(\mathbf{r}, t) + \varepsilon′(\partial \mathbf{E}(\mathbf{r}, t)/\partial t)$. The effects of eddy currents become relevant with an increase in the device volume and, since we will deal with structures with relative small volumes, this contribution is generally disregarded (Martinez et al. 2004).

It has to be noted that both $\mathbf{H}_{Amp}$ and $\mathbf{H}_{mc}$ are nonuniform fields so that their effects cannot be correctly evaluated by using *macrospin* models (by "macrospin" we mean a particle that remains homogeneously magnetized independently of external conditions). Figures 8.8 and 8.9 show examples of the spatial configuration of these fields for a structure having circular and rectangular cross-sections, respectively.

The last contribution to the effective field arises from the effect of thermal fluctuations (or noise) on the magnetization

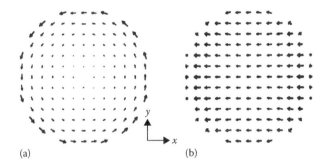

FIGURE 8.8 Spatial distribution of (a) Ampère field and (b) magnetostatic coupling field for a device having a circular cross-section.

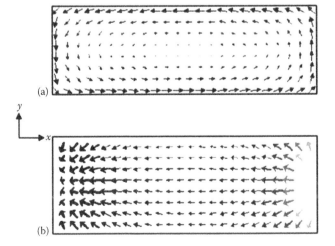

FIGURE 8.9 Spatial distribution of (a) Ampère field and (b) magnetostatic coupling field for a device having a rectangular cross-section.

dynamics of nanoscale devices. Considering that experimental data are well described by the current dependent activation barriers that agree with the prediction of the LLG-based models (Garcia-Palacios and Lazaro 1998, Krivorotov et al. 2004, Li and Zhang 2004), it is possible to include in a micromagnetic framework a thermal field as an additive random field to the deterministic effective field for each cell. It is done by starting from the formalism of Brownian motion (Gardiner 1985, Kloeden and Platen 1999).

In this case, the previous deterministic equation become a stochastic one, named the *Langevin equation*. The fluctuating term, which appears in the Langevin equation presents an integral structure that needs to be analyzed by using the theory of stochastic processes (Brown 1963, 1979, Garcia-Palacios and Lazaro 1998).

From the physical point of view (Garcia-Palacios and Lazaro 1998), the more appropriate interpretation of those stochastic integrals is called the *Stratonovich interpretation*. In this sense, Brown (1963, 1979) developed the *Fokker–Planck equation* to study the temporal evolution of nonequilibrium probability distributions. Several techniques have been developed to solve the Fokker–Planck equation. Some of those allow the direct computation of the probability distribution of magnetization, while others numerically compute the eigenvalues and amplitudes of the most relevant dynamics modes. Instead of solving the Fokker–Planck equation, an approximation is to build the solutions of the Langevin equation (Garcia-Palacios and Lazaro 1998). The Langevin dynamics gives origin to the stochastic trajectories of the system variables (time-evolution of magnetization) and, starting from this approach, we can compute the statistic behavior by averaging over many realizations (for "realizations" we mean a particular statistic trajectory obtained by starting from a given sequence of aleatory numbers for the statistic system variables).

From the computational point of view, we should take into account that thermal fluctuations in magnetic media occur at correlation times much shorter than the typical response time of magnetic systems (Brown 1963, 1979). In other words, thermal fluctuations arise from high-frequency perturbations. The thermal field that is generally used to simulate thermal effects can be therefore represented as a *Wiener stochastic process* (Van Kampen 1987). Moreover, because fluctuations arise from the interaction between magnetization and an enormous number of microscopic degrees of freedom (phonons, conduction electrons, nuclear spins, etc.) having equivalent stochastic properties, it is generally assumed that the above stochastic process can be treated as a *Gaussian white noise* (Brown 1963, Garcia-Palacios and Lazaro 1998). The same microscopic degrees of freedom are responsible for the relaxation of the magnetization precession, because fluctuations and dissipation are phenomena strictly related each other (*Fluctuation-Dissipation Theorem*). Because of those degrees of freedom, the thermal field $\mathbf{H}_{th}$, to be added to the deterministic effective field, has to satisfy the properties of a Gaussian stochastic vector process.

This leads to a definition of the stochastic Langevin–Landau–Lifshitz–Gilbert (LLLG) equation. In order to also take into account the spin-torque term within this new formulation, the main hypotheses are: (1) the spin torque does not contain a fluctuating field, (2) the fluctuating field does not depend on spin torque, and (3) the magnetization configuration of the PL does not depend on temperature.

The thermal field, $\mathbf{H}_{th}$, is hence included as a random fluctuating three-dimensional vector quantity given by

$$\mathbf{H}_{th} = \xi \sqrt{2 \frac{\alpha}{1+\alpha^2} \frac{k_B T}{\mu_0 \gamma' \Delta V M_S \Delta t}} \qquad (8.37)$$

where

k_B is the Boltzmann constant
ΔV is the volume of the computational cubic cell
Δt is the simulation time-step
T is the temperature of the sample
$\gamma' = \gamma/(1 + \alpha^2)$
ξ is a Gaussian stochastic process

The thermal field satisfies the following statistical properties:

$$\begin{cases} \langle H_{th,k}(t) \rangle = 0 \\ \langle H_{th,k}(t) H_{th,l}(t') \rangle = D \delta_{kl} \delta(t - t') \end{cases} \qquad (8.38)$$

where k and l represent the Cartesian coordinates x, y, and z. According to this, each component of $\mathbf{H}_{th}$ is a space and time independent random Gaussian distributed number (Wiener process) with zero mean value. The constant D measures the strength of thermal fluctuations and its value is obtained from the Fokker–Planck equation.

Finally, the micromagnetic framework has to be completed with proper boundary conditions, which describe the behavior of the magnetization vector at the computational boundaries. The formulation of boundary conditions is therefore dependent on the geometry under investigation. Some examples will be given in Section 8.3.3.

8.3.2 Spin-Transfer Torque and Giant Magnetoresistance: Dual Phenomena

In Section 8.3.1, we described the spin-transfer torque mechanism and showed the dynamic equation of motion for the magnetization which includes this effect. We described that, if a spin-polarized current presents proper sign and amplitude, it may destabilize the equilibrium configuration of the magnetization of the thinner FL leading it to either a reversal process or a persistent precession. Lots of questions could immediately follow such analysis: are these phenomena so interesting to deserve potential applications in the field of information technology? If yes, how do we observe and detect experimentally these magnetization dynamics?

The answer lies in the existence of another phenomenon, called giant magnetoresistance (GMR), which can be

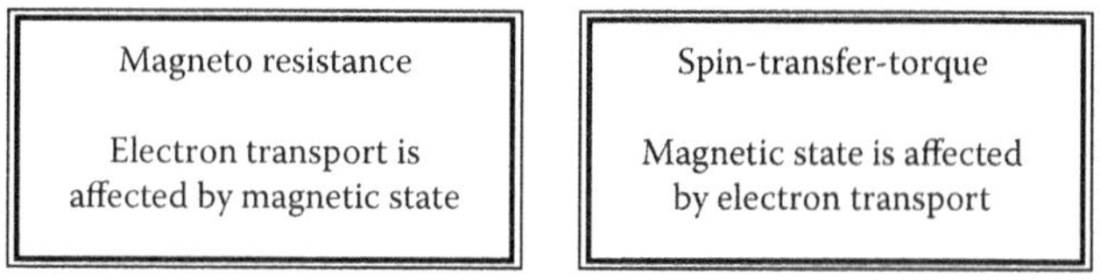

FIGURE 8.10 STT and GMR effects as dual mechanisms.

considered the dual mechanism of the spin-transfer torque (STT). Observing the phenomena from a more general viewpoint, we can describe the STT as the mechanism by which we account for the modifications of the magnetic state of a system induced by an electric current. For duality, the GMR effect (Baibich et al, 1988, Parkin et al. 1990, 1991, Schad et al. 1994) accounts for the variations of the electrical state of a system induced by the magnetic state (see Figure 8.10). For the discovery of the GMR effect, Fert and Gruenberg shared the Nobel Prize for Physics in 2007.

The GMR is a quantum-mechanical effect observed in thin film structures composed of alternating ferromagnetic and nonmagnetic metal layers and arises from the spin-dependent scattering of conduction electrons by spins in ferromagnetic layers. When the spin of ferromagnetic layers is anti-parallel to the conduction electrons, then the scattering is increased. Therefore, the magnetoresistivity will be varied by the relative orientation of magnetization in neighboring ferromagnetic layers separated by a nonmagnetic metallic thin layer. In detail, the GMR effect manifests itself as a significant decrease in resistance from the zero-field state, when the magnetization of adjacent ferromagnetic layers are antiparallel due to a weak anti-ferromagnetic coupling between layers, to a lower level of resistance when the magnetization of the adjacent layers align due to an applied external field. The spin of the electrons of the nonmagnetic metal align parallel or antiparallel with an applied magnetic field in equal numbers, and therefore suffer less magnetic scattering when the magnetizations of the ferromagnetic layers are parallel.

In the analyzed case of the three-layers system, it behaves as a spin-valve for electron transport from one ferromagnetic layer to the other. The relative alignment of the magnetization vectors of the two ferromagnetic layers can be easily detected by measuring the system resistance (see Figure 8.11). At the same time, it is possible to associate two different logic states to the different electric configurations (e.g., bit 0 = low-resistance state, bit 1 = high-resistance state). This is the basic working principle of magnetoresistive random access memories (MRAM). On the other hand, if the magnetization undergoes a persistent precession, the output signal might manifest itself as a periodic variation of the resistance of the device, which could be used to build up nanoscale oscillators.

8.3.3 Nanoscale Devices: Pillar vs. Point-Contact

Within the class of nanoscale devices, the two most common experimental geometries are the so-called nano-pillar and nano-contact (or point-contact).

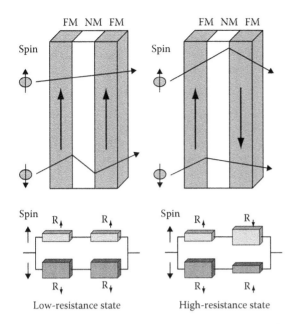

FIGURE 8.11 A schematic of a GMR-based spin-valve GMR showing a higher resistance state in the anti-parallel configuration with respect to the parallel one.

The main differences existing between them involve both the lateral sizes and the current density distribution. In fact, nano-contact devices generally present larger lateral dimensions (in the order of tenths of microns) with respect to those of nano-pillar (hundreds of nanometers). In addition, the perpendicular-to-plane applied current involves the whole physical (cross-section) area of pillar devices, whereas only a reduced region (the contact area, typically of the radius of 10–80 nm) is involved in nano-contact devices (see Figure 8.12).

The nano-pillar structure is widely used as an elementary memory cell where spin-polarized current drives the process of magnetization reversal (or switching), even in the absence of an external bias magnetic field.

On the other hand, the nano-contact geometry becomes particularly suitable to design microwave spin-transfer oscillators

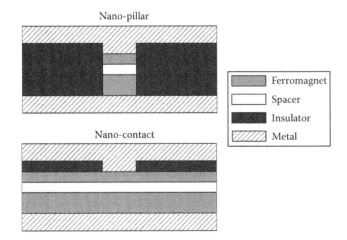

FIGURE 8.12 Nano-pillar and nano-contact spin-valve geometries.

with very narrow linewidths because it combines the advantages of having a larger effective volume (with respect to pillar devices) (Kim et al. 2008) and a negligible influence of edge roughness and defects (Rippard et al. 2004). In fact, in the case of extended nano-contact geometry (also referred to as "nonconfined" systems), the FL is not bounded in the plane; therefore, the magnetic oscillations induced by spin-polarized current can excite spin-wave modes, which propagate through the FL material (in analogy to the radiation of electromagnetic waves via a transmitting antenna). On the contrary, the finite sizes of a nano-pillar device make it more similar to a magnetic resonator because of its reflecting boundaries. In fact, the spin-wave eigenmodes of such resonators, even though they could be spatially nonuniform, can have only a discrete frequency spectrum determined by the finite in-plane sizes of the device.

The structural differences between these two geometries imply not only different technological applications, but also require different approaches in the corresponding numerical investigations.

In fact, the micromagnetic study of the dynamics involved in a nano-pillar device can be carried out by directly applying the hints given in Sections 8.3.1 and 8.3.2 with the inclusion of normal boundary conditions at the computational edges (which also correspond to the physical edges, because of the finite dimensions):

$$\frac{\partial \mathbf{m}}{\partial \mathbf{n}}\bigg|_{\partial\Omega} = 0 \qquad (8.39)$$

(**n** is the versor of the normal to the boundary $\partial\Omega$). We should notice that the normal derivative is different from zero only if surface anisotropy effects are taken into account.

On the other hand, the analysis of dynamics involved in nano-contact geometry requires a more sophisticated approach. In fact, one of the difficulties of FD methods lies in the fact that the investigated motion equation has to be solved in a discretized domain which, if applied to extended areas, would require prohibitive memory allocation and computational times. This is the case of nano-contact devices.

Nevertheless, theoretically boundless space extension problems have been solved by using special conditions at the boundaries of the reduced computational domain, in order to absorb the outgoing waves. Such a need for free-space simulation happens in a lot of physical problems and especially in wave-structure interactions.

Following the previous discussions, it is clear that FD techniques applied to extended geometries impose the numerical implementation of computational areas smaller than the physical ones. Therefore, it is necessary to develop a method for the waves absorption at the artificial boundaries to prevent the reflections of the propagating spin wave modes from these boundaries that, in a medium with relatively low dissipation (as Permalloy), might occur otherwise. If proper absorbing boundary conditions (ABC) are not considered, a spurious (purely computational) interference occurs which, in turn, can introduce a substantial distortion in the computed picture of the phenomenon and can seriously affect the properties of the excited spin-waves. As it will be shown below, due to the impossibility of implementing some sort of analytical operators within a micromagnetic framework for point-contact geometries, absorbing boundary conditions based on spatial-dependent damping functions have to be introduced.

The problem of finding the exact analytical formulation of the perfectly absorbing conditions for spin waves at the edges of the computational region has not been solved so far. The analytical formulation of ABC generally requires the knowledge of the exact wave solution (Enquist and Majda 1977, Renaut 1992, Alpert et al. 2002). In fact, the most general first-order analytical formulation for local ABC was introduced by Higdon (1987) as follows:

$$\left[\cos\theta\,\frac{\partial}{\partial t} - c\frac{\partial}{\partial x}\right]U = 0 \qquad (8.40)$$

where the Higdon operator is enclosed in brackets and it is used to completely absorb plane waves U propagating with speed c at the angles $\pm\theta$ with respect to the x axis. Unfortunately, from the numerical point of view, the lack of knowledge of the exact spin-wave velocity c does not allow the implementation of this formalism within a micromagnetic framework. Moreover, if we restrict the analysis to the case of perpendicular incidence (where $\cos\theta = 0$), the previous condition becomes the common boundary condition (see Equation 3.22) used in the modeling of the finite specimen $\partial U/\partial x = 0$, which cannot assure the waves absorption at each boundary.

Consequently, we need an alternative numerical strategy for the waves absorption at the computational boundaries. The attempts to find such conditions in numerical simulations have been undertaken by assuming that magnetic dissipation in the magnetic medium of the FL increases near the borders of the computational region according to a certain chosen empirical dependence on coordinates (Berkov and Gorn 2006, Consolo et al. 2007a,c). Following this idea, a micromagnetic approach could be generalized by including a site-dependent damping function in the model.

A possible scheme was first introduced by Berkov et al. in (Berkov and Gorn 2006) where the damping parameter was considered to vary between its physical value (α_O) and a higher artificial one ($\alpha_O + 2\Delta\alpha$) in a *smoothed* way, which starts at the distance from the center $r \approx (R_O - \sigma_\alpha)$ and involves a ring with a width of $\approx 2\sigma_\alpha$:

$$\alpha(r) = \alpha_O + \Delta\alpha\left(1 + \tanh\frac{r - R_O}{\sigma_\alpha}\right) \qquad (8.41)$$

The ABC introduced by Berkov are based on the following three main assumptions:

1. The dissipation within and nearby the contact area must be equal to its physical value.
2. Dissipation far from the contact area and close to the simulated boundaries must be large enough to ensure the wave energy absorption.
3. The spatial variation between these two values must be smooth enough to prevent wave reflection at the boundary.

Leaving the first two Berkov's assumptions unchanged, an alternative approach has been proposed in Consolo et al. (2007c) for perpendicular-to-plane magnetized films, and is based on the following requirements:

1. Creating a nonphysical absorbing medium surrounding the computational area
2. Considering a damping coefficient two orders of magnitude greater than the physical one just in the boundary cells
3. Entirely preserving the physical properties of the computational area

It has to be pointed out that, in spite of the formal equivalence between these empiric procedures and of the effectiveness of both procedures to absorb sufficiently well outgoing waves, the latter proposal, differently from the first one, allows the preservation of the physical properties in a larger computational area since it is based on the association of a localized *abrupt* change of the energy absorption at the computational boundaries

$$\begin{cases} \alpha(r) = \alpha_O & \text{for } r < R \\ \alpha(r) = m \cdot \alpha_O & \text{for } r = R \end{cases} \qquad (8.42)$$

where

R is the radius of the computational area
m is the factor that amplifies the dissipation at the boundaries

Finally, as briefly introduced above, we need to mathematically formulate the difference existing in the spatial distribution of current for pillar and nano-contact devices, which is expressed by a proper choice of the dimensionless function $f(r)$ in Equation 8.15.

In particular, for pillar devices, by assuming a uniform current density distribution, it is quite straightforward to consider it as

$$\begin{cases} f(r) = 1 & \text{for } r \le R \\ f(r) = 0 & \text{for } r > R \end{cases} \qquad (8.43)$$

which simply implies that all the computational cells are traversed by the same current value I.

On the other hand, for nanocontact geometry, it has to be taken into account that the current flow involves only the contact region of radius R_C so that, under the same condition of uniformity, we consider

$$\begin{cases} f(r) = 1 & \text{for } r \le R_C \\ f(r) = 0 & \text{for } r > R_C \end{cases} \qquad (8.44)$$

which implies that only the computational cells within the contact area are traversed by the current I, whereas there exists an abrupt cut-off of the current outside the contact. While such an approximation is physically unrealistic, it has been demonstrated that it is able to capture most of the dynamics observed in laboratory experiments. However, more sophisticated shape functions can be derived by numerically solving the *Poisson equation* to get the corresponding current density distribution through the whole computational area.

8.4 Conclusions and Future Perspective

In this contribution, we presented a brief overview of a micromagnetic technique that is used to investigate the behavior of nanoscale spintronic devices. It is based on the numerical integration of the nonlinear LLGS equation. After developing a full-scale FD micromagnetic tool, a question might be asked. Is it possible to develop a framework simpler than the micromagnetic one but that allows for a gain in the equivalent understanding of the complex magnetization dynamics? Actually, magnetization dynamics might be most easily investigated using *macrospin* (less properly called "single-domain") approximation. The macrospin approximation assumes that the magnetization of a sample stays spatially uniform throughout its motion and can be treated as a single macroscopic spin. Since the spatial variation of the magnetization is frozen out, exploring the dynamics of magnetic systems is much more tractable using the macrospin approximation than it is using full micromagnetic simulations. The macrospin model makes it easy to explore the phase space of different torque models, and it has been a very useful tool for gaining a *zeroth-order* understanding of spin-torque physics. On the other hand, it obviously suffers from some intrinsic limitations (e.g., the impossibility of reproducing the nonuniform spatial distribution of magnetization and fields), which sometimes prevents the possibility of mimicking experimental data. It is possible to estimate qualitatively the critical size for which a macrospin approximation could be considered still sufficiently appropriate. It mainly depends on the competition between exchange and magnetostatic energy: the former favoring collinear (uniform) magnetic state, the latter favoring closed-flux configurations. As the exchange energy density of the closed-loop configuration obviously increases with decreasing particle size (because magnetization gradients become larger), it leads to the result that only a collinear magnetization state is energetically stable below a certain critical size. It could be demonstrated that such critical length-scale is about—four to eight times the exchange length (about 20–40 nm for usual soft magnetic materials).

On the other hand, micromagnetic frameworks offer a powerful tool for investigating magnetization phenomena occurring at the mesoscopic scale because of the accurate spatial resolution (few nanometers) and of the theoretically infinite bandwidth (restricted only by the integration time step). Because of that, it is also possible to gain additional information on the investigated problem, which could not be acquired from a laboratory experiment.

How can we summarize this issue between an accurate but time-consuming micromagnetic approach and a fast and often oversimplified macrospin model?

Let us recall that full-scale simulations are supposed to include more (and not fewer) known features of the investigated

system than macrospin models and, at the same time, use fewer free (adjustable) parameters. Because of that, if a simplified macrospin approach were able to produce a better agreement with experiment than a micromagnetic approach, it would imply an incomplete comprehension of some crucial properties of the system under study. In other words, it does not necessarily mean that a wrong choice of the values of parameters has been considered, but rather it should recall for further studies to gain a better understanding of the problem, which in turn should bring some corrections to the model. Moreover, because full-scale simulations are quite time-consuming, it is necessary to get an accurate knowledge of the system parameters, together with their spatial dependence (especially near critical regions, such as close to the edges) from independent sources, such as a high-quality experiment. Such information is of crucial importance for the modeling of nano-scale devices due to their extremely small sizes.

What about present and future applications of micromagnetic framework?

So far, full-micromagnetic frameworks have been able to mimic the experimental behavior of most spintronic devices (pillar, nano-contact, MTJ, exchange-bias systems, phase-locked nano-contacts) as well as to perform predictions on new high-performance setups (Finocchio et al. 2006a,b, 2007, 2008, Choi et al. 2007, Consolo et al. 2007d, 2008, Hrkac 2008). At the same time, micromagnetic tools are successfully used to validate analytical theories as well. For example, the analysis of the disagreement reported between an analytical theory about the excitation of the nonlinear evanescent spin-wave "bullet" mode in in-plane magnetized nanocontact devices (Slavin and Tiberkevich 2005) and earlier results of micromagnetic simulations for the same geometry was intriguing and stimulating (Berkov and Gorn 2006). In fact, while the approximated theory was able to capture the underlying physics and consequently reproduce most of the experimental observations, micromagnetic simulations failed in that attempt. The problem was solved by using a new numerical strategy based on the application of decreasing currents starting from a large supercritical regime (Consolo et al. 2007b), which contains a lot of nonlinear "seeds." Apart from validating the theory, the micromagnetic approach has also been able to attribute the additional "subcritically-unstable" nature of these modes, enlarging the knowledge on the spin-wave modes supported by nano-contact devices.

From the experimental point of view, there is also an increasing interest in the dynamics (both switching and precession) involving magnetic tunnel junctions because of the discovery of very large TMR values (TMR is the difference in resistance between parallel and antiparallel orientation for the electrode magnetizations of a magnetic tunnel junction) at room temperature.

One reason for the interest in spin-torque effects in tunnel junctions is that these devices are better-suited than metallic magnetic multilayers for many types of applications. Tunnel junctions have higher resistances that can often be better impedance-matched to silicon-based electronics, and TMR values can now be made larger than the GMR values in metallic devices.

Devices based on those effects have already found very widespread applications as magnetic-field sensors in the read heads of magnetic hard disk drives as well as nonvolatile random access memory based on magnetic tunnel junctions.

Applications of spin transfer torques are so fascinating as well. Magnetic switching driven by the spin transfer effect can be much more efficient and more industrially intriguing than the usage of static magnetic fields or current-induced magnetic fields. This may enable the production of magnetic memory devices with much lower switching currents and hence greater energy efficiency as well as a larger integration density. The steady-state magnetic precession mode that can be excited by spin transfer is under investigation for a number of high-frequency applications, for example, nanometer-scale microwave sources (tunable by both magnetic field and current), detectors, mixers, modulators, phase shifters, and arrays (or matrixes) of phase-locked devices (whose coupling is used to increase the output power). One potential area of use is short-range chip-to-chip or even within-chip communications.

References

Alpert, B., Greengard, L., and Hagstrom T. 2002. Nonreflecting boundary conditions for the time-dependent wave equation. *J. Comput. Phys.* 180: 270–296.

Baibich, M.N. et al. 1988. Giant magnetoresistance of (001)Fe/(001) Cr magnetic superlattices. *Phys. Rev. Lett.* 61: 2472–2475.

Berger, L. 1996. Emission of spin waves by a magnetic multilayer traversed by a current. *Phys. Rev. B* 54: 9353–9358.

Berkov, D.V. and Gorn, N. 2006. Micromagnetic simulations of the magnetization precession induced by a spin-polarized current in a point-contact geometry. *J. Appl. Phys.* 99: 08Q701.

Berkov, D.V. and Miltat, J. 2008. Spin-torque driven magnetization dynamics: Micromagnetic modelling, *J. Magn. Magn. Mater.* 320: 1238–1259.

Bertotti, G. 1998. *Hysteresis in Magnetism.* Academic Press, Boston, MA.

Brown, W.F. 1940. Theory of the approach to magnetic saturation. *Phys. Rev.* 58: 736–743.

Brown, W.F. Jr. 1962. *Magnetostatic Principles in Ferromagnetism,* North-Holland Publishing Company, Amsterdam, the Netherlands.

Brown, W.F. 1963. *Micromagnetics.* Wiley, New York.

Brown, W.F. 1979. Thermal fluctuations in fine magnetic particles. *IEEE Trans. Magn.* MAG-15(5): 1196–1208.

Choi, S. et al. 2007. Double-contact spin-torque nano-oscillator with optimized spin-wave coupling: Micromagnetic modelling. *Appl. Phys. Lett.* 90: 083114.

Consolo, G. et al. 2007a. Boundary conditions for spin-wave absorption based on different site-dependent damping functions. *IEEE Trans. Magn.* 43: 2974–2976.

Consolo, G. et al. 2007b. Excitation of self-localized spin-wave bullets by spin-polarized current in in-plane magnetized magnetic nanocontacts: A micromagnetic study. *Phys. Rev. B* 76: 144410.

Consolo, G. et al. 2007c. Magnetization dynamics in nanocontact current controlled oscillators. *Phys. Rev. B* 75: 214428.

Consolo, G. et al. 2007d. Nanocontact spin-transfer oscillators based on perpendicular anisotropy in the free layer. *Appl. Phys. Lett.* 91: 162506.

Consolo, G. et al. 2008. Micromagnetic study of the above-threshold generation regime in a spin-torque oscillator based on a magnetic nano-contact magnetized at an arbitrary angle. *Phys. Rev. B* 78: 014420.

Enquist, B. and Majda, A. 1977. Absorbing boundary conditions for the numerical simulation of waves. *Math. Comput.* 31: 629–651.

Finocchio, G. et al. 2006a. Magnetization dynamics driven by the combined action of AC magnetic field and DC spin-polarized current. *J. Appl. Phys.* 99: 08G507.

Finocchio, G. et al. 2006b. Trends in spin-transfer driven magnetization dynamics of CoFe/AlO/Py and CoFe/MgO/Py magnetic tunnel junctions. *Appl. Phys. Lett.* 89: 262509.

Finocchio, G. et al. 2007. Magnetization reversal driven by spin-polarized current in exchange-biased nanoscale spin valves. *Phys. Rev. B* 76: 174408.

Finocchio, G. et al. 2008. Numerical study of the magnetization reversal driven by spin-polarized current in MgO based magnetic tunnel junctions. *Physica B* 403: 364–367.

Garcia-Palacios, J.L. and Lazaro, F.J. 1998. Langevin-dynamics study of the dynamical properties of small magnetic particles. *Phys. Rev. B* 58: 14937–14958.

Gardiner, C.W. 1985. *Handbook of Stochastic Methods.* Springer, Berlin, Germany.

Gilbert, T.L. 2004. A phenomenological theory of damping in ferromagnetic materials. *IEEE Trans. Magn.* 40: 3443–3449.

Higdon, R.L. 1987. Numerical absorbing boundary conditions for the wave equation. *Math. Comput.* 49: 65–90.

Hrkac, G. 2008. Mutual phase locking in high-frequency microwave nano-oscillators as a function of field angle. *J. Magn. Magn. Mater.* 320: L111–L115.

Kim, J.V., Tiberkevich, V., and Slavin, A. 2008. Generation linewidth of an auto-oscillator with a nonlinear frequency shift: Spin-torque nano-oscillator. *Phys. Rev. Lett.* 100: 017207.

Kloeden, P.E. and Platen E. 1999. *Numerical Solution of Stochastic Differential Equations.* Springer, Berlin, Germany.

Krivorotov, I.N. et al. 2004. Temperature dependence of spin-transfer-induced switching of nanomagnets. *Phys. Rev. Lett.* 93: 166603.

Landau, L.D. and Lifshitz, E.M. 1935. On the theory of the dispersion of magnetic permeability in ferromagnetic bodies. *Phys. Z. Sowjetunion* 8: 153–169.

Li, Z. and Zhang, S. 2004. Thermally assisted magnetization reversal in the presence of a spin-transfer torque. *Phys. Rev. B* 69: 134416.

Martinez, E., Torres, L., and Lopez-Diaz, L. 2004. Computing solenoidal fields in micromagnetic simulations. *IEEE Trans. Magn.* 40: 3240–3243.

Newell, A.J., Williams, W., and Dunlop, D.J. 1993. A generalization of the demagnetizing tensor for nonuniform magnetization. *J. Geophys. Res.* 98: 9551–9555.

Parkin, S.S., More, N., and Roche, K.P. 1990. Oscillations in exchange coupling and magnetoresistance in metallic superlattice structures: Co/Ru, Co/Cr, and Fe/Cr. *Phys. Rev. Lett.* 64: 2304–2307.

Parkin, S.S., Bhadra, R., and Roche, K.P. 1991. Oscillatory magnetic exchange coupling through thin copper layers. *Phys. Rev. Lett.* 66: 2152–2155.

Preisach, F. 1929. Investigations on the Barkhausen effect. *Ann. Physik* 3: 737–799.

Ralph, D.C. and Stiles, M.D. 2008. Spin-transfer torques. *J. Magn. Magn. Mater.* 320: 1190–1216.

Renaut, R.A. 1992. Absorbing boundary conditions, difference operators and stability. *J. Comput. Phys.* 102: 236–251.

Rippard, W.H. et al. 2004. Direct-current induced dynamics in $Co_{90}Fe_{10}/Ni_{80}Fe_{20}$ point contacts. *Phys. Rev. Lett.* 92: 027201.

Schad, R. et al. 1994. Giant magnetoresistance in Fe/Cr superlattices with very thin Fe layers. *Appl. Phys. Lett.* 64: 3500–3502.

Slavin, A.N. and Kabos, P. 2005. Approximate theory of microwave generation in a current-driven magnetic nanocontact magnetized in an arbitrary direction. *IEEE Trans. Magn.* 41: 1264–1273.

Slavin, A. and Tiberkevich, V. 2005. Spin-wave mode excited by spin-polarized current in a magnetic nanocontact is a standing self-localized wave bullet. *Phys. Rev. Lett.* 95: 237201.

Slavin, A. and Tiberkevich, V. 2008. Excitation of spin waves by spin-polarized current in magnetic nano-structures. *IEEE Trans. Magn.* 44: 1916–1927.

Slonczewski, J.C. 1996. Current-driven excitation of magnetic multilayers. *J. Magn. Magn. Mater.* 159: L1–L7.

Slonczewski, J.C. 1999. Excitation of spin waves by an electric current. *J. Magn. Magn. Mater.* 195: L261–L268.

Stiles, M.D. and Miltat J. 2006. Spin-transfer torque and dynamics. In *Spin Dynamics in Confined Magnetic Structures III.* Eds. B. Hillebrands and A. Thiaville, pp. 225–308. Springer, Berlin/Heidelberg, Germany.

Van Kampen, N.G. 1987. *Stochastic Processes in Physics and Chemistry.* North-Holland, Amsterdam, the Netherlands.

Quantum Spin Tunneling in Molecular Nanomagnets

Gabriel González
University of Central Florida

Michael N. Leuenberger
University of Central Florida

9.1 Introduction

Molecular magnets have attracted considerable interest recently among researchers because they are considered to be ideal systems to probe the interface between classical and quantum physics as well as to study decoherence in nanoscale systems. The advances in nanoscience during the last decade have made it possible to design and fabricate a wide variety of nanosize objects, ranging from a couple of micrometers all the way down to a few tens of nanometers, where quantum effects become important and give rise to new properties. One of the goals of miniaturization was the possibility of observing quantum tunneling effects in mesoscopic systems. One source for such a task is the so-called single-molecule magnets (SMM) and antiferromagnetic molecular wheels, in which the spin state of the molecule is known to behave quantum mechanically at low temperatures.

During the last decade, a tremendous progress in the experimental methods for contacting single molecules and measuring the electrical current through them has been achieved. The current through single magnetic molecules like Mn_{12} and Fe_8 has been measured and magnetic excited states have been identified (Jo et al., 2006; Henderson et al., 2007). In a three-terminal molecular single electron transistor, the current can flow between the source and drain leads via a sequential tunneling process through the molecular charge levels, thereby bringing the whole field of Coulomb-blockade physics to molecular systems.

Single-molecule magnets are mainly organic molecules containing multiple transition-metal ions bridged by organic ligands. These ions are strongly coupled by exchange interaction, yielding a large magnetic moment per molecule. The large spin combined with the large magnetic anisotropy provides an energy barrier for magnetization reversal. These systems provided for the first time evidence of quantum tunneling of the magnetization and interference effects as well as oscillations of the tunnel splitting.

Ferromagnetic molecular magnets such as Mn_{12} and Fe_8 show incoherent tunneling of the magnetization (Korenblit and Shender, 1978; Enz and Schilling, 1986; van Hemmen and Süto, 1986; Chudnovsky and Gunther, 1988) and allow one to study the interplay of thermally activated processes and quantum tunneling. The spin tunneling leads to two effects. First, the magnetization relaxation is accelerated whenever spin states of opposite direction become degenerate due to the variation of the external longitudinal magnetic field (Friedman et al., 1996; Thomas et al., 1996; Leuenberger and Loss, 1999, 2000a; Leuenberger and Loss). Second, the spin acquires a Berry phase during the tunneling process, which leads to oscillations of the tunnel splitting as a function of the external transverse magnetic field (Wernsdorfer and Sessoli, 1996; Wernsdorfer et al., 2000; Leuenberger and Loss, 2000b, 2001a).

Antiferromagnetic molecular wheels are another type of molecular magnets where an even number of transition metal ions with spin form a closed ring in which the exchange interaction between neighbors gives rise to a strong spin quantum dynamics. Antiferromagnetic molecular magnets such as ferric wheels belong to the most promising candidates for the observation of coherent quantum tunneling on the mesoscopic scale (Chiolero and Loss, 1998; Meier and Loss, 2000, 2001). In contrast to incoherent tunneling, in quantum coherent tunneling spins tunnel back and forth between energetically degenerate configurations at a tunneling rate which is *large* compared to the decoherence rate. The detection of quantum behavior is more challenging in antiferromagnetic molecular magnets than in ferromagnetic systems, but is feasible with present day experimental techniques.

Understanding the properties of molecular magnets is only a first step toward achieving technological applications in data storage, data processing, and quantum technologies. A possible next step will be the preparation and control of a well-defined single-spin quantum state of a molecular cluster. Although challenging, this task appears feasible with present-day experiments and would allow one to carry out quantum computing with molecular magnets (Leuenberger and Loss, 2001b).

9.2 Spin Tunneling in Molecular Nanomagnets

An effective spin Hamiltonian that captures the basic physics of a molecular nanomagnet is given by the following form:

$$\mathcal{H} = \mathcal{H}_0(\vec{S}) + \mathcal{H}_Z + \mathcal{H}_T + \mathcal{H}_{sb}. \tag{9.1}$$

We will now analyze each term in the Hamiltonian given in Equation 9.1 separately. The first term in Equation 9.1 is the dominant term of the Hamiltonian with spin quantum number $S = |\vec{S}| \gg 1$ and is called the uniaxial anisotropy, which results from the spin–orbit interactions and generates and easy magnetic axis for the magnetic moment of the nanomagnet. The assumption is that we have a single domain magnetic molecule in which the very strong exchange interactions align the microscopic electronic spins into a parallel or antiparallel configuration, resulting in a total spin $\vec{S} = \sum_j \vec{s}_j$ for the ferromagnetic moment or the Neél vector with $\vec{S} = \sum_j (-1)^j \vec{s}_j$, respectively.

For a nanomagnet like Fe_8, $\mathcal{H}_0(\vec{S}) = -AS_z^2$ where $A/k_B \approx 0.275\,K$ and generates an energy barrier that separates opposite spin projections (see Figure 9.1).

Quantum mechanics specifies that a particle in a symmetric double well potential undergoes a tunnel effect that makes the particle go back and forth from one well to the other with a frequency ω_T. This implies that the eigenfunctions of the Hamiltonian are delocalized, which means that there is a probability to find the particle either in one well or the other. In a similar fashion, there is a probability for the spin of the nanomagnet to change direction, i.e., if the spin of the nanomagnet points up there is a probability that at a later time the spin will be pointing down. At equilibrium and finite temperature, the probability that the spin is in state $|m\rangle$, where $S_z|m\rangle = m|m\rangle$ for $m = -s, -s+1, \ldots, s-1, s$, is given by

$$P_m = \frac{e^{-E_m/k_B T}}{Z} \tag{9.2}$$

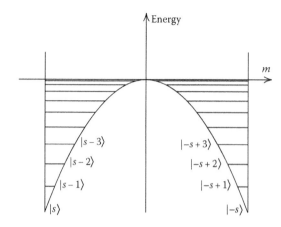

FIGURE 9.1 The graph shows $\mathcal{H}_0$ vs. S_z where the horizontal lines indicate the energies of the quantum states.

where
    Z is the partition function
    $E_m = -Dm^2$

For the case when $T \to 0$ the probability for the spin to go from one well to the other satisfies the following equation:

$$\frac{dP(t)}{dt} \approx -\Gamma P(t) < 0. \tag{9.3}$$

Therefore $P(t) \propto e^{-\Gamma t}$, where Γ is the rate of decay and Γ^{-1} is known as the lifetime or characteristic time of the particle.

The second term in Equation 9.1 is the Zeeman energy resulting from the interaction of the spin with an external magnetic field, i.e., $\mathcal{H}_Z = \mu_B g \vec{S} \cdot \vec{H}$. A magnetic field applied along the easy axis of the molecule will tilt the potential well and for certain values of the magnetic field ($\delta H_z = (m + m')|A|/g\mu_B$), the energy levels on both sides of the anisotropy barrier coincide. Therefore the Hamiltonian given by $\mathcal{H} = -AS_z^2 + g\mu_B \delta H_z S_z$ is still doubly degenerate (see Figure 9.2).

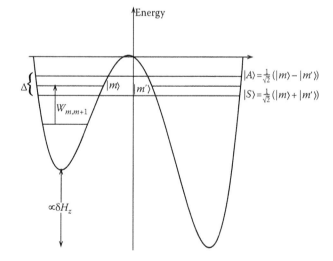

FIGURE 9.2 The graph shows $(\mathcal{H}_0 + \mathcal{H}_Z)$ vs. S_z in a constant magnetic field parallel to the easy axis. Δ denotes the tunnel splitting between states $|m\rangle$ and $|m'\rangle$.

These degeneracies can be removed by the term $\mathcal{H}_T$, which corresponds to the transverse anisotropies. For the nanomagnet Fe$_8$, the transverse anisotropy term is given by $\mathcal{H}_T = E(S_x^2 - S_y^2)$, where $(E/k_B \approx 0.046\,\text{K})$. The anisotropy term lifts the degeneracy, thereby creating an energy gap between the spin states $|m\rangle$ and $|m'\rangle$, denoted by $\Delta_{mm'}$. If the Hamiltonian possesses transverse terms, the pairwise degenerate states get split by $\Delta_{mm'}$ amount of energy. If the coupling of our system to an external bath is assumed to be zero, each pair of the degenerate states can be described by means of a two-state Hamiltonian:

$$\mathcal{H} = \begin{bmatrix} hm & \dfrac{\Delta_{mm'}}{2} \\[2mm] \dfrac{\Delta_{mm'}}{2} & hm' \end{bmatrix} = \frac{h}{2}(m+m')\begin{bmatrix} 1 & 0 \\ 0 & 1 \end{bmatrix}$$
$$+ \frac{h}{2}(m-m')\begin{bmatrix} 1 & \tan\theta \\ \tan\theta & -1 \end{bmatrix}, \quad (9.4)$$

where $\tan\theta = \Delta_{mm'}/h(m-m')$, $0 \leq \theta < \pi$ and h is the longitudinal magnetic field with the coupling constant g and Bohr magneton μ_B absorbed, which makes the double well asymmetric, and $\Delta_{mm'}$ is the tunnel splitting between the states $|m\rangle$ and $|m'\rangle$. The eigenstates of Equation 9.4 read

$$|\psi_+\rangle = \cos\frac{\theta}{2}|m\rangle + \sin\frac{\theta}{2}|m'\rangle, \quad (9.5)$$

$$|\psi_-\rangle = -\sin\frac{\theta}{2}|m\rangle + \cos\frac{\theta}{2}|m'\rangle, \quad (9.6)$$

with eigenvalues

$$E_\pm = \frac{h}{2}\left[(m+m') \pm \frac{(m-m')}{\cos\theta}\right] = \frac{1}{2}\left[h(m+m') \pm \sqrt{h^2(m-m')^2 + \Delta_{mm'}^2}\right]. \quad (9.7)$$

$|\psi_+\rangle$ and $|\psi_-\rangle$ are delocalized as long as $h(m-m') \leq \Delta_{mm'}$. It is only in this regime that a spin state can tunnel from one side of the barrier to the other, which is also relevant for the Zener tunneling (see Section 9.5). In the limit $h \to 0$, the eigenstates $|\psi_+\rangle$ and $|\psi_-\rangle$ are the symmetrical and antisymmetrical combinations of $|m\rangle$ and $|m'\rangle$, i.e., $|\psi_+\rangle = |S\rangle = \frac{1}{\sqrt{2}}(|m\rangle + |m'\rangle)$ and $|\psi_-\rangle = |A\rangle = \frac{1}{\sqrt{2}}(|m\rangle - |m'\rangle)$. For $\Delta_{mm'} \to 0$ we are left with $|\psi_+\rangle = |m\rangle$ and $|\psi_-\rangle = |m'\rangle$ for $h > 0$ and $|\psi_-\rangle = |m\rangle$ and $|\psi_+\rangle = |m'\rangle$ for $h < 0$ (Figure 9.3).

Let us now assume that the system starts in the state

$$|\psi(t=0)\rangle = |m\rangle = \cos\frac{\theta}{2}|\psi_+\rangle - \sin\frac{\theta}{2}|\psi_-\rangle, \quad (9.8)$$

from which one obtains immediately the time evolution

$$|\psi(t)\rangle = \cos\frac{\theta}{2}e^{-iE_+t/\hbar}|\psi_+\rangle - \sin\frac{\theta}{2}e^{-iE_-t/\hbar}|\psi_-\rangle. \quad (9.9)$$

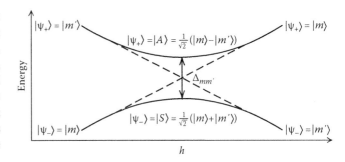

FIGURE 9.3 Anticrossing of the adiabatic eigenvalues $E_\pm = \frac{1}{2}\left[h(m+m') \pm \sqrt{h^2(m-m')^2 + \Delta_{mm'}^2}\right]$. The Zener tunneling transition of the state $|\psi_+\rangle$ is adiabatic (see Section 9.5).

Then one can calculate the probability that the state $|m\rangle$ has tunneled to the state $|m'\rangle$ after time t:

$$\left|\langle m' | \psi(t)\rangle\right|^2 = \left|\sin\frac{\theta}{2}\langle\psi_+ | \psi(t)\rangle + \cos\frac{\theta}{2}\langle\psi_- | \psi(t)\rangle\right|^2$$

$$= \left|\sin\frac{\theta}{2}\cos\frac{\theta}{2}(e^{-iE_+t/\hbar} - e^{-iE_-t/\hbar})\right|^2$$

$$= \sin^2\theta\sin^2\left(\frac{\sqrt{h^2(m-m')^2 + \Delta_{mm'}^2}}{2\hbar}t\right)$$

$$= \frac{\Delta_{mm'}^2}{h^2(m-m')^2 + \Delta_{mm'}^2}\sin^2\left(\frac{\sqrt{h^2(m-m')^2 + \Delta_{mm'}^2}}{2\hbar}t\right)$$
$$\hspace{10cm}(9.10)$$

This means that both the tunneling behavior and also the two-level spin resonance behavior are described by a Rabi oscillation.

In reality, the spin of the SMM is interacting with the environment and this interaction is described by the last term in Equation 9.1. The interaction of the molecular magnet with the environment is due to the hyperfine interaction, dipole interaction, spin-phonon interaction, the interaction between the molecule and two contact leads, etc. When the spin interacts with its environment, the wavefunction loses the memory of its phase. This phenomenon is known as decoherence. In this case, the wavefunction is not longer appropriate to describe the system. The problem of a system that interacts with the environment is formulated by means of the density matrix formalism. The spin has a certain probability to be in state $|m\rangle$ at time t, which is described by the density matrix $\rho_m(t)$. Due to the interaction with the environment, the spin undergoes incoherent transitions from a state $|m\rangle$ to another state $|m'\rangle$ at a rate $W_{m'm}$ until $\rho_m(t)$, $-s \leq m \leq s$ reach their equilibrium value. This process is known as relaxation. If these transitions are independent of each other, the density matrix evolves according to the Pauli equation or master equation

$$\frac{d\rho_m(t)}{dt} = \sum_{m'}\left[W_{mm'}\rho_{m'}(t) - W_{m'm}\rho_m(t)\right] \quad (9.11)$$

where $W_{m'm}$ is the transition probability from state $|m\rangle$ to state $|m'\rangle$. The master equation (Equation 9.11) can also be written as

$$\frac{d\rho_m(t)}{dt} = \sum_{m'} A_{mm'}\rho_{m'}(t) \tag{9.12}$$

where

$$A_{mm'} = \begin{cases} W_{mm'} & \text{if } m \neq m' \\ -\sum_{m''}' W_{m''m'} & \text{if } m' = m, \end{cases} \tag{9.13}$$

the prime indicates that the term $m'' = m$ is to be omitted in the summation. A formal solution to Equation 9.11 can be obtained in the following manner, let

$$\rho_m(t) = p_m^{(n)} e^{-\Gamma_n t} \tag{9.14}$$

where Γ_n^{-1} is the characteristic time labeled by the index $n = 0$, $1, \ldots, 2s$. Substituting Equation 9.14 into Equation 9.11, one obtains

$$-\Gamma_n p_m^{(n)} = \sum_{m'}\left[W_{mm'} - \delta_{mm'}\sum_{m''}W_{mm''}\right]p_{m'}^{(n)} = \sum_{m'}A_{mm'}p_{m'}^{(n)} \tag{9.15}$$

which implies that $-\Gamma_n$ is an eigenvalue of the master matrix defined in Equation 9.13. The master matrix is a square matrix $(2s + 1) \times (2s + 1)$, which is not Hermitian. However, the eigenvalue problem can be written in terms of the Hermitian real matrix by making the following transformation

$$B_{mm'} = A_{mm'}\exp\left[\frac{\beta(E_m - E_{m'})}{2}\right] = B_{m'm}. \tag{9.16}$$

It follows that the general solution of Equation 9.11 is then

$$\rho_m(t) = \exp\left[\frac{-\beta E_m}{2}\right]\sum_n c_n \mathcal{P}_m^{(n)} e^{-\Gamma_n t} \tag{9.17}$$

where $(\mathcal{P}_m^{(n)}, -\Gamma_n)$ are the $(2s + 1)$ orthonormal set of eigenvectors with the corresponding eigenvalues of the matrix (9.16) and $\mathcal{P}_m^{(n)} = p_m^{(n)}\exp(\beta E_m/2)$. The c_n are the constants related to the initial distribution of ρ_m by

$$c_n = \sum_k \rho_k(0)\mathcal{P}_k^{(n)}\exp\left[\frac{\beta E_k}{2}\right]. \tag{9.18}$$

Equation 9.11 is only valid for times where the correlation time τ_c in the heat bath is much smaller than the relaxation time of the spin system, necessary for the establishment of irreversibility in a macroscopic system. In fact, in order to describe quantum tunneling, it is necessary to use the *generalized master equation*.

The generalized master equation that describes the relaxation of the spin due to phonon-assisted transitions including resonances due to tunneling is given by

$$\dot{\rho}_{mm'} = \frac{i}{\hbar}[\rho, \mathcal{H}_0]_{mm'} + \delta_{mm'}\sum_{n\neq m}\rho_n W_{mn} - \gamma_{mm'}\rho_{mm'}. \tag{9.19}$$

The difference to the usual master equation is that Equation 9.19 takes also off-diagonal elements of the density matrix $\rho(t)$ into account. This is essential to describe tunneling of the magnetization, which is caused by the overlap of the S_z states.

Let us consider the two-state system $\{|m\rangle, |m'\rangle\}$, which yields the two-state Hamiltonian in the presence of a bias field given in Equation 9.4. Next we insert the two-state Hamiltonian into the generalized master equation (Equation 9.19), which yields for the diagonal elements of the density matrix

$$\dot{\rho}_m = \frac{i\Delta_{mm'}}{2\hbar}(\rho_{mm'} - \rho_{m'm}) - W_m\rho_m + \sum_{n\neq m, m'}W_{mn}\rho_n \tag{9.20}$$

and for the off-diagonal elements

$$\dot{\rho}_{mm'} = -\left(\frac{i}{\hbar}\xi_{mm'} + \gamma_{mm'}\right)\rho_{mm'} + \frac{i\Delta_{mm'}}{2\hbar}(\rho_m - \rho_{m'}), \tag{9.21}$$

with $\xi_m = -Am^2 + g\mu_B\delta H_z m$ and $\xi_{mm'} = \xi_m - \xi_{m'}$, similarly for $m \leftrightarrow m'$. Ultimately, we are interested in the overall relaxation time τ of the quantity $\rho_s - \rho_{-s}$ (see Section 9.7) due to phonon-induced transitions. This τ turns out to be much longer than $\tau_d = 1/\gamma_{mm'}$, which is the decoherence time of the decay of the off-diagonal elements $\rho_{mm'} \propto e^{-t/\tau_d}$ of the density matrix ρ. Thus, we can neglect the time dependence of the off-diagonal elements, i.e., $\dot{\rho}_{mm'} \approx 0$. Physically this means that we deal with incoherent tunneling for times $t > \tau_d$. Inserting then the stationary solution of Equation 9.21 into Equation 9.20, which leads to the complete master equation including resonant as well as nonresonant levels, we get

$$\dot{\rho}_m = -W_m\rho_m + \sum_{n\neq m, m'}W_{mn}\rho_n + \Gamma_{mm'}(\rho_{m'} - \rho_m), \tag{9.22}$$

where

$$\Gamma_{mm'} = \Delta_{mm'}^2\frac{W_m + W_{m'}}{4\xi_{mm'}^2 + \hbar^2(W_m + W_{m'})^2} \tag{9.23}$$

is the transition rate from m to m' (induced by tunneling) in the presence of phonon damping. Note that Equation 9.22 is now of the usual form of a master equation, i.e., only diagonal elements of the density matrix $\rho(t)$ occur. For levels $k \neq m, m'$, Equation 9.22 reduces to

$$\dot{\rho}_k = -W_k\rho_k + \sum_n W_{kn}\rho_n. \tag{9.24}$$

We note that $\Gamma_{mm'}$ has a Lorentzian shape with respect to the external magnetic field δH_z occurring in $\xi_{mm'}$. It is thus this $\Gamma_{mm'}$ that will determine the peak shape of the magnetization resonances.

9.3 Phonon-Assisted Spin Tunneling in Mn₁₂ Acetate

The magnetization relaxation of crystals and powders made of molecular magnets Mn_{12} has attracted much recent interest since several experiments (Sessoli et al., 1993; Novak and Sessoli, 1995; Novak et al., 1995; Paulsen and Park, 1995; Paulsen et al., 1995) have indicated unusually long relaxation times as well as increased relaxation rates (Friedman et al., 1996; Hernández et al., 1996; Thomas et al., 1996) whenever two spin states become degenerate in response to a varying longitudinal magnetic field H_z. According to earlier suggestions (Barbara et al., 1995), this phenomenon has been interpreted as a manifestation of incoherent macroscopic quantum tunneling (MQT) of the spin.

As long as the external magnetic field H_z is much smaller than the internal exchange interactions between the Mn ions of the Mn_{12} cluster, the Mn_{12} cluster behaves like a large single spin $\vec{S}$ of length $|\vec{S}| = 10$. For temperatures $1 \leq T \leq 10\,K$, its spin dynamics can be described by the spin Hamiltonian given in Equation 9.1, including the coupling between this large spin and the phonons in the crystal (Villain et al., 1994; Garanin and Chudnovsky, 1997; Hernández et al., 1997; Fort et al., 1998; Luis et al., 1998; Leuenberger and Loss, 1999, 2000a; Leuenberger and Loss).

The most general spin–phonon coupling reads

$$\mathcal{H}_{sp} = g_1(\epsilon_{xx} - \epsilon_{yy}) \otimes (S_x^2 - S_y^2) + \frac{1}{2} g_2 \epsilon_{xy} \otimes \{S_x, S_y\}$$

$$+ \frac{1}{2} g_3(\epsilon_{xz} \otimes \{S_x, S_z\} + \epsilon_{yz} \otimes \{S_y, S_z\})$$

$$+ \frac{1}{2} g_4(\omega_{xz} \otimes \{S_x, S_z\} + \omega_{yz} \otimes \{S_y, S_z\}), \qquad (9.25)$$

where

g_i are the spin–phonon coupling constants
$\epsilon_{\alpha\beta}(\omega_{\alpha\beta})$ is the (anti-)symmetric part of the strain tensor

From the comparison between experimental data (Friedman et al., 1996; Thomas et al., 1996) and calculation, it turns out that the constants $g_i \approx A\ \forall i$ (Leuenberger and Loss, 1999, 2000a; Leuenberger and Loss).

9.4 Interference between Spin Tunneling Paths in Molecular Nanomagnets

Recent experiments have pointed out the importance of the interference between spin tunneling paths in molecules. For instance, measurements of the magnetization in bulk Fe_8 have observed oscillations in the tunnel splitting $\Delta_{m,-m}$ between states $S_z = m$ and $-m$ as a function of a transverse magnetic field at temperatures between 0.05 and 0.7 K (see Landau, 1932). This effect can be explained by the interference between Berry phases associated to spin tunneling paths of opposite windings (DiVincenzo et al., 1992; van Delft and Henley, 1992).

Usually the Berry phase effect arises in systems that undergo an adiabatic cyclic evolution, and which occurs in such diverse fields as atomic, condensed matter, nuclear and elementary particle physics, and optics. The basic assumption for the derivation of the Berry phase is that, for a slowly varying time-dependent Hamiltonian $H(\vec{R}(t))$ that depends on parameters $R_1(t)$, $R_2(t)$, ..., $R_N(t)$ components of a vector $\vec{R}$, the eigenstate evolves in time according to

$$\left| \psi_i(\vec{R}(0)) \right\rangle \rightarrow e^{i\varphi_i(t)} \left| \psi_i(\vec{R}(t)) \right\rangle \exp\left[-\frac{i}{\hbar} \int_0^t E_i(\vec{R}) dt \right]. \qquad (9.26)$$

Plugging Equation 9.26 into Schrödinger's equation we have

$$\frac{d\varphi}{dt} = i\left\langle \psi_i \left| \nabla_{\vec{R}} \right| \psi_i \right\rangle \cdot \frac{d\vec{R}}{dt}, \qquad (9.27)$$

thus, for $t = 0$ to $t = \tau$ we end up with

$$\varphi(\tau) - \varphi(0) = \int_{\vec{R}(0)}^{\vec{R}(\tau)} \left\langle \psi_i \left| i\nabla_{\vec{R}} \right| \psi_i \right\rangle d\vec{R}. \qquad (9.28)$$

The term of Equation 9.28 is known as Berry's phase. The overall phase of a quantum state is not observable but the relative phases for a quantum system, which undergoes a coherent evolution, can be detected experimentally. Consider for example two paths $\vec{R}$ and $\vec{R}'$ with the same end points $\vec{R}(0) = \vec{R}'(0)$ and $\vec{R}(\tau) = \vec{R}'(\tau)$, the net Berry phase change is given by

$$\Delta\varphi = \oint \left\langle \psi_i \left| i\nabla_{\vec{R}} \right| \psi_i \right\rangle \cdot d\vec{R}. \qquad (9.29)$$

This is a line integral around a closed loop in parameter space, which is nonzero in general. The classic example of Berry's phase is an electron at rest subjected to a time-dependent magnetic field of constant magnitude but changing direction. The Hamiltonian for this system is given by

$$\mathcal{H} = -\mu_B \vec{\sigma} \cdot \vec{H} = -\mu_B \begin{bmatrix} H_z & H_x - iH_y \\ H_x + iH_y & -H_z \end{bmatrix}, \qquad (9.30)$$

where $\vec{H}(t) = H(\sin\theta(t)\cos(\phi(t)), \sin(\theta(t))\sin(\phi(t)), \cos(\theta(t)))$ and μ_B is Bohr magneton. It is easy to show that the eigenstate representing spin up along $\vec{H}(t)$ for Equation 9.30 has the form

$$|\uparrow\rangle = \begin{pmatrix} e^{-i\phi/2}\cos(\theta/2) \\ e^{i\phi/2}\sin(\theta/2) \end{pmatrix}. \qquad (9.31)$$

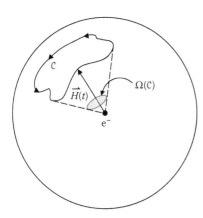

FIGURE 9.4 The figure shows an electron in a constant magnetic field which sweeps around a closed curve $\mathcal{C}$ and subtends a solid angle Ω.

For the case when the magnetic field sweeps out an arbitrary closed circuit $\mathcal{C}$ (see Figure 9.4) we can apply Equation 9.29 to get the Berry phase

$$\Delta\varphi = -\frac{1}{2}\Omega(\mathcal{C}), \qquad (9.32)$$

where $\Omega(\mathcal{C})$ is the solid angle described by the field $\vec{H}$ along the circuit $\mathcal{C}$.

9.4.1 Berry's Phase and Path Integrals

An alternative approach for quantum mechanics and hence for the calculation of the tunneling frequency is by means of path integrals which were introduced by Richard Feynman (Feynman, 1948). In this section, we will briefly explain the path integral approach to calculate the Berry phase in single molecular magnets.

Feynman showed that the probability amplitude $\langle x' | \exp[i(t' - t)\mathcal{H}/\hbar] | x \rangle$ for a particle that is at point x' at time t' if it was at point x at a time t can be expressed as the path integral

$$U(x', t' \mid x, t) = \int_{x}^{x'} \mathcal{D}[x''] \exp\left[\frac{iS_E}{\hbar}\right], \qquad (9.33)$$

where

$$S_E = \int_{t}^{t'} dt\, L[x(t), \dot{x}(t)], \qquad (9.34)$$

is the *Euclidean action*, which involves the classical Lagrangian $L[x(t), \dot{x}(t)]$ and $\int \mathcal{D}[x'']$ is an integral over all paths from x to x' defined as

$$\int_{x}^{x'} \mathcal{D}[x''] \equiv \lim_{N \to \infty} \left(\frac{m}{2\pi i\hbar\Delta t}\right)^{N/2} \prod_{n=1}^{N-1} \int dx_n, \qquad (9.35)$$

where $x_n = x(t_n)$ and $t_n \in [t, t_1, \ldots, t_{N-1}, t']$. For purposes of actual evaluation of quantities using path-integral methods,

it is common to get rid of the factor i in the exponential of Equation 9.33 by analytic continuation to imaginary times, through $\beta = it/\hbar$. Then, the path integral becomes equivalent to the partition function Z by noting $\beta = 1/k_B T$.

Quantum tunneling of a spin is often described with the terminology of a path integral. Consider a general single-spin Hamiltonian $\mathcal{H}_{z,n} = -AS_z^2 + B_n(S_+^n + S_-^n)$, with easy-axis ($A$) and transverse ($B_n$) anisotropy constants satisfying $A \gg B_n > 0$. Here, n is an even integer, i.e., $n = 2, 4, 6, \ldots$, so that $\mathcal{H}_z$ is invariant under time reversal. Such Hamiltonians are relevant for molecular magnets such as Mn_{12} (Fort et al., 1998; Leuenberger and Loss, 1999, 2000a) and Fe_8 (Leuenberger and Loss, 2000b). The corresponding classical anisotropy energy $E_{z,n}(\theta, \phi) = -As^2 \cos^2\theta + B_n s^n \sin^n\theta \cos(n\phi)$ has the shape of a double-well potential, with the easy axis pointing along the z direction. It is obvious that the anisotropy energy remains invariant under rotations around the z-axis by multiples of the angle $\eta = 2\pi/n$. For the following calculation, it proves favorable to choose the easy axis along the y direction where the contour path does not include the south pole. Then the spin Hamiltonian and the anisotropy energy are changed into

$$\mathcal{H}_{y,n} = -AS_y^2 + B_n(S_+^n + S_-^n), \qquad (9.36)$$

where now $S_\pm = S_z \pm iS_x$, and

$$E_{y,n}(\theta, \phi) = -As^2 \sin^2\theta \sin^2\phi + B_n s^n$$
$$\times \left[(\cos\theta + i\sin\theta\cos\phi)^n + (\cos\theta - i\sin\theta\cos\phi)^n\right]. \qquad (9.37)$$

We are interested in the tunneling between the eigenstates $|m\rangle$ and $|-m\rangle$ of $-AS_y^2$, corresponding to the global minimum points $(\theta = \pi/2, \phi = -\pi/2)$ and $(\theta = \pi/2, \phi = +\pi/2)$ of $E_{y,n}$. For this, we evaluate the imaginary time transition amplitude between these points. For the ground-state tunneling ($m = s \gg 1$), this can be done by means of the coherent spin-state path integral (DiVincenzo et al., 1992)

$$\left\langle -\frac{\pi}{2} \middle| e^{-\beta\mathcal{H}} \middle| +\frac{\pi}{2} \right\rangle = \int_{\pi/2,-\pi/2}^{\pi/2,+\pi/2} \mathcal{D}\Omega\, e^{-S_E}, \qquad (9.38)$$

where $\beta = 1/k_B T$ is the inverse temperature, $\mathcal{D}\Omega = \Pi_\tau d\Omega_\tau$, $d\Omega_\tau = [4\pi/(2s + 1)]d(\cos\theta_\tau)d\phi_\tau$ the Haar measure of the S^2 sphere, and $S_E = \int_0^\beta d\tau[is\dot{\phi}(1 - \cos\theta) + E_{y,n}]$ the Euclidean action, where the first term in S_E defines the Wess–Zumino (or Berry phase) term, which gives rise to topological interference effects for spin tunneling (DiVincenzo et al., 1992; von Delft and Henley, 1992). We can divide the S^2 surface area of integration in Equation 9.38 into n equally shaped subareas $\mathcal{A}_n$, so that we can factor out the following sum over Berry phase terms without the need of

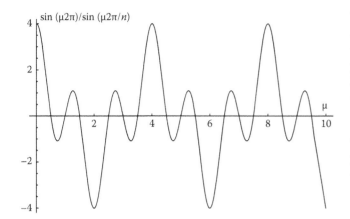

FIGURE 9.5 Sum over Berry phase terms for $s = 10$ and $n = 4$.

evaluating the dynamical part of the path integral (i.e., the following result is valid for all m),

$$\left\langle -\frac{\pi}{2} \Big| e^{-\beta \mathcal{H}_{y,n}} \Big| +\frac{\pi}{2} \right\rangle \propto \sum_{k=1}^{n} e^{i(2\pi/n)(2k-1)m} = \frac{\sin(2\pi m)}{\sin(2\pi m/n)}, \quad (9.39)$$

which vanishes whenever n is not a divisor of $2m$. Thus, the tunnel splitting energy $\Delta_{m,-m}$ between the states $|m\rangle$ and $|-m\rangle$ vanishes if $2m/n \notin Z$. However, if $2m/n \in Z$, the variable m must be extended to real numbers, i.e., $m \to \mu \in R$, in order to calculate the limit $\mu \to m$ of the ratio of the sine functions, which is plotted in Figure 9.5 for a special case. In order to visualize the interference between the Berry phases in Equation 9.39, we select one representative path of each subarea $\mathcal{A}_n$. Then the vanishing of the amplitude in Equation 9.39 for the case $n = 6$ (see Figure 9.6) can be thought of as a destructive interference

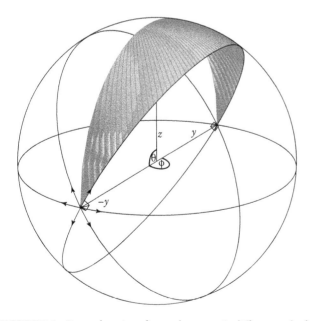

FIGURE 9.6 Berry phase interference between six different paths for $s = 10$ and $n = 6$.

between six different paths. Also, it is important to note that from the semiclassical point of view, the total classical energy of the spin system E must be conserved during the tunneling process. Therefore, the semiclassical paths do not follow the local minima of E with respect to θ alone, except for the case of purely quadratic (i.e., $n = 2$) anisotropies (DiVincenzo et al., 1992; Garg, 1993).

From the above treatment, we conclude that tunneling between two degenerate spin states $|m\rangle$ and $|m'\rangle = |-m\rangle$ is topologically suppressed (i.e., the twofold degeneracy is not lifted) whenever n is *not* a divisor of $2m$. Since n is even this excludes immediately tunneling for all half-odd integer spins s (for all m and n), in accordance with Kramers degeneracy. For s integer, however, tunneling can be either allowed or suppressed, depending on the ratio $2m/n$. In the latter case, the twofold degeneracy of spin states is not lifted by the anisotropy, and we can view this result as a generalization of the Kramers theorem to integer spins.

9.5 Incoherent Zener Tunneling in Fe$_8$

Besides Mn$_{12}$, there have been several experiments on the molecular magnet Fe$_8$ that revealed macroscopic quantum tunneling of the spin (Barra et al., 1996; Wernsdorfer and Sessoli, 1996; Sangregorio et al., 1997; Ohm et al., 1998; Wernsdorfer et al., 2000). In particular, recent measurements on Fe$_8$ (Wernsdorfer and Sessoli, 1996; Wernsdorfer et al., 2000) lead to the development of the concept of the incoherent Zener tunneling (Leuenberger and Loss, 2000b; Leuenberger et al., 2003). The resulting Zener tunneling probability P_{inc} exhibits Berry phase oscillations as a function of the external transverse field H_x.

For many physical systems, the Landau–Zener model (Landau, 1932; Zener, 1932) has become an important tool for studying tunneling transitions (Crothers and Huges, 1977; Garga et al., 1985; Shimshoni and Gefen, 1991; Averin and Bardas, 1995). It must be noted that all quantum systems to which the Zener model (Landau, 1932) is applicable can be described by *pure* states and their *coherent* time evolution. In particular, the theory presented in Leuenberger and Loss (2000b) agrees well with recent measurements of $P_{\text{inc}}(H_x)$ for various temperatures in Fe$_8$ (Wernsdorfer and Sessoli, 1996; Wernsdorfer et al., 2000).

For the Zener transition, usually only the asymptotic limit is of interest. Therefore it is required that the range over which $\varepsilon_{mm'}(t) = \varepsilon_m - \varepsilon_{m'}$ is swept is much larger than the tunnel splitting $\Delta_{mm'}$ and the decoherence rate $\hbar\gamma_{mm'}$. In addition, the evolution of the spin system is restricted to times t that are much longer than the decoherence time $\tau_d = 1/\gamma_{mm'}$. In this case, tunneling transitions between pairs of degenerate excited states are incoherent. This tunneling is only observable if the temperature T is kept well below the activation energy of the potential barrier. Accordingly, one is interested only in times t that are larger than the relaxation times of the excited states. It was shown by Leuenberger and Loss (2000b) that the Zener tunneling can be described by Equation 9.22, where

$$\Gamma_{mm'}(t) = \frac{\Delta^2_{mm'}}{2} \frac{\gamma_{mm'}}{\varepsilon^2_{mm'}(t) + \hbar^2 \gamma^2_{mm'}} \qquad (9.40)$$

is time dependent, in contrast to Equation 9.23. As usual, the abbreviations $\gamma_{mm'} = (W_m + W_{m'})/2$ and $W_m = \Sigma_n W_{nm}$ are used, where W_{nm} denotes the approximately time-independent transition rate from $|m\rangle$ to $|n\rangle$, which can be obtained via Fermi's golden rule (Leuenberger and Loss, 1999, 2000a). The tunnel splitting (Leuenberger and Loss, 1999, 2000a) is given by

$$\Delta_{mm'} = 2 \left| \sum_{\substack{m_1, \dots, m_N \\ m_i \neq m, m'}} \frac{V_{m,m_1}}{\varepsilon_m - \varepsilon_{m_1}} \prod_{i=1}^{N-1} \frac{V_{m_i, m_{i+1}}}{\varepsilon_m - \varepsilon_{m_{i+1}}} V_{m_N, m'} \right|. \qquad (9.41)$$

V_{m_i, m_j} denote off-diagonal matrix elements of the total Hamiltonian $\mathcal{H}_{\text{tot}}$.

Since all resonances n lead to similar results, Equation 9.22 is solved only in the unbiased case—corresponding to $n = 0$ (see below)—where the ground states $|s\rangle$, $|-s\rangle$ and the excited states $|m\rangle$, $|-m\rangle$, $m \in [-s + 1, s - 1]$ of the spin system with spin s are pairwise degenerate. In addition, it is assumed that the excited states are already in their stationary state, i.e., $\dot{\rho}_m = 0 \; \forall m \neq s, -s$. Equation 9.22 leads then to

$$1 - P_{\text{inc}} \equiv \Delta\rho(t) = \exp\left\{ -\int_{t_0}^{t} dt' \Gamma_{\text{tot}}(t') \right\}, \qquad (9.42)$$

where $\Delta\rho(t) = \rho_s - \rho_{-s}$, which satisfies the initial condition $\Delta\rho(t = t_0) = 1$, and thus $P_{\text{inc}}(t = t_0) = 0$. The total time-dependent relaxation rate is given by $\Gamma_{\text{tot}} = 2[\Gamma_{s,-s} + \Gamma_{\text{th}}]$, where the thermal rate Γ_{th}, which determines the incoherent relaxation via the excited states, is evaluated by means of relaxation diagrams (Leuenberger and Loss, 1999, 2000a).

Assuming linear time dependence, i.e., $\varepsilon_{mm'}(t) = \alpha_m^{m'} t$, in the transition region (Landau, 1932), and with $|\varepsilon_{mm'}^{<,>}| \gg \hbar\gamma_{mm'}$ one obtains from Equation 9.42

$$\Delta\rho = \exp\left\{ -\frac{2\Delta E^2_{s,-s}}{\hbar\alpha_s^{-s}} \arctan\left(\frac{\alpha_s^{-s}}{\hbar\gamma_{s,-s}} t \right) - \int_{-t}^{t} dt' \Gamma_{\text{th}} \right\}$$

$$\approx \exp\left\{ -\frac{\pi \Delta E^2_{s,-s}}{\hbar\alpha_s^{-s}} - \int_{-t}^{t} dt' \Gamma_{\text{th}} \right\}, \qquad (9.43)$$

where $t_0 = -t$. In the low-temperature limit $T \to 0$ the excited states are not populated anymore and thus Γ_{th}, which consists of intermediate rates that are weighted by Boltzmann factors b_m (Leuenberger and Loss, 1999, 2000a), vanishes. Consequently, Equation 9.43 simplifies to

$$\Delta\rho = \exp\left\{ -\frac{\pi E^2_{s,-s}}{\hbar\alpha_s^{-s}} \right\} = \exp\left\{ -\frac{\pi E^2_{s,-s}}{\hbar |\dot{\varepsilon}_{s,-s}(0)|} \right\}. \qquad (9.44)$$

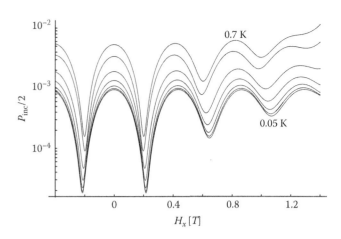

FIGURE 9.7 Zener transition probability $P_{\text{inc}}(H_x)$ for temperatures $T = 0.7, 0.65, 0.6, 0.55, 0.5, 0.45,$ and $0.05\,\text{K}$. The fit agrees well with data. Note that P_{inc} is equal to $2P$. (From Wernsdorfer, W. et al., *Europhys. Lett.*, 50, 552, 2000.)

The exponent in Equation 9.44 differs by a factor of 2 from the Zener exponent (Landau, 1932). This is not surprising since Γ_{tot} is the relaxation rate of $\Delta\rho$, where both ρ_s and ρ_{-s} are changed in time by the same amount, and *not* an escape rate like in the case of coherent Zener transition, where only the population of the initial state is changed in time. Equation 9.44 implies $P_{\text{inc}} = 1$ for $|\dot{\varepsilon}_{s,-s}(0)| \to 0$ (adiabatic limit) and $P_{\text{inc}} = 0$ for $|\dot{\varepsilon}_{s,-s}(0)| \to \infty$ (sudden limit). After fitting the parameters the incoherent Zener theory is in excellent agreement with experiments (Wernsdorfer and Sessoli, 1996; Wernsdorfer et al., 2000) for the temperature range $0.05\,\text{K} \leq T \leq 0.7\,\text{K}$ if the states $|\pm10\rangle$, $|\pm9\rangle$, and $|\pm8\rangle$ are taken into account. In particular, the path leading through $|\pm8\rangle$ gives a non-negligible contribution for $T \geq 0.6\,\text{K}$. One obtains the following from Equation 9.43 for Fe_8 in the case $n = 0$:

$$\Gamma_{\text{tot}} = 2\left(\Gamma_{10}^{-10} + \sum_{n=9}^{8} \frac{b_n}{\dfrac{2}{W_{10,n}} + \dfrac{1}{\Gamma_{n,-n}}} \right),$$

$$\Delta\rho = \exp\left\{ -\frac{\pi E^2_{10,-10}}{\hbar\alpha_{10}^{-10}} - \sum_{n=9}^{8} \frac{\pi E^2_{n,-n} W_{10,n} b_n}{\alpha_n^{-n} \sqrt{E^2_{n,-n} + \hbar^2 W^2_{10,n}}} \right\}, \qquad (9.45)$$

where the approximation $\gamma_{n,-n} \approx W_{10,n}$ and $|\varepsilon_{mm'}^{<,>}| \gg E_{n,-n}, \gamma_{n,-n}$ is used. $P_{\text{inc}} = 1 - \Delta\rho$, which is plotted in Figure 9.7, is in excellent agreement with the measurements (Wernsdorfer et al., 2000).

9.6 Coherent Néel Vector Tunneling in Antiferromagnetic Molecular Wheels

Antiferromagnetic molecular clusters are the most promising candidates for the observation of coherent quantum tunneling on the mesoscopic scale currently available (Chiolero and Loss,

1998). Several systems in which an even number N of antiferromagnetically coupled ions is arranged on a ring have been synthesized to date (Papaefthymiou et al., 1994; Fabretti et al., 1996; Schromm et al., 2001; Gatteschi et al., 2002). These systems are well described by the spin Hamiltonian

$$\hat{\mathcal{H}} = J\sum_{i=1}^{N} \hat{\mathbf{s}}_i \cdot \hat{\mathbf{s}}_{i+1} + g\mu_B \mathbf{H} \cdot \sum_{i=1}^{N} \hat{\mathbf{s}}_i - k_z \sum_{i=1}^{N} \hat{s}_{i,z}^2, \quad (9.46)$$

where

$\hat{\mathbf{s}}_i$ is the spin operator at site i with spin quantum number s, $\hat{\mathbf{s}}_{N+1} \equiv \hat{\mathbf{s}}_1$

J is the nearest neighbor exchange

$\mathbf{H}$ is the magnetic field

$k_z > 0$ the single ion anisotropy directed along the ring axis

The parameters J and k_z have been well established both for various ferric wheels (Papaefthymiou et al., 1994; Fabretti et al., 1996; Kelemen et al., 1998; Cornia et al., 1999; Jansen et al. 1999; Koch et al., 1999; Schromm et al., 2001; Zotos et al., 2001) with $N = 6, 8, 10$, and, more recently, also for a Cr wheel (Gatteschi et al., 2002). For $\mathbf{H} = 0$, the classical ground-state spin configuration of the wheel shows alternating (Néel) order with the spins pointing along $\pm \mathbf{e}_z$. The two states with the Néel vector $\mathbf{n}$ along $\pm \mathbf{e}_z$ (Figure 9.8), labeled $|\uparrow\rangle$ and $|\downarrow\rangle$, are energetically degenerate and separated by an energy barrier of height $N k_z s^2$. Because antiferromagnetic exchange induces dynamics of Néel-ordered spins, the states $|\uparrow\rangle$ and $|\downarrow\rangle$ are not energy eigenstates. Rather, a molecule prepared in spin state $|\uparrow\rangle$ would tunnel coherently between $|\uparrow\rangle$ and $|\downarrow\rangle$ at a rate Δ/h, where Δ is the tunnel splitting (Barbara and Chudnovsky, 1990; Krive and Zaslavskii, 1990). This tunneling of the Néel vector corresponds to a simultaneous tunneling of all N spins within the wheel through a potential barrier governed by the easy axis anisotropy. Within the framework of coherent state spin path integrals, an explicit expression for the tunnel splitting Δ as a function of magnetic field $\mathbf{H}$ has been derived (Chiolero and Loss, 1998). A magnetic field applied in the ring plane, H_x, gives rise to a Berry phase acquired by the spins during tunneling (DiVincenzo et al., 1992; van Delft and Henley, 1992; Garg, 1993; Leuenberger et al., 2003). The resulting interference

of different tunneling paths leads to a sinusoidal dependence of Δ on H_x, which allows one to continuously tune the tunnel splitting from 0 to a maximum value which is of order of some Kelvin for the antiferromagnetic wheels synthesized to date.

The tunnel splitting Δ also enters the energy spectrum of the antiferromagnetic wheel as level spacing between the ground and first excited state. Thus, Δ can be experimentally determined from various quantities such as magnetization, static susceptibility, and specific heat. Even more information on the physical properties of antiferromagnetic wheels (Equation 9.46) can be obtained from a theoretical and experimental investigation of dynamical quantities, such as the correlation functions of the total spin $\hat{\mathbf{S}} = \sum_{i=1}^{N} \hat{\mathbf{s}}_i$ or of single spins within the antiferromagnetic wheel (Meier and Loss, 2000, 2001). By symmetry arguments, it follows that the correlation function of total spin, $\langle \hat{S}_\alpha(t)\hat{S}_\alpha(0)\rangle$, which is experimentally accessible via measurement of the alternating current (AC) susceptibility does not contain a component, which oscillates with the tunnel frequency Δ/h (Meier and Loss, 2000, 2001). Hence, neither the tunnel splitting nor the decoherence rate of Néel vector tunneling can be obtained by experimental techniques which couple to the *total* spin of the wheel. In contrast, the correlation function of a single spin

$$\langle \hat{s}_{i,z}(t)\hat{s}_{i,z}(0)\rangle \simeq s^2 \left(\frac{e^{-\beta\Delta/2}}{2\cosh(\beta\Delta/2)} e^{i\Delta t/h} + \frac{e^{\beta\Delta/2}}{2\cosh(\beta\Delta/2)} e^{-i\Delta t/h} \right)$$
$$(9.47)$$

exhibits the time dependence characteristic of coherent tunneling of the quantity $\hat{S}_{i,z}$ with a tunneling rate Δ/h (Meier and Loss, 2001). We conclude that *local* spin probes are required for the observation of the Néel vector dynamics. Nuclear spins, which couple (predominantly) to a given single spin $\hat{\mathbf{s}}_i$ are ideal candidates for such probes (Meier and Loss, 2001) and have already been used to study spin cross-relaxation between electron and nuclear spins in ferric wheels (Lascialfari et al., 1999; Pini et al., 2000).

For simplicity, we consider a single nuclear spin $\hat{\mathbf{I}}$, $I = 1/2$, coupled to one-electron spin by a hyperfine contact interaction $\hat{\mathcal{H}}' = A\hat{\mathbf{s}}_1 \cdot \hat{\mathbf{I}}$. According to Equation 9.47, the tunneling electron spin $\hat{\mathbf{s}}_1$ produces a rapidly oscillating hyperfine field $As \cos(\Delta t/\hbar)$ at the site of the nucleus.

Signatures of the coherent electron spin tunneling can thus also be found in the nuclear susceptibility. For a static magnetic field applied in the plane of the ring, H_x, it can be shown that the nuclear susceptibility

$$\chi''_{I,yy}(\omega) \simeq \frac{\pi}{4}\left[\tanh\left(\frac{\beta\gamma_I H_x}{2}\right)\delta(\omega - \gamma_I H_x/\hbar) \right.$$

$$\left. + \left(\frac{As}{\Delta}\right)^2 \tanh\left(\frac{\beta\Delta}{2}\right)\delta(\omega - \Delta/\hbar) \right] - [\omega \to -\omega] \quad (9.48)$$

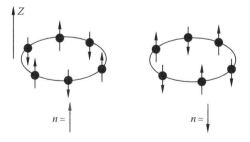

FIGURE 9.8 The two degenerate classical ground state spin configurations of an antiferromagnetic molecular wheel with easy axis anisotropy.

exhibits a *satellite resonance at the tunnel splitting* Δ *of the electron spin system* (Meier and Loss, 2001). Here, $\gamma_I H_x$ is the Lamor frequency of the nuclear spin and the first term in Equation 9.48 corresponds to the transition between the Zeeman split energy levels of I. Because typically $As \simeq 1\,\mathrm{mK}$ and $\Delta \leq 2\,\mathrm{K}$ in Fe$_{10}$, the spectral weight of the satellite peak is small compared to the one of the first term in Equation 9.48 unless the magnetic field is tuned such that Δ is significantly reduced compared to its maximum value. The observation of the satellite peak in Equation 9.48 is challenging, but possible with current experimental techniques (Meier and Loss, 2001). The experiment must be conducted with single crystals of an antiferromagnetic molecular wheel with sufficiently large anisotropy $k_z > 2J/(Ns)^2$ at high, tunable fields (10 T) and low temperatures (2 K). Moreover, because the tunnel splitting $\Delta(\mathbf{H})$ depends sensitively on the relative orientation of $\mathbf{H}$ and the easy axis (Jansen et al., 1999; Zotos et al., 2001), careful field sweeps are necessary to ensure that the satellite peak in Equation 9.48 has a large spectral weight.

The need for local spin probes such as NMR or inelastic neutron scattering to detect coherent Néel vector tunneling can be traced back to the translation symmetry of the spin Hamiltonian $\hat{\mathcal{H}}$ (Meier and Loss, 2000). If this symmetry is broken, e.g., by doping of the wheel, ESR also provides an adequate technique for the detection of coherent Néel vector tunneling. If one of the original Fe or Cr ions of the wheel with spin $s = 5/2$ or $s = 3/2$, respectively, is replaced by an ion with different spin $s' \neq s$, this will in general also result in a different exchange constant J' and single ion anisotropy k_z' at the dopand site, i.e.,

$$\hat{\mathcal{H}} = J \sum_{i=2}^{N-1} \hat{\mathbf{s}}_i \cdot \hat{\mathbf{s}}_{i+1} + J'(\hat{\mathbf{s}}_1 \cdot \hat{\mathbf{s}}_2 + \hat{\mathbf{s}}_1 \cdot \hat{\mathbf{s}}_N)$$

$$+ g\mu_B \mathbf{H} \cdot \sum_{i=1}^{N} \hat{\mathbf{s}}_i - \left(k_z' \hat{s}_{1,z}^2 + k_z \sum_{i=2}^{N} \hat{s}_{i,z}^2 \right). \qquad (9.49)$$

Although thermodynamic quantities, such as magnetization, of the doped wheel may differ significantly from the ones of

the undoped wheel, the picture of spin tunneling in antiferromagnetic molecular systems (Barbara and Chudnovsky, 1990; Krive and Zaslavskii, 1990) remains valid (Meier and Loss, 2000). However, due to unequal sublattice spins, a net total spin remains even in the Néel ordered state of the doped wheel (Figure 9.9). This allows one to distinguish the configurations sketched in Figure 9.8 according to their total spin. The dynamics of the total spin $\hat{s}$ is coupled to the one of the Néel vector (DiVincenzo et al., 1992), and coherent tunneling of the Néel vector results in a coherent oscillation of the total spin. Coherent Néel vector tunneling in doped wheels can hence also be probed by ESR. The AC susceptibility shows a resonance peak at the tunnel splitting Δ,

$$\chi_{zz}''(\omega \simeq \Delta / \hbar) = \pi(g\mu_B)^2 \left| \langle e | \hat{S}_z | g \rangle \right|^2 \tanh\left(\frac{\beta\Delta}{2} \right) \delta(\omega - \Delta/\hbar).$$

$$(9.50)$$

with a transition matrix element between the ground state $|g\rangle$ and first excited state $|e\rangle$,

$$\left| \langle e | \hat{S}_z | g \rangle \right| \simeq |s' - s| \frac{8Jk_z s^2}{(g\mu_B H_x)^2} \qquad (9.51)$$

for $g\mu_B H_x \gg s\sqrt{8Jk_z}$. The matrix element in Equation 9.51 determines the spectral weight of the absorption peak in the ESR spectrum. The analytical dependence has been determined within a semiclassical framework and is in good agreement with numerical results obtained from exact diagonalization of small systems (Figure 9.9).

In conclusion, several antiferromagnetic molecular wheels synthesized recently are promising candidates for the observation of coherent Néel vector tunneling. Although the observation of this phenomenon is experimentally challenging, nuclear magnetic resonance, inelastic neutron scattering, and ESR on doped wheels are adequate experimental techniques for the observation of coherent Néel vector tunneling in antiferromagnetic molecular wheels.

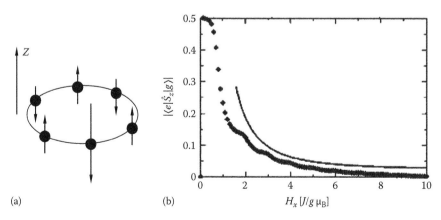

(a)   (b)

FIGURE 9.9 The doped antiferromagnetic molecular wheel acquires a tracer spin which follows the Néel vector dynamics (a). Comparison of results obtained for the matrix element $|\langle e | \hat{S}_z | g \rangle|$ with a coherent state spin path integral formalism (solid line) and by numerical exact diagonalization (symbols) for $N = 4$, $s = 5/2$, $s' = 2$, $J' = J$, $k_z = k_z' = 0.0055J$ (b).

9.7 Berry-Phase Blockade in Single-Molecule Magnet Transistors

Single-electron devices show many promising characteristics such as ultimate low power consumption, down scalability to atomic dimensions, and high switching speed. In addition, these devices are supposed to be good candidates for applications in quantum computation and quantum information technologies. Recently there has been a huge interest in using magnetic molecules with large spin in single-electron devices to uncover the effects that a large spin has upon electron transport through single-molecule magnets (SMM) (Sangregorio et al., 1997). During the past few years, recent experiments demonstrated the possibility to place individual molecules between the source and drain leads allowing electron transport measurements (Jo et al., 2006; Henderson et al., 2007). In a three terminal molecular single-electron transistor (SET), the molecule is situated between the source and drain leads with an insulated gate electrode underneath. The insulating ligands on the periphery of the molecule act as isolating barriers and current can flow between the source and drain leads via a sequential tunneling process through the molecular energy levels, which are tuned by the gate electrode (see Figure 9.10). Several experimental and theoretical groups have been trying to predict and prove the effects on electron transport through an individual molecular nanomagnet SET. Heersche et al. reported Coulomb blockade and conduction excitation characteristics in a Mn_{12} individual molecular nanomagnet SET. Negative differential conductance and current suppression effects were explained with a model that combines spin properties of the molecule with standard sequential tunneling theory.

Recently, experiments have pointed out the importance of the interference between spin tunneling paths in molecules and its effects on electron transport scenarios involving SMMs. These results indicate that the Berry-phase interference plays an important role in the transport properties of SMMs in SET devices. Last year, the authors of this book chapter published an article

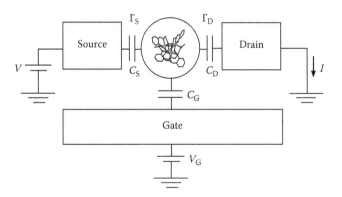

FIGURE 9.10 The sketch shows a three terminal molecular single-electron transistor. The ligands on the periphery of the molecule act as isolating barriers and electrons can flow between source and drain leads by sequential or cotunneling processes through the discrete energy levels of the molecule.

(González and Leuenberger, 2007) in which an effect called the *Berry-phase blockade* was explained. The effect we described is a quantum interference effect that can be detected experimentally by measuring the current through a single molecular electron transistor with oppositely polarized leads and by applying a transverse magnetic field along specific angles. In the following we summarize our results.

For weak coupling between the leads and the SMM we used the generalized master equation describing the electronic spin states of the SMM (see Section 9.2). The sequential tunneling rates for absorption of an electron in Equation 9.19 for ground states with spin s and s' in the case of low temperatures are given by

$$W_{s',s} = \sum_l W_{s',s}^{(l)}, \quad W_{s',s}^{(l)} = w_\downarrow^{(l)} f_l(\Delta_{s',s}),$$

$$W_{-s',-s} = \sum_l W_{-s',-s}^{(l)}, \quad W_{-s',-s}^{(l)} = w_\uparrow^{(l)} f_l(\Delta_{-s',-s}),$$

(9.52)

and the tunneling rates for the emission of an electron are given by

$$W_{s,s'} = \sum_l W_{s,s'}^{(l)}, \quad W_{s,s'}^{(l)} = w_\downarrow^{(l)}[1 - f_l(\Delta_{s,s'})],$$

$$W_{-s,-s'} = \sum_l W_{-s,-s'}^{(l)}, \quad W_{-s,-s'}^{(l)} = w_\uparrow^{(l)}[1 - f_l(\Delta_{-s,-s'})],$$

(9.53)

where $f_l(\Delta_{s',s}) = [1 + e^{(\Delta_{s',s} - \mu_l)/kT}]^{-1}$ is the Fermi function. $w_{\downarrow\uparrow}^{(l)}$ represents the spin-dependent transition rate from the $l \in$ [Left, Right] lead to the SMM and are defined in Fermi's golden rule approximation by $w_\downarrow^{(l)} = 2\pi D v_\downarrow^{(l)} |t_\downarrow^{(l)}|^2/\hbar$ and $w_\uparrow^{(l)} = 2\pi D v_\uparrow^{(l)} |t_\uparrow^{(l)}|^2/\hbar$, respectively, where D is the density of states and $v_\uparrow^{(l)}$ and $v_\downarrow^{(l)}$ are fractions of the number of spins polarized up and down of lead l such that $v_\downarrow^{(l)} + v_\uparrow^{(l)} = 1$. $t_\uparrow^{(l)}$ and $t_\downarrow^{(l)}$ are the tunneling amplitudes of lead l, respectively. Typical values for the tunneling rate of the electron range from around $w = 10^6$ s^{-1} to $w = 10^{10}$ s^{-1} (see González and Leuenberger, 2007, and references therein).

Solving the generalized master equation for the stationary limit, we obtained the coupled differential equations

$$\dot{\rho}_s = \left(\frac{\Delta_{s,-s}}{2\hbar}\right)^2 \frac{2\gamma_{s,-s}}{(g\mu_B H_z(s-(-s)))^2/\hbar^2 + \gamma_{s,-s}^2}$$
$$\times (\rho_{-s} - \rho_s) + W_{s,s'}\rho_{s'} - W_{s',s}\rho_s,$$

(9.54)

$$\dot{\rho}_{-s} = \left(\frac{\Delta_{s,-s}}{2\hbar}\right)^2 \frac{2\gamma_{s,-s}}{(g\mu_B H_z(s-(-s)))^2/\hbar^2 + \gamma_{s,-s}^2}$$
$$\times (\rho_s - \rho_{-s}) + W_{-s,-s'}\rho_{-s'} - W_{-s',-s}\rho_{-s}.$$

(9.55)

The other two differential equations are obtained by just replacing $s \leftrightarrow s'$ in the above equations. Solving the set of differential equations for ρ_s, ρ_{-s}, $\rho_{s'}$ and $\rho_{-s'}$ in the stationary case ($t \gg 1/W_{m,n}$), we obtain

$$\rho_s = (W_{s,s'}(W_{-s',-s} + \Gamma_{s,-s})\Gamma_{s',-s'} + W_{-s,-s'}(W_{s',s} + \Gamma_{s',-s'})\Gamma_{s,-s})/\eta,$$

$$\rho_{-s} = (W_{s,s'}(W_{-s,-s'} + \Gamma_{s',-s'})\Gamma_{s,-s} + W_{-s,-s'}(W_{s',s} + \Gamma_{s,-s})\Gamma_{s',-s'})/\eta,$$

$$\rho_{s'} = (W_{s',s}(W_{-s,-s'} + \Gamma_{s',-s'})\Gamma_{s,-s} + W_{-s',-s}(W_{s,s'} + \Gamma_{s,-s})\Gamma_{s',-s'})/\eta,$$

$$\rho_{-s'} = (W_{s',s}(W_{-s',-s} + \Gamma_{s,-s})\Gamma_{s',-s'} + W_{-s',-s}(W_{s,s'} + \Gamma_{s',-s'})\Gamma_{s,-s})/\eta,$$

$$(9.56)$$

where η is a normalization factor such that $\sum_n \rho_n = 1$. The incoherent tunneling rate is

$$\Gamma_{s,-s} = \left(\frac{\Delta_{s,-s}}{2\hbar}\right)^2 \frac{2\gamma_{s,-s}}{\left(g\mu_B H_z\left(s - (-s)\right)\right)^2 / \hbar^2 + \gamma_{s,-s}^2}. \quad (9.57)$$

We now proceed to define the current through the SMM in terms of the density matrix for the case of the single molecule magnet Ni_4. In the case of Ni_4 we have $s = 4$ and $s' = 7/2$, therefore the current reads

$$I = e(W_{4,7/2}\rho_{7/2} + W_{-4,-7/2}\rho_{-7/2}). \quad (9.58)$$

In the case of leads that are fully polarized in opposite directions, i.e., $v_\uparrow^{(L)} = v_\downarrow^{(R)} = 1$ or $v_\downarrow^{(L)} = v_\uparrow^{(R)} = 1$, we get one of the two following conditions for the transition rates, respectively:

$$W_{-4,-7/2} = W_{7/2,4} = 0 \quad \text{or} \quad W_{4,7/2} = W_{-7/2,-4} = 0. \quad (9.59)$$

Choosing the case $v_\uparrow^{(L)} = v_\downarrow^{(R)} = 1$ and using the condition $W_{-4,-7/2} = W_{7/2,4} = 0$, we obtain from Equation 9.58

$$\frac{e}{I} = \frac{2}{W_{-7/2,-4}} + \frac{1}{\Gamma_{4,-4}} + \frac{2}{W_{4,7/2}} + \frac{1}{\Gamma_{7/2,-7/2}}. \quad (9.60)$$

Equation 9.60 reflects the fact that the current through the SMM depends on the tunnel splittings. The transitions that contribute to the current through the SMM in the case of fully polarized leads $v_\uparrow^{(L)} = v_\downarrow^{(R)} = 1$ are $4 \rightarrow 7/2 \rightarrow -7/2 \rightarrow -4$. Figure 9.11 shows the current as a function of $H_\perp$ for fully polarized leads. If the tunnel splitting $\Delta_{4,-4}$ or $\Delta_{7/2,-7/2}$ is topologically quenched (see Figure 9.11), $\Gamma_{4,-4}$ or $\Gamma_{7/2,-7/2}$ vanishes (see Equation 9.57), which leads to complete current suppression according to Equation 9.60. Since this current blockade is a consequence of the topologically quenched tunnel splitting, we call it *Berry-phase blockade*. Note that the current can also be suppressed by applying H_z, which follows immediately from Equations 9.57 and 9.60.

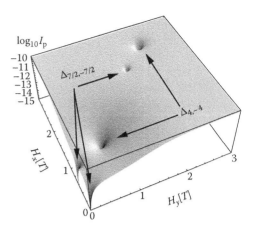

FIGURE 9.11 The graph shows the $\log_{10} I$ vs. $H_\perp$. The scale varies from $I = 0.1$ nA to 1 fA. From the figure we see that the current is completely suppressed at the zeros of the tunnel splitting, i.e., $\Delta_{4,-4} = \Delta_{7/2,-7/2} = 0$.

9.8 Concluding Remarks

We have seen how molecular nanomagnets offer an interesting platform to explore the quantum dynamics of mesoscopic systems. The theoretical advantage of using molecular nanomagnets is that molecules are all identical to each other, allowing the performance of experiments on large assemblies of identical particles and detect quantum effects. Molecules can be investigated in solutions or solids depending on the experiment you want to perform, allowing accurate measurements. We have paid special attention to the quantum tunneling of the magnetization of molecular nanomagnets and the different approaches that exist to describe this phenomenon. In particular we have placed emphasis on the interference effects of the quantum tunneling of the magnetization via Berry's phase and we have included some of our recent results in this field (González and Leuenberger, 2007).

Molecular nanomagnets are also the subject of intense research in molecular electronics. Single-electron devices have taken advantage of the small size of nanomagnets to create electronic structures of atomic dimensions. These devices are supposed to be good candidates for applications in quantum computation and quantum information technologies. There has been also a lot of research on single-electron transistors (SET) made of molecular nanomagnets to study quantum effects such as the Kondo effect in this kind of structures (Leuenberger and Mucciolo, 2006). The Kondo effect is a very well known and studied phenomenon in condensed matter physics. This effect arises when a magnetic impurity is placed into a conductor, which causes a dramatic increase in the resistivity of the metal at low temperatures. The Kondo effect can be seen also in a single-electron transistor where the molecular nanomagnet plays the role of the magnetic impurity and the leads of the SET mimics the bulk metal. It has been shown that the Hamiltonian for a molecular nanomagnet placed in a SET device can be mapped onto the Kondo Hamiltonian (González et al., 2008). This result is of great interest because one can apply the poor man's scaling theory to describe the Kondo physics at low temperatures and see how the Berry-phase oscillations become temperature dependent. This fact allows us to conclude that the scaling equations can be checked experimentally by measuring the renormalized zero points of the Berry phase.

The field of molecular nanomagnets is growing very fast and chemistry is playing a major role concerning the growth of these nanostructures. New nanomagnets with novel and interesting magnetic properties are being produced, which demand more sophisticated theories to explain and describe the dynamics of these magnetic mesoscopic systems. One direction which remains a major area of investigation lately is related to storing and decoding information on a single spin state to be able to create a quantum computer. The implementation of Grover's algorithm with molecular nanomagnets has already been proposed, however, it is required to perform simultaneous manipulation of many spin phases without losing quantum coherence, which remains still an experimental challenge.

Acknowledgment

We acknowledge support from NSF-ECCS 0725514, the DARPA/MTO Young Faculty Award HR0011-08-1-0059, 0901784, and AFOSR FA 9550-09-1-0450.

References

Averin, D. and A. Bardas. AC Josephson effect in a single quantum channel. *Phys. Rev. Lett.*, 75:1831–1834, 1995.

Barbara, B. and E.M. Chudnovsky. Macroscopic quantum tunneling in antiferromagnets. *Phys. Lett. A*, 145:205–208, 1990.

Barbara, B., W. Wernsdorfer, L.C. Sampaio, J.G. Park, C. Paulsen, M.A. Novak, R. Ferré, D. Mailly, R. Sessoli, A. Caneschi, K. Hasselback, A. Benoit, and L. Thomas. Mesoscopic quantum tunneling of the magnetization. *J. Magn. Magn. Mater*, 140–144:1825–1828, 1995.

Barra, A.L., P. Debrunner, D. Gatteschi, C.E. Schulz, and R. Sessoli. Superparamagnetic-like behavior in an octanuclear iron cluster. *Europhys. Lett.*, 35:133–138, 1996.

Chiolero, A. and D. Loss. Macroscopic quantum coherence in molecular magnets. *Phys. Rev. Lett.*, 80:169–172, 1998.

Chudnovsky, E.M. and L. Gunther. Quantum tunneling of magnetization in small ferromagnetic particles. *Phys. Rev. Lett.*, 60:661–664, 1988.

Cornia, A., M. Affronte, J.C. Lasjaunias, and A. Caneschi. Low-temperature specific heat of Fe6 and Fe10 molecular magnets. *Phys. Rev. B*, 60:1161–1166, 1999.

Crothers, D.S. and J.G. Huges. Stueckelberg close-curve-crossing phases. *J. Phys. B*, 10:L557–L560, 1977.

DiVincenzo, D.P., D. Loss, and G. Grinstein. Suppression of tunneling by interference in half-integer-spin particles. *Phys. Rev. Lett.*, 69:3232–3235, 1992.

Enz, M. and R. Schilling. Magnetic field dependence of the tunnelling splitting of quantum spins. *J. Phys. C: Solid State Phys.*, 19:L711–L715, 1986.

Fabretti, A.C., S. Foner, D. Gatteschi, R. Grandi, A. Caneshi, A. Cornia, and L. Schenetti. Synthesis, crystal structure, magnetism, and magnetic anisotropy of cyclic clusters comprising six iron(iii) ions and entrapping alkaline ions. *Chem. Eur. J.*, 2:1379–1387, 1996.

Feynman R.P. Space-Time Approach to Non-Relativistic Quantum Mechanics. *Rev. Mod. Phys.*, 20:367–387, 1948.

Fort, A., A. Rettori, J. Villain, D. Gatteschi, and R. Sessoli. Mixed quantum-thermal relaxation in Mn12 acetate molecules. *Phys. Rev. Lett.*, 80:612–615, 1998.

Friedman, J.R., M.P. Sarachik, J. Tejada, and R. Ziolo. Macroscopic measurement of resonant magnetization tunneling in high-spin molecules. *Phys. Rev. Lett*, 76:3830–3833, 1996.

Garanin, D.A. and E.M. Chudnovsky. Thermally activated resonant magnetization tunneling in molecular magnets: Mn12ac and others. *Phys. Rev. B*, 56:11102–11118, 1997.

Garg, A. Topologically quenched tunnel splitting in spin systems without kramers' degeneracy. *Europhys. Lett.*, 22:205–210, 1993.

Garga, A., N.J. Onuchi, and V. Ambegaokar. Effect of friction on electron transfer in bio-molecules. *J. Chem. Phys.*, 83:4491–4503, 1985.

Gatteschi, D., A.A. Smith, M. Helliwell, R.E.P. Winpenny, A. Cornia, A.L. Barra, A.G.M. Jansen, E. Rentschler, J. van Slageren, R. Sessoli, and G.A. Timco. Magnetic anisotropy of the antiferromagnetic ring [cr8f8piv16]. *Chem. Eur. J.*, 8:277–285, 2002.

González, G. and M.N. Leuenberger. Berry-phase blockade in single-molecule magnets. *Phys. Rev. Lett.*, 98:256804-4, 2007.

González, G., M.N. Leuenberger, and E.R. Mucciolo. Kondo effect in single-molecule magnet transistors. *Phys. Rev. B*, 78:054445-12, 2008.

Henderson, J.J., C.M. Ramsey, E. del Barco, A. Mishra, and G. Christou. Fabrication of nano-gapped single-electron transistors for transport studies of individual single-molecule magnets. *J. Appl. Phys.*, 101:09E102, 2007.

Hernández, J.M., X.X. Zhang, F. Luis, J. Bartolomé, J. Tejada, J.R. Friedman, M.P. Sarachik, and R. Ziolo. Field tuning of thermally activated magnetic quantum tunnelling in mn12—ac molecules. *Europhys. Lett.*, 35:301–306, 1996.

Hernández, J.M., X.X. Zhang, J. Tejada, F. Luis, and R. Ziolo. Evidence for resonant tunneling of magnetization in Mn12sacetate complex. *Phys. Rev. B*, 55:5858–5865, 1997.

Jansen, A.G.M., A. Cornia, and M. Affronte. Magnetic anisotropy of Fe6 and Fe10 molecular rings by cantilever torque magnetometry in high magnetic fields. *Phys. Rev. B*, 60:12177–12183, 1999.

Jo, M.-H., J.E. Grose, K. Baheti, M.M. Deshmukh, J.J. Sokol, E.M. Rumberger, D.N. Hendrickson, J.R. Long, H. Park, and D.C. Ralph. Signatures of molecular magnetism in single-molecule transport spectroscopy. *Nano Lett.*, 6:2014–2020, 2006.

Kelemen, M.T., M. Weickenmeier, B. Pilawa, R. Desquiotz, and A. Geisselmann. Magnetic properties of new Fe6(triethanolaminate(3-))6 spin-clusters. *J. Magn. Magn. Mat.*, 177–181:748–749, 1998.

Koch, R., P. Müller, I. Bernt, R.W. Saalfrank, H.P. Andres, H.U. Güdel, O. Waldmann, J. Schülein, and O. Allenspach. Magnetic anisotropy of two cyclic hexanuclear Fe(iii) clusters entrapping alkaline ions. *Inorg. Chem.*, 38:5879–5886, 1999.

Korenblit, I.Ya. and E.F. Shender. Low-temperature properties of amorphous magnetic materials with random axis of anisotropy. *Sov. Phys. JETP*, 48:937–942, 1978.

Krive, I.V. and O.B. Zaslavskii. Macroscopic quantum tunnelling in antiferromagnets. *J. Phys. Condens. Matter*, 2:9457–9462, 1990.

Landau, L.D. On the theory of transfer of energy at collisions. *Phys. Z. Sowjetunion*, 2:46, 1932.

Lascialfari, A., F. Borsa, M. Horvati, A. Caneschi, M.H. Julien, Z.H. Jang, and D. Gatteschi. Proton NMR for measuring quantum level crossing in the magnetic molecular ring Fe10. *Phys. Rev. Lett.*, 83:227–230, 1999.

Leuenberger, M.N. and D. Loss. Spin relaxation in Mn12-acetate. *Europhys. Lett.*, 46:692–698, 1999.

Leuenberger, M.N. and D. Loss. Spin tunneling and phonon-assisted relaxation in Mn12-acetate. *Phys. Rev. B*, 61:1286–1302, 2000a.

Leuenberger, M.N. and D. Loss. Incoherent zener tunneling and its application to molecular magnets. *Phys. Rev. B*, 61:12200–12203, 2000b.

Leuenberger, M.N. and D. Loss. Spin tunneling and topological selection rules for integer spins. *Phys. Rev. B*, 63:054414-4, 2001a.

Leuenberger, M.N. and D. Loss. Quantum computing in molecular magnets. *Nature*, 410:789–793, 2001b.

Leuenberger, M.N. and D. Loss. Reply to the comment of E. M. Chudnovsky and D. A. Garanin on Spin relaxation in Mn_{12}-acetate. *Europhys. Lett.*, 52: 247–248, 2000.

Leuenberger, M.N., F. Meier, and D. Loss. Quantum spin dynamics in molecular magnets. *Monatshefte fuer Chemie*, 134:217–233, 2003.

Leuenberger, M.N. and E.R. Mucciolo. Berry-phase oscillation of the Kondo effect in single-molecule magnets. *Phys. Rev. Lett.*, 97:126601, 2006.

Luis, F., J. Bartolomé, and F. Fernández. Resonant magnetic quantum tunneling through thermally activated states. *Phys. Rev. B*, 57:505–513, 1998.

Meier, F. and D. Loss. Thermodynamics and spin-tunneling dynamics in ferric wheels with excess spin. *Phys. Rev. B*, 64:224411–224414, 2000.

Meier, F. and D. Loss. Electron and nuclear spin dynamics in antiferromagnetic molecular rings. *Phys. Rev. Lett.*, 86:5373–5376, 2001.

Novak, M.A. and R. Sessoli. *Quantum Tunneling of Magnetization.* L. Gunther and B. Barbara (eds.), Kluwer, Dordrecht, the Netherlands, 1995.

Novak, M.A., R. Sessoli, A. Caneschi, and D. Gatteschi. Magnetic properties of a Mn cluster organic compound. *J. Magn. Magn. Mater*, 146:211–213, 1995.

Ohm, T., C. Sangregorio, and C. Paulsen. Local field dynamics in a resonant quantum tunneling system of magnetic molecules. *Europhys. J. B*, 6:195–199, 1998.

Papaefthymiou, G.C., S. Foner, D. Gatteschi, K.L. Taft, C.D. Delfs, and S.J. Lippard. [Fe(OMe)2(O2CCH2Cl)]10, a molecular ferric wheel. *J. Am. Chem. Soc.*, 116:823–832, 1994.

Paulsen, C. and J.G. Park. *Quantum Tunneling of Magnetization.* L. Gunther and B. Barbara (eds.), Kluwer, Dordrecht, the Netherlands, 1995.

Paulsen, C., J.G. Park, B. Barbara, R. Sessoli, and A. Caneschi. Novel features in the relaxation times of Mn12Ac. *J. Magn. Magn. Mater*, 140–144:379–380, 1995.

Pini, M.G., A. Cornia, A. Fort, and A. Rettori. Low-temperature theory of proton NMR in the molecular antiferromagnetic ring Fe10. *Europhys. Lett.*, 50:88–93, 2000.

Sangregorio, C., T. Ohm, C. Paulsen, R. Sessoli, and D. Gatteschi. Quantum tunneling of the magnetization in an iron cluster nanomagnet. *Phys. Rev. Lett.*, 78:4645–4648, 1997.

Schromm, S., J. Schülein, P. Müller, I. Bernt, R.W. Saalfrank, O. Waldmann, R. Koch, and F. Hampel. Magnetic anisotropy of a cyclic octanuclear Fe(iii) cluster and magneto-structural correlations in molecular ferric wheels. *Inorg. Chem.*, 40:2986–2995, 2001.

Sessoli, R., D. Gatteschi, A. Caneschi, and M.A. Novak. Magnetic bistability in a metal-ion cluster. *Nature*, 365:141–143, 1993.

Shimshoni, E. and Y. Gefen. Onset of dissipation in zener dynamics: Relaxation vs. dephasing. *Ann. Phys.*, 210:16–80, 1991.

Thomas, L., F. Lionti, R. Ballou, D. Gatteschi, R. Sessoli, and B. Barbara. Macroscopic quantum tunnelling of magnetization in a single crystal of nanomagnets. *Nature*, 383:145–147, 1996.

von Delft, J. and C.L. Henley. Destructive quantum interference in spin tunneling problems. *Phys. Rev. Lett.*, 69:3236–3239, 1992.

van Hemmen, J.L. and A. Süto. Tunnelling of quantum spins. *Europhys. Lett.*, 1:481–490, 1986.

Villain, J., F. Hartmann-Boutron, R. Sessoli, and A. Rettori. Magnetic relaxation in big magnetic molecules. *Europhys. Lett.*, 27:159–164, 1994.

Wernsdorfer, W. and R. Sessoli. Quantum phase interference and parity effects in magnetic molecular clusters. *Science*, 284:133–135, 1996.

Wernsdorfer, W., R. Sessoli, A. Caneshi, D. Gatteschi, and A. Cornia. Nonadiabatic Landau-Zener tunneling in Fe8 molecular nanomagnets. *Europhys. Lett.*, 50:552–558, 2000.

Zener, C. Non-adiabatic crossing of energy levels. *Proc. R. Soc. Lond. A*, 137:696, 1932.

Zotos, X., B. Normand, X. Wang, and D. Loss. Magnetization in molecular iron rings. *Phys. Rev. B*, 63:184409, 2001.

10

Inelastic Electron Transport through Molecular Junctions

Natalya A. Zimbovskaya
University of Puerto Rico

10.1 Introduction

Molecular electronics is known to be one of the most promising developments in nanoelectronics, and the past decade has seen extraordinary progress in this field (Aviram et al. 2002, Cuniberti et al. 2005, Nitzan 2001). Present activities on molecular electronics reflect the convergence of two trends in the fabrication of nanodevices, namely, the top-down device miniaturization through the lithographic methods and bottom-up device manufacturing through the atom-engineering and the self-assembly approaches. The key element and the basic building block of molecular electronics is a junction including two electrodes (leads) linked by a molecule, as schematically shown in Figure 10.1. Usually, the electrodes are microscopic large but macroscopic small contacts that could be connected to a battery to provide the bias voltage across the junction. Such a junction may be treated as a quantum dot coupled to the charge reservoirs. The discrete character of energy levels on the dot (molecule) is combined with nearly continuous energy spectra on the reservoirs (leads) occurring due to their comparatively large size.

When the voltage is applied, an electric current flows through the junction. Successful transport experiments with molecular junctions (Ho 2002, Lortscher et al. 2007, Park et al. 2000, Poot et al. 2006, Reichert et al. 2002, Smit et al. 2002, Yu et al. 2004) confirm their significance as active elements in nanodevices. These include applications as rectifiers (molecular diodes), field-effect transistors (molecular triodes), switches, memory elements and sensors. Also, these experiments emphasize the importance of a thorough analysis of the physics underlying electron transport through molecular junctions. A detailed understanding of electron transport at the molecular scale is a key step to future device operations. A theory of electron transport in molecular junctions is being developed since the last two decades, and main transport mechanisms are currently elucidated in general terms (Datta 2005, Imry and Landauer 1999). However, a progress of the experimental capabilities in the field of molecular electronics brings new theoretical challenges causing a further development of the theory.

Speaking of transport mechanisms, it is useful to make a distinction between elastic electron transport when the electron energy remains the same as it travels through the junction, and inelastic transport processes when the electron undergoes energy changes due to its interactions with the environment. There are several kinds of processes bringing inelasticity in the electron transport in mesoscopic systems including molecular junctions. Chief among these are electron–electron and electron–phonon scattering processes. These processes may bring about significant inelastic effects modifying the transport properties of molecular devices and charging, desorption, and chemical reactions as well. To keep this chapter at a reasonable length, we concentrate on the inelastic effects originating from electron–phonon interactions.

In practical molecular junctions, the electron transport is always accompanied by nuclear motions in the environment. Therefore, the conduction process is influenced by the coupling between the electronic and the vibrational degrees of freedom. Nuclear motions underlie the interplay between the coherent electron tunneling through the junction and the inelastic thermally assisted hopping transport

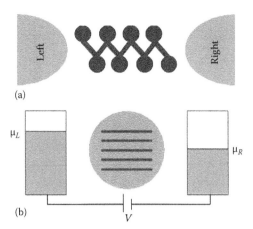

(a)

μ_L　　　　　　　μ_R

(b)　　　　V

FIGURE 10.1 Schematic drawing of a junction including two electrodes and a molecule in between (a). When the voltage is applied across the junction, electrochemical potentials μ_L and μ_R differ, and the conduction window opens up (b).

(Nitzan 2001). Also, electron–phonon interactions may result in polaronic conduction (Galperin et al. 2005, Gutierrez et al. 2005, Kubatkin et al. 2003, Ryndyk et al. 2008), and they are directly related to the junction heating (Segal et al. 2003) and to some specific effects such as alterations in both shape of the molecule and its position with respect to the leads (Komeda et al. 2002, Mitra et al. 2004, Stipe et al. 1999). The effects of electron–phonon interactions may be manifested in the inelastic tunneling spectrum (IETS) that presents the second derivative of the current in the junction d^2I/dV versus the applied voltage V. The inelastic electron tunneling microscopy has proven to be a valuable method for the identification of molecular species within the conduction region, especially when employed in combination with scanning tunneling microscopy and/or spectroscopy (Galperin et al. 2004).

Inelasticity in the electron transport through molecular junctions is closely related to the dephasing effects. One may say that incoherent electron transport always includes an inelastic contribution with the possible exception of the low temperature range. The general approach to theoretically analyzing electron transport through molecular junctions in the presence of dissipative/phase-breaking processes in both electronic and nuclear degrees of freedom is based on advanced formalisms (Segal et al. 2000, Skourtis and Mukamel 1995, Wingreen et al. 1989, 1993). These microscopic computational approaches have the advantages of being capable of providing detailed dynamics information. However, while considering stationary electron transport through molecular junctions, one may turn to the less time-consuming approach based on scattering matrix formalism (Buttiker 1986, Li and Yan 2001a,b), as discussed below.

10.2 Coherent Transport

To better show the effects of dissipation/dephasing on the electron transport through molecular junctions, it seems reasonable to start from the case where these effects do not occur. So, we consider a molecule (presented as a set of energy levels) placed in between two electrodes with nearly continuous energy spectra. While there is no bias voltage applied across the junction, the latter remains in equilibrium characterized by the equilibrium Fermi energy E_F, and there is no current flowing through it. When the bias voltage is applied, it keeps the left and right electrodes at different electrochemical potentials μ_L and μ_R. Then the electric current appears in the junction, and the molecular energy levels located in between the electrochemical potentials μ_L and μ_R play a major part in maintaining this current. Electrons from occupied molecular states tunnel to the electrodes in accordance with the voltage polarity, and the electrons from one electrode travel to another one using unoccupied molecular levels as intermediate states for tunneling. Usually, the electron transport in molecular junctions occurs via the highest occupied (HOMO) and the lowest unoccupied (LUMO) molecular orbitals that work as channels for electron transmission. Obviously, the current through the junction depends on the quality of contacts between the leads and the molecule ends. However, there also exists the limit for the conductance in the channels. As was theoretically shown (Landauer 1970), the maximum conductance of a channel with a single-spin degenerate-energy level equals:

$$G_0 = \frac{e^2}{\pi\hbar} = (25.8 \text{ k}\Omega)^{-1} \qquad (10.1)$$

where
　e is the electron charge
　$\hbar$ is Planck's constant

This is a truly remarkable result for it proves that the minimum resistance $R_0 = G_0^{-1}$ of a molecular junction cannot become zero. In another words, one never can short-circuit a device operating with quantum channels. Also, the expression (10.1) shows that the conductance is a quantized quantity.

Conductance g in practical quantum channels associated with molecular orbitals can take on values significantly smaller that G_0, depending on the delocalization in the molecular orbitals participating in the electron transport. In molecular junctions it also strongly depends on the molecule coupling to the leads (quality of contacts), as was remarked before. The total resistance $r = g^{-1}$ includes contributions from the contact and the molecular resistances, and could be written as (Wingreen et al. 1993)

$$r = \frac{1}{G_0}\left(1 + \frac{1-T}{T}\right). \qquad (10.2)$$

Here, T is the electron transmission coefficient that generally takes on values less than unity.

The general expression for the electric current flowing through the molecular junction could be obtained if one calculates the total probability for an electron to travel between two electrodes at a certain tunnel energy E and then integrates the latter over the whole energy range (Datta 1995). This results in the well-known Landauer expression:

$$I = \frac{e}{\pi\hbar} \int T(E)\big[f_L(E) - f_R(E)\big]dE. \qquad (10.3)$$

Here, $f_{L,R}(E)$ are Fermi distribution functions for the electrodes with chemical potentials $\mu_{L,R}$, respectively. The values of $\mu_{L,R}$ differ from the equilibrium energy E_F, and they are determined by the voltage distribution inside the system. Assuming that coherent tunneling predominates in electron transport, the electron transmission function is given by (Datta et al. 1997, Samanta et al. 1996):

$$T(E) = 2Tr\big\{\Delta_L G \Delta_R G^+\big\}, \qquad (10.4)$$

where

the matrices $\Delta_{L,R}$ represent the imaginary parts of the self-energy terms $\Sigma_{L,R}$ describing the coupling of the molecule to the electrodes

G is the Green's function matrix for the molecule whose matrix elements between the molecular states $\langle i|$ and $|j\rangle$ have the form

$$G_{ij} = \langle i|E - H|j\rangle. \qquad (10.5)$$

Here, H is the molecular Hamiltonian including the self-energy parts $\Sigma_{L,R}$.

When a molecule contacts the surface of electrodes, this results in a charge transfer between the molecule and the electrodes, and in a modification of the molecule energy states due to the redistribution of the electrostatic potential within the molecule. Besides, the external voltage applied across the junction brings additional changes to the electrostatic potential further modifying the molecular orbitals. The coupling of the molecule to the leads may also depend on the voltage distribution. So, generally, the electron transmission T inserted in Equation 10.3 and the electrochemical potentials $\mu_{L,R}$ depend on the electrostatic potential distribution in the system. To find the correct distribution of the electric field inside the junction one must simultaneously solve the Schrodinger equation for the molecule and the Poisson equation for the charge density, following a self-consistent converging procedure. This is a nontrivial and complicated task, and significant effort was applied to study the effect of electrostatic potential distribution on the electron transport through molecules (Damle et al. 2001, Di Ventra et al. 2000, Galperin et al. 2006, Lang and Avouris 2000a,b, Mujica et al. 2000, Xue and Ratner 2003a,b, 2004, Xue et al. 2001). Here, we put these detailed considerations aside, and we use the simplified expression for the electrochemical potentials:

$$\mu_L = E_F + \eta|e|V; \quad \mu_R = E_F - (1-\eta)|e|V, \qquad (10.6)$$

where the parameter η indicates how the bias voltage is distributed between the electrodes. Also, we assume that inside the molecule, the external electrostatic field is screened due to the charge redistribution, and the electron transmission is not sensitive to the changes in the voltage V. Although very simple, this model allows one to analyze the main characteristics of the electron transport through molecular junctions. Within this model one may write down the following expression for the self-energy parts (D'Amato and Pastawski 1990):

$$(\Sigma_\beta)_{ij} = \sum_k \frac{\tau^*_{ik,\beta}\tau_{kj,\beta}}{E - \varepsilon_{k,\beta} + is}. \qquad (10.7)$$

Here,

$\beta \in L, R, \tau_{ik,\beta}$ is the coupling strength between the ith molecular state and the kth state on the left/right lead

$\varepsilon_{k,\beta}$ are the energy levels on the electrodes

s is a positive infinitesimal parameter

Assuming that the molecule is reduced to a single orbital with the energy E_0 (a single-site bridge), Green's function accepts the form

$$G(E) = \frac{1}{E - E_0 - \Sigma_L - \Sigma_R}. \qquad (10.8)$$

Accordingly, in this case, one may simplify the expression (10.4) for the electron transmission:

$$T(E) = \frac{4\Delta_L\Delta_R}{(E - E_0)^2 + (\Delta_L + \Delta_R)^2}. \qquad (10.9)$$

To elucidate the main features of electron transport through molecular junctions we consider a few examples. In the first example, we mimic a molecule as a one-dimensional chain consisting of N identical hydrogen-like atoms with the nearest neighbors interaction. We assume that there is one state per isolated atom with the energy E_0, and that the coupling between the neighboring sites in the chain is characterized by the parameter b. Such a model was theoretically analyzed by D'Amato and Pastawski and in some other works (see e.g., Mujica et al. 1994). Based on Equations 10.4 and 10.5, it could be shown that for a single-site chain ($N = 1$) the electron transmission reveals a well-distinguished peak at $E = E_0$, shown in Figure 10.2. The height of this peak is determined by the coupling of the bridge site to the electrodes. The peak in the electron transmission arises because the molecular orbital $E = E_0$ works as the channel/bridge for electron transport between the leads. Similar peaks appear in the conductance $g = dI/dV$. Assuming the symmetric voltage distribution ($\eta = 1/2$), the peak in the conductance is located at $V = \pm 2E_0$. As for the current voltage characteristics, they display step-like shapes with the steps at $V = \pm 2E_0$. When the chain includes several sites, we obtain a set of states (orbitals) for our bridge instead of the single state $E = E_0$, and their number equals the number of

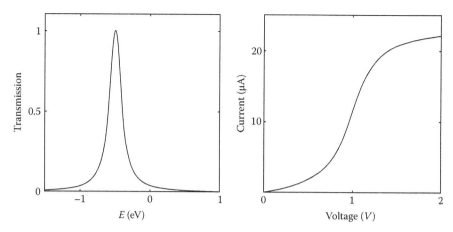

FIGURE 10.2 Coherent electron transmission (left panel) and current (right panel) versus bias voltage applied across a molecular junction where the molecule is simulated by a single electronic state. The curves are plotted assuming $\Delta_L = \Delta_R = 0.1$ eV, $E_0 = -0.5$ eV, $T = 30$ K.

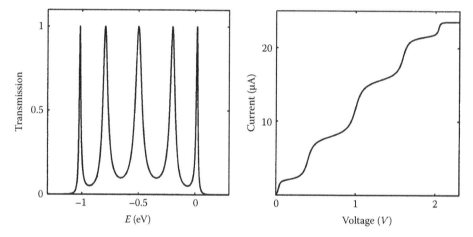

FIGURE 10.3 Coherent electron transmission (left panel) and current (right panel) through a junction with a five electronic states bridge. The curves are plotted for $\Delta_L = \Delta_R = 0.1$ eV, $b = 0.3$ eV, $E_0 = -0.5$ eV, $T = 30$ K.

sites in the chain. All these states are the channels for the electron transport. Correspondingly, the transmission reveals a set of peaks as presented in Figure 10.3. The peaks are located within the energy range with the width $4b$ around $E = E_0$. The coupling of the chain ends to the electrodes affects the transmission, especially near $E = E_0$. As the coupling strengthens, the transmission minimum values increase. Now, the current voltage curves exhibit a sequence of steps. The longer the chain, the more energy levels it possesses, and the more the number of steps in the $I-V$ curves.

The second example concerns the electron transport through a carbon chain placed between copper electrodes. In this case, as well as for practical molecules, a preliminary step in transport calculations is to compute the relevant molecular energy levels and wave functions. Usually, these computations are carried out employing quantum chemistry software packages (e.g., Gaussian) or density functional-based software. Also, a proper treatment of the molecular coupling to the electrodes is necessary for it brings changes into the molecular energy states. For this purpose, one may use the concept of an

"extended molecule" proposed by Xue et al. (2001). The point of this concept is that only a few atoms on the surface of the metallic electrode are significantly disturbed when the molecule is attached to the latter. These atoms are located in the immediate vicinity of the molecule end. Therefore, one may form a system consisting of the molecule itself and the atoms from the electrode surfaces perturbed by the molecular presence. This system is called the extended molecule and treated as such while computing the molecular orbitals. In the example considered, the extended molecule included four copper atoms on each side of the carbon chain. The results for electron transport are shown in Figure 10.4. Again, we observe a comb-like shape of the electron transmission corresponding to the set of transport channels provided by the molecular orbitals, and the stepwise $I-V$ curve originating from the latter. Transport calculations similar to those described above were repeatedly carried out in the last two decades for various practical molecules (see e.g., Galperin et al. 2006, Xue and Ratner 2003a,b, Xue et al. 2001, Zimbovskaya 2003, 2008, Zimbovskaya and Gumbs 2002).

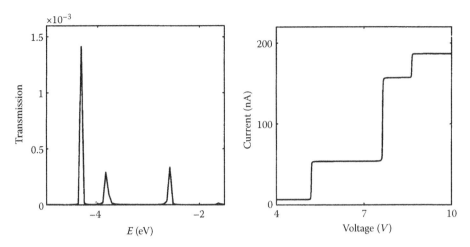

FIGURE 10.4 Coherent electron transmission (left panel) and current (right panel) through a carbon chain coupled to the copper leads at $T = 30\,\text{K}$.

10.3 Buttiker Model for Inelastic Transport

An important advantage of the phenomenological model for the incoherent/inelastic quantum transport proposed by Buttiker (1986) is that this model could easily be adapted to analyze various inelastic effects in electron transport through molecules (and some other mesoscopic systems) avoiding complicated and time-consuming advanced methods, such as those based on the nonequilibrium Green's functions formalism (NEGF).

Here, we present the Buttiker model for a simple junction including two electrodes linked by a single-site molecular bridge. The bridge is attached to a phase-randomizing electron reservoir, as shown in Figure 10.5. Electrons tunnel from the electrodes to the bridge and vice versa via the channels 1 and 2. While on the bridge, an electron could be scattered into the channels 3 and 4 with a certain probability ε. Such an electron arrives at the reservoir where it undergoes inelastic scattering accompanied by phase-breaking and then the reservoir reemits it back to the channels 3 and 4 with the same probability. So, within the Buttiker model, the electron transport through the junction is treated as the combination of tunnelings through the barriers separating the molecule from the electrodes and the interaction with the phase-breaking electron reservoir coupled to the bridge site. The key parameter of the model is the probability ε that is closely related to the coupling strength between the bridge site and the reservoir. When $\varepsilon = 0$ the reservoir is detached from the bridge, and the electron transport is completely coherent and elastic. Within the opposite limit ($\varepsilon = 1$), electrons are certainly scattered into the reservoir that results in the overall phase randomization and inelastic transport.

Within the Buttiker model, the particle fluxes outgoing from the junctions J_i' could be presented as the linear combinations of the incoming fluxes J_k where the indexes i, k label the channels for the transport: $1 \leq i, k \leq 4$.

$$J_i' = \sum_k T_{ik} J_k. \tag{10.10}$$

The coefficients T_{ik} in these linear combinations are matrix elements of the transmission matrix that are related to the elements of the scattering matrix S, namely, $T_{ik} = |S_{ik}|^2$. The matrix S expresses the outgoing wave amplitudes b_1', b_2', a_3', a_4' in terms of the incident ones b_1, b_2, a_3, a_4. To provide the charge conservation in the system, the net current in the channels 3 and 4 linking the system with the dephasing reservoir must be zero, so we may write

$$J_3 + J_4 - J_3' - J_4' = 0. \tag{10.11}$$

The transmission for quantum transport could be defined as the ratio of the particle flux outgoing from the system and that one incoming to the latter. Solving Equations 10.10 and 10.11 we obtain

$$T(E) = \frac{J_2'}{J_1} = T_{21} + \frac{K_1 \cdot K_2}{2R}, \tag{10.12}$$

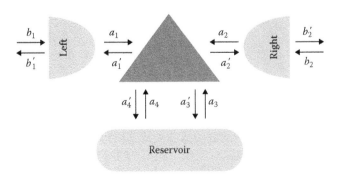

FIGURE 10.5 Schematic drawing illustrating inelastic electron transport through a molecular junction within the Buttiker model.

where

$$K_1 = T_{31} + T_{41}; \quad K_2 = T_{23} + T_{24};$$
$$R = T_{33} + T_{44} + T_{43} + T_{34}. \tag{10.13}$$

For the junction including the single-site bridge the scattering matrix S has the form (Buttiker 1986)

$$S = \frac{1}{Z} \begin{pmatrix} r_1 + \alpha^2 r_2 & \alpha t_1 t_2 & \beta t_1 & \alpha \beta t_1 r_2 \\ \alpha t_1 t_2 & r_2 + \alpha^2 r_1 & \alpha \beta r_1 t_2 & \beta t_2 \\ \beta t_1 & -\beta r_1 t_2 & \beta^2 r_1 & \alpha r_1 r_2 - \alpha \\ \alpha \beta t_1 r_2 & \beta t_2 & \alpha r_1 r_2 - \alpha & \beta^2 r_2 \end{pmatrix}. \tag{10.14}$$

Here, $Z = 1 - \alpha^2 r_1 r_2$, $\alpha = \sqrt{1-\varepsilon}$, $\beta = \sqrt{\varepsilon}$, and $r_{1,2}$ and $t_{1,2}$ are the amplitude reflection and the transmission coefficients for the two barriers. Later, the expression for this matrix suitable for the case of multisite bridges including several inelastic scatterers was derived (Li and Yan 2001a,b).

Assuming for certainty the charge flow from the left to the right, we may write down the following expression (Zimbovskaya 2005):

$$T(E) = \frac{g(E)(1+\alpha^2)\left[g(E)(1+\alpha^2) + 1 - \alpha^2\right]}{\left[g(E)(1-\alpha^2) + 1 + \alpha^2\right]^2}, \tag{10.15}$$

where

$$g(E) = 2\sqrt{\frac{\Delta_L \Delta_R}{(E - E_0)^2 (\Delta_L + \Delta_R)^2}}. \tag{10.16}$$

Now, the electron transmission strongly depends on the dephasing strength ε. As shown in Figure 10.6, coherent transmission ($\varepsilon = 0$) exhibits a sharp peak at $E = E_0$ that gives a step-like shape to the volt–ampere curve, as was discussed in the previous section. In the presence of dephasing, the peak gets eroded. When the ε value approaches 1 the *I–V* curve becomes linear, corroborating the ohmic law for the inelastic transport.

Within Buttiker's model, ε is introduced as a phenomenological parameter whose relation to the microscopic characteristics of the dissipative processes affecting electron transport through molecular junctions remains uncertain. To further advance this model one should find out how to express ε in terms of the relevant microscopic characteristics for various transport mechanisms. This should open the way to a making of a link between the phenomenological Buttiker model and the NEGF. Such an attempt has been carried out in recent works (Zimbovskaya 2005, 2008) where the effect of stochastic nuclear motion on electron transport through molecules was analyzed.

10.4 Vibration-Induced Inelastic Effects

The interaction of electrons with molecular vibrations is known to be an important source of inelastic contribution to the electron transport through molecules. Theoretical studies of vibrationally inelastic electron transport through molecules and other similar nanosystems (e.g., carbon nanotubes) have been carried out over the past few years by a large number of researchers (Cornaglia et al. 2004, 2005, Donarini et al. 2006, Egger and Gogolin 2008, Galperin et al. 2007, Gutirrez et al. 2006, Kushmerick et al. 2004, Mii et al. 2003, Ryndyk and Cuniberti 2007, Siddiqui et al. 2007, Tikhodeev and Ueba 2004, Troisi and Ratner 2006, Zazunov and Martin 2007, Zazunov et al. 2006a,b, Zimmerman et al. 2008). Also, manifestations of the electron–vibron interactions were experimentally observed (Agrait et al. 2003, Djukic et al. 2005, Lorente et al. 2001, Qin et al. 2004, Repp et al. 2005a,b, Segal 2001, Smit et al. 2004, Tsutsui et al. 2006, Wang et al. 2004, Wu et al. 2004, Zhitenev et al. 2002). To analyze vibration induced effects on electron transport through molecular bridges, one must assume that molecular orbitals are coupled

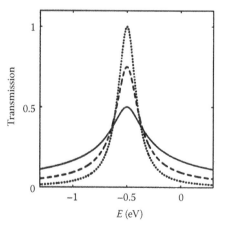
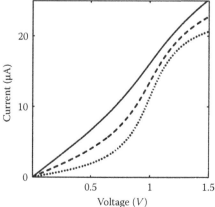

FIGURE 10.6 Electron transmission (left panel) and current (right panel) computed within the Buttiker model at various values of the dephasing parameter ε, namely, $\varepsilon = 0$ (dotted lines), $\varepsilon = 0.5$ (dashed lines), and $\varepsilon = 1$ (solid lines). The curves are plotted assuming that the molecule is simulated by a single orbital with $E_0 = -0.5\,\text{eV}$ at $\Delta_L = \Delta_R = 0.1\,\text{eV}$, $T = 30\,\text{K}$.

to the phonons describing vibrations. While on the bridge, electrons may participate in the events generated by their interactions with vibrational phonons. These events involve a virtual phonon emission and absorption. For rather strong electron–phonon interaction this leads to the appearance of metastable electron levels that could participate in the electron transport through the junctions, bringing an inelastic component to the current. As a result, vibration induced features occur in the differential molecular conductance dI/dV and in the IETS d^2I/dV^2. This was observed in experiments (see e.g., Qin et al. 2004, Zhitenev et al. 2002). Sometimes, even current voltage curves themselves exhibit an extra step originating from the electron-vibron interactions (Djukic et al. 2005).

Particular manifestations of electron–vibron interaction effects in the transport characteristics are determined by the relation of three relevant energies. These are the coupling strengths of the molecule to the electrodes $\Delta_{L,R}$, the electron–phonon coupling strength λ, and the thermal energy kT (k is the Boltzmann constant). When the molecule is weakly coupled to the electrodes ($\Delta_{L,R} \ll \lambda, \hbar\Omega$) and the temperature is low ($kT \ll \Delta_{L,R}$), the electron transfer through the junction may give rise to a strong vibrational excitation, and one may expect a pronounced vibrational resonance structure to appear in the electron transmission. Correspondingly, extra steps should occur in the I–V curves. A proper theoretical consideration of the electron transport in this regime could be carried out within the approach proposed by Wingreen et al. (1989). Here, we employ a very simple semi-quantitative approximation that, nevertheless, allows the qualitative description of this structure including the effect of higher phonon harmonics on the transport characteristics.

We consider a junction including a single-site bridge that is coupled to a single vibrational mode with the frequency Ω. An electron on the bridge may virtually absorb several phonons, which results in the creation of a set of metastable states with the energies $E_n = \tilde{E}_0 + n\,\hbar\Omega$ ($n = 0, 1, 2, \ldots$). Here, the energy $\tilde{E}_0$ is shifted with respect to E_0 due to the electron–phonon interaction. The difference in these energies E_p is called a polaronic

shift and could be estimated as $E_p = \lambda^2/\hbar\Omega$ (Gutirrez et al. 2006, Ryndyk and Cuniberti 2007, Wingreen et al. 1989). At a weak coupling of the bridge state to the electrodes, the lifetime of these metastable states is long enough for them to serve as channels for the electron transmission. Therefore, one may roughly approximate the transmission as a sum of contributions from all these channels. The terms in the sum have a form similar to the well-known expression for coherent transmission (see Equation 10.9). However, every term includes the factor $P(n)$ that corresponds to the probability of the appearance of the metastable state. So, we obtain

$$T(E) = 2\Delta_L\Delta_R \sum_n \frac{P(n)}{(E - \tilde{E}_0 - n\hbar\Omega)^2 + (\Delta_L + \Delta_R)^2}. \tag{10.17}$$

Here (Cizek et al. 2004):

$$P(n) = \frac{1}{n!}\left(\frac{\lambda^2}{2\hbar^2\Omega^2}\right)^n \exp\left(-\frac{\lambda^2}{2\hbar^2\Omega^2}\right). \tag{10.18}$$

The phonon-induced peaks in the transmission are displayed in Figure 10.7 along with the transmission peak for coherent transport through a single-site bridge. As expected, the coupling of the electronic degrees of freedom to the vibrational motion splits the single peak in the coherent transmission into a set of smaller peaks associated with vibrational levels. The peaks could be resolved when $\Delta_{L,R} \ll \hbar\Omega$. This agrees with the results of the earlier theoretical works (Wingreen et al. 1989), as well as with the experiments (Qin et al. 2004, Zhitenev et al. 2002). Phonon-induced peaks in the transmission give rise to the steps in the I–V curves and rather sharp features (peaks and dips) in the IETS. The latter are shown in Figure 10.8 (left panel) and they resemble those obtained using proper NEGF-based calculations (Galperin et al. 2004).

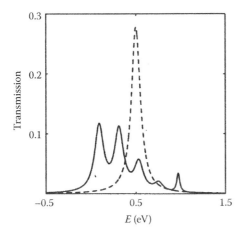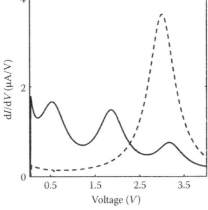

FIGURE 10.7 The electron transmission versus energy (left panel) and the conductance versus voltage (right panel) for a junction with the molecular bridge simulated by a single electronic state weakly coupled to the leads: $\Delta_L = \Delta_R = 0.01\,\text{eV}$, $E_0 = 0.5\,\text{eV}$. Solid lines are plotted assuming that the bridge is coupled to a single phonon mode ($\hbar\Omega = 0.22\,\text{eV}$, $\lambda = 0.3\,\text{eV}$). Dashed lines correspond to the coherent electron transport.

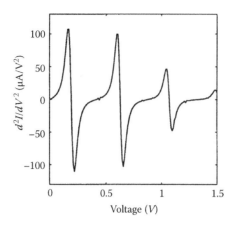
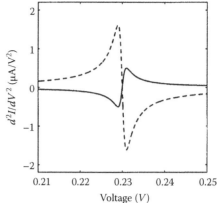

FIGURE 10.8 The inelastic electron tunneling spectrum plotted against the bias voltage at $\hbar\Omega = 0.22\,\text{eV}$, $\lambda = 0.3\,\text{eV}$. Left panel: $\Delta_L = \Delta_R = 0.01\,\text{eV}$, $E_0 = 0.5\,\text{eV}$. Right panel: $\Delta_L = \Delta_R = 0.5\,\text{eV}$, $E_0 = 0.5\,\text{eV}$ (solid line), $E_0 = -0.5\,\text{eV}$ (dashed line).

When the molecule is strongly coupled to the electrodes ($\Delta_{L,R} \gg \lambda$) and the temperature is still low ($kT \ll \lambda$, $\hbar\Omega$), electron–vibron interaction effects are less pronounced. Both the current and the conductance are weakly affected by the electron–phonon coupling (Galperin et al. 2007, Tikhodeev and Ueba 2004). However, the IETS features remain distinguishable. These features appear at the threshold $V = \hbar\Omega/|e|$ that corresponds to the opening of a channel for inelastic transport. To analyze IETS in the simplest way one may use the result for electron transmission derived within the Buttiker model where the dephasing parameter ε is expressed in terms of the relevant energies, namely,

$$\varepsilon = \frac{\Gamma_{ph}}{2(\Delta_L + \Delta_R) + \Gamma_{ph}}, \tag{10.19}$$

where $\Gamma_{ph} = 2\text{Im}(\Sigma_{ph})$, and Σ_{ph} is the self-energy term originating from the electron–phonon interaction. Based on the nonequilibrium Green's function formalism, the expression for Γ_{ph} was derived in the form (Mii et al. 2003)

$$\Gamma_{ph}(E) = 2\pi\lambda^2 \int d\omega\rho(\omega)\big\{N(\omega)[\rho_{el}(E - \hbar\omega)$$
$$+ \rho_{el}(E + \hbar\omega)] + \big(1 + N(\omega)\big)\big(\big[1 - n(E - \hbar\omega)\big]\rho_{el}(E - \hbar\omega)$$
$$+ n(E + \hbar\omega)\rho_{el}(E + \hbar\omega)\big) + N(\omega)\big(\big[1 - n(E + \hbar\omega)\big]\rho_{el}(E + \hbar\omega)$$
$$+ n(E - \hbar\omega)\rho_{el}(E - \hbar\omega)\big)\big\}. \tag{10.20}$$

Here, $\rho_{el}(E)$ and $\rho_{ph}(\omega)$ are the phonon and the electron densities of states, respectively; $N(\omega)$ is the Bose–Einstein distribution function at the temperature T, and

$$n(E) = \frac{1}{2}\big[f_L(E) + f_R(E)\big]. \tag{10.21}$$

While considering a junction with a single-site bridge state coupled to the single vibrational mode, we can write the following expressions for ρ_{ph} and ρ_{el}:

$$\rho_{ph}(\omega) = \frac{1}{\pi\hbar}\frac{\gamma}{(\omega - \Omega)^2 + \gamma^2}, \tag{10.22}$$

$$\rho_{el}(E) = \frac{1}{\pi}\frac{\Delta_L + \Delta_R}{(E - E_0)^2 + (\Delta_L + \Delta_R)^2}, \tag{10.23}$$

where the polaron shift is neglected, and the parameter γ characterizes the broadening of the maximum in ρ_{ph} at $\omega = \Omega$ due to the interaction of the vibrionic mode with the environment. At low temperatures, we may significantly simplify the expression for Γ_{ph}. Within the conduction window $\mu_R < E < \mu_L$, we get

$$\Gamma_{ph}(E) \approx \pi\lambda^2\Bigg\{\int_0^{(\mu_L - E)/\hbar} d\omega\rho_{ph}(\omega)\rho_{el}(E + \hbar\omega)$$
$$+ \int_0^{(E - \mu_R)/\hbar} d\omega\rho_{ph}(\omega)\rho_{el}(E - \hbar\omega)\Bigg\}. \tag{10.24}$$

Omitting from consideration the coupling of the phonon mode to the environment ($\gamma \to 0$), we may easily carry out the integration over ω in Equation 10.24, and we arrive at the result

$$\Gamma_{ph}(E) \approx \pi\lambda^2\{\rho_{el}(E + \hbar\Omega)\theta(\mu_L - \hbar\Omega - E)$$
$$+ \rho_{el}(E - \hbar\omega)\theta(E - \mu_R - \hbar\Omega)\}, \tag{10.25}$$

where $\theta(x)$ is the step function.

Substituting the approximation for Γ_{ph} given by Equations 10.24 and 10.25 into the expression (10.19) for the dephasing parameter $\int$ and employing the earlier result (10.15) for the electron transmission, we may calculate the transport characteristics. The adopted simplified approach, as well as the NEGF-based calculations, shows that the IETS signal appears at the threshold determined by the frequency of the vibrational mode (see Figure 10.8, right panel). At first glance, one may expect the net current through the junction to increase at the threshold. Indeed, the inelastic contribution to the current increases from zero to a certain nonzero value at this threshold for the channel for inelastic transport to open up. However, more thorough studies show that both elastic and nonelastic contributions to the net current undergo changes at the inelastic tunneling threshold, and the elastic current could decrease there, as was first shown by Persson and Baratoff (1987). Moreover, this decrease in the elastic current may overweigh the contribution coming from the inelastic channel. Depending on the relative value of the elastic and the inelastic contributions to the net current near threshold, the IETS reveals a peak or a dip at the corresponding voltage. Experiments corroborate the variety in the IETS taken for molecular junctions (Djukic et al. 2005, Hahn et al. 2000, Wang et al. 2004, Zhitenev et al. 2002). The shape of the signal is very sensitive to the characteristics of the junction, such as the position of the electronic state, the electron–phonon and the molecule-to-leads coupling strengths, and the vibron frequency. For instance, at a very strong coupling of the molecular bridge to the leads, it could so happen that the backscattering by the negatively biased electrode electrons whose energies belong to the conduction window between μ_L and μ_R, are locally depleted near the junction. In this situation, the opening of the inelastic channel may cause the increased reflection (otherwise forbidden by the Fermi exclusion) leading to the decrease in the conduction. Consequently, the described scenario should result in the dip in the IETS signal. Also, as concluded in the recent work of Ryndyk and Cuniberti (2007, the sideband phonon-induced features in the electron spectral density discussed above could give rise to the corresponding features in the differential conductance dI/dV and IETS assuming that the molecule coupling to the leads is not too strong. Contributions from these sideband features may be responsible for the shape of the IETS signal at the threshold of the inelastic tunneling channel. These contributions could produce an extra inelastic signal, as well. The latter appears as an additional peak or a dip in the differential conductance.

The question of the current decrease/increase at the phonon excitation threshold that corresponds to the peak/dip in the IETS is not completely answered so far, and the appropriate theory is being developed (Balseiro et al. 2006, Egger and Gogolin 2008). Nevertheless, it is presently understood that three relevant energies, namely, molecule–electrode couplings $\Delta_{L,R}$, electron–phonon coupling strength λ, and the phonon energy $\hbar\Omega$ play a very important part in determining the shape of the IETS signal. Varying these parameters one may convert the IETS signal from a peak to a dip and vice versa.

At finite temperatures, molecular vibrations always occur in the presence of stochastic nuclear motions. These motions could be described as a phonon thermal bath. Coupling of vibrational modes to this bath further affects the electron transport causing energy dissipation. The dissipative processes must be taken into account to properly analyze the effects of electron–phonon interactions in the electron transport. Also, the displacements of ions involved in the molecular vibrations are accompanied by the changes in the electrostatic field inside the molecule. This could give rise to polaronic effects in electron transport. We discuss these issues in the next sections.

10.5 Dissipative Transport

Electron transport through molecular junctions is always accompanied by stochastic nuclear/ion motions in the close environment. Interactions of traveling electrons with these environmental fluctuations cause energy dissipation. The importance of dissipative effects depends on several factors. Among these factors the temperature, the size, and the complexity of the molecular bridge predominate. The temperature determines the intensity of the nuclear motions, and the size of the molecule determines the so-called contact time, that is, the time for an electron to travel through the junction, and in consequence, to contact the environment. It was shown by Buttiker and Landauer (1985) that the contact time is proportional to the number of sites (subunits) in the molecule providing intermediate states for the electron tunneling. For small molecules at low temperatures, the contact time is shorter than the characteristic times for fluctuations in the environment, so, the effect of the latter on the electron transport is not very significant. One may expect a small broadening of the molecule energy states to occur that brings a moderate erosion of the peaks in the electron transmission and the steps in the *I–V* characteristics.

On the contrary, in large-sized molecules such as proteins and DNA, electron transport is accompanied by strong energy dissipation. The significance of the system-environment interactions in macromolecules was recognized long ago in studies of long-range electron-transfer reactions. In these reactions, electrons travel between distant sites on the molecule called a donor and an acceptor. It was established that when an electron initially localized on the donor site moves to the acceptor site with a lower energy, the energy difference must be dissipated to the environment to provide the irreversibility of the transfer (Garg et al. 1985).

A usual way to theoretically analyze dissipative effects in the electron transport through molecules is to introduce a phonon bath representing the random motions in the environment. In general, there is no one/several dominating modes in the bath. Instead, the bath is characterized by a continuous spectral function of all relevant phonon modes $\rho_{ph}(\omega)$. The electrons are supposed to be coupled to the phonon bath, and this coupling is specified by the spectral function. The particular form of $\rho_{ph}(\omega)$ may be found based on molecular dynamics simulations. However, to qualitatively study the effect of dissipation on electron transport one can employ the expression (Mahan 2000)

$$\rho_{ph}(\omega) = \lambda \frac{\omega}{\omega_c} \exp\left(-\frac{\omega}{\omega_c}\right), \qquad (10.26)$$

where
- the parameter λ characterizes the electron–phonon coupling strength
- ω_c is the cutoff frequency for the bath related to its thermal relaxation time $\tau_c = \omega_c^{-1}$

To illustrate the possible effects of dissipation on the electron transport we return to our simple model where the molecular bridge is represented by a single state. Now, we assume that this state is coupled to the phonon bath. This model is hardly appropriate to properly analyze the dissipative effects in the electron transport through practical molecules for the molecule length is very important for the dissipative effects to be pronounced. Nevertheless, it still could serve to basically outline the main features of the dissipative electron transport through molecular junctions. Also, the proposed model could be useful to analyze electron transport in doped polyacetylene/polyaniline–polyethylene oxide nanofibers (Zimbovskaya 2008). These conducting polymers could be treated as some kind of granular metals where metallic-like regions (grains) are embedded in the poorly conducting medium of the disorderly arranged polymer chains (MacDiarmid 2001). While in the metallic state, the intergrain electron transport in these nanofibers is mostly provided by electron tunneling through the intermediate states on the polymer chains between the grains (Prigodin and Epstein 2002). In this case, the contact time could be long enough for the effects of dissipation to be well manifested that justifies the adoption of the above model.

Again, one may carry out transport calculations using Equation 10.15 for the electron transmission, and by expressing ε in terms of the relevant energies. Substituting Equation 10.26 in the expression (10.20), we may calculate $\Gamma_{ph}(E)$. The energy dissipation effects are more distinctly pronounced at moderately high temperatures, so we assume $kT \gg \hbar\omega_c$. Then the main contribution to the integral over ω in Equation 10.20 originates from the low frequency region $\omega \ll \omega_c$; and we obtain the following approximation:

$$\Gamma_{ph}(E) = \frac{8kT\lambda\Gamma}{(E - E_0)^2 + \Gamma^2}, \qquad (10.27)$$

where $\Gamma = \Delta_L + \Delta_R + (1/2)\Gamma_{ph}(E)$. Solving the obtained equation for Γ_{ph} and using Equation 10.19, we obtain (Zimbovskaya 2008)

$$\varepsilon = \frac{1}{2} \frac{\rho^2 \left(1 + \sqrt{1 + \rho^2}\right)}{4\left(\dfrac{E - E_0}{\Delta_L + \Delta_R}\right)^2 + \dfrac{1}{2}\left(1 + \sqrt{1 + \rho^2}\right)^3}, \qquad (10.28)$$

where $\rho^2 = 32kT\lambda/(\Delta_L + \Delta_R)^2$.

Voltage dependencies of the conductance computed based on Equations 10.3, 10.15, and 10.28 are presented in Figure 10.10 (left panel). One may see that at low values of the bias voltage, the electrons coupling to the phonon bath bring an enhancement in the conduction. The effect becomes reversed as the voltage grows above a certain value. This happens because the phonon-induced broadening of the molecular level (the bridge) assists the electron transport at a small bias voltage. When the voltage rises, this effect is surpassed by the scattering effect of phonons that resist electron transport. The significance of the electron–phonon interactions is determined by the ratio of the coupling constant λ and the self-energy terms describing the bridge coupling to the electrodes. The phonon bath makes a significant effect on the transport characteristics when $\lambda > \Delta_{L,R}$.

Dissipative electron transport through large DNA molecules was studied both theoretically and experimentally (Gutierrez et al. 2005, Xu et al. 2004). Theoretical studies were based on a model where the molecule was simulated by a tight-binding chain of sites linking the electrodes and the attached side chains. Electrons were allowed to travel along the bridge chain and to hop to the nearby side chains. These chains were coupled to the phonon bath providing the energy dissipation (see Figure 10.9). Although proposed for specific kinds of poly (dG)-poly (dC) molecules, this model seems to be quite generic and useful for a larger class of macromolecules.

Several coupling regimes to the bath may be analyzed. The most preferred regime for dissipative effects to appear is the strong-coupling limit defined by the condition $\lambda/\omega_c > 1$. Within this regime, the characteristic time for the electron bath interactions is much shorter than the typical electron time scales. Consequently, the bath makes a significant impact on the molecule electronic structure. New bath-induced states appear in the molecular spectrum inside the HOMO–LUMO gap. However, these states are strongly damped due to the dissipative action of the bath (Gutierrez et al. 2005). As a result, a small finite density of phonon-induced states appears inside the gap supporting electron transport at a low bias voltage. So again, the environment induces incoherent phonon-assisted transport through molecular bridges. For illustration, we show here the results of

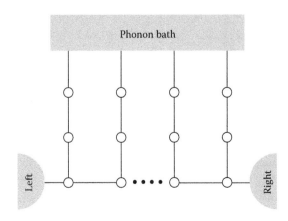

FIGURE 10.9 Schematic drawing of a molecular junction where the molecular bridge is coupled to the phonon bath via side chains.

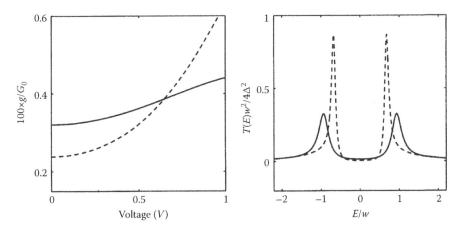

FIGURE 10.10 Left panel: The electron conductance versus voltage for a junction with a single electronic state bridge directly coupled to the phonon thermal bath. The curves are plotted assuming $E_0 = 0.4\,\text{eV}$, $\Delta_L = \Delta_R = 0.1\,\text{eV}$, $T = 30\,\text{K}$, $\lambda = 0.3\,\text{eV}$ (solid line), $\lambda = 0.05\,\text{eV}$ (dashed line). Right panel: Electron transmission through the junction in the case when the bridge state interacts with the phonon bath via the side chain coupled to the bridge state with the coupling parameter w. The curves are plotted assuming $\Delta_L = \Delta_R = \Delta$, $w/\Delta = 20$, $E_0 = 0$, $\lambda = 0.3\,\text{eV}$ (solid line), $\lambda = 0.05\,\text{eV}$ (dashed line).

calculations carried out for a toy model with a single-site bridge with a side chain attached to the latter. The side chain is supposed to be coupled to the phonon bath. The results for the electron transmission are displayed in Figure 10.10 (right panel). We see that the original bridge state at $E = 0$ is completely damped but two new phonon-induced states emerge nearby that could support electron transport.

An important characteristics of the dissipative electron transport through molecular junctions is the power loss in the junction, that is, the energy flux from the electronic into the phononic system. Assuming the current flow from the left to the right, this quantity may be estimated as the sum of the energy fluxes $Q_{L,R}$ at the left and the right terminals (leads):

$$P = Q_L + Q_R. \tag{10.29}$$

One may express the energy fluxes in terms of renormalized currents at the electrodes $\tilde{I}_{L,R}(E)$ that are defined as follows (Datta 2005):

$$I_{L,R} = \frac{e}{\pi\hbar} \int \tilde{I}_{L,R}(E)\,dE. \tag{10.30}$$

Then $Q_{L,R}$ may be presented in the form

$$Q_{L,R} = \frac{1}{\pi\hbar} \int E\tilde{I}_{L,R}(E)\,dE. \tag{10.31}$$

We remark that the current $I_{L,R}$ in Equation 10.30 are related to the corresponding leads, and their signs are accordingly defined. An outgoing current is supposed to be positive whereas an incoming one is negative for each lead. To provide an electric charge conservation in the junction one must require that $I = I_L = -I_R$ for the chosen direction of the current flow. Therefore, the energy fluxes also have different signs. As for the $Q_{L,R}$ magnitudes, they

may differ only if the renormalized currents $\tilde{I}_L(E)$ and $\tilde{I}_R(E)$ are distributed over energies in different ways. This cannot happen in the case of elastic transport, for in this case, $\tilde{I}_L(E) = -\tilde{I}_R(E)$. However, if the transport process is accompanied with the energy dissipation, the energy distributions for $\tilde{I}_L$ and $\tilde{I}_R$ may differ. In this case, electrons lose some energy while moving through the junction and this gives rise to the differences in the renormalized currents' energy distributions. For instance, in the case when the electrodes are linked with a single-state bridge, the energy distribution of $\tilde{I}_L(E)$ has a single maximum whose position is determined by the site energy E_0 and the applied bias voltage V. Assuming that the average energy loss due to dissipation could be estimated as ΔE, the maximum in the $\tilde{I}_R(E)$ distribution is shifted by this quantity, so the current $\tilde{I}_R(E)$ flows at lower energies compared to $\tilde{I}_L(E)$. This results in the power loss and the Joule heating in the junction (Segal et al. 2003).

10.6 Polaron Effects: Hysteresis, Switching, and Negative Differential Resistance

While studying electron transport through molecular junctions, hysteresis in the current–voltage characteristics was reported in some systems (Li et al. 2003). Multistability and stochastic switching were reported in single-molecule junctions (Lortscher et al. 2007) and in single metal atoms coupled to a metallic substrate through a thin ionic insulating film (Olsson et al. 2007, Repp et al. 2004).

The coupling of an electron belonging to a certain atomic energy level to the displacements of ions in the film brings a possibility of polaron formation in there. This leads to the polaron shift in the electron energy. It was noticed that multistability and hysteresis in molecular junctions mostly occurred when the molecular bridges included centers of long-living charged electronic states (redox centers). On these grounds, it was suggested that hysteresis in the *I–V* curves observed in molecular junctions appear due to the formation of polarons on the molecules (Galperin et al. 2005).

The presence of the polaron shift in the energy of a charged (occupied) electron state creates a difference between the latter and the energy of the same state while it remains unoccupied. Assuming, for simplicity, a single-state model for the molecular bridge coupled to a single optical phonon mode, we may write the following expression for the renormalized energy:

$$\tilde{E}_0(n_0) = E_0 - \frac{\lambda^2 n_0}{\hbar\Omega}, \tag{10.32}$$

where the electronic population on the bridge n_0 is given by

$$n_0 = \frac{1}{\pi}\int dE \frac{f_L(E)\Delta_L + f_R(E)\Delta_R}{\left[E - \tilde{E}_0(n_0)\right]^2 + \left(\Delta_L + \Delta_R\right)^2}. \tag{10.33}$$

So, as follows from Equation 10.32, the polaron shift depends on the bridge occupation n_0, and the latter is related to $\tilde{E}_0$ by Equation 10.33. Therefore, the derivation of an explicit expression for $\tilde{E}_0(n_0)$ is a nontrivial task even within the chosen simple model. Nevertheless, it could be shown that two local minima emerge in the dependence of the potential energy of the molecular junction including two electrodes linked by the molecular bridge, of the occupation number n_0. These minima are located near $n_0 = 0$ and 1, and they correspond to the neutral (unoccupied) and charged (occupied) states, respectively. This is illustrated in Figure 10.11 (left panel). These states are metastable, and their lifetime could be limited by the quantum switching (Mitra et al. 2005, Mozyrsky et al. 2006). When the switching time between the two states is longer than the characteristic time for the external voltage sweeping, one may expect the hysteresis to appear, for the states of interest live long enough to maintain it. Within the opposite limit, the average

washes out the hysteresis. When the states are especially short-lived this could even result in a telegraph noise at a finite bias voltage that replaces the controlled switching.

Further, we consider long-lived metastable states and we concentrate on the *I–V* behavior. Let us for certainty assume that the bridge state at zero bias voltage is situated above the Fermi energy of the system and remains empty. As the voltage increases, one of the electrode's chemical potentials crosses the bridge level position, and the current starts to flow through the system. The *I–V* curve reveals a step at the bias voltage value corresponding to the crossing of the unoccupied bridge level with the energy E_0 by the chemical potential. However, while the current flows through the bridge, the level becomes occupied and, consequently, shifted due to the polaron formation. When the bias voltage is reversed, the current continues to flow through the shifted bridge state of the energy $\tilde{E}_0$, until the recrossing happens. Due to the difference in the energies of the neutral and the charged states, the step in the *I–V* curves appears at different values of the voltage, and this is the reason for the hysteresis loop to appear as shown on the right panel of Figure 10.11. One could also trace the hysteresis in the *I–V* characteristics starting from the filled (and shifted) bridge state.

Again, the hysteresis loop in the *I–V* curves may occur when both occupied and unoccupied states are rather stable, which means that the potential barrier separating the corresponding minima in the potential energy profile is high enough, so that quantum switching between the states is unlikely. This happens when the bridge is weakly coupled to the electrodes ($\Delta_{L,R} \ll \lambda^2/\hbar\Omega$), which is an obvious requirement for the involved states (neutral or charged) to the distinguishable. In other words, the broadening of the relevant levels due to the coupling to the electrodes must be much smaller that the polaron shift. As was recently shown (Ryndyk et al. 2008), the stronger the electron-phonon coupling, the less becomes the probability of the switching. At large values of λ, the tunneling between the charged and

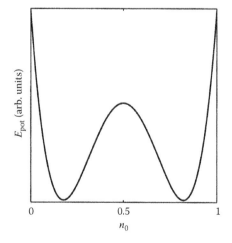
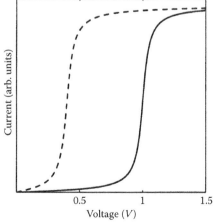

FIGURE 10.11 Left panel: Schematic of the potential energy of the molecular junction versus the occupation number n_0. Right panel: Hysteresis in the current–voltage characteristics. The solid line corresponds to the transport via the unoccupied bridge electronic state, and the dashed line corresponds to the transport via the occupied state shifted due to the polaron formation.

the neutral states is exponentially suppressed. Also, it was shown that the symmetry/asymmetry in the coupling to the electrodes could significantly affect the hysteresis behavior (D'Amico et al. 2008, Ryndyk et al. 2008). For asymmetric junctions ($\Delta_L \neq \Delta_R$), two nearly stable states exist at the zero bias voltage. When the asymmetry is very strong, both states could appear stable at one bias voltage polarity and unstable when the polarity is reversed. It was suggested that under an appropriate choice of parameters, one may create a situation where the instability regions for the involved states do not overlap. These properties give grounds to conjecture that such strongly asymmetric junctions could reveal memory functionalities, that make them potentially useful in the fabrication of nanodevices.

Among the various potentially important properties of the electron transport through metal–molecule junctions, one may separate the negative differential resistance (NDR), that is, the decrease in the current I while the bias voltage across the molecule increases. The NDR effect was originally observed in tunneling semiconducting diodes (Kastner 1992). Later, it was viewed in quantum dots (Grobis et al. 2005, Repp et al. 2005a,b). Several possible scenarios were proposed to explain the NDR occurrence in electron transport through molecules. The effect could originate from the alignment and the subsequent disalignment of the Fermi levels of the electrodes with the molecular orbitals that happens as the bias voltage varies (Xue et al. 1999). Also, the NDR could appear as a Coulomb blockade-induced effect (Simonian et al. 2007) and/or due to some other reasons (Zimbovskaya and Pederson 2008). It is likely that different mechanisms could play a major part in the NDR appearance in different molecular junctions where it was observed so far.

Here, we discuss the NDR features in the current–voltage characteristics that originate from the electron coupling with the molecular vibrational modes. As was shown (Galperin et al. 2005, Yeganeh et al. 2007a,b), the polaron formation could give rise to the NDR. This may happen if the polaron shift in the energy of the occupied state moves the energy level away from the conduction window. As the shift depends on the electronic population, and the latter changes as the bias voltage increases, the occupied state falls out from the window between the chemical potentials of the electrodes at a certain value of the voltage. If the transport is being conducted via this occupied state, then it stops when this voltage value is reached. Correspondingly, the current value drops to zero revealing a distinctive NDR feature. This is a realistic scenario based on the main features of the electron transport through molecules under polaron formation.

In conclusion, electron–vibron coupling could lead to the formation of a polaron that results in the energy differences between the occupied and the unoccupied states on the molecular bridge. At a weak interaction between the bridge and the electrodes and the strong electron–phonon coupling these charged and neutral states are metastable and could serve for electron conduction. This results in such interesting and potentially useful effects as the hysteresis and the NDR features in the current–voltage characteristics.

10.7 Molecular Junction Conductance and Long-Range Electron-Transfer Reactions

Long-range electron-transfer reactions play an important part in many biological processes such as photosynthesis and enzyme catalyses (Kuznetsov and Ulstrup 1999). Theoretical and experimental studies of these processes have lasted for more than four decades but they still remain within a very active research area. In the intramolecular electron-transfer reactions, the electric charge moves from one section of the molecule to another section of the same molecule. Long-range transfer typically occurs in large molecules such as proteins and/or DNA, so that these two sections are situated far apart from each other. A common setup for the transfer reactions includes a donor, a bridge, and an acceptor. Due to the large distances between the donor (where an electron leaves) and the acceptor (where it arrives) typical for the long-range transfer reactions, a direct coupling between the two is negligible. Therefore, the electron participating in the transfer needs a molecular bridge providing a set of intermediate states for electron transport.

Essentially, intramolecular electron transfer is a combination of nuclear environment fluctuations and electron tunneling. Electron-transfer reactions result from the response of a molecule to environmental polarization fluctuations that accompany nuclear fluctuations. The molecule responds by redistribution of the electronic density thus establishing opportunities for the charge transfer to occur. The main characteristic of the electron-transfer processes is the transfer rate K_{et} that is the inversed time of the reaction. The transfer rate for the electron tunneling in the molecule interior from the donor to the acceptor depends both on the electronic transmission amplitudes and the vibrionic coupling. The latter provides the energy exchange between the electronic and the nuclear systems.

Viewing fast electronic motions in the background of the slowly moving nuclei (nonadiabatic electron transfer) and applying the Fermi golden rule of perturbation theory, it was shown that K_{et} could be written in the form first suggested by Marcus (1965)

$$K_{et} = \frac{2\pi}{\hbar} |H_{DA}|^2 (FC). \qquad (10.34)$$

Here, the first cofactor is the electron transmission coefficient, H_{DA}, which is the effective matrix element between the donor and the acceptor. The second term is the density of states weighted Franck–Condon factor that describes the effect of nuclear motions in the environment.

There exists a noticeable resemblance between the long-range electron transfer and the molecular conductance, and this resemblance was analyzed in several theoretical works (Dahnovsky 2006, Mujica et al. 1994, Nitzan 2001, Yeganeh et al. 2007a,b, Zimbovskaya 2003, Zimbovskaya and Gumbs 2002). The fundamental part in both molecular conduction and electron-transfer reactions is taken by the electron tunneling inside the molecule via the set of intermediate states. In large molecules, the electron involved in the transfer reaction or contributing to the conductance, with a high probability follows a few pathways that work as molecular bridges. The rest of the molecules do not significantly participate in the transfer/conduction and may be omitted from consideration in calculations of electron transmission (Jones et al. 2002; Beratan et al. 1992). On these grounds, it was suggested that the original molecule could be reduced to a simpler chain-like structure, and the latter may replace the former in calculations of molecular conductance and/or transfer rates (Daizadeh et al. 1997). Such simplification significantly eases computations of the molecular conductance and the electron-transfer rates in macromolecules.

The resemblance between the molecular conductance and the long-range electron transfer does not mean that these processes are nearly identical. Along with the similarities there are substantial differences between the former and the latter. For instance, the continuum of states causing the charge transport arises from the multitude of electronic states on the electrodes in the case of molecular conduction, and from the vibrations and fluctuations in the environment in the case of electron-transfer reactions. Also, the driving force that puts electrons in motion originates from the external bias voltage applied across the junction in the case of molecular conduction. In the electron-transfer situation, this force appears due to the electron–vibronic interactions in the system, or the electron-transfer reaction starts as a result of a photoexcitation of the donor part of the molecule. The observables, namely, the molecular conductance g and/or current I, and the transfer rate K_{et} differ as well.

Notwithstanding these differences, the electron transport through molecular junctions and the electron transfer could be theoretically analyzed using the common formalism as recently proposed by Yeganeh et al. (2007b). Further, we follow their approach. The starting point is that one may simulate the donor–bridge–acceptor system as some kind of molecular junction, shown in Figure 10.12. Here, we represent the donor as a single state $|i\rangle$, and this state mimics the left lead (assuming the transport is from the left to the right). The donor is coupled to the bridge and the latter is coupled to the continuum of the final states $|f\rangle$ simulating the right lead. These couplings are described by the self-energy terms $\Delta_{i,f}$. For simplicity, we assume that there exists a single state on the bridge, and this state is coupled to a vibronic mode of the frequency Ω. The latter mimics environment fluctuations. We assume that the

"leads" are weakly coupled to the bridge ($\Delta_{L,R} \ll \lambda$), which is typical for electron-transfer situations. Also, the representation of the environment motions by the single vibronic mode is justified only at low temperatures when the thermal energy kT is much smaller than the electron–vibrion coupling parameter λ.

Now, we can write the Landauer expression for the current flowing through the "junction" using Equations 10.3 and 10.4. Assuming that the bridge includes only one state, we simplify Equation 10.4:

$$T(E) = 2T_r\left\{\Delta_i G \Delta_f G^\dagger\right\} = -2\frac{\Delta_i \Delta_f}{\Delta_i + \Delta_f}\mathrm{Im}(G). \quad (10.35)$$

Here, the subscripts i/f label the initial and the final states. Within the Fermi golden rule regime that allows the introduction of the transfer rate, the bridge must be coupled to the final states much more strongly than to the initial state ($\Delta_f \gg \Delta_i$). Therefore

$$\frac{\Delta_i \Delta_f}{\Delta_i + \Delta_f} \approx \Delta_i \quad (10.36)$$

and the coupling to the final states falls out of the expressions for electron transmission and current. Within the chosen model, this is a physically reasonable result for the final states reservoir (the right "electrode" in our junction) was merely introduced to impose a continuum of states maintaining the transfer process at a steady-state rate. Also, considering the current flow, we may suppose that the initial state is always filled [$f_i(E) = 1$], and the final states are empty [$f_f(E) = 0$]. Therefore, the current flow through the "junction" accepts the form

$$I = \frac{2e}{\pi\hbar}\int dE\,\Delta_i\,\mathrm{Im}(G). \quad (10.37)$$

Both the current and the transfer rate are fluxes closely related to each other, namely,

$K_{et} = I/e$. So, we may write

$$K_{et} = -\frac{2}{\pi\hbar}\int dE\,\Delta_i\,\mathrm{Im}(G). \quad (10.38)$$

Now, Δ_i could be computed using the expression (10.7) for the corresponding self-energy term. Keeping in mind that the "left electrode" includes a single state with the certain energy $\int_i$ we obtain

$$\Delta_i = \mathrm{Im}\left(\frac{|\tau_i|^2}{E - \varepsilon_i + is}\right) = \pi|\tau_i|^2\delta(E - \varepsilon_i). \quad (10.39)$$

Accordingly, the expression (10.38) for the transfer rate may be reduced to the form:

$$K_{et} = -\frac{2}{\hbar}|\tau_i|^2\,\mathrm{Im}\left[G(\varepsilon_i)\right], \quad (10.40)$$

FIGURE 10.12 Schematic of the model system used in the transfer rate calculations. The initial/final reservoirs correspond to the left/right leads in the molecular junction.

where τ_i represents the coupling between the donor and the molecule bridge. It must be stressed that within the chosen model, τ_i is the only term representing the relevant state coupling that may be identified with the electronic transmission coefficient H_{DA} in the general expression (10.34) for K_{et}. This leaves us with the following expression for the Franck–Condon factor:

$$(FC) = -\frac{1}{\pi}\text{Im}\left[G(\varepsilon_i)\right]. \tag{10.41}$$

As discussed in Section 10.4, at a weak coupling of the bridge to the leads the electron–vibronic interaction opens the set of metastable channels for the electron transport at the energies $E_n = \tilde{E}_0 + n\hbar\Omega$ ($n = 0, 1, 2\ldots$) where $\tilde{E}_0$ is the energy of the bridge state with the polaronic shift included. Green's function may be approximated as a weighted sum of contributions from these channels:

$$G(\varepsilon_i) = \sum_{n=0}^{\infty} P(n)\left[\varepsilon_i - \tilde{E}_0 - n\hbar\Omega + is\right]^{-1}, \tag{10.42}$$

where $s \to 0^+$, and the coefficients $P(n)$ are probabilities for the channels to appear given by Equation 10.18.

Substituting Equation 10.42 into Equation 10.41 we get

$$(FC) = \sum_{n=0}^{\infty} P(n)\delta(\Delta F - n\hbar\Omega). \tag{10.43}$$

Here, $\Delta F = \int_i - \tilde{E}_0 \equiv \int_i - E_0 + \lambda^2/(\hbar\Omega)$ is the exergicity of the transfer reaction, that is, the free energy change originating from the nuclear displacements accompanied by polarization fluctuations. The effect of the latter is inserted via the reorganization term $\lambda^2/(\hbar\Omega)$ related to polaron formation. The exergicity in the transfer reaction takes on a part similar to that of the bias voltage in the electron transport through molecular junctions. It gives rise to the electron motion through the molecules. In the particular case when the voltage drops between the initial state (left electrode) and the molecular bridge these two quantities are directly related by $|e|V = \Delta F$.

Usually, the long-range electron transfer is observed at moderately high (room) temperatures, so the low temperature approximation (10.43) for the Franck–Condon factor cannot be employed. However, the expression (10.41) remains valid at finite temperatures, only the expression for Green's function must be modified to include the thermal effects. It is shown (Yeganeh et al. 2007a,b) that within the high-temperature limit ($kT > \hbar\Omega$) the expression for the (FC) may be converted to the well-known form first proposed by Marcus:

$$(FC) = \frac{1}{\sqrt{4\pi E_p kT}}\exp\left[-\frac{(\Delta F - E)_p^2}{4E_p kT}\right], \tag{10.44}$$

where $E_p = \lambda^2/(\hbar\Omega)$ is the reorganization energy.

While studying the electron-transfer reactions in practical macromolecules, one keeps in mind that both the donor and the acceptor subsystems in the standard donor–bridge–acceptor triad are usually complex structures including multiple sites coupled to the bridge. Correspondingly, the bridge has a set of entrances and a set of exits that an electron can employ. At different values of the tunnel energy different sites of the donor and/or acceptor subsystems can give predominant contributions to the transfer. Consequently, an electron involved in the transfer arrives at the bridge and leaves from it via different entrances/exits, and it follows different pathways while on the bridge.

Also, nuclear vibrations in the environment could strongly affect the electron transmission destroying the pathways and providing a transition to a completely incoherent sequential hopping mechanism of the electron transfer. All this means that a proper computation of the electron transmission factor H_{DA} for practical macromolecules is a very complicated and nontrivial task. The strong resemblance between the electron-transfer reactions and the electron transport through molecules gives grounds to believe that studies of molecular conduction can provide important information concerning the quantum dynamics of electrons participating in the transfer reactions. One may expect that some intrinsic characteristics of the intramolecular electron transfer, such as the pathways of the tunneling electrons and the distinctive features of the donor/acceptor coupling to the bridge could be obtained in experiments on the electron transport through molecules. For instance, it was recently suggested to characterize electron pathways in molecules using the inelastic electron tunneling spectroscopy, and other advances in this area are to be expected.

10.8 Concluding Remarks

Presently, the electron transport through molecular-scale systems is being intensively studied both theoretically and experimentally. Largely, unceasing efforts of the research community to further advance these studies are motivated by important application potentials of single molecules as active elements of various nanodevices intended to compliment current silicon-based electronics. An elucidation of the physics underlying electron transport through molecules is necessary in designing and operating molecular-based nanodevices. Elastic mechanisms for the electron transport through metal–molecule–metal junctions are currently understood and described fairly well. However, while moving through the molecular bridge, an electron is usually affected by the environment, and it results in a change of its energy. So, inelastic effects appear, and they may bring noticeable changes in the electron transport characteristics. Here, we concentrated on the inelastic and the dissipative effects originating from the molecular bridge vibrations and thermally activated stochastic fluctuations. For simplicity and also to keep this chapter within a reasonable length, we avoided a detailed description of computational formalisms commonly used to theoretically analyze electron transport through molecular junctions. These formalisms are described elsewhere. We mostly focus on the physics of the inelastic effects in the electron transport. Therefore, we employ very simple models and techniques.

We remark that along with nonelasticity originating from electron–vibron interactions, which is the subject of the present review, there exist inelastic effects of different kinds. For instance, inelastic effects arise due to electron–photon interactions. Photoassisted transport through molecular junctions was demonstrated and theoretically addressed using several techniques. It is known that optical pumping could give rise to the charge flow in an unbiased metal–molecule–metal junction, and light emission in biased current-carrying junctions could occur. Also, there is the issue of molecular geometry. There are grounds to conjecture that in some cases the geometry of a molecule included in the junction may change as the current flows through the latter for bonds can break with enough amount of current. Obviously, this should bring a strong inelastic component to the transport, consequently affecting observables.

It is common knowledge that electron–electron interactions may significantly influence molecular conductance leading to a Coulomb blockade and a Kondo effect. To properly treat electron transport through molecular junctions one must take these interactions into consideration. Corresponding studies were carried out omitting electron–phonon interactions. However, a full treatment of the problem including both electron–electron and electron–phonon interactions has not been completed so far. There exist other theoretical challenges, such as the effect of bipolaron formation that originates from an effective electron-electron attraction via phonons.

Finally, practical molecular junctions are complex systems, and a significant effort is necessary to bring electron transport calculations to a result that could be successfully compared with the experimental data. For this purpose, one needs to compute molecular orbitals and the voltage distribution inside the junctions to get sufficient information on the vibronic spectrum of the molecule, the electron–phonon coupling strengths, and electron–electron interactions. One needs a good quantitative computational scheme for transport calculations where all this information can be accounted for. Currently, there remain some challenges that have not been properly addressed by theory. Therefore, a comparison between the theoretical and the experimental results on the molecular conductance sometimes does not bring satisfactory results. However, there are firm grounds to believe that further efforts of the research community will result in a detailed understanding of all the important aspects of molecular conductance including inelastic and dissipative effects. Such an understanding is paramount to the conversion of molecular electronics into a viable technology.

References

Agrait, N.; Yeati, A. L., and van Ruitenbeek, L. M. 2003. Quantum properties of atomic-sized conductors. *Phys. Rep.* **377**: 81–279.

Aviram, A.; Ratner, M. A., and Mujica, V. (Eds). 2002. *Molecular Electronics II*. New York: Annals of the New York Academy of Sciences.

Balseiro, C.A.; Cornaglia, P. S., and Grempel, D. R. 2006. Electron–phonon correlation effects in molecular transistors. *Phys. Rev. B* **74**: 235409.

Beratan, D. N.; Onuchic, J. N.; Winkler, J. R. et al. 1992. Electron-tunneling pathways in proteins. *Science* **258**: 1740–1741.

Buttiker, M. 1986. Role of quantum coherence in series resistors. *Phys. Rev. B* **33**: 3020–3026.

Buttiker, M. and Landauer, R. 1985. Traversal time for tunneling. *Phys. Scri.* **32**: 429–434.

Cizek, M.; Thoss, M., and Domcke, W. 2004. Theory of vibrationally inelastic electron transport through molecular bridges. *Phys. Rev. B* **70**: 125406.

Cornaglia, P. S.; Ness, H., and Grempel, D. R. 2004. Many-body effects on the transport properties of single-molecule devices. *Phys. Rev. Lett.* **93**: 147201.

Cornaglia, P. S.; Grempel, D. R., and Ness, H. 2005. Quantum transport through a deformable molecular transistor. *Phys. Rev. B* **71**: 075320.

Cuniberti, G.; Fagas, G., and Richter, K. (Eds). 2005. *Introductory Molecular Electronics: A Brief Overview*, Lecture Notes in Physics, Volume **680**. Berlin, Germany: Springer.

Dahnovsky, Yu. 2006. Modulating electron dynamics: Modified spin-boson approach. *Phys. Rev. B* **73**: 144303.

Daizadeh, I.; Gehlen, J. N., and Stuchebrukhov, A. A. 1997. Calculation of electronic tunneling matrix element in proteins: Comparison of exact and approximate one-electron methods for Ru-modified azurin. *J. Chem. Phys.* **106**: 5658–5666.

D'Amato, J. L. and Pastawski, H. M. 1990. Conductance of a disordered linear chain including inelastic scattering events. *Phys. Rev. B* **41**: 7411–7420.

D'Amico, P.; Ryndyk, D. A.; Cuniberti, G. et al. 2008. Charge-memory effect in a polaron model: Equation-of-motion method for Green functions. *New J. Phys.* **10**: 085002.

Damle, P. S.; Ghosh, A. W., and Datta, S. 2001. Unified description of molecular conduction: From molecules to metallic wires. *Phys. Rev. B* **64**: 201403.

Datta, S. 1997. *Electron Transport in Mesoscopic Systems*. Cambridge, U. K.: Cambridge University Press.

Datta, S. 2005. *Quantum Transport: Atom to Transistor*. Cambridge, U. K.: Cambridge University Press.

Di Ventra, M.; Pantelides, S. T., and Lang, N. D. 2000. First-principles calculations of transport properties of a molecular device. *Phys. Rev. Lett.* **84**: 979–982.

Djukic, D.; Thygesen, K. S.; Untiedt, C. et al. 2005. Stretching dependence of the vibration modes of a single-molecule Pt–H_2–Pt bridge. *Phys. Rev. B* **71**: 161402(R).

Donarini, A.; Grifoni, M., and Richter, K. 2006. Dynamical symmetry breaking in transport through molecules. *Phys. Rev. Lett.* **97**: 166801.

Egger, R. and Gogolin, A. O. 2008. Vibration-induced correction to the current through a single molecule. *Phys. Rev. B* **77**: 1098.

Galperin, M.; Ratner, M. A., and Nitzan, A. 2004. Inelastic electron tunneling spectroscopy in molecular junctions: Peaks and dips. *J. Chem. Phys.* **121**: 11965–11979.

Galperin, M.; Ratner, M. A., and Nitzan, A. 2005. Hysteresis, switching, and negative differential resistance in molecular junctions: A polaron model. *Nano Lett.* **5**: 125–130.

Galperin, M.; Nitzan, A., and Ratner, M. A. 2006. Molecular transport junction: Current from electronic excitations in the leads. *Phys. Rev. Lett.* **96**: 166803.

Galperin, M.; Ratner, M. A., and Nitzan, A. 2007. Topical review: Molecular transport junctions: Vibrational effects. *J. Phys. Condens. Matter* **19**: 103201.

Garg, A.; Onuchic, J. N., and Ambegaokar, V. 1985. Effect of friction on electron transfer in biomolecules. *J. Chem. Phys.* **83**: 4491–4503.

Grobis, M.; Wachowiak, A.; Yamachika, M. F. et al. 2005. Tuning negative differential resistance in a molecular film. *Appl. Phys. Lett.* **86**: 204102.

Gutierrez, R.; Mandal, S., and Cuniberti, G. 2005. Dissipative effects in the electronic transport through DNA molecular wires. *Phys. Rev. B* **71**: 235116.

Gutirrez, R.; Mohapatra, S.; Cohen, H. et al. 2006. Inelastic quantum transport in a ladder model: Implications for DNA conduction and comparison to experiments on suspended DNA oligomers. *Phys. Rev. B* **74**: 235105.

Hahn, J. R.; Lee, H. J., and Ho, W. 2000. Electronic resonance and symmetry in single-molecule inelastic electron tunneling. *Phys. Rev. Lett.* **85**: 1914–1917.

Ho, J. W. 2002. Single molecule chemistry. *J. Chem. Phys.* **117**: 11033–11061.

Imry, Y. and Landauer, R. 1999. Conductance viewed as transmission. *Rev. Mod. Phys.* **71**: S306–S312.

Jones, M. L.; Kurnikov I. V., and Beratan, D. N. 2002. The nature of tunneling pathway and average packing density model for protein-mediated electron transfer. *J. Phys. Chem. A* **106**: 200206.

Kastner, M. A. 1992. The single-electron transistor. *Rev. Mod. Phys.* **64**: 849–858.

Komeda, T.; Kim, Y.; Kawai, M. et al. 2002. Lateral hopping of molecules induced by excitation of internal vibration mode. *Science* **295**: 2055–2058.

Kubatkin, S.; Danilov, A.; Hjort, M. et al. 2003. Single-electron transistor of a single organic molecule with access to several redox states. *Nature* **425**: 698–701.

Kushmerick, J. G.; Lazorcik, J.; Patterson, C. H. et al. 2004. Vibronic contributions to charge transport across molecular junctions. *Nano Lett.* **4**: 63942.

Kuznetsov, A. M. and Ulstrup, I. 1999. *Electron Transfer in Physics and Biology*. Chichester, U. K.: John Wiley.

Landauer, R. 1970. Electrical resistance of disordered one-dimensional lattices. *Philos. Mag.* **21**: 863–867.

Lang, N. D. and Avouris, Ph. 2000a. Carbon-atom wires: Charge-transfer doping, voltage drop and the effect of distortions. *Phys. Rev. Lett.* **84**: 358–361.

Lang, N. D. and Avouris, Ph. 2000b. Electrical conductance of parallel atomic wires. *Phys. Rev. B* **62**: 7325–7329.

Li, X.-Q. and Yan, Y.-J. 2001a. Electrical transport through individual DNA molecules. *Appl. Phys. Lett.* **79**: 2190–2192.

Li, X.-Q. and Yan, Y.-J. 2001b. Scattering matrix approach to electronic dephasing in longrange electron transfer. *J. Chem. Phys.* **115**: 4169–4174.

Li, C.; Zhang, D.; Liu, X. et al. 2003. In$_2$O$_3$ nanowires as chemical sensors. *Appl. Phys. Lett.* **82**: 1613.

Lorente, N.; Persson, M.; Lauhon, L. J. et al. 2001. Symmetry selection rules for vibrationally inelastic tunneling. *Phys. Rev. Lett.* **86**: 2593–2596.

Lortscher, E.; Weber, H. B., and Riel, H. 2007. Statistical approach to investigating transport through single molecules. *Phys. Rev. Lett.* **98**: 176807.

MacDiarmid, A. G. 2001. Nobel lecture: "Synthetic metals": A novel role for organic polymers. *Rev. Mod. Phys.* **73**: 701–712.

Mahan, G. D. 2000. *Many-Particle Physics*. New York: Plenum.

Marcus, R. A. 1965. On the theory of electron-transfer reactions. VI. Unified treatment for homogeneous and electrode reactions. *J. Chem. Phys.* **43**: 679–701.

Mii, T.; Tikhodeev, S. G., and Ueba, H. 2003. Spectral features of inelastic electron transport via a localized state. *Phys. Rev. B* **68**: 205406.

Mitra, A.; Aleiner I., and Millis, A. J. 2004. Phonon effects in molecular transistors: Quantal and classical treatment. *Phys. Rev. B* **69**: 245302.

Mitra, A.; Aleiner I., and Millis, A. J. 2005. Semiclassical analysis of the nonequilibrium local polaron. *Phys. Rev. Lett.* **94**: 076404.

Mozyrsky, D.; Hastings, M. B., and Martin, I. 2006. Intermittent polaron dynamics: Born–Oppenheimer approximation out of equilibrium. *Phys. Rev. B* **73**: 035104.

Mujica, V.; Kemp, M., and Ratner, M. A. 1994. Electron conduction in molecular wires. I. A scattering formalism. *J. Chem. Phys.* **101**: 6849–6855.

Mujica, V.; Roitberg, A. E., and Ratner, M. A. 2000. Molecular wire conductance: Electrostatic potential spatial profile. *J. Chem. Phys.* **112**: 6834–6839.

Nitzan, A. 2001. Electron transmission through molecules and molecular interfaces. *Ann. Rev. Phys. Chem.* **52**: 681–750.

Olsson, F. E.; Paavilainen, S.; Persson, M. et al. 2007. Multiple charge states of Ag atoms on ultrathin NaCl films. *Phys. Rev. Lett.* **17**: 176803.

Park, H.; Park, J.; Lim, A. K. L. et al. 2000. Nanomechanical oscillations in a single-C$_{60}$ transistor. *Nature* **407**: 57–60.

Persson, B. N. J. and Baratoff, A. 1987. Inelastic electron tunneling from a metal tip: The contribution from resonant processes. *Phys. Rev. Lett.* **59**: 339–342.

Prigodin, V. N. and Epstein, A. J. 2002. Nature of insulator–metal transition and novel mechanism of charge transport in highly doped electronic polymers. *Synth. Met.* **125**: 43–53.

Poot, M.; Osorio, E.; O'Neil, K. et al. 2006. Temperature dependence of three-terminal molecular junctions with sulfur end-functionalized tercyclohexylidenes. *Nano Lett.* **6**: 1031–1035.

Qin, X. H.; Nazin G. V., and Ho, W. 2004. Vibronic states in single molecule electron transport. *Phys. Rev. Lett.* **92**: 206102.

Reichert, J.; Ochs, R.; Beckmann, D. et al. 2002. Driving current through single organic molecules. *Phys. Rev. Lett.* **88**: 176804.

Repp, J.; Meyer, G.; Olsson, F. E. et al. 2004. Controlling the charge state of individual gold adatoms. *Science* **305**: 493–495.

Repp, J.; Meyer, G.; Stojkovic, S. M. et al. 2005a. Molecules on insulating films: Scanning-tunneling microscopy imaging of individual molecular orbitals. *Phys. Rev. Lett.* **94**: 026803.

Repp, J.; Meyer, G.; Paavilainen, S. et al. 2005b. Scanning tunneling spectroscopy of Cl vacancies in NaCl films: Strong electron–phonon coupling in double-barrier tunneling junctions. *Phys. Rev. Lett.* **95**: 225503.

Ryndyk, D. A. and Cuniberti, G. 2007. Nonequilibrium resonant spectroscopy of molecular vibrons. *Phys. Rev. B* **76**: 155430.

Ryndyk, D. A.; D'Amico, P.; Cuniberty, G. et al. 2008. Charge-memory polaron effect in molecular junctions. *Phys. Rev. B* **78**: 085409.

Samanta, M. P.; Tian, W.; Datta, S. et al. 1996. Electronic conduction through organic molecules. *Phys. Rev. B* **53**: R7626–R7629.

Segal, D.; Nitzan, A.; Davis, W. B. et al. 2000. Electron transfer rates in bridged molecular systems. 2. A steady-state analysis of coherent tunneling and thermal transitions. *J. Phys. Chem. B* **104**: 3817–3829.

Segal, D.; Nitzan, A., and Hnggi, P. 2003. Thermal conductance through molecular wires. *J. Chem. Phys.* **119**: 6840–6855.

Siddiqui, L.; Ghosh, A. W., and Datta, S. 2007. Phonon runaway in nanotube quantum dots. *Phys. Rev. B* **76**: 085433.

Simonian, N.; Li, J., and Likharev, K. 2007. Negative differential resistance at sequential single-electron tunnelling through atoms and molecules. *Nanotechnology* **18**: 424006.

Skourtis, S. S. and Mukamel, S. 1995. Superexchange versus sequential long range electron transfer: Density matrix pathways in Liouville space. *Chem. Phys.* **197**: 367–388.

Smit, R. H. M.; Noat, Y.; Untiedt, C. et al. 2002. Measurement of the conductance of a hydrogen molecule. *Nature* **419**: 906–909.

Smit, R. H. M.; Untiedt, C., and van Ruitenbeek, J. M. 2004. The high-bias stability of monatomic chains. *Nanotechnology* **15**: S472–S478.

Stipe, B. C.; Rezaei, M. A., and Ho, W. 1999. Localization of inelastic tunneling and the determination of atomic-scale structure with chemical specificity. *Phys. Rev. Lett.* **82**: 1724–1727.

Tikhodeev, S. G. and Ueba, H. 2004. Relation between inelastic electron tunneling and vibrational excitation of single adsorbateson metal surfaces. *Phys. Rev. B* **70**: 125414.

Troisi, A. and Ratner, M. A. 2006. Molecular transport junctions: Propensity rules for inelastic tunneling. *Nano Lett.* **6**: 1784–1788.

Tsutsui, M.; Kurokawa, S., and Sakai, A. 2006. Bias-induced local heating in Au atom-sized contacts. *Nanotechnology* **17**: 5334–5338.

Wang, W.; Lee, T.; Kretzchmar, I. et al. 2004. Inelastic electron tunneling spectroscopy of an alkanedithiol self-assembled monolayer. *Nano Lett.* **4**: 643–646.

Wingreen, N. S.; Jacobsen, K. W., and Wilkins, J. W. 1989. Inelastic scattering in resonant tunneling. *Phys. Rev. B* **40**: 11834–11850.

Wingreen, N. S.; Jauho, A.-P., and Meir, Y. 1993. Time-dependent transport through a mesoscopic structure. *Phys. Rev. B* **48**: 8487–8490.

Wu, S. W.; Nazin, G. V.; Chen, X. et al. 2004. Control of relative tunneling rates in single molecule bipolar electron transport. *Phys. Rev. Lett.* **93**: 236802.

Xu, B.; Zhang, P.; Li, X. et al. 2004. Direct conductance measurement of single DNA molecules in aqueous solution. *Nano Lett.* **4**: 1105–1108.

Xue, Y. and Ratner, M. A. 2003a. Microscopic study of electrical transport through individual molecules with metallic contacts. I. Band lineup, voltage drop, and high-field transport. *Phys. Rev. B* **68**: 115406.

Xue, Y. and Ratner, M. A. 2003b. Microscopic study of electrical transport through individual molecules with molecular contacts. II. Effect of interface structure. *Phys. Rev. B* **68**: 115407.

Xue, Y. and Ratner, M. A. 2004. End group effect on electrical transport through individual molecules: A microscopic study. *Phys. Rev. B* **69**: 085403.

Xue, Y.; Datta, S.; Hong, S. et al, 1999. Negative differential resistance in the scanning-tunneling spectroscopy of organic molecules. *Phys. Rev. B* **59**: R7852–R7855.

Xue, Y.; Datta, S., and Ratner, M. A. 2001. Charge transfer and "band lineup" in molecular electronic devices: A chemical and numerical interpretation. *J. Chem. Phys.* **115**: 4292–4299.

Yeganeh, S.; Galperin, M., and Ratner M. A. 2007a. Switching in molecular transport junctions: Polarization response. *J. Am. Chem. Soc.* **129**: 13313–13320.

Yeganeh, S.; Ratner, M. A., and Mujica, V. 2007b. Dynamics of charge transfer: Rate processes formulated with nonequilibrium Green's functions. *J. Chem. Phys.* **126**: 161103.

Yu, L. H.; Keane, Z. K.; Ciszek, J. W. et al. 2004. Inelastic electron tunneling via molecular vibrations in single-molecule transistors. *Phys. Rev. Lett.* **93**: 266802.

Zazunov, A. and Martin, T. 2007. Transport through a molecular quantum dot in the polaron crossover regime. *Phys. Rev. B* **76**: 033417.

Zazunov, A.; Feinberg, D., and Martin, T. 2006a. Phonon squeezing in a superconducting molecular transistor. *Phys. Rev. Lett.* **97**: 196801.

Zazunov, A.; Feinberg, D., and Martin, T. 2006b. Phonon-mediated negative differential conductance in molecular quantum dots. *Phys. Rev. B* **73**: 115405.

Zhitenev, N. B.; Meng, H., and Bao, Z. 2002. Conductance of small molecular junctions. *Phys. Rev. Lett.* **88**: 226801.

Zimbovskaya, N. A. 2003. Low temperature electronic transport and electron transfer through organic macromolecules. *J. Chem. Phys.* **118**: 4–7.

Zimbovskaya, N. A. 2005. Low temperature electronic transport through macromolecules and characteristics of intramolecular electron transfer. *J. Chem. Phys.* **123**: 114708.

Zimbovskaya, N. A. 2008. Inelastic electron transport in polymer nanofibers. *J. Chem. Phys.* **123**: 114705.

Zimbovskaya, N. A. and Gumbs, G. 2002. Long-range electron transfer and electronic transport through macromolecules. *Appl. Phys. Lett.* **81**: 1518–1520.

Zimbovskaya, N. A. and Pederson, M. R. 2008. Negative differential resistance in molecular junction: The effect of electrodes electronic structure. *Phys. Rev. B* **78**: 153105.

Zimmerman, J.; Pavone, P., and Cuniberti, G. 2008. Vibrational modes and low-temperature thermal properties of graphene and carbon nanotubes: Minimal force constant model. *Phys. Rev. B* **78**: 045410.

11

Bridging Biomolecules with Nanoelectronics

Kien Wen Sun
National Chiao Tung University

Chia-Ching Chang
National Chiao Tung University

11.1 Introduction and Background

The field of nanostructures has grown out of the lithographic technology developed for integrated circuits, but is now much more than simply making smaller transistors. In the early 1980s, microstructures became small enough to observe interesting quantum effects. These structures were smaller than the inelastic scattering length of an electron so that the electrons could remain coherent as they traversed them, giving rise to interference phenomena. Studies on the Aharonov–Bohm effect and universal conductance fluctuations led to the field of "mesoscale physics"—between macroscopic classical systems and fully quantized ones.

Now the size of the structures that can be produced is approaching the de Broglie wavelength of the electrons in the solids, leading to stronger quantum effects. In addition to interesting new physics, this drive toward smaller length scales has important practical consequences. When semiconductor devices reach about 100 nm, the essentially classical models of their behavior will no longer be valid. It is not yet clear how to make devices and circuits that will operate properly on these smaller scales. The replacement for the transistor, which must carry the technology to well below 100 nm, has not been identified. It is anticipated that the semiconductor industry will run up against this "wall" within about 10 years.

The current very large-scale integrated circuit paradigm based on complementary metal oxide semiconductor (CMOS) technology cannot be extended into a region with features smaller than 10 nm.[1] With a gate length well below 10 nm, the sensitivity of

the silicon field-effect transistor parameters may grow exponentially due to the inevitable random variations in the device size. Therefore, an alternative nanodevice concept of molecular circuits was proposed, which was a radical paradigm shift from the pure CMOS technology to the hybrid semiconductor.[2] The concept combines the advantages of nanoscale components, such as the reliability of CMOS circuits, and the advantages of patterning techniques, which include the flexibility of traditional photolithography and the potentially low cost of nanoimprinting and chemically directed self-assembly. The major attraction of this concept is the incorporation of the richness of organic chemistry with the versatilities of semiconductor science and technology. However, before this, one needs to bring directed self-assembly from the present level of single-layer growth on smooth substrates to the reliable placement of three-terminal molecules on patterned semiconductor structures.

While physics and electrical engineering have been evolving from microstructures, to mesoscopic structures, and now to nanostructures, molecular biologists have always worked with objects of a few nanometers or less. A DNA molecule, for example, is very long (when stretched out), but it is about 2.5 nm wide with base pairs separated by 0.34 nm. Since the technology has not existed to directly fabricate and manipulate objects this small, various chemical techniques have been developed for cutting, tagging, and sorting large biological molecules. While enormously successful, these methods utilize batch processing of huge numbers of the molecules and rely on statistical interpretations. It is not possible, in principle, to sequence one particular DNA molecule with these techniques, for example. Ideally,

$$\vec{J} = -D\nabla n \tag{11.2}$$

where

n denotes the concentration of the denaturant that is dissociated from the protein

D denotes the diffusive constant

vector J denotes the flux, respectively, of the solute

According to the Einstein relation

$$D = \frac{kT}{6\pi\eta R_{\mathrm{H}}} \tag{11.3}$$

where

k is the Boltzmann constant

T is the temperature in Kelvin

η is the viscosity of the solvent

R_{H} is the hydration radius of the solutes

Due to the intrinsic diffusion process, the solute exchange processes are not synchronous for all protein molecules. Therefore, the folding rate of protein may not be measured directly by a simple spectral technique, that is, the stopped-flow CD,[45] continuous-flow CD,[46] or fluorescence.[47] However, the reaction interval of protein folding can be revealed by the autocorrelation of reaction time from these direct measurements. The detailed mechanism and an example will be discussed later.

If we look at the energy landscape funnel model of protein folding,[48] it appears that proteins can be trapped in a multitude of local minima of the potential well in a complicated protein system. The native state, though, is at the lowest energy level. When thermal equilibrium is reached, most of the protein molecules are located in the lowest energy state, with a population ratio as low as $e^{-\Delta E/kT}$, according to the Maxwell–Boltzmann distribution in thermodynamics. The ΔE denotes the energy difference between the native state and a local minimum; k and T denote the Boltzmann constant and temperature in Kelvin, respectively. At high concentration (>0.1 mg/mL), however, considerable amount of insoluble protein has been observed in protein folding,[31,33–37,48] indicating that insoluble proteins are at an even lower energy state than the native protein. Therefore, by considering the intermolecular interactions during the protein folding process, the reaction energy landscape may be expressed as a three-well model (Figure 11.2). As shown in Figure 11.2, the unfolded protein (U) is in the highest energy state; the native protein (N) is in the lower energy state. However, the intermediate (I) that may cause further protein aggregation/precipitation is in the lowest energy state. Although the energy state of the intermediate/aggresome is the lowest energy state, the conformational energy of the individual proteins composing the aggresome may not be lower than the native protein. Namely, in single molecular simulation, this extra potential well of intermediate (I) is nonexistent. Therefore, in the conventional energy landscape model (the single molecule simulation model),

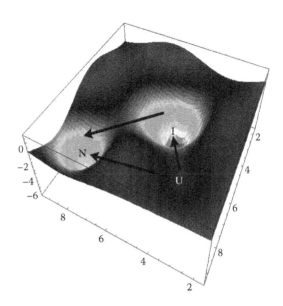

FIGURE 11.2 Three well model of multi-protein molecules folding reaction. The U denotes the unfolded state. N denotes the native state and I denotes the protein–protein complex (aggresome) intermediate.

the lowest energy state "I" cannot be observed. According to the Zwanzig's definition of state, the protein molecules in the intermediate (I) belong to an unfolded state.[13] Hence, in a direct folding reaction, the soluble (N) and the insoluble parts (U) can coexist and they can be observed simultaneously, which is similar to the situation where the phase transition line is crossed in reactions congruent to the "first-order phase transition" model. Therefore, we named the protein folding reaction as "first-order like state transition model" (as shown in Figure 11.3).

The $\Phi(n_1, n_2,...)$ in Figure 11.3 denotes the folding status of protein, where $n_1, n_2,...$ represent the variables affecting the folding status, such as, temperature, concentration of denaturants, etc. The reaction curve indicates an overcritical reaction path of a quasi-static folding reaction. The gray area in Figure 11.3 indicates the state transition boundary of protein folding. The gray line and dash line indicate the reaction path of direct folding. By combining the three-well model (Figure 11.2) and the direct

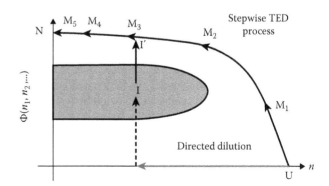

FIGURE 11.3 The protein folding phase diagram, where the $\Phi(n_1, n_2,...)$ denotes the folding status (the order parameter) of protein. The $n, n_1, n_2,...$ denote the variables that affect the folding status such as temperature, concentration of denaturants, etc.

folding reaction of the "first-order like state transition model" (Figure 11.3), we realized that those folded protein molecules along the direct folding path might fold spontaneously or form aggregates. Spontaneous folding may be driven by enthalpy–entropy compensation.

As indicated previously, the conformation of protein changed with changes of the solvent environment. It seems that the protein may fold spontaneously, such as in Anfinsen's experiment[44] and direct folding reactions. The protein folding reaction, similar to all chemical reactions, reaches its equilibrium by following the fundamental laws of thermodynamics. Although protein folding has been studied extensively in certain model systems for over 40 years, the driving force at the molecular level remained unclear until recently.

It is known that polymers and macromolecules may self-assemble/self-organize into a wide range of highly ordered phases/states at thermal equilibrium.[49–52] In a condensed solvent environment, large molecules may self-organize to reduce their effective volume. Meanwhile, the number of the allowed states (Ω) of small molecules, such as buffer salt and other counter-ions in solution, increases considerably. Therefore, the entropy of the system, $\Delta S = R \ln(\Omega_f/\Omega_i)$, becomes large, where i and f denote the initial and final states, respectively. Meanwhile, the enthalpy change (ΔH) between the unfolded and native protein is around hundreds kcal/mol.[53] Therefore, the Gibbs free energy of the system, $\Delta G = \Delta H - TdS$, becomes more negative in this system when the large molecules self-organize.[54] A similar entropy–enthalpy compensation mechanism has been used to solve the reaction of colloidal crystals that self-assemble spontaneously.[55,56]

According to our studies,[31,33–37] the effective diameter of the unfolded protein is about 1.7–2.5 fold larger than the folded protein. Therefore, with the same mechanism, those macromolecules (proteins) may tend to reduce their effective volumes and increase the system entropy when thermal equilibrium is reached. The increase in entropy may compensate for the change of the enthalpy of the system and enable the reaction to take place spontaneously. This may be the reaction molecular mechanism of spontaneous protein folding reactions. Meanwhile, a similar mechanism can be adopted into the self-assembly process of magnetic protein in nanopore arrays.

11.2.3 Quasi-Static Thermal Equilibrium Dialysis for Magnetic Protein Folding

Due to the intrinsic diffusion process, the solvent exchange rate is slow and thus the variation of λ is slow and can be thought of as quasi-static. Therefore, we named this buffer exchanging process as a quasi-static process. We manipulated the reaction direction of the protein folding through this process. Meanwhile, we can obtain stable intermediates in each thermal equilibrium state. These intermediates may help us reveal the molecular folding mechanism of protein that is to be discussed in Section 11.3. The following is an example of the stepwise folding method,[31,33–37] and the buffers used were described in these studies.

Step 1: The unfolded protein (U) was obtained by treating the precipitate or inclusion body with denaturing/unfolding buffer to make it 10 mg/mL in concentration. This solution was left at room temperature for 1 h. This process was meant to relax the protein structure by urea and pH (acidic or basic) environments. The disulfide bridges were reduced to SH groups and the protein was unfolded completely.

Step 2: The unfolded protein (U) in the denature/unfolding buffer was dialyzed against the folding buffer 1 for 72 h to dilute the urea concentration to 2 M, producing intermediate 1, or M_1.

Step 3: M_2 was obtained by dialyzing M_1 against the folding buffer 2 for 24 h to dilute urea concentration to 1 M.

Step 4: M_3, an intermediate without denaturant (urea) in solution, was then obtained by dialyzing M_2 against the folding buffer 3 for 24 h.

Step 5: M_3 was further dialyzed against the folding buffer 4 for 24 h, and the pH changed from 11 to 8.8 to produce M_4.

Step 6: Finally, the chemical chaperonin mannitol was removed by dialyzing M_4 against the native buffer for 8 h to yield M_5.

It should be noted that all the equilibrium time of each step is longer than the conventional dialysis time. In general, for the free solvent case, the solute may exchange with the buffer completely within hours. However, it is known that the denaturant molecules interact with protein, similar to the Donna effect, and the solute exchange may be slow and needs more time for the system to reach thermal equilibrium, especially for the first refolding stage. The folding time of each process is relatively longer than the regular solvent exchange process. Therefore, we can obtain the magnetic protein that follows a similar process. Protein microenvironment protects the net electron spin of molecules from thermal fluctuation.

The bridging ligands (i.e., sulfur atom, S) between the magnetic ions may be responsible for aligning the electron spin of magnetic ions. As indicated in Figure 11.4, the valance bonding electrons of the bridging Cys may hop between the bonded metal ions, such as Mn^{2+} and Cd^{2+}; whereas the Cd^{2+} in the β metal cluster is rather important in restraining the orientation of the electron spins of the bridging sulfurs and in aligning the spins of Mn^{2+} in the metal binding clusters. Therefore, this electron hopping effect may turn the Mn,Cd-MT into a magnetic molecule. However, the protein backbone surrounding the β metal cluster may provide a strong restraining effect to overcome the thermal fluctuations from the environment. Therefore, the magnetization can be observed in room temperature. However, the geometrical symmetry of the spin arrangement in all Mn-MT may cause partial or complete cancellation of detectable magnetization. These results also indicated that the threshold temperature of the molecular magnet might rise to room temperature if the proper prosthetic environment, such as protein backbone, can be linked against the thermal fluctuation of the temperature.

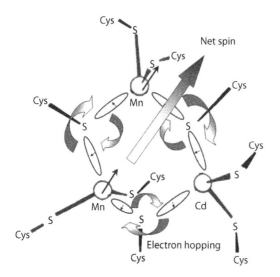

FIGURE 11.4 Proposed electron spin model of Mn^{2+} in β metal binding cluster of Mn,Cd-MT-2.

Therefore, we have successfully constructed a molecular magnet, Mn,Cd-MT, that is stable from 10 to 330 K. The observed magnetic moment can be explained by the highly ordered alignment of $(Mn_2CdS_9)^{3-}$ clusters embedded in the β-domain in which sulfur atoms serve as key bridging ligands. The discovery of mMT may allude to new schemes in constructing a completely different category of molecular magnets.

11.3 Nanostructured Semiconductor Templates: Nanofabrication and Patterning

The rapidly developing of interdisciplinary activity in nanostructuring is truly exciting. The intersections between the various disciplines are where much of the novel activity resides, and this activity is growing in importance. The basis of the field is any type of material (metal, ceramic, polymer, semiconductor, glass, and composite) created from nanoscale building blocks (clusters of nanoparticles, nanotube, nanolayers, etc.) that are themselves synthesized from atoms and molecules. Thus, the controlled synthesis of those building blocks and their subsequent assembly into nanostructures is one fundamental theme of this field. This theme draws upon all of the material-related disciplines from physics to chemistry to biology and to essentially all of the engineering disciplines as well.

The second and most fundamental important theme in this field is that the nanoscale building blocks, because of their size being below about 100 nm, impart to the nanostructures that are created from them new and improved properties and functionalities that are still unavailable in conventional materials and devices. The reason for this is that the materials in this size range can exhibit fundamentally new behavior when their sizes fall below the critical length scale associated with

any given property. Thus, essentially any material property can be dramatically changed and engineered through the controlled size-selective synthesis and assembly of nanoscale building blocks. The present juncture is important in the fields of nanoscale solid-state physics, nanoelectronics, and molecular biology. The length scales and their associated physics and fabrication technology are all converging to the nanometer range.

The ability to fabricate structures with nanometer precision is of fundamental importance for any exploitation of nanotechnology. In particular, cost effective methods that are able to fabricate complex structures over large areas will be required. One of the main goals in the nanofabrication area is to develop general techniques for rapidly patterning large areas (a square centimeter, or more) with structures of nanometer sizes. Presently, electron-beam (e-beam) lithography is capable of defining patterns that are less than 10 nm. These patterns can then be transferred to a substrate using various ion milling/etching techniques. However, these are "heroic" experiments, and can only be made over a very limited area—typically a few thousand square microns, at most. While such areas are immediately useful for investigating the physics of nanostructures, the applications we would like to pursue will eventually require a faster writing scheme and much larger areas. The time and area constraints are determined by the direct e-beam writing. It is a "serial" process, defining single small regions at a time. Furthermore, the field of view for the e-beam system is typically less than 100 nm when defining the nanometer-scale structures. One simply cannot position the electron beam with nm precision over larger areas.

While e-beam lithographic methods are very general, in that essentially any shape can be written, we will also make use of "natural lithography" (tricks). Similarly, advances in the knowledge of the DNA structure has recently been applied to the fabrication of self-organized nm surface structures. There are many such "tricks" that could prove crucial to the success of the projects in allied fields. A list of current methods toward nanofabrication is given below.

11.3.1 Patterned Self-Assembly for Pattern Replication

By exploiting e-beam and focused ion beam lithography, self-assembled monolayers can be patterned into 10–20 nm features that can be functionalized with single molecules or small molecular groupings. These patterned areas will then be used as templates to direct the vertical assembly of stacks of molecules or to direct the growth of polymeric molecules. Schematics of the processes and the possible templates used are shown in Figure 11.5. The initial 2D pattern will thus be translated into 3D nanosized objects. With the capability of the full control over the interfacial properties, it will be possible to release the objects from the templates and transfer them to another substrate, after which the nanopatterned surface can be used again to provide an inexpensive replication technique.

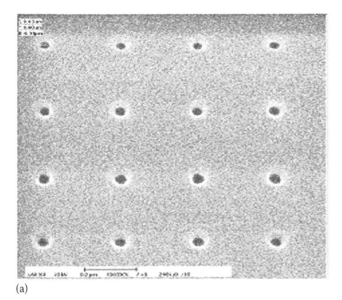

(a)

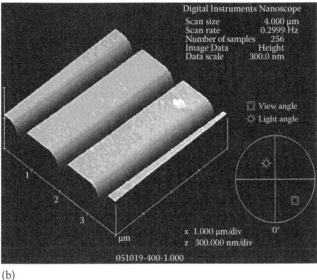

(b)

FIGURE 11.5 E-beam lithography defined (a) nanopores and (b) nanotrenches on silicon templates. (From Chang, C.-C. et al., *Biomaterials*, 28, 1941, 2007. With permission. Elsevier.)

11.3.2 Fabrication by Direct Inkjet and Mold Imprinting

This part of the technique will draw on recent advances in printing techniques for direct patterning of surfaces. This involves the use of inkjet printing to deliver a functional material (semiconductor or metal) to a substrate, which is then controlled by patterning in the surface free energy of the substrate, to allow very accurate patterning of the printed material. Currently, devices show channel lengths down to 5 μm, but indications are that geometries can be reduced to submicron dimensions. The method will be concerned with the limits to resolution that can be achieved by this process, and also the structure and the associated electronic structure at polymer–polymer interfaces, such as that between semiconductor and insulator layers in the field-effect device.

FIGURE 11.6 Nanogratings on silicon templates by using thermal nano-imprinting technique.

In 1995, Professor Stephen Y. Chou of Princeton University invented a new fabrication method in the field of semiconductor fabrication. It is called nano-imprint lithography (NIL).[57] Briefly speaking, this technique was demonstrated by pressing the patterned mold to contact with the polymer resist directly. The patterns on the mold will transfer to the polymer resist without any exposure source. Therefore, the diffraction effect of light can be ignored, and the limitation is dependent only on the pattern size of the mold rather than on the wavelength of the exposure light. In Figure 11.6 we show a nano-grating structure made by using the thermal imprinting technique.

NIL technology is a physical deformation process and is very different from conventional optical lithography. This technology provides a different way to fabricate nanostructures with easy processes, high throughput, and low cost. Currently, there are three main NIL techniques under investigation, namely, hot-embossing nano-imprint lithography (H-NIL[57]), ultraviolet nano-imprint lithography (UV-NIL[58]), and soft lithography.[59] Those NIL technologies can be applied to many different research fields, including nano-electric devices,[60] bio-chips,[61] micro-optic devices,[62] micro-fluidic channels,[63] etc.

11.3.3 Nanopatterned SAMs as 2D Templates for 3D Fabrication

Modern lithographic techniques (e-beam or focused ion beam (FIB), SNOM lithography) are able to generate topographic (relief) patterns in the 10–50 nm size ranges. As a further step toward more complicated and functional 3D structures, chemical functionalization at a similar size scale is necessary. By exploiting e-beam, FIB, and near-field optical lithography, it is possible to pattern SAMs directly. Using lithography, the SAMs can either be locally destroyed and refilled with other molecules, or the surface of the SAMs can be activated to allow further chemical reactions. The resulting patterned surfaces will be chemically patterned. Patterns can be introduced to incorporate H-bonding, pi–pi stacking, or chemical reactivity. These patterned areas will be used as templates to direct the vertical assembly of stacks of molecules or to direct the growth of polymeric molecules. Large, extended aromatic molecules prefer to stack on top of each other due to pi–pi stacking. These molecules have interesting electronic properties as molecular wires. When surfaces can be patterned to incorporate "seeds" for the large aromatic molecules, the stacking can be directed away from the surface. The initial 2D pattern will thus be translated into 3D nanosized objects. With the full control over the interfacial properties, it will be possible to release the objects from the templates and transfer them to another substrate, after which the nanopatterned surface can be used again to provide an inexpensive replication technique.

11.4 Self-Assembling Growth of Molecules on the Patterned Templates

The self-assembling growth of the MT-2 proteins is demonstrated as follows. One mg/mL magnetic MT in Tris. HCL buffer solution was placed onto the patterned surface, and an electric field with an intensity of 100 V/cm was then applied for 5 min to drive the MT molecules into the nanopores. The sample was then washed with DI water twice to remove the unbounded MT molecules and salts on the surface (the schematic of the process is also shown in Figure 11.7a. Figure 11.8 shows the atomic force microscopy (AFM) image of the template surface with 40 nm nanopores after they were filled by the MT-molecules. Keep in mind that most of the Si surface was still protected by photoresist after the etching processes, which has prevented the MT-molecules from forming strong OH bonds with the Si surface underneath. Therefore, the electrical field-driven MT molecules were all anchored on those areas that were not covered with photoresist. The molecules landing in each pore were then self-assembly grown vertically from the bottom of the pore into the shape of a rod (as shown in Figure 11.8). These molecular nanorods have an average height of ~120 nm above the template surface and a diameter equal to the size of the nanopore.

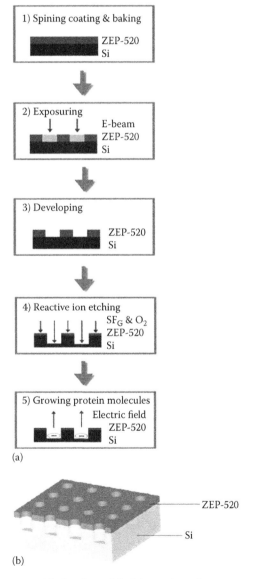

(a)

(b)

FIGURE 11.7 (a) Flowchart of the lithography, etching processes, and growth of protein molecules, (b) schematics of the patterned templates with nanopores. (From Chang, C.-C. et al., *Biomaterials*, 28, 1941, 2007. With permission. Elsevier.)

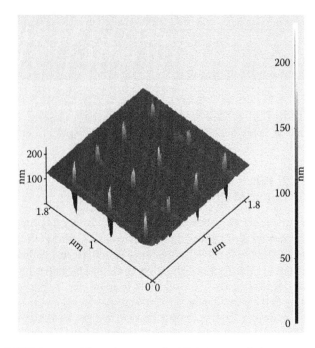

FIGURE 11.8 Three-dimensional AFM image of the patterned magnetic molecules. The molecules have self-assembled to grow into a rod shape. (From Chang, C.-C. et al., *Biomaterials*, 28, 1941, 2007. With permission. Elsevier.)

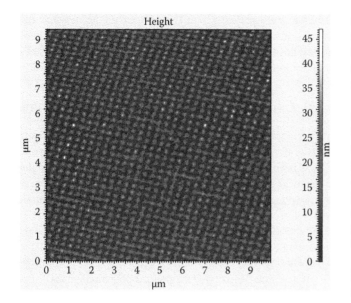

FIGURE 11.9 Two-dimensional AFM images of the patterned MT-molecules on the template with pore size of 130 nm and pitch size of 300 nm. (From Chang, C.-C. et al., *Biomaterials*, 28, 1941, 2007. With permission. Elsevier.)

However, experiments on the templates with pore sizes larger than 100 nm gave quite different results. Figure 11.9 shows the two-dimensional AFM image of the template surface with larger pores, where we can see that the molecules did not grow vertically above the template surface. Therefore, we were not able to generate 3D images of this type of template. However, judging from the AFM phase images, the MT molecules did form

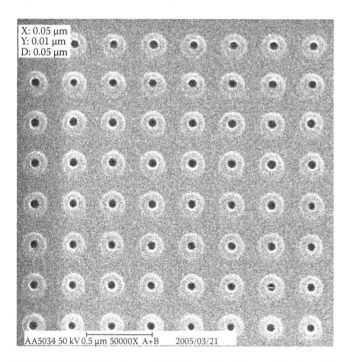

FIGURE 11.10 SEM image of the Si template shows a ring shape Si exposed area around the circumference of nanopores. (From Chang, C.-C. et al., *Biomaterials*, 28, 1941, 2007. With permission. Elsevier.)

a more dense structure in the larger pores compared with the case of the smaller pores. On templates with thinner photoresist and smaller pitch sizes (less than 600 nm), we also found that the molecules anchored in the pore can grow laterally toward the neighboring pores (data not shown).

By increasing both the e-beam exposure and dry etching time on the Si surface covered with a thin photoresist layer with a thickness of less than 150 nm, we were able to create a ring-type area with an exposed Si surface along the periphery of the nanopores. The SEM image of this type of template is shown in Figure 11.10. On this particular template, the molecules not only independently grew inside the pores, but they also grew along the circumference of the pores to form molecular rings on the template. Figure 11.11 shows the two-dimensional (2D) and three-dimensional (3D) AFM images of such molecular rings.

In order to gain better control of the formation of molecular nanostructures, it is important to uncover the underlying self-assembling growth mechanism. Molecular self-assembly

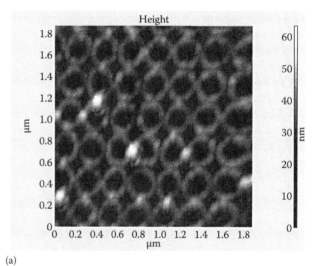

(a)

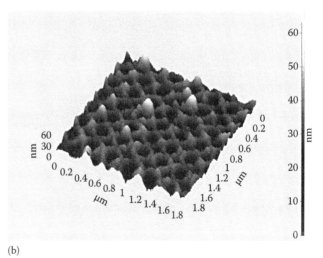

(b)

FIGURE 11.11 (a) 2D and (b) 3D AFM images of the patterned MT-molecules. (From Chang, C.-W. et al., *Appl. Phys. Lett.*, 88(26), 263104, 2006. With permission. AIP.)

can be mediated by weak, noncovalent bonds—notably hydrogen bonds, ionic bonds (electrostatic interactions), hydrophobic interactions, van der Waals interactions, and water-mediated hydrogen bonds. Although these bonds are relatively insignificant in isolation, when combined together as a whole, they govern the structural conformation of all biological macromolecules and influence their interaction with other molecules. The water-mediated hydrogen bond is especially important for living systems, as all biological materials interact with water. We believe that the first layer of proteins anchored inside the nanopores was bonded with the Si surface dangling bonds. They have provided building blocks for proteins that arrived later. With the assistance of spatial confinement from the patterned nanostructures, the rest of the proteins are able to self-assemble via the van der Waals interactions and perform molecular self-assembly.

11.5 Magnetic Properties of Molecular Nanostructures

The magnetic properties of the self-assembled molecular nanorods were investigated with magnetic force microscopy (MFM). We monitored the change of contour of a particular nanorod on the template when an external magnetic field was applied. Figure 11.12a shows the MFM image of the nanorod without the external magnetic field. In Figure 11.12b, a magnetic field of 500 Oe was applied during the measurement with a field direction from the right to left. The strength of the field was kept at a minimum so as not to perturb the magnetic tip on the instrument. In Figure

11.12 we can see clearly that the contour of the nanorod has changed in shape as compared to the case with no applied field. It indicates that the molecular self-assembly carries a magnetic dipole moment that interacts with the external magnetic field.

11.6 Conclusion and Future Perspectives

Success in the synthesis of the magnetic molecules produced from metallothionein (MT-2) by replacing the Zn atoms with Mn and Cd has been demonstrated in this chapter. Hysteresis behavior in the magnetic dipole momentum measurements was observed over a wide range of temperatures when an external magnetic field was scanned. These magnetic MT molecules were also found to self-assemble into nanostructures with various shapes depending on the nanostructures patterned on the Si templates. Data from the MFM measurements indicate that these molecular self-assemblies also carry magnetic dipole momentum. Since the pore size, spacing and shape can easily and precisely be controlled by lithography and etching techniques, this work should open up a new path toward an entire class of new biomaterials that can be easily designed and prepared. The techniques developed in this particular work promise to facilitate the creation of many bio-related nanodevices and spintronics. Magnetic molecular self-assembly may find its use in data storage or magnetic recording systems, as an example. They can also act as spin biosensors and be placed at the gate of the semiconductor spin valve to control the spin current from the source to the drain. More importantly, this work should not be limited to MT-2 molecules

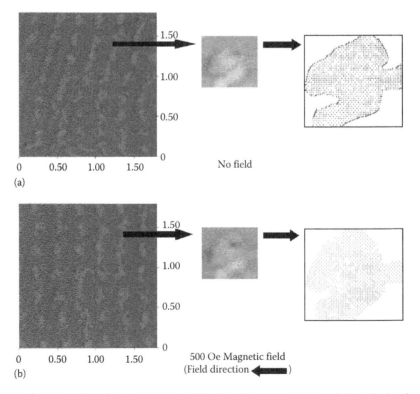

FIGURE 11.12 MFM images of nanorods (a) without the magnetic field (b) with a 500 Oe magnetic field applied with a field direction from right to left. (From Chang, C.-C. et al., *Biomaterials*, 28, 1941, 2007. With permission. Elsevier.)

and should be extended to other type of molecules and proteins as well. As mentioned in the beginning of this chapter, the various surface patterning techniques developed over the years to interface organic or biological materials with semiconductors have not only provided new tools for controlled 2D and 3D self-organized assemblies, but have also been essential to the creation and emergence of new semiconductor-molecular nanoelectronics as recently discussed by Likharev.[64]

In the future, it is important to develop techniques for growing and characterizing molecular self-assembly, single nanostructure, and molecule on semiconductor templates to bring a measure of control to the density, order, and size distribution of these molecular nanostructures. Using self-assembly techniques, one can routinely make molecular assembly with precise distances between them. Recent explorations of molecular self-assembly have sought to provide transverse dimensions on the mesoscopic nanometer scale. As a general—although not inviolate—rule, these attempts have led to very good local ordering (e.g., nearest neighbors).

We can anticipate, (1) the development of new (supra) molecular nanostructures via the self-assembling method (bottom up technique) and immobilization of single nanostructure or molecule; (2) the design of methods to functionalize molecular self-assembly and devices; (3) an integration of bottom-up and top-down procedures for the nano- and microfabrication of molecularly driven sensors, actuators, amplifiers, and switches; and (4) an increased understanding and appreciation of the science and engineering that lie behind nanoscale processes. All this and more is in the nature of the nanotechnology bonds as it impacts on biology and beyond. In the final analysis, however, the practice of biological synthesis that relies on molecular recognition and self-assembling processes within a very much more catholic framework than is currently being contemplated by most researchers that will dictate the pace of progress in synthesis.

The final goal is to understand this whole notion of what self-assembly is. One needs to really learn how to make use of the methods of organizing structures in more complicated ways than we can do now. On a molecular scale, the accurate and controlled application of intermolecular forces can lead to new and previously unachievable nanostructures. This is why molecular self-assembly (MSA) is a highly topical and promising field of research in nanotechnology today. MSA encompasses all structures formed by molecules selectively binding to a molecular site without external influence. With many complex examples all around us in nature (ourselves included), MSA is a widely observed phenomenon that has yet to be fully understood. Being more a physical principle than a single quantifiable property, it appears in engineering, physics, chemistry, and biochemistry, and is therefore truly interdisciplinary.

References

1. Sessoli R., Gatteschi D., Caneschi A., and Novak M.A. (1993), *Nature* (*London*), **365**, 141–143.
2. Fonticelli M., Azzaroni O., Benitez G., Martins M.E., Carro P., and Salvarezza R.C. (2004), *J. Phys. Chem. B*, **108**, 1898.
3. Park M., Harrison C., Chaikin P.M., Register R.A., and Adamson D.H. (1997), *Science* (*Washington, DC*), **276**, 1401–1404.
4. Li R.R., Dapkus P.D., Thompson M.E., Jeong W.G., Harrison C., Chaikin P.M., Register R.A., and Adamson D.H. (2000), *Appl. Phys. Lett.*, **76**, 1689–1691.
5. Thurn-Albrecht T., Schotter J., Kästle G.A., Emley N., Shibauchi T., Krusin-Elbaum L., Guarini K., Black C.T., Tuominen M.T., and Russell T.P. (2000), *Science* (*Washington, DC*), **290**, 2126–2129.
6. Kim H.C. et al. (2001), *Adv. Mater.*, **13**, 795–797.
7. Lopes W.A. and Jaeger H.M. (2001), *Nature*, **414**, 735–738.
8. Cheng J.Y. et al. (2001), *Adv. Mater.*, **13**, 1174–1178.
9. Kim S.O., Solak H.H., Stoykovich M.P., Ferrier N.J., de Pablo J.J., and Nealey P.F. (2003), *Nature*, **424**, 411–414.
10. Demers L.M., Ginger D.S., Park S.J., Li Z., Chung S.W., and Mirkin C.A. (2002), *Science* (*Washington, DC*), **296**, 1836–1838.
11. Hodneland C.D., Lee Y.S., Min D.H., and Mrksich M. (2002), *Proc. Natl. Acad. Sci.*, **99**, 5048–5052.
12. Houseman B.T. and Mrksich M. (2002), *Chem. Biol.*, **9**, 443–454.
13. Chen C.S., Mrksich M., Huang S., Whitesides G.M., and Ingber D.E. (1997), *Science* (*Washington, DC*), **276**, 1345–1347.
14. Mrksich M., Dike L.E., Tien J.Y., Ingber D.E., and Whitesides G.M. (1997), *Exp. Cell. Res.*, **235**, 305–313.
15. Chen C.S., Mrksich M., Huang S., Whitesides G.M., and Ingber D.E. (1998), *Biotech. Prog.*, **14**, 356–363.
16. Pitters J.L., Piva P.G., Tong X., and Wolkow R.A. (2003), *Nano Lett.*, **3**, 1431.
17. Wacaser B.A., Maughan M.J., Mowat I.A., Niederhauser T.L., Linford M.R., and Davis R.C. (2003), *Appl. Phys. Lett.*, **82**, 808.
18. Eaton D.L. (1985), *Toxicol. Appl. Pharmacol.*, **78**, 158–162.
19. Robbins A.H., McRee D.E., Williamson M., Collett S.A., Xuong N.H., Furey W.F., Wang B.C., and Stout C.D. (1991), *J. Mol. Biol.*, **221**, 1269–1293.
20. Messerle B.A., Schaffer A., Vasak M., Kagi J.H.R., and Wuthrich K. (1992), *J. Mol. Biol.* **225**, 433–443.
21. Boulanger Y., Goodman C.M., Forte C.P., Fesik S.W., and Armitage I.M. (1983), *Proc. Natl. Acad. Sci.*, **80**, 1501–1505.
22. Chang C.C. and Huang P.C. (1996), *Protein Eng.*, **9**, 1165–1172.
23. Wei S.H. and Zunger A. (1986), *Phys. Rev. Lett.*, **56**, 2391–2394.
24. Christou G., Gatteschi D., Hendrickson D.N., and Sessoli R. (2000), *MRS Bull.*, **25**, 66–71.
25. Sessoli R., Tsai H.L., Schake A.R., Wang S., Vincent J.B., Folting K., Gatteschi D., Christou G., and Hendrickson D.N. (1993), *J. Am. Chem. Soc.*, **115**, 1804–1816.
26. Aubin S.M.J., Dilley N.R., Pardi L., Kryzystek J., Wemple M.W., Brunel L.C., Maple M.B., Christou G., and Hendrickson D.N. (1998), *J. Am. Chem. Soc.*, **120**, 4991.
27. Friedman J.R., Sarachik M.P., Tejada J., and Ziolo R. (1996), *Phys. Rev. Lett.*, **76**, 3830–3833.

28. Bushby R.J. and Pailland J.-L. (1995), Molecular magnets, in *Introduction to Molecular Electronics*, M.C. Petty, M.R. Bryce, and D. Bloor (eds.), Edward Arnold, Hodder Headline, London, U.K.

29. Wolynes P.G. (1997), *Proc. Natl. Acad. Sci. U.S.A.*, **94**, 6170–6175.

30. Zwanzig R. (1997), *Proc. Natl. Acad. Sci. U.S.A.*, **94**, 148–150.

31. Chang C.-C., Yeh X.-C., Lee H.-T., Lin P.-Y., and Kan L.S. (2004), *Phys. Rev. E*, **70**, 011904.

32. Welker E., Wedemeyer W.J., Narayan M., and Scheraga H.A. (2001), *Biochemistry*, **40**, 9059–9064.

33. Chang C.-C., Su Y.-C., Cheng M.-S., and Kan L.S. (2002), *Phys. Rev. E*, **66**, 021903.

34. Chang C.-C., Tsai C.T., and Chang C.Y. (2002), *Protein Eng.*, **5**, 437–441.

35. Chang C.-C., Su Y.-C., Cheng M.-S., and Kan L.S. (2003), *J. Biomol. Struct. Dyn.*, **21**, 247–256.

36. Chang C.-C. and Kan L.-S. (2007), *Chin. J. Phys.*, **45**, 693–702.

37. Liu Y.-L., Lee H.-T., Chang C.-C., and Kan L.S. (2003), *Biochem. Biophys. Res. Commun.*, **306**, 59–63.

38. Levy Y., Jortner J., and Becker O. (2001), *Proc. Natl. Acad. Sci. U.S.A.*, **98**, 2188–2193.

39. Creighton T.E. (1993), *Proteins*, W.H. Freeman, New York.

40. Elcock A.H. (1999), *J. Mol. Biol.*, **294**, 1051–1062.

41. Ulrih N.P., Anderluh G., Macek P., and Chalikian, T.V. (2004), *Biochemistry*, **43**, 9536–9545.

42. Despa F., Fernandez A., and Berry R.S. (2004), *Phys. Rev. Lett.*, **93**, 228104.

43. Seefeldt M.B., Ouyang J., Froland W.A., Carpenter J.F., and Randolph T.W. (2004), *Protein Sci.*, **13**, 2639–2650.

44. Anfinsen C.B., Haber E., Sela M., and White F.H. Jr. (1961), *Proc. Natl. Acad. Sci.*, **47**, 1309–1314.

45. Su Z.D., Arooz M.T., Chen H.M., Gross C.J., and Tsong T.Y. (1996), *Proc. Natl. Acad. Sci.*, **93**, 2539–2544.

46. Roder H., Maki K., Cheng H., and Shastry M.C. (2004), *Methods*, **34**, 15–27.

47. Royer C.A. (1995), *Methods Mol. Biol.*, **40**, 65–89.

48. Misawa S. and Kumagai I. (1999), *Biopolymers*, **51**, 297–307.

49. Dinsmore A.D., Crocker J.C., and Yodh A.G. (1998), *Curr. Opin. Colloid Interface Sci.*, **3**, 5–11.

50. Gast A.P. and Russel W.B. (1998), *Phys. Today*, **51**, 24–30.

51. Pusey P.N. and van Megan W. (1986), *Nature*, **320**, 340–341.

52. Ackerson B.J. and Pusey P.N. (1988), *Phys. Rev. Lett.*, **61**, 1033–1036.

53. Tsong T.Y., Hearn R.P., Warthal D.P., and Sturtevant J.M. (1970), *Biochemistry*, **9**, 2666–2677.

54. Tokuriki N., Kinjo M., Negi S., Hoshino M., Goto Y., Urabe I., and Yomo T. (2004), *Protein Sci.*, **13**, 125–133.

55. Lin K.-H., Crocker J.C., Prasad V., Schofield A., Weitz D.A., Lubensky T.C., and Yodh A.G. (2000), *Phys. Rev. Lett.*, **85**, 1770–1773.

56. Lau A.W.C., Lin K.-H., and Yodh A.G. (2002), *Phys. Rev. E*, **66**, 020401.

57. Stephen Y.C., Peter R.K., and Preston J. R. (1995), *Appl. Phys. Lett.*, **67**, 3114–3116.

58. Bender M., Otto M., Hadam B., Spangenberg B., and Kurz H. (2000), *Microelectron. Eng.*, **53**, 233–236.

59. Xia Y. and Whitesides G.M. (1998), *Angew. Chem. Int.*, **37**, 550–575.

60. Lingjie G., Peter R.K., and Stephen Y.C. (1997), *Appl. Phys. Lett.*, **71**, 1881–1883.

61. Pépin A., Youinou P., Studer V., Lebib A., and Chen Y. (2002), *Microelectron. Eng.*, **61–62**, 927–932.

62. Li M., Tan H., Chen L., Wang J., and Chou S.Y., (2003), *J. Vac. Sci. Technol. B*, **21**, 660–663.

63. Cho Y.H., Lee S.W., Kim B.J., and Fujii T. (2007), *Nanotechnology*, **18**, 465303.

64. Likharev K.K. (2003), Hybrid semiconductor–molecular nanoelectronics. *The Industrial Physicist*, June/July 2003, Forum: 20–23.

65. Chang C.-C., Sun K.W., Lee S.F., and Kan L.S. (2007), *Biomaterials*, **28**, 1941–1947.

66. Chang C.-C., Sun, K.W., Kan, L.-S., and Kuan, C.-H. (2006), *Appl. Phys. Lett.*, **88**, 263104.

II

Nanoscale Transistors

Transistor Structures for Nanoelectronics

Jean-Pierre Colinge
University College Cork

Jim Greer
University College Cork

12.1 Introduction

In 1965, Gordon Moore predicted that the number of transistors that could be placed on a silicon chip would double every 18 months. This prediction became the official roadmap of the semiconductor industry and it still is the yardstick by which the progress of microelectronic devices and circuits is measured (Moore 1965). The MOS transistor, also called MOSFET (metal-oxide-semiconductor field-effect transistor) is the workhorse of the semiconductor industry and is at the heart of every digital circuit. Without the MOSFET, there would be no computer industry, no digital telecommunication systems, no video games, no pocket calculators, and no digital wristwatches.

Figure 12.1 shows the evolution of the number of MOS transistor per chip vs. calendar year, known as "Moore's law." Note that the vertical axis has a logarithmic scale. The effective length of the transistor, L, is the distance between the source and the drain in a region called the "channel region" where the electrostatics and the current flow are controlled by the gate. When the effective gate length becomes too small, the gate loses the control of that region and the so-called short-channel effects appear. These effects increase the OFF current of the device and render the current dependent of the drain voltage. In the most extreme cases, the transistor can no longer be turned off. The effective length of MOSFETs has shrunken from 10 µm in

1971 (Intel® 4004 processor) to 1 µm in the early 1980s, 0.1 µm in 2000, and 10 nm dimensions should be reached around 2015. Short-channel effects become so important in classical, planar MOSFETs with dimensions below 5 nm that new device structures that improve the electrostatic control of the channel by the gate are being explored. These devices are called multi-gate MOSFETs. They basically are silicon nanowires with a gate electrode wrapped around the channel region. This device architecture maximizes the control of the electrostatics in the channel region and allows one to reduce the effective channel length to sub-10 nm dimensions. At those dimensions, quantum transport effects start to kick in, and device simulators need to incorporate the Schrödinger equation to accurately model the transistor's electrical characteristics. Of course, the reduction of transistor size goes hand in hand with the increase of transistors on a chip. The first 32 Gb flash memory chips made using a 40 nm technology were announced in 2006. Each of these chips contains over 32 billion transistors. The number of transistor per chip is literally reaching astronomical proportions, considering that the 400 Gb mark will be reached around the year 2020. When this milestone is reached, there will be as many transistors on a single chip as there are stars in the Milky Way (Figure 12.1).

In 2008, the effective gate length of transistors that are used to fabricate microprocessors is 22 nm, and the thickness of the gate insulator is approximately 1 nm. Fully functional

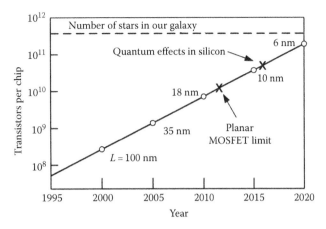

FIGURE 12.1 Moore's law: number of transistors per chip vs. calendar year.

"gate-all-around" transistors made in a silicon nanowire with a diameter of only 3 nm, that is, roughly the same diameter as a carbon nanotube, were reported in 2006 (Singh et al. 2006). Although these transistors are not formally called "nanoelectronic devices," they are clearly of nanometer dimensions. Strictly speaking, the nanoelectronics era has already begun.

Section 12.2 describes the evolution of the MOS transistor as its structure reaches nanometer dimensions: new materials are used to optimize electrical conductivity and dielectric constant, semiconductor alloying and the introduction of stress are used to increase carrier mobility, and novel transistor geometries are employed to increase the electrostatic control of the channel by the gate.

12.2 Evolution of Silicon Processing: Technology Boosters

In the 1980s, only a handful of elements were used in silicon chips: boron, phosphorus, arsenic, and antimony for doping the silicon, oxygen, and nitrogen for growing or depositing insulators, and aluminum for making interconnections. A few elements, such as hydrogen, argon, chlorine, and fluorine are used during processing in the form of etching plasmas or oxidation-enhancing agents. Sulfur is used in sulfuric acid to clean wafers. In the 1990s, a few more elements were added to the list, such as titanium, cobalt, and nickel, used to form low-resistivity metal silicides, tungsten, used to form vertical interconnects known as "plugs," and bromine, used in a plasma form to etch silicon. The 2000s saw an explosion in the number of elements used in silicon processing: lanthanide (rare earth) metals are being used to form oxides with high dielectric constants (high-k dielectrics), carbon and germanium are used to change the lattice parameter and induce mechanical stresses in silicon, fluorides of noble gases are used in excimer laser lithography, and a variety of metals are used to synthesize compounds that have desirable work functions or Schottky characteristics. Virtually all elements are used, with the notable exception of alkaline metals, which create mobile charges in oxides, and, of course, radioactive elements (Figure 12.2).

The use of new elements to obtain new desirable properties is a technology booster that has made it possible to extend the life of CMOS and reduce dimensions beyond barriers that were previously considered insurmountable. For instance, the reduction

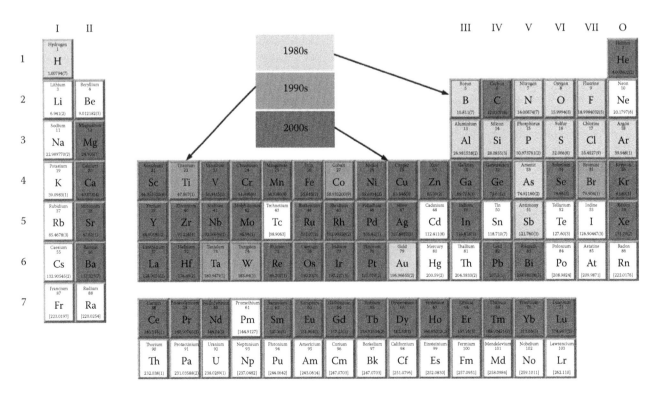

FIGURE 12.2 Elements used in silicon processing.

of gate oxide thickness below 1.5 nm leads to a gate tunnel current that quickly becomes prohibitively high. Replacing silicon dioxide by a high-k dielectrics such as hafnium oxide (HfO_2), which has a dielectric constant of 22 (vs. 3.9 for SiO_2), allows one to increase the thickness of the gate dielectric by a factor 22/3.9 = 5.5 without reducing the gate capacitance (i.e., the current drive of the transistor). The use of new gate dielectrics gave rise to the notion of equivalent oxide thickness, EOT, which is defined by the relationship $EOT = t_d(\varepsilon_{ox}/\varepsilon_d)$, where t_d is the thickness of the dielectric layer, and ε_{ox} and ε_d are the permittivity of silicon dioxide and the dielectric material, respectively. For example, a 4 nm thick layer of HfO_2 is electrically equivalent to an SiO_2 layer of 0.7 nm.

To improve the properties of transistors, another technology booster is commonly used: stress. Compressive stress increases hole mobility in (110) silicon (i.e., in the direction of current flow of most transistors), while tensile stress increases electron mobility. Mobility can also be modified by using Si:Ge or Si:Ge:C alloys. Compressive stress can be induced in the channel region of a transistor by introducing germanium in the source and drain. The resulting "swelling" of the silicon in the source and drain compresses the channel region situated between them. Tensile stress can readily be obtained by depositing a silicon nitride contact-etch stop layer on top of the device. Mobility (and thus speed) improvement in excess of 50% can be obtained using stress techniques.

The third technology booster deals with the physical geometry of the transistor and is worth being described in detail. As the dimensions of the transistors are shrunk, the close proximity between the source and the drain reduces the ability of the gate electrode to control the potential distribution and the flow of current in the channel region, and undesirable effects, called the "short-channel effects" render MOSFETs inoperable. For all practical purposes, it seems impossible to scale the dimensions of classical "bulk" MOSFETs below 15 nm. If that limitation cannot be overcome, Moore's law would reach an end around year 2012.

Short-channel effects arise when electric field lines from source and drain affect the control of the channel region by the gate. These reduce the threshold voltage V_{TH} according to the expression $V_{TH} = V_{TH\infty} - \Delta V_{TH} - DIBL$ where $V_{TH\infty}$ is the threshold voltage of a long-channel device, ΔV_{TH} is the "threshold voltage roll-off," due to the sharing of the space charge region underneath the gate between the gate, and the unbiased source and drain junctions, and DIBL is the "drain barrier lowering" which results from the increase of the share of the space charge region related to the drain junction when the drain voltage is increased. Short-channel effects can be minimized by reducing the junction depth and the gate oxide thickness. They can also be minimized by reducing the depletion depth through an increase in channel doping concentration. In modern devices, however, practical limits on the scaling of junction depth and gate oxide thickness lead to a significant increase of short channel effects and excessively large values of DIBL can quickly be reached.

12.3 Nonplanar Multi-Gate Transistors

The field lines in MOS transistors are illustrated graphically in Figure 12.3. In a bulk device (Figure 12.3A), the electric field lines propagate through the depletion regions associated with the junctions. Their influence on the channel can be reduced by increasing the doping concentration in the channel region. In very small devices, unfortunately, the doping concentration becomes very high ($>10^{19}$ cm^{-3}), which degrades carrier mobility. The situation can be improved by making the transistor in a thin film of silicon atop of an insulator. The silicon film is thin enough that it is fully depleted of majority carriers when a gate voltage is applied. Such a device is called fully depleted silicon-on-insulator (FDSOI) MOSFET. In FDSOI devices, most of the field lines propagate through the buried oxide (BOX) before reaching the channel region (Figure 12.3B). Short-channel effects can be further reduced by using a thin buried oxide and an underlying ground plane. In that case, most of the electric field lines from the source and drain terminate on the buried ground plane instead of the channel region (Figure 12.3C). This approach, however, has the inconvenience of increased junction capacitance and body effect (Xiong et al. 2002).

A much more efficient device configuration is obtained by using the double-gate transistor structure. This device structure was first proposed by Sekigawa and Hayashi in 1984 and was shown to reduce threshold voltage roll-off in short-channel devices (Sekigawa and Hayashi 1984). In a double-gate device, both gates are connected together. The electric field lines from source and drain underneath the device terminate

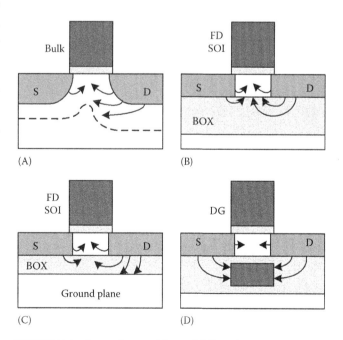

FIGURE 12.3 Encroachment of electric field lines from source and drain on the channel region in different types of MOSFETs: (A) Bulk MOSFET, (B) fully depleted SOI MOSFET, (C) fully depleted SOI MOSFET with thin buried oxide and ground plane, (D) double-gate MOSFET.

on the bottom gate electrode and, therefore, cannot reach the channel region (Figure 12.3D). Only the field lines that propagate through the silicon film itself can encroach on the channel region and degrade short-channel characteristics. This encroachment can be reduced by reducing the silicon film thickness.

Further improvement of short-channel effect control can be achieved by using a gate electrode that is wrapped around three sides of the channel region. The resulting device is known as a triple-gate FET, a term that includes the quantum-wire silicon-on-insulator (SOI) MOSFET (Baie et al. 1995) and the tri-gate MOSFET (Doyle et al. 2003, Kavalieros et al. 2006). The electrostatic integrity of triple-gate MOSFETs can be improved by extending the sidewall portions of the gate electrode to some depth in the buried oxide and underneath the channel region, which results in the formation of a Π-gate MOSFET (Park et al. 2001) or an Ω-gate FET (Yang et al. 2002, Ritzenthaler et al. 2006). The best possible electrostatic of the gate is obtained using "four" gates, or a gate that is wrapped around all sides of the channel region, thereby forming a gate-all-around silicon nanowire transistor (Singh et al. 2006). Using a combination of technology boosters, that is, using a tri-gate device in conjunction with strained silicon, a metal gate and/or high-k dielectric as gate insulator can be used to further enhance the performances of the device (Krivokapic et al. 2003, Andrieu et al. 2006, Kavalieros et al. 2006).

The quality of the electrostatic gate control over the channel can be expressed using a simple expression that can be derived from Poisson's equation (Colinge 2007a,b). It is found that short channel effects are basically nonexistent if the effective channel length is five times larger than that of a value called the "natural length" of the device, and noted λ. The natural length is given by

$$\lambda \cong \sqrt{\frac{\varepsilon_{si}}{n\varepsilon_{ox}} t_{si} t_{ox}}$$

where

- t_{si} and t_{ox} are the thickness of the silicon film and the gate dielectric
- ε_{si} and ε_{ox} is the permittivity of the silicon and the gate dielectric, respectively
- n is the equivalent number of gates (ENG)

The ENG is basically equal to the physical number of sides of the devices covered by a gate and it expresses how efficiently the gate controls the electrostatics in the channel, assuming a square cross section ($W_{si} = t_{si}$). In a single gate device, n is equal to unity, $n = 2$ in a double-gate device, $n = 3$ in a tri-gate configuration, $n = 4$ in a gate-all-around (4 gates device), and $n \approx \pi$ in a π-gate device (Lee et al. 2007). The Ω-gate structure is a π-gate configuration with a larger lateral extension of the gate electrode underneath the channel region. The Ω-gate MOSFET has an effective number of gates larger than a π-gate,

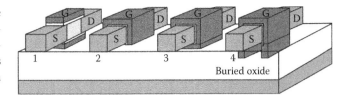

FIGURE 12.4 Different multi-gate MOSFET configurations. 1, double-gate; 2, triple-gate; 3, quadruple-gate or gate-all-around; 4, Π or Ω gate.

but smaller than a 4-gate device. The different configurations are shown in Figure 12.4.

Short-channel effects are absent if the effective gate length is larger than seven to eight times the natural length, λ. This behavior is common to all gate configurations. Acceptable level of short-channel effects (DIBL < 50 mV and subthreshold slope <75 mV/decade) are obtained if the effective gate length, L, is smaller than 4λ (Figure 12.5) (Lee et al. 2007). Thus, for instance, to have the same short-channel behavior than a gate-all-around device, a single-gate SOI transistor needs to be made in a silicon film that is twice as thin.

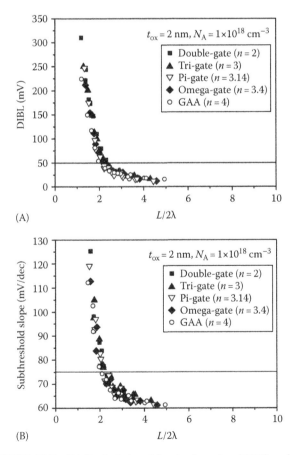

FIGURE 12.5 (A) Drain-induced barrier lowering (DIBL) and (B) subthreshold slope as a function of $L/2\lambda$ for different gate configurations. The gate oxide thickness is equal to 2 nm and the p-type channel doping concentration, N_A, is equal to 10^{18} cm^{-3}.

12.4 The Rise of Quantum Effects

As long as the cross section of the transistor stays above 10 nm classical models using the Poisson equation and drift-diffusion can be used. In smaller devices, however, quantum confinement of the electrons in two directions (perpendicular to the current flow) introduces some fundamental changes in the electronic properties of silicon. It is now necessary to solve the Schrödinger equation in conjunction with the Poisson equation. The density of states (DoS) for electrons in the conduction band in a three-dimensional crystal is a square root function of the electron energy: $DoS \approx \sqrt{E - E_{co}}$, where E_{co} is the minimum energy in the conduction band. In a two-dimensional crystal, the electrons are confined in one direction of space and free to move in the two other directions. As a result, energy subbands are formed within the conduction band. Each subband corresponds to a discrete energy level $E_{nz} = \hbar^2/2m(\pi n/t_{si})^2$ in the direction of confinement, z, and free motion in the two other directions of space (x and y). In this relationship, m is the mass of the electron, t_{si} is the crystal thickness in the z direction, and $n = 1, 2, 3,\dots$. Within each subband the density of states is constant and thus independent of the electron energy: $DoS = (m/\hbar^2 t_{si})$ (Ando et al. 1982). In a one-dimensional crystal, where the electrons are confined in two directions of space and free to move in the

third direction, energy subbands are formed within the conduction band. Each subband corresponds to a discrete energy level $E_{mnyz} = \hbar^2/2m[(\pi m/W_{si})^2 + (\pi n/t_{si})^2]$ in the directions of confinement, y and z, and free motion in the third directions, x. In this relationship, t_{si} is the crystal thickness in the z direction, W_{si} is the crystal width in the y direction and m and $n = 1, 2, 3,\dots$. Within each subband the density of states is shaped as a peak as a function of the electron energy: $DoS \approx (E - E_{nmyz})^{-1/2}$ (Davies 1998). This is illustrated in Figure 12.6, which shows the DoS as a function of energy in a tri-gate transistor with a thickness and width of 5 nm × 5 nm (Figure 12.6A) and a similar device with a thickness and width of 100 nm × 5 nm (Figure 12.6B). The density of states in 3D silicon is shown as well for comparison. In the device with the smaller section the DoS shows a series of peaks, which is characteristic of a one-dimensional system. In the taller device, the peaks gather to form steps of a staircase. Each step of the staircase corresponds to an energy subband of a 2D system. An ideal 2D device would be infinitely tall, while the one simulated here has a thickness of only 100 nm. One nonetheless can observe that the curve tends to a series of steps having each a constant DoS value, which is characteristic of an ideal 2D system (Colinge 2007a,b).

The confinement of carriers also modifies the spatial distribution of the electrons in the transistor. Let us take the example of a pi-gate device. Classical simulation tools predict that inversion channels form preferentially along the corner edges of the device (Figure 12.7A). A simulation of the same device

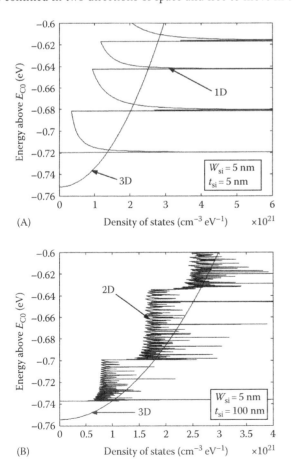

(A)

(B)

FIGURE 12.6 Density of states (DoS) vs. electron energy in a tri-gate MOSFET with (A) $W_{si} = t_{si} = 5$ nm and (B) $W_{si} = 100$ nm and $t_{si} = 5$ nm.

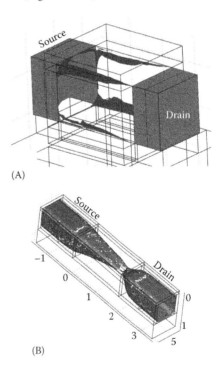

(A)

(B)

FIGURE 12.7 Electron isoconcentration contours in an n-channel pi-gate MOSFET with $t_{si} = W_{si} = 7$ nm and $L = 10$ nm. $V_{GS} = 0.7$ V and $V_{DS} = 0.5$ V. (A) Classical simulation using Poisson and drift-diffusion equations; (B) quantum simulation using Poisson and Schrödinger equations with nonequilibrium Green function formalism.

that includes Schrödinger's equation paints a very different figure: in this case, the electrons are confined to one or two channels located midway between the center of the device and the top corners (Figure 12.7B). Under the particular bias conditions used in Figure 12.7B the device is near saturation and the channel emanating from the source splits into two channels near the drain. Quite clearly, quantum confinement has profound influences on the electron distribution and, therefore, on scattering and transport mechanisms (Kim and Lundstrom 2008).

12.5 Carbon Nanotube Transistors

Carbon (C), silicon (Si), and germanium (Ge) are group IV elements, a description used to indicate that these elements share the same column in the periodic table of the elements. Another way of stating that all three are group IV elements is to say that they share the same configurations for their outer shell electrons. One consequence is that C, Si, and Ge can occur naturally in the diamond structure; see Figure 12.8A. Carbon is also commonly found in the graphite structure consisting of layers of carbon sheets. Each sheet is formed by a monolayer of carbon atoms arranged in a close packed hexagonal pattern. A single layer or carbon sheet in graphite is known as graphene. Carbon nanotubes (CNTs) are graphene strips rolled up into hollow columns or tubes. Carbon in three of its naturally occurring forms, diamond, graphene, and nanotube, is depicted within Figure 12.8. The CNT shown in Figure 12.8C is formed by wrapping

a single graphene sheet into itself and onto a cylindrical shape. Nanotubes formed in this way are referred to as single walled carbon nanotubes (SWCNT). Although not difficult to produce, SWCNTs were not observed experimentally until 1991 (Iijima 1991, Bethune et al. 1993). Under typical growth conditions, many concentric tubes can form and this structure is referred to as multi-walled carbon nanotubes (MWCNT). Graphene can be wrapped onto a cylinder in different ways, referred to as the chirality or "handedness" of the nanotube. Specific chiralities give rise to different structures that are categorized as "armchair," "zig-zag," or simply "chiral." The specific structures give rise to different electronic properties for a specific CNT, which may be insulating, semiconducting, or metallic. Selecting the chirality, and hence controlling a CNT's electronic properties, remains a challenge to the growth and subsequent processing for these materials (Ding et al. 2008).

SWCNTs can have diameters of less than 1 nm but they are typically found within a range of 1–3 nm diameters, whereas their lengths can be on the order of centimeters. For a given diameter and chirality, the electronic, thermal, and mechanical properties between two SWCNT are well matched as the atomic structure of a tube is essentially defect free. This perfect atomic structure results in near ballistic transport, i.e., no scattering along the tube length, and is an attractive feature suggesting the use of CNTs for nanoelectronics. Also, the nanotubes are predicted to be able to carry current densities of up to three orders of magnitude larger (×1000) than typical conductors such as copper and aluminium making them attractive for applications

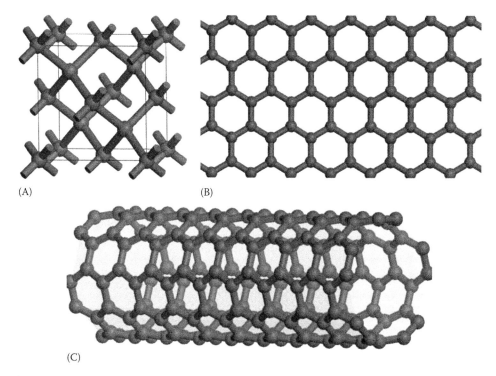

(A)   (B)

(C)

FIGURE 12.8 Carbon in three of its forms: (A) diamond structure, (B) graphene sheet, (C) an example of a carbon nanotube. Note that other forms of carbon exist including amorphous carbon and hollow spheres known as the fullerenes.

in nanoelectronic interconnects. Initially, use for CNTs in electronics focuses on two applications: as "vias" between different levels of metallization (the "wiring" in an integrated circuit) and as the channel material in field effect transistors.

The use of CNT's as a material to fill vias is motivated by the need to have a highly conducting interconnection between two metallization layers, and the fact as circuit dimensions shrink, the cross sections of the vias also decreases. Extremely narrow vias are difficult to fill with conventional material deposition methods. One alternative is to deposit carbon in molecular form, for example, in the form of hollow carbon spheres C_{60} or "buckminsterfullerene." The radius of the C_{60} molecule is on the order of 1 nm and can readily fill vias with cross sections of only a few square nanometers. As the deposited carbon is heated with an appropriate catalytic metal at the base of the via, CNTs grow. The resulting lengths of the CNTs are easily able to span between two metallization layers. However, as mentioned previously, it is presently not possible to control during growth whether a tube will be insulating, semiconducting, or metallic. Hence it is required to grow MWCNTs or bundles of SWCNTs to insure that an appropriate number of metallic tubes are contained within the via. This imposes a constraint on the minimum allowable cross section and introduces a statistical element to via resistances.

CNT field effect transistors (CNTFETs) are transistors fabricated with a CNT as the channel. In the first demonstrations of transistor with a CNT channel, a nanotube is placed from solution or other means onto an oxidized silicon substrate, and a metal layer is deposited and patterned as two metal electrodes. The resulting structure is a CNT spanning the gap between a metal source and drain (Martel et al. 1998, Tans et al. 1998). A cross section of the resulting structure is depicted in Figure 12.9. These original structures required the CNTs to be deposited and located, followed by the deposition of the electrodes. The silicon substrate is used as the gate electrode in a so-called back gate configuration, limiting the performance of the transistors

fabricated in this manner. Typical metals used to form the source and drain contacts include but are not limited to Ti, Ni, Al, and Au. For semiconducting nanotubes, the metal/CNT junctions form a Schottky barrier and it is found that holes are injected into the nanotubes resulting in a p-channel MOS transistor. It has been determined that thermal treatments of the structures can be used to generate a shift in the source/drain Fermi levels, believed to be caused by driving oxygen out of the metal, with the result that n-channel CNTFET devices can be fabricated (Appenzeller et al. 2002).

There are now more sophisticated procedures for fabricating CNTFETs including deposited gate oxides and gate electrodes. CNTFETs have been fabricated and compared to Si MOSFETs and it is found that the CNTFETs can have lower switching delays for devices with similar on–off ratios (Javey et al. 2004, Seidel et al. 2005). CNTFET device layout and fabrication has not been fully optimized for high-frequency behavior, but the theoretical predictions for the high velocities achievable for the charge carriers in the CNTs and the measured results for high frequency performance on non-optimized structures indicate that ballistically limited CNTFETS should outperform ballistically limited Si FETs (Guo et al. 2005). However, in order to achieve the benefits of CNTs in large-scale integration schemes, progress is needed in the placement and controlled growth of nanotubes with preselected electronic character.

12.6 Nonclassical Transistor Structures

Classical transistors are based on the drift and diffusion of electrons and holes due to the presence of the electric fields and carrier concentration gradients. Novel transistor concepts based on quantum effects such as tunneling through a barrier, intraband tunneling, or spin polarization, open the possibility of improved performance and new functionality.

12.7 Graphene Ribbon Nanotransistors

As mentioned, one of the obstacles to the use of CNTs in nanoelectronics is the ability to grow nanotubes with a specific electronic character, that is, insulating, semiconducting, or conducting. Recently, the ability to produce single sheets of graphene with linear dimensions of micrometers (or "microns") and with thicknesses of a single carbon atom has been achieved (Novoselov et al. 2004, 2005). In this sense, graphene is a truly "two-dimensional" material and displays unique electron properties—it is the only two-dimensional material with atomic thickness to be isolated to date. Graphene exhibits high conductivity and is being considered for use in high-speed electronic transistors. Due to the unique electronic structure of graphene, charge carriers move through a layer with zero mass and constant velocity. In common with CNTs, there is little scattering from defects. However, an ideal graphene sheet has a zero bandgap limiting its use in conventional

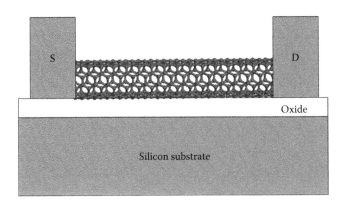

FIGURE 12.9 A CNT transistor in a back gated configuration. In this configuration, the CNT is connected between source and drain contacts and sits on an insulating oxide layer. The silicon substrate is used as a "back gate."

electronic devices, although the unusual behavior of charge carriers in the material suggests it may be useful for nonclassical electronics. On the other hand, it is known that interactions of graphene with a semiconducting substrate can induce a bandgap of a fraction of an electron volt resulting in a narrow bandgap semiconductor.

Alternatively, for very narrow sheets or "nanoribbons" of graphene, the sheet edges can be chemically modified or "terminated" by different atoms or chemical groups. This results in a tailoring of the nanoribbon's electronic structure, and the resulting structure is a two-dimensional confined electron gas with lateral confinement lengths of only a few nanometers and a layer confinement length given by a single atom thickness. The material properties of graphene are only now beginning to be explored and transistor structures allowing for the introduction of local gate-tunable potential barriers within a graphene plane have been developed to explore charge transport mechanisms (Huard et al. 2007).

12.8 Sub-kT/q Switch

The power dissipated by a switching device is equal to $W = fC(V^2/2)$, where f is the switching frequency, C is the capacitance of the device, and V is the supply voltage. Quite clearly, reducing the supply voltage is the best way to reduce power consumption. Unfortunately, transistors, unlike mechanical switches and reed relays, cannot switch "instantly," and require a finite input voltage variation to switch from the off state to the on state. This behavior is fixed by the energy distribution of the electrons, and is a function of temperature T. The thermal energy is kT/q, where k is the Boltzmann constant, and q is the electron charge. At room temperature, the current in the best possible transistor needs an increase of input voltage of 60 mV to increase a decade (a 10-fold increase). This limit is referred to as the 60 mV/decade subthreshold slope. Thus, if one uses a supply voltage of 60 mV, the on/off current ratio of the transistor will only be equal to 10. In a chip that contains a billion of transistors, it is mandatory to reduce the off current to the sub-nanoampere level, which requires on/off current ratios in the order of 10^6 and thus a supply voltage of at very least 600 mV (in practice 0.8–1 V to get a decent current drive). Apart from cooling the devices in liquid nitrogen (where they require only 150 mV to achieve an on/off current ratio of 10^6), there is no real practical solution around the problem. Impact ionization in silicon-on-insulator (SOI) transistors has been long known to be capable of reducing the subthreshold slope below 60 mV/decade (Sundaresan and Chen 1990), but impact ionization requires large supply voltages, which defeats the purpose of realizing a low-voltage switch. More promising are devices based on band-to-band tunneling. Here, the variation of current is controlled by the variation of a tunnel barrier rather than by a change in electron concentration, which overcomes the kT/q limit. Such devices have been simulated and made in thin SOI films (Wang et al. 2004) and in carbon nanotubes

(Appenzeller et al. 2004) but present technological challenges such as realizing extremely high doping concentrations in very thin films.

12.9 Single-Electron Transistor

The single-electron transistor (SET) comprises a source, a drain, a gate, and a channel region isolated from the source and drain by tunnel junctions. Since this region is usually very small, we will call it a "dot." When a small voltage is applied across a tunnel junction, it behaves like a capacitor, but electrons can tunnel through the barrier. If a constant voltage is applied to a tunnel junction, periodic current pulses, called "Coulomb oscillations," are produced. If the average current through the structure is I, the frequency of the Coulomb oscillations is equal to $f = I/q$, each pulse caused by the tunneling of a single electron through the barrier. Let us now consider two identical tunnel barriers in series (Figure 12.10A) with initial conditions $V_A = V_D = 0$ V. If voltage V_A is decreased, V_D is equal to $V_A/2$, since $C_1 = C_2 \equiv C$, but no current flows through the structure as long as the potential across C_1 is small enough. At the precise moment where the voltage across C_1 becomes large enough for an electron to tunnel through the junction, the voltage across C_1 is equal to $V_{At} - V_{At}/2 = V_{At}/2$. Note that the voltage across C_2 is also equal to $V_{At}/2$. When an electron tunnels into the dot, V_D jumps from $V_{At}/2$ to $V_{At}/2 - qC_2$ (V_A and V_D are negative values). At that moment, the voltage across C_1 drops to $V_{At}/2 + qC$ that is smaller than $V_{At}/2$ in absolute value, which inhibits the tunneling of another electron from the source into the dot. The voltage across C_2, on the other hand, is equal to $V_{At}/2 - qC$ that is larger than $V_{At}/2$ in absolute value, and the electron that was "stored" in the dot can now tunnel through C_2, giving rise to a drain current. In that process, the potential across C_1 is now increased above the critical value for tunneling, and the process repeats itself. The device is thus conducting if the applied voltage is larger, in absolute value, than a critical value, V_{At}. A more detailed analysis concludes that $V_{At} = -q/2C$ (Wasshuber et al. 1997). The region where no current flow is permitted ($V_{At} < V_A < -V_{At}$ with $V_{At} < 0$) is called the "Coulomb gap" (Figure 12.10B).

If we add a gate electrode that is capacitively coupled with the dot, we obtain a single-electron transistor. The dot potential, and thus the current flow, can now be modulated by the gate voltage. Depending on the gate voltage a discrete number of electrons

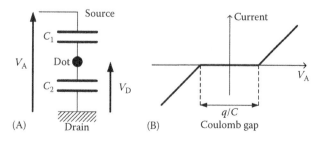

FIGURE 12.10 (A) Two tunnel barriers, C_1 and C_2, in series. (B) I–V characteristics of a double tunnel barrier structure.

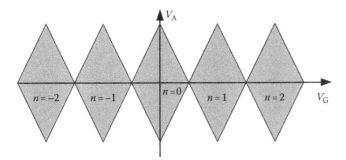

FIGURE 12.11 Current flow in an SET in the V_G–V_A plane. The shadowed areas indicate regions of the V_G–V_A plane where Coulomb blockade prevents current flow and $n = -2, -1, 0, 1, 2$ indicates the number of electrons stored in the dot. Outside these regions, the current increases with the gate voltage.

(or holes, $n = ...-2, -1, 0, 1, 2,...$) can be stored in the dot, which produces a series of Coulomb gaps. These can be observed in the V_G–V_A plane as a series of rhombus-shaped regions where no current flows through the device (Figure 12.11).

12.10 Spin Transistor

Another nonclassical device is based on the manipulation of information not by a transfer of charge, but by a transfer or modification of electron spin. "Spintronics," in analogy with "electronics" groups a series of devices that use electron spin instead of charge flow to carry information. One advantage of spin over charge is that spin can be easily manipulated by externally applied magnetic fields, a property that has been utilized in magnetic storage technology for many years (Das Sarma 2001). Spin transistors involve the injection of spin-polarized electrons from an emitter made out of a ferroelectric material into a semiconductor. A ferroelectric gate is used to either conserve the spin or reverse it as the electrons flow through toward the collector. The collector is ferroelectric, with the same spin as the source. The electrons will reach the collector if their original spin has been conserved, but if they are no longer aligned with the direction of magnetization of the collector, no current can pass.

12.11 Molecular Tunnel Junctions

The idea using single molecules as circuit elements along with a proposal for building a molecular rectifier dates back over 30 years (Aviram and Ratner 1974). However, experimental developments to explore the use of single molecules in electronics did not significantly advance until the introduction of scanning probe microscopy (SPM) and the use of break junctions to measure current voltage characteristics for nanometer scale objects. Measurement of currents across single molecules has been investigated for just over a decade (Reed et al. 1997). State of the art photo- and e-beam lithography is not able to reliably produce metal lines with gap spacing of less than 1 nm as needed to contact single molecules. SPM and break junction methods allow for

molecules to span gaps of a nanometer or less formed between two metal electrodes. The motivation for using SPM and break junctions to form molecular tunnel junctions is similar, and we briefly describe the formation of a break junction gap used in the investigation of molecular electronics.

In Figure 12.12, the idea behind a break junction is illustrated. Two triangular metal regions are patterned on a polymer substrate and they meet at the "apex" formed by each of the triangular regions. A piezoelectrically activated piston stretches the polymer layer leading to the "breaking" of the metal junction. Since piezoelectric actuation can be controlled on an Ångström-scale, a gap spacing of a few atomic lengths can separate the metallic contacts formed. Many ingenious methods for trapping molecules in the resulting gaps have been devised, but it remains true that the detailed structure of the resulting molecular junctions remains largely undetermined. Given the ability to trap single molecules in the gaps and to measure the current–voltage characteristics for simple tunnel junctions, many experiments and calculations of the currents across molecules have been performed (Cuniberti et al. 2005). A typical atomistic view of a single molecule bonded between two metal electrodes is given in Figure 12.13.

Although there remain discrepancies between different measurements, different theoretical treatments, and between theory and experiment, there is convergence and improvement in the

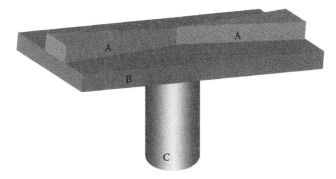

FIGURE 12.12 Schematic of a break junction experiment: Two triangular metal patterns A are deposited on a polymer substrate B and meet at a "point." The piston C under the substrate is piezoelectrically actuated causing the polymer layer to stretch resulting in the metal contacts to separate producing gaps of less than 1 nm.

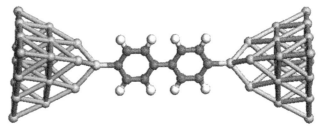

FIGURE 12.13 A "ball and stick" representation of the atomic positions and bonding structure of a 4,4′-biphenyldithiol (BPD) molecule bonded between two metal electrodes. Electrode atoms are metals, often gold is used. The "apex" of the electrode is a "linker", in this case a sulphur atom. The two molecular rings are formed by carbon atoms and "saturated" by hydrogen bonds.

reproducibility of the measurements of current–voltage characterisitics in the experimental measurements on, for example, simple prototypical molecular junctions such as benzene dithiol (BDT) molecule, which is a benzene ring attached to the metal contacts through sulfur end groups (Reed et al. 1997, Dadosh et al. 2005, Lörtscher et al. 2007, Martin et al. 2008). Current experimental estimates for the resistance for a BDT molecule bonded between gold electrodes range between several MΩ to tens of MΩ. Early theoretical predictions often predicted currents orders of magnitude higher than those experimentally observed, and at least some of the reasons for these discrepancies have been identified and resolved (Delaney and Greer 2004, Lindsay and Ratner 2007). Theoretical predictions are now capable of predicting currents across molecular tunnel junctions within experimental uncertainties (Fagas and Greer 2007).

In Figure 12.14A, a representation of molecular tunnel junction is given in terms of an energy level diagram. On the left and right of the junction, the electrodes are represented as a set of dense energy levels characteristic of metals. Between the two electrodes, a few widely spaced energy levels represent the molecular region. For typical molecules being investigated in tunnel junctions, the molecular energy level spacings are on the order of an electron volt. If we take the Fermi level as the highest energy level in the electrodes, it is seen that in Figure 12.14A, one of the molecular levels lies above the Fermi energy and the other lies below. Hence,

the highest energy level on the molecule is unoccupied and the lower level is filled. If a small voltage is applied across the junction, and if it is assumed that little current can flow through the lower occupied level (the electrons on the left and right "block" each other and current cannot flow), then it is seen that electrons leaving the electrodes near the Fermi level have no electronic states to flow across. The molecule in this case represents a barrier to the electrons, which must tunnel from the electrode at a lower voltage to that of a higher voltage. In this case, the current flow is described as nonresonant tunneling. As the voltage V across the tunnel is increased, the energy levels in the left and right electrodes are shifted up and down by $\varepsilon + eV/2$ and $\varepsilon - eV/2$, respectively, with the voltage converted to energy by multiplication of the electronic charge e. This leads to a situation where the unoccupied level on the molecule becomes "resonant" or aligned with an electrode energy level on one side of the molecular junction allowing current to flow freely between the electrodes as shown schematically in Figure 12.14B. In this case, the current across the molecule is due to resonant tunneling. These considerations can be quantified by what is commonly referred to as Landauer's formula (Datta 1997, Imry and Landauer 1999)

$$I = \frac{2e}{h} \int T(\varepsilon)[f_{\mathrm{L}}(\varepsilon) - f_{\mathrm{R}}(\varepsilon)]d\varepsilon,$$

with h Planck's constant. In this picture, states occupied in both the left and right electrodes "block" the current flow. This is described as the difference between the Fermi occupations $[f_{\mathrm{L}}(\varepsilon) - f_{\mathrm{R}}(\varepsilon)]$ for each state of energy ε in the left and right electrodes, respectively. For states that are not blocked, current can flow with transmission $T(\varepsilon)$. In Figure 12.14C, a typical current–voltage (IV) characteristic for a molecular junction is shown. At low voltages, the electrode states on one side of the junction become "unblocked" and current can flow, but there is no molecular state in the voltage window that allows electrons to traverse the molecule. The nonresonant tunneling behavior of the junction then implies a low value for the transmission coefficient and a corresponding small current. As a molecular state enters the voltage window as in Figure 12.14B, the transmission coefficient at energy ε increases substantially and results in a sudden increase in the current as the voltage increases. In Figure 12.14C, the onset of the resonant tunneling regime is clearly marked at the sudden increase in current above and below ±2 V.

Much of the recent work in molecular electronics is focused on describing the physics of current flow on a nanometer scale. There is effort needed to understand how to build switching or logic functions into a single molecule at a level that meets the stringent electronics requirements in terms of reproducibility, integration, and reliability, or indeed to pursue new functions for molecular electronics (Galperin et al. 2008). To explore the use of molecular electronics, efficient simulations of the behavior of molecules embedded into electronic circuits needs to be developed in a manner that allows the device models to be incorporated into circuit design methods (Fransson et al. 2006), but

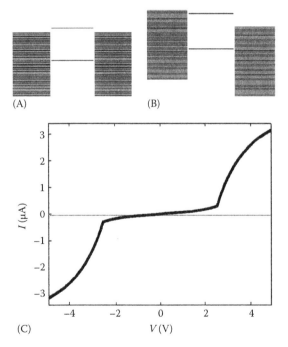

(A)   (B)

(C)

FIGURE 12.14 (A) An energy level diagram of a molecular tunnel junction. The dense set of electron states on the left and right represent the electrodes; the discrete set of widely separated states represents the molecule in the junction. (B) As a voltage is applied, the electrode states are shifted and the unoccupied molecular energy level enters the voltage window. (C) A representative current–voltage characteristic for a molecular tunnel junction. (After Delaney, P. and Greer, J.C., *Phys. Rev. Lett.*, 93, 036805-1, 2004.)

this work can be characterized as in its infancy. The goal in these approaches is to develop analytical models combining mesoscopic quantum transport equations with information about the molecular junctions determined from electronic structure calculations. Unknown physical parameters are then extracted by fitting to measurements of electron transport across molecular tunnel junctions, for example, for the prototypical case of benzene-1,4-dithiol bonded between gold electrodes. The coupling strength of the molecule attached to the metal electrodes is difficult to measure or calculate, so this parameter is extracted by minimizing the difference between the model predictions and measured data. The resulting model can be used to reproduce the *IV* characteristics for a molecule in a manner similar to that used for conventional compact models of transistors.

12.12 Point Contacts and Conductance Quantum

There is a remarkable consequence of Landauer's formula. In Figure 12.15, a pictorial representation of a break junction is given with a single atom bridging between two electrodes. It has been demonstrated using high-resolution transmission electron microscopy that a single atom can be reproducibly trapped in the break junction just prior to breaking of the junction (Rodrigues et al. 2000). For these junctions, a single atomic state can be in resonance with the electrodes and the transmission can be near ideal with $T \approx 1$ within the applied voltage window. Landauer's formula predicts the conductance in this case to be $g = \partial I/\partial V \approx 2e^2/h$, corresponding to a resistance of approximately $12.9\,k\Omega$. This result indicates for a single quantum "channel" with perfectly ballistic transmission $T = 1$, there is an intrinsic resistance known as the "contact resistance." This resistance cannot be eliminated for a quantum conduction channel and in this sense, resistance is quantized per ballistic conducting channel. For metals with many parallel conducting channels, this intrinsically quantum effect is not noticeable. However for quantum scale conductors with a few or only one conducting channel, the contact resistance is significant and has been observed both in low-dimensional semiconductor devices (van Wees et al. 1988, Wharam et al. 1988) and for atomic point contacts (Scheer et al. 1998).

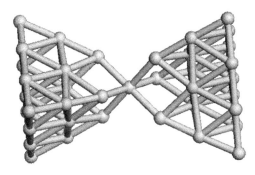

FIGURE 12.15 A "ball and stick" representation of quantum point contact. In this case, a single atom is located at the center of the junction.

12.13 Atomic-Scale Technology Computer-Aided Design

The continued down scaling of MOSFETs and the exploration of nonclassical transistor structures impose new requirements for technology computer-aided design (TCAD). For the "end-of-the-roadmap" silicon devices and for transistors relying on quantum mechanical effects for their operation, explicit atomic-scale structures of the devices and the ability to predict quantum behavior needs to be included in TCAD tools. Methods from computational chemistry and physics are being coupled to TCAD tools for design at the atomic and molecular length scales, and this represents a true change in simulation philosophy. Microelectronics device design was incremental for much of the period 1960–2000; few revolutionary technology changes were needed. Hence, parameterizations and simulations optimized for a production technology generation were adequate to predict properties for the design of the next technology generation. As the new technology generation was developed, re-parameterizations of the models were then made to refine and optimize transistors and circuits. In this way, technology design and manufacture leapfrogged one another forward. Now, the fields of molecular electronics, quantum computation, as well as classical alternatives are actively under investigation as means to overcome anticipated technology limitations. Hence, the nature of TCAD simulation is changing: instead of predicting incremental changes to existing technologies, new TCAD tools must be able to assess proposed candidates from a large set of replacement technologies in order to remain useful. And, useful has a rigorous definition in this field—TCAD simulation must be capable of reliably assessing new technologies, be cheaper and faster than design-by-trial strategies, with the constraints including design cycle time, cost, and accuracy.

The technology simulation community is not starting from scratch in terms of atomic scale and quantum computations. The development of algorithms and programs dates back to early days of quantum theory (Hartree 1928). However, the development of algorithms and programs were not written with nanoelectronic TCAD needs in mind and with means for treating open quantum systems, that is, systems that exchange charge or energy with their environment as needed to describe transistors and other circuit elements driven by voltage sources are currently under development.

In Figure 12.16, a simple graphical representation of Moore's law deduced from the minimum transistor feature size is compared to the system sizes accessible to atomistic and ab initio calculations. Clearly increased computing power and memory resources have allowed for computation and simulation at larger and larger sizes. In the diagram, we have differentiated between ab initio calculations, which directly solve the electronic Schrödinger equation, and much less accurate but computationally cheaper atomistic simulations. Atomistic simulations rely on simple empirical models of atom–atom interactions, but cannot deliver any information about the electronic structure of materials. They require, however, many orders of magnitude

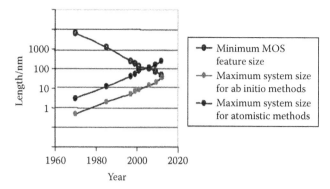

FIGURE 12.16 Convergence of atomistic and quantum mechanical simulations and nanoelectronics length scales.

less computational power for their description of matter and their use in the pharmaceutical and chemical industries is widespread. As a crude rule of thumb, accurate ab initio electronic structure codes can treat systems with a few hundred atoms to thousands of atoms, whereas atomistic methods can handle system sizes up to several tens of millions of atoms. Within a multiscale approach, atomistic simulations can be used to model aspects of the growth and processing of nanoscale structures. The electronic structure and charge transport in these systems can then be determined from ab initio techniques.

12.14 Conclusion

The size of transistors produced during the last 30 years by the microelectronics industry has shrunk by a factor of a hundred, from a few micrometers to a few tens of nanometers. As the limits of classical planar architecture are being reached, new transistor structures based on nanowire geometries are used to push the scaling down to nanometer dimensions. Quantum effects then begin to emerge, and the electrical characteristics of transistors no longer follow the predictions of classical simulation tools. The new behaviors imposed by quantum mechanics, on the other hand, pave the way for a new generation of devices with quantized current transport and nonconventional switching mechanisms.

References

Ando, T., Fowler, A.B., and Stern, F. 1982. Electronic properties of two-dimensional systems. *Review of Modern Physics* 54:437–672.

Andrieu, F., Dupré, C., Rochette, F., Faynot, O., Tosti, L., Buj, C., Rouchouze, E., Cassé, M., Ghyselen, B., Cayrefourcq, I., Brévard, L., Allain, F., Barbé, J.C., Cluzel, J., Vandooren, A., Denorme, S., Ernst, T., Fenouillet-Béranger, C., Jahan, C., Lafond, D., Dansas, H., Previtali, B., Colonna, J.P., Grampeix, H., Gaud, P., Mazuré, C., and Deleonibus, S. 2006. 25 nm short and narrow strained FDSOI with TiN/HfO₂ gate stack. *Symposium on VLSI Technology*, 16.4, Honolulu, HI.

Appenzeller, J., Knoch, J., Martel, R., Derycke, V., Wind, S.J., and Avouris, P. 2002. Carbon nanotube electronics. *IEEE Transactions on Nanotechnology* 1(4):84–89.

Appenzeller, J., Lin, Y.M., Knoch, J., and Avouris, P. 2004. Band-to-band tunneling in carbon nanotube field-effect transistors. *Physical Review Letters* 93(19):196805-1–196805-4.

Aviram, A. and Ratner, M.A. 1974. Molecular rectifiers. *Chemical Physics Letters* 29:277–283.

Baie, X., Colinge, J.P., Bayot, V., and Grivei, E. 1995. Quantum-wire effects in thin and narrow SOI MOSFETs. *Proceedings of the IEEE International SOI Conference*, Tucson, AZ, pp. 66–67.

Bethune, D.S., Klang, M.S., de Vries, M.S., Gorman, G., Savoy, R., Vazquez, J., and Beyers, R. 1993. Cobalt-catalysed growth of carbon nanotubes with single-atomic-layer walls. *Nature* 363:605–607.

Colinge, J.P. 2007a. Quantum-wire effects in trigate SOI MOSFETs. *Solid-State Electronics* 51(9):1153–1160.

Colinge, J.P. 2007b. *FinFETs and Other Multi-Gate Transistors*. Springer, Berlin, Germany.

Cuniberti, G., Fagas, G., and Richter, K. (eds.) 2005. *Introducing Molecular Electronics*, Lecture Notes in Physics. Springer, Berlin, Germany.

Dadosh, T., Gordin, Y., Krahne, R., Khivrich, I., Mahalu, D., Frydman, V., Sperling, J., Yacoby, A., and Bar-Joseph, I. 2005. Measurement of the conductance of single conjugated molecules. *Nature* 436:677–680.

Das Sarma, S. 2001. Spintronics. *American Scientist* 89:516–523.

Datta, S. 1997. *Electronic Transport in Mesoscopic Systems*. Cambridge University Press, Cambridge, U.K.

Davies, J.H. 1998. *The Physics of Low-dimensional Devices*. Cambridge University Press, Cambridge, U.K.

Delaney, P. and Greer, J.C. 2004. Correlated electron transport for molecular electronics. *Physical Review Letters* 93:036805-1–036805-4.

Ding, F., Larsson, P., Larsson, J.A., Ahuja, R., Duan, H., Rosen, A., and Bolton, K. 2008. The importance of strong carbon-metal adhesion for catalytic nucleation of single-wall carbon nanotubes. *Nano Letters* 8:463–468.

Doyle, B.S., Datta, S., Doczy, M., Jin, B., Kavalieros, J., Linton, T., Murthy, A., Rios, R., and Chau, R. 2003. High performance fully-depleted tri-gate CMOS transistors. *IEEE Electron Device Letters* 24(4):263–265.

Fagas, G. and Greer, J.C. 2007. Tunnelling in alkanes anchored to gold electrodes via amine end groups. *Nanotechnology* 18:424010-1–424010-4.

Fransson, J., Bengone, O.M., Larsson, J.A., and Greer, J.C. 2006. A physical compact model for electron transport across single molecules. *IEEE Nanotechnology* 5(6):745–749.

Galperin, M., Ratner, M.A., Nitzan, A., and Troisi, A. 2008. Nuclear coupling and polarization in molecular transport: Beyond tunneling to function. *Science* 319(5866):1056–1060.

Guo, J., Hasan, S., Javey, A., Bosman, G., and Lundstrom, M. 2005. Assessment of high frequency performance of carbon nanotube transistors. *IEEE Transactions on Nanotechnology* 4(6):715–721.

Hartree, D.R. 1928. The wave mechanics of an atom with a non-Coulomb central field. *Proceedings of the Cambridge Philosophical Society* 24:89–132.

Huard, B., Sulpizio, J.A., Stander, N., Todd, K., Yang, B., and Goldhaber-Gordon, D. 2007. Transport measurements across a tunable potential barrier in graphene. *Physical Review Letters* 98:236803-1–236803-4.

Iijima, S. 1991. Helical microtubules of graphitic carbon. *Nature* 354(6348):56–58.

Imry, Y. and Landauer, R. 1999. Conductance viewed as transmission. *Review of Modern Physics* 71:S306–S312.

Javey, A., Guo, J., Farmer, D.B., Wang, Q., Wang, D., Gordon, R.G., Lundstrom, M., and Daih. 2004. Carbon nanotube field-effect transistors with integrated ohmic contacts and high-? gate dielectrics. *Nano Letters* 4(3):447–450.

Kavalieros, J., Doyle, B., Datta, S., Dewey, G., Doczy, M., Jin, B., Lionberger, D., Metz, M., Rachmady, W., Radosavljevic, M., Shah, U., Zelick, N., and Chau, R. 2006. Tri-gate transistor architecture with high-k gate dielectrics, metal gates and strain engineering. *Symposium on VLSI Technology* 7.1, Honolulu, HI.

Kim, R. and Lundstrom, M. 2008. Characteristic features of 1-D ballistic transport in nanowire MOSFETs. *IEEE Transactions on Nanotechnology* 7(6):787–794.

Krivokapic, Z., Tabery, C., Maszara, W., Xiang, Q., and Lin, M.R. 2003. High-.performance 45-nm CMOS technology with 20-nm multi-gate devices. *Extended Abstracts of the International Conference on Solid State Devices and Materials (SSDM)*, Nagoya, Japan, pp. 760–761.

Lee, C.W., Yun, S.R.N., Yu, C.G., Park, J.T., and Colinge, J.P. 2007. Device design guidelines for nano-scale MuGFETs. *Solid-State Electronics* 51(3):505–510.

Lindsay, S.M. and Ratner, M.A. 2007. Molecular transport junctions: Clearing mists. *Advanced Materials* 19:23–31.

Lörtscher, E., Weber, H.B., and Riel, H. 2007. Statistical approach to investigating transport through single molecules. *Physical Review Letters* 98:176807-1–176807-4.

Martel, R., Schmidt, T., Shea, H.R., Hertel, T., and Avouris, P. 1998. Single- and multi-wall carbon nanotube field-effect transistors. *Applied Physics Letters* 73:2447–2449.

Martin, C.A., Ding, D., van der Zant, H.S.J., and van Ruitenbeek, J.M. 2008. Lithographic mechanical break junctions for single-molecule measurements in vacuum: Possibilities and limitations. *New Journal of Physics* 10:065008-1–065008-18.

Moore, G. 1965. Cramming more components onto integrated circuits. *Electronics* 38(8):114–117.

Novoselov, K.S., Geim, A.K., Morozov, S.V., Jiang, D., Zhang, Y., Dubonos, S.V., Grigorieva, I.V., and Firsov, A.A. 2004. Electric field effect in atomically thin carbon films. *Science* 306:666–669.

Novoselov, K.S., Jiang, D., Schedin, F., Booth, T.J., Khotkevich, V.V., Morozov, S.V., and Geim, A.K. 2005. Two-dimensional atomic crystals. *Proceedings of the National Academy of Sciences USA* 102:10451–10453.

Park, J.T., Colinge, J.P., and Diaz, C.H. 2001. Pi-gate SOI MOSFET. *IEEE Electron Device Letters* 22:405–406.

Reed, M.A., Zhou, C., Muller, C.J., Burgin, T.P., and Tour, J.M. 1997. Conductance of a molecular junction. *Science* 278:252–254.

Ritzenthaler, R., Dupré, C., Mescot, X., Faynot, O., Ernst, T., Barbé, J.C., Jahan, C., Brévard, L., Andrieu, F., Deleonibus, S., and Cristoloveanu, S. 2006. Mobility behavior in narrow Ω-gate FET devices. *Proceedings of the IEEE International SOI Conference*, Niagara falls, ON, pp. 77–78.

Rodrigues, V., Fuhrer, T., and Ugarte, D. 2000. Signature of atomic structure in the quantum conductance of gold nanowires. *Physical Review Letters* 85:4124–4127.

Scheer, E., Agraït, N., Cuevas, J.C., Yeyati, A.L., Ludoph, B., Martín-Rodero, A., Bollinger, G.R., van Ruitenbeek, J.M., and Urbina, C. 1998. The signature of chemical valence in the electrical conduction through a single atom contact. *Nature* 394:154–157.

Seidel, R.V., Gvaham, A.P., Kretz, J., Rajasekharan, B., Duesberg, G.S., Liebau, M., Unger, E., Kreupl, F., and Hoenlein, W. 2005. Sub-20nm short channel carbon nanotube transistors. *Nano Letters* 5(1):147–150.

Sekigawa, T. and Hayashi, Y. 1984. Calculated threshold-voltage characteristics of an XMOS transistor having an additional bottom gate. *Solid-State Electronics* 27:827–829.

Singh, N., Agarwal, A., Bera, L.K., Liow, T.Y., Yang, R., Rustagi, S.C., Tung, C.H., Kumar, R., Lo, G.Q., Balasubramanian, N., and Kwong, D.L. 2006. High-performance fully depleted silicon nanowire (diameter<5 nm) gate-all-around CMOS devices. *IEEE Electron Device Letters* 27(5):383–386.

Sundaresan, R. and Chen, C.E. 1990. SIMOX devices and circuits. *Electrochemical Society Proceedings* 90(6):437–454.

Tans, S.J., Verschueren, A.R.M., and Dekker, C. 1998. Room-temperature transistor based on a single carbon nanotube. *Nature* 393:49–52.

Van Wees, B.J., van Houten, H., Beenakker, C.W.J., Williamson, J.G., Kouwenhoven, L.P., van der Marel, D., and Foxon, C.T. 1988. Quantized conductance of point contacts in a two-dimensional electron gas. *Physical Review Letters* 60:848–850.

Wang, P.F., Hilsenbeck, K., Nirschl, T., Oswald, M., Stepper, C., Weis, M., Schmitt-Landsiedel, D., and Hansch, W. 2004. Complementary tunneling transistor for low power application. *Solid-State Electronics* 48:2281–2286.

Wasshuber, C., Kosina, H., and Selberherr, S. 1997. SIMON-A simulator for single-electron tunnel devices and circuits. *IEEE Transactions on Computer-Aided Design of Integrated Circuits and Systems* 16(9):937–944.

Wharam, D.A., Thornton, T.J., Newbury, R., Pepper, M., Ahmed, H., Frost, J.E.F., Hasko, D.G., Peacock, D.C., Ritchie, D.A., and Jones, G.A.C. 1988. One-dimensional transport and the quantisation of the ballistic resistance. *Journal of Physics C* 21:L209–L214.

Xiong, W., Ramkumar, K., Jamg, S.J., Park, J.T., and Colinge, J.P. 2002. Self-aligned ground-plane FDSOI MOSFET. *Proceedings of the IEEE International SOI Conference*, San Jose, CA, pp. 23–24.

Yang, F.L., Chen, H.Y., Cheng, F.C., Huang, C.C., Chang, C.Y., Chiu, H.K., Lee, C.C., Chen, C.C., Huang, H.T., Chen, C.J., Tao, H.J., Yeo, Y.C., Liang, M.S., and Hu, C. 2002. 25 nm CMOS Omega FETs. *Technical Digest of IEDM*, New York, pp. 255–258.

13

Metal Nanolayer-Base Transistor

André Avelino Pasa
Universidade Federal de Santa Catarina

13.1 Introduction

A metal nanolayer-base transistor is a transistor with a metallic layer with nanometric thickness sandwiched between two semiconductor layers. The metal forms the base and the semiconductors the emitter and the collector. This device was proposed and demonstrated at the beginning of the 1960s by Rose (1960), Geppert (1962), and Atalla and Kahng (1962). The main characteristic of the device for applied purposes was the expected high operation frequency when compared to bipolar transistors with low resistance and short transit times in the thin metallic nanolayer base.

13.2 Band Diagram

The mechanism of the metal-base transistor can be understood by drawing the energy band diagram of the sandwich structure with emphasis on the interfaces. The contact between a metal and a semiconductor forms a metal–semiconductor (MS) junction that has a special electric characteristic described independently by Schottky (1938) and Mott (1938). These junctions have a rectifying electrical behavior due to the presence of a potential barrier at the interface between the two materials. In Figure 13.1, we illustrate two different ways of performing MS contacts. The most common one at the time when metal-base transistors (MBT) emerged was by pressing a metallic tip against the surface of the semiconductor (a), known as the point contact, and the most reliable one used currently (b), obtained by depositing a metallic layer on top of a semiconductor substrate.

The band diagram of an MS contact is displayed in Figure 13.2a. The metal, represented by the energy of the Fermi level (E_{Fm}) and the semiconductor, represented by the bottom of the conduction band (E_C), and the top of the valence band (E_V), are in contact and under equilibrium conditions. The Fermi level in the semiconductor side (E_{FS}) is close to E_C, since we are considering an n-type material. In this case, the barrier at the interface ($q\phi_b$), named the Schottky barrier, is due to the flow of electrons from the semiconductor to the metal to attain the equilibrium conditions, i.e., $E_{FM} - E_{FS}$ along the two materials corresponding in solid-state physics to the condition of the same chemical potential through the whole system. Electrons flow to the metal instead of from the metal to the semiconductor, since we are considering the work function of the metal ($q\phi_m$) to be higher than the semiconductor affinity ($q\chi_s$). The band banding in the semiconductor is then a consequence of the flow of electrons and leads to a depletion region and the appearance of an electric field close to the interface. The field is contrary to that flow and reaches its maximum intensity in the equilibrium condition. Figure 13.2b and c display the band banding conditions under direct and reversed bias, respectively. Electrons at E_{FM} have to overcome the Schottky barrier with a height of $q\phi_b$, which is independent of the applied voltage, to cross the interface. On the other hand, electrons at the conduction band of the semiconductor are favored to flow to the metal by applying a forward bias as shown in Figure 13.2b. Due to the fact that the flow from the semiconductor to the metal is facilitated by the external voltage *V*, the MS interface acts as a diode rectifying the current, and the device is named Schottky diode. From the point of view of fundamental physics, this device is known as a hot-carrier diode. The carriers are injected with energy higher than the energy of the Fermi level. In the case of silicon, the most used semiconductor in the microelectronic industry, the height of the Schottky barrier varies in the range of 0.5 and 1.0 eV, this being the energy in excess of the carriers entering the metal. In Figure 13.2c, the diode is a reverse bias and no flux of charge is expected from both sides of the barrier. However, in real devices, a reverse current is always observed with a magnitude depending on the thermal ejection of carriers over the barrier, and most important, on the existence of defects that locally control the transport mechanism. Band diagrams similar to the ones shown in Figure 13.2 can also be drawn for p-type semiconductors.

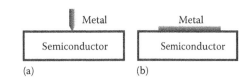

FIGURE 13.1 (a) Metal semiconductor contact obtained by pressing a metallic tip against the semiconductor surface (point contact). (b) MS contact formed by depositing a metallic layer on the surface of a semiconducting substrate.

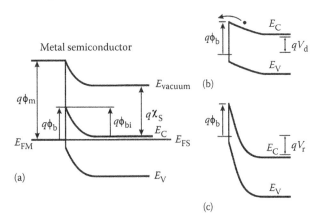

FIGURE 13.2 (a) Energy band diagram of the MS contact with a Schottky barrier at the interface. (b) and (c) Shift of conduction and valence bands of the semiconductor under influence of an external bias. The quantities indicated in the figure are: E_{FM}, Fermi level of the metal; $q\phi_b$, Schottky barrier; $q\chi_s$, electron affinity; $q\psi_{bi}$, built in potential; E_{FS}, Fermi level of the semiconductor; E_C, energy of the bottom of the conduction band; E_V, energy of the top of the valence band; E_{vacuum}, energy of the vacuum level; and q is just the electronic charge.

The current density J across an MS interface under the application of the external bias voltage V is described by the thermionic emission (TE) theory (Rhoderick and Williams 1988), which depicts the transport of carriers injected over the interface barrier $q\phi_b$, given by

$$J = J_o e^{\frac{qV}{nkT}} \left[1 - e^{\frac{-qV}{kT}} \right],$$ (13.1)

where
  n is the ideality factor
  k Boltzmann's constant
  T the absolute temperature
  J_o the saturation current density defined as

$$J_o = A^* T^2 e^{\frac{-q\phi_b}{kT}},$$ (13.2)

where A^* is Richardson's constant, given by

$$A^* = \frac{4\pi q m^* k^2}{h^3},$$ (13.3)

where
  m^* is the effective mass
  h is Planck's constant

For silicon, A^* is equal to 32 A/cm² K² for the p-type and 110 A/cm² K² for the n-type. The ideality factor n, equal to 1 corresponds to the case when the TE theory ideally describes the transport of charge over the potential barrier. For $n > 1$ are represented situations with deviations from the ideal case and are originated by the presence of, for example, inhomogeneities and interfacial layers.

Figure 13.3a shows the characteristic J–V curve of a Co/Si-n Schottky diode. The Co layer with a thickness of 80 nm was electrodeposited and the contact area was 0.5 cm². The curve follows the expected behavior of the TE theory and barrier heights and ideality factors of about 0.6 eV and 1.2, respectively, were obtained for this system (Zandonay et al. 2008). Figure 13.3b shows the characteristic of the diode when the logarithm function is applied to absolute values of current, i.e., the logarithm function was applied to the positive current values in the positive voltage range and to the negative current values, multiplied by −1, in the negative voltage range. This typical representation of the diode response is easily obtained through the observation of the magnitude of the reverse current and the linear interval of the forward current that follows the exponential dependence on the voltage predicted by the TE theory.

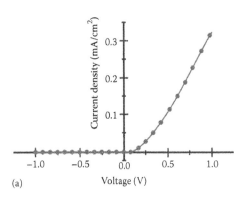

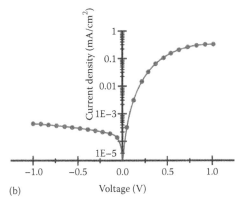

FIGURE 13.3 Linear (a) and logarithm (b) representation of the J–V electric characteristic of a Co/Si-n electrodeposited Schottky diode with a thickness of Co of about 80 nm and $q\phi_b = 0.60$ eV and $n = 1.2$. (After Zandonay, R. et al., *ECS Trans.*, 14, 359, 2008.)

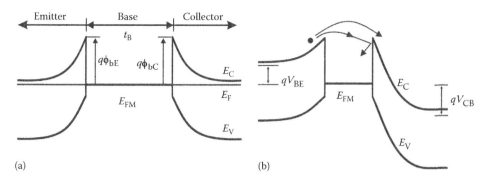

FIGURE 13.4 Band diagram of the MBT consisting of a semiconductor-metal-semiconductor planar trilayer structure under equilibrium conditions (a) and under external bias (b). Emitter-base diode forward biased and collector-base diode reverse biased. V_{CB} and V_{EB} are the applied voltages at the collector and the emitter relative to the base, $q\phi_{bE}$ and $q\phi_{bC}$ are the Schottky barrier heights at the emitter-base and collector-base interfaces, respectively, and t_B is the thickness of the base.

In an MBT, the metal layer of the base forms two Schottky barriers, as shown in Figure 13.4a, one with the emitter, and one with the collector. The barriers are placed face-to-face and can be polarized independently by applying voltages to the emitter and collector terminals relative to the base terminal. The forward operation mode of the transistor consists in applying a forward voltage at the emitter-base diode, to inject carriers into the base, and a reverse bias at the collector-base diode, in order to assist the collection of carriers crossing the base, as displayed in Figure 13.4b.

For the case of n-type semiconductors, as illustrated in Figure 13.4, the electrons ejected by the emitter have to have enough energy to cross the base and overcome the barrier at collector side. In order to have current gains close to 1, the collector barrier height should be lower than the barrier height at the emitter side and the base thickness should be relatively thin. In other words, the number of electrons that reaches the collector is reduced by processes occurring when traveling from the emitter to the collector through the base, which could be expressed in general as

$$J_C = \alpha_T J_E, \tag{13.4}$$

where α_T is the total transport factor of hot electrons that takes into account a series of individual factors, such as electron scattering in the metal base, quantum-mechanical reflection at the base-collector interface, and emission and collection efficiency due to interaction in the emitter and the collector interfaces (Sze 1969). All these processes contribute toward reducing the current that is measured in the collector-base circuit. Specifically, the base-transport factor, α_B, responsible for the scattering of the electrons in the base by collisions with electrons, phonons, impurities, and defects, can be written as

$$\alpha_B = e^{-\frac{t_B}{\lambda}}, \tag{13.5}$$

where λ is the mean free path of the injected hot electrons into the base. This dependency indicates that α_B, and consequently

α_T, is strongly affected by the base thickness. Typical values for λ are a few tens of nanometers as $\lambda = 22$ nm observed for electrons with a 1 eV crossing Au layer sandwiched between Si-n and Ge-n at room temperature (Sze 1969).

13.3 *J–V* Characteristic

Similarly to bipolar transistors, the MBTs can be characterized electrically by measuring the performance of the device in the three different configurations: common-base, common-emitter, and common-collector. The common-base configuration is the most commonly used to obtain the electrical response of the transistor and to determine the common-base current gain α_o, defined by the ratio between collector– emitter current densities, i.e.,

$$\alpha_o = \frac{J_C}{J_E} = \alpha_T - \frac{J_{reverse}}{J_E}, \tag{13.6}$$

where $J_{reverse}$ is the density of the current of the reverse biased collector-base diode. When the ratio between the reverse and the emitter current densities is considerable low, it can be neglected and $\alpha_o \approx \alpha_T$. In the common-base configuration, a fixed current is applied to the emitter and the collector-current density J_C is measured as a function of the collector-to-base voltage (V_{CB}). Figure 13.5 shows the vertical structure of the MBT, with emitter, base, and collector, in planar technology, and the circuit used for common-base measurements.

The transistor action for the common-base measurement corresponds to collector currents weakly dependent on the applied voltage between the collector and the base and strongly dependent on the emitter current. Figure 13.6 shows the typical response of a metal-base transistor for four different emitter currents densities, J_{E1} to J_{E4} with increasing values, showing a small increase of the collector current with the increase of the collector-to-base voltage, and the defined stages for the collector current that are determined by the intensity of the emitter current. The weak dependence of J_C on V_{CB} is attributed to a barrier height lowering with the reverse applied voltage at the collector Schottky diode

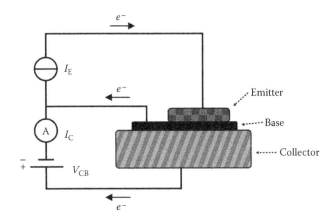

FIGURE 13.5 Schematic representation of a vertical structure with the emitter-base-collector of an MBT in planar technology. The bias circuit for common-base characterization is also shown. I_E is the current source connected in series to the emitter-base circuit and A is the current meter connected in series to the collector-base circuit that measures I_C, the collector current, as a function of voltage V_{CB} between the collector and the base.

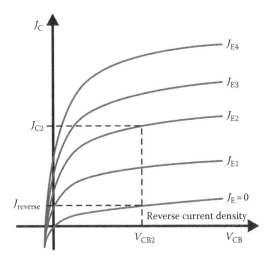

FIGURE 13.6 Typical common-base characteristic of an n-metal-n transistor with collector current levels determined by the intensity of the emitter current and weakly dependent on the collector-to-base reverse voltage.

(Sze 1969), and for this reason, the current gain has to be calculated for a fixed voltage as shown in the figure, where

$$\alpha_o = \frac{J_{C2}}{J_{E2}} - \frac{J_{reverse}}{J_{E2}}. \qquad (13.7)$$

13.4 Fabrication

The main concern related to metal-base transistors is the fabrication process. It is necessary to prepare a structure with metal-semiconductor interfaces that have Schottky barriers and reduced losses for the electrons ejected by the emitter. Electronic devices are usually based on single crystal semiconductors and

the growth of semiconductor epitaxial layers on top of metals is a difficult issue, requiring sophisticated methods. Epitaxial growth is a process whereby the atoms reaching the surface are deposited following the same atomic structure (order) of the crystalline substrate. Molecular beam epitaxy (MBE), liquid phase epitaxy (LPE), and chemical vapor deposition (CVD) are techniques that allow the growth of monolithic structures. This kind of process increases the production cost and reduces possibilities of large-scale integration (circuits with a large density of devices), requiring ultrahigh vacuum deposition or high temperature fabrication steps.

The pioneering work of Geppert (1962) and Atalla and Kahng (1962) were performed by depositing metal layers (base) on top of single crystal semiconductor substrates (collector). The emitter was a crystalline semiconductor piece with a tip or a blunt corner pressed against the metallic layer (point contact). Geppert electrodeposited Cu on n-Ge and contacted with an n-Ge probe. Atalla and Kahng deposited Au thin films with thicknesses of 10 nm on n-Ge and the point contact was n-Si. A few years later, Kanô (1964) used a vacuum-evaporated emitter, CdSe, an Au thin layer as base, and Ge as the collector to fabricate a structure with a Schottky barrier height higher at the emitter (0.75 eV) and lower at the collector (0.45 eV). However, in all these cases described above, the achieved current gains were not more than 0.5, and others factors, such as reproducibility and rough interfaces limited the performance of the device for applications.

To improve the quality of the interfaces, especially since one of the main issues to be worked out was the point contacts with a low yield of reproducibility, different approaches were tried. Brodsky and Deneuville (Deneuville and Brodsky 1978) grew amorphous semiconductor layers to surmount the need for monolithic structures. The glow discharge-deposition process was used for depositing hydrogenated silicon carbide (a-CSi:H) as the emitter, and hydrogenated silicon (a-Si:H) as the collector. The base was a Pt layer. The barrier heights at the emitter and the collector were 0.96 and 0.75 eV, respectively. Though the fact that the heights are adequate for the collection process, the common-base current gain measured was not higher than 0.1.

Moreover, no significant increase in the transfer ratio was observed by growing the semiconductor-metal-semiconductor (SMS) structure using the epitaxy techniques mentioned above. Rosencher et al. (1984) and Hensel et al. (1985) grew monolithic Si-CoSi$_2$-Si structures and obtained transistors with characteristic similar to the one shown in Figure 13.6. The monolithic atomic structure obtained by these authors is shown in Figure 13.7. CoSi$_2$ is a metallic disilicide with a resistivity comparable to metals and with a small lattice mismatch (~1.4%) with Si allowing the epitaxial growth.

Tung et al. (1986) observed high values of α_o, as high as 0.95, with MBTs grown by MBE. However, they also detected the existence of pinholes in the base that directly connected the emitter to the collector. The electron transfer was in this case mainly through the small openings in the base layer, where the barrier height is lower than the Schottky height. This mechanism of transport is the operation principle behind the permeable-base

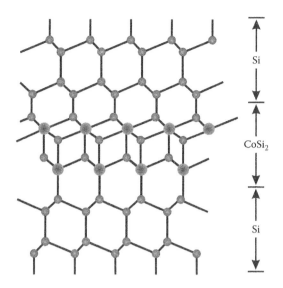

FIGURE 13.7 Atomic arrangement of the monocrystalline Si-CoSi$_2$-Si structure.

transistors (PMT), which incorporate a metal grid as base between the emitter and the collector (Bozler and Alley 1980). Even though the Si/CoSi$_2$/Si MBT showed a useful and a controllable current gain, its fabrication remained a challenge, requiring ultrahigh vacuum deposition and annealing.

The presence of pinholes in the base needs to be checked in all cases in order to see if the measured α_o is not a misleading interpretation of the direct contact between the emitter and the collector through the pinholes. A checking procedure was suggested by Rosencher et al. (1986) where the voltage between the emitter and the base is measured as a function of the collector-to-base voltage. The absence of pinholes leads to an independent voltage at the emitter-base junction, which is screened by the compact metal of the base.

A very significant innovation concerning metal nanolayer-base transistors was achieved with the advent of the spin transistor. Monsma et al. (1995) fabricated a metal-base transistor with a magnetic multilayer at the base. Magnetic multilayers are composed of ferromagnetic (F) and non-ferromagnetic (N) layers,

with individual layer thicknesses of a few nanometers, and show electrical resistance depending on the applied magnetic field, a phenomenon known as giant magnetoresistance (GMR) (Baibich et al. 1988). Figure 13.8a shows the usual band structure of a metal-base transistor with the magnetic multilayer (MM) represented at the base. The difference is that in this case, the base-transport factor will also depend on the applied magnetic field, i.e., $\alpha_B = \alpha_B(H)$. For a ferromagnetic configuration of the multilayer, represented by ($\uparrow\uparrow$) and indicating the same magnetization direction (parallel) for the ferromagnetic layers, the magnetoresistance (MR) is low, and for anti-ferromagnetic, with the non-ferromagnetic layers aligned anti-parallel ($\uparrow\downarrow$), the MR is high. These two MR levels correspond to distinct α_B, $\alpha_B(\uparrow\uparrow) < \alpha_B(\uparrow\uparrow)$, since lower values of MR correspond to less scattering of the electrons when crossing the base, and consequently higher values of collector current. Figure 13.8b shows a common-base characteristic curve of a spin transistor for both configurations of the MM, with the excess current for the ferromagnetic configuration named magnetocurrent, quantified as the difference $\Delta J = J(\uparrow\uparrow) - J(\uparrow\downarrow)$. The magnetocurrent effect is calculated as MC(%) = 100 [$(J(\uparrow\uparrow) - J(\uparrow\downarrow)/J(\uparrow\downarrow)$)]. The device demonstrated by Monsma et al. (1995) had initially a Co/Cu/Co multilayer at the base, with Co as ferromagnetic and Cu as non-ferromagnetic materials, and posteriorly a Pt layer was added to the emitter interface to tailor the barrier height relative to the collector interface (Monsma et al. 1998).

The magnetoresistance phenomenon is an effect observed at nanometer dimensions. It was first described in metallic nanolayers (Baibich et al. 1988) and subsequently in granular systems (Berkowitz et al. 1992, Xiao et al. 1992). It can be understood by assuming the model proposed by Mott (1964) that considers two channels for the conduction of electrons in ferromagnetic metals, one for spin up and one for spin down. The electrons transversally crossing a magnetic multilayer are assumed to suffer scattering in the layer with a magnetization contrary to their spin orientation. At low magnetic fields, when the magnetic multilayer is supposed to be in the anti-ferromagnetic alignment, the electron of both spin orientations will suffer scattering in alternate layers, and a higher MR will be measured. For high

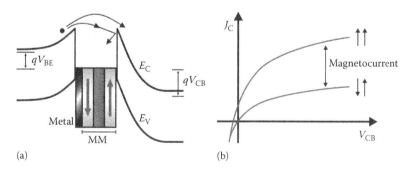

FIGURE 13.8 (a) Band diagram of an MBT with an F/N/F magnetic multilayer (MM) in the base. A metal was added at the emitter interface to tailor the barrier height in the emitter relative to the collector. The arrows indicate the direction of the magnetization of the layers. (b) Common-base characteristic of the spin transistor collector current for both configurations of the multilayer, ferromagnetic ($\uparrow\uparrow$), and antiferromagnetic ($\uparrow\downarrow$). (After Monsma, D.J. et al., *Phys. Rev. Lett.*, 74, 5260, 1995; Monsma, D.J. et al., *Science*, 281, 407, 1998.)

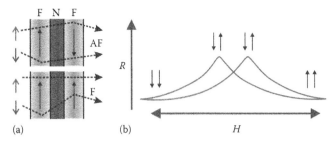

FIGURE 13.9 (a) Model of two channels and antiferromagnetic ($\uparrow\downarrow$) and ferromagnetic ($\uparrow\uparrow$) ordering in magnetic multilayers. (b) Typical curve for the magnetoresistance versus the applied magnetic field in magnetic multilayers.

magnetic fields, when the MM is ferromagnetic ordered, the electrons in the channel with the same orientation of the magnetization of the F layers will not suffer scattering by crossing the structure, and the electrons in the other channel will be scattered at each F layer. Since the channels are parallel, the total resistance will be reduced by the existence of the channel aligned with the magnetization of the ferromagnetic ordering of the layers. The alignment of the MM will have lower resistance, and the difference between both magnetic ordering is the magnetoresistive effect. Similarly to the magnetocurrent effect defined above, the MR effect can be written as MR(%) = 100 [($I(\uparrow\downarrow) - I(\uparrow\uparrow)/I(\uparrow\uparrow)$)]. Figure 13.9a illustrates the scattering of electrons through the multilayered structure for both magnetic alignments, and in Figure 13.9b, the corresponding magnetoresistance curve when the field is varied from negative to positive values is shown.

The main disadvantages of the spin transistor proposed by Momsma et al. (1995) were related to the fabrication process that requires a vacuum bonding in order to sandwich the ferromagnetic multilayer between two-device quality silicon substrates, and the low transport factor observed (<10^{-4}). However, an easy method of fabricating MBTs with high current gains remained elusive.

A significant progress toward the fabrication process of TBMs was achieved by Meruvia et al. (2004) by using an organic

semiconductor as the top layer in the vertical structure shown in Figure 13.5. A current gain as high as 1 was observed in the MBT with fullerene (C_{60}) as the emitter, Au thin layers as the base, and n-type silicon as the collector. The device is easy to fabricate since the organic layers can be prepared by regular methods such as electrodeposition, spin coating, and evaporation, and the approach opened the field of MBTs to hybrid structures with conducting polymers as the emitter and also as a pseudo-metallic base (Meruvia et al. 2005). The energy-band diagram of organic semiconductors is analogous to the one shown in Figure 13.4 for inorganic MBTs with barriers at interfaces and for the injection of hot carriers.

Most of the work expended on metal nanolayer-base transistors was in n-type devices. A p-type MBT was achieved by electrodepositing cuprous oxide (Cu_2O) (Delatorre et al. 2006). This semiconducting oxide is a p-type material with a 2.1 eV band-gap, a factor 2 higher than the band-gap of silicon of 1.1 eV. Figure 13.10 shows characteristic common-base curves obtained with the Cu_2O/Co/Si metal-base transistors with Co nanolayers of (a) 29 nm and (b) 33 nm, which display the transistor action in the third quadrant of the graphic as expected for p-metal-p devices.

The differences observed in Figure 13.10 by comparing the family of curves for the two base thicknesses used, 29 and 33 nm, indicate that the current gain is not only dependent on the thickness of the metallic layer but also that different transport mechanism are acting. For the 29 nm case, the collector current saturates at levels that are proportional to the emitter current, corresponding to a constant current gain. However, for 33 nm the I_C plateaus are not linear with I_E. When the test suggested in reference Rosencher et al. (1986) and described above to detect the presence of pinholes was applied to this p-metal-p transistor, a dependence of V_{EB} on V_{CB} was found. As can be seen in Figure 13.11a, for the thinner base layer, a significant dependence of V_{EB} on V_{CB} is observed. This dependency is a signature of the presence of pinholes in the base, which were also confirmed by the atomic force microscopy image shown in Figure 13.11b. The dark regions correspond to areas where no electrodeposition occurred, acting then as pinholes, allowing a direct contact of

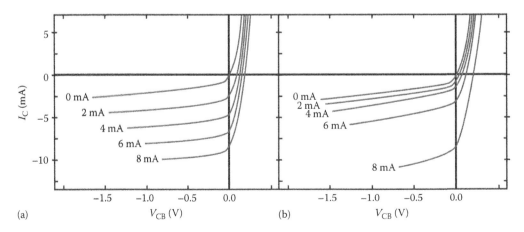

FIGURE 13.10 Common-base characteristics of a p-type metal transistor with Cu_2O as emitter, p-type silicon as collector, and Co with thicknesses of (a) 29 nm and (b) 33 nm as base. (After Delatorre, R.G. et al., *Appl. Phys. Lett.*, 88, 233504-1, 2006. With permission.)

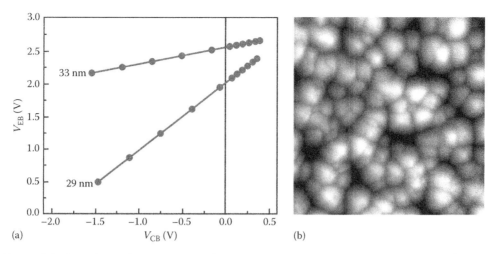

FIGURE 13.11 (a) Test of the dependency of V_{EB} on V_{CB} to check for the existence of pinholes in the base. (b) Atomic force microscopy of the Co metallic base with dark areas corresponding to regions not covered by the metal. (After Delatorre, R.G. et al., *Appl. Phys. Lett.*, 88, 233504-1, 2006. With permission.)

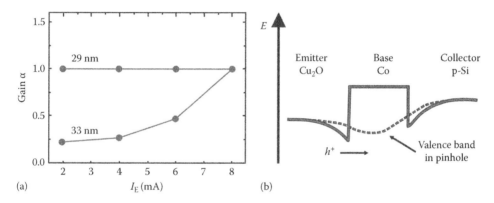

FIGURE 13.12 (a) Dependency of the current gain in the common-base configuration for two different Co thicknesses (29 and 33 nm). (b) Band diagram for the p-type metal base transistor with the two Schottky barriers and the expected barrier (dotted line) in the pinholes where the emitter and the collector are in contact, as expected for a permeable-base transistor. (After Delatorre, R.G. et al., *Appl. Phys. Lett.*, 88, 233504-1, 2006. With permission.)

the emitter with the collector. The bright areas are the Co grains. By increasing the base thickness, a considerable reduction in the pinholes occurred, since a lower dependence of V_{EB} on V_{CB} is observed for 33 nm in Figure 13.11a.

Figure 13.12a shows the dependency of the gain as a function of the emitter current for the two thicknesses of Co. The gain is equal to 1 and is independent of the emitter current for the case where the transistor has a higher density of pinholes. For a lower density of pinholes, the dependency of the current gain with the emitter current indicates that the gain increases with the energy of the ejected electron, i.e., with an increase of the emitter current, which is compatible with the view of electrons with higher energies being collected more easily at the collector interface. Figure 13.12b depicts the expected energy barriers for the p-type metal-base transistor in the case of two Schottky interfaces in the Cu_2O/Co/Si structure and the gradual variation of the barrier in the pinholes where the Cu_2O is in contact with p-Si. With an increasing base thicknesses, the MBT action dominates the PMT one, and on increasing the emitter current, higher gains

should be observed since the energy of the electrons is higher, favoring the overcome of the collector to the base barrier.

References

Atalla, M. M. and Kahng, D. 1962. A new hot electron triode structure with semiconductor metal emitter. *IEEE Transactions on Electron Devices* ED-9: 507–508.

Baibich, M. N., Broto, J. M., Fert, A., Nguyen van Dau, F., Petroff, F., Eitenne, P., Creuset, G., Friederich, A., and Chazelas, J. 1988. Giant magnetoresistance of (001)Fe/(001)Cr magnetic superlattices. *Physical Review Letters* 61: 2472–2475.

Berkowitz, A. E., Mitchell, J. R., Carey, M. J., Young, A. P., Zhang, S., Spada, F. E., Parker, F. T., Hutten, A., and Thomas, G. 1992. Giant magnetoresistance in heterogeneous Cu-Co alloys. *Physical Review Letters* 68: 3745–3748.

Bozler, C. O. and Alley, G. D. 1980. Fabrication and numerical simulation of the permeable base transistor. *IEEE Transactions on Electron Devices* ED-27: 1128–1141.

Delatorre, R. G., Munford, M. L., Zandonay, R., Zoldan, V. C., Pasa, A. A., Schwarzacher, W., Meruvia, M. S., and Hümmelgen, I. A. 2006. p-Type metal-base transistor. *Applied Physics Letters* 88: 233504-1–233504-3.

Deneuville, A. and Brodsky, M. H. 1978. Thin-film metal base transistor structure with amorphous silicon. *Thin Solid Films* 55: 137–141.

Geppert, D. V. 1962. The metal base transistor. *Proceedings of the IRE* 50: 1527.

Hensel, J. C., Levi, A. F. J., Tung, R. T., and Gibson, J. M. 1985. Transistor action in Si/CoSi$_2$/Si heterostructures. *Applied Physics Letters* 47: 151–153.

Kanô, G. 1964. Schottky emitter type metal base transistor. *Japanese Journal of Applied Physics* 3: 363–364.

Meruvia, M. S., Hummelgen, I. A., Sartorelli, M. L., Pasa, A. A., and Schwarzacher, W. 2004. Organic-metal-semiconductor transistor with high gain. *Applied Physics Letters* 84: 3978–3980.

Meruvia, M. S., Benvenho, A. R. V., Hümmelgen, I. A., Pasa, A. A., and Schwarzacher, W. 2005. Pseudo-metal-base transistor with high gain. *Applied Physics Letters* 86: 263504-1–263504-3.

Monsma, D. J., Lodder, J. C., Popma, T. J. A., and Dieny, B. 1995. Perpendicular hot electron spin-valve effect in a new magnetic field sensor: The spin-valve transistor. *Physical Review Letters* 74: 5260–5263.

Monsma, D. J., Vlutters, R., and Lodder, J. C. 1998. Room temperature operating spin-valve transistors formed by vacuum bonding. *Science* 281: 407–409.

Mott, N. F. 1938. Note on the contact between a metal and an insulator on semiconductor. *Proceedings of the Cambridge Philosophical Society* 34: 568–572.

Mott, N. F. 1964. Electrons in transition metals. *Advances in Physics* 13: 325.

Rhoderick, E. H. and Williams, R. H. 1988. *Metal-Semiconductor Contacts*, 2nd edn. Oxford, U.K.: Clarendon.

Rose, A. 1960 (June). RCA Interim report 6A.

Rosencher, E., Delage, S., Campidelli, Y., and Arnaud D'Avitaya, F. 1984. Transistor effect in monolithic Si/CoSi$_2$/Si epitaxial structures. *Electronics Letters* 20: 762–764.

Rosencher, E., Badoz, P. A., Pfister, J. C., Arnaud d'Avitaya, F., Vincent, G., and Delage, S. 1986. Study of ballistic transport in Si-CoSi$_2$-Si metal base transistors. *Applied Physics Letters* 49: 271–273.

Schottky, W. 1938. Halbleitertheorie der sperrschicht. *Naturwissenschaften* 26: 843.

Sze, S. M. 1969. *Physics of Semiconductor Devices*. New York: Wiley.

Tung, R. T., Levi, A. F. J., and Gibson, J. M. 1986. Control of a natural permeable CoSi$_2$ base transistor. *Applied Physics Letters* 48: 635–637.

Xiao, J. Q., Jiang, J. S., and Chien, C. L. 1992. Giant magnetoresistance in nonmultilayer magnetic systems. *Physical Review Letters* 68: 3749–3752.

Zandonay, R., Delatorre, R. G., and Pasa, A. A. 2008. Electrical characterization of electrodeposited Co/p-Si Schottky diodes. *ECS Transactions* 14: 359–363.

14

ZnO Nanowire Field-Effect Transistors

Woong-Ki Hong
Gwangju Institute of Science and Technology

Gunho Jo
Gwangju Institute of Science and Technology

Sunghoon Song
Gwangju Institute of Science and Technology

Jongsun Maeng
Gwangju Institute of Science and Technology

Takhee Lee
Gwangju Institute of Science and Technology

14.1 Introduction

The top-down fabrication techniques, which involve photolithographic and subsequent etching techniques, have been used as conventional methods for device scaling in semiconductor industry for integrated electronic systems. However, due to the further downscaling of device size, the top-down approach faces the limitations of photolithographic/etching techniques as well as the increasing costs associated with lithography equipment (Lu and Lieber 2007). One promising method to overcome the limitations with the top-down methods is the bottom-up approach, which is a unique technique for fabricating functional devices, without the use of lithography techniques, through the chemical synthesis, self-assembly, and the manipulation of nanoscale elements such as small molecules, nanoparticles, nanotubes, and nanowires (Agarwal 2008). Recently, for the realization of integrated electronic systems through the bottom-up approach, these nanomaterials have attracted considerable interest due to their unique properties such as large surface-to-volume ratio, carrier and photon confinement, and high sensitivity. In particular, various semiconductor nanowires offer great potential as building blocks for nanoscale electronic and photonic device applications.

Among many semiconductor nanowires, ZnO with the preferentially hexagonal wurzite-type structure has a wide range of properties, from metallic to insulating conductivities (including n- and p-type conductivities), a direct wide band gap (~3.4 eV), large exciton-binding energy (~60 meV), radiation hardness, high transparency, piezoelectricity, room-temperature ferromagnetism, and chemical-sensing effects (Fan and Lu 2005, Klingshirn 2007, Schmidt-Mende and MacManus-Driscoll 2007). In particular, the advantage of large exciton-binding energy and a variety of nanostructures with relative cost-effectiveness provides excellent optical and electrical properties as a promising material for blue and ultraviolet optical devices, compared with GaN and GaN-based materials. Moreover, since ZnO has stronger radiation hardness than other common semiconductor materials such as Si, GaAs, CdS, and GaN, ZnO materials are promising for space applications (Look et al. 1999). In recent decades, due to these remarkable properties of ZnO materials and the demand for further miniaturization by semiconductor technology, tremendous efforts have been devoted to the growth and characterization of one-dimensional semiconductor ZnO nanostructures (or ZnO nanowires) as well as their versatile applications as a potential material for future nanoelectronics. For example, field-effect transistors (FETs) (Goldberger et al. 2005), sensors (Dorfman et al. 2006), light-emitting devices (Bao et al. 2006), solar cells (Law et al. 2005), as well as logic circuits (Park et al. 2005, Yeom et al. 2008) have been extensively investigated and demonstrated.

In particular, since nanowire-based FETs are the fundamental element for nanoelectronics among these versatile applications, ZnO nanowire-based FETs have been fabricated and intensively investigated in a typical back-gate configuration due to fabrication simplicity (Goldberger et al. 2005, Cha et al. 2006, Chang et al. 2006). The early studies of ZnO nanowire FETs have focused on their device performance (Cha et al. 2006, Chang et al. 2006), gas-sensing application (Fan and Lu 2006, Hsueh et al. 2007), photodetection (Kind et al. 2002, Fan et al. 2004a,b, Soci et al. 2007), and chemisorption/photodesorption (Li et al. 2005, Suehiro et al. 2006). However, in order to fabricate the functional devices from the ZnO nanowires, their fundamental properties must be carefully investigated because the presence of a large surface and interfacial area in nanowires can profoundly alter their performance (Hong et al. 2007, Xiong et al. 2007). Recently, there are a few studies on the role of geometric properties, surface states, and passivation on the transport properties of ZnO nanowire FETs, specifically on the influence of the presence of surface trap states at the interfaces associated with nanowire size and surface roughness (Hong et al. 2007, 2008a,b,c). The interfaces and interfacial states related to interface roughness play an important role not only in nanowire heterostructures, but also in the electronic transport properties of nanoscale transistors as fundamental elements for nanoelectronics (Hastas et al. 2002, Wang et al. 2005, Jung et al. 2007, Mieszawska et al. 2007). In addition, it is well known that metal-oxide semiconductors including ZnO nanostructures are strongly affected by the chemical adsorptions of ambient gases (Lagowski et al. 1977, Kolmakov and Moskovits 2004). Therefore, it is essential to understand the surface-roughness- and size-dependent effects on the electronic transport properties of nanowire transistors. Moreover, since ZnO nanowires are strongly influenced by chemical environments due to large surface-to-volume ratio (Goldberger et al. 2005, Li et al. 2005, Fan and Lu 2006, Suehiro et al. 2006, Hsueh et al. 2007), it is important to study the passivation effects of dielectric layer on the transport properties correlated to geometry and adsorbed species on the ZnO nanowires.

In this chapter, we review the basic electrical properties of FETs made from ZnO nanowires that were grown by a vapor transport method. We demonstrate that the transport properties of ZnO nanowire FETs are associated with the surface roughness, nanowire size, surface states, and/or defects, as well as the surface chemistry in surrounding environments. These explorations will give insights not only in understanding the transport properties but also in developing practically useful applications of nanowire-based devices.

14.2 Background

14.2.1 Basic Physical Parameters of ZnO

The ZnO in wurzite structure, which can be considered as a hexagonal close-packed lattice of zinc atoms tetrahedrally coordinated with four oxygen atoms, has the primitive translation vectors $\mathbf{a}_0$, $\mathbf{b}_0$, and $\mathbf{c}_0$ in which $\mathbf{a}_0$ and $\mathbf{b}_0$ lay in the x–y plane and are of equal length, whereas $\mathbf{c}_0$ is parallel to the z-axis

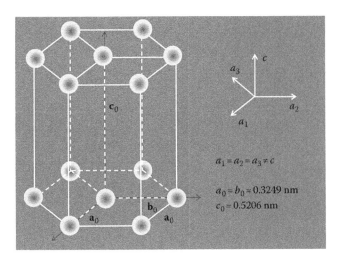

FIGURE 14.1   Crystal structure of ZnO.

TABLE 14.1   Basic Physical Parameters of ZnO

Property	Value
Lattice constants (at $T = 300\,\mathrm{K}$)	
a_0	0.3249 nm
c_0	0.5206 nm
a_0/c_0	1.602 (ideal structure: 1.633)
Density (ρ)	5.606 g/cm³
Melting point	1975°C (2248 K)
Dielectric constant (ε)	8.66
Energy gap (E_g)	3.4 eV
Exciton-binding energy	60 meV
Electron effective mass ($m_e^\star$)	0.24
Electron mobility (μ_e)	200 cm²/V s
Hole effective mass ($m_h^\star$)	0.59
Hole mobility (μ_h) (at $T = 300\,\mathrm{K}$)	5–50 cm²/V s

(Figure 14.1). The values of the primitive translation vectors are $a_0 = b_0 = 0.3249\,\mathrm{nm}$ and $c_0 = 0.5206\,\mathrm{nm}$ at room temperature. The ratio c_0/a_0 of 1.602 deviates slightly from the value $c_0/a_0 = 1.633$ of the ideal hexagonal structure. The basic physical parameters for ZnO are summarized in Table 14.1 (Fan and Lu 2005, Klingshirn 2007, Schmidt-Mende and MacManus-Driscoll 2007). Based on these fundamental physical properties, intense research by many groups has been focused on the growth and characterization of ZnO nanostructures and their versatile device applications.

In the following we explain briefly the growth by a vapor transport method and the characterization of ZnO nanowires and the detail fundamental properties of the FETs made from ZnO nanowires are explained in Section 14.3.

14.2.2 General Growth Methods of ZnO Nanowires

Semiconductor nanowires including ZnO nanowires have attracted considerable attention due to their unique properties, including their high anisotropic geometry and large surface-to-volume ratio. In order to utilize semiconductor nanowires for nanoelectronic devices, it is essential to rationally grow high-quality nanowires with tunable and modulated chemical composition, size, and morphology, and to precisely control doping concentration with both n- and p-type dopants. To this end, there are many efforts to control the nanowire properties, and numerous growth techniques have been developed (Dai et al. 2003, Law et al. 2004, Schmidt-Mende and MacManus-Driscoll 2007). In recent years, the interest in ZnO nanostructures has increased drastically and many researchers have intensively investigated on them. In general, ZnO nanostructures can be grown by either vapor-phase growth or aqueous solution growth methods. In particular, vapor-phase growth methods use gas-phase species as the initial starting reactants for the nanowire formation, and numerous growth techniques such as laser-assisted growth (Sun et al. 2006), chemical vapor deposition (Yang et al. 2002, Park et al. 2003), and thermal evaporation (Dai et al. 2003) have been developed to prepare gas-phase species. In these gas-phase production methods, nanowires are commonly grown via vapor–liquid–solid (VLS) or vapor–solid (VS) mechanisms. VLS process can be regarded as nanowire nucleation and growth via Zn accumulation in liquid catalyst particle (often Au), supersaturation, precipitation, and oxidation, whereas VS process can be described as one-dimensional nanostructure formation from vapor phase precursors in the absence of metal catalysts (Law et al. 2004).

14.2.3 Field-Effect Transistor

The FET, which is often called a unipolar transistor due to a majority carrier device, uses either electrons or holes for conduction. The FET is a three-terminal (source, drain, and gate) device in which the current through two terminals between source and drain is controlled by the gate (Sze 1981, Streetman and Banerjee 2000). The conductivity is varied by the electric field that is produced when the gate–source voltage (V_G) is increased; the drain–source current (I_{DS}) increases exponentially for V_G below threshold. The FETs are divided into several forms: a junction FET (JFET), a metal-semiconductor FET (MESFET), and a metal-insulator semiconductor FET (MISFET) or metal-oxide semiconductor FET (MOSFET) (Figure 14.2) (Sze 1981, Streetman and Banerjee 2000). In the JFET, the control (gate) voltage varies the depletion width of a reverse biased p–n junction with the channel between the source and drain (Figure 14.2a). Similarly, the MESFET is the JFET in which the reverse biased p–n junction is replaced by a metal-semiconductor Schottky junction (Figure 14.2b) The MOSFET is used to amplify or switch electronic signals and generally is consisting of three terminals (source, drain, and gate) and substrate (or body) (Figure 14.2c through f). In particular, since a gate electrode is separated from the semiconductor (substrate) by an oxide, there is effectively no dc gate current, and the channel is capacitively coupled to the gate via the electric field in the oxide. Furthermore, the FETs are

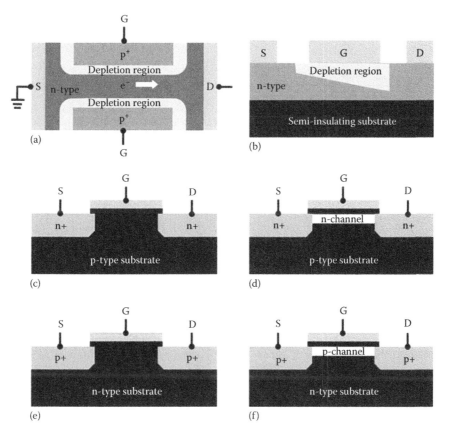

FIGURE 14.2 Basic classification of FETs. (a) JFET, (b) MESFET, and (c)–(f) four different types of MOSFET. (c) n-channel E-mode MOSFET and (d) n-channel D-mode MOSFET. (e) p-channel E-mode MOSFET and (f) p-channel D-mode MOSFET.

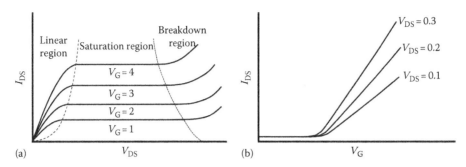

FIGURE 14.3 Basic current–voltage characteristics of FETs. (a) Typical transistor output characteristics (I_{DS}–V_{DS}). (b) Typical transistor transfer characteristics (I_{DS}–V_G).

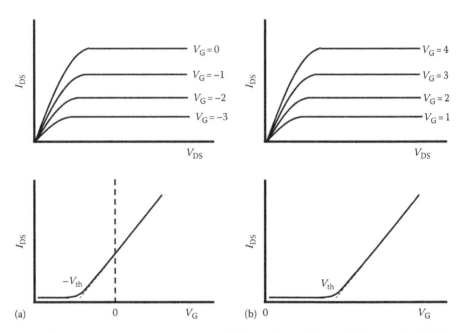

FIGURE 14.4 Comparison of basic current–voltage characteristics of (a) "normally on" FET and (b) "normally off" FET.

distinguished as depletion-mode (D-mode) and enhancement-mode (E-mode) types in terms of the turn-on/off position and the polarity of the threshold voltage. The D-mode FETs exhibit the turn-on of the channel at zero gate bias and negative threshold voltages (Figure 14.2d and f), whereas the E-mode FETs exhibit the turn-off at zero gate bias and positive threshold voltages (Figure 14.2c and e). Therefore, D- and E-mode types are often called "normally on" and "normally off," respectively.

The basic current–voltage (*I–V*) characteristics of FETs are shown in Figure 14.3, where the complete set of I_{DS}–V_{DS} (drain–source current versus voltage) curves are called "output characteristics" (Figure 14.3a) and I_{DS}–V_G (drain–source current versus gate voltage) curves are called "transfer characteristics" (Figure 14.3b). In particular, the output characteristics are plotted as a function of the drain–source voltage (V_{DS}) for various gate voltages (V_G) and can be divided into three regions: (1) the linear region, where I_{DS} increase linearly with V_{DS} for a given $V_G(>V_{th})$; (2) the saturation region, where I_{DS} no longer increases as V_{DS} increases, i.e., it remains essentially constant; and (3) the breakdown region, where

I_{DS} suddenly increases with the further increase of V_{DS} beyond saturation. Figure 14.4 displays the comparison of *I–V* characteristics between "normally on" (D-mode) FET (Figure 4.4a) and "normally off" (E-mode) FET (Figure 14.4b). The main difference is the position of the threshold voltage (V_{th}) along the V_G.

14.3 Results and Discussion

14.3.1 Growth by a Vapor Transport Method and Characterization of ZnO Nanowires

ZnO nanostructures can be simply grown using a thermal furnace system consisting of quartz tube, gas flow controller, and vacuum system. Figure 14.5 shows a schematic of furnace system and a series of the field emission scanning electron microscopy (FESEM) images of various ZnO nanostructures with distinctive geometrical morphologies formed by the vapor transport method using a furnace system, which is the most common method to grow ZnO nanowires. The shape and size of these nanostructures are strongly dependent on the various growth

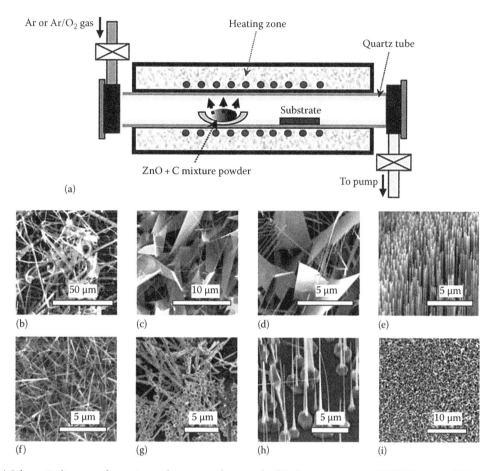

FIGURE 14.5 (a) Schematic diagram of experimental apparatus for growth of ZnO nanostructures. FESEM images of (b) nanoribbons or nanoblets, (c) nanosheets, (d) nanocombs, (e) vertically well-aligned nanowire/nanorods, (f) random-oriented nanowire/nanorods, (g) corrugated nanowires, (h) nanotips or nanoneedles, and (i) nanowalls.

conditions such as temperature, gas flow rate, substrate itself, and the position of substrate. In general, the size, surface morphology, and the crystal structure of ZnO nanowires are commonly characterized using FESEM and transmission electron microscopy (TEM). In order to study the optical properties of ZnO nanowires, the photoluminescence (PL) spectra are measured. In addition, to investigate the electrical properties of ZnO nanowires, the FET-based device structures are fabricated as shown in Figure 14.6e and f and characterized.

The vertically well-aligned ZnO nanowires were grown on the sapphire substrates or ZnO-film-coated sapphire substrates (Figure 14.6a). Figure 14.6b shows the micro-photoluminescence (μPL) spectra of ZnO nanowires grown on the two different substrates. A He–Cd laser (325 nm) was used as an excitation source in the PL measurements. The μPL spectra were measured on the ZnO nanowires after they were transferred from the growth substrates to a silicon wafer in order to eliminate the signals coming from the ZnO film substrate itself. The ZnO nanowires grown on an Au-catalyst-free ZnO film substrate show a stronger defect emission in the PL spectrum than those grown on an Au-coated sapphire substrate. Since it is generally agreed that the defect emission is a surface-related process (Wang et al. 2006), the PL spectra suggest that the ZnO nanowires grown on an Au-catalyst-free ZnO film have a significantly larger number of surface defect sites. Figure 14.6c and d shows typical high-resolution TEM (HRTEM) images of ZnO nanowires grown on an Au-coated sapphire substrate and on an Au-catalyst-free ZnO film, respectively. The upper insets of Figure 14.6c and d are computed fast Fourier transform patterns obtained from the lattice fringes of ZnO nanowires, indicating that the ZnO nanowires grown on both substrates are single crystalline with a preferred growth direction of [0001]. The lower insets of Figure 14.6c and d are the low-magnification TEM images of ZnO nanowires showing the overall surface roughness along the nanowires. Both the low-magnification TEM images and the HRTEM images allow comparison between the surface structures of the ZnO nanowires grown on the two different types of substrates. Compared to the ZnO nanowires grown on an Au-coated sapphire substrate (Figure 14.6c), the ZnO nanowires grown on an Au-catalyst-free ZnO film (Figure 14.6d) are seen to be significantly rougher across the surfaces parallel to the growth direction, as is apparent in the HRTEM images, and as is indicated by the circled regions in the low-magnification TEM images of the lower inset in Figure 14.6d. Figure 14.6e and f show the schematic of nanowire FET device structure and the SEM image of a single ZnO nanowire connecting the source and drain electrodes.

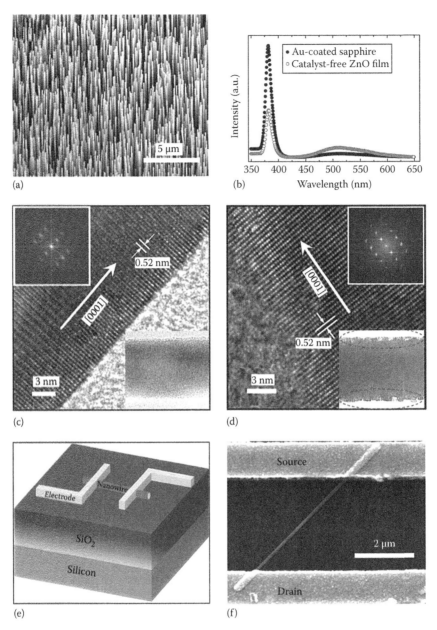

FIGURE 14.6 (a) FESEM images of vertically well-aligned ZnO nanowires grown on Au-coated sapphire substrates or ZnO buffer film-coated substrates. (b) The μPL spectra of ZnO nanowires on the two substrates showing emissions at approximately 378 nm. HRTEM images of ZnO NWs grown on (c) an Au-coated sapphire substrate and (d) an Au-catalyst-free ZnO film. Insets are computed fast Fourier transform patterns (upper) and low-magnification TEM images (lower). (e) A schematic of the FET device structures. (f) A FESEM image of a single ZnO nanowire connected between source and drain electrodes in FET. (Adapted from Hong, W.-K. et al., *Appl. Phys. Lett.*, 90, 243103, 2007. With permission.)

14.3.2 Fabrication and Characterization of ZnO Nanowire Field-Effect Transistors

ZnO nanowire FETs with different gate structures have been fabricated and extensively investigated in the past several years. For example, there are four different types of gate configurations: back gate (Goldberger et al. 2005, Hong et al. 2007), top gate (Huang et al. 2001, Yeom et al. 2008), side gate (Cha et al. 2006), and surrounding gate (Ng et al. 2004, Wade et al. 2007). In particular, the FETs with back-gate and top-gate configurations

have been intensively explored due to the fabrication simplicity. To fabricate ZnO nanowire FET device, the grown ZnO nanowires are first transferred from the growth substrate to a silicon wafer with thermally grown oxide or a bare silicon wafer by dropping and drying a nanowire suspension (Figure 14.7). The nanowire suspension is made by briefly sonicating the growth substrate of ZnO nanowires in ethanol or isopropyl alcohol (IPA) for tens of seconds. Metal electrodes forming Ohmic contacts between metal and ZnO nanowires are deposited by an evaporator or a sputter and defined as source and drain electrodes by

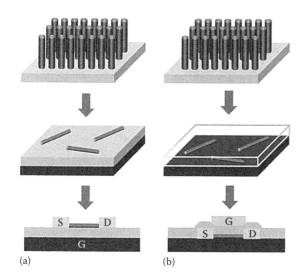

FIGURE 14.7 Fabrication process of ZnO nanowire FETs with (a) back gate and (b) top gate configurations.

lithography techniques and liftoff process (Figure 14.7). In back-gated FET configuration (Figure 14.7a), a pair of metal electrodes patterned on the ZnO nanowire channel serves as the source and drain and a degenerately doped silicon substrate serves the gate electrode. In top-gated FET configuration (Figure 14.7b), a metal gate electrode is formed above a gate dielectric layer on nanowires. Recently, tremendous efforts have been devoted to the fabrication and characterization of nanowire-based transistors. However, if functional FET devices with electrical contacts are to be made from nanowires, their fundamental properties must be carefully investigated due to the presence of the different interfaces and interfacial states in nanowire-based device. The different interfaces exist between the nanowire and passivation layer, nanowire and gate dielectrics, nanowire and electrodes, and nanowire and ambient gases.

In particular, since the semiconductor surfaces are strongly affected by the chemical adsorption of ambient gases (Wolkenstein 1991), the influence of chemisorptions of ambient gases in the case of semiconducting nanowires including ZnO nanowires are much more sensitive than their bulk counter part due to large surface-to-volume ratio (Lagowski et al. 1977, Kolmakov and Moskovits 2004). For example, surface defects such as oxygen vacancies act as the adsorption sites of O_2 molecules and form oxygen ions (O^-, O^{2-}, or O_2^-), and then the chemisorption of O_2 molecules depletes the surface electron and reduces the channel conductivity (Li et al. 2004, Zhang et al. 2004). Therefore, the passivation of ZnO nanowire FET devices is significantly important to remove the influence of water or gas molecules in ambient air on nanowire and to improve the FET performance by enhancing the gate-coupling effects (Hong et al. 2008a,b,c). In addition, other characterization conditions such as gate bias sweep rate can also influence the electronic transport in nanowire transistors due to the presence of surface states (Maeng et al. 2008).

In this section, we review the electrical properties of the surface-tailored smooth and rough ZnO nanowire FETs with and without surface passivation using poly(methyl metahacrylate)

(PMMA), which is a polymeric resist commonly used in nano-lithographic processes involving electron beam, deep UV, or x-ray radiation (Hong et al. 2008a,b,c). Furthermore, a detailed study of the passivation and gate sweep rate effects on the ZnO nanowire FETs under different oxygen environments is presented.

14.3.2.1 Passivation and Surface-Roughness Effects

The electrical characteristics before and after the passivation of the FET devices made from the surface-tailored smooth and rough ZnO nanowires are summarized in Figure 14.8. Figure 14.8a and b shows the output characteristics (source–drain current versus voltage, I_{DS}–V_{DS}) and transfer characteristics (source–drain current versus gate voltage, I_{DS}–V_G) for a FET made from smooth ZnO nanowires, respectively. The FETs made from smooth ZnO nanowires exhibit more well-defined saturation and pinch-off characteristics after passivation in comparison with the FET devices before passivation. In particular, the FET devices before passivation show decreasing separation in current between I_{DS} curves at larger currents in I_{DS}–V_{DS} curves (Figure 14.8a), which is attributed to either an electron injection barrier at the source electrode or to mobility degradation associated with the interface roughness scattering of channel electrons at the channel–insulator interface with increasing gate voltage (Dehuff et al. 2005). The I_{DS}–V_G curves in Figure 14.8b show the threshold voltages (V_{th}) of −4.16 and −2.25 V before and after passivation, respectively. The threshold voltage after passivation shifts toward the positive gate bias direction due to surface-depletion-induced channel-narrowing effect (Fan et al. 2004a,b, Ju et al. 2006, 2007). The current on/off ratios (I_{on}/I_{off}) both before and after passivation for this FET device of smooth ZnO nanowire exhibited 10^4–10^5 (insets of Figure 14.8b).

In contrast, Figure 14.8c and d shows the I_{DS}–V_{DS} and I_{DS}–V_G curves for a FET made from rough ZnO nanowires, respectively. The unpassivated FET devices exhibited the poor electrical characteristics, as the I_{DS}–V_{DS} curves shown in Figure 14.8c. The I_{DS} suddenly decreased with the further increase of V_{DS} and V_G beyond saturation and also the current between I_{DS} curves was not clearly separated. However, the passivated FET devices exhibited well-defined linear regions at low biases and saturation regions at high biases as typical transistors, indicating clear pinch-off behavior. Unlike the FET devices made from smooth ZnO nanowires, the threshold voltage for the FET devices made from rough ZnO nanowires exhibits a considerable shift toward the positive gate bias direction after passivation. The threshold voltage of the particular FET device shown in Figure 14.8d shifted from −12.5 to 8.9 V, indicating transition from D-mode to E-mode operations. Most of the passivated FET devices made from both smooth and rough ZnO nanowires exhibit superior electrical performance than the unpassivated FET devices made from rough ZnO nanowires. These results are in good agreement with the superior characteristics of the FETs passivated by SiO_2/Si_3N_4 or PMMA layer (Kim et al. 2003, Chang et al. 2006, Hong et al. 2008a,b,c). The current on/off ratios (I_{on}/I_{off}) both before

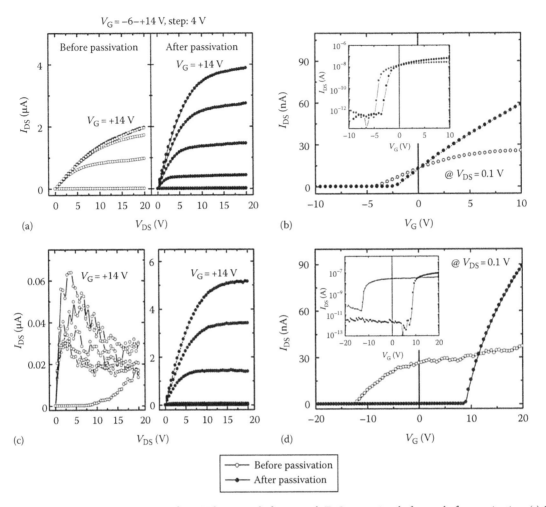

FIGURE 14.8 (a) I_{DS}–V_{DS} and (b) I_{DS}–V_G curves of a FET device made from rough ZnO nanowires before and after passivation. (c) I_{DS}–V_{DS} and (d) I_{DS}–V_G curves of a FET device made from rough ZnO nanowires before and after passivation. (Adapted from Hong, W.-K. et al., *Appl. Surf. Sci.*, 254, 7559, 2008c. With permission.)

and after passivation for the FET device of rough ZnO nanowire exhibited approximately 10^4–10^5 (insets of Figure 14.8d).

Furthermore, the electrical properties of the ZnO nanowire FETs have also been investigated and compared before and after the PMMA passivation under different environments of ambient air (20% O_2), dry O_2, and vacuum (10^{-3} Torr) environments (Song et al. 2008). Figure 14.9 shows a series of the I_{DS}–V_{DS} and the I_{DS}–V_G curves for a ZnO nanowire FET measured at different gate voltages under various oxygen environments before and after passivation, respectively. Figure 14.9a and b displays the I_{DS}–V_{DS} curves for the ZnO nanowire FET before and after passivation, respectively. As shown in Figure 14.9a, the current level is significantly affected by the environmental condition with O_2 molecules. The currents decrease significantly when the ZnO nanowire FET is exposed to dry O_2, as compared to the case in ambient air. When the oxygen is evacuated, the currents increase again. On the contrary, the I_{DS}–V_{DS} curves in Figure 14.9b show a negligible effect on the passivated ZnO nanowire FET by oxygen molecules in different oxygen environments. This is because the PMMA passivation layer prevents oxygen molecules from being adsorbed onto the nanowire. One can also notice that the ZnO

nanowire FETs after passivation have a better electrical performance in terms of well-defined linear and saturation regions in I_{DS}–V_{DS} curves (Figure 14.9b). It has been reported that the passivation can improve the FET performance by enhancing the gate-coupling effect (Kim et al. 2006, Wunnicke 2006, Hong et al. 2008a,b,c). This passivation effect is more prominent in transfer characteristics, as shown in Figure 14.9c and d. A series of plots in Figure 14.9c and d shows the I_{DS}–V_G curves for the ZnO nanowire FET before and after passivation, respectively. The FET device was measured at a fixed source–drain voltage (V_{DS} = 0.1 V) under the same different environments of ambient air (20% O_2), dry O_2, and vacuum (~10^{-3} Torr). The I_{DS}–V_G plots in the semilogarithmic scale (insets of Figure 14.9c and d) show an on/off current ratio as large as 10^4–10^5 for both before and after passivation. Before passivation (Figure 14.9c), the threshold voltages of ZnO nanowire FET shifted from −1.73 (in ambient air) to 6.05 V (in dry O_2) in the positive gate bias direction. This is because more electrons are captured by the oxygen as the oxygen pressure is raised. Since the charge density is reduced after being exposed to dry O_2, more positive gate bias is needed to make currents flow in the nanowire channel, and thus the

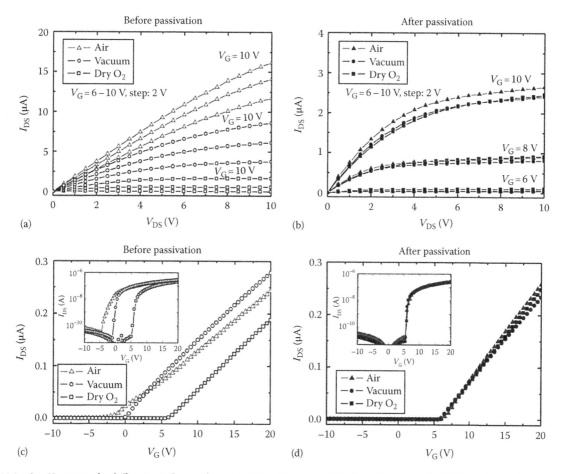

FIGURE 14.9 I_{DS}–V_{DS} curves for different gate biases (from 6 to 10 V, with a step of 2 V) for a ZnO nanowire FET, acquired under ambient air, dry O_2, and vacuum conditions (a) before and (b) after passivation. I_{DS}–V_G curves at V_{DS} = 0.1 V for the ZnO nanowire FET, acquired under ambient air, dry O_2, and vacuum conditions (c) before and (d) after passivation. The insets in (c and d) show the semilogarithmic plot of the I_{DS}–V_G curves. (Adapted from Song, S. et al., *Appl. Phys. Lett.*, 92, 263109, 2008. With permission.)

threshold voltage shifts to the positive gate bias direction. When oxygen gas was evacuated to a vacuum level of 10^{-3} Torr, the threshold voltages of the nanowire FET shifted back from 6.05 (in dry O_2) to −0.10 V (in a vacuum) in the negative gate bias direction, as compared when they were exposed to a dry O_2 environment. This is due to the reduction of the trapping effects of oxygen in a vacuum. On the contrary, when the FET was passivated with PMMA, the threshold voltages were ~6 V and shifted very little under different oxygen environments, as shown in Figure 14.9d. Note that the threshold voltage is defined as the gate voltage obtained by extrapolating the linear portion of the I_{DS}–V_G curve from the point of maximum slope to zero drain current, in which the point of maximum slope is the point where transconductance (dI_{DS}/dV_G) is at a maximum (Arora 1993).

As a result, these phenomena can be explained by oxygen effects. The ambient oxygen partial pressure has a considerable effect on the electrical properties of ZnO nanowire FETs (Fan et al. 2004a,b). Particularly, the adsorbed oxygen molecules deplete the electrons in the ZnO nanowire and form oxygen ions (O^-, O^{2-}, or O_2^-) (Li et al. 2004). Then, the electrons in the ZnO nanowires are trapped by the adsorbed oxygen molecules,

and thus the surface depletion region of ZnO nanowire can be formed, leading to current reduction and threshold voltage shift to the positive gate bias direction (Figure 14.9c). As the oxygen pressure is raised, more electrons are captured by the oxygen molecules at the nanowire surface. Correspondingly, the depletion region is widened and the carrier density in the ZnO nanowire is decreased even more when the FET is exposed to dry O_2 environment.

14.3.2.2 Sweep Rate Effect of Gate Bias

As aforementioned, the corresponding characteristics of nanowire FETs are sensitively dependent on the surrounding environments to which the devices are exposed, and whether or not the devices are passivated (Park et al. 2004, Chang et al. 2006, Song et al. 2008). Other characterization conditions can also influence electrically the interactive absorption of oxygen molecules on the nanowire surface. Particularly, the effect of gate bias sweep rate on the electronic transport properties of ZnO nanowire FETs has been investigated (Maeng et al. 2008). The electrical properties of ZnO nanowire FETs were measured as a function of gate bias sweep rate in different measurement environments.

Figure 14.8 show a series of I_{DS}–V_G curves measured at different gate bias sweep rates (2500, 250, 130, 100, 12, 6, 1.2, 0.3, 0.2, 0.1 V/s) at $V_{DS} = 0.5$ V under various environments. The insets in Figure 14.10 show the I_{DS}–V_G plots on a semilogarithmic scale. A ZnO nanowire FET device was systematically measured in ambient air, in an N_2-filled glove box, and in ambient air after the same device was passivated with a PMMA layer. Figure 14.10a shows the case of a ZnO nanowire FET (unpassivated) measured in ambient air. As the gate bias sweep rate was decreased from 2500 to 0.1 V/s, the current decreased and the threshold voltage shifted to the positive gate bias direction. These phenomena can

be explained by the depletion of electrons by oxygen adsorption on the surface of the ZnO nanowire. As previously mentioned, the increase of oxygen concentration causes the change of conductivity by surface depletion in the nanowire channel, i.e., the adsorbed oxygen molecules that bind the electrons localized at the nanowire surface become oxygen ions in the forms of O^-, O^{2-}, or O_2^-, resulting in depletion of electrons and thus lowering the conductivity (Takata et al. 1976, Li et al. 2004, Sadek et al. 2007). Oxygen adsorption is sustained at an equilibrium condition in ambient air at zero gate bias. However, the gate bias induces more adsorption of oxygen ions due to the applied positive gate bias. Therefore, when the nanowire FET is applied with a slower gate bias rate, more electrons on the ZnO nanowire surface are trapped by the adsorbed oxygen ions, which results in a reduction of nanowire conduction channel region by the more depleted region. Thus, the current decreases, and the threshold voltage shifts to the positive gate bias direction. The effect of the gate bias sweep rate on the ZnO nanowire FET was investigated under different oxygen environments. The same ZnO nanowire FET device measured in ambient air (Figure 14.10a) was placed into an N_2-filled glove box after holding at a vacuum (~10^{-5} Torr) for 24 h. The I_{DS}–V_G curves measured in this N_2 environment are shown in Figure 14.10b. The conductivity of the ZnO nanowire FET device in the N_2 environment increased by about an order of magnitude, compared with the case of ambient air. In the N_2 environment, the nanowire FET has fewer oxygen ions on the surface, thus less surface depletion of electrons, resulting in an increase of the current. Unlike the case of the nanowire measured in ambient air, the nanowire device (unpassivated) in the N_2 environment was not influenced by the gate bias sweep rate, as is shown in Figure 14.10b. The current did not change and the threshold voltage did not shift. This is because of the relative absence of oxygen in the N_2-filled glove box; therefore, there is a negligible effect of oxygen absorption even at different gate bias sweep rates. In addition, the effect of the gate bias sweep rate on the same ZnO nanowire FET device after the passivation of PMMA layer was explored. The resulting I_{DS}–V_G curves after the passivation, measured in ambient air, are shown in Figure 14.10c. Although the change and shift are not as significant, the I_{DS}–V_G curves exhibited somewhat similar trends of the current change and threshold voltage shift for different gate bias sweep rates as the case of Figure 14.10a. Therefore, the effect of gate bias sweep rate on the nanowire channel can be explained as the following. Slower gate bias sweep rates imply longer gate biasing time. Captured oxygen ions are strongly bonded by positive gate bias, so that oxygen ions would not be easily detached. Thus, slow gate bias sweep rates will cause more oxygen absorption than fast gate bias sweep rates. More adsorbed oxygen molecules extend the surface depletion region. Consequently, a larger depletion region in the nanowire occurs by slower gate bias sweep rates and longer gate-biasing time, which results in the reduction of current and carrier density and a shift of gate bias in the positive gate bias direction. On the contrary, faster gate bias sweep rates are applied in relatively shorter gate biasing time. Therefore, fast gate bias sweep rates lead to less oxygen absorption.

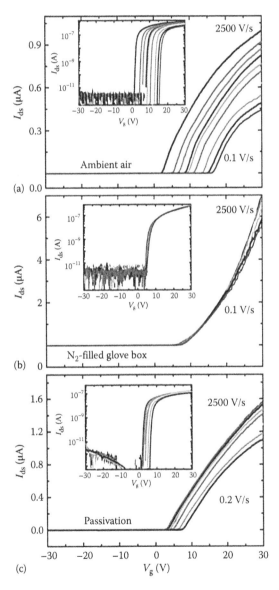

FIGURE 14.10 I_{DS}–V_G curves for a ZnO nanowire FET as a function of gate bias sweep rate (2500, 250, 130, 100, 12, 6, 1.2, 0.3, 0.2, and 0.1 V/s) measured at $V_{DS} = 0.5$ V (a) under ambient air, (b) N_2-filled glove box, and (c) ambient air after PMMA passivation. The semilogarithmic plots of I_{DS}–V_G curves are shown in the insets. (Adapted from Maeng, J. et al., *Appl. Phys. Lett.*, 92, 233120, 2008. With permission.)

The current, threshold voltage, and carrier density were sensitively dependent on the gate bias sweep rate. These phenomena became more pronounced under environments with increased oxygen. The results of this study may offer how to characterize the surface effects of ZnO nanowire devices or other materials.

14.3.3 N-Channel Depletion-Mode and Enhancement-Mode ZnO Nanowire FETs

To date, although FETs using ZnO nanowires as active channels have been extensively investigated, most of the fabricated ZnO nanowire FETs were typically normally on type, n-channel depletion-mode (D-mode) behavior, which exhibited nonzero current at zero gate bias and negative threshold voltages. However, for a wide number of the applications of nanowire FETs in logic circuits, both D-mode and enhancement-mode (E-mode) FETs are required (Sedra and Smith 1991, Fortunato et al. 2004, Lee et al. 2006, Ma et al. 2007).

The electrical characteristics of D-mode and E-mode ZnO nanowire FETs are summarized in Figure 14.11. All the fabricated ZnO nanowire FETs were passivated by PMMA layer to remove the influence of water or gas molecules in ambient air on nanowire and to improve the FET performance by enhancing the gate-coupling effects. Note that although the structure of ZnO nanowire FET device is different from the conventional metal-oxide-semiconductor transistors, the operation mode of ZnO nanowire FETs is distinguished as D-mode or E-mode in terms of the polarity of the threshold voltage (Hong et al 2007). ZnO nanowire FETs with negative threshold voltages are normally on type, n-channel D-mode transistors, indicating that more negative gate bias should be applied to deplete carriers in the channel to reduce channel conductance, since the n-channel for current flow already exists at zero gate bias. In contrast, ZnO nanowire FETs with positive threshold voltages are normally off type, n-channel E-mode transistors, indicating that more positive gate bias is needed to make the channel, since the channel current does not flow at zero gate bias.

Figure 14.11a and b shows the electrical characteristics for n-channel D-mode ZnO nanowire FETs, whereas Figure 14.11c and d shows those for n-channel E-mode FETs. The I_{DS}-V_{DS} curves of both D-mode and E-mode FETs have well-defined linear regimes at low biases and saturation regimes at high biases. Figure 14.11b shows that the threshold voltage V_{th} is −4.14 V, indicating n-channel D-mode behavior. The I_{DS}-V_G plot in the semilogarithmic scale displays an on/off current ratio as large as 10^5 (inset of Figure 14.11b). In contrast, in the I_{DS}-V_G curves of Figure 14.11d, the threshold voltage V_{th} is +10.85 V, indicating n-channel E-mode behavior. The I_{DS}-V_G plot in the semilogarithmic scale shows an on/off current ratio as large as 10^6 (inset of Figure 14.11d).

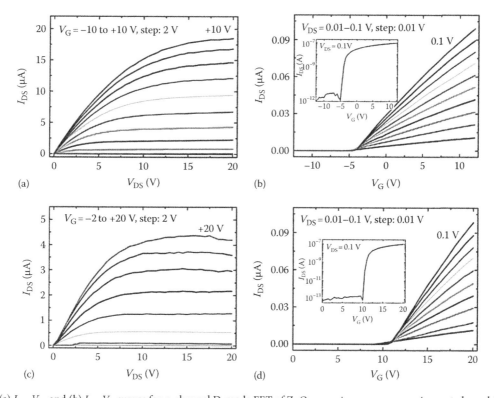

FIGURE 14.11 (a) I_{DS}-V_{DS} and (b) I_{DS}-V_G curves for n-channel D-mode FET of ZnO nanowires grown on an Au-coated sapphire substrate. The inset shows the semilogarithmic plot of the I_{DS}-V_G curve at V_{DS} = 0.1 V. (c) I_{DS}-V_{DS} and (d) I_{DS}-V_G curves for n-channel E-mode FET of ZnO nanowires grown on an Au-catalyst-free ZnO film. The inset shows the semilogarithmic plot of the I_{DS}-V_G curve at V_{DS} = 0.1 V. (Adapted from Hong, W.-K. et al., *Appl. Phys. Lett.*, 90, 243103, 2007. With permission.)

14.3.4 Electronic Transport in Surface-Architecture-Controlled ZnO Nanowire FETs

It is well-known that interface roughness plays an important role in electronic transport for transistor devices (Hastas et al. 2002, Wang et al. 2005). In particular, the electronic transport in nanowire transistors can be influenced by the surface morphology of nanowires associated with the interface roughness, including the presence of surface trap states at the interfaces between the nanowires and the dielectric layers of nanowire FETs (Wang et al. 2004, Dayeh et al. 2007a,b). In order to investigate the influence of surface states and/or defects (associated with surface roughness and nanowire size) on the electronic transport properties in nanowire transistors, the surface morphology- and size-controlled ZnO nanowires were grown on the three sets of sapphire substrates with ZnO buffer films: (1) an undoped ZnO film with Au catalyst (denoted as Au–ZnO film) or without Au catalyst (denoted as ZnO film), (2) a gallium-doped ZnO film with Au catalyst (denoted as Au–GZO film) or without Au catalyst (denoted as GZO film), (3) an aluminum-doped ZnO film with Au catalyst (denoted as Au–AZO film) or without Au catalyst (denoted as AZO film). In addition, an Au-coated sapphire (denoted as Au-sapphire) substrate was also used for comparison with the ZnO buffer film-coated sapphire substrates. ZnO buffer films with approximately 1 μm thickness were grown on the sapphire substrates by a radio frequency (rf) sputtering system using a commercially sintered ZnO target, a

Ga$_2$O$_3$ (1 wt%)-doped ZnO target, and an Al$_2$O$_3$ (1 wt%)-doped ZnO target. The Au thin film (~3 nm) was deposited on some substrates using an e-beam evaporator to form Au catalysts for the growth of nanowires.

Figure 14.12 shows the I_{DS}–V_G curves at V_{DS} = 0.1 V of the surface-architecture-controlled ZnO nanowire FETs, which exhibit different electronic transport characteristics. Insets in Figure 14.10 show the low-magnification TEM images of the surface-architecture-controlled ZnO nanowires grown on different substrates. The ZnO nanowires with relatively smaller diameters were found to have rougher surfaces than those with relatively larger diameters. The rough ZnO nanowires can have a larger surface-area-to-volume ratio than smooth ZnO nanowires. The FET devices made from smooth ZnO nanowires (nanowires grown on GZO, Au–AZO films, and Au-sapphire) have negative threshold voltages, indicating n-channel D-mode behavior that exhibits nonzero current at the zero gate bias. In contrast, the FETs made from rough ZnO nanowires (nanowires grown on Au–ZnO, ZnO, Au–GZO, and AZO films) have positive threshold voltages, indicating n-channel E-mode behavior that exhibits off-current status at zero gate bias.

Figure 14.13 shows cross-sectional schematics (Figure 14.13a and c) across the electrodes, ZnO nanowire, and dielectric layers with the corresponding equilibrium energy band diagram (Figure 14.13b and d) of the ZnO nanowire FETs at V_G = 0 V. Figure 14.13a and b shows the case for smooth ZnO nanowires, and Figure 14.13c and d depicts the case for rough ZnO

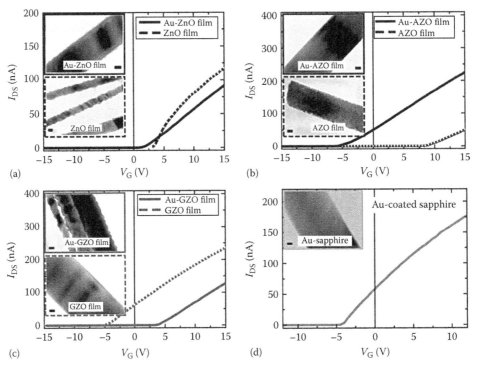

FIGURE 14.12 I_{DS}–V_G curves (measured at V_{DS} = 0.1 V) of FETs made from ZnO nanowires grown on (a) Au–ZnO and ZnO films, (b) Au–AZO and AZO films, (c) Au–GZO and GZO films, and (d) as Au-sapphire substrate. Labels are explained in the text. The insets show the low magnification TEM images of the surface-architecture-controlled ZnO nanowires. (Adapted from Hong, W.-K. et al., *Nano Lett.* 8, 950, 2008b. With permission.)

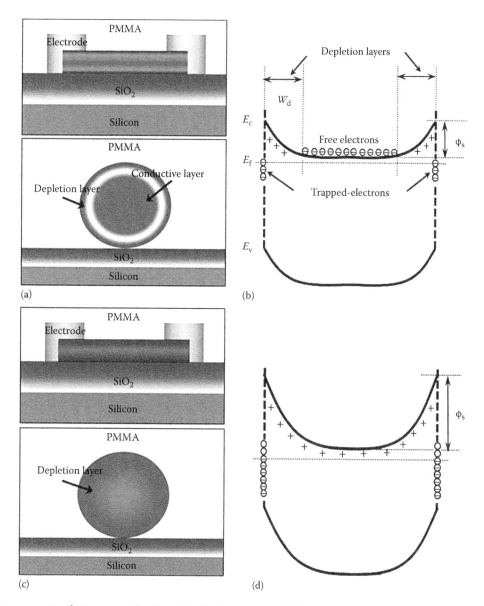

FIGURE 14.13 A cross-sectional view across the electrodes, ZnO nanowire, and dielectric layers (a and c) and the corresponding equilibrium energy band diagrams of nanowire FETs at $V_G = 0\,V$ (b and d) for smooth ZnO nanowires (a and b) and rough ZnO nanowires (c and d). (Adapted from Hong, W.-K. et al., *Nano Lett.*, 8, 950, 2008b. With permission.)

nanowires. For single crystalline semiconductor materials, band bending due to Fermi-level pinning can occur at surfaces and/or interfaces because the surface states and/or defects on single crystalline nanowires can induce trap energy levels at the interfaces (Wang et al. 2004, Calarco et al. 2005, Jones et al. 2007, Liao et al. 2007). The trapping of carrier electrons in the trap states can cause electron depletion in the channel, resulting in a threshold voltage shift and a conductance modulation (Ikeda 2002, Hossain et al. 2003, Fan et al. 2004a,b, Ju et al. 2007, Yoon et al. 2007). Correspondingly, the different electronic transport behavior between smooth and rough ZnO nanowires can be explained by considering the depletion of electron carriers due to surface band bending at the PMMA/ZnO nanowire and/or ZnO nanowire–SiO$_2$ interfaces as shown in Figure 14.13. Note that for the convenience of discussion and simplification of charge

transport mechanism, it is assumed that the uniform charge and gate potential distribution for nanowires is with circular cross sections. However, the typical back-gate nanowire FET geometry can be significantly influenced by the cross-sectional shapes of nanowires and whether the nanowire is embedded or not (Wunnicke 2006). In addition, the back-gate nanowire FET geometry has the limitation in accuracy for calculating the gate-nanowire capacitance as well as the complication for the experimental extraction of transport parameters under the presence of depletion region and trap states (Dayeh et al. 2007a,b, Khanal and Wu 2007).

The depletion region by surface band bending and the effective conduction region (or effective diameter) can be estimated as (Taur and Ning 1998, Yu and Cardona 2001, Wang et al. 2004, Goldberger et al. 2005)

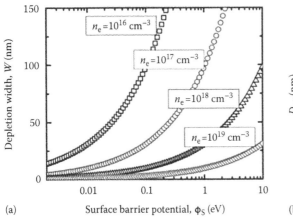
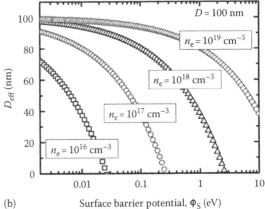

FIGURE 14.14 (a) Depletion width (W_d) and (b) effective diameter (D_{eff}) of ZnO nanowires as a function of surface barrier potential of ZnO nanowires at $N_D = 10^{16}$–10^{19} cm^{-3}.

$$W_d = \left(\frac{2\varepsilon_{ZnO}\, \phi_S}{e\, N_D} \right)^{1/2} \qquad (14.1)$$

where

ϕ_S is the surface barrier potential
e is the electronic charge
N_D is the doping density
ε_{ZnO} is the dielectric constant of ZnO

Then by using the depletion approximation and charge neutrality condition (Taur and Ning 1998, Yu and Cardona 2001, Hossain et al. 2003), the surface trap density (N_t) can be estimated as $N_t = 2N_D W_d$. Thus, the effective diameter (D_{eff}) of the conduction layer in the nanowire channel can be expressed as

$$D_{eff} = D - \left(\frac{N_t}{N_D} \right) \qquad (14.2)$$

where D is the geometrical diameter of nanowire. Therefore, by assuming that the $N_D = 10^{17}$/cm^3, $\varepsilon_{ZnO} = 8.66$ (Fan and Lu 2005), $\phi_S = 0.3$ eV, and the nanowire diameter (D) is 100 nm, we obtain $W_d = 54$ nm and $D < 2W_d$ from Equations 14.1 and 14.2 as shown in Figure 14.14.

14.4 Summary

In order to fabricate the functional devices from the ZnO nanowires, their fundamental properties must be carefully investigated because the presence of a large surface and interfacial area in nanowires can profoundly alter their performance. Therefore, it is essential to understand the surface-roughness- and size-dependent effects on the electronic transport properties of nanowire transistors. Furthermore, since ZnO nanowires are strongly influenced by chemical environments due to large surface-to-volume ratio, it is also important to study on the passivation effects of dielectric layer on the transport properties correlated to geometry and adsorbed species on the ZnO

nanowires. In this chapter, we reviewed the electrical properties of FETs made from the surface-tailored smooth and rough ZnO nanowires that were grown by a vapor transport method. The transport properties of ZnO nanowire FETs are associated with the surface roughness, nanowire size, surface states, and/or defects as well as the surface chemistry in surrounding environments. The characteristics show that the electrical properties of ZnO nanowire FETs must be carefully investigated due to the various effects of nanowire surface and surrounding environments as well as interfaces. In the field of nanowire-based electronics, these explorations are important in understanding the transport properties and developing into practically useful device applications.

In future, in order to enable ZnO nanowire FETs to function as building blocks for the next-generation electronic circuitry, the ability to rationally grow nanowires with precisely the control of chemical composition, size, morphology, structure, and doping concentration will be required. Furthermore, the implementation of assembling any kind of functional nanodevices into highly integrated nanosystems must be intensively investigated.

Acknowledgment

This work was supported in part by the National Research Laboratory (NRL) Program, the National Core Research Center (NCRC) grant, and the World Clan University (WCU) program by the Korean Ministry of Education, Science and Technology (MEST).

References

Agarwal, R. 2008. Heterointerfaces in semiconductor nanowires. *Small* 4: 1872–1893.

Arora, N. 1993. *MOSFET Models for VLSI Circuit Simulation*. New York: Springer, Wien.

Bao, J., Zimmler, M. A., and Capasso, F. 2006. Broadband ZnO single-nanowire light-emitting diode. *Nano Lett.* 6: 1719–1722.

Calarco, R., Marso, M., Richter, T. et al. 2005. Size-dependent photoconductivity in MBE-grown GaN-nanowires. *Nano Lett.* 5: 981–984.

Cha, S. N, Jang, J. E., Choi, Y. et al. 2006. High performance ZnO nanowire field effect transistor using self-aligned nanogap gate electrodes. *Appl. Phys. Lett.* 89: 263102 (3 pp).

Chang, P.-C., Fan, Z., Chien, C.-J. et al. 2006. High-performance ZnO nanowire field effect transistors. *Appl. Phys. Lett.* 89: 133113 (3 pp).

Dai, Z. R., Pan, Z. W., and Wang, Z. L. 2003. Novel nanostructures of functional oxides synthesized by thermal evaporation. *Adv. Funct. Mater.* 13: 9–24.

Dayeh, S. A., Soci, C., Yu, P. K. L., Yu, E. T., and Wang, D. 2007a. Influence of surface states on the extraction of transport parameters from InAs nanowire field effect transistors. *Appl. Phys. Lett.* 90: 162112 (3 pp).

Dayeh, S. A., Soci, C., Yu, P. K. L., Yu, E. T., and Wang, D. 2007b. Transport properties of InAs nanowire field effect transistors: The effects of surface states. *J. Vac. Sic. Technol. B* 25: 1432–1436.

Dehuff, N. L., Kettenring, E. S., Hong, D. et al. 2005. Transparent thin-film transistors with zinc indium oxide channel layer. *J. Appl. Phys.* 97: 064505 (5 pp).

Dorfman, A., Kumar, N., and Hahm, J. 2006. Nanoscale ZnO-enhanced fluorescence detection of protein interactions. *Adv. Mater.* 19: 2685–2690.

Fan, Z. and Lu, J. G. 2005. Zinc oxide nanostructure: Synthesis and properties. *J. Nanosci. Nanotechnol.* 5: 1561–1573.

Fan, Z. and Lu, J. G. 2006. Chemical sensing with ZnO nanowire field-effect transistor. *IEEE Trans. Nanotechnol.* 5: 393–396.

Fan, Z., Chang, P.-C., Lu, J. G. et al. 2004a. Photoluminescence and polarized photodetection of single ZnO nanowires. *Appl. Phys. Lett.* 85: 6128–6130.

Fan, Z., Wang, D., Chang, P.-C., Tseng, W.-Y., and Lu, J. G. 2004b. ZnO nanowire field-effect transistor and oxygen sensing property. *Appl. Phys. Lett.* 85: 5923–5925.

Fortunato, E. M. C., Barquinha, P. M. C., Pimentel, C. C. M. B. G. et al. 2004. Wide-bandgap high-mobility ZnO thin-film transistors produced at room temperature. *Appl. Phys. Lett.* 85: 2541–2543.

Goldberger, J., Sirbuly, D. J., Law, M., and Yang, P. 2005. ZnO nanowire transistors. *J. Phys. Chem. B* 109: 9–14.

Hastas, N. A., Dimitriadis, C. A., and Kamarinos, G. 2002. Effect of interface roughness on gate bias instability of polycrystalline silicon thin-film transistors. *J. Appl. Phys.* 92: 4741–4745.

Hong, W.-K., Hwang, D.-K., Park, I.-K. et al. 2007. Realization of highly reproducible ZnO nanowire field effect transistors with n-channel depletion and enhancement modes. *Appl. Phys. Lett.* 90: 243103 (3 pp).

Hong, W.-K., Kim, B.-J., Kim, T.-W. et al. 2008a. Electrical properties of ZnO nanowire field effect transistors by surface passivation. *Colloids Surf. A* 313–314: 378–382.

Hong, W.-K., Sohn, J. I., Hwang, D.-K. et al. 2008b. Tunable electronic transport characteristics of surface-architecture-controlled ZnO nanowire field effect transistors. *Nano Lett.* 8: 950–956.

Hong, W.-K., Song, S., Hwang, D.-K. et al. 2008c. Effects of surface roughness on the electrical characteristics of ZnO nanowire field effect transistors. *Appl. Surf. Sci.* 254: 7559–7564.

Hossain, F. M., Nishii, J., Takagi, S. et al. 2003. Modeling and simulation of polycrystalline ZnO thin-film transistors. *J. Appl. Phys.* 94: 7768–7777.

Hsueh, T.-J., Chang, S.-J., Hsu, C.-L., Lin, Y.-R., and Chen, I.-C. 2007. Highly sensitive ZnO nanowire ethanol sensor with Pd adsorption. *Appl. Phys. Lett.* 91: 053111 (3 pp).

Huang, Y., Duan, X. F., Cui, Y. et al. 2001. Logic gates and computation from assembled nanowire building blocks. *Science* 294: 1313–1317.

Ikeda, H. 2002. Evaluation of grain boundary trap states in polycrystalline–silicon thin-film transistors by mobility and capacitance measurements. *J. Appl. Phys.* 91: 4637–4645.

Jones, F., Léonard, F., Talin, A. A., and Bell, N. S. 2007. Electrical conduction and photoluminescence properties of solution-grown ZnO nanowires. *J. Appl. Phys.* 102: 014305 (7 pp).

Ju, S., Lee, K., Janes, D. B. et al. 2006. ZnO nanowire field-effect transistors: Ozone-induced threshold voltage shift and multiple nanowire effects. In *Proceedings of the 6th IEEE Conference on Nanotechnology,* Cincinnati, OH, pp. 445–448.

Ju, S., Lee, K., Yoon, M.-H. et al. 2007. High performance ZnO nanowire field effect transistors with organic gate nano-dielectrics: Effects of metal contacts and ozone treatment. *Nanotechnology* 18: 155201 (7 pp).

Jung, Y., Ko, D.-K., and Agarwal, R. 2007. Synthesis and structural characterization of single-crystalline branched nanowire heterostructures. *Nano Lett.* 7: 264–268.

Khanal, D. R. and Wu, J. 2007. Gate coupling and charge distribution in nanowire field effect transistors. *Nano Lett.* 7: 2778–2783.

Kim, W., Javey, A., Vermesh, O. et al. 2003. Hysteresis caused by water molecules in carbon nanotube field-effect transistors. *Nano Lett.* 3: 193–198.

Kim, H.-J., Lee, C.-H., Kim, D.-W., and Yi, G.-C. 2006. Fabrication and electrical characteristics of dual-gate ZnO nano-rod metal-oxide semiconductor field-effect transistors. *Nanotechnology* 17: S327–S331.

Kind, H., Yan, H., Messer, B., Law, M., and Yang, P. 2002. Nanowire ultraviolet photodetectors and optical switches. *Adv. Mater.* 14: 158–160.

Klingshirn, C. 2007. ZnO: From basis towards applications. *Phys. Stat. Sol. B* 244: 3027–3073.

Kolmakov, A. and Moskovits, M. 2004. Chemical sensing and catalysis by one-dimensional metal-oxide nanostructures. *Annu. Rev. Mater. Res.* 34: 151–180.

Lagowski, J., Sproles, E. S. Jr., and Gatos, H. C. 1977. Quantitative study of the charge transfer in chemisorption; oxygen chemisorption on ZnO. *J. Appl. Phys.* 48: 3566–3575.

Law, M., Goldberger, J., and Yang, P. 2004. Semiconductor nanowires and nanotubes. *Annu. Rev. Mater. Res.* 34: 83–122.

Law, M., Greene, L. E., Johnson, J. C., Saykally, R., and Yang, P. 2005. Nanowire dye-sensitized solar cells. *Nat. Mater.* 4: 455–459.

Lee, C. A., Jin, S. H., Jung, K. D., Lee, J. D., and Park, B.-G. 2006. Full-swing pentacene organic inverter with enhancement-mode driver and depletion-mode load. *Solid State Electron.* 50: 1216–1218.

Li, Q. H., Liang, Y. X., Wan, Q. et al. 2004. Oxygen sensing characteristics of individual ZnO nanowire transistors. *Appl. Phys. Lett.* 85: 6389–6391.

Li, Q. H., Gao, T., Wang, Y. G., and Wang, T. H. 2005. Adsorption and desorption of oxygen probed from ZnO nanowire films by photocurrent measurements. *Appl. Phys. Lett.* 86: 123117 (3 pp).

Liao, Z.-M., Liu, K.-J., Zhang, J.-M., Xu, J., and Yu, D.-P. 2007. Effect of surface states on electron transport in individual ZnO nanowires. *Phys. Lett. A* 367: 207–210.

Look, D. C., Reynolds, D. C., Hemsky, J. W., Jones, R. L., and Sizelove, J. R. 1999. Production and annealing of electron irradiation damage in ZnO. *Appl. Phys. Lett.* 75: 811–813.

Lu, W. and Lieber, C. M. 2007. Nanoelectroncis from the bottom up. *Nat. Mater.* 6: 841–850.

Ma, R. M., Dai, L., and Qin, G. G. 2007. Enhancement-mode metal-semiconductor field-effect transistors based on single n-CdS nanowires. *Appl. Phys. Lett.* 90: 093109 (3 pp).

Maeng, J., Jo, G., Kwon, S.-S. et al. 2008. Effect of gate bias sweep rate on the electronic properties of ZnO nanowire field-effect transistors under different environments. *Appl. Phys. Lett.* 92: 233120 (3 pp).

Mieszawska, A. J., Jalilian, R., Sumanasekera, G. U., and Zamborini, F. P. 2007. The synthesis and fabrication of one-dimensional nanoscale heterojunctions. *Small* 3: 722–756.

Ng, H. T., Han, J., Yamada, T. et al. 2004. Single crystal nanowire vertical surrounded-gate field-effect transistor. *Nano Lett.* 4: 1247–1252.

Park, W. I., Yi, G.-C., Miyoung K., and Pennycook, S. J. 2003. Quantum confinement observed in ZnO/ZnMgO nanorod heterostructures. *Adv. Mater.* 15: 526–529.

Park, W. I, Kim, J. S., Yi, G.-C., Bae, M. H., and Lee, H.-J. 2004. Fabrication and electrical characteristics of high-performance ZnO nanorod field-effect transistors. *Appl. Phys. Lett.* 85: 5052–5054.

Park, W. I., Kim, J. S., Yi, G.-C., and Lee, H.-J. 2005. ZnO nanorod logic circuits. *Adv. Mater.* 17: 1393–1397.

Sadek, A. Z., Choopun, S., Wlodarski, W. et al. 2007. Characterization of ZnO nanoblet-based gas sensor for H_2, NO_2, and hydrocarbon sensing. *IEEE Sens. J.* 7: 919–924.

Schmidt-Mende, L. and MacManus-Driscoll, J. L. 2007. ZnO-nanostructures, defects, and devices. *Mater. Today* 10: 40–48.

Sedra, A. S. and Smith, K. C. 1991. *Microelectronic Circuits*, 3rd edn. Orlando, FL: Harcourt Brace College Publishers.

Soci, C., Zhang, A., Xiang, B. et al. 2007. ZnO nanowire UV photodetectors with high internal gain. *Nano Lett.* 7: 1003–1009.

Song, S., Hong, W.-K., Kwon, S.-S., and Lee, T. 2008. Passivation effects of ZnO nanowire field effect transistors under oxygen, ambient, and vacuum environments. *Appl. Phys. Lett.* 92: 263109 (3 pp).

Streetman, B. G. and Banerjee, S. 2000. *Solid State Electronic Devices*. Upper Saddle River, NJ: Prentice Hall.

Suehiro, J., Nakagawa, N., Hidaka, S.-I. et al. 2006. Dielectrophoretic fabrication and characterization of a ZnO nanowire-based UV photosensor. *Nanotechnology* 17: 2567–2573.

Sun, Y., Fuge, G. M., and Ashfold, M. N. R. 2006. Growth mechanisms for ZnO nanorods formed by pulsed laser deposition. *Superlattice. Microstruet.* 39: 33–40.

Sze, S. M. 1981. *Physics of Semiconductor Devices*. New York: John Wiley & Sons.

Takata, M., Tsubone, D., and Yanagida, H. 1976. Dependence of electrical conductivity of ZnO on degree of sintering. *J. Am. Ceram. Soc.* 59: 4–8.

Taur, Y. and Ning, T. H. 1998. *Fundamentals of Modern VLSI Devices*. Cambridge, U.K.: Cambridge University Press.

Wade, T. L., Hoffer, X., Mohammed, A. D. et al. 2007. Nanoporous alumina wire templates for surrounding-gate nanowire transistors. *Nanotechnology* 18: 125201 (4 pp).

Wang, D., Chang, Y.-L., Wang, Q. et al. 2004. Surface chemistry and electrical properties of germanium nanowires. *J. Am. Chem. Soc.* 126: 11602–11611.

Wang, J., Polizzi, E., Ghosh, A., Datta, S., and Lundstrom, M. 2005. Theoretical investigation of surface roughness scattering in silicon nanowire transistors. *Appl. Phys. Lett.* 87: 043101 (3 pp).

Wang, D., Seo, H. W., Tin, C.-C. et al. 2006. Effects of postgrowth annealing treatment on the pthotoluminescence of zinc oxide nanorods. *J. Appl. Phys.* 99: 113509 (3 pp).

Wolkenstein, T. 1991. *Electronic Processes on Semiconductor Surfaces during Chemisorptions*, (ed.) R. Morrison. New York: Consultants Bureau.

Wunnicke, O. 2006. Gate capacitance of back-gated nanowire field-effect transistors. *Appl. Phys. Lett.* 89: 083102 (3 pp).

Xiong, H. D., Wang, W., Li, Q. et al. 2007. Random telegraph signals in n-type ZnO nanowire field effect transistors at low temperature. *Appl. Phys. Lett.* 91: 053107 (3 pp).

Yang, P., Yan, H., Mao, S. et al. 2002. Controlled growth of ZnO nanowires and their optical properties. *Adv. Funct. Mater.* 12: 323–331.

Yeom, D., Keem, K., Kang, J. et al. 2008. NOT and NAND logic circuits composed of top-gate ZnO nanowire field-effect transistors with high-k Al_2O_3 gate layers. *Nanotechnology* 19: 265202 (5 pp).

Yoon, Y., Lin, J., Pearton, S. J., and Guo, J. 2007. Role of grain boundaries in ZnO nanowire field-effect transistors. *J. Appl. Phys.* 101: 024301 (5 pp).

Yu, P. Y. and Cardona, M. 2001. *Fundamentals of Semiconductors: Physics and Materials Properties*. Berlin, Germany: Springer-Verlag.

Zhang, Y., Kolmakov, A., Chretien, S. et al. 2004. Control of catalytic reactions at the surface of a metal oxide nanowire by manipulating electron density inside it. *Nano Lett.* 4: 403–407.

15

C$_{60}$ Field Effect Transistors

Akihiro Hashimoto
University of Fukui

15.1 Introduction

15.1.1 Fullerene (C$_{60}$)

Fullerene is a new carbon material system closely related with carbon micro-clusters and/or carbon ultrafine particles. In general, *micro-clusters* are defined as atomic clusters consisting of less than 1000 atoms. On the other hand, *ultrafine particles* are usually defined as particles consisting of atoms above the upper limit of micro-clusters. The discovery of fullerene was achieved in the process of researching of carbon micro-clusters and/or carbon ultrafine particles. Fullerene is the name of a family of carbon clusters such as C$_{60}$, C$_{70}$, and so on. In this chapter, we will use the word "fullerene" mainly as the name of the C$_{60}$ clusters.

C$_{60}$ clusters were first physically observed as a magic number cluster in the mass spectrum of cluster molecular beams, although their theoretical prediction had already existed in the pioneering research of these clusters. Furthermore, several large-scale production methods for C$_{60}$ clusters were found in a study on the formation process of carbon fine particles from the resistive heating for carbon rods. In a sense, the discovery of the fullerene was due to serendipity. Further information will be provided in Section 15.2.

15.1.2 Solid C$_{60}$ Crystals

High-purity solid C$_{60}$ molecular crystals can be obtained from high-purity C$_{60}$ powders by using the sublimation method or the vacuum evaporation method in an ultrahigh vacuum (UHV) environment similar to the molecular beam epitaxy (MBE). This method is preferable to the solution growth from a solvent for the separation of the C$_{60}$ molecules because of the reduction of residual impurities during growth. The solid C$_{60}$ crystal has a face-centered cubic structure (fcc, $Fm\overline{3}m$ symmetry) above 260 K, and the lattice constant is 1.4154 nm at 270 K. It is well known that each of the C$_{60}$ molecules in the solid C$_{60}$ crystal rotates continuously very fast with a rotation correlation time on the order of 10 ps. The C$_{60}$ molecules behave as a sphere with a diameter of 0.71 nm. Therefore, the solid C$_{60}$ crystal is called "the plastic crystal" with "a rotational diffusion" of the C$_{60}$ molecules. The rotational diffusion means that the degrees of freedom of the molecular rotation are softened before the ones of the translational displacements. The electronic structure of the solid C$_{60}$ has been extensively studied by many groups, and its semiconductor properties such as an energy band gap of about 1.6 eV are commonly known. Moreover, one of the alkali-doped C$_{60}$s, RbCs$_{2}$C$_{60}$, shows superconductivity with a critical temperature of 33 K. In addition, it is well known that all alkali-doped C_{60} systems, in general, show superconductivity. The epitaxy of the solid C_{60} layer has also been extensively studied by many groups known as the van der Waals epitaxy. The concept of lattice mismatches of the epitaxial layers to the substrate is applicable as they are similar to the cases of the conventional semiconductor epitaxy. Therefore, it is a little difficult to obtain high-quality solid C_{60} layers without any dislocations and/or any boundary for high lattice-mismatch combinations even in the van der Waals epitaxy. Further information will be provided in Sections 15.3 and 15.4.

15.1.3 C_{60} Field Effect Transistors (C_{60} FETs)

Fullerene field effect transistors (C_{60} FETs) are typical of the family of the organic FETs (OFETs). It is well known that solid C_{60} FETs have the highest electron mobility in the organic FET family, although most OFETs show the *p*-type, as shown in Figure 15.1 [1]. The various types of the OFET have been extensively studied in recent years, because of their large area fabrication, flexibility, light weight, and low cost. The most essential reason for their extensive research may be due to a resource problem in conventional Si and compound semiconductors. Carbon materials exist almost infinitely on earth and the modification of carbon compounds is very easy to obtain, because of vast chemical knowledge known about carbon compounds. Essentially, organic materials used in the OFETs are high-resistance materials without doping and, moreover, rather unique electronic properties appear even in the doped organic materials due to a strong electron–lattice interaction compared with conventional covalent semiconductors such as Si and GaAs. The doping materials in organic semiconductors, in general, do not form impurity levels

in the mid-gap of the band gap. However, the electrons of highest isolated occupied molecular orbital (HOMO) levels of the doping materials can transfer directly to the lowest unoccupied molecular orbital (LUMO) levels of the organic semiconductor. In general, such charge transfer process is often accompanied with strong lattice distortion. Consequently, the mid-gap level or the impurity band never forms in conductive organic materials. Almost all conductive organic materials have similar electronic properties to intrinsic semiconductor systems even in impurity doping. The conductivity in the organic materials essentially comes from the properties of the electrode metals for organic materials. In that sense, the conventional Shockley–Bardeen theory cannot be applied to OFET operations. The situation for general OFETs is also common for solid C_{60} FETs. Further information will be provided in Sections 15.5 and 15.6.

15.1.4 Bell Laboratory Misconduct

In 2001, a research group in the Bell Laboratory had reported about the high-temperature superconductivity in the solid C_{60} FETs, and consequently an intensive study of C_{60} FET superconductivity has been performed by many research groups. However, none of the groups could show similar high-temperature superconductivity such as presented in the Bell Laboratory's presentation. This story is the beginning of infamous misconduct by Bell Laboratory. The investigation team for misconduct was immediately formed and they concluded that all reports for the solid C_{60} FET superconductivity from the Bell Laboratory were false [2]. One of the causes of the misconduct may be due to the severe competitive circumstances in the U.S. research society based on the market-driven principle in the research field. From a scientific viewpoint, we have not succeeded in forming

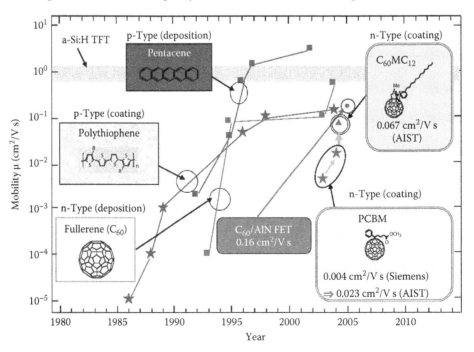

FIGURE 15.1 Improvement of mobility in OFET research field.

an ideal interface between the solid C$_{60}$ and the insulator layers in the FET structure until now. In that sense, the superconductivity problem in the solid C$_{60}$ FET remains until now an open question, in spite of the Bell Laboratory misconduct.

15.1.5 Interfaces between Solid C$_{60}$ and Insulator Films

The interface between the solid C$_{60}$ layer and the insulator layer such as SiO$_2$ is formed mainly by two different ways. The first way is a conventional FET fabrication process technique for C$_{60}$ bulk. The second is an epitaxy of the solid C$_{60}$ layer on a patterned insulator layer with electrode metals. In both ways, there are several problems in forming an ideal interface. Namely, there are damage problems accompanied with the formation process of the insulator layer on the C$_{60}$ bulk surface in the first way, and lattice mismatch problems for the solid C$_{60}$ epitaxy in the second way. The interface problem is very essential for electron transport; therefore, the characteristics of the C$_{60}$ FETs such as *I–V* characteristics are strongly dependent on the interfacial states. It is well known that the *I–V* characteristics depend on adsorption molecules such as OH$^-$ ions during the *I–V* measurements under the atmosphere. The conventional sputtering method to form the insulator layers on the solid C$_{60}$ surface easily results in physical plasma damages from ions and/or the sputtered elements from the target during the deposition of the layers. Therefore, we can only obtain a rather disordered interface like the interface found between the solid C$_{60}$ and the SiO$_2$ layers. On the other hand, even in the second way, we can also only obtain the amorphous C$_{60}$ interface when the solid C$_{60}$ layers are grown on the amorphous surfaces like SiO$_2$. The author's research group has proposed a new insulator layer made of epitaxial single crystalline AlN grown on the SiC and/or the Si substrates for single crystalline solid C$_{60}$ growth. In principle, we can obtain a single crystal epitaxial layer of the solid C$_{60}$ on the single crystalline AlN insulator layer. We will discuss about this new structure and the FET characteristics in the later sections. In these sections, we will see that we can obtain a higher performance C$_{60}$ FET in comparison with the FET structures consisting of conventional amorphous solid C$_{60}$/SiO$_2$. Further information will be found in Sections 15.5 through 15.7.

15.1.6 Field Effect Doping

It is well known that chemical doping in solid C$_{60}$ is not as controllable as with conventional semiconductors such as Si. This issue is the common problem for all the organic materials, including solid C$_{60}$. The key point of the problem is that almost all of organic materials behave as ideal intrinsic semiconductors with low conductivity and without mid-gap states. The carriers come from the source and the drain electrodes into the channel layer induced by an applied gate voltage. The carrier concentrations of the channel layer can be varied by controlling the supply of the carriers from the source and the drain electrodes through the modulation of the applied gate bias voltages. Certainly, in

high-performance OFET, the am-bipolar transistor operation has been observed as direct proof of field effect doping, rather than by conventional chemical doping. Namely, this doping technique is completely different to chemical doping in conventional FET, in which the carriers come from bulk conduction bands due to the band bending at the interface, except for the similarity of the device structures. An important advantage of the field effect doping technique is am-bipolar carrier injections by controlling the polarity of the gate bias voltage. The operation polarity is determined by the combination of metals and organic materials. For example, when gold, Au, is used as an electrode, the channel layer of the C$_{60}$ FET shows n-type conduction and it is said that the solid C$_{60}$ is n-type material.

15.1.7 Operation Principles of C$_{60}$ FETs

When the channel layer is formed by applying the gate bias voltage, the modulation of the gate bias voltage produces modulation of the channel conductance between the source and the drain. The typical characteristics of the drain current and the drain-source voltage (I_D–V_{SD} characteristics) with some gate voltage (as shown in Figure 15.2 as a schematic illustration) consist of three parts: the sub-threshold, the linear, and the saturation regions. In the sub-threshold region, the drain current gradually increases with the drain-source voltage due to the hopping transport mechanism. In the linear region, the uniform channel layer is formed by the applied gate voltages. Therefore, the drain current is proportional to the drain-source voltage just before the "pinch-off" voltage, according to the conventional FET theory. The channel layer just disappears due to a reverse bias voltage near the drain electrode at the pinch-off voltage. After the formation of depletion regions near the drain electrode at the pinch-off voltage, the external excess applied voltage of the drain-source electrodes is used as an extension of the depletion region in the channel layer. Therefore, in this region, the drain current saturates as a function of the applied drain-source voltage. These pictures are commonly observed for conventional semiconductor FET performances, except for the sub-threshold regions. However, in principle, the solid C$_{60}$ FET, and in general, also high-performance OFETs are expected to result in the am-bipolar operation, which is completely different from the

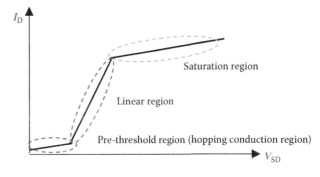

FIGURE 15.2 Schematic illustration of typical *I–V* characteristics of OFET.

conventional semiconductor FET, because of the difference in the origin of the induced channel carriers. In solid C_{60} FETs, we can inject both the electrons and the holes into the channel layers. In principle, we can control the p- and/or the n-channels of FETs by using the polarity of the external applied gate voltage.

15.1.8 Superconductivity

In the future, it is expected that the C_{60} FET will possibly have superconductivity with a considerably high critical temperature, in spite of the Bell Laboratory misconduct. If we can realize the ideal interface between the high-quality intrinsic solid C_{60} layer and the insulator layer, then the am-bipolar operation will be observed with very high mobility. This will be possible, though the present stage for C_{60} FETs is very far from achieving the above-ideal situation even with the highest performances. Therefore, the superconductivity of the C_{60} FET still now remains an open question. It is a very interesting issue, especially related to the holes' injection in the FET structure. We have already proposed a new type of C_{60}/single crystalline AlN FET structures on SiC and/or Si substrates for investigating the possibility of superconductivity with a considerably high critical temperature. However, so far, the preliminary C_{60} FET performance is not at all so high.

15.1.9 Guides for This Chapter

In this chapter, we discuss in detail the basic and the advanced topics about C_{60} FETs. In Section 15.2, we describe the fundamental physical properties of C_{60} molecules, including a brief historical review, a large-scale production synthesis method, an electronic structure of C_{60} molecules, and the significant aspects of C_{60} FETs. In Section 15.3, we describe the various physical properties of solid C_{60} crystals. In Section 15.4, we describe the van der Waals epitaxy of the solid C_{60} layer to be used in a realistic application of the FET structures. In Section 15.5, the basic concept of organic FETs, including the C_{60} FET, will be described in order to understand the C_{60} FET operation. In Section 15.6, the realistic fabrication process of the C_{60} FET is discussed in detail. In Section 15.7, the characterizations of the C_{60} FET performance are described and we discuss about a strategy for the realization of high-performance C_{60} FETs. Finally, we summarize this article in Section 15.8.

15.2 Fullerene (C_{60})

In this section, we describe the minimal knowledge about fullerene, in particular, the C_{60} molecule, to help in understanding the following sections. For those who want to know more details than presented in every section, look in the review and textbook references at the end of this chapter.

15.2.1 Discovery of Fullerene Family

The discovery of fullerene certainly gave a heavy shock to the scientific society, especially in carbon chemistry. In 1985, Curl and coworkers discovered fullerene, the C_{60} molecules, at the Sussex and the Rice Universities [3]. It is well known that the name of "fullerene" is named after Richard Buckminster Fuller who is one of the representative modern architects in the United States. The existence of the C_{60} molecule had already theoretically been predicted as a "corannulene" molecule by Eiji Osawa in 1970. He noticed that the structure of a corannulene molecule was a subset of a soccer-ball shape and, therefore, he predicted that a full ball shape could also exist. Unfortunately, he reported his idea only in a Japanese journal, but news of his discovery did not reach Europe and America. Before the discovery of the fullerene, the only known stable structures of the carbon system were diamond, graphite, and amorphous carbon. Therefore, most people thought that it would be difficult for a new structure of the carbon system to exist, because of the long research history of carbon systems by the chemistry. However, in 1985, Harold Kroto, who was studying the origin of interstellar molecules, was interested in interstellar molecules that have curious absorption and emission spectra. He thought there was a possibility that some carbon systems, produced in the supernova process, were strongly related with the origin of the emission spectra. Initially, Kroto did not think about the C_{60} molecular structure at the beginning of the 1980s, but he thought about some carbon molecules with strange chain structures. In 1984, Kroto visited the Richard Smalley Lab in Rice University to discuss a new proposal for "a laser-vaporization cluster beam mass spectroscopy" (see Figure 15.3) in which a similar condition to the supernova process for the carbon cluster formation seems to have been generated [4]. The experimental results were unexpected. The mass spectra for the laser-ablated carbon systems showed strong intensity in the C_{60} molecule [5–7]. However, Kroto, Smalley,

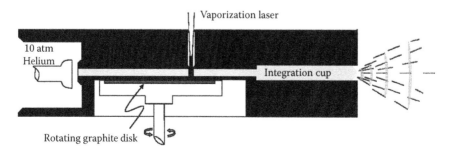

FIGURE 15.3   Laser-vaporization cluster beam mass spectroscopy.

and Robert Curl could not explain at all the stable structure of the C_{60} molecule when they first observed the strong mass spectra from it. It is said that they had a sudden inspiration about the stable structure of the C_{60} molecule from a geodesic dome designed by a famous modern American architect, Buckminster Fuller. The geodesic dome has a mechanically stable structure due to the proper dispersion of the internal force of the dome. They reported their ideas about the stable structure of the C_{60} in an academic paper with several other scientific data. However, many carbon researchers did not pay so much attention to the paper. The researchers needed direct proof such as optical spectroscopic data of the C_{60} for the discussions about the structure of C_{60} molecule to be taken seriously. However, the amounts of C_{60} molecules produced by the laser-vaporization cluster beam mass spectroscopy were not enough at all for an optical spectroscopic measurement in spite of the high generation rate of the C_{60} molecules. It was difficult to construct an apparatus large enough for the laser-vapor evaporation method of the large-scale production of the C_{60} molecules to occur. Therefore, the fullerene proposal by Kroto et al. had to wait until large-scale production methods for the fullerene had improved.

15.2.2 Synthesis of C_{60} Molecules

In 1990, the American and German cooperative research groups found a new "gas evaporation method" for the large-scale production of fullerene [8]. In the new method, a highly purified carbon rod was directly heated by an electric current under the He gas ambient of 100 Torr. The yield of C_{60} molecules by using this method was about 10%. Moreover, Smalley's group also developed another large-scale production method named the "contact arc method," as shown in Figure 15.4, in which the

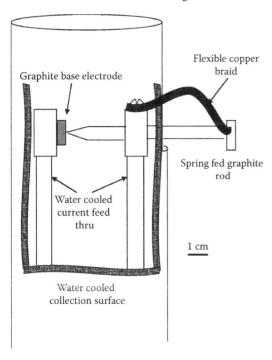

Graphite base electrode

Flexible copper braid

Spring fed graphite rod

Water cooled current feed thru

1 cm

Water cooled collection surface

FIGURE 15.4 Contact arc method.

maximum yield of C_{60} molecules reached at about 13% [9]. Some other large-scale production methods such as the RF-heating method [10] and the gas burning method [11] have been developed by now. They also have been used for the commercial large-scale production of C_{60} molecules.

Typical carbon soot produced by the contact arc method, for example, usually contains 0%–15% fullerene, in which C_{60} molecules consist of 70%–85% of the weight, 10%–15% of the weight is C_{70}, and other higher-order residual fullerene molecules. Krätschmer et al. discovered abstraction methods for obtaining C_{60} molecules from carbon soot produced by large-scale production methods like the gas evaporation method. They found out that fullerene such as C_{60} molecules can be easily dissolved into an organic solvent, while the rest of the carbon cannot be dissolved. They also observed that the sublimation temperatures of C_{60} and C_{70} molecules in UHV are different from each other: 300°C and 350°C for the C_{60} and C_{70} molecules, respectively [8]. The high purity abstraction rate of more than 99.97% of the C_{60} molecules from the soot was achieved by this molecular distillation method using the sublimation apparatus with a constant temperature gradient [12]. Furthermore, very recently, it has been reported that a purity of more than 99.99999% classes has been achieved by a N_2 atmospheric train-sublimation method by Laudise et al. (see Figure 15.5) [13].

On the other hand, liquid chromatography such as open-column chromatography for C_{60} molecules [14] and high-performance liquid chromatography (HPLC) [15] are also very popular and useful methods for the abstraction of C_{60} and higher-order fullerene like the C_{70} molecules. However, in general, the purity of the fullerene obtained by these methods cannot be as high compared with the molecular distillation method because of residues of the organic solvent.

In any case, the C_{60} molecular powder with a purity of 99.98% can be easily bought commercially. For the van der Waals epitaxy of solid C_{60} for the FETs, in general, the C_{60} molecular powder with the purity more than 99.98% has been used.

15.2.3 Properties of C_{60} Molecule

The hypothesis for the structure of the Buckminster fullerene C_{60} molecule proposed by Kroto and Smalley in 1985 was researched for the large-scale production of C_{60} samples using various types of spectroscopy. In particular, there were several important theoretical predictions about the lattice-vibration spectrum, for example, for the infrared absorption (IR) spectrum [16–19]. Theoretical investigations predicted that only the first four-order vibration modes in the IR spectrum were active among 174 degrees of freedom in the lattice vibrations of the C_{60} molecules, when the structure was a soccer-ball shape. Indeed, Krätchmer and Huffman used the four IR modes in the IR spectra to identify the C_{60} molecules in the carbon soot produced by their gas evaporation method, as mentioned in their 1990 academic paper on the large-scale production of fullerene.

According to their theoretical predictions, the C_{60} molecules with I_h symmetry have 46 vibration modes, including many

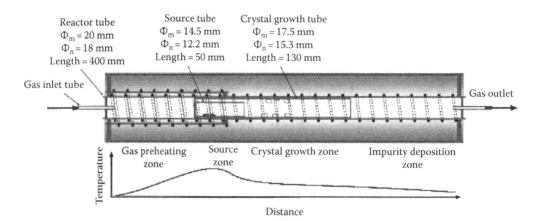

FIGURE 15.5 N_2 atmospheric train-sublimation method.

degenerated modes with 174 degrees of freedom in lattice vibrations. The 46 inner-molecule lattice-vibration modes of the C_{60} molecule were classified by the I_h symmetry as follows [20]:

$$\Gamma(C_{60}) = 2A_g + 3F_{1g} + 4F_{2g} + 6G_g + 8H_g + A_u$$
$$+ 4F_{1u} + 5F_{2u} + 6G_u + 7H_u \qquad (15.1)$$

where the $4F_{1u}$ modes are only allowed for the IR active mode. On the other hand, the 10 modes of the $2A_g$ and the $8H_g$ modes are also allowed as Raman active modes. The IR (A) and the Raman (B) spectra from the solid C_{60} layers deposited on the Si substrate are shown in Figure 15.6. The 4 IR and the 10 Raman peaks in the spectra are clearly observed [21].

Other strong experimental data from the soccer-ball shape of C_{60} molecules can be obtained by ^{13}C nuclear magnetic resonance (^{13}C-NMR) measurements. Indeed, the Kroto research group at Sussex University and the Bethune research group at IBM (Almaden) obtained only one NMR signal due to the high I_h symmetry in the ^{13}C-NMR measurements [14,22]. Moreover, the x-ray diffraction for the $C_{60}(OsO_4)$ single crystal [23] and the neutron diffraction for the low-temperature solid C_{60} crystal succeeded in obtaining more direct proof of the soccer-ball shape structure of

C_{60} molecules after stopping of the high-speed rotation of the C_{60} molecules by using various scientific techniques [24].

Buckminster fullerene consists of 12 pentagons and 20 hexagons that is a geometrical stable combination due to Euler's theorem as follows:

$$V + F = E + 2 \qquad (15.2)$$

where V, F, and E are the numbers of the vertex, the face, and the side of a polyhedron, respectively. Euler's theorem is very helpful for the discussion about the geometrical stability of the higher-order fullerene structures. However, because the chemical stability is not equal to the geometrical one, we have to take into account the electronic structure of Buckminster fullerene. In the simplest calculation using the Hückel approximation, we can estimate the electron energy levels in the C_{60} molecule. When the electron pairs, in which each electron has the opposite spin from each other, occupy the lowest energy level and the electron pairs finally form a closed electron shell, the Buckminster fullerene becomes chemically stable, in addition to being geometrically stable. This is known as the "Hückel rule," and it is used for describing the chemical stability of the fullerene family.

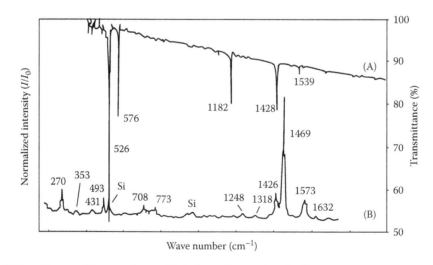

FIGURE 15.6 Infrared (A) and Raman (B) spectra from solid C_{60} layers deposited on Si substrate.

The electronic structure of the C_{60} molecule is very unique because it has many degenerate levels due to high I_h symmetry. In the Hückel approximation, it is estimated that the HOMO level, h_u, is five-folded degenerated and the LUMO level, t_{1u}, is three-folded, and that the energy gap between the HOMO and the LUMO levels is about 1.5 eV. Therefore, it is assumed that the solid C_{60} crystal becomes a semiconductor [25–28]. A scanning tunneling microprobe (STM) observation under the UHV showed that the C_{60} molecules on a well-defined solid surface like Cu (111) or Si (111) 7 × 7 do not spin due to the interaction between the surface and the C_{60} molecules induced by a charge transfer from surface to the C_{60} molecules. Many other interesting properties of the C_{60} molecules are referred to another chapter in this handbook series.

15.3 Solid C_{60}

A solid C_{60} single crystal without any residual organic solvent can be obtained by using the sublimation method or a vacuum evaporation method like MBE, while solid C_{60} crystals grown by solution growth can easily include residual solvent.

15.3.1 Crystal Structure

The structure of the solid C_{60} bulk crystal grown by using the sublimation method has two kinds of structural phases. They are a high-temperature and a low-temperature phase above and below 260 K, respectively [29]. In the high-temperature phase above 260 K, the structure of the solid C_{60} is a fcc lattice with ($Fm\bar{3}m$) symmetry in the space group and the lattice constant is 1.4145 nm. The four C_{60} molecules in the unit cell of the solid C_{60} crystal look like balls with a diameter of 0.71 nm due to the high-speed spin rotation in the x-ray diffraction measurements. Each C_{60} molecule in the solid C_{60} crystal rotates very fast continuously with a rotation correlation time on the order of 10 ps, estimated from the line width of the ^{13}C-NMR spectra [30]. The rotational diffusion property of the C_{60} molecules in the solid C_{60} bulk was shown estimating the neutron inelastic scattering and the rotation potential of the C_{60} molecules that have many local minima with the thermal activation energy of 35 meV [31]. Therefore, the solid C_{60} crystal is called "the plastic crystal" with a rotational diffusion of C_{60} molecules, in which the degrees of freedom of molecular rotation are softened before the translational displacement.

15.3.2 van der Waals Interaction of C_{60}

The lattice vibration of solid C_{60} consists of two kinds of vibration modes, an inner-molecular and an intermolecular vibration. The inner-molecular vibration mode is due to the vibration in the degree of freedom in the C_{60} inner molecule. Therefore, the spectrum of the inner-molecule vibration is very similar to the isolated C_{60} molecule, described in Section 15.2. On the other hand, the intermolecular vibration modes are the proper vibration modes of solid C_{60}, and there are three kinds of modes: acoustic, optical, and libration modes. The phonon frequency of

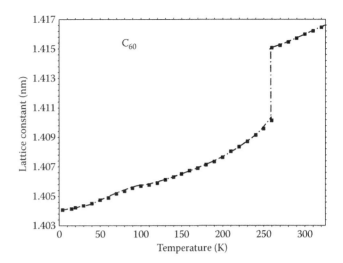

FIGURE 15.7 Lattice constant of solid C_{60} as a function of temperature.

all modes (0–50 cm^{-1}) is lower than the vibration modes in conventional covalent crystal modes such as Si (0–530 cm^{-1}), because of the large mass of the fullerene molecules and the weak van der Waals interaction between the C_{60} molecules. The innermolecular vibration modes have a considerable higher frequency than the intermolecular vibration modes [32].

15.3.3 Structural Phase Transition

The crystal structure of solid C_{60} suddenly changes from a fcc structure to a simple cubic (sc) structure when below 270 K in the first-order phase transition, because of the discontinuity in the variations of the lattice constant, as shown in Figure 15.7 [33].

15.3.4 Electronic Structures

The electronic structure of solid C_{60} has been extensively studied by many groups. They have already revealed certain aspects of its semiconductor nature, for example, the energy band gap of about 1.5 eV. Figure 15.8 shows the band structure of the fcc C_{60} crystal calculated by the density functional method, accompanied with π electron levels from an isolated C_{60} molecule calculated by the Hückel method, as presented by Saito and Oshiyama [28]. The electronic structure has a rather small dispersion and a narrow bandwidth (0.4 eV). They are just the origin of the band gap of fcc solid C_{60}. The solid C_{60} is a direct band-gap semiconductor in which the gap is the difference between the bottom of the LUMO and the top of the HOMO at X point.

The photo-emission spectroscopy (PES) and the inversion photo-emission spectroscopy (IPES) are very useful in directly observing the electronic structures of solid C_{60}. The HOMO–LUMO gap (3.7 eV) estimated from the PES and IPES spectra is larger than the gap estimated by the other techniques such as the absorption measurements. The middle energy points of the HOMO and the LUMO bands are above 2.25 eV and below 1.15 eV from the Fermi level, respectively. It is said that the large band gap in the PES and the IPES is due to the electron correlation effect [34].

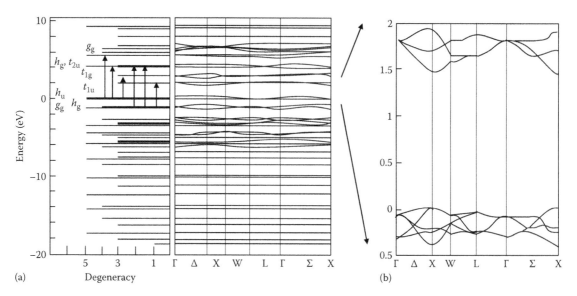

FIGURE 15.8   Band structure of fcc C_{60} crystal calculated by density functional theory.

The HOMO and the HOMO-1 bands, which come from the h_u and the h_g-g_g, respectively, are π electron-like states. These bands are 5- and the 9-degenerated electron states. On the other hand, the empty states consist of the LUMO and the LUMO+1 bands that originate from the t_u and the t_g. They form 3-degenerated states in the empty bands.

The electrical conductivity σ of solid C_{60} is extremely low, that is, from 10^{-8} to 10^{-14} (Ω cm)$^{-1}$. Therefore, solid C_{60} is like an insulator [35,36]. The fluctuation in electrical conductivity is due to impurities and defects. Furthermore, there are difficulties in measuring high-resistance materials. The relative dielectric constant of solid C_{60} is 4.4 ± 0.2 [37]. The relative deviation of the dielectric constant (Δε/ε) is due slightly to the temperature with a small discontinuity at 260 K. This variation is due to the phase transition from the fcc to the sc structures and a broad step-like dependence around 165 K in the low-temperature phase which occurs due to a disorder in the molecule's axis in the pentagon and hexagon directions [38].

15.3.5 Optical Properties

The absorption and luminescence spectra for solid C_{60} deposited on a quartz substrate are shown in Figure 15.9 [39]. Three strong absorption peaks are observed at 221, 271, and 347 nm. These peaks come from the π–π* transition of the C_{60} molecules, because the peaks are also observed from C_{60} molecules dispersed in an organic solvent. The broad and weak bands, observed at around 600 nm, come from the forbidden transition between the t_{1u} and h_u levels. In contrast to these absorption peaks, the rather strong band at 430–520 nm comes from the proper absorption of solid C_{60}. A broad photoluminescence spectrum with peak energy at 737 nm is observed above 10 K, as shown in Figure 15.9b. However, the photoluminescence spectrum at 1.2 K, as shown in Figure 15.10, shows two luminescence bands excited by

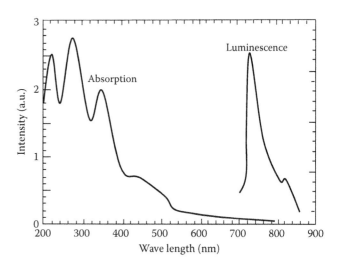

FIGURE 15.9   Absorption and luminescence spectra for solid C_{60} deposited on a quartz substrate.

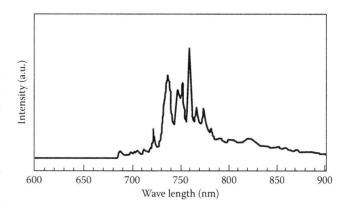

FIGURE 15.10   Photoluminescence spectrum from solid C_{60} at 1.2 K.

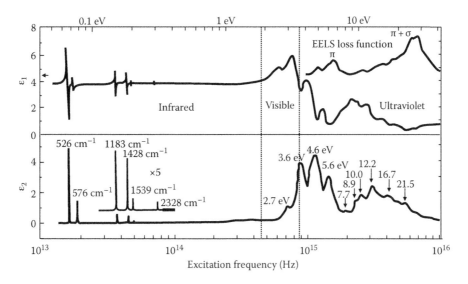

FIGURE 15.11 Optical measurements results from infrared to ultraviolet wavelengths for $\varepsilon_1(\omega)$ and/or the $\varepsilon_2(\omega)$.

734 and 684 nm. The 734 and the 684 nm excited bands consist of the electronic transition and a vibronic transition [40].

The dielectric function, $\varepsilon(\omega)$, consists of the real part $\varepsilon_1(\omega)$ and the imaginary part $\varepsilon_2(\omega)$, as follows:

$$\varepsilon(\omega) = \varepsilon_1(\omega) + i\varepsilon_2(\omega) \qquad (15.3)$$

The complex refractive index, $n_c(\omega)$, is the root of the dielectric function:

$$n_c(\omega)^2 = \varepsilon(\omega) \qquad (15.4)$$

in which $n_c(\omega)$ consists of the real part, the refractive index $n(\omega)$, and the imaginary part, the extinction coefficient $\kappa(\omega)$. The $\varepsilon_1(\omega)$ and the $\varepsilon_2(\omega)$ are connected with the $n(\omega)$ and the $\kappa(\omega)$, as follows:

$$\varepsilon_1(\omega) = n(\omega)^2 - \kappa(\omega)^2$$
$$\varepsilon_2(\omega) = 2n(\omega)\kappa(\omega) \qquad (15.5)$$

These optical constants can be obtained from the absorption and the reflection spectra. The dielectric function is the most important and basic physical quantity concerning the optical properties of the solid C_{60} layers. The typical results of the optical measurements results from the infrared to the ultraviolet wavelengths for the $\varepsilon_1(\omega)$ and/or the $\varepsilon_2(\omega)$ are shown in Figure 15.11 [41]. In the measurements, the solid 100 nm thick C_{60} layers were deposited on proper substrates such as Si and glass by using the vacuum evaporation method. Many coupled modes were observed in the thick solid C_{60} spectra in addition to inner-molecular F_{1u} vibration modes (526, 576, 1183, and 1428 cm^{-1}) and their two coupled modes (1539 and 2328 cm^{-1}), roughly below 0.3 eV. A solid C_{60} film is also assumed to be a nonlinear optical system due to the π-electrons in the molecules. Third-harmonic generations were reported by Kajzar. The third nonlinear susceptibility as a function of the wavelength is shown in Figure 15.12 [42].

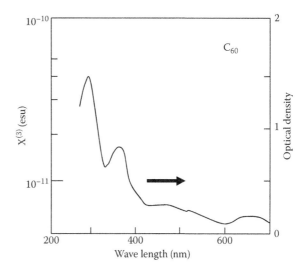

FIGURE 15.12 Third nonlinear susceptibility as a function of the wavelength.

The photo-conducting spectra of the solid C_{60} deposited on glass substrates are shown in Figure 15.13 [39]. The photo-conducting spectrum from the thin C_{60} layer just corresponds to the absorption spectrum. However, the behavior of the photo-conductivity is out of phase with the absorption spectrum, as shown in Figure 15.13. The photo-conducting spectra show that the photo-active region of solid C_{60} is only limited to around the thin surface region and, therefore, the diffusion length of the photo-produced electrons and/or holes is very short. The electron-hole pairs can be produced within the thin small region, because the photons cannot penetrate deeply into the solid C_{60} layers. There is a decrease in one order of magnitude of the photo-conductance due to oxygen exposure, because the oxygen molecules absorbed into the solid C_{60} layer work as recombination centers for the carriers, decreasing their lifetime and mobility. The absorption of the oxygen molecules into the solid C_{60} layers

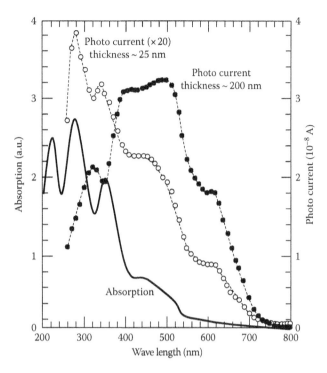

FIGURE 15.13 Photo-conducting spectra of solid C_{60} deposited on glass substrates.

is a reversible process. In addition, photo-conductance recovering accompanied with oxygen adsorption is observed by the 180°C temperature mark annealing under a vacuum. However, a large irreversible deterioration of photo-conductance is induced by white light irradiation exposure to oxygen in the solid C_{60} layers, because of a chemical reaction between the oxygen atoms and the C_{60} molecules induced by white light irradiation.

The polymerization of solid C_{60} is induced by ultraviolet strong light irradiation and/or a low-energy (3–1500 eV) electron beam. We can check the polymerization in several ways; for example, by using the Raman frequency shift. The polymerization by ultraviolet light and the low-energy electron beam is due to an excitation of the electrons in the double bonds of the hexagons and the following reconstruction of new bonds structures between the C_{60} molecules [21,43]. However, the oxygen contamination in solid C_{60} layers has a suppression effect in the polymerization process, because the oxygen atoms work as the killer centers of the excited electrons before the formation of new bonds. Furthermore, the transition from dimmer to monomers is also induced by a thermal annealing at 200°C under an UHV environment. The activation energy of this process is estimated at about 1.25 eV [44].

15.4 van der Waals Epitaxy of C_{60}

The epitaxial growth of solid C_{60} on several proper substrates offers very attractive potential due to its new electrical and optical properties, provided we can obtain the ideal interface between solid C_{60} and the insulator layers. The intermolecular

distance on the fcc solid C_{60} film is about 1 nm. This is several times longer than the inter-atomic distance on the top surface of semiconductor substrates such as GaN. The atomic arrangement of the GaN (0001) surface has a sixfold symmetry similar to a fcc solid C_{60} (111) surface. The inter-atomic distance GaN (0001) is about 0.319 nm, which is nearly one-third of the intermolecular distance of the C_{60} molecules in a fcc solid C_{60} (111) surface. If we define the lattice-mismatch as the difference between the intermolecular distances on the C_{60} (111) surface [45] and the distance is almost equal to three times the inter-atomic one on the GaN (0001) surface, then the lattice-mismatch becomes about 4.5% which is not so serious of a difference in the lattice-mismatch of solid C_{60} epitaxial growth. Additionally, it is well known that nitride is rather stable surface for chemical reactions. Therefore, this surface is assumed to be a chemically inactive plane. This plane is necessary for the epitaxy of a van der Waals molecular crystal without any covalent bonds to substrates [46]. It is equally important in the formation of a van der Waals interface in layered materials such as Mica [47] and MoS_2 [48]. Consequently, the solid C_{60} epitaxial growth on the proper substrates can be expected to appear under proper growth conditions. We call the epitaxy like this (solid C_{60} epitaxial growth without any covalent bonds to the substrates, except for the potential periodicity), "the van der Waals epitaxy." In this section, the basic knowledge about the epitaxial growth of a solid C_{60} single crystal layer is described.

15.4.1 Molecular Beam Epitaxy

Conventional solid source MBE is named after the epitaxial growth using molecular beams such as the C_{60} beam produced by conventional Knudsen thermal cells (K-cells) under an UHV environment below a background pressure of lower than 10^{-7} Pa. The conventional MBE system usually consists of a main chamber for epitaxial growth, a preparation chamber for the surface cleaning of the substrate, and an introduction chamber for the substrate charge from the atmosphere, as shown in Figure 15.14. Each chamber has a pumping system with several vacuum

FIGURE 15.14 Conventional MBE system.

monitoring systems such as a nude ion gauge in order to keep the UHV environment, and all chambers are connected via gate valves with each other. The main chamber has a sample holder with a sample heating system and a monitoring system for epitaxy. For example, a reflection of high-energy electron diffraction (RHEED) system and a quadrupole mass spectroscopic (Q-Mass) system are used for monitoring the growth surface and the background ambient during epitaxy, respectively. The main chamber should maintain the UHV environment as much as it is possible in order to suppress contamination during growth, allowing a cooling bag using liquid nitrogen or a water circular system to settle inside the chamber. The preparation chamber has a thermal heating system for sample cleaning and, in many cases, other in-situ characterization systems such as the Auger electron spectroscopic (AES) system in order to characterize the sample surface before and/or after epitaxy. The substrate prepared by the proper chemical cleaning treatment is installed into the introduction chamber from the atmosphere to the high vacuum environment. Then, the substrate is transferred from the introduction chamber to the preparation chamber through the gate valve to clean the surface and, occasionally, to perform some in-situ characterization under a UHV environment before growth. After the surface cleaning, the substrate is transferred into the main chamber for growth. In the main chamber, the K-cell and/or the substrate heating systems should fully do the out-gas procedure before the installation of the substrate. The beam flux also should be estimated by a nude ion gauge monitor before and after growth, if possible. The details of the general information of MBE growth and systems can be obtained from the proper reference books.

15.4.2 van der Waals Epitaxy

van der Waals epitaxy is an important technique in overcoming the restrictions in large lattice-mismatches in the crystal growth. According to Koma, the van der Waals interface can drastically relax the lattice-matched growth conditions, because of a weak interaction between the films and the substrates [46]. There has been much interest in using single crystalline C_{60} thin films grown by van der Waals epitaxy on hydrogen-terminated (H-terminated) Si substrates. These newly functional films have a great potential in many areas due to the discovery of the large-scale production synthesis methods of C_{60} molecules. However, it has still been difficult to obtain single-domain C_{60} thin solid films with the fcc structure on Si substrates, though single-domain C_{60} films can be grown on layered material substrates such as MoS_2 [48], GaSe [49], and Mica [47]. The growth of C_{60} films on H-terminated Si (111) substrates with quasi van der Waals interfaces shows a three-dimensional island growth mode. In addition, the film's grown structure contains a fcc double domain. Therefore, a large fcc single-domain film has not been obtained yet to the best of our knowledge.

The most popular device structures of C_{60} FETs is a bottom contact structure with a SiO_2 insulator layer formed on a conductive Si substrate as a gate electrode, and the details will be described in the later sections. In the bottom contact structure with the SiO_2 insulator layer, the C_{60} layer becomes an amorphous or a polycrystalline structure due to the amorphous surface of the SiO_2 layer. If we want to use the single crystalline solid C_{60} surface, then we have to apply the top contact structure to the single crystalline solid C_{60} bulk. The solid C_{60} bulk can only be obtained on the inside of the furnace wall by methods such as train sublimation. However, the obtained single crystal is usually very small. Although it is not impossible to make C_{60} FETs using such tiny C_{60} bulks, interface problems such as the ion damage always occur in the formation process of the insulator layer when using such a sputtering method. Therefore, in the current state-of-the-art technology, it is very difficult to obtain an ideal interface between solid C_{60} and insulator layer in both FET structures. There is a lot of discussion about the use of the bottom contact structure in the C_{60} FET because of the easy fabrication process, even for amorphous-like interfaces.

15.4.3 Initial Growth Surface

Here, we will discuss the relationship between the van der Waals epitaxy and the initial growth surface. We have already pointed out that the interaction between the C_{60} molecules and the initial surface plays an important role in the epitaxial growth of the solid C_{60} layer. Therefore, we discuss several types of surfaces such as the H-terminated Si (111) surface and/or the single crystalline AlN surface.

15.4.3.1 H-Terminated Si (111) Surfaces

In Figure 15.15, typical atomic force microscope (AFM) image and RHEED pattern of C_{60} epitaxy on an H-terminated Si (111) surface are shown. Petal-like islands with considerable low density are formed on the H-terminated Si (111) surface. The results strongly indicate that the van der Waals interaction between C_{60} molecules and the surface is not so strong and, therefore, that the migration length of the C_{60} molecules on a quasi van der Waals surface like the H-terminated Si (111) surface is considerably longer than conventional epitaxy such as with Si layers. However, we can control the island density and size by using a two-step growth technique, as shown in Figure 15.15. The two-step growth technique is well known as one of the most important techniques in achieving a high-quality single crystalline for high lattice-mismatch hetero-epitaxial systems. According to the two-step growth procedure, the C_{60} layers are grown at 150°C to enhance the surface migration of C_{60} molecules, after the formation of low-temperature buffer layer growth at 30°C successively annealed at 150°C to form a high nucleus density, respectively. The typical beam equivalent pressure of the C_{60} molecular beam (purity; >99.99%) was 5.0×10^{-9} Torr during growth, and the beam pressure corresponds to a growth rate of 0.8 nm/min [50].

A RHEED pattern from the grown layer by using the two-step growth technique showed a fcc single-domain structure, as shown in Figure 15.16. However, the RHEED pattern from the directly grown layer showed a double-domain fcc structure in

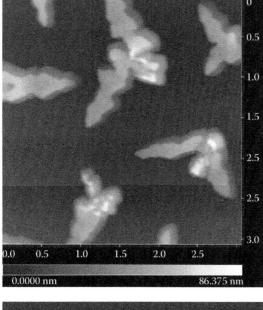

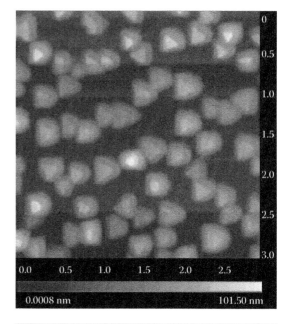

FIGURE 15.15 Typical AFM image and RHEED pattern of C_{60} epitaxy on an H-terminated Si (111) surface.

FIGURE 15.16 Typical AFM image and RHEED pattern of C_{60} grown layer by using two-step growth technique.

which the normal and the rotated domains on the Si substrate mixed with each other, as shown in Figure 15.16. According to Saito, in the case of C_{60} epitaxy on NaCl substrates, not only hexagonal (or truncated triangular) plates (which are single crystals) but also particles with irregular shapes which look like petals are observed. Such petal-like particles are rich in planar defects such as parallel and cyclic twining [47]. Therefore, the triangle islands formed in the two-step growth have a single-domain fcc structure. The petal-like islands formed in the direct growth have a double-domain fcc structure. These well-arranged dense islands on the surface obtained by the two-step growth may have a potential to form thick C_{60} films with the fcc single-domain structure.

15.4.3.2 Hexagonal GaN (0001) Surfaces with Ga-Polarity

The epitaxial growth of solid C_{60} on semiconductor substrates offers a very attractive potential due to its newly discovered

electrical and optical properties. The atomic arrangement of the hexagonal GaN (0001) surface has a sixfold symmetry similar to the fcc solid C_{60} (111) surface. Its inter-atomic distance is about 0.319 nm, which is near one-third of the intermolecular distance of the C_{60} molecules in fcc solid C_{60} (111) surfaces. An extended lattice-mismatch in the van der Waals epitaxy can be defined as the difference between the intermolecular distance in the C_{60} (111) surface and the distance equal to three times the inter-atomic one in the h-GaN (0001) surface. The extended lattice-mismatch then becomes 4.5% which is the lowest among conventional semiconductor substrates in solid C_{60} epitaxial growth. Additionally, it is well known that the h-GaN surface is rather stable for chemical reactions. Therefore, the surface is expected to act as a chemically inactive plane which is necessary for van der Waals epitaxy as well as the van der Waals interface in layered materials such as Mica and MoS_2 [51].

FIGURE 15.17 Typical RHEED patterns from solid C$_{60}$ layer grown on flat h-GaN (0001) surface at 100°C for 60 min.

The GaN (0001) surfaces formed on the α-Al$_2$O$_3$ substrates are prepared by a metal organic vapor phase epitaxy. Typical RHEED patterns from a solid C$_{60}$ layer grown on a flat h-GaN (0001) surface at 100°C for 60 min are shown in Figure 15.17. The directions of the incident electron beam are parallel to (a) [1120] and (b) [11$\bar{1}$0] crystal axes of the h-GaN (0001) surface. The in-situ RHEED observation revealed that the C$_{60}$ film growth did not occur at 150°C because of the re-evaporation of the C$_{60}$ molecules from the surface. On the other hand, sharp streaks from the fcc (111) surface can be observed in the RHEED patterns, as shown in Figure 15.17. The results imply that the lattice-structure of the grown layer is mainly the fcc structure, and the relationship of the crystal axes between the epitaxial grown layers and the h-GaN (0001) substrates is C$_{60}$ (111)‖h-GaN (0001), C$_{60}$ [10$\bar{1}$]‖h-GaN [11$\bar{2}$0], C$_{60}$ [11$\bar{2}$]‖h-GaN [01$\bar{1}$0], as shown in Figure 15.18. Figure 15.19 shows a typical AFM image of the C$_{60}$ layer on a flat h-GaN (0001) surface.

FIGURE 15.19 Typical AFM image of C$_{60}$ layer on flat h-GaN (0001) surface.

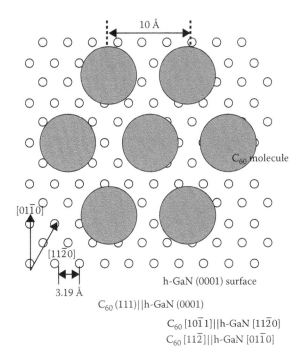

FIGURE 15.18 Relationship of crystal axes between epitaxial grown layers and h-GaN (0001) substrates.

The whole surface of the substrate was completely covered with highly dense C$_{60}$ islands. The size of the islands was on the order of a sub-micron meter. Figure 15.20 shows the growth temperature dependency on grain size and density estimated from AFM images, as shown in Figure 15.19, in C$_{60}$ growth on h-GaN and H-terminated Si. In the case of C$_{60}$ growth on the h-GaN, the grain size slowly increases and the grain density rapidly decreases as a function of the growth temperature in comparison with the H-terminated Si substrate. The results indicate that the interaction between the C$_{60}$ molecule and the h-GaN surface is extremely weak compared with C$_{60}$ film growth on a H-terminated Si (111) surface. The evaporation of the C$_{60}$ molecules on the GaN surface may very well easily occur and, consequently, it leads to a reduction of the diffusion length of C$_{60}$ molecules during growth. Short diffusion and/or migration lengths may induce highly dense small grains, as shown in Figure 15.20.

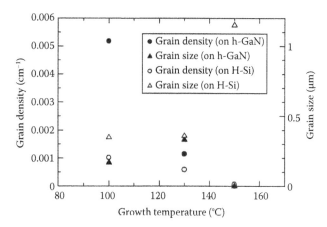

FIGURE 15.20 Growth temperature dependency on grain size and density estimated from AFM images, as shown in Figure 15.19.

The experimental results for C_{60} direct growth on rough h-GaN (0001) surfaces show that a polycrystalline C_{60} layer can only be obtained. Furthermore, it suggests that the surface migration of the C_{60} molecules may be suppressed by the existence of various planes on the rough surface. The fcc C_{60} layers can be grown epitaxially even on a rough h-GaN (0001) surface by using the two-step growth technique, though only the polycrystalline C_{60} layers are obtained by direct growth. All experimental results strongly indicate that the periodicity of the h-GaN surface potential for the van der Waals epitaxy is easily and strongly modified by surface roughness. Therefore, the flatness of GaN substrate is also a very important factor in obtaining a high-quality C_{60} epitaxial layer.

15.4.3.3 Hexagonal AlN (0001) Surface with N-Polarity

The AlN single crystal layer can be grown on 6H-SiC [53,54], Si [55,56], and so on. The single crystal AlN layer has large bandgap energy (6.2 eV) and enough high dielectric breakdown voltage for C_{60} FET applications. Lattice-mismatch between the C_{60} fcc structure (111) and the h-AlN (0001) is 7.3% in C_{60} [10$\bar{1}$]∥AlN

[11$\bar{2}$0] (the 0° orientation) and 7.0% in C_{60} [11$\bar{2}$]∥AlN [11$\bar{2}$0] (the 30° orientation), as shown in Figure 15.21. These values are acceptable under C_{60} van der Waals epitaxy [45]. Therefore, the AlN layer is assumed to be the substrate for single crystal C_{60} layers. However, the lattice-mismatch in both orientations, shown in Figure 15.21, is almost the same. Because the C_{60} van der Waals epitaxy has a character that the growth orientation on the smooth surface strongly depends on the lattice mismatch [45], it is a little difficult to predict the C_{60} orientations on the AlN surface. If the single crystal C_{60} layers can be grown on a AlN surface, a more perfect interface can be obtained than by an interface formed by the conventional sputtering method.

Figure 15.22 shows typical RHEED patterns from a solid C_{60} layer grown on a smooth AlN surface (RMS value: 0.9 nm) for 2 h at 100°C–160°C. All the RHEED patterns in Figure 15.22 indicate a fcc (111) structure. The ring pattern appears on a spot pattern below a growth temperature of 100°C. Both the RHEED patterns with the azimuth of electron beam paralleled to AlN [10$\bar{1}$0] and AlN [11$\bar{2}$0] at 100°C–130°C show the inclusion of C_{60} [10$\bar{1}$] and C_{60} [11$\bar{2}$] patterns, as shown in Figure 15.22. Moreover, twin patterns are also observed. On the other hand, the C_{60} [10$\bar{1}$] pattern is only observed with a AlN [11$\bar{2}$0] azimuth of the electron beam at 160°C, and the C_{60} [11$\bar{2}$] pattern is only observed with a AlN [10$\bar{1}$0] azimuth of the electron beam at 160°C, respectively, although the twin patterns are still observed at 160°C.

From the RHEED results we can summarize that the C_{60} layers that are grown at 100°C–130°C contain four kinds of grains, which are the C_{60} [10$\bar{1}$]∥AlN [11$\bar{2}$0] (the 0° orientation) and its twin grains and the C_{60} [11$\bar{2}$]∥AlN [11$\bar{2}$0] (the 30° orientation) and its twin grains, as shown in Figure 15.23. However, the RHEED intensity from the 0° orientation is always stronger than that from the 30° orientation. Figure 15.24 shows the intensity ratio of the 0° orientation grain as a function of the growth temperature. In this figure, the 0° orientation grains are grown more than the 30° orientation grains, and its ratio is about 7:3. The C_{60} layers that are grown at 160°C contain only the 0° orientation grains, although the twin grains still remain. The results strongly

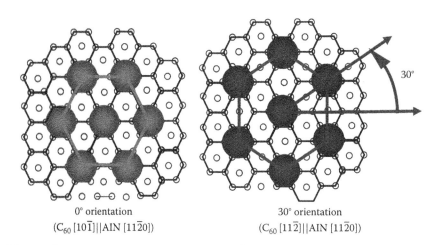

0° orientation
(C_{60} [10$\bar{1}$]∥AlN [11$\bar{2}$0])

30° orientation
(C_{60} [11$\bar{2}$]∥AlN [11$\bar{2}$0])

FIGURE 15.21 C_{60} [10$\bar{1}$]∥AlN [11$\bar{2}$0] (the 0° orientation) and 7.0% in C_{60} [11$\bar{2}$]∥AlN [11$\bar{2}$0] (the 30° orientation).

• The RHEED patterns from the solid C_{60} layer on smooth AlN surface

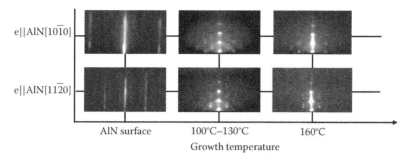

FIGURE 15.22 Typical RHEED patterns from solid C_{60} layer grown on smooth AlN surface (RMS value: 0.9 nm) for 2 h at 100°C–160°C.

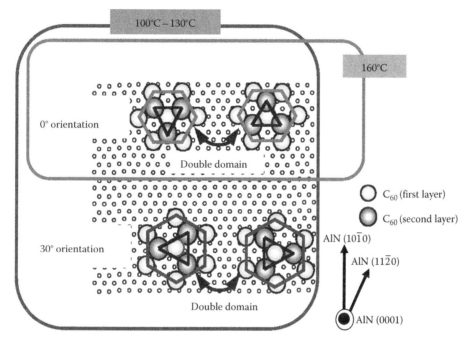

FIGURE 15.23 Summarized illustration of epitaxial relationship between solid C_{60} and AlN (0001) surface.

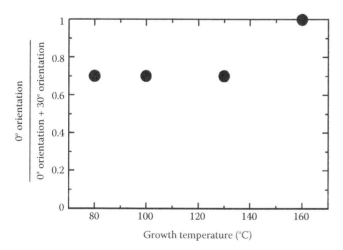

FIGURE 15.24 RHEED intensity ratio of 0° orientation grain as a function of growth temperature.

indicate that the C_{60} grains with the 0° orientations are stable on a smooth AlN surface in high-temperature growth, although the lattice-mismatch for each orientation is almost equal.

The solid C_{60} grains grown on the smooth AlN surface for 5 h at 100°C–160°C showed a three-dimensional island growth. Their grain density and grain size as a function of growth temperature show a trade-off relation which is strongly connected to the migration of C_{60} molecules and the covering of C_{60} layers on the AlN surface. The RHEED pattern for the solid C_{60} layer grown on the rough AlN (RMS value: 9.5 nm) surface at the 130°C shows a mixed pattern with a spot and a ring. The RHEED result indicates that the orientation of the grains is not controlled at all. Such random orientation can be improved by thermal annealing at 130°C after the low temperature growth at 30°C. However, the obtained C_{60} layers have multi-domain structures. The results indicate that the roughness of the AlN surface is also an important factor for C_{60} epitaxy. The RHEED pattern from the solid C_{60} layer grown on the polycrystalline AlN (RMS value: 0.7 nm) at 130°C shows a full ring pattern, which means the polycrystalline C_{60} layer is grown on the surface. It strongly indicates that the periodic atomic arrangement of the surface is also an important factor in C_{60} epitaxy as well as the surface roughness [52].

15.5 Principles of Organic FET Functions

The first report on the mobility measurement of an organic semiconductor material, merocyanine, using the field effect was published in the *Japanese Journal of Applied Physics* in 1984 by Kudo et al. [57]. Just after the first report on the field effect of the organic semiconductors, Koezuka et al. described the first OFET using polythiophene in 1987 [58]. After these innovations by Japanese researchers, the study of organic FETs has been extensively developed through the last two decades. The mobility of the carrier has achieved the same level as the ploy Si FETs in the late 1990s, as shown in Figure 15.1. Now, it has been reported that the highest mobility of organic FETs is 40 cm²/V s in a single crystalline rubrene FET [59]. According to Haddon et al., the solid C_{60} layers were utilized at the active element in thin film FETs [60]. However, the mobility of the FET was found to be rather low (10^{-4} cm²/V s) by Tang et al. [61]. Very recently, a group from the University of Tokyo has reported the highest electron mobility of 6.0 cm²/V s in the C_{60} FET structures at a low temperature [62]. The developments of the OFETs performance have been rapid in recent years like other organic devices such as organic electro-luminescence (OEL) [63]. In this section, we discuss about the general principles of OFETs, including C_{60} FETs for a deeper understanding of C_{60} FET operation in the later sections.

15.5.1 Device Structures of OFETs

Organic transistor materials are classified mainly into two categories: low- and high-molecular compound systems such as rubrene and polythiophene, respectively. In general, low-molecular compound systems have a rather high mobility. Consequently, the research on the OFETs using low-molecular systems has been intense in recent years. Therefore, in this section, we will discuss about FETs using low-molecular compound systems. The various types of FET structures such as vertical structure [64] and the Schottky gate structure [65] have already been reported. However, in this section we will discuss metal insulator semiconductor (MIS) structures, which are the most popular FET structures. In the OFET structure, mainly two types of structures are usually used in research: (a) bottom and (b) top contact structures, as shown in Figure 15.25a and b, respectively. The bottom contact structure is the most popular structure in OFETs, because a fine-patterning technique for the short channel is easily applicable just after the deposition of the electrode metal, as shown in Figure 15.25a. The organic active

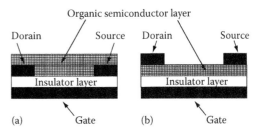

FIGURE 15.25 Schematics of (a) bottom and (b) top contact FET structures.

elements are deposited on the source and drain electrodes and the insulator layer after the fine-patterning of the electrodes. In general, in the bottom contact structure, the considerably large height difference between the electrodes and the insulator surface becomes an origin of high contact resistance between the active organic element and the metal electrodes. The discontinuity between the electrode and the organic materials at the edge of the electrodes becomes an origin of fluctuation in the FET performance such as in the *I–V* curves. In contrast, active organic elements are deposited on a flat insulator surface before the electrode formation in the top contact structure. However, achieving the fine-patterning of the organic semiconductor material systems is considerably difficult. Therefore, the metal electrodes formation is usually performed by using a metal-mask vacuum evaporation method. Moreover, a high contact resistance between the top metal electrodes and the interface of the insulator–organic layer in the active channel region cannot be essentially avoided. In practice, the contact resistance is lower than the case of the bottom contact FET, because the metal atoms of the electrodes may diffuse into the organic layer. In general, SiO_2 layers thermally grown on a Si substrate are used as the insulator layer and gold is used as source and drain electrode. The interface between metal and organic semiconductor becomes a Schottky junction because of the large difference in the carrier density between the metal and organic materials without doping. Therefore, the off-current is very low due to the large contact resistance in the OFET operations.

15.5.2 Metal/Organic Layer Interface

Organic semiconductors show intrinsic semiconductor nature and have very low impurity states in the band gap. Therefore, carriers excited by thermal energy play an essential role in the current conduction of OFETs. Both the electrons and the holes contribute to the carrier transport in the same manner as in ambipolar operations. However, many experimental reports have been published on *p*-type materials such as rubrene and *n*-type materials such as solid C_{60} layers. Such a material dependency on the conducting types in spite of an essentially am-bipolar nature in organic semiconductors had been a large mystery in the OFET research for many years. However, very recently, Yasuda et al. have reported about the am-bipolar operations in OFETs such as pentacene or Cu-ftalocianina which had been considered a *p*-type OFET [66]. According to them, the am-bipolar operation can be realized by using a potassium electrode with a low work function, instead of the Au with a considerably high work function. Moreover, Nishikawa et al. have reported about the am-bipolar operation of OFETs using the work function as a control by using a modification technique that uses self-organized mono-layered molecules on the surface of Au electrodes [67]. These results strongly indicate that the formation of the carrier accumulation layer at the interface of OFETs is completely different from that of conventional semiconductor FETs. Namely, although the carrier accumulation layer for conventional semiconductor materials is formed by impurity doping into bulk, the carrier injection from the electrode plays an essential and important role for the formation of the carrier accumulation layer in OFET structures. According to Tamura et al., the formation of the accumulation layer in OFETs can be explained consistently using by the Maxwell–Wagner model for dielectric materials. In other words, the formation of the carrier accumulation layer can only be realized by carrier injection from the electrode with a low injection barrier height. However, the carrier accumulation layer cannot be formed for a high injection barrier [68]. The results of the analysis of the formation of accumulation layers indicate that the type of carriers of the accumulation layer can be determined by the relationship between the work function of the electrode metals and the HOMO–LUMO levels of the organic semiconductor materials, that is, the Schottky barrier between the electrode metals and organic semiconductor surfaces.

15.5.3 Field Effect Doping for Solid C_{60} Layers

A field effect doping technique has been proposed as a new type carrier doping that is applied to the interface between solid C_{60} and insulator layers by applying an external voltage to it in a similar manner to conventional field effect transistor (FET) [60]. In this doping technique, the solid C_{60} layer is treated as a semiconductor that was theoretically predicted by Saito and Oshiyama [28]. Injected career density and/or polarity through the interactions between the electrodes metals and the solid C_{60} layer can be controlled by the value and the polarity of external gate voltage in this technique. This point is a great advantage in the field effect doping for C_{60} device fabrication techniques in comparison with conventional chemical doping. However, several serious problems need to be resolved before there are realistic device applications using the field effect doping technique. The most important problem to be fixed is how to form a well-controlled interface between the insulator and the solid C_{60} layers. The channel region is formed in several mono layers from the interface. In the fabrication of the FET structure, an excellent C_{60} interface without any damage can be formed by the deposition of a C_{60} vacuum on a single crystal insulator layer such as the AlN layer (as shown in Figure 15.26) because the AlN single crystal layer can be grown on 6H-SiC [53,54], Si [55,56], and so on. In the conventional sputtering method for insulator film deposition, it is easy to damage the interface region. Therefore, it is a little difficult to form a reproducible and excellent interface. On the other hand, the single crystalline AlN

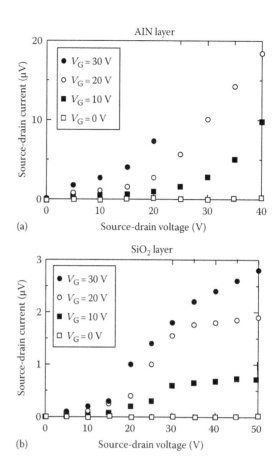

FIGURE 15.26 Typical I_D–V_{SD} characteristics of fabricated C_{60} FETs with (a) conventional SiO_2 layers and with (b) newly proposed AlN layers.

layer has the large band-gap energy (6.2 eV). In addition, it also has enough high dielectric breakdown and dielectric constant for C_{60} FET applications. Consequently, it is expected that the interface between the single crystalline solid C_{60} layers grown on a single crystalline AlN insulator layer will become one of the most promising structures for field effect doping.

15.5.4 Am-Bipolar OFET Operations

In general, organic semiconductors such as solid C_{60} films are essentially intrinsic semiconductors, with the characteristics of low carrier density and an absence of impurity levels in the band gap. Therefore, if we can remove the extrinsic factors such as the absorption of gas and the carrier traps and can choose the proper electrode metals, an am-bipolar transistor can be easily realized in contrast to conventional semiconductors. There are many ways to realize a practical am-bipolar transistor such as fabricating the device under a vacuum or using inert gas ambient to prevent the absorption gas effect. Another typical idea is the usage of an asymmetric Schottky barrier source–drain electrode structure using different metals with large or small work functions from each other. Moreover, organic polymers such as polymethyl methacrylate (PMMA) can be applied as insulator layers to remove the electron traps on the SiO_2 surface. The electrons and the holes can be injected simultaneously under am-bipolar operation. Therefore, it is expected that a pseudo-PN junction will result in light emission due to a recombination of simultaneously injected electrons and holes in the channel. Am-bipolar light-emitting transistors have been reported in the carbon nanotubes, the organic polymers, and the tetracene channel FETs [69–71]. Unfortunately, however, am-bipolar operations have not been observed in solid C_{60} FETs until now for no essential reason.

15.6 Fabrication of Fullerene Field Effect Transistors (C_{60} FET)

The most important problem in the fabrication of the C_{60} FETs is how to form an effective interface between the insulator and the solid C_{60} layer. In this section, we discuss a new structure of the C_{60} FETs, which consists of an epitaxial AlN layer as the insulator and a solid C_{60} epitaxial layer grown on a AlN (0001) surface to realize the ideal interface, in comparison with the most popular structure of C_{60}/SiO_2 FETs. First of all, the solid C_{60} growth on the AlN (0001) surface by using the MBE technique is described again, although we have already described it in Section 15.4. Furthermore, how C_{60} islands with a fcc structure can be obtained on a smooth single crystalline AlN surface has been explained. The experimental results show that four types of C_{60} islands, in general, are obtained: those with 0° of orientation and its twin ones, those with 30° of orientation and its twin ones. However, only the islands with 0° orientation can be obtained by controlling of the solid C_{60} growth temperature. Such results strongly indicate that an effective interface between the AlN insulator and the solid C_{60} layer can be formed for device applications. The fabrication of polycrystalline solid C_{60}/AlN FET structures and the conductive characteristic of C_{60} FETs as a function of the C_{60} grain size have also been described. The C_{60} grain size grown on the polycrystalline AlN insulator layer is always larger than on SiO_2 insulator layers. Moreover, the electron mobility increases with the C_{60} grain size. It is also shown that the activation energy in the hopping conduction below the threshold voltage decreases along with the C_{60} grain size. These results all strongly indicate that the C_{60}/AlN FET structure has a great potential in achieving an outstanding performance in C_{60} FETs [72].

15.6.1 Substrates: SiO_2 or AlN

The most popular substrate is a conductive Si substrate with a thermal oxidation layer, SiO_2. The SiO_2 layer has some excellent properties as an insulator layer, and it is well known in the Si-MOSFET world. However, the fatal shortcoming in the SiO_2 layer as the insulator layer of the solid C_{60} FET is that the amorphous surface structure prevents single crystalline solid C_{60} epitaxy. On the other hand, AlN layers grown on conductive Si (or SiC) substrates have similar excellent properties as the insulator films of C_{60} FETs to SiO_2 films, except that they have a crystal surface. The crystal surface of the insulator film is necessary in forming an ideal interface between solid C_{60} layers and the

insulator layers. For van der Waals epitaxy of solid C_{60} layer to form the ideal interface, however, there are also other necessary conditions such as lattice-matched conditions and the surface roughness of the insulator layer to be met. Therefore, the AlN layer grown on the Si (or SiC) substrates has a great potential for the insulator layer of solid C_{60} FETs.

15.6.2 Device Process

The drain and source electrodes of gold are usually fabricated on the insulator layer by conventional lithography techniques. The channel length (L) and width (W) of the fabricated FET described here are 7.5 and 175 μm, respectively. After the lithography of the electrodes, the C_{60} layers are grown at 30°C–160°C by a conventional solid source MBE system.

15.6.3 C_{60} Epitaxy

The lattice mismatches between the solid C_{60} (111) surface with the fcc structure and the AlN (0001) surface with the wurtzite structure (h-AlN) are 7.3% in C_{60} [10$\bar{1}$]‖AlN [11$\bar{2}$0] (0° orientation) and 7.0% in C_{60} [11$\bar{2}$]‖AlN [11$\bar{2}$0] (30° orientation), as shown in Figure 15.21. The lattice mismatch can be permitted in the van der Waals epitaxy of the solid C_{60} layer [45]. It has been reported that a fcc C_{60} layer can be grown on h-GaN (0001) which is chemically stable just like AlN [51]. Therefore, AlN can be expected to be a substrate for the single crystal C_{60} layers. However, the lattice-mismatch in both orientations (shown in Figure 15.21) is almost the same. Because the C_{60} van der Waals epitaxy growth orientation on the smooth surface strongly depends on the lattice mismatch [45], it is a little difficult to predict the C_{60} orientations on the AlN surface by using the above considerations. If single crystalline C_{60} layers can be grown on the AlN surface, a more effective interface can be obtained than from an interface formed by the conventional sputtering method. However, no one has succeeded in forming a single crystalline C_{60} layer grown on the AlN surface so far. Therefore, we have to investigate the characteristics of FETs with polycrystalline solid C_{60} layers on the AlN surface while at the same time studying the single C_{60} layer growth on single AlN layers. In particular, it is very interesting that the size of the C_{60} grains may be strongly related to conductive characteristic such as the mobility and/or the activation energy of hopping conduction in FET performance.

A smooth single crystalline AlN surface (root mean square: RMS value: 0.9 nm), a rough single crystalline AlN surface (RMS value: 9.5 nm), and a smooth polycrystalline AlN surface (RMS value: 0.7 nm) can be formed by the nitridation of Al_2O_3 (0001) substrate or by the AlN growth on them by using radio-frequency plasma-assisted molecular beam epitaxy (RF-MBE). The C_{60} layers described here can also be grown on a AlN surface for 2–5 h at 60°C–160°C by a conventional solid-source MBE system under a base pressure of ~10^{-7} Pa. The monitoring of the growth process is usually performed by RHEED. The surface morphology of the solid C_{60} grown layer can be characterized by using the AFM. The AlN layers should be grown on a *p*-type

Si (111) substrate by using the RF-MBE system at about 500 nm (RMS value: 1.7 nm) in order to fabricate the bottom contact FET structure. For the C$_{60}$/SiO$_2$ FET structure, for example, the p-type Si (100) substrate is ordinarily used as a substrate covered with a SiO$_2$ insulator layer thermally formed at roughly 250 nm (RMS value: 0.2 nm).

Figure 15.22 shows typical RHEED patterns from solid C$_{60}$ layers grown on a smooth AlN surface (RMS value: 0.9 nm) for 2 h at 100°C–160°C. All RHEED patterns in Figure 15.22 show ones from the fcc (111) structure. Ring patterns appear in spot patterns below a growth temperature of 100°C. Both the RHEED patterns, the one with azimuth electron beam paralleled to the AlN [10$\bar{1}$0] and the AlN [11$\bar{2}$0] crystal directions from samples grown at 100°C–130°C, show the inclusion of both C$_{60}$ [10$\bar{1}$] and C$_{60}$ [11$\bar{2}$] patterns, as shown in Figure 15.22. Moreover, twin patterns are also observed. On the other hand, the C$_{60}$ [10$\bar{1}$] and the C$_{60}$ [11$\bar{2}$] patterns are only observed with the AlN [11–20] and AlN [10$\bar{1}$0] azimuths from samples grown at 160°C, respectively. However, the twin patterns are still observed even at 160°C.

The RHEED results can be summarized as follows: the C$_{60}$ layers grown at 100°C–130°C contain four kinds of islands which are in the C$_{60}$ [10$\bar{1}$]∥AlN [11] (the 0° orientation) epitaxial relation and its twin islands. They are also in the C$_{60}$ [11$\bar{2}$]∥AlN [11$\bar{2}$0] (the 30° orientation) epitaxial relation and its twin islands, as shown in Figure 15.23. However, the RHEED intensity from the islands with the 0° orientation is always stronger than from the islands with the 30° orientation. Figure 15.24 shows the RHEED intensity ratio for both islands in the epitaxial relation with the 0° and the 30° orientations as a function of the growth temperature. The islands with 0° orientation are grown much larger than ones with 30° orientation. The ratio is about 7:3. The C$_{60}$ layer grown at 160°C contains only islands with 0° orientation. However, the twin islands still remain. The results strongly indicate that the C$_{60}$ islands with the 0° orientation are stable on a smooth AlN surface in high-temperature growth, while the lattice-mismatch for each orientation is almost the same.

The C$_{60}$ islands grown on the smooth AlN surface at 100°C–160°C show a three-dimensional island growth. The density and the size of the islands, as a function of growth temperature, show a trade-off relation which is strongly connected to the migration of the C$_{60}$ molecules and the covering of the C$_{60}$ layers on the AlN surface.

The RHEED pattern for a solid C$_{60}$ layer grown on a single AlN with a rough surface (RMS value: 9.5 nm) at 130°C shows a mixed pattern with a spot and a ring. The RHEED results indicate that the orientation of the grains is not controlled at all. Such random orientation can be improved by the thermal annealing at 130°C after a low-temperature growth at 30°C. However, the obtained C$_{60}$ layers have multi-domain structures. The results indicate that the roughness of a single crystalline AlN surface is also one of the important factors for solid C$_{60}$ epitaxy.

The RHEED pattern from a solid C$_{60}$ layer grown on a polycrystalline AlN (RMS value: 1.7 nm) at 130°C shows a complete ring pattern. This means that a polycrystalline C$_{60}$ layer is grown on the surface. Furthermore, it strongly indicates that

the periodic atomic arrangement on the AlN surface is also an important factor in solid crystalline C$_{60}$ epitaxy as well as in surface roughness [52].

The RHEED observations show that polycrystalline C$_{60}$ layers can be grown on both AlN and SiO$_2$ layers. By observing the AFM in surface morphology, it was revealed that the grain size of the polycrystalline C$_{60}$ layers grown on polycrystalline AlN layers was larger than those grown on SiO$_2$ layers under the same conditions of C$_{60}$ growth. The results indicate that a better interface due to the obtained larger C$_{60}$ grains can be made on AlN layers rather than on SiO$_2$ layers.

15.6.4 Measurements of FET Performance

A drain current versus a source–drain voltage (I_D–V_{SD}) characteristic of FETs are usually measured by a conventional three probe current-voltage measuring system in an ambient N$_2$ atmosphere to avoid oxidation.

15.7 Characterization of C$_{60}$ FETs on SiO$_2$ or on AlN

15.7.1 I–V Characteristics

Typical I_D–V_{SD} characteristics of fabricated C$_{60}$ FETs with (a) conventional SiO$_2$ layers and with (b) newly proposed AlN layers are shown in Figure 15.26a and b, respectively. The typical I_D–V_{SD} characteristics consist of three regions: (1) the sub-threshold, (2) the linear, and (3) the saturation regions, similar to the I_D–V_{SD} characteristics of polycrystalline Si thin-film transistors (TFTs) [73]. According to the standard explanation of polycrystalline Si TFTs, those three regions can be explained by the differences in the conduction mechanisms. The linear and/or the saturation regions in the I_D–V_{SD} curves can be understood by normal band conduction. The electron mobility can be estimated from the I_D–V_{SD} curves. The electron mobility of such C$_{60}$ FETs with (a) SiO$_2$ and with (b) AlN layers are estimated from the I_D–V_{SD} curves, as shown in Figure 15.26a and b and are 2.0×10^{-2} and 9.0×10^{-2} cm^2/V s, respectively. It is well known that the grain size is closely related to the mobility in conventional polycrystalline Si TFTs. The electron mobility estimated from I_D–V_{SD} curves as a function of the C$_{60}$ grain size is shown in Figure 15.27. The electron mobility increases with grain size similar to conventional polycrystalline Si TFTs.

15.7.2 Sub-Threshold Region

The sub-threshold region at a low V_{SD} can be understood as hopping a conduction region [74]. According to many reports on organic FETs, a hopping conduction is commonly observed in them, and it strongly depends on the temperature and the electric-field-strength in the channel [75]. The activation energy of carrier hopping in C$_{60}$ FETs, which is estimated from the temperature dependence of the I_D–V_{SD} curves in the range of 200–300 K by a gate voltage of 7.5 V, is shown in Figure 15.28. The

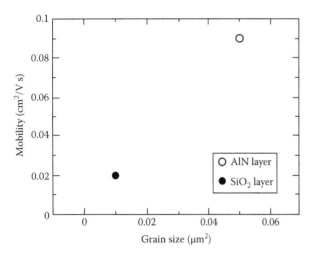

FIGURE 15.27 Electron mobility estimated from I_D–V_{SD} curves as a function of C_{60} grain size.

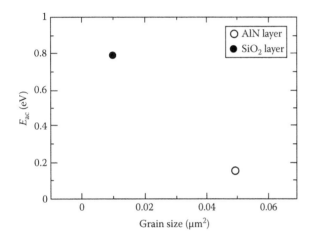

FIGURE 15.28 Activation energy of carrier hopping in C_{60} FETs with AlN and SiO_2 insulator layers.

results clearly indicate that the activation energy decreases with grain size in a manner similar to organic FET cases [76].

Moreover, a disorder model for a hopping conduction is given by the following equation:

$$\mu = \mu_0 \exp\left[-\left(\frac{2}{3}\hat{\sigma}\right)^2\right] \exp\left[C(\hat{\sigma}^2 - \Sigma^2)E^{\frac{1}{2}}\right] \quad \hat{\sigma} = \frac{\sigma}{kT} \quad (15.6)$$

where μ_0, σ, and Σ are the pre-exponential factors, the parameters of diagonal disorder, and the parameter of off-diagonal disorders, respectively [75]. The diagonal and the off-diagonal disorders mean the energetic and the positional disorders, respectively. When the Σ is larger than the σ, the electron mobility in a hopping conduction has a negative dependence on the electric-field-strength. The electron mobility in the hopping region is shown in Figure 15.29 as a function of the electrical field in the interface. The permittivity and the thickness of the AlN and the SiO_2 layers are different from each other. Consequently, the number of carriers induced by the gate voltage should be estimated as a function

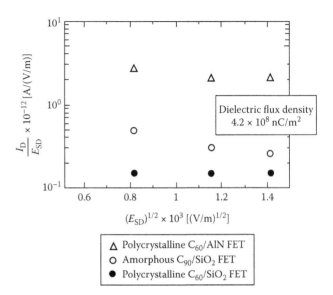

FIGURE 15.29 Electron mobility in hopping region as a function of electrical field.

of the dielectric flux density. As shown in Figure 15.29, the electron mobility in a polycrystalline C_{60}/SiO_2 FET does not depend on the electric-field-strength. However, the electron mobility in the amorphous C_{60}/SiO_2 FET and the polycrystalline $C_{60}/$AlN FET decreases with the electric-field-strength. Therefore, such a negative slope in the electron mobility as a function of the electric-field-strength in the hopping region means that a positional disorder is dominant at the interface of those FETs. That is, a route for the conductive current without any relation to the electric-field-strength must be formed in polycrystalline $C_{60}/$AlN FETs and amorphous C_{60}/SiO_2 FETs. Such large positional disorders can be understood by looking at the geometrical interface roughness which is due to a rough surface morphology of the AlN layers in the case of $C_{60}/$AlN FETs and by the structural interface roughness due to the poor amorphous structure of the C_{60} layers in C_{60}/SiO_2 FETs.

15.8 Summary

We have described solid C_{60} FETs starting from a basic explanation of C_{60} molecules to FET applications, through an introductory discussion about van der Waals epitaxy and organic FET operations. The most essential issue in C_{60} FET research at the present stage is the formation of an ideal interface between solid C_{60} and the insulator layers. To accomplish this, we need a new combination of solid C_{60} and insulator layers. In this chapter, we discussed that one of the possible strong candidates is a combination of solid C_{60} and single-crystalline AlN grown on either SiC or Si substrates. However, this problem still remains open to debate.

In the near future, it is expected that rather high-temperature superconductivity will be obtained to form the ideal interface in some solid C_{60} FET structures. The reasons of this expectation are based on the potential of hole-injection in field effect doping and high-energy inner-cluster phonon (~1000 K) which will assist in the superconductivity in the bulk of solid C_{60} [77,78]. Furthermore, the various combinations of carbon materials such as nanotubes and graphene with solid C_{60} FET structures will bring new applications like new molecular devices to reality. The C_{60} FET structures must play essential and important roles in carbon nano-electronics in the research beyond CMOS in the future.

References

1. http://www.aist.go.jp/aist e/latest reserach/2004/20041118/ 20041118.html: Latest researches, "n-Type Organic Thin Film Transistor Prepared by Printing Method."
2. http://www.lucent.com/news: Bell Lab Research Review Report, Report of the investigation committee on the possibility of scientific misconduct in the work of Hendrik Schön and Coauthors, 2006/11/29.
3. H. W. Kroto, J. R. Heath, S. C. O'Brien, R. F. Curl, and R. E. Smally, *Nature*, *318*, 162 (1985).
4. E. A. Rohlfing, D. M. Cox, and A. Kaldon, *J. Chem. Phys.*, *81*, 3322 (1984).
5. H. W. Kroto, *Angew. Chem. Int. Ed. Engl.*, *31*, 111 (1992).
6. R. E. Smally, *The Science*, March/April, p. 22 (1991).
7. J. Baggott, *Perfect Symmetry*, Oxford University Press, Oxford, U.K. (1994).
8. W. Kratchmer, L. D. Lamb, K. Fostiropoulos, and D. R. Huffman, *Nature*, *347*, 354 (1990).
9. R. E. Haufler et al., *J. Phys. Chem.*, *94*, 8634 (9990).
10. G. Peters and M. Jansen, *Angew. Chem. Int. Ed. Engl.*, *31*, 223 (1992).
11. J. B. Howard, J. T. McKinnon, Y. Markarousky, A. L. Lafleur, and M. E. Johnson, *Nature*, *352*, 139 (1991).
12. R. D. Averitt, J. M. Alford, and N. J. Halas, *Appl. Phys. Lett.*, *65*, 374 (1994).
13. R. A. Laudise, Ch. Kloc, P. G. Sinpkins, and T. Siegrist, *J. Cryst. Growth* *187*, 449 (1998).
14. R. Taylor, J. P. Hare, A. K. Abdul-Sada, and H. W. Kroto, *J. Chem. Soc. Chem. Commun.*, *20*, 1423–1425 (1990).
15. C. J. Welch and H. Pirkle, *J. Chromatogr.*, *609*, 89 (1992).
16. S. Larsson, A. Volosov, and A. Rosen, *Chem. Phys. Lett.*, *137*, 501 (1987).
17. D. E. Weeks and W. G. Harter, *J. Chem. Phys.*, *90*, 4744 (1989).
18. R. E. Stanton and M. D. Newton, *J. Phys. Chem.*, *92*, 2141 (1988).
19. S. J. Cyvin, E. Brendsdal, B. N. Cyvin, and J. Brunroll, *Chem. Phys. Lett.*, *143*, 377 (1988).
20. M. S. Dresselhaus, G. Dresselhaus, and P. C. Eklund, *Science of Fullerenes and Carbon Nanotubes*, Chap.11, Academic Press, New York (1996).
21. A. M. Rao et al., *Science*, *259*, 955 (1993).
22. R. D. Johnson, G. Meijer, and D. S. Bethune, *J. Am. Chem. Soc.*, *112*, 8983 (1990).
23. J. M. Hawkins, A. L. Meyer, A. Timothy, S. Loren, and F. J. Salam, *J. Phys. Chem.*, *95*, 9 (9991).
24. W. I. F. David et al., *Nature*, *353*, 147 (1991).
25. R. C. Haddon, *Acc. Chem. Res.*, *25*, 127 (1992).
26. R. C. Haddon, L. E. Brus, and K. Raghavachari, *Chem. Phys. Lett.*, *125*, 459 (1996).
27. A. D. J. Haymet, *Chem. Phys. Lett.*, *122*, 421 (1985).
28. S. Saito and A. Oshiyama, *Phys. Rev. Lett.*, *66*, 2637–2640 (1991).
29. W. I. F. David, R. M. Ibberson, and T. Matsuo, *Proc. R. Soc. Lond.*, *A442*, 129 (1993).
30. R. Tycko, R. C. Haddon, G. Dabbagh, S. H. Giarum, D. C. Douglassc, and A. M. Muujsce, *J. Phys. Chem.*, *95*, 518 (1991).
31. D. A. Neumann et al., *Phys. Rev. Lett.*, *67*, 3808 (1991).
32. A. F. Hebard, *Phys. Today*, *45*, 26 (November 1992).
33. K. Prassides, H. W. Kroto, R. Taylor, D. H. M. Walton, W. I. F. David, J. Tomkinson, R. C. Haddon, M. J. Rosseinsky, and D. W. Murphy, *Carbon*, *30*, 1277 (1992).
34. J. H. Weaver, *J. Phys. Chem. Solids*, *53*, 1433 (1993).
35. T. Arai, Y. Murakami, H. Suenatsu, K. Kikuchi, Y. Achiba, and I. Ikemoto, *Solid State Commun.*, *84*, 827 (1992).
36. J. Mort et al., *Chem. Phys. Lett.*, *186*, 284 (1992).
37. A. F. Hebard, R. C. Haddon, R. M. Fleming, and A. R. Kortan, *Appl. Phys. Lett.*, *59*, 2109 (1991).
38. G. B. Alers, B. Golding, A. R. Kortan, R. C. Haddon, and F. A. Thiel, *Science*, *257*, 511 (1992).
39. S. Kazaoui, R. Ross, and N. Minami, *Solid State Commun.*, *90*, 623 (1994).
40. D. J. van den Heuvel et al., *Chem. Phys. Lett.*, *233*, 284 (1995).
41. B. Pevzner, A. F. Hebard, R. C. Haddon, S. D. Senturia, and M. D. Dresselhaus, *Mater. Res. Soc. Symp. Proc.*, *359*, 423 (1995).
42. F. Kajzar, *Synth. Met.*, *54*, 21 (1993).
43. Y. B. Zhao, D. M. Poirier, R. J. Pachman, and J. H. Weaver, *Appl. Phys. Lett.*, *64*, 577 (1994).
44. P. Zhou et al., *Appl. Phys. Lett.*, *60*, 2871 (1992).
45. K. Tanigaki, S. Kuroshima, and T. W. Ebbesen, *Thin Solid Film*, *257*, 154 (1995).
46. A. Koma, *J. Cryst. Growth*, *201/202*, 236 (1999).
47. Y. Saito, *Mater. Sci. Eng. B*, *18*, 229 (1993).
48. M. Sakurai, H. Tada, and A. Koma, *Jpn. J. Appl.*, *30*, L1892 (1991).
49. G. Genstlnblum, L. M. Yu, J. J. Pireaux, P. A. Thiry, R. Caudano, J. M. Themlin, S. Bouzidi, F. Coletti, and J. M. Debever, *Appl. Phys.*, *A56*, 175 (1993).
50. H. Takashima, M. Nakaya, A. Yamamoto, and A. Hashimoto, *J. Cryst. Growth*, *227/228*, 825 (2001).
51. H. Takashima, M. Nakaya, A. Yamamoto, and A. Hashimoto, *J. Cryst. Growth*, *227/228*, 829 (2001).
52. D. Yokoyama, H. Nojiri, A. Yamamoto, and A. Hashimoto, *Phys. Stat. Sol. (a)*, *195*, 3 (2003).
53. N. Teraguchi, A. Suzuki, Y. Saito, T. Yamaguchi, T. Araki, and Y. Nanishi, *J. Cryst. Growth*, *230*, 392 (2001).

54. N. Onojima, J. Suda, and H. Matsunami, *J. Cryst. Growth*, **237–239**, 1012 (2002).

55. V. Lebedev, B. Schroter, G. Kipshidze, and W. Richter, *J. Cryst. Growth* **207**, 266 (1999).

56. U. Kaiser, P. D. Brown, I. Khodos, C. J. Humphreys, H. P. D. Schenk, and W. Richter, *J. Mater. Res.*, **14**, 2036 (1999).

57. K. Kudo, M. Yamashina, and T. Moriizumi, *Jpn. J. Appl. Phys.*, **23**, 130 (1984).

58. H. Koezuka, A. Tsumura, and T. Ando, *Synth. Met.*, **18**, 699 (1987).

59. J. Takeya, M. Yamagishi, Y. Tominari, R. Hirahara, Y. Nakazawa, T. Nishikawa, T. Kawase, T. Shimoda, and S. Ogawa, *Appl. Phys. Lett.*, **90**, 102120 (2007).

60. R. C. Haddon, A. S. Perei, R. C. Morris, T. T. M. Palsta, A. F. Hebard, and R. M. Fleming, *Appl. Phys. Lett.*, **67**, 121 (1995).

61. J. Tang, G. Xing, Y. Zhao, L. Jing, H. Yuan, F. Zhao, X. Gao, H. Qian, R. Su, K. Ibrahim, W. Chu, L. Zhang, and K. Tanigaki, *J. Phys. Chem.*, **B111**, 11929 (2007).

62. M. Kitamura, S. Aomori, J. Ho Na, and Y. Arakawa, *Appl. Phys. Lett.*, **93**, 033313 (2008).

63. J. Kido, M. Kimura, and K. Nagai, *Science*, **267**, 1332 (1995).

64. D. X. Wang, Y. Tanaka, M. Iizuka, S. Kuniyoshi, K. Kudo, and K. Tanaka, *Jpn. J. Appl. Phys.*, **38**, 256 (1999).

65. Y. Ohmori, K. Muro, M. Onoda, and K. Yoshino, *Jpn. J. Appl. Phys.*, **31**, L646 (1992).

66. T. Yasuda, T. Goto, K. Fujita, and T. Tsutsui, *Appl. Phys. Lett.*, **85**, 2098 (2004).

67. T. Nishikawa, S. Kobayashi, T. Nakanowatari, T. Mitani, T. Shimoda, Y. Kubozono, G. Yamamoto, H. Ishii, M. Niwano, and Y. Iwasa, *J. Appl. Phys.*, **97**, 104509 (2005).

68. R. Tamura, E. Lim, T. Manaka, and M. Iwamoto, *J. Appl. Phys.*, **100**, 114515 (2006).

69. J. A. Misewich, R. Martel, Ph. Avouris, J. C. Tsang, S. Heinze, and J. Tersoff, *Science*, **300**, 783 (2003).

70. J. Zaumseil, R. H. Friend, and H, Sirringhaus, *Nat. Mater.*, **5**, 69, (2006).

71. T. Takahashi, T. Takenobu, J. Takeya, and Y. Iwasa, *Adv. Funct. Mater.*, **17**, 1623 (2007).

72. H. Nojiri, D. Yokoyama, A. Yamamoto, and A. Hashimoto, *Diamond Relat. Mater.*, **14**, 518 (2005).

73. Y. G. Yoon, G. B. Kim, B. I. Lee, and S. K. Joo, *Thin Solid Films*, **466**, 303–306 (2004).

74. C. P. Jarrett, K. Pichler, R. Newbould, and R. H. Friend, *Synth. Met.*, **77**, 35–38 (1996).

75. H. Bassler, *Phys. Stat. Sol.*, (b) **175**, 15–56 (1993).

76. K. Horiuchi, S. Uchino, K. Nakada, N. Aoki, M. Shimizu, and Y. Ochiai, *Physica B*, **329–333**, 1538–1539 (2003).

77. T. W. Ebbesen, J. S. Tsai, K. Tanigaki, J. Tabuchi, Y. Shimakawa, Y. Kubo, I. Hirosawa, and J. Mizuki, *Nature*, **355**, 620 (1992).

78. T. W. Ebbesen, J. S. Tsai, K. Tanigaki, H. Hiura, Y. Shimakawa, Y. Kubo, I. Hirosawa, and J. Mizuki, *Physica C*, **203**, 163 (1992).

16

The Cooper-Pair Transistor

José Aumentado
National Institute of
Standards and Technology

16.1 Introduction

The Cooper-pair transistor (CPT) is a three-terminal superconducting device composed of a mesoscale superconducting island connected to drain and source leads via ultrasmall (~100 nm) Josephson junctions (see Figure 16.1). Due to its small size, the energy to add a single Cooper pair to the island can be large compared with the typical temperatures at which similar quantum circuits are operated, $T < 100$ mK. Because of this, the transport properties of the device can be strongly dependent on the polarization charge presented by a capacitively coupled gate electrode and operated as a transistor with high input impedance. As an electrometer, it is useful as it can be operated with very little dissipation and should minimally influence the systems that it is measuring. As we will see, its similarity to another quantum circuit, the Cooper-pair box (CPB), marks it as a useful device to study decoherence in superconducting quantum bits (qubits) due to unpaired electrons.

While the CPT is topologically identical to the normal-state single-electron transistor (SET), it is operated in the superconducting state, and the superconducting nature of the device modifies the transport properties in several important ways. For our purposes, we also distinguish the CPT from the superconducting single-electron transistor (SSET) as a mode of operating this particular double-junction device. As a CPT, it is operated on the supercurrent branch near zero voltage in the coherent Cooper-pair transport regime, while as an SSET, it is operated at finite voltage bias and can include several incoherent hybrid quasiparticle/Cooper-pair tunneling processes.

16.2 Theory of Operation

As shown in Figure 16.1a, the device is composed of two Josephson junctions and can, at its core, be regarded as a single Josephson-like element with an added internal charge degree of freedom. This modification is important as it allows the CPT to act as a tuneable Josephson junction that is sensitive to charge. In this spirit, it is important that we begin with a conceptual understanding of a single Josephson junction.

16.2.1 The Single Josephson Junction

A superconducting Josephson junction is composed of a thin insulating barrier connecting two superconducting electrodes. In 1962, Brian Josephson made the remarkable statement that these junctions should support a supercurrent, that is, a dissipationless current at zero voltage (Josephson, 1962). This is a consequence of the finite quantum mechanical tunneling probability that arises from the spatial overlap in the superconducting order parameter on either side of a thin junction barrier. Josephson encapsulated this effect in the following equations that relate the current and the voltage to the dynamics of the phase difference across the junction*

$$I = I_0 \sin \delta \qquad (16.1)$$

$$V = \varphi_0 \dot{\delta}. \qquad (16.2)$$

---
* We use the notational shorthand $\varphi_0 \equiv \hbar/2e$. This is simply the superconducting flux quantum $\Phi_0 = h/2e$ divided by 2π.

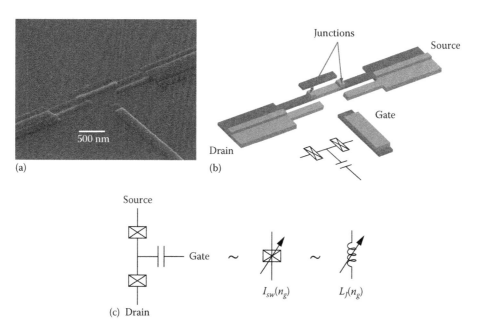

FIGURE 16.1 (a) Field emission micrograph of typical CPT. (b) 3D equivalent model showing two-angle/double image deposition (see Section 16.3) and location of junctions at the overlap. (c) The CPT is effectively a Josephson junction with a critical current and inductance that modulate with gate polarization charge/voltage $n_g \equiv C_g V_g / e$.

Taken at face value, these equations are the constitutive equations defining the Josephson junction as an inductor-like nonlinear circuit element. I_0 is a "critical current" beyond which the junction cannot support a supercurrent and the junction switches to a finite voltage branch in the current–voltage (I–V) characteristic. It can be computed from the low-temperature form of the Ambegaokar–Baratoff relation* (Ambegaokar and Baratoff, 1963).

$$I_0 = \frac{2}{eR_N} \frac{\Delta_1 \Delta_2}{\Delta_1 + \Delta_2} K\left(\frac{|\Delta_1 - \Delta_2|}{\Delta_1 + \Delta_2} \right), \quad (16.3)$$

where

Δ_1 and Δ_2 are the superconducting gap energies on either side of the insulating barrier

K is the complete elliptic integral of the first kind

If $\Delta_1 \sim \Delta_2$, $K \to \pi/2$ and

$$I_0 \simeq \frac{\pi}{eR_N} \frac{\Delta_1 \Delta_2}{\Delta_1 + \Delta_2}. \quad (16.4)$$

The most commonly used superconductor for CPTs (and SETs) is aluminum, Δ_{Al} = 180–250 μeV. Typical junction sizes are ~100 nm × 100 nm that yield junction resistances in the range of

10 kΩ < R_N < 100 kΩ. With these parameters, the critical current ranges from 1 to 100 nA.

The presence of the phase δ across the junction in the equations marks it as an inductor-like element that can store magnetic flux. Following this reasoning, one can derive a Josephson inductance by finding the scaling coefficient between the voltage and the time derivative of the current,

$$L_J(I) = \frac{L_{J0}}{|\cos \delta|} = \frac{L_{J0}}{\sqrt{1 - (I/I_0)^2}}, \quad \text{where } L_{J0} \equiv \frac{\varphi_0}{I_0}. \quad (16.5)$$

One can also define a Josephson energy that is a function of the phase difference δ,

$$E_J = \varphi_0 I_0 |\cos \delta| = E_{J0} \sqrt{1 - \left(\frac{I}{I_0} \right)^2}, \quad \text{where } E_{J0} \equiv \varphi_0 I_0. \quad (16.6)$$

This can be interpreted as the magnetic energy that is stored in the element. This energy has a deeper significance as the coupling energy between Cooper-pair charge states on either side of the junction and will be important in calculating the energy bands of the CPT. For CPT Josephson junctions, the equivalent inductances are typically ~10–30 nH and the energies 20–80 μeV (0.2–1 K in temperature units).

16.2.1.1 The *I–V* Curve and Phase Dynamics

16.2.1.1.1 The *I–V* Curve

The CPT is operated on the so-called "supercurrent branch" of the *I–V* curve at (or near) zero voltage. The two measurement modes that we discuss in this chapter involve measuring the

---

* The form of the Ambegaokar–Baratoff relation shown here is uncommon in that it acknowledges the difference in gap energies that seems to be typical in evaporated aluminum junctions. This gap energy difference is usually ignored in the early SSET and CPT literature but has become important in more recent CPT experiments.

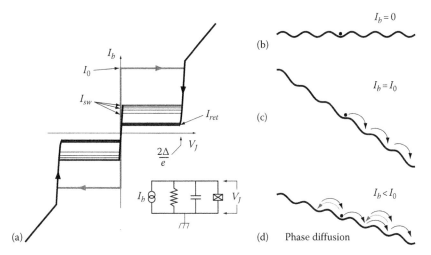

FIGURE 16.2 (a) Schematic I–V curve/characteristic for an ultrasmall Josephson junction. The ideal $T = 0$ I–V is shown in light gray, switching to the voltage state at $I_b = I_0$. For $T > 0$, thermal fluctuations can drive the system to the voltage state at much lower bias currents in ultrasmall ($A_{jn} < 0.01\,\mu m^2$) junctions, as shown in black. (b) $I_b < I_0$. The phase particle is trapped in a local minima of a washboard potential and $\langle \dot{\delta} \rangle = 0$ and no voltage is developed across the junction. (c) $I_b = I_0$. The washboard potential tilted at exactly the critical slope necessary to "release" the phase particle from its local minimum and enter a free-running state at finite voltage, $V (= \varphi_0 \langle \dot{\delta} \rangle > 0)$. (d) $I_b < I_0$. Phase diffusion. In ultrasmall junctions, thermal fluctuations can drive the phase to evolve in a 1D random walk of phase steps $\Delta\phi \approx 2\pi$. Since the washboard is tilted, this results in a slow diffusion in one direction, $V (= \varphi_0 \langle \dot{\delta} \rangle) > 0$, although it is still considered to be on the "supercurrent branch."

switching current I_{sw}, at which the CPT switches from the super-current branch to the finite-voltage branch as well as measuring the effective inductance L_J of the CPT when biased at zero current. While the small size of the junctions and internal charge degree of freedom will affect these parameters significantly, it is worthwhile to first look at the single junction case and understand how the dynamics of the phase determine the structure of the I–V curve (see Figure 16.2a).

16.2.1.1.2 The Resistively and Capacitively Shunted Junction (RCSJ) Model

A more realistic model of a single Josephson junction includes a parallel shunt resistance and capacitance (see inset in Figure 16.2a). The shunt resistance R encapsulates the resistance of the junction to normal/quasiparticle currents through the junction, while the capacitance C is simply the physical capacitance that arises from having two planar electrodes overlapping with an insulating barrier in between. In the junctions comprising a CPT, the intrinsic $R > 10\,k\Omega$ and is considered effectively "unshunted" since R is much bigger than the impedance presented by the Josephson inductance. As noted previously, the Josephson tunnel element is a nonlinear inductance but, for small phase excursions, it looks like a linear inductor. The circuit model presented then looks very much like a parallel LCR circuit or damped simple harmonic oscillator with a natural frequency

$$\omega_p = \frac{1}{\sqrt{L_{J0}C}} = \sqrt{\frac{2eI_0}{\hbar C}}. \tag{16.7}$$

This is usually called the "plasma frequency" and has the significance of simply being the frequency of small oscillations for a Josephson junction.

Although this model can be used to describe the intrinsic Josephson junction dynamics, it is also useful in determining the dynamics when the junction is placed in an arbitrary measurement circuit. If, for instance, we want to current bias the junction, we might add a voltage source V_b in series with a resistor R_b and construct a Norton equivalent that is simply an ideal current bias $I_b = V_b/R_b$ in parallel with an output impedance R_b.[*] In this case, we can roll R_b into the shunt R in our *RSJ* model. With this simple circuit equivalent, we can derive a set of first-order differential equations defining the equations of motion for the phase by writing down Kirchoff's equations for the currents in each branch. We can write this system of equations as a single second-order differential equation

$$\ddot{\delta} + \frac{1}{RC}\dot{\delta} + \omega_p^2\left(\sin\delta - \left(\frac{I_b}{I_0}\right)\right) = 0. \tag{16.8}$$

For small δ at zero current bias, this describes the motion of a particle with coordinate δ oscillating in a quadratic potential with a friction term, viz., it's a damped simple harmonic oscillator (and is consistent with our earlier statement that L_J is linear for small oscillations). At finite current bias, however, this equation describes the motion of the particle in a tilted sinusoidal (or *washboard*) potential.[†] We know from Equation 16.2 that in order for the junction to support a finite dc voltage across

---

[*] This vast oversimplification has its caveats and must be amended to account for stray capacitance in the bias lines and any other reactances relevant in the frequency range of our junction dynamics (typically $\omega < 50\,GHz$ for small aluminum junctions).

[†] More in depth discussions of the RCSJ model can be found in Tinkham (2004) and Clarke and Braginski (2004).

it, the particle must have some net motion in the phase, that is, $\langle \dot{\delta} \rangle \neq 0$. Since the particle is trapped at a fixed phase when $I_b < I_0$, we have a well-defined phase and the junction remains in the supercurrent state (see Figure 16.2b). By applying a bias current, we tilt the washboard and, when $I_b = I_0$, the particle is tipped out of its well and begins to run away (Figure 16.2c). In this free-running state, the time derivative of the phase has both an ac and a dc component corresponding to ac and dc voltages across the junction. In a more realistic treatment, I_b includes a stochastic contribution from thermal fluctuations due to the dissipation in the environment and so the current at which the junction switches to the free-running state is lower than I_0 since thermal excitations can drive the phase particle over the potential barrier prematurely. Because of this, the measured current is distributed according to the statistics of a thermally excited (Kramers) escape process (Fulton and Dunkleberger, 1974). We will differentiate this switching current I_{sw} from the ideal Ambegaokar–Baratoff value I_0 and remember that it is actually a statistical variable.

If we reduce the current bias in the free-running, finite voltage state, the particle velocity begins to slow down due to the dissipation in the circuit. Once the dissipation rate can compensate for the ac power generated by the free-running voltage oscillations, the particle "retraps" into a minima in the washboard potential and the junction returns to the zero-voltage supercurrent branch. The current at which this happens, I_{ret}, the "return current" is an indication of the amount of damping or dissipation in the environment at the free-running oscillation frequency (Tinkham, 2004). Roughly, the closer to zero I_{ret} is, the less damping there is. Although the CPT is designed and operated in the unshunted, hysteretic regime, in many practical dc SQUID magnetometers, the junctions are intentionally shunted with damping resistors to eliminate hysteresis in the I–V measurement permitting stable finite voltage operation.

16.2.1.1.3 *Concerns Specific to Small Junctions*

While the picture above is the standard way to look at relatively large junctions (area $\gtrsim 1\,\mu m^2$), the junctions in CPTs are usually about a factor of 100 smaller and have much smaller capacitances. The main consequence of this is that the observed switching currents can easily be smaller than I_0 by an order of magnitude. One way to understand this is to return to the harmonic oscillator analogy. We can see that $\omega_p^2 \propto 1/C$, so C is an effective mass for our "phase particle." In CPT junctions, this means that the equivalent phase particle can be relatively light compared with its larger junction cousins and so thermal fluctuations, even at 100 mK, can have a more significant effect, kicking the particle into the free-running state much more easily. This has the end effect of reducing I_{sw} far below I_0 and also widening the distribution of I_{sw} (Figure 16.2a).

A more problematic effect of the light phase particle mass is the fact that the same thermal fluctuations can "kick" the phase particle from well to well in the washboard potential in a process called "phase diffusion" (Kautz and Martinis, 1990).

At zero current bias, this results in a one-dimensional random walk in the phase with zero mean. At finite current, the tilt in the washboard biases this diffusive motion in one direction, such that the mean phase velocity $\langle \dot{\delta} \rangle$ can be positive (negative) at positive (negative) bias (see Figure 16.2d). The result of all of this is that at finite currents below the switching current, we often see a small finite voltage. This slope is fairly uniform near zero current and gives the supercurrent branch a slight resistive contribution that can correspond to 10 s or even 100 s of Ωs. The downside of this behavior is that the junction or CPT is not truly dissipationless.

For electrometer operation, we want I_{sw} to have as narrow a distribution as possible since it is the quantity that will vary with applied gate voltage/polarization charge. It has been demonstrated, however, that one can narrow this distribution and also increase I_{sw}, so that it is closer to I_0 by fabricating a larger capacitance (a few picofarad) in the leads shunting the junction to "weigh down" the phase particle (Joyez et al., 1994; Joyez, 1995).

16.2.2 Coulomb Blockade

While the above discussion of a single junction will eventually be relevant to the CPT measurements we are interested in, we have not yet seriously considered the role of the internal charge degree of freedom in determining the overall critical current and Josephson inductance in this device. This is, afterall, the knob that turns the CPT into a transistor.

We can qualitatively recognize the role of the island charge by first determining the energy required to charge the CPT island with a single electron charge e,

$$E_C = \frac{e^2}{2C_\Sigma}, \tag{16.9}$$

where

$$C_\Sigma \equiv C_{J1} + C_{J2} + C_g + C_{stray}. \tag{16.10}$$

Here, C_Σ is the total capacitance as seen by the island, and while it includes the junction and gate capacitances, it also includes a "catch-all" term, C_{stray}, that is meant to include any contributions from unintentional coupling to ground. In practice, $C_{stray} \ll C_\Sigma$ and can usually be ignored. The energy E_C is called the single-electron Coulomb blockade energy (Devoret and Grabert, 1992).

For typical junction capacitances $C_{J1,J2} \lesssim 1\,\text{fF}$ and gate capacitances of 0.01–1 fF, E_C can be on the order of ~100–400 μeV. In temperature units, this temperature is 1–5 K and is therefore a significant energy scale compared with typical operating temperatures. As noted previously, the Josephson energies of these junctions are roughly on par with the Coulomb blockade energies, so we can qualitatively assume that both energy scales will play a role in the transport.

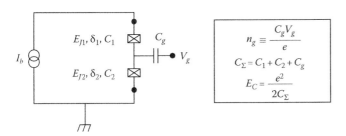

FIGURE 16.3 Simplified CPT schematic. Each ultrasmall junction has an associated Josephson energy and phase along with an overall Coulomb blockade energy that defines the energy required to place discrete charge on the island. Voltage applied to the gate injects "gate polarization charge" into modulating the Coulomb or ionization energy. This modulates its Josephson-like behavior through charge-phase duality (see text).

16.2.3 Band Structure of the CPT

Since we have two junctions in series (Figure 16.3), we must define separate phases for each junction δ_1 and δ_2.* We can now be more precise about the role of the charging energy by looking at the Hamiltonian for the CPT and how the contributions couple to the phases and the island charge n,

$$H = H_{J1} + H_{J2} + H_{CB} + H_{QP}. \qquad (16.11)$$

$H_{J1,J2}$ are the contributions due to the Josephson energy coupling the charge states on either side of the junctions, while H_{CB} is the Coulomb blockade contribution. Explicitly, we have

$$H_{J1,J2} = -E_J \cos \delta_{1,2} \qquad (16.12)$$

$$H_{CB} = \frac{E_C}{4}(n - n_g)^2. \qquad (16.13)$$

Here we have used the reduced gate voltage $n_g \equiv C_g V_g/e$ that is the polarization charge applied by the gate in electron units. H_{QP} is the contribution from quasiparticle excitations in the leads and on the island. Physically, they correspond to unpaired electrons/holes that can be introduced to the system via thermal excitations or some nonequilibrium process that can break Cooper pairs. While this is an important contribution, we ignore it now to facilitate the discussion.

We can write down a matrix equivalent of Equation 16.11 in the basis of island charge states $|n\rangle$, noting that the Josephson coupling terms that link Cooper-pair charge states on either state of the junction also link "adjacent" Cooper-pair charge states $|n\rangle \to |n \pm 2\rangle$.† Defining "external" and "internal" phases $\delta = \delta_1 + \delta_2$ and $\theta = \delta_1 - \delta_2$ and setting $\theta = 0$ (equivalent to assum-

ing that the island has negligible inductance), we can write H in the charge basis. For example, the Hamiltonian spanning the Cooper-pair island charge states $|-2\rangle$, $|0\rangle$, and $|2\rangle$ is

$$H = \begin{bmatrix} E_C(-2 - n_g)^2 & -\frac{1}{2}(E_{J1}e^{-i\delta/2} + E_{J2}e^{+i\delta/2}) & 0 \\ -\frac{1}{2}(E_{J1}e^{+i\delta/2} + E_{J2}e^{-i\delta/2}) & E_C(-n_g)^2 - \frac{1}{2} & (E_{J1}e^{-i\delta/2} + E_{J2}e^{+i\delta/2}) \\ 0 & -\frac{1}{2}(E_{J1}e^{+i\delta/2} + E_{J2}e^{-i\delta/2}) & E_C(2 - n_g)^2 \end{bmatrix}.$$

$$(16.14)$$

The eigenvalues for this matrix represent the energy bands and are functions of n_g and external phase δ. The choice of basis size (how big a matrix one really needs to write down) is dictated by the ratio of $E_{J1,2}/E_C$. If this ratio is small, charge is easily localized and the corresponding external phase fluctuations are large, so the device has a weak effective Josephson energy. In this case, the charge basis may be described well by a small subspace of $|n\rangle$ spanning only a few charge states near $|0\rangle$. If $E_{J1,2}/E_C \gtrsim 1$, this approximation fails and one must write down a larger matrix spanning a bigger subspace of $|n\rangle$. In this limit, the charge on the island is no longer as well defined as its quantum mechanical state is now described by a linear superposition of several charge states. Likewise, fluctuations in the CPT external phase are then small and the CPT begins to look like a single Josephson junction.

16.2.3.1 The Uncertainty Principle at Work

We can numerically diagonalize Equation 16.14 to calculate the eigenvalues at various n_g and δ. Figure 16.4 shows the resulting eigenenergy surfaces corresponding to the ground and first excited state energies, ϵ_0 and ϵ_1. The quantities that we want to measure, $I_{sw}(n_g)$ and $L_J(n_g)$, are defined by the first and second derivatives of these surfaces in the phase. The size of this modulation is a direct consequence of how well the CPT island localizes Cooper pairs, $n2e$, through the uncertainty relation between charge number and phase, $\Delta n \Delta \delta \geq 1/2$. In this sense, the CPT is a strange charge-based device because it relies on the competition between number and phase uncertainty. This is in contrast to the SSET or SET where E_C is entirely dominant. In the CPT, we would like E_C to be big but would also like E_J to be comparable in magnitude, so that the switching current is not so small that it is difficult to measure well. Similarly, rf measurements of the Josephson inductance (described later in this review) become easier at bigger E_J as larger signal powers can be used while biased on the supercurrent branch.

16.2.3.2 Calculation of the Critical Current Modulation

The critical current of the CPT can now be determined from the eigenenergies calculated above. By restricting to the ground state band $\epsilon(n_g, \delta)$, we can compute the critical current (Joyez, 1995)

$$I_0(n_g) = \frac{1}{\varphi_0} \frac{\partial \epsilon_0(n_g, \delta)}{\partial \delta}\bigg|_{\delta_{max}}, \qquad (16.15)$$

---

* The phases δ_1 and δ_2 bear a strict conjugate correspondence to the number of Cooper pairs having "flown" through junctions 1 and 2. This discussion closely follows that given by Joyez (1995) and a more extensive discussion of choosing appropriate quantum variables appears there.

† For consistency with Joyez (1995), n in $|n\rangle$ refers to the single electron number and *not* Cooper pair number.

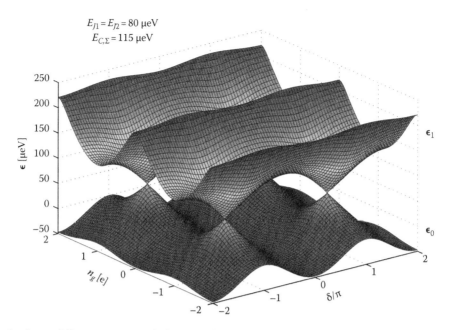

$E_{J1} = E_{J2} = 80\,\mu eV$
$E_{C,\Sigma} = 115\,\mu eV$

FIGURE 16.4 (**See color insert following page 21-4.**) The ground (ϵ_0) and first excited (ϵ_1) states calculated by numerically diagonalizing Equation 16.11 with the energies indicated in the figure.

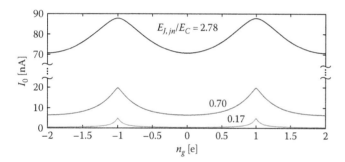

FIGURE 16.5 Numerically calculated CPT critical currents for various ratios of $E_{J,jn}/E_C$. In this example, E_C is fixed at $115\,\mu eV$, while $E_{J,jn}$ is varied.

where δ_{max} is the phase maximizing the derivative.* In Figure 16.5, we show the calculated critical currents for several ratios of E_J/E_C. Note that the calculated currents are $2e$ periodic in the applied gate charge which is the result of the Josephson coupling of Cooper-pair charge states. In practice, measured switching currents in CPTs are usually significantly smaller than the calculated value due to small-junction effects noted above. However, the total magnitude of the modulation can be on the order of 10 nA for practical aluminum device parameters and is easily measured.

16.2.3.3 Calculation of the Effective Inductance Modulation

The ground state energy surface $\epsilon_0(n_g, \delta)$ corresponds to an effective Josephson energy that is a function of charge as well as phase. When the ratio $E_J/E_C \lesssim 1$, the phase dependence, while

---

* Incidentally, one can also take derivatives of the excited state eigenenergies and compute critical currents for those bands. Interested readers can see Flees et al. (1997) for further details.

still 2π periodic, gets "peaky" near odd-integer n_g. Despite this, one may still define a Josephson inductance in a similar manner (Sillanpää et al., 2004; Naaman and Aumentado, 2006b), that is,

$$L_J(n_g) = \frac{1}{\varphi_0^2} \frac{\partial^2 \epsilon_0(n_g, \delta)}{\partial \delta^2}\bigg|_{\delta=0}. \quad (16.16)$$

This is similar to Equation 16.5, but since the $\epsilon_0(n_g, \delta)$ is not strictly sinusoidal, we do not get the exactly the same result. For typical aluminum device parameters, the inductance modulation can be 10–100 nH and, like the switching current, is easily measured.

16.2.4 Quasiparticle Poisoning

Although the "quasiparticle" excitations that are talked about in superconductivity do not explicitly bear a well-defined charge, they do represent the screened excitations presented by an unpaired electron that might exist due to the breaking of a Cooper pair. A quasiparticle can tunnel onto a CPT island by "undressing" itself of its screening cloud and tunneling through a junction barrier alone. At this point, this additional charge presents a full electron charge offset to the CPT island. Since the CPT switching current and inductance are nominally $2e$ periodic in the gate charge, the addition of an extra electron can present a significant change in the way the CPT is operated, since it fluctuates the charge by e. In the literature, this problem was first discussed in terms of island "parity," but more recent work has favored the more colorful term "quasiparticle poisoning." The most important thing about quasiparticle poisoning for electrometry is that it is a stochastic process whose dynamics can happen on timescales comparable to our measurement time, so its effect on the resulting measurement must be well understood. That being said, quasiparticle poisoning has proven to be an interesting problem in itself, and the CPT has been very useful in its study.

To see how quasiparticles enter the picture, we now include the Hamiltonian H_{QP} in our description of the CPT,

$$H_{QP} = \sum_j \epsilon_j \gamma_j \gamma_j^\dagger. \tag{16.17}$$

The γ_j, $\gamma_j^\dagger$ are annihilation/creation operators for quasiparticle excitations in the superconductor (Tinkham, 2004), while ϵ_j is the energy of the excitation. In our case, this can be Δ_1 or Δ_2, remembering that the island and leads can have different gap energies. These quasiparticles are usually taken to be thermally generated and therefore have an exponentially small probability of existing at low temperatures. The expression for the quasiparticle density is (cf., Shaw et al., 2008),

$$n_{qp} = D(\epsilon_F)\sqrt{2\pi\Delta_j k_B T} \exp\left(\frac{-\Delta_j}{k_B T}\right). \tag{16.18}$$

$D(\epsilon_F)$ ($=2.3 \times 10^4$ μm^{-3} J^{-1}) is the density of states at the Fermi energy and Δ_j is the superconducting gap energy in either film 1 or 2. Plugging in $\Delta \sim 200\,\mu\text{eV}$ (aluminum) and $T = 100\,\text{mK}$ gives us 2×10^{-4} μm^{-3}. This is a very low number considering the volume of a generic CPT island and leads; yet we know, experimentally, that the actual quasiparticle density is typically 10–$1000\,\mu\text{m}^{-3}$ (Mazin, 2004; Shaw et al., 2008). The disparity in the thermal prediction versus the experimentally measured numbers is the first indication that the source for the quasiparticles we see at low temperatures is distinctly nonthermal in nature. As yet, no one has determined the source of these *nonequilibrium* quasiparticles and we are stuck with the problem of understanding them.

16.2.4.1 Energetics of the Nonequilibrium Quasiparticle Poisoning Process

Since we are forced to work in an environment filled with nonequilibrium quasiparticles, we must figure out how to include them in the band picture that we constructed above. We have three states of interest (see Figure 16.6) (Aumentado et al., 2004):

State 0 No quasiparticles in or near the CPT island. *Even parity.*

State ℓ A quasiparticle is in the leads, in the vicinity of the CPT island. No quasiparticles on the island. *Even parity.*

State i A quasiparticle is on the CPT island. No quasiparticles in the leads near the island. *Odd parity.*

These states can all be described with the band structure derived in the previous section by offsetting the modulated energy surfaces by the superconducting gap energy, corresponding to where the quasiparticle lives, viz.,

$$\begin{aligned} E_0(n_g) &\equiv \epsilon_0(n_g, \delta = 0), \\ E_\ell(n_g) &\equiv \epsilon_0(n_g, \delta = 0) + \Delta_\ell, \\ E_i(n_g) &\equiv \epsilon_0(n_g + 1, \delta = 0) + \Delta_i. \end{aligned} \tag{16.19}$$

Here, we include the possibility of different gap energies in the leads and island, Δ_ℓ and Δ_i, respectively, and confine ourselves to $\delta = 0$. In principle, phase diffusion will smear these levels somewhat, but the average phase on the supercurrent branch will always be localized in the bottom of a potential well in the phase when biased near $I_b = 0$.

In Figure 16.7, we show the energy bands for two different pairs of island + gap energies using typical Coulomb, Josephson, and gap energies for an aluminum device. In the "type H" device (Figure 16.7a), the island gap is greater than the lead gap, and in the "type L" device, the reverse is true. Assuming that nonequilibrium quasiparticles are present, we can limit our discussion to E_ℓ and E_i. In the $T = 0$ limit, we expect the system to relax to the lowest energy state and the problem is reduced to whether the ℓ or i state energy is smallest (denoted in Figure 16.7a and d by the black dotted trace). For this purpose, we introduce the energy difference:[*]

$$\delta E_{\ell i}(n_g) \equiv E_i(n_g) - E_\ell(n_g). \tag{16.20}$$

At $T = 0$, the sign of this quantity determines what state we are in, that is,

$$\text{sgn}[\delta E_{\ell i}] = \begin{cases} +1 & \text{even parity} \\ -1 & \text{odd parity} \end{cases} \tag{16.21}$$

---

[*] *A brief history detour.* In early treatments of quasiparticle poisoning, all quasiparticles were assumed to be thermal in nature, so the important energy in the problem is the free energy required for transition from 0 to i state (Tuominen et al., 1992, 1993; Amar et al., 1994; Joyez et al., 1994; Tinkham et al., 1995). This is basically the process of breaking a Cooper pair with thermal fluctuations. Experimentally, this model was more or less verified early on, but the situation was complicated by the anecdotal evidence that this picture failed to explain the numerous unpublished experiments that showed poisoning at low temperatures. In the absence of nonequilibrium quasiparticles, the early free energy theories are still valid and, in any case, when the system is heated sufficiently $T > 250\,\text{mK}$, the thermal quasiparticle generation rate dominates over the nonequilibrium rate. For a good review of the early theory, see Tinkham (2004).

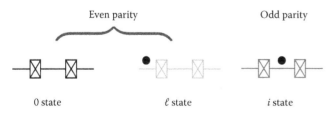

FIGURE 16.6 Three state nonequilibrium quasiparticle model. "0" state: no quasiparticles in vicinity of CPT, even parity. "ℓ" state: quasiparticle in leads even parity. "i" quasiparticle on island, odd parity.

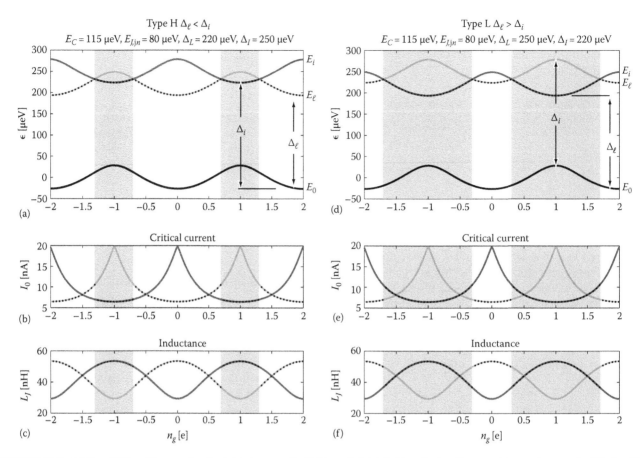

FIGURE 16.7 (a) Type H ($\Delta_\ell < \Delta_i$) energy bands. ℓ state/even parity (light gray), i state/odd parity (medium gray), 0 state/even parity (black solid). Minimum energy state is denoted as the black dotted trace. Corresponding even (light gray) and odd (medium gray) state (b) critical current and (c) effective inductance for a type H device. (d) Type L ($\Delta_\ell > \Delta_i$) energy bands. Corresponding even (light gray) and odd (medium gray) state (e) critical currents and (f) effective inductances for a type L device. Gray areas mark range in n_g where the CPT island is "trap-like," that is, a potential well for quasiparticles and the parity state is bimodal and can rapidly switch when coupled to thermal excitations.

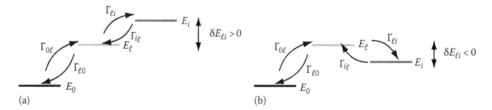

FIGURE 16.8 (a) Barrier-like configuration of levels. (b) Trap-like configuration of levels.

$\delta E_{\ell i}$ is an effective potential barrier height for quasiparticles. When $\delta E_{\ell i}$ is positive, the CPT island looks like a barrier, and when it is negative, it looks like a trap (Figure 16.8). In reality, the system is at finite temperature, so this qualitative picture must be amended to include the possibility of thermal excitations (phonons) coupling to the quasiparticles and exciting them out of the trap. Since this is a thermal escape process, we must characterize the system in terms of the average lifetimes of the poisoned and unpoisoned states, τ_o and τ_e.* Likewise, it is also useful to talk about these lifetimes as rates, $\Gamma_{eo} = \Gamma_{0\ell} + \Gamma_{\ell i} = 1/\tau_o$

and $\Gamma_{oe} = \Gamma_{i\ell} = 1/\tau_e$. Γ_{eo} is known as the "poisoning rate" and Γ_{oe} is the "ejection rate."

Experimentally, at $T \lesssim 250\,\mathrm{mK}$, the poisoning rate Γ_{eo} has little temperature dependence and can be anywhere between 10^3 and 10^5 s^{-1} (Aumentado et al., 2004; Ferguson et al., 2006; Naaman and Aumentado, 2006b; Court et al., 2008b). The fact that this rate is constant in this temperature range points to the notion that the source is nonthermal in nature and the rate for Cooper-pair breaking in the leads determines the poisoning rate, $\Gamma_{eo} \sim \Gamma_{0\ell} \gg \Gamma_{\ell i}$. The ejection rate Γ_{oe} does, however, have a temperature dependence at these low temperatures since the process of kicking a quasiparticle out of a potential well is a thermal escape process. One can derive the probability of the

---

* The "e" and "o" refer to "even" and "odd" parity states. Odd parity corresponds to a single excess quasiparticle on the CPT island.

even parity state (no quasiparticle on the island) using detailed balance arguments (Aumentado et al., 2004),

$$P_e = P_0 + P_\ell = \frac{1}{1 + \beta_{0\ell} e^{-\delta E_{\ell i}/k_B T}}. \qquad (16.22)$$

The factor, $\beta_{0\ell} \equiv \Gamma_{0\ell}/(\Gamma_{0\ell} + \Gamma_{\ell 0})$, accounts for the generation and recombination rates for nonequilibrium quasiparticles in the leads. What this all means is that there is not really any strictly quasiparticle-free zone for typical energies that we would use in practice.* In Figure 16.7, we denote the places where the CPT looks trap-like and it is apparent that both the L and H devices are affected. Even with the higher island gap, the H device can trap quasiparticles, albeit in a shallower potential, $\delta E_{\ell i}$ than presented by the L device and also over a smaller range in n_g.

It is interesting to note that if there is no source for nonequilibrium quasiparticles, $\Gamma_{0\ell} = 0$ ($\beta_{0\ell} = 0$) and $P_e = 1$ whether or not we have a barrier-like or trap-like profile. In other words, there would be no poisoning regardless of the relative gap energies. The fact that one sees poisoning easily in trap-like devices at low temperatures is the strongest indication that the quasiparticles in these systems are generated by some nonequilibrium process. Surprisingly, the temperature dependence above also predicts that the even state probability can actually be enhanced since, even though the poisoning rate might be fixed, the ejection rate can be increased by heating the CPT and giving quasiparticles energy to escape (Aumentado et al., 2004; Palmer et al., 2007).

This model of nonequilibrium quasiparticle poisoning was first roughly outlined and tested in Aumentado et al. (2004) using CPTs with engineered gap energy profiles and measured with the ramped current technique (see below). It was subsequently confirmed in several later experiments in both CPTs and CPBs using rf and dc techniques (Gunnarsson et al., 2004; Yamamoto et al., 2006; Palmer et al., 2007; Savin et al., 2007; Court et al., 2008b; Shaw et al., 2008).

16.3 Fabrication

16.3.1 Electron-Beam Lithography and Two-Angle Deposition

The majority of CPTs (and single-charge tunneling devices in general) are fabricated from evaporated aluminum using conventional electron beam lithography and electron-gun (or thermal) deposition techniques.

The most conventional method of fabrication is to first define the island and lead pattern of the CPT in a special double-layer e-beam resist stack that is spun onto a silicon substrate. The double-layer is constructed such that the upper "image" layer defines

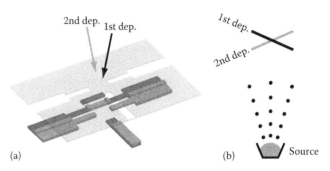

FIGURE 16.9 Angle deposition for the CPT. A mask is defined in electron-beam sensitive resist, usually in a bilayer configuration such that the underlayer is overdeveloped and the top (image) layer pattern can cast a shadow on the surface of the substrate from several angles. In this example, the leads for the device are deposited in the first deposition (dark gray) at an angle θ_1. After this deposition, oxygen is admitted into the deposition chamber forming an oxide on the surface of the metal. Finally, a second evaporation (light gray) is performed at θ_2 depositing the CPT island. In the process, Josephson junctions are formed at the overlap between the island and oxidized leads. (a) Shadow deposition through the top layer "image" resist. (b) Schematic of tilt configuration with respect to source.

the pattern outline, while the lower "ballast" layer is meant to provide a vast "undercut" region underneath the image. This is usually achieved using resist for the lower layer that is much more sensitive to the e-beam exposure than the upper layer. When the pattern is written, the dosage required to generate the image overexposes the lower layer resist such that the pattern is much wider in the lower layer resist, forming an "undercut" region when the sample is developed (cf., Cord et al., 2006).

The undercut region allows us to tilt the substrate so that the impinging aluminum atoms can deposit an image through the image resist mask at an angle (Figure 16.9). After the first aluminum deposition, oxygen is released into the chamber that grows a thin (<1 nm) insulating Al_xO_y layer on the surface of the aluminum. The substrate is then rotated to another angle and a second image is deposited onto the substrate such that it overlaps with the first oxidized image. In this manner, insulating junctions can be formed *in situ* without exposing the device to air. Subsequent processing removes the resist, so that it can be bonded up in a suitable measurement circuit. This process is called the "Dolan bridge" or "shadow deposition" technique (Dolan, 1977) and is used in almost all single-charge tunneling devices as well as most superconducting qubit designs. There are many variations on this technique but, surprisingly, the basic method has remained unchanged for more than 30 years.

16.3.2 Aluminum and Gap Engineering

Aluminum is regarded as the material of choice for these devices because of the general ease with which one can grow a reliable oxide layer at room temperature within the deposition chamber. Aluminum is easily deposited using both thermal and e-beam deposition techniques that are common to most clean rooms and

---

* The reason for this is not obvious. Basically, the difference in aluminum gap energies that one can achieve practically is ~50 μeV. In order to see any level of charge localization at $T = 100$ mK, we need $E_C \sim 100$ μeV. If we want useful switching current modulation (~10 nA) and inductance modulation (~10 nH), then we are stuck with $E_{J,n} \sim 50$ μeV. For these values, it is difficult to get the E_ℓ and E_i bands to separate.

requires very little in the way of special equipment. That being said, the CPT is a superconducting device whose operation is somewhat dependent on the quality of the superconducting gap.* In most general-purpose metal evaporators, the quality of the aluminum is completely beholden to the deposition hygiene of the community that uses the machine. For instance, the gap energy of aluminum is notoriously sensitive to the disorder and impurities that are present in the as-deposited film. For instance, two films deposited at identical rates but at different base pressures can yield gap energies that differ by 10s of μeV. This was done intentionally by Aumentado et al. (2004) to fabricate type L and H devices, but the same effect can be achieved by simply changing the thickness of the aluminum. It turns out that the gap energy can be enhanced substantially by making the film thinner (Townsend et al., 1972), and subsequent CPT experiments demonstrated that this method can be used to make CPTs (Yamamoto et al., 2006; Court et al., 2008a).

For reference, typical cryopumped e-gun deposition systems can have base pressures in the low 10^{-8} Torr range, especially if they are dedicated to aluminum and a few other well-behaved metals and employ a load-lock system for exchanging samples. These systems usually surpass more general use machines in film quality and oxide reproducibility. The author prefers e-gun deposition out of a bare water-cooled copper hearth rather than using thermal deposition out of a tungsten boat. Tungsten has a tendency to alloy with aluminum and repeated depositions from the same source yield films with inconsistent quality (resistivity and gap quality) and must be replaced on every pumpdown.

We note in passing that CPTs have been successfully fabricated from niobium with aluminum oxide junctions, but in none of the published work has anyone been able to reduce quasiparticle poisoning in any significant way. While this is awful for making charge-based qubits, these devices can still be used as electrometers, but the fabrication techniques can prove to be much more difficult since niobium is a refractory metal and difficult to evaporate. At its evaporating temperature, most practical deposition systems outgas so much that the quality of the film can be compromised and the integrity of the resist can also become an issue (Dolata et al., 2002, 2005). Others have used a sputtered niobium/aluminum oxide/niobium trilayer (ubiquitous in SQUID fabrication) and patterned using focused ion beam (Watanabe et al., 2004).

16.4 Practical Operation and Performance

The Coulomb, Josephson, and superconducting gap energies are all 10–$300\,\mu eV$, so practical measurements are performed well below 1 K in a dilution refrigerator, usually at $T \lesssim 100\,mK$. There are currently two generic modes of operation of the CPT: current

switching electrometry and rf electrometry. The main advantage of the current switching electrometry is its cost. rf electrometry can require relatively expensive microwave amplifiers, generators, and fast digitizers, while current switching electrometry can be achieved with simple function generators and voltage preamps operating with less than 1 MHz of bandwidth.

16.4.1 Current Switching Electrometry

One of the earliest methods for measuring the CPT was to measure the modulation of the switching current I_{sw} as a function of an applied gate voltage. This measurement is performed by cycling the current bias between the supercurrent and voltage branches in the I–V characteristic. We can track the current at which the CPT switches to the voltage state by tracking the voltage across the device and using a simple threshold trigger (see Figure 16.10). We can subdivide this measurement approach into ramped and

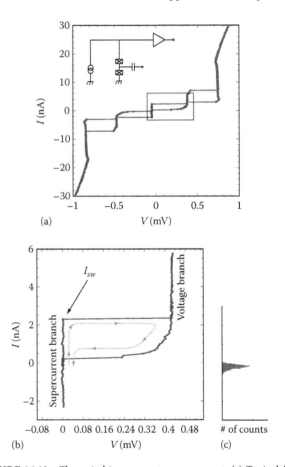

FIGURE 16.10 The switching current measurement. (a) Typical I–V curve for CPT device. The feature at ~$0.4\,mV$ ($\Delta_i + \Delta_\ell$) is the so-called Josephson quasiparticle peak or JQP and is due to more a complex Cooper-pair quasiparticle tunneling cycle. The device switches to the true voltage branch at $\geq 2(\Delta_i + \Delta_\ell)$. Inset: Simplified measurement circuit schematic. (b) Expanded view of switching current cycle. The current is ramped along the supercurrent branch until the voltage across the CPT switches to the finite valued voltage branch. The current at which this happens is recorded, and the results of many ramp cycles are recorded in a switching current histogram as in (c).

---

* If the gap edge is not well defined and there is some finite density of states in the subgap, then the gap engineering arguments that we made previously lose their applicability. This is mostly only an issue in "dirtier" deposition systems.

pulsed current bias techniques. Each has its own merits, but the experimental literature is largely dominated by the ramped technique.

16.4.1.1 Ramped Switching Current Measurement

As the name suggests, the bias current through the CPT is ramped linearly in time such that the current exceeds the maximum switching current and is then returned to zero bias for the next cycle. The probability of switching at a particular current can be backed out from the histogram of switching currents accumulated over many cycles (see Figure 16.10a). This ramped bias current measurement was originally done in the late 1970s in relatively large junctions (Fulton and Dunkleberger, 1974), but the methodology is also suitable to small Josephson junction devices such as the CPT (Joyez, 1995).

This method can be used for electrometry since it yields a switching current histogram whose mean value changes with gate, but its power lies in the fact that one can obtain the switching or "escape" rate $\gamma_{sw}(I_p, n_g)$ through a straightforward transformation of the whole histogram as in the original work by Fulton and Dunkleberger (1974). This is useful since the escape rate contains information about the electron temperature and how well the system is isolated from external noise (Devoret et al., 1987).

16.4.1.2 Quasiparticle Poisoning in the Ramped Current Measurement

If quasiparticle poisoning is present, the effective Josephson energy of the CPT flickers between two different values corresponding to even and odd parities. Each of these energies has its own escape rate $\gamma_{sw,e}$ and $\gamma_{sw,o}$, and the observed switching current in any given ramp cycle is determined by the instantaneous state of the system as the current is ramped. If the ramp rate is fast

enough such that it can span the distance separating the odd and even switching currents faster than the odd/even lifetimes, then the ramped measurement becomes a snapshot of the parity state and we see that the switching currents group around two distinct values as shown in type L device histograms in Figure 16.12.

In our example, the modulation of the switching current histograms is similar to that of the type H device, except that it shows a distinct 1e shifted image corresponding to the presence of the odd state, particularly where we predict the island potential to be trap-like. This bimodal behavior is really only evident if the measurement is faster than the state lifetimes. If the current bias ramp rate were slowed down significantly, then the system could flicker back and forth between parity states many times and we would only see switching distributions grouped around whichever parity state had the lowest switching current. That is, we would see a purely 1e modulation in the switching current. In the early literature, the available dc measurement techniques were unable to capture the dynamical nature of the quasiparticle parity states, and the periodicity (1e versus 2e) of the modulation was the only handle with which to gauge whether poisoning was present. Because of the dynamics of the poisoning process, this correlation can be misleading.

In the type L example given here, the device is definitely poisoned, particularly where we expect the island potential to look trap-like. However, if we examine the type H device's switching current modulation, it is distinctly 2e periodic (Figure 16.11). The early picture of poisoning would naively assume that the poisoning is not present when, in fact, it is happening much faster than a simple analysis would indicate. We know, for instance, that the island potential is trap-like over a narrower range and that the trap potential is shallow compared with the operating temperature (this is due to the gap engineering for the H device). Therefore, quasiparticles that get trapped on the island

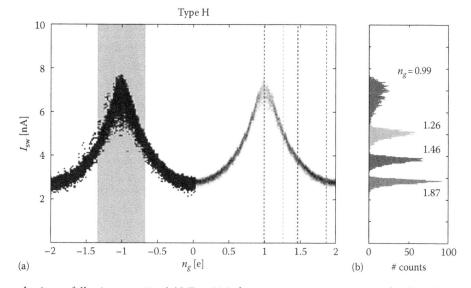

FIGURE 16.11 (See color insert following page 21-4.) (a) Type H I_{sw} histograms versus n_g. Histogram height is displayed in grayscale on the right-hand side, whereas all counts are displayed equally on the left-hand side. As in Figure 16.7, the gray box in (a) denotes regions where the island potential is trap like for quasiparticles. (b) Selected histograms corresponding to several gate voltages. Device parameters: $\Delta_i = 246\,\mu eV$, $\Delta_\ell = 205\,\mu eV$, $E_C \approx 115\,\mu eV$, and $E_{J1} = E_{J2} \approx 82\,\mu eV$.

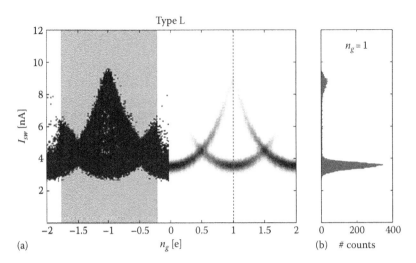

FIGURE 16.12 (a) Type L I_{sw} histograms versus n_g. Histogram height is displayed in grayscale on the right-hand side, whereas all counts are displayed equally on the left-hand side. As in Figure 16.7, the gray box in (a) denotes regions where the island potential is trap-like for quasiparticles. (b) Histogram at $n_g = 1$. Bimodal structure indicates relatively long-lived even and odd states. Device parameters: $\Delta_i = 205\,\mu eV$, $\Delta_\ell = 246\,\mu eV$, $E_C \approx 115\,\mu eV$, and $E_{J1} = E_{J2} \approx 78\,\mu eV$.

are rapidly ejected by thermal fluctuations. In this case, quasiparticles can jump in and out of the island before the CPT has time to latch into the free-running voltage state. In other words, the switching current measurement is bandwidth limited and is not able to see short-lived odd-parity events. While the average probability of being in the even state might be close to unity, quasiparticles may constantly be getting trapped and ejected at very rapid rates. In terms of the poisoning and ejection rates,

$$P_e = \frac{\Gamma_{oe}}{\Gamma_{eo} + \Gamma_{oe}}. \qquad (16.23)$$

If the ejection rate Γ_{oe} is big compared with the poisoning rate Γ_{eo}, then P_e approaches one even though Γ_{eo} may be arbitrarily large itself. This is an insidious effect since it was previously common for many groups working in CPB qubits to assess their quasiparticle situation from what appeared to be clean $2e$ modulation characteristics. In fact, $2e$ modulation is observed in the presence of these fast quasiparticle dynamics when the measurement technique is slow in comparison. Since quasiparticles might have lifetimes far shorter than $1\,\mu s$, most available methods (dc and rf alike) can have problems with this.

In CPTs with deeper trap potentials, such as the type L devices we show here, the ejection rate can be slowed down significantly. Since the measurement shown in Figures 16.12 and 16.13c,d was performed with a ramp rate that was comparable to the quasiparticle ejection rate, it was possible to obtain a bimodal histogram of I_{sw}. In this case, it is easy to correlate the derived escape function with meaningful rates. The two slopes in Figure 16.13d are the even and odd state escape rates, $\gamma_{sw,e}$ and $\gamma_{sw,o}$, respectively. Thus, the bias current ramp is not infinitely fast, and the parity can flip from even to odd in the time it takes to ramp between the two current values corresponding to the odd and even state switching currents resulting in several switching events in the region between the peaks. These correspond to quasiparticles

jumping into the CPT island during the bias ramp. Since the plateau that connects the even and odd switching rates is derived from the histogram counts between the even and odd peaks, it is a direct indication of the poisoning rate, Γ_{eo}. We can apply similar reasoning to the type H data in Figure 16.13a,b and notice that the escape rate for $n_g = 0.99$ is a little curvy, although the other escape rates shown look quite linear (on a semilogarithmic scale). While we do not develop a distinct plateau in this case, the deformation in the form of the escape rate is still an indication of very fast poisoning. This should not be a surprise since we expect the island to look trap-like at this gate voltage as we noted above. Thus, the lesson to be learned here is that despite appearances, the data shown in Figure 16.11 masks the fact that the type H device is poisoned.

16.4.1.3 Single Shot Measurement

In a single shot operation (Cottet et al., 2001), we pulse the current bias up from zero to some value I_p for a time τ_p. The probability of switching to the voltage state is given by

$$P_{sw}(n_g) = 1 - e^{-\gamma_{sw}(I_p, n_g)\tau}, \qquad (16.24)$$

where $\gamma_{sw}(I_p, n_g)$ is the rate at which the CPT phase escapes to the free-running voltage state when instantaneously biased at I_p and n_g. If we are biased at n_g and wish to determine whether the polarization charge has shifted by δn_g, we need to figure out the likelihood of seeing a switching event that is actually due to the shift in switching probability versus just the probability of switching with no charge shift at all,

$$\Delta P_{sw}(n_g, \delta n_g) = e^{-\gamma_{sw}(I_p, n_g)\tau} - e^{-\gamma_{sw}(I_p, n_g + \delta n_g)\tau}. \qquad (16.25)$$

If we want to measure this change in polarization charge in a single measurement, then we require that $\Delta P_{sw} = 1$. That is, if we pulse the bias current and see the CPT switch to the voltage state,

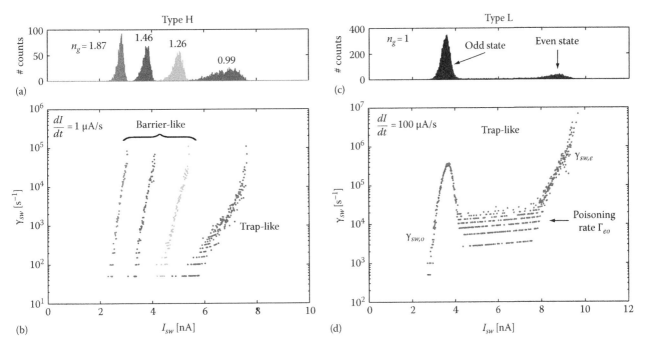

FIGURE 16.13 **(See color insert following page 21-4.)** (a,c) Switching current (I_{sw}) histograms and (b,d) derived switching/escape rates for a type H (barrier-like) and type L (trap-like) CPTs. For the type L device, the quasiparticle trapping behavior is evident in the bimodal I_{sw} distribution. In this case, the poisoning rate Γ_{eo} can be read directly from the derived escape rate in (d) as shown. Although the type H device is barrier-like for most n_g, it still looks like a trap near $n_g = 1$ (see Figure 16.7). This is apparent in the "curvy" structure of the escape curve for $n_g = 0.99$ as compared with the escape rates at other n_g in (b).

then the polarization charge has shifted by δn_g with certainty. This defines a minimum charge shift $\delta n_{g,ss}$ or voltage change $\delta V_{g,ss} (= \delta n_{g,ss} e/C_g)$ on the gate that can be detected with one measurement. If we want to detect a smaller change in gate voltage, then we would have to take multiple measurements until the uncertainty in the shift of $\langle I_{sw} \rangle$ is satisfactory. In many experiments, this is a completely reasonable approach. It is possible, however, to attempt to reduce the width of the switching current distribution by increasing the shunt capacitance across the CPT (increasing the effective mass of our imaginary phase particle) (Joyez, 1995).

In the end, pulsed single shot measurements have never been very popular. This may be because their chief use would have been in charge-based qubits, and other significant measurement schemes were shown to outperform it (Vion et al., 2002). However, it might still be a viable measurement method outside of superconducting quantum computing.

16.4.2 Zero-Biased rf Electrometry

The initial aim of the rf inductance measurements was to demonstrate a dissipationless (or near dissipationless) electrometer that could be used in charge-based quantum circuits such as the CPB, but the most useful application in recent years seems to have been to study the dynamics of quasiparticle tunneling.

Initial measurements of the Josephson (or "quantum") inductance of a CPT were performed by Sillanpää et al. (2004) and soon thereafter by Naaman and Aumentado (2006b). rf measurements of CPTs were also performed by Court and coworkers

(Ferguson et al., 2006) as well, but these measurements focused on a modulation of the dissipation and were not strictly confined to the supercurrent branch.*

16.4.2.1 The rf Measurement Setup

Figure 16.14a shows a typical microwave circuit used in rf-CPT electrometry. In this scheme, the CPT is embedded in a tank circuit composed of the parasitic capacitance provided by the leads, the Josephson inductance of the CPT, and an extra surface mount inductor soldered to a printed circuit board near the chip to lower the tank resonance to the range of the microwave amplifier.†

Following the circuit path in Figure 16.14a, we reflect an incoming microwave signal (the *carrier*) off of the CPT resonator circuit and measure the amplitude and phase of the outgoing signal, that is, we measure the scattering parameter S_{11}. As shown

---

* It's important to note that measurement of the Josephson inductance (the second derivative of the ground state energy in phase) is conceptually linked to the "quantum capacitance" (the second derivative in charge). rf measurements of the capacitance have recently been shown to be an extremely useful measurement of charge-based qubits such as the CPB (Wallraff et al., 2004; Duty et al., 2005) and the transmon (Schuster, 2007).

† The word "parasitic" should be a tip-off to the reader that this scheme might be a little hit-or-miss and, indeed, it can be frustrating to target the operating frequency within 10% (this is the typical bandwidth for sub-1 GHz low-noise cryogenic amplifiers). While this setup is very similar to that used in typical rf-SET operation (Schoelkopf et al., 1998), it is not necessarily trivial to implement.

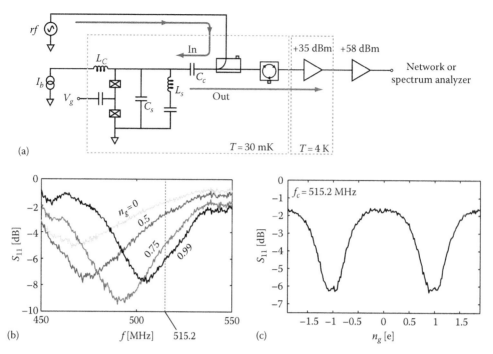

FIGURE 16.14 (a) Simplified rf-CPT measurement schematic. Incoming microwave power is directed toward CPT+ resonator through a directional coupler. The reflected wave is then amplified and measured at room temperature. In this schematic, several bandpass filters and attenuators have been omitted for simplicity. (b) Reflected power versus frequency at various gate voltages. (c) Reflected power modulation as a function of gate voltage for a fixed carrier frequency, $f_c = 515.2$ MHz. (Adapted from Naaman, O. and Aumentado, J., *Phys. Rev. B*, 73, 172504, 2006b.)

earlier, the gate-dependent Josephson inductance can be many 10s of nano-Henries and, since this inductance provides much of the total inductance available to the resonator, the frequency shift can be on the order of the bandwidth (typical Qs ~ 20–30). While a purely dissipationless circuit will reflect all of the signal ($|S_{11}|^2 = 1$) (Pozar, 2004), there are, in fact, a number of sources of dissipation, including the lossy traces on the pc board and leads on chip, the losses in the surface mount inductor and wire bonds, and, finally, the intrinsic CPT losses from any phase-diffusion resistance present. The end result is that there is a visible resonance dip in the reflected power amplitude $|S_{11}|^2$ (Figure 16.14b,c). For our purposes then, the frequency shift due to the modulation of the CPT Josephson inductance can be inferred from the reflected amplitude modulation.* That is, since the CPT inductance is a function of island charge, the final output microwave power amplitude and phase modulate with the gate-induced polarization charge.

Like the rf-SET, one can characterize the ultimate noise performance of this device as an electrometer in terms of an effective charge resolution in a 1 Hz bandwidth. In the rf measurement of the CPT inductance, the best number that has been reported is $\delta Q \sim 50\,\mu e/\sqrt{\text{Hz}}$ (Naaman and Aumentado, 2006b). That is, if we integrate the reflected power at the end of our measurement chain for 1 s, we can resolve a change in polarization charge at

the gate electrode $C_g V_g$ of 5.2×10^{-5} electrons. If we compare this with the best rf-SSET numbers $\delta Q < 5\,\mu e/\sqrt{\text{Hz}}$ (Brenning et al., 2006), the rf-CPT is more than 100 times slower at resolving the polarization charge. At first blush, this seems to put the rf-CPT at a disadvantage, but the rf-SSET charge resolution comes at the expense of using a relatively complex charge transport cycle that involves *both* quasiparticles and Cooper pairs. The back-acting voltage fluctuations of this cycle presented at the device they are measuring are equally nontrivial and have even been used to *cool* an electromechanical system that it was intended to measure (Naik et al., 2006). While some would consider this a feature of the rf-SSET, it seems to get away from the notion of simply wanting to use the device as a noninvasive electrometer. In fact, quasiparticle poisoning the rf-CPT has been used as a method to measure the effect of nearby SSETs, demonstrating that the latter emits nontrivial microwave power into its environment when voltage-biased (as they would be when operated as electrometers) (Naaman and Aumentado, 2007).

16.4.2.2 Operation Beyond 1 GHz

Although all of the published CPT experiments were operated with carrier frequencies below 1 GHz, this is not a fundamental requirement. In fact, this limitation seems to be determined by parasitic self-resonances in the surface-mount inductors, often used in these experiments. In principle, the operating frequency can be raised to many gigahertz using coplanar resonator techniques. The advantage of moving to higher frequencies is that wideband, low-noise HEMT amplifiers are now readily available in the 4–8 GHz range and all of the associated passive

---

* In principle, this information is also accessible through the phase of the outgoing signal and can be measured using an IQ mixer. Since loss was present in many of these experiments and the frequency shift was usually very obvious, measurement of the amplitude modulation has ended up being the simplest method.

components (directional couplers and isolators) are smaller and much more common (less expensive).

16.4.2.3 Quasiparticle Poisoning in the rf-CPT

As in the switching current measurement, quasiparticle poisoning also has a dramatic effect on rf measurement. Unlike the switching current measurement, a qualitative understanding of the measurement response is very straightforward. Since the inductance of the CPT can switch instantaneously between two different parity states when the gate is biased into a trap-like regime, the reflected power also switches between two different amplitudes. This kind of response is more commonly known as "telegraph noise." A typical example is shown in Figure 16.15a for a type L device. We note that this is the same bimodal behavior that we observe in the switching current measurements except that we can sit at zero bias and watch quasiparticles jumping in and out of the CPT island. In Figure 16.15c, we histogram the telegraph time traces as a function of gate voltage, we get data that, unsurprisingly, are very similar to our bimodal switching current histograms in Figure 16.12.

The rates Γ_{eo} and Γ_{oe} can be derived from an analysis of these time-domain traces but requires a careful characterization of the system measurement bandwidth (Naaman and Aumentado, 2006a). Although it is possible to extract the poisoning rate Γ_{eo} from a switching current measurement, it is far more difficult to pull out an ejection rate. This is a direct consequence of the time-ordered nature of the current bias ramp. In contrast, the rf measurement is not burdened by this kind of time ordering in any obvious way, and both the poisoning and ejection rates are available.

The availability of the poisoning and ejection rates has provided further verification of the model presented in Aumentado et al. (2004), while confirming that nonequilibrium quasiparticles are generated in the leads at a constant rate below ~250 mK. In addition, the ejection rate Γ_{oe} has been used to validate the notion that biased SSETs emit nontrivial levels of microwave power into their environment (Naaman and Aumentado, 2007). This is important as for a long time rf-SSET electrometry had been considered a viable method of measuring CPB qubit charge states.

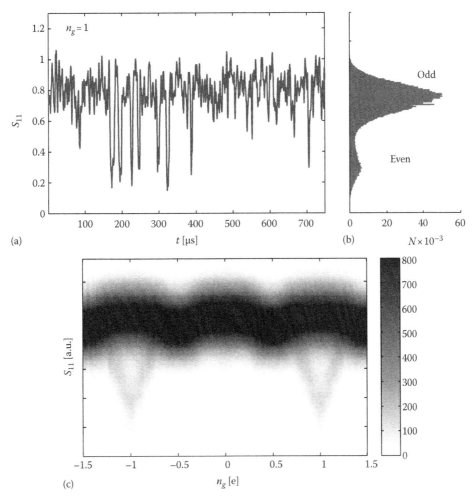

FIGURE 16.15 (a) Reflected power (linear units) of type L CPT at $n_g = 1$ in the time domain. The even (upper) and odd (lower) state levels are evident in telegraph-noise time traces. (b) Histogram of the full time trace. (c) Histograms versus gate voltage. (Adapted from Naaman, O. and Aumentado, J., *Phys. Rev. B*, 73, 172504, 2006b.)

16.4.2.4 Relation to the Cooper-Pair Box

Earlier, we alluded to the fact that the CPT was related to the CPB. In fact, when the CPT is biased at zero, it is identical to the CPB if the biasing circuit series impedance is small at the relevant parallel-junction plasma frequency. Since the problem of quasiparticle poisoning is also important to charge-based qubits such as the CPB (Nakamura et al., 1999; Wallraff et al., 2004), quantronium (Vion et al., 2002), and transmon (Koch et al., 2007), the fact that we can study it with relatively high instantaneous bandwidth in the CPT motivated several experiments aimed at studying the dynamics of the poisoning process in detail (Aumentado et al., 2004; Ferguson et al., 2006; Naaman and Aumentado, 2006b; Court et al., 2008b). Ultimately, it was realized that quantum capacitance measurements could achieve the same objective measuring CPBs directly and have yielded the most detailed quasiparticle poisoning studies to date (Lutchyn and Glazman, 2007; Shaw et al., 2008).

16.5 Present Status and Future Directions

The CPT is the simplest device in which dc transport characteristics can be correlated to the duality between charge and phase. Although the most prominent recent CPT experiments have focused on the problem of nonequilibrium quasiparticle poisoning, the CPT is also a promising general-purpose low-temperature, low-backaction electrometer. Since the most reliable techniques for fabricating these devices involve aluminum fabrication with gap, Josephson, and Coulomb blockade energies in the ~100–300 µeV range, operation is restricted to dilution refrigeration, so that the systems that one attaches it to must also be cold. While this seems restrictive at present, there are already several mesoscopic condensed matter systems that might benefit from minimally invasive fast electrometry.

Acknowledgment

The author wishes to acknowledge several important conceptual discussions with Ofer Naaman, Michel H. Devoret, and John M. Martinis.

References

Amar, A., Song, D., Lobb, C. J., and Wellstood, F. C. (1994). 2e and e periodic pair currents in superconducting coulomb-blockade electrometers. *Physical Review Letters*, 72(20):3234.

Ambegaokar, V. and Baratoff, A. (1963). Tunneling between superconductors. *Physical Review Letters*, 10(11):486.

Aumentado, J., Keller, M. W., Martinis, J. M., and Devoret, M. H. (2004). Nonequilibrium quasiparticles and 2e periodicity in single-cooper-pair transistors. *Physical Review Letters*, 92(6):066802.

Brenning, H., Kafanov, S., Duty, T., Kubatkin, S., and Delsing, P. (2006). An ultrasensitive radio-frequency single-electron transistor working up to 4.2 k. *Journal of Applied Physics*, 100:114321.

Clarke, J. and Braginski, A. I. (eds.). (2004). *The SQUID Handbook: Volume 1: Fundamentals and Technology of SQUIDs and SQUID Systems*. Wiley-VCH, Weinheim, Germany.

Cord, B., Dames, C., and Bergren, K. K. (2006). Robust shadow-mask evaporation via lithographically defined undercut. *Journal of Vacuum Science and Technology B*, 24(6):3139.

Cottet, A., Steinbach, A., Joyez, P., Vion, D., and Pothier, H. (2001). *Macroscopic Quantum Coherence and Quantum Computing*, p. 111. Kluwer Academic, Plenum Publishers, New York.

Court, N. A., Ferguson, A. J., and Clark, R. G. (2008a). Energy gap measurement of nanostructured aluminium thin films for single cooper-pair devices. *Superconductor Science and Technology*, 21(1):015013.

Court, N. A., Ferguson, A. J., Lutchyn, R., and Clark, R. G. (2008b). Quantitative study of quasiparticle traps using the single-cooper-pair transistor. *Physical Review B*, 77(10):100501.

Devoret, M. H. and Grabert, H. (1992). Introduction to single-charge tunneling. *Single Charge Tunneling: Coulomb Blockade Phenomena in Nanostructures (NATO Science Series: B)*, p. 1. Springer, New York.

Devoret, M. H., Esteve, D., Martinis, J. M., Cleland, A., and Clarke, J. (1987). Resonant activation of a brownian particle out of a potential well: Microwave-enhanced escape from the zero-voltage state of a Josephson junction. *Physical Review B*, 36(1):58.

Dolan, G. (1977). Offset masks for lift-off photoprocessing. *Applied Physics Letters*, 31(5):333.

Dolata, R., Scherer, H., Zorin, A., and Niemeyer, J. (2002). Single electron transistors with high-quality superconducting niobium islands. *Applied Physics Letters*, 80(15):2776.

Dolata, R., Scherer, H., Zorin, A., and Niemeyer, J. (2005). Single-charge devices with ultrasmall nb/alo/nb trilayer Josephson junctions. *Journal of Applied Physics*, 97:054501.

Duty, T., Johansson, G., Bladh, K., Gunnarsson, D., Wilson, C., and Delsing, P. (2005). Observation of quantum capacitance in the cooper-pair transistor. *Physical Review Letters*, 95(20):206807.

Ferguson, A. J., Court, N. A., Hudson, F. E., and Clark, R. G. (2006). Microsecond resolution of quasiparticle tunneling in the single-cooper-pair transistor. *Physical Review Letters*, 97(10):106603.

Flees, D., Han, S., and Lukens, J. (1997). Interband transitions and band gap measurements in bloch transistors. *Physical Review Letters*, 78(25):4817.

Fulton, T. and Dunkleberger, L. (1974). Lifetime of the zero-voltage state in Josephson tunnel junctions. *Physical Review B*, 9(11):4760.

Gunnarsson, D., Duty, T., Bladh, K., and Delsing, P. (2004). Tunability of a 2e periodic single cooper pair box. *Physical Review B*, 70(22):224523.

Josephson, B. (1962). Possible new effects in superconductive tunnelling. *Physics Letters*, 1(7):251.

Joyez, P. (1995). Le Transistor a une Paire de Cooper: un Systeme Quantique Macro-scopique. PhD thesis, L'Universite Paris, Paris, France.

Joyez, P., Lafarge, P., Filipe, A., Esteve, D., and Devoret, M. H. (1994). Observation of parity-induced suppression of Josephson tunneling in the superconducting single.... *Physical Review Letters*, 72(15):2458.

Kautz, R. and Martinis, J. (1990). Noise-affected iv curves in small hysteretic Josephson junctions. *Physical Review B*, 42(16):9903.

Koch, J., Yu, T. M., Gambetta, J. M., Houck, A. A., Schuster, D. I., Majer, J., Blais, A., Devoret, M. H., Girvin, S. M., and Schoelkopf, R. J. (2007). Charge insensitive qubit design from optimizing the cooper-pair box. *Physical Review A*, 74(4):042319.

Lutchyn, R. M. and Glazman, L. I. (2007). Kinetics of quasiparticle trapping in a cooper-pair box. *Physical Review B*, 75(18):184520.

Mazin, B. A. (2004). Microwave kinetic inductance detectors. PhD thesis, California Institute of Technology, Pasadena, CA.

Naaman, O. and Aumentado, J. (2006a). Poisson transition rates from time-domain measurements with a finite bandwidth. *Physical Review Letters*, 96(10):100201.

Naaman, O. and Aumentado, J. (2006b). Time-domain measurements of quasiparticle tunneling rates in a single-cooper-pair transistor. *Physical Review B*, 73(17):172504.

Naaman, O. and Aumentado, J. (2007). Narrow-band microwave radiation from a biased single-cooper-pair transistor. *Physical Review Letters*, 98(22):227001.

Naik, A., Buu, O., LaHaye, M. D., Armour, A. D., Clerk, A. A., Blencowe, M. P., and Schwab, K. C. (2006). Cooling a nanomechanical resonator with quantum back-action. *Nature*, 443:193.

Nakamura, Y., Pashkin, Y. A., and Tsai, J. S. (1999). Coherent control of macroscopic quantum states in a single-cooper-pair box. *Nature*, 398:786.

Palmer, B. S., Sanchez, C. A., Naik, A., Manheimer, M. A., Schneiderman, J. F., Echternach, P. M., and Wellstood, F. C. (2007). Steady-state thermodynamics of nonequilibrium quasiparticles in a cooper-pair box. *Physical Review B*, 76(5):054501.

Pozar, D. M. (2004). *Microwave Engineering*, 3rd edn. Wiley, New York.

Savin, A. M., Meschke, M., Pekola, J. P., Pashkin, Y. A., Li, T. F., Im, H., and Tsai, J. S. (2007). Parity effect in Al and Nb single electron transistors in a tunable environment. *Applied Physics Letters*, 91:063512.

Schoelkopf, R. J., Wahlgren, P., Kozhevnikov, A. A., Delsing, P., and Prober, D. E. (1998). The radio-frequency single-electron transistor (rf-set): A fast and ultrasensitive electrometer. *Science*, 280:1238.

Schuster, D. I. (2007). Circuit quantum electrodynamics. PhD thesis, Yale University, New Haven, CT.

Shaw, M. D., Lutchyn, R. M., Delsing, P., and Echternach, P. M. (2008). Kinetics of nonequilibrium quasiparticle tunneling in superconducting charge qubits. *Physical Review B*, 78(2):024503.

Sillanpää, M., Roschier, L., and Hakonen, P. (2004). Inductive single-electron transistor. *Physical Review Letters*, 93(6):066805.

Tinkham, M. (2004). *Introduction to Superconductivity*, 2nd edn. Dover, New York.

Tinkham, M., Hergenrother, J. M., and Lu, J. G. (1995). Temperature dependence of even-odd electron-number effects in the single-electron transistor. *Physical Review B*, 51(18):12649.

Townsend, P., Taylor, R., and Gregory, S. (1972). Superconducting behavior of thin-films and small particles of aluminum. *Physical Review B*, 5(1):54.

Tuominen, M. T., Hergenrother, J. M., Tighe, T. S., and Tinkham, M. (1992). Experimental evidence for parity-based 2e periodicity in a superconducting single-electron tunneling transistor. *Physical Review Letters*, 69(13):1997.

Tuominen, M. T., Hergenrother, J. M., Tighe, T. S., and Tinkham, M. (1993). Even-odd electron number effects in a small superconducting island: Magnetic-field dependence. *Physical Review B*, 47(17):11599.

Vion, D., Aassime, A., Cottet, A., Joyez, P., Pothier, H., Urbina, C., Esteve, D., and Devoret, M. H. (2002). Manipulating the quantum state of an electrical circuit. *Science*, 296:886.

Wallraff, A., Schuster, D. I., Blais, A., Frunzio, L., Huang, R.-S., Majer, J., Kumar, S., Girvin, S. M., and Schoelkopf, R. J. (2004). Strong coupling of a single photon to a superconducting qubit using circuit quantum electrodynamics. *Nature*, 431:162.

Watanabe, M., Nakamura, Y., and Tsai, J. (2004). Circuit with small-capacitance high-quality Nb Josephson junctions. *Applied Physics Letters*, 84(3):410.

Yamamoto, T., Nakamura, Y., Pashkin, Y. A., Astafiev, O., and Tsai, J. S. (2006). Parity effect in superconducting aluminum single electron transistors with spatial gap profile controlled by film thickness. *Applied Physics Letters*, 88(21):212509.

III

Nanolithography

17

Multispacer Patterning: A Technology for the Nano Era

Gianfranco Cerofolini
University of Milano-Bicocca

Elisabetta Romano
University of Milano-Bicocca

Paolo Amato
*Numonyx and University
of Milano-Bicacca*

17.1 Introduction

The evolution of integrated circuits (ICs) has been dominated by the idea of scaling down its basic constituent—the metal-oxide-semiconductor (MOS) field-effect transistor (FET). In turn, this has required the development of suitable lithographic techniques for its definition on smaller and smaller length scales. There are several generations of lithographic techniques, usually classified according to the technology required for the definition of the wanted features on photo- or electro-sensitive materials (resists): standard photolithography (436 nm, Hg *g*-line; refractive optics), deep ultraviolet (DUV) photolithography (193 nm, ArF excimer laser; refractive optics), immersion DUV photolithography (refractive optics), extreme ultraviolet (EUV) photolithography (13.5 nm, plasma-light source; reflective optics), and electron beam (EB) lithography (electron wavelength controlled by the energy, typically in the interval 10^{-3} to 10^{-2} nm).

The industrial system has succeeded in that, but the cost of ownership has in the meanwhile dramatically increased, because of either the required investment per machine,

$$\text{DUV} \ll \text{immersion DUV} \ll \text{EUV},$$

or the throughput,

$$\text{EB} \ll \text{DUV}.$$

It is just the dramatic cost escalation that is necessary for the reduction of the feature size that casts doubts on the possibility of continuing the current increase of IC density beyond the next 10 years.

Entirely new revolutionary technological device platforms, overcoming the complementary MOS (CMOS) paradigm, must likely be developed to enable the economical feasibility and scalability of electronic circuits to the tera scale intergation (TSI).

On another side, the preparation of ICs with bit density as high as 10^{11} cm^{-2} seems now possible, with modest changes in the current production process and marginal investment for the fabrication facility, within a different paradigm. The new paradigm is based on a structure, where a crossbar embodies in each of its cross-points a functional material able to perform by itself the functions of a memory cell [1].

The crossbar is indeed producible (via nonconventional lithography or even without any lithographic method) with geometry on the 10 nm length scale. Although the crossbar is not yet a circuit, it may nonetheless become a circuit if each cross-point contains a memory cell and each of them can be addressed, written, and read—that requires an external circuitry for addressing, power supply, and sensing. The best architecture for satisfying those functions is manifestly achieved embedding the crossbar in a CMOS circuitry [2]:

TSI IC = submicro CMOS IC∪
      nano crossbar ∪
      nanoscopic cells.

This architecture reduces the problem of preparing TSI ICs to that of producing nanoscopic memory cells and inserting them into the cross-points of a crossbar structure. It would be a mere declaration of will were it not for the fact that molecules by themselves able to behave as memory cells have been not only designed [3–5] and synthesized [6,7], but also inserted in hybrid devices [8–12]. This fact opens immediately the possibility of a *hybrid route* to TSI ICs:

Hybrid TSI IC = submicro CMOS IC∪
      nano crossbar ∪
      grafted functional molecules.   (17.1)

In this approach, the transport properties of programmable molecules are exploited for the preparation of externally accessible circuits. Because of this, it is usually referred to as molecular electronics.

The hypothesized TSI IC has thus a hybrid structure, formed by a nanometer-sized kernel (the functionalized crossbar), linked to a conventional submicrometer-sized CMOS control circuitry (producible with currently achievable technologies) and hosting molecular devices (whose production is left to chemistry).

17.2 The Crossbar Process

That self-assembled monolayers on preformed gold contacts may behave as nanoscale memory elements was demonstrated for thiol-terminated π-conjugated molecules containing amino or nitro groups [9]. The first demonstration of nonvolatile molecular crossbar memories employed self-assembled monolayers of thiol-terminated rotaxanes as reprogrammable cells [10]. The molecules were embedded between the metal layers forming the crossbar via a process that can be summarized as follows:

XB^1, deposition and definition of the first-level ("bottom")
     wire array
XB^2, deposition of the active reconfigurable molecules, work-
     ing also as vertical spacer separating lower and upper
     arrays
XB^3, deposition and definition of the second-level ("top")
     wire array

Figure 17.1 sketches the XB process.

Although potentially revolutionary, the XB approach, with double metal strips, has been found to have serious limits:

- The organic active element is incompatible with high-temperature processing, so that the top layer must be deposited in XB^3 at room or slightly higher temperature. This need implies a preparation based on physical vapor

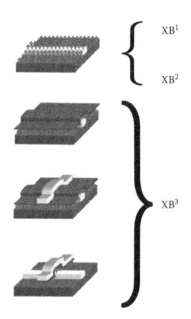

FIGURE 17.1 The first proposed crossbar-architecture fabrication steps. (Reprinted from Cerofolini, G., *Nanoscale devices*, Springer, Berlin, Germany, 2009. With permission.)

deposition, where the metallic electrode results from the condensation of metal *atoms* on the outer surface of the deposited organic films. This process, however, poses severe problems of compatibility, because isolated metal atoms, quite irrespective of their chemical nature, are mobile and decorate the molecule, rather than being held as a film at its outer extremity [13–15].

- A safe determination of the conductance state of bistable molecules requires the application of a voltage V appreciably larger than $k_B T/e$ (with k_B being the Boltzmann constant, T the absolute temperature, and $-e$ the electron charge), say $V = 0.1$–0.2 V. Applied to molecules with typical length around 3 nm, this potential sustains an electric field, of the order of 5×10^5 V cm^{-1}, sufficiently high to produce metal electromigration along the molecules [16].

- The energy barrier for metal-to-molecule electron transfer is controlled by the polarity of the contact, in turn increasing with the electronegativity difference along the bond linking metal and molecule [17]. The use of thiol terminations for the molecule, as implicit in the XB approach, is expected to be responsible for high-energy barriers because of the relatively high electronegativity of sulfur.

Even though the first difficulty can in principle be removed by slight sophistication of the process (for instance, as follows: spin-coating the organic monolayer with a dispersion of metal nanoparticles in a volatile solvent, evaporating the solvent, forming a relatively compact layer via coalescence of the metal particles, and compacting the resulting film by means of an additional amount of PVD metal), other difficulties are more fundamental in nature and require different materials.

A solution to the electromigration problem can be achieved preparing the bottom electrodes in the form of silicon wires

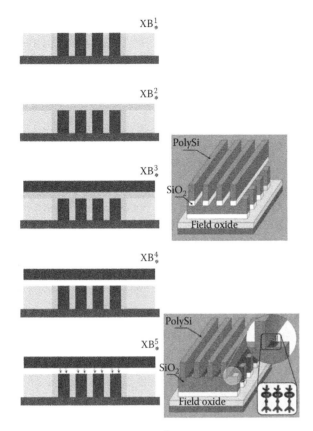

FIGURE 17.2 The basic idea of XB*: preparing the crossbar before its functionalization.

(as done in [12]), and the top electrodes in the form conducting π-conjugated polymers (as suggested in [18]):

XB$_+^1$, deposition and definition of the bottom array of poly-silicon wires

XB$_+^2$, deposition of the active (reconfigurable) element, working also as vertical spacer separating lower and upper arrays

XB$_+^3$, deposition and definition of the top array of conducting π-conjugated polymers

The use of poly-silicon as material for the top array too seems impossible because it is prepared almost uniquely via chemical vapor deposition at incompatible temperatures with organic molecules.* The only way to overcome this difficulty consists thus in a process, XB*, where the two poly-silicon arrays defining the crossbar matrix are prepared *before* the insertion of the organic element [11]. Preserving a constant separation on the nanometer length scale is possible only via the growth of a sacrificial thin film on the first array before the deposition of the second one [21]:

XB$_*^1$, preparation of a bottom array of poly-silicon wires

XB$_*^2$, deposition of a sacrificial layer as vertical spacer separating lower and upper arrays

---

* It is instead possible if the spacer is an inorganic film. This case, although not for molecular electronics, is interesting for memories based on phase-change materials [19] or mimicking the memristor [20].

XB$_*^3$, preparation of a top array of poly-silicon wires crossing the first-floor array

XB$_*^4$, selective chemical etching of the spacer

XB$_*^5$, insertion of the reprogrammable molecules in a way to link upper and lower wires in each cross-point

The basic idea of process XB*, of inserting the functional molecules after the preparation of the crossbar, is sketched in Figure 17.2.

Of the three considered processes (XB, XB$_+$, and XB*), the one based on double-silicon strips is certainly the most conservative one and is thus expected to be of easiest integration in IC processing. For this reason (and for the possibility of using three-terminal molecules, see Section 17.4.3), our attention will be concentrated on the XB* route.

17.3 Nonlithographic Preparation of Nanowires

The preparation of a crossbar requires the use of simple geometries—essentially arrays of dielectrically insulated conductive wires. What is especially interesting is that wire arrays with pitch on the nanometer length scale are producible via nonlithographic techniques (NLTs). Not only is this preparation possible, but also the wire linear density already achieved with the NLTs described in the following is smaller than the one achievable via the most advanced EUV or EB lithographies. Such NLTs exploit the following features:

(V), the "vertical" control of film thickness, possible down to the subnanometer length scale provided that the film is sufficiently homogeneous.

(V-to-H), the transformation of films with "vertical" thickness t into patterns with "horizontal" width w:

$$t \xrightarrow{\text{NLT}} w.$$

These techniques are *imprint lithography* and two variants of the *multispacer patterning technique*.

17.3.1 Imprint Lithography

Imprint lithography (IL) is a contact lithography where properties (V) and (V-to-H) are exploited for the preparation of the mask. The process is essentially based on the sequential alternate deposition of two films, A and B, characterized by the existence of a preferential etching for one (say A) of them. After cutting at 90°, polishing, and controlled etching of A, one eventually gets a mask formed by nanometer-sized trenches running parallel to one another at a distance fixed by the thickness of B [22,23]. For instance, a contact mask for imprint lithography with pitch of 16 nm was prepared by growing on a substrate a quantum well via molecular beam epitaxy, cutting the sample perpendicularly to the surface, polishing the newly exposed surface, and etching selectively the different strata of the well [23]. The potentials of the preparation method based on superlattice nanowire pattern transfer (SNAP) are reviewed in Ref. [24].

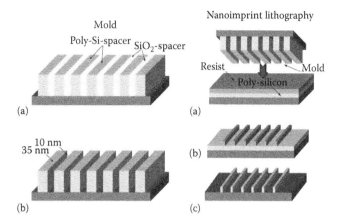

FIGURE 17.3 Preparation of mold for imprint lithography (left) and its use as contact mask (right); the multilayer has been supposed to be produced with cycles of sequential depositions of silicon and SiO_2. (Reprinted from Cerofolini, G., *Nanoscale devices*, Springer, Berlin, Germany, 2009. With permission.)

Actually a number of variants for transferring the pattern to the surface have been developed: molding, embossing, and stamping are the ones most frequently considered [25]. In one of them (molding), after filling the trenches with a suitable polymer, the mask is used as a stamp, pressing it onto the surface; if the polymer has a higher affinity for the surface than for the mask, the pattern is transferred to the surface when the mask is eventually removed [25]. The transfer of the polymer to the surface is possible without loss of geometry only if the trench is sufficiently shallow; this implies that the polymer must sustain a subsequent process where it is used as mask for the definition (via directional etching) of the underlying structure with a high aspect ratio. Another method (embossing), sketched in the right-hand side of Figure 17.3, involves the pressure-induced transfer of the pattern from the mask to a plastic film and its subsequent polymerization.

Imprint lithography is generally believed to have potential advantages over conventional lithography because it can be carried out in ordinary laboratory with no special equipments [26]. This situation is expected to make it easy to run along the learning curve to a mature technology. However, very little is known about the overall yield, eventually resulting in production cost, of this process (preparation of mask and stamp, imprint, etching) when the geometries are on the length scale of tens of nanometers.

Imprint lithography has been the matter of extended investigation (see for instance, Refs. [25,26]) and will not be discussed here. Rather, this chapter is devoted to describe the multispacer patterning technique and to compare its two variants.

17.3.2 Spacer Patterning Technique

The multispacer patterning technique (S^nPT) is essentially based on the repetition of the spacer patterning technique (SPT). In turn, the SPT is an age-old technology originally developed for the dielectric insulation of metal electrodes contacting source and drain from the gate of metal-oxide-semiconductor (MOS) transistors.

The SPT involves the following steps:

SPT^0, the *lithographic definition* of a seed with sharp edge and high aspect ratio

SPT^1, the *conformal deposition* on this feature of a film of uniform thickness

SPT^2, the *directional etching* of the film until the original seed surface is exposed

If the process is stopped at this stage, it results in the formation of sidewalls of the original seed; otherwise, if

SPT^3, the original seed is removed via a *selective etching*

what remains is constituted only by the walls of the seed edges. Figure 17.4 sketches the various stages of SPT.

This technique has been demonstrated to be suitable for the preparation of features with minimum size of 7 nm [27,28] and has already achieved a high level of maturity, succeeding in the definition, with yield very close to unity, of nanoscopic bars with high aspect ratio.

The SPT may be sophisticated via the deposition of a multilayered film; Figure 17.5 shows the sidewalls resulting from the deposition of a multilayer and compares it with what is really done in the original application of SPT—the insulation of the source and drain electrodes from the gate [29].

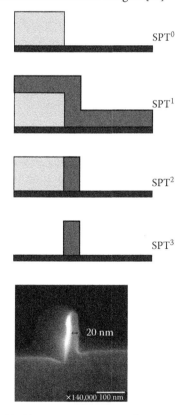

FIGURE 17.4 Up: The spacer patterning technique: SPT^0, definition of a pattern with sharp edges; SPT^1, conformal deposition of a uniform film; SPT^2, directional etching of the deposited film up to the appearance of the original seed; and SPT^3, selective etching of the original feature. Down: Cross-section of a wire produced via SPT. (Reprinted from Cerofolini, G., *Nanoscale devices*, Springer, Berlin, Germany, 2009. With permission.)

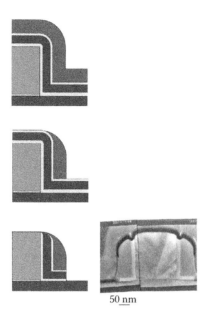

FIGURE 17.5 The original application of the SPT in microelectronics—dielectric insulation of the gate from source and drain electrodes. (Reprinted from Cerofolini, G., *Nanoscale devices*, Springer, Berlin, Germany, 2009. With permission.)

17.4 Multispacer Patterning Techniques

Two S^nPT routes have been considered: the additive (S^nPT_+) and multiplicative ($S^nPT_\times$) routes.

The S^nPT_+ is recent and was proposed having in mind the preparation of crossbars for molecular electronics [30–32]. The $S^nPT_\times$ is instead much older: the first demonstrators were developed for the generation of gratings with sub-lithographic period [33]; recently, however, this technique has been used for the preparation of wire arrays in biochips also [28]. Since no detailed comparison of the limits and relative advantages of these techniques is known, the following part will try to provide an understanding about them on the basis of fundamental considerations.

17.4.1 Additive Route—S^nPT_+

The S^nPT_+ is substantially based on n STP repetitions where *the original seed is not removed and each free wall of newly grown bars is used as a seed for the subsequent STP*. Each SPT_+ cycle starts from an assigned seed and proceeds with the following steps:

$S^nPT_+^1$, conformal deposition of a conductive material
$S^nPT_+^2$, directional etching of this material up to the exposure of the original seed
$S^nPT_+^3$, conformal deposition of an insulating material
$S^nPT_+^4$, directional etching of this material up to the exposure of the original seed

The basic idea of the S^nPT_+ is shown in Figure 17.6: the upper part sketches the process; the lower part shows instead how

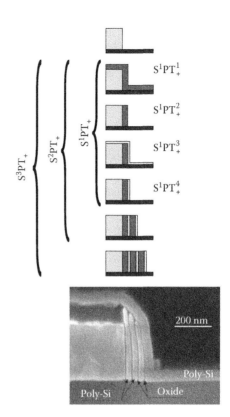

FIGURE 17.6 Up: the additive multispacer patterning technique. Down an example of S^3PT_+ multispacer (with pitch of 35 nm and formed by a double layer poly-Si|SiO$_2$) resulting after three repetitions of the SPT$_+$. (Reprinted from Cerofolini, G., *Nanoscale devices*, Springer, Berlin, Germany, 2009. With permission.)

poly-silicon arrays separated by SiO$_2$ dielectrics with sub-lithographic pitch (35 nm) can indeed be produced [30–32].

The sketch in Figure 17.6 shows a process in which lines are additively generated onto a progressively growing seed, preserving the original lithographic feature along the repetitions of the unit process. The unit process is based on two conformal depositions of uniform layers (poly-silicon and SiO$_2$) each followed by a directional etching.

Figure 17.7 shows, however, that a similar structure could be obtained by a cycle formed by

S^nPT_+', conformal deposition of a bilayer film (formed by an insulating layer deposited before the conductive one—the order of deposition is fundamental)
S^nPT_+'', directional etching of this film up to the exposure of the original seed

Consider an array with pitch P of lithographic seeds, each with width W (and thus separated from one another by a distance $P-W$). Denote with the same symbols in lower case, p and w, the corresponding sub-lithographic quantities. Starting from the said array of lithographic seeds, after n repetitions of SPT$_+$ any seed is surrounded by $2n$ lines (an example with $n = 4$ is shown in Figure 17.8), so that the corresponding effective linear density K_n of spacer bars is given by

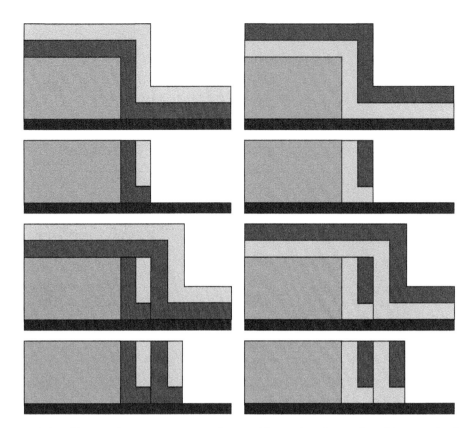

FIGURE 17.7 A variant of the additive multispacer patterning technique able to reduce the number of directional etching by a factor of 2 via sequential deposition of a bilayered film. Two possibilities are considered, consisting in the deposition first of either poly-silicon and then of SiO₂ (left), or the same layers in reverse order (right). (Reprinted from Cerofolini, G., *Nanoscale devices*, Springer, Berlin, Germany, 2009. With permission.)

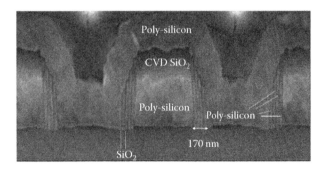

FIGURE 17.8 An example of S⁴PT₊, showing the construction of four silicon bars per side of the seed. (Reprinted from Cerofolini, G., *Nanoscale devices*, Springer, Berlin, Germany, 2009. With permission.)

$$K_n = \frac{2n}{P}.$$

The example of Figure 17.8 shows also that the process can be tuned to preserve the constancy with n of wire width w_n, $w_n = w$, and pitch p_n, $p_n = p$. On the contrary, the spacer heights s_n decrease almost linearly with n,

$$s_n \simeq s_0 - \tau n \qquad (17.2)$$

(with τ the spacer height loss per SPT cycle), at least for n lower than a characteristic value n^{max}.

This unavoidable decrease is ultimately due to the fact that the conformal coverage of a feature with high aspect ratio results necessarily in a rounding off of the edge shape with curvature radius equal the film thickness and that the subsequent directional etching produces a nonplanar surface. Figure 17.9 explains the reasons for the shape of the resulting bars: even assuming a perfect directional attack, the resulting spacer is not flat and there is a loss of height τ not smaller than t: $\tau \geq t$.

Figure 17.6 shows that the process can actually be controlled to have τ coinciding, within error, with its minimum theoretical value, $\tau = 1.0t$. Figure 17.8 shows however that τ depends on the process, and this can be tuned to have $\tau = 3.2t$.

Although the loss of height may seem a disadvantage, in Section 17.5.1 it will be shown how the controlled decrease of s_n with n may be usefully exploited.

Assuming the validity of Equation 17.2 until s_n vanishes, the maximum number n^{max} of SPT₊ repetitions is given by $n^{max} = s_0/\tau$; after n^{max} SPT₊ repetitions, the seed is lost and the process cannot continue further. In view of the availability of techniques for the production of deep trenches with very high aspect ratios, in this analysis s_0 (and hence n^{max}) may be regarded as almost a free parameter.

The optimum distance allowing the complete filling of the void regions separating the original lithographic seeds is therefore given by $P - W = 2n^{max}p$. Hence, the maximum number of cross-points that can be arranged in any square of side P is given

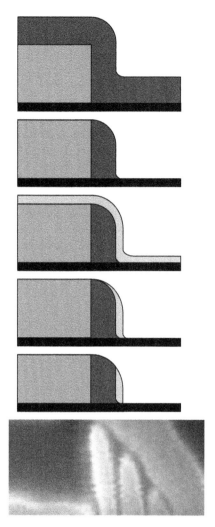

FIGURE 17.9 Shape of a sidewall resulting after ideal conformal deposition and directional etching (top; five sketches) and an image of how it results in practice (bottom; magnification of the spacers shown in Fig. 6). (Reprinted from Cerofolini, G., *Nanoscale devices*, Springer, Berlin, Germany, 2009. With permission.)

by $(2n^{max})^2$, and the maximum effective cross-point density δ_+^{max} achievable with the S^nPT_+ is given by

$$
\delta_+^{max} = \left(\frac{2n^{max}}{P} \right)^2
$$

$$
= \frac{1}{p^2} \left(\frac{1}{1 + W/2n^{max}p} \right)^2. \tag{17.3}
$$

Equation 17.3 shows that δ_+^{max} depends on the lithography (through W) and on the sub-lithographic technique S^nPT_+ (through n^{max} and p).

Just to give an idea of the maximum obtainable density, Figure 17.6 shows that $p = 35\,\text{nm}$ has already been achieved and $n^{max} \simeq 10$ is at the reach of the S^nPT_+; for $W = 0.1\,\mu\text{m}$ (characteristic value for IC high volume production) and $p = 30\,\text{nm}$,

Equation 17.3 gives $\delta_+^{max} \simeq 8 \times 10^{10}\ \text{cm}^{-2}$. The comparison of this prediction with the lithographically achievable cross-point density (currently of about $2 \times 10^9\ \text{cm}^{-2}$), shows that S^nPT_+ allows the cross-point density to be magnified by a factor of about 40. This is however achieved only with the construction of 10 consecutive spacers per (bottom and top) layer. The spacer technology is a mature technology with yields close to unity also when employed in more complex geometries than single lines. The increase of processing cost implied by its repeated application in IC processing has been discussed in Ref. [34]. Although this increase is moderate, the integration of so many SPT_+ cycles may, however, be not trivial and passes through the development of dedicated cluster tools.

It is however noted that the basic idea sketched in Figure 17.7 can be extended to remove this difficulty at least partially: the conformal deposition of a slab with n poly-silicon|insulator bilayers (the insulator being SiO_2, Al_2O_3,…) followed by its directional etching would indeed result in the formation of $2n$ dielectrically insulated poly-silicon wires. Figure 17.10 sketches the process.

The figure however shows that this dramatic simplification of the process can be done only for the deposition of the bottom array; its use for the top array would result in a distance of the top wire from the lower one varying with the order of poly-silicon layer.

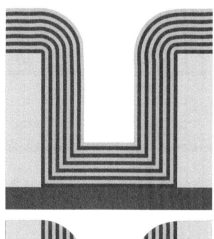

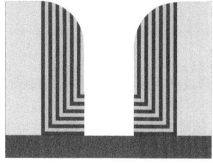

FIGURE 17.10 Structure resulting after the deposition of a slab of six poly-silicon|SiO_2 bilayers (top) in one shot followed by the conformal attack stopped with the exposure of the original lithographic seed, resulting in an array of 12 sub-lithographic wires (bottom). (Reprinted from Cerofolini, G., *Nanoscale devices*, Springer, Berlin, Germany, 2009. With permission.)

17.4.2 Multiplicative Route—S^nPT_x

The SPT allows, starting from one seed, the preparation of *two* spacers [27]; in principle, this fact allows another, multiplicative, growth technique—S^nPT_x. The multiplicative generation requires that both sides of each newly grown spacer are used as seeds for the subsequent growth—that is possible only if the original seed is etched away at the end of any cycle. In S^nPT_x each multiplicative SPT_x cycle involves therefore the following steps:

$S^nPT_x^1$, conformal deposition of a film on the seed
$S^nPT_x^2$, directional etching of the newly deposited film up to the exposure of the seed
$S^nPT_x^3$, selective etching of the original seed

Figure 17.11 sketches two S^nPT_x repetitions.

Assume that the process starts from a seed formed by an array with pitch P of lithographically defined seeds (lines) each of width W, the linear density K_0 of lines being thus given by $K_0 = P^{-1}$. If lower and upper arrays have the same linear density K_0, the lithographic cross-point density is K_0^2 and the repetition of n (bottom) plus n (top) S^nPT_x results in a sub-lithographic cross-point density δ_x given by

$$\delta_x = 2^{2n} K_0. \tag{17.4}$$

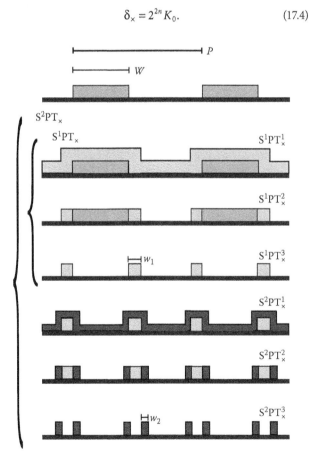

FIGURE 17.11 Two STP_x steps for the formation of a sub-lithographic wire array starting from a lithographic seed array. (Reprinted from Cerofolini, G., *Nanoscale devices*, Springer, Berlin, Germany, 2009. With permission.)

For any assigned n, this density is optimized maximizing K_0, i.e., minimizing P. The minimum value of P is determined by the considered lithography, while W is adjusted to the wanted value controlling exposure, etching, etc. The appropriate ratio W/P is obtained with the following considerations.

Let the seeds be formed by a given material A (to be concrete we shall think of it as SiO_2) and the SPT_x be carried out depositing another material B (again for concreteness, we shall think of B as poly-silicon) forming a conformal layer of thickness $t_1 = \varrho_1 W$, with $\varrho_1 < 1$. After completion of the multiplicative SPT cycle, the surface will thus be covered by an array of $2K_0$ wires per unit length each of width $w_1 = t_1 = \varrho_1 W$.

Let the process proceed with the deposition of a film of A with thickness $t_2 = \varrho_2 w_1 = \varrho_1 \varrho_2 W$, with $\varrho_2 < 1$ (in the considered example, this process could be the oxidation of poly-silicon to an SiO_2 thickness t_2). After completion of the second SPT_x cycle, the surface will be covered by a spacer array of linear density $2^2 K_0$ each of width $w_2 = t_2 = \varrho_1 \varrho_2 W$. It is noted that in S^nPT_x the seed material at the end of each SPT_x is inverted from A to B or vice versa, so that the material of the original seed must be chosen in the relation to the parity (even or odd) of n.

In the following, the focus is on the search of the mask geometry that maximizes the spacer density. After n reiterations of the SPT_x, the spacers will extend both beyond and beneath the original lithographic feature. The zone containing the spacers extends from the edge of the original lithographic feature both into the region separating them and into the region beneath the original feature by amounts l_{out}^n and l_{in}^n given by

$$l_{out}^n = w_1 + w_2 + \cdots + w_n$$

$$= W \sum_{k=1}^{n} \prod_{j=1}^{k} \varrho_j, \tag{17.5}$$

$$l_{in}^n = w_2 + \cdots + w_n$$

$$= W \sum_{k=2}^{n} \prod_{j=1}^{k} \varrho_j. \tag{17.6}$$

The estimate of l_{out}^n and l_{in}^n requires knowledge of various ϱ_k values; at this stage, it is impossible to state anything about them. Without pretending to describe the actual technology, but simply to have quantitative (although presumably correct at the order of magnitude) estimates, we assume ϱ_k independent of k, $\forall k(\varrho_k = \varrho)$. With this assumption Equations 17.5 and 17.6 become

$$l_{out}^n = W \sum_{k=1}^{n} \varrho^k$$

$$= W \frac{\varrho}{1-\varrho}(1 - \varrho^n), \tag{17.7}$$

$$l_{in}^n = W \sum_{k=2}^{n} \varrho^k$$

$$= W \frac{\varrho^2}{1-\varrho}(1-\varrho^{n-1}). \tag{17.8}$$

The least upper bounds of l_{out} and l_{in} are obtained taking the limit for $n \to +\infty$ in Equations 17.7 and 17.8: $l_{out} = \varrho/(1-\varrho)$ and $l_{in} = \varrho^2/(1-\varrho)$; moreover, already for relatively low values of n both ϱ^n and ϱ^{n-1} are negligible with respect to 1 so that we can reasonably assume

$$l_{out}^n \simeq W\varrho/(1-\varrho), \tag{17.9}$$

$$l_{in}^n \simeq W\varrho^2/(1-\varrho). \tag{17.10}$$

For any W, the optimum ϱ is obtained imposing the condition that all the region beneath the original lithographic feature is filled with nonoverlapping spacers: $2l_{in} = W$. Inserting this condition into Equation 17.10 gives

$$\varrho \simeq \frac{1}{2}. \tag{17.11}$$

Similarly, the optimum size of the outer region is given by the following condition: $2l_{out} = P - W$. Inserting this condition into Equation 17.9 gives

$$P \simeq 3W. \tag{17.12}$$

Choi et al. [28] have demonstrated that three SPT_x repetitions on a lithographically defined seed result in nanowire arrays of device quality and suggest that very long wires can indeed be produced with a high yield. However, even accepting that the process have a yield so high as to allow the preparation of non-interrupted wires over a length (on the centimeter length scale) comparable with the chip size, if the lines are used as conductive wires of the crossbar its length is so high to have a series resistance larger than the resistance of the molecules forming the memory cell. This problem was considered in Ref. [2], where it was shown that the crossbar memory can conveniently be organized in modules each hosting a sub-memory of size 1–4 kbits. This implies that each module must be framed in a region sufficiently large to allow the addressing of the memory cells. In this way the density calculated with Equation 17.4 is an upper value to the exploitable density.

17.4.3 Three-Terminal Molecules

The use of molecules in molecular electronics is essentially due to the fact that they embody in themselves the electrical characteristics of existing devices. The characteristics of nonlinear resistors, diodes, and Schmitt triggers have been reported for two-terminal molecules; their use as nonvolatile memory

cells is possible thanks to the stabilization of a metastable state excited by the application of a high voltage (thus behaving as a kind of virtual third terminal). Three-terminal molecules offer more application perspectives not only because they can mimic transistors but also because they could exploit genuine quantum phenomena like the Aharonov–Bohm effect [35].

The application potentials of three-terminal molecules can however be really exploited only if all terminals can be contacted singularly. This is manifestly impossible using the XB or XB_+ routes, but is possible in the XB_* framework. The major advantage of poly-silicon in the XB_* route is that it does not pose the problem of metal electromigration. However, the multispacer technology can also be adapted for the preparation of nanowire arrays of poly-silicon and metals in arrangements that

- Allow the use of three-terminal molecules
- Facilitate the self-assembly of functional molecules
- Avoid the problem of metal electromigration

Assume that the top array defining the crossbar is formed by poly-silicon nanowires whereas the bottom array has a more complicate structure. Assume, as sketched in Figure 17.12a, that each conformal deposition is formed by the following multilayer:

$$SiO_2 \mid polysilicon \mid SiO_2 \mid metal \mid SiO_2 \mid polysilicon,$$

where the metal might be, for instance, platinum obtained via CVD from PtF_6 precursor.

After an SPT, one gets the structure sketched in Figure 17.12b; a subsequent time-controlled selective etch of the metal electrode will result in the formation of a recessed region, as sketched in Figure 17.12c.

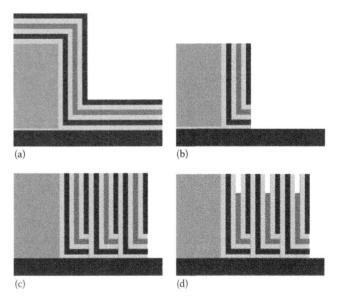

(a)

(b)

(c)

(d)

FIGURE 17.12 The structure resulting after (a) conformal deposition of a multilayer, (b) its directional etching, (c) three repetitions of the above processes, and (d) the time-limited preferential attack of the metal. (Reprinted from Cerofolini, G., *Nanoscale devices*, Springer, Berlin, Germany, 2009. With permission.)

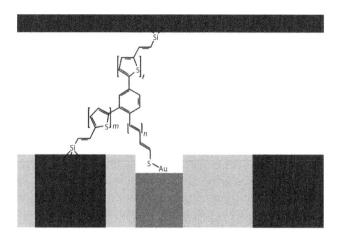

FIGURE 17.13 The three-terminal molecule after the formation of a self-assembled monolayer on the metal and the grafting to the silicon. The larger separation between metal and top poly-silicon electrode than between top and bottom poly-silicon electrodes suggests that the metal surface is subjected to an electric stress much lower than that at the bottom silicon surface.

Now, observe that thiol-terminated molecules self-assemble spontaneously on many metals (like platinum, gold, etc.) forming closely packed monolayers. Therefore, if the considered three-terminal molecules contain two alkyne and one thiol terminations, they can arrange in the cross-points in an ordered way, allowing their covalent grafting by simple heat treatment, as sketched in Figure 17.13. This figure also explains why the electric field at the metal–sulfur interface may be significantly lower than at the silicon–carbon interface.

17.5 Influence of Technology on Architecture

Device architecture and preparation procedure are strongly interlocked. This will become especially clear considering that crossbars obtained via S^nPT_+ may be linked to the external world via methods that cannot be extended to S^nPT_x.

17.5.1 Addressing

If the availability of nanofabrication techniques is fundamental in establishing a nanotechnology, not less vital is the integration of the nanostructures with higher-level structures: once the crossbar structure is formed, it is necessary to link it to the conventional silicon circuitry. This is especially difficult because the nanoworld is not directly accessible by means of standard lithographic methods—"the difficulties in communication between the nanoworld and the macroworld represent a central issue in the development of nanotechnology" [36].

The importance of addressing nanoscale elements in arrays goes beyond the area of memories and will be critical to the realization of other integrated nanosystems such as chemical or biological sensors, electrically driven nanophotonics, or even quantum computers.

In the following the attention will however be limited to the problem of addressing cross-points in a nanoscopic crossbar structure by means of externally accessible lithographic contacts.

Several strategies have been adopted to attack this problem: many of them involve materials and methods quite far from, if not orthogonal to, those of the planar technology [37–43].

The consistent strategies with the planar technology are discussed in Ref. [2]. Of them, one can be applied to all crossbars irrespective of their preparation methods. According to this strategy, each line defining the crossbar extends beyond the crossing region and in this zone it is used for addressing. This region is then covered with a protecting cap, which is etched away along a narrow (sub-lithographic) line misoriented with respect to the array by a small angle α. In this way the zones where the bars are not covered are separated by a distance that diverges for $\alpha \to 0$; thus, if α is sufficiently small, the separation between the zones no longer protected makes them accessible to conventional lithography and suitable for contacting the CMOS circuitry. In this method, each line is linked separately from the others to the external circuitry—addressing n^2 cross-points requires therefore $2n$ contacts.

The multispacer technique (in particular, the S^nPT_+) permits, however, novel strategies for the nano-to-litho link in addition to the ones suitable for crossbar prepared with other techniques.

In the first of such strategies, the original mask defining the seed is shaped with n indentations with size so scaled that (1) the first indentation is filled, with the fusion of the wires, after the first deposition; (2) the second indentation is filled, with the fusion of the wires, after the second deposition; (3)...; and (4) n, the nth indentation is filled, with the fusion of the wires, after the nth deposition.

Taking into account that the minimum distance between the centers of two adjacent fused layers is $W + 3p$ (say 150 nm) and that each contact requires the definition of a hole in a region with side $2p$ (say 70 nm), this technique allows the nano-to-litho link. Figure 17.14 shows the cross section demonstrating

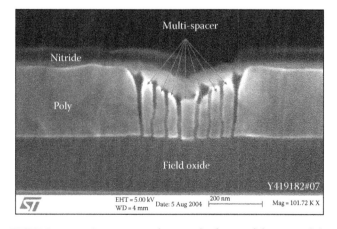

FIGURE 17.14 Cross-section showing the fusion of the arms of the fourth wire grown on two sides of the indentation. (Reprinted from Cerofolini, G., *Nanoscale devices*, Springer, Berlin, Germany, 2009. With permission.)

how filling indentation with the fusion of the central wires may render them accessible to lithography. Similarly to the strategy considered above, in this method, each line is linked separately from the others to the external circuitry, so that addressing n^2 cross-points requires therefore $2n$ contacts.

The S^nPT_+ technique results in wires with different heights. This fact can be exploited to inhibit or enable them by means of dielectrically insulated lithographically defined electrodes in the geometry described in Ref. [2]. A crossbar with $n \times n$ cross-points can therefore be addressed by controlling the conduction along the wires from one Ohmic contact to the other by means of 2 + 2 electrodes. Therefore, for the addressing of n^2 cross-points, this method needs 3 + 3 electrodes only; however, as discussed in Ref. [2], this architecture requires a complex elaboration of the information involving analog-to-digital conversion, with subsequent analysis and elaboration of data.

17.5.2 Comparing Crossbars Prepared with Additive or Multiplicative Routes

While the repetition in additive way of n SPTs per (bottom and top) layers magnifies the lithographically achievable cross-point density K_0^2 by a factor of $(2n)^2$, the repetition in multiplicative way gives a magnification of 2^{2n}. The magnification factor increases quadratically for the additive way and exponentially with the multiplicative way.

To estimate numerically the process simplifications offered by S^nPT_x over SPT_+, consider for instance the case of the 3 SPT_x repetitions per layer. This would produce a magnification of the lithographic cross-point density by a factor of $2^3 \times 2^3$. Taking $W = 0.1\,\mu m$, after 3 SPT_x repetitions, the spacer width should be of 12.5 nm, with minimum separation of 25 nm. Taking into account Equation 17.12, the cross-point density achievable with the repetition of 2×3 SPT_x would thus be almost the same as that obtainable with the repetition of 2×10 SPT_+ (7×10^{10} cm^{-2} vs. 8×10^{10} cm^{-2}).

Figure 17.15 shows in plan view a comparison between the following crossbars:

(a) A 2×2 crossbar obtained by crossing lithographically defined lines.
(b) A 16×16 crossbar obtained via S^8PT_+ starting from lithographically defined seeds separated by a distance allowing the optimal arrangement of the wire arrays.
(c) An 16×16 crossbar obtained via S^3PT_x starting from lithographically defined seeds separated by a distance satisfying Equation 17.12.

The figure has been drawn in the following hypotheses:

- The lithographic lines in (a) and (b) have width at the current limit for large-volume production, say $W = 65$ nm.
- The height loss τ is such that the maximum number of repetitions in the additive route is 8, and the sub-lithographic pitch is the same as shown in Figure 17.6.

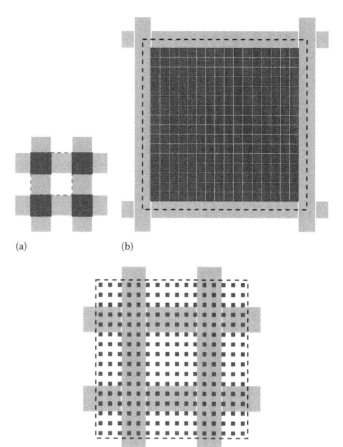

(a)　(b)

(c)

FIGURE 17.15 Plan-view comparison of the crossbars obtained (a) crossing lithographically defined lines, (b) using the lithographically defined lines above as seeds for S^8PT_+, and (c) using the lithographically defined lines above as seeds for S^3PT_x. In each structure the square with dashed sides denotes a unit cell suitable for the complete surface tiling. (Reprinted from Cerofolini, G., *Nanoscale devices*, Springer, Berlin, Germany, 2009. With permission.)

- The lithographic width of (c) is chosen to allow the minimum pitch to be consistent with the one obtained with the additive route ($W = 100$ nm); in this way, producing sub-lithographic wires with width (12.5 nm) has been proved to be producible [28].

For $n \geq 3$, this comparison is so favorable to S^nPT_x to suggest its practical application. The following factors, however, would make S^nPT_+ preferable to S^nPT_x:

1. If addressing the wires defining the crossbar are used also as addressing lines, they cannot run along the entire plane; their interruption for addressing reduces the available area.
2. In S^nPT_x all wires are produced collectively and have the same height and material characteristics. On the contrary, in S^nPT_+ the wires are produced sequentially, each SPT_+ repetition produces wires of decreasing height, and the characteristics of the material (or even the materials themselves) may vary in a controlled way from one cycle to another.

Whereas at the first glance all the S^nPT_+ features described the second item may seem detrimental, the analysis of Section 17.5.1 has clarified that they may be usefully exploited for cross-point addressing.

Deciding which route, between the additive and multiplicative ones, is actually more convenient for the preparation of hybrid devices depends on the particular application and circuit architecture. In fact, the multiplicative route (certainly less demanding for what concerns the preparation of the crossbar but more expensive for what concerns the nano-to-litho link) is presumably suitable for random access memories; on the contrary, the additive route (more demanding for the crossbar preparation, but much heavier for the elaboration of the signal) seems consistent with nonvolatile memories.

17.5.3 Applications—Not Only Nanoelectronics

The proposal of the S^nPT is quite recent [30,31], so that it has had only short time to be tested.

17.5.3.1 Electronics

Concerning the preparation of crossbars, the attention has mainly been concentrated on one side on the verification of the possibility of scaling the S^nPT_+ to large values of n (the productions of arrays with $n = 3$ [30,31], 4 [44], and 6 [45] have been reported) and on the other on the architectural impact of this technology [46–48]. In the middle, a demonstrator has been prepared showing the feasibility, without stressing the technology, of crossbars with cross-point density of 10^{10} cm^{-2} [45].

The electrical characterization of silicon nanowires has been the subject of only a few studies: Ref. [24], addressed to the nanonowires produced via the SNAP technique and Ref. [45], devoted to nanowires built with the S^nPT_+. The latter paper demonstrates that the S^nPT_+ is already suitable for the preparation of ultrahigh density memory.

The interest of a technology for the cheap production of nanowires is not limited to nanoelectronics but might extend to energetics.

17.5.3.2 Energetics

The large-scale availability of devices able to transform low-enthalpy heat (which would otherwise be dispersed into the environment) into electrical energy without complicate mechanical systems might impact the energy problem on a global scale. In principle the Seebeck effect provides a way for that. The ability of a material to operate as a Seebeck generator is contained in a parameter, ZT—a function of the Seebeck coefficient and of the electrical and thermal conductivities. The possibility of practical application is related to the occurrence of $ZT \gtrsim 1$.

From the technical point of view, Seebeck devices are already known, but materials with high Seebeck coefficient (multicomponent nanostructured thermoelectrics, such as Bi_2Te_3/Sb_2Te_3 thin-film superlattices, or embedded PbSeTe quantum dot super-lattices) are expensive. This fact has allowed the application of direct thermoelectric generators only to situations (e.g., space) where cost is not important or due to other factors (e.g., weight in space application).

This state of affairs would significantly change only in the presence of highly efficient, low-cost materials. A new avenue to progress in such a direction has been the claim, by two independent collaborations [49,50], that silicon nanowires with width of 20–30 nm and rough surfaces having $ZT \simeq 1$. Although silicon is certainly a cheap material and a lot of technologies are known for its controlled deposition, this result is of potential practical interest only if the production of nanowires does not involve electron beam or extreme ultraviolet lithographies—hence the relevance of S^nPT.

It is also noted that the researchers of Refs. [49,50] employed wires with comparable height and width, thus characterized with a very high resistance. The S^nPT allows the preparation of wires with much higher aspect ratio (same width and much higher height—actually nanosheets). If it were possible to impart a sufficient roughness to the sidewalls of these sheets, it would result in much more efficient Seebeck generators.

17.6 Fractal Nanotechnology

There is an application where the multiplicative route is manifestly superior: the preparation of fractal structures on the sublithographic length scale.

17.6.1 Fractals in Nature

The volume V of any body with regular shape (spheric, cubic, …) varies with its area A as

$$V = g_3 A^{3/2}, \tag{17.13}$$

with g_3 being a coefficient related to the shape (for instance, $g_3 = (1/6)^{3/2}$ for cubes, $g_3 = 1/6\sqrt{\pi}$ for spheres, etc.; a well-known variational properties of the sphere guarantees that for all bodies $g_3 \leq 1/6\sqrt{\pi}$). For bodies with regular shape the ratio A/V diverges for small V and vanishes for large V.

Life can be preserved against the second law of thermodynamics only in the presence of a production of negative entropy in proportion to the total mass of the organism. This is achieved via the establishment of a diffusion field sustained by the metabolism inside the organism [51]. Since the consumption of energy and neg-entropy of living systems increases in proportion to V whereas the exchange of matter increases A, small unicellular organisms (like bacteria) have no metabolic problem due to their shape and can grow satisfying Equation 17.13 preserving highly symmetric shapes (the smallest living bacterium, the pleuropneumonia like organism, with diameter 0.1–0.2 μm, is spherical). On the contrary, larger organisms (even procaryotic, like the amoeba) may survive only adapting their shapes to have an energy uptake coinciding with what is required to preserve living functions [52,53]:

$$V = g_2 A, \tag{17.14}$$

with g_2 being a coefficient characteristic of two-dimensional growth. Moreover, the need to adapt itself to the variable environmental conditions is satisfied only thanks to the existence of an inner organellum, the endoplasmatic reticulum [54], which can fuse to the external membrane, thus allowing an efficient change of area at constant volume [52,53]. This mechanism, however, can sustain the metabolism of unicellular organisms (like the amoeba) only for diameter of at most a few tens of micrometers.

Larger organisms require the organization of the constituting cells in tissues specialized to single functions. This specialization, however, requires the formation of a vascular network (to transport catabolites to, and anabolites from, each constituting cell) whose space-filling nature implies a fractal character [55]. Nature prefers to manifest itself with fractal shapes in other situations too, like for the mammalian lung (to allow a better O_2–CO_2 exchange) [56], the dendritic links of the neuron (to allow a high interconnection degree [57]).

That smoothness is not a mandatory feature of the way how nature expresses itself, but rather fractality is of ubiquitous occurrence in a large class of phenomena even in the mineral kingdom has become clear with Mandelbrot's question about the length of the Great Britain coast [58].

The fractality of nature is manifestly approximate, the lowest length scale being ultimately limited by the atomic nature of matter. That surfaces may continue to have a scale invariance down to the atomic size became however clear only after Avnir, Farin, and Pfeifer's study of the adsorption behavior of porous adsorbents [59–61].

Surprisingly enough, the artificial production of self-similar structures has remained largely unexplored. This is somewhat disappointing because the use of arrays or matrices with self-similar structures is potentially interesting even for applications. For instance, a matrix formed by the Cartesian product of two Cantor sets would provide a way for sensing with infinite probes a surface, leaving it almost completely uncovered. Although the minimum length scale is actually larger than the atomic one, that ideal case suggests the usefulness of the idea.

The lowest length scale of biological fractals is determined by cell size, say 10^4 nm. The preparation of self-similar structures on length scales between 10^2 and 10^4 nm is relatively easy via photolithography, but this scale seems inadequate for interesting applications like the highly parallel probing of single cells (as required, for instance, by the needs of systems biology [62]). Rather, for that application the appropriate length scale seems the one characteristic of nanotechnology, 1–10^2 nm.

17.6.2 Producing Nanoscale Fractals via $S^n PT_x$

Imagine for a moment that, in spite of the atomistic nature of matter and of the inherent technological difficulties, the multiplicative route can be repeated indefinitely. Remembering that the $(n + 1)$th step generates a set S_{n+1}, which is, nothing but the one at the nth, S_n, at a lower scale, the sequence $\{S_0, \ldots, S_n, \ldots\}$

defines a fractal. This fractal, referred to as multispacer fractal set, is self-similar only if the height of each spacer varies with n as 2^{-n}, otherwise, the fractal is self-affine [63]. As mentioned above, the "spontaneous" decrease of height with n, Equation 17.2 renders the fractal self-affine. A self-similar fractal can be obtained at the end of process planarizing the whole structure with a resist and sputter-etching in a nonselective way the composite film until the thickness is reduced to $s_0/2^n$.

It is however noted that even assuming our ability to scale down the fabrication technology, the atomistic structure of matter limits anyway the above considerations to an interval of 1–2 orders of magnitude, ranging from few atomic layers to the lower limits of standard lithography.

Having clarified in which limits the set S_n may be considered a fractal, it is interesting to compare it with other fractal sets. The prototype of such sets, and certainly the most interesting from the speculative point of view, is the Cantor middle-excluded set. Figure 17.16 compares sequences of three process steps eventually leading to the multispacer fractal set S and to the Cantor set C. The comparison shows interesting analogies: consider a multiplicative multispacer with $P = 2W$; if $w_n = (1/3)w_{n-1}$, the measure at each step of multifractal set coincides with that of the Cantor set. This implies that the multispacer fractal set has null measure. Similarly, it can be argued that the multispacer set, considered as a subset of the unit interval, has the same fractal dimension as the Cantor middle-excluded set—$\ln(2)/\ln(3)$ [63]. At each step, the multispacer fractal set is characterized by a more uniform distribution of single intervals than the Cantor set; this makes the former more interesting for potential applications than the latter.

Once one has one-dimensional fractal sets, two-dimensional fractal sets can be constructed taking their Cartesian products. Although the mixed product $C \times S$ is possible, in the following, we concentrate on a comparison of the Cantor and multispacer fractal crossbars. Figure 17.17 compares their plan views showing

(A) A 16×16 crossbar obtained via S^3PT_x starting from lithographically defined seeds separated by a distance satisfying Equation 17.12

(B) A 16×16 crossbar obtained via S^3PT_x starting from lithographically defined seeds and arranging the process to generate the Cantor middle-excluded set

Although the potential applications of the Cantor set are quite far, trying to reproduce it on the nanometer length scale seems of a certain interest. This is compatible with existing technologies; a possible process would involve

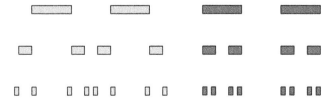

FIGURE 17.16 Generation of the multispacer set (left) and of the Cantor middle-excluded set (right). (Reprinted from Cerofolini, G., *Nanoscale devices*, Springer, Berlin, Germany, 2009. With permission.)

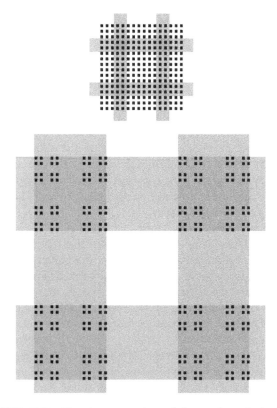

FIGURE 17.17 Plan-view comparison of the crossbars obtained via S^3PT_x, and a corresponding Cantor middle-excluded set; the width of the lithographic seed in S^3PT_x has been assumed to be at the technology forefront, whereas in the Cantor set it has been assumed sufficiently large to allow three reiterations of the process. (Reprinted from Cerofolini, G., *Nanoscale devices*, Springer, Berlin, Germany, 2009. With permission.)

(C1) the lithographic definition of the seed (formed, for instance, by poly-silicon) generating the Cantor set

(C2) its planarization (for instance, via the deposition of a low viscosity glass and its reflow upon heating)

(C3) the etching of this film to a thickness controlled by the exposure of the original seed

(C4) the selective etching of the original film

(C5) the conformal deposition of a film of the same material as the original seed (poly-silicon, in the considered example) and of thickness equal to 1/3 of its width

(C6) its directional etching

(C7) the selective etching of the space seed (glass, in the considered example)

Figure 17.18 sketches the overall process [64].

The preparation of fractal structures may appear at a first sight nothing but a mere exercise of technology stressing. The following examples suggest however the usefulness of a fractal technology:

1. The highly parallel and real-time sensing of single cells (as required, for instance, by the needs of systems biology [62] of by label-free immunodetection [65]) could be done with a minimum of perturbation (i.e., leaving almost completely uncovered the cell surface) by a matrix formed by the Cartesian product of two Cantor sets [63].

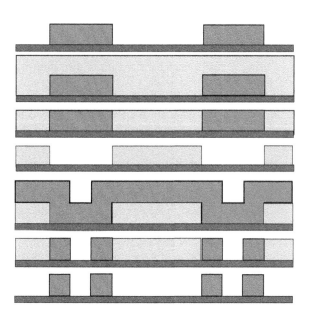

FIGURE 17.18 A process for the generation of Cantor's middle-excluded set. (Reprinted from Cerofolini, G., *Nanoscale devices*, Springer, Berlin, Germany, 2009. With permission.)

2. Superhydrophobic surfaces may be prepared controlling roughness and surface tension of nonwetting surfaces [66]. Whereas surface tension is a material property, roughness can be controlled by the preparation. For instance, roughness may be achieved by imparting a suitable (fractal) relief on the surface.

3. If the S^nPT is used for the preparation of crossbar structures for molecular electronics, the functionalization with organic molecules of the cross-points can only be done after the preparation of the hosting structure. According to the analysis of Ref. [44], this requires an accurate control of the rheological and diffusion properties in a medium embedded in a domain of complex geometry. Understanding how such properties change when the size is scaled and clarifying to which extent the domain can indeed be viewed as a fractal (so allowing the analysis on fractals [67] to be used for their description) may be a key point for the actual exploitation of already producible nanometer-sized wire arrays in molecular electronics.

Appendixes

17.A Abstract Technology

The processes considered in this chapter (conformal deposition, directional etching, and selective etching) have been introduced without specifying exactly what they are and which materials they involve. Although the reader certainly has a built-in idea of them, we however believe useful to provide their rigorous definition.

Once defined in rigorous mathematical terms, the processes of interest for this work are the building blocks of a nanotechnology that may be seen as the equivalent of Euclid's geometry, where the construction with straightedge and compass is replaced by the construction via conformal, directional or selective, deposition or etchings. Of course, the properties of a figure built with any real straightedge and compass are not exactly the same as those of the same figure constructed with the ideal straightedge and compass. Similarly, the use of real materials and processes produces structures that differ from those achievable via the ideal materials and processes. Nonetheless, it is convenient to develop the theory in abstract terms.

17.A.1 Bodies and Surfaces

The minimal characterization of a body is in terms of its geometry and constituting materials.

Definition 17.1 (Simple body)   *A simple body $\mathcal{B}_X$ is a three-dimensional closed connected subset B of the Euclidean space $\mathbb{E}^3$ filled with a material X:*

$$\mathcal{B}_X = (B, X).$$

What is a material is considered here a primitive concept. It may be a substance, a mixture, a solution, or an alloy. The emphasis is on homogeneity—whichever region, however small, of B is considered, the constituting material is the same. When not necessary, the index denoting the constituting material will not be specified.

In view of the granular nature of matter, considering the limit for vanishing size is not relevant. Here and in the following when dealing with the concept of vanishing length (as it implicitly happens when considering the frontier B^* of B) we mean that the property holds true on the *ultimate length scale*. Although the ultimate length scale is an operative concept related to the probe used for observing the body, we however have in mind the molecular one. Analogously, we consider a body as indefinitely extended when its size is much bigger than the size variations induced by the considered processes. In this sense, a *wafer* is a simple body indefinitely extended over two dimensions.

Definition 17.2 (Composite body)   *Consider the N-tuple of simple body $(\mathcal{B}_{X^1}, \mathcal{B}_{X^2}, ..., \mathcal{B}_{X^N})$ formed by nonoverlapping sets $(B_1, B_2, ..., B_N)$, with*

$$\forall I, J\left(B_i \cap B_j = B_i^* \cap B_j^*\right),$$

and let

$$B = \bigcup_{i=1}^{N} B_i.$$

If B is connected and the simple bodies $(\mathcal{B}_{X^1}, \mathcal{B}_{X^2}, ..., \mathcal{B}_{X^N})$ contain at least two different materials X^I and X^J, then the N-tuple is a composite body.

Definition 17.3 (Interface)   *Let a composite body $\mathcal{B}$ be formed by simple bodies $\mathcal{B}_{X^1}$ and $\mathcal{B}_{X^2}$ with $X^1 \neq X^2$; the region*

$$F_{1|2} = B_1^* \cap B_2^*$$

defines the interface between $\mathcal{B}_{X^1}$ and $\mathcal{B}_{X^2}$.

For any $\mathbf{x}$ in $F_{1|2}$ all neighborhoods, however small, contain materials X^1 and X^2.

Definition 17.4 (Total surface)   *The set*

$$S_{tot} = B^*$$

is the total surface of $\mathcal{B}$.

Proposition 17.1   *Let a composite body $\mathcal{B}$ be formed by the N-tuple of simple body $(\mathcal{B}_{X^1}, \mathcal{B}_{X^2}, ..., \mathcal{B}_{X^N})$. Then*

$$S_{tot} = \bigcup_i B_i^* \setminus \bigcup_{j,k} F_{j/k}.$$

Definition 17.5 (Surface)   *Let $\mathcal{B}$ be a composite body, and $B_{\bullet}$ be the smallest simply-connected (i.e., without holes) set containing B. Then*

$$S = B_{\bullet}^*$$

is the outer surface (or simply surface) of $\mathcal{B}$.

Definition 17.6 (Inner surface)   *Let $\mathcal{B}$ be a composite body. Then*

$$S_{in} = S_{tot}/S$$

is the inner surface of $\mathcal{B}$.

17.A.2 Conformal Deposition and Isotropic Etching

Definition 17.7 (Delta coverage)   *For any body $\mathcal{B} \leftarrow$ with surface S, the additive delta coverage $\overline{D}_{+\delta}$ of thickness δ is the set*

$$\overline{D}_{+\delta} = \left\{ \mathbf{x} : |\mathbf{x} - \mathbf{y}| \leq \delta \wedge \mathbf{y} \in S \wedge \mathbf{x} \notin B_{\bullet} \setminus S \right\}.$$

The delta coverage can be imagined to result from the application of an operator a_δ to B:

$$\overline{D}_{+\delta} = a_\delta(B).$$

The delta coverage $\overline{D}_{+\delta}$ is the "pod" of thickness δ covering the set B. The set obtained as union of B with its delta coverage is referred to a $B_{+\delta}$:

$$B_{+\delta} = B \cup a_\delta(B).$$

This operation can be reiterated, and the final set obtained after n iteration is referred to as $B_{+\delta}^n$:

$$B_{+\delta}^n = B_{+\delta}^{n-1} \cup a_\delta\left(B_{+\delta}^{n-1}\right),$$

where $B_{+\delta}^1 = B_{+\delta}$ and $B_{+\delta}^0 = B$.

The reiteration of a_δ is the base for the following definition:

Definition 17.8 (Conformal coverage)   *A conformal coverage $C_{+\delta}$ of thickness δ is the following limit**:

$$C_{+\delta} = \lim_{n\to+\infty} \bigcup_{i=0}^{n-1} a_\delta\left(B_{+\delta/n}^i\right).$$

In other words, the conformal coverage process can be thought of as obtained by applying infinitely many delta coverages; one after another, and each one of infinitesimal thickness. As shown in Figure 17.19, in general conformal coverage and delta coverage of the same thickness can lead to different bodies. Moreover from the same figure it is clear that these transformations are not topological invariants.

Definition 17.9 (Conformal deposition)   *A conformal deposition of a film of material Z of thickness δ is the pair $(C_{+\delta}, Z)$.*

Note that the material Z may be different from the materials X^i forming the original body.

The following theorem is trivial:

Theorem 17.1 (Planarization)   *Let d_S the diameter of $\mathcal{B}$. Irrespective of the shape of $\mathcal{B} \leftarrow$ the conformal deposition of a layer for which $\delta \gg d_S$ produces a body whose shape becomes progressively closer and closer to the spherical one as δ increases. If $\mathcal{B} \leftarrow$ is indefinitely extended in two directions, it undergoes a progressive planarization.*

Isotropic etching can be defined in a way similar to that used for conformal coverage by introducing an operator s_δ, characterized by the following definitions:

---

* Such a limit is naively clear, but its rigorous specification should require to elaborate many technical details that we prefer to skip in this appendix. The same considerations hold true anytime we introduce limits of this kind.

FIGURE 17.19   The different behavior of conformal coverage (left) and delta coverage (right). (Reprinted from Cerofolini, G., *Nanoscale devices*, Springer, Berlin, Germany, 2009. With permission.)

$$s_\delta(B) = a_\delta(\mathbb{E}^3 \setminus B_\bullet),$$

$$B_{-\delta}^n = B_{-\delta}^{n-1} \setminus s_\delta\left(B_{-\delta}^{n-1}\right).$$

In general, the operator s_δ is not the inverse of a_δ. In fact, $s_\delta(B) = a_{-\delta}(B)$ only if B is a simply connected set. We refer to s_δ as *delta depletion*.

Definition 17.10 (Conformal depletion)   *A conformal depletion $E_{-\delta}$ of thickness δ is the following limit:*

$$E_{-\delta} = \lim_{n\to+\infty} \bigcup_{i=0}^{n-1} s_\delta\left(B_{-\delta/n}^i\right).$$

Also in this case, it is true that in general the conformal depletion and delta depletion s_δ of the same thickness can lead to different bodies (see Figure 17.20).

Definition 17.11 (Isotropic etching)   *An isotropic (nonselective) etching is a conformal depletion of thickness δ applied to a body $\mathcal{B} \leftarrow$ regardless of its constituting materials.*

The etching process just defined is nonselective. However the most part of etching are selective (see Section 17.A.4).

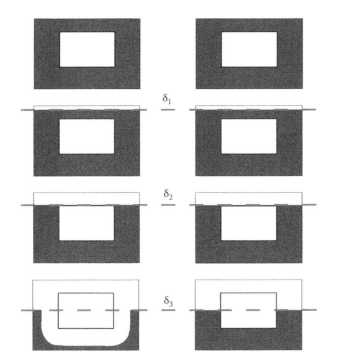

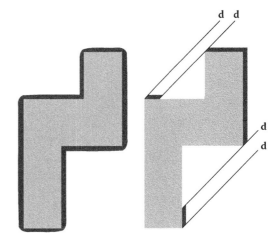

FIGURE 17.21　Comparison between delta coverage (left) and directional delta coverage along **d** (right) of the same thickness δ. (Reprinted from Cerofolini, G., *Nanoscale devices*, Springer, Berlin, Germany, 2009. With permission.)

FIGURE 17.20　The different behavior of conformal depletion (left) and delta depletion (right). (Reprinted from Cerofolini, G., *Nanoscale devices*, Springer, Berlin, Germany, 2009. With permission.)

17.A.3　Directional Processes

Definition 17.12 (Shadowed surface along d)　*Let* **d** *be a unit vector in* $\mathbb{E}^3$. *For any point s belonging to the surface S consider the straight line* l_s *starting from s and oriented as* **d**. *If* l_s *intersects S, then s is said to be shadowed along* **d**. *The set all shadowed points of S is referred to as the shadowed surface* $S_{\mathbf{d}}$ *along* **d** *of the body* $\mathcal{B}$.

Definition 17.13 (Directional delta coverage along d)　*The directional delta coverage along* **d** *is the set of points lying on the straight lines directed along* **d** *and going from the exposed surface to the same surface shifted along* **d** *by an amount* δ:

$$\overline{D}_{+\delta,+\mathbf{d}} = \{\mathbf{x} : \mathbf{x} = \mathbf{y} + a\mathbf{d} \wedge a \leq \delta \wedge \mathbf{y} \in S \setminus S_{\mathbf{d}}\}.$$

Figure 17.21 shows the results of applying to a given set the delta coverage and the directional delta coverage along **d**.

By using the directional delta coverage along δ instead of the delta coverage, it is possible to define the directional deposition.

Definition 17.14 (Directional deposition)　*A directional deposition* $C_{+\delta,+\mathbf{d}}$ *of thickness* δ *along* **d** *is the following limit:*

$$C_{+\delta,+\mathbf{d}} = \lim_{n \to +\infty} \bigcup_{i=0}^{n-1} a_{\delta,+\mathbf{d}}\left(B^i_{+\delta/n,+\mathbf{d}}\right),$$

where $a_{\delta,+\mathbf{d}}$ *is the operator associated with the directional delta coverage and* $B^n_{+\delta,+\mathbf{d}} = B^{n-1}_{+\delta,+\mathbf{d}} \cup a_{\delta,+\mathbf{d}}(B^{n-1}_{+\delta,+\mathbf{d}})$.

FIGURE 17.22　An example of directional deposition. (Reprinted from Cerofolini, G., *Nanoscale devices*, Springer, Berlin, Germany, 2009. With permission.)

Figure 17.22 shows an example of directional deposition. This process (as shown in the figure) can create holes inside the bodies.

Analogously, to define the directional depletion we introduce the directional delta depletion:

Definition 17.15 (Directional delta depletion along d)

$$\overline{D}_{+\delta,-\mathbf{d}} = \left\{\mathbf{x} : \mathbf{x} = \mathbf{y} - a\mathbf{d} \wedge a \leq \delta \wedge \mathbf{y} \in S \setminus S_{\mathbf{d}}\right\}.$$

Definition 17.16 (Directional etching along d)   *A directional etching $E_{-\delta,-\mathbf{d}}$ of thickness δ is the following limit:*

$$E_{-\delta,-\mathbf{d}} = \lim_{n\to+\infty} \bigcup_{i=0}^{n-1} s_{\delta,-\mathbf{d}}\left(B_{-\delta/n,-\mathbf{d}}^i\right),$$

where $s_{\delta,-\mathbf{d}}$ is the operator associated to the directional delta depletion and $B_{-\delta,+\mathbf{d}}^n = B_{-\delta,-\mathbf{d}}^{n-1}\backslash s_{\delta,-\mathbf{d}}\left(B_{-\delta,-\mathbf{d}}^{n-1}\right)$.

17.A.4 Selective Processes

No indication has hitherto been given about the dependence of the same unit process on the material to which it is applied. In general the effect depends on the material, and this difference is referred to in terms of selectivity.

Before introducing selectivity, we note that the surface S of a body $B \leftarrow$ is, in general, composed by different materials. In particular, by using Proposition 1 it is possible to define $S_{\text{tot},X}$, the subset of S_{tot} containing material X:

$$S_{\text{tot},X} = \bigcup_{X^i=X} B_i^* \Big\backslash \bigcup_{j,k} F_{j|k},$$

and then the (outer) surface S_X composed by material X:

$$S_X = S_{\text{tot},X} \cap S.$$

Definition 17.17 (Selective delta depletion)   *For any body $B \leftarrow$ with surface S, the selective delta depletion $s_{\delta,X}(B)$ of thickness δ is the set*

$$s_{\delta,X}(B) = \Big\{\mathbf{x} : |\,\mathbf{x}-\mathbf{y}\,| \le \delta \wedge \mathbf{y} \in S_X \wedge \mathbf{x} \notin B_\bullet\backslash S_X\Big\}.$$

In other words, the selective delta depletion is a delta depletion that affects S_X only, not the whole surface S. We remark that, in general, $s_{\delta,X}(B)$ is not a connected set.

Definition 17.18 (Selectivity)   *Any process is said to be selective with respect to materials X when, applied to a composite body B with frontier S, affects S_X only.*

Definition 17.19 (Selective etching)   *A selective etching $E_{-\delta,X}$ of thickness δ is the following limit (in a sense to be defined):*

$$E_{-\delta,X} = \lim_{n\to+\infty} \bigcup_{i=0}^{n-1} s_{\delta,X}\left(B_{-\delta/n,X}^i\right),$$

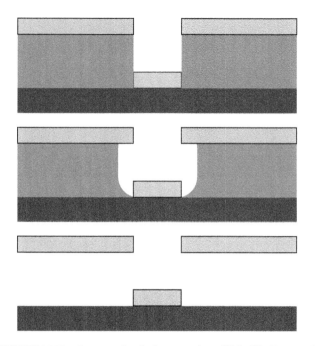

FIGURE 17.23   An example of selective etching ("liftoff"). (Reprinted from Cerofolini, G., *Nanoscale devices*, Springer, Berlin, Germany, 2009. With permission.)

where

$$B_{-\delta/n,X} = \left(B\backslash s_{\delta,X}(B)\right) \cup \left(\bigcup_{X^j \neq X} B_j\right),$$

and the iterative process is defined as in the previous cases.

As example of selective etching is given in Figure 17.23.

17.B Concrete Technology

Together with the above operations, one could certainly define other, more complicate (or perhaps even formally more interesting), operations. What is of uppermost importance here is that *there already exist combinations of materials and processes of the silicon technology mimicking the ideal behaviors described above.*

The materials considered in this chapter are well known in the silicon technology: a practical model for wafer is any body extending in two directions exceedingly more than the change of thickness resulting from the considered processes and such that only the processes are actually carried out on only one of its major surfaces. Before undergoing any operations, such wafers have generally homogeneous chemical composition and high flatness and are referred to as substrates. In the typical situations considered herein the substrates are slices of single crystalline silicon. The other materials are polycrystalline silicon (poly-Si), SiO_2, Si_3N_4, and various metals (Al, Ti, Pt, Au,…).

The typical processes involved in the silicon technology are lithography (for the definition of geometries), wet or

TABLE 17.1 Shape Resulting after Etching or Growth Processes

	Shape	
Process	Conformal	Directional
Attack	Wet etching	Sputter etching
	← Plasma etching	
		Reactive ion etching →
Growth	CVD	PVD

gas-phase etchings, chemical or physical vapor deposition, planarization, doping (typically via ion implantation), and diffusion. A special class of materials is formed by resists, i.e., photoactive materials undergoing polymerization (or depolymerization) under illumination. The processes of interest here are the following:

Lithography requires a preliminary planarization with a resist, its patterning (i.e., the definition of a geometry) via the exposure through a mask to light, the selective etching of the exposed (or unexposed) resist, the selective etching first of the region not protected by the patterned resist, and eventually of this material.

Wet etchings are usually isotropic and are used for their selectivity: HF_{aq} etches isotropically SiO_2 leaving unchanged silicon and Si_3N_4; H_3PO_4 etches isotropically Si_3N_4 leaving unchanged silicon and SiO_2; $HF_{aq} + HNO_{3\,aq}$ etches Si leaving unchanged Si_3N_4 (but has poor selectivity with respect to SiO_2).

Sputter etching is produced by momentum transfer from a beam to a target and results typically in nonselective directional etching. Selectivity can be imparted exploiting reactive ion etching.

Plasma etching can be tuned to the situation: via a suitable choice of the atmosphere it can be used for the isotropic selective etching of Si, SiO_2 or Si_3N_4; it becomes progressively more directional and less selective applying a bias to the body ("target") with respect to the plasma.

Chemical vapor deposition is the typical way for the conformal deposition that occurs when the growth is controlled by reactions occurring at the growing surface. Poly-Si grows well on SiO_2 but requires an SiO_2 buffer layer for the growth on Si_3N_4; conversely Si_3N_4 is easily deposited on SiO_2. Silicon can be conformally covered by extremely uniform layers of SiO_2 with controlled thickness on the subnanometer range via *thermal oxidation*.

Physical vapor deposition is the typical way for the deposition of metal films. The corresponding growth mode is the positive counterpart of directional etching. The conformal deposition of a metal layer can only be done via CVD from volatile precursors (like metal carbonyls or metalorganic monomers).

Table 17.1 summarizes the shape (conformal or directional) resulting from the above processes.

References

1. Heath J. R., Kuekes P. J., Snider G. S., Williams R. S., A defect-tolerant computer architecture: Opportunities for nanotechnology. *Science, 280* (1998) 1716–1721.
2. Cerofolini G. F., Realistic limits to computation. II. The technological side. *Appl. Phys. A, 86* (2007) 31–42.
3. Aviram A., Ratner M., Molecular rectifiers. *Chem. Phys. Lett., 29* (1974) 277–283.
4. Joachim C., Gimzewski J. K., Aviram A., Electronics using hybrid-molecular and mono-molecular devices. *Nature, 408* (2000) 541–548.
5. Joachim C., Ratner M. A., Molecular electronics: Some views on transport junctions and beyond. *Proc. Natl. Acad. Sci. USA, 102* (2005) 8801–8808.
6. Tour J. M., Rawlett A. M., Kozaki M., Yao Y., Jagessar R. C., Dirk S. M., Price D. W., et al., Synthesis and preliminary testing of molecular wires and devices. *Chem. Eur. J., 7* (2001) 5118–5134.
7. Mendes P. M., Flood A. H., Stoddart J. F., Nanoelectronic devices from self-organized molecular switches. *Appl. Phys. A, 80* (2005) 1197–1209.
8. Chen J., Reed M. A., Rawlett A. M., Tour J. M., Large on-off ratios and negative differential resistance in a molecular electronic device. *Science, 286* (1999) 1550–1552.
9. Reed M. A., Chen J., Rawlett A. M., Price D. W., Tour J. M., Molecular random access memory cell. *Appl. Phys. Lett., 78* (2001) 3735–3737.
10. Luo Y., Collier C. P., Jeppesen J. O., Nielsen K. A., Delonno E., Ho G., Perkins J., et al, Two-dimensional molecular electronics circuits. *Chem. Phys. Chem., 3* (2002) 519–525.
11. Cerofolini G. F., Ferla G., Toward a hybrid micro-nanoelectronics. *J. Nanoparticle Res., 4* (2002) 185–191.
12. Green J. E., Choi J. W., Boukai A., Bunimovich Y., Johnston-Halperin E., Delonno E., Luo Y., et al., A 160-kilobit molecular electronic memory patterned at 10^{11} bits per square centimetre. *Nature, 445* (2007) 414–417.
13. Service R. F., Next-generation technology hits an early midlife crisis. *Science, 302* (2003) 556–559.
14. Stewart D. R., Ohlberg D. A. A., Beck P., Chen Y., Williams R. S., Jeppesen J. O., Nielsen K. A., Stoddart J. F., Molecule-independent electrical switching in Pt/organic monolayer/Ti devices. *Nano Lett., 4* (2004) 133–136.
15. Lau C. N., Stewart D. R., Williams R. S., Bockrath D., Direct observation of nanoscale switching centers in metal/molecule/metal structures. *Nano Lett., 4* (2004) 569–572.
16. Zhitenev N. B., Jiang W., Erbe A., Bao Z., Garfunkel E., Tennant D. M., Cirelli R. A., Control of topography, stress and diffusion at molecule metal interfaces. *Nanotechnology, 17* (2006) 1272–1277.
17. Stewart M. P., Maya F., Kosynkin D. V., Dirk S. M., Stapleton J. J., McGuiness C. L., Allara D. L., Tour J. M., Direct covalent grafting of conjugated molecules onto Si, GaAs and Pd surfaces from aryldiazonium salts. *J. Am. Chem. Soc., 126* (2004) 370–378.

18. Akkerman H. B., Blom P. W. M., de Leeuw D. M., de Boer B., Towards molecular electronics with large-area molecular junctions. *Nature, 441* (2006) 69–71.

19. Lankhorst M. H. R., Ketelaars B. W. S. M. M., Wolters R. A. M., Low-cost and nanoscale non-volatile memory concept for future silicon chips. *Nature Mater. 4* (2005) 347–352.

20. Strukov D. B., Snider G. S., Stewart D. R., Williams R. S., The missing memristor found. *Nature, 453* (2008) 80–83.

21. Cerofolini G. F., Romano E., Molecular electronics in silico. *Appl. Phys. A, 91* (2008) 181–210.

22. Natelson D., Willett R. L., West K. W., Pfeiffer L. N., Fabrication of extremely narrow metal wires. *Appl. Phys. Lett., 77* (2000) 1991–1993.

23. Melosh N. A., Boukai A., Diana F., Gerardot B., Badolato A., Heath J. R., Ultrahigh-density nanowire lattices and circuits. *Science, 300* (2003) 112–115.

24. Wang D., Sheriff B. A., McAlpine M., Heath J. R., Development of ultra-high density silicon nanowire arrays for electronics applications. *Nano Res., 1* (2008) 9–21.

25. Gates D. B., Xu Q. B., Stewart M., Ryan D., Willson C. G., Whitesides G. M., New approaches to nanofabrication. *Chem. Rev., 105* (2005) 1171–1196.

26. Whitesides G. M., Love J. C., The art of building small. *Sci. Am. Rep., 17* (2007) 12–21.

27. Choi Y.-K., Zhu J., Grunes J., Bokor J., Somorjai G. A., Fabrication of sub-10-nm silicon nanowire arrays by size reduction lithography. *J. Phys. Chem. B, 107* (2003) 3340–3343.

28. Choi Y.-K., Lee J. S., Zhu J., Somorjai G. A., Lee L. P., Bokor J., Sub-lithographic nanofabrication technology for nanocatalysts and DNA chips. *J. Vac. Sci. Technol. B, 21* (2003) 2951–2955.

29. Augendre E., Rooyackers R., de Potter de ten Broeck M., Kunnen E., Beckx S., Mannaert G., Vrancken C., et al., Thin L-shaped spacers for CMOS devices. *Eur. Solid-State Device Res., 2003. ESSDERC'03,* 2003, pp. 219–222.

30. Cerofolini G. F., Arena G., Camalleri M., Galati C., Reina S., Renna L., Mascolo D., Nosik V., Strategies for nanoelectronics. *Microelectron. Eng., 81* (2005) 405–419.

31. Cerofolini G. F., Arena G., Camalleri M., Galati C., Reina S., Renna L., Mascolo D., A hybrid approach to nanoelectronics. *Nanotechnology, 16* (2005) 1040–1047.

32. Cerofolini G. F., An extension of microelectronic technology to nanoelectronics. *Nanotechnol. E-Newslett., 7* (2005) 5–6.

33. Flanders D. C., Efremow N. N., Generation of <50 nm period gratings using edge defined techniques. *J. Vac. Sci. Technol. B, 1* (1983) 1105.

34. Cerofolini G. F., Mascolo D., A hybrid micro-nano-molecular route for nonvolatile memories. *Semicond. Sci. Technol., 21* (2006) 1315–1325.

35. Cardamone D. M., Stafford C. A., Mazumdar S., Controlling quantum transport through a single molecule. *Nano Lett., 6* (2006) 2422–2426.

36. Roukes M., Plenty of room indeed. *Sci. Am. Rep., 17,* 3 (2007) 4–11.

37. Huang Y., Duan X., Cui Y., Lauhon L. J., Kim K.-H., Lieber C. M., Logic gates and computation from assembled nanowire building blocks. *Science, 294* (2001) 1313–1317.

38. Zhong Z., Wang D., Cui Y., Bockrath M. W., Lieber C. M., Nanowire crossbar arrays as address decoders for integrated nanosystems. *Science, 302* (2003) 1377–1379.

39. DeHon A., Lincoln P., Savage J. E., Stochastic assembly of sublithographic nanoscale interfaces. *IEEE Trans. Nanotechnol., 2* (2003) 165–174.

40. Likharev K. K., Strukov D. B., CMOL: Devices, circuits, and architectures. In: *Introducing Molecular Electronics,* Cuniberti G., Fagas G., Richter K. Berlin, Germany: Springer, 2005, pp. 447–477.

41. Strukov D. B., Likharev K. K., Prospects for terabit-scale nanoelectronic memories. *Nanotechnology, 16* (2005) 137–148.

42. Beckman R., Johnston-Halperin E., Luo Y., Green J. E., Heath J. R., Bridging dimensions: Demultiplexing ultrahigh-density nanowire circuits. *Science, 310* (2005) 465–468.

43. Aswal D. K., Lenfant S., Guerin D., Yakhmi J. V., Vuillaume D., Self assembled monolayers on silicon for molecular electronics. *Anal. Chim. Acta, 568* (2006) 84–108.

44. Cerofolini G. F., Casuscelli V., Cimmino A., Di Matteo A., Di Palma V., Mascolo D., Romanelli E., Volpe M. V., Romano E., Steps farther toward micro-nano-mole integration via the multispacer patterning technique. *Semicond. Sci. Technol., 22* (2007) 1053–1060.

45. Ben Jamaa M. H., Cerofolini G., Leblebici Y., De Micheli G., Cost-efficient and CMOS-compatible fabrication of high-density poly-silicon nanowire crossbars with the multispacer patterning technique. In: *2009 VLSI Technology Symposium,* Kyoto, Japan, 2009.

46. Ben Jamaa M. H., Moselund K. E., Atienza D., Bouvet D., Ionescu A. M, Leblebici Y., De Micheli G., Fault-tolerant multi-level logic decoder for nanoscale crossbar memory arrays. In: *Proceedings of the 2007 IEEE/ACM International Conference on Computer-Aided Design,* IEEE Press, Piscataway, NJ, 2007, pp. 765–772.

47. Ben Jamaa M. H., Moselund K. E., Atienza D., Bouvet D., Ionescu A. M, Leblebici Y., De Micheli G., Variability-aware design of multilevel logic decoders for nanoscale crossbar memories. *IEEE Trans. Comput. Aided Des. Integr. Circuits Syst., 27* (2008) 2053–2067.

48. Ben Jamaa M. H., Atienza D., Leblebici Y., De Micheli G., A stochastic perturbative approach to design a defect-aware thresholder in the sense amplifier of crossbar memories. In: *14th Asia and South Pacific Design Automation Conference,* Yokohama, Japan, 2009.

49. Hochbaum A. I., Chen R. K., Delgado R. D., Liang W., Garnett E. C., Najarian M., Majumdar A., Yang P., Enhanced thermoelectric performance of rough silicon nanowires. *Nature, 451* (2008) 163–167.

50. Boukai A. I., Bunimovich Y., Tahir-Kheli J., Yu J.-K., Goddard III, W. A., Heath J. R., Silicon nanowires as efficient thermoelectric materials. *Nature, 451* (2008) 168–171.

51. Rashevsky N., *Mathematical Biophysics*, 3rd edn. New York: Dover, 1960.

52. Cerofolini G. F., Size, shape, growth and reproduction—Towards a physical morphology. *Solid Films, 79* (1981) 277–299.

53. Cerofolini G. F., The biomedium. Adsorbed water as a model for the aqueous medium, supporting life functions. *Adv. Colloid Interface Sci., 19* (1983) 103–136.

54. Ploegh H. L., A lipid-based model for the creation of an escape hatch from the endoplasmic reticulum. *Nature, 448* (2007) 435–438.

55. Gazit Y., Berk D. A., Leunig M., Baxter L. T., Jain R. K., Scale-invariant behavior and vascular network formation in normal and tumor tissue. *Phys. Rev. Lett., 75* (1995) 2428–2431.

56. Shlesinger M. F., West B. J., Complex fractal dimension of the bronchial tree. *Phys. Rev. Lett., 67* (1991) 2106–2108.

57. Simons M. J., Pellionisz A. J., Genomics, morphogenesis and biophysics: Triangulation of Purkinje cell development. *Cerebellum, 4* (2005) 1–9.

58. Mandelbrot B. B., *The Fractal Geometry of Nature*. New York: Freeman, 1982.

59. Pfeifer P., Avnir D., Chemistry in noninteger dimensions between two and three. I. Fractal theory of heterogeneous surfaces. *J. Chem. Phys., 79* (1983) 3558–3565; Erratum *80* (1984) 4573.

60. Avnir D., Farin D., Pfeifer P., Chemistry in noninteger dimensions between two and three. II. Fractal surfaces of adsorbents. *J. Chem. Phys., 79* (1983) 3566–3571.

61. Avnir D., Farin D., Pfeifer P., Molecular fractal surfaces. *Nature, 308* (1984) 261–263.

62. Hood L., Heath J. R., Phelps M. E., Lin B., Systems biology and new technologies enable predictive and preventative medicine. *Science, 306* (2004) 640–643.

63. Falconer K., *Fractal Geometry: Mathematical Foundations and Applications*, 2nd edn. New York: Wiley, 2003.

64. Cerofolini G. F., Narducci D., Amato P., Romano E., Fractal nanotechnology. *Nanoscale Res. Lett., 3* (2008) 381–383.

65. Stern E., Klemic J. F., Routenberg D. A., Wyrembak P. N., Turner-Evans D. B., Hamilton A. D., LaVan D. A., Fahmy T. M., Reed M. A., Label-free immunodetection with CMOS-compatible semiconducting nanowires. *Nature, 445* (2007) 519–522.

66. Tuteja A., Choi W., Ma M., Mabry J. M., Mazzella S. A., Rutledge G. C., McKinley G. H., Cohen R. E., Designing superoleophobic surfaces. *Science, 318* (2007) 1618–1622.

67. Kigami J., *Analysis on Fractals*. Cambridge, U.K.: Cambridge University Press, 2001.

68. Cerofolini, G., *Nanoscale Devices*. Berlin, Germany: Springer, 2009.

Patterning and Ordering with Nanoimprint Lithography

Zhijun Hu
Soochow University

Alain M. Jonas
Catholic University of Louvain

18.1 Introduction

Photolithography, which consists of etching locally a photosensitive polymer film spin-cast on a flat substrate (Figure 18.1, left column), is central to top-down micro- and nanofabrication methodologies. The image of a drawing is first projected on the film through a mask and possibly a set of lenses; subsequent development of the exposed regions, that is, removal of the exposed regions when a positive tone resist is used, results in the etching of the drawing in the polymer film (Rai-Choudhury 1997). This topographically patterned film then serves for further operations such as localized deposition of metals, etching of the underlying substrate, or ion implantation, which are basic processes for the fabrication of integrated circuits (IC) (Campbell 2001).

The need to increase the density of integration has resulted in a continuously decreasing size of the features to be lithographied. It is well known that a wave of wavelength λ is diffracted when passing through an opening of size a, and that the width of the central diffraction lobe is proportional to λ/a in the far-field limit (Born and Wolf 1980). Therefore, it does not come as a surprise that the resolution limit of photolithography is directly linked to the wavelength of the radiation used as well. Other factors also come into play (Campbell 2001) but need not interest us here. As a result, progress in downsizing photolithography led to a continuous decrease of λ, resulting simultaneously in rapidly rising costs (Resnick 2007). Starting from visible light, the trend thus went to deep UV, extreme UV (Wurm and Gwyn 2007), and soft x-ray lithography (Ueno 2007). In parallel, electron beam lithography (EBL) was developed to benefit from the short wavelength of accelerated electrons (Owen 1985, Suzuki 2007). However, EBL consists of scanning a beam of electrons over a film of an electrosensitive resist, which is essentially

a serial process much slower than the all-at-once projection of a drawing. Therefore, EBL is restricted to niche uses such as research, low-volume production, or the fabrication of masks for photolithography.

The photolithographic process is actually more related to the etching technique used by artists to reproduce drawings than to the original lithographic technique. In etching, a film of wax is deposited on the surface of a metal plate. The artist removes locally the wax using a sharp stylus, and the metal plate is subsequently etched with acid where the wax is missing. The resulting metal plate is then used to reprint the drawing (bottom-left of Figure 18.1). Historically, other methods have been devised by mankind to reproduce drawings. Probably the oldest method is stamping, which can be traced back to prehistoric times when men and women of all continents inked their hands to impress the walls of caves. Traditional ink seals such as used in East Asia, known as 印章 in China, are remote descendants of this method. The stamping technique was also translated to micro- and nanotechnologies, in which it is currently known as microcontact printing (μCP) (Kumar and Whitesides 1993), the most prominent of the soft lithographic methods (Xia and Whitesides 1998, Xia et al. 1999, Gates et al. 2005). In μCP, a soft elastomer stamp is inked with a solution of a molecule capable to chemisorb on a substrate (Figure 18.1, central column). By pressing the stamp on the substrate, local grafting of the molecules ensues. The stamp softness is important to ensure conformal contact with the surface; however, it also results in undesirable sagging and deformation that usually limit μCP to the micrometer scale, even though higher resolutions were reported in favorable circumstances (Biebyuck et al. 1997, Odom et al. 2002). But more importantly, stamping methods only transfer a monolayer of ink on the surface of the substrate. The resulting drawing is thus

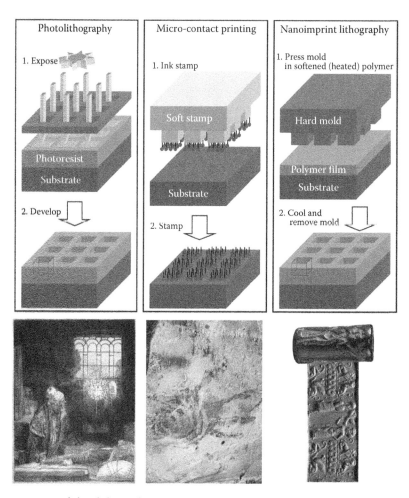

FIGURE 18.1 Schematic presentation of photolithography, microcontact printing, and nanoimprint lithography. The art drawings obtained by the corresponding historical drawing method are shown at the bottom. (Left) Doctor Faust, etching from Rembrandt, 1652; (center) a paleo-Indian handprint from Pedra Pintada, Brazil; (right) a Mesopotamian seal and its printout, *ca.* 1720–1650 BC, Syria.

essentially two-dimensional, as opposed to the three-dimensional nature of the topographically patterned polymer films used for integration technology. Therefore, µCP is more suited to create micrometer-scale chemical patterns that are used as platforms for bottom-up assembly processes (Mahalingam et al. 2004) or for the fabrication of arrays of biosensors (Kane et al. 1999), rather than for nanolithography.

Another ancient method of reproducing drawings is the embossing of soft materials such as paper, clay, or wax by hard seals. Historical records trace back to the kingdoms of Ancient Egypt and Mesopotamia, and there is evidence that seals were used as signatures in official Chinese documents before 220 BC. The method of embossing found its way in nanotechnology 5000 years after its initial discovery, where it is currently known as nanoimprint lithography (NIL) (Chou et al. 1995), hot embossing (Heyderman et al. 2000), or nanoembossing (Studer et al. 2002). NIL is a newcomer in the lithography field, since it was initially proposed in the mid-1990s only by Chou and coworkers (Chou et al. 1995, 1996); however, the method has rapidly attracted strong interest and is currently considered as one of the possible industrial options for reaching feature

sizes below 30 nm (ITRS 2007). An overview of the method that consists of transferring a drawing in a polymer film by pressing a topographically patterned hard mold into the film (Figure 18.1, right column) is given in Section 18.2. Because NIL as such has already been reviewed extensively (Torres 2003, Guo 2004, Stewart and Willson 2005, Guo 2007, Resnick 2007, Schift and Kristensen 2007, Schift 2008), this section concentrates only on the most important features of the method required to understand the rest of the chapter. Note that NIL, like all other techniques mentioned above, leaves aside the issue of the fabrication of the master mold (or stamp for µCP, or mask for photolithography). Actually, the idea behind most advanced lithographic techniques is to fabricate masters by a slow, expensive but efficient technique such as EBL, and to replicate them in large numbers by a less expensive, more rapid method. This is not basically different from printing or stencil techniques, which also create one template from which many copies are produced. Obviously, the wear of the master is a potentially serious problem that is certainly more significant for methods involving a direct contact between the master and the film to be patterned, as in NIL or µCP.

The main advantages of NIL as a lithographic technique are the relatively limited cost of the method, its high throughput, and its capacity to reach very small feature sizes. But there is intrinsically more to NIL than simply the fabrication of three-dimensionally patterned polymer films for integration technology. NIL is basically a molding technique acting at the nanometer scale, which can thus be used to shape functional materials into nanowires, nanopillars, or other useful nanoobjects. This new trend is reviewed in Section 18.3. Even more interestingly, NIL can also be used as a way to control ordering processes (such as crystallization) at the nanometer scale. In this case, not only is the shape of the nanoobjects controlled during embossing but also their internal structure, which may give rise to enhanced performance compared with macroscopic materials. Recent examples of this promising emerging trend are given in Section 18.4. NIL is thus progressively moving from a purely lithographic method to a real polymer nanoprocessing tool, which might well be the niche where NIL will thrive in the future.

Before closing this section, it is useful to mention an interesting parallelism between the historical development of microscopy and lithography. Microscopy began with optical microscopes that are limited by diffraction to a spatial resolution of about $\lambda/2$ (Bergmann and Schaefer 1999), except when special setups are used in combination with fluorescent molecules (Klar et al. 2000). Higher resolutions are attained by decreasing the wavelength, leading to electron microscopes and soft x-ray microscopes (Chao et al. 2005). This clearly parallels the evolution of photolithography. However, the diffraction limit can also be beaten by changing entirely the principle of operation of the microscope. This is the case of scanning probe microscopes, in which a sharp tip is placed in contact or close to contact with a surface. In atomic force microscopy (AFM), for instance, the local force experienced by the tip is translated into an image in a way that depends on the specific mode of the operation selected (Kalinin and Gruverman 2007). The main point in the present context is that images of very high resolution can be obtained by switching from a wave-based imaging process to a contact-mode imaging process. This is not unlike the switch from a photolithography-based process to an embossing-based process.

The comparison is even stronger when one realizes that AFM tips can be used as mechanical tools to orient polymer molecules (Leung and Goh 1992), just as NIL is shown in Section 18.4.

18.2 Nanoimprint Lithography

Contrarily to photolithography and EBL, NIL replicates a relief pattern in a heated thermoplastic or in a thermosetting polymer by mechanical deformation, rather than by a local photo- or electrochemical modification of the resist. The pattern is then frozen by cooling down or cross-linking the polymer. A schematic drawing of the original NIL process that was proposed for thermoplastic polymers is shown in Figure 18.2a (Chou et al. 1997). A thin film of an amorphous thermoplastic polymer is cast onto a rigid surface. The polymer film is softened by heating above its glass transition temperature. A rigid mold bearing nanoscale features is then pressed into the polymer film until the polymer flows into the recesses of the mold. The polymer film is finally hardened by decreasing the temperature below the glass transition, before the mold is removed. A thickness contrast is thus created in the polymer film after imprinting. A thin residual layer of polymer is most often intentionally left between the protrusions of the mold and the substrate in order to prevent direct impact of the hard mold on the substrate and thereby protect the fragile protrusions of the mold (Guo 2007). This residual layer is then removed by an anisotropic O_2 plasma etching process, which completes the pattern replication. However, Schulz et al. (2003) have concluded from a study of local mass transport and its effects on pattern replication that for optimal performance, the initial film thickness should be selected so as to have less than 100% filling of the mold cavities, which corresponds to nearly zero residual layer thickness.

As can be realized, a very attractive aspect of NIL is its intrinsic simplicity: the basic elements needed for NIL are limited to (1) a mold with a relief comprising a large number of nanofeatures, (2) a suitable resist such as a thermoplastic polymer that can be softened during imprinting and hardened after imprinting, and (3) a printing equipment setup allowing some control of process parameters such as temperature and

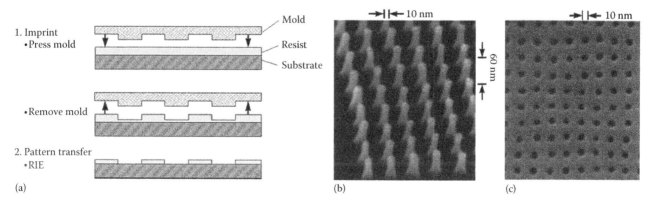

FIGURE 18.2 (a) Schematic drawing of the originally proposed NIL method. (b) SEM image of a mold with pillars of 10 nm lateral size and 60 nm height. (c) SEM image of the corresponding array of holes imprinted in a PMMA film. (Reprinted from Chou, S.Y. et al., *J. Vac. Sci. Technol. B*, 15, 2897, 1997. With permission.)

pressure. NIL is thus cost-efficient compared with other lithographic methods, while at the same time offering a good performance with respect to resolution and throughput. Unlike the development of photolithography and EBL, the lack of expensive optical components and light sources makes it much easier to construct an imprint setup. To meet the requirements of an NIL process, one needs only a well-controlled heating and cooling system and a setup for applying pressures. There are now many different companies selling NIL equipment, such as Obducat, EVGroup, or Süss MicroTec in the European Union, and Nanonex in the United States.

The mold used in NIL can essentially be any type of solid material that has a good mechanical strength and reasonable durability, and is usually made of silicon, dielectric materials such as silicon dioxide or silicon nitride, metals such as nickel, or even polymeric materials that are harder than the resist (Guo 2007). The desired nano- or microscale features defined on the molds are produced by any other lithographic processing such as photolithography, EBL, or even NIL itself, followed by localized metal deposition, reactive ion etching of metal-uncovered regions, and metal removal. In order to facilitate the separation of the mold from the substrate after imprinting, an anti-adhesion layer is normally coated on the mold surface. The requirements for this layer are that its interfacial tension with the imprinted material be large, whereas its surface tension in air be low. This favors the separation of the mold and imprinted material for thermodynamic reasons. Therefore, the most widely adopted approach is to form a self-assembled monolayer of a partially perfluorinated silane (e.g., $1H,1H,2H,2H$-perfluorodecyltrichlorosilane) on the mold surface, preferably by a vapor-phase reaction (Jung et al. 2005). Mixtures of mono- and trichloro-perfluorinated silanes were also proposed (Schift et al. 2005). Perfluorinated organic materials are interesting because of their low surface tension in air and known incompatibility with hydrocarbons that favors a high interfacial tension with most polymers.

The imprinting resists used in NIL are typically thermoplastic amorphous homopolymers such as poly(methyl methacrylate) (PMMA), polycarbonate (PC), and polystyrene (PS). One common feature of these thermoplastic materials is that their viscosity and elastic modulus are significantly reduced by increasing the temperature above their glass transition temperature, T_g. Hence, the polymer viscoelastic fluid can be forced to flow into the recesses of the mold under a pressure of a few tens of bar and to conform exactly to the surface relief of the mold. As a rule of thumb, the pressure ranges usually from 10 to 100 bar and the imprint temperature is most often ~70°C–90°C above the T_g of the imprinted polymers. The polymer fluid can be easily hardened by cooling back to below T_g to freeze the imprinted nanostructures.

One key parameter for the selected polymer is its molar mass M that controls physical properties such as the glass transition temperature, the stiffness, the strength, and the viscosity (Gedde 1995). For instance, the viscosity scales as $M^{3.4}$ above some critical M_e, below which entanglements between chains do not exist in the melt (Rubinstein and Colby 2003). The pressure, imprinting

times, and/or imprinting temperatures have therefore to be significantly increased for polymers of higher molar mass. Both printing temperature and pressure should be kept as low as possible in view of the time needed for temperature and pressure cycling. It is thus favorable to work with polymers of lower M, hence lower viscosity. However, because amorphous polymers of molar mass lower than M_e are brittle, they are not suitable for nanoimprinting. Typically, M_e is in the 10,000 g mol⁻¹ range for styrenic and methacrylate amorphous polymers (Rubinstein and Colby 2003). Note also that the distribution of molar mass and the topology of the polymer (e.g., presence of long chain branches or cyclic chains) are important parameters with respect to rheological properties; these may thus be tuned for optimal throughput.

One unique advantage of NIL is that its resolution is not limited by factors such as light diffraction, scattering, and interference in the resist. The resolution limit of NIL can be smaller than the radius of gyration of polymer chains that is on the level of a few or a few tens of nanometers depending on molar mass. Figure 18.2b and c shows scanning electron microscopy (SEM) images of a mold with 10 nm diameter pillars and the replicated 10 nm hole array in PMMA (Chou et al. 1997), showing that indeed NIL can be used as a high-resolution nanolithographic technique.

Since the first report on NIL, the method has attracted strong interest from both industry and academia due to its merits such as high-resolution, low-cost, and inherent high-throughput resulting from parallelized processing. For lithography purposes, however, one of the drawbacks of the originally proposed NIL process is the thermal cycle that limits the throughput for semiconductor IC manufacturing. A variation of the NIL technique that uses a transparent mold and low-viscosity UV-curable monomers that cross-link upon UV exposure (step and flash imprint lithography, SFIL) was developed soon after (Ruchhoeft et al. 1999), allowing the process to be carried out at lower pressures and room temperature. Since the thermal cycle is avoided, SFIL is much faster than thermal NIL which makes it very attractive for the manufacturing of semiconductor IC devices (Resnick et al. 2003). Other variations include solvent-assisted NIL, in which the polymer is softened by a solvent vapor, and imprinting is performed at room temperature (Khang et al. 2001, Voicu et al. 2007).

In order to optimize the NIL process or widen its applications, understanding the cavity-filling process during imprinting is essential. At first sight, the initial step in the process of NIL is very similar to the technique of hot embossing, in which a polymer layer is thermally deformed using a rigid mold. However, with ever smaller feature sizes and increasingly complex mold geometries, the behavior of the thin polymer film during both molding and demolding becomes more complicated. The cavity fill process can be understood in first approximation as a squeeze flow, as shown in Figure 18.3a (Heyderman et al. 2000). Initially, the polymer flows on the cavity walls and wets them. The material in the center of the cavity is under compressive stress, causing a convex profile to form. The polymer surface thus raises and eventually touches the top of the cavity. Because squeeze flow, electrostatic interactions between the mold and polymer, capillary forces, van der Waals forces between the mold and the substrate

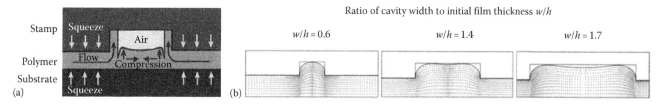

FIGURE 18.3   (a) Scheme of the forces acting on a polymer film during the filling of a cavity. (Reprinted from Heyderman, L.J. et al., *Microelectron. Eng.*, 54, 229, 2000. With permission.) (b) Cavity fill process depending on the ratio of cavity width to film thickness. (Reprinted from Rowland, H.D. et al., *J. Micromech. Microeng.*, 15, 2414, 2005. With permission.)

across the polymer fluid, and surface tensions appear in the balance of forces (Schift et al. 2001), a variety of self-organization phenomena can be triggered during imprinting, resulting sometimes in the formation of different undesirable patterns such as pillars, fractal-like patterns, and viscous fingers. Rowland et al. (2005) have investigated by numerical simulation the impact of polymer properties, mold geometry, and process conditions on the cavity-filling process. When the polymer flows vertically into an open cavity during imprinting, the polymer can deform either as a single peak centered in the cavity or as a dual peak where each peak remains close to the vertical sidewalls, depending upon geometry, as shown in Figure 18.3b. The ratio of cavity width to film thickness determines single versus dual-peak cavity filling, regardless of the absolute size of the features and also of the pressure or temperature applied during embossing.

18.3 Nanoshaping Functional Materials by Nanoimprint Lithography

It is important to note that the cavity fill process of NIL only moves the macromolecules of the imprinted resist from one place to another. The chemical properties of the imprinted resist are thus kept after imprinting, contrary to conventional lithography where a pattern is generally created by the irradiation-induced reaction of organic materials. Therefore, NIL can be adapted directly to pattern or shape of a wide range of relatively fragile materials such as functional polymers and biological macromolecules instead of photo- or electroresists. In this respect, NIL is attracting increasing attention for its ability to shape directly functional materials into nanostructures. NIL can therefore be used as an efficient nanoprocessing tool to shape a material for applications, requiring a synergy between its topography and its functional properties. NIL was therefore used

1. *To generate nanopores in materials*: The demand for low dielectric constant (low k) materials in the microelectronics industry has recently led to considerable interest in porous materials, especially nanoporous materials (Maier 2001). Imprinting can be used to create pores into a thin film of a dielectric material and therefore decreases its dielectric constant. For instance, Soles et al. demonstrated the direct patterning by NIL of sub-100 nm features into porous spin-on organosilicate glass materials (Ro et al. 2007, 2008).

2. *To shape the interface between two functional materials*: In organic heterojunction solar cells, the interface between donor and acceptor materials is crucial. This interface could be tailored by NIL. A thin film of conjugated polymer (serving as acceptor) was first imprinted in a controlled atmosphere (low vacuum or inert gas). Well-defined nanoscale structures were thus created in the acceptor layer. The other component (donor) of the system was then poured in the created trenches (Kim et al. 2007, Cheyns et al. 2008 and Aryal et al. 2008). The fill factor and power conversion efficiency were correspondingly increased.

3. *To create optical gratings in light-emitting materials*: Within recent years, organic dye-doped or conjugated polymer laser devices have received much attention as convenient and compact light sources. One of the important issues to make polymer lasers is the fabrication of a periodic surface relief in the conjugated or dye-doped polymers, with a period close to the wavelength of light. The periodic relief can modify the generation and propagation of light in these materials by interference effects. Chou et al. were the first to apply NIL to directly shape organic light-emitting materials, such as Alq3 (8-hydroxyquinoline aluminum) doped with DSMII dye molecules, or their blends with PMMA (Wang et al. 1999). Figure 18.4a shows AFM images of the resulting Alq3/DSMII gratings of 300 nm period. It was demonstrated that no degradation of the light emission efficiency is caused by the NIL process (Figure 18.4b). In addition, because the small dye molecules embedded in the PMMA matrix are slightly aligned by the imprinting process (see Section 18.4), an anisotropic photoluminescence behavior was observed (Wang et al. 2000).

Subsequently, Guo et al. (2002) successfully imprinted two-dimensional photonic crystal nanostructures in nonlinear optical dye-doped polymers. Pisignano et al. (2003, 2004) and Mele et al. (2005, 2006) showed that conjugated polymers and oligomers can be directly imprinted at room temperature under a very high pressure, based on the plastic deformation of the polymer. Figure 18.5a is an AFM image of a room temperature-imprinted organic semiconductor thin film. A study of the optical properties before and after imprinting (absorbance, luminescence, and quantum yield) again ruled out the possible degradation of the active materials even when imprinting was performed in air (Figure 18.5b). The imprinting of

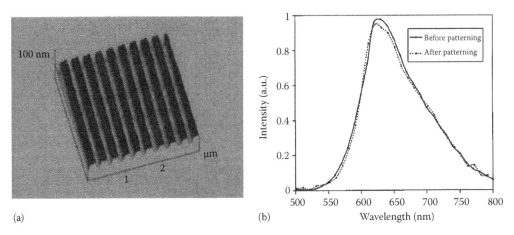

(a)   (b)

FIGURE 18.4 Optical grating imprinted by NIL in a light-emitting material. (a) AFM height image of Alq3/DMSII gratings with a period of 300 nm. (b) The measured luminescence spectral intensity before and after imprinting, demonstrating the absence of sample degradation. (Reprinted from Wang, J. et al., *Appl. Phys. Lett.*, 75, 2767, 1999. With permission.)

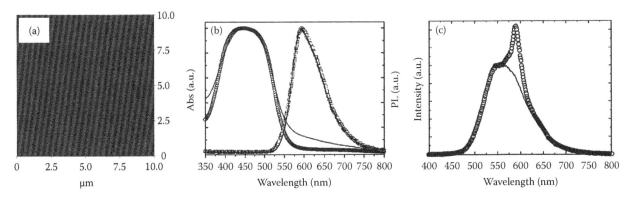

FIGURE 18.5 Optical grating imprinted by NIL at room temperature in a light-emitting organic semiconductor. (a) AFM height image of the surface grating with a period of 400 nm. Image side: 10 μm. (b) Absorbance and integrated photoluminescence spectra before (line) and after (circles) imprinting, showing the absence of degradation. (c) Photoluminescence spectra at a selected emission angle, showing the presence of an interference peak after imprinting (circles) compared with the uniform film (line). The interference peak is due to the surface grating. (Reprinted from Pisignano, D. et al., *Adv. Mater.*, 16, 525, 2004. With permission.)

a surface periodic grating with wavelength-scale periodicity enhances the light emitted in specific directions due to interference effects (Figure 18.5c), and reduces the effective length covered by the photons in the material and thus the self-absorption inside the organic slab. These interference effects were subsequently used to build a distributed feedback (DFB) laser, wherein the surface grating serves to produce a narrow band of wavelengths by interference. Figure 18.6 displays the photoluminescence spectrum of such a DFB laser obtained by NIL using the conjugated light-emitting polymer (poly[2-methoxy-5-(2-ethylhexyloxy)-1,4-phenylenevinylene]) (MEH-PPV), showing a narrow emission at about 630 nm wavelength.

4. *To create antireflective structures on the surface of optical materials*: Two-dimensional antireflection subwavelength structures were also created by NIL on the surface of thin films in order to suppress Fresnel reflection. Fresnel reflection is a drawback for many optical

systems, since it reduces the fraction of transmitted light, deteriorates the contrast of displays, and generates ghost images in imaging systems (Ting et al. 2008a,b). Surface relief gratings with periods smaller than the wavelength of light, named subwavelength gratings, may behave as antireflection surfaces. Figure 18.7a is an SEM image of antireflective PMMA nanostructures fabricated by NIL. The optical properties of a plain PMMA sheet with the ones of a patterned PMMA sheet are shown in Figure 18.7b, showing a significant reduction of glare after imprinting.

5. *To shape biomaterials*: Ohtake et al. (2004) reported that DNA can be directly patterned by NIL; interestingly, the patterned DNA retains its activity, showing that these biomacromolecules can withstand the relatively harsh processing conditions of NIL. Park et al. (2007) showed that chitosan, a derivative of a natural polysaccharide, can also be directly patterned by NIL at low temperature

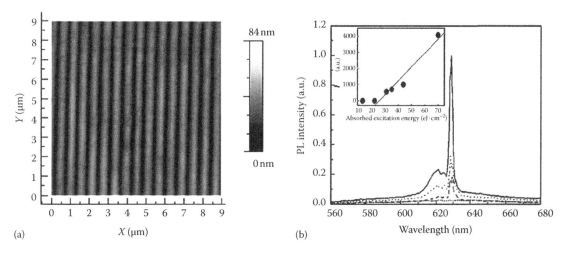

FIGURE 18.6 (a) AFM height image of MEH-PPV gratings with a period of 600 nm patterned directly by NIL. (b) Distributed feedback emission spectra under pulsed excitations of increasing dose. The narrow band of wavelengths emitted at about 630 nm results from interference from the surface grating. Inset: the influence of excitation dose on the DFB emission intensity. (Reprinted from Mele, E. et al., *Appl. Phys. Lett.*, 89, 131109, 2006. With permission.)

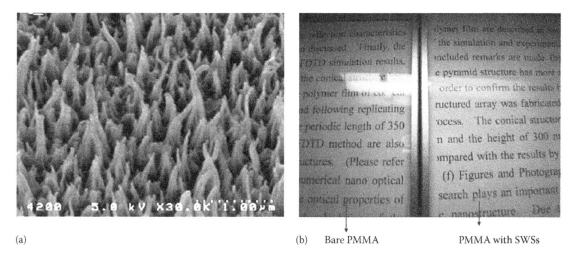

FIGURE 18.7 (a) SEM image of PMMA films with imprinted antireflective nanostructures. (b) Comparison between a bare PMMA sheet and the imprinted PMMA sheet, showing an obvious damping of backlight glare. (Reprinted from Ting, C.-J. et al., *Nanotechnology*, 19, 205301, 2008a. With permission.)

and low pressure. The nanoimprinted chitosan patterns contain a large amount of primary amino groups for further chemical modification, and can be used, for instance, to graft proteins and DNA strands for biosensing purposes.

Interestingly, nanoshaping of functional *inorganic* materials is also possible by NIL. For instance, Grigoropoulos et al. (Ko et al. 2007, Park et al. 2008) demonstrated the direct nanoimprinting of thin films of gold nanoparticles. Because of their small size (2–3 nm), the nanoparticles start to melt at ~130°C–140°C, well below the melting point of bulk gold. A variety of gold nanostructures were fabricated, such as nanodots, nanowires, and serpentine nanowire arrays, showing the potential of the method for the direct fabrication of electrodes for nanoscopic devices. Imprinting of soft precursors of inorganic crystals and glasses

was also reported, showing the possibility to use NIL to fabricate ceramic or inorganic glass nanostructures (Harnagea et al. 2003, Okinaka et al. 2006, Peroz et al. 2007). Finally, direct imprinting of silicon wafers was also achieved via laser-induced transient liquefaction of the imprinted material (Chou et al. 2002).

18.4 Controlling Ordering and Assembly Processes by Nanoimprint Lithography

As mentioned in Section 18.1, the application of NIL to organic materials such as semicrystalline polymers or liquid crystalline oligomers is capable to control the internal structure of the resulting nanoobjects. Consider semicrystalline polymers as an example. These materials crystallize in ~10 nm thick lamellar-shaped crystals, in which the chains are perpendicular or slightly inclined to the lamellar surface, and when emerging from the crystal surface either fold back in the same lamella or reenter a neighboring quasi-parallel lamella (Figure 18.8) (Gedde 1995, Reiter and Strobl 2006). The lateral sizes of polymer lamellae are in the micrometer range, and the lamellae are elongated in the preferred growth direction of the crystal. Successive lamellae are separated by amorphous interlayers, which accommodate the folds, loose loops, tie segments, and dangling ends of the chains (Mandelkern 1979, Gautam et al. 2000). The crystallization process occurs through nucleation from a central seed, most often an impurity (Wunderlich 1976), from which a complex assembly of lamellae forms by crystal growth, branching, and splaying (Bassett 1984). The resulting so-called spherulite has a radius in the micrometer range, and the preferred growth direction of the lamellae is essentially radial within the spherulites. Hence, at the supraspherulitic scale, semicrystalline polymers are isotropic, and consist of an intricate mixture of crystallized and amorphous nanoregions. Therefore, their properties are averages of different phases over all possible orientations, and include components resulting from the numerous crystal–amorphous interfaces present in a spherulite. This significantly reduces the performance of such materials.

Consider now the possible effects of NIL on such a complex morphology. Embossing is performed in the molten state, but the polymer crystallizes in the mold during cooling. First, the nucleation process could be affected, because NIL splits the melt into a very large number of small cavities. Since the probability to find a heterogeneous nucleating seed in a cavity depends on the product between the volume of each nanocavity and the number density of nucleating seeds in the melt, this probability may be well below 1.

If so, crystallization in the nanocavity will either require the spontaneous formation of a nucleating germ in the melt, a process called homogeneous nucleation rarely encountered in practice when crystallizing bulk polymers, or the propagation of a crystal from the outside into the nanocavity (if the cavity is not totally isolated), which will only be possible when a nearby crystal grows in the proper orientation. The situation is similar to the one encountered when crystallizing polymers in microdroplets (Cormia et al. 1962, Massa and Dalnoki-Veress 2004), in microdomains of phase-separated block copolymers (Reiter et al. 2001, Loo et al. 2002, Muller et al. 2002), or in the nanopores of membranes (Steinhart et al. 2006, Woo et al. 2007), cases for which retarded crystallization and preferred orientation are often reported. One can thus expect identical effects to occur in NIL as well.

Second, the crystal growth process might also be affected. It was demonstrated for the first time in Russia in 1972 (Sheftal and Bouzynin 1972), and rediscovered later in the United States (Smith and Flanders 1978), that topographically patterned amorphous surfaces may induce preferential orientation of growing crystals, provided the symmetry of the topography be adapted to the habit of the growing crystals. This phenomenon is known as "graphoepitaxy" or "artificial epitaxy" (Givargizov 1994, 2008), with reference to the epitaxy of crystals on crystalline substrates, where matching between the lattice planes of the growing crystal and of the substrate provides for preferential orientation. In graphoepitaxy, however, the matching occurs at a higher level and does not require a crystalline substrate. Key thermodynamic parameters are the shape of the growing entity and the symmetry of the substrate topography, as well as the surface tensions and surface tension anisotropies of the crystal–substrate and melt–substrate interfaces, which dictate the energy balance of the growing system (Mouthuy et al. 2007). On the kinetic side, the growth rates of the crystal along specific crystalline directions are also important parameters because they control the crystal shape, hence the graphoepitaxial matching. In the present context, the topography of the NIL mold may thus control graphoepitaxially the crystallization process of the polymer, resulting in a preferential alignment of specific crystallographic directions with respect to the nanofeatures of the mold.

A third factor that could affect the crystallization process in a NIL experiment is the degree of chain alignment that results from the melt embossing process. Polymers are viscoelastic fluids, which may keep for a long time a partial memory of the considerable squeeze flow experienced during the imprinting. In addition, polymer chains tend to align along flow lines during processing. This favors oriented crystallization (Miller 1979, Kumaraswamy et al. 1999, Somani et al. 2000) and the emergence of specific morphologies such as "shish-kebabs," in which long crystalline nanofibers made of aligned chains (the "shishes") serve as nuclei for oriented lamellar crystals (the "kebabs") (Eder et al. 1989, Kimita et al. 2007). In NIL, the residual degree of chain alignment will depend on the molar mass of the polymer, on its previous processing history, and on its annealing time in the melt before crystallization. Note that chain orientation under flow is not easy to predict at the nanometer scale, since molecular confinement of the chains in a thin film leads to a substantially perturbed rheological behavior (Rowland et al. 2008).

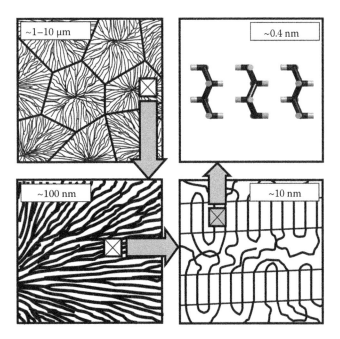

FIGURE 18.8 Scheme of the morphology of semicrystalline polymers seen at different length scales, showing spherulites (top-left image), stacks of lamellae (bottom-left image), lamellar crystals and amorphous interlayers (bottom-right image), and crystalline stems (top-right image).

A fourth factor is confinement itself. Most macromolecules have radii of gyration in the 10–50 nm range, close to the thickness of films used in NIL and to mold feature sizes, resulting in strong confinement of the chains. There is currently an ongoing discussion in the literature regarding the way confinement affects polymer properties such as glass transition and segmental mobility (Ellison and Torkelson 2003, Alcoutlabi and McKenna 2005) or whole chain motion (Rowland et al. 2008). Conflicting results have been published, and it is not possible to review them fairly in the present chapter. Instead, one should keep in mind that the conformation and dynamic properties of confined polymers may be substantially different in a nanocavity than in the bulk, which is likely to modify their crystallization behavior as well. Think only of the need that chains fold back and forth to form lamellar crystals (Figure 18.8). If this occurs in a space only slightly larger than the intrinsic lamellar thickness, significant perturbations of the crystallization process are clearly to be expected.

What is true for polymer crystallization should also be valid for other supramolecular assembly processes, such as the formation of liquid crystalline domains or the ordering of block copolymer mesophases. In the sequel, we will review the results obtained in this emerging field, for a range of systems such as amorphous rigid polymers, liquid crystalline polymers, semicrystalline polymers, and phase-separated block copolymers. NIL will be shown to be unique in its ability to orient crystals and chains in specific directions, resulting in favorable cases in a significant improvement of the performance of nanodevices. The results will be divided in two classes: NIL under either full or partial confinement. In the first case, the starting film thickness and the height of the protrusions on the mold are such that the mold can touch the surface of the substrate if pressed strongly enough (Figure 18.9). For a mold having a protruding surface fraction σ, each protrusion being of height h, full confinement will only be possible when the starting film thickness $d < (1 - \sigma)h$. Under this circumstance, each recess of the mold will be effectively isolated from other recesses. In contrast, when $d > (1 - \sigma)h$, a continuous residual film always remains between the substrate and the protruding features of the mold, no matter how hard the mold is pressed in the film (Schift and Heyderman

2003). In this case, information on the ordering process may propagate from recess to recess, which will have significant consequences on the global ordering of the sample.

18.4.1 Ordering under Partial Confinement

The first experimental demonstration of molecular alignment by NIL was obtained for chromophores dissolved in a standard PMMA resist, using line gratings as molds (Wang et al. 2000). Polarized photoluminescence and absorption measurements indicated that the dye molecules are aligned after imprint with their long axis along the grating lines, most probably due to the effect of flow during processing. The fluorescence anisotropy, $R = (I_{//} - I_{\perp})/(I_{//} + 2I_{\perp})$, quantifies the degree of alignment by comparing the fluorescence emission polarized parallel to the grating lines ($I_{//}$) and perpendicular to them ($I_{\perp}$). R values of 1 correspond to complete alignment, whereas $R = 0$ for isotropic systems. From the reported dichroic ratios $DR = I_{//}/I_{\perp}$, R values of 0.1–0.2 can be computed for this specific example, which shows that the polymer flow is able to align the dye molecules only moderately.

For semicrystalline polymers crystallized in partial confinement, no or limited preferential orientation of polymer crystals occurs when line gratings are used as molds (Hu et al. 2005, Okerberg et al. 2007). Thin films of poly(vinylidene fluoride) (PVDF, α-phase) crystallize in a normal spherulitic morphology and appear to be insensitive to the presence of the grooves of the NIL mold (Hu et al. 2005). The crystalline lamellae propagate from groove to groove, and the electron diffraction patterns do not show preferred orientation relative to the line grating direction (Figure 18.10). This can be ascribed to fast crystal growth in the thin residual film below the protrusions of the mold, from which crystallization propagates into the linear grooves of the mold. The final morphology is thus not dictated by the mold, but mainly by the processes occurring in the residual film. Similar results were obtained for poly(ethylene oxide) (PEO) (Okerberg et al. 2007), although partial elongation of the crystalline lamellae along the grooves of the mold was sometimes observed. As is readily apparent in Figure 18.11a, the global spherulitic shape is not affected by

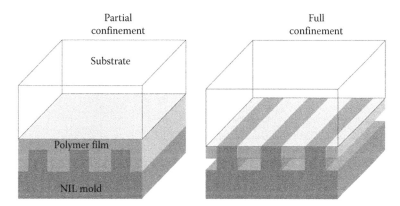

FIGURE 18.9 Two configurations of NIL leading to different microstructures after imprint. For clarity, the mold was placed at the bottom of the drawing.

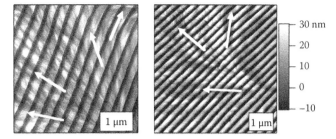

FIGURE 18.10 Microstructure of PVDF (α-phase) crystallized in NIL line grating molds under partial confinement (left: transmission electron micrograph; right: atomic force microscopy topograph). No preferential alignment is obtained, as shown by the arrows that highlight the direction of a few lamellar crystals. (Adapted from Hu, Z. et al., *Nano Lett.*, 5, 1738, 2005.)

the NIL mold, and lamellae are found growing either perpendicular to the grooves (zone 1 in Figure 18.11a and b) or parallel to them (zone 2 in Figure 18.11a and c), depending on the orientation of the radius of the spherulite with respect to the line grating direction. These results clearly indicate that chain

alignment by the flow is not sufficient to induce oriented crystal growth and that graphoepitaxy of semicrystalline polymers does not occur in partial confinement.

The situation is somewhat different when imprinting a low molar mass liquid crystalline conjugated polymer, poly(9,9-dioctylfluorene-*co*-benzothiadiazole) (F8BT), in its liquid crystalline nematic mesophase (Zheng et al. 2007, Schmid et al. 2008). This green-light-emitting polymer displayed after imprinting *R* values as large as 0.97, indicating that most polymer chains are aligned parallel to the line grating direction. The order parameter *R* was found to increase for thinner films, suggesting that the ordering starts from the surface of the mold and progressively vanishes when going down into the residual film. Polymer-light-emitting diodes (PLEDs) were successfully built based on nanoimprinted F8BT, giving rise to the emission of preferentially linearly polarized light (Figure 18.12). A field-effect transistor was also realized, and the hole mobility was 10–15 times larger parallel to the grating lines than perpendicular to them. These results directly illustrate the beneficial impact of aligning molecules, as far as charge mobility or emission polarization is concerned. It should,

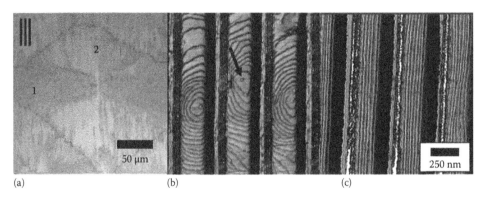

FIGURE 18.11 Microstructure of PEO crystallized in NIL grating molds under partial confinement. Panel (a) is an optical micrograph displaying a spherulite; the vertical stripes indicate the direction of the lines of the grating used to imprint the sample. Panels (b) and (c) are AFM topographic images corresponding to the regions 1 and 2 of panel (a), respectively, and show different relative orientations of the lamellae compared with the line grating direction. (Reprinted from Okerberg, B.C. et al., *Macromolecules*, 40, 2968, 2007. With permission.)

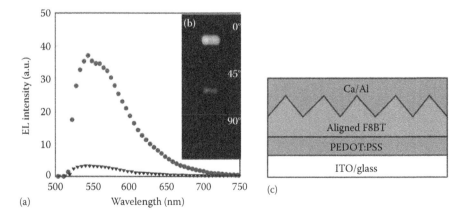

FIGURE 18.12 (a) Electroluminescence of a PLED made of F8BT imprinted in partial confinement with a line grating. The luminescence is measured through a polarizer parallel (circles) or perpendicular (triangles) to the grating lines. (b) Optical images of the electroluminescent device seen through a polarizer oriented with respect to the grating lines as indicated in the figure. (c) Schematic drawing of the PLED structure. (Reprinted from Zheng, Z. et al., *Nano Lett.*, 7, 987, 2007. With permission.)

however, be noted that NIL failed to align a similar light-emitting conjugated polymer, poly(9,9-dioctylfluorene) (F8) also displaying a liquid crystalline nematic mesophase (Song et al. 2008). This was tentatively ascribed to the higher molar mass (hence larger viscosity) of this polymer compared with F8BT. Clearly, NIL under partial confinement is able to align liquid crystalline polymers in favorable circumstances; however, further work is required to fully understand the factors which control this alignment that are probably similar to the ones driving the alignment of liquid crystals on buffing layers (Toney et al. 1995).

18.4.2 Ordering under Full Confinement

When the starting film thickness is small enough for the NIL mold to enter in contact with the substrate, full confinement arises (Figure 18.9b). In this case, the propagation of information from groove to groove is suppressed, and complete graphoepitaxial alignment happens as is demonstrated below. It is critical to realize that graphoepitaxial alignment occurs at some stage during the formation of a structure, depending on the details of the processing history. It may thus orient a growing crystal, or a liquid crystalline domain, or even a single molecule, depending on the basic structural element that interacts with the mold. The notion of basic structural element is thus key to understand and reconcile the results published in the literature, and will serve us as guide in the sequel.

When a *semicrystalline* polymer crystallizes from the isotropic melt in a NIL line grating mold, the basic structural element to consider is the *nascent crystal*. Because polymer crystals are usually elongated along their fast growth axis direction, alignment of this crystal axis along the grating lines is favored for geometrical reasons. This is, for instance, the case for PVDF crystallized in complete confinement in its orthorhombic α-form (Hu et al. 2005), whose fast b-axis is found by electron diffraction to align parallel to the grooves of the NIL mold (Figure 18.13a). Interestingly, when NIL molds containing curved grooves are used, the b-axis follows these curved tracks (Hu and Jonas, unpublished). This can only be possible if the resulting curved nanowires are polycrystalline. Note that graphoepitaxy in linear channels only aligns one crystal axis, leaving open the issue of the rotation of the crystal about this axis. It was found for NIL-imprinted α-PVDF that the chain c-axis is perpendicular to the substrate for films below about 100 nm (flat-on lamellae), whereas in thicker films, the crystals have their chain c-axis parallel to the substrate (edge-on lamellae) (Hu and Jonas, unpublished). In both cases, however, the b-axis remains parallel to the grooves. The variation from a flat-on to an edge-on orientation upon increasing film thickness is frequent with polymers and was tentatively rationalized recently (Wang et al. 2008).

For polymers imprinted in a *liquid crystalline phase*, then cooled into their crystalline phase, the basic structural element to consider differs depending on the detailed microstructure of the liquid crystalline phase. Poly(9,9-dioctylfluorene) (F8) was imprinted in its nematic liquid crystalline phase, in which chain axes tend to orient parallel to a director with no other form of long-range ordering (Hu et al. 2007). Thus, the basic structural element

is simply a *rod-like chain*, which for entropic and geometrical reasons aligns parallel to the grooves. After cooling and crystallization, the preferential ordering is maintained, and the chain axis (which is the c-axis in the low-temperature crystal phase) remains parallel to the grooves, hence to the axis of the resulting nanowire (Figure 18.13b). The preferential alignment translates into polarized light emission, with an R factor of 0.65–0.7 testifying for a high degree of ordering. This is in stark contrast with the absence of preferential ordering reported when F8 was imprinted in partial confinement (Song et al. 2008), and shows the importance of full confinement for NIL-induced graphoepitaxy.

The semiconducting poly(3,3‴-didodecyl-quaterthiophene) (PQT) was also imprinted in its liquid crystalline phase under full confinement (Hu et al. 2007). PQT chains form in the liquid crystalline phase supramolecular rod-like nanostructures, in which extended chains pack with their π-stacking direction along the rod axis. The basic structural element is thus the *rod-like nanostructure*, whose long axis aligns parallel to the grooves of the mold. As a consequence, the π-stacking direction is parallel to the axis of the imprinted nanowire, and this orientation is kept when the polymer crystallizes on cooling. Therefore, the b-axis of the crystal, which is the π-stacking direction, is similarly aligned at the end of the process (Figure 18.13c), whereas the chain c-axis direction is perpendicular to the nanowire axis. The situation is therefore entirely different from F8, where the chain axis was aligned parallel to the nanowire axis. In the present case, this specific crystal setting is interesting for applications, because the PQT b-axis is the axis of high carrier mobility. A nanowire-based field-effect transistor was thus realized by nanoimprint, and it was demonstrated that the hole mobility is increased by a factor of 1.7 along the nanowire axis compared with an isotropic PQT film. Essentially identical results were reported later for nanoimprinted poly (3-hexylthiophene) (P3HT), which adopts a similar morphology as PQT (Aryal et al. 2009).

For rigid conjugated *amorphous* polymers, which cannot crystallize and do not adopt a liquid crystalline phase, the basic structural element to consider during imprinting is simply their *Kuhn segment*. The flexibility of a polymer is quantified by its persistence length L_p (Grosberg and Khokhlov 1994), which is the average distance over which the local orientation of the chain persists. Real chains can be modeled as equivalent freely jointed chains having rigid (Kuhn) segments of length $b_K = 2L_p$ connected by freely rotating junctions (Rubinstein and Colby 2003). Conjugated polymers have large persistence lengths, and therefore long Kuhn segments. These will tend to align parallel to the groove direction during imprinting. The resulting nanowire thus contains chains whose axes are parallel to the nanowire axis, which is favorable for conductivity. For polypyrrole (PPy), for instance, the conductivity was improved by a factor of 1.7 in the imprinted nanowires compared with the isotropic film (Hu et al. 2007), and this was correlated to the preferential chain alignment observed by birefringence measurements (Figure 18.13d).

As a final example of graphoepitaxial alignment by NIL in full confinement, the supramolecular assembly of a phase-separated

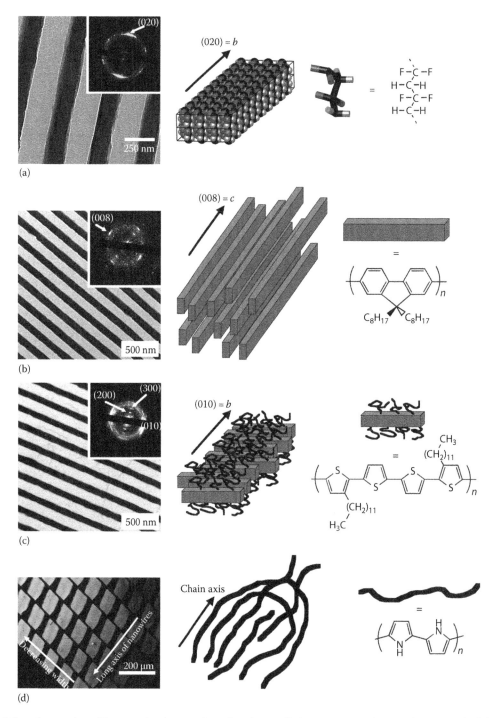

FIGURE 18.13 Selected examples of functional polymers aligned preferentially by NIL in complete confinement. The left column contains transmission electron microscopy images of (a) nanowires of PVDF, (b) electroluminescent F8, (c) semiconducting PQT, and (d) a polarized microscopy image of birefringent arrays of nanowires of conducting PPy (the nanowire axes are at 45° with respect to the polarizer and analyzer directions). Insets in (a)–(c) are the corresponding electron diffraction, showing the crystallographic axis aligned parallel to the line grating direction of the mold. The right column presents schematic drawings of the basic structural elements aligned during imprinting (a nascent crystal for semicrystalline PVDF, rod-like chains of the nematic liquid crystalline phase of F8, a supramolecular rod of π-stacked chains in the liquid crystalline phase of PQT, and chain segments of the amorphous PPy). (Adapted from Hu, Z. et al., *Nano Lett.*, 5, 1738, 2005; Hu, Z. et al., *Nano Lett.*, 7, 3639, 2007.)

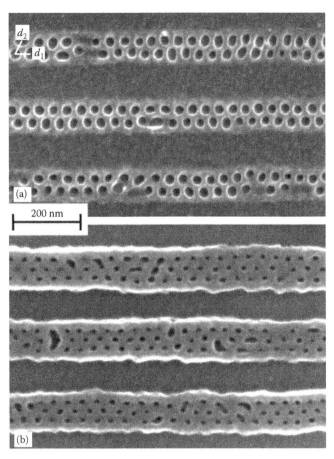

FIGURE 18.14 SEM images of a (PS-*b*-PMMA) diblock copolymer imprinted in full confinement in its hexagonal mesophase. The PMMA cylinders were etched away after mold removal, showing the graphoepitaxial alignment of the mesophase. (Reprinted from Li, H.-W. and Huck, W.T.S., *Nano Lett.*, 4, 1633, 2004. With permission.)

block copolymer is presented (Li and Huck 2004). The selected poly(styrene)-*block*-poly(methyl methacrylate) (PS-*b*-PMMA) copolymer consists of flexible amorphous blocks, which do not align by themselves. However, due to the microphase separation of the two incompatible blocks, the PMMA regions form nanocylinders located on the nodes of a hexagonal lattice, within a continuous PS matrix (Hamley 1998). Such microphase-separated copolymers are known to form graphoepitaxially oriented supramolecular crystals (Cheng et al. 2006, Bita et al. 2008). They should thus also align in NIL. Here, the proper structural element is the unit cell of the supramolecular crystal, whose unit cell parameters are in the 10 nm range and above. The axes of the block copolymer unit cell were indeed found to be parallel to the direction of the grooves of the mold (Figure 18.14), although which axis is parallel to the grooves depends on the thickness of the starting block copolymer film (Li and Huck 2004). However, the regularity of the packing was limited, most probably because the dimensions of the grooves were not exactly adapted to the natural period of the block copolymer (Li and Huck 2004). Nevertheless, this example illustrates nicely how one can combine two lithographic techniques together, namely, NIL and block copolymer lithography, since the PMMA can be etched away with UV radiation after imprinting, giving rise to a second level of patterning.

So far, all examples given above referred to line gratings as molds. Other feature shapes can be selected, but the success or failure of graphoepitaxy will depend on the match between the intrinsic size of the basic structural element and the shape of the nanofeatures of the mold (Givargizov 1994). When square cavities are selected instead of grooves, it is obvious that one cannot expect anymore preferential alignment in the plane of the film. However, this does not preclude ordering effects from appearing. This was reported for a statistical copolymer poly(vinylidene fluoride-*stat*-trifluoroethylene) [P(VDF-TrFE)], which was imprinted in its paraelectric liquid crystalline phase, then crystallized by cooling into the ferroelectric pseudohexagonal β-phase (Hu et al. 2009). The mold consisted of square nanocavities about 100 nm in lateral size, allowing to shape the polymer film into a dense array of nanosquares (Figure 18.15a). As expected, there is no preferential orientation in the plane of the film after imprint; however, the polar *b*-axis is aligned almost vertically, and the crystalline perfection is improved. This results in a much lower voltage than usually required to switch the direction of the electrical dipole moment. Figure 18.15b is a map of local polarization acquired by piezoresponse force microscopy (PFM), after writing the word "FeRAM" with a positive or negative bias (5 V) in the array of P(VDF-TrFE) imprinted nanostructures. Interestingly,

(a)

(b)

FIGURE 18.15 Array of ferroelectric P(VDF-TrFE) nanostructures nanoimprinted in complete confinement. The orientation of the polar *b*-axis is close to vertical. (a) AFM topography. (b) Piezoresponse amplitude of the same region after local poling with a positive (bright letters) or negative (dark letters) 5 V voltage. The piezoresponse amplitude is proportional to the local electric dipole moment, and was measured by PFM. (Adapted from Hu, Z. et al., *Nat. Mater.*, 8, 62, 2009.)

when nanoimprinting is performed with micro- instead of nanometer-sized cavities, no preferential orientation of the *b*-axis is noted, and the performance of the device is not improved compared with the macroscopic film (Zhang et al. 2007). This underlines the importance of confining crystallization to dimensions comparable to the intrinsic crystal size for optimal results. This example also illustrates that the combination of shaping and crystalline improvement afforded by NIL is extremely useful for device development. Indeed, high-density arrays of ferroelectric polymer crystals exhibiting a low switching voltage are attractive for the development of cheap organic random-access memories. Such systems are considered to be superior to electrically erasable and programmable read-only memories and to Flash memories in terms of write-access time and power consumption (Sheikholeslami and Gulak 2000).

18.5 Conclusions and Perspectives

NIL is by now a well-established lithographic technique, expected to play a significant role in the development of semiconductor technology. Attractive features are the parallelism afforded by the

methodology, its potential low-cost, its high resolution, its versatility toward a large range of materials, the possibility to mix it with photoresist technology, and its compatibility with integration technology. However, the potential of NIL as a method to control the shape and internal order of functional materials is only being realized currently. NIL was shown in this chapter to be able to shape functional materials into gratings, arrays, nanowires, and other interesting nanoobjects for applications in optics or microelectronics. Far from being limited to polymeric materials, NIL is further opening its application range to biomacromolecules, sol–gel precursors of inorganic ceramics and glasses, and even hard crystalline solids such as silicon. When applied to materials capable to self-organize or crystallize, NIL is also capable to orient graphoepitaxially the basic structural elements of the material, provided it be performed under full confinement. In much rarer cases such as for liquid crystals of low molar mass, partial confinement may also lead to preferential orientation. The control over the internal structure of nanomolded materials translates into improved device performance. Thus, it was shown that NIL-induced graphoepitaxy results in polarized light emission for electroluminescent polymers, increased mobility of charge carriers in semiconducting polymers, increased conductivity for conjugated conducting polymers, or easier switching of electrical dipole moments for ferroelectric materials. It was also demonstrated in this chapter that the principles governing graphoepitaxial orientation can be rationalized based on the knowledge of the structure of the material and on the notion of basic structural element. This should help device designers to predict the effects of NIL on the microstructure of materials, and therefore to use fully the power of this promising nanoprocessing method.

Acknowledgments

The authors acknowledge the generous financial support from the Wallonia Region (Nanolitho, Nanosens, and Nanotic projects), the French Community of Belgium (ARC Nanorg and Dynanomove), the Belgian Federal Science Policy (IUAP SC² and FS²), the Belgian National Fund for Scientific Research, the Solvay company through the Fondation Louvain, and the European Commission (STREP Metamos, NoE FAME), which permitted part of the research mentioned here to be performed. Access to the WinFab clean rooms is also gratefully acknowledged, as well as stimulating discussions and collaborations with colleagues and students in UCLouvain and elsewhere, too numerous to be all cited here.

References

Alcoutlabi, M. and McKenna, G. B. 2005. Effects of confinement on material behaviour at the nanometre size scale. *J. Phys. Condens. Matter* 17: R461–R524.

Aryal, M., Buyukserin, F., Mielczarek, K., Zhao, X.-M., Gao, J., Zhakidov, A., and Hu, W. W. 2008. Imprinted larger-scale high density polymer nanopillars for organic solar cells. *J. Vac. Sci. Technol. B* 26: 2562–2566.

Aryal, M., Trivedi, K., Hu, W. W. 2009. Nano-confinement induced chain alignment in ordered P3HT nonstructures defined by nanoimprint lithography. *ACS Nano* 3: 3085–3090.

Bassett, D. C. 1984. Electron microscopy and spherulitic organization in polymers. *CRC Crit. Rev. Solid State Mater. Sci.* 12: 97–163.

Bergmann, L. and Schaefer, C. 1999. *Optics of Waves and Particles*. Berlin, Germany: Walter de Gruyter.

Biebyuck, H. A., Larsen, N. B., Delamarche, E., and Michel, B. 1997. Lithography beyond light: Microcontact printing with monolayer resists. *IBM J. Res. Dev.* 41: 159–170.

Bita, I., Yang, J. K. W., Jung, Y. S., Ross, C. A., Thomas, E. L., and Berggren, K. K. 2008. Graphoepitaxy of self-assembled block copolymers on two-dimensional periodic patterned templates. *Science* 321: 939–943.

Born, M. and Wolf, E. 1980. *Principles of Optics*, 6th edition. Oxford, NY: Pergamon Press.

Campbell, S. A. 2001. *The Science and Engineering of Microelectronic Fabrication*. New York: Oxford University Press.

Chao, W. L., Harteneck, B. D., Liddle, J. A., Anderson, E. H., and Attwood, D. T. 2005. Soft X-ray microscopy at a spatial resolution better than 15 nm. *Nature* 435: 1210–1213.

Cheng, J. Y., Zhang, F., Smith, H. I., Vancso, G. J., and Ross, C. A. 2006. Pattern registration between spherical block-copolymer domains and topographical templates. *Adv. Mater.* 18: 597–601.

Cheyns, D., Vasseur, K., Rolin, C., Genoe, J., Poortmans, J., and Heremans, P. 2008. Nanoimprinted semiconducting polymer films with 50 nm features and their application to organic heterojunction solar cells. *Nanotechnology* 19: 424016.

Chou, S. Y., Krauss, P. R., and Renstrom, P. J. 1995. Imprint of sub-25 nm vias and trenches in polymers. *Appl. Phys. Lett.* 67: 3114–3116.

Chou, S. Y., Krauss, P. R., and Renstrom, P. J. 1996. Imprint lithography with 25 nm resolution. *Science* 252: 85–87.

Chou, S. Y., Krauss, P. R., Zhang, W., Guo, L. J., and Zhuang, L. 1997. Sub-10 nm imprint lithography and applications. *J. Vac. Sci. Technol. B* 15: 2897–2904.

Chou, S. Y., Keimel, C., and Gu, J. 2002. Ultrafast and direct imprint of nanostructures in silicon. *Nature* 417: 835–837.

Cormia, R. L., Price, F. P., and Turnbull, D. 1962. Kinetics of crystal nucleation in polyethylene. *J. Chem. Phys.* 37: 1333–1340.

Eder, G., Janeschitz-Kriegel, H., and Krobath, G. 1989. Shear induced crystallization, a relaxation phenomenon in polymer melts. *Prog. Coll. Polym. Sci.* 80: 1–7.

Ellison, C. J. and Torkelson, J. M. 2003. The distribution of glass-transition temperatures in nanoscopically confined glass formers. *Nat. Mater.* 2: 695–700.

Gates, B. D., Xu, Q. B., Stewart, M., Ryan, D., Wilson, C. G., and Whitesides, G. M. 2005. New approaches to nanofabrication: Molding, printing, and other techniques. *Chem. Rev.* 105: 1171–1196.

Gautam, S., Balijepalli, S., and Rutledge, G. C. 2000. Molecular simulations of the interlamellar phase in polymers: Effect of chain tilt. *Macromolecules* 33: 9136–9145.

Gedde, U. W. 1995. *Polymer Physics*. London, U.K.: Chapman & Hall.

Givargizov, E. I. 1994. Artificial epitaxy (Graphoepitaxy). In *Handbook of Crystal Growth*, Vol. 3, D. T. J. Hurle (Ed.), pp. 941–995. Amsterdam, the Netherlands: Elsevier.

Givargizov, E. I. 2008. Graphoepitaxy as an approach to oriented crystallization on amorphous substrates. *J. Cryst. Growth* 310: 1686–1690.

Grosberg, A. Y. and Khokhlov, A. R. 1994. *Statistical Physics of Macromolecules*. New York: American Institute of Physics.

Guo, L. J. 2004. Recent progress in nanoimprint technology and its applications. *J. Phys. D Appl. Phys.* 37: R123–R141.

Guo, L. J. 2007. Nanoimprint lithography: Methods and material requirements. *Adv. Mater.* 19: 495–513.

Guo, L. J., Cheng, X., and Chao, C. Y. 2002. Fabrication of photonic nanostructures in nonlinear optical polymers. *J. Mod. Opt.* 49: 663–673.

Hamley, I. W. 1998. *The Physics of Block Copolymers*. Oxford, NY: Oxford University Press.

Harnagea, C., Alexe, M., Schilling, J., Choi, J., Wehrspohn, R. B., Hesse, D., and Gosele, U. 2003. Mesoscopic ferroelectric cell arrays prepared by imprint lithography. *Appl. Phys. Lett.* 83: 1827–1829.

Heyderman, L. J., Schift, H., David, C., Gobrecht, J., and Schweizer, T. 2000. Flow behavior of thin polymer films used for hot embossing lithography. *Microelectron. Eng.* 54: 229–245.

Hu, Z., Baralia, G., Bayot, V., Gohy, J.-F., and Jonas, A. M. 2005. Nanoscale control of polymer crystallization by nanoimprint lithography. *Nano Lett.* 5: 1738–1743.

Hu, Z., Muls, B., Gence, L., Serban, D. A., Hofkens, J., Melinte, S., Nysten, B., Demoustier-Champagne, S., and Jonas, A. M. 2007. High-throughput fabrication of organic nanowire devices with preferential internal alignment and improved performance. *Nano Lett.* 7: 3639–3644.

Hu, Z., Tian, M., Nysten, B., and Jonas, A. M. 2009. Regular arrays of highly ordered ferroelectric polymer nanostructures for non-volatile low-voltage memories. *Nat. Mater.* 8: 62–67.

ITRS (International Technology Roadmap for Semiconductors), 2007 edition (http://www.itrs.net).

Jung, G.-Y., Li, Z., Wu, W., Chen, Y., Olynick, D. L., Wang, S.-Y., Tong, W. M., and Williams, R. S. 2005. Vapor-phase self-assembled monolayer for improved mold release in nanoimprint lithography. *Langmuir* 21: 1158–1161.

Kalinin, S. and Gruverman, A. 2007. *Scanning Probe Microscopy: Electrical and Electromechanical Phenomena at the Nanoscale*. New York: Springer.

Kane, R. S., Takayama, S., Ostuni, E., Ingber, D. E., and Whitesides, G. M. 1999. Patterning proteins and cells using soft lithography. *Biomaterials* 20: 2363–2376.

Khang, D.-Y., Yoon, H., and Lee, H. H. 2001. Room-temperature imprint lithography. *Adv. Mater.* 13: 749–752.

Kim, M.-S., Kim, J.-S., Cho, J. C., Shtein, M., Guo, L. J., and Kim, J. 2007. Flexible conjugated polymer photovoltaic cells with controlled heterojunctions fabricated using nanoimprint lithography. *Appl. Phys. Lett.* 90: 123113.

Kimita, S., Sakurai, T., Nozue, Y., Kasahara, T., Yamaguchi, N., Karino, T., Shibayama, M., and Kornfiled, J. A. 2007. Molecular basis of the shish-kebab morphology in polymer crystallization. *Science* 316: 1014–1017.

Klar, T. A., Jakobs, S., Dyba, M., Egner, A., and Hell, S. W. 2000. Fluorescence microscopy with diffraction resolution barrier broken by stimulated emission. *Proc. Natl. Acad. Sci. U.S.A.* 97: 8206–8210.

Ko, S. H., Park, I., Pan, H., Grigoropoulos, C. P., Pisano, A. P., Luscombe, C. K., and Fréchet, J. M. J. 2007. Direct Nanoimprinting of metal nanoparticles for nanoscale electronics fabrication. *Nano Lett.* 7: 1869–1877.

Kumar, A. and Whitesides, G. M. 1993. Features of gold having micrometer to centimeter dimensions can be formed through a combination of stamping with an elastomeric stamp and an alkanethiol ink followed by chemical etching. *Appl. Phys. Lett.* 63: 2002–2004.

Kumaraswamy, G., Issaian, A. M., and Kornfield, J. A. 1999. Shear-enhanced crystallization in isotactic polypropylene. 1. Correspondence between in situ rheo-optics and ex situ structure determination. *Macromolecules* 32: 7537–7547.

Leung, O. M. and Goh, M. C. 1992. Orientational ordering of polymers by atomic force microscope tip-surface interaction. *Science* 255: 64–66.

Li, H.-W. and Huck, W. T. S. 2004. Ordered block-copolymer assembly using nanoimprint lithography. *Nano Lett.* 4: 1633–1636.

Loo, Y. L., Register, R. A., and Ryan, A. J. 2002. Modes of crystallization in block copolymer microdomains: Breakout, templated, and confined. *Macromolecules* 35: 2365–2374.

Mahalingam, V., Onclin, S., Peter, M., Ravoo, B. J., Huskens, J., and Reinhoudt, D. N. 2004. Directed self-assembly of functionalized silica nanoparticles on molecular printboards through multivalent supramolecular interactions. *Langmuir* 20: 11756–11762.

Maier, G. 2001. Low dielectric constant polymers for microelectronics. *Prog. Polym. Sci.* 26: 3–65.

Mandelkern, L. 1979. Relation between properties and molecular morphology of semicrystalline polymers. *Faraday Discuss.* 68: 310–319.

Massa, M. V. and Dalnoki-Veress, K. 2004. Homogeneous crystallization of poly(ethylene oxide) confined to droplets: The dependence of the crystal nucleation rate on length scale and temperature. *Phys. Rev. Lett.* 92: 255509.

Mele, E., Di Benedetto, F., Persano, L., Cingolani, R., and Pisignano, D. 2005. Multilevel, room temperature nanoimprint lithography for conjugated polymer-based photonics. *Nano Lett.* 5: 1915–1919.

Mele, E., Camposeo, A., Stabile, R., Del Carro, P., Di Benedetto, F., Persano, L., Cingolani, R., and Pisignano, D. 2006. Polymeric distributed feedback lasers by room-temperature nanoimprint lithography. *Appl. Phys. Lett.* 89: 131109.

Miller, R. L. (Ed.). 1979. *Flow-Induced Crystallization in Polymer Systems*. New York: Gordon & Breach.

Mouthuy, P.-O., Melinte, S., Geerts, Y. H., and Jonas, A. M. 2007. Uniaxial alignment of nanoconfined columnar mesophases. *Nano Lett.* 7: 2627–2632.

Muller, A. J., Balsamo, V., Arnal, M. L., Jakob, T., Schmalz, H., and Abetz, V. 2002. Homogeneous nucleation and fractionated crystallization in block copolymers. *Macromolecules* 35: 3048–3058.

Odom, T. W., Love, J. C., Wolfe, D. B., Paul, K. E., and Whitesides, G. M. 2002. Improved pattern transfer in soft lithography using composite stamps. *Langmuir* 18: 5314–5320.

Ohtake, T., Nakamatsu, K., Matsui, S., Tabata, H., and Kawai, T. 2004. DNA nanopatterning with self-organization by using nanoimprint. *J. Vac. Sci. Technol. B* 22: 3275–3278.

Okerberg, B. C., Soles, C. L., Douglas, J. F., Ro, H. W., Karim, A., and Hines, D. R. 2007. Crystallization of poly(ethylene oxide) patterned by nanoimprint lithography. *Macromolecules* 40: 2968–2970.

Okinaka, M., Tsukagoshi, K., and Aoyagi, Y. 2006. Direct nanoimprint of inorganic-organic hybrid glass. *J. Vac. Sci. Technol. B* 24: 1402–1404.

Owen G. 1985. Electron lithography for the fabrication of microelectronic devices. *Rep. Prog. Phys.* 48: 795–851.

Park, I., Cheng, J., Pusano, A. P., Lee, E.-S., and Jeong, J.-H. 2007. Low temperature, low pressure nanoimprinting of chitosan as a biomaterial for bionanotechnology applications. *Appl. Phys. Lett.* 90: 093902.

Park, I., Ko, S. H., Pan, H., Grigoropoulos, C. P., Pisano, A. P., Fréchet, J. M. J., Lee, E.-S., and Jeong, J.-H. 2008. Nanoscale patterning and electronics on flexible substrate by direct nanoimprinting of metallic nanoparticles. *Adv. Mater.* 20: 489–496.

Peroz, C., Heitz, C., Barthel, E., Sondergard, E., and Goletto, V. 2007. Glass nanostructures fabricated by soft thermal nanoimprint. *J. Vac. Sci. Technol. B* 25: L27–L30.

Pisignano, D., Persano, L., Visconti, P., Cingolani, R., Gigli, G., Barbarella, G., and Favaretto, L. 2003. Oligomer-based organic distributed feedback lasers by room-temperature nanoimprint lithography. *Appl. Phys. Lett.* 83: 2545–2547.

Pisignano, D., Persano, L., Raganato M. F., Visconti, P., Cingolani, R., Barbarella, G., Favaretto, L., and Gigli, G. 2004. Room temperature nanoimprint lithography of non-thermoplastic organic film. *Adv. Mater.* 16: 525–529.

Rai-Choudhury, P. 1997. *Handbook of Microlithography, Micromachining, and Microfabrication: Vol. 1: Microlithography*. Bellingham, WA: SPIE Optical Engineering Press.

Reiter, G. and Strobl, G. R. (Eds.). 2006. *Progress in Understanding of Polymer Crystallization*. Lecture Notes in Physics, Vol. 714. Berlin, Germany: Springer.

Reiter, G., Castelein, G., Sommer, J. U., Rottele, A., and Thurn-Albrecht, T. 2001. Direct visualization of random crystallization and melting in arrays of nanometer-size polymer crystals. *Phys. Rev. Lett.* 87: 226101.

Resnick, D. J. 2007. Imprint lithography. In *Microlithography Science and Technology*, K. Suzuki and B. W. Smith (Eds.), pp. 465–499. Boca Raton, FL: Taylor & Francis.

Resnick, D. J., Dauksher, W. J., Mancini, D., Nordquist, K. J., Bailey, T. C., Johnson, S., Stacey, N., Ekerdt, J. G., Willson, C. G., Sreenivasan, S. V., and Schumaker, N. 2003. Imprint lithography for integrated circuit fabrication. *J. Vac. Sci. Technol. B* 21: 2624–2631.

Ro, H. W., Jones, R. L., Peng, H., Hines, D. R., Lee, H.-J., Lin, E. K., Karim, A., Yoon, D. Y., Gidley, D. W., and Soles, C. L. 2007. The direct patterning of nanoporous interlayer dielectric insulator films by nanoimprint lithography. *Adv. Mater.* 19: 2919–2924.

Ro, H. W., Peng, H., Niihara, K., Lee, H.-J., Lin, E. K., Karim, A., Gidley, D. W., Jinnai, H., Yoon, D. Y., Gidley, D. W., and Soles, C. L. 2008. Self-sealing of nanoporous low dielectric constant patterns fabricated by nanoimprint lithography. *Adv. Mater.* 20: 1934–1939.

Rowland, H. D., Sun, A. C., Schunk, P. R., and King, W. P. 2005. Impact of polymer film thickness and cavity size on polymer flow during embossing: Toward process design rules for nanoimprint lithography. *J. Micromech. Microeng.* 15: 2414–2425.

Rowland, H. D., King, W. P., Pethica, J. B., and Cross, G. L. W. 2008. Molecular confinement accelerates deformation of entangled polymers during squeeze flow. *Science* 322: 720–724.

Rubinstein, M. and Colby, R. H. 2003. *Polymer Physics*. Oxford, NY: Oxford University Press.

Ruchhoeft, P., Colburn, M., Choi, B., Nounu, H., Johnson, S., Bailey, T., Damle, S., Stewart, M., Ekerdt, J., Sreenivasan, S. V., Wolfe, J. C., and Willson, C. G. 1999. Patterning curved surfaces: Template generation by ion beam proximity lithography and relief transfer by step and flash imprint lithography. *J. Vac. Sci. Technol. B* 17: 2965–2969.

Schift, H. 2008. Nanoimprint lithography: An old story in modern times? A review. *J. Vac. Sci. Technol. B* 26: 458–480.

Schift, H. and Heyderman, L. J. 2003. Nanorheology. Squeeze flow in hot embossing of thin films. In *Alternative Lithography. Unleashing the Potentials of Nanotechnology*, C. M. Sotomayor Torres (Ed.), pp. 47–76. New York: Kluwer Academic/Plenum.

Schift, H. and Kristensen, A. 2007. Nanoimprint lithography. In *Springer Handbook of Nanotechnology*, 2nd edition, B. Bushan (Ed.), pp. 239–278. Berlin, Germany: Springer.

Schift, H., Heyderman, L. J., Auf der Maur, M., and Gobrecht, J. 2001. Pattern formation in hot embossing of thin polymer films. *Nanotechnology* 12: 173–177.

Schift, H., Saxer, S., Park, S., Padeste, C., Peiles, U., and Gobrecht, J. 2005. Controlled co-evaporation of silanes for nanoimprint stamps. *Nanotechnology* 16: S171–S175.

Schmid, S. A., Yim, K. H., Chang, M. H., Zheng, Z., Huck, W. T. S., Friend, R. H., Kim, J. S., and Herz, L. M. 2008. Polarization anisotropy dynamics for thin films of a conjugated polymer aligned by nanoimprinting. *Phys. Rev. B* 77: 115338.

Schulz, H., Wissen, M., and Scheer, H.-C. 2003. Local mass transport and its effect on global pattern replication during hot embossing. *Microelectron. Eng.* 67–68: 657–663.

Sheftal, N. N. and Bouzynin, N. A. 1972. Preferred orientation of crystallites on the substrate and effect of scratches. *Vestn. Mosk. Univ. Ser. Geol.* 27: 102–104.

Sheikholeslami, A. and Gulak, P. G. 2000. A survey of circuit innovations in ferroelectric random-access memories. *Proc. IEEE* 88: 667–688.

Smith, H. I. and Flanders, D. C. 1978. Oriented crystal growth on amorphous substrates using artificial surface-relief gratings. *Appl. Phys. Lett.* 32: 349–350.

Somani, R. H., Hsiao, B. S., Nogales, A., Srinivas, S., Tsou, A. H., Sics, I., Balta-Calleja, F. J., and Ezquerra, T. A. 2000. Structure development during shear flow-induced crystallization of i-PP: In-situ small-angle x-ray scattering study. *Macromolecules* 33: 9385–9394.

Song, M. H., Wenger, B., and Friend, R. H. 2008. Tuning the wavelength of lasing emission in organic semiconducting laser by the orientation of liquid crystalline conjugated polymer. *J. Appl. Phys.* 104: 033107.

Steinhart, M., Göring, P., Dernaika, H., Prabhukaran, M., Gösele, U., Hempel, E., and Thurn-Albrecht, T. 2006. Coherent kinetic control over crystal orientation in macroscopic ensembles of polymer nanorods and nanotubes. *Phys. Rev. Lett.* 97: 027801.

Stewart, M. D. and Willson, C. G. 2005. Imprint materials for nanoscale devices. *MRS Bull.* 30: 947–951.

Studer, V., Pepin, A., and Chen, Y. 2002. Nanoembossing of thermoplastic polymers for microfluidic applications. *Appl. Phys. Lett.* 80: 3614–3616.

Suzuki, K. 2007. Electron beam lithography systems. In *Microlithography Science and Technology*, K. Suzuki and B. W. Smith (Eds.), pp. 465–499. Boca Raton, FL: Taylor & Francis.

Ting, C.-J., Huang, M.-C., Tsai, H.-Y., Chou, C.-P., and Fu, C.-C. 2008a. Low cost fabrication of the large-area anti-reflection films from polymer by nanoimprint/hot-embossing technology. *Nanotechnology* 19: 205301.

Ting, C.-J., Chang, F.-Y., Chen, C.-F., and Chou, C.-P. 2008b. Fabrication of an antireflective polymer optical film with subwavelength structures using a roll-to-roll micro-replication process. *J. Micromech. Microeng.* 18: 075001.

Toney, M. F., Russell, T. P., Logan, J. A., Kikuchi, H., Sands, J. M., and Kumar, S. K. 1995. Near-surface alignment of polymers in rubbed films. *Nature* 374: 709–711.

Torres, C. M. (Ed.). 2003. *Alternative Lithography. Unleashing the Potentials of Nanotechnology*. New York: Kluwer Academic/Plenum.

Ueno, T. 2007. X-ray lithography. In *Microlithography Science and Technology*, K. Suzuki and B. W. Smith (Eds.), pp. 361–382. Boca Raton, FL: Taylor & Francis.

Voicu, N. E., Ludwigs, S., Crossland, E. J. W., Andrew, P., and Steiner, U. 2007. Solvent-vapor-assisted imprint lithography. *Adv. Mater.* 19: 757–767.

Wang, J., Sun, X., Chen, L., and Chou, S. Y. 1999. Direct nanoimprint of submicron organic light-emitting structures. *Appl. Phys. Lett.* 75: 2767–2769.

Wang, J., Sun, X., Chen, L., Zhuang, L., and Chou, S. Y. 2000. Molecular alignment in submicron patterned polymer matrix using nanoimprint lithography. *Appl. Phys. Lett.* 77: 166–168.

Wang, Y., Chan, C. M., Ng, K. M., and Li, L. 2008. What controls the lamellar orientation at the surface of polymer films during crystallization? *Macromolecules* 41: 2548–2553.

Woo, E., Huh, J., Jeong, Y. G., and Shin, K. 2007. From homogeneous to heterogeneous nucleation of chain molecules under nanoscopic cylindrical confinement. *Phys. Rev. Lett.* 98: 136103.

Wunderlich, B. 1976. *Macromolecular Physics. Vol. 2. Crystal Nucleation, Growth, Annealing*. New York: Academic Press.

Wurm, S. and Gwyn, C. 2007. EUV lithography. In *Microlithography Science and Technology*, K. Suzuki and B. W. Smith (Eds.), pp. 383–464. Boca Raton, FL: Taylor & Francis.

Xia, Y. N. and Whitesides, G. M. 1998. Soft lithography. *Annu. Rev. Mater. Sci.* 28: 153–184.

Xia, Y. N., Rogers, J. A., Paul, K. E., and Whitesides, G. M. 1999. Unconventional methods for fabricating and patterning nanostructures. *Chem. Rev.* 99: 1823–1848.

Zhang, L., Ducharme, S., and Li, J. 2007. Microimprinting and ferroelectric properties of poly(vinylidene fluoride-trifluoroethylene) copolymer films. *Appl. Phys. Lett.* 91: 172906.

Zheng, Z., Yim, K.-H., Saifullah, M. S. M., Welland, M. E., Friend, R. H., Kim, J.-S., and Huck, W. T. S. 2007. Uniaxial alignment of liquid-crystalline conjugated polymers by nanoconfinement. *Nano Lett.* 7: 987–992.

19

Nanoelectronics Lithography

Stephen Knight
National Institute of Standards and Technology

Vivek M. Prabhu
National Institute of Standards and Technology

John H. Burnett
National Institute of Standards and Technology

James Alexander Liddle
National Institute of Standards and Technology

Christopher L. Soles
National Institute of Standards and Technology

Alain C. Diebold
University at Albany

19.1 Introduction

The modern integrated circuit (IC), comprising memory, logic processors and analog function devices, are multicomponent and multilevel nanostructures prepared by a series of patterning and pattern-transfer steps. Figure 19.1 shows the cross-sectional hierarchical structure that starts from the smallest feature, the transistor, to dielectrics and metal contacts that are each well defined and must precisely overlay the previous layer. This three-dimensional nanoelectronics structure is manufactured by a rapid patterning process called lithography.

Since the invention of the transistor in 1947 by Bell Labs and Intel's first microprocessor in 1971, modern lithography has enabled the semiconductor industry to shrink device dimensions. The early progress was first quantified by Gordon Moore in 1965 and is now known as Moore's law.[1] While progress in all process steps are required to achieve the continual shrinking of circuit elements, the advancements in lithography have been the overwhelming driving force. Figure 19.2 shows how Moore's law has guided the industry's systematic increase in the number of transistors per chip since the introduction of the IC. Indeed, now and in the future, the industry will achieve productivity improvement primarily by feature reduction.[2]

While a wide variety of lithography technologies have been developed, optical step and repeat lithography technologies are the predominant methods of printing features on the semiconductor surfaces and overlying the interconnect structures. A number of competing technologies have been explored, such as x-ray[3,4] and flood electron beam (SCALPEL),[5,6] but optical lithography has dominated through vigorous optical technology and tool developments that have left it the lowest cost solution with highest throughput. Direct-write electron-beam technology provides the highest resolution and remains the leading technique for manufacturing the masks used by optical lithography. A novel way of comparing different lithography strategies is by a plot of resolution versus throughput, also known as "Tennant's Law,"[7,8] as plotted in Figure 19.3. In this plot, direct-write approaches, such as single-atom placement by scanning tunneling microscopy (STM), atomic-force microscopy (AFM) tip-induced oxidation, and electron-beam

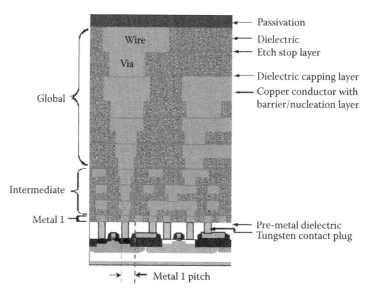

FIGURE 19.1 Cross section of an integrated circuit, showing the active semiconductor and multilayer interconnect levels. (Reproduced from *International Technology Roadmap for Semiconductors*, 2007 edition, SEMATECH, Austin, TX, Figure INTC2, 2007. With permission.)

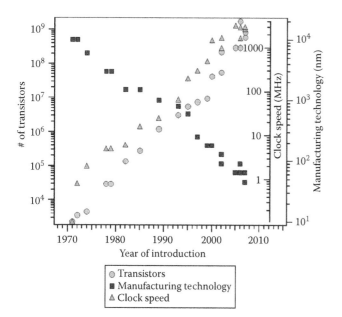

FIGURE 19.2 A composite plot of the scaling of the number of transistors, clock speed, and manufacturing technology versus the year of introduction of Intel Processors. (Data from www.intel.com/technology/timeline.pdf.)

lithography are compared with optical step and repeat lithography that replicate the features of a mask. The continued advancements in optical lithography have pushed the resolution to smaller features at higher pixel throughput, as analyzed by Brunner.[9]

The fundamental guide for optical step and repeat lithography is the Rayleigh equation, which provides the scaling criteria for predicting the smallest optically definable image[2]

$$R = k_1 \frac{\lambda}{\mathrm{NA}} = k_1 \frac{\lambda}{n \sin \theta}; \tag{19.1}$$

where

λ is the source wavelength

n is the medium refractive index

$n \sin \theta$ is the numerical aperture (NA) with the incidence angle (θ)

k_1 is a process dependent factor

While the Rayleigh equation defines the resolution, the ability of the photoresist to replicate the mask features is of critical importance. All approaches and manufacturing tools have taken advantage of the Rayleigh equation to predict transitions from each manufacturing technology node (smallest feature size) up to the diffraction limit. Optical lithography started with the near *ultra*violet (UV) g-line (436 nm) and i-lines (365 nm) of mercury-arc lamps and made way for laser sources in the deep *ultra*violet (DUV) from KrF (248 nm) and ArF (193 nm) excimers. A significant extension, immersion lithography, decreases the 193 nm wavelength by using water as the immersion fluid and is on track to produce features down to 32 nm. The next-generation photoresist materials, described in Section 19.2, for sub 22 nm features expect to image 13.4 nm radiation in extreme *ultra*violet (EUV) with all reflective lithography imaging tools.

While most of this chapter focuses on the leading-edge and next-generation technologies used for defining the finest features in nanoelectronics, it is important to recognize that for most applications it is necessary to connect the nano-world (transistor) to the macro-world (a computer motherboard). Figure 19.1 shows the cross section of an IC. Note that the interconnect sizes increase at subsequent higher levels. Therefore, the previous-generation tools continue to play crucial roles by migrating to the higher levels of interconnect.

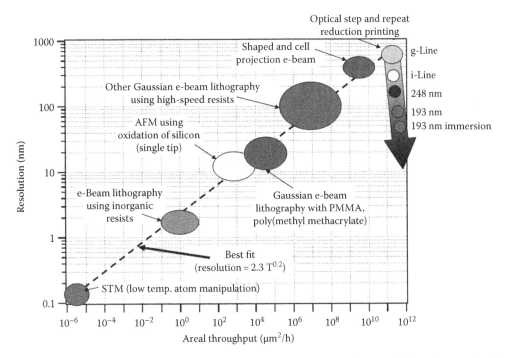

FIGURE 19.3 The progress of optical step and repeat lithography has driven to higher resolution and throughput as described by "Tennant's Law," whereby the resolution (R) versus areal throughput (T) scales as $R = 2.3\,T^{0.2}$ (Adapted from Tennant, D., Limits of conventional lithography, in Timp, G.M. (ed.), *Nanotechnology*, AIP Press, New York, Chapter 4, p. 161, 1999. With permission.)

Microelectronics and now nanoelectronics technology have become a collaborative and competitive worldwide effort, as exemplified by the *International Technology Roadmap for Semiconductors* (ITRS), available to the public at http://www.itrs.net. The ITRS is updated every two years by subject matter experts from the semiconductor manufacturing industry, the tool and materials supplier industries, the factory automation infrastructure, academia, and government agencies. An important guiding aspect of the ITRS roadmap is the identification of the status of the technology nodes and guidance of the phases of research, development, and pre-production. In particular, the 2007 edition identifies potential lithographic solutions out to 2022, with a predicted dynamic random access memory half pitch of 11 nm and FLASH memory half pitch of 9 nm as shown in Figure 19.4.

In this roadmap, several emerging lithography approaches are highlighted. Double patterning with 193 nm DUV water immersion is expected to extend to the 32 nm half pitch era with high-volume production in 2013. Alternatively, contending technologies for 22 nm half pitch are EUV Lithography, 193 nm DUV immersion with higher index fluid and lens materials, maskless lithography (ML2), and nanoimprint lithography (NIL).[10,11] NIL is rapidly emerging as a low-cost, high-resolution, and versatile alternative to optical lithography. Below 22 nm half-pitch, the likely technology solutions are less clear. All the technologies mentioned above, along with new contenders, such as directed self assembly, have credible paths. However, each would have to surmount numerous technical difficulties while limiting exorbitant cost and loss in throughput.

In the rest of this chapter, we describe some of the challenges facing the photoresist materials (Section 19.2) used in optical lithography including some crucial aspects facing these materials as the feature dimensions are reduced to the length scale of the basic photoresist polymers. In Section 19.3, DUV lithography, the basic optics, advancements in steppers, and approaches to extend to higher-resolution, denser features are described. In Section 19.4, electron-beam lithography is covered with respect to the metrics of resolution, throughput, overlay requirements, and cost. Section 19.5 highlights a non-optical lithography approach, NIL. In this section, several variations of NIL for transferring mask features into a photoresist are described. Finally, in Section 19.6, we end with an overview of the metrology requirements for nanolithography. Figure 19.1 is a reminder that one must print and measure a wide-range of feature dimensions, from the nanoscale to the macroscale. Lastly, a separate chapter in this handbook is dedicated to EUV lithography.

19.2 Photoresist Technology

The optical image projected from a mask upon a thin film at the semiconductor wafer plane is the first step of photolithography.[10,12] The thin films that replicate the mask features are called photoresists with basic process as shown in Figure 19.5. The high sensitivity of photoresists to radiation have consistently met the challenges of smaller, high-fidelity features with increased throughput driven by memory and processor chip performance to feature size gains (Moore's law). The etch resistance of the

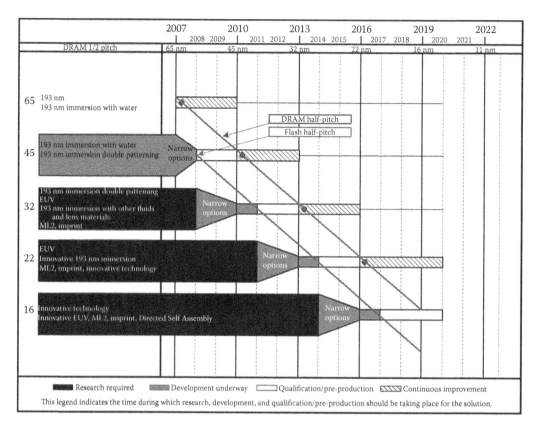

FIGURE 19.4 Roadmap for half pitch scaling. (Reproduced from *International Technology Roadmap for Semiconductors*, 2007 edition, SEMATECH, Austin, TX, Figure LITH5, 2007. With permission.)

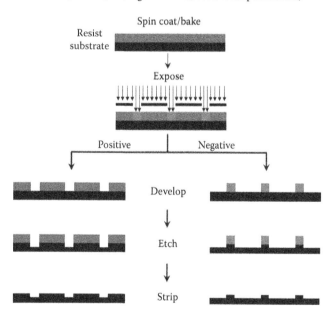

FIGURE 19.5 Schematic of the lithographic process for positive and negative tone resist.

photoresist allows pattern transfer into the underlying semiconductor wafer.

Test structures, such as a line and space pattern as shown in Figure 19.6, must meet criteria as defined by the ITRS roadmap for critical dimension (CD). The CD is the feature size,

for example, in a 32 nm half-pitch 1:1 dense line, a line with a width of 32 nm is followed by a space of equal size. The Rayleigh equation defines the resolution; however, the ability of the photoresist to perfectly replicate the mask features is of critical importance. The line-width variations called line-width roughness (LWR) and line-edge roughness (LER) must be reduced to less than 2 nm for 32 nm half pitch (HP) as they impact device performance.[13,14] Therefore, the photoresist materials chemistry plays a substantial role for both resolution and LER. Methods to extend resist resolution by double patterning methods and new photoresist architectures must be leveraged against meeting the LER requirements for sub-22 nm lithography.

19.2.1 Fundamentals

The requirements for advanced photoresists are discussed in the 2007 ITRS roadmap.[14] Specifically, the "Lithography" chapter highlights difficult challenges:

Resist materials at <32 nm indicates three issues (1) Resist and antireflective coating materials composed of alternatives to PFAS [perfluoroalkylsulfonate] compounds, (2) Limits of chemically amplified resist sensitivity for <32 nm half-pitch due to acid diffusion length, and (3) materials with improved dimensional and LWR control.

The details of the photoresist materials chemistry are at the heart of turning the optically defined patterns into three-dimensional

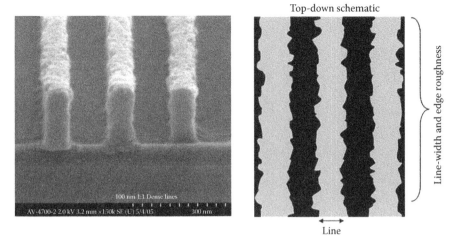

FIGURE 19.6 Example of a cross-sectional SEM image of a 100 nm 1:1 line-space lithographic line. The fluctuations in the feature critical dimension along the line are quantified by line-edge roughness and line-width roughness metrics as indicated in the schematic.

nanoscale features. Chemically amplified photoresists are formulations of an acid-sensitive polymer film mixed with photoacid generators (PAGs) and other additives, such as base quenchers. Upon exposure through a mask, strong acids are formed by photolysis of the molecularly mixed PAG within the thin film. A post-exposure bake is then applied and the acidic protons (photoacids) diffuse along with the counter-anions and catalyze a deprotection reaction on the acid-sensitive polymer to change the local solubility for development in an aqueous hydroxide solution as shown in Figure 19.7. The photoacids are true catalysts as they are regenerated by each deprotection reaction;[15] hence, the term chemical amplification refers to the cascade of reactions that occur within the photoresist induced by a single photon. Therefore, these photoresists may be used at low exposure doses.

The chemical reaction-diffusion that forms the latent image and development must be understood and controlled at the nanometer length scale. Chemically amplified photoresists are also deposited onto bottom anti-reflection coatings (BARC) to eliminate the effects of standing waves. Interactions and component transport between the BARC and resist layer can lead to loss of profile control or pattern collapse and therefore must also be understood. Detailed studies of these interactions and transport mechanisms are needed to design materials for the successful fabrication of sub-32 nm structures.

As feature sizes are reduced to below 32 nm, a general problem of simultaneously minimizing dose sensitivity, CD, and LER was observed experimentally and theoretically. These observations suggest only two of these metrics may be met at the sacrifice of

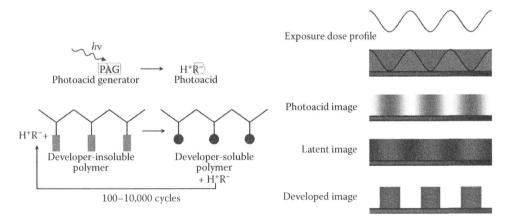

FIGURE 19.7 Schematic of the photolysis process of a photoacid generator that forms a strong acid (H^+R^-) and the subsequent acid catalytic reaction that changes the solubility of the polymer in a developer solution. Schematic of the mask-defined dose profile, subsequent profile in the photoresist film that leads to a photoacid image and chemical latent image formed during the post-exposure bake. The final developed image is formed after selective dissolution in a developer solution as determined by the extent of chemical reaction in the nominally exposed regions.

the third. Several theoretical models[16–18] have been proposed, such as those by Bristol (2007)

$$\text{LER} \approx \frac{1}{\sqrt{q\alpha D_e}}\sqrt{1 + \frac{1}{\varepsilon}\frac{1}{\text{LILS}\cdot\delta}}, \qquad (19.2)$$

where

    q is the number of photons/nm^2
    D_e is the dose at the line-edge (threshold for development)
    α is the fraction of light absorbed
    ε is the quantum efficiency defined by the (molecules of acids produced)/(number of absorbed photons)
    δ is the effective photoacid diffusion length
    LILS is the latent-image log slope

As shown in the schematic of Figure 19.7, the initial photoacid image is followed by a post-exposure bake step whereby the photoacid diffuses and reacts. The extent of the reaction will define a chemical latent image in the photoresist. The local slope of the chemical transformation, or latent-image log slope (LILS), at the point of development is critical to maximize. Qualitatively, a large exposure dose leads to a lower LER at fixed LILS and δ. However, notice that once smaller features are desired ($\delta <$ CD), then the photoacid diffusion length should be reduced; hence at fixed D_e, a smaller δ may increase LER. The full problem is a nonlinear reaction-diffusion equation and typically requires additional parameterization to include the effects of photoacid loss, photoacid trapping, and amine quenchers that react and diffuse. However, approaches such as Equation 19.2 provide qualitative insights into the resist problem. In cases when the material specifics were not known, the form of material constant = (half-pitch)^3 × (LER)^2 × (Sensitivity) has been used.[19]

Such resolution limits are linked to the fundamentals of photoacid generation (via D_e, α, ε) and chemical reaction-diffusion during the post-exposure bake that defines the LILS, choice of photoresist chemistry, PAG size, and processing time and the temperature that ultimately determines the photoacid diffusion length and development step that resolves the final feature. Experimental methods have been developed to quantitatively measure many of these parameters. More recently, parameters such as the photoacid diffusion length were estimated by modeling of the lithographic feature power spectrum.[16]

The efficient generation of photoacids is also crucial to photoresist technology. Analytical methods have been developed to measure photoacid generation and quantum efficiency.[20–23] Many of these techniques rely on a well-defined dose upon photoresist thin films followed by analytical approaches to determine the acid concentration via titration with a standard base. More recently, quantifying the quantum efficiency and acid generation mechanisms in EUV photoresists has taken a more important role as increasing the acid concentration or efficiency (Equation 19.2) is a method for reducing LER. For 248 and 193 nm lithography, photoresist polymers were designed for low absorbance, while the PAG absorbs strongly for photolysis. Since organic photoresist polymers absorb strongly in the EUV, the photolysis induced by the direct exposure of the 92 eV photon and secondary electrons are equally important. In fact, the ε for EUV exposure is greater than 1 due to the non-negligible contribution from lower energy secondary electrons. In both DUV and EUV lithography, the photoacid image plays the important role of changing the local solubility of the photoresist polymer in the nominally exposed regions (Figure 19.7).

During the post-exposure bake, the latent image (Figure 19.7) formed is a composition profile of the photoresist within the line:space feature. The true shape of this profile is due to the photoacid catalyzed reaction-diffusion process that cleaves (deprotection) a nonpolar side-group of the photoresist polymer. The influence of the optical image quality on the final developed feature may be probed by varying the aerial image contrast. Hinsberg et al. elegantly showed that the initial photoacid distribution, controlled through the exposure quality by interferometric lithography, significantly affects the printed feature quality.[24] Furthermore, using an image-fading technique, Pawloski et al. identified an apparent resolution limit in the final feature of 193 nm resists as quantified by LER versus the image-log slope (ILS).[25] These changes in the feature quality are also controlled by the reaction-diffusion of the photoacid[12] into the unexposed regions that leads to image spreading or blurring.[26] However, even with an ideal step-exposure condition, mimicked by forming a bilayer stack, the reaction-diffusion process induces an image blur. Controlling the photoacid reaction-diffusion (via LILS and δ) remains an important strategy for improving feature quality.

The transport properties of the photoacid are also influenced by the changes in the local chemical composition that occur during the deprotection reaction.[27] Houle et al. demonstrated that the evolving resist polymer chemistry plays a crucial role in lithographic imaging.[26] An increased photoacid size decreased the apparent diffusion length,[28,29] but also quantitatively increased the LILS.[30,31] However, the catalytic efficiency of the photoacid proton can dominate, such that by increasing the size of the photoacid counter-anion, image blur occurs primarily due to the local proton mobility, not diffusion of the acid-counter-anion pair.[32] Neutron reflectivity methods were developed to characterize the shape of the reaction-diffusion front with nanometer resolution.[14,30,31,33–37] In these studies, the influence of the evolving chemical composition on the spatial-extent of the reaction profile was directly measured.

Base quencher additives are also used in photoresist formulations to limit the reaction and diffusion of the photoacid catalyst into unexposed regions.[12,38] The influence of these neutralizing species on the reaction-diffusion process is complex.[39] The simplistic view of the quenchers solely acting to neutralize photoacid, thus decreasing the acid concentration, is not always correct.[40] The quencher appears to partially neutralize the photoacid less than stoichiometrically, influence the dissolution either as promoters or inhibitors, and increase the development induction time.[39] The partial neutralization initially reduces the amount of acid available after photolysis proportional to the base concentration.[40–42] The influence of quenchers on the chemical gradients in the film were suspected to control LER.[43] Direct measurements of the effect of amine quenchers on the reaction-diffusion profile shape (LILS) and spatial extent (δ) were determined by neutron reflectivity.[35]

A central assumption in these resolution limit models is a direct transfer of the chemical deprotection reaction-diffusion heterogeneity on the feature quality; the details of the development process are not considered. However, the bulk of photoresist thin films dissolve via reactive dissolution kinetics involving a well-defined steady-state swollen layer.[44–46] This swollen layer must approach the nominally unreacted and unexposed zone as bulk development ceases.[47] This crucial transition zone results from the initial deprotection latent image that can be controlled by aerial images, polymer chemistry, photoacid generators and base additives, and post-exposure bake conditions.[30,35] The mechanism of how the advancing swelling dissolution front faces the transition of soluble to insoluble species (solubility switch) is crucial for the understanding of resist resolution limits. This was addressed by neutron reflectivity techniques that directly measured the developer penetration and extent of line-edge swelling.[48–50] The residual swelling fraction at the feature edge remains diffuse over length scales (>10 nm) far exceeding the polymer chain dimensions during hydroxide development and water rinse step; the swelling layer eventually collapses upon drying.

Alternative development approaches to control this residual swelling fraction may be needed to smooth and reduce LER.

19.2.2 Advances by Material Structure

In a typical polymer photoresist, the PAG is dissolved along with the photoresist polymer and spin cast to form a thin film mixture, or binary blend. Polymers provide flexible platforms to change functional groups in order to meet etch resistance, optical transparency, refractive index requirements, and a variety of acid-sensitive protecting groups. The high glass transition temperatures (typically, $T_g > 140°C$) provide dimensional stability and wide latitude in post-exposure bake temperature that increases the rates of reaction and photoacid diffusivity. Lastly, polymers have a large degree of lipophilicity in an aqueous base developer that provides a high-development contrast.[12]

Driven by the CD requirements, it was considered that reducing the photoacid diffusion length would enable smaller CD. The PAG and polymer blend approach may not be the most effective route, since the photoacid could diffuse to lengths longer than the CD. In order to address this viewpoint, an alternative resist structure was devised that covalently bonds the photoacid generator to the polymer as shown in Figure 19.8a.[51–54] With this approach, after exposure, the photoacid counter-anion remains covalently bound to the polymer thereby restricting the acidic

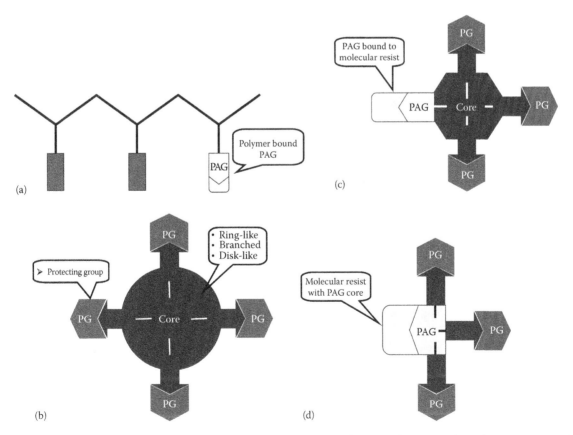

FIGURE 19.8 Cartoons of photoresist architecture alternative to polymer and PAG blends: (a) Photoresist polymer bound to the PAG, (b) molecular glass resist with variable core structure, (c) molecular glass resist bound to the PAG, and (d) PAG-core molecular resist.

proton diffusion length. Such approaches remain promising especially to increase the dose sensitivity.

Since the CD and LER metrics are approaching the characteristic dimensions of the photoresist polymers, alternative architectures were considered to extend photolithography by using lower molar mass molecules.[55] These molecular glass (MG) resists, while smaller, may also improve the uniformity of blends with PAG and other additives since miscibility of polymer blends decreases with increasing molar mass.[56] In general, the molecular glass resist has a well-defined small-molecule core that bears protected base-soluble groups (such as hydroxyls and carboxyls) as shown in Figure 19.8b. With this approach, the core chemistry can vary from calix[4]resorcinarenes (ring-like),[57–59] branched phenolic groups,[60,61] and hexaphenolic groups (disk-like).[62,63] Early approaches with MG led to low glass transition temperatures, however, such problems were resolved by increased hydrogen bonding functionality and the design of the core structure. The MG resists may also benefit from a more uniform development due to the lack of chain entanglements and reduced swelling, when compared with polymers; these are active areas of research. Experimental data using the quartz crystal microbalance method demonstrate that swelling appears during development even with molecular glass resists.[64] Most of these alternative resist structures adhere to the chemical amplification strategy. However, nonchemically amplified photoresists are also being considered as they do not contain photoacid generators and hence do not suffer from photoacid diffusion length constraints.[65]

Two other novel MG variants are a PAG covalently bound to the molecular resist (Figure 19.8c) and the core of the molecule serving as the PAG[66] (Figure 19.8d). As designed, there would be no need for the blending of PAG with such resist systems. In the case of Figure 19.8d, photolysis produces a photoacid, which then would deprotect the unexposed acid-sensitive protecting groups of the PAG-core molecular glass. These two approaches (Figure 19.8c and d) are smaller pixel sizes and true one-component systems that in principle eliminate surface segregation and phase separation in cast films. These alternate resist structures, however, typically require additives such as amine base quenchers to limit the diffusion of the photoacid catalyst into unexposed regions.

19.2.3 Progress in Resists for EUV

There has been progress in designing photoresists for EUV lithography in anticipation of the 22 nm nodes. The testing and development of new materials typically relies on direct lithographic testing to screen formulations. However, due to the lack of widely available EUV exposure tools, micro-field exposure tools (MET), such as the 0.3 NA SEMATECH Berkeley MET have an important role for resist testing. Substantial progress[67] in reaching CD challenges with commercial chemically amplified photoresist were reported with 20 nm half-pitch with an EUV dose sensitivity of (12.7–15.2) mJ/cm², which is near the theoretical Rayleigh resolution limit of the 0.3-NA system with

a k_1 of 0.45. While progress in resolution and sensitivity were observed, LER remains a challenge. In fact, the fidelity of the resist feature may have non-negligible LER contributions from the EUV mask and optical trains as ascertained by modeling. After subtracting an estimated mask contribution, resist LER values appear to approach the 2 nm level. Additionally, a pattern collapse of sub-20 nm dense features suggests the intrinsic resolution could be better than expected. In fact, unoptimized model EUV polymers and MG photoresists clearly show 20 nm features as quantified by the latent image and developed image roughness.[68] Alternative development approaches[69] may provide new directions to break the resolution limits (Equation 19.2), which currently do not directly consider development effects, such as swelling and swelling layer collapse.

19.2.4 Progress in Resists for 193 nm Immersion Lithography

By inspection, the Rayleigh equation (Equation 19.1) directs that the smallest feature may be achieved by employing higher refractive index media (photoresist and immersion fluids). Currently, highly purified water with $n = 1.44$ is used as the immersion fluid and typical 193 nm polymers have $n \approx 1.7$. The development of photoresists and immersion fluids with higher n are needed. A strategy to increase the refractive index is by incorporating more polarizable heteroatoms, such as sulfur,[70–73] into the polymer structure. Halides such as Cl, Br, and I would increase the refractive index at the expense of absorption and possible photo-induced side reactions. Developer-soluble topcoat barriers[74–77] or engineered surface-segregating barrier-layer additives[78–83] are typically used to reduce or mitigate the leaching of critical components (PAG and quenchers) into the immersion fluid.

19.2.5 Concluding Remarks

Sub-32 nm critical dimensions come with many resist challenges. The challenges of reducing feature size and decreasing LER and sensitivity are inter-related. Current attempts to surpass resolution challenges include novel resist process strategies such as double patterning and double exposure as well as novel architectures. However, it is clear that the quality of the final patterned structures is inter-dependent upon the spatial distribution of the photoacid (optical image quality), the spatial extent of the reaction-diffusion process (latent image), and the development mechanisms. The guidance of the resist resolution metrics in cooperation with improved materials fundamentals will help photoresist materials achieve ever smaller features as suggested by the Rayleigh equation.

19.3 Deep Ultraviolet Lithography

Deep ultraviolet (DUV) lithography, using KrF or ArF excimer lasers with illumination wavelengths of 248 or 193 nm, respectively, is now the predominant technology for producing critical

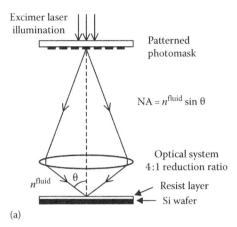

(a) (b)

FIGURE 19.9 (a) Schematic of the DUV photolithography exposure process. The numerical aperture is defined by NA = $n^{\text{fluid}} \sin \theta$, where θ is the half angle of the marginal rays and n^{fluid} is the index of the fluid (gas or liquid) between the final lens element and the resist. (b) Illustration of a commercial 193 nm immersion lithography system (ASML TWINSCAN XT:1950i), showing the cylindrical optics barrel in the center, the mask above the optics system, and an illuminated wafer below. (Courtesy of ASML, Veldhoven, the Netherlands. This illustration is based on an artist impression. No rights can be derived from it.)

layers of leading-edge, mass-market semiconductor logic and memory ICs. The technology is based on illuminating a patterned fused silica mask with diffused excimer laser radiation and imaging the transmitted light patterns with 4:1 image reduction onto a photosensitive-resist-coated Si wafer, using a diffraction-limited optical system. The process is indicated schematically in Figure 19.9a. The illuminated wafer is removed from the optical system for chemical processing steps to convert the exposed pattern (latent image) in the resist to a patterned structure on the wafer and put back into the optical system multiple times for further exposure/processing steps. State-of-the-art ICs are built up from about 20–30 layers, a significant fraction of which involve DUV exposure steps. An illustration of a leading-edge 193 nm immersion lithography exposure tool for volume production is shown in Figure 19.9b.

DUV lithography is an evolution of optical-projection lithography developed in the late 1960s and the 1970s using the strong spectral lines from mercury discharge lamps to illuminate first at 436 nm (g-line) and later at 365 nm (i-line). In these early systems, the pattern on the mask for the entire circuit was illuminated and imaged onto the wafer for a specified exposure time, and then the wafer was "stepped" to an adjacent unexposed position for another imaging of the circuit pattern. This was repeated until the wafer was filled with as many identical exposures as the wafer size allowed. These wafer "step-and-repeat" systems were termed "steppers."

In modern leading-edge DUV lithography systems, the full circuit illumination of the mask is replaced by a "scanned" illumination. A fraction of the circuit pattern on the mask is illuminated in a narrow strip across its width and the mask is scanned under this strip, while its image is projected with a 4:1 reduction ratio onto a wafer scanning in the opposite direction, at a speed relative to the mask reduced by the factor of 4. When the full circuit pattern has been exposed, the wafer is stepped to the next position and the process repeated. The exposure image is

stepped-and-scanned across the wafer, producing typically over 50 full chip exposures on a 300 mm diameter wafer at a rate giving about 100 wafers per hour. These wafer "step-and-scan" systems are referred to as "scanners," though they are often referred to as "steppers" as well.

The first commercially available production optical-projection lithography system, the DSW4800, was introduced by GCA in 1978. It achieved a minimum feature resolution of a little over 1 μm and a slightly larger depth of focus.[10] At the time, it was fully understood that the resolution of the optical-projection lithography approach was fundamentally restricted by the diffraction limit given roughly by the Rayleigh resolution criterion in Equation 19.1. In DUV lithography, the process-dependent factor k_1 is of order 1 Ref. [10,84]. Actually, Equation 19.1 is intended to capture both the limiting effects of diffraction and the impact of the resist processing. Considering just the diffraction effects, Equation 19.1 would refer to the aerial image at the image plane, and k_1 would be determined only by the profile of the light intensity. However, the actual size of a feature ultimately created by the lithography process also depends on the chemical processing of the exposed resist as described in Section 19.2. For an isolated feature, this size can be any value, i.e., k_1 has no fundamental restrictions. On the other hand, for the dense structures of real circuits, e.g., modeled by a periodic structure, the separation between printed features (the pitch) is fundamentally limited. With the minimum feature size in Equation 19.1 defined as half the pitch for a single exposure, k_1 can be rigorously shown to have a minimum value of $k_1 = 0.25$ Ref. [10,84]. This limit holds for any single-exposure process, as long as the process has a linear dose response.

Because a functional circuit must have some topology and because imaging control is imperfect, the resist exposure process requires some finite depth of focus (DOF). This also has a diffraction and geometric limitation, which is characterized by Equation 19.3

$$\text{Depth of focus} = \frac{\lambda}{2n\left(1-\sqrt{1-\dfrac{NA^2}{n^2}}\right)} \xrightarrow{\text{Small NA}} k_2\frac{n\lambda}{NA^2}$$

$$= k_2\frac{\lambda}{n\sin^2\theta}, \tag{19.3}$$

where, for small NAs, a process-dependent prefactor k_2 is conventionally used.[10,84] For large NAs, the exact form of the equation conventionally incorporates a different process-dependent prefactor $k_3 \times 4$.[85]

In the early 1980s, reasonable considerations about the prospect of significantly altering the parameters in Equations 19.1 and 19.3 to improve the resolution and DOF led to the conclusion that optical-projection lithography had an ultimate dense feature size resolution of about 0.5 μm. Consequently, it was assumed then that progress in high-volume-production ICs would require switching within a decade or so to alternative lithography techniques with more extendibility potential. These included electron-beam direct-write lithography (with multiple parallel beams), electron-projection lithography, ion-projection lithography, x-ray-proximity lithography, extreme-ultraviolet lithography, and nanoimprint lithography.[10] Few people, if anyone, publicly predicted then that the factors in Equations 19.1 and 19.3 would be relentlessly driven to enable optical-projection lithography to out-complete all the alternative technologies mentioned above, in cost and performance capabilities for volume production, at least into the second decade of this century.

19.3.1 DUV Lithography Steppers

A primary driver for this progress has been the wavelength factor in Equation 19.1. This can be seen in Figure 19.10, which shows a log-linear plot of critical feature sizes of leading-edge ICs as a function of year introduced into large-scale production, along with the illumination wavelength used. Wavelength reduction has been aggressively pursued in part because, as is clear from Equations 19.1 and 19.3, feature size reduction by decreasing wavelength reduces DOF less than that by increasing the NA. However, the switch from Hg-arc lamp illuminators with wavelengths at 436 and 365 nm to the DUV excimer laser sources with wavelengths at 248 nm (KrF) around 1995 and 193 nm (ArF) around 2000 was a very challenging one. Hg-arc lamp sources are compact, relatively inexpensive, reliable, and operate continuously. KrF and ArF excimer lasers are much more complex, are substantially more expensive to purchase and operate reliably, have undesirable laser coherence properties, and the illumination is pulsed. This last characteristic has been particularly troublesome because obtaining the required exposure doses fast enough for acceptable wafer throughputs requires very high pulse peak intensities. This puts very stringent requirements on the durability of lens and window materials in the optical system. Further, the shorter UV wavelengths

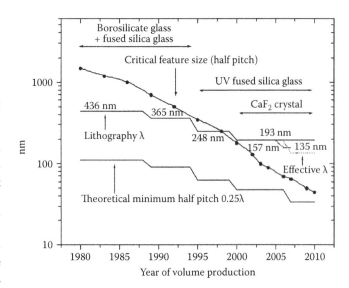

FIGURE 19.10 Plot of critical feature size (half pitch) of leading-edge semiconductor integrated circuits versus year of volume production, on a log-linear plot. Also included is the lithography wavelength used and the theoretical minimum half pitch for single exposures (0.25λ) for this wavelength. 157 nm technology has not been brought into production. Effective λ is λ/n for 193 nm immersion lithography. Optical materials used for lenses at each λ are also indicated.

substantially limit material options for these optical elements. In particular, at the time of the introduction of 248 nm lithography, UV-fused silica glass was the only UV-transmitting material that could meet the tight optical properties specifications for the large lenses required. For 193 nm systems, in addition to UV-fused silica, a second high-quality material, crystalline calcium fluoride (CaF_2), had to be developed for lenses to correct for chromatic aberrations in the optical system due to the wavelength dispersion of the index of fused silica (see Figure 19.10).

Fused silica glass is an amorphous form of SiO_2, which minimizes the large birefringence effects of the uniaxial crystal structure of crystalline SiO_2 (quartz). However, the amorphous structure makes fused silica a thermodynamically metastable material and it suffers structural changes under exposure to high 193 nm laser intensities. These changes result in both volume compaction and rarefaction effects with different intensity dependencies.[86] The resulting refractive index and lens geometry changes can substantially degrade lens performance. To minimize these effects, maximum intensities have to be limited throughout the stepper optics, and the more durable CaF_2 lens elements generally must be used in the positions with highest intensities, for example, at the final lens position.

For wavelengths much below 193 nm, UV-fused silica glass has poor transmission, leaving only the cubic structure Group II fluorides as practical lens materials. Of these, only CaF_2 has been made with lithography-grade optical quality. Hence, wavelength extension to 157 nm with F_2 excimer lasers would require the lenses to be made only from CaF_2, or possibly from other related fluorides, such as BaF_2, if the optical quality could be significantly improved.[87] The availability of lithography-quality CaF_2 for lens material has been a key issue for the development of both 193 and 157 nm lithography technologies, and building a CaF_2 manufacturing infrastructure to support the needs of both of these technologies has been a major challenge. A principal difficulty with CaF_2 production is that its relatively low thermal conductivity and high thermal expansion coefficient requires it to be cooled very slowly from the melt in order to obtain low-strain, single-crystal material of the sufficiently large sizes needed for lenses. Even many weeks of controlled cooling in elaborate temperature-controlled furnaces gives a low yield of lens blanks meeting specifications. One result is that 193 nm stepper systems have been designed to use the minimum number of CaF_2 lens elements possible, though CaF_2 has not been designed out entirely.

A further unexpected complication from the use of CaF_2 lens elements in 193 and 157 nm lithography systems resulted from a faulty assumption that the cubic crystalline structure of CaF_2 would ensure isotropic- and polarization-independent optical properties (for high-quality crystals), as could be "demonstrated" by naive symmetry arguments and measurements at longer wavelengths. In fact, CaF_2 turned out to have substantial index anisotropy and birefringence at the short wavelengths of 193 and 157 nm.[88] This is due to the symmetry-breaking effects of the finite photon momentum at these wavelengths, giving rise to a "spatial-dispersion-induced" or "intrinsic" birefringence.[89] Fortunately, the symmetric nature of this effect has enabled it to be minimized by judiciously orienting the crystal axes of several lens elements to substantially cancel the effects.

Beyond 157 nm, a few shorter wavelengths have been considered for further resolution extension: 126 nm from Ar_2 excimer lasers[90] and 121.6 nm from hydrogen Lyman-α discharge sources.[91] At these wavelengths, the only transmissive optical materials known are LiF, which has high extrinsic absorption and poor exposure durability, and MgF_2, which has high natural birefringence due to its noncubic crystal structure. These shorter wavelength technologies have not been developed beyond feasibility studies.[90,91] Below about 100 nm, no practical material is transparent and refractive optics are not possible. For reasons primarily associated with the development of immersion lithography discussed later, 157 nm lithography, though demonstrated to be technically feasible,[87] has been dropped off of technology roadmaps.[14] It now appears likely that the shortest wavelength that will be used for production lithography with refractive optics is 193 nm.

As of 2008, 193 nm steppers are the primary tools for producing critical layers of high-volume, leading-edge ICs. Their ArF excimer laser sources are pulsed ($\approx$6 kHz, 10 mJ/pulse) and line narrowed ($\leq$0.25 pm spectral bandwidth).[92] The transmission photomasks are made from fused silica with Cr absorbing features and have thin-membrane pellicles to keep uncontrolled particles from collecting on the mask and imaging onto the wafer. The stepper optics are made of fused silica spherical and aspheric lenses, calcium fluoride lenses with clocked crystal axes, mirror elements with aspheric surfaces, and immersion fluids between the last lens element and the wafer, as discussed below.

While the wavelength was progressing down to 193 nm, the other two factors in Equation 19.1, k_1 and NA, were also being pushed towards their limits. A number of resolution enhancement techniques have been used to drive the k_1 factor down to achieve IC structures with half pitches corresponding to a value of k_1 approaching the limiting value of $k_1 = 0.25$.[93] Nearly all of these have been taken over from established optical techniques in other fields such as microscopy. These have included (1) illumination methods (off-axis illumination and partial-coherence control), (2) mask modifications to engineer desired wavefronts (phase shift masks, sub-resolution mask structures, and other optical proximity corrections), and (3) resist contrast improvements. For the 45 nm technology node, logic manufacturers are expected to operate with k_1 factors down to about 0.31,[94] and further extension with k_1 factors down to 0.29 is considered feasible.[95] Unfortunately, accompanying these gains, process latitudes have been shrinking to marginally tolerable levels. Clearly, k_1 factor improvements for single exposures are nearly tapped out.

The last factor left in Equation 19.1 is the numerical aperture, $NA = n \sin \theta$. For air between the final lens element and the wafer, the liming value is $NA = 1$. The GCA 4800 DSW stepper in 1978 had an NA of 0.28.[96] The NA has been increased steadily in newer designs by increasing the size and complexity, and consequently the cost of the lens systems. To contain the number and size of lens elements as the NA was increased, the lens system designs had to incorporate aspheric lenses and off-axis mirror elements (catadioptric designs). Further, as the NAs approached 0.9, the polarization effects could no longer be ignored and illumination polarization control became an essential part of the lithography process, further complicating the lithography tools and processes.[84,97] State-of-the-art dry 193 nm stepper optics have an NA of about 0.93, contain about 30 lenses, with a path length through lens material of about 1 m, and weigh 500 kg or more.[10] With dry NA limit of 1.0, there is very little possible gain left to justify the vastly increased cost and complexity needed for improvement—an exponential increase in cost for an asymptotic gain. The $\sin \theta$ factor is now also essentially tapped out.

Figure 19.10 shows, along with the critical feature size (half pitch), the limiting value of 0.25λ for each wavelength. Clearly, by 2006, the critical feature size 65 nm had approached its limit for 193 or 157 nm. In the 1990s, this projection was the reason why it was nearly universally assumed that alternative technologies, such as EUV lithography, would have to take over below this feature size. This has turned out to be incorrect primarily for two reasons: the NA is not limited to 1 if immersion fluids are included and the effective k_1 factor can be decreased below 0.25 with multiple exposures and processing on the same layer.

19.3.2 193 nm Immersion Lithography

Equation 19.1 is valid at the image plane, which of course must be in the resist. In principle, the $NA = n \sin \theta$ should be evaluated in the resist, which typically has a 193 nm index in the range

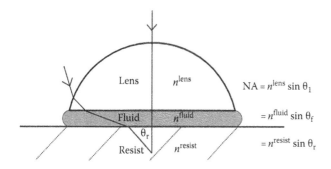

FIGURE 19.11 Schematic of an on-axis ray and a marginal ray through the last elements of an immersion lithography optical system. The angles θ_l, θ_f, and 0_r are the angles of the marginal ray with respect to the surface normal in the lens, fluid, and resist, respectively. For the case shown here, the refractive index of the fluid is less than that of the lens and resist, and n^{fluid} is then the maximum possible NA of the system.

$n = 1.6–1.7$. However, by Snell's law ($n_1 \sin \theta_1 = n_2 \sin \theta_2$), NA = $n \sin \theta$ can equally well be evaluated above the resist. Until the recent introduction of the immersion stepper, the space between the final lens element and the wafer was filled with air or N_2 gas, with an index near $n = 1.0$. A consequence is that in spite of the much higher index of the resist, the NA has a maximum value of 1.0. However, as microscopists have known and exploited for centuries, adding a fluid between the image plane and the final lens element allows the maximum NA to be increased.[98–100] For a planer final lens element, the NA can theoretically be increased up to the lowest index of the resist, the fluid, or the lens, as indicated in Equation 19.4 and Figure 19.11:

$$NA_{max} \leq (n^{resist}, n^{fluid}, n^{lens}) \qquad (19.4)$$

A nearly ideal immersion fluid for 193 nm lithography turned out to be purified water. It has sufficient transparency at this wavelength, is relatively innocuous to the resist and lens materials, is compatible with resist processing, has low enough viscosity to enable rapid wafer scanning, and is inexpensive. These properties enabled remarkably rapid development and implementation of immersion technology.[101] From the inception of substantial 193 nm immersion lithography efforts in late 2002,[102–105] it took only about 4 years for production-worthy water immersion systems to be built and the process brought into production. Immersion imaging does introduce some new issues, however, including bubble formation in the fluid, evaporation residue defects, resist-immersion fluid interactions, and fluid thermal effects. Polarization effects, already issues for dry systems, have been exacerbated at the extreme NAs of immersion lithography, requiring careful polarization control for differently oriented structures, which puts some restrictions on IC design.[106]

With water as the immersion fluid, having a 193 nm index of $n^{water} = 1.437$, the theoretical minimum half pitch (HP^{min}) for a 193 nm exposure tool decreases to HP^{min} = 0.25 × 193.4 nm/1.437 = 33.6 nm. Furthermore, at HPs achievable for dry systems, Equations 19.1 and 19.3 show that for the same

HP, the DOF is larger for the immersion approach. Note that the minimum HP for 157 nm dry systems is HP^{min} = 0.25 × 157.6 nm/1.0 = 39.4 nm, higher than that for 193 nm immersion systems. Attempts were made to identify high-index 157 nm immersion fluids, but only relatively low-index (n [157 nm] ≈ 1.35) fluorocarbon liquids were found to have any practical transparency at 157 nm,[107] and consequently 157 nm lithography technology was dropped from technology roadmaps. As of 2008, water-based 193 nm immersion lithography systems are being operated at an NA in the range of 1.30–1.35 to satisfy the 45 nm technology node with acceptable process latitudes.[14]

Approaches to extend 193 nm immersion lithography technology further with higher-index fluids have been pursued. Practical 193 nm transmitting organic fluids with 193 nm indices near $n = 1.65$ nm (second generation fluids) have been developed.[108,109] However, the last stepper lens element is made of calcium fluoride with a 193 nm index of $n = 1.50$, or fused silica, with an index of $n = 1.57$. By Equation 19.4, these materials are the bottlenecks for resolution extension. Thus, significant NA gain from higher-index fluids requires higher-index last lens materials as well. Higher-index UV lens materials have been explored for this purpose. Key practical materials identified include $Lu_3Al_5O_{12}$ (LuAG) [$n = 2.21$] and polycrystalline $MgAl_2O_4$ (ceramic spinel) [$n = 1.93$].[110] LuAG has the highest index, but efforts have not yet succeeded in improving the 193 nm transmission to the specifications, which are stringent due to its high thermo-optic coefficient, dn/dT. Also, the large value of the intrinsic birefringence for this material is difficult to compensate for. Ceramic spinel solves this problem with its polycrystalline structure, but the polycrystalline nature also results in a high degree of scattering. A further possible high-index lens material being considered is α-Al_2O_3 (crystalline sapphire) [$n = 1.92$]. Its large natural birefringence due to its uniaxial crystal structure has limited its use in precision optics. However, with its crystal optic axis oriented along the optical axis of the system and with the polarization of all rays arranged to be oriented perpendicular to this axis, a sapphire last lens element could be manageable. As of this writing, LuAG is considered the most promising high-index lens material candidate,[111] and the industry is still considering whether the practical NA gain to about 1.50 is worth the development effort.

If a third-generation immersion fluid with an index near or above $n = 1.80$ could be developed, then an NA increase to near NA ≈ 1.7 could provide enough incentive to justify the substantial development effort required to implement it, because this would enable the 32 nm half-pitch technology node. Pure organic fluids with indices this high cannot have practical 193 nm transmissions. However, the indices of the fluids could in principle be increased to this level by loading the liquids with approximately (2–10) nanometer-size suspended particles of high-index oxide crystals, e.g., HfO_2 or LuAG, while maintaining acceptable transmission, scattering levels, and viscosities.[112–114] This approach is being explored, but it is regarded at best as complementary to multiple patterning approaches to resolution extension discussed next.

19.3.3 Double Patterning

The process-dependent k_1 factor has been driven down to near the hard limit of $k_1 = 0.25$ for single exposures. However, for multiple exposures and processing steps on the same layer, structures with half pitches corresponding to lower effective k_1 values are possible. A number of these multiple patterning approaches are now being pursued and these are expected to satisfy the requirements for the 32 nm half-pitch technology node and possibly the 22 nm node and below. These approaches differ by the number and sequence of exposure and processing steps used to achieve the desired features. Four basic types are illustrated schematically in Figure 19.12.[14] They all use the fact that while the structure pitch created by a single exposure is limited by diffraction, as characterized by Equation 19.1, processing can be used to tailor the line/space ratios in each period, and subsequent exposure and/or processing steps can interleave further structures to increase the structure density. In the Litho-Etch-Litho-Etch process, Figure 19.12a, two Litho-Etch processes are done in sequence with the second process shifted by half the period to give a pitch doubling. This general process can be used to create arbitrary structures with the half pitch shrunk by a factor of 2 below the diffraction limit, though this scale shrink requires tighter tolerances on line edge control and overlay, among other issues.

A serious difficulty with this process is that it requires that the wafer be taken out of the stepper track for etching and then realigned before the second exposure, creating overlay challenges and decreasing the throughput for the layer by a factor of about 2. A preferred approach would have the two exposures done in sequence with just one etch step at the end: Litho-Litho-Etch, shown in Figure 19.12b. Unfortunately, it can be shown that for a resist system with a linear dose response, as has been universal in DUV lithography, two exposures cannot generate patterns with pitches below the diffraction limit. However, with a nonlinear dose response system, this is possible. A number of nonlinear response approaches are being pursued, including contrast enhancement layers, 2-photon resists, and positive-and negative-tone threshold response resists.[115] One version, Litho-Freeze-Litho-Etch, shown in Figure 19.12c, is, particularly promising.[116] In this approach, the latent image captured in the resist from the first exposure is "frozen" by chemical treatment. The first frozen image region is protected from any photo response to a second exposure, then a single etch step can create the pitch-doubled structures.

A different approach, known as side wall (or spacer assist) double patterning, requires only one exposure step to create pitch doubling.[117] As illustrated in Figure 19.12d, the first exposure and development step is only to create sidewalls at the desired positions. A subsequent thin-film deposition on the sidewalls, chemical-mechanical polishing (CMP) to split the deposited structure,

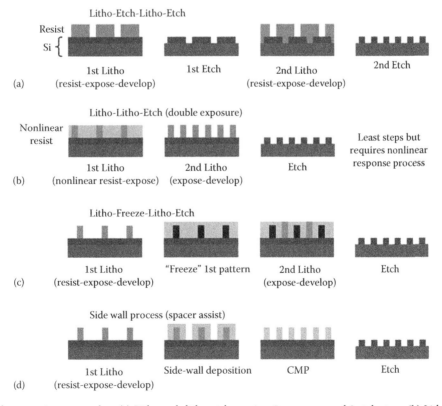

FIGURE 19.12 Double patterning approaches: (a) Litho-etch-litho-etch requires 2 exposures and 2 etch steps. (b) Litho-litho-etch (or double exposure) requires 2 exposures and only one etch step, but needs a resist process with a nonlinear dose response. (c) Litho-freeze-litho-etch has intermediate complexity, requiring 2 exposures, a chemical freezing process after the first exposure, and one etch step. (d) Side wall process (spacer assist) requires one exposure and one etch. It uses side wall deposition and chemical-mechanical polishing (CMP) to achieve doubled pitch.

and elimination of the original pattern enables the desired pitch doubled structures. This approach takes advantage of the well-understood and highly controllable thin-film-deposition technology to obtain the desired linewidths, independent of diffraction limits. Furthermore, the sidewall-derived structures are automatically self-aligned by the original structure. A difficulty with this approach is that restriction to patterning along sidewalls makes design layout much more difficult, and not all structures are topologically possible with single exposures. At least a further exposure/processing step is generally required for arbitrary structures. Still, this method is inherently extendable, and pitch quadrupling, etc. is possible in principle.

19.3.4 Ultimate Resolution Limits of DUV Lithography

Double patterning methods are already being used in IC production, and they are the declared technology solutions, using 193 nm water immersion tools, for several volume semiconductor manufacturers for the 32 nm half-pitch technology node projected for about 2013.[14] These approaches appear feasible for further resolution shrinking as well, e.g., the 22 nm half-pitch technology node. Redesign of 193 nm immersion steppers with high-index fluids and lens materials, in combination with double patterning techniques, offer the potential for further resolution extension and relaxation of the k_1 factor for a given resolution. When double patterning becomes routine, consideration of triple patterning, quadruple patterning, etc. may become tempting. However, this would likely have to come at the cost of much tighter process control requirements, such as CD and overlay control, than is presently attainable. Improvements to meet the requirements, along with the increased complexity (more exposures and processing steps per layer) will surely drive up costs substantially. There is little doubt that these DUV lithography extension approaches can be made to work technically. It is just a matter of whether they can operate at commercially viable costs and whether any reliable alternatives can be implemented at less cost. For example, rapidly increasing mask costs have made the numerous maskless technologies, such as multi electron–beam direct write schemes, very attractive, at least for low-volume production.

For high-volume leading-edge ICs, it is generally agreed that for the 22 nm half-pitch technology node and below, a reliable single-exposure EUV solution, with its larger depth of focus and more natural extendibility, would be preferred. However, formidable challenges, repeated introduction delays, and especially rapidly rising projected costs, have made this solution not so inevitable as it once seemed at the beginning of the decade. Cost is the ultimate driver. If or when DUV lithography is displaced, and what its ultimate practical resolution limits are, is now less clear than ever. The technology has thrived on remarkable innovativeness and resourcefulness and its development history does not suggest that these will cease. Previous predictions of its limits and of its imminent replacement have always been wrong.

19.4 Electron-Beam Lithography

Electron-beam lithography has demonstrated its utility over several decades as both a primary pattern generation tool for the semiconductor industry and as the premiere means of patterning small structures for advanced device development and research in myriad fields. Like any lithographic technology, its suitability for a given application depends on its performance with respect to the four metrics of resolution, throughput, pattern placement, and cost-of-ownership. As we shall see in the following sections, electron-beam systems excel with respect to resolution and pattern placement, but suffer from fundamental limits in terms of throughput.

19.4.1 Resolution

Resolution in a lithographic process is a rather ill-defined quantity since, by virtue of the fact that it is a process involving many steps, a large number of variables are involved. These can frequently be manipulated to produce isolated features far smaller than might, at first sight, be judged possible. The true test of the process is the minimum pitch of the features that can be fabricated. In what follows, we will consider primarily those factors that are unique to electron-beam exposure and neglect those that are common across the various exposure techniques—those are covered elsewhere in this chapter.

The ultimate resolution that can be obtained using electron-beam exposure is determined by the nature of the interactions of the electrons with the material being exposed. There are two parts to be considered: first, the trajectories that the incident, or primary, electrons follow within the material and second, the processes by which the energy deposited by the primary electrons is translated into developable chemistry in the exposed material.

The electron trajectories are determined by the combination of incident electron energy and resist/substrate atomic number.[118,119] At low energies/high atomic numbers, the electrons scatter strongly within the solid, which leads to substantial broadening of the beam as it penetrates the material. Since most resist materials are organic, their average atomic numbers are not dissimilar from that of carbon, so the beam broadening, or forward scattering in the resist, is determined principally by the electron energy. For this reason, most high-resolution systems operate at energies of 50–100 keV. In addition, because the beam broadening increases progressively as the electrons undergo additional elastic scattering events on their way through the resist, thin resists yield higher resolution images. Very low (<500 eV) incident energies have also been proposed as a means of generating high-resolution images.[120–122] In this case, the beam broadening is on the order of and constrained by the very short total range of the electrons within the resist. The resist thickness is therefore limited to that range, which may be insufficient for subsequent processing needs, and the resolution can only be improved by reducing the energy (and the resist thickness) to impractically small values.

The incident electron energy, E_0, has a dramatic impact on the dose needed to expose resist, which, for energies above about 5 keV, varies linearly with E_0. This is a reflection of how the inelastic scattering cross section varies with electron energy. The effect is captured by the Bethe continuous slowing-down approximation[123] that describes the average rate at which fast electrons lose energy along their trajectories. At high energies and resist thicknesses that are small relative to the electron range, the feature width is determined not by the deviations in the trajectories of the primary electrons due to elastic scattering, but by the volume in which they deposit energy. Although the primary electrons can cause chemical changes in the resist directly, it is normally assumed that the so-called fast secondary electrons generated by inelastic scattering of the primary beam are responsible for the majority of broken bonds in the resist.[124] This is because the low-energy secondary electrons have a much larger interaction cross section than those of the primary beam.[118] Conceptually, each primary electron trajectory can be thought of as surrounded by a cylinder of material 10–20 nm in diameter exposed by the secondaries that it generates. At very low incident electron energies, the stopping power increases significantly and the electrons deposit all their energy efficiently in the resist film, rather than carrying on through to the substrate.[125]

The resolution that can be achieved in a resist can also be estimated by considering the impact parameter for energy-loss events for the primary electrons. An electron traveling with velocity v, interacting with a stationary electron at a distance b delivers an impulse having a duration of approximately b/v and contains frequency components, ω, up to v/b. These correspond to a maximum energy transfer, ΔE, of $h\omega/2\pi$.[126,127] For 100 kV electrons and ΔEs of 5, 50, and 500 eV, b is 10, 1, and 0.1 nm, respectively. Normally, the energy scale relevant to resist exposure is on the order of a few electron volts (3.6 eV for the C–C bond), leading to an interaction distance of a few nanometers but in a few cases the energy needed is much higher, corresponding to a core-shell excitation, for example. In systems such as alkali metal halides, in which the exposure mechanism depends on the dissociation of the halide ions from the lattice, resolutions of 1–2 nm are achievable.[128-130] Unfortunately, the doses required are in the C/cm^2 range, which is not practical for manufacturing. Currently, the highest resolution material that is suitable for nanofabrication is hydrogen silsesquioxane (HSQ), which has demonstrated feature sizes as small as 7 nm half-pitch.[131] This is an inorganic, small-molecule resist that essentially forms silica upon exposure and development. Its increased mechanical strength relative to organic resist materials is one reason for its ability to produce small features, while its higher average atomic number may help reduce the range of the fast secondaries, contributing to its high resolution.

Electron–solid interactions also affect the resolution in electron-beam systems through the proximity effect. This is the additional exposure of the resist by electrons backscattered from the substrate and it leads to a loss in contrast and thus to a loss in resolution. At higher voltages and on low-atomic number substrates (e.g., Si or SiO_2), the backscattered electrons have a relatively large range ($\propto V^{5/3}$) and the contribution of the forward and backscattered electron doses can be captured by the two-Gaussian model[132]

$$D = D_0 \left[e^{x^2/\alpha^2} + \eta e^{x^2/\beta^2} \right] \tag{19.5}$$

where

D_0 is the incident dose
α is the width of the forward-scattering distribution
β is the width of the backscattered distribution
η is the effective dose contributed by the backscattered electrons

As we have discussed above, α is typically on the order of 10 nm, while β is on the order of micrometers. η can be quite large, depending on the substrate's atomic number and is approximately 0.5 for Si at 100 kV.[133] This model is applicable if the forward and backscattered dose distributions can be readily separated. In this case, small, isolated features receive a dose of D_0, the centers of large features receive a dose of $D_0(1 + \eta)$, and all other types of features receive a pattern-density dependent dose that varies between those values. As long as the feature size is small compared with β, this dose variation can be corrected for either by adjusting the dose delivered to individual features,[134] applying the inverse of the background dose to produce a flat background,[135,136] or adjusting the size of the features[137,138] by using a pattern density averaged over a length scale on the order of β. However, if either α or β approach the feature size, then much more computationally intensive techniques must be used to perform an iterative correction procedure, as the correction applied to each feature affects its neighbor and vice versa.[139] At low energies and in high atomic number substrates such as GaAs or InP, the deposited energy distribution is more complex[140,141] and the proximity effect becomes much more difficult to correct. It is, however, important to point out that these phenomena are analogous to the flare and optical proximity effects encountered in advanced optical lithography, for which correction techniques are at an advanced stage of development.

So far in our discussion, we have implicitly assumed an ideal point source of electrons incident upon the resist. A close approximation can be achieved in practice with low-current, focused probes in electron microscopes. However, for practical applications, we need high beam currents, and this has a dramatic impact on the design and operation of electron-beam lithography tools and leads to systems in which the intrinsic resolution of the resist and process is often better than that of the electron optics.

19.4.2 Throughput

In the design of most optical systems, including electron microscopes, there is a trade-off between blur due to diffraction, which scales as $1/NA$, where NA is the numerical aperture of the system, and aberrations that generally scale as some power of NA (a spherical aberration is often a dominant term and varies as NA^3). In electron-beam lithography systems, throughput is critical to their cost of ownership and ultimate utility. Unlike photons in a light-optical system, the electrons in an electron-optical system affect one another strongly through unscreened Coulomb repulsion. The interactions give rise to three principal effects: (1) the forces between electrons along the electron-optical axis lead to the speeding up or slowing down of individual electrons, which is equivalent to an energy spread in the beam, known as the Boersch effect; (2) the average force due to all the other electrons in the beam leads to deterministic changes in electron trajectories and leads, to first order, to an overall defocusing of the beam, known as the global space-charge effect; and (3) the interactions between individual electrons lead to random changes in individual electron trajectories, known as the stochastic space-charge effect. The severity of the Boersch effect depends on the chromatic aberration in the system. The global space-charge effect can be thought of as creating a negative lens in the system whose effects can be compensated for by refocusing the beam. However, if the current density is not uniform across the beam, that lens becomes highly aberrated and its effects cannot be compensated. The stochastic space-charge effect is not correctable. As might be expected, the space-charge effects are made worse when the electrons are forced closer together or can interact for longer periods and increase with beam current, I, and column length, L, and are mitigated when the electrons are spread farther apart or interact for shorter times and decrease with increasing numerical aperture α, the demagnification factor, M, and the accelerating voltage, V. The detailed dependence of beam blur on I, L, V, M, and α is particular to a given system and has been studied in depth for both single-column, probe-forming,[142] and projection systems.[143,144] Note that improvements realized by increasing the voltage are offset by the need to increase the beam current to compensate for the linear decrease in resist sensitivity with voltage.

The net result of these effects is that rather than balancing diffraction and aberration blurs, the system optimization must be performed by balancing space-charge and aberration blurs. This means that for any given beam current in a system, there is an optimum numerical aperture that must be used and a corresponding minimum blur that can be achieved. Any attempt to increase the beam current leads to an increase in beam blur. Electron-beam lithography systems for nanofabrication research are generally designed to provide the optimum resolution, with throughput as an important, but secondary goal. However, the semiconductor industry's continuing demand for ever-smaller features and higher throughputs has resulted in the disappearance of electron-beam lithography from the direct IC fabrication process because of the conflict between the need for increased beam currents to increase throughput and the need for the decreased beam currents necessary for improving resolution cannot be resolved at the feature sizes now required.

In the future, the throughput of electron-beam lithography, even for mask making, is likely to become worse. To understand why this is the case, it is necessary to examine how resist sensitivity scales with feature size.

If we consider a feature of CD L and allow it to vary by no more than $L/10$, then we can define an image pixel as $L/10 \times L/10$. In order for each pixel to be faithfully reproduced, we need the resist to be able to clearly distinguish between exposed and unexposed pixels. Even if the dose over a large area is fixed at a precise value, the arrival of the electrons at any pixel is a statistical process, with the number of electrons delivered following a Poisson distribution. This effect is frequently referred to as shot noise and has been extensively studied.[145–149] We can, therefore, define a minimum number, n, of electrons per pixel that is necessary to ensure that the error rate of improperly exposed pixels is suitably small. The total dose needed per unit area therefore becomes $100n/L^2$, i.e., it scales as $1/L^2$, and the beam current necessary to maintain a given real exposure rate must therefore scale in the same fashion.

19.4.3 Overlay

The fact that magnetic and electric fields can deflect electrons precisely and at high speed is one of the great strengths of electron-beam systems. In optical lithography systems, accurate and precise registration and overlay can only be achieved by ensuring nanometer-scale mechanical alignment between mask and wafer stages—physically massive systems that can only be controlled with bandwidths of a few tens of Hz. In electron-beam tools, it is not necessary to control the stage precisely (offsets of several micrometers between the actual and planned stage coordinates are common), because the position of the electron beam can be adjusted, at speeds of up to 100 kHz, to make up the difference, provided a suitable signal from the stage interferometers is available. There are generally differences in the coordinate system that are generated by the stage and the substrate. Normally, these are accounted for by periodically acquiring alignment mark information from the substrate. This procedure is susceptible to errors arising from drift occurring between mark measurements and schemes have been developed that enable registration data to be generated from the substrate continuously.[150] The ability to adjust the position of the beam, as well as characteristics such as focus and stigmation, in real time, makes it possible to correct for a large variety of errors, such as those caused by variations in substrate height. Along with the benefits it brings, the sensitivity to fields is also a source of difficulty associated with electron-beam systems: interference from dynamic fields must be reduced as far as possible, while that from static ones, such as those due to the small magnetic inhomogeneities of the stage, must be carefully mapped and calibrated out.

Another source of error peculiar to electron-beam tools, particularly at high voltages, is substrate heating. The total amount of energy deposited in the substrate scales as the square of the accelerating voltage because of the combination of the increasing energy per electron and the increasing resist dose required. The concomitant substrate heating can affect both pattern placement/overlay, as a result of thermal expansion,[151] and CD control, as resist sensitivity may vary with temperature.[152] Temperature increases are more severe in SiO_2 substrates because of silica's poor thermal conductivity, while Si substrates, with their higher coefficient of thermal expansion, are liable to greater placement errors.

19.4.4 Cost of Ownership

As we have seen, although electron-beam systems provided exceptional resolution, and are capable of precise pattern placement,[153] they are fundamentally limited in throughput. Historically, for relatively large feature sizes, the limiting factor in throughput was the serial nature of the writing process in probe-forming, or Gaussian-beam, systems and this led to the development of schemes such as shaped-beam,[154] cell-projection,[155] large-area,[6,156] and full-field projection[157] that sought to deliver more than one pixel's worth of information at a time. Unfortunately, as feature sizes decreased and the resolution requirements became more demanding, they all ran into space-charge limits. This has relegated electron-beam systems for production purposes to the fabrication of small numbers of high-value items such as photomasks, CD/DVD masters, and nanoimprint templates. There are two routes out of this dilemma: the first is to accept the limited throughput, but use the essential addressability of electron beams to create flexible systems—maskless lithography tools—that can create ICs without the need for masks and their associated costs;[158,159] the second is to avoid the space-charge limits of single columns by dividing and conquering—using multiple small columns to keep the electrons far enough apart so that their interactions are insignificant.[160,161] In the first case, the cost of ownership is competitive with photolithography for short-production run devices where only a few wafers worth of devices are produced for each mask set. In the second case, providing that issues associated with source uniformity, control, and brightness can be overcome, then it may be possible to have the best of both worlds—flexibility in patterning together with high throughput—making it directly competitive with optical lithography. Research and development are active in both areas.

19.5 Nanoimprint Lithography

Nanoimprint lithography (NIL) is rapidly emerging as a low-cost, high-resolution, and versatile alternative to optical lithography as a patterning technology for semiconductor fabrication and other applications that incorporate nanoelectronic devices.[162,163] Optical lithographic techniques, including state-of-the-art tools based on 193 nm radiation[98] or next-generation potential solutions, including extreme ultraviolet (EUV, 13.5 nm radiation) lithography,[164–166] rely upon selectively exposing a photoactive film to radiation via a photomask that defines the lithographic pattern of interest. A chemical reaction occurs in the exposed regions of the film that renders them soluble in an aqueous base, creating a physical pattern defined by the unexposed regions in the film. It is becoming increasingly difficult to control the resolution of this patterning process.

NIL offers a potentially simplified alternative to chemically amplified resists. All NIL techniques rely on a simple squeeze-flow mechanism whereby a resist flows into a well-defined physical cavity of nanoscale dimensions. It is a molding or replication technique where the ultimate resolution is defined by the physical dimensions of the imprint mold cavities.[167–170] If high-resolution masters can be created, high-resolution copies can be generated. It is also a repetitive stamping or replication technique. The well-defined patterns in the imprint master can be used to create thousands of replicas, analogous to how the metallic master for an LP record were used to stamp thousands of inexpensive vinyl copies for consumer use. Compared with optical lithography for high-volume nanomanufacturing, there is an important distinction worth highlighting. With optical lithography, every printing or copy of the nanostructured surface requires extremely tight control of the environmental conditions to maintain the sensitive balance of the complicated reaction, diffusion, and dissolution process for the chemically amplified resists.[171] This is precisely why state-of-the-art optical lithography tools cost upwards of $30 M apiece, providing a 65 nm patterning resolution, and semiconductor fabrication lines cost billions of dollars to construct.

The resolution of NIL is largely controlled by a simple squeeze-flow process.[167–170] As long as high-quality, high-resolution imprint masters can be fabricated, it is relatively easy to stamp out copies of the mold. In optical lithography, the ultimate patterning resolution is intrinsically coupled with the high volume manufacturing process making process control measures extremely important. Every single copy manufactured requires stringent control over the reaction, diffusion, and dissolution processes. With NIL, the resolution is less coupled with the high-volume manufacturing process and more dependent on the mold or imprint master fabrication. For these reasons, NIL tools tend to be significantly less expensive. Research and development grade imprint tools can be purchased for about $100 K, while manufacturing grade tools cost approximately $2 M, both with a sub 10 nm resolution.

Both optical and nanoimprint lithography are dependent on high-resolution masks or masters. In optical lithography, the mask sets required to fabricate a semiconductor device can cost several millions of dollars to produce, comprising a significant fraction of the manufacturing costs. These are phase-shifting optical masks in which blanket UV radiation is exposed to one side, generating selectively exposed regions in the resist film on the other side. This exposure is also done through a series of lenses where the features in the mask are approximately four times larger than the image at the wafer plane. In NIL, these mask costs are likely to increase. The squeeze-flow is a direct write process where the features in the mold are the same size as the features in the wafer plane; there is no longer this 4x reduction. This means that high-resolution and costly electron-beam lithography is needed to fabricate the imprint masks, compared with optical lithography where the mask resolution requirements are somewhat reduced because of the lens system. There are no lenses in NIL and the imprint masks may very well be more costly to produce. Furthermore, these expensive NIL mask sets must be mechanically pressed into the resist film. In optical lithography, the masks are isolated from the resist behind a transparent and protective pellicle. There are no protective pellicles in NIL and one must make direct mechanical contact of the expensive mask with the resist film.

In this section, we provide a general overview of the NIL technology. Our discussion will be largely limited to the field of nanoelectronics, including complementary metal-oxide semiconductor (CMOS) devices, although NIL is also widely considered to be a manufacturing alternative for several other forms of nanotechnology. We will elaborate on the differences between NIL and existing optical lithography techniques and discuss some of the unique materials and metrology challenges that NIL processes introduce. We will also provide a brief overview of some of the recent developments in applying NIL to nanoelectronics technologies.

19.5.1 Variations of NIL

19.5.1.1 Thermal NIL

NIL generically refers to pressing a hard mold or template decorated with nanoscale patterns into a softer material to lithographically define well-controlled nanostructures.[167–170] In most cases, pattern formation is accomplished by a squeeze-flow mechanism whereby the material flows into the mold in a liquid-like state (Figure 19.13). Of course, this requires

"setting" the pattern in the liquid material. In the thermal embossing form of NIL, the resist is typically a glassy polymer.[167,168] During the imprint process, the film and the imprint mold are heated to above the glass transition of the polymer while applying an external pressure. Under elevated pressure and temperature, the polymer flows into the cavities of the mold. The pattern is set by cooling the system to below the glass transition of the polymer, where the material vitrifies into a rigid glass. When the mold is separated from the substrate, the surface is left with nanoscale patterns that are relief images of the mold cavities. A cavity in the mold becomes a polymeric pattern on the surface.

The thermal embossing form of NIL is similar in concept to many of the bulk molding or forming processes used to manufacture large scale plastic components. High molecular mass polymers, even at temperatures far above their glass transition, tend to be high viscosity fluids. Flow does not always readily occur under quiescent conditions meaning that pressures must be applied to squeeze the material into the mold. Most NIL tools designed for thermal embossing are capable of applying pressure as high as 5 MPa. This is combined with maximum temperatures of typically 200°C–300°C, which means that most tools are capable of exceeding the rubbery plateau modulus for most polymer melts and able to induce material flow.

There are several different options when it comes to applying the heat and pressure for thermal NIL. One class of tools is largely based on the photomask aligner and wafer bonding systems widely used in the semiconductor industry. In many instances, an imprint module can be purchased as an add-on for an existing tool. In these systems, mechanical platens with heating elements are used to apply the temperature and pressure to the imprint mold and substrate. Their advantages are that they provide uniform heating through the large thermal mass of the platens and they often have a built-in capability for optical overlay alignment between the imprint mold and the substrate. The primary disadvantage is that maintaining parallel contact between the platens can be difficult. Misalignment of just a few μm can prevent conformal contact, which makes it very difficult to imprint nanoscale features over large areas. To overcome the conformal contact issues, some imprint tools utilize hydrostatic gas pressure to squeeze the imprint mold into the resist film.[172] This pressure is applied through what is essentially a deformable gas bladder. Since the NIL molds or masters are usually etched into thin Si or quartz wafers and imprinted onto similar thin Si substrates, the molds and wafers are able to bend or flex to a certain extent and increase the degree of conformal contact. In some tools, the pressure is applied through the top through the gas bladder while the substrate rests on a rigid metallic platen that provides the heating. These systems provide an optimal balance between uniform heating and conformal contact. There are also tools that utilize an entirely soft press technology, where pressure is applied through both the top and bottom through deformable gas bladders. In these cases, the bladders are transparent and the heating of the substrate is provided through infrared heat lamps. This soft press technology maximizes the pressure uniformity

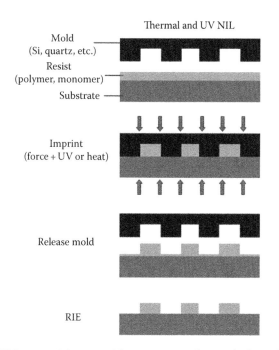

FIGURE 19.13 Schematic of the NIL process, showing both the thermal and UV forms.

across the mold and substrate while the reduced thermal mass (no platens) allows for rapid heating and cooling rates. However, the noncontact optical heating means that temperature uniformity and stability are more difficult to achieve.

19.5.1.2 UV NIL

One disadvantage of thermal NIL is that it requires heating and applying pressure to the substrate. For certain applications, the patterned substrate may or may not be able to withstand the thermal budget associated with this process. Likewise, large temperature changes can induce thermal distortions, which make overlay and alignment difficult. UV NIL offers a low temperature alternative to thermal NIL. In UV NIL, the resist is a liquid-like (or low viscosity) monomer that readily flows into the mold cavities at room temperature with no heating and very little pressure.[169,170] In many instances, the liquid monomer is actually wicked into the mold via capillary forces, minimizing distortions of the substrate or the template (Figure 19.13). In the mold, the monomer undergoes a cross-linking reaction upon exposure to UV radiation, creating a hard glassy pattern that retains its shape after separating the mold from the substrate. This means that either the NIL mold or the supporting substrate must be transparent to the appropriate UV radiation. For this reason, the molds are typically fabricated in quartz wafers, whereas silicon and nickel are more common materials in thermal NIL. An additional requirement is that UV NIL is limited to patterning materials that cross-link under UV radiation. Thermal NIL can, in principle, be applied to any material in which flow can be induced.

19.5.1.3 Transfer Printing

Both the thermal and UV forms of NIL described thus far are similar to optical lithography in that a pattern is created in a resist film that has been previously deposited on the substrate of interest. It is also possible to employ the NIL technologies in a material transfer mode that is similar to most industrial forms of printing. In transfer printing, the material that is to be patterned is first applied to a patterned stamp or mold and then transferred via mechanical contact to the target substrate. This requires tuning of the differential adhesion between the pattern mold and the substrate such that the material adheres to the substrate more strongly than the patterned mold. There are two primary modes in which transfer printing can occur[173] (Figure 19.14). The first is a whole layer transfer where a continuous film with a nanoscale texture defined by the mold topology is transferred to the substrate. The second is a discrete transfer or inking mode where material is selectively transferred to the substrate from the protrusions of the mold; any material in the cavities of the mold is not transferred over. Whole layer transfer is used primarily to create topological patterns of a given material on the target substrate. Examples of this technology are being used to create nano- or microfluidic channels or devices. The discrete transfer mode can also be used to create physical patterns, but it also has the ability to create chemical patterns of a given surface energy or chemical functionality by essentially inking monolayers of a

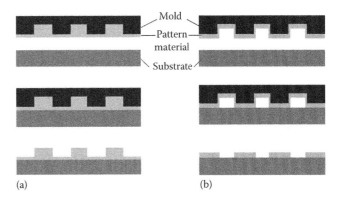

FIGURE 19.14 Schematic of two modes of transfer printing: (a) whole-layer transfer and (b) inking.

chemical compound. In this last case, the pattern may, to a first approximation, lack topology and not alter the smooth surface of the substrate. Chemical patterns like these are commonly considered in sensor-type applications.

19.5.1.4 Functional Materials

One of the biggest distinctions between NIL and competing optical lithography methods is rooted in the generality of the patterning process. In nearly all forms, optical lithography focuses on creating high-resolution patterns in sacrificial photoresist formulations. As mentioned previously, photoresists are highly complicated materials systems that balance chemical reactivity, acid diffusivity, and their dissolution behavior to optimize their patterning resolution. Once the pattern is formed, the resist becomes a sacrificial component for subsequent additive or subtractive processes to transfer the pattern into the functional materials of interest. On the contrary, the mechanical stamping and squeeze-flow processes used in NIL are generic mechanisms applicable to a wide range of materials. This means that a wide range of materials can be patterned directly by NIL, including functional materials. The resolution of NIL processes is controlled by the mold or mask fabrication process more so than optical lithography. The ability to directly stamp nanoscale structures into functional materials is a highly attractive way to simplify the number of parameters impacting patterning resolution, reduce the number of nanofabrication steps, and thereby reduce manufacturing costs.[174]

19.5.1.5 Tool Types

NIL is a low-cost nanopatterning alternative to optical lithography with the potential for high-volume manufacturability. Mechanical stamping processes have a long history in manufacturing environments and the same appears to be bearing true for the emerging field of nanomanufacturing. Stamping processes are intrinsically compatible with manufacturing assembly lines. However, NIL is still in its infancy period and most of the tools that are commercially available today are geared for R&D or the technology development. There are examples of high-throughput tools in the market, but most of the activity is still in the lab as opposed to pre-production. In this section, we review the different types of tools with respect to patterning throughput.

The most versatile tools on the market today also have the lowest throughput and least applicability for high volume nanofabrication. These account for most of the imprint tools sold today from companies like Nanonex, Obducat, or Jenoptik that utilize gas pressure through some type of deformable bladder to apply a very uniform pressure to the mold and the substrate. Typically, these tools are very flexible and provide the user with a wide variety of process control, at the sacrifice of throughput. The user has the freedom to place an arbitrary mold onto an arbitrary substrate, sandwiched between one or two sheets of deformable rubber or foil. This stack is then placed inside a movable gas chamber. The gas chamber closes down on the deformable bladders, creating pressure reservoirs on the top and/or bottom of the sample. A vacuum pumping system is also usually involved to evacuate the sample region between the bladders. Nitrogen or some other inert gas is then pumped into the chamber, generating pressures as high as approximately 5 MPa to press the mold into the substrate. Heat is then applied through either the optical heat lamps or thermal sample block to enable flow of the material into the mold cavity. After a prescribed heating cycle of a few minutes, the temperature is cooled back to room temperature and gas chambers are vented to release the pressure. The sample and the mold are then manually removed from the chamber and manually separated to reveal the patterned substrate. This process obviously describes the thermal form of NIL. The same general process applies for UV NIL with these tools with the exception that lower pressure is required and UV radiation is applied instead of heat. In either case, UV or NIL, these single stamp tools are capable of generating one patterned substrate approximately every 10–15 min with manual input from the user. For most applications, this throughput is not suitable for manufacturing. Rather they are designed for flexibility and ease of use. The sample chambers typically vary from approximately 5–20 cm in diameter and allow arbitrary form factors or piece parts to be imprinted. This is primarily useful for technology development. One notable exception is a large area single imprint tool being developed at the Korean Institute of Machinery and Materials.[175] This is a low pressure UV imprint tool capable of imprinting an entire 45 cm (diagonal) flat panel display uniformly in a single imprint. Cycle times of 4–5 min here would probably be more acceptable given the limited demand for such high-end displays. These cycle times would be totally unacceptable for the high volume demand of bit patterned magnetic media for data storage hard drives.

There are efforts to increase the throughput of imprinting tools. The next class of tools generically utilizes a stamp and repeat technology to generate multiple copies of a single pattern across a large substrate. This form of stamp or step and repeat technology is naturally compatible with the form factor of multiple dies (or devices) per wafer that is currently used in semiconductor fabrication.[170] So these types of tools tend to be targeted

for CMOS devices. Molecular Imprints, Inc. is the pioneer in developing step and repeat tools, although other companies including EV Group and Nanonex now offer tools with similar capabilities. Throughput or manufacturability tends to be the primary driver for going from a single imprint to a step and repeat tool. For this reason, most step and repeat tools are based on the UV NIL process, which is usually faster because it lacks the heating and cooling cycles associated with thermal NIL. Furthermore, locally heating a selected region for imprinting on a larger wafer or substrate is difficult. It is much easier and faster to locally expose the imprinted region to UV radiation. Lastly, these step and repeat tools are usually optimized for throughput that comes at the price of the flexibility in the patterning processes or conditions. The general use single imprint tools typically offer greater flexibility. For example, single imprint tools are usually capable of performing either UV or thermal NIL on a variety of materials whereas step and repeat tools tend to be limited to UV NIL patterning of a given resist formulation that has been optimized for the system.

The NIL tools discussed thus far have largely been designed as drop-in replacement technologies for optical lithography, with the target applications in CMOS. However, there have been recent advances in building roll-to-roll NIL tools that are capable of patterning continuous sheets or strips of a nanopatterned substrate.[176,177] Roll-to-roll processes are capable of generating patterned media at a rate at least several orders of magnitude faster than traditional optical lithography, enabling high volume nanofabrication. However, roll-to-roll processes tend to provide very poor overlay capabilities, practically limiting the advantages to single-layer devices. For these reasons, the target applications migrate away from CMOS to emerging markets where nanopatterning has yet to be realized. Examples might include antifouling biological films, polarizing or anti-reflective optical coatings,[178] brightness enhancing films for large area displays, or flexible electronic devices like radio frequency ID tags.[179] In the roll-to-roll form of NIL, the mold cavities are fabricated onto either a circular drum or a tractor tread-like device. A continuous film is fed in between a pair of rollers that rotate in opposite directions, squeezing their textured surface(s) into the film. Roll-to-roll processes can be implemented in many different forms. The thermal form of NIL can be used to emboss nanoscale topology on the film or substrate by using heated rollers. Likewise, there are examples of UV NIL being implemented to create textured surfaces with roll-to-roll devices. The traditional forms of gravure or flexography can also be adapted to create roll-to-roll NIL transfer printing processes.

19.5.2 Fundamental Issues

There are a few commercial products being made with imprint today, proof that the technology is inherently manufacturable. Some of the companies operating in this space include Nano Opto, Heptagon, Omron, OVD Kingram, LG Electronics, Reflexite, MacDermid, Wavefront Technologies, and Spectratek.

Dr. Michael P.C. Watts is the founder and president of Impattern Solutions, an independent consulting company in the field of NIL. The company's website tracks the progress and commercialization of imprint-related technologies and provides updates at www.impattern.com. Most of the commercial success stories in the imprint field have thus far been related to optics or light. These include, in order of decreasing critical dimensions, refractive, diffractive, and sub-wavelength optical devices. However, even with the sub-wavelength devices, the minimum dimensions of the structures being fabricated commercially are typically on the order of 100 nm or larger, which by many definitions, is not within the nanoscale domain. The community has yet to see the realization of devices fabricated by NIL with dimensions strongly into the sub-100 nm regime. There are many examples of nanoelectronic technologies in the R&D phase that appear to have strong potential with respect to NIL, but several fundamental issues need to be resolved before these ideas can successfully make the transition from the lab to manufacturing. In this section, we briefly summarize some of the remaining technical roadblocks to the commercialization of nanoelectronics via NIL techniques.

19.5.2.1 Nanoimprint Materials Development

There are several examples of high resolution patterning being achieved with NIL. For years, Professor Chou's fabrication of 10 nm vias was the ultimate demonstration of high resolution pattering with NIL.[168] More recently, Professor John Roger's group demonstrated that the fabrication of 2 nm replicas of carbon nanotubes randomly distributed across a rigid substrate could be created within a polydimethylsiloxane (PDMS) resin using NIL processes.[180] Professor Chou's group has gone on to report the patterning periodic line-space patterns with a 6 nm half-pitch,[181] indicating that NIL has the potential resolution and patterning control that is commensurate with nanoelectronics (Figure 19.15).

However, there is still a need for improving NIL materials or resists. Resolution is only part of the picture and a NIL resist must also satisfy a number of different properties. First, the material must be able to readily flow into the imprint mold. If harsh conditions that plastically deform the patterned material are required, the NIL processes may induce residual stress into the patterns.[182–184] It is desirable that the patterned media readily flows into the imprint mold. Likewise, the imprinted patterns must have the mechanical strength to resist pattern collapse. There are several examples of the critical aspect ratio for pattern collapse with decreasing feature size.[185] NIL is capable of generating very small patterns that make pattern collapse a significant concern. Furthermore, it is critical that the mold separates easily from the imprinted structure without destroying the sample. This requires both low surface energy mold treatment strategies that minimize adhesion between the imprint and the substrate as well as imprint materials with the cohesive strength to withstand the shear stress generated during mold separation.[186] Since external forces are required to pry the mold from the substrate, a mechanical load is placed on the

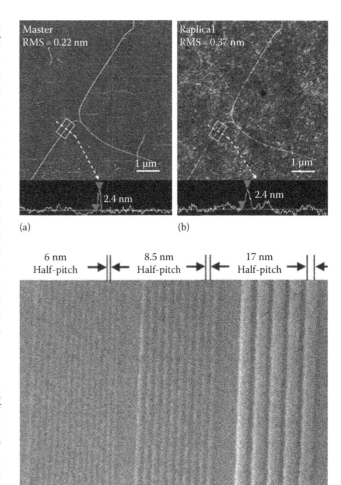

FIGURE 19.15 Examples of high resolution patterning with NIL. Panels (a) shows a few carbon nanotubes with a diameter of 2 nm on a flat surface that served as the model imprint mold while panel (b) shows the resulting patterns that were replicated in PDMS. (Reprinted from Hua, F. et al., *Nano Lett.*, 4(12), 2467, 2004. With permission. © 2004, American Chemical Society.) Panel (c) shows well regular line-space patterns that have been imprinted with critical dimension to a 6 nm half-pitch. (Reprinted from Austin, M.D. et al., *Nanotechnology*, 16(8), 1058, 2005. With permission.)

patterns. It is a significant materials science challenge to maximize these properties simultaneously especially with functional nanoimprint materials.

19.5.2.2 Template Fabrication and Availability

The fabrication and availability of high quality templates or molds are crucial for NIL patterning technologies. With an ultimate resolution approaching just a few nm, even the slightest flaw in the master has the potential to be replicated into hundreds or thousands of imprinted copies. This means that significant efforts must be made to produce high quality imprint masters. Electron-beam (e-beam) lithography is the industry standard for producing high-resolution patterns with a resolution on the order of 10–20 nm. However, e-beam lithography is not practical for manufacturing since it is a slow, serial writing process and is very costly. One of the greatest attributes of NIL is that it can be considered as an e-beam replication technique. Once a costly pattern is fabricated by e-beam lithography into a mold or master material, thousands of high-resolution copies of the pattern can be replicated via NIL processes at a fraction of the initial cost.[187] This is a very attractive proposition, but it also means that a high quality imprint master is critical. Any mistake will be easily propagated through all of the copies.

19.5.2.3 Defect Inspection and Dimensional Metrology

The need to have high quality imprint templates has been articulated in the preceding section. Hand-in-hand with this requirement is the need to evaluate, quantify, and certify the pattern quality in both the imprint master and the resulting imprints. High-resolution patterning demands high-resolution pattern metrology. We have already mentioned the excessive costs for fabricating a NIL master or mold with large areas patterned with nanoscale features. Quantitative measurements to qualify that the imprint master meets the design specifications in terms of pattern shape are absolutely critical for making this investment. Likewise, fast and efficient high-resolution pattern metrology methods are also needed to monitor the high volume manufacturing processes. It is critical to have quantitative measures in place to determine if the manufacturing processes drift out of tolerance. For high-resolution patterning, this means high-resolution metrology. Even at the current 65 nm technology node, the "Metrology" chapter of the ITRS Roadmap contains roadblocks in terms of high-resolution CD metrologies.[188] The leap from 65 nm optical lithography to NIL processes that offer better than a 5 nm resolution raise significant concerns regarding our ability to quantitatively evaluate the patterns that we fabricate. This is not that big of an issue in the R&D phase of technology development, where one usually deals with highly optimized prototypes. However, high volume manufacturing processes demand quantitative process monitoring capabilities. With respect to pattern shape metrology, the ability to pattern with NIL greatly exceeds the limits of the current inspection tools.

NIL is essentially a direct pattern transfer process whereby the patterns in the imprint have the potential to be a mirror image of the patterns in the mold. This brings up the fidelity of the pattern transfer concept—how closely do the imprinted features resemble the mold. To achieve the greatest fidelity of pattern transfer, one must be able to minimize the shrinkage of the imprinted material in the mold (either up cross-linking or cooling) or any distortions of the pattern after imprinting. The fidelity of the pattern transfer concept is fundamentally unique to NIL. However, in NIL one can directly compare the dimensions of the mold cavity to the pattern to quantify the fidelity of the pattern transfer. This means that dimensional metrologies must be able to quantify both the mold and the imprinted patterns.[189]

19.5.2.4 Residual Layer Control

The objective in most imprint processes is typically to create isolated patterns on the substrate. However, there is always a continuous residual layer of resist material between the substrate and the patterns that needs to be removed. Simple volume filling arguments dictate the thickness of the residual layer, comparing the volume of material per unit area in the smooth film to the volume of the cavities in the mold. Even if there is less material in the film than there is in the mold cavities, the protrusions of the mold cannot be completely pressed into direct contact with the substrate; there is always a finite residual layer on the order of a few nanometers thick. While this residual layer is easily removed through reactive ion etching (RIE) processes, it is critical to achieve both a minimized and uniform residual layer. The minimized residual layer is important to maintain CD control. RIE processes are highly anisotropic, etching primarily vertically, however, there is a small amount of lateral trimming that occurs (Figure 19.16). If one has to etch through a thick residual layer, then the extent of lateral trimming can be significant; minimizing the residual layer minimizes the trimming. Furthermore, it is important to achieve a uniform residual layer across the entire field of the imprint. If there are thick and thin regions of residual layer, the RIE will break through to the substrate first in the thinner regions. When this happens, a micro loading effect occurs in the etch process where the concentration of plasma increases upon breakthrough (the plasma etch chemistries are biased to consume resist over the substrate). The result of micro-loading is an increased lateral etch rate, i.e., a localized enhancement in the lateral trimming. This nonuniform lateral trimming leads to a distribution of the pattern dimensions (loss of CD control).

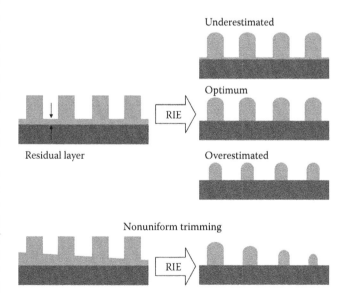

FIGURE 19.16 A schematic illustrating how an accurate knowledge and control of the residual layer thickness is critical to controlling lateral trimming and critical dimension control. Overestimating the thickness of the residual layer can lead to excessive lateral trimming. A non-uniform residual layer can lead to a distribution of linewidths and a loss of CD control through a micro-loading effect.

Controlling the residual layer thickness is straight forward for imprint molds with uniform pattern fields. Simple volume filling calculations can be used to adjust the thickness of the film to minimize the residual layer, assuming that a uniform pressure is applied.[190] However, residual layer control becomes difficult when the pattern field contains a distribution of different patterns with different sizes and density. In this case, the volume filling calculations are different for the different regions of the mold.[191] Resist films are usually applied via spin coating techniques, which leads to a film of uniform thickness. In these situations, it is difficult to control the residual layer. The Molecular Imprints series of tools use a GDS-based drop on demand technology to deposit an optimized amount of resist material in each location to properly fill the mold. This technique can be very effective for minimizing the residual layer.[192,193]

19.5.3 Overlay Accuracy and Control

Depending on the application, it may or may not be important to align the imprint within the lateral directions with some features or fiduca markers on the substrate. There are many applications, such as anti-reflective optical coatings, where pattern overlay is not required. However, for nanoelectronic applications, one could imagine it being very important to align the imprint with circuit elements in the substrate. Most imprint tools offer some type of alignment capability. The alignment is usually achieved by fabricating a series of vernier gratings in both the imprint mold and the patterning substrate and use interferometers to sense very small changes in the in-plane dimensions.[194] Piezoelectric actuators can be incorporated into the imprint head to generate very small lateral position corrections that help precisely align the mold with the substrate. Most general use imprint tools are designed for molds and substrates that are not transparent, meaning that the comparison of the interference pattern from the two verniers is achieved through a series of mirrors and reflective light. With these types of tools, one is able to achieve overlay accuracies on the order of ±200 nm in the in-plane directions. Given the 5 nm or better patterning resolution of NIL, the accuracy with which a feature can be placed within the plane of the film is significantly worse. There are tools with higher overlay accuracy. Molecular Imprints produces highly specialized tools that are optimized for patterning with UV radiation only, using quartz molds that are optically transparent.[195] When it comes to overlay alignment, systems that are dedicated to patterning with transparent molds have a significant advantage because through wafer comparison of the optical verniers is possible. Eliminating the series of mirrors and reflective optics reduces the error in the overlay alignment, enabling sub-15 nm 3σ overlay accuracy. This is more commensurate with the ultimate patterning resolution of a few nm.

19.5.4 Technology Examples

There are a growing number of examples in the literature of nanoelectronic devices being fabricated by NIL processes, proving that the technique has potential (Figure 19.17). The most common examples are the cross-bar memory structures where a semiconducting material is sandwiched between orthogonally situated parallel line-space electrodes to be used as either data storage or logic devices.[196] Here, NIL processes are used to fabricate the parallel line-space electrode arrays. These devices can have half-pitch values as small as 17 nm for the electrode arrays and memory densities as high as 100 Gbit/cm^2. On the CMOS front, a recent joint effort from Toshiba and Molecular Imprints have demonstrated the fabrication of a functional sub-32 nm logic circuit using NIL.[197] Likewise, IBM and SEMATECH have reported the fabrication of 27 nm functional FINFET structures using NIL technologies.[198] Another sector where NIL processes are being widely adapted is for magnetic data storage media. The equivalent Moore's law for the hard disc industry is even more aggressive than that of the semiconductor industry, doubling in data storage capacity ever 12 months. To keep pace with this trend, the industry is moving towards bit patterned media. NIL processes are being developed to fabricate 25 nm posts of magnetic media on a pitch of approximately 40 nm.[199] These are just a few examples related to nanoelectronics where NIL technologies are being actively pursued. There are also a number of other applications in the optical and biotechnology fields where NIL methods are poised to have an impact. Two of the most prominent examples include wire grid polarizer elements for screen and display technologies[200,201] and photonic structures to facilitate light extraction in high brightness LEDs.[187]

19.5.5 Future Perspectives

In some respects, NIL is just one of a long list of next-generation lithography techniques that has been developed as an alternative to deep UV optical lithography for the semiconductor industry. However, none of these alternate technologies have even come close to supplanting optical lithography as the high volume lithography of choice for the electronics industry. While a debate over the merits of NIL to replace optical lithography for CMOS fabrication is beyond the scope of this review, it is important to realize that NIL is the first next generation lithography candidate to have a viable life outside of CMOS, and there are already success stories for the patterning technology in non-CMOS technologies. This means that the success or failure of NIL is not tied to the semiconductor industry; it has already succeeded. As microelectronics moves beyond CMOS into nanoelectronics with novel device architectures, dramatically different principals of operation, and the need for nanoscale patterning, NIL is on-pace to be a well-established generic nanoscale patterning technology of relevance. Next generation lithography techniques tied directly to the CMOS community (193 nm immersion, double patterning, EUV) are solely dependent on this community for success or failure. These patterning technologies will likely go away if they are not adapted by the semiconductor community. The same is not true for NIL.

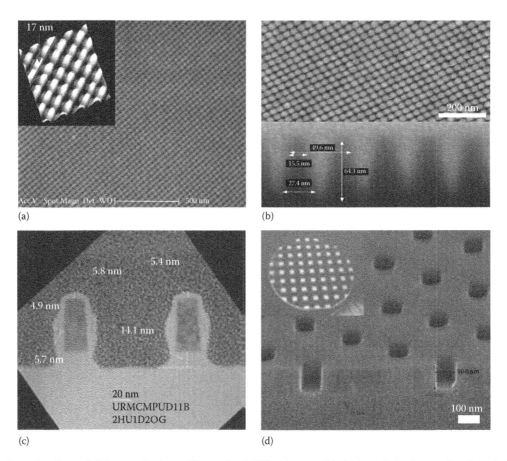

FIGURE 19.17 Examples of several different technologies fabricated with NIL techniques: (a) a high-resolution image taken from the 17 nm half-pitch cross-bar nanowire circuit arrays made by SNAP. (Reprinted from Jung, G.Y. et al., *Nano Lett.*, 6(3), 351, 2006. With permission. © 2006, American Chemical Society.) (b) Bit patterned magnetic media for high density data storage showing 25 nm magnetic posts fabricated by NIL. (Reprinted from Dobisz, E.A. et al., *Proc. IEEE*, 96(11), 1836, 2008. With permission. © 2008, IEEE.) (c) A cross-sectional view of a 27 nm FinFET structure whose critical dimension was patterned by Step-and-Flash Imprint Lithography at Molecular Imprints, Austin, TX. Nominally identical structures led to functional FinFET devices. (From IBM Almaden Research Center, San Jose, CA. With permission.) (d) An array of photonic devices patterned uniformly over an entire 6 nm wafer for high brightness LED devices. (Reprinted from Miller, M. et al., *Proc. SPIE*, 6883, 68830D, 2008. With permission.)

19.6 Metrology for Nanolithography

19.6.1 Advanced Lithographic Processes

As optical lithography reaches its limits, new processes such as double patterning and double exposure are being explored as means of extending 193 nm immersion-based lithography. The goal of these patterning methods is patterning at dimensions that are less than those that can be printed in a single lithographic exposure step. Each of these process-based methods poses a unique challenge to all aspects of process control including metrology. These processes are summarized in Figures 19.18 through 19.20.[188]

Double exposure uses two reticles that must be precisely aligned so that the pattern placement is one pitch across the printed area. One photoresist layer is exposed twice with specially designed reticles resulting in densely spaced lines. This pattern can then be etched into the underlying films patterning the wafer with features having a ½ pitch less than can be printed using one reticle. This process is shown in Figure 19.18.

Double patterning is also done using two reticles that must be aligned with accuracy, so that the pattern placement has one pitch and close to one linewidth across the printed area. The first photoresist layer is exposed and then the film stack is etched leaving the pattern in the hard mask layer. Then the wafer is coated with BARC and resist and the second set of lines is patterned into the second BARC/resist layer. The resultant double pattern structure is etched into the poly silicon. Figure 19.19 provides great detail for double patterning processes for lines. Similar process steps are used for contacts and trenches.

Spacer double patterning is believed to be already in use in manufacturing. A slightly different film stack is used for spacer double patterning. First, a sacrificial poly silicon layer is patterned, spacer oxide is deposited, and the sacrificial poly silicon layer is removed resulting in a spacer pattern. This pattern is etched into the poly silicon layer. In Figure 19.20, this single reticle process is more fully described. The origin of the two sets of line shapes is due to the different shapes of each side of a spacer. The two sets of critical dimensions are due to the nature of the spacer deposition process.

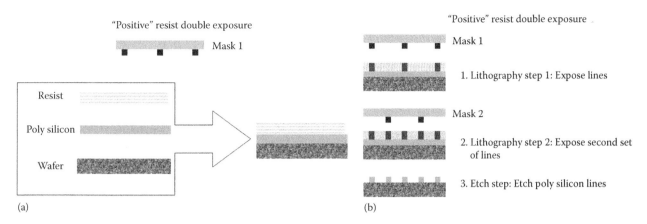

FIGURE 19.18 The double exposure method of patterning lines is shown. Lines are patterned using positive resist. (a) shows the film stack, and (b) shows the patterning process.

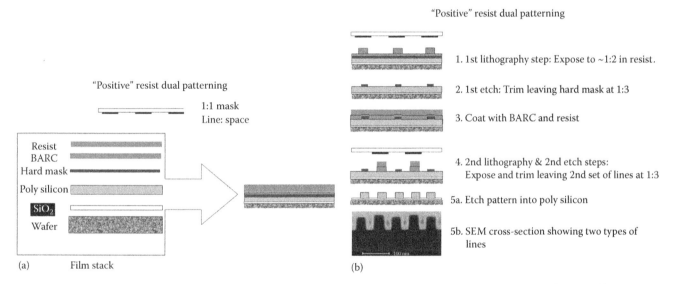

FIGURE 19.19 The double patterning method of patterning lines is shown. Lines are patterned using positive resist. (a) shows the film stack, and (b) shows the patterning process.

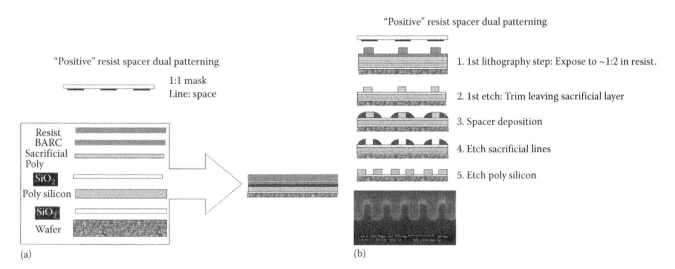

FIGURE 19.20 The spacer double patterning method is shown. Lines are patterned using positive resist. (a) shows the film stack, and (b) shows the patterning process.

Overlay plays an important role in double pattering, spacer patterning, and double exposure pattering of lines. Overlay must be near perfect to achieve a uniform pitch. If the second exposure is shifted, every other space will be different. This impacts further processing, the measurement of CD and lineshape, and device performance. Etch is known to be sensitive to lineshape and the distance between lines. Etch uniformity can be reduced when the resist pattern is etched into the layer below.

19.6.2 Metrology for Advanced Lithography Processes

Measurement needs start with control of feature shape and placement on the mask (reticle), and follow through until the features are etched into the active layer of the device or interconnect structure.[188] Since optical masks are glass with chrome features, scanning electron microscope (SEM)-based measurements must overcome the charging inherent in the measurement of an insulator. The chrome features must have the correct linewidth and shape and must be placed on the mask in the correct position relative to each other. The recent introduction of "environmental" SEM allows for charge neutralization during the SEM-based measurement of CD and line shape on the mask.

The advanced lithographic process challenge metrology mainly through the two distributions of critical dimensions and line shapes especially the side wall angles of the lines. Two methods of CD measurement are used during manufacturing, namely CD-scanning electron microscopy (CD-SEM) and scatterometry. The term scatterometry typically refers to one of two optical methods of measuring critical dimensions using a grating test structure. One example of a grating test structure is an array of lines with a constant pitch and linewidth. Others include an array of contact holes. Scatterometry measurements can be done in a spectroscopic ellipsometer; the main challenge is analyzing the changes in the optical response that the grating structure imparts versus the response of an un-patterned film stack. The other approach to scatterometry is to measure the diffraction pattern from a single wavelength across a range of angles. Both methods of scatterometry require that an extensive database of simulated ellipsometric responses be placed into a library. Some libraries contain tens to hundreds of thousands of simulated structures for a single process step. Data from a sample is matched to this set of simulated responses. Accuracy requires that ellipsometry be sensitive to the small differences in optical response.

Ellipsometry measures the change in polarization of light after reflection from the sample surface. Spectroscopic ellipsometry-based scatterometry relies on the changes in polarization due to changes in the grating test structure as a function of wavelength. Advanced lithographic processes pose two challenges to scatterometry. One is shrinking feature size and the second is the two sets of distributions of CD and feature shape. Sensitivity to small changes in dimension is becoming increasingly difficult as feature size decreases below 50 nm for both CD-SEM and scatterometry. The previous method of single exposure and etch patterning required that scatterometry measure the average CD and line shape of a uniform grating structure. Double patterning and double exposure requires that scatterometry produce the average CD and line shape of two distributions of line shape. In situations where overlay errors result in a nonuniform pitch, the number of simulated data sets is greatly increased. Small differences in optical response make it very difficult to distinguish changes in CD and line shape. Recent commercial instrumentation extended the wavelength range of scatterometry further into the UV down to 150 nm. In general, UV wavelengths have improved the ability of scatterometry to measure CD and line shape for features with sub 50 nm dimensions.

The specifics of ellipsometry-based scatterometry are useful in understanding how critical dimensions are determined from a method associated with the determination of film thickness for unpatterned layers. The premise of ellipsometry is that light polarized in-plane of reflection (P) reflects differently than light polarized perpendicular to the plane of reflection (S). Ellipsometry determines the change in polarization of light in terms of Δ, which is the difference in phase change $= \Delta_P - \Delta_S$ and Ψ, which is related to the ratio of the change in intensity upon reflection through $\tan^{-1}(|R_P|/|R_S|)$, where R_P and R_S are the complex reflectivity's parallel and perpendicular to the reflectance plane. The reflectivity is determined by the dielectric function of the film stack. Reflection from a single film on a substrate depends on the complex dielectric function (refractive index) of the materials and the Fresnel reflection coefficients r_{12} between (1) air and (2) the film and r_{23} between (2) film and (3) the substrate. The reflection from an unpatterned film is determined by well-understood reflection coefficients[202]

$$R_P = \frac{(r_{12}^P + r_{23}^P\, e^{-i2\beta})}{(1 + r_{12}^P\, r_{23}^P\, e^{-i2\beta})} \tag{19.6}$$

$$R_S = \frac{(r_{12}^S + r_{23}^S\, e^{-i2\beta})}{(1 + r_{12}^S\, r_{23}^S\, e^{-i2\beta})} \tag{19.7}$$

where

$$\beta = 2\pi\left(\frac{d}{\lambda}\right) N_2 \cos\phi \tag{19.8}$$

The film thickness dependence comes from the exponential term β where d is the film thickness, λ is the wavelength of the light, and N_2 is the complex refractive index of the film. The measurement of a patterned film stack requires that Maxwell's equations be solved. Typically, this is done using a rigorous coupled wave approximation (RCWA), and the RCWA equations must be solved at each wavelength for each structure. Changes in Ψ and Δ vs wavelength require solving the RCWA equations at enough wavelengths to observe subtle changes due to CD and line shape.[202]

The CD-SEM measurement of critical dimensions is also challenged by double patterning and double exposure. Measuring patterned photoresist requires the calibration of the rate of shrinkage of the resist vs. electron beam exposure time. This calibration requires the collection of enough data to provide statistically significant shrinkage rates. This must be done for each resist material. It is important to note that different resists are used for the fabrication of line and vias. Many IC manufacturers hold the resist formulas to be proprietary meaning that each resist formula will have different responses to electron beam exposure. Recent unpublished studies show that resist lines with a CD less than 25 nm shrink more rapidly than lines with a greater CD. Scatterometry measurements may also impact resist lines especially when measurements are done in UV.

There are important differences between CD-SEM and scatterometry in terms of the nature of information each provides. Advanced CD-SEM is capable of measuring the CD of two or more lines simultaneously. Future CD-SEMs may be able to measure several lines in a test area providing the average CD as well as the range of the CD in the test area. Scatterometry provides the average CD in the test area.

The measurement of line edge roughness in resist is critical for the control of device performance including leakage current. Industry standards organizations such as SEMI have driven consensus documents describing measurement methods for quantifying line edge roughness.[203] Line edge roughness on a nm scale produces gate CD width roughness that results in an increase in leakage current. Line edge roughness at longer scales produces an increase in carrier surface scattering that increases line resistivity. At this time, line edge roughness measurements can be done using CD-SEM or CD-AFM.

Overlay metrology is considerably more challenging for advanced lithography processes. Overlay measurements are typically done using specially designed optical microscopes and target structures that are patterned into both the upper and lower layers. Double patterning of trenches for interconnect lines have additional complications for metrology. If the overlay is shifted, then the trench CD also changes. Overlay measurement precision for all double patterning methods becomes at best a factor of 0.7 of the overlay required for a single exposure patterning of all the trenches. This has driven the development of new overlay test structures that include scattering bars that diffract light. SEM may provide an alternate means of measuring overlay.

19.6.3 Proposed New CD Metrology Methods

Many have speculated about the demise of CD-SEM and scatterometry. Even optical microscopy-based CD measurement is still done for features have larger CDs such as microelectromechanical systems (MEMS). It is possible that CD-SEM and scatterometry can not extend beyond the 22 nm half pitch. Two methods are being considered as potential replacements for scatterometry and CD-SEM. These are small angle x-ray scattering, which is known as CD-SAXS,[204] and He ion microscopy.[205] In CD-SAXS, a focused, monochromatic x-ray beam is transmitted through a grating test structure, as shown in

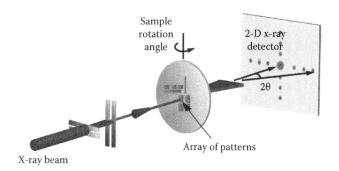

FIGURE 19.21 CD SAXS schematic of using collimated x-ray beams with energy that are transmissive through a patterned silicon wafer resulting in scattering from an array of patterns.

Figure 19.21, that is similar to the ones used for scatterometry measurement. It is important to note that CD-SAXS can measure line edge roughness, average CD, individual CD, and line shape for the lines in the test structure.[206–209] At this time (2009), evaluation of the capabilities of He ion microscopy are in the initial stages.

Abbreviations

α	Fraction of light absorbed, when referring to resist technology
α	Width of the forward-scattering distribution, when referring to electron-beam lithography
β	Width of the backscattered distribution, when referring to electron-beam lithography
δ	Photoacid diffusion length, when referring to resist technology
ε	Photoacid generation efficiency, when referring to resist technology
η	The effective dose contributed by the backscattered electrons, when referring to electron-beam lithography
BARC	Bottom anti-reflective coating
CD	Critical dimension
CD-SEM	Critical dimension scanning electron microscope
CD-SAXS	Critical dimension small-angle x-ray scattering
CD-AFM	Critical dimension atomic force microscopy
CMOS	Complementary metal-oxide semiconductor
CMP	Chemical mechanical polishing
De	Dose at line-edge, when referring to resist technology
D_0	Incident dose, when referring to electron-beam lithography
DOF	Depth of focus
DUV	Deep-ultraviolet
E_0	Incident electron energy, when referring to electron-beam lithography
EUV	Extreme-ultraviolet
FINFET	Nonplanar, double-gate field effect transistor, "fin" refers to shape of semiconductor
HP	Half-pitch

HSQ	Hydrogen silsesquioxane
IC	Integrated circuit
ITRS	*International Technology Roadmap for Semiconductors*
LILS	Latent-image log slope
LED	Light emitting diode
LER	Line-edge roughness
LWR	Line-width roughness
MEMS	Microelectromechanical systems
MG	Molecular glass
ML2	Maskless lithography
NA	Numerical aperture
NIL	Nanoimprint lithography
PAG	Photoacid generator
PDMS	Polydimethylsiloxane
q	Number of photons/nm², when referring to resist technology
RIE	Reactive ion etch
SCALPEL	Scattering with angular limitation in projection electron-beam lithography
STM	Scanning tunneling microscopy
UV	Ultraviolet

Official contribution of the National Institute of Standards and Technology; not subject to copyright in the United States.

Certain commercial equipment and materials are identified in this paper in order to specify adequately the experimental procedure. In no case does such identification imply recommendations by the National Institute of Standards and Technology nor does it imply that the material or equipment identified is necessarily the best available for this purpose.

References

1. Moore, G. E., Cramming more components onto integrated circuits, *Electronics*, **38**, 114–117 (1965).
2. Huff, H., *Into the Nano Era, Springer Series in Materials Science*, Springer-Verlag, Berlin, Germany, **106**, (2009).
3. Maydan, D., Coquin, G. A., Maldonado, J. R. et al., High-speed replication of submicron features on large areas by x-ray lithography, *IEEE Transactions on Electron Devices*, **ED22**(7), 429 (1975).
4. Silverman, J. P., X-ray lithography: Status, challenges, and outlook for 0.13 mu m, *Journal of Vacuum Science & Technology B*, **15**(6), 2117 (1997).
5. Harriott, L. R., Berger, S. D., Biddick, C. et al., The SCALPEL proof of concept system, *Microelectronic Engineering*, **35**(1–4), 477 (1997).
6. Liddle, J. A., Berger, S. D., Biddick, C. J. et al., The scattering with angular limitation in projection electron-beam lithography (SCALPEL) system, *Japanese Journal of Applied Physics Part 1—Regular Papers Short Notes & Review Papers*, **34**(12B), 6663 (1995).
7. Tennant, D., Limits of conventional lithography, in G. M. Timp (ed.), *Nanotechnology*, AIP/Springer, New York, Chap. 4, p. 161 (1999).
8. Marrian, C. R. K. and Tennant, D. M., Nanofabrication, *Journal of Vacuum Science & Technology A*, **21**(5), S207–S215 (2003).
9. Brunner, T. A., Why optical lithography will live forever, *Journal of Vacuum Science & Technology B*, **21**(6), 2632 (2003).
10. Levinson, H. J., *Principles of Lithography*, 2nd edn., SPIE Press, Bellingham, WA, (2005).
11. Brillouet, M., An introduction to ultimate lithography, *Comptes Rendus Physique*, **7**(8), 837 (2006).
12. Ito, H., Chemical amplification resists for microlithography, *Advances in Polymer Science*, **172**, 37 (2005).
13. Chandhok, M., Datta, S., Lionberger, D., and Vesecky, S., Impact of line-width roughness on Intel's 65-nm process devices, *Advances in Resist Materials and Processing Technology XXIV, SPIE*, **6519**, 65191A (2007).
14. *International Technology Roadmap for Semiconductors*, 2007 edition, http://www.itrs.net, Lithography, p. 10. SEMATECH, Austin, TX.
15. Wallraff, G. M. and Hinsberg, W. D., Lithographic imaging techniques for the formation of nanoscopic features, *Chemical Reviews*, **99**(7), 1801 (1999).
16. Gallatin, G. M., Naulleau, P., Niakoula, D. et al., Resolution, LER, and sensitivity limitations of photoresists, *Emerging Lithographic Technologies XII, SPIE*, **6921**, 69211E (2008).
17. Bristol, R. L., The tri-lateral challenge of resolution, photospeed, and LER: Scaling below 50 nm? *Advances in Resist Materials and Processing Technology XXIV, SPIE*, **6519**, 65190W (2007).
18. Van Steenwinckel, D., Gronheid, R., Lammers, J. H. et al., A novel method for characterizing resist performance, *Advances in Resist Materials and Processing Technology XXIV, SPIE*, **6519**, 65190V (2007).
19. Wallow, T., Higgins, C., Brainard, R. et al., Evaluation of EUV resist materials for use at the 32 nm half-pitch node, *Emerging Lithographic Technologies XII, SPIE*, **6921**, 69211F (2008).
20. Fedynyshyn, T. H., Goodman, R. B., and Roberts, J., Polymer matrix effects on acid generation, *Advances in Resist Materials and Processing Technology XXV, SPIE*, **6923**, 692319 (2008).
21. Pawloski, A. R., Christian, and Nealey, P. F., A standard addition technique to quantify photoacid generation in chemically amplified photoresist, *Chemistry of Materials*, **13**(11), 4154 (2001).
22. Szmanda, C. R., Brainard, R. L., Mackevich, J. F. et al., Measuring acid generation efficiency in chemically amplified resists with all three beams, *Journal of Vacuum Science & Technology B*, **17**(6), 3356 (1999).
23. Brainard, R., Hassanein, E., Li, J. et al., Photons, electrons, and acid yields in EUV photoresists: A progress report, *Advances in Resist Materials and Processing Technology XXV, SPIE*, **6923**, 692325 (2008).

24. Hinsberg, W., Houle, F. A., Hoffnagle, J. et al., Deep-ultraviolet interferometric lithography as a tool for assessment of chemically amplified photoresist performance, *Journal of Vacuum Science & Technology B*, **16**(6), 3689 (1998).

25. Pawloski, A., Acheta, A., Lalovic, I., LaFontaine, B., and Levinson, H., Characterization of line edge roughness in photoresist using an image fading technique, *Proceedings of the SPIE, Advances in Resist Technology and Processing XXI*, **5376**, 414 (2004).

26. Houle, F. A., Hinsberg, W. D., Sanchez, M. I., and Hoffnagle, J. A., Influence of resist components on image blur in a patterned positive-tone chemically amplified photoresist, *Journal of Vacuum Science & Technology B*, **20**(3), 924 (2002).

27. Houle, F. A., Hinsberg, W. D., Morrison, M. et al., Determination of coupled acid catalysis-diffusion processes in a positive-tone chemically amplified photoresist, *Journal of Vacuum Science & Technology B*, **18**(4), 1874 (2000).

28. Croffie, E., Yuan, L., Cheng, M. S. et al., Modeling influence of structural changes in photoacid generators an 193 nm single layer resist imaging, *Journal of Vacuum Science & Technology B*, **18**(6), 3340 (2000).

29. Itani, T., Yoshino, H., Fujimoto, M., and Kasama, K., Photoacid bulkiness on dissolution kinetics in chemically amplified deep ultraviolet resists, *Journal of Vacuum Science & Technology B*, **13**(6), 3026 (1995).

30. Vogt, B. D., Kang, S., Prabhu, V. M. et al., Measurements of the reaction-diffusion front of model chemically amplified photoresists with varying photoacid size, *Macromolecules*, **39**(24), 8311 (2006).

31. Vogt, B. D., Kang, S., Prabhu, V. M. et al., The deprotection reaction front profile in model 193 nm methacrylate-based chemically amplified photoresists, *Advances in Resist Technology and Processing XXIII, Proceedings of SPIE*, **6153**, 615316 (2006).

32. Shi, X. L., Effect of Coulomb interaction and pKa on acid diffusion in chemically amplified resists, *Journal of Vacuum Science & Technology B*, **17**(2), 350 (1999).

33. Goldfarb, D. L., Angelopoulos, M., Lin, E. K. et al., Confinement effects on the spatial extent of the reaction front in ultrathin chemically amplified photoresists, *Journal of Vacuum Science & Technology B*, **19**(6), 2699 (2001).

34. Lin, E. K., Soles, C. L., Goldfarb, D. L. et al., Direct measurement of the reaction front in chemically amplified photoresists, *Science*, **297**(5580), 372 (2002).

35. Vogt, B. D., Kang, S., Prabhu, V. M. et al., Influence of base additives on the reaction diffusion front of model chemically amplified photoresists, *Journal of Vacuum Science & Technology B*, **25**, 175 (2006).

36. Wu, W. L., Prabhu, V. M., and Lin, E. K., Identifying materials limits of chemically amplified photoresists, *Advances in Resist Materials and Processing Technology XXIV, SPIE*, **6519**, 651902 (2007).

37. Lavery, K. A., Choi, K. W., Vogt, B. D. et al., Fundamentals of the reaction-diffusion process in model EUV photoresists, *Advances in Resist Technology and Processing XXIII, SPIE*, **6153**, 615313 (2006).

38. Houlihan, F., Person, D., Nalamasu, O. et al., Study of base additives for use in a single layer 193 nm resist based upon poly(norbornene/maleic anhydride/acrylic acid/tert-butyl acrylate), *Proceedings of the SPIE—The International Society for Optical Engineering*, **4345**, 67 (2001).

39. Pawloski, A. R., Christian, and Nealey, P. F., The multifunctional role of base quenchers in chemically amplified photoresists, *Chemistry of Materials*, **14**(10), 4192 (2002).

40. Houle, F. A., Hinsberg, W. D., and Sanchez, M. I., Acid-base reactions in a positive tone chemically amplified photoresist and their effect on imaging, *Journal of Vacuum Science & Technology B*, **22**(2), 747 (2004).

41. Hinsberg, W. D., Houle, F. A., Sanchez, M. I., and Wallraff, G. M., Chemical and physical aspects of the post-exposure baking process used for positive-tone chemically amplified resists, *IBM Journal of Research and Development*, **45**(5), 667 (2001).

42. Hinsberg, W., Houle, F., Sanchez, M. et al., Effect of resist components on image spreading during postexposure bake of chemically amplified resists, *Proceedings of the SPIE—The International Society for Optical Engineering*, **3999**, 148 (2000).

43. Michaelson, T. B., Pawloski, A. R., Acheta, A., Nishimura, Y., and Willson, C. G., The effects of chemical gradients and photoresist composition on lithographically generated line edge roughness, *Proceedings of SPIE*, **5753**, 368 (2005).

44. Hinsberg, W., Houle, F. A., Lee, S. W., Ito, H., and Kanazawa, K., Characterization of reactive dissolution and swelling of polymer films using a quartz crystal microbalance and visible and infrared reflectance spectroscopy, *Macromolecules*, **38**(5), 1882 (2005).

45. Hinsberg, W. D., Houle, F. A., Ito, H., Kanazawa, K., and Lee, S. W., Kinetics of reactive dissolution of lithographic polymers, *Proceedings of the 13th International Conference on Photopolymers, Advances in Imaging Materials and Processes, RETEC* 2003, 193 (2003).

46. Hinsberg, W. D., Houle, F. A., and Ito, H., Reactive dissolution of lithographic copolymers, *Proceedings of the SPIE, Advances in Resist Technology and Processing XXI*, **5376**, 352 (2004).

47. Houle, F. A., Hinsberg, W. D., and Sanchez, M. I., Kinetic model for positive tone resist dissolution and roughening, *Macromolecules*, **35**(22), 8591 (2002).

48. Prabhu, V. M., Vogt, B. D., Kang, S. et al., Direct measurement of the spatial extent of the in situ developed latent image by neutron reflectivity, *Journal of Vacuum Science & Technology B*, **25**(6), 2514 (2007).

49. Prabhu, V. M., Vogt, B. D., Kang, S. et al., Direct measurement of the in situ developed latent image: The residual swelling fraction, *Proceedings of SPIE*, **6519**, 651910 (2007).

50. Prabhu, V. M., Rao, A., Kang, S., Lin, E. K., and Satija, S. K., Manipulation of the asymmetric swelling fronts of photoresist polyelectrolyte gradient thin films, *The Journal of Physical Chemistry B*, **112**(49), 15628 (2008).

51. Wang, M. X., Gonsalves, K. E., Rabinovich, M., Yueh, W., and Roberts, J. M., Novel anionic photoacid generators (PAGs) and corresponding PAG bound polymers for sub-50 nm EUV lithography, *Journal of Materials Chemistry*, **17**(17), 1699 (2007).

52. Wang, M. X., Lee, C. T., Henderson, C. L. et al., Incorporation of ionic photoacid generator (PAG) and base quencher into the resist polymer main chain for sub-50 nm resolution patterning, *Journal of Materials Chemistry*, **18**(23), 2704 (2008).

53. Fukushima, Y., Watanabe, T., Ohnishi, R. et al., Optimization of photoacid generator in photoacid generation-bonded resist, *Japanese Journal of Applied Physics*, **47**(8), 6293 (2008).

54. Wu, H. P. and Gonsalves, K. E., Preparation of a photoacid generating monomer and its application in lithography, *Advanced Functional Materials*, **11**(4), 271 (2001).

55. Hirayama, T., Shiono, D., Hada, H., and Onodera, J., New photoresist based on amorphous low molecular weight polyphenols, *Journal of Photopolymer Science and Technology*, **17**(3), 435 (2004).

56. VanderHart, D. L., De Silva, A., Felix, N., Prabhu, V. M., and Ober, C. K., The effect of EUV molecular glass architecture on the bulk dispersion of a photo-acid generator, *Advances in Resist Materials and Processing Technology XXV, SPIE*, **6923**, 69231M (2008).

57. Young-Gil, K., Kim, J. B., Fujigaya, T., Shibasaki, Y., and Ueda, M., A positive-working alkaline developable photoresist based on partially tert-Boc-protected calix[4]resorcinarene and a photoacid generator, *Journal of Materials Chemistry*, **12**(1), 53 (2002).

58. Chang, S. W., Ayothi, R., Bratton, D. et al., Sub-50 nm feature sizes using positive tone molecular glass resists for EUV lithography, *Journal of Materials Chemistry*, **16**(15), 1470 (2006).

59. Yang, D., Chang, S. W., and Ober, C. K., Molecular glass photoresists for advanced lithography, *Journal of Materials Chemistry*, **16**(18), 1693 (2006).

60. De Silva, A., Lee, J. K., Andre, X. et al., Study of the structure-properties relationship of phenolic molecular glass resists for next generation photolithography, *Chemistry of Materials*, **20**(4), 1606 (2008).

61. Felix, N. M., De Silva, A., Luk, C. M. Y., and Ober, C. K., Dissolution phenomena of phenolic molecular glass photoresist films in supercritical CO2, *Journal of Materials Chemistry*, **17**(43), 4598 (2007).

62. De Silva, A., Felix, N., Sha, J., Lee, J. K., and Ober, C. K., Molecular glass resists for next generation lithography, *Advances in Resist Materials and Processing Technology XXV, SPIE*, **6923**, 69231L (2008).

63. De Silva, A., Felix, N., Forman, D., Sha, J., and Ober, C. K., New architectures for high resolution patterning, *Advances in Resist Materials and Processing Technology XXV, SPIE*, **6923**, 69230O (2008).

64. Toriumi, M., Santillan, J., Itani, T., Kozawa, T., and Tagawa, S., Dissolution characteristics and reaction kinetics of molecular resists for extreme-ultraviolet lithography, *Journal of Vacuum Science & Technology B*, **25**(6), 2486 (2007).

65. Nishimura, I., Heath, W. H., Matsumoto, K. et al., Non-chemically amplified resists for 193 nm lithography, *Advances in Resist Materials and Processing Technology XXV, SPIE*, **6923**, 69231C (2008).

66. Lawson, R. A., Lee, C. T., Yueh, W., Tolbert, L., and Henderson, C. L., Single molecule chemically amplified resists based on ionic and non-ionic PAGs, *Advances in Resist Materials and Processing Technology XXV, SPIE*, **6923**, 69230K (2008).

67. Naulleau, P. P., Anderson, C. N., Chiu, J. et al., Latest results from the SEMATECH Berkeley extreme ultraviolet microfield exposure tool, *Journal of Vacuum Science & Technology B* **27**, 66 (2009).

68. Woodward, J. T., Choi, K. W., Prabhu, V. M. et al., Characterization of the latent image to developed image in model EUV photoresists, *Advances in Resist Materials and Processing Technology XXV, SPIE*, **6923**, 69232B (2008).

69. Choi, K. W., Prabhu, V. M., Lavery, K. A. et al., Effect of photo-acid generator concentration and developer strength on the patterning capabilities of a model EUV photoresist, *Proceedings of SPIE*, **6519**, 651943 (2007).

70. Zimmerman, P. A., Byers, J., Piscani, E. et al., Development of an operational high refractive index resist for 193 nm immersion lithography, *Advances in Resist Materials and Processing Technology XXV, SPIE*, **6923**, 692306 (2008).

71. Matsumoto, K., Costner, E., Nishimura, I., Ueda, M., and Willson, C. G., High-index resist for 193-nm immersion lithography, *Advances in Resist Materials and Processing Technology XXV, SPIE*, **6923**, 692305 (2008).

72. Gonsalves, K. E., Wang, M., and Pujari, N. S., High refractive-index resists composed of anionic photoacid generator (PAG) bound polymers for 193 nm immersion lithography, *Advances in Resist Materials and Processing Technology XXV, SPIE*, **6923**, 69231P (2008).

73. Liu, H. P., Blakey, I., Conley, W. E. et al., Application of quantitative structure property relationship to the design of high refractive index 193i resist, *Journal of Micro-Nanolithography MEMS and MOEMS*, **7**(2), 023001 (2008).

74. Allen, R. D., Brock, P. J., Sundberg, L. et al., Design of protective topcoats for immersion lithography, *Journal of Photopolymer Science and Technology*, **18**(5), 615 (2005).

75. Terai, M., Kumada, T., Ishibashi, T. et al., Mechanism of immersion specific defects with high receding-angle topcoat, *Advances in Resist Materials and Processing Technology XXIV, SPIE*, **6519**, 65191S (2007).

76. Takebe, Y., Shirota, N., Sasaki, T., and Yokokoji, O., Development of top coat materials for ArF immersion lithography, *Advances in Resist Materials and Processing Technology XXIV, SPIE*, **6519**, 65191Y (2007).

77. Nakagawa, H., Goto, K., Shima, M. et al., Process optimization for developer soluble immersion topcoat material, *Advances in Resist Materials and Processing Technology XXIV, SPIE*, **6519**, 651923 (2007).

78. Sundberg, L. K., Sanders, D. P., Sooriyakumaran, R., Brock, P. J., and Allen, R. D., Contact angles and structure/surface property relationships of immersion materials, *Advances in Resist Materials and Processing Technology XXIV, SPIE,* **6519**, 65191Q (2007).

79. Shirota, N., Takebe, Y., Wang, S. Z., Sasaki, T., and Yokokoji, O., Development of non-topcoat resist polymers for 193-nm immersion lithography, *Advances in Resist Materials and Processing Technology XXIV, SPIE,* **6519**, 651905 (2007).

80. Wada, K., Kanna, S., and Kanda, H., Novel materials design for immersion lithography, *Advances in Resist Materials and Processing Technology XXIV, SPIE,* **6519**, 651908 (2007).

81. Wu, S., Tseng, A., Lin, B. et al., Non-topcoat resist design for immersion process at 32-nm node, *Advances in Resist Materials and Processing Technology XXV, SPIE,* **6923**, 692307 (2008).

82. Sanders, D. P., Sundberg, L. K., Brock, P. J. et al., Self-segregating materials for immersion lithography, *Advances in Resist Materials and Processing Technology XXV, SPIE,* **6923**, 692309 (2008).

83. Wang, D., Caporale, S., Andes, C. et al., Design consideration for immersion 193: Embedded barrier layer and pattern collapse margin, *Journal of Photopolymer Science and Technology,* **20**(5), 687 (2007).

84. Smith, B. W., Optics for photolithography, in K. Suzuki and B. W. Smith (eds.), *Microlithography: Science and Technology,* 2nd edn., CRC Press, Boca Raton, FL, (2007).

85. Lin, B. J., The k_3 coefficient in nonparaxial λ/NA scaling equations for resolution, depth of focus, and immersion lithography, *Journal of Microlithography Microfabrication and Microsystems,* **1**(1), 7 (2002).

86. Oldham, W. G. and Schenker, R. E., 193-nm lithographic system lifetimes as limited by UV compaction, *Solid State Technology,* **40**(4), 95–102 (1997).

87. Sewell, H., McClay, J., Jenkins, P. et al., 157 nm lithography—Window of opportunity, *Journal of Photopolymer Science and Technology,* **15**(4), 569 (2002).

88. Burnett, J. H., Levine, Z. H., and Shirley, E. L., Intrinsic birefringence in calcium fluoride and barium fluoride, *Physical Review B,* **64**(24), 241102 (2001).

89. Burnett, J. H., Levine, Z. H., Shirley, E. L., and Bruning, J. H., Symmetry of spatial-dispersion-induced birefringence and its implications for CaF_2 ultraviolet optics, *Journal of Microlithography Microfabrication and Microsystems,* **1**(3), 213 (2002).

90. Kang, H., Bourov, A., and Smith, B. W., Optical lithography at a 126-nm wavelength, *Emerging Lithographic Technologies V, SPIE,* **4343**, 797 (2001).

91. Liberman, V., Rothschild, M., Murphy, P. G., and Palmacci, S. T., Prospects for photolithography at 121 nm, *Journal of Vacuum Science & Technology B,* **20**(6), 2567 (2002).

92. Trintchouk, F., Ishihara, T., Gillespie, W. et al., XLA-300: The fourth-generation ArF MOPA light source for immersion lithography, *Optical Microlithography XIX, SPIE,* **6154**, 615423 (2006).

93. Wong, A. K. K., *Resolution Enhancement Techniques in Optical Lithography,* SPIE Press, Bellingham, WA, (2001).

94. Smayling, M. C., Liu, H. Y., and Cai, L., Low k_1 logic design using gridded design rules, *Design for Manufacturability through Design-Process Integration II, SPIE,* **6925**, 69250B (2008).

95. Borodovsky, Y., Marching to the beat of Moore's Law, *Advances in Resist Technology and Processing XXIII, SPIE,* **6153**, 615301 (2006).

96. Bruning, J. H., Optical lithography: 40 years and holding, *Optical Microlithography XX, SPIE,* **6520**, 652004 (2007).

97. Matsuyama, T., Ohmura, Y., and Williamson, D. M., The lithographic lens: Its history and evolution, *Optical Microlithography XIX, SPIE,* **6154**, 615403 (2006).

98. Kawata, H., Carter, J., Yen, A., and Smith, H. I., Optical projection lithography using lenses with numerical apertures greater than unity, *Microelectronic Engineering,* **9**, 31 (1989).

99. Mulkens, J., Flagello, D., Streefkerk, B., and Graeupner, P., Benefits and limitations of immersion lithography, *Journal of Microlithography Microfabrication and Microsystems,* **3**(1), 104 (2004).

100. Owen, G., Pease, R. F. W., Markle, D. A. et al., 1/8 μm optical lithography, *Journal of Vacuum Science & Technology B,* **10**(6), 3032 (1992).

101. de Klerk, J., Wagner, C., Droste, R. et al., Performance of a 1.35NA ArF immersion lithography system for 40-nm applications, *Optical Microlithography XX, SPIE,* **6520**, 65201Y (2007).

102. Lin, B., Drivers, prospects, and challenges for immersion lithography, *Third International Symposium on 157 nm Lithography,* Antwerp, Belgium (2002).

103. Owa, S., Shiraishi, N., Tanaka, I. et al., Nikon F2 exposure tool, *Third International Symposium on 157 nm Lithography,* Antwerp, Belgium (2002).

104. Smith, B. W., Kang, H., Bourov, A., and Cropanese, F., Extreme-NA water immersion lithography for 35–65 nm technology, *Third International Symposium on 157 nm Lithography,* Antwerp, Belgium (2002).

105. Switkes, M. and Rothschild, M., Resolution enhancement of 157-nm lithography by liquid immersion, *Optical Microlithography XV, SPIE,* **4691**, 459 (2002).

106. Adam, K. and Maurer, W., Polarization effects in immersion lithography, *Journal of Microlithography Microfabrication and Microsystems,* **4**(3), 031106 (2005).

107. Kunz, R. R., Switkes, M., Sinta, R. et al., Transparent fluids for 157-nm immersion lithography, *Journal of Microlithography Microfabrication and Microsystems,* **3**(1), 73 (2004).

108. French, R. H., Qiu, W., Yang, M. K. et al., Second generation fluids for 193 nm immersion lithography, *Optical Microlithography XIX, SPIE,* **6154**, 615415 (2006).

109. Furukawa, T., Hieda, K., Wang, Y. et al., High refractive index fluid for next generation ArF immersion lithography, *Journal of Photopolymer Science and Technology*, **19**(5), 641 (2006).

110. Burnett, J. H., Kaplan, S. G., Shirley, E. L. et al., High-index optical materials for 193 nm immersion lithography, *Optical Microlithography XIX, SPIE,* **6154**, 615418 (2006).

111. Parthier, L., Wehrhan, G., Seifert, F. et al., Development update of high index lens material LuAG for ArF Hyper NA immersion systems, *Fifth International Symposium on Immersion Lithography Extensions*, The Hague, the Netherlands, 22–25 September 2008.

112. Jahromi, S., Bremer, L., Tuinier, R., and Liebregts, S., Development of third generation immersion fluids based on dispersion of nanoparticles, *Fifth International Symposium on Immersion Lithography Extensions*, The Hague, the Netherlands, 22–25 September 2008.

113. Zimmerman, P., Rice, B., Rodriguez, R. et al., High index fluids to enable 1.55 and higher NA 193 nm immersion imaging, *Fifth International Symposium on Immersion Lithography Extensions*, The Hague, the Netherlands, 22–25 September 2008.

114. Zimmerman, P. A., Rice, B., Rodriguez, R. et al., The use of nanocomposite materials for high refractive index immersion lithography, *Journal of Photopolymer Science and Technology*, **21**(5), 621 (2008).

115. Byers, J., Lee, S., Jeri, K. et al., Double exposure materials: Simulation study of feasibility, *Journal of Photopolymer Science and Technology*, **20**(5), 707 (2007).

116. Hori, M., Nagai, T., Nakamura, A. et al., Sub-40-nm half-pitch double patterning with resist freezing process, *Advances in Resist Materials and Processing Technology XXV, SPIE,* **6923**, 69230H (2008).

117. Jung, W. Y., Kim, C. D., Eom, J. D. et al., Patterning with spacer for expanding the resolution limit of current lithography tool, *Design and Process Integration for Microelectronic Manufacturing IV, SPIE,* **6156**, 61561J (2006).

118. Reimer, L., *Transmission Electron Microscopy: Physics of Image Formation and Microanalysis*, 4th edn., Springer-Verlag, Berlin, Germany, (1985).

119. Bolorizadeh, M. and Joy, D. C., Effects of fast secondary electrons to low-voltage electron beam lithography, *Journal of Micro-Nanolithography MEMS and MOEMS*, **6**(2), 023004 (2007).

120. Baylor, L. R., Lowndes, D. H., Simpson, M. L. et al., Digital electrostatic electron-beam array lithography, *Journal of Vacuum Science & Technology B*, **20**(6), 2646 (2002).

121. Utsumi, T., Low-energy e-beam proximity lithography (LEEPL): Is the simplest the best? *Japanese Journal of Applied Physics Part 1-Regular Papers Short Notes & Review Papers*, **38**(12B), 7046 (1999).

122. Utsumi, T., Present status and future prospects of LEEPL, *Microelectronic Engineering*, **83**(4–9), 738 (2006).

123. Bethe, H., The theory of the passage of rapid neutron radiation through matter, *Annalen Der Physik*, **5**(3), 325 (1930).

124. Broers, A. N., Resolution limits of PMMA resist for exposure with 50 kV electrons, *Journal of the Electrochemical Society*, **128**(1), 166 (1981).

125. Tanuma, S., Powell, C. J., and Penn, D. R., Calculations of stopping powers of 100 eV to 30 keV electrons in 10 elemental solids, *Surface and Interface Analysis*, **37**(11), 978 (2005).

126. Han, G., Khan, M., Fang, Y. H., and Cerrina, F., Comprehensive model of electron energy deposition, *Journal of Vacuum Science & Technology B*, **20**(6), 2666 (2002).

127. Jackson, J. D., *Classical Electrodynamics*, 3rd edn., John Wiley & Sons Inc., New York, (1999).

128. Salisbury, I. G., Timsit, R. S., Berger, S. D., and Humphreys, C. J., Nanometer scale electron-beam lithography in inorganic materials, *Applied Physics Letters*, **45**(12), 1289 (1984).

129. Langheinrich, W. and Beneking, H., The resolution of the inorganic electron-beam resist LiF(AlF(3)), *Microelectronic Engineering*, **23**(1–4), 287 (1994).

130. Muray, A., Isaacson, M., and Adesida, I., AiF3—A new very high-resolution electron-beam resist, *Applied Physics Letters*, **45**(5), 589 (1984).

131. Yang, J. K. W. and Berggren, K. K., Using high-contrast salty development of hydrogen silsesquioxane for sub-10-nm half-pitch lithography, *Journal of Vacuum Science & Technology B*, **25**(6), 2025 (2007).

132. Parikh, M. and Kyser, D. F., Energy deposition functions in electron resist films on substrates, *Journal of Applied Physics*, **50**(2), 1104 (1979).

133. Watson, G. P., Berger, S. D., Liddle, J. A. et al., Precise measurement of the effective backscatter coefficient for 100-keV electron-beam lithography on Si, *Journal of Vacuum Science & Technology B*, **13**(6), 2535 (1995).

134. Watson, G. P., Fetter, L. A., and Liddle, J. A., Dose modification proximity effect correction scheme with inherent forward scattering corrections, *Journal of Vacuum Science & Technology B*, **15**(6), 2309 (1997).

135. Owen, G. and Rissman, P., Proximity effect correction for electron-beam lithography by equalization of background dose, *Journal of Applied Physics*, **54**(6), 3573 (1983).

136. Watson, G. P., Berger, S. D., Liddle, J. A., and Waskiewicz, W. K., A background dose proximity effect correction technique for scattering with angular limitation projection electron lithography implemented in hardware, *Journal of Vacuum Science & Technology B*, **13**(6), 2504 (1995).

137. Groves, T. R., Efficiency of electron-beam proximity effect correction, *Journal of Vacuum Science & Technology B*, **11**(6), 2746 (1993).

138. Seo, E. and Kim, O., Dose and shape modification proximity effect correction for forward-scattering range scale features in electron beam lithography, *Japanese Journal of Applied Physics Part 1-Regular Papers Short Notes & Review Papers*, **39**(12B), 6827 (2000).

139. Lee, S. Y., Jacob, J. C., Chen, C. M., Mcmillan, J. A., and Macdonald, N. C., Proximity effect correction in electron-beam lithography—A hierarchical rule-based scheme-pyramid, *Journal of Vacuum Science & Technology B*, **9**(6), 3048 (1991).

140. Patrick, W. and Vettiger, P., Optimization of the proximity parameters for the electron-beam exposure of nanometer gate-length GaAs metal-semiconductor field-effect transistors, *Journal of Vacuum Science & Technology B*, **6**(6), 2037 (1988).

141. Wuest, R., Strasser, P., Jungo, M. et al., An efficient proximity-effect correction method for electron-beam patterning of photonic-crystal devices, *Microelectronic Engineering*, **67–8**, 182 (2003).

142. Jansen, G. H., *Coulomb Interactions in Particle Beams*, Academic Press, New York, (1990).

143. Mkrtchyan, M. M., Liddle, J. A., Berger, S. D. et al., Stochastic scattering in charged particle projection systems: A nearest neighbor approach, *Journal of Applied Physics*, **78**(12), 6888 (1995).

144. Stickel, W., Simulation of Coulomb interactions in electron beam lithography systems—A comparison of theoretical models, *Journal of Vacuum Science & Technology B*, **16**(6), 3211 (1998).

145. Gallatin, G. M. and Liddle, J. A., Analytical model of the "Shot Noise" effect in photoresist, *Microelectronic Engineering*, **46**(1–4), 365 (1999).

146. Gallatin, G. M., Continuum model of shot noise and line edge roughness, *Lithography for Semiconductor Manufacturing II, SPIE*, **4404**, 123 (2001).

147. Kruit, P., Steenbrink, S., and Wieland, M., Predicted effect of shot noise on contact hole dimension in e-beam lithography, *Journal of Vacuum Science & Technology B*, **24**(6), 2931 (2006).

148. Neureuther, A. R., Pease, R. F. W., Yuan, L. et al., Shot noise models for sequential processes and the role of lateral mixing, *Journal of Vacuum Science & Technology B*, **24**(4), 1902 (2006).

149. Smith, H. I., A statistical-analysis of ultraviolet, x-ray, and charged-particle lithographies, *Journal of Vacuum Science & Technology B*, **4**(1), 148 (1986).

150. Ferrera, J., Wong, V. V., Rishton, S. et al., Spatial-phase-locked electron-beam lithography—Initial test-results, *Journal of Vacuum Science & Technology B*, **11**(6), 2342 (1993).

151. Fares, N., Stanton, S., Liddle, J., and Gallatin, G., Analytical-based solutions for SCALPEL wafer heating, *Journal of Vacuum Science & Technology B*, **18**(6), 3115 (2000).

152. Babin, S. and Kuzmin, I. Y., Experimental verification of the TEMPTATION (temperature simulation) software tool, *Journal of Vacuum Science & Technology B*, **16**(6), 3241 (1998).

153. Chao, W. L., Harteneck, B. D., Liddle, J. A., Anderson, E. H., and Attwood, D. T., Soft x-ray microscopy at a spatial resolution better than 15 nm, *Nature*, **435**(7046), 1210 (2005).

154. Sturans, M. A., Hartley, J. G., Pfeiffer, H. C. et al., EL5: One tool for advanced x-ray and chrome on glass mask making, *Journal of Vacuum Science & Technology B*, **16**(6), 3164 (1998).

155. Nakayama, Y., Okazaki, S., Saitou, N., and Wakabayashi, H., Electron-beam cell projection lithography—A new high-throughput electron-beam direct-writing technology using a specially tailored Si aperture, *Journal of Vacuum Science & Technology B*, **8**(6), 1836 (1990).

156. Pfeiffer, H. C., Projection reduction exposure with variable axis immersion lenses (PREVAIL)-A high throughput e-beam projection approach for next generation lithography, *Japanese Journal of Applied Physics Part 1-Regular Papers Short Notes & Review Papers*, **38**(12B), 7022 (1999).

157. Heritage, M. B., Electron-projection microfabrication system, *Journal of Vacuum Science & Technology*, **12**(6), 1135 (1975).

158. http://www.darpa.mil/MTO/Programs/nanowriter/index.html, 2009.

159. Kruit, P., High throughput electron lithography with the multiple aperture pixel by pixel enhancement of resolution concept, *Journal of Vacuum Science & Technology B*, **16**(6), 3177 (1998).

160. Chang, T. H. P., Thomson, M. G. R., Kratschmer, E. et al., Electron-beam microcolumns for lithography and related applications, *Journal of Vacuum Science & Technology B*, **14**(6), 3774 (1996).

161. Groves, T. R. and Kendall, R. A., Distributed, multiple variable shaped electron beam column for high through-put maskless lithography, *Journal of Vacuum Science & Technology B*, **16**(6), 3168 (1998).

162. Pease, R. F. and Chou, S. Y., Lithography and other patterning techniques for future electronics, *Proceedings of IEEE*, **96**(2), 248 (2008).

163. Sreenivasan, S. V., Nanoscale manufacturing enabled by imprint lithography, *MRS Bulletin*, **33**(9), 854 (2008).

164. Barty, A. and Goldberg, K. A., The effects of radiation induced carbon contamination on the performance of an EUV lithographic optic, *Proceedings of SPIE*, **5037**, 450 (2003).

165. Brainard, R. L., Henderson, C., Cobb, J. et al., Comparison of the lithographic properties of positive resists upon exposure to deep- and extreme-ultraviolet radiation, *Journal of Vacuum Science & Technology B*, **17**, 3384 (1999).

166. Brainard, R. L., Barclay, G. G., Anderson, E. H., and Ocola, L. E., Resists for next generation lithography, *Microelectronic Engineering*, **62**(707), 715 (2002).

167. Chou, S. Y., Krauss, P. R., and Renstrom, P. J., Imprint of sub-25 Nm vias and trenches in polymers, *Applied Physics Letters*, **67**(21), 3114 (1995).

168. Chou, S. Y., Krauss, P. R., and Renstrom, P. J., Imprint lithography with 25-nanometer resolution, *Science*, **272**(5258), 85 (1996).

169. Haisma, J., Verheijen, M., vandenHeuvel, K., and van den Berg, J., Mold-assisted nanolithography: A process for reliable pattern replication, *Journal of Vacuum Science & Technology B*, **14**(6), 4124 (1996).

170. Colburn, M., Johnson, S., Stewart, M. et al., Step and flash imprint lithography: A new approach to high-resolution patterning, *Proceedings of SPIE*, **3676**, 379 (1999).

171. Bratton, D., Yang, D., Dai, J. Y., and Ober, C. K., Recent progress in high resolution lithography, *Polymers for Advanced Technologies*, **17**(2), 94 (2006).

172. Gao, H., Tan, H., Zhang, W., Morton, K., and Chou, S. Y., Air cushion press for excellent uniformity, high yield, and fast nanoimprint across a 100 mm field, *Nano Letters*, **6**(11), 2438 (2006).

173. Huang, X. D., Bao, L. R., Cheng, X. et al., Reversal imprinting by transferring polymer from mold to substrate, *Journal of Vacuum Science & Technology B*, **20**(6), 2872 (2002).

174. Schmid, G. M., Stewart, M. D., Wetzel, J. et al., Implementation of an imprint damascene process for interconnect fabrication, *Journal of Vacuum Science & Technology B*, **24**(3), 1283 (2006).

175. Kim, K., Jeong, J., Park, S. et al., Development of a very large-area ultraviolet imprint lithography process, *Microelectronic Engineering*, **86**, 1983 (2009).

176. Ahn, S. H. and Guo, L. J., High-speed roll-to-roll nanoimprint lithography on flexible plastic substrates, *Advanced Materials*, **20**(11), 2044 (2008).

177. Tan, H., Gilbertson, A., and Chou, S. Y., Roller nanoimprint lithography, *Journal of Vacuum Science & Technology B*, **16**(6), 3926 (1998).

178. Ahn, S. H., Kim, J. S., and Guo, L. J., Bilayer metal wire-grid polarizer fabricated by roll-to-roll nanoimprint lithography on flexible plastic substrate, *Journal of Vacuum Science & Technology B*, **25**(6), 2388 (2007).

179. PolyIC. http://www.polyic.com, 2009.

180. Hua, F., Sun, Y. G., Gaur, A. et al., Polymer imprint lithography with molecular-scale resolution, *Nano Letters*, **4**(12), 2467 (2004).

181. Austin, M. D., Zhang, W., Ge, H. X. et al., 6 nm half-pitch lines and 0.04 mu m(2) static random access memory patterns by nanoimprint lithography, *Nanotechnology*, **16**(8), 1058 (2005).

182. Ding, Y. F., Ro, H. W., Germer, T. A. et al., Relaxation behavior of polymer structures fabricated by nanoimprint lithography, *ACS Nano*, **1**(2), 84 (2007).

183. Ding, Y. F., Ro, H. W., Alvine, K. J. et al., Nanoimprint lithography and the role of viscoelasticity in the generation of residual stress in model polystyrene patterns, *Advanced Functional Materials*, **18**(12), 1854 (2008).

184. Jones, R. L., Hu, T. J., Soles, C. L. et al., Real-time shape evolution of nanoimprinted polymer structures during thermal annealing, *Nano Letters*, **6**(8), 1723 (2006).

185. Hirai, Y., Yoshida, S., and Takagi, N., Defect analysis in thermal nanoimprint lithography, *Journal of Vacuum Science & Technology B*, **21**(6), 2765 (2003).

186. Jung, G. Y., Li, Z. Y., Wu, W. et al., Vapor-phase self-assembled monolayer for improved mold release in nanoimprint lithography, *Langmuir*, **21**(4), 1158 (2005).

187. Miller, M., Brooks, C., Lentz, D. et al., Step and flash imprint process integration techniques for photonic crystal patterning: Template replication through wafer patterning irrespective of tone, *Proceedings of SPIE*, **6883**, 68830D (2008).

188. *International Technology Roadmap for Semiconductors*, 2007 edition, http://www.itrs.net, Metrology. SEMATECH, Austin, TX

189. Li, M. T., Chen, L., Zhang, W., and Chou, S. Y., Pattern transfer fidelity of nanoimprint lithography on six-inch wafers, *Nanotechnology*, **14**(1), 33 (2003).

190. Lee, H. J., Ro, H. W., Soles, C. L. et al., Effect of initial resist thickness on residual layer thickness of nanoimprinted structures, *Journal of Vacuum Science & Technology B*, **23**, 3023 (2005).

191. Schulz, H., Wissen, M., and Scheer, H. C., Local mass transport and its effect on global pattern replication during hot embossing, *Microelectronic Engineering*, **67–68**, 657 (2003).

192. Kim, K. D., Jeong, J. H., Sim, Y. S., and Lee, E. S., Minimization of residual layer thickness by using the optimized dispensing method in S-FIL (TM) process, *Microelectronic Engineering*, **83**, 847 (2006).

193. Melliar-Smith, M., Lithography beyond 32 nm—A role for imprint? *Proceedings of SPIE*, **6520**, 652001 (2007).

194. Zhang, W. and Chou, S. Y., Multilevel nanoimprint lithography with submicron alignment over 4 in. Si wafers, *Applied Physics Letters*, **79**(6), 845 (2001).

195. Resnick, D. J., Dauksher, W. J., Mancini, D. et al., Imprint lithography for integrated circuit fabrication, *Journal of Vacuum Science & Technology B*, **21**(6), 2624 (2003).

196. Jung, G. Y., Johnston-Halperin, E., Wu, W. et al., Circuit fabrication at 17 nm half-pitch by nanoimprint lithography, *Nano Letters*, **6**(3), 351 (2006).

197. Hand, A., *Toshiba Validates Imprint Lithography for < 32 nm*, Semiconductor International, October, (2007).

198. Hart, M. W. IBM Almaden Research Center, DARPA Presentation. 2007.

199. Dobisz, E. A., Bandic, Z. Z., Tsai-Wei, W., and Albrecht, T., Patterned media: Nanofabrication challenges of future disk drives, *Proceedings of IEEE*, **96**(11), 1836 (2008).

200. Ahn, S. W., Lee, K. D., Kim, J. S. et al., Fabrication of a 50 nm half-pitch wire grid polarizer using nanoimprint lithography, *Nanotechnology*, **16**(9), 1874 (2005).

201. Yu, Z. N., Deshpande, P., Wu, W., Wang, J., and Chou, S. Y., Reflective polarizer based on a stacked double-layer sub-wavelength metal grating structure fabricated using nanoimprint lithography, *Applied Physics Letters*, **77**(7), 927 (2000).

202. Raymond, C., Scatterometry for semiconductor metrology, in A. C. Diebold (ed.), *Handbook of Silicon Semiconductor Metrology*, Dekker, New York, p. 477 (2001).

203. SEMI Standard. Test method for evaluation of line edge roughness and line width roughness, SEMI P47-0307.

204. Jones, R. L., Lin, E. K., Lin, Q. et al., Cross-section and critical dimension metrology in dense high aspect ratio patterns with CD-SAXS, *AIP Conference Proceedings*, **788**, 403 (2005).

205. Notte, J., Ward, B., Economou, N. et al., An introduction to the helium ion microscope, *AIP Conference Proceedings*, **931**, 489 (2007).

206. Wang, C. Q., Jones, R. L., Lin, E. K. et al., Characterization of correlated line edge roughness of nanoscale line gratings using small angle x-ray scattering, *Journal of Applied Physics*, **102**(2), 024901 (2007).

207. Wang, C. Q., Jones, R. L., Lin, E. K., Wu, W. L., and Leu, J., Small angle x-ray scattering measurements of lithographic patterns with sidewall roughness from vertical standing waves, *Applied Physics Letters*, **90**(19), 193122-3 (2007).

208. Hu, T. J., Jones, R. L., Wu, W. L. et al., Small angle x-ray scattering metrology for sidewall angle and cross section of nanometer scale line gratings, *Journal of Applied Physics*, **96**(4), 1983 (2004).

209. Jones, R. L., Hu, T., Lin, E. K. et al., Small angle x-ray scattering for sub-100 nm pattern characterization, *Applied Physics Letters*, **83**(19), 4059 (2003).

Extreme Ultraviolet Lithography

Obert R. Wood II
GLOBALFOUNDRIES

20.1 Introduction

Extreme ultraviolet (EUV) lithography extends conventional optical lithography to higher resolution because it utilizes an exceedingly short imaging wavelength (λ = 13.5 nm) and provides a larger depth of focus because it employs a small numerical aperture (NA = 0.25) imaging system. Although it has a number of similarities with visible, deep ultraviolet (DUV), and 193 nm lithography, EUV lithography (EUVL) presents several unique technical challenges. For example, because EUV radiation is not transmitted through ambient air, an EUVL tool must be operated in a vacuum environment. Since all solid materials strongly absorb EUV radiation, conventional refractive optics is not an option and reflective reticles and optics must be employed. Since no naturally occurring material provides more than about 1% reflectivity at normal incidence in the EUV spectral region, EUV mirrors must be coated with special multilayer (ML) reflective coatings. Since EUV sources are inefficient, a high-temperature plasma source requiring extremely high input power and careful management of waste heat must be used. Furthermore, because hydrocarbons and water vapor are cracked by EUV radiation, depositing carbon on mirror surfaces and oxidizing their reflective coatings, hydrocarbon and water vapor levels in an EUV exposure tool must be carefully controlled. While the development of EUVL has been and continues to be challenging, the successful insertion of EUVL into semiconductor manufacturing will bring many benefits, among which are a much wider process window and an ability to print linewidths at 22 nm and smaller.

The key components of a system for EUVL include (1) an illumination system consisting of a plasma EUV radiation source, a condenser to collect the light, and an illuminator; (2) a patterned reflective mask; (3) a 4×-reduction reflective imaging system comprised of six or more highly precise ML-coated aspheric mirrors; and (4) an EUV-sensitive resist. A diagram illustrating the key components in an EUVL exposure tool is shown in Figure 20.1.

Although a variety of sources can generate EUV radiation, the source illustrated in Figure 20.1 is a high-temperature plasma of the kind that can be produced by focusing a high-power pulsed laser on a target of lithium, tin, or xenon. EUV radiation reflected from a patterned reticle located on the mask stage is imaged by 4×-reduction projection optics onto a resist-coated silicon wafer located on the wafer stage. EUVL exposure tools use the same step and scan architecture employed in today's DUV and 193 nm exposure tools and will ultimately operate at similar throughputs, approximately 80–100 wafers/h.

The development of EUV imaging systems from the early 1980s to the present is summarized in Section 20.2, which also describes the key technology developments that made EUVL possible: ML reflective coatings, highly precise aspheric mirrors, and phase-shifting point-diffraction interferometers. The current state of the art in EUVL is summarized in Section 20.3. The current status of EUVL infrastructure development in sources, masks, resists, mask handling, and optics and mask contamination is described in Section 20.4. The elements of EUV technology at highest risk of not being ready when needed and the timing of EUV insertion into manufacturing are discussed in Section 20.5, and an extensive list of references is given at the end.

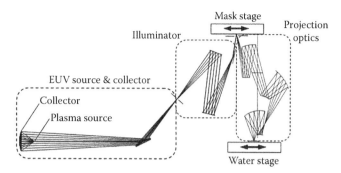

FIGURE 20.1 Sketch showing the major subsystems in an exposure tool for extreme ultraviolet lithography.

20.2 Historical Background

20.2.1 EUV Imaging System Development

The most important developments in EUV imaging systems that took place in Europe, Japan, and the United States from 1985 to the present are illustrated in Figure 20.2.

The surface of each optical element in an EUV imaging system must be provided with a ML coating that reflects EUV radiation at near-normal angles of incidence. The use of normal-incidence reflective optics for x-ray microscopes and x-ray telescopes was proposed first in the early 1980s,[1-3] but efficient ML reflective coatings were not sufficiently advanced at that time to be of much use and the results of the early soft x-ray imaging experiments were not very encouraging. The earliest EUV imaging experiments were carried out using projection optics consisting of two spherical mirrors, a convex primary, and a concave secondary, arranged in a Schwarzschild form. In fact, projection optics of this type were used by Hiroo

Kinoshita of NTT in 1985 in the very first EUVL imaging experiments. Initially, the imaging system mirrors consisted of tungsten–carbon MLs deposited on SiC substrates.[4] Later, an imaging system with reflective coatings designed to work near the absorption edge of Si was assembled at the High Energy Physics Laboratory in Tsukuba, Japan, but the alignment accuracy of the optics was poor and all of the replicated patterns were severely distorted. In 1984, an image of a 4 μm line and space pattern was finally produced and in 1986, at the Autumn Meeting of the Japan Society of Applied Physics, the results of the early imaging experiments were first made public.[5] Other early experiments that employed two-spherical-mirror projection optics were the demonstration by Bjorkholm et al. of AT&T in 1990 of diffraction-limited imaging near 14 nm wavelength using 20× reduction 0.08 NA Mo/Si ML-coated Schwarzschild optics to print features as small as 0.05 μm in poly (methyl metacrylate) resist,[6] and the development by Tichenor et al. of Sandia National Laboratory (SNL) in 1995 of the first EUVL laboratory tool capable of precise overlay combining 10× reduction, 0.088 NA Schwarzschild optics with accurate mask and wafer stages, and an integrated through-the-lens alignment system.[7]

The second stage in the evolution of EUVL imaging systems involved the development of two-aspherical-mirror projection optics. In 1993, Kinoshita et al. of NTT fabricated a 5× reduction, 0.07 NA two-aspherical-mirror imaging system with a 20 × 0.4 mm exposure area[8] based on a design first described by Kurihara et al.[9] and demonstrated the imaging of features as small as 0.25 μm across a 10 × 0.6 mm area and imaging of 0.15 μm features over a portion of a 2 × 0.6 mm ring-shaped field. In 1996, Haga et al. succeeded in expanding the exposure area of the two-aspherical-mirror imaging system to 20 × 25 mm by

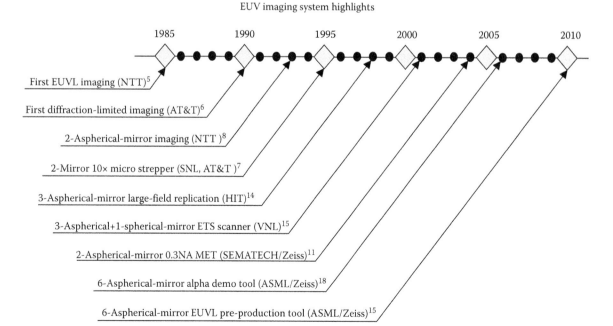

FIGURE 20.2 Major developments in EUVL imaging systems from 1985 to present.

utilizing new critical-illumination optics to illuminate a 20 × 0.6 mm ring-shaped area of the mask and by scanning the mask and wafer stages in synchronism.[10] In 2004, Naulleau et al. of Lawrence Berkeley National Laboratory (LBNL) demonstrated the printing capabilities of a 0.3 NA EUV micro-exposure tool (MET) at the Advanced Light Source in Berkeley.[11] The MET's 0.3 NA 5×-reduction two-aspherical-mirror imaging system (with maximum aspheric departure of 3.82 μm on the primary and 5.61 μm on the secondary), manufactured by Carl Zeiss in Germany, had an annular pupil with a central obscuration radius equal to 30% of the full pupil, a 1 × 3 mm field of view at the reticle plane (200 × 600 μm at the wafer plane), and provided ~22.5 nm resolution with conventional illumination (Rayleigh $k_1 = 0.5$) and smaller than 16 nm resolution with dipole illumination (Rayleigh $k_1 = 0.35$).[12]

The third stage in the evolution of EUV imaging systems involved the development of projection optics, which could support the larger exposure fields needed for device demonstrations. In 1996, Murakami et al. of Nikon Corporation described the design of a three-aspherical-mirror ring-field imaging system.[13] In 1998, Kinoshita et al. of Himeji Institute of Technology (Hyogo University today) described a three-aspherical mirror system for EUVL that included illumination optics, a scanning and alignment mechanism, a 0.1 NA 5× reduction imaging system that utilized three aspherical mirrors and one plane mirror, which supported 0.1 μm resolution printing by scanning a 1 mm wide by 30 mm long ring-shaped image field across a wafer, and a load-lock chamber for bringing wafers into and out of vacuum.[14] In 2001, Tichenor et al. of the Virtual National Laboratory, with personnel from LBNL, Lawrence Livermore National Laboratory (LLNL), and SNL, described system integration and performance of the EUV Engineering Test Stand (ETS) developed for the EUVLLC,[15] an Intel-led consortium of semiconductor manufacturers formed in 1997 to accelerate the commercialization of EUVL. The ETS's 0.1 NA 4× reduction projection optics, consisting of three aspherical mirrors (with max aspheric departures less than 10 μm) and one spherical mirror, provided a ±0.5 μm depth of focus (DOF), a scanned field size of 24 × 32.8 mm, and 100 nm resolution imaging. The extensive experimental work on the ETS and the involvement of personnel from six semiconductor companies in the EUVLLC project resulted in a high level of confidence that EUVL was ready for commercial development.

The most recent stage in the evolution of EUV imaging systems includes the development of six-aspherical-mirror ring-field projection optics by each of the three major suppliers of commercial lithographic exposure tools. The 0.25 NA 4× reduction imaging system in ASML's full-field scanning EUV Alpha Demo Tool (ADT), two of which were delivered to customers in 2007, is designed to print 40 nm dense lines and spaces over a 26 × 33 mm field and has recently been used to print 28 nm dense lines and spaces in a state-of-the-art resist.[16] The 0.25 NA 4× reduction six-aspherical-mirror imaging system in Nikon's EUV1 process development tool is designed to print 45 nm dense lines over a 26 × 23 mm field, and a set of EUV1 projection optics has recently demonstrated an extraordinarily low 0.4 nm root mean square (rms)

wavefront error over a 26 × 2 mm static exposure field.[17] ASML's Model NXE: 3100 EUV preproduction tool, which is expected to be delivered to customers in 2010, will include a 0.25 NA six-aspherical-mirror imaging system with even lower wavefront error and with reduced imaging system flare.[16]

20.2.2 EUVL-Enabling Technologies

Four scientific and technical breakthroughs critical to the development of EUVL occurred between 1981 and 1996. These four major advances were the first demonstrations of normal-incidence imaging in the soft x-ray spectral region, the development of efficient ML reflective coatings, the fabrication and characterization of precision aspheres, and the improvement of phase-measuring interferometry (PMI).

20.2.2.1 Multilayer Reflective Coatings

Below 30 nm wavelength, no naturally occurring material provides more than about 1% normal incidence reflectance because the index of refraction of all materials in this region is nearly unity (actually slightly less than one). In 1972, Eberhard Spiller of IBM showed that ML films composed of two materials that have nearly identical index of refraction, but that have widely different values of EUV absorption, can lead to constructive interference and, hence, enhanced reflection in a narrow wavelength band peaked at the Bragg wavelength, $\lambda = 2\Lambda \sin(\theta)$, where Λ is the ML period and θ is the angle of incidence.[18] In 1985, Barbee et al.[19] demonstrated that the particular material combination of molybdenum and silicon results in an exceptionally high normal-incidence reflectivity at wavelengths just longer than the Si L absorption edge at 12.4 nm. Today, ML coatings can provide a usable normal incidence reflectance in narrow bands of EUV wavelength from 4.5 to 30 nm.[20] The normal incidence reflectance of the best modern Mo/Si ML, an example of which is shown in Figure 20.3, now approaches 70% in a narrow band of wavelengths near 13.5 nm.[21]

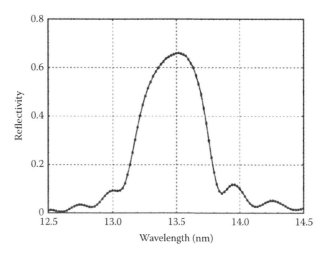

FIGURE 20.3 Reflectance of a Mo/Si ML measured with a commercial EUV reflectometer. (Reprinted from Wood, O., *AIP Conf. Proc.*, 931, 375, 2007. With permission.)

To maximize tool throughput, each of the mirrors in EUVL projection optics must have the highest possible reflectivity at 13.5 nm. To maximize imaging performance, the central wavelength of the reflectivity must be controlled to better than 0.01 nm. To support diffraction-limited printing, the mirror surfaces must have surface figure errors of the order of about 0.1 nm rms, and the ML coatings must not contribute more than approximately one-quarter of the total error, a requirement that leads to strict requirements on the thickness uniformity. The processing conditions for the deposition of Mo/Si ML coatings have now been optimized to obtain high reflectivity, low internal stress, and graded thickness simultaneously, and ML coatings are now available commercially at a number of institutions around the globe. Nikon Corporation recently succeeded in fabricating an aspheric mirror for an EUV imaging system with as little as 60 pm rms surface figure change before and after ML coating (small enough so that its effect on wavefront error can be ignored).[17]

20.2.2.2 Fabrication and Characterization of Aspheric Mirrors

The progress that has taken place in the fabrication and characterization of aspheric mirrors since the early 1990s is impressive. In the beginning, only low-departure aspheres could be fabricated and these were found to be only marginally usable in two-mirror EUV imaging systems. Polishing processes for aspheric mirrors have since been developed that can simultaneously reduce low spatial frequency roughness (LSFR), mid-spatial frequency roughness (MSFR), and high spatial frequency roughness (HSFR). Today, aspheres that meet the stringent requirements of a high-NA six-mirror EUV lithographic imaging system, less than 0.1 nm rms for LSRF, MSFR, and HSFR, are available commercially. Figure 20.4 shows some recent data from Nikon Corporation from a polished aspheric mirror for its EUV1 projection optics on which Nikon achieved an LSFR of 38 pm rms, an MSFR of 80 nm rms, and a HSFR of 68 nm rms.[17]

Surface figure data at low spatial frequencies, obtained from visible light interferometry, are needed to ensure diffraction-limited imaging. Surface finish data at high- and mid-spatial frequencies, obtained with an atomic force microscope and a white-light interferometric microscope, are needed to predict the distribution of scattered radiation in the imaging system. Surface roughness at high-spatial frequencies results in a loss of EUV reflectivity and, hence, system throughput. Surface roughness at mid-spatial frequencies can scatter light into the image plane and result in a loss of image contrast. For this reason, MSFR and the imaging system flare it causes is a serious concern for EUVL even today. A complete set of power spectral density (PSD) surface height data from an asphere for Nikon's EUV1 projection optics is also shown in Figure 20.4.[17] The smooth PSD curve shows that a high degree of precision has been successfully achieved in each of the important spatial wavelength regions.

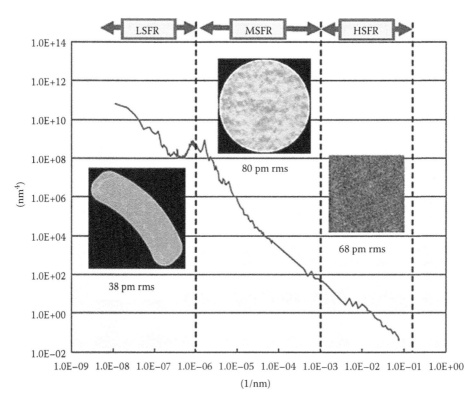

FIGURE 20.4 Plot of surface height data from a polished aspheric mirror for the projection optics of Nikon's EUV1 projection optics expressed as power spectral density (nm⁴) versus spatial frequency (nm⁻¹). (Reprinted from Murakami, K. et al., *Proc. SPIE*, 6921, 69210Q, 2008. With permission.)

20.2.2.3 The Phase-Shifting Point- Diffraction Interferometer

To achieve the highest possible resolution (diffraction-limited printing), the wavefront error of an imaging system must be less than $\lambda/14$, where λ is the operating wavelength (less than 1 nm when $\lambda = 13.5$ nm). For a six-mirror imaging system, this means that the error contributions from each individual mirror can be no larger than about 0.1 nm rms (assuming that the errors are uncorrelated and taking into account the doubling of the error that occurs on reflection). Visible-light phase-measuring interferometry (PMI) is the metrology most commonly used to provide the feedback needed for figuring a mirror surface. Since PMI is a comparative technique, its accuracy is limited by the quality of the reference. Recognizing this, Gary Sommargren (LLNL) in 1996 developed a new type of PMI, called phase-shifting diffraction interferometry (PSDI) that utilized a reference based on diffraction.[22] An experimental arrangement for PSDI configured to measure the surface figure of a concave off-axis aspheric mirror is illustrated in Figure 20.5.

Light leaving the end of a single-mode optical fiber diffracts, forming a spherical wavefront. Part of this wavefront is incident on the mirror being tested and is reflected back toward the fiber. This aberrated wavefront reflects from a semitransparent metal film on the face of the fiber and interferes with part of the original spherical wavefront to produce an interference pattern on a charge-coupled-device camera. Subsequent work has shown that PSDI can result in measurement accuracies approaching 0.10 nm rms.[23]

The total wavefront error of an EUV imaging system depends not only on the surface figure and alignment of each mirror but also on the spatially varying properties of their ML coatings. EUV interferometry, with phase-shifting point-diffraction interferometry (PS/PDI) and various forms of lateral-shearing interferometry, has been used to test and align a number of EUV imaging systems.[24] In fact, the 0.3 NA SEMATECH Berkeley MET was interferometrically aligned at EUV wavelengths using the PS/PDI technique and its wavefront error was shown to be about 0.8 nm rms in a 37-term annular Zernike polynomial dominated by higher order spherical aberration.[25] Because the wavefront error measured with visible light is not exactly the same as that measured with EUV radiation due to the phase change in reflection that occurs at a ML, to ensure that the six-mirror projection optics in future EUV exposure tools have wavefront errors below 1 nm, EUV interferometry has been or is being developed by each of the major suppliers of commercial lithography tools.

20.3 Current State of the Art

EUVL development has accelerated significantly in the last few years. Alpha-class EUV exposure tools are now in operation in Europe, Japan, and the United States. Significant improvements in both line/space and contact-hole performance of EUV-resist materials have been made due in large part to the availability of EUV MET tools that provide high-quality and high-contrast aerial images. The first EUV device integration exercises have been carried out using full-field EUVL exposure tools in Albany, New York and in Leuven, Belgium. This section presents the current state of the art in EUV exposure tools and a brief summary of the results of the device integrations exercise.

20.3.1 EUVL Exposure Tools

Two full-field EUVL research and development exposure tools were delivered by ASML in 2006, one to the College of Nanoscale Science and Engineering in Albany, New York and the other to the IMEC microelectronics research laboratory in Leuven, Belgium.[26] Another full-field EUVL R&D tool was delivered by Nikon Corporation to the Semiconductor Leading

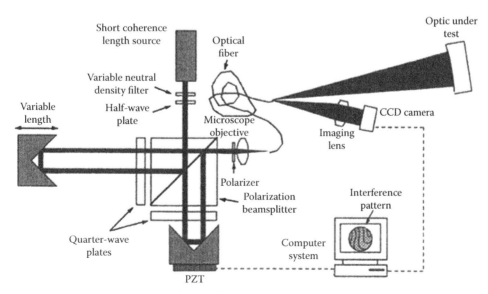

FIGURE 20.5 Experimental arrangement for a phase-shifting diffraction interferometer configured to measure the surface figure of a concave off-axis aspheric mirror. (Reprinted from Sommargren, G., *OSA Trends Opt. Photon.*, 4, 108, 1996. With permission.)

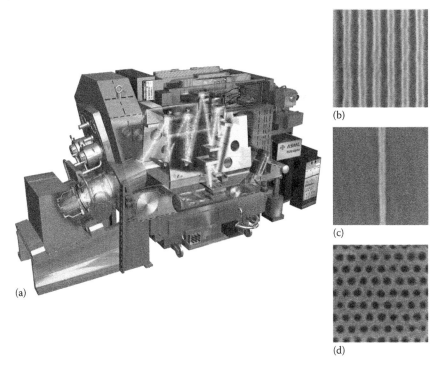

FIGURE 20.6   (See color insert following page 21-4.) (a) Artist illustration of ASML's EUV ADT showing a DPP EUV source on the left, illumination optics in the middle, and mask, projection optics, and wafer on the right. (Courtesy of ASML, Veldhoven, the Netherlands.) (b) Image of 28 nm lines and spaces printed in a 50 nm thick layer of a chemically amplified resist using an ASML ADT. (Reprinted from Meiling, H. et al., *Proc. SPIE*, **7271**, 727102, 2009. With permission.) (c) Image of 27 nm isolated line in 80 nm thick EUV-65 resist printed with an ASML ADT. (Reprinted from Pierson, B. et al., EUV resist performance on the ASML ADT and LBNL MET, *International EUVL Symposium*, Lake Tahoe, CA, 2008. With permission.) (d) Image of 28 nm diameter contact holes in 80 nm thick film of a chemically amplified resist printed at 45 mJ/cm^2 dose using an ASML ADT. (Reprinted from Koh, C. et al., *Proc. SPIE*, **7271**, 727124, 2009. With permission.)

Edge Technologies (Selete) consortium in Tsukuba, Japan in 2007.[27] An artist's rendering of an ASML EUV ADT is shown in Figure 20.6a.

The EUV source for the ADT is a gas discharge–produced plasma fueled by Sn vapor. The ADT utilizes a reflective illuminator to supply partially coherent illumination ($\sigma = 0.5$) to a 6″ × 6″ × 0.25″ patterned reflective mask. A 4× reduction 0.25 NA reflective imaging system, comprised of six highly precise ML-coated aspheric mirrors, produces an image of the mask on resist-coated 300 mm diameter wafers. The ADT utilizes the same step and scan architecture employed in today's DUV and 193 nm exposure tools, although its throughput is limited to a few wafers per hour by the available source power. The specifications for the ADT include 40 nm resolution for dense lines and spaces, 30 nm for isolated lines, and 55 nm for contact holes, a field size of 26 × 33 mm, an overlay error of 12 nm mean +3σ, and an imaging system flare of 16%.[28] Nikon's EUV1 exposure tool includes a 4× reduction 0.25 NA six-mirror imaging system with about 10% flare and an illumination system that provides $\sigma = 0.8$ conventional illumination, and was designed for use in process development at the 45 nm half-pitch node. The specifications for the EUV1 include 45 nm resolution for dense lines and spaces and a field size of 26 × 33 mm.[29] Using 0.25 NA EUV ADTs, ASML recently demonstrated 28 nm resolution for dense lines and spaces in a 50 nm thick layer of a state-of-the-art chemically amplified resist (CAR) (illustrated in Figure 20.6b).[15] ASML in joint work with AMD, IBM, and Toshiba demonstrated 27 nm resolution for isolated lines at a DOF of 200 nm (illustrated in Figure 20.6c),[30] and SEMATECH demonstrated 28 nm resolution for dense contact holes in 80 nm thick films of resist at a DOF of 150 nm (illustrated in Figure 20.6d).[31]

Micro-field exposure tools have played and continue to play a key role in the development of EUV resists. This is because the relative simplicity of these tools enables them to provide a higher aerial image contrast than the early alpha-class EUVL exposure tools, since they utilize higher NA projection optics (0.3 NA instead of 0.25 NA) and support more sophisticated illumination modes, and their owners have been more willing to risk contaminating the projection optics by agreeing to print with resist materials that have high outgassing rates. Two MET tools, the Canon 0.3 NA small-field exposure tool at Selete in Tsukuba, Japan[32] and the SEMATECH Berkeley MET at the Advanced Light Source in California[11] use synchrotron radiation as their EUV source. The other two, an Exitech MS-13 at Intel Corporation in Oregon and another at the SEMATECH Resist Test Center in Albany, New York use a Xe discharge–produced plasma source. A CAD model of the SEMATECH Berkeley MET illustrating the major components of the exposure system as well as the EUV beam path[33] is shown in Figure 20.7a.

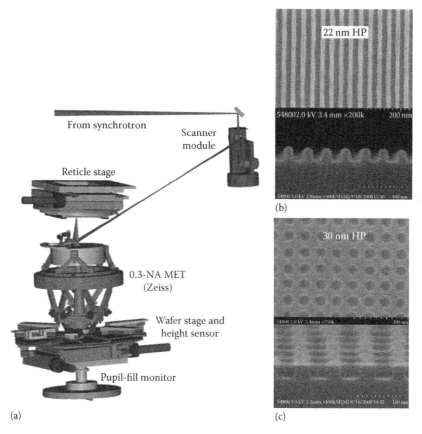

FIGURE 20.7 (a) CAD model of major exposure station components and beam path of the SEMATECH Berkeley MET. (b) Patterning of 1:1 dense lines and spaces in a 50 nm thick film of a chemically amplified resist at a sizing dose of 15.2 mJ/cm² using the SEMATECH Berkeley MET using rotated dipole illumination. (c) Patterning of 1:1 dense contact holes in a chemically amplified resist using the SEMATECH Berkeley MET using annular illumination. (Reprinted from Naulleau, P. et al., *Proc. SPIE*, 7271, 72710W, 2009. With permission.)

As mentioned in Section 20.2.1, the 0.3 NA MET optics support a resolution of 22.5 nm at a Rayleigh k_1 factor of 0.5 and a resolution of 16 nm at a Rayleigh k_1 factor of 0.35. Although the SEMATECH Berkeley MET utilizes the same projection optics design as the commercial METs, its unique lossless programmable-coherence illuminator allows it to print at considerably lower Rayleigh k_1 factor (that is, it supports better resolution) than the other MET tools. Another significant benefit of the SEMATECH Berkeley MET is that it is the only MET to have been interferometry-aligned at EUV wavelengths after integration into the tool at the operational location. Impressive resist performance improvements have been demonstrated for both line/space and contact-hole printing during the last year using the SEMATECH Berkeley MET. Figure 20.7b shows the printing of 22 nm half-pitch lines/spaces in a 50 nm thick resist film at about 20 mJ/cm² dose using 45° rotated dipole illumination[33]; Figure 20.7c shows the printing of 30 nm diameter contact holes at 48.6 mJ/cm² dose using 0.35/0.55 annular illumination.[33]

EUV interference lithography (IL) tools, like the ones at Paul Scherrer Institute (PSI) in Switzerland and at the University of Wisconsin in Madison, are exploring the ultimate resolution limits of resist materials and screening new candidate resist materials for lithographic performance before subjecting the expensive projection optics in the full-field EUV exposure tools to the risk of contamination from resist outgas products. The EUV-IL tool at PSI uses a spatially coherent EUV beam from the XIL beamline at the Swiss Light Source, with a central wavelength of 13.4 nm and a spectral width of about 2%–3%, to illuminate a transmission-diffraction-grating-based interferometer like the one illustrated in Figure 20.8a. The period of the interference fringes formed by this interferometer is equal to half the period of the diffraction gratings, which are patterned on thin silicon nitride membranes using electron beam lithography. Although the image contrast of EUV-IL can be influenced by a number of factors (transverse coherence length of the source, a difference in intensity between the two interfering EUV beams, mechanical instabilities, etc.), EUV-IL offers the advantage of nearly perfect aerial image contrast, performance that is difficult to achieve using projection or proximity lithography methods. EUV-IL at PSI has demonstrated resolution below 12.5 nm half-pitch in several electron beam resist materials. For example, a cross-sectional scanning electron microscope (SEM) image of 16.25 nm half-pitch lines and spaces in a 25 nm thick film of calixarene resist (*p*-chloromethyl-methoxy-calix[4]arene) is shown in Figure 20.8b[34] and a cross-sectional SEM image of 20 nm half-pitch lines and spaces in a 35 nm thick film of HSQ resist (hydrogen silsesquioxane) is shown in Figure 20.8c.[35]

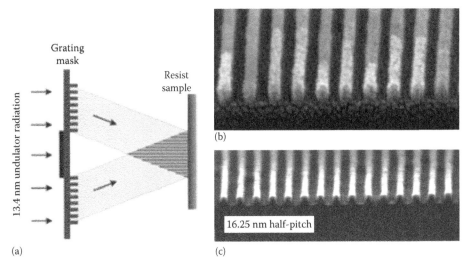

FIGURE 20.8 (a) Sketch of system for EUV interference lithography produced by EUV radiation from two transmission gratings illuminated with EUV radiation from an undulator beamline at a synchrotron light source. (b) Cross-sectional SEM image of HSQ resist patterns with half-pitch of 20 nm produced by the EUV interference lithography at the PSI. (Reprinted from Ekinci, Y. et al., *Microelectron. Eng.*, 84, 700, 2007. With permission from Elsevier.) (c) Cross-sectional SEM image of calixarene patterns with half-pitch of 16.25 nm produced by the EUV interference lithography at PSI. (Reprinted from Solak, H. et al., *J. Vac. Sci. Technol. B*, 25, 91, 2007. With permission.)

20.3.2 EUV Device Integration

Now that the first generation of alpha-class EUVL exposure tools has been deployed and major exposure tool suppliers have established detailed roadmaps for the ramp-up of EUVL technology, one of the main challenges facing the semiconductor industry is to begin integrating EUVL into the semiconductor fabrication process so the technology will be ready to support high-volume manufacturing (HVM) when needed. EUV device integration exercises not only provide a realistic test of technology readiness but also highlight any remaining technical issues. Two EUV device integration exercises were carried out in 2008 and a number of others are currently underway. In 2008, the ASML EUV ADT in Albany was integrated into a standard semiconductor manufacturing flow by the AMD/IBM/Toshiba Alliance and used to pattern the first metal interconnect level of AMD test chips fabricated using 45 nm logic design rules,[36] and the ADT at IMEC in Belgium was used to pattern the contact hole level of a 32 nm

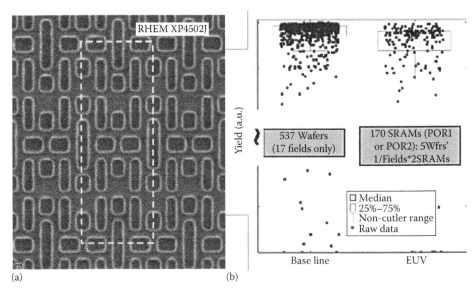

FIGURE 20.9 (a) EUV resist image of an area of the mask containing SRAM patterns. The unit cell is defined by the dashed rectangle. Rohm and Haas resist XP4502J with a film thickness of 150 nm was used. (Reprinted from La Fontaine, B. et al., *Proc. SPIE*, 6921, 69210P, 2008. With permission from SPIE.) (b) Comparison of yield from a baseline 45 nm process and a process using EUVL to print the M1 level. Each point in the baseline statistics represents a full wafer, whereas each point in the EUV statistics represents a 256 kbit SRAM area. (Reprinted from La Fontaine, B. et al., The future of EUV lithography, in Yamamoto, N. (ed), *Proceedings of SEMI Technology Symposium 2008*, SEMI Japan, Chiba, Japan, 2008. With permission.)

six-transistor SRAM cell based on FinFET technology.[37] More recently, EUV Alpha Demo Tools were used by the IBM/AMD/Toshiba Alliance in Albany to pattern two levels of a 22 nm test chip[38] by Samsung to pattern a contract hole level at the 4× DRAM node,[39] and by Hynix to compare the patterning fidelity of 193 nm immersion and EUVL at the 4×, 3×, and 2× DRAM nodes.[40]

Some results from the AMD/IBM/Toshiba Alliance EUV device integration exercise at the 45 nm logic node are shown in Figure 20.9. Figure 20.9a shows an EUVL image of memory structure (SRAM) patterns in a 150 nm thick film of Rohm and Haas XP4502J resist.[41] The SRAM unit cell is defined by the dashed rectangle. Once EUVL of the M1 layer was completed, the patterns were transferred to the underlying dielectric layer using reactive ion etching. The resulting trenches were filled with Cu using a damascene process and then polished. Electrical tests at the M1 level showed that individual transistors connected in this way had electrical characteristics very similar to those patterned using standard 193 nm immersion lithography. After further processing of the 45 nm node test chips to the fifth interconnect level, 256 kbit SRAMs were analyzed electrically for yield. The results in Figure 20.9b demonstrate the potential for EUVL to produce yield in extended devices when used to pattern one level. The yield data shown for the baseline process was calculated on full wafers, whereas the yield data for the EUV case was based on individual 256 kbit SRAM areas on five wafers.[41] Some of the best yielding 26 Mbit SRAM blocks had nearly 100% yield but did contain two failing cells. The location of the failing cells coincided with the location of a hard defect on the mask discovered during subsequent wafer and reticle inspection.

20.4 EUVL Infrastructure Status

Even after more than two decades of development, a number of technical challenges still need to be met before EUVL can be used in HVM. The steering committee of the International EUVL Symposium, with representatives from leading semiconductor manufacturers, EUVL Consortia in Europe, Japan, and the United States, and commercial EUVL exposure tool suppliers, has identified and ranked the list of critical technical issues on a yearly basis since 2002. The current list of critical issues, which was last updated and re-ranked in October 2008, is shown in Table 20.1. Even though this year's list is shorter than ever, the

TABLE 20.1 2008 EUV Focus Areas

Rank	Key Focus Areas
1	Long-term source operation with 100 W at IF and delivery of 5 MJ/day
2	Defect-free masks through lifecycle and inspection/review infrastructure
3	Resist resolution, sensitivity, and LEF met simultaneously
Areas of Significant Concern	
•	Reticle protection during storage, handling, and use
•	Projection/illuminator optics and mask lifetime

top three issues (focus areas, sources, masks, and resists) have remained at or near the top of each year's critical issue list and must be completely resolved before EUVL can be used in HVM.

20.4.1 EUV Sources

The wavelength and spectral bandwidth of the source for an EUVL exposure tool, a 2% wavelength band around 13.5 nm, are set by the ML reflective coatings on the imaging system mirrors. The source power is set by the required exposure tool throughput, which to first order is determined by the transmission of the optical train (collector efficiency and illuminator, reticle, and projection optic reflectivities) and the resist sensitivity. Lithography exposure tool manufacturers have agreed on the following top-level specifications for an HVM EUV source: 115 W of collected source power at a plane behind the first source collector optic called the intermediate focus (IF), an etendue limit of 3 mm² sr, and a repetition rate of 10 kHz or higher.[42] The cost of ownership (COO) for an EUV source is likely to be determined by the lifetime of the source collector, which in turn will be limited by the effectiveness of the source debris mitigation system. The current EUV source requirements are very severe and significant progress needs to be made to achieve the performance required for HVM while still maintaining an acceptable COO. For the later stages of HVM, EUV sources with up to 400 W of IF power are likely to be needed. All sources under consideration for EUVL are based on hot plasmas created by an electrical discharge or by laser irradiation. The current status of the two leading EUV source technologies, discharge-produced plasma (DPP) and laser-produced plasma (LPP) are described in the next two sections.

20.4.1.1 Discharge-Produced Plasma EUV Sources

The major components of a DPP EUV source-collector module, the DPP source, the debris mitigation system, and the collector optic, are illustrated in Figure 20.10a.

A number of discharge configurations have been evaluated for use in a DPP EUV source. What these discharge configurations have in common is the azimuthal magnetic field generated by the large axial discharge current between the discharge electrodes, which compresses the source fuel (typically Xe gas or Sn vapor) to the hot plasma (about 30 eV electron temperature) required for efficient generation of EUV radiation at 13.5 nm. The debris mitigation system for a DPP EUV source must protect the collector against (1) sputtering by high-energy ions and atoms, (2) contamination by tin fuel, and (3) contamination by electrode material. A reduction in the collector sputter rate by a factor of approximately 50,000 has been demonstrated through improvements in debris mitigation, and the lifetime of the collector in a Sn DPP source today is limited only by contamination. When the limits of active protection of the collector against contamination are reached, *in situ* cleaning will be needed to further prolong the collector lifetime. The collector used with DPP sources usually consists of a number of nested shells (in a so-called Wolter design) that are fabricated using a nickel electroforming process and coated with a thick layer of ruthenium

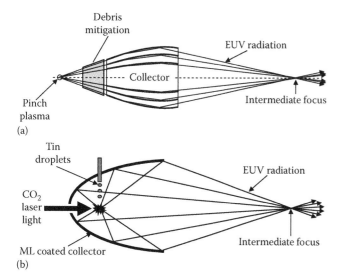

FIGURE 20.10 (a) Schematic diagram of an EUV source-collector module containing an EUV Sn DPP source. (b) Schematic diagram of an EUV source-collector module for a Sn droplet source powered by a carbon dioxide laser.

to provide high grazing-incidence reflectance of EUV radiation. At SPIE Advanced Lithography 2009, Philips Extreme UV in a joint Philips/Xtreme solution reported achieving 8 W of in-band power at IF using 170 W/2πsr Sn DPP EUV source, a foil-trap-based debris mitigation system, and a six-shell Ru-coated grazing-incidence collector.[43]

The advantages of a DPP EUV source are its lower cost of manufacturing, its lower level of complexity, and its use of a lower cost grazing-incidence collector optic. DPP EUV source technology is more mature than LPP EUV source technology and DPP EUV sources are commercially available today. The disadvantages of DPP EUV sources are the large thermal load and short lifetime of the debris mitigation system and collector optic and the difficulty of scaling to higher powers without sacrificing component lifetime and reliability.

20.4.1.2 Laser-Produced Plasma EUV Sources

The major components of an LPP EUV source-collector module, an LPP source consisting of high-power pulsed laser, a droplet target generator, and a ML-coated normal-incidence collector, are shown in Figure 20.10b. In the early 1990s, a number of experimental and theoretical studies showed that LPP EUV sources could achieve conversion efficiencies from laser energy to EUV energy in a 2% bandwidth near a 13 nm wavelength of about 1%–2% depending on target material, laser intensity, laser pulse width, and laser wavelength. It was apparent from the very earliest experiments that debris ejected from the plasma posed a major threat to the lifetime of the collector due to (1) deposition of debris (particulates) from the source hardware, (2) deposition of atomic vapor from the source plasma, and (3) erosion by high-energy ions. Since that time, a great deal of effort has been put into designing and testing a variety of debris-mitigation schemes (gas curtains, electrostatic

repeller fields, etc.) for LPP EUV sources. The only alternative to debris mitigation is to minimize debris production using mass-limited targets that supply only the amount of target material needed for efficient production of EUV radiation. One of the most promising methods of producing a mass-limited target is to introduce a stream of tiny droplets of target material into the laser focus, as shown in Figure 20.10b. At SPIE Advanced Lithography 2009, Cymer Inc. reported achieving about 45 W of EUV power at 50 kHz repetition rate in 400 ms-long bursts from a LPP EUV source operating at 50 kHz repetition rate using an approximately 12 kW CO_2 laser to ionize a succession of tiny Sn droplet targets.[44] Cymer has recently been able to accumulate nearly 1 MJ of EUV dose by operating a LPP Sn droplet source over a 3 day period for about 8 h per day.[44]

The advantages of an LPP EUV source are a lower rate of debris production due to the use of a mass-limited target and a larger standoff distance between the hot plasma and the collector optic, a higher radiation-collection efficiency and better thermal management due to the use of a normal-incidence collector, and the possibility of power scaling through spatial, temporal, and spectral multiplexing. The disadvantages of an LPP EUV source are its potentially higher capital cost and greater complexity (a high-power pulsed laser system, a droplet generator, and an ML-coated collector are required) and the relative immaturity of LPP technology (even after many years of development, the measured power at IF from LPP EUV sources has only recently surpassed that from DPP EUV sources).

20.4.2 EUV Masks

The second most critical issue for EUVL is the inability to fabricate a defect-free mask blank and the lack of the infrastructure needed for defect inspection and review. A reflective mask for EUVL consists of buffer and absorber (e.g., Cr or TaN) layers deposited on a ruthenium-capped Mo/Si ML-coated blank. The performance requirements for an EUV mask blank can be found in the SEMI P38-1103 Specification for Absorber Film Stacks and Multilayers on Extreme Ultraviolet Mask Blanks.[45] EUV masks are similar to masks for conventional optical lithography, except they are coated with 40 layer pairs of Mo (2.8 nm thick) and Si (4.1 nm thick) instead of a single layer of chrome, and the 6″ × 6″ × 0.25″ mask substrate is a low thermal expansion material instead of quartz, as shown in Figure 20.11.[46] The performance specifications for EUV mask substrates can be found in the SEMI P27-1102 Specifications for Extreme Ultraviolet Lithography Mask Substrates.[47] A patterned EUV mask, like the one shown in Figure 20.11, was fabricated from an ML-coated blank by selectively etching the absorber and buffer layers.

20.4.2.1 EUV Mask Blanks

Fabrication of a defect-free EUV mask blank is exceedingly difficult. First, it requires an almost perfectly flat (front-side flatness no more than 32 nm peak to valley for the 32 nm technology node) and defect-free substrate with an extremely low coefficient of thermal expansion (0 ± 12 ppb/K in the temperature

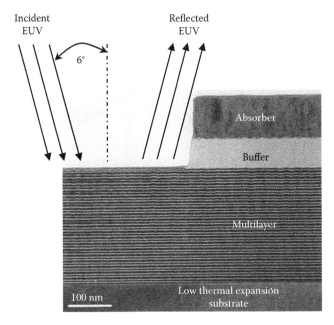

FIGURE 20.11 Cross-sectional TEM image showing EUV reflective mask architecture. (Reprinted from Levinson, H.J., *Principles of Lithography* 2nd edn., SPIE Press, Bellingham, WA, 2005. With permission.)

range from 12°C to 25°C). Then, it requires a coating technology that can deposit Mo/Si MLs with high reflectivity (65%–70%), good thickness uniformity, and nearly zero defect density (the defectivity specification for the 32 nm technology node is a defect density less than 0.003 defects/cm² at defect sizes greater than 25 nm). EUV mask blanks meeting the SEMI P38-1103 Class A production specifications for reflectance and uniformity are available today; deposition technologies and processes for zero-added–defect Mo/Si ML deposition are not.

In an attempt to accelerate the development of the zero-defect-adder Mo/Si deposition process needed to produce defect-free mask blanks, SEMATECH established the Mask Blank Development Center (MBDC) in Albany, New York in 2003. During the MBDC ramp-up, SEMATECH, in partnership with Veeco Instruments, successfully transferred technology developed at LLNL specifically for depositing low-defect ML coating on EUV blanks. A record of the MBDC's progress in reducing the defectivity on EUV blanks is shown in Figure 20.12.[48]

The champion plate from a recent deposition run exhibited only eight defects in the quality area of the plate (132 × 132 mm), corresponding to a defect density of 0.04 defects/cm² at defect sizes greater than 53 nm (polystyrene latex sphere equivalent size). At the present time, the defect density on the best blanks are dominated by defects that originate from substrate pits and bumps. The rate of progress in blank defect reduction is currently limited by the performance of available defect inspection tools. According to recent data from Intel Corporation, the most sensitive defect inspection tools have a 98% capture rate for 45 nm (SiO_2 sphere equivalent) defects on quartz and for 50 nm defects on MLs.[49] Further progress in defect-free EUVL mask development will require defect inspection tools with sensitivities down to approximately 25 nm. Whether this level of performance can be achieved by incremental improvements to existing tools or will require the development of an entirely new defect inspection tool platform is still an open question.

20.4.2.2 EUV Mask Defectivity

As with chrome-on-quartz masks for optical lithography today, the absorbing pattern on an EUV mask can be inspected using a commercially available DUV mask inspection tool and repaired as needed using focused-ion or electron-beam techniques. While the repair of defects that occur during the absorber patterning process is reasonably straightforward, repair of defects

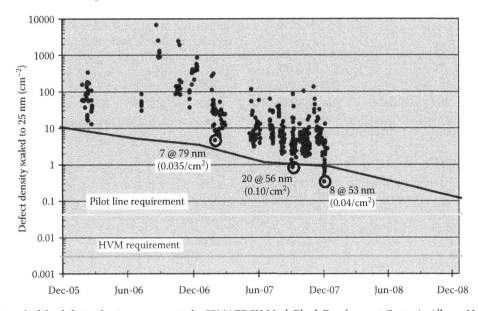

FIGURE 20.12 Record of the defect reduction progress at the SEMATECH Mask Blank Development Center in Albany, New York. (Reprinted from Kearney, P. et al., *Proc. SPIE*, 6921, 69211X, 2008. With permission.)

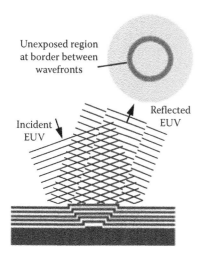

Incident
EUV

Reflected
EUV

Unexposed region
at border between
wavefronts

FIGURE 20.13 Illustration of a multilayer phase defect that could result from a growth discontinuity or replicated substrate imperfection, which leads to an abrupt phase shift and, if resolved, a printable defect.

in the ML coating or on the substrate below the reflective coating is problematic. Examples of these "phase defects" were noticed during the course of EUV printing experiments as early as 1992 and were studied systematically by Nguyen et al. in 1994.[50] The illustration in Figure 20.13 shows how a small height difference in a ML coating can lead to cancellation of the reflective waves and an unexpected, high-contrast feature in the aerial image.

Recent results on EUV mask defectivity, determined by printing an EUV mask on resist-coated wafers and inspecting those wafers in a KLA-Tencor 2800 wafer inspection system operating in cell-to-cell mode, showed that only a few defects originated from the mask.[46] The resulting mask blank defect density inferred from these measurements (about 1 defect/cm²) is approximately 30 times smaller than expected based on mask blank inspection statistics (30–60 defects/cm²) and masks fabricated from state-of-the-art mask blanks should provide even better defectivity performance.

20.4.3 EUV Resists

The development of an EUV resist that can simultaneously meet the resolution, line-edge roughness (LER), and sensitivity requirements for insertion of EUVL into manufacturing is now ranked by the International EUVL Symposium Technical Steering Committee as the third highest risk to the commercialization of EUVL.

20.4.3.1 Current State of the Art

Because of the limited power available from current EUV sources, state-of-the-art EUV resists must be chemically amplified. The important properties of a chemically amplified resist (CAR) are due mainly to its photo acid generator and its polymer type, which in most cases is the same as those used in resists for 248 and 193 nm lithographies. The chief advantages of a CAR is its high sensitivity and resolution; its drawbacks are high LER, tendency towards pattern collapse, and high EUV light absorption. Pattern collapse depends on the aspect ratio of the resist pattern and, for features less than 30 nm in width, the critical aspect ratio is

approximately 2. Assuming aspect ratio scaling continues to hold true, patterning for the 22 nm half-pitch node will require film thicknesses near 40 nm, although the situation can be improved somewhat using surfactant rinses, bilayer resist technologies, and supercritical CO_2 drying. The mechanism responsible for EUV light absorption in EUV resists differs from that in optical resists because the absorption results from the atomic nature of the excitation rather than from the making or breaking of molecular bonds. The high light absorption of all resist material in the EUV spectral region leads to resist patterns with sloping (non-vertical) sidewall profiles that, like pattern collapse, can be improved by using an ultrathin imaging layer over a hard mask in a bilayer or trilayer approach. The resolution performance of state-of-the-art EUV CAR resists printed with the SEMATECH Berkeley MET with 45° rotated dipole illumination is shown in Figure 20.7b, and with 0.35/0.55 annular illumination in Figure 20.7c.[33]

20.4.3.2 The RLS Challenge

The key lithographic performance parameters of CARs, resolution (R), line-edge roughness (L), and sensitivity (S), are highly interdependent—a fact that is often called the RLS challenge or the RLS trade-off. The correlation between resist sensitivity and resolution for a collection of EUV resists evaluated with the SEMATECH Berkeley MET in 2007[51] is shown in Figure 20.14a. Clearly, the materials that exhibit the highest resolution tend to be the least sensitive. The correlation between line edge roughness and sensitivity for a large selection of EUV resists, also tested on the SEMATECH Berkeley MET,[52] is shown in Figure 20.14b. Several attempts to model the interdependences among the performance parameters R, L, and S have been made. Among these, the lithographic uncertainty principle (K_{LUP}) described by van Steenwinckel et al.[53] and the Z-constant proposed by Wallow et al.[51] where [Z ~ (half-pitch)³·LER²·sensitivity] have proven to be the most useful. While no single material simultaneously meets resolution, LER, and sensitivity targets for the 22 nm half-pitch node, a number of materials have been developed that meet both the resolution and sensitivity targets. LER remains the single most challenging performance parameter and, at this point, satisfactory values can only be obtained with a high aerial image contrast imaging system and at an unacceptably high exposure dose. The curve in Figure 20.14b is the prediction of a simple model for the shot noise-limited LER. Because of its short exposure wavelength, it is possible that EUVL could have an additional contribution to LER due to shot noise; for the same exposure energy, a EUV image will have a smaller number of photons per pixel. For the most sensitive resist formulations shown in Figure 20.14b, LER tends to be dominated by shot noise, whereas the LER in slower resists appears to be dominated by other effects.

20.4.4 EUV Mask Handling

Because EUV radiation is highly absorbed by all materials, a membrane pellicle for an EUVL mask is not generally considered to be practical and the current strategy for maintaining a defect-free EUV mask is to utilize a removable pellicle to protect the mask

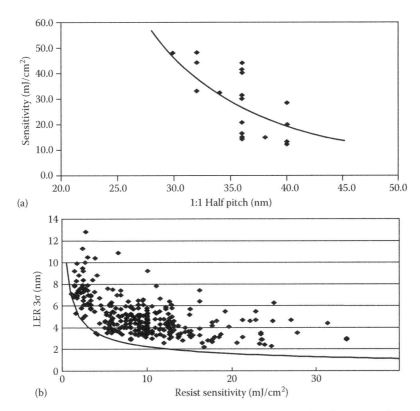

(a)

(b)

FIGURE 20.14 (a) Observed correlation between resist sensitivity (dose to size at 40 nm half-pitch) and limiting resolution for a collection of EUV resist evaluated using the SEMATECH Berkeley MET. (Reprinted from Wallow, T. et al., *Proc. SPIE*, 6921, 69211F, 2008. With permission from SPIE.) (b) LER and sensitivity for a large selection of EUV resists evaluated using the SEMATECH Berkeley MET. The curve represents a simple model estimate for shot-noise-limited LER. (Reprinted from La Fontaine, B. et al., *J. Vac. Sci. Technol. B*, 25, 2089, 2007. With permission.)

during shipping, handling, and storage and to remove the pellicle only when the mask is being inspected or when it is in the vacuum environment of an EUV exposure tool and is ready for exposure. An EUV-removable pellicle can be as simple as a protective, flat face plate that shields the patterned side of the mask from particulate contamination. Variants of the simple face plate are the Fala/Intel Bracket, which adds an additional level of protection by shielding the sides of the mask, the ASML frame, and storage box, which is used as the reticle carrier in their ADTs, and the Nikon, Canon, and Entegris inner pod, which includes a backside lid to form the enclosure shown in the inset of Figure 20.15. The results of four recent tests of a SEMI (draft) 4466-compliant inner pod fabricated by SEMATECH (a so-called S-pod) are shown in Figure 20.15. The data shows, on average, about 0.1 particle added per cycle of use, where cycle is defined as a round-trip shipment, vacuum pump vent, and accelerated storage test at 53 nm PSL equivalent inspection capability.[54]

20.4.5 EUV Optics and Mask Contamination

Another area of significant concern is the shortening of the lifetime of the projection and illuminator optics and the reflective mask due to surface contamination. Water vapor and hydrocarbons are the two major contaminants present in the vacuum environment of an EUVL exposure tool that are known to reduce

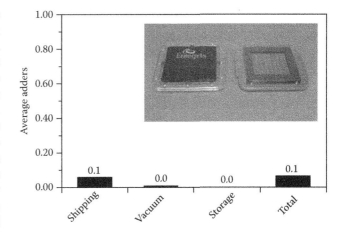

FIGURE 20.15 s-Pod EUV reticle carrier test results for round-trip shipping, vacuum pumping and venting, and accelerated storage. Inset: photograph of reticle pod. (Reprinted from Gomei, Y. et al., *Proc. SPIE*, 6517, 65170W, 2008. With permission from SPIE; He, L. et al., *Proc. SPIE*, 6921, 69211Z, 2008. With permission.)

the lifetime of the reflecting surfaces. Of the two contaminants, carbon contamination is generally considered to be less damaging to optics' lifetimes because carbon films can be removed by hydrogen or oxygen cleaning. A particularly egregious example of carbon contamination is depicted in Figure 20.16,

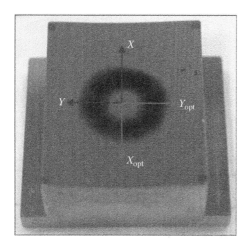

FIGURE 20.16 EUV illuminator mirror showing strong signs of contamination. (Reprinted from La Fontaine, B. et al., *J. Vac. Sci. Technol. B*, 25, 2089, 2007. With permission.)

which shows a photograph of the first illuminator mirror from the SEMATECH EUV MET in Albany with an annular-shaped zone of darkening where the EUV radiation impinged on the mirror surface.

Because an oxide layer strongly absorbs EUV radiation (a 0.3 nm increase in oxide thickness leads to a more than 1% EUV reflectance loss) and is more difficult to remove, oxidation is viewed as a more serious threat to optics' lifetimes. The vacuum environment in an EUVL exposure tool can be modified (by introducing the appropriate gases) to strike a balance between oxidation and carbon growth. This method of contamination control was utilized in 2001 in the EUV Engineering Test Stand[55] (discussed in Section 20.2.1). Oxidation control involving the use of long-life anti-oxidative capping layers comprised of the metal oxides of V, Cr, or W together with carbon-film suppression and removal using EUV + O_2 *in situ* cleaning is a contamination-control strategy currently being explored in Japan.[56] At the 2008 EUVL Symposium, ASML presented data showing that the lifetime of the projection optics in both of its ADTs is at least 4.5×10^9 source pulses (about 1.5 years).[57] For an acceptable COO, the operational lifetime of the imaging system in an EUV exposure tool for HVM must be greater than 5 years (or 30,000 light-on hours). Since the models that have been developed to predict how contamination scales with EUV power density and environmental conditions have not yet reached the level of sophistication needed to accurately predict what the lifetime will be in an HVM tool, the integration of a moderately high-power EUV source with a full-field EUV scanner has become a critical need.

The masks in the EUV METs currently in operation around the world have now received a sufficiently large cumulative dose of EUV radiation that carbon contamination on their surfaces is becoming an issue. Attempts are being made to characterize mask contamination so the lifetime of an EUV mask can be predicted and a safe and effective mask-cleaning process can

be developed. Preliminary results from Auger emission spectroscopy of the mask surface show clear signals of carbon and oxygen in the contaminated areas. Preliminary results from a print-based study of a contaminated mask showed that the primary effect of the contamination is an increase in dose to size with increasing contamination.[41] An analysis of wafers printed with the same mask after cleaning show that the original dose to size was recovered with only a slight reduction in DOF. The results also suggest that the increase in dose to size could be used as a metric to trigger a mask clean.

20.5 Future Perspectives

The two EUVL technology elements that are at highest risk of not being ready when needed are source power at intermediate focus (IF) and mask blank defectivity. Source power at IF, which is critically important to exposure tool throughput, currently falls far short of what will be needed for an EUVL pilot production tool. However, according to Cymer Inc.'s LPP source roadmap, a LPP EUV source with beta-level performance will be delivered to ASML in 1Q09 and a LPP EUV source with 100 W at IF will be available before the end of the 2009 calendar year.[57] If Cymer's LPP EUV source delivery schedule slips, ASML's EUV preproduction tool schedule will also slip and—because of this—source power at IF is generally regarded as the technology element that will determine the date of EUVL manufacturing insertion. Mask blank defectivity, which is critically important to the yield of electrically functional devices, also falls far short of what will be needed for HVM, and given the current rate of progress in blank defectivity reduction, it now seems likely that completely defect-free blanks will not be available in the foreseeable future. However, several of the blank-defect workarounds that are now being actively pursued, such as defect compensation and fiducials for blank disposition, could allow functional devices to be manufactured even at current blank defect levels. Most of the other EUVL technology elements are already able to satisfy 15 nm pilot line requirements. Mask handling, optics quality, resist resolution, and sensitivity are already in specification. The only other performance metric that is of concern is resist LER. The current best LER, about 3.0 nm (3σ) at 32 nm HP and 15 mJ/cm² dose, is much higher than the International Technology Roadmap for Semiconductors specification for the 15 nm hp node, 1.1 nm (3σ)[31]; however, the use of smoothing under-layers, special rinse liquids, and/or smoothing during etch transfer is expected to help close this gap.

EUVL provides a wider manufacturing process window, has a potentially higher throughput, and is the most easily extendible of all the lithographic technologies that have been evaluated as possible replacements for optical projection lithography. The wider process window and easier extendibility for EUVL can be seen most clearly by rearranging the Rayleigh equation for resolution (RES = $k_1\lambda$/NA, where RES is resolution, λ is wavelength,

NA is numerical aperture, and k_1 is a process-dependent constant) to give an expression for k_1 in terms of resolution:

$$k_1 = \left[\frac{NA}{\lambda}\right] RES = \left[\frac{NA}{\lambda}\right] HP \qquad (20.1)$$

where resolution (RES) has been replaced with the period, or half-pitch, of dense line and space patterns. Plots of k_1 versus HP (nm) for a 193 nm immersion exposure tool with a 1.30 NA imaging system and for EUV exposure tools with 0.25, 0.35, and 0.45 NA imaging systems are shown in Figure 20.17.

Also shown in Figure 20.17 are some dates from the roadmaps of the most aggressive semiconductor chip makers for the production ramps at the 32, 22, and 15 nm technology nodes (i.e., at which the logic HP is assumed to undergo an approximate 0.7 shrink). In the not-too-distant past, k_1 values of 0.6 and larger were needed for HVM. More recently, it has been found possible to work at lower values of k_1, at the cost of tighter process control, by employing a variety of resolution enhancement techniques such as off-axis illumination, phase-shifting masks, and application of optical proximity correction to the patterns on the mask. The limiting value for k_1 is 0.25, below which the image contrast for periodic line and space patterns is zero. The calculated values plotted in Figure 20.17 show that EUVL, even with the smallest NA imaging system, will allow a return to manufacturing process windows that seems impossibly large by today's standards. By utilizing EUV imaging systems with larger and larger NAs, the route to higher resolution that optical lithography took in the past, EUVL should be able to print dense features at HP well below 20 nm. In contrast, the process window for single-exposure 193 nm immersion lithography, with water as the immersion fluid, will shrink to zero at an HP approximately equal to 39 nm (solid line at $k_1 = 0.25$ in Figure 20.17). Even if major advances in high-index immersion fluids, high-index lens materials, and high-index resists are made, SE 193 nm immersion lithography is unlikely to be extended much below 32 nm HP, and double-exposure double-etch (DE^2) 193 nm immersion lithography is unlikely to be extended much below 22 nm HP (dashed line at $k_1 = 0.18$ in Figure 20.17).

Several full-field EUVL pilot production tools will be available in 2011. According to ASML's EUVL tool roadmap,[15] two EUVL exposure tool models will be available in 2011: the NXE 3100 preproduction tool (illustrated in Figure 20.18), which has a 0.25 NA imaging system, will support 4 nm overlay, and provide a throughput of about 60 wafers per hour; and a first-generation HVM tool, which will have a 0.32 NA imaging system, will support 3 nm overlay, and provide a throughput of about 150 wafers per hour. According to Canon's EUV tool development plan,[58] a six-mirror 0.3 NA imaging system will be available in 2011 for in-house testing purposes, but the first of Canon's VS2 HVM EUV exposure tools will not be available until late in 2015. According to Nikon's EUVL tool development scenario,[56] a 0.25 NA EUV2

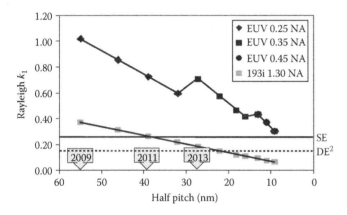

FIGURE 20.17 A plot of the Rayleigh k_1 parameter vs. the half-pitch of dense 1:1 lines and spaces for a 193 nm immersion imaging system with 1.30 NA and 13.5 nm EUV imaging systems with 0.25, 0.35, and 0.45 NA.

FIGURE 20.18 **(See color insert following page 21-4.)** Artist illustration of ASML NXE: 3100 preproduction EUV exposure tool showing the LPP EUV source on the right, the illuminator optics in the center, and the mask and the projection optics on the right. (Courtesy of ASML, Veldhoven, the Netherlands.)

process development tool with lower wavefront error and lower flare imaging optics will be available at its factory early in 2011 and a greater than 0.3 NA EUV3 HVM EUVL exposure tools will be available early in 2012.

Abbreviations

ADT	Alpha Demo Tool
CAR	chemically amplified resist
COO	cost of ownership
DOF	depth of focus
DPP	discharge-produced plasma
DRAM	dynamic random access memory
DUV	deep ultraviolet
ETS	Engineering Test Stand
EUV	extreme ultraviolet
EUVL	extreme ultraviolet lithography
FET	field-effect transistor
HP	half-pitch
HSFR	high-spatial frequency roughness
HVM	high-volume manufacturing
IF	intermediate focus
IL	interference lithography
LBNL	Lawrence Berkeley National Laboratory
LER	line-edge roughness
LLNL	Lawrence Livermore National Laboratory
LPP	laser-produced plasma
LSRF	low spatial frequency roughness
LSI	lateral-shearing interferometry
MBDC	Mask Blank Development Center
MET	micro-exposure tool
ML	multilayer
MSRF	mid-spatial frequency roughness
NA	numerical aperture
PMI	phase-measuring interferometry
PSD	power spectral density
PSDI	phase-shifting diffraction interferometry
PSI	Paul Scherrer Institute
PSL	polystyrene latex (sphere)
PS/PDI	phase-shifting point-diffraction interferometry
SEM	scanning electron microscope
SNL	Sandia National Laboratory
SRAM	static random access memory
TEM	transmission electron microscope

References

1. Henry, J., Spiller, E., and Weisskopf, M., Imaging performance of a normal incidence x-ray telescope measured at 0.18 KeV, *Proc. SPIE* **316**, 166–168 (1981).

2. Lovas, I., Stanty, W., Spiller, E. et al., Design and assembly of a high resolution Schwarzschild microscope for soft-x-rays, *Proc. SPIE* **316**, 90–97 (1981).

3. Underwood, J.H. and Barbee, T.W. Jr., Soft x-ray imaging with a normal incidence mirror, *Nature* **294**, 429–431 (1981).

4. Takei, K. and Maeda, Y., Preparation of multi-layered tungsten-carbon films by ion beam sputtering, *Jpn. J. Appl. Phys.* **24**, 1366–1367 (1985).

5. Kinoshita, H., Kaneko, T., Takei, H. et al., Study on x-ray reduction projection lithography, *Presentation at the 47th Autumn Meeting of the Japan Society of Applied Physics*, Sapporo, Japan, Paper No. 28-ZF-15 (1986).

6. Bjorkholm, J., Bokor, J., Eichner, L. et al., Reduction imaging using multilayer coated optics: Printing of features smaller than 0.1 microns, *J. Vac. Sci. Technol. B* **8**, 1509–1513 (1990).

7. Tichenor, D.A., Kubiak, G.D., Haney, S.J. et al., Recent results in the development of an integrated EUVL laboratory tool, *Proc. SPIE* **2437**, 292–307 (1995).

8. Kinoshita, H., Kurihara, K., Mizota, T. et al., Large-area, high-resolution pattern replication by the use of a two-aspherical-mirror system, *Appl. Opt.* **32**, 7079–7083 (1993).

9. Kurihara, K., Kinoshita, H., Mizota, T. et al., Two-mirror telecentric optics for soft x-ray reduction lithography, *J. Vac. Sci. Technol. B* **9**, 3189–3192 (1991).

10. Haga, T., Tinone, M., Takenaka, H., and Kinoshita, H., Large-field (>20 × 25 mm^2) replication by EUV lithography, *Microelectron. Eng.* **30**, 179–182 (1996).

11. Naulleau, P., Goldberg, K.A., Anderson, E. et al., Status of EUV microexposure capabilities at the ALS using the 0.3-NA MET optic, *Proc. SPIE* **5374**, 881–891 (2004).

12. Naulleau, P., Anderson, C., Chiu, J. et al., Advanced extreme ultraviolet resist testing using the SEMATECH Berkeley 0.3-NA micro field exposure tool, *Proc. SPIE* **6921**, 69213N1–69213N11 (2008).

13. Murakami, K., Oshino, T., Shimizu, S. et al., Basic technologies for extreme ultraviolet lithography, *OSA Trends Opt. Photon.* **4**, 16–20 (1996).

14. Kinoshita, H., Watanabe, T., Niibe, M. et al., Three-aspherical mirror system for EUV lithography, *Proc. SPIE* **3331**, 20–31 (1998).

15. Tichenor, D.A., Ray-Chaudhuri, A.K., Replogle, W.C. et al., System integration and performance of the EUV engineering test stand, *Proc. SPIE* **4343**, 19–37 (2001).

16. Meiling, H., Buzing, N., Cummings, K. et al., EUV system—Moving towards production, *Proc. SPIE* **7271**, 7271021–72710215 (2009).

17. Murakami, K., Oshino, T., Kondo, H. et al., Development status of projection optics and illumination optics for EUV1, *Proc. SPIE* **6921**, 69210Q1–69210Q7 (2008).

18. Spiller, E., Multilayer optics for x-rays, in P. Dhez and C. Weisbuch (eds.), *Physics, Fabrication and Applications of Multilayer Structures*, Plenum, New York (1987), pp. 271–309.

19. Barbee, T.W. Jr., Mrowka, S., and Hettrick, M., Molybedenum-silicon multilayer mirrors for the extreme ultraviolet, *Appl. Opt.* **24**, 883–886 (1985).

20. Stearns, D.G., Rosen, R.S., and Vernon, S.P., Multilayer mirror technology for soft-x-ray projection lithography, *Appl. Opt.* **32**, 6952–6960 (1993).

21. Perera, R.C.C. and Underwood, J.H., Results from our recently delivered automated reflectometer for measurement of reflectivity of EUV lithographic masks, *Poster Presentation at the 2006 International EUVL Symposium*, Barcelona, Spain, 15–18 October 2006.

22. Sommargren, G., Phase shifting diffraction interferometry for measuring extreme ultraviolet optics, *OSA Trends Opt. Photon.* **4**, 108–112 (1996).

23. Sommargren, G., Phillion, D., Johnson, M. et al., 100-picometer interferometry for EUVL, *Proc. SPIE* **4688**, 316–328 (2002).

24. Goldberg, K., Naulleau, P., Denham, P. et al., EUV interferometry of the 0.3 NA MET optics, *Proc. SPIE* **5037**, 69–74 (2003).

25. Goldberg, K., Naulleau, P., Denham, P. et al., EUV interferometric testing and alignment of the 0.3 NA MET optic, *Proc. SPIE* **5373**, 64–73 (2004).

26. Harned, N., Goethals, M., Groeneveld, R. et al., EUV lithography with the Alpha Demo Tools: Status and challenges, *Proc. SPIE* **6517**, 6517061–65170612 (2007).

27. Tawarayama, K., Aoyama, H., Magoshi, S. et al., Recent progress of EUV full-field exposure tools in Selete, *Proc. SPIE* **7271**, 7271181–7271188 (2009).

28. Meiling, H., Meyer, H., Banine, V. et al., First performance results of the ASML alpha demo tool, *Proc SPIE* **6151**, 6151081–61510812 (2006).

29. Miura, T., Murakami, K. Suzuki, K. et al., Nikon EUVL development progress update, *Proc. SPIE* **6517**, 6517071–65170710 (2007).

30. Pierson, B., Wallow, T., Mizuno, H. et al., EUV resist performance on the ASML ADT and LBNL MET, *Oral Presentation at the 2008 International EUVL Symposium*, Lake Tahoe, CA, 28 September 2008.

31. Koh, C., Ren, L., Georger, J. et al., Assessment of EUV resist readiness for 32 nm hp manufacturing, and extendibility study of EUV ADT using state-of-the-art resist, *Proc. SPIE* **7271**, 7271241–72712411 (2009).

32. Uzawa, S., Kubo, H., Miwa, Y. et al., Path to the HVM in EUVL through the development and evaluation of the SFET, *Proc. SPIE* **6517**, 6517081–65170811 (2007).

33. Naulleau, P., Anderson, E., Baclea-an, L.-M. et al., The SEMATECH Berkeley microfield exposure tool: Learning at the 22 nm node and beyond, *Proc. SPIE* **7271**, 72710W1–72710W11 (2009).

34. Solak, H.H., Ekinci, Y., Kaser, P., and Park, S., Photon-beam lithography reaches 12.5 nm half-pitch resolution, *J. Vac. Sci. Technol. B* **25**, 91–95 (2007).

35. Ekinci, Y., Solak, H.H., Padeste, C. et al., 20 nm line/space patterns in HSQ fabricated by EUV interference lithography, *Microelectron. Eng.* **84**, 700–704 (2007).

36. La Fontaine, B., Deng, Y., and Kim, R.-H., The use of EUV lithography to produced demonstration devices, *Proc. SPIE* **6921**, 69210P1–69210P10 (2008).

37. Lorusso, G.F., Hermans, J., Goethals, A.M. et al., Imaging performance of the EUV alpha demo tool at IMEC, *Proc. SPIE* **6921**, 69210O1–69210O11 (2008).

38. Wood, O., Koay, C.-S., Petrillo, K. et al., Integration of EUV lithography in the fabrication of 22-nm node devices, *Proc. SPIE* **7271**, 7271041–72710411 (2008).

39. Park, J.-O., Koh, C., Goo, D. et al., The application of EUV lithography for 40 nm node DRAM device and beyond, *Proc. SPIE* **7271**, 7271141–7271149 (2009).

40. Eom, T.-S., Park, S., and Park, J.-T., Comparative study of DRAM cell patterning between ArF immersion and EUV lithography, *Proc. SPIE* **7271**, 7271151–72711511 (2009).

41. La Fontaine, B., Wood, O., and Medeiros, D., The future of EUV lithography, N. Yamamoto (ed.), *Proceedings of SEMI Technology Symposium*, SEMI Japan, Chiba, Japan, 2008.

42. Ota., K., Watanabe, Y., Banine, V., and Frankin, H., EUV source requirements for EUV lithography, in Vivek Bakshi (ed.), *EUV Sources for Lithography*, SPIE Press, Bellingham, WA (2006), pp. 27–43.

43. Corthout, M., Apetz, R., Bruederman, J. et al., Sn DPP source-collector modules: Status of Alpha sources, Beta developments, and HVM experiments, *Proc. SPIE* **7271**, 72710A1–72710A10 (2009).

44. Brandt, D.C., Fomenkov, I.V., Ershov, A.I. et al., LPP source system development for HVM, *Proc. SPIE* **7271**, 727031–72710311 (2009).

45. SEMI P38-1103, Specifications for absorbing film stacks and multilayers on extreme ultraviolet lithography blanks, Semiconductor Equipment and Materials International, San Jose, CA (2003).

46. Levinson, H.J., *Principles of Lithography*, 2nd edn, SPIE Press, Bellingham, WA (2005).

47. SEMI P37-1102, Specifications for extreme ultraviolet lithography mask substrates, Semiconductor Equipment and Materials International, San Jose, CA (2003).

48. Kearney, P., Lin, C.C., Sugiyama, T. et al., Ion beam deposition for defect-free EUVL mask blanks, *Proc. SPIE* **6921**, 69211X1–69211X7 (2008).

49. Ma, A., Liang, T., Park, S.-J. et al., EUVL blank defect inspection capability at Intel, *2008 International EUVL Symposium*, Lake Tahoe, CA, September 29, 2008.

50. Nguyen, K., Attwood, D., Mizota, T. et al., Imaging of EUV lithography masks with programmed defects, *OSA Proc. Extreme Ultraviolet Lithography* **23**, 193–203 (1994).

51. Wallow, T., Higgins, C., Brainard, R. et al., Evaluation of EUV resist materials for use at the 32 nm half-pitch node, *Proc. SPIE* **6921**, 69211F1–69211F11 (2008).

52. La Fontaine, B., Deng, Y., Kim, R.-H. et al., Extreme ultraviolet lithography: From research to manufacturing, *J. Vac. Sci. Technol. B* **25**, 2089–2093 (2007).

53. Van Steenwinckel, D., Gronheid, T., Lammers, J.H. et al., A novel method for characterizing resist performance, *Proc. SPIE* **6519**, 65190V1–65190V12 (2007).

54. He, L., Wurm, S., Seidel, P. et al., Status of EUV reticle handling solutions in meeting 32 nm HP EUV lithography, *Proc. SPIE* **6921**, 69211Z1–69211Z10 (2008).

55. Malinowski, M., Grunow, P., Steinhaus, C. et al., Use of molecular oxygen to reduce EUV-induced carbon contamination of optics, *Proc. SPIE* **4343**, 347–356 (2001).

56. Miura, T., Murakami, K., Kawai, H. et al., Nikon EUVL development progress update, *Proc. SPIE* **7271**, 72711X1–72711X11 (2009).

57. Meiling, H., Lok, B., Hultermans, B. et al., EUV alpha demo tools—Stepping stones towards volume production, *2008 International EUVL Symposium*, Lake Tahoe, CA, 28 September 2008.

58. Hasegawa, T., Uzawa, S., Honda, T. et al., Development status of Canon's full-field EUVL tool, *Proc. SPIE* **7271**, 72711Y1–72711Y11 (2009).

IV

Optics of Nanomaterials

Cathodoluminescence of Nanomaterials

Naoki Yamamoto
Tokyo Institute of Technology

21.1 Introduction

Dimensionality and size are two factors that introduce new properties into semiconductor materials and enable us to provide new functionality in semiconductor nanoelectrics and photonics. Semiconductor quantum wires have been intensively studied, on one hand, for the theoretical interest in physical properties due to quasi-one-dimensional quantum confinement of an electronic system (Sakaki, 1980; Ogawa and Takagahara, 1991; Tanatar et al., 1998) and, on the other hand, for practical application to future optoelectronic devices such as low-threshold lasers (Arakawa and Sakaki, 1982) and polarization-sensitive devices (Wang et al. 2001). One-dimensional nanostructures have been primarily synthesized by lithography and an epitaxial technique on semiconductor substrates (Arakawa and Sakaki, 1982; Ils et al., 1994; Someya et al., 1995). V- (Kapon et al., 1989) and T-shaped quantum wires (Pfeiffer et al., 1990) were fabricated using epitaxial growth techniques (Wang and Voliotis, 2006), namely, molecular beam epitaxy (MBE) and metalorganic vapor phase epitaxy (MOVPE). However, excitons in the quantum wires were observed to be localized in monolayer step-induced islands at low temperatures (Guilet et al., 2003). The fluctuation of confinement potential along a quantum wire is a serious issue for realizing the characteristic property of a one-dimensional structure.

The free-standing nanowires of many semiconductors are recently produced by the growth technique using nanosized liquid droplets of the metal solvent (Hiruma et al., 1995; Gudiksen et al., 2002; Wu et al., 2002). This technique has an advantage to produce heterogeneous structure along the wire such as p–n junctions (Gudiksen et al., 2002) and superlattices (Wu et al., 2002). The optical properties of the nanowires have been studied by photoluminescence (PL) and absorption spectroscopy, which showed the blue shift of the peak energy due to the quantum confinement effect and giant polarization anisotropy in PL and optical absorption. Those properties will provide a potentiality for new application of the semiconductor nanowires. In spite of many works for synthesizing nanowires, there have been very few experimental works for studying the optical properties of self-standing nanowires, especially for a single nanowire, because of the limitation in spatial resolution. An isolated single nanowire of InP was studied by PL (Wang et al., 2001), which showed giant polarization anisotropy in PL and photoconductivity measurement. The giant polarization anisotropy is partly due to the excitation process by light, because the electromagnetic field in the nanowire, which generates carriers, depends on the polarization direction of incident light. Polarization anisotropy can also be expected in the light emission process, as in the case of electroluminescence (EL) and cathodoluminescence (CL).

We studied the optical properties of semiconductor nanowires, observing cathodoluminescence (CL) spectra and polarized monochromatic CL images of nanowires by using a transmission electron microscope (TEM) combined with a CL detection system (TEM-CL). The TEM-CL technique has an advantage in high spatial resolution and ability to characterize nanoscale structures (Yamamoto, 1984; Mitsui et al., 1996; Yamamoto et al., 2003, 2008; Ino and Yamamoto, 2008). In Section 21.2, a principle of TEM-CL technique is described based on the fundamental knowledge of radiative recombination processes and property of exciton. Luminescence in semiconductor nanowires shows characteristic properties in emission energy and polarization, because the optical transition between electronic states is affected by the quantum confinement effect. For understanding these properties, the theory of the band structure in semiconductors and optical transitions are described in Section 21.3. Quantum confinement effect on the optical properties of semiconductor nanowires and dielectric effect on polarization properties are described in Section 21.4. Finally, the experimental results of the TEM-CL technique for the InP and GaAs nanowires, which show the quantum confinement effect, and for the ZnO nanowires, which show the dielectric effect, are described in Section 21.5.

21.2 Fundamentals of Cathodoluminescence

21.2.1 Radiative Recombination

Luminescence in semiconductors is light emission phenomenon due to the recombination of excited electrons in the conduction band with holes in the valence band (Voos et al., 1980). It is called photoluminescence (PL) when those electrons and holes, the carriers, are excited by light, and called cathodoluminescence (CL) when they are excited by fast electrons. The band structures are classified into two types, direct gap type and indirect gap type. In the former type, the conduction band minimum is located at the Γ point ($k = 0$) in the k-space, whereas it is shifted from the Γ point in the latter type. The valence band maximum is located at the Γ point. Then in the case of the direct gap band structure, the radiative interband transition from the electron states in the conduction band to the hole states in the valence band occur at the Γ point accompanying with emission of photon at high recombination probability. GaAs and InP have band structure of this type. Whereas in the indirect gap type,

the transition occurs with the association of phonon, and the recombination rate is rather small. Si and GaP have band structure of this type.

Typical recombination processes in semiconductors are depicted in Figure 21.1 (Pankove, 1971). First group is radiative recombination processes, i.e., (a) interband transition from conduction band to valence band, (b)–(d) annihilation of free exciton (FX) composed of a pair of electron and hole, and related processes of bound excitons (BX), (e)–(g) transitions of carriers trapped by impurity states such as donor and acceptor states. Second group is nonradiative recombinations, i.e., (h) multiphonon scattering, and (i) Auger process.

Photon energies in the recombination processes in Figure 21.1 are slightly lower than the band gap energy E_g, though their values are different from each other. The emission intensity by the interband transition from conduction band to valence band is generally small compared with the other processes. Photon energies in the impurity-associated recombination processes (Figure 21.1e and f) are given as follows:

$$(D,h): E = E_g - E_D \tag{21.1}$$

$$(e,A): E = E_g - E_A, \tag{21.2}$$

where E_D and E_A are binding energies of the donor level and acceptor level, respectively. Photon energy associated with the donor–acceptor pair transition (D,A) is given by

$$(D,A): E = E_g - E_D - E_A + \frac{e^2}{4\pi\varepsilon R}, \tag{21.3}$$

where

R is a distance between donor and acceptor
ε is the dielectric constant of the material
e is the elemental charge

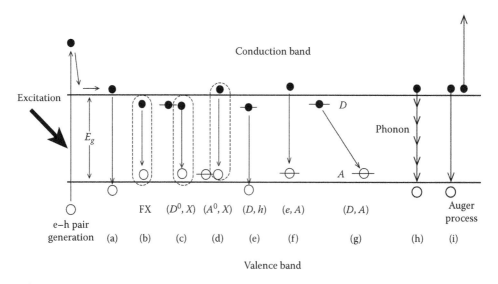

FIGURE 21.1 Recombination processes in semiconductors.

TABLE 21.1 Basic Parameters on Exciton in Typical Semiconductors

	Structure	Band Type	Band Gap (eV) E_g (4 K)	E_g (300 K)	Dielectric Constant	E_x (meV)	FX Radius (nm)
Si	Diamond	Indirect	1.1698	1.11	11.4	14.7	403
Ge	Diamond	Indirect	0.7454	0.669	15.36	4.15	11
GaAs	Zinc-blend	Direct	1.519	1.43	13.1	4.21	13
InP	Zinc-blend	Direct	1.4236	1.34	12.61	5.12	11
GaN	Wurtzite	Direct	3.507	3.39	10.4 ($\parallel_c$) 9.5 ($\perp_c$)	23	3
ZnO	Wurtzite	Direct	3.436	3.37	8.84 ($\parallel_c$) 8.47 ($\perp_c$)	59	1.4

Note: The data were gathered from many literatures. The values of FX radius are calculated from the data in the other columns.

The last term indicates the Coulomb energy of the electron and hole pair. The impurities locate at lattice sites and R takes many different values. Then the emission peak due to (D,A) is composed of the multiple peaks to become a single broad peak, and dependence of the peak position on temperature and excitation rate is different from that of the other luminescence peaks of a single recombination process.

21.2.2 Exciton

An electron and a hole in semiconductors attract each other to form a FX. FX emission dominates in the radiative recombination process at low temperatures. The binding energy of exciton can be obtained by solving Schrödinger equation for electron and hole system attracted by Coulomb potential (Basu, 1997; Yu and Cardona, 1999). The binding energy of the lowest energy level and Bohr radius of FX in a bulk crystal are expressed by the hydrogen atom model as

$$E_X = 13.6 \left(\frac{\mu}{m_0} \right) \frac{1}{\varepsilon^2} \, [\text{eV}] \qquad (21.4)$$

$$a_X = \varepsilon \left(\frac{m_0}{\mu} \right) a_B, \qquad (21.5)$$

where

ε is a relative dielectric constant
m_0 is the electron mass
μ is the reduced effective mass $((1/\mu = (1/m_e^\star) + (1/m_h^\star))$
a_B is Bohr radius (0.0529 nm), respectively

Photon energy of the FX emission is given by

$$E = E_g - E_X. \qquad (21.6)$$

Some of excitons are trapped by impurities to form BX. The recombinations of excitons bounded by a neutral donor and ionized donor are expressed by (D^0, X) and (D, X), respectively. Similarly those of excitons bounded by a neutral acceptor and ionized acceptor are expressed by (A^0, X) and (A, X), respectively. The photon energy of the recombination is given by

$$E = E_g - E_X - E_b, \qquad (21.7)$$

where E_b is a binding energy for trapping exciton at impurities. The magnitude of E_b is about 1/3 of E_D and E_A or less.

Basic parameters are listed in Table 21.1 for typical semiconductors. The values of the band gap energy and exciton binding energy are taken after several literatures (Pankove, 1971; Vugaftman et al., 2001). Exciton radius in the last column is calculated by (21.4) and (21.5) using the relative dielectric constants and exciton binding energies listed in the table. The values give only a measure of the exciton radius, because the differences among the reported values for the listed parameters are fairly large.

21.2.3 TEM-Cathodoluminescence Technique

TEM-cathodoluminescence (TEM-CL) is a technique to observe luminescence spectra and scanning images using a converged electron beam having a probe size of nm order (Yamamoto, 2002, 2008). The characteristics of TEM-CL are as follows: (1) Inner structures can be observed by TEM using thin samples and comparison between TEM and CL images is possible and (2) high spatial resolution of about 100 nm can be achieved. In CL, incident electrons are scattered in a specimen from the incident beam direction, as shown in Figure 21.2. The scattered electrons produce secondary electrons and plasmons, which generate electron-hole pairs in the hatched region (generation region). The excited carriers diffuse into surroundings and some of them recombine radiatively in a surrounding region (diffusion region). The diffusion of the excess minority carriers is important, because majority carriers exist everywhere. The resolution of CL is given by the diameter of the diffusion region, which is determined by the following factors: (1) the electron beam diameter, (2) size of the generation region, and (3) the diffusion length of the minority carrier. The diameter of the electron beam is less than 10 nm, and is negligible compared with the other factors.

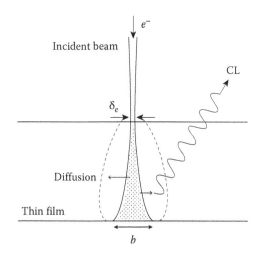

FIGURE 21.2 Carrier generation and light emission process in TEM-CL.

The diameter of the beam spread due to elastic scattering in a thin specimen of thickness t (μm) is given by

$$b = 6.25 \left(\frac{Z}{E}\right)\left(\frac{\rho}{A}\right)^{1/2} t^{3/2} \; [\mu m], \qquad (21.8)$$

where

Z is the atomic number
A is the atomic mass
E (keV) is an incident electron energy
ρ (g/cm³) is the density (Goldstein et al., 1977)

For example, assuming that for GaAs, $Z_{av} = 32$, $A_{av} = 72.3$, $\rho = 5.3\,g/cm^3$, and a specimen thickness is 0.5 μm, we obtain $b = 0.24$ μm. However, Monte Carlo simulation of the electron

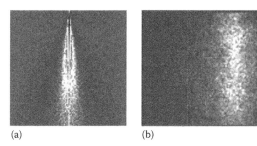

(a)   (b)

FIGURE 21.3 Monte Carlo simulation of (a) carrier generation by incident electrons and (b) carrier diffusion.

scattering showed that the electron density is concentrated on the center of the beam, so the effective beam spread is smaller than that given by Equation 21.8 (see Figure 21.3).

Equation 21.8 indicates that the higher accelerating voltage and thinner specimen thickness are advantageous for realizing high spatial resolution. However, when the accelerating voltage exceeds 100 kV, CL intensity starts to decrease because of the formation of specimen damage such as vacancies. TEM is typically operated at an accelerating voltage of 80 kV for CL measurements.

Diffusion length is an effective parameter for determining spatial resolution in CL of semiconductors. In insulators we can ignore carrier diffusion, but in semiconductors, carrier diffusion occurs with diffusion length of about 100 nm for GaN and 1 μm for GaAs. The diffusion length depends on dopant density and temperature, and reduces by surface recombination effect in thin specimens (Nakaji et al., 2005; Ino and Yamamoto, 2008). In quantum structures of semiconductors, carriers are confined in the localized structures smaller than the diffusion length, and individual nanostructures can be observed in CL images with high resolution as in the case of self-standing nanowires.

TEM-CL system based on a 200 kV transmission electron microscope (JEM-2000FX) is illustrated in Figure 21.4. Samples

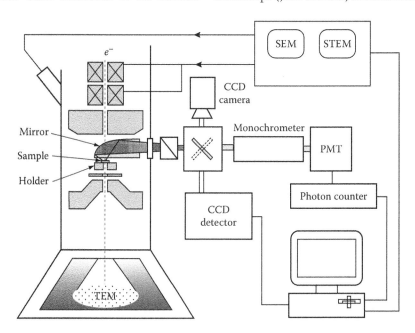

FIGURE 21.4 TEM-CL system.

are examined in the TEM, both in the stationary beam illumination mode and the scanning beam illumination mode. The scanning electron microscopy (SEM) imaging using a secondary electron detector and the scanning transmission electron microscopy (STEM) imaging using a transmitted electron beam detector are both possible in the TEM. A beam size of 10 nm in diameter and a probe current of 1 nA was conventionally used. The temperature of a sample can be varied from 20 K to room temperature using a liquid-He cooling holder and liquid-N$_2$ cooling holder.

Light emitted from a sample is collected by an ellipsoidal mirror or parabolic mirror, being guided through a polarizer and a monochromator, and is detected by a photomultiplier tube (PMT) with a GaAs photocathode for visible light and an InGaAs photocathode for infrared light. A CCD detector is also used for measuring spectra with high sensitivity. Monochromatic CL images can be obtained by scanning the electron beam in the SEM mode. This imaging technique is similar to SEM or EDX mapping, because we use the light intensity collected by the mirror, instead of secondary electron yields or x-ray intensities, as in the case of the SEM. An image is composed of, for example, 100 × 100 pixels, so that the focused electron beam is scanned across the specimen and stays 0.1 s for each pixel. Therefore, it takes 1,000 s to obtain one complete image. The CL images in this paper are shown with the absolute intensity.

In the CL detection system, a linear polarizer is located between the ellipsoidal mirror and monochrometer to select linearly polarized components of the emitted light; here *p*-polarized (*s*-polarized) light, which is detected with the polarization direction parallel (perpendicular) to the monochrometer slit, mainly involves linearly polarized light component parallel (perpendicular) to the longer axis of the ellipsoidal mirror or the axis of the parabola. A mask with a small hole is set in front of the CCD detector for the angular resolved measurement using the parabolic mirror. The mask is movable using an *X-Y* stage in a plane normal to the light path and selects an emission angle.

21.3 Optical Properties of Semiconductor Quantum Structures

21.3.1 Band Structure of Semiconductor

The wavefunction of a carrier in a semiconductor crystal of volume *V* is of a form of Bloch function,

$$\psi_{nk} = \frac{1}{\sqrt{V}} e^{ik \cdot r} u_{nk}(r), \qquad (21.9)$$

which is a product of a periodic function $u_{nk}(r)$ of the crystal lattice and a slowly varying function $\exp(ik \cdot r)$. Subscripts *n* and *k* indicate a number of the Bloch band and wavevector, respectively. In many semiconductors of diamond and zinc-blend structures, the carrier wavefunction at the bottom of the conduction band has the *s*-like orbital and that at the top of the

valence band has the *p*-like orbital (Basu, 1997; Yu and Cardona, 1999). The function $u_{n0}(r)$ has a form of the atomic orbital in the vicinity of the atoms. The band structure of semiconductor has been treated by the $k \cdot p$ perturbation theory (Luttinger and Kohn, 1955). The wavefunctions of electron states at the Γ point in the conduction band are given by the product of the atomic *s*-orbital function and spin orbital function as

$$u^c \frac{1}{2} = |s\uparrow\rangle \quad \text{and} \quad u^c\left(-\frac{1}{2}\right) = |s\downarrow\rangle \qquad (21.10)$$

Here, $|\uparrow\rangle$ and $|\downarrow\rangle$ are the spin orbital functions with up and down spins, respectively. The wavefunctions of the hole states at Γ point in the valence band are given by the $k \cdot p$ theory using the atomic *p*-orbital functions $|X\rangle$, $|Y\rangle$, and $|Z\rangle$. Assuming that k is parallel to the *z* direction, the following functions are used as the valence band wavefunctions,

$$
\begin{aligned}
u_{\frac{3}{2},\frac{3}{2}} &= -\frac{1}{\sqrt{2}} |(X+iY)\uparrow\rangle \\
u_{\frac{3}{2},-\frac{3}{2}} &= \frac{1}{\sqrt{2}} |(X-iY)\downarrow\rangle \\
u_{\frac{3}{2},\frac{1}{2}} &= -\frac{1}{\sqrt{6}} \Big[|(X+iY)\downarrow\rangle - |2Z\uparrow\rangle\Big] \\
u_{\frac{3}{2},-\frac{1}{2}} &= \frac{1}{\sqrt{6}} \Big[|(X-iY)\uparrow\rangle + |2Z\downarrow\rangle\Big] \\
u_{\frac{1}{2},\frac{1}{2}} &= -\frac{1}{\sqrt{3}} \Big[|(X+iY)\downarrow\rangle + |Z\uparrow\rangle\Big] \\
u_{\frac{1}{2},-\frac{1}{2}} &= -\frac{1}{\sqrt{3}} \Big[|(X-iY)\uparrow\rangle - |Z\downarrow\rangle\Big]
\end{aligned}
\qquad (21.11)
$$

where subscripts *j* and m_j in u_{j,m_j} are quantum number of total angular momentum J and the *z* component J_z, respectively. The functions $u_{\frac{3}{2},\frac{3}{2}}$ and $u_{\frac{3}{2},-\frac{3}{2}}$ express the heavy hole (hh) wavefunctions, $u_{\frac{3}{2},\frac{1}{2}}$ and $u_{\frac{3}{2},-\frac{1}{2}}$ the light hole (lh) wavefunctions, and $u_{\frac{1}{2},\frac{1}{2}}$ and $u_{\frac{1}{2},-\frac{1}{2}}$ are those of the split-off band due to the spin–orbit interaction. Band structure, energy level, and dispersion can be calculated by orthogonalizing the Luttinger–Kohn Hamiltonian containing 2nd order $k \cdot p$ perturbation terms (Basu, 1997). The band structure of the direct gap type is schematically illustrated in Figure 21.5a. In the valence band of the zinc-blend structure, the hh and lh bands are degenerated at the Γ point, as seen in the figure, and the split-off band locates below them. Whereas in the wurtzite structure this degeneracy is dissolved, and the A, B, C bands exist separately (Figure 21.5b). The effective mass of hole is determined by the curvature of the dispersion curve at $k = 0$, i.e., the small curvature gives a large effective mass.

If the wavevector of hole k_h is inclined from the *z*-axis, the hole wavefunctions of the valence band change from (21.11). Here we

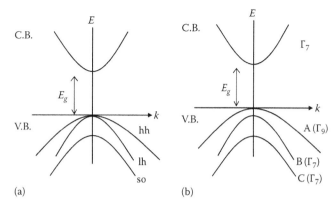

FIGURE 21.5 Band structures near the direct gap of (a) zinc-blend structure and (b) wurtzite structure.

take a z' axis parallel to $\boldsymbol{k}_h$, and x' and y' axes to be normal to z', as shown in Figure 21.6. The Cartesian coordinates (x, y, z) are fixed in space, and the coordinates (x', y', z') are related to (x, y, z) by the rotation of angle θ around the y-axis and successive rotation of angle ϕ around the z-axis. The conversion matrix between the wavefunctions in the two coordinates is expressed as

$$T = \begin{pmatrix} \cos\theta\cos\phi & -\sin\phi & \sin\theta\cos\phi \\ \cos\theta\sin\phi & \cos\phi & \sin\theta\sin\phi \\ -\sin\theta & 0 & \cos\theta \end{pmatrix}. \quad (21.12)$$

Namely, the wavefunction of the heavy hole is expressed in each coordinates as

$$-\frac{1}{\sqrt{2}} \left| \left(X' + iY' \right) \uparrow \right\rangle = -\frac{1}{\sqrt{2}} \{ (\cos\theta\cos\phi - i\sin\phi) \left| X \right\rangle$$
$$+ (\cos\theta\sin\phi + i\cos\phi) \left| Y \right\rangle - \sin\theta \left| Z \right\rangle \} \left| \uparrow \right\rangle$$

$$(21.13)$$

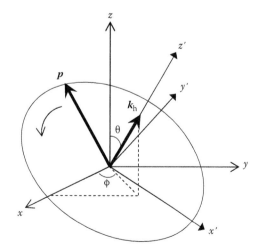

FIGURE 21.6 Relation between the coordinates fixed in space and those associated with the wavevector of hole $\boldsymbol{k}_h$.

and that of the light hole is expressed as

$$\frac{1}{\sqrt{6}} \left| (X' - iY') \uparrow \right\rangle + \sqrt{\frac{2}{3}} \left| Z' \downarrow \right\rangle = \frac{1}{\sqrt{6}} \{ (\cos\theta\cos\phi + i\sin\phi) \left| X \right\rangle$$
$$+ (\cos\theta\sin\phi - i\cos\phi) \left| Y \right\rangle - \sin\theta \left| Z \right\rangle \} \left| \uparrow \right\rangle$$
$$+ \sqrt{\frac{2}{3}} \{ \sin\theta\cos\phi \left| X \right\rangle + \sin\theta\sin\phi \left| Y \right\rangle + \cos\theta \left| Z \right\rangle \} \left| \downarrow \right\rangle \quad (21.14)$$

21.3.2 Theory of Light Emission

Photon is emitted when an electron in the higher energy state transfers to the lower state. The transition probability can be given by the Fermi Golden rule,

$$W = \frac{2\pi}{\hbar} \sum_{\lambda} \sum_{f,i} |\langle f | H' | i \rangle|^2 \, \delta(E_{f\lambda} - E_{i\lambda}). \quad (21.15)$$

Here, H' is the perturbation term due to the interaction between electron and photon appearing in the Hamiltonian, and is expressed as $H' = -(e/m)\boldsymbol{A} \cdot \boldsymbol{p}$ using a vector potential $\boldsymbol{A}$ and momentum $\boldsymbol{p}$. The wavefunctions of the initial and final states are denoted by $|i\rangle$ and $|f\rangle$, which are given by a product of wavefunctions of electron and photon. The summation over λ in (21.15) is taken over modes of photons. Substituting the above expression for the perturbation term, (21.15) can be rewritten as

$$W = \frac{2\pi}{\hbar} \sum_{\lambda} \sum_{f,i} \left(\frac{eA_0}{m} \right)^2 |\langle \psi_f | (\boldsymbol{e} \cdot \boldsymbol{p}) e^{-i\boldsymbol{K} \cdot \boldsymbol{r}} | \psi_i \rangle|^2$$
$$\times (n_{\lambda} + 1) \delta(E_i - E_f - \hbar\omega). \quad (21.16)$$

where $\boldsymbol{K}$ is a wavevector of photon. The first term in $(n_{\lambda} + 1)$ expresses the stimulated emission and the second term expresses the spontaneous emission. In the following, we are only concerned with the spontaneous emission.

The rate of the spontaneous emission at a frequency ω in the transition from electron state in the conduction band to the hole state in the valence band is given by

$$R(\omega) = \frac{2\pi}{\hbar} \left(\frac{eA_0}{m} \right)^2 \sum_{c,v} |M|^2 G(\hbar\omega) f_c(E_c) \left[1 - f_v(E_v) \right] \delta(E_c - E_v - \hbar\omega). \quad (21.17)$$

The summation over the photon modes is replaced by the optical density of states G, which is derived from the van Roosbroeck–Shockly relation as

$$G(\hbar\omega) = \frac{n^3}{\pi^2 c^3 \hbar^3} (\hbar\omega)^2, \quad (21.18)$$

where n is the refractive index of the material. $f_c(E_c)$ and $f_v(E_v)$ are Fermi distribution functions. It is noted that we should use the pseudo-Fermi energy instead of the equilibrium one, when carriers are excited by outer sources. The product of $f_c(1-f_v)$ in (21.17) produces a sharp peak near the band gap energy ($E_c - E_v$) in luminescence spectra. This is clear difference from the absorption spectrum which shows multiple peaks due to $|M(\omega)|^2$ above the band gap energy.

The term M is an optical transition matrix element, giving emission intensity of the transition, and is written as

$$M = \langle \psi_v | e^{-iK \cdot r}(e \cdot p) | \psi_c \rangle \qquad (21.19)$$

where e is a unit vector parallel to the electric field of light, i.e., the polarization direction. Using the momentum operator $p = -i\hbar\nabla$ and wavefunctions of the conduction and valence bands, (21.19) can be rewritten as

$$M = -i\frac{\hbar}{V}\left[\int e^{i(k_c - K - k_v)\cdot r}\left\{ u_v^*(e \cdot \nabla u_c) + i e \cdot k_c u_v^* u_c \right\} dr\right] \quad (21.20)$$

The second term is zero because of orthogonality of the wavefunctions. The following equation is valid for any periodic function of the lattice translation, $f(r) = f(r + R_n)$,

$$\int f(r)e^{i(k_c-K-k_v)\cdot r}\,dr = \sum_n e^{i(k_c-K-k_v)\cdot R_n}$$
$$\times \int_{V_c} f(r)e^{i(k_c-K-k_v)\cdot r}\,dr \propto \delta(k_c - K - k_v)$$
$$(21.21)$$

Using this, the integral in (21.20) is found to be non-zero only when $k_c = k_v + K$, indicating the conservation of wavevector in the transition. The wavelength of light with band gap energy is relatively long, i.e., $K \ll k_c, k_v$ so we may approximate as $k_c \approx k_v$. As a result we obtain

$$M = \langle u_v | e \cdot p | u_c \rangle \delta_{k_v, k_c} \qquad (21.22)$$

Here the condition $k_c \approx k_v$ indicates the "vertical" interband transition.

21.3.3 Polarization

The vector e in the optical transition matrix element indicates polarization direction of the electric field of light. Thus, the polarization in luminescence derives from the property of the matrix element. Here we consider

$$\langle u_v | p | u_c \rangle$$

as a momentum vector. The vector p reflects the polarization of the emitted light; when e is perpendicular to p, M is zero, which means that the polarization component perpendicular to p is

zero. In other words, the x component of the electric field, E_x, is proportional to $\langle u_v | p_x | u_c \rangle$. From symmetry, only the following matrix elements of

$$\langle X | p_x | s \rangle = \langle Y | p_y | s \rangle = \langle Z | p_z | s \rangle = -i\frac{\hbar}{m}P = p_{cv} \qquad (21.23)$$

are non-zero, where P is the factor introduced by Kane (1957).

For example, in the transition from the electron state of $|s\uparrow\rangle$ in the conduction band to the heavy hole state, the matrix elements can be written as

$$\left\langle \frac{3}{2},\frac{3}{2}\middle| p_x \middle| s\uparrow \right\rangle = -\frac{1}{\sqrt{2}}\langle X|p_x|s\rangle = -\frac{1}{\sqrt{2}}p_{cv}$$

$$\left\langle \frac{3}{2},\frac{3}{2}\middle| p_y \middle| s\uparrow \right\rangle = -\frac{i}{\sqrt{2}}\langle Y|p_y|s\rangle = -\frac{i}{\sqrt{2}}p_{cv} \qquad (21.24)$$

$$\left\langle \frac{3}{2},\frac{3}{2}\middle| p_z \middle| s\uparrow \right\rangle = 0$$

From these, the relation among the electric field components can be written as

$$\frac{E_y}{E_x} = e^{i\frac{\pi}{2}}, \quad E_z = 0 \qquad (21.25)$$

This means that the light emitted toward the z direction has a right-handed circular polarization. This situation is illustrated in Figure 21.7, where we express $|X\rangle$ and $|Y\rangle$ in terms of pure angular momentum states. The wavevector k_h of the heavy hole is pointing to the z direction. The wavefunction of the heavy hole is extended parallel to the x-y plane, and its momentum vector p rotates on the same plane. The wavevector k_e of the electron in the conduction band is parallel to the z-axis, and the wavefunction has a spherical symmetry. In this recombination process of the electron and heavy hole, the angular momentum of the heavy hole is preserved as that of the circular polarized light.

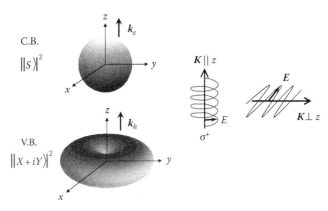

FIGURE 21.7 Wavefunctions of the electron in the conduction band and the heavy hole in the valence band. Circularly polarized and linearly polarized light are emitted by the optical transition between them.

Similarly in the transition from the electron state $|s\uparrow\rangle$ in the conduction band to the light hole state $\left|\dfrac{3}{2},-\dfrac{1}{2}\right\rangle$, the matrix elements can be written as

$$\left\langle\frac{3}{2},-\frac{1}{2}\right|p_x\left|s\uparrow\right\rangle=\frac{1}{\sqrt{6}}\langle X|p_x|s\rangle=\frac{1}{\sqrt{6}}p_{cv}$$

$$\left\langle\frac{3}{2},-\frac{1}{2}\right|p_y\left|s\uparrow\right\rangle=-\frac{i}{\sqrt{6}}\langle Y|p_y|s\rangle=-\frac{i}{\sqrt{6}}p_{cv} \qquad (21.26)$$

$$\left\langle\frac{3}{2},-\frac{1}{2}\right|p_z\left|s\uparrow\right\rangle=0$$

Then, light with a left-handed circular polarization is emitted. Whereas in the transition from the electron state $|s\downarrow\rangle$ to the light hole state $\left|\dfrac{3}{2},-\dfrac{1}{2}\right\rangle$, the matrix elements are written as

$$\left\langle\frac{3}{2},-\frac{1}{2}\right|p_x\left|s\downarrow\right\rangle=0$$

$$\left\langle\frac{3}{2},-\frac{1}{2}\right|p_y\left|s\downarrow\right\rangle=0 \qquad (21.27)$$

$$\left\langle\frac{3}{2},-\frac{1}{2}\right|p_z\left|s\downarrow\right\rangle=\sqrt{\frac{2}{3}}\langle Z|p_z|s\rangle=\sqrt{\frac{2}{3}}p_{cv}$$

In this case, light is linearly polarized parallel to the z-axis.

The emission intensity of CL is proportional to a mean square of the electric field vector. Therefore, in the transition from $|s\uparrow\rangle$ in the conduction band to the heavy hole state in the valence band, the emission intensity of light polarized in each direction is expressed as

$$I_x\propto\frac{1}{2}p_{cv}^2,\quad I_y\propto\frac{1}{2}p_{cv}^2,\quad I_z\propto 0 \qquad (21.28)$$

Similarly in the transition from $|s\uparrow\rangle$ in conduction band to the light hole state $\left|\dfrac{3}{2},-\dfrac{1}{2}\right\rangle$ in the valence band,

$$I_x\propto\frac{1}{6}p_{cv}^2,\quad I_y\propto\frac{1}{6}p_{cv}^2,\quad I_z\propto 0 \qquad (21.29)$$

And in the transition from $|s\downarrow\rangle$ in the conduction band to the light hole state $\left|\dfrac{3}{2},-\dfrac{1}{2}\right\rangle$ in the valence band,

$$I_x\propto 0,\quad I_y\propto 0,\quad I_z\propto\frac{2}{3}p_{cv}^2 \qquad (21.30)$$

The relative intensity and polarization of the emission associated with the transition between the conduction band and the valence band states are schematically depicted in Figure 21.8.

The semiconductors of zinc-blend structure are crystallographically isotropic, and any direction can be taken as the z-axis. Thus,

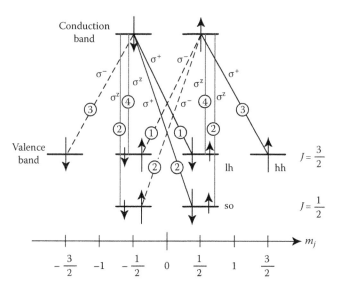

FIGURE 21.8 Relative intensity and polarization of the emission associated with the transition between the conduction band and the valence band states.

the luminescence from the bulk crystal is not polarized, because the polarization of each recombination process cancels out as a whole. Typical cases in which polarized emission occurs are luminescence from quantum structures and that from semiconductors of wurtzite structure. In the case of quantum well structures, a unique axis appears normal to the well, and the heavy and light hole states have anisotropic distribution which causes dissociation of the band degeneration (Yu and Cardona, 1999). Similarly in semiconductors of wurtzite structure, the c-axis is a unique axis, and the bands of the heavy and light hole states are dissolved into the A and B bands. If the highest energy state in the valence band is heavy hole as in the case of GaN, the optical transition to the heavy hole state mainly occurs, and the luminescence is polarized perpendicular to the c-axis. In addition, the polarization property of luminescence changes by a strain field in semiconductor crystals, because a lattice strain causes an anisotropic deformation potential and reconstructs the band structure (Hensel and Feher, 1963; Pollak and Cardona, 1968).

21.4 Quantum Structures

When carriers are confined in small-scaled semiconductor structures, carrier wavefunctions form standing waves and start to show quantum character. The wavelength and wave number of a carrier in a one-dimensional potential well of an infinite height with a width of L are given as

$$\lambda=\frac{2L}{n},\quad k=\frac{n\pi}{L} \qquad (21.31)$$

where n is an integer number. The wave number k can only take discrete values in this direction corresponding to $n = 1, 2, 3, \ldots$, and has continuous value in the other direction. Kinetic energy in this direction takes discrete values given by

$$\Delta E = \frac{\hbar^2}{2m}k^2 = \frac{\hbar^2}{2m}\left(\frac{n\pi}{L}\right)^2 \qquad (21.32)$$

This anisotropy in k gives rise to anisotropy in optical transition and polarization of luminescence. In the following section, we start with a quantum well structure because of simplicity in theoretical treatment and next move to a quantum wire structure which shows strong polarization property.

21.4.1 Quantum Well

We consider carriers confined in a thin semiconductor plate of a width L_z in the z direction. The wavefunction of electrons in the conduction band and holes in the valence band are expressed using the Bloch function $u_{c,v}(\boldsymbol{r})$ at $\boldsymbol{k} = 0$ under the effective mass approximation as

$$\Psi^e(\boldsymbol{r}) = \frac{1}{\sqrt{V}}\sin\left(\frac{n_c\pi}{L_z}z\right)e^{i(k_x x + k_y y)}u_c(\boldsymbol{r}) = \phi_e(z)e^{i(k_x x + k_y y)}u_c(\boldsymbol{r})$$
$$(21.33)$$

$$\Psi^v(\boldsymbol{r}) = \frac{1}{\sqrt{V}}\sin\left(\frac{n_v\pi}{L_z}z\right)e^{i(k_x x + k_y y)}u_v(\boldsymbol{r}) = \phi_h(z)e^{i(k_x x + k_y y)}u_v(\boldsymbol{r})$$
$$(21.34)$$

where ϕ_e and ϕ_h are envelop functions of the electron and hole, respectively, showing standing waves in the confined direction. The energy of the electron and hole measured from the top of the valence band are given by

$$E_e(n) = \frac{\hbar^2}{2m_e}\left(\frac{n_c\pi}{L_z}\right)^2 + \frac{\hbar^2}{2m_e}(k_x^2 + k_y^2) + E_g \qquad (21.35)$$

$$E_h(n) = \frac{\hbar^2}{2m_h}\left(\frac{n_v\pi}{L_z}\right)^2 + \frac{\hbar^2}{2m_h}(k_x^2 + k_y^2) \qquad (21.36)$$

where

- m_e and m_h are the effective mass of electron and hole, respectively
- n_c and n_v are their quantum numbers

Since wavevector components, k_x and k_y, take continuous values, mini bands are formed for each quantum number, as shown in Figure 21.9. It should be noted that (21.36) is a rough approximation because the band mixing should be considered for the valence band in the quantum structure, as shown in Figure 21.9. However, this gives a simple picture for the band splitting in the valence band. In the quantum well structure, the degenerated bands at the Γ point are dissolved, and the highest energy state is the heavy hole. Consequently, the transition from the electron state to the heavy hole state becomes dominant. Since the energy

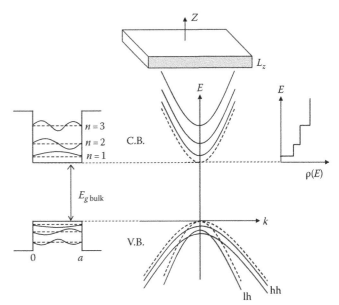

FIGURE 21.9 Wave functions, band structure, and density of states in a quantum well.

shift of the valence band is relatively small compared to that of the conduction band, the emission energy is roughly estimated from (21.35) and (21.36).

The probability of interband transition from the conduction band to the heavy hole state at the band edge (BE) is given by the following matrix element,

$$M = \langle \Psi^v | \boldsymbol{e}\cdot\boldsymbol{p} | \Psi^c \rangle = \langle \phi_h(z) | \phi_e(z) \rangle \left\langle u_{\frac{3}{2},\frac{3}{2}} \middle| \boldsymbol{e}\cdot\boldsymbol{p} \middle| u_c \right\rangle \qquad (21.37)$$

The product of the electron and hole envelop functions is non-zero only when

$$n_c = n_v \qquad (21.38)$$

If M is zero, the corresponding transition is inhibited. This gives the selection rule of the transition, and means that the optical interband transition only occurs between the electron and hole states of the same quantum number. The emission energy is given by

$$E(n) = E_e(n) - E_h(n) = \frac{\hbar^2}{2\mu}\left(\frac{n_c\pi}{L_z}\right)^2 + \frac{\hbar^2}{2\mu}(k_x^2 + k_y^2) + E_g \qquad (21.39)$$

where μ is the reduced effective mass. The lowest energy transition occurs for $n_c = n_v = n = 1$. However, if the barrier height is finite, other transitions become permitted.

The polarization of emission from the quantum well is deduced from (21.37). Here we take the z-axis normal to the well. In the transition between the lowest mini bands at the BE, the wavevector of carrier is $\boldsymbol{k}_c = \boldsymbol{k}_v = (0, 0, \pi/L_z)$, which points to the z direction. Therefore, the result of (21.24) can be applied for this case, so the emission toward the z direction is a right-handed circular polarized light. Whereas light emitted along the well is linearly polarized parallel to the well.

In the interband transition between the electronic states with a finite $k_\perp$ component normal to the z-axis, the wavevector of the carriers is inclined from the z-axis, and the wavefunction of the valence band changes according to (21.13). The emission energy increases according to (21.39). In order to find the polarization of emitted light with this energy, the transition matrix element should be integrated over equivalent transitions of the same energy. Firstly we consider the transition from the $|s\uparrow\rangle$ state in the conduction band to the heavy hole state in the valence band. The wavevector component parallel to the z-axis is π/L_z, and then $\boldsymbol{k} = (k\sin\theta\cos\phi, k\sin\theta\sin\phi, \pi/L_z)$. The wavefunction of the heavy hole is expressed by (21.13). The x and y directions are equivalent, so here we consider the emission propagating in the x direction with the electric field parallel to the y direction (TE wave in Figure 21.10). After integrating over the azimuth angle ϕ, the averaged transition matrix element is given as

$$\langle |M|^2 \rangle_{hh,TE} = \frac{1}{2\pi} \int_0^{2\pi} \left| \left\langle \frac{3}{2}, \frac{3}{2} \middle| p_y \middle| s \right\rangle \right|^2 d\phi$$

$$= \frac{1}{2} \cdot \frac{1}{2\pi} \int_0^{2\pi} (\cos^2\theta\sin^2\phi + \cos^2\phi)\, d\phi = \frac{1}{4} p_{cv}^2 (1 + \cos^2\theta)$$

$$(21.40)$$

Similarly for the electric field parallel to the z direction (TM wave), the averaged transition matrix element is given as

$$\langle |M|^2 \rangle_{hh,TM} = \frac{1}{2\pi} \int_0^{2\pi} \left| \left\langle \frac{3}{2}, \frac{3}{2} \middle| p_z \middle| s \right\rangle \right|^2 d\phi$$

$$= \frac{1}{2} \cdot \frac{1}{2\pi} \int_0^{2\pi} \sin^2\theta\, d\phi = \frac{1}{2} p_{cv}^2 \sin^2\theta \qquad (21.41)$$

If the $k_\perp$ component is zero ($k_x = k_y = 0$), i.e., $\theta = 0$ at the mini-BE, the TM component is zero and the polarization of the emission is parallel to the well. As the tilted angle θ from the z-axis increases, the emission energy increases with increasing $\boldsymbol{k}$ and the polarization direction tilts from the well plane.

21.4.2 Quantum Wire

21.4.2.1 Energy Levels in Quantum Wire

A semiconductor quantum wire of an infinite length along the z-axis with a rectangular cross section of the widths of L_x and L_y confines a carrier in an infinite barrier height. Electron wavefunction in the quantum nanowire is expressed in the effective mass approximation by a product of the Bloch function at the BE and the envelop function in the x-y plane normal to the wire axis. For simplicity we consider the electron state at the BE of the conduction band in semiconductors with a cubic symmetry and a direct band structure. The electron wavefunction is expressed as

$$\Psi^e(\boldsymbol{r}) = \phi^c(x,y)\, e^{ik_z z}\, u_c(\boldsymbol{r}) \qquad (21.42)$$

The envelop function of the electron is written as

$$\phi^c(x,y) = \frac{1}{\sqrt{L_x L_y}} \sin\left(\frac{\pi n_c}{L_x} x\right) \sin\left(\frac{\pi m_c}{L_y} y\right) \qquad (21.43)$$

where n_c and m_c are integer. The energy of the electron in the (n_c, m_c) band is given by

$$E_e(n_c, m_c) = \frac{\hbar^2}{2m_e}\left\{\left(\frac{\pi n_c}{L_x}\right)^2 + \left(\frac{\pi m_c}{L_y}\right)^2\right\} + \frac{\hbar^2}{2m_e} k_z^2 + E_g \qquad (21.44)$$

where m_e is the electron effective mass. The lowest energy band corresponds to the quantum numbers of $n_c = m_c = 1$. Mini bands are formed from (21.44) in the conduction band, as shown in Figure 21.11. The density of states as a function of energy has sharp peaks at the mini-BE given by the first term, and shows a sawtooth shape.

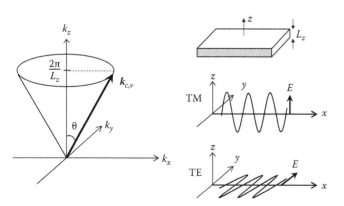

FIGURE 21.10 Wavevectors of the same energy states in the miniband. Light emission of both the TM mode and TE mode occurs from a quantum well.

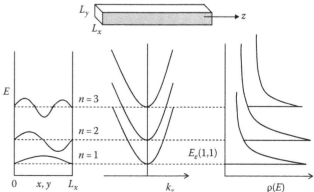

FIGURE 21.11 Wave functions, band structure, and density of states in a quantum wire with a square cross section.

On the other hand, electrons in a quantum wire with a circular cross section of radius R has the following envelop functions expressed using the cylindrical coordinates (ρ, ϕ) as

$$\phi_e(\rho,\phi) \propto J_\ell\left(\frac{j_{n,\ell}\rho}{R}\right)\cos(\ell\phi) \quad \text{or} \quad J_\ell\left(\frac{j_{n,\ell}\rho}{R}\right)\sin(\ell\phi) \quad (21.45)$$

where

J_l is the Bessel function of the ℓth order
$j_{l,n}$ is the 0 point value of nth order given by the condition that $J_l(j_{l,n}) = 0$

The lowest values are given as $j_{0,1} = 2.405$, $j_{1,1} = 3.832$, and $j_{0,2} = 5.520$. A set of energy levels are derived as

$$E_e(\ell,n) = \frac{\hbar^2 j_{\ell,n}^2}{2m_e R^2} + \frac{\hbar^2}{2m_e}k_z^2 + E_g \quad (21.46)$$

All the energy states indicated by the quantum number (ℓ, n) are doubly degenerated, except for $\ell = 0$.

Change in electron energy levels in the conduction band due to the quantum confinement effect is large, whereas those of holes in the valence band are relatively small because of small curvature of the bands or large effective mass of hole. However, the band mixing occurs, and then the hole wavefunctions are formed by a mixture of the bulk wavefunctions in (21.11). This affects the property of the optical transitions. The hole wavefunctions are expressed by a linear combination of the bulk wavefunctions as

$$\Psi^\nu(\mathbf{r}) = e^{ik_z z}\sum_j \phi_j^\nu(x,y)u_j(\mathbf{r}) \quad \left(j = \frac{3}{2}, \frac{1}{2}, -\frac{1}{2}, -\frac{3}{2}\right) \quad (21.47)$$

The envelop functions are written as

$$\phi_j^\nu(x,y) = \frac{1}{\sqrt{L_x L_y}}\sum_{n_\nu, m_\nu} c_j(n_\nu, m_\nu)\sin\left(\frac{\pi n_\nu}{L_x}x\right)\sin\left(\frac{\pi m_\nu}{L_y}y\right) \quad (21.48)$$

Here, we ignore the contribution of the split-off band to the mixing. The wavefunctions and energies in the valence band of the quantum wire can be derived from the theoretical calculation using the Luttinger–Kohn Hamiltonian in the effective mass approximation (Sercel and Vahala, 1990; Bockelmann and Bastard, 1992; McIntyre and Sham, 1992).

Theoretical treatments were firstly applied to the quantum wires manufactured from the quantum well structures, in which the wire has an axis parallel to the growth surface and rectangular shape in cross section (Tanaka and Sakaki, 1989; Tsuchiya et al., 1989). Because of the technological development for the fabrication of self-standing nanowire, energy band calculations have been done for self-standing wires with circular and rectangular cross sections of Si (Zhao et al., 2004), GaAs (Persson and Xu, 2004, Redlinski and Peeters, 2008),

and InP (Pellegrini et al., 2005). The band dispersion of camel back structure appears in some of the high energy bands of the valence band. The character of the band structure such as the sequence of the levels does not change with the diameter of the quantum wire. The highest energy band in the valence band has a character of the light hole, and the second highest energy band has a character of the heavy hole. The difference in band structure between the nanowires grown in the [100] and [111] directions was shown in the GaAs nanowires (Redlinski and Peeters, 2008).

The luminescence due to exciton is generally dominant at low temperatures. The emission energy of the FX is given by the band gap energy minus the exciton binding energy as in (21.6). If the carrier confinement in quantum wires is perfect, the exciton binding energy becomes infinite in the limit of small wire size (Basu, 1997). A real nanowire has a finite size in diameter, the divergence can be avoided using a following potential for the Coulomb interaction between the electron-hole pair (Ando et al., 1991),

$$V(z,z_0) = -\frac{e^2}{4\pi\varepsilon}\frac{1}{\left(|z| + z_0\right)} \quad (21.49)$$

where z_0 is a parameter for the size effect of nanowire. Quantum confinement effect is large enough to be detected, when the diameter of the wire becomes comparable to or less than the exciton radius. The size dependence of the emission energy is approximately proportional to R^{-2}, as in (21.46). The theoretical calculations, including other effects, have been carried out for the size dependence of the emission energy; for example, the dielectric mismatch effect results in the $R^{-1.8}$ dependence (Redlinski and Peeters, 2008).

The barrier height of the carrier confinement is finite in real when a quantum wire is surrounded by another semiconductor or even by air. In such cases, the energy levels decrease from those given by (21.44) and (21.46), and the distances between the energy levels decrease, because they must exist below the barrier height level. When the diameter of wire decreases with the barrier height fixed, the energy levels move upward to gather near the barrier height level, and their total number decreases. However, at least one level is left at the barrier height in the symmetrical potential well in the limit of small size.

21.4.2.2 Luminescence from Quantum Wire

Luminescence from bulk crystals with cubic symmetry such as Si and GaAs is generally unpolarized, whereas that with hexagonal symmetry such as GaN and ZnO shows anisotropy in polarization because the c-axis is a unique axis. The luminescence from nanowires has been frequently observed to be polarized even for semiconductors with cubic symmetry. There are two reasons for the appearance of the polarization anisotropy in luminescence from nanowire; one is the anisotropy originated from the quantum effect, and another is the anisotropic response of nanowire to the electromagnetic fields parallel or perpendicular to the wire axis (Ruda and Shik, 2005, 2006).

The former effect appears when a diameter of nanowire becomes smaller than the exciton radius, which gives rise to the polarization anisotropy in the luminescence of excitons in the nanowire. The latter effect appears even in nanowires with diameter larger than the exciton radius.

21.4.2.2.1 Quantum Confinement Effect

The quantum confinement of carriers in thin nanowires affect on the polarization of luminescence. Here we treat a nanowire of a square cross section with a dimension of a and an infinite length. The wire is isotropic in the cross-sectional plane and the confinement potential has an infinite barrier height for carriers in the wire. At first we ignore the band mixing in the valence band, and use the wavefunctions of the one-band model expressed as follows:

$$\Psi^v(\mathbf{r}) = e^{ik_z z}\phi_j^v(x, y)u_j(\mathbf{r}) \tag{21.50}$$

When we take the heavy hole state for j, the transition matrix element is written as

$$M = \langle \psi_v | e^{-i\mathbf{K}\cdot\mathbf{r}}(\mathbf{e}\cdot\mathbf{p}) | \psi_c \rangle = \sum_{k_i} \langle \phi_j^v | \phi_e \rangle \langle u_j(\mathbf{k}_i) | \mathbf{e}\cdot\mathbf{p} | u_c \rangle \tag{21.51}$$

Electron wavefunctions of the lowest energy states in the conduction band are expressed by (21.42) and (21.43) with $n_c = m_c = \pm 1$. The transition occurs under the selection rule of $n_c = n_v$ and $m_c = m_v$ in (21.43) and (21.48). There are four corresponding heavy hole states of the same energy at the BE ($k_z = 0$), having the following wavevectors:

$$\mathbf{k}_i = \left(\pm\frac{\pi}{a}, \pm\frac{\pi}{a}, 0\right) : i = 1 \sim 4 \tag{21.52}$$

The wavefunctions of the heavy hole states with these wavevectors are given by substituting $\theta = 90°$ in (21.13) as

$$\left|\frac{3}{2}, \frac{3}{2}\right\rangle_{k_i} = -\frac{1}{\sqrt{2}}\left\{-i\sin\phi_i |X\rangle + i\cos\phi_i |Y\rangle - |Z\rangle\right\}|\uparrow\rangle \tag{21.53}$$

Four azimuth angles of

$$\phi_1 = \frac{\pi}{4}, \quad \phi_2 = \frac{3\pi}{4}, \quad \phi_3 = \frac{5\pi}{4}, \quad \phi_4 = \frac{7\pi}{4} \tag{21.54}$$

correspond to the four wavevectors. Similarly, the wavefunctions of the light hole states with the same wavevectors are written as

$$\left|\frac{3}{2}, -\frac{1}{2}\right\rangle = \frac{1}{\sqrt{6}}\left\{i\sin\phi_i |X\rangle - i\cos\phi_i |Y\rangle - |Z\rangle\right\}|\uparrow\rangle$$

$$+ \sqrt{\frac{2}{3}}\left\{\cos\phi_i |X\rangle + \sin\phi_i |Y\rangle\right\}|\downarrow\rangle \tag{21.55}$$

If the transitions from the conduction band to those valence band states are assumed to occur independently, mean square of the transition matrix element is given as

$$|M|_{e-hh}^2 = \frac{1}{4}\sum_{k_i}\left|\left\langle u_{\frac{3}{2},\frac{3}{2}}(\mathbf{k}_i)\middle|\mathbf{e}\cdot\mathbf{p}\middle|s\uparrow\right\rangle\right|^2 \tag{21.56}$$

Here, the term from $|s\downarrow\rangle$ to $\left|u_{\frac{3}{2},-\frac{3}{2}}(\mathbf{k}_i)\right\rangle$ is ignored, because it gives the same value as above. If we take the z direction along the wire axis, it follows that $\mathbf{e}\cdot\mathbf{p} = p_z$ for the polarization of emitted light parallel to the wire axis and $\mathbf{e}\cdot\mathbf{p} = p_x$ for that normal to the wire axis. Then the square of the transition matrix element is calculated using (21.52), and the two components parallel and perpendicular to the z-axis are obtained as

$$|M|_{e-hh\parallel}^2 = \frac{1}{2}p_{ev}^2, \quad |M|_{e-hh\perp}^2 = \frac{1}{4}p_{ev}^2 \tag{21.57}$$

The component parallel to the wire axis is larger than the perpendicular one because the transformed wavefunction in (21.52) involves a large $|Z\rangle$ component. When the polarization of light is inclined from the wire axis by an angle θ (Figure 21.12), the square of the transition matrix element can be written as

$$|M|_{e-hh}^2 = \frac{1}{2}p_{ev}^2\cos^2\theta + \frac{1}{4}p_{ev}^2\sin^2\theta \tag{21.58}$$

The transition matrix element from the conduction band to the light hole state is calculated in the same way using (21.54),

$$|M|_{e-\ell h}^2 = \frac{1}{6}p_{ev}^2\cos^2\theta + \frac{5}{12}p_{ev}^2\sin^2\theta \tag{21.59}$$

In this transition, the component perpendicular to the wire axis is larger than the parallel one. The two hole states locate at different energy levels, so the emission energy of this transition is

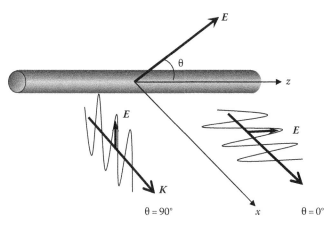

FIGURE 21.12 Light emissions from a nanowire polarized in directions inclined from the wire axis by angles of $\theta = 0°$ and $90°$.

also different. If we use the relation $p_{cv}^2 = 3M^2$ in which M is the averaged value in the isotropic bulk crystal, the ratio between these transition matrix elements can be written as

$$\frac{|M|_{e-\ell h}^2}{|M|_{e-hh}^2} = \frac{\frac{1}{2}\cos^2\theta + \frac{5}{4}\sin^2\theta}{\frac{3}{2}\cos^2\theta + \frac{3}{4}\sin^2\theta} \qquad (21.60)$$

This result was derived by several authors on the basis of the one-band model (Tanaka and Sakaki, 1989; Tsuchiya et al., 1989), though this treatment was claimed to be too simple, and later more reliable one was proposed on the basis of band mixing in the valence band.

In the photoluminescence excitation (PLE) spectrum of the quantum wires formed in the GaAs quantum well structure, two peaks were observed to appear near the BE energy, as shown in Figure 21.13 (Tsuchiya et al., 1989). On the other hand in the PL spectrum, only the lower energy peak appeared. The lower energy peak was assigned to be due to the transition to the heavy hole state, and the higher peak due to the transition to the light hole state, respectively. It appeared to be consistent with the result of (21.57). However, it was suggested that the one-band model cannot explain the polarization property correctly, and the theoretical treatment taking account of the band mixing in the valence band should be necessary. Sercel and Vahala (1990) showed that if the coherent combination of the hole states with the four wavevectors is considered in the one-band model, Equation 21.57 has a constant value (=1/3) independent of θ. They calculated the polarization dependence of the optical transition using the mixed states of the heavy and light holes in a cylindrical wire. As a result, the transition

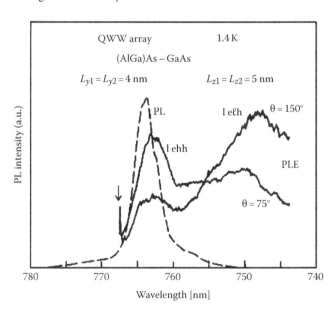

FIGURE 21.13 Photoluminescence (PL) and photoluminescence excitation (PLE) spectra of the quantum wires formed in the GaAs quantum well structure. (From Tsuchiya, M. et al., *Phys. Rev.*, 62, 455, 1989. With permission.)

matrix elements associated with the two highest hole states were obtained as

$$|M|_{C-V\left(\frac{1}{2}\right)}^2 = \left(\frac{2}{3}\cos^2\theta + \frac{1}{6}\sin^2\theta\right) p_{ev}^2 I_{C-V\left(\frac{1}{2}\right)}^2 \qquad (21.61)$$

$$|M|_{C-V\left(\frac{3}{2}\right)}^2 = \frac{1}{2}\sin^2\theta \cdot p_{ev}^2 I_{C-V\left(\frac{3}{2}\right)}^2 \qquad (21.62)$$

where I is an overlap integral. The optical transition of the lowest energy is $C-V(1/2)$. The $V(1/2)$ state involves only 41% of the heavy hole state. The single peak in the PL spectrum is due to this transition. The polarization component parallel to the wire axis is dominant in the emission, and the polarization ratio deduced from (21.60) is 60%. The polarization ratio becomes smaller as the barrier height decreases. The optical transition of the next lowest energy is $C-V(3/2)$, and the $V(3/2)$ state involves 67% of the heavy hole state. The polarization direction of the corresponding emission is perpendicular to the wire axis. The property of the polarization in these two transitions is well consistent with those of the two emission peaks in the PLE spectrum in Figure 21.13.

Similar results were obtained in the calculation of optical absorption for self-standing GaAs nanowires (Redlinski and Peeters, 2008). The highest energy state in the valence band has a character of the light hole. The optical transition matrix element associated with this hole state has both polarization components parallel and perpendicular to the wire axis, and the former is much larger than the latter. The fine splitting of the lowest energy peak in the absorption spectrum appears in the [111] wire because of the anisotropy in the cross-sectional plane. The next highest energy state in the valence band has a character of the heavy hole, and the absorption occurs only for the polarization direction perpendicular to the wire axis. Emission energies in these two transitions are slightly different, so in PL and CL, the transition to the band having the light hole character is predominant.

21.4.2.2.2 Dielectric Effect

The second factor to affect the polarization of luminescence from semiconductor nanowires is the dielectric effect (Ruda and Shik, 2005). When a spatially homogeneous electric field is applied to a nanowire of cylindrical shape with the electric field vector parallel to the wire axis, magnitude and direction are the same both in and outside of the wire. Whereas the electric field is applied perpendicular to the wire axis, its magnitude is reduced by a factor of $(2\varepsilon_0/\varepsilon + \varepsilon_0)$, where ε and ε_0 are relative dielectric constants of semiconductor and surrounding material, respectively (Landau et al., 1984), i.e.,

$$E_{\parallel} = E_{0\parallel}$$

$$E_{\perp} = \frac{2\varepsilon_0}{\varepsilon + \varepsilon_0} E_{0\perp} \qquad (21.63)$$

This is true for the oscillating electric field of light if the diameter of the wire is sufficiently smaller than the wavelength of light. From this, the polarization dependence of the optical absorption is expressed by

$$\frac{k_{\parallel}}{k_{\perp}} = \left| \frac{\varepsilon + \varepsilon_0}{2\varepsilon_0} \right|^2 \tag{21.64}$$

This anisotropy appears in PL and light-induced current excited by polarized light.

As for the dielectric effect on the polarization of luminescence, we consider an oscillating electric dipole with a frequency of ω and a dipole moment of $\boldsymbol{d}_0 = (d_{0x}, d_{0y}, d_{0z})$ in a cylindrical nanowire. We take the z-axis parallel to the wire axis and the x-axis parallel to the propagating direction of emitted light which is normal to the wire axis. Here we consider an effective dipole moment $\boldsymbol{d} = (d_x, d_y, d_z)$ of which the electric potential in a free space is equivalent to that produced by $\boldsymbol{d}_0$ outside the nanowire. The relation between $\boldsymbol{d}$ and $\boldsymbol{d}_0$ can be written as

$$d_{x,y} = \frac{2\varepsilon_0}{\varepsilon + \varepsilon_0} d_{0x,y}, \quad d_z = d_{0z} \tag{21.65}$$

The magnitude of the pointing vector of light at the position far from the center of wire by x_0 is given by

$$S_{\parallel} = \frac{a\omega^4}{6\pi\varepsilon_0^2 c^3 x_0} (d_x^2 + 2d_z^2) \tag{21.66}$$

$$S_{\perp} = \frac{a\omega^4}{2\pi\varepsilon_0^2 c^3 x_0} d_y^2$$

for the components parallel and perpendicular to the wire axis, respectively. Here, c is the speed of light and a is a radius of the wire. Therefore, when the semiconductor is isotropic and $d_{0x} = d_{0y} = d_{0z}$, the ratio of intensity between the emissions polarized parallel and perpendicular to the wire axis is given as

$$\frac{I_{\parallel}}{I_{\perp}} = \frac{S_{\parallel}}{S_{\perp}} = \frac{d_x^2 + 2d_z^2}{3d_x^2} = \frac{(\varepsilon + \varepsilon_0)^2 + 2\varepsilon_0^2}{6\varepsilon_0^2} \tag{21.67}$$

The polarization ratio is given by

$$P = \frac{I_{\parallel} - I_{\perp}}{I_{\parallel} + I_{\perp}} = \frac{(\varepsilon + \varepsilon_0)^2 - 4\varepsilon_0^2}{(\varepsilon + \varepsilon_0)^2 + 8\varepsilon_0^2} \tag{21.68}$$

This indicates that P does not depend on the wire size. Since the dielectric constants of semiconductors are much larger than 1, the dielectric effect affects seriously on the polarization anisotropy. In case that the internal emission is not isotropic such as in the case of the wurtzite-type semiconductors and quantum wires, we have to consider that $d_{0x} = d_{0y} \neq d_{0z}$. In such cases, (21.68) changes to

$$P = \frac{I_{\parallel} - I_{\perp}}{I_{\parallel} + I_{\perp}} = \frac{(\varepsilon + \varepsilon_0)^2 p - 4\varepsilon_0^2}{(\varepsilon + \varepsilon_0)^2 p + 8\varepsilon_0^2} \tag{21.69}$$

where we define

$$p = \frac{d_{0z}^2}{d_{0x}^2} = \frac{|M_{\parallel}|^2}{|M_{\perp}|^2} \tag{21.70}$$

as a measure of the anisotropic internal emission.

For large nanowires, the above treatment becomes invalid when the diameter of the wire is comparable with the wavelength of emitted light, and the polarization depends on wire size. In this case, the retardation effect should be involved, and we have to solve the Helmholtz equation to find the electromagnetic field in the nanowires instead of the Laplace equation (Ruda and Shik, 2006). Consequently, the following equations are derived corresponding to (21.65)

$$d_x = \sqrt{\varepsilon} \frac{J_1'(ka)H_1^{(1)}(ka) - J_1(ka)H_1^{(1)'}(ka)}{\sqrt{\varepsilon_0}J_1'(ka)H_1^{(1)}(k_0a) - \sqrt{\varepsilon}J_1(ka)H_1^{(1)'}(k_0a)} d_{0x} \equiv A_x d_{0x}$$

$$d_z = \sqrt{\varepsilon} \frac{J_1(ka)H_0^{(1)}(ka) - J_0(ka)H_1^{(1)}(ka)}{\sqrt{\varepsilon} J_1(ka)H_0^{(1)}(k_0a) - \sqrt{\varepsilon_0}J_0(ka)H_1^{(1)}(k_0a)} d_{0z} \equiv A_z d_{0z}$$

$$\tag{21.71}$$

where J and H are the Bessel and first-kind Hankel functions

$$k = \sqrt{\varepsilon} \frac{\omega}{c}$$

$$k_0 = \sqrt{\varepsilon_0} \frac{\omega}{c}$$

The coefficient A expresses the effect of image force. It is noted that ε_0 and ε are the relative dielectric constants at the frequency ω, and different from the static values listed in Table 21.1. The relation of the wave numbers of light in the nanowire and surrounding material is written as $k = k_0\sqrt{\varepsilon/\varepsilon_0} = nk_0$, using the refractive index n of the material of the wire. The polarization ratio is given as

$$P = \frac{A_z^2 p - A_x^2}{A_z^2 p + 2A_x^2} \tag{21.72}$$

Figure 21.14 shows polarization ratio P as a function of $k_0a(=(\omega/c)a)$, calculated for various values of ε and p. The polarization is seen to strongly oscillate with k_0a. In the case that $\varepsilon = 4$ and $p = 1$, the points where $P = 0$ are given by $k_0a = 1.21, 1.91, 2.72, 3.5, \ldots$. It is found that the polarization ratio has a value calculated from (21.68), when the radius a is sufficiently smaller than the wavelength of light, λ_0. Then, the polarization ratio P decreases

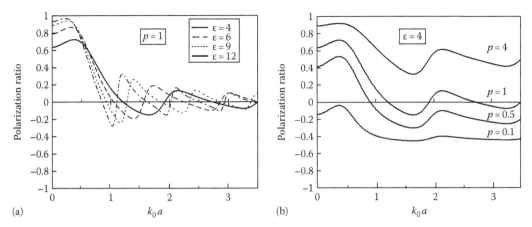

(a) $k_0 a$ (b) $k_0 a$

FIGURE 21.14 Polarization ratio as a function of $k_0 a (=(\omega/c)a)$ calculated by (21.72) for (a) various values of ε with $p = 1$, and (b) various values of p with $\varepsilon = 4$.

with increasing a, and is equal to zero at $a_1 = 0.193\lambda_0$. With further increasing a, P gradually approaches zero with oscillation. If p is not equal to 1, the asymptotic value of P is to be given by (21.69) for $k_0 a = 0$ and $(p - 1)/(p + 2)$ in the limit of $k_0 a \to \infty$.

21.5 Application of TEM-CL

21.5.1 InP Nanowires

InP has a direct band gap structure with a gap energy of 1.42 eV at 4 K. Exciton Bohr radius is about 11 nm, so the quantum confinement effects, energy shift of the near-edge emission, and polarization characteristics are expected to appear for nanowires with a diameter smaller than this radius. Exciton binding energy is 5 meV. The emission intensity of exciton in the bulk crystal is weak at higher temperatures, but it can be observed even at room temperature if the quantum effect is strong enough. The size dependence of emission energy by quantum confinement effect has been measured in PL and absorption spectroscopy (Yu et al., 2003), and theoretically treated for cylindrical nanowires (Pellegrini et al., 2005). The PL study by Wang et al. (2001) using a polarized light excitation showed extremely large degree of polarization ($P = 0.96$) along the wire axis in the luminescence from single InP nanowires. The peak shift in PL was small, and the anisotropic polarization was considered to be due to the dielectric effect. Mishra et al. (2007) showed that the PL emission from the InP nanowires of 80 nm in diameter having zinc-blend and wurtzite structures are polarized parallel and perpendicular to the wire axis, respectively.

InP nanowires were grown on InP substrates by the metal-organic vapor phase epitaxy (MOCVPE) technique with gold nanoparticles as the catalyst (Figure 21.15). From the TEM observation, the InP nanowire has the zinc-blend structure of cubic lattice. The wire axis is parallel to the [111] axis, and the nanowire contains many twin boundaries normal to the wire axis (Bhunia et al., 2003). The diameter of the nanowire is distributed in the range from 5 to 30 nm, and the most frequently observed size is 20 nm. The length can be controlled by the

FIGURE 21.15 A SEM image of InP nanowires grown on InP.

growth conditions such as group-III source gas flow rate and total growth time, and that of the present sample is 700 nm. Cross-sectional samples were made by cleaving the sample crystal normal to the (111) substrate surface, and were examined by TEM-CL (Yamamoto et al., 2006).

Figure 21.15a shows a SEM image of InP nanowires standing straight on the InP substrate with various diameters. The image was taken at a tilt angle of 40°. Figure 21.16a shows a SEM image of a cross-sectional sample where the InP nanowires are standing perpendicular to the incident electron beam. The CL spectra from the nanowires and a substrate region are shown in Figure 21.16b with a common vertical scale, which were taken at room temperature with an electron beam scanning over an 0.5 μm × 0.5 μm area in each region. A broad peak appears in the CL spectrum from the nanowire region at peak energy of 1.50 eV. The band gap energy of the bulk InP is 1.34 eV at room temperature, and this peak energy is much higher than that expected from the quantum confinement effect for the nanowire of 20 nm in diameter. The intensity of the nanowire emission is one order stronger than that of the InP substrate. A monochromatic CL image in Figure 21.16c clearly shows that an intense emission comes from the

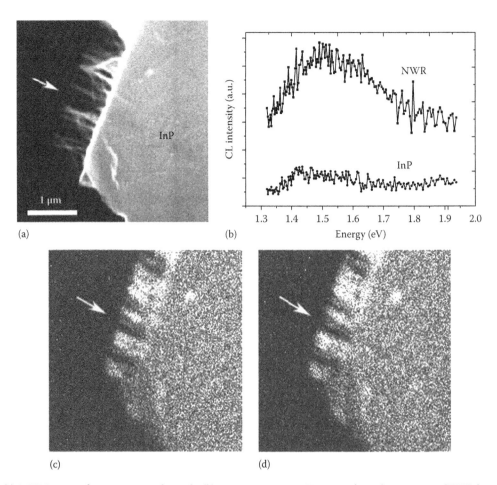

FIGURE 21.16 (a) A SEM image of a cross-sectional sample, (b) room temperature CL spectra from the nanowires (NWRs) and substrate region with a common vertical scale. (c) and (d) are monochromatic CL images taken at photon energies of 1.419 and 1.473 eV, respectively.

nanowires rather than from the substrate at room temperature. Individual nanowire can be resolved in the CL image, as indicated by an arrow. The broad emission peak from the nanowires is considered to be due to the size distribution in wire diameter because the electron confinement energy of nanowire changes with wire diameter. Then the emissions of different energies are expected to come from the nanowires with different diameters. Figure 21.16c and d shows monochromatic CL images taken at photon energies of 1.419 and 1.473 eV, respectively. The images should reveal the distribution of nanowires which produce the emissions of these energies. However, the intensity distribution is almost the same in those images, and the nanowires with different diameters cannot be distinguished. This fact indicates that each nanowire gives the broad peak in the emission spectrum.

The temperature dependence of the emission spectrum from the nanowire is shown in Figure 21.17. The emission intensity of the broad peak does not increase with decreasing temperature from room temperature to 40 K, while that from the InP substrate strongly increases with cooling. The broad peak blueshifts with decreasing temperature and the behavior of the shift follows the Varshni's law. The deviation of the broad peak from the exciton peak in the bulk InP is about 170 meV. For the

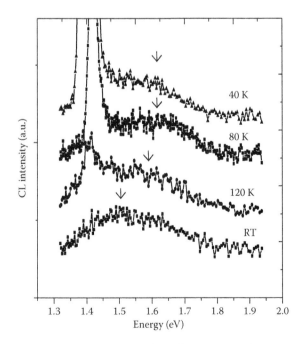

FIGURE 21.17 Emission spectra from the nanowires taken at various temperatures.

recombination of carriers in the ground states ($l = 0$ and $n = 1$), the emission energy shift from the bulk one is roughly estimated from (21.46) to be $\Delta E_{0,1} = 31$ meV for the InP wire of $R = 10$ nm with $\mu = 0.071$ m and $j_{0,1} = 2.405$. From the comparison, the observed value is much larger than the value expected from the average diameter of 20 nm. The diameter of the InP quantum wire with the energy shift of 170 meV is estimated to be about 5 nm from the previous data by (Yu et al., 2003). It can be considered that there exists a depletion region near the surface or an oxide layer on the nanowire sidewalls (Dionizio et al., 2008), which decreases a practical size of carrier confinement region.

Figure 21.18a shows CL spectra from the nanowire region and substrate region taken at 80 K, with an electron beam scanning over an 0.5 μm × 0.5 μm area. The emission peak at 1.412 eV comes from the InP substrate with a strong intensity, and the emission from the nanowires shows a strong peak at 1.417 eV. The deviation from the InP peak is 5 meV. This peak shift is much smaller than the value estimated from the quantum confinement effect. It is more likely to be considered that the emission comes from the strained InP substrate region near the surface, because a part of the carriers generated in the nanowires can drift into the substrate.

Figure 21.18b shows CL spectra in the wide energy range extended to more than 1.5 eV. In this spectral range, the CL intensity from the nanowire region is much stronger than that from the InP substrate. A broad peak appears in the spectrum from the nanowire region at 1.60 eV with a full-width half-maximum (FWHM) of 150 meV. Figure 21.18c is a SEM image of the nanowire region, and a corresponding panchromatic CL image taken at 80 K is shown in Figure 21.18d. The InP substrate shows a brighter contrast than the nanowires in this image. On the other hand, in monochromatic CL images taken at the photon energy of 1.55 and 1.632 eV (Figure 21.18e and f), the nanowires show much brighter contrast than the InP substrate. It is noticed that the intensity distributions in Figure 21.18e and f are similar to each other in spite of the existence of the size distribution in the nanowires. This means that the emission peak from each nanowire is sufficiently broad, even at low temperatures. If the broadening of the emission peak due to the phonon scattering process is dominant, the FWHM of the peak should become narrow with decreasing sample temperature, and split peaks from the nanowires with different diameters are expected to be observed. However, any split from the broad peak has not been observed, even for the measurement under a converged electron beam illumination at the nanowire region. One possible

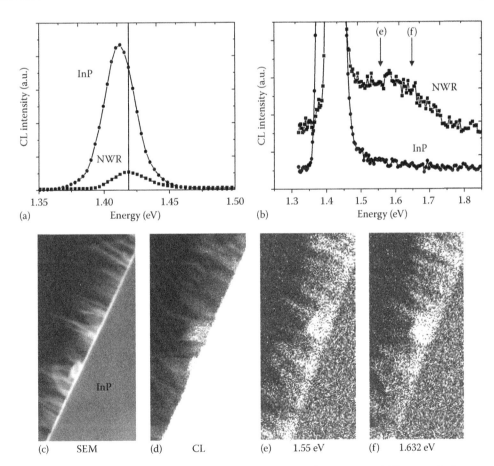

(a) Energy (eV)   (b) Energy (eV)

(c) SEM   (d) CL   (e) 1.55 eV   (f) 1.632 eV

FIGURE 21.18 (a) A CL spectrum from the nanowire region and substrate region taken at 80 K, and (b) a magnified spectrum of (a). (c) A SEM image of the nanowire region and (d) a corresponding panchromatic CL image. (e) and (f) are monochromatic CL images taken at photon energy of 1.55 and 1.632 eV, respectively.

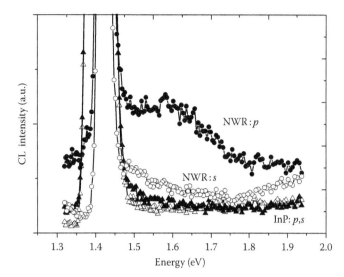

FIGURE 21.19 CL spectra from the nanowire region and substrate region taken for the p (solid spots) and s (open spots) polarization directions.

explanation is that the broadening of the luminescence peak occurs due to carrier scattering by the impurities or defects such as twin boundaries involved in the nanowire during the growth. In such case, the FWHM is independent of temperature. However, the origin of the broadening is not yet clear.

The polarization dependence of the CL spectrum was measured, as shown in Figure 21.19, where $p(s)$ polarization corresponds to the electric field oscillation of light parallel (normal) to the nanowire direction. It is found that the luminescent light emitted from the nanowires is highly polarized parallel to the wire direction, while that from the substrate is not polarized. The polarization ratio at the peak energy of 1.6 eV is 0.50. The value expected from the dielectric effect is $P = 0.93$ calculated in (21.68) with $\varepsilon = 12.0$, where the size dependence of the dielectric effect is neglected. In addition, the quantum confinement effect should be involved, i.e., (21.69) is used with $p(>1)$, the expected value must be larger than 0.93. The observed value is much smaller than the expected one. This is partly because the polarization mixing occurs due to reflection by the ellipsoidal mirror which corrects light emitted to a large solid angle. Wang et al. (2001) reported a large degree of polarization ($P = 0.96$) in the photoluminescence from InP nanowires. The PL was excited by a polarized light, so the polarization anisotropy is enhanced in the absorption process. Whereas in the CL experiment, the excitation is performed by the incident electrons, there is no anisotropy in excitation of carriers. Thus, the degree of polarization in the CL measurement is smaller than that in the polarized PL. The CL detection is recently improved by introducing the angler-resolved measurement system, as shown in the later section of ZnO nanowires.

21.5.2 GaAs Nanowire

Gallium arsenide (GaAs) has a zinc-blend structure, with a direct band gap of 1.519 eV at 0 K. Exciton radius is rather large,

about 15 nm, so the quantum confinement effect on photon energy and polarization of the FX emission is expected for thin nanowires. The binding energy of FX in the bulk is about 4 meV, and the emission intensity of FX becomes weak at higher temperatures above 200 K, even in thin nanowires. Recently, self-standing nanowires of many semiconductors are produced by a growth technique using nanosized liquid droplets of a metal solvent (Hiruma et al., 1995; Gudiksen et al., 2002; Wu et al., 2002), and optical properties were theoretically studied (Persson and Xu, 2004; Redlinski and Peeters, 2008).

Figure 21.20a shows a SEM image of a GaAs nanowire sample covered by an $Al_{0.4}Ga_{0.6}As$ layer, self-standing on a Si(111) substrate. The average outer dimensions of the nanowires are 80 nm in diameter and 700 nm in height, and the average density of the nanowires is about $10\,\mu m^{-2}$. The sample was fabricated by a metalorganic chemical vapor phase epitaxy (MOCVPE) technique using Au nanoparticles as the catalyst (Bhunia et al., 2003; Tateno et al., 2004). First, an $Al_{0.3}Ga_{0.7}As$ nanowire was grown to 100 nm in height on the Si substrate, and then GaAs nanowire and $Al_{0.3}Ga_{0.7}As$ nanowires were successively grown to 100 and 180 nm, respectively. Finally, $Al_{0.4}Ga_{0.6}As$ was deposited as a capping layer, with a thickness of 100 nm to reduce the nonradiative recombination on the nanowire surface. Au dots with a radius of about 20 nm can be observed on top of the nanowires in Figure. 21.20a. Figure 21.20b schematically represents a diagram of the sample structure. The materials are direct-gap

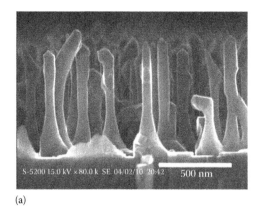

(a)

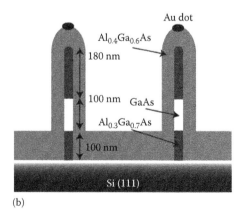

(b)

FIGURE 21.20 (a) A SEM image of GaAs/AlGaAs nanowires with a core-shell structure, and (b) schematic diagram of the sample structure.

semiconductors, with energy gaps of 1.52 eV for GaAs, 1.96 eV for $Al_{0.3}Ga_{0.7}As$, and 2.11 eV for $Al_{0.4}Ga_{0.6}As$ at 0 K (Pavesi and Guzzi, 1994). The $Al_{0.4}Ga_{0.6}As$ capping layer acts as a potential barrier, so a stronger (weaker) confinement structure is formed in the GaAs ($Al_{0.3}Ga_{0.7}As$) wire region. Carriers excited in the wire by an electron beam are expected to flow into the GaAs wire region, and recombine at the bottom of the band. Therefore, CL only from GaAs nanowires is expected to have sharp emission peaks in the CL spectrum corresponding to their diameters.

Figure 21.21a shows a SEM image of the sample observed from the direction normal to the surface. The image was observed in the SEM mode using the TEM operated at an accelerating voltage of 80 kV with a scanning area of 1 μm² (Ishikawa et al., 2008). Figure 21.21b shows a panchromatic CL image of the same area as that shown in Figure 21.20a taken at 60 K. The localized CL intensity is distributed at the nanowires, and we can specify the emission of each nanowire from the comparison between the CL image and the SEM image. A spectrum from this area is shown

in Figure 21.21c, where two broad peaks with a complex shape appear, reflecting the size distribution of the nanowires existing in this area. The emission spectra of similar shapes were frequently observed when the scanning area is of the order of 1 μm², although the peak energies are different. The spectrum taken from the wider area becomes a single broad peak extending over the energy range from 1.7 to 2.0 eV. The broadening of the CL peak is due to the superposition of many CL peaks from the nanowires of different diameters. To determine the effect of size distribution, we used a focused electron beam illuminating individual nanowires, A, B, and C indicated by circles in Figure 21.21a. As shown in Figure 21.21d, sharp peaks were observed to appear at different energies because of the difference in diameter of the wire. It is noted that some nanowires such as A and C show double emission peaks.

As for the diameters of the nanowires, it is difficult to determine them from the SEM image since the wires are covered by the AlGaAs layer with outer diameters of about 100 nm. The

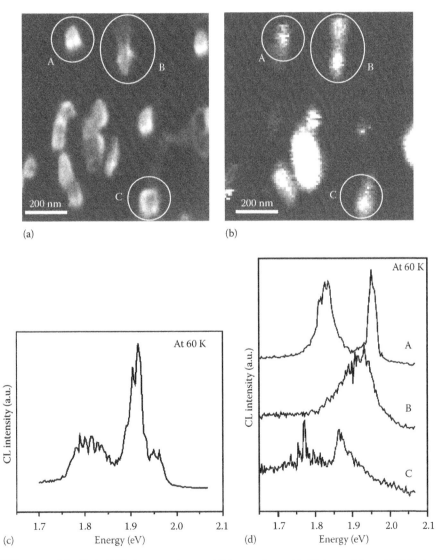

(a)

(b)

(c)

(d)

FIGURE 21.21 (a) A SEM image of the GaAs/AlGaAs nanowires, and (b) a panchromatic CL image taken at 130 K. A scan area is 1 μm × 1 μm. (c) A CL spectrum taken from the 1 μm × 1 μm area in (b). (d) CL spectra taken from the individual nanowires marked A, B, and C in (a) and (b).

energy levels were previously theoretically studied by Persson and Xu (2004) for GaAs nanowhisker grown in the [111] direction, on the basis of a tight binding approach. According to the calculation of the transition at the Γ point, the peak energies from 1.7 to 2.0 eV in the CL spectrum from the wide area correspond to the diameters from 5 to 2 nm. These values are much smaller than that expected from a Au particle size of 20 nm. This has been similarly observed in InP self-standing nanowires (Yamamoto et al., 2006), where the peak shift is much larger than that expected from the outer dimensions of the wire. This means that a practical size of carrier confinement region is much smaller than the apparent diameter. Such narrowing of the confinement region may result from the inner stress near the GaAs/AlGaAs interface and interdiffusion of Al into the wire.

The temperature dependence of the CL spectra from the single nanowire A shown in Figure 21.21a is shown in Figure 21.22a. The lower energy peak (P1) appears below 140 K and the higher energy peak (P2) appears below 100 K. Their intensities increase with decreasing temperature. The intensity of the P2 peak increases rapidly at lower temperatures, and exceeds the P1 peak intensity at about 60 K. As shown in Figure 21.22b, the P1 peak energy shows a blue shift as temperature decreases, following the band gap shift described by Varshni's law (solid lines). The P1 emission can be attributed to the radiative recombination of excitons at the ground state of the GaAs wire because the emission peak predominates at low temperatures. The luminescence of excitons remains at rather high

temperatures compared with that in bulk GaAs because the binding energy of excitons becomes large in nanowires owing to the quantum confinement effect (Basu, 1997). The appearance of the two peaks can be explained, assuming that the two parts having different energy states coexist along a single nanowire. The inhomogeneous intensity distribution was frequently observed along the tilted wires in the CL images and the polarization of the emitted light was randomly distributed with respect to the wire axis (Ishikawa et al., 2008). These facts indicate that the quantum-dot-like potential structure is formed along the wire, which does not emit a polarized light parallel to the wire axis.

Figure 21.23a shows a SEM image of a GaAs/AlGaAs nanowire with a core-shell structure, as illustrated in Figure 21.20, and Figure 21.23b shows a corresponding panchromatic CL image taken at 130 K. In the CL image, discontinuous contrast is seen to appear along the wire, which suggests the existence of a quantum dot-like structure in the GaAs core. Figure 21.23c is a spectrum image taken with the electron beam scanning along the wire axis. A peak line at a wavelength of 830 nm is attributed to the bulk luminescence of GaAs. This indicates that the carriers excited in the wire diffuse into the GaAs substrate because this peak vanishes for the nanowire separated from the substrate. It is noted that the emissions from those dots have a common spectral shape in the wavelength range from 700 to 800 nm, and the anisotropy in polarization is very small. We still need more consideration about the property of this luminescence.

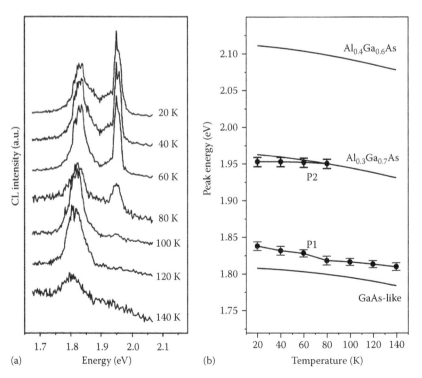

FIGURE 21.22 (a) Temperature dependence of the CL spectrum from the nanowire A in Figure 21.21. (b) Temperature dependence of the peak energies of the two peaks. Solid lines show the temperature dependence of the band gaps of GaAs, $Al_{0.3}Ga_{0.7}As$, and $Al_{0.4}Ga_{0.6}As$, respectively, given by the Varshni's law. The solid curve of GaAs is shifted for easy comparison with the observed one.

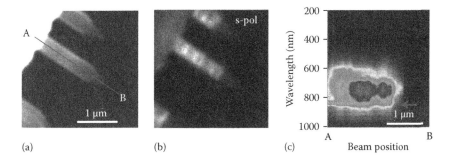

FIGURE 21.23 (a) A SEM image of GaAs/AlGaAs nanowires and (b) a panchromatic CL image of the same area (3 μm × 3 μm) taken at 130 K. (c) A spectrum image taken with the electron beam scanning along a line A-B through the central nanowire in (a).

21.5.3 ZnO Nanowires

Zinc oxide (ZnO) is a wide gap semiconductor of wurtzite structure, with a gap energy of 3.44 eV at 4 K. The valence band structure of bulk ZnO near the BE is composed of three bands (A, B, and C band), and the character of the band has been controversy for many years. The highest energy band (the A band) of many wurtzite-type semiconductors such as GaN and CdSe has a character of heavy hole (the Γ_9 symmetry, as shown in Figure 21.5b), whereas many authors assume the Γ_7 symmetry character for ZnO (Meyer et al., 2004). The polarization studies showed that the inner emission due to the transition from the conduction band to the A band contain both polarization components parallel and perpendicular to the c-axis. The energy difference between the A and B band was reported to be very small, 4.9 meV. The exciton binding energy is about 60 meV which is much larger than the thermal energy at room temperature ($k_BT = 25$ meV), so the band-edge emission can be observed even at room temperature. However, the exciton radius is very small ($a_X = 1.3$ nm), and it is difficult to fabricate a ZnO quantum wire of such a small diameter. The quantum effect of ZnO nanowires has rarely been reported so far.

Recently, the optical properties of ZnO nanowires have been studied using self-standing nanowires grown parallel to the c-axis by MOCVD technique. Hsu et al. (2004) measured PL from narrow nanowires at room temperature and observed BE emission and green emission originated from defects. They showed that the BE emission is polarized parallel to the c-axis with the intensity ratio of $I_{\parallel} : I_{\perp} = 1 : 0.7$, whereas the green emission is polarized perpendicular to the c-axis with the ratio of $I_{\parallel} : I_{\perp} = 0.85 : 1$. The polarization dependence of photoconductivity was studied by Fan et al. (2004) using nanowires of 30–150 nm in diameter. The photoconductivity had a maximum value for polarization direction parallel to the c-axis. Yu et al. (2003) measured temperature dependence of PL spectrum from ZnO nanowires, showing peaks of BX emission, the P emission, and their phonon replica. The nanowires of wurtzite-type semiconductors such as GaN (Zhang and Xia, 2006; Chen et al., 2008) and CdSe (Lan et al., 2008) have also been studied. Chen et al. (2008) measured the size dependence in polarization of the BE emission from GaN nanowires, and explained the results from

the dielectric effect and the anisotropy in the inner emission. The polarization ratio changed with temperature; the BE emission is polarized parallel to the c-axis at room temperature, and is unpolarized at 20 K. They explained this from the change in anisotropy of the internal emission.

In the TEM-CL study, ZnO smoke particles were made by burning a Zn wire in air, and were collected by a copper mesh coated by a carbon film with micrometer-size holes. The smoke particle has a tetrapod-like shape, with four legs growing along the c-axis. Figure 21.24 shows CL spectra taken from a thick leg of the ZnO smoke particle at various temperatures. Two peaks at wavelengths of 368 nm (3.370 eV) and 375 nm (3.307 eV) appear in the spectrum at 20 K. The peak at 368 nm can be attributed to the ionized donor BX emission (Meyer et al., 2004), and the peak

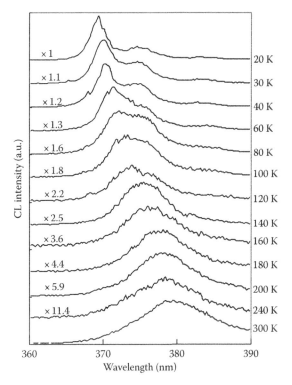

FIGURE 21.24 CL spectra taken from a thick leg of a ZnO smoke particle at various temperatures.

at 375 nm is due to the *P* emission, which is generated by Auger-like process associated with two excitons (Hvam, 1973; Li et al., 2008). A small peak at 393 nm (3.155 eV) is a phonon replica of the *P* emission. The photon energies of those peaks shift to lower energies with temperature, following the Varshni's law. A broad single peak (a BE emission) at room temperature can be recognized as a mixed peak composed of the BX emission, *P* emission, and their phonon replica.

Figure 21.25a shows a SEM image of a ZnO smoke particle with thick legs, and Figure 21.25b and c are monochromatic CL images taken at the peak wavelength at room temperature for the *p* and *s* polarizations, respectively. The BE emission from a leg marked A extending along the *y* direction is strong in the *p* polarization, whereas that from a leg marked B extending along

the *x* direction is strong in the *s* polarization. These facts indicate that the BE emission is polarized perpendicular to the *c*-axis. The central part of the particle shows a dark contrast because the emission peak from this part is slightly shifted to the longer wavelength. The central part showed a bright contrast in the monochromatic CL image taken at 395 nm (Figure 21.25d). This peak shift may be caused by the band gap narrowing due to strain in the multiple twin structure at the central part.

The polarization of the BE emission from ZnO nanowires depends on the diameter of the wire. CL light emitted from a single wire with a large diameter (300 nm) elongating perpendicular to the axis of the parabolic mirror was measured with changing polarization angle θ defined in Figure 21.12. The CL intensity as a function of θ is shown in Figure 21.26a. The light was emitted in a direction perpendicular to the thick wire axis. The solid angle of the detection is about 0.05 str. The intensity of the BE emission at 383 nm has a minimum at $\theta = 0°$, and shows $\cos^2 \theta$-dependence. Polarization ratio is estimated to be -0.43 from the intensity ratio of $I_{\parallel c}/I_{\perp c} = 0.40$, which is deduced using the fitting function of

$$I(\theta) = (I_{\parallel} - I_{\perp})\cos^2\theta + I_{\perp} \qquad (21.73)$$

Whereas the BE emission from a thin nanowire with a diameter of 60 nm presented an opposite behavior in polarization, as shown in Figure 21.26b; the intensity has a maximum at $\theta = 0°$. The polarization ratio is estimated to be 0.31 from the intensity ratio of $I_{\parallel c}/I_{\perp c} = 1.9$.

The quantum confinement effect cannot affect the polarization property of the CL emission from the ZnO nanowires because the diameters of those wires are much larger than the exciton radius. Therefore, the parameter *p* related to the anisotropic internal emission is fixed regardless of the diameter of the nanowires, and should be given by the character of the optical transition in the bulk crystal. If the intensity ratio $I_{\parallel c}/I_{\perp c} = 0.40$ is considered as a bulk one, the parameter *p* can be calculated from (21.67) and (21.70) as follows:

$$p = \frac{d_{0z}^2}{d_{0x}^2} = \frac{1}{2}\left\{3\left(\frac{I_{\parallel}}{I_{\perp}}\right) - 1\right\} = 0.1 \qquad (21.74)$$

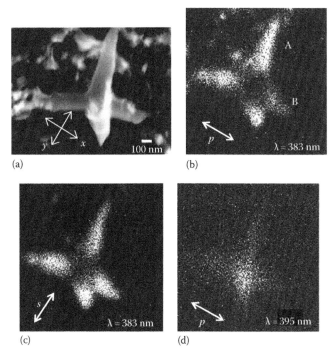

(a)

(c)   (d)

FIGURE 21.25 (a) A SEM image of a ZnO smoke particle with thick legs, and (b) and (c) are monochromatic CL images taken at the peak wavelength (383 nm) at room temperature for the *p* and *s* polarizations, respectively. (d) A monochromatic CL image taken at 395 nm.

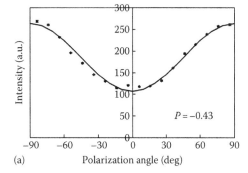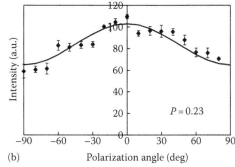

(a)   (b)

FIGURE 21.26 CL intensity from (a) a thick wire (diameter of 300 nm) and (b) a thin wire (60 nm) of ZnO measured by changing an angle θ.

If the highest energy state in the valence band has a character of the light hole, $p = 4$ is expected from (21.58) for the transition from the conduction band to valence band. The above value is too small compared to the expected one. In other words, the emission due to the transition to the light hole state should be mainly polarized parallel to the c-axis, but the observed one is rather polarized perpendicular to the c-axis. This fact indicates that the transition to the heavy hole state component in the A band considerably contributes to the CL emission. It was also

observed that the polarization ratio does not change with temperature from 20 to 300 K.

The CL emission from the thick nanowire is polarized perpendicular to the wire axis (the c-axis), whereas that from the thin nanowire is polarized parallel to the wire axis. We should consider the dielectric effect for explaining this size dependence of the polarization. Figure 21.27 shows the dependence of the polarization ratio on nanowire diameter calculated from (21.71) and (21.72) with the dielectric constant at the peak wavelength to be 4.0 and with $p = 0.25$. This result well explains the behavior that the polarization ratio P is positive for thin nanowires, and turns to be negative for thick nanowires. The experimental values measured from ZnO nanowires with different diameters are plotted in Figure 21.27. The polarization ratio changes sign around a diameter of 100 nm, being fitted with the calculated curve.

The spatial variation of polarization can be measured from spectral images taken with scanning electron beam. Figure 21.28a and b shows a SEM image of a ZnO smoke particle with thick legs and a panchromatic CL image, respectively. Change in CL intensity appears along the leg. Figure 21.28c and d are p- and s-polarized spectral images taken with scanning electron beam along the line A-B in Figure 21.28a. The peak wavelength is 383 nm at room temperature. Polarization ratio calculated from the division of the peak intensities in the images of Figure 21.28c and d is shown in Figure 21.28e. The polarization ratio is seen to be −0.45 at a thinner part ($D = 150$ nm) and −0.25 at a thicker part along the line. One of the legs is standing vertically near the position B, and the polarization is vanished there.

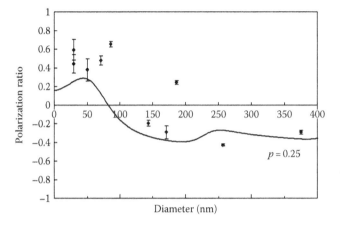

FIGURE 21.27 Dependence of the polarization ratio on nanowire diameter measured from several nanowires with various diameters. Solid line is a calculated curve using the dielectric constant at the peak wavelength of 4.0 and $p = 0.25$.

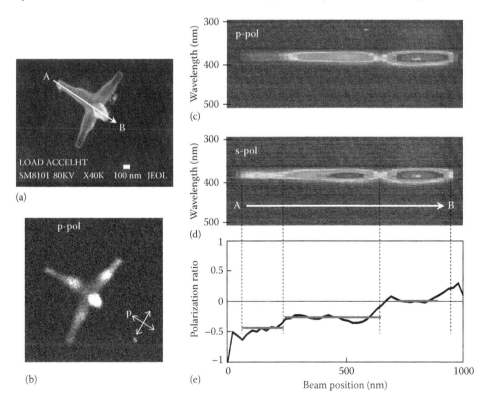

FIGURE 21.28 (a) A SEM image of a ZnO smoke particle and (b) a panchromatic CL image, respectively. (c) and (d) are p- and s-polarized spectral images taken with scanning electron beam along the line A-B in (a). (e) Spatial variation of the polarization ratio along the line.

This behavior is qualitatively explained by the theoretical curve in Figure 21.27.

Other types of emission appear under a high excitation condition, when a thin nanowire is illuminated by a strong electron beam. The *P* emission is one of such types. Emission peak due to the exciton molecules appears near the BE peak, and shifts to the lower energies with increasing electron beam density. The emission peak due to the electron-hole plasma also appears at the lower energy side of the BE peak, and becomes predominant under the high excitation rate. The effect of shape and size of nanowires on the property of these emissions such as threshold condition of lasing has been studied recently (Johnson et al., 2003).

Acknowledgment

The author acknowledges Y. Watanabe, K. Tateno, G. Salviati, and L. Lazzarini for the supply of nanowire samples. This work has been supported by the Grant-in-Aid for Scientific Research from the Ministry of Education, Culture, Sports, Science and Technology of Japan.

References

Ando, H, Oohashi, H, and Kanbe, H, 1991, *J. Appl. Phys.* 70, 7024.

Arakawa, Y and Sakaki, H, 1982, *Appl. Phys. Lett.* **40**, 939.

Basu, PK, 1997, *Theory of Optical Processes in Semiconductors*, Oxford University Press, New York.

Bockelmann, U and Bastard, G, 1992, *Phys. Rev. B* **45**, 1688.

Bhunia, S, Kawamura, T, Watanabe, Y, Fujikawa, S, and Tokushima, K, 2003, *Appl. Phys. Lett.* **83**, 3371.

Chen, H-Y, Yang, Y-C, Lin, H-W, Chang, S-C, and Gwo, S, 2008, *Opt. Express* 16(17), 13465.

Dionizio, M, Venezuela, P, and Schmidt, TM, 2008, *Nanotechnology* 19, 065203.

Fan, Z, Chang, P-C, Lu, JG, Walter, EC, Penner, RM, Lin, C-H, and Lee, HP, 2004, *Appl. Phys. Lett.* 85(25), 6128.

Goldstein, JI, Costley, JL, Lorimer, GW, and Reed, SJB, 1977, *Scanning Electron Microscopy*, Vol. 1, pp. 315, IITRI, Chicago, IL.

Gudiksen, MS, Lauhon, LJ, Wang, J, Smith, DC, and Lieber, CM, 2002, *Nature* **415**, 617.

Guilet, T, Grousson, R, Voliotis, V, Wang, X-L, and Ogura, M, 2003, *Phys. Rev. B* **68**, 045319.

Hiruma, K, Yazawa, M, Katsuyama, T, Ogawa, K, Haraguchi, K, Koguchi, M, and Kakibayashi, H, 1995, *J. Appl. Phys.* **77**, 447.

Hsu, NE, Hung, WK, and Chen, YF, 2004, *J. Appl. Phys.* **96**(8), 4671.

Hvan, JM, 1973, *Solid State Com.* **12**, 95.

Hensel, JC and Feher, G, 1963, *Phys. Rev.* **129**, 1041.

Ils, P, Michel, M, Forchel, A, Gyuro, I, Klenk, M, and Zielinski, E, 1994, *Appl. Phys. Lett.* **64**, 496.

Ino, N and Yamamoto, N, 2008, *Appl. Phys. Lett.* **93**, 232103.

Ishikawa, K, Yamamoto, N, Tateno, K, and Watanabe, Y, 2008, *Jpn. J. Appl. Phys.* **47**(8), 6596.

Johnson, JC, Yan, H, Yang, P, and Saykally, RJ, 2003, *J. Phys. Chem. B* **107**, 8816.

Kane, EO, 1957, *J. Phys. Chem. Solids.* **1**, 249.

Kapon, E, Hwang, DM, and Bhat, R, 1989, *Phys. Rev. Lett.* **63**, 430.

Lan, A, Giblin, J, Protasenko, V, and Kuno, M, 2008, *Appl. Phys. Lett.* **92**, 183110.

Landau, LD, Lifshitz, EM, and Pitaevskii, LP, 1984, *Electrodynamics of Continuous Media*, Pergamon, Oxford, U.K.

Li, W, Gao, M, Chen, R, Zhang, X, Xie, S, and Peng, L-M, 2008, *Appl. Phys. Lett.* **93**, 023117.

Luttinger, JM and Kohn, W, 1955, *Phys. Rev.* **97**, 869.

McIntyre, CR and Sham, LJ, 1992, *Phys. Rev.* **45**, 9443.

Meyer, BK, Alves, H, Hofmann, DM et al., 2004, *Phys. Stat. Sol. (b)* **241**(2), 231.

Mishra, A, Titova, LV, Hoang, TB et al., 2007, *Appl. Phys. Lett.* **91**, 263104.

Mitsui, T, Yamamoto, N, Tadokoro, T, and Ohta, S, 1996, *J. Appl. Phys.* **80**, 6972.

Nakaji, D, Grillo, V, Yananoto, N, and Mukai, T, 2005, *J. Elect. Microscopy.* **54**, 223.

Ogawa, T and Takagahara, T, 1991, *Phys. Rev. B* **43**, 14325.

Pankove, JI, 1971, *Optical Processes in Semiconductors*, Dover Pub. Inc., New York.

Pavesi, L and Guzzi, M, 1994, *J. Appl. Phys.* **75**, 4779.

Pellegrini, G, Mattei, G, and Mazzoldi, P, 2005, *J. Appl. Phys.* **97**, 073706.

Persson, MP and Xu, HQ, 2004, *Nano Lett.* **4**, 2409.

Pfeiffer, L, West, KW, Stormer, HL, Eisenstein, JP, Baldwin, KW, Gershoni, D, and Spector, J, 1990, *Appl. Phys. Lett.* **56** 1697.

Pollak, FH and Cardona, M, 1968, *Phys. Rev.* **172**, 816.

Redlinski, P and Peeters, FM, 2008, *Phys. Rev. B* **77**, 075329.

Ruda, HE and Shik, A, 2005, *Phys. Rev. B* **72**, 115308.

Ruda, HE and Shik, A, 2006, *J. Appl. Phys.* **100**, 024314.

Sakaki, H, 1980, *Jpn. J. Appl. Phys.* **19**, L735.

Sercel, PC and Vahala, KJ, 1990, *Appl. Phys. Lett.* **57**, 545.

Someya, T, Akiyama, H, and Sakaki, H, 1995, *Phys. Rev. Lett.* **74**, 3664.

Tanaka, M and Sakaki, H, 1989, *Appl. Phys. Lett.* **54**, 1326.

Tanatar, B, Al-Hayek, I, and Tomak, M, 1998, *Phys. Rev. B* **58**, 9886.

Tateno, K, Gotoh, H, and Watanabe, Y, 2004, *Appl. Phys. Lett.* **85**, 1808.

Tsuchiya, M, Gaines, JM, Yan, RH, Simes, RJ, Holtz, PO, Coldren, LA, and Petroff, PM, 1989, *Phys. Rev.* **62**, 455.

Voos, M, Leheny, RF, and Shah, J 1980, Radiative recombination, in *Handbook on Semiconductors: Vol. 2 Optical Properties of Solids*, M. Balkanski (ed.), North-Holland, Amsterdam, the Netherlands.

Vugaftman, I, Meyer, JR, and Ram-Mohan, LR, 2001, *Appl. Phys. Rev.* **89**, 5815.

Wang, J, Gudiksen, MS, Duan, X, Cui, Y, and Lieber, CM, 2001, *Science* **293**, 1455.

Wang, X-L and Voliotis, V, 2006, *J. Appl. Phys.* **99**, 121301.

Wu, Y, Fan, R, and Yang, P, 2002, *Nano Lett.* **2**, 83.

Yananoto, N, Spence, JCH, and Fatty, D, 1984, *Phil. Mag. A* **49**, 609.

Yamamoto, N 2002, Development of CL for semiconductor research I: EM-CL study of microstructures and defects in semiconductor epilayers, in *Nanoscale Spectroscopy and Its Applications to Semiconductor Research*, Watanabe Y et al. (eds.), *Lecture Notes in Physics*, Vol. 588, Springer Verlag, Berlin, Germany.

Yamamoto, N, 2008, TEM-Cathodoluminescence study of semiconductor quantum dots and quantum wires, in *Beam Injection Based Nanocharacterization of Advanced Materials*, G. Salviati et al. (eds.), Research Signpost, Kerala, India.

Yamamoto, N, Itoh, H, Grillo, V, et al., 2003, *J. Appl. Phys.* **94**, 4315.

Yamamoto, N, Bhunia, S, and Watanabe, Y, 2006, *Appl. Phys. Lett.* **88**, 153106.

Yamamoto, N, Ishikawa, K, Akiba, K, Bhunia, S, Tateno, K, and Watanabe, Y, 2008, TEM-Cathodoluminescence study of semiconductor quantum dots and quantum wires, in *Beam Injection Based Nanocharacterization of Advanced Materials*, G. Salviati et al. (eds.), Research Signpost, Kerala, India.

Yu, H, Li, J, Loomis, RA, Wang L-W, and Buhro, WE, 2003, *Nat. Mater.* **2**, 517.

Yu, PY and Cardona, M 1999, *Fundamentals of Semiconductors*, Springer Verlag, Berlin, Germany.

Zhang, XW and Xia, JB, 2006, *J. Phys. Condens. Matter* **18**, 3107.

Zhao, X, Wei, CM, Yang, L, and Chou, MY, 2004, *Phys. Rev. Lett.* **92**(23), 236805.

Optical Spectroscopy of Nanomaterials

Yoshihiko Kanemitsu
Kyoto University

22.1 Introduction

Over the past two decades, there have been extensive studies on the optical properties of semiconductor nanomaterials from the fundamental physics viewpoint and from the interest in the application to functional devices, because they exhibit unique size-dependent quantum properties [1–11]. In this chapter, we discuss optical properties of semiconductor nanomaterials of zero-dimensional (0D) nanoparticle quantum dots and one-dimensional (1D) carbon nanotubes. In optical studies of nanoparticle quantum dots and carbon nanotubes, we would like to point out two important reports opening new active fields: the discovery of room-temperature-visible luminescence from porous silicon in 1990 [12] and the discovery of efficient luminescence from isolated carbon nanotubes in 2002 [13]. These observations of efficient luminescence clearly show that nanoparticles and carbon nanotubes are high-quality crystalline semiconductors. Many different fabrication methods have been developed to obtain stable and efficient luminescence from nanoparticles and carbon nanotubes, e.g., core/shell nanoparticles, suspended isolated nanotubes, and so on [14–20]. These nanomaterials become new materials for optoelectronic devices such as wavelength-tunable light-emitting diodes and lasers, quantum light sources, and solar cell applications.

When semiconductor nanoparticles of sizes are comparable to or smaller than the exciton Bohr radius in bulk crystals, the excited state energies and optical properties are very sensitive to their sizes [1,21,22]. Usually, nanoparticle samples are an inhomogeneous system in the sense that they have a distribution of size and shape, and variations of surface structures and surrounding environments [2]. A large nanoparticle has small band-gap energy and a small nanoparticle has large band-gap energy. Furthermore, the nanoparticles have large surface-to-volume

ratios, and then the optical properties of nanoparticles are also sensitive to surrounding environments. The exciton band-gap energy of semiconducting carbon nanotubes is also sensitive to the nanotube diameter and the chiral index. Then, we need to study the intrinsic optical processes in nanoparticle quantum dots and carbon nanotubes hidden by sample inhomogeneity using sophisticated optical spectroscopy.

If the laser light excites all nanoparticles or all nanotubes in the sample, the sample shows broad luminescence, reflecting size or diameter distributions. This "global" photoluminescence (PL) or nonresonantly excited luminescence contains contributions from all nanoparticles or all nanotubes in the sample, and the PL spectrum is inhomogeneously broadened, as shown in Figure 22.1a. These sample inhomogeneities are the origin of nonexponential PL decay. In inhomogeneously broadened systems, resonant excitation spectroscopy is a powerful method to obtain intrinsic information from broadened optical spectra. Under resonant excitation at energies within the global PL band, we can observe fine structures in PL spectra at low temperatures, as shown in Figure 22.1b. Resonant excitation at energies within the luminescence band results in a single zero-phonon PL line or a well-resolved phonon progression in PL spectra at low temperatures. In this case, we suppress the inhomogeneous broadening of the luminescence by selectively exciting a narrow subset of nanoparticles. Resonant excitation results in fluorescence line narrowing (FLN) in nanoparticle samples [23–27]. The resonantly excited PL spectra are sensitive to the nature of the band-edge structure and the surface structure [27–29]. Moreover, luminescence hole-burning (LHB) spectroscopy is another resonant excitation spectroscopy. In the LHB experiments, the sample is excited by intense laser at the energy within the PL band [30–32]. After prolonged laser irradiation (burning laser excitation), a spectral hole is formed near the burning laser energy in the luminescence

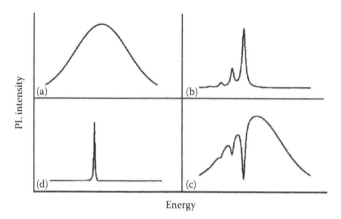

FIGURE 22.1 Luminescence spectra of semiconductor nanoparticles: (a) global luminescence spectrum, (b) resonantly excited luminescence spectrum, (c) LHB spectrum, and (d) single nanoparticle luminescence spectrum.

band. We obtain similar optical information from the FLN and LHB experiments. However, in the resonantly excited PL or FLN spectra, it is difficult to obtain the detailed spectrum near the excitation energy because of the scattering of the excitation laser light. In the LHB experiment, it is comparatively easy to observe the spectral change just at the excitation energy [32].

Single molecule (or nanomaterial) spectroscopy is the most powerful tool for understanding intrinsic optical properties of isolated molecules and nanomaterials [5,33]. Single nanoparticle spectroscopy makes it possible to probe inherent and novel optical properties in nondoped and impurity-doped semiconductor nanoparticles hidden by inhomogeneity, such as size distributions and surrounding environment variations [34–37]. As an example of unique optical phenomena, PL blinking (or PL intermittency) is revealed by single nanoparticle spectroscopy [35]. We have developed many different types of luminescence imaging microscopes. Our systems cover a wide spectral region and wide response times. The spatial resolution is typically about 1 μm in our confocal microscopes and about 50–100 nm in home-built near-field optical microscopes. In these space-resolved optical measurements, the most important point is the fabrication of stable and isolated nanomaterial samples. The nanomaterials should be isolated from each other, and the number density is dilute (about $1/\mu m^2$ or less). The sample should be stable against oxidation and strong light illumination. Then, chemically synthesized materials are good samples for our optical studies. Here, we discuss the luminescence properties and the mechanism of spectral diffusion and PL blinking of single carbon nanotubes and nanoparticle quantum dots.

22.2 Carbon Nanotubes

A single-walled carbon nanotube (SWNT) with about 1 nm diameter and a length greater than several hundred nanometers is a prototype of 1D structures. The recent discovery of efficient PL from semiconducting SWNTs [13,38] has stimulated considerable efforts in understanding optical properties of SWNTs. The

semiconducting SWNTs are 1D direct-gap band structures [8]. Because of the extremely strong electron–hole interactions (excitonic effects) in 1D materials, unique optical properties of SWNTs are determined by the dynamics of 1D excitons [8,39]. In addition, the electronic structure and the PL energy of SWNTs strongly depend on the diameter and the chiral index [38]. The SWNT samples are also inhomogeneous systems, similar to the nanoparticle samples, because many different species of nanotubes exist in the sample. The inhomogeneous broadening and the spectral overlapping of PL spectra cause the complicated PL dynamics of SWNTs. Single nanotube spectroscopy reveals the intrinsic excitonic properties of SWNTs [40–46], such as exciton energy, bright and dark exciton structures, exciton–phonon interaction, and so on.

For single nanotube spectroscopy, we synthesized spatially isolated carbon nanotubes on Si substrates using an alcohol catalytic chemical vapor deposition method [45,46]. In our experiments, the Si or SiO_2 substrates were patterned with parallel grooves, typically 500 nm wide and 500 nm deep using an electron-beam lithography technique. The isolated SWNTs grow from one side toward the opposite side of the groove. We used these SWNT samples without matrix and surfactant around the nanotubes to reduce the local environmental fluctuation effect. We show a typical PL spectrum of a single carbon nanotube suspended on the groove [assigned chiral index: (7,6)] at about 40 K in Figure 22.2a [46]. Very broad PL bands are observed in the ensemble-averaged spectrum of micelle-wrapped SWNTs dispersed in gelatin. The PL spectral shape of a single carbon

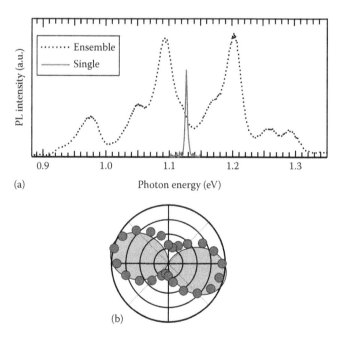

FIGURE 22.2 (a) PL spectrum of a typical suspended single SWNT [assigned chiral index (7,6)] in comparison with the ensemble-averaged spectrum of micelle-wrapped SWNTs dispersed in gelatin. (b) Polar plot of the PL intensity of a typical single SWNT versus the polarized direction of the excitation laser. The PL data (circle) were fitted using $\cos^2\theta$ (solid line). (Reprinted from Matsunaga, R. et al., *Phys. Rev. Lett.*, 101, 147404, 2008. With permission.)

nanotube is given by a Lorentzian function, and its linewidth of a few meV reflects homogeneous broadening. Figure 22.2b shows a polar plot of the PL intensity of a typical single carbon nanotube versus the polarization direction of the excitation laser light [46]. Since a 1D dipole moment exists, strong optical absorption occurs when the polarization of the excitation light parallels the nanotube axis. This PL anisotropy is useful for determining the direction of the observed nanotube axis for single nanotube spectroscopy and modulation spectroscopy.

The sharp luminescence spectra provide detailed information on the exciton fine structures. We studied the PL fine structure of single SWNTs under magnetic fields at low temperatures. A single sharp PL spectrum arising from bright exciton recombination is observed at zero magnetic field, as shown in Figure 22.2a. When the magnetic field is parallel to the nanotube axis, a new peak appears below the bright exciton peak. Figure 22.3a shows PL spectra of a single carbon nanotube under magnetic fields [46]. These PL spectra are fit well by two Lorentzian functions. With the magnetic field, the lower energy peak shows a redshift and the lower-energy peak intensity increases. We cannot observe these changes when the magnetic field is perpendicular to a single nanotube axis, as shown in Figure 22.3b [46]. The splitting of the PL peak occurs due to the magnetic flux parallel to the nanotube axis. These splitting and magnetic field dependence can be explained by the Aharonov–Bohm splitting of excitons based on the Ajiki and Ando model [8,47]. The singlet

exciton states split into the bonding and antibonding exciton states, and this due to the short-range Coulomb interaction. The bonding state is odd parity (bright) and the antibonding is even (dark). The energy difference between the bright and dark exciton states also depends on the diameter of SWNTs. These experimental observations are consistent with the theoretical calculation. The dark exciton state exists about several meV below the bright exciton state. Studies of 1D dark excitons influencing optical responses of carbon nanotubes [48–52] are very important for optical device applications.

The diameter dependence of the exciton energy in single carbon nanotubes is also revealed by single nanotube spectroscopy. At room temperature, the experimentally obtained PL spectra can be approximately reproduced by single Lorentzian functions. The observed PL peaks correspond to the zero-phonon lines of free excitons, and the spectral linewidth of the PL spectra is determined by the homogeneous broadening. We obtained PL spectra from many different isolated SWNTs with a variety of chiral indices. Figure 22.4a shows a distribution of the PL peak energies for the single SWNTs, indicated by diamonds [45]. In Figure 22.4b, we show some of the PL spectra from isolated SWNTs with various emission energies [45]. Only a single sharp peak can be seen in each spectrum. The PL linewidth clearly becomes broader as the diameter decreases. This shows that the exciton–phonon interaction is stronger in smaller diameter tubes.

The lowest exciton has fine structures (bright and dark excitons), and the fine structures will determine optical responses and cause unique phenomena. At low temperatures, we observe an interesting phenomenon, spectral diffusion. A few ten percents

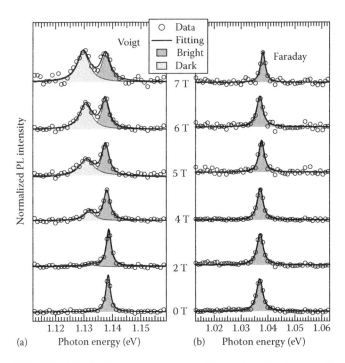

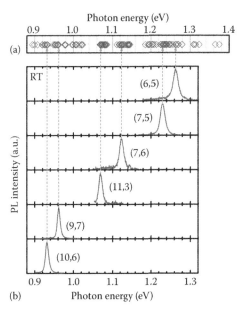

FIGURE 22.3 (a) Normalized magneto-PL spectra of a single (9,4) carbon nanotube at 20 K in the Voigt geometry. The split PL spectra are fit by two Lorentzian functions. (b) The normalized magneto-PL spectra of a single (9,5) carbon nanotube at 20 K in the Faraday geometry. The PL spectra are fit by a Lorentzian function. (Reprinted from Matsunaga, R. et al., *Phys. Rev. Lett.*, 101, 147404, 2008. With permission.)

FIGURE 22.4 (a) PL peak energy distribution of obtained PL spectra from about 180 different isolated SWNTs. (b) PL spectra for several species of single SWNTs at room temperature. SWNTs with higher PL emission energy tend to have a larger spectral linewidth. (Reprinted from Inoue, T. et al., *Phys. Rev. B*, 73, 233401, 2006. With permission.)

of the nanotubes show spectral diffusion. We consider that spectral diffusion is related to the exciton fine structure, bright and dark excitons. The PL fluctuation due to spectral diffusion is clearly observed at low temperatures. Spectral fluctuation occurs very slowly, the order of several seconds. Figure 22.5a shows a typical temporal evolution of the PL spectrum of SWNT showing spectral fluctuations at 40 K [53]. During spectral diffusion, the PL spectra clearly show two peaks. The lower energy is fitted by Gaussian and the higher energy is Lorentzian. From spectral fitting, we can determine the peak positions and linewidth of both peaks of the PL spectra. Temporal changes in two PL peak energies and linewidths of the lower energy PL band are shown in Figure 22.5b [53]. The higher energy peak is almost constant, but the lower energy peak fluctuates. We find a good correlation between the PL peak energy and the PL linewidth of the lower energy band. When the PL peak shows a low energy, the linewidth becomes wider. To clarify the origin

of the spectral diffusion in the lower energy peak, the PL linewidth of the lower energy peaks is plotted as a function of the emission energy. We obtain the square root dependence of the linewidth on the emission energy [53]. This dependence suggests the quantum-confined Stark effect [54]. The spectral diffusion can be explained by the fluctuation of local electric field. The Stark effect causes a redshift in the exciton energy. A small, fast, local electric field fluctuation results from surface charge oscillations. These observations show that the energy splitting between the bright and the dark exciton states is estimated to be about a few meV [53]. This conclusion is well consistent with the magneto-optic results as mentioned before. Detailed understanding of fine structures of the lowest excitons is important for the optoelectronic applications of carbon nanotubes.

22.3 Nanoparticle Quantum Dots

Similar spectral diffusion and PL blinking phenomena are also observed in nanoparticle quantum dots. PL blinking phenomenon is quite enhanced in 0D nanoparticles rather than 1D carbon nanotubes. PL blinking in single nanoparticles is caused by a random switching between light-emitting "on" and non-light-emitting "off" states under continuous-wave (cw) laser excitation. Since the first blinking observation in nanoparticles [35], the mechanism of nanoparticles PL blinking has been extensively discussed [55–70]. Ionization of nanoparticles due to nonradiative Auger recombination plays an essential role in PL blinking of single nanoparticles [55]. It is well accepted in this field that PL blinking originates from the photoionization and neutralization of nanocrystals under cw light illumination. In CdSe nanocrystals, for example, the non-light-emitting off-time state is due to positively charged nanoparticles, and the light-emitting on-time state is due to the neutral nanoparticle [55,70]. Spectral fluctuations during on-time suggest that both electrons and holes trapped on the nanoparticle surface cause transient and local electric field fluctuations in the light-emitting neutral nanoparticle. In neutral nanoparticles, excitons recombine radiatively. In ionized (or charged) nanoparticles, the photogenerated excitons and excess holes recombine nonradiatively through fast three-body Auger recombination. The PL blinking means that a nanoparticle repeats the neutralization and ionization under cw laser excitation.

PL blinking behavior is very sensitive to the local surrounding environments. Figure 22.6 shows the PL intensity time traces of a single CdSe nanoparticle on different substrates: glass, rough, and flat Au surfaces [66]. The samples are excited by cw laser at room temperature. The on-off PL blinking behavior is clearly observed on the glass substrate. The time distribution of the on and off states can be characterized by power law functions [57]. The power law distributions suggest that the PL blinking is caused by a very complicated process. On the metal surfaces, on the other hand, the PL off-time is drastically suppressed. The PL blinking suppression indicates that the very rapid neutralization of ionized nanoparticles occurs through the fast energy transfer. However, the enhancement

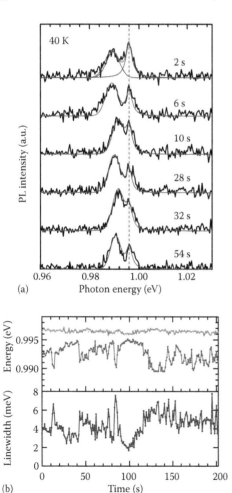

FIGURE 22.5 (a) Temporal evolutions of the PL spectrum of a single SWNT showing spectral fluctuations at 40 K. The solid curves indicate the results of fitting analysis assuming Gaussian and Lorentzian functions. (b) Temporal trace of the higher and lower energy peaks of the PL spectra of a single SWNT at 40 K. The temporal trace of the linewidth (FWHM) of the lower energy peaks. (Reprinted from Matsuda, K. et al., *Phys. Rev. B*, 77, 193405, 2008. With permission.)

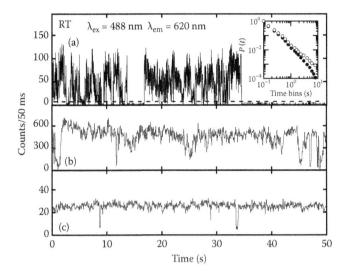

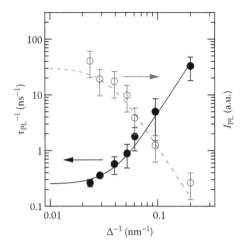

FIGURE 22.6 PL intensity time-traces of single CdSe/ZnS nanoparticles on the glass (a), the rough Au surface (b), and the flat Au surface (c). The inset of (a) is the histogram of the on-time (solid circles) and off-time (open circles) duration of the PL intensity time trace on the glass. (Reprinted from Matsunaga, R. et al., *Phys. Rev. Lett.*, 101, 147404, 2008. With permission.)

FIGURE 22.7 PL lifetime and PL intensity as a function of the distance between the CdSe nanoparticles monolayer and the Au film. (Reprinted from Ueda, A. et al., *Appl. Phys. Lett.*, 92, 133118, 2008. With permission.)

and quenching of the PL intensity depend on the roughness of the metal surface. On the rough surface, the PL intensity increases. In this experiment, rough Au surfaces are composed of an assembly of hemispherical particles with lateral sizes of 20–50 nm and peaks and valleys of roughly 15 nm. On the flat surface, however, the PL intensity decreases. The enhancement of the PL intensity depends on the excitation laser wavelength. The enhancement spectrum agrees well with the absorption spectrum for localized plasmon resonance [66].

Even on rough Au surfaces, we can observe both enhancement and quenching phenomena of the PL intensity. PL intensity on a rough Au surface depends on the polarization angle of the excitation laser [71]. Electric field enhancement depends on the polarization direction. The PL intensity of a single nanoparticle on glass does not change with the polarization angle. We obtain the microscopic structure of semiconductor nanoparticles and rough surfaces from the polarization and wavelength dependences. Thus, we conclude that the PL intensity enhancement is related to the electric field enhancement due to localized plasmon excitation. Studying PL blinking behaviors is a way to understand energy transfer processes between nanoparticles and surrounding environments.

Close-packed nanoparticle films or nanoparticle arrays show unique exciton energy transfer and charge carrier transport beyond isolated nanoparticles [72–79]. Many different types of closely packed nanoparticle films, arrayed nanoparticle solids, and nanoparticle suprasolids have been prepared [72–79]. In order to control energy transfer between excitons in semiconductors and plasmons in metals, we fabricate metal-semiconductor hetero-nanostructures and their PL spectrum and dynamics. We fabricated two types of semiconductor-metal nanoparticle heterostructures using the Langmuir–Blodgett technique: close-packed CdSe nanoparticle monolayers on Au substrates [80]

and mixed CdSe and Au nanoparticle monolayers [81]. In close-packed CdSe nanoparticle monolayers on Au surfaces, the inert polymer thin film was inserted between the nanoparticle monolayer and the Au substrate. The distance between the excitons and plasmons, Δ, is controlled by the polymer thickness. Figure 22.7 summarizes the distance dependence of the PL lifetime and the time-integrated PL intensity [80]. There is a good correlation between the PL decay rate and the PL intensity. PL quenching only occurs when the distance between excitons and plasmons is less than 30 nm. In large distance samples, the PL decay rates in close-packed monolayers are much larger than those in isolated nanoparticles in solutions. The PL lifetime in the CdSe monolayers on the glass is governed by the nonradiative recombination of excitons in nanoparticles and the energy transfer from small to large CdSe nanoparticles, which have lower exciton energies. Furthermore, the PL decay increases with a decrease of the distance between the Au surface and the CdSe nanoparticle monolayer. This can be attributed to energy transfer from nanoparticles to surface plasmons of the Au surfaces. The reduction of both the PL lifetime and the PL intensity simultaneously occurs. PL quenching only occurs in the CdSe nanoparticle monolayer in close proximity to the Au films [80].

Close-packed monolayer films composed of CdSe and Au nanoparticles have simple two-dimensional hexagonal lattices [81]. The PL and optical density of the sample films depend on the Au nanoparticle concentration in the film, because of the spectral overlap between the exciton luminescence of CdSe nanoparticles and the plasmon absorption in Au nanoparticle. In the CdSe and Au nanoparticle mixed monolayer samples, the PL decay curves can be reproduced successfully using three exponential decay components. Three kinds of decay channel of excited states exist in the close-packed CdSe and Au nanoparticle monolayer. In mixed monolayer samples, the decay times are classified into three components: 0.2, 1, and 10 ns, as summarized in Figure 22.8a [81]. These decay times are almost independent of

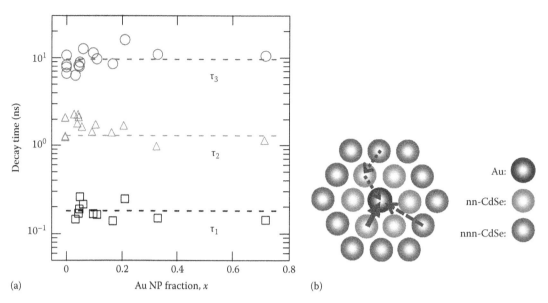

FIGURE 22.8 (a) Decay time obtained by fitting using the three exponential decays as a function of the Au nanoparticle fraction. The broken lines are guides. (b) Schematic illustration of the configurations of the metal and semiconductor nanoparticles. Direct energy transfer from a nearest-neighbor (nn) CdSe nanoparticle (solid arrow) and from a next-nearest-neighbor (nnn) CdSe nanoparticle [broken arrow]. Stepwise CdSe → CdSe → Au nanoparticle energy transfer [dotted arrow]. (Reprinted from Hosoki, K. et al., *Phys. Rev. Lett.*, 100, 207404, 2008. With permission.)

the Au nanoparticle mixing ratio in the film. The 10-ns decay time originates from radiative recombination within CdSe nanoparticles in the films. As the Au nanoparticle fraction increases, the amplitudes of two 1- and 10-ns components decrease. The amplitude of the fast 0.2-ns decay component becomes dominant. These results indicate that the fast PL quenching is caused by the energy transfer to Au nanoparticles for the CdSe nanoparticles in contact with Au nanoparticles. Here, note that the decay component of about 1 ns is a unique characteristic of close-packed mixed nanoparticles solids. Energy transfer between the nearest-neighbor CdSe nanoparticles takes part in the slow PL quenching process of 1 ns in the mixed film. The 1-ns PL quenching process is the stepwise energy transfer from a CdSe nanoparticle to a CdSe nanoparticle to a Au nanoparticle, as illustrated in Figure 22.8b [81]. Therefore, we conclude that the PL dynamics are explained by three kinds of decay channel. The energy relaxation rate in semiconductor nanoparticles is controlled by changing local surrounding environments. These close-packed nanoparticle heterostructures will show unique exciton energy transfer, and charge carrier transport beyond isolated nanoparticles opens new application fields.

22.4 Multiexciton Generation

Finally, we discuss unique exciton–exciton interactions in nanoparticles and nanotubes. Nanoparticle quantum dots and carbon nanotubes provide an excellent stage for experimental studies of many-body effects of excitons or electrons on optical processes in semiconductors [82–84]. The reduced dielectric screening and the relaxation of the energy–momentum conservation rule in nanostructures enhance the Coulomb carrier–carrier interactions,

leading to multi-carrier processes such as the quantized Auger recombination, multiple exciton generation (MEG), carrier multiplication (CM), and so on [7,85]. The achievement of efficient CM in semiconductors makes it possible to produce highly efficient solar cells with conversion efficiencies that exceed the Shockley–Queisser limit of 32% [86]. Strongly confined electrons and excitons in nanomaterials show unique nonlinear optical properties, compared to semiconductor bulk crystals.

Strong interactions between carriers or between excitons cause fast nonradiative Auger recombination of multiple excitons or carriers [82]. Intense interest in Auger recombination in nanoparticles has been stimulated by investigations and searches for new laser and solar cell materials [87–89]. In laser and solar cell applications, nonradiative Auger recombination dominates both the carrier density and the carrier lifetime, determining the device performance. Moreover, in transient absorption, PL, and terahertz conductivity experiments, fast Auger recombination has also been used as a probe in MEG and CM processes [89–96]. The CM efficiencies of nanoparticles are not clear, and the CM mechanism is under discussion. In SWNTs, for example, strong Coulomb interactions enhance the many-body effects of excitons. The fast Auger recombination of excitons has been observed by means of exciton homogeneous linewidth [97] and pump-probe measurements [98–100]. We studied MEG processes in SWNTs at room temperature by temporal change in the carrier density [101]. The fast-decay component grows at increasing excitation intensity. When the photon energy is three times larger than the band-gap energy, Auger recombination occurs efficiently even in the weak intensity region. In our experiment, CM is estimated to be about 1.3 under 4.65 eV excitation [101]. We pointed out that a possible mechanism of CM in carbon nanotubes is the impact ionization. Strong interactions between carriers or between excitons cause unique optical processes in semiconductor nanomaterials. Highly excited semiconductor nanomaterials show new optical functionalities.

22.5 Summary

We briefly discussed luminescence properties of carbon nanotubes and nanoparticle quantum dots by means of single molecular spectroscopy and time-resolved optical spectroscopy. These semiconductor nanomaterials show unique luminescence properties such as spectral diffusion and luminescence blinking. Energy transfer between nanomaterials and surrounding environments affects the PL spectra and dynamics of nanomaterials. Although this chapter is written as a review-type survey of our recent studies, we hope that discussions and many references cited are useful for the readers.

Acknowledgments

The author would like to thank many colleagues and graduate students for their contributions and discussions. In particular, Prof. K. Matsuda deserves mention here. Part of this work was supported by a Grant-in-Aid for Scientific Research on Innovative Area "Optical Science of Dynamically Correlated Electrons (DYCE)" (No. 20104006) from MEXT, Japan, and a Grant-in-Aid for Scientific Research (No. 21340084) from Japan Society for Promotion of Science (JSPS).

References

1. L. Brus: *J. Phys. Chem.* 90, 2555 (1986).
2. Y. Kanemitsu: *Phys. Rep.* 263, 1 (1995).
3. A. P. Alivisatos: *J. Phys. Chem.* 100, 13226 (1996).
4. A. G. Cullis, L. T. Canham, and P. D. J. Calcott: *J. Appl. Phys.* 82, 909 (1997).
5. S. A. Empedocles, R. Neuhauser, K. Shimizu, and M. G. Bawendi: *Adv. Mater.* 11, 1243 (1999).
6. A. P. Alivisatos: *Nat. Biotechnol.* 22, 47 (2004).
7. V. I. Klimov: *J. Phys. Chem. B* 110, 16827 (2006).
8. T. Ando: *J. Phys. Soc. Jpn.* 75, 024707 (2006).
9. D. J. Norris, Al. L. Efros, and S. C. Erwin: *Science*, 319, 776 (2008).
10. Ph. Avouris, M. Freitag, and V. Perebeinos: *Nat. Photon.* 2, 341 (2008).
11. R. Beaulac, P. I. Archer, S. T. Ochsenbein, and D. R. Gamelin: *Adv. Funct. Mater.* 18, 3873 (2008).
12. L. T. Canham: *Appl. Phys. Lett.* 57, 1046 (1990).
13. M. J. O'Connell, S. M. Bachilo, X. B. Huffman, V. C. Moore, M. S. Strano, E. H. Haroz, K. L. Rialon, P. J. Boul, W. H. Noon, C. Kittrell, J. Ma, R. H. Hauge, R. B. Weisman, and R. E. Smalley: *Science* 297, 593 (2002).
14. Y. Kanemitsu, T. Ogawa, K. Shiraishi, and K. Takeda: *Phys. Rev. B* 48, 4883 (1993).
15. C. B. Murray, D. J. Norris, and M. G. Bawendi: *J. Am. Chem. Soc.* 115, 8706 (1993).
16. M. A. Hines and P. Guyot-Sionnest: *J. Phys. Chem.* 100, 468 (1996).
17. X. Peng, M. C. Schlamp, A. V. Kadavanich, and A. P. Alivisatos: *J. Am. Chem. Soc.* 119, 7019 (1997).
18. Z. A. Peng and X. G. Peng: *J. Am. Chem. Soc.* 123, 183 (2001).
19. J. Lefebvre, Y. Homma, and P. Finnie: *Phys. Rev. Lett.* 90, 217401 (2003).
20. A. Nish, J.-Y Hwang, J. Doig, and R. J. Nicholas, *Nat Nanotechnol.* 2, 640 (2007).
21. Y. Kayanuma: *Phys. Rev. B* 38, 9797 (1988).
22. S. V. Gaponenko: *Optical Properties of Semiconductor Nanoparticles* (Cambridge University, Cambridge, U.K., 1998).
23. P. D. J. Calcott, K. J. Nash, L. T. Canham, M. J. Kane, and D. Brumhead: *J. Phys. Condens. Matter* 5, L91 (1993); *J. Lumin.* 57, 257 (1993).
24. T. Itoh, M. Nishijima, A. I. Ekimov, C. Gourdon, Al. L. Efros, and M. Rosen: *Phys. Rev. Lett.* 74, 1645 (1995).
25. N. Nirmal, D. J. Norris, M. Kuno, M. G. Bawendi, Al. L. Efros, and M. Rosen: *Phys. Rev. Lett.* 75, 3728 (1995).
26. Al. L. Efros, M. Rosen, M. Kuno, M. Nirmal, D. J. Norris, and M. Bawendi: *Phys. Rev. B* 54, 4843 (1995).
27. Y. Kanemitsu: *J. Lumin.* 100, 209 (2002).
28. Y. Kanemitsu, S. Okamoto, M. Otobe, and S. Oda: *Phys. Rev. B* 55, R7375 (1997).
29. Y. Kanemitsu and S. Okamoto: *Phys. Rev. B* 58, 9652 (1998).
30. T. Kawazoe and Y. Masumoto: *Phys. Rev. Lett.* 77, 4942 (1996).
31. D. Kovalev, H. Heckler, B. Averboukh, M. Ben-Chorin, M. Schwartzkopff, and F. Koch: *Phys. Rev. B* 57, 3741 (1998).
32. Y. Kanemitsu, H. Tanaka, Y. Fukunishi, T. Kushida, K. S. Min, and H. A. Atwater: *Phys. Rev. B* 62, 5100 (2000).
33. W. E. Moerner: *J. Phys. Chem. B* 106, 910 (2002).
34. S. A. Empedocles, D. J. Norris, and M. G. Bawendi: *Phys. Rev. Lett.* 77, 3873 (1996).
35. N. Nirmal, B. O. Dabbousi, M. G. Bawendi, J. I. Macklin, J. K. Trautman, T. D. Harris, and L. E. Brus: *Nature* 383, 802 (1996).
36. A. Ishizumi, K. Matsuda, T. Saiki, C. W. White, and Y. Kanemitsu: *Appl. Phys. Lett.* 87, 133104 (2005).
37. A. Ishizumi, C. W. White, and Y. Kanemitsu: *Appl. Phys. Lett.* 84, 2397 (2004).
38. S. M. Bachilo, M. S. Strano, C. Kittrell, R. H. Hauge, R. E. Smalley, and R. B. Weisman: *Science* 298, 2361 (2002).
39. F. Wang, G. Dukovic, L. E. Brus, and T. F. Heinz: *Science* 308, 838 (2005).
40. A. Hartschuh, H. N. Pedrosa, L. Novotny, and T. D. Krauss: *Science* 301, 1354 (2003).
41. J. Lefebvre, J. M. Fraser, P. Finnie, and Y. Homma: *Phys. Rev. B* 69, 075403 (2004).
42. K. Matsuda, Y. Kanemitsu, K. Irie, T. Saiki, T. Someya, Y. Miyauchi, and S. Maruyama: *Appl. Phys. Lett.* 86, 123116 (2005).

43. J. Lefebvre, D. G. Austing, J. Bond, and P. Finnie: *Nano Lett.* 6, 1603 (2006).

44. H. Htoon, M. J. O'Connell, P. J. Cox, S. K. Doorn, and V. I. Klimov: *Phys. Rev. Lett.* 93, 027401 (2004).

45. T. Inoue, K. Matsuda, Y. Murakami, S. Maruyama, and Y. Kanemitsu: *Phys. Rev. B* 73, 233401 (2006).

46. R. Matsunaga, K. Matsuda, and Y. Kanemitsu: *Phys. Rev. Lett.* 101, 147404 (2008).

47. H. Ajiki and T. Ando: *J. Phys. Soc. Jpn.* 62, 1255 (1993).

48. S. Zaric, G. N. Ostojic, J. Kono, J. Shaver, V. C. Moore, M. S. Strano, R. H. Hauge, R. E. Smalley, and X. Wei: *Science* 304, 1129 (2004).

49. S. Zaric, G. N. Ostojic, J. Shaver, J. Kono, O. Portugall, P. H. Frings, G. L. J. A. Rikken, M. Furis, S. A. Crooker, X. Wei, V. C. Moore, R. H. Hauge, and R. E. Smalley: *Phys. Rev. Lett.* 96, 016406 (2006).

50. H. Hirori, K. Matsuda, Y. Miyauchi, S. Maruyama, and Y. Kanemitsu: *Phys. Rev. Lett.* 97, 257401 (2006).

51. I. B. Mortimer and R. J. Nicholas: *Phys. Rev. Lett.* 98, 027404 (2007).

52. A. Srivastava, H. Htoon, V. I. Klimov, and J. Kono: *Phys. Rev. Lett.* 101, 087402 (2008).

53. K. Matsuda, T. Inoue, Y. Murakami, S. Maruyama, and Y. Kanemitsu: *Phys. Rev. B* 77, 193405 (2008).

54. S. A. Empedocles and M. G. Bawendi: *Science* 278, 2114 (1997); *J. Phys. Chem. B* 103, 1826 (1999).

55. Al. L. Efros and M. Rosen: *Phys. Rev. Lett.* 78, 1110 (1997).

56. R. G. Neuhauser, K. T. Shimizu, W. K. Woo, S. A. Empedocles, and M. G. Bawendi: *Phys. Rev. Lett.* 85, 3301 (2000).

57. M. Kuno, D. P. Fromm, H. F. Hamann, A. Gallagher, and D. J. Nesbitt: *J. Chem. Phys.* 115, 1028 (2001).

58. F. Koberling, A. Mews, and T. Basché: *Adv. Mater.* 13, 672 (2001).

59. K. T. Shimizu, W. K. Woo, B. R. Fisher, H. J. Eisler, and M. G. Bawendi: *Phys. Rev. Lett.* 89, 117401 (2002).

60. W. G. J. H. M. van Sark, P. L. T. M. Frederix, A. A. Bol, H. C. Gerritsen, and A. Meijerink: *Chem. Phys. Chem.* 3, 871 (2002).

61. M. Kuno, D. P. Fromm, S. T. Johnson, A. Gallagher, and D. J. Nesbitt: *Phys. Rev. B* 67, 125304 (2003).

62. J. Muller, J. M. Lupton, A. L. Rogach, J. Feldmann, D. V. Talapin, and H. Weller: *Appl. Phys. Lett.* 85, 381 (2004).

63. A. Issac, C. von Borczyskowski, and F. Cichos: *Phys. Rev. B* 71, 161302(R) (2005).

64. D. E. Gómez, J. van Embden, and P. Mulvaney: *Appl. Phys. Lett.* 88, 154106 (2006).

65. K. Zhang, H. Chang, A. Fu, A. P. Alivisatos, and H. Yang: *Nano Lett.* 6, 843 (2006).

66. Y. Ito, K. Matsuda and Y. Kanemitsu: *Phys. Rev. B* 75, 033309 (2007).

67. Y. Ito, K. Matsuda, and Y. Kanemitsu: *J. Phys. Soc. Jpn.* 77, 103713 (2008).

68. P. Frantsuzov, M. Kuno, B. Janko, and R. A. Marcus: *Nat. Phys.* 4, 519 (2008).

69. F. Stefani, J. Hoogenboom, and E. Barkal: *Phys. Today* 62(2), 34 (2009).

70. T. D. Krauss and L. E. Brus: *Phys. Rev. Lett.* 83, 4840 (1999).

71. K. Matsuda, Y. Ito, and Y. Kanemitsu: *Appl. Phys. Lett.* 92, 211911 (2008).

72. C. M. Murray, C. R. Kagan, and M. G. Bawendi: *Science* 270, 1966 (1995).

73. C. A. Leatherdale, C. R. Kagan, N. Y. Morgan, S. A. Empedocles, M. A. Kastner, and M. G. Bawendi: *Phys. Rev. B* 62, 2669 (2000).

74. M. P. Pileni: *J. Phys. Chem. B* 105, 3358 (2001).

75. C. R. Kagan, C. B. Murray, M. Nirmal, and M. G. Bawendi: *Phys. Rev. Lett.* 76, 1517 (1996).

76. C. R. Kagan, C. B. Murray, and M. G. Bawendi: *Phys. Rev. B* 54, 8633 (1996).

77. S. A. Crooker, J. A. Hollingsworth, S. Tretiak, and V. I. Klimov: *Phys. Rev. Lett.* 89, 186802 (2002).

78. M. Achermann, M. A. Petruska, S. A. Crooker, and V. I. Klimov: *J. Phys. Chem. B* 107, 13782 (2003).

79. O. I. Mićić, K. M. Jones, A. Cahill, and A. J. Nozik: *J. Phys. Chem. B* 102, 9791 (1998).

80. A. Ueda, T. Tayagaki, and Y. Kanemitsu: *Appl. Phys. Lett.* 92, 133118 (2008).

81. K. Hosoki, S. Yamamoto, T. Tayagaki, K. Matsuda, and Y. Kanemitsu: *Phys. Rev. Lett.* 100, 207404 (2008).

82. V. I. Klimov, A. A. Mikhailovsky, D. W. McBranch, C. A. Leatherdale, and M. G. Bawendi: *Science* 287, 1011 (2000).

83. V. I. Klimov, A. A. Mikhailovsky, S. Xu, A. Malko, J. A. Hollingsworth, C. A. Leatherdale, H. J. Eisler, and M. G. Bawendi: *Science* 290, 314 (2000).

84. Y. Kanemitsu, T. J. Inagaki, M. Ando, K. Matsuda, T. Saiki, and C. W. White: *Appl. Phys. Lett.* 81, 141 (2002).

85. A. J. Nozik: *Chem. Phys. Lett.* 457, 3 (2008).

86. W. Shockley and H. J. Queisser: *J. Appl. Phys.* 32, 510 (1961).

87. V. I. Klimov, S. A. Ivanov, J. Nanda, M. Achermann, I. Bezel, J. A. McGuire, and A. Piryatinski: *Nature* 447, 441 (2007).

88. T. Tayagaki, S. Fukatsu, and Y. Kanemitsu: *Phys. Rev. B* 79, 041301(R) (2009).

89. R. D. Schaller and V. I. Klimov: *Phys. Rev. Lett.* 92, 186601 (2004).

90. R. Ellingson, M. C. Beard, J. C. Johnson, P. Yu, O. I. Micic, A. J. Nozik, A. Shabaev, and A. L. Efros: *Nano Lett.* 5, 865 (2005).

91. M. C. Beard, K. P. Knutsen, P. Yu, J. M. Luther, Q. Song, W. K. Metzger, R. J. Ellingson, and A. J. Nozik: *Nano Lett.* 7, 2506 (2007).

92. G. Nair and M. G. Bawendi: *Phys. Rev. B* 76, 081304(R) (2007).

93. V. I. Klimov, J. A. McGuire, R. D. Schaller, and V. I. Rupasov: *Phys. Rev. B* 77, 195324 (2008).

94. G. Nair, S. M. Geyer, L. Y. Chang, and M. G. Bawendi: *Phys. Rev. B* 78, 125325 (2008).

95. J. J. H. Pijpers, E. Hendry, M. T. W. Milder, R. Fanciulli, J. Savolainen, J. L. Herek, D. Vanmaekelbergh, S. Ruhman, D. Mocatta, D. Oron, A. Aharoni, U. Banin, and M. Bonn: *J. Phys. Chem. C* 111, 4146 (2007); 112, 4783 (2008).

96. M. T. Trinh, A. J. Houtepen, J. M. Schins, T. Hanrath, J. Piris, W. Knulst, A. P. L. M. Goossens, and L. D. A. Siebbeles: *Nano Lett.* 8, 1713 (2008).

97. K. Matsuda, T. Inoue, Y. Murakami, S. Maruyama, and Y. Kanemitsu: *Phys. Rev. B* 77, 033406 (2008).

98. F. Wang, G. Dukovic, E. Knoesel, L. E. Brus, and T. F. Heinz: *Phys. Rev. B* 70, 241403(R) (2004).

99. F. Wang, Y. Wu, M. S. Hybertsen, and T. F. Heinz: *Phys. Rev. B* 73, 245424 (2006).

100. Y.-Z. Ma, L. Valkunas, S. L. Dexheimer, S. M. Bachilo, and G. R. Fleming: *Phys. Rev. Lett.* 94, 157402 (2005).

101. A. Ueda, K. Matsuda, T. Tayagaki, and Y. Kanemitsu: *Appl. Phys. Lett.* 92, 233105 (2008).

23
Nanoscale Excitons and Semiconductor Quantum Dots

Vanessa M. Huxter
University of Toronto

Jun He
University of Toronto

Gregory D. Scholes
University of Toronto

23.1 Introduction to Nanoscale Excitons

The optical properties of nanoscale excitons are of interest to researchers because their size tunability makes them of practical use; however, understanding the nature of these delocalized electronic states is a challenge (Scholes and Rumbles 2006, 2008). Nanoscale excitons are the electronic excited states formed by the absorption of light by nanoscale systems. A nanoscale system is any chemical or physical system that is in the nanometer-size regime. The kinds of systems that are the subject of this chapter tend to have properties lying between those of molecules and those of bulk semiconductors. Unlike excitons in bulk semiconductors, the energies of nanoscale excitons can often be changed by the size of the system. That property is envisioned, for example, to allow the color of solid-state lasers to be tuned.

Well-studied examples of nanoscale systems include semiconductor nanocrystals, carbon nanotubes (CNTs), organic conjugated polymers, and molecular aggregates, shown in Figure 23.1. The excitons in quantum dots and CNTs are closely related to molecular excited states; however, the size of the nanoscale systems has forced researchers to make severe approximations in their quantum-mechanical descriptions. Nonetheless, the excited states of nanoscale systems tend to be more amenable to approximate descriptions than molecules because the wavefunction delocalization reduces the importance of the electron correlation for an accurate calculation of the energies of electronic states. Such electronic excited states are often described as Wannier–Mott excitons (Banyai and Koch 1993; Basu 1997; Gaponenko 1998; Jorio et al. 2008). Other systems, like molecular aggregates, crystals, and certain proteins—namely, those involved in photosynthetic energy transduction—have optical properties that are better described with reference to the molecular building blocks of the aggregate. The lowest electronic excited states of these aggregates are called Frenkel excitons (Kasha 1976). Conjugated polymers, now used in organic light-emitting diodes and displays (OLEDs), have excited states somewhere between these limits (Sariciftci 1997; Hadziioannou and Malliaras 2006). This relationship is illustrated in Figure 23.2.

In the limiting case of Wannier–Mott excitons, it is assumed that the electronic interaction between the building blocks of the system—the atoms or unit cells—is large, akin to a chemical bond. In this case, orbitals that are delocalized over an entire system are a good starting point for describing one-electron states. Photoexcitation introduces an electron into the conduction orbitals, leaving a "hole" in the valence orbitals. The Wannier–Mott description assumes that there are many electrons in the system, so the atomic centers are highly screened from the outermost electrons, that is, the dielectric constant is high. In that case, it is considered reasonable to assume that the electron and the hole move freely in the background dielectric continuum and are weakly bound. The strength of this electron–hole attraction determines the "binding" of the lower energy, optically allowed states compared to the dense manifold of charge carrier states that lie higher in energy. In this model, the electron and the hole move under their mutual attraction in a dielectric continuum, and then the exciton energy levels are found as a series analogous to the states of the hydrogen atom. Note that this model can be useful, but it must be realized that it is highly approximate (Scholes 2008b), and it is only able to give limited insights. The conceptual advantage of this model is that it converges to the free carrier limit where the electron-hole attraction is negligible compared to the thermal energies. It can thereby be seen how the photoexcitation of a bulk semiconductor exciton efficiently produces charge carriers, which is how a typical semiconductor solar cell works.

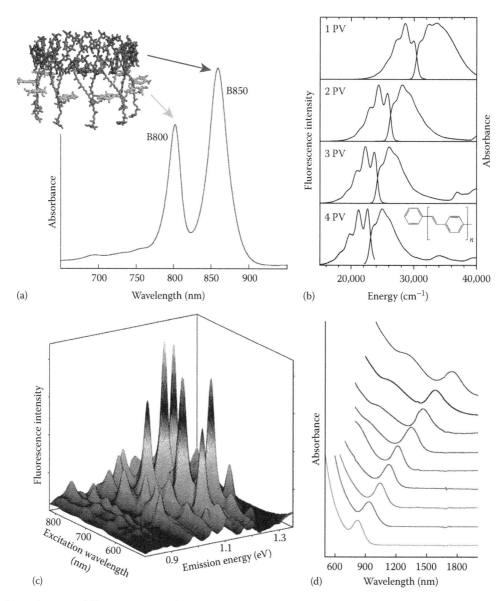

FIGURE 23.1 **(See color insert following page 21-4.)** Excitons and structural size variations on the nanometer length scale. (a) The photosynthetic antenna of purple bacteria, LH2, is an example of a molecular exciton. The absorption spectrum clearly shows the dramatic distinction between the B800 absorption band, arising from essentially "monomeric" bacteriochlorophyll-a (Bchl) molecules, and the redshifted B850 band that is attributed to the optically bright lower exciton states of the 18 electronically coupled Bchl molecules. (b) The size-scaling of polyene properties, for example, oligophenylenevinylene oligomers, derives from the size-limited delocalization of the molecular orbitals. However, as the length of the chains increases, disorder in the chain conformation impacts the picture for exciton dynamics. Absorption and fluorescence spectra are shown as a function of the number of repeat units. (c) SWCNT size and "wrapping" determine the exciton energies. Samples contain many different kinds of tubes, therefore optical spectra are markedly inhomogeneously broadened. By scanning excitation wavelengths and recording a map of fluorescence spectra, the emission bands from various different CNTs can be discerned, as shown. (Courtesy of Dr. M. Jones). (d) Rather than thinking in terms of delocalizing the wavefunction of a semiconductor through interactions between the unit cells, the small size of the nanocrystal confines the exciton relative to the bulk. Size-dependent absorption spectra of PbS quantum dots are shown. (Adapted from Scholes, G. D. and Rumbles, G., *Nat. Mater.*, 5, 683, 2006.)

In order to further understand the formation of excitons in nanoscale systems, we will examine the case of semiconductors as a model system. The optical properties of semiconductor nanostructures lie in an intermediate regime between a molecular and a bulk description. In a bulk semiconductor, the dimensions of the system are practically infinite compared to the dimensions of the carriers (electrons and holes). In this case, the wavefunctions, which are standing waves in the material, are spread over an infinite number of unit cells (the repeat unit in a crystal), and the carriers (electrons and holes) are free particles. The density of states for the bulk material is continuous, as shown in Figure 23.3a. Note that in the case of a bulk semiconductor, the density of states for both the hole and the electron are continuous. These two continuous regions are separated by

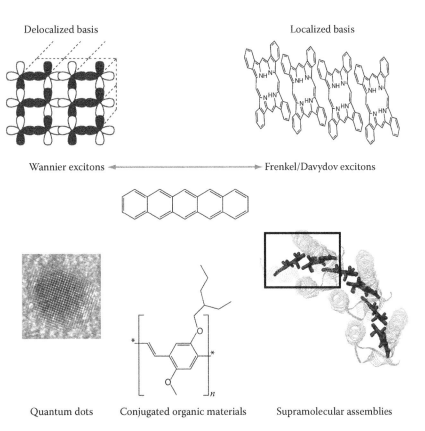

FIGURE 23.2 Collective properties of nanoscale materials modify the optical properties, such as the wavelength and the dipole strength for light absorption. The elementary excitations are known as excitons. Excitons are formed through the collective absorption of light by two or more repeat units in a crystal, a molecular assembly, or a macromolecule. Wannier excitons are typical of atomic crystals, semiconductor quantum dots, aromatic molecules, CNTs, conjugated polymers, and so on. Supramolecular assemblies, including J-aggregates and photosynthetic light-harvesting antennae, typify Frenkel excitons—excitons in which the repeat units retain their identity to a significant degree.

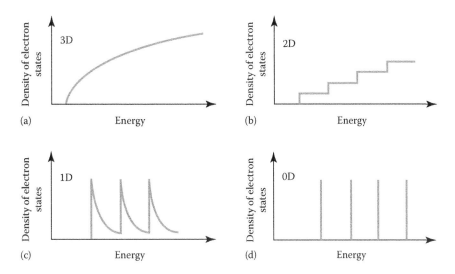

FIGURE 23.3 Idealized representation of the density of electron states for (a) a bulk semiconductor system (3D) and a semiconductor system confined in (b) one dimension (a 2D system), (c) confined in two dimensions (a 1D system) and (d) confined in all three dimensions (a 0D system). In the bulk case, the energy levels form a continuous band. With increasing confinement, the density of states becomes more discrete, resulting in the delta functional form of the density of states in the 0D case.

the bandgap in which the density of states goes to zero for an ideal system. The bulk system is infinite in all three dimensions. The effect of spatially confining the electrons and holes in one, two, or three of those dimensions is to change the density of states. The spatial confinement of the electrons and holes in one dimension means that the wavefunctions of those particles are quantized in that same one dimension. However, the electrons and holes are not confined and can move freely in the other two dimensions. This is similar to the concept of a plane as a two-dimensional object in a three-dimensional space. Examples of these two-dimensional systems with a spatial confinement in one dimension are nanometer-height thin films and quantum wells. The quantization of the electrons and holes in one dimension changes the density of states from the bulk continuum to a step-type function, as shown in Figure 23.3b. Similarly, spatial confinement of the electrons and holes in two dimensions results in the quantization of the particles in those same two dimensions. In these systems, called quantum wires, the electrons and the holes can only move freely in one dimension. The quantization in two of three dimensions results in a further change in the density of states, which becomes more discrete with individual peaks as shown in Figure 23.3c. Finally, the spatial confinement of the electron and the hole in all three dimensions results in quantization in all three dimensions. In this case, the density of states is discrete and takes the form of delta functions, as shown in Figure 23.3d. These systems are called quantum dots and are discussed in more detail later in this chapter.

The physical confinement in one, two, or three dimensions changes the boundary conditions imposed on the wavefunctions and quantizes the behavior of the electrons and holes in one, two, or three dimensions. The boundary conditions associated with the spatial confinement of the wavefunctions, in turn, modify the density of states. The resulting change in the density of states from a continuous band in the bulk to discrete levels in a quantum dot is analogous to the particle in a box model in quantum mechanics. The physical dimensions required to spatially confine the electron or the hole depend on the size of the electron and hole wavefunctions. These are determined by the physical properties of the specific material and are usually on the order of one to tens of nanometers.

The quantum confinement of the electron and the hole wavefunctions modifies their interaction. The spatially overlapped electron and hole wavefunctions can associate through an energy-lowering Coulomb interaction due to their relative negative and positive charges, forming an exciton. This association can be described in analogy to a hydrogen atom and, as such, the size of the exciton is described using a Bohr radius. Just like the hydrogen atom, the lowest energy set of states is comprised of closely bound electron-hole pair configurations. These are the bound exciton states, which are the optically active states of the system (those that absorb and emit light). They tend to be clearly distinguished from a band of many higher energy states, which do not absorb light, but are the nanoscale-free carrier states (Scholes 2008b).

The energy separation between the lowest energy-exciton states and the onset of the carrier states is called the exciton-binding energy.

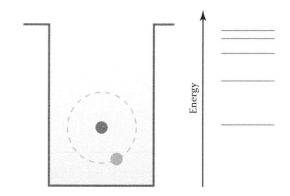

FIGURE 23.4 Confining the electron and the hole in a potential results in the formation of a hydrogen-like exciton and bound exciton states. The binding of the exciton is mediated through the Coulomb interaction.

energy. We can think of the exciton-binding energy as the energy required to ionize an exciton. Notably, this ionization energy is significantly reduced compared to small, molecular systems because of the many different ways that the electron and the hole can be separated. The exciton-binding energy, mediated by an attractive Coulomb interaction, can also be thought of as the energy reduction associated with forming an exciton as compared to the energies of the free electron and hole confined in the three-dimensional potential. This exciton-binding energy shifts the positions of the energy levels. The exciton in a three-dimensional potential and the associated exciton levels are shown in Figure 23.4.

The binding energy of excitons is a topic of widespread interest, especially because of its relevance to photovoltaic science. In high dielectric constant bulk-semiconductor materials the exciton-binding energy is typically small: 27 meV for CdS, 15 meV for CdSe, 5.1 meV for InP, and 4.9 meV for GaAs. The small exciton-binding energies of these materials make them well-suited for photovoltaic applications because the optically active exciton states absorb light, then the ambient thermal energy (~25 meV) is sufficient to convert the exciton to free carriers. Thus, the potential energy of the absorbed photon is converted to an electrical potential. On the other hand, in molecular materials, the electron-hole Coulomb interaction is substantial—usually a few eV. In nanoscale materials, we find a middle ground where exciton-binding energies are significant in magnitude—that is, excitons are important.

As an example, the valence and the conduction orbitals of a single-wall carbon nanotube (SWCNT) are shown in Figure 23.5. These are typical 1D densities-of-states as were introduced in Figure 23.3. If the electrons were non-interacting, then the lowest excitation energy of a SWCNT would be the same as the energy gap between the highest valence orbital and the lowest conduction orbital. In fact, this is the energy corresponding to the onset of carrier formation. Suitable quantum-chemical calculations introduce interactions between the electrons in these various orbitals, which captures the electron-hole attraction and thus leads to a significant stabilization of the excited states relative to

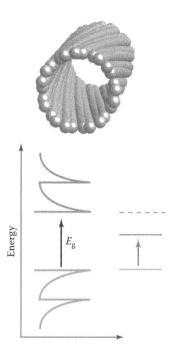

FIGURE 23.5 The attractive interaction of electrons and holes in a CNT (represented by the model at the top of the figure) results in the formation of excitons. This stabilizes the excited states relative to the orbital energy difference, lowering the energy of the optical gap.

those estimated with the more primitive model of orbital energy differences. That gives the correct energy of the optical absorption bands. The energy difference between the orbital energy difference and the optical gap is the exciton-binding energy. It has been predicted for SWCNTs to be ~0.3–0.5 eV (Zhao and Mazumdar 2004), which was later confirmed by an experiment (Ma et al. 2005; Wang et al. 2005).

As a result of the nature of excitons, the optical properties of nanoscale systems provide an interesting link between the properties of extended "bulk" systems and those of molecules. The study of excitons provides the opportunity for new insights into the behavior of nanoscale systems. For the rest of this chapter, we will focus on the properties of a particular nanoscale system: semiconductor quantum dots.

23.2 Nonlinear Optical Properties of Semiconductor Quantum Dots

Quantum-confined semiconductor nanocrystals, or quantum dots, have been the focus of intense study over the past three decades due to their size-tunable optical properties and unique physical characteristics. This area of research was founded in the early 1980s when researchers at Bell Laboratories (Rossetti et al. 1983) in the United States and the Yoffe Institute (Ekimov et al. 1980, Ekimov and Onushchenko 1982; Efros and Efros 1982) in Russia (then the U.S.S.R) independently described the properties of nanometer-sized semiconductor quantum dots. This work was quickly followed by studies on colloidal samples (Spanhel et al. 1987), leading to a further understanding of the optical and

physical properties of quantum dots. Within less than a decade, a basic theoretical framework to describe the observed properties had been proposed and work was underway to explore the fundamental physics of these materials.

Much of the early work on quantum dots focused on semiconductor-doped glasses. These materials are characterized by broad size distributions and offer no possibility of control over the shape or the interface characteristics of the particles. The limitations of these doped glasses, particularly the broad size distribution that results in a large static inhomogeneity, masked many of the fundamental physical processes that were occurring in the quantum dots. However, the study of quantum dot systems underwent a revolution in 1993 when it was discovered that nucleation and a controlled growth of colloidal semiconductors could be achieved by injecting highly reactive organometallic precursors into a solvent system that coordinates to the colloid surface (Murray et al. 1993). This coordinating solvent, trioctylphosphine oxide, was an important discovery because it serves to arrest growth and stabilize the nanocrystals. This method allowed a simple and reproducible synthesis of high quality, nearly monodisperse cadmium chalcogenide nanocrystal samples. This development led to an explosion in the field and opened new avenues for research and technology including increased processibility, the possibility of mass production, and chemical manipulation for tailored shape control. In particular, this reliable method of making high-quality samples allowed the exploration of phenomena that had been previously unobservable due to static inhomogeneity associated with broad size distributions.

In addition to doped glasses and colloidal samples, there are epitaxially grown quantum dots. While colloidal nanocrystals tend to be sized in the range of 1–10 nm in diameter, epitaxial dots, which are grown on solid substrates, may be a couple of nanometers in height but with lateral dimensions of tens of nanometers. The physics of these materials differ from colloidal samples in some fundamental ways; however, the basic properties associated with quantum-confined nanocrystals apply to all three types of quantum dots. Different aspects of these properties have been explored in many comprehensive reviews. For instance, a review of the electronic properties was presented by Yoffe (1993, 2001). The optical nonlinearities of semiconductor nanocrystals were reviewed by Banfi's group (Banfi et al. 1998). Research on semiconductor quantum dots has evolved from fundamental science (Alivisatos 1996; Empedocles and Bawendi 1997; Klimov 2000; Klimov et al. 2000a,b) to lasing and amplification (Klimov 2006; Klimov et al. 2007), optical power limiting (He et al. 2007a,b), biological imaging (Bruchez et al. 1998; Dubertret et al. 2002; Larson et al. 2003; Michalet et al. 2005) and labeling (Seydack 2005), sensitization (Dayal and Burda 2008), and optical switching (Etienne et al. 2005; He et al. 2005a).

One of the defining features of a semiconductor is the bandgap, which separates the conduction band and the valence band. When a semiconductor material absorbs light, an electron is promoted from the valence to the conduction band. The wavelength of light absorbed and emitted from a semiconductor material is determined by the width of the bandgap. In semiconductors, when an electron moves from the valence to the conduction band, the resulting gap left in the valence band is called a hole. An exciton can be formed through an energy-lowering Coulomb interaction between the negative electron and the positive hole. In analogy with the hydrogen atom, the spatial extent of the exciton wavefunction is described by a quantity called the exciton Bohr radius. As a result of the three-dimensional spatial confinement of the exciton wavefunction, the density of states becomes discrete, as described earlier in the chapter. The position of these states, and therefore the energy of the gap, depends on the spatial confinement of the exciton, which is determined by the physical size of the nanocrystal. To a first approximation, this quantum-confinement effect can be described using the particle in a box or a simple quantum box model (Efros and Efros 1982; Brus 1984), in which the electron motion is restricted in all three dimensions by impenetrable walls. For a spherical nanocrystal with radius R, the quantum box model predicts that the size-dependence of the energy gap is proportional to $1/R^2$, indicating that the energy of the lowest transition increases as the nanocrystal size decreases. In addition, as described above, quantum confinement changes the continuous energy bands of a bulk semiconductor into discrete exciton energy levels. The exciton energy levels produce peaks in the absorption spectrum of quantum dots, which is in contrast to the continuous absorption spectrum of a bulk semiconductor (Alivisatos 1996).

The colloidal quantum dots discussed here are composed of a semiconductor core surrounded by a shell of organic ligand molecules (Murray et al. 1993). The organic capping prevents the uncontrolled growth and agglomeration of the nanoparticles. It also allows quantum dots to be chemically manipulated as if they were large molecules, with solubility and chemical reactivity determined by the identity of the organic molecules. The capping also provides an electronic passivation of the nanocrystals by terminating the dangling bonds on the surface. The unterminated dangling bonds can affect the emission efficiency of the quantum dots because they lead to a loss mechanism where electrons are rapidly trapped at the semiconductor surface before they have a chance to emit photons. Using colloidal chemical synthesis, one can prepare nanocrystals with nearly atomic precision with diameters ranging from nanometers to tens of nanometers and a size dispersion as narrow as 5% standard deviation. Because of the quantum-confinement effect, the ability to tune the size or shape of the nanocrystals translates into a means of controlling their optical properties, such as the absorption and emission wavelengths (Scholes and Rumbles 2006; Scholes 2008a).

In a quantum dot system, the three-dimensional confinement modifies the Hamiltonian, adding a potential that restricts how far apart the electron and the hole can be. This forced spatial overlap changes the density of states as described earlier in the chapter. The interaction between the confined electron and the hole, mediated by a Coulomb potential, leads to the formation of an exciton, which can be described in analogy to a hydrogen atom where an electron interacts with a nucleus. As a result of the discrete character of the density of states in quantum dots, as shown in Figure 23.3d, the oscillator strength is concentrated into those few transitions instead of being spread over a continuum of states. This means that the oscillator strength of the states in quantum dots is significantly enhanced. The concentration of the oscillator strength into a few transitions also enhances the nonlinear optical properties of quantum dots as compared to the bulk (Shalaev et al. 1996). As a result of their enhanced nonlinear optical properties, synthetically controllable size tunability and photostability, quantum dots continue to be of great interest both for fundamental research and device applications. One of these areas of research involves nonlinear two-photon absorption (2PA).

23.3 Two-Photon Absorption in Semiconductor Quantum Dots

Two-photon absorption in semiconductors is the simultaneous absorption of two photons of identical or different frequencies required to move an electron from the valence to the conduction band. As opposed to the linear intensity dependence of one-photon absorption (1PA), 2PA depends on the square of the light intensity. Therefore, 2PA is a third-order nonlinear optical process and many orders of magnitude weaker than 1PA. The 2PA coefficient, β, is directly related to the imaginary part of the third-order nonlinear susceptibility, $\chi^{(3)}$, by $\beta = 3\pi \mathrm{Im}\chi^{(3)}/(\lambda n_0^2 c \varepsilon_0)$, where n_0 is the linear refractive index, ε_0 is the dielectric constant in vacuum, λ the laser wavelength, and c the speed of light in vacuum (Sutherland 2003). Compared to 1PA, 2PA is associated with different selection rules for dipole transitions. For example, in CdSe quantum dots, one-photon transitions satisfy selection rules $\Delta L = 0, \pm 2$ and $\Delta F = 0, \pm 1$ while two-photon transitions satisfy $\Delta L = \pm 1, \pm 3$ and $\Delta F = 0, \pm 1, \pm 2$, where L is the orbital angular momentum of the envelope wavefunction and F is the total angular momentum (Schmidt et al. 1996). Therefore, the comparison between 1PA and 2PA spectra allows a more detailed optical analysis of electronic states in semiconductor quantum dots (Schmidt et al. 1996). For direct bandgap semiconductor quantum dots, a simple theory based on the effective mass approximation was proposed for both 1PA and 2PA (Fedorov et al. 1996). Although this simple model does not consider the mixing between the heavy and light holes bands, it works quite well for describing degenerate 2PA. Recently, this theory has been extended to obtain the analytical expressions for both degenerate and nondegenerate 2PA spectra of semiconductor quantum dots with the parabolic band approximation and $\vec{k} \cdot \vec{p}$ theory (Padilha et al. 2007).

As the incident laser intensity increases, the 2PA in semiconductor quantum dots will be saturated due to the limited density of states of the quantized energy levels (He et al. 2005b). If the allowed 2PA transitions are assumed to occur between the one atomic-like energy level in the conduction band and the other in the valence band, the ensemble of quantum dots can be approximately treated as a two-level system. Since the semiconductor quantum dots are not uniform in size, the saturation of 2PA for such an inhomogeneous system may be derived as $\beta(I) = \beta / \sqrt{1 + I_0^2/I_{s,2PA}^2}$ (Sutherland 2003). The saturation intensity, $I_{s,2PA}$, in quantum dots can be described quantitatively by an inhomogeneously broadened, saturated 2PA model as (He et al. 2005b)

$$I_{s,2PA}^2 = \frac{\hbar\omega\pi\Delta\omega p(2\omega)N_0}{\tau_p\sqrt{\pi/2}\left(1 + \frac{g_k}{g_n}\right)\beta} \frac{g_k}{g_n} \tag{23.1}$$

where

$g_k(g_n)$ is the electronic degeneracy of the upper (lower) state
τ_p is the half width at the maximum of the femtosecond laser pulse
$\hbar\omega$ is the photon energy
N_0 is the number density of the quantum dots in the sample material
$p(2\omega)$ is the probability of a homogeneous class of absorbers with a central frequency of 2ω

The quantity $\Delta\omega$ is related to the dephasing time (T_2) of the excitation, i.e., the width of the homogeneous line shape.

Two-photon transition does not involve a real intermediate state. The single photon energy is less than that of the quantum dots bandgap. So 2PA is a non-resonant optical nonlinearity. The 2PA coefficients, β, can be determined by two-photon excited fluorescence, nonlinear transmission, Z scan, and pump-probe techniques (Sutherland 2003). For example, the differential equation describing the intensity change in a one-photon transparent but two-photon absorbing material is given by $dI_0/dz = -\beta I_0^2$, where I_0 is the peak intensity of the input laser beam inside the sample. The nonlinear transmission in such a material, excited by a focused continuous-wave (cw) Gaussian beam, can be expressed as follows (Sutherland 2003):

$$T(I_0) = \frac{\left[\ln(1 + \beta l I_0)\right]}{(\beta l I_0)} \tag{23.2}$$

where l is the thickness of the quantum dot sample. A comparison of the expressions for the energy transmittance of a cw top-hat, pulsed Gaussian and sech^2 beam are available in Sutherland (2003). From Equation 23.2 one can see that at a given input intensity (I_0) level, if the nonlinear transmission value is measured, the β value can be readily determined. The 2PA coefficient β is a macroscopic parameter characterizing the quantum dot composite material. The intrinsic 2PA coefficient of quantum dots, β_{QD}, can be derived as $\beta_{QD} = \beta_{composite}n_{0,composite}^2/(n_{0,QD}^2 f_v |f|^4)$, where f_v is the volume fraction of the quantum dots in the matrix and $f = 3n_{0,matrix}^2/(n_{0,QD}^2 + 2n_{0,matrix}^2)$ is the local field correction that depends on the dielectric constant of the quantum dots and the matrix material. In addition, the 2PA cross section can be calculated by the use of the definition: $\sigma_2 = \beta\hbar\omega/N_0$.

Two-photon absorption in semiconductor quantum dots has many important potential applications, some of which are discussed in more detail below.

23.3.1 Biological Imaging

Using quantum dot labeling, 2PA provides a possible way of performing biological imaging that is not possible by traditional one-photon methods as visible wavelengths cannot penetrate human tissue. As tagging materials, semiconductor quantum dots have advantages over fluorescent dyes, such as broad excitation and narrow emission bands, emission wavelength tunability, photostability, and enhanced brightness. The 2PA in robust water-soluble CdSe/ZnS core-shell quantum dots was found to be well-suited for use as fluorescent labels in multiphoton microscopy for biological imaging (Larson et al. 2003). The near-infrared two-photon excitation of such quantum dots in the "tissue optical window" (0.7–1.1 μm), in which water and hemoglobin absorb very little light, allows the extensive imaging of living systems, making further developments in medical diagnostics possible.

23.3.2 Amplified Stimulated Emission and Lasing

Quantum dot lasers have potential advantages, such as a temperature-insensitive lasing threshold and wide-range color tunability. Due to the quantum-size confinement, the 2PA cross-sections in semiconductor quantum dots are enhanced compared to the corresponding bulk material. This allows for a lower lasing threshold with a 2PA pumping mechanism. In addition, quantum dot lasers with 2PA excitation in the "tissue optical window" have important application prospects on laser-assisted biological-medical diagnostics and therapy. Recently, upconverted laser emission from a solution-processed CdSe/CdS/ZnS quantum dot waveguide-resonant cavity has been successfully demonstrated with femtosecond excitation at a wavelength of 800 nm (Zhang et al. 2008).

23.3.3 Optical Power Limiting

Optical power limiting and stabilization can be used to protect sensitive equipment or to control noise in laser beams (He et al. 2008). In this case, the output laser intensity approaches a constant value when the input intensity increases beyond a certain threshold, limiting the amount of optical power entering a system. Compared to optical-limiting materials based on organic chromophores, crystals, and polymers, semiconductor quantum dots have a large 2PA cross section and better photostability.

23.3.4 Two-Photon Sensitizer

Semiconductor quantum dots are excellent sensitizers for near-infrared 2PA due to their large and size-tunable 2PA cross sections. For example, quantum dots linked to phthalocyanines

(Pcs) can be excited using the 2PA of the quantum dots without any significant direct excitation of Pc molecules. In this case, Pcs are electron acceptor molecules, and a near-infrared excitation of the Pc molecules within the spectral therapeutic window (0.7–1.2 μm) can be realized via a two-photon sensitization of the quantum dot followed by an energy transfer to the Pc (Dayal and Burda 2008).

23.4 Conclusion

Excitons in nanoscale systems, particularly in semiconductor quantum dots, provide a link between the bulk and the molecular regimes. In quantum dots, the discrete density of states associated with the exciton results in an increased oscillator strength in the optically active levels and an increased nonlinear cross section. The increased nonlinear optical properties of quantum dots provide an opportunity for novel devices and experimental applications including those associated with 2PA. The unique properties of nanoscale excitons and the materials associated with them will provide inspiration and direction to future research.

References

Alivisatos, A. P. 1996. *Science* 271: 933.

Banfi, G. P., Degiorgio, V., and Ricard, D. 1998. *Adv. Phys.* 47: 447.

Banyai, L. and Koch, S. W. 1993. *Semiconductor Quantum Dots*. River Edge, NJ: World Scientific.

Basu, P. K. 1997. *Theory of Optical Processes in Semiconductors: Bulk and Microstructures*. New York: Oxford University Press.

Bruchez, M., Moronne, M., Gin, P., Weiss, S., and Alivisatos, A. P. 1998. *Science* 281: 2013.

Brus, L. E. 1984. *J. Chem. Phys.* 80: 4403.

Dayal, S. and Burda, C. 2008. *J. Am. Chem. Soc.* 130: 2890.

Dubertret, B., Skourides, P., Norris, D. J. et al. 2002. *Science* 298: 1759.

Efros, Al. L. and Efros, A. L. 1982. *Semiconductors* 16: 1209–1214.

Ekimov, A. I. and Onushchenko, A. A. 1982. *Semiconductors* 16: 1215–1219.

Ekimov, A. I., Onushchenko, A. A., and Tsekhomskii, V. A. 1980. *Sov. Glass Phys. Chem.* 6: 511–512.

Empedocles, S. A. and Bawendi, M. G. 1997. *Science* 278: 2114.

Etienne, M., Biney, A., Walser, A. D. et al. 2005. *Appl. Phys. Lett.* 87: 181913.

Fedorov, A. V., Baranv, A. V., and Inoue, K. 1996. *Phys. Rev. B* 54: 8627.

Gaponenko, S. V. 1998. *Optical Properties of Semiconductor Nanocrystals*. New York: Cambridge University Press.

Hadziioannou, G. and Malliaras, G. G. eds. 2006. *Semiconducting Polymers: Chemistry, Physics and Engineering*, Weinheim, Germany: Wiley-VCH.

He, J., Ji, W., Ma, G. H. et al. 2005a. *J. Phys. Chem. B* 109: 4373.

He, J., Mi, J., Li, H. P., and Ji, W. 2005b. *J. Phys. Chem. B* 109: 19184.

He, G. S., Yong, K. T., Zheng, Q. et al. 2007a. *Opt. Express* 15: 12818.

He, G. S., Zheng, Q., Yong, K. T. et al. 2007b. *Appl. Phys. Lett.* 90: 181108.

He, G. S., Tan, L., Zheng, Q., and Prasad, P. N. 2008. *Chem. Rev.* 108: 1245.

Jorio, A., Dresselhaus, M. S., and Dresselhaus, G. 2008. *Carbon Nanotubes: Advanced Topics in Synthesis, Structure, Properties and Applications*. New York: Springer.

Kasha, M. 1976. Molecular excitons in small aggregates. In *Spectroscopy of the Excited State*, ed. B. DiBartolo. New York: Plenum Press.

Klimov, V. I. 2000. *J. Phys. Chem. B* 104: 6112.

Klimov, V. I. 2006. *J. Phys. Chem. B* 110: 16827.

Klimov, V. I., Mikhailovsky, A. A., McBranch, D. W., Leatherdale, C. A., and Bawendi, M. G. 2000a. *Science* 287: 1011.

Klimov, V. I., Mikhailovsky, A. A., Xu, S. et al. 2000b. *Science* 290: 314.

Klimov, V. I., Ivanov, S. A., Nanda, J. et al. 2007. *Nature* 447: 441.

Larson, D. R., Zipfel, W. R., Williams, R. M. et al. 2003. *Science* 300: 1434.

Ma, Y. Z., Valkunas, L., Bachilo, S. M., and Fleming, G. R. 2005. *J. Phys. Chem. B* 109: 15671–15674.

Michalet, X., Pinaud, F. F., Bentolila, L. A. et al. 2005. *Science* 307: 538.

Murray, C. B., Norris, D. J., and Bawendi, M. G. 1993. *J. Am. Chem. Soc.* 115: 8706–8715.

Padilha, L. A., Fu, J., Hagan, D. J. et al. 2007. *Phys. Rev. B* 75: 075325.

Rossetti, R., Nakahara, S., and Brus, L. E. 1983. *J. Chem. Phys.* 79: 1086–1088.

Sariciftci, N. S. 1997. *Primary Photoexcitations in Conjugated Polymers: Molecular Exciton versus Semiconductor Band Model*. Singapore: World Scientific.

Schmidt, M. E., Blanton, S. A., Hines, M. A., and Guyot-Sionnest, P. 1996. *Phys. Rev. B* 53: 12629.

Scholes, G. D. 2008a. *Adv. Funct. Mater.* 18: 1157.

Scholes, G. D. 2008b. *ACS Nano* 2: 523–537.

Scholes, G. D. and Rumbles, G. 2006. *Nat. Mater.* 5: 683.

Scholes, G. D. and Rumbles, G. 2008. Excitons in nanoscale systems: Fundamentals and applications. In *Annual Review of Nano Research*, eds. G. Cao and C. J. Brinker. Hackensack, NJ: World Scientific.

Seydack, M. 2005. *Biosens. Bioelectron.* 20: 2454.

Shalaev, V. M., Poliakov, E. Y., and Markel, V. A. 1996. *Phys. Rev. B* 53: 2437.

Spanhel, L., Haase, M., Weller, H., and Henglein, A. 1987. *J. Am. Chem. Soc.* 109: 5649–5655.

Sutherland, R. L. 2003. *Handbook of Nonlinear Optics*. New York: Marcel Dekker.

Wang, F., Dukovic, G., Brus, L. E., and Heinz, T. F. 2005. *Science* 308: 838–841.

Yoffe, A. D. 1993. *Adv. Phys.* 42: 173.

Yoffe, A. D. 2001. *Adv. Phys.* 50: 208.

Zhang, C. F., Zhang, F., Zhu, T. et al. 2008. *Opt. Lett.* 33: 2437.

Zhao, H. B. and Mazumdar, S. 2004. *Phys. Rev. Lett.* 93: 157402.

Optical Properties of Metal Clusters and Nanoparticles

Emmanuel Cottancin
Université de Lyon

Michel Broyer
Université de Lyon

Jean Lermé
Université de Lyon

Michel Pellarin
Université de Lyon

24.1 Introduction

The discovery of fullerenes (Kroto et al. 1985) and the experimental evidence of the so-called magic cluster sizes (Knight et al. 1984, de Heer 1993), more than 25 years ago, were at the origin of unceasing investigations on small clusters and nanoparticles and have contributed to the emergence of nanosciences, at the crossing point of several branches such as physics, chemistry, or even biology. In view of their high surface-to-volume ratio, clusters possess properties, different from those of bulk matter, that are very sensitive to their size and shape, rendering them very attractive from both the fundamental and technological points of view. In particular, the metallic species of a few nanometers in diameter disclose electronic properties intermediate between those of molecular systems and those of bulk matter. Whereas the sparse energy levels are quantized in atoms and molecules and continuously distributed in the energy bands of the crystal (Ashcroft and Mermin 1976), they tend to bunch together in metallic clusters and thus pattern the so-called electronic shells. This shell structure was evidenced experimentally (magic sizes) in the early 1980s in several metals (de Heer 1993) and nicely interpreted in the frame of the *jellium* model (Brack 1993). However, it was shown that magic sizes can also be correlated with atomic shell closures (Martin 1996) and even that both electronic and geometric structures may compete with each other, depending on the temperature (Martin et al. 1990).

The optical properties of metallic clusters being directly underlain by their electronic structure, their optical study is, in this respect, of fundamental interest. For the rest, the fascinating colors of glasses doped with metallic powders have been known for ages even if their origin long remained mysterious. The synthesis of such materials is utilized since antiquity in the art of making jewels, ornamental glassware, or stained glass in the Middle Ages. One can quote the famous cup of Licurgus (Barber and Freestone 1990) from the fourth century AC (appearing red in transmission and green in reflection) or the stained-glass windows of the Chartres Cathedral in France. The colorful shade of these materials (such as glass ruby) was discovered empirically and various colors were obtained during the seventeenth century by alchemists, metallurgists, or glassworkers. For instance, Glauber mentions that *purple can be obtained by precipitating gold from its solution in aqua regia with a solution of tin compound* (Hunt 1976). It was however only in the nineteenth century that their optical properties started to become more systematically studied by Faraday who succeeded in producing gold colloids by reducing gold salts (Faraday 1857b) and showed that color effects are intimately correlated to the size and morphology of colloids. In the Bakerian lecture (Faraday 1857a), Faraday spoke about the experimental relations of gold to light in the following terms: "Light has a relation to the matter which it meets with in its course, and is affected by it, being reflected, deflected, transmitted, refracted, absorbed, &c. by particles very minute in their dimensions." In the lineage of Faraday, Zsigmondy worked on colloidal suspensions and set up the ultra-microscope based on the scattering of particles observed in dark field by illuminating the material to be viewed with a light source placed at right angle to the plane of the objective (Zsigmondy 1926). In 1925, Zsigmondy was awarded the Nobel Prize for Chemistry for his work on the heterogeneous nature of colloidal solutions.

On a theoretical point of view, the pioneering experiments by Faraday have been interpreted later by Mie (1908) who solved Maxwell's equations of a metallic sphere in a homogeneous medium submitted to an external electromagnetic field (plane wave). Mie showed that the original properties of metallic nanoparticles are a consequence of the "dielectric confinement" (limited volume of the material) of particles whose sizes are smaller or of the same order as the wavelength of the excitation field. At the same time, Maxwell-Garnett developed an effective field model (Maxwell-Garnett 1904) in order to describe the optical properties of a medium containing *minute metal spheres.* The main feature in the optical extinction spectra of small metallic clusters is the emergence of a giant resonance in the near UV–visible range, called surface plasmon resonance (SPR) that is related to the collective motion of the conduction electrons induced by the applied field. This resonance is clearly noticeable only in the case of simple metal (alkali, trivalent) and noble metal (gold, silver, and copper) clusters. Its spectral position and width depend on the morphology of the particles (size, shape, and internal structure for alloyed systems), but also on their dielectric environment (medium in which the particles are embedded, local neighborhood) (Kreibig and Vollmer 1995). The development of numerous cluster sources (Sattler et al. 1980, Smalley 1983, Milani and de Heer 1990, Siekmann et al. 1991) in the 1980s enabled to probe the intrinsic properties of the clusters of very small size for which quantum size effects were expected. From a fundamental point of view, alkali clusters constitute a perfect model owing to their simple electronic structure and have been widely studied in the gas phase (Pedersen et al. 1991, Blanc et al. 1992, Bréchignac et al. 1992, de Heer 1993). However, they are immediately oxidized in contact with air once deposited on a surface, and thus are not suitable for applications. In spite of their more complex electronic structure, noble metal clusters in solution or embedded in a transparent matrix are more promising for potential applications because they are more robust toward oxidization. By varying the morphology, structure, or environment of these clusters, their optical response may be more or less controlled, making them attractive in several areas (linear and nonlinear optics [Kreibig and Genzel 1985], nano-materials, nano-photonics, plasmonics, biosensors [Raschke et al. 2003]). Conversely, as the optical response is closely linked with the electronic structure, the SPR can also be used as a probe of the structure of metallic clusters, especially bimetallic systems or nanoalloys. Furthermore, the exaltation of the electromagnetic field in the vicinity of the particle can be exploited to increase the coupling of molecules with light, for developing biological markers for instance (McFarland and Duyne 2003).

Until the past few years, most of the experiments in this field were performed on cluster assemblies, and except the investigation of size-selected clusters in the gas phase, the results are blurred by averaging effects due to the unavoidable cluster size and shape dispersions in samples (Kreibig and Vollmer 1995, Cottancin et al. 2006). To overcome this drawback, new methods of spectroscopy have been developed within the past 10 years, in order to study a single nanoparticle (Tamaru et al. 2002, Raschke et al. 2003, Arbouet et al. 2004, Dijk et al. 2006, Billaud et al. 2007). This has opened up a new field of research, allowing for instance to investigate reliably more complex

nano-objects or to study in detail shape effects on the optical response best than ever.

The aim of this chapter is to give some keys to understand the linear optical properties (absorption, scattering, and extinction) of metallic clusters and nanoparticles,* and to present the state of the art of the research in this field. Section 24.2 deals with the theoretical description of the optical response of a metallic particle submitted to an external field. After a brief focus on the optical response of a bulk metal, particular attention is paid to the dipolar approximation that is appropriate for clusters sizes much smaller than the wavelength of excitation. Size, shape, and structural effects that can be expected are then fully discussed. The case of small clusters, for which quantum finite size effects are expected, is concisely sketched, disregarding the case of very small clusters (below 100 atoms per cluster) for which ab initio calculations are necessary (Rubio et al. 1997, Bonacic-Koutecky et al. 2001, Harb et al. 2008). Finally, the broad outlines of the Mie theory required to describe the optical response of larger nanoparticles are given.

Section 24.3 is divided into three subsections. The first one briefly sets out various methods for producing clusters together with spectroscopic techniques for probing their optical properties. The second one gives an overview of the major results obtained on simple and noble metal clusters concerning size, shape, and multipolar effects in clusters and nanoparticles. Illustrations are taken equally from results obtained for cluster assemblies or single nanoparticles. Finally, the case of bimetallic clusters is described to illustrate the possibility of using the SPR as a structural probe.

24.2 Theoretical Description

24.2.1 General Considerations on Bulk Metals

In the bulk phase, the close packing of atoms involves an overlap of their outer atomic orbitals that strongly interact. This leads to a broadening of the discrete levels of the free atoms into bands (a continuum of electronic states), of very high density of states in metals. In this last case, the highest occupied energy band is called the conduction band and is filled with electrons originating from the outermost atomic orbitals. These electrons weakly interact with the ionic cores, and can be considered as quasi-free particles. They are delocalized in the metal and responsible for most of electrical and thermal transport properties. The highest occupied energy level in the conduction band (not completely filled in metals) is the Fermi level ε_F for which the electron velocity is the Fermi velocity v_F. Lower-energy atomic orbitals will give rise to deeper energy bands (valence bands) in the solid. In noble metals for instance, the highest of these bands originates from the nd atomic orbitals ($n = 3, 4, 5$ for Cu, Ag, and Au, respectively). It is narrow owing to its weak hybridization with

---

* The term "cluster" is generally reserved for sizes lower than a few nanometers in diameter, whereas the term "nanoparticle" is used in the range between a few nanometers to a few hundreds of nanometers.

the conduction band (Figure 24.1). When a metal interacts with light, it may absorb a photon of energy $h\nu$ promoting an electron from an occupied state to an unoccupied state, the latter being located in the incompletely filled conduction band. As for the occupied state, it belongs either to the conduction band or to the valence band. In the first case, the corresponding transition is called "intraband" transition and occurs in the infrared (IR)–visible range. The second case corresponds to "interband" transitions that can occur only if the photon energy is larger than the limiting value $\hbar\Omega_{ib}$ corresponding to the energy threshold required for reaching the Fermi level from the top of the valence band (see Figure 24.1). In simple metals like alkali or trivalents, this threshold is out of the optical domain (near UV–visible–near IR). Thus, only optical transitions within the conduction band may occur. For noble metals, it lies in the UV–visible range and interband transitions may happen in the visible-UV range (see Figure 24.1). The energy threshold occurs in the UV range for silver ($\hbar\Omega_{ib} \sim 4\,\mathrm{eV}$, $\lambda_{ib} \sim 310\,\mathrm{nm}$) and in the visible range for gold and copper ($\hbar\Omega_{ib} \sim 1.9\,\mathrm{eV}$, $\lambda_{ib} \sim 650\,\mathrm{nm}$).

The optical response of the metal can be described entirely by its dielectric function $\varepsilon(\omega)$, which reflects its electronic structure. If a metal is submitted to an external electromagnetic (EM) field $\vec{E} = \vec{E}_o \cos(\omega t) = \Re(\vec{E}_o e^{-i\omega t})$, it is polarized such that its polarization (defined as the dipolar moment per volume unit) is $\vec{P} = \varepsilon_o \chi \vec{E}$, where ε_o is the vacuum permittivity and χ the dielectric susceptibility. The displacement vector can then be written as $\vec{D} = \varepsilon_o \vec{E} + \vec{P} = \varepsilon_o \varepsilon(\omega)\vec{E}$, where $\varepsilon(\omega) = 1 + \chi(\omega)$ denotes the relative dielectric function which is characteristic of the metal. A part of the response may be in phase with the exciting field ($\alpha \cos(\omega t)$), whereas the absorption effects induce a response in quadrature phase ($\alpha \sin(\omega t)$). Therefore, the dielectric function can be decomposed into a complex form $\varepsilon = \varepsilon_1 + i\varepsilon_2$. Let us recall that it is directly correlated to the optical index of the medium, $n_{opt} = n + i\kappa = \sqrt{\varepsilon}$, in which n is the refractive index and κ the extinction coefficient (Fox 2001). n_{opt} is real if the medium is non-absorbing.

In the case of simple metals, the Drude model (Drude 1900a,b, Ashcroft and Mermin 1976), developed originally to explain why metals are good conductors of heat and electricity, remains successful to interpret their optical properties (such as the fact that metals are good reflectors for frequencies lower than a threshold frequency, called plasma frequency [Fox 2001]). This model makes the basic assumption that most of the metal properties can be explained in first approximation by those of the conduction electrons if they are considered as independent and quasi-free. In this frame, when an oscillating electric field is applied, the free electrons oscillate and undergo collisions with other particles (electrons, ions, defects) with a characteristic scattering time $\tau = (1/\gamma_o)$, where γ_o is the average collision rate of electrons. Since the electrons are independent, their global response is the sum of all individual responses. By applying the principle of dynamics for an electron of effective mass m_e and charge $-e$, one obtains:

$$m_e \frac{d^2\vec{r}}{dt^2} = -\gamma_o m_e \frac{d\vec{r}}{dt} - e\vec{E}_o e^{-i\omega t}, \qquad (24.1)$$

where $\vec{r}$ is the complex position vector of the electron. One can easily solve this equation and deduce the polarization in the metal:

$$\vec{P} = -\rho e\vec{r} = \frac{\varepsilon_o \omega_P^2}{m_e \omega(\omega + i\gamma_o)} \qquad (24.2)$$

where ρ is the number of electrons per volume unit (electronic density) and

$$\omega_P = \sqrt{\frac{\rho e^2}{m_e \varepsilon_o}} \qquad (24.3)$$

the plasma angular frequency that is introduced to interpret the high reflectivity of metals (Fox 2001) for frequencies lower than ω_P.

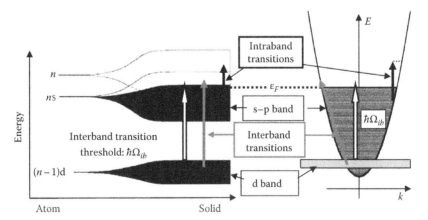

FIGURE 24.1 (Left) Schematic diagram of the transition from discrete electronic levels in atoms to electronic bands in the solid for noble metals. (Right) Schematic density of states for the $(n-1)$d and ns–p bands (left) of a noble metal. The ns–p band corresponding to quasi-free electrons in the metal is almost parabolic. As the atoms are brought closer together, their outer orbitals begin to overlap with each other and their strong interaction involves the formation of energy bands. Optical transitions inside the s–p band take place in the IR–visible range whereas interband transitions between the d-band and the conduction band may occur in the visible–near UV range.

where $\vec{j}(t) = -\rho e \vec{v} = \dfrac{\partial \vec{P}}{\partial t}$ and finally the absorption cross section takes the following form:

$$\sigma_{abs}(\omega) = \frac{9\omega \varepsilon_m^{3/2}}{c} \left(\frac{4}{3} \pi R^3 \right) \frac{\varepsilon_2(\omega)}{(\varepsilon_1(\omega) + 2\varepsilon_m)^2 + \varepsilon_2(\omega)^2}, \quad (24.6)$$

where c denotes the velocity of light.

In general $\varepsilon_2(\omega)$ is relatively small in the case of simple and silver metals in the UV–visible range and $\sigma_{abs}(\omega)$ is strongly enhanced when

$$\varepsilon_1(\omega) + 2\varepsilon_m = 0 \quad (24.7)$$

This condition named resonance condition may be fulfilled for metals for which the real part of the dielectric function $\varepsilon_1(\omega)$ is negative (see Figure 24.2). The correlative resonance phenomenon is known as the SPR or Mie resonance and corresponds to the collective oscillation of the conduction electrons relative to the ionic background. By using Equations 24.4 and 24.5, the spectral position of the SPR can be approximately deduced from the resonance condition Equation 24.7. One obtains

$$\omega_{SPR} \simeq \frac{\omega_p}{\sqrt{\varepsilon_1^{IB}(\omega_{RPS}) + 2\varepsilon_m}}. \quad (24.8)$$

This equation indicates that the frequency resonance depends both on the material through ω_p and ε_1^{IB}, and on the medium in which the particle is embedded through ε_m. In alkali metals, as the IB transitions are negligible, one gets in vacuum ($\varepsilon_m = 1$) the simple relation $\omega_{SPR} = \omega_p/\sqrt{3}$, which generally holds for a perfectly conducting metallic sphere.

It can be underlined that, in the frame of the quasi-static approximation, the extinction and scattering cross sections can be deduced directly from the Mie theory in the small-size range (Bohren and Huffman 1983). They take the following form at the lowest order:

$$\sigma_{ext}(\omega) = \frac{9\omega \varepsilon_m^{3/2}}{c} \left(\frac{4}{3} \pi R^3 \right) \frac{\varepsilon_2(\omega)}{(\varepsilon_1(\omega) + 2\varepsilon_m)^2 + \varepsilon_2(\omega)^2} \quad (24.9)$$

$$\sigma_{sca}(\omega) = \frac{3\omega^4 \varepsilon_m^2}{2\pi c^4} \left(\frac{4}{3} \pi R^3 \right)^2 \frac{(\varepsilon_1(\omega) - \varepsilon_m)^2 + \varepsilon_2(\omega)^2}{(\varepsilon_1(\omega) + 2\varepsilon_m)^2 + \varepsilon_2(\omega)^2} \quad (24.10)$$

σ_{sca} varies with λ as $1/\lambda^4$ (Rayleigh-scattering) ($\lambda = (2\pi c/\omega)$ being the wavelength in vacuum) and exhibits the same resonance condition as the absorption cross section. For a given nanoparticle of diameter R, $(\sigma_{sca}/\sigma_{ext}) \propto (R/\lambda)^3$. In the dipolar approximation ($\phi \ll \lambda$) the extinction is clearly dominated by absorption ($\sigma_{sca} \ll \sigma_{abs} = \sigma_{ext}$ [comparison of Equations 24.6 and 24.9]).

24.2.2.3 Intrinsic Size Effects

24.2.2.3.1 Dielectric Function of a Confined Metal

Except a mere volume factor, it is clear from Equation 24.6 that no size effects are predicted in the dipolar approximation.

However, in this description, the dielectric function of the particle was assumed to be the one of the bulk. Actually, some modifications of the metallic dielectric function are expected because of confinement. Indeed, in a confined system, the collision rate of electrons is modified because of surface scattering that reduces the mean free path of electrons (Kreibig and Fragstein 1969, Kreibig and Genzel 1985, Kreibig and Vollmer 1995). The extra-scattering rate scales at v_F/ℓ, where v_F is the Fermi electron velocity and ℓ the limited mean free path for electron motion (of the order of the particle radius). For a spherical particle of radius R, the collision rate can be written as

$$\gamma(R) = \gamma_o + g \frac{v_F}{R}, \quad (24.11)$$

where g is the surface-scattering coefficient whose value is about unity (Kreibig and Vollmer 1995). The modification of the collision rate with size induces a modification of $\varepsilon^D(\omega, R)$. The dielectric function is then $\varepsilon(\omega, R) = \varepsilon^D(\omega, R) + \varepsilon^{IB}(\omega, \infty) - 1$, where $\varepsilon^{IB}(\omega, \infty)$ refers to as the interband dielectric function.

In the dipolar approximation, this size dependence of the dielectric function induces a size variation of the absorption cross section, as illustrated Figure 24.5 in the case of copper, silver, and gold clusters embedded in alumina. The main features in the spectra are a damping and broadening of the SPR with decreasing size. For gold and copper, a slight red-shift of the resonance is expected within the frame of this model, which is not observed experimentally due to quantum size effects (Cottancin et al. 2006) as will be shown in the following section.

This phenomenological size evolution of $\varepsilon(\omega, R)$ may be recovered from a simple quantum approach of confined electrons in an infinite potential well (Kubo model [Kawabata and Kubo 1966]). Nevertheless this simple model is too crude to give an interpretation of the size effects observed in small alkali and noble metal clusters (except silver) because an infinite potential well cannot lead to the well-known spillout of electrons beyond the classical radius of clusters.

24.2.2.3.2 Quantum Size Effects

In the classical Mie theory, the electronic density $\rho_e(r)$ perfectly matches the positive ionic charge, but quantum calculations (Brack 1993) predict a significant overflow of electrons beyond the classical radius* of the cluster $R_C = r_S N_e^{1/3}$. Because of the finite depth of the potential well, the electronic density does not vanish beyond the cluster radius R_C (exponential decay). This spillout corresponds to an increase of a few angstroms of the effective cluster radius (conduction electron cloud) that becomes significant in small clusters ($\phi < 5\,nm$). In particular, it has a great influence on the optical properties of metallic clusters, especially on the size evolution of the SPR spectral position.

---

* Keeping in mind that the Wigner–Seitz radius r_S is correlated with the volume occupied by a conduction electron ($V_e = (4/3)\pi r_S^3$) in the bulk, one can easily deduce that the classical radius of a cluster containing N_e conduction electrons is: $R_C = r_S N_e^{1/3}$.

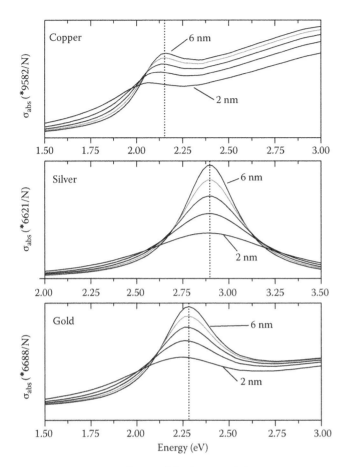

FIGURE 24.5 Size evolution of the theoretical absorption cross section σ_{abs} versus energy within the classical Mie theory in the dipolar approximation, taking into account the reduction of the mean free path in the particle, for copper, silver, and gold clusters embedded in porous alumina ($\varepsilon_m \simeq 2.7$). (From Cottancin, E. et al., *Theor. Chem. Acc.*, 116, 514, 2006.)

On a quantum basis, the metallic clusters of a few nanometers in diameter are usually described by the standard jellium model (Brack 1993) in which the conduction electrons of a spherical cluster in vacuum are quantum mechanically treated, whereas the granular ionic structure is replaced by a spherical, homogeneous, positive-charge distribution: the jellium (of total positive charge Q). The problem to be solved is then the one of N_e electrons interacting together and with the potential created by the positively charged jellium. For $r < R_{N_e}$, the interaction potential energy between an electron and the jellium is equal to

$$V_{jel}(r) = \frac{1}{2} m (\omega_{RPS}^{cl})^2 r^2 - \frac{3}{8\pi\varepsilon_o} \frac{qQ}{R_C} \quad \text{where } \omega_{RPS}^{cl} = \frac{\omega_P}{\sqrt{3}}$$

The first term corresponds to the potential of a harmonic oscillator of angular frequency ω_{RPS}^{cl} (the SPR frequency in vacuum) in the case of simple metals for which $\varepsilon^{IB} = 0$ (see Equation 24.8).

The resolution of the problem may be achieved in the frame of the density functional theory (DFT) by solving self-consistently the Kohn–Sham equations (Hohenberg and Kohn 1964,

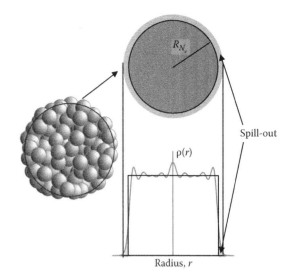

FIGURE 24.6 Schematic view of the jellium model: the ionic background is replaced by a homogeneous positive charge distribution with which the conduction electrons interact. (Right, top) Geometrical aspect of the jellium (dark gray) and the electron spillout (light gray). (Bottom) Corresponding evolution of both ionic (black line) and electronic (gray line) densities versus r. The electronic density overflows beyond the classical cluster radius: this is the so-called spillout.

Kohn and Sham 1965). When convergence is reached, the system can be considered formally as equivalent to a system of N_e-independent electrons moving in an effective central potential $V_{eff}(r)$ of finite depth. The most relevant results obtained in this framework are the electronic shell and supershell structures explaining the famous magic sizes, the beating patterns (Pedersen et al. 1991, Pellarin et al. 1995) and the rise of the spillout (see Figure 24.6) whose profile is almost independent of cluster size (Brack 1993). The interaction of a cluster with light may be investigated in the frame of the random phase approximation (Yannouleas et al. 1989) (RPA) or in the frame of the time-dependent local-density approximation (Eckardt 1985) (TD LDA), both using the jellium model. These formalisms account for the quantum size effects ignored in the classical approach. Figure 24.7 shows for instance results obtained with the TD LDA for the magic sizes of sodium clusters in vacuum. For small clusters ($N < 100$), the absorption spectrum is broad and highly structured. This fragmentation of the plasmon band is due to a phenomenon known as Landau damping and can be explained in a simple manner in terms of a discrete state (the plasmon) coupled to a continuum (Cohen-Tannoudji et al. 1996) (one-electron excitations). When the cluster size increases, the fragmentation persists, but the oscillator strength is concentrated in an ever-narrower spectral range around the plasmon frequency, adopting a quasi-Lorentzian distribution. A red-shift of the SPR with decreasing size is also observed, in qualitative agreement with experimental spectra obtained for alkali clusters as will be shown in the following section.

This red-shift of the resonance obtained through quantum calculations is intimately linked to the increasing spillout

24.2.2.4.2 Influence of Chemical Structure

The optical response of metallic nanoparticles not only depends on their size and shape but also on their internal structure that can be heterogeneous. This is, for example, the case of mixed clusters made of several species (two for simplicity) in various proportions. Consider two chemical structures: a more or less homogeneous bimetallic alloying or a full spatial segregation of both metals (core/shell structure for instance). In the case of alloyed structures, the main difficulty is the knowledge of an effective dielectric function to use as an input for calculating the optical response. If the dielectric function has not been previously determined on corresponding alloys in the bulk phase (by optical measurements for instance), it is usually taken as the weighted average of the dielectric function of each constituent. But this hypothesis is crude because it assumes that the constituents are randomly segregated in nanodomains, each of them being described by a dielectric function taken as the one of the bulk phase. For alloyed structures at an atomic level, this approach may fail because the effective dielectric function of the alloy may be very different from those of each component.

The only structure for which calculations are currently feasible is the segregated core/shell structure that may develop in some particular systems. The problem can be solved exactly in the dipolar approximation for spherical or ellipsoidal shapes, and may be generalized to multi-shell structures. By solving Maxwell's equations and applying the successive boundary conditions on each interface, one can deduce the extinction and scattering cross sections. For a spherical core/shell cluster of core radius R_{core} and total radius R composed of two materials (with core and shell complex dielectric functions ε_c and ε_{sh}, respectively), the cross sections take the following form (Bohren and Huffman 1983):

$$\begin{cases} \sigma_{ext}(\omega) \simeq \sigma_{abs}(\omega) = \dfrac{3\omega\varepsilon_m^{1/2}}{c}\left(\dfrac{4}{3}\pi R^3\right)\Im\left(\dfrac{(\varepsilon_{sh}-\varepsilon_m)(\varepsilon_c+2\varepsilon_{sh})+f_v(\varepsilon_c-\varepsilon_{sh})(\varepsilon_m+2\varepsilon_{sh})}{(\varepsilon_{sh}+2\varepsilon_m)(\varepsilon_c+2\varepsilon_{sh})+2f_v(\varepsilon_{sh}-\varepsilon_m)(\varepsilon_c-\varepsilon_{sh})}\right) \\[4mm] \sigma_{sca}(\omega) = \dfrac{3\omega^4\varepsilon_m^2}{2\pi c^4}\left(\dfrac{4}{3}\pi R^3\right)^2\Im\left|\dfrac{(\varepsilon_{sh}-\varepsilon_m)(\varepsilon_c+2\varepsilon_{sh})+f_v(\varepsilon_c-\varepsilon_{sh})(\varepsilon_m+2\varepsilon_{sh})}{(\varepsilon_{sh}+2\varepsilon_m)(\varepsilon_c+2\varepsilon_{sh})+2f_v(\varepsilon_{sh}-\varepsilon_m)(\varepsilon_c-\varepsilon_{sh})}\right|^2 \end{cases}$$

where $f_v = (R_{core}/R)^3$ is the volume ratio between the core and the cluster.

Such calculations can be performed on various bimetallic clusters or in the case of nanohybrid systems (dielectric–metal or semiconductor–metal). The reduction of the mean free path in the core and in the shell may be also taken into account (Granqvist and Hunderi 1978). As an illustration the optical response of a nanoshell containing a water core and a silver or copper shell embedded in water is displayed in Figure 24.9 for various shell thicknesses. This calculation, made in the frame of the dipolar approximation, is valid only for relatively small particles (Bohren and Huffman 1983). The most important feature in the spectra is a large red-shift of the resonance when the thickness of the metal decreases. In the case of copper, the SPR, damped and broadened in the full metal cluster, is progressively uncoupled from interband transitions (and thus enhanced) when the ratio between the thickness and the total

radius decreases. A very similar evolution is expected for a gold nanoshell (Jain and El-Sayed 2007). One can also notice that for small shell thicknesses the resonance occurs at the same spectral range independently of the material (silver, copper, or gold). Such systems are very interesting because the position of the plasmon resonance can be easily tuned in the visible range by adjusting the shell thickness, which is very convenient for potential applications in medicine (Weissleder 2001).

24.2.2.5 Beyond the Dipolar Approximation: The Mie Theory

As emphasized before, when the size of the particle is very small compared to the wavelength of the exciting field, light scattering is negligible and the absorption cross section can be appropriately calculated by assuming the quasi-static (or dipolar) approximation. On the contrary, for large sizes as compared to the wavelength, the spatial distribution of the field amplitude in the particle volume cannot be disregarded (see the larger particle in Figure 24.4). One has to resort to the Mie theory for calculating the optical properties of particles subject to an incoming plane wave of angular frequency ω. This theory allows to determine the absorption and scattering cross sections, as well as the angular scattering pattern, of homogeneous spherical particles, of any radius, embedded in a homogeneous non-absorbing medium of dielectric function ε_m.

The theory (Bohren and Huffman 1983, Lermé et al. 2008) may be shortly sketched as follows. In a homogeneous medium the electric and magnetic field, $\vec{E}$ and $\vec{H}$ respectively, both satisfy the equations $\nabla \vec{V} = 0$ (zero divergence) and $\nabla^2 \vec{V} + k^2\vec{V} = 0$ (vector Helmholtz wave equation), where $k = [\omega\varepsilon(\omega)\mu_0]^{1/2}$ is the wave vector in the medium and $\vec{V}$ stands for $\vec{E}$ or $\vec{H}$ ($\varepsilon(\omega) = \varepsilon_{metal}(\omega)$ in the particle and $\varepsilon(\omega) = \varepsilon_m$ outside the particle). In the Mie theory the incident, internal and scattered fields are expanded on the two sets of appropriate vector functions $\vec{M}_{nm}$ and $\vec{N}_{nm}$ (referred to as "the vector spherical harmonics") satisfying both previous Maxwell equations in the relevant homogeneous medium. These vector functions can be expressed in spherical coordinates (the origin being the particle center), and written as follows:

$$\vec{M}_{nm}(\vec{r},k) = K_{nm}\nabla \times \left[\vec{r}z_n(kr)P_n^m[\cos(\theta)]\exp(im\varphi)\right]$$

$$\vec{N}_{nm}(\vec{r},k) = \frac{1}{k}\nabla \times \vec{M}_{nm}(\vec{r}),$$

where

the $P_n^m(\theta)$'s are associated Legendre polynomials ($n = 1, 2, 3, \ldots$; $-n \leq m \leq n$)

$z_n(kr)$ are either spherical Bessel functions of the first kind $j_n(kr)$ (involved in the incident and internal electromagnetic field expansions), or spherical Hankel functions of the first kind $h_n^+(kr)$ (involved in the scattered field expansion, for ensuring the outgoing wave radiation condition, $\vec{M}_{nm}$ and $\vec{N}_{nm}$ are then denoted $\vec{M}_{nm}^+$ and $\vec{N}_{nm}^+$)

K_{nm} is a numerical factor

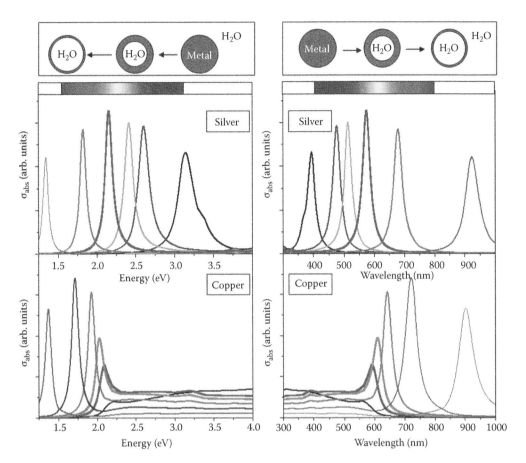

FIGURE 24.9 (See color insert following page 21-4.) Evolution of the absorption cross section of a nanoshell of silver (top) or copper (bottom) in the dipolar approximation versus energy (left) or versus wavelength (right) for various thicknesses of the shell. The core is filled with water and the external medium is also water ($n = 1.33$). The dielectric functions of copper and silver have been extracted from Palik (1985–1991). The correspondence between the energy in eV and the wavelength in nm is E (eV) = 1239.85/λ (nm). The total radius of the cluster is always 15 nm and the thickness takes the following values: $e = R - R_c = 5$; 4; 3; 2; 1 nm corresponding to ratios between the shell thickness and the total cluster radius: (e/R) = 0.33; 0.27; 0.2; 0.13; 0.07. The spectra in black correspond to the fully homogeneous cluster.

The three involved fields (incident, scattering, and internal fields) can then be expressed as follows:

$$
\begin{cases}
\vec{E}_i(\vec{r}) = \sum_{n=1}^{\infty} \sum_{m=-n}^{n} [p_{nm}\vec{M}_{nm} + q_{nm}\vec{N}_{nm}] \\
\vec{E}_{sca}(\vec{r}) = \sum_{n=1}^{\infty} \sum_{m=-n}^{n} [a_{nm}\vec{M}_{nm}^+ + b_{nm}\vec{N}_{nm}^+] \\
\vec{E}_{int}(\vec{r}) = \sum_{n=1}^{\infty} \sum_{m=-n}^{n} [c_{nm}\vec{M}_{nm} + d_{nm}\vec{N}_{nm}]
\end{cases}
$$

Owing to the orthogonality properties of the vector spherical harmonics, the four boundary conditions expressing the tangential continuity of the electric and magnetic fields at the surface of the sphere lead to a one-to-one relationship between the six various expansion coefficients corresponding to a given (n, m)-index set. For a given index set, they result in four linear algebraic equations allowing the expansion coefficients of the internal and scattered fields to be determined as a function of those characterizing the incident field. A mere proportionality

relationship is obtained first between p_{nm}, a_{nm}, and c_{nm}, and second between q_{nm}, b_{nm}, and d_{nm}, the coefficients depending on $\varepsilon(\omega)$, ε_m, R, $j_n(x)$, and $h_n^+(x)$ (and their derivatives) for $x = kR$ or $x = k_m R$. Finally, the integration of the flux of the Poynting vector over a sphere enclosing the particle allows one to obtain the energy dissipated per second by scattering and absorption and thus extinction, and consequently the cross sections. All these observables can be expressed in terms of the expansion coefficients of the three involved fields:

$$
\begin{cases}
\sigma_{sca} = \frac{4\pi}{k^2} \sum_{n,m} \left\{ |a_{nm}|^2 + |b_{nm}|^2 \right\} \frac{1}{E_o^2} \\
\sigma_{abs} = -\frac{4\pi}{k^2} \sum_{n,m} \text{Re} \left\{ a_{nm} p_{nm}^* + b_{nm} q_{nm}^* \right\} \frac{1}{E_o^2} - \sigma_{sca} \\
\sigma_{ext} = \sigma_{abs} + \sigma_{sca}
\end{cases}
$$

The vector spherical harmonics are defined according to Appendix A in reference Lermé et al. (2008).

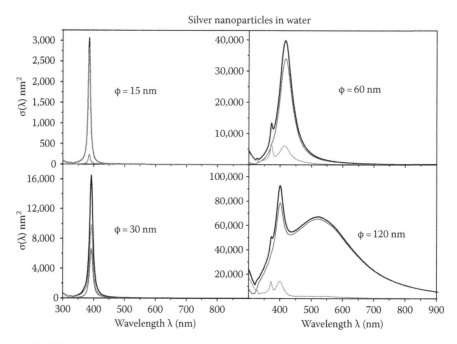

FIGURE 24.10 Extinction (bold black line), absorption (gray line), and scattering (thin black line) cross sections calculated for silver clusters of diameter φ in water ($n = 1.33$) in the frame of the Mie theory. The dielectric function of silver is extracted from Johnson and Christy (1972). For small sizes extinction is dominated by absorption (for φ = 15 nm absorption and extinction are identical), whereas scattering becomes predominant for larger sizes.

Figure 24.10 gives the results of the Mie theory for silver nanoparticles of various sizes embedded in water ($n = 1.33$). A red-shift of the resonance with increasing size, which is correlated to multipolar effects, is clearly evidenced. Such size effects are sometimes referred to as extrinsic size effects (Kreibig and Vollmer 1995). The resonance is very different from the one expected in the quasi-static approximation and scattering is seen to be fastly dominant in the optical response for larger sizes. Notice that the spectra for smaller sizes do not involve surface effects and are thinner than the ones obtained previously since the modification of the collision rate at the cluster surface has not been included here (Section 24.2.2.3). For larger sizes, a shoulder appears at about 375 nm, which corresponds to the quadrupolar mode. This quadrupolar mode only appears in the absorption cross section for a diameter of about 60 nm. Experimentally, it has been observed in silver nanoparticles embedded in silica glass (p. 289 from reference Kreibig and Vollmer (1995)), but it is easier to observe on single nanoparticles (Billaud et al. 2007).

The Mie theory can be generalized to spherical core/shell structures and extended to several interacting spherical nanoparticles excited by an incident plane wave. In this latter case, the field felt by the particle is then the sum of the applied field and the fields scattered by the other particles. The optical response can be auto-coherently resolved. It should be emphasized that in addition to the analytical approaches, numerous numerical methods exist to deal with the particles of any shape or size (Kahnert 2003) or with pairs of nanoparticles in interaction (Romero, 2006). This subject arouses a large interest but is out of the range of this chapter.

24.3 Experimental Results: State of the Art

This section gives a brief outline of the main results obtained in this field of investigation through some illustrations on size, shape, and multipolar effects observed in simple and noble metal clusters. But first, the most common methods for producing clusters and some techniques for performing their optical spectroscopy are described.

24.3.1 Nanoparticle Synthesis

Clusters and nanoparticles can be synthesized either by chemical or physical processes. The former have been known for a long time and allow producing clusters in solution, in matrices or deposited on various substrates. The latter have been developed for almost 30 years.

24.3.1.1 Chemical Methods

Chemical methods can be classified into two main routes: the formation of clusters in glasses or the colloidal synthesis. The former has been worked out from antiquity and consists in including metallic powders and some reducing agent in a melting glass, which is then fastly cooled down in order to obtain a homogeneous glass doped with metal in an oxidized or ionic state (Doremus et al. 1992, Kreibig and Vollmer 1995, Shiomi and Umehara 2000). Various thermal or irradiation treatments are further applied to induce metallic reduction and the growth of metallic nanoparticles embedded in a transparent glass. This technique is nevertheless limited to noble metals with a low concentration (10^{-6} to 10^{-3}), but this is the only one giving volumic nanostructured glasses. The solgel technique (Brinker and Scherer 1990) is an alternative to elaborate such composite materials but only in the form of thin films. It consists in transforming a solution of precursors (solid particles in suspension: the sol) into a polymerized gel containing metallic ions on various substrates (by dip-coating, spin-coating, or pulverization). After drying, the films are thermally treated (annealed under reducing atmosphere, UV, or laser irradiation) allowing the reduction in metallic nanoparticles. This technique enables to prepare noble and transition metal clusters (Nogami et al. 1995, De et al. 1996, Canut et al. 2007) of sizes between a few nanometers to a few tens of nanometers embedded in various matrices with concentration ranging from 10^{-3} to 10^{-1}.

The second route is the colloidal technique (Link and El-Sayed 1999), which has known a large development within the past decades for elaborating self-organized nanostructures. The standard way consists in reducing metallic salts in aqueous solution at intermediate temperatures (300–400 K) thanks to reducing agents. The addition of organic compounds adsorbing on cluster surfaces will stabilize and control the growth reaction. The final cluster diameter is controlled by the nature and the concentration of the reagents (organic compounds and reducing agents): it can vary from a few nanometers to a few hundreds of nanometers. By this way, bimetallic clusters can also be prepared in the form of alloyed or segregated structures (Papavassiliou 1976, Torigoe et al. 1993, Liz-Marzan and Philipse 1995, Link et al. 1999). The optical properties of colloids are investigated either in solution or once they have been deposited on a substrate. However, they can be also transferred into the gas phase, with an electrospray source for instance, and then analyzed through mass spectrometry. They can even be deposited after an accurate selection of their size (Rauschenbach et al. 2006, Böttger et al. 2007). More recently, new methods have emerged allowing a reduction of the size dispersion that is a prerequisite for cluster self-assembling. By those ways clusters are generally passivated with ligands or surface-active molecules having an hydrophilic head and an hydrophobic chain (surfactants). Depending on their shape, related to the ratio between the head size and chain length, these molecules may form direct or reverse dynamic micelles (Pileni 1989) that can be viewed as molds in which nanoparticles can be chemically synthesized. For instance, the reverse micelles form aqueous droplets in nonaqueous solution that serve as nanoreactors for the cluster growth (Lisiecki and Pileni 1995, Lisiecki et al. 1996). By adding anions that adsorb on clusters, the final cluster size may be accurately controlled. Moreover, the shape may be also monitored by taking advantage of a selective adsorption on the various faces of the cluster during the growth. The mean size of the so-produced clusters is of a few nanometers in diameter, and nanorods, nanodisks, or nanowires may be obtained as well. This process reduces considerably the size dispersion and is thus a suitable way to generate 2D or 3D superlattices (Schmitt et al. 1997, Courty et al. 2001) that are model systems for investigating the influence of the cluster interactions on their optical or magnetic properties.

24.3.1.2 Physical Methods

An alternative way to generate clusters is the physical route. Two main approaches are usually considered. The first starts with the production under vacuum of an atomic vapor (thermal heating, laser ablation, etc.) that is further deposited on a suited substrate (de Heer 1993, Perez et al. 1997, 2001). Taking advantage of their diffusion on the surface, atoms can meet or be trapped on defects so as to form nucleation seeds that further grow under the form of clusters by the subsequent capture of other mobile atoms. This gives rise to 2D assemblies of cluster islands with size and shape dispersions that are generally difficult to control accurately. The atoms can also be deposited simultaneously with a dielectric matrix in which they are trapped and isolated. A further thermal treatment may then promote diffusion in the matrix volume so as to induce their grouping and the formation of embedded clusters. In this way, thin nanostructured films can be obtained (3D arrangement), but the cluster size and concentration are hardly controlled independently. It must be stressed that such samples can also be obtained, before annealing, by implanting high-energy metallic ions in a preformed thin film. A second method allows an independent and easier control of the cluster size and volumic concentration. It consists in directly producing clusters in the gas phase and depositing them on a substrate simultaneously with an embedding matrix. The available sources generating cluster beams work generally on the following principle (de Heer 1993): a metallic vapor obtained for instance by heating a metal in an oven is mixed with an inert gas to induce the cluster nucleation and growth. Then, the mixture undergoes an expansion through a nozzle giving birth to a continuous cluster beam. Depending on the temperature and pressure conditions in the source, the size range may strongly vary from a source to another one (seeded supersonic beam sources [Larsen et al. 1974], Sattler sources [Sattler et al. 1980], etc.). Instead of a metallic vapor, a plasma can be produced by lasers (pulsed laser vaporization sources [Smalley 1983, Milani and de Heer 1990]), by an electric discharge (PACIS sources [Siekmann et al. 1991]), or by magnetron sputtering (Hahn and Averback 1990). Cluster growth takes place during the high cooling of the plasma with an inert gas and the so-formed clusters may be charged or neutral. The main advantage is that clusters of refractory metals or alloys may be produced. Moreover, the cluster size distribution can be varied by controlling the inert gas pressure in the source (Richtsmeier et al. 1985).

The co-deposition technique of preformed clusters by the so-called low-energy cluster beam deposition (LECBD) has proven to be one of the most efficient technique for elaborating nanostructured thin films for optical as well as for other investigations. Here we mainly focus on the optical studies of supported nanoparticles, but it must be emphasized that many experiments have been carried out on free cluster beams produced by sources such as those mentioned before. This will be outlined in the following.

24.3.2 Techniques of Optical Spectroscopy

Metallic clusters may be optically investigated either in the gas phase or in nanostructured films. In the former, several methods are available to probe their optical response such as photodissociation, multiphoton ionization (MPI) or photo-depletion spectroscopy (Kreibig and Vollmer 1995). They are of particular importance as they allow to get information about their electronic structure. The principle of such methods is the following: a cluster absorbs a given number of photons of known energy, and its desexcitation leads to the appearance of at least two reaction products (electrons, ions, or neutral fragments) that are detected and counted by mass spectrometry. The nature of these products and the wavelength dependence of their counting rates give insight into the electronic structure of the primary cluster. MPI is rather devoted to very small clusters (of a few atoms), whereas beam depletion spectroscopy can be extended to larger clusters. By using size-selected cluster ion beams and detecting the fragment-size distribution, it is also possible to deduce photo-absorption cross sections (Bréchignac et al. 1992, 1993).

Concerning embedded clusters (samples of nanostructured films or of colloids in solution), the general setup to investigate their optical properties is composed of a light source directed through the sample and detectors for the measurements of the incident, transmitted, reflected, and scattered light that may be analyzed with a grating monochromator (Fox 2001) (see Figure 24.11). The transmittivity T yields the extinction of light correlated to absorption and scattering losses.

For small clusters, the scattering being negligible with respect to absorption, the transmittivity directly reflects absorption (if the thin film reflectivity R can be neglected) through the Beer–Lambert law: $T = (I_t / I_o) = e^{-K_{obs}e}$, where $K_{abs} = (4\pi/\lambda)\kappa(\omega)$, κ being the imaginary part of the effective optical index of the composite sample. This effective optical index may be estimated through effective medium theories, for instance the Maxwell–Garnett theory developed in the frame of the dipolar approximation for small embedded spheres and valid for low-volumic particle concentrations (<10%). Note that the particles can be considered almost isolated from each other if the volumic metal proportions are lower than a few percents. The optical response can then be compared to the expected response of a single cluster in a transparent medium. Of course, this requires an accurate characterization of the matrix in which the clusters are embedded. On the other hand, in the case of weakly absorbing thin films, in particular when the resonance is strongly damped as is observed in small clusters), the reflectivity can no longer be neglected and multilayer effects (Fabry–Pérot fringes) may blur the intrinsic response of the embedded clusters. To get rid of this spurious effect, a simple trick consists in using p-polarized light under Brewster incidence when measuring transmission (Cottancin et al. 2003).

All these measurements are carried out on cluster assemblies and the optical response corresponds to the average response of all the probed clusters. The unavoidable geometry fluctuations in the samples make delicate a precise determination of their specific properties. To overcome this difficulty the clusters may be size selected (Issendorff and Palmer 1999, Alayan et al. 2004) before co-deposition, but shape and environment fluctuations will persist. Another approach developed within the past 10 years aims at investigating a single deposited or embedded metallic nanoparticle. It requires highly sensitive spectroscopic methods since the optical density of the sample is considerably lower than in conventional nanostrutured films. Such investigations may be achieved through near-field or far-field microscopy. Indeed, when a particle is excited by an incident field, the diffracted field gives rise to evanescent waves in the vicinity of the particle and propagating waves far from the particle. In near-field microscopy an optical fiber tip is used as a probe of the field near the particle. By scanning the position of the tip on a sample, a map of the surface covered with deposited nanoparticles may be recorded, but the resolution is still limited by the tip size. In this way, gold nanoparticles of 20 nm in diameter have been detected (Klar et al. 1998). However, the interaction between the tip and the nanoparticles prevents a straightforward deduction of the

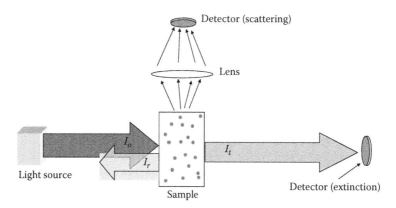

FIGURE 24.11 Schematic experimental setup for conventional scattering and extinction spectroscopy. Light sources and detectors may be combined with monochromators.

particle intrinsic properties. Far-field is more convenient as the obtained response is easier to compare with theory. To "bypass" the Rayleigh criterion, the basic idea is to isolate nanoparticles from each other by depositing them with very low concentration (less than one particle per square micrometer) to be sure that only one particle will be shined by a highly focused light beam. The most widespread technique is the dark-field confocal microscopy allowing the measurement of single-nanoparticle scattering and which can be viewed as an improvement of the ultramicroscope of Zsigmondy. The high sensitivity is obtained by using high numerical aperture refractive objectives with a mask (Dijk et al. 2006) or reflective objectives, both generating a shadow cone in which the scattered light can be collected with a negligible background signal. Nevertheless, as the scattering varies as V^2, such investigations remain restricted to particles larger than 20 nm in diameter (Tamaru et al. 2002, Raschke et al. 2003). On the other hand, interferometric methods have been also developed to detect smaller nanoparticles (Lindfors et al. 2004, Berciaud et al. 2005).

Besides scattering experiments, it is also possible to detect the overall extinction of light by a nanoparticle in a transmission measurement configuration, just as in conventional spectrophotometers. Using tightly focused light beams in a confocal geometry, an increased sensitivity may be reached thanks to a spatial modulation of the sample and a lock-in detection (Arbouet et al. 2004, Muskens et al. 2006, Billaud et al. 2007). The main advantage of this technique is that it enables the absolute measurement of the extinction cross-section if the incident field distribution in the focal plane is accurately characterized. Finally, it should be noticed that scattering or extinction measurements on single nanoparticles may be correlated to their electron microscopy images by using substrates compatible with optical measurements and electron microscopy (Tamaru et al. 2002, Billaud et al. 2008).

24.3.3 Experimental Results

This last section draws up some of the major experimental results relative to size and shape effects on the optical properties of metal clusters, emphasizing the role of the SPR as a probe of the internal structure of the cluster material. Because of the wealth of publications in this field for 20 years, the relevant state of the art will not be exhaustive. A more complete bibliography may be found in the appendix of Kreibig and Vollmer (1995) as well as in the references list cited below.

24.3.3.1 Size and Shape Effects in the Optical Response of Metal Clusters

24.3.3.1.1 Alkali Clusters

Let us first examine the experimental results concerning the SPR in alkali clusters. As mentioned before, alkalis are the simplest metals and their clusters represent ideal model systems that were widely studied during the 1990s. Because of their high sensitivity toward oxidization, most of the corresponding experiments were usually performed on cluster beams in the gas phase through photo-evaporation or depletion spectroscopy (Bréchignac et al. 1992, 1993, Borggreen et al. 1993, Pedersen et al. 1993). In all

cases (lithium, sodium, potassium), the optical spectra are dominated from very small sizes (~8 atoms) by a broad absorption peak (the SPR) qualitatively similar to the classical predictions of the Drude model. Nevertheless, a red-shift of the resonance with decreasing size, correlated to the increasing spillout influence, is patent (see Figure 24.12a) whatever the element. The signature of the spillout effect has been also observed in the measurements of the static polarizability of sodium and lithium clusters (Knight et al. 1985, Bénichou et al. 1999). Indeed we have already underlined that because of the electronic spillout, the sphere containing the conduction electrons (of effective radius R_e) is larger than the geometrical volume enclosing the ionic cores (sphere of radius R_C). The main consequence is a correlated increase of the static cluster polarizability since it is proportional to the effective volume $(4/3)\pi R_e^3$. The calculations in the frame of the time-dependent local-density-approximation formalism (TDLDA) (within the simple jellium model) that have been shown in Figure 24.7 reproduce qualitatively the experimental results, in particular the red-shift of the resonance with decreasing size. Nevertheless, it should be emphasized that these calculations slightly overestimate the absolute value of the SPR in sodium clusters (the SPR occurs generally around 2.5–2.8 eV for sizes smaller than 100 atoms per cluster, whereas the calculations give values around 3 eV). In lithium the disagreement is larger. These discrepancies show that various ingredients, not included in the standard jellium model, have a noticeable influence. For instance the ionic-core polarization, which is disregarded in this framework, may induce a shift of the resonance toward smaller energies. This shift can be viewed classically through Equation 24.8 in which ε_1^{IB} would be chosen different from zero. Moreover, the profile of the effective potential confining the conduction s-electrons and hence also the electron density are known to be softened when electron–ion interaction pseudo-potentials (non-coulombic) are taken into consideration (Rubio et al. 1993, Lermé et al. 1995). The surface softness increases the electron spillout relative to the standard jellium model in which the electron–ion interaction is assumed to be coulombic. At last, the nonlocal character of the electron–ion pseudo-potential results in an increase of the effective conduction electron mass (for instance, in lithium the effective mass is $m_{eff} = 1.4\,m_e$). Both these effects lead to a decrease of the bulk plasma angular frequency ω_p, and hence of ω_{SPR}. On the other hand, the Drude parametrization of the dielectric function also leads to a larger value of the SPR frequency (~3.5 eV) than that deduced from experimental measurements on thin alkali films (~3.1 eV), showing similarly the crudeness of the model for the bulk phase. This shows that the conduction electrons do not really constitute a free electron gas in the bulk. If the details of the cluster electronic structure are obviously of great importance to obtain an accurate quantitative description of the optical properties of alkali clusters, the possible deviation of their shape from the ideal case of a sphere relative to the ideal case of a sphere deserves to be considered with care. The one-peak resonance observed for Na_{21}^+ and Na_{41}^+ (see Figure 24.12b) indicates that these clusters adopt an almost spherical shape since these magic sizes correspond to exact electronic closures in

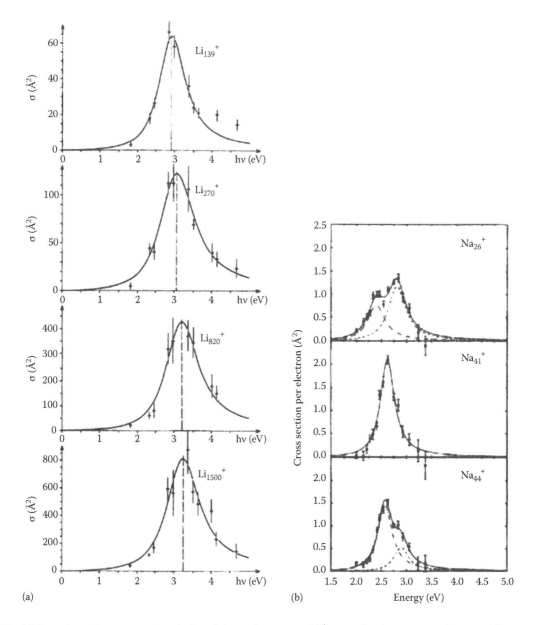

FIGURE 24.12 (a) Photoabsorption cross sections for large lithium clusters ions Li_n^+ versus the photon energy. (From Bréchignac, C. et al., *Phys. Rev. Lett.*, 70, 2036, 1993. With permission.) (b) Photoabsorption cross sections per electron versus the photon energy, for various sizes of sodium clusters. (From Pedersen et al., *Z. Phys. D*, 26, 281, 1993. With permission.) In both figures, the lines are Lorentzian fits of the data.

the spherical jellium model. For non-magic sizes, the resonance is asymmetric, reflecting the existence of at least two resonance modes corresponding to plasmon excitations along the different principal axes of deformation of the cluster (see Figure 24.12b). One can thus say that the profile of resonance is an efficient probe of the cluster shape. Nevertheless, for very small sizes the multi-peak absorption spectra also reflect the electronic-level quantization.

At last, potassium and sodium clusters have been also investigated once deposited under ultrahigh vacuum (UHV) and exposed to various gases (Iline et al. 1999). Since the gases modify their surface tension after being adsorbed, clusters may experience a change in shape that can be detected through variations

in absorption spectra. Nevertheless, the influence of the gas is not so easy to interpret as the size was not well controlled.

24.3.3.1.2 Noble Metal Clusters

In noble metal clusters, as the valence d-band and the conduction s–p band are energetically close to each other, the SPR emerges more or less easily depending on the element and is strongly influenced by the d-electrons. The simplest metal is silver because the threshold for interband transitions in the bulk is quite high (around 4 eV). The SPR emerges for very small sizes, in the range of less than 20 atoms, and it was measured both for ionic clusters (Tiggesbäumker et al. 1993) and for size-selected neutral clusters deposited in a rare gas matrix (Fedrigo et al. 1993). Nevertheless, the core electrons (d electrons mainly) play an important role. The screening they induce is responsible for a global red-shift of the SPR frequency as compared to the frequency expected from the standard jellium model (Pollack et al. 1991, de Heer 1993, Liebsch 1993a,b, Rubio and Serra 1993, Kreibig and Vollmer 1995, Lermé et al. 1998a,b, Palpant et al. 1998). For gold, the threshold for interband transitions is much lower (of the order of 2 eV). This results in a stronger red-shift of the SPR frequency and, in addition, a strong damping and broadening of the resonance band, especially for small particles (Kreibig and Vollmer 1995, Alvarez et al. 1997, Palpant et al. 1998, Gaudry et al. 2001). For copper, the interband threshold is similar to that of gold, of the order of 2 eV. The SPR occurs in the same frequency range as gold, but experiments show that the resonance band is much more damped and even disappears at low sizes for particles of typically 2–3 nm in diameter (few hundred of atoms) depending of the observation conditions (Lisiecki and Pileni 1995, Celep et al. 2004), e.g., on the dielectric constant of the matrix in which the particles are embedded. This specific behavior of copper clusters will be explained below.

Figure 24.13 illustrates the size evolution of the SPR in noble metal clusters in a size range below 10 nm in diameter for which quantum size effects are expected. The upper part of the figure

displays experimental results on the size evolution of the SPR on the samples of noble metal clusters (copper, silver, and gold) embedded in alumina matrix (Lermé et al. 1998b, Palpant et al. 1998, Lermé 2000, Celep et al. 2004, Cottancin et al. 2006) at low concentrations (the matrix and the cluster size distributions have been carefully characterized). As the samples were elaborated with the same protocol and in the same matrix, a comparative study can be achieved. In gold clusters, the SPR band occurs around 2.3–2.4 eV in the same spectral region than the threshold of the interband transitions (2 eV), which continue, as for them, to induce adsorption in the UV range. Moreover, a blue shift and a damping of the resonance are observed when the cluster size decreases. In silver clusters, the SPR band peaks at about 2.9–3.0 eV, well below 4 eV the threshold of the interband transitions. Only a very small blue-shift of the resonance is measured when the cluster size decreases. For copper, the SPR band emerges hardly at about 2.15 eV from the steep increase of the interband transitions. Actually in copper the interband threshold is close to 2 eV (as in gold), but the corresponding absorption increases much more rapidly above 2 eV. The damping and the broadening of the SPR band, with decreasing sizes, are the most important features of the experimental observations, rather than a clear blue-shift as in gold. At very small sizes, the SPR

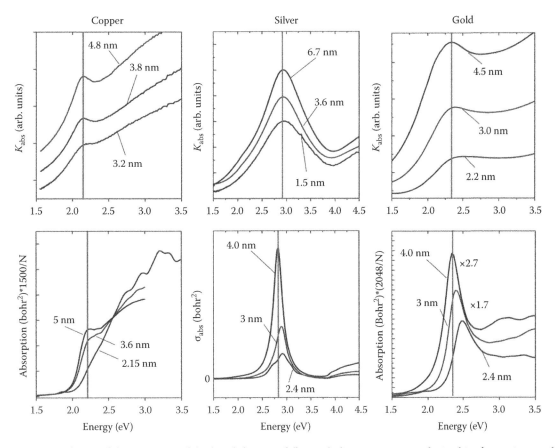

FIGURE 24.13 Size evolution of the experimental (top) and theoretical (bottom) absorption spectra obtained in the semi-quantal model for noble metal clusters of copper, silver and gold embedded in alumina (Cottancin et al. 2006). The volumic concentrations of metallic clusters embedded in the transparent matrix are in all cases rather low (<5%–6%) allowing to consider these systems as isolated clusters randomly dispersed. (From Cottancin, E. et al., *Theor. Chem. Acc.*, 116, 514, 2006.)

band tends to vanish in the steep rising edge of the interband transitions, but in the present experiments, it was not possible to produce clusters as small as 2 nm.

In the present size range (2 nm < φ < 5 nm), the Mie theory is not suited to explain the size evolution of the SPR, even if a size dependence of the scattering rate is assumed (see Equation 24.11). Indeed, the size evolution obtained from such calculations (Figure 24.5) shows a red-shift of the resonance in gold and copper in contradiction with experimental results. For a better interpretation of the experimental results, and as long as large particles are involved, a structureless jellium-type model approach seems quite appropriate and reasonable for investigating mean-size trends over broad-size domains. The calculations are based on the TDLDA within a jellium model including phenomenologically the absorption and screening properties of both the ionic-core background and the surrounding matrix through their dielectric functions $\varepsilon^{IB}(\omega)$ and ε_m (Lermé 2000). A first ingredient added in the model, well established in surface physics and correlated to the spatial localization of d-electrons wavefunctions, is a skin of a vanishing polarizability of thickness d of a few angstroms in which the polarization of the d-core electrons is ineffective ($\varepsilon^{IB} = 1$ in this shell). This skin of ineffective polarizability was firstly introduced by Liebsch for silver surfaces (Liebsch 1993a) and by Kresin for clusters (Kresin 1995). Moreover, a vacuum ring surrounding the particle, of thickness d_m, is introduced in the model in order to mimic the local matrix porosity estimated experimentally (see Figure 24.14).

Figure 24.13 shows the comparison between experiment and this semi-quantal theory. The theoretical calculations well reproduce the experimental trends when the cluster size decreases, namely the strong blue-shift of the SPR for gold, the very small blue-shift for silver, and the gradual disappearance of the SPR in the rising edge of the interband transitions for copper. This blue-shift is mainly due to the skin of ineffective screening by the d-electrons, which results in an all the more important blue-shift that the sizes are small. This blue-shift competes with the red-shift associated with the spillout of the s-electrons, that is a pure quantum effect. In gold, the blue-shift is predominant as compared to the red-shift. In silver, both are close to be compensated, inducing a rough quenching of the size effects. This different behavior of gold and silver clusters is easy to understand because the real part of ε^{IB} (which quantifies the screening) is much smaller in silver than in gold (of the order of 4 and 11, respectively, in the spectral range of the SPR [Cottancin et al. 2006]). The difference between gold and copper may be understood with regard to their respective dielectric functions. The main difference lies in the steeper increase of the imaginary part of ε^{IB} for copper, just above the interband threshold, leading to a stronger damping of the resonance. Actually, the vanishing of the resonance in copper is a signature of a blue-shift of the SPR, resulting in an increasing coupling with IB transitions.

Finally, the semi-quantal calculations appear to be the best compromise between completely ab initio methods limited to relatively small sizes and completely classical calculations that disregard the quantum size effects. These semi-quantal calculations, with a very limited number of semiempirical parameters, are able to predict the optical properties of the three noble metals in a size range from 1.5 to 10 nm in diameter. Above 10 nm, the quantum size effects are negligible and the Mie theory is well adapted.

As for shape effects, experiments have been performed on small silver cluster beams showing similar results as in alkalis (Tiggesbäumker et al. 1992, Tiggesbäumker et al. 1993). However the shape influence on the optical response of noble metal clusters and nanoparticles (essentially gold and silver) has been mostly investigated within the past 10 years on single nanoparticles. Indeed, the tremendous advances in the synthesis of various nano-objects (such as rods, prisms, chains, core/shell structures, pairs of nanoparticles, etc.) opened up a new field of investigations on their optical properties, in a size range where the multipolar effects are of major importance. For instance, silver nanoparticles of various shapes have been studied using dark-field optical microscopy and spectroscopy, in correlation with their electron microscopy images (Mock et al.

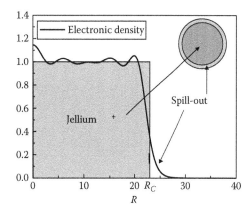
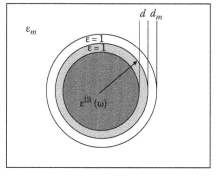

FIGURE 24.14 (Left) *Jellium* model: ionic (gray) and electronic (black line) densities. (Right) Semi-quantal model: cluster of radius $R_C = r_S N_e^{1/3}$ in matrix of dielectric function ε_m (d et d_m denote respectively the thicknesses of the shell of reduced polarizability and the vacuum ring surrounding the particle).

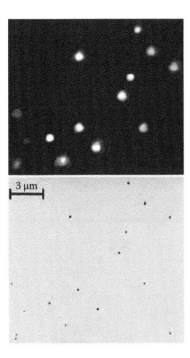

FIGURE 24.15 (See color insert following page 21-4.) (Top) Color image of a typical sample of silver nanoparticles as viewed under the dark-field microscope. The brightness of the particles increases from blue to red due to both the intrinsic optical scattering cross section and the spectral output of the light source (the red particle is overexposed). This image is correlated to its electron microscopy image (Bottom). (Reprinted from Mock, J.J. et al., *J. Chem. Phys.*, 116, 6755, 2002. With permission.)

2002). The spectral position of the resonance is so much shape dependent that spheres appear blue, pentagons green, and triangles mostly red (see Figure 24.15). A significant dispersion of the SPR peak position for triangular particles is observed. This phenomenon is probably due to the high sensitivity of the resonance to the height, the edge length, and the apex shapes (sharp or truncated) of the triangles as has been evidenced through numerical calculations based on the discrete dipole approximation (Sherry et al. 2006).

The main conclusion concerning all the numerous studies for which the state of the art is out of the topic of this chapter is that the SPR may be tailored by varying the shape or the structure of the nano-objects. Consequently, the attachment of specific molecules (such as proteins or antibodies) on the nanoparticle surface can be detected by an induced shift of the SPR (biosensors) (McFarland and Duyne 2003).

24.3.3.2 The SPR as a Probe of the Internal Structure of Bimetallic Clusters

Bimetallic particles are of double interest since they disclose size- and shape-dependent properties and mirror the original features of bulk alloys (a complete bibliography about nanoalloys can be found in Ferrando et al. (2008)). Depending on the thermodynamic properties of both components, mixed particles can form either homogeneous alloys or segregated core/shell structures. An insight into the bulk properties of both constituent

materials (Wigner–Seitz radius (r_s), surface energies) and their corresponding binary phase diagrams brings indications about the possible structure of the composite clusters. In general, one expects the metal component with the lower surface energy to accumulate preferentially at the surface. However, in the case of systems in solution like colloids, one has to be careful with such simple rules, because the kinetics of reduction as well as the surface-adsorbed ions and the surrounding medium also come into play. Thermodynamical equilibrium conditions are not necessary reached and such particle architectures may be only metastable in this respect.

The dielectric function of mixed clusters being governed by its overall electronic structure, the optical response of bimetallic systems will be modified by varying the relative proportion of both constituents (Kelly et al. 2003, Hubenthal et al. 2005). For instance, in the case of gold–silver nanoparticles, by far the most-studied binary system (Morriss and Collins 1964, Papavassiliou 1976, Teo et al. 1987, Mulvaney et al. 1993, Liz-Marzan and Philipse 1995, Link et al. 1999), the SPR falls between those of gold and silver for alloyed systems. On the contrary, in core/shell geometries (that can be essentially obtained by chemical ways), the optical response displays two resonance peaks for sufficiently large particles.

As an illustration, Figure 24.16 shows the evolution of the optical response with varying compositions of aqueous solutions of alloyed gold–silver nanoparticles of diameters ranging from 15 to 50 nm (Russier-Antoine et al. 2008). The spectra exhibit a single SPR whose wavelength is found to move regularly, with increasing gold content, from the silver SPR to the gold one. This observation thus rules out the possibility of a core/shell structure for which two resonances would be expected.

However, the experimental spectra cannot be correctly reproduced by the Mie theory if the dielectric function of the alloy is taken as the volume-weighted average of the pure material dielectric functions. Such an hypothesis is physically relevant if the particle can be considered as built by the stacking of pure nanodomains that individually retain the electronic structure (dielectric function) of their corresponding bulk phase. Its failure indicates that the effective dielectric function is closely related to an alloying at a much lower scale (atomic level). It cannot be simply inferred from the dielectric function of each component but must be either obtained from available experimental data on similar bulk alloys or calculated from solid-state theory assuming an alloying at an atomic scale. In this system, measurements on annealed alloys (Nilsson 1970) have been performed and the corresponding dielectric functions exhibit a regular shift of the interband transition threshold with the gold composition that is not reproduced in the average dielectric function evoked just before. If this IB threshold is taken into account, i.e., if the dielectric function reproduces correctly the one measured experimentally, the optical spectra are in agreement with the theory, confirming thus that the nanoparticles adopt an alloyed structure at an atomic level. Similar works on smaller bimetallic gold–silver clusters produced by laser vaporization and embedded in alumina had lead to the

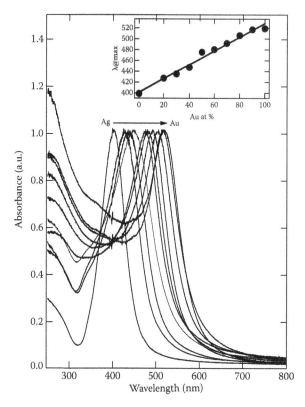

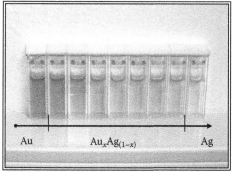

FIGURE 24.16 (See color insert following page 21-4.) (Left) Normalized UV–visible spectra of Au–Ag alloy nanoparticles with varying composition. (Insert) Location of the SPR maximum as a function of the gold content. (Right) Corresponding solutions whose colors vary from the red (pure gold nanoparticles) to the yellow (pure silver nanoparticle). (From Russier-Antoine, I. et al., *Phys. Rev. B*, 78, 35436, 2008. With permission.)

same conclusions about the internal structure of clusters, near from the one of annealed alloys (Gaudry 2001). Such a homogeneous alloyed structure is the likely one (for clusters produced by physical ways) in view of both their relative surface energies and Wigner–Seitz radii and regarding the bulk phase diagram of the Au–Ag system in which homogeneous solutions are thermodynamically possible for any relative composition (Massalski et al. 1986).

In the same way, the internal structure of mixed Ni–Ag and Pt–Ag clusters of a few nanometers in diameter, produced by laser vaporization and embedded in an alumina matrix (Cottancin et al. 2003), has been investigated. Concerning the bulk properties of these materials, the phase diagram of Ni–Ag shows that Ni and Ag are immiscible over the entire composition range (Massalski et al. 1986). One can suppose that the silver atoms will move onto the cluster surface for both systems since the surface energy of silver is the lowest one and that the resulting structure will be a core/shell one. As for the Pt–Ag phase diagram, several alloys are present (Massalski et al. 1986), so the assumption of a core/shell structure has to be taken with caution. Moreover, the atomic Wigner–Seitz radii of Ag and Ni are noticeably different (r_s(Ag) = 3.02 a.u. and r_s(Ni) = 2.6 a.u.) and a pronounced lattice mismatch is expected, whereas they are closer for Pt and Ag (r_s(Pt) = 2.9 a.u.).

The experimental spectra of both systems are displayed in Figure 24.17. In the case of the Ni–Ag system, the optical properties are found to be intermediate between those of pure silver and pure nickel clusters, with an SPR enlarged and blue-shifted compared to pure silver clusters. This feature can be understood within the classical Mie theory assuming a segregation between both metals in a core/shell geometry and including the reduction of the mean free path in the silver shell. Moreover, low energy ion scattering (LEIS) measurements that consist in probing the cluster surface provide the evidence that the surface of the clusters is mainly covered with silver atoms. Such results permit to conclude that the core/shell geometry with silver on the surface is the most likely structure and that the dielectric function of the bimetallic system remains closely connected with the ones of its constituents. As for Pt–Ag clusters, LEIS measurements lead also to the conclusion that the cluster surface is mainly composed of silver; nevertheless, the absorption spectra do not exhibit a marked SPR, in contradiction with the calculated spectra for which a clear resonance is expected as in Ni–Ag clusters. Calculations assuming a dielectric function equal to the volumic-weighted average of the constituent ones predict also a large resonance band. Those results suggest that the Pt–Ag particles do not form a segregated core/shell structure, but rather an alloyed core with a surface enriched with silver atoms. Monte Carlo simulations (Calvo et al. 2008) performed on these systems show the same trends as from small sizes. In particular at zero temperature, silver and platinum seem to be segregated, whereas an increase in the temperature

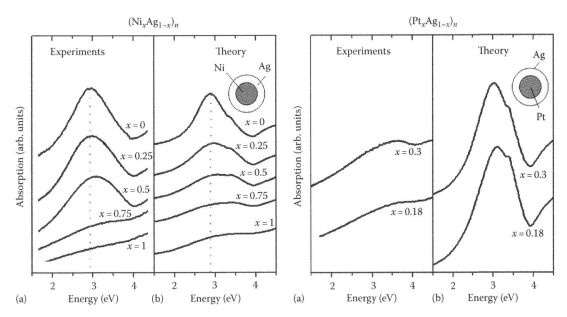

FIGURE 24.17 (Left) (a) Evolution of the optical absorption spectra of Ni–Ag clusters (optical diameter of about 2.6–2.7 nm) with the nickel proportion. (b) Theoretical absorption K_{abs} within the classical core/shell model when the reduction of the mean free path in the silver shell is taken into account. (Right) (a) Evolution of the optical absorption spectra of Pt–Ag clusters (optical diameter of about 4 nm) with the platinum proportion. (b) Theoretical absorption K_{abs} within the classical core/shell model.

induces a better alloying with the surface remaining rich in silver. Finally, one can conclude that the lack of the SPR in experimental spectra shows in this case that the dielectric function of the bimetallic system is not simply correlated to those of its constituents as it is the case for Ni–Ag.

24.4 Conclusion and Outlooks

This chapter intended to shed light on the main ingredients involved in the linear optical properties of metallic clusters and nanoparticles such as size, shape, environment, and the chemical structure of particles. The optical studies of nanoparticles are naturally a wide-ranging field including dynamical processes or nonlinear optics that have not been developed in this chapter.

It has been shown that the colors of samples containing clusters or nanoparticles originate from the strong absorption and scattering (SPR) of the metal nanoparticles when the frequency of the exciting electromagnetic field is coherent with the collective conduction electron motion.

In small particles (ϕ < 20–30 nm), the optical response can be in a first approximation described in the frame of the dipolar approximation. In addition, quantum size effects have to be taken into account for cluster sizes below 5–10 nm in diameter, involving a red-shift of the SPR in alkali clusters because of the increasing effect of the electronic spillout. In noble metals, the polarization of the core electrons (d electrons mainly) and their ineffective influence near the cluster surface result in a blue-shift that competes with the red-shift associated with the spillout effect. The size evolution of the SPR depends then on the element (Cu, Ag, or Au) constituting the clusters.

For larger sizes (ϕ > 20–30 nm), the Mie theory accounts well for the optical response of spherical nanoparticles in a homogeneous medium. When the size increases, higher resonance modes become dominant and the SPR is red-shifted and spectrally broadened.

The synthesis of new particles of various sizes and shapes in the past decades and the development of highly sensitive methods for probing single nano-objects have opened a new field of investigations showing that the SPR may be tailored by tuning the shape or the composition of the metal particles. The understanding of their behavior with the help of computational methods and experiments on single particles gives high hopes for potential applications. For instance, the high sensitivity of the SPR toward the local neighborhood of the particle or the chemical nature of its surface can be used for making biosensors, giving information about surface changes of the particles, as for instance the red-shift observed in silver nanoparticles after several molecules have adsorbed (McFarland and Duyne 2003). Moreover the large local field enhancement that occurs near the particle surface is being used in Surface-Enhanced Raman Spectroscopy (Campion and Kambhampati 1998) or Plasmon Enhanced Fluorescence (Lee et al. 2004). The strong absorption of metal nanoparticles can even be exploited in plasmonic photothermal therapy (Dickerson et al. 2008, El-Sayed et al. 2006).

Acknowledgments

The authors are grateful to F. Calvo for the reading of the chapter and for his fruitful remarks.

References

Alayan, R., L. Arnaud, M. Broyer et al. (2004), Application of a static quadrupole deviator to the deposition of size-selected cluster ions from a laser vaporization source, *Rev. Sci. Instrum.*, **75**, 2461.

Alvarez, M. M., J. T. Khoury, T. G. Schaaff et al. (1997), Optical absorption spectra of nanocrystal gold molecules, *J. Phys. Chem. B*, **101**, 3706.

Arbouet, A., D. Christofilos, N. Del Fatti et al. (2004), Direct measurement of the single-metal-cluster optical absorption, *Phys. Rev. Lett.*, **93**, 127401.

Ashcroft, N. W. and N. D. Mermin (1976), *Solid State Physics*, International Edition, Saunders College, Philadelphia, PA.

Barber, D. J. and I. C. Freestone (1990), An investigation of the origin of the colour of the Licurgus cup by analytical Transmission Electron Microscopy, *Archaeometry*, **32**, 33–45.

Bénichou, E., R. Antoine, D. Rayane et al. (1999), Measurement of static electric dipole polarizabilities of lithium clusters: Consistency with measured dynamic polarizabilities, *Phys. Rev. A*, **59**, R1–R4.

Berciaud, S., L. Cognet and B. Lounis (2005), Photothermal absorption spectroscopy of individual semiconductor nanocrystals, *Nanoletters*, **5**, 2160.

Berthier, S. (1993), *Optique des Milieux Composites*, Polytechnica, Paris, France.

Billaud, P., J.-R. Huntzinger, E. Cottancin et al. (2007), Optical extinction spectroscopy of single silver nanoparticles, *Eur. Phys. J. D*, **43**, 271–274.

Billaud, P., S. Marhaba, E. Cottancin et al. (2008), Correlation between the extinction spectrum of a single metal nanoparticle ant its electron microscopy image, *J. Phys. Chem. C*, **112**, 978–982.

Blanc, J., V. Bonacic-Koutecky, M. Broyer et al. (1992), Evolution of the electronic structure of lithium clusters between four and eight atoms, *J. Chem. Phys.*, **96**, 1793.

Bohren, C. F. and D. P. Huffman (1983), *Absorption and Scattering of Light by Small Particles*, Wiley, New York.

Bonacic-Koutecky, V., P. Fantucci, and J. Koutecky (1991), Quantum chemistry of small clusters of elements of groups Ia, Ib, and IIa: Fundamental concepts, predictions, and interpretation of experiments, *Chem. Rev.*, **91**, 1035–1108.

Bonacic-Koutecky, V., V. Veyret, and R. Miltric (2001), Ab initio study of the absorption spectra of Ag_n (n = 5–8) clusters, *J. Chem. Phys.*, **115**, 10450.

Borggreen, J., P. Chowdhury, N. Kebaïli et al. (1993), Plasma excitations in charged sodium clusters, *Phys. Rev. B*, **48**, 17507.

Böttger, P. H. M., Z. Bi, D. Adolph et al. (2007), Electrospraying of colloidal nanoparticles for seeding of nanostructure growth, *Nanotechnology*, **18**, 105304.

Brack, M. (1993), The physics of simple metal clusters: Self-consistent jellium model and semiclassical approaches, *Rev. Mod. Phys.*, **65**, 677–731.

Bréchignac, C., P. Cahuzac, N. Kebaïli, J. Leygnier, and A. Sarfati (1992), Collective resonance in large free potassium cluster ions, *Phys. Rev. Lett.*, **68**, 3916.

Bréchignac, C., P. Cahuzac, J. Leygnier, and A. Sarfati (1993), Optical response of large lithium clusters: Evolution toward the bulk, *Phys. Rev. Lett.*, **70**, 2036–2039.

Brinker, C. J. and G. W. Scherer (1990), *Sol-Gel Science*, Academic Press, San Diego, CA.

Calvo, F., E. Cottancin, and M. Broyer (2008), Segregation, core alloying, and shape transitions in bimetallic nanoclusters: Monte Carlo simulations, *Phys. Rev. B*, **77**, 121406(R).

Campion, A. and P. Kambhampati (1998), Surface-enhanced Raman scattering, *Chem. Soc. Rev.*, **27**, 241–250.

Canut, B., M. G. Blanchin, V. Teodorescu, and A. Traverse (2007), Structure of Ni/SiO_2 films prepared by sol-gel dip coating, *J. Non Cryst. Solids*, **353**, 2646–2653.

Celep, G., E. Cottancin, J. Lermé et al. (2004), Optical properties of copper clusters embedded in alumina: An experimental and theoretical study of size dependence, *Phys. Rev. B*, **70**, 165409.

Clemenger, K. (1985), Ellipsoidal shell structure in free-electron metal clusters, *Phys. Rev. B*, **32**, 1359–1362.

Cohen-Tannoudji, C., B. Diu, and F. Laloë (1996) *Mécanique Quantique, Tome II*, Hermann, Paris, France.

Cottancin, E., G. Celep, J. Lermé et al. (2006), Optical properties of noble metal clusters as a function of the size: Comparison between experiments and a semi-quantal theory, *Theor. Chem. Acc.*, **116**, 514.

Cottancin, E., M. Gaudry, M. Pellarin, et al. (2003), Optical properties of mixed clusters: Comparative study of Ni/Ag and Pt/Ag clusters, *Eur. Phys. J. D*, **24**, 111–114.

Courty, A., C. Fermon, and M.-P. Pileni (2001), "Supra Crystals" made of nanocrystals, *Adv. Mater.*, **13**, 254.

De, G., M. Gusso, L. Tapfer et al. (1996), Annealing behavior of silver, copper and silver-copper nanoclusters in a silica matrix synthesized by the sol-gel technique, *J. Appl. Phys.*, **80**, 6734.

de Heer, W. A. (1993), The physics of simple metal clusters: Experimental aspects and simple models, *Rev. Mod. Phys.*, **65**, 611–675.

Dickerson, E. B., E. C. Dreaden, X. Huang et al. (2008), Gold nanorod assisted near-infrared plasmonic photothermal therapy (PPTT) of squamous cell carcinoma in mice, *Cancer Lett.*, **269**, 57–66.

Dijk, M. A. V., A. L. Tchebotareva, M. Orrit et al. (2006), Absorption and scattering microscopy of single nanoparticles, *Phys. Chem. Chem. Phys.*, **8**, 3486.

Doremus, R., S. C. Kao, and R. Garcia (1992), Optical absorption of small copper particles and the optical properties of copper, *Appl. Opt.*, **31**, 5773.

Drude, P. (1900a), Zur Elektronentheorie der Metalle, *Ann. Phys.*, **306**, 566–613.

Drude, P. (1900b), Zur Elektronentheorie der Metalle; II. Teil. Galvanomagnetische und thermomagnetische Effecte, *Ann. Phys.*, **308**, 369–402.

Eckardt, W. (1985), Size-dependent photoabsorption and photo-emission of small metal particles, *Phys. Rev. B*, **31**, 6360.

El-Sayed, I. H., X. Hunang, and M. A. El-Sayed (2006), Selective laser photo-thermal therapy of epithelial carcinoma using anti-EGFR antibody conjugated gold nanoparticles, *Cancer Lett.*, **239**, 129–135.

Faraday, M. (1857a), The Bakerian lecture: Experimental relations of gold (and other metals) to light, *Philos. Trans. R. Soc. Lond.*, **147**, 145–181.

Faraday, M. (1857b), Experimental relations of gold (and other metals) to light, *Proc. R. Soc. Lond.*, 145–181.

Fedrigo, S., W. Harbich, and J. Buttet (1993), Collective dipole oscillations in small silver clusters embedded in rare-gas matrices, *Phys. Rev. B*, **47**, 10706.

Ferrando, R., J. Jellinek, and R. L. Johnston (2008), Nanoalloys: From theory to applications of alloys clusters and nanoparticles, *Chem. Rev.*, **108**, 845.

Fox, M. (2001), *Optical Properties of Solids*, Oxford University Press, New York.

Gaudry, M., J. Lermé, E. Cottancin et al. (2001), Optical properties of $(Au_xAg_{1-x})_n$ clusters embedded in alumina: Evolution with size and stoichiometry, *Phys. Rev. B*, **64**, 085407.

Granqvist, C. and O. Hunderi (1978), Optical absorption of ultrafine metal spheres with dielectric cores, *Z. Phys. B: Condens. Matter*, **30**, 47–51.

Hahn, H. and R. S. Averback (1990), The production of nanocrystalline powders by magnetron sputtering, *J. Appl. Phys.*, **67**, 1113.

Harb, M., F. Rabilloud, D. Simon et al. (2008), Optical absorption of small silver clusters: Ag_n, (n = 4–22), *J. Chem. Phys.*, **129**, 194108.

Hohenberg, P. and W. Kohn (1964), Inhomogeneous electron gas, *Phys. Rev.*, **136**, B864.

Hubenthal, F., T. Ziegler, C. Hendrich, M. Alschinger, and F. Träger (2005), Tuning the surface plasmon resonance by preparation of gold-core/silver-shell and alloy nanoparticles, *Eur. Phys. J. D*, **34**, 165–168.

Hunt, L. B. (1976), The true story of purple of cassius. The birth of gold-based glass and enamel colours, *Gold Bull.*, **9**, 134–139.

Iline, A., M. Simon, F. Stietz, and F. Träger (1999), Adsorption of molecules on the surface of small metal particles studied by optical spectroscopy, *Surf. Sci.*, **436**, 51–62.

Issendorff, B. V. and R. E. Palmer (1999), A new high transmission infinite range mass selector for cluster and nanoparticle beams, *Rev. Sci. Instrum.*, **70**, 4497.

Jain, P. K. and M. A. El-Sayed (2007), Universal scaling of plasmon coupling in metal nanostructures: Extension from particle pairs to nanoshells, *Nano Lett.*, **7**, 2854–2858.

Johnson, P. B. and R. W. Christy (1972), Optical constants of the noble metals, *Phys. Rev. B*, **6**, 4370–4379.

Kahnert, N. M. (2003), Numerical methods in electromagnetic scattering theory, *J. Quant. Spectrosc. Radiative Trans.*, **79–80**, 775–824.

Kawabata, A. and R. Kubo (1966), Electronic properties of fine metallic particles, *J. Phys. Soc. Jpn.*, **21**, 1767.

Kelly, K. L., E. Coronado, L. L. Zhao, and G. C. Schatz (2003), The optical properties of metal nanoparticles: The influence of size, shape and dielectric environment, *J. Phys. Chem. B*, **107**, 668–677.

Klar, T., M. Perner, S. Grosse et al. (1998), Surface plasmon resonance in single metallic nanoparticles, *Phys. Rev. Lett.*, **80**, 4249.

Knight, W. D., K. Clemenger, W. A. de Heer et al. (1984), Electronic shell structure and abundances of sodium clusters, *Phys. Rev. Lett.*, **52**, 2141 LP–2143.

Knight, W. D., K. Clemenger, W. A. de Heer, and W. A. Saunders (1985), Polarizability of alkali clusters, *Phys. Rev. B*, **31**, 2539–2540.

Kohn, W. and L. J. Sham (1965), Self-consistent equations including exchange and correlation effects, *Phys. Rev.*, **140**, A1133.

Kreibig, U. and C. V. Fragstein (1969), The limitation of electron mean free path in small silver particles, *Z. Phys.*, **224**, 307.

Kreibig, U. and L. Genzel (1985), Optical absorption of small metallic particles, *Surf. Sci.*, **156**, 678–700.

Kreibig, U. and M. Vollmer (1995) *Optical Properties of Metal Clusters*, Springer, Berlin, Germany.

Kresin, V. V. (1995), Collective resonances in silver clusters: Role of d electrons and the polarization-free surface layer, *Phys. Rev. B*, **51**, 1844.

Kroto, H. W., J. R. Heath, S. C. O'Brien, R. F. Curl, and R. E. Smalley (1985), C-60—Buckminsterfullerene, *Nature*, **318**, 162.

Larsen, R. A., S. K. Neoh, and D. R. Herschbach (1974), Seeded supersonic alkali atom beams, *Rev. Sci. Instrum.*, **45**, 1511.

Lee, J., A. O. Govorov, J. Dulka, and N. A. Kotov (2004), Bioconjugates of CdTe nanowires and Au nanoparticles: Plasmon-exciton interactions, luminescence enhancement, and collective effects, *Nano Lett.*, **4**, 2323.

Lermé, J. (2000), Introduction of quantum finite-size effects in the Mie's theory for a multilayered metal sphere in the dipolar approximation: Application to free and matrix-embedded noble metal clusters, *Eur. Phys. J. D*, **10**, 265.

Lermé, J., G. Bachelier, P. Billaud et al. (2008), Optical response of a single spherical particle in a tightly focused light beam: Application to the spatial modulation spectroscopy technique, *J. Opt. Soc. Am. A*, **25**, 493.

Lermé, J., B. Palpant, B. Prével et al. (1998a), Optical properties of gold metal clusters: A time dependent local density approximation investigation, *Eur. Phys. J. D*, **4**, 95–108.

Lermé, J., B. Palpant, B. Prével et al. (1998b), Quenching of the size effects in free and matrix-embedded silver clusters, *Phys. Rev. Lett.*, **80**, 5105.

Lermé, J., M. Pellarin, J. L. Vialle, and M. Broyer (1995), Effects of nonlocal ion pseudopotential on the electronic shell structure of metal clusters, *Phys. Rev. B*, **52**, 2868–2877.

Liebsch, A. (1993a), Surface plasmon dispersion and size dependence of Mie resonance: Silver versus simple metals, *Phys. Rev. B*, **48**, 11317.

Liebsch, A. (1993b), Surface plasmon dispersion of Ag, *Phys. Rev. Lett.*, **71**, 145.

Lindfors, K., T. Kalkbrenner, P. Stoller, and V. Sandoghdar (2004), Detection and spectroscopy of gold nanoparticles using supercontinuum white light confocal microscopy, *Phys. Rev. Lett.*, **93**, 037401.

Link, S. and M. A. El-Sayed (1999), Size and temperature dependance of the plasmon absorption of colloidal gold nanoparticles, *J. Phys. Chem. B*, **103**, 4212–4217.

Link, S., Z. L. Wang, and M. A. El-Sayed (1999), Alloy formation of gold-silver nanoparticles and the dependence on their absorption, *J. Phys. Chem. B*, **103**, 3529–3533.

Lisiecki, I. and M. P. Pileni (1995), Copper metallic particles synthesized "in situ" in reverse micelles: Influence of various parameters on the size of the particles, *J. Phys. Chem.*, **99**, 5077.

Lisiecki, I., F. Billoudet, and M. P. Pileni (1996), Control of the shape and the size of copper metallic clusters, *J. Phys. Chem.*, **100**, 4160–4166.

Liz-Marzan, L. M. and A. P. Philipse (1995), Stable hydrosols of bimetallic nanoparticles immobilized on imogolite fibers, *J. Phys. Chem.*, **99**, 15120–15128.

Martin, T. P. (1996), Shells of atoms, *Phys. Rep.*, **273**, 199–241.

Martin, T. P., T. Bergmann, H. Göhlich, and T. Lange (1990), Observation of electronic shells and shells of atoms in large Na clusters, *Chem. Phys. Lett.*, **172**, 209.

Massalski, T. B., J. L. Murray, L. H. Bernett, and H. Baker (1986), *Binary Alloy Phase Diagrams* (Vol. 1), American Society for Metals, Metals Park, OH.

Maxwell-Garnett, J. C. (1904), Colours in metal glasses and in metallic films, *Philos. Trans. R. Soc. Lond.*, **203**, 385.

McFarland, A. D. and R. P. V. Duyne (2003), Single silver nanoparticles as real-time optical sensors with zeptomole sensitivity, *Nano Lett.*, **3**, 1057–1062.

Mie, G. (1908), Beiträge zur Optik trüber Medien, speziell kolloidaler Metallösungen, *Ann. Phys.*, **25**, 377.

Milani, P. and W. A. de Heer (1990), Improved pulsed laser vaporization source for production of intense beams of neutral and ionized clusters, *Rev. Sci. Instrum*, **61**, 1835.

Mock, J. J., M. Barbic, D. R. Smith, D. A. Schultz, and S. Schultz (2002), Shape effects in plasmon resonance of individual colloidal silver nanoparticles, *J. Chem. Phys.*, **116**, 6755.

Morriss, R. H. and L. F. Collins (1964), Optical properties of multilayer colloids, *J. Chem. Phys.*, **41**, 3357–3363.

Mulvaney, P., M. Giersig, and A. Henglein (1993), Electrochemistry of multilayer colloids: Preparation and absorption spectrum of gold-coated silver particles, *J. Chem. Phys.*, **97**, 7061–7064.

Muskens, O. L., N. Del Fatti, F. Vallée et al. (2006), Single metal nanoparticle absorption spectroscopy and optical characterization, *Appl. Phys. Lett.*, **88**, 063109.

Nilsson, P. O. (1970), Electronic structure of disordered alloys: Optical and photoemission measurements on Ag-Au and Cu-Au alloys, *Phys. Kondens. Mater.*, **11**, 1–18.

Nogami, M., Y. Abe, and A. Nakamura (1995), Cu microcrystals in sol-gel derived glass, *J. Mater. Res.*, **10**, 2648.

Palik, E. D. (1985–1991), *Handbook of Optical Constants of Solids*, Academic Press, New York.

Palpant, B., B. Prével, J. Lermé et al. (1998), Optical properties of gold clusters in the size range 2–4 nm, *Phys. Rev. B.*, **57**, 1963.

Papavassiliou, G. C. (1976), Surface plasmons in small Au-Ag alloy particles, *J. Phys. F*, **6**, L103–L105.

Pedersen, J., S. Bjørnholm, J. Borggreen et al. (1991), Observation of quantum supershells in clusters of sodium atoms, *Nature*, **353**, 733–735.

Pedersen, J., J. Borggreen, P. Chowdhury et al. (1993), Plasmon profiles and shapes of sodium cluster ions, *Z. Phys. D*, **26**, 281–283.

Pellarin, M., E. Cottancin, B. Baguenard et al. (1995), Observation of two successive quantum supershells in a 15000-electron fermionic system, *Phys. Rev. B*, **52**, 16807.

Perez, A., P. Mélinon, V. Dupuis et al. (1997), Cluster assembled materials: A novel class of nanostructured solids with original structures and properties, *J. Phys. D*, **30**, 709.

Perez, A., P. Mélinon, V. Dupuis et al. (2001), Nanostructured materials from clusters: synthesis and properties, *Mater. Trans.*, **42**, 1460.

Perez, A., P. Mélinon, J. Lermé, and P.-F. Brevet (2007), Nanoscience (Chapter 7), *Clusters and Colloids*, Springer Verlag, Berlin/Heidelberg, Germany.

Pileni, M. P. (1989), *Structure and Reactivity in Reverse Micelles*, Elsevier Science, Amsterdam, the Netherlands.

Pollack, S. P., C. R. C. Wang, and M. M. Kappes (1991), On the optical response of Na$_{20}$ and its relation to computational prediction, *J. Chem. Phys.*, **94**, 2496.

Raschke, G., S. Kowarik, T. Franzl et al. (2003), Biomolecular recognition based on single gold nanoparticle light scattering, *Nanoletters*, **3**, 935.

Rauschenbach, S., F. L. Stadler, E. Lunedei et al. (2006), Electrospray ion beam deposition of Clusters and Biomolecules, *Small*, **2**, 540–547.

Richtsmeier, S. C., E. K. Parks, K. Liu, L. G. Pobo, and S. J. Riley (1985), Gas phase reactions of iron clusters with hydrogen. I. Kinetics, *J. Chem. Phys.*, **82**, 3659.

Romero, I., J. Aizpurua, G. W. Bryant, and J. G. de Abajo (2006), Plasmons in nearly touching metallic nanoparticles: Singular response in the limit of touching dimers, *Opt. Express*, **14**, 9988.

Rubio, A. and L. Serra (1993), Dielectric screening effects on the photoabsorption cross section of embedded metallic clusters, *Phys. Rev. B*, **48**, 18222.

Rubio, A., J. A. Alonso, X. Blase, and S. G. Louie (1997), Theoretical models for the optical properties of clusters and nanostructures, *Int. J. Mod. Phys. B*, **11**, 2727–2776.

Rubio, A., L. C. Balbás, and J. A. Alonso (1993), Photoabsorption cross sections of sodium clusters: Electronic and geometrical effects, *Z. Phys. D*, **26**, 284–286.

Russier-Antoine, I., G. Bachelier, V. Sablonière et al. (2008), Surface heterogeneity in Au-Ag nanoparticles probed by hyper-Rayleigh scattering, *Phys. Rev. B*, **78**, 35436.

Sattler, K., J. Mühlbach, and E. Recknagel (1980), Generation of metal clusters containing from 2 to 500 atoms, *Phys. Rev. Lett.*, **45**, 821.

Schmitt, J., G. Decher, W. J. Dressick et al. (1997), Metal nanoparticle/polymer superlattice films: Fabrication and control of layer structure, *Adv. Mater.*, **9**, 61–65.

Sherry, L. J., R. Jin, C. A. Mirkin, G. C. Schatz, and R. P. V. Duyne (2006), Localized surface plasmon resonance spectroscopy of single silver triangular nanoprisms, *Nano Lett.*, **6**, 2060–2065.

Shiomi, H. and K. Umehara (2000), Preparation of SnO_2-Glass composite containing Cu particles reduced from copper ions in glass matrix—Effect of glass particle size on microstructure and electrical property, *Mater. Sci. Res. Int.*, **6**, 144–149.

Siekmann, H. R., C. Lüder, J. Faehrmann, H. O. Lutz, and K. H. Meiwes-Broer (1991), The pulsed arc cluster ion source (PACIS), *Z. Phys. D*, **20**, 417.

Smalley, R. E. (1983), Laser studies of metal cluster beams, *Laser Chem.*, **2**, 167.

Tamaru, H., H. Kuwata, H. Miyazaki, and K. Miyano (2002), Resonant light scattering from individual Ag nanoparticle and particle pairs, *Appl. Phys. Lett.*, **80**, 1826–1828.

Teo, B. K., K. Keating, and Y. H. Kao (1987), Observation of plasmon frequency in the optical spectrum of $Au_{18}Ag_{20}$ clusters: The beginning of the collective phenomenon characteristic of the bulk?, *J. Am. Chem. Soc.*, **109**, 3494–3495.

Tiggesbäumker, J., L. Köller, H. O. Lutz, and K. H. Meiwes-Broer (1992), Giant resonances in silver-cluster photofragmentation, *Chem. Phys. Lett.*, **190**, 42.

Tiggesbäumker, J., L. Köller, K. H. Meiwes-Broer, and A. Liebsch (1993), Blue shift of the Mie plasma frequency in Ag clusters and particles, *Phys. Rev. A*, **48**, R1749.

Torigoe, K., Y. Nakajima, and K. Esumi (1993), Preparation and characterization of colloidal silver platinum alloys, *J. Phys. Chem.*, **97**, 8304–8309.

Weissleder, R. (2001), A clearer vision for in vivo imaging, *Nat. Biotechnol.*, **19**, 316–317.

Yannouleas, C., R. A. Broglia, M. Brack, and P. F. Bortignon (1989), Fragmentation of the photoabsorption strength in neutral and charged metal microclusters, *Phys. Rev. Lett.*, **63**, 255–259.

Zsigmondy, R. A. (1926), Properties of colloids, *Nobel Lecture*.

25

Photoluminescence from Silicon Nanostructures

Amir Sa'ar
The Hebrew University
of Jerusalem

25.1 Introduction

Silicon, the second-most abundant element on earth (after oxygen) having superior mechanical and electronic properties, has become the principal material of the semiconductors industry from the dawn of microelectronics, and apparently will remain dominant in the foreseen future. Silicon is a semiconductor whose electrical conductivity can be controlled over a wide range, either dynamically or permanently. Its oxidized state (SiO_2) is one of the best and most stable electrical insulator and its superior chemical and mechanical properties make silicon the ideal material for advanced materials processing. For all these reasons, silicon became essentially the sole player in electronic integrated circuits, being the basic building block of most electronic devices, from transistors and diodes to microprocessors, solar cells, wireless communication devices, and more.[1] Yet, silicon is not a good choice for photonic applications where optically active elements are required due to its indirect energy bandgap characteristics where energy's minima of the conduction and the valence bands do not fall at the same wavevector (i.e., the crystal Bloch momentum normalized to $\hbar$). The situation is schematically illustrated in Figure 25.1 where the energy-band diagram of silicon is shown and compared to that of GaAs, which is a direct bandgap semiconductor.

In a steady state, electrons (open circles) occupy the lowest energy states of the conduction band while holes (filled circles) occupy the upper states of the valence band. The emission of a photon takes place once the electron and the hole recombine radiatively, namely, conduction electrons drop down to empty states of holes in the valence band and, releasing their energy to photons of energy, $\hbar\omega = E_C - E_V \cong E_g$, where E_g is the bandgap energy of the semiconductor and $\hbar\omega$ is the photon energy. Besides energy, momentum should also be conserved during the process. However, as the photon wavelength ($\lambda = 2\pi/k \sim 1\,\mu m$ for silicon) is about three orders of magnitude larger than the de Broglie wavelength of the electrons (which is of the order of the lattice constant of the semiconductors ~5 Å), the photon momentum can be neglected relative to that of the electrons and the holes. In direct bandgap semiconductors like GaAs, both the electrons and the holes have the same momentum at the center of the Brillouin zone (Γ-point in Figure 25.1) and vertical radiative recombination can take place. In silicon, however, the large momentum mismatch between electrons and holes does not permit direct radiative recombination unless another entity, a phonon, for example, is involved in the recombination process. As a result, silicon is a poor emitter of light and cannot be utilized for applications where active light sources are required. In many papers and even textbooks this property of silicon is considered to be a disadvantage. However, one should remember that radiative recombination in direct bandgap semiconductors is a fast process (usually of the order of a few nanoseconds in direct bandgap semiconductors with $E_g \sim 1$–2 eV) that limits the lifetime of the carriers (mainly that of the minority carriers in the semiconductor). In silicon, the slow radiative lifetime (of the order of few milliseconds in pure silicon[1]) allows the minority carriers to diffuse over macroscopically large distances (a few hundreds of micrometers and more), a highly favorable property for electronic applications.

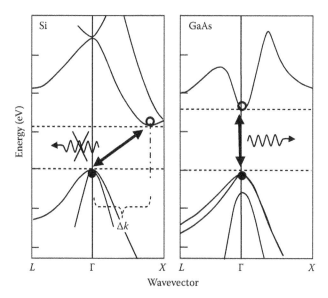

FIGURE 25.1 Energy-band diagrams for silicon and GaAs where the dispersion curves are along the [100] direction (*X*-point) and the [111] direction (*L*-point). Conduction electrons (open circles) and valence holes (filled circles) occupy the band's minima with Δ*k* (momentum/ℏ) being the wavevector's difference for silicon. Photon emission is not allowed for silicon.

The situation is substantially different when dealing with silicon nanostructures. Generally speaking, semiconductor nanostructures[2,3] are artificially made semiconductor objects whose dimensions have been shrunk down to the nanometer length scales. They are classified according to their dimensionality, e.g., two-dimensional (2D) quantum wells, one-dimensional (1D) nanowires (or quantum wires), and zero-dimensional (0D) nanocrystals (or quantum dots[4]), where the classification is according to the number of unconfined dimensions along which carriers (electrons and holes) are still free to move. At the nanometric length scales, quantum confinement (QC) phenomena become important. For example, one should anticipate the momentum conservation law to relax with the decreasing size of the object. In case of silicon nanostructures, for example, a 0D silicon nanocrystal of radius *R*, we may use the Heisenberg uncertainty principle to estimate the wavevector relaxation: Δ*k* ~ 1/*R*. As the relaxation of the wavevector conservation law increases with the decreasing size of the nanostructure, one may expect the radiative recombination rate to considerably increase once the wavevector mismatch, shown in Figure 25.1, becomes comparable to Δ*k*, raising the question: Can we generate active photonic elements[5,6] from silicon nanostructures?

The above question has turned to be a practical subject in 1990 when Canham[7] reported on efficient red light emission from porous silicon (PS) under UV light illumination. PS is a nanometric random network of pores and silicon prepared by electrochemical etching of a silicon substrate.

Figure 25.2 shows a typical spectrum of the photoluminescence (PL) and a photograph of red light coming out from a PS layer. As the efficiency of the PL is comparable to that obtained from direct bandgap semiconductors, Canham in his pioneering

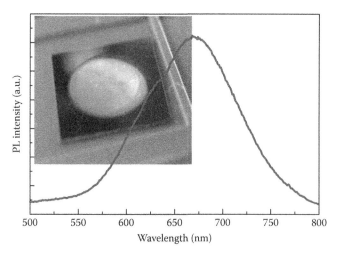

FIGURE 25.2 **(See color insert following page 21-4.)** The PL spectrum from porous silicon with a maximum PL at a wavelength of about 675 nm. The photograph at the inset demonstrates the red color of the emitted PL from a circular layer of porous silicon (the sample has been illuminated with a UV lamp).

work has suggested that quantum size effects might be responsible for efficient PL. Two complementary results seem to support this conclusion. The first is the presence of fairly small crystalline silicon nanostructures, in the form of nanocrystals and undulating nanowires, in the PS medium.[8] Secondly, the considerable blueshift of the maximum PL energy, and accordingly the energy bandgap, from about 1.12 eV (of bulk silicon crystals) to ~1.8 eV in PS, is another manifestation of QC in small semiconductor nanostructures.[2] Following Canham's discovery, a very extensive investigation has been conducted by many groups, aimed at verifying the QC model. Surprisingly, while many investigations provided additional support to the QC model,[9] a considerable number of works have reported results and properties of the PL that cannot be explained by the QC picture.[10] This puzzle has led numerous researchers to propose alternative models and theories that do not rely on QC, the more notable ones suggesting that surface phenomena are responsible for the PL.[11,12] In this picture, radiative transitions take place on the surface of the nanostructures either due to surface bonds, surface defects, imperfections, or even surface molecular species. In particular, the fact that certain properties of the PL depend on "surface chemistry," for example, the specific way that silicon surface bonds are terminated, have led certain researchers to suggest that surface phenomena are responsible for the luminescence from silicon nanostructures.[13]

Let us emphasize that surface phenomena are expected to play a major role in small nanostructures. To follow this, let us estimate the surface-to-volume (STV) ratio for few nanostructures of different dimensionality but having a simple geometrical form such as spheres (0D), cylinders (1D), and slabs (2D), as schematically shown in Figure 25.3.

In order to define a "volume" of the surface, we may estimate the surface's thickness of a given nanostructure, *d*, to be about 1–3 monolayers thick. For silicon (with a lattice constant of ~0.54 nm),

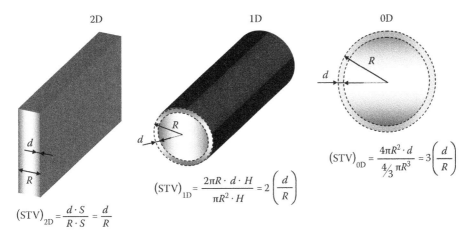

2D   1D   0D

$$(\text{STV})_{\text{2D}} = \frac{d \cdot S}{R \cdot S} = \frac{d}{R}$$

$$(\text{STV})_{\text{1D}} = \frac{2\pi R \cdot d \cdot H}{\pi R^2 \cdot H} = 2\left(\frac{d}{R}\right)$$

$$(\text{STV})_{\text{0D}} = \frac{4\pi R^2 \cdot d}{\frac{4}{3}\pi R^3} = 3\left(\frac{d}{R}\right)$$

FIGURE 25.3 The surface-to-volume (STV) ratio for a 2D slab, 1D cylinder, and 0D sphere where R is the thickness of the slab, radius of the cylinder, and radius of the sphere, respectively. In all three cases, d represents the thickness of the surface.

we may take, $d \sim 1\,\text{nm}$, to be an estimate of the surface's thickness. Hence, for a 0D spherical nanocrystal we find, STV = $3(d/R)$, where R is the radius of the sphere, so that for $R = 6\,\text{nm}$ nanocrystal we get, STV ~ 0.5, meaning that 50% of the silicon atoms belong to the surface while for $R = 10\,\text{nm}$, 33% of the silicon atoms sit on the surface of the nanocrystal. We conclude that surface phenomena become more and more appreciable with the decreasing size of the nanostructures and, on the nanoscales not only quantum size effects can affect the electronic properties of the nanostructures but also surface phenomena should carefully be considered as a possible source of interactions that may affect electronic properties of the nanostructures. This is particularly true for silicon, which is known to be material sensitive to surface termination and usually requires a special treatment for passivation.

Unfortunately, PS is not the ideal medium for the study of quantum size effects and surface phenomena at the nanoscale. PS is a random network of electrochemically etched pores and silicon that is characterized by a broad size and shapes distribution, it is chemically and mechanically unstable and it tends to change its properties with time (aging effects).[11] Hence, to a great extent, the puzzles and discrepancies between the QC and the surface chemistry (SC) models could not be resolved despite of the extensive investigation on PS. Yet, over the last decade, alternative techniques to fabricate silicon nanostructures, with better capabilities to control their size, shape, and the host matrix in which they are embedded, have emerged.[14] With these improved techniques, it is now possible to investigate the evolution of optical and electrical properties of silicon nanostructures versus size and dimensionality. Recent experimental results together with refined theories indicate that none of the above models alone, for example, the QC and the SC models, can explain the entire optical properties of silicon nanostructures.[15–17] Instead, a refined comprehensive model, which takes into account both quantum size effects and surface phenomena has to be developed.[18,19] The purpose of this chapter is to describe these recent developments in the field of silicon nanostructures, particularly those experiments and models that provide a support and verification to the mutual role played by QC and SC in shaping the optical

properties of silicon nanostructures.[19] The interested readers are referred to earlier reviews and the rich literature in the field discussing each individual model (QC and SC) and the supporting/contradicting experimental results.[5–12]

25.2 Synthesis of Silicon Nanostructures

In this section, we briefly review some of the most popular techniques to synthesize light-emitting silicon nanostructures. The simplest and the less expensive method is based on PS, which is fabricated by electrochemical etching of crystalline silicon substrates using hydrofluoric acid (HF)–based solutions. The specific conditions to obtain light-emitting PS (like the PS sample shown in Figure 25.2)[20,21] are reviewed in detail in Refs. [22,23]. Light-emitting PS has a random, nanometric sponge-like structure with a fairly large surface area that can easily be accessed and chemically modified due to its porous characteristics. While this property of PS is of great advantage for certain applications such as chemical[24,25] and biochemical sensing,[26,27] it represents a major limitation for a consistent study of quantum size effects as the size, shape, and even the dimensionality of the nanocrystalline silicon objects are not well defined and can vary between samples and preparation techniques. Furthermore, as the PS matrix may contain various species such as SiO_x, amorphous silicon (a:Si), and other amorphous derivatives of silicon, the exact surface termination of the nanocrystalline silicon objects cannot be determined accurately. Aging effects can be quite significant in PS and may affect both optical properties and transport phenomena. For all these reasons and despite the extensive literature on luminescence from PS, the author of this chapter is in the opinion that in practice, it is impossible to draw consistent conclusions about nanoscale phenomena from PS alone.

The first experimental report on room-temperature PL from silicon nanostructures other than PS is related to silicon nanocrystals (SiNCs) fabricated by ion implantation of silicon into SiO_2 matrix.[28] Recently after, other methods to produce SiNCs embedded in SiO_2 matrices emerged,[14] including chemical vapor deposition (CVD)[29] of sub-stoichiometric oxide (SiO_x), RF magnetron sputtering,[30] reactive evaporation,[31] and plasma-enhanced CVD (PECVD).[32–34] In principle, all these methods require high-temperature annealing of the deposited films to produce phase separation of the excess silicon from the SiO_2 matrix followed by crystallization of the silicon into nano-crystalline particles. The ion implantation method[35–37] is quite popular due to its compatibility with the standard silicon CMOS technology (where it is routinely used to create doped silicon regions). In this technique, silicon ions are extracted from the plasma, accelerated toward the SiO_2 substrate by an electric field and losing their energy after traveling a given depth in the substrate. The thickness and the profile of the implanted layer depend mainly on the implantation dose and the ion energy. In principle, the lower the ion energy the narrower the implanted zone, however, implantation dose also decreases at lower ion energies, setting up a practical limit on the ion energy, which is typically in the range of 1–10 keV.

A similar result of forming oxide layers with excess silicon concentration has been obtained by RF magnetron sputtering. The sputtering process (sometime referred as physical vapor deposition ≡ PVD) involves a bombardment of silicon and SiO_2 solid targets by energetic ions (usually argon ions), removing atoms and molecules from the targets that are deposited on the substrate. As silicon dioxide is an electric insulator, an RF electric field is used to create the plasma between the substrate and the targets. A popular method to produce SiNCs is the co-sputtering technique[15,38] in which, the two targets (usually pure silicon and quartz) are simultaneously exposed to the ions, producing a mixed layer of silicon and oxide with the excess silicon being used to create SiNCs after high-temperature annealing. The thickness and the amount of excess silicon concentration are determined by the RF power as well as by the geometry of the sputtering chamber. Usually, in order to improve uniformity, the substrate is rotated and placed far enough from the targets. In certain cases however, particularly when size-dependent phenomena are to be investigated, it is desirable to put the substrate near the targets without rotation. In this case, the excess silicon concentration along the deposited film continuously varies from one edge of the substrate to the other edge,[15] as schematically illustrated in Figure 25.4.

The amount of excess silicon concentration can be determined by measuring the volume ratio of silicon and SiO_2 obtained from reference depositions of the single components (Si and SiO_2) under the same deposition conditions and is measured in unit of % excess silicon content in a given volume (x). In the experiments reported in Refs. [15,16], x varied from about 10% (at the edge

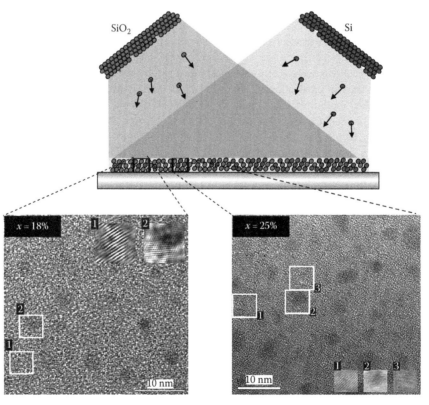

FIGURE 25.4 Schematic view of the co-sputtering method with the two targets of silicon and SiO_2 simultaneously exposed to energetic ions for producing a mixed layer of silicon and SiO_2. The amount of excess silicon decreases from the silicon target edge toward the SiO_2 target edge. Isolated silicon nanocrystals (and PL) have been observed in the range of $10\% \leq x \leq 35\%$. The two cross-section TEM images were taken from two different locations along the substrate with $x = 18\%$ and $x = 25\%$. A few nanocrystals are marked by white squares where the insets are the corresponding images obtained after spatial filtering.

closer to the SiO$_2$ target) up to about 80% (at the edge closer to the Si target). Luminescent SiNCs are obtained after high-temperature annealing from a limited range of 10% ≤ x ≤ 35% where isolated nanocrystals can be observed. At higher excess silicon content, a connected network of silicon clusters is formed, the PL diminishes, and both dark- and photoconductivities from the deposited layers appear, leading Balberg et al.[39] to propose that a formation of percolating connected network of silicon clusters gives rise to mutual exclusion of the PL and the transport.[40] Annealing of the as-deposited films at high temperatures of about 1150°C–1200°C in a controlled environment (Ar or N gas) leads to phase separation and crystallization of the silicon monomers. In Figure 25.4, we present a few examples of cross-section transmission electron microscope (TEM) micrographs obtained after thinning the samples by ion milling to allow passing of the electron beam through the specimen. The images shown in Figure 25.4 are related to two different values of excess silicon content, x ≈ 18% and 25%. Nearly spherical SiNCs can clearly be seen in both images, revealing typical silicon lattice fringes of mainly the {111} silicon crystal planes (and sometime also the {220} crystal planes). The insets to each figure show the silicon crystallographic planes obtained after spatial filtering of the images in the frequency domain for selected SiNCs marked by solid white squares. The diameter of the nanocrystals is not constant but rather varies in the range of 2–10 nm. The histograms shown in Figure 25.5 were obtained after statistical processing of numerous TEM images, identifying SiNCs and measuring their diameters for each value of the excess silicon volume content. The profile of the histograms can be fitted to a Gaussian shape size distribution with a full width at half maximum (FWHM) of approximately 2 nm at the lower content (x ≈ 10%) and 3 nm at the higher silicon volume content (x ≈ 30%), reflecting the inhomogeneous characteristics of the samples, which are typical to almost all fabrication techniques. The average diameter of the nanocrystals increases with increasing silicon content and can be fitted to the following power law (solid line in Figure 25.5b):

$$d\,(\text{nm}) \cong 2(x - x_0)^{1/3} \qquad (25.1)$$

where

d is the average diameter of the SiNCs
$x_0 = (8 \pm 2)\%$

The power exponent of 1/3 stems from the relationship between the diameter and the volume of the nanocrystals, reflecting the spherical characteristics of the SiNCs. Let us emphasize that SiNCs with a diameter smaller than 2 nm have not been observed so far, independent of the fabrication method that has been utilized, and Equation 25.1 is rather limited to the range of 10% ≤ x ≤ 35%.

Another elegant method to produce continuous size variation of SiNCs in a single deposition run is the laser pyrolysis of silane in a gas-flow reactor, which has been reported by Ledoux et al.[41] In this technique, a pulsed CO$_2$ laser produces a molecular beam of silicon nanoparticles having a size distribution of 3–7 nm

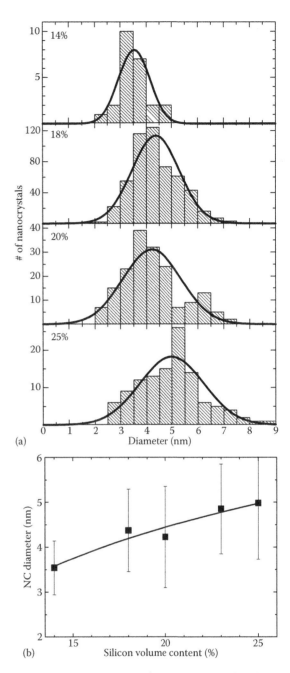

FIGURE 25.5 (a) Histograms showing the size distribution for selected samples having different excess silicon volume contents. The histograms have been obtained after statistical analysis of numerous TEM images, similar to those presented in Figure 25.4, identifying the nanocrystals and measuring their diameters. The solid lines represent the best fit to a Gaussian distribution function. (b) The variation of the average diameter of the nanocrystals versus the excess silicon volume content. The solid line represents the best fit to Equation 25.1.

with the larger nanoparticles moving slower than the smaller ones. Using a rotating mechanical chopper synchronized with the laser pulses, size-selected nanoparticles are transmitted and deposited at different locations across the substrate. The PL image shown in Figure 25.6 reveals a color variation of the PL

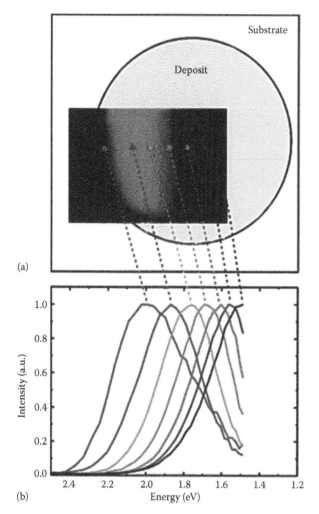

(a)

(b)

FIGURE 25.6 **(See color insert following page 21-4.)** (a) A photograph showing the PL variation along the substrate from silicon nanocrystals deposited by the laser pyrolysis technique. (b) The normalized PL spectra from different positions along the substrate. (Reprinted from Ledoux, G. et al., *Appl. Phys. Lett.*, 80, 4834, 2002. With permission.)

across the substrate that is correlated with the average size of the silicon nanoparticles.

A slightly narrower size distribution but a much better control of the position and the density of SiNCs has been achieved by synthesizing Si/SiO$_2$ superlattices. This method has been introduced by Lockwood et al.[42] who have used silicon molecular beam epitaxy (MBE) system combined with ex situ UV ozone oxidation for growing alternating nanolayers of amorphous silicon (a:Si) and SiO$_2$. Later, this approach has been adopted to create alternating layers of SiNCs and SiO$_2$ using PECVD,[43] low-pressure CVD (LPCVD)[44,45] and RF magnetron sputtering in the serial sputtering mode.[46] A key factor in all these methods is a good control of the layer's thicknesses down to the nanometric length scales and good uniformity, particularly of the interfaces between layers. A similar approach, based on depositing alternating SiO$_x$/SiO$_2$ layers with $1 \leq x < 2$, has also been introduced where phase separation during annealing gives rise to a formation of a:Si layers that later on crystallize to create SiNCs.[47]

Results obtained by serial sputtering of silicon/SiO$_2$ superlattices are shown in Figure 25.7.[48]

In this method, individual layers are deposited one on top of the other at deposition rates of about 2–4 nm/min to achieve a good thickness control. Cross-section HRTEM images, shown in Figure 25.7, reveal the presence of 2D alternating layers of SiNCs separated by amorphous SiO$_2$ layers. The order and the high quality of the interfaces can be appreciated from the low-magnification TEM image at the inset to Figure 25.7(a) while the higher magnification HRTEM micrograph of Figure 25.7(b) reveals 2D layers of SiNCs with average diameter of about 4.3 nm and a statistical size distribution similar to those presented in Figure 25.5. Figure 25.7(a) presents a STEM (scanning TEM) micrograph obtained after introducing high angle annular dark field (HAADF) detector into the microscope for collecting

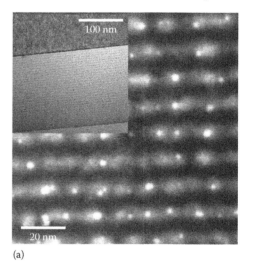

(a)

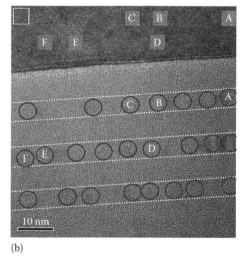

(b)

FIGURE 25.7 High-resolution, cross-section TEM micrographs of serially sputtered SiNCs embedded in SiO$_2$ matrices. (a) STEM image (high z-contrast) revealing nearly spherical SiNCs (bright spots) with average diameter of 3 nm, separated by 10 nm of SiO$_2$ layers. The inset shows a lower magnification image of the superlattice. (b) HRTEM image of selected layers of SiNCs (marked by circles) having average diameter of 4.3 nm.

electrons that undergo high angle inelastic scattering to yield high z-contrast (where z is the atomic number of the element). In this image, the average diameter of the SiNCs is about 3 nm.

25.3 Luminescence Properties of Silicon Nanostructures

25.3.1 Optical Characterizations and Luminescence Bands

In general, several luminescence bands can be excited from silicon nanostructures, but most of them are not directly related to the nanostructures but rather to defects and imperfections in the host matrix. A relatively strong but misleading band is the so-called "F"-band (Fast-band) which has been observed for oxidized PS and for most silicon nanostructures embedded in SiO$_2$ matrices.[11,49,50] The emission spectrum of this band extends over the blue-green (400–550 nm) range of the visible spectrum and is characterized by a fast PL decay time of several nanoseconds. This band has been observed from fully oxidized PS and from silicon nanostructures, such as silicon nanowires,[51,52] with fairly large diameters and therefore, cannot be associated with the nanostructures themselves but with luminescence oxide and interface defects. Other luminescence bands that have been reported but do not relate to quantum size effects are UV bands (~350 nm) and near-IR bands of bulk silicon, unsaturated silicon bonds, and rare-earth heavy ions in the silicon matrix.[11,12]

The size-dependent luminescence band, which shows a clear correlation with quantum size effects in silicon nanostructures, is called the "S"-band (Slow-band) and has been the subject of extensive investigation over the past two decades.[5–13] The emission spectrum from this band can be tuned over the green–red–near IR spectral ranges (500–900 nm) and is characterized by significantly slower PL decay times (compared to the F-band) in the range of few microseconds. In a typical PL experiment, carriers (electrons and holes) are optically excited via absorption of photons of energies exceeding the energy bandgap of the nanostructures, followed by nonradiative relaxation of the photocarriers into the lowest energy levels of the nanostructures and finally, radiative recombination of the carriers that generates PL photons (see Figure 25.8).

Several complementary methods including direct optical absorption,[53] ellipsometry,[54–56] and PLE[11,57] (PL excitation) have been utilized for revealing the absorption spectra from silicon nanostructures. In general, the absorption is substantially blueshifted relative to the PL spectrum with a weak tail below 3 eV and increasing absorption above 3 eV that resembles some of the crystalline and the amorphous silicon characteristics. Therefore, it is not simple to correlate absorption data with the nanostructures themselves as contributions from silicon species, defects, and other absorbing centers in the host matrix cannot be distinguished. The photoexcited carriers undergo a fast relaxation, on timescales of several picoseconds, releasing their energy via nonradiative processes (such as phonons emission) and are trapped in the lower energy levels of the nanostructures. Hence, the

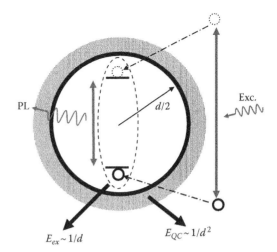

FIGURE 25.8 Schematic of the optical excitation process (solid line) generating electron (dotted circle)–hole (solid circle) pairs, followed by nonradiative relaxation into the lowest energy levels of the nanocrystal (dashed-dotted lines) and finally, radiative recombination of the electron–hole pair (exciton) to generate a photon (gray arrow). The scaling laws for the exciton energy and the bandgap energy (according to the QC model) are shown at the bottom of the figure.

luminescence spectrum, particularly the luminescence from the S-band, can directly be correlated with the silicon nanostructures to provide a direct tool for studying the size effects in these nanostructures. Two kinds of PL experiments can be performed. In continuous wave (cw) PL experiments, the system approaches a steady state where all relaxation and transient phenomena have already disappeared and a steady state PL spectrum is measured. Time-resolved PL experiments allow studying dynamical aspects of the nanocrystals where, the investigated dynamics depends on the temporal resolution of the measurement system, particularly the time resolution of the optical detection system that follows the decay of the PL signal shortly after the excitation pulse is switched off. In silicon nanostructures, the dynamics associated with interband (e.g., conduction to valence) radiative and nonradiative relaxation processes is on timescales of few microseconds and therefore, the excitation source can be either a pulsed laser or a beam of a cw laser modulated by an external light modulator (such as acousto-optical modulator).

25.3.2 CW Photoluminescence Experiments

One of the most convincing evidences for quantum size effects in silicon nanostructures comes from cw PL experiments on series of SiNCs having different diameters. The room temperature, normalized PL spectra shown in Figure 25.6 (SiNCs fabricated by laser pyrolysis of silane[41]) and Figure 25.9 (SiNCs fabricated by co-sputtering[15]) demonstrate this phenomenon where all PL spectra are blueshifted to higher energies for smaller nanocrystals. This behavior of the PL, e.g., a blueshift of the PL peak energy with the decreasing size of the nanocrystals, has also been observed for PS[9–11] and for SiNCs fabricated by other methods.[14] The relatively large bandwidth of all PL spectra (FWHM ≈ 200 meV) has been

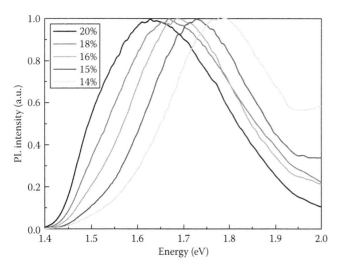

FIGURE 25.9 The normalized PL spectra for co-sputtered SiNCs of different excess silicon volume content.

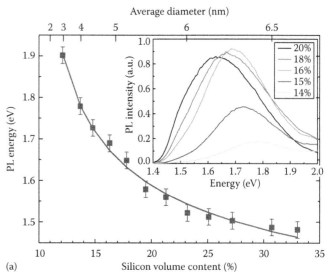

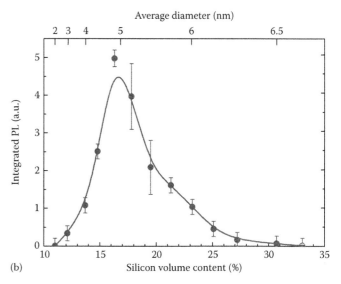

FIGURE 25.10 The variation of (a) the PL peak energy and (b) the integrated PL versus the excess silicon volume content (bottom axis) and the average diameter of the nanocrystals (top axis). The inset shows the non-normalized PL spectra at room temperature.

assigned to the inhomogeneous characteristics of the samples, for example, the broad size distributions shown in Figure 25.5 for co-sputtered SiNCs.

Figure 25.10 (top) presents the variation of the PL peak energy with the increasing excess silicon volume content, x, in the range of $10\% \leq x \leq 35\%$ (bottom axis of Figure 25.10) and accordingly, with the decreasing average diameter of the co-sputtered SiNCs (upper axis of Figure 25.10). The solid line in this figure represents the best fit of the experimental results to an inverse power law as follows:

$$E - E_G = \frac{b}{(x - x_0)^{\delta/3}} \qquad (25.2)$$

where

$\delta/3 \cong 0.45 \pm 0.1$ is the power exponent

$E_G = 1.12\,\text{eV}$ is the energy bandgap of bulk silicon

$x_0 \cong (9 \pm 2)\%$ (very close to the experimental value of x_0 obtained for the variation of the average diameter with x; see Equation 25.1)

Using Equations 25.1 and 25.2, we find a relationship between the PL energy shift (relative to bulk silicon) and the diameter of the SiNCs:

$$\Delta E = E - E_G = \frac{A}{d^\delta} \qquad (25.3)$$

where

$\delta \cong 1.35 \pm 0.3$

ΔE is called the "confinement energy" of the nanocrystals

A very similar inverse power law with essentially the same power exponent has been found for SiNCs fabricated by laser pyrolysis of silane,[41] by PECVD,[34] and by other techniques. However, this power exponent deviates from the exponent predicted from a simple QC model, e.g., a particle confined into a spherical quantum dot having infinite potential barrier. In the latter case, the quantized energy levels of the dot are given by[2] $E_{n,l} = 2\hbar^2 z_{n,l}^2 /(m^\star d^2) \sim 1/d^2$, where $z_{n,l}$ are the lth root of the nth order spherical Bessel function and $m^\star$ is the effective mass of the carriers in the bulk medium. Therefore, in such a simple effective mass model, one would expect the confinement energy to vary as $\Delta E \sim 1/d^2$. Several factors may account for the deviation of δ from this simple model. A major factor, which has been ignored so far, is the Coulomb interaction between electrons and holes to create excitons. Excitons are bound states of electron–hole pairs that are created by the Coulomb attraction between the charged carriers. In a bulk silicon crystal, the Bohr radius of the exciton is given by $a_{ex} = \hbar^2 \varepsilon_{Si}/\mu e^2 \approx 5-7$ nm, where ε_{Si} is the dielectric constant of silicon and μ is the reduced mass of the electron–hole pairs. QC modifies this picture, particularly when the size of the nanocrystals becomes comparable to the

Bohr radius of the exciton. In this case ($d/2 \leq a_{ex}$), which is called the strong confinement regime, the electrostatic Coulomb energy is scaled as $E_{ex} \approx e^2/(\varepsilon_{Si}d/2) \sim 1/d$ as opposed to $\sim 1/d^2$ of the effective mass model; see Figure 25.8. Thus, we can expect the power exponent to vary in the range of $1 \leq \delta \leq 2$ with $\delta \approx 2$ being the weak confinement limit for big nanocrystals. Indeed, recent first-principle calculations (meaning that the many-body problem of having a finite number of silicon atoms in a dot, including the specific form of surface termination bonds, has numerically been solved) such as the LCAO (linear combination of atomic orbitals)[58] and the pseudo-potential[59] methods, have found a power exponent of $\delta \cong 1.4$, in a very good agreement with the experimental results, thus providing a solid support to the QC model.

Yet, cw PL experiments provide additional information about SiNCs that cannot be easily interpreted by the QC model. Figure 25.10b shows the variation of the integrated PL intensity, which is obtained by integrating the area below the non-normalized PL spectra[15] (inset to Figure 25.10a) versus the silicon volume content (and the average diameter of the nanocrystals). In order to obtain the actual PL yield, one should take into account the varying density of SiNCs as lower x means lower density of SiNCs. The corrected PL yield, obtained after dividing the integrated PL by x, is shown in Figure 25.11. In this presentation, the precise number of photons absorbed per each unit area could only be estimated. While this quantity has not been measured directly, in several other experiments[41,60–63] aimed at measuring the absolute value of the PL yield, it has been found that the PL yield for SiNCs having a diameter of 4–5 nm is fairly high in the range of 60%–80%. Similar values have also been found for PS.[64] The maximum yield of 30%–50% shown in Figure 25.11 (for nanocrystals emitting their light in the 650–750 nm, red-orange spectral range) should be considered as a lower limit. However, the general trend of the

PL yield, e.g., sharp increase with the decreasing diameter up to a maximum for 4–5 nm in diameter nanocrystals and a decrease of the yield for smaller nanocrystals, has been confirmed by numerous groups.[41,63] This finding cannot simply be explained by the QC model. According to this model, the smaller the nanocrystals the more pronounced is the relaxation of the momentum conservation rule (as confirmed experimentally[65]) and accordingly, a larger PL yield should be measured.[66] Further implications of the high PL yield and its relation to the rates of radiative and nonradiative processes in SiNCs is discussed in Section 25.3.3.

Among the first experiments indicating the mutual role of QC and SC is the one reported by Wolkin et al.[17] In this experiment, a series of PS samples having different porosities have been kept under controlled environment to avoid oxidation and to ensure that the surface of the silicon nanostructures is passivated by Si–H chemical bonds. The state of passivation has been monitored by FTIR infrared spectroscopy to exclude the presence of Si–O and Si–OH bonds. A second group of similar PS samples were exposed to air for 24 h, allowing surface oxidation and the presence of Si–O bonds. Results of room-temperature PL measurements from both series of samples are presented in Figure 25.12. Despite that the actual average size of the silicon nanostructures cannot accurately be measured for PS, it is well known that smaller silicon features appear for larger porosities of the samples.[67] The PL spectra, shown in Figure 25.12a, show indeed a consistent blueshift with the increasing porosity for hydrogen-terminated PS samples. On the other hand, oxygen-terminated samples (Figure 25.12b) show, at first, a blueshift of the PL spectra up to the yellow (~600 nm) and then, no blueshift and even a redshift for the blue-green emitting PS samples. Clearly, this phenomenon cannot be explained by the QC model alone and the role of SC, particularly the passivation of the PS surface must be taken into account. The model proposed by Wolkin et al.[17] assumes that some of the dangling bonds, on the silicon oxide interface, are passivated by the Si=O double bond, which is known to be more stable than other forms of surface passivation.[68] Based on this assumption, the authors simulated the electronic band structure of SiNCs having Si=O double bonds at the silicon oxide interface. The results, shown in Figure 25.13, suggest that the QC model is valid for hydrogen-terminated SiNCs as radiative recombinations are via free excitons (e.g., excitons of the bulk nanocrystalline silicon core) for all sizes of the nanocrystals. However, for Si=O passivated SiNCs, both electrons and holes can be localized (or trapped) by the Si=O bonds where, in zone I of Figure 25.13 none of the carriers are trapped, in zone II only electrons are trapped, and in zone III both electrons and holes are trapped to form surface-trapped excitons. This model provides a nice and elegant explanation to the experimental results shown in Figure 25.12, particularly for kinetic properties associated with the joint contribution of SC and QC to the shift of the energy bandgap. Yet, as pointed out by these authors, the intensity of the PL (which is proportional to the integrated PL) increases by few orders of magnitude up to the orange emitting samples and then, diminishes for higher porosity samples emitting in the blue-green; quite similar to the PL yield presented in Figure 25.11 for co-sputtered

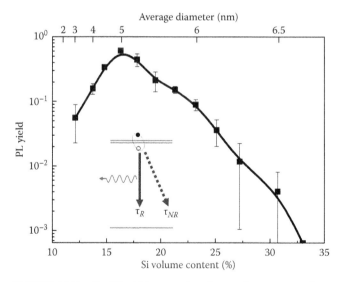

FIGURE 25.11 The PL yield versus the silicon volume content (bottom axis) and the average diameter of the nanocrystals (top axis). The inset illustrates schematically the two relaxation channels of radiative recombination (followed by photon emission) and nonradiative relaxation of the exciton.

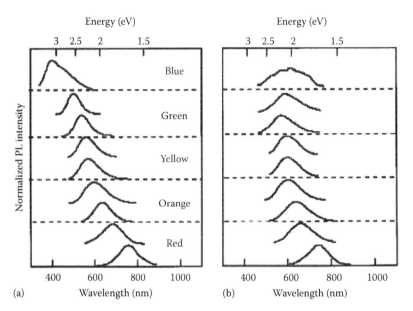

FIGURE 25.12 Room-temperature PL spectra from PS with different porosities (a) kept in Ar atmosphere, (b) after exposure to air. (Reprinted from Wolkin, M.V. et al., *Phys. Rev. Lett.*, 82, 197, 1999. With permission.)

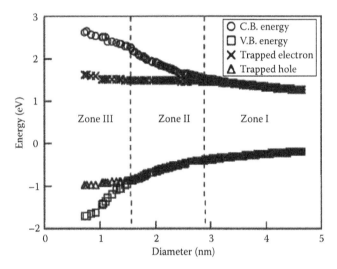

FIGURE 25.13 Electronic states of SiNCs as a function of the cluster size and surface passivation. (Reprinted from Wolkin, M.V. et al., *Phys. Rev. Lett.*, 82, 197, 1999. With permission.)

SiNCs. Hence, dynamical characteristics of the nanocrystals still require sufficient explanations, particularly as the PL decay time from the blue-green PS samples have been reported[17] to be much faster (0.07 μs in the blue) than those of the red samples (~2 μs) and therefore, according to the QC model, should exhibit higher PL yield. Further discussion of this topic will be given in Section 25.4.2 after presenting the vibron model.

25.3.3 Time-Resolved Photoluminescence Experiments

The PL decay process from silicon nanostructures has been intensively investigated over the recent years, both from PS[11] and from SiNCs embedded in SiO$_2$ matrices.[69] In a typical setup,

a train of short laser pulses excites the carriers to their lowest excitonic state from which they radiatively recombine to generate PL photons. The PL decay time is measured shortly after the pump laser is switched off using, for example, a gated photon-counting system that is mostly suitable for SiNCs since the relevant dynamics occurs on timescales of sub-microseconds up to few milliseconds.

A typical room-temperature, time-resolved PL decay curve from co-sputtered SiNCs with x = 18% (average diameter of ~4.5 nm) and at a PL energy of 1.65 eV (PL wavelength of ~750 nm), is shown in Figure 25.14. The temporal behavior of the

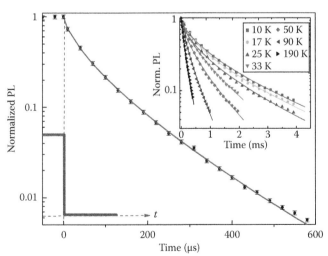

FIGURE 25.14 Room temperature, time-resolved PL decay normalized to the initial PL intensity taken at PL photon energy of 1.65 eV for the x ~ 18% sample (average diameter of 4.5 nm). The solid line represents the best fit of the experimental data (symbols) to the stretched-exponential decay function (Equation 25.4). The inset shows the PL decay curves at various low temperatures.

PL does not follow a simple exponential decay function [e.g., $I/I_0 = \exp(-t/\tau)$; notice the semilogarithmic scale of Figure 25.14], but rather a stretched-exponential decay function of the form:

$$I(t) = I_0 \exp\left[-\left(\frac{t}{\tau}\right)^{\beta} \right] \qquad (25.4)$$

where

    τ is the characteristic PL lifetime

    $0 < \beta \leq 1$ is the dispersion exponent of the PL decay function

This form of the PL decay function has been observed for many classes of silicon nanostructures[37,70] and is frequently assigned to disorder characteristics of systems where dispersive transport take place, particularly migration[71,72] and trapping–detrapping (or release) of photoexcited carriers.[73] This topic is directly related to transport phenomena in PS and SiNCs where dispersive transport has been observed and discussed,[21,74] but will not be reviewed here. The solid line in Figure 25.14 represents the best fit of the experimental data to Equation 25.4, yielding a PL lifetime of about ~40 μs and $\beta \sim 0.85$. As opposed to the dispersion exponent that is approximately independent on the PL energy, the PL decay time decreases with the increasing PL energy, from about 100 μs at 1.4 eV to few μs at 2 eV. This is rather important result as it indicates, again, about the inhomogeneous nature of the PL spectrum, in accordance with the inhomogeneous size distributions of the nanocrystals shown in Figure 25.5. According to the QC model, the smaller the SiNCs (and therefore, the breakdown of the k-conservation rule), the larger the radiative transitions rate and the faster the radiative lifetime. We will follow this picture, assuming that each wavelength in the PL spectrum probes a different size of the nanocrystals, as suggested by the QC model. In addition, let us denote the substantial dependence of the PL decay time (at a given PL energy) on temperature, shown at the inset to Figure 25.14. The lower is the sample temperature the longer is the PL decay time, approaching fairly slow PL lifetimes of about few milliseconds at low temperatures (below 50 K). The variation of the PL lifetime with temperature, for several PL energies, is presented in the Arrhenius plot (semilogarithmic scale versus the inverse temperature) of Figure 25.15.

We can identify two distinct temperature regimes in Figure 25.15: a low-temperature regime (below 50–60 K) where the PL lifetime is essentially independent of temperature and a high-temperature regime (above 60 K) where the PL lifetime gets shorter with the increasing temperature. To follow these characteristics of the PL lifetime let us briefly describe the exchange-splitting model that has originally been proposed by Calcott et al.[75] for PS, and later on adopted for other classes of silicon nanostructures.[69] In typical semiconductors the lowest excitonic state is composed of conduction electrons having a total angular momentum of $J = 1/2$ (the sum of $L = 0$ orbital angular momentum of the conduction band and spin 1/2 of the electrons) and heavy holes states having $J = 3/2$ ($L = 1$ of the valence band and spin 1/2). Hence, the total angular momentum of the exciton can be either 1 or 2. Of course, QC makes this picture much more

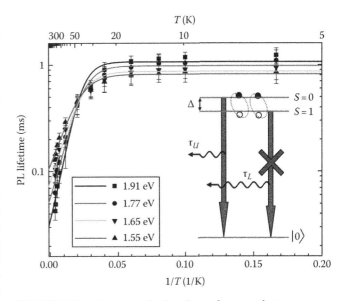

FIGURE 25.15 Arrhenius plot (semilog scale versus the inverse temperature) of the PL lifetime, from the $x \sim 18\%$ sample, for several photon energies. The solid lines represent the best fit of the experimental data to the exchange-splitting model shown schematically at the inset with τ_U, τ_L, and Δ being the upper state (singlet) lifetime, the lower state (triplet) lifetime, and the splitting energy, respectively.

complicated as different bands can be mixed and the spherical symmetry of the ideal dot can be distorted (in fact, the spherical symmetry of the dot is always slightly distorted when constructed from a cubic crystal). Therefore, Calcott et al.[75] treated the exciton as composed of two spin 1/2 particles assuming that only the spin components of the total angular momentum are conserved. In this case, the total spin of the exciton can be either $S = 1$ (triplet) or $S = 0$ (singlet). The Coulomb exchange interaction between electrons and holes lifts the singlet–triplet degeneracy. To follow the role of the exchange interaction, let us point out that, according to the Pauli principle, a singlet state (which is antisymmetric with respect to exchange of the electron–hole spins) has to be symmetric with respect to exchange of the electron–hole orbital states, while the spin-triplet state has to be orbitally antisymmetric with respect to this exchange. Hence, the Coulomb energy associated with the exchange interaction is always larger for the spin-singlet state ($S = 0$) relative to the spin-triplet state ($S = 1$). The situation is schematically illustrated at the inset to Figure 25.15 where Δ is the singlet–triplet exchange splitting energy. In a bulk silicon crystal, the splitting energy is fairly small, of the order of ~150 μeV; however, since the exchange interaction is proportional to the overlap between the electron and the hole states, it can significantly be enhanced by confining the exciton into small nanostructures. This picture explains very well the behavior of the PL lifetime. Denoting the lifetimes of the upper singlet and the lower triplet states by τ_U and τ_L, respectively, and taking the population of the two states to be in thermal equilibrium (which is a reasonable assumption as relaxation processes between these neighboring states should very fast, of the order of few picoseconds), we find the following expression for the PL radiative lifetime:

$$\frac{1}{\tau_R} = \frac{g/\tau_L + (1/\tau_U)e^{-\Delta/kT}}{g + e^{-\Delta/kT}} \qquad (25.5)$$

where

$1/\tau_L$ is the rate of radiative transitions from the lower triplet state

$1/\tau_U$ is the rate of radiative transitions from the upper singlet state

$g = g_L/g_U = 3$ is the level's degeneracy ratio* and the Bolzmann factor, $\exp(-\Delta/kT)$, takes into account the relative (thermal) population of the states

The solid lines in Figure 25.15 represent the best fit of the PL lifetimes to Equation 25.5. The appearance of two temperature regimes can be understood now as follows. At low temperatures (where $kT < \Delta$), only the lower triplet state of the exciton is populated and the PL decay time is dominated by the long lifetime of the triplet state. At higher temperatures, the upper singlet state becomes populated. If the radiative lifetime of this state is much faster than that of the lower triplet state, it will dominate the PL decay time at high temperatures, giving rise to exponentially faster lifetimes (versus the inverse temperature), which is in a very good agreement with the results shown in Figure 25.15. The upper state lifetime is expected to be (relatively) short since the spin singlet is an optically active state (e.g., radiative recombination of the exciton into the ground state of no exciton is allowed) as opposed to the lower triplet state, which is optically forbidden state according to the spin selection rules. The spin–orbit interaction[11] can mix the singlet–triplet states, making the triplet state weakly allowed. However, this interaction is fairly weak in silicon so that the lifetime of the triplet state is expected to be quite long.

Before discussing the application of this analysis to the various SiNCs samples (of variable silicon volume content), let us denote the contribution of nonradiative relaxation processes to the measured PL decay time. In principle, the measured (inverse) PL decay time is the sum of radiative and nonradiative relaxation rates, $\tau^{-1} = \tau_R^{-1} + \tau_{NR}^{-1}$, where τ_{NR} is the nonradiative relaxation time. Yet, as will be further explained hereafter, at high temperatures the measured PL decay time is dominated by radiative processes and we may take, $\tau_U \cong \tau_S$ (where τ_S is the singlet lifetime). At low temperatures, the contribution of nonradiative processes cannot be neglected; thus we have $\tau_L^{-1} = \tau_T^{-1} + \tau_{NR}^{-1}$, with τ_T being the triplet (radiative) lifetime.

The above analysis, which yields the upper level lifetime, the lower level lifetime, and the exchange-splitting energy for each of the PL energies, has been exploited to the entire set of SiNC samples having different average diameters.[15,16] The results are presented in Figure 25.16 for selected samples. Let us emphasize again that each of these samples is characterized by inhomogeneous, approximately Gaussian, size distribution of the nanocrystals (see Figure 25.5). The PL energy probes a given

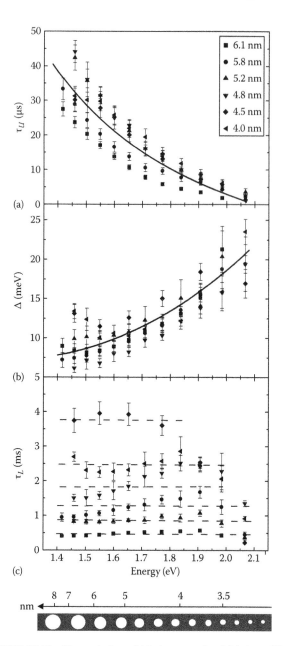

FIGURE 25.16 The variation of (a) the upper level lifetime, (b) the triplet–singlet exchange-splitting energy and (c) the lower level lifetime, as a function of the PL energy for various samples having different average diameters. The lower diameter scale correlates the PL energy to the diameter of the nanocrystals according to Equation 25.3. The solid lines in (a and b) represent a universal dependence of the upper level lifetime and the splitting energy on the diameter of the nanocrystals, while the dashed lines in (c) show the dependence of the lower level lifetime on the average size of the samples.

size (or diameter) of the nanocrystals within this distribution. This is schematically illustrated at the bottom of Figure 25.16, where a smaller PL energy is linked to a larger nanocrystal with a scale of diameters (based on Equation 25.3—the relationship between the PL energy and the diameter of the nanocrystals). For a given size distribution, the PL intensity varies across the PL spectrum yielding more photons near the average size of the

---

* The degeneracy ratio depends on the assumption regarding the total angular momentum of the states. However, the lifetime analysis is almost insensitive to the specific value of this ratio.

SiNCs. Nevertheless, according to the QC model the PL lifetime should be independent of the size distribution but rather characteristic of a given nanocrystal's diameter. In other words, according to the QC model, the PL lifetime should vary with the PL energy but should be *independent of the size distribution of the nanocrystals*. The results shown in Figure 25.16a for the upper state (singlet) lifetime and in Figure 25.16b for the exchange-splitting energy, remarkably follow the QC model. The upper state lifetime, τ_U, decreases by an order of magnitude (from about 40–50 μs down to 2–3 μs) with increasing PL energy, in agreement with the QC model where larger relaxation of the *k*-conservation rule is expected for smaller nanocrystals. Furthermore, all the results for τ_U from different samples having different average diameters, collapse into a single line (the blue line in Figure 25.16a) independent of the size distributions of the samples. The PL energy probes a lifetime of nanocrystals having the same diameter, which should be independent of how many nanocrystals of that size appear at a given distribution. A similar conclusion holds for the exchange-splitting energy. The smaller the nanocrystal, the larger the exciton confinement and the bigger the splitting energy. Here again, as the splitting energy is a size-dependent property of the nanocrystals, this energy should be independent of the size distribution and the results for all samples should collapse into a single line.

So far, we have found the upper state (singlet) lifetime and the exchange-splitting energy to provide a solid support to the QC model. However, a completely different behavior appears for the lower state (triplet) lifetime, τ_L. The lower state lifetime is essentially independent of the PL energy and therefore, is not a size-dependent property of the nanocrystals. In contrast, τ_L depends on the excess silicon volume content (x), and accordingly, on the average diameter of the SiNCs. Thus, the lower state lifetime cannot be assigned to radiative relaxation from the triplet state but rather to nonradiative relaxation processes that depend on the environment of the SiNCs, e.g., the amount of excess silicon in the host matrix and, as will be explained later, on the SC (or passivation) of the SiNCs. These findings suggest that *QC is responsible for the relatively fast radiative transitions rates from SiNCs, while SC is responsible for the fairly slow nonradiative relaxation processes in these nanocrystals*. Yet, both processes are responsible for the PL yield, η (see Section 25.3.2), which is derived from the competition between radiative and nonradiative recombination rates, as schematically illustrated at the inset to Figure 25.11, and is given by the following expression:

$$\eta = \frac{\tau_R^{-1}}{\tau_R^{-1} + \tau_{NR}^{-1}} = \frac{\tau_{NR}}{\tau_R + \tau_{NR}} = \frac{\tau}{\tau_R} \quad (25.6)$$

where $\tau^{-1} = \tau_R^{-1} + \tau_{NR}^{-1}$ is the total (measured) PL decay time. For high-yield samples (average diameter in the range of 4.5–5.5 nm), the PL decay time varies by more than two orders of magnitude over the 50–300 K temperature range (see Figure 25.15), while the integrated PL is approximately constant over the same temperature range (it varies by less than a factor of 2; see Figure 25.17).

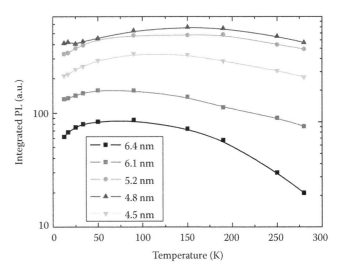

FIGURE 25.17 The variation of the integrated PL as a function of temperature for various samples having different average diameters.

Both experiments (Figures 25.16 and 25.17) suggest that over the 50–300 K temperature range, the PL decay time is dominated by *radiative transitions* of the exciton where, $\tau_R \ll \tau_{NR}$. In this range, we find $\eta \cong$ constant ~1, in a good agreement with the high yield measured for these SiNCs. At this temperature range, radiative transitions are dominated by the upper-state lifetimes, $\tau \cong \tau_U \sim 2$–50 μs, which should be compared to the nonradiative lifetimes, $\tau_{NR} \sim 1$–3 ms. Notice that nonradiative rates have been measured at low temperatures and, in general, one should expect faster nonradiative lifetimes at higher temperatures. Nevertheless, all experimental results indicate that $\tau_R \ll \tau_{NR}$ over the entire temperature range, up to room temperature.

Based on the assignment of radiative lifetime to the upper (singlet) state, we can estimate now the oscillator strength for optical transitions from this state. The oscillator strength, $f_{i \to j}$, is a dimensionless parameter that measures the strength of a dipole-allowed optical transition from a given *i*th state to another *j*th state, relative to all other dipole-allowed transitions from the same (*i*th) state. The oscillator strength is normalized so that the sum over all oscillator strengths originated from the same level, is equal to 1, a property which is known as the *f*-sum rule[76] (or the Thomas–Reiche–Kuhn sum rule). The relationship between the oscillator strength and the (spontaneous) radiative lifetime of the singlet state is given by[18]

$$f_{OS} = \frac{2\pi m c^3 \hbar^2 \varepsilon_0}{E^2 e^2 n} \frac{1}{\tau_U} \quad (25.7)$$

where

m is the electron mass
n is the refractive index of the medium
E is the photon energy

In Figure 25.18, we have plotted the oscillator strength for optical transitions, from the upper singlet state, versus the confinement energy, ΔE. Here again, all results from samples having

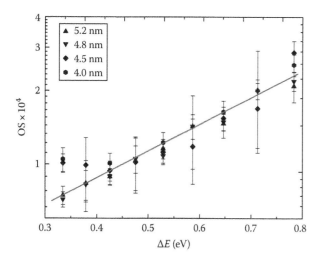

FIGURE 25.18 Semilogarithmic plot of the upper level oscillator strength versus the confinement energy, for various samples having different average diameters. The red line represents the best fit to Equation 25.8.

different average diameters collapse into a single function that, taking into account the limited range of variation, can be fitted into exponential form:*

$$f_{OS} = f_0 \exp\left(\frac{\Delta E}{E_{OS}}\right) \qquad (25.8)$$

where

$E_{OS} \cong 0.4 \pm 0.1\,\text{eV}$ is the characteristic oscillator strength energy

$f_0 \cong (3.5 \pm 1) \times 10^{-5}$ is the oscillator strength of the bulk crystal (no confinement)

The above analysis of the PL characteristics seems to adequately explain most of the experimental findings but, at the same time, open up other fundamental questions concerning the origin of the PL. Let us briefly summarize these findings and the most relevant questions[19]:

1. *Radiative Processes*: QC plays a major role here. For small silicon nanostructures, QC gives rise to a blueshift of the bandgap energy, up to the visible range of the spectrum, according to the power law of Equation 25.3. At the same time, the increasing bandgap energy gives rise to shorter radiative lifetimes since $\tau_U \sim 1/E^2$ (without taking into account the effect of confinement on the oscillator strength, which will be discussed hereafter). A second contribution of QC to radiative processes comes from the breakdown of the *k*- (or momentum) conservation rule. However, the effect of confinement on the rate of radiative transitions has been overestimated by many researchers over the years. Our estimate of the oscillator strength for the case of no confinement is, $f_0 \sim 10^{-5}$, in a fairly good

agreement with the poor optical emission and the fairly slow radiative lifetimes ($\geq 10\,\text{ms}$) from bulk silicon. QC enhances the oscillator strength for small nanocrystals, but only to the order of $\sim(2-3) \times 10^{-4}$ and, apparently up to 10^{-3} for blue-emitting nanocrystals. For a comparison, in direct bandgap semiconductors the oscillator strength for interband transitions is ~ 1, about 3–4 orders of magnitude larger than the oscillator strength of our silicon nanostructures. A direct consequence of this result is the fairly slow radiative lifetimes in silicon nanostructures relative to direct bandgap semiconductors. For example, the radiative lifetime in GaAs (with energy bandgap of $\sim 1.42\,\text{eV}$) is about few nanoseconds compared to few microseconds in silicon nanostructures; a result which is a direct manifestation of the 3–4 orders of magnitude differences in the oscillator strength. Some researchers might assign this conclusion to the so-called quasi-direct bandgap of silicon nanostructures, meaning that the oscillator strength for dipole allowed optical transitions is substantially smaller compared to direct bandgap semiconductors. However, the author of this chapter prefers not to talk about "bands" in small silicon nanostructures as we actually deal with a discrete set of levels rather than "bands."

2. *Nonradiative Processes*: As opposed to radiative processes, QC does not play a major role here. The efficient light emission from SiNCs is due to exceptionally slow nonradiative relaxation processes in SiNCs, of the order of few milliseconds. While this result explains very well the high quantum yield from SiNCs, it raises another fundamental question. The measured nonradiative lifetimes of few milliseconds are typical for the purest silicon wafers, with impurity levels of about 10^{13}–$10^{14}\,\text{cm}^{-3}$ (e.g., one impurity atom per more than a billion of silicon atoms).[1] It is well known that higher density of impurities enhances nonradiative relaxation rates, for example, in p-type silicon crystals with boron concentration of about $10^{16}\,\text{cm}^{-3}$ the minority carriers' lifetime is about $0.1–1\,\mu\text{s}$; adding $10^{17}\,\text{cm}^{-3}$ Au impurities into the crystal gives rise to nonradiative lifetimes as short as $10^{-10}\,\text{s}$. The high PL yield and the slow nonradiative relaxation times in SiNCs seem to be a universal property of SiNCs, essentially independent of the method of fabrication (see Section 25.2). Yet, these fabrication methods are not necessarily "clean" and free of imperfections, defects, and impurities. In fact, the quality of interfaces between SiNCs and the host matrices cannot be compared to those obtained by epitaxial growth methods; their shape and symmetry are less regular than nanocrystals synthesized by colloidal chemistry and the level of impurities and defects is far from those obtained by advanced silicon crystal growth techniques. Considering all these facts, one would expect to find much faster nonradiative relaxation times in SiNCs, at least a considerable variation of the PL yield among the various preparation techniques. It seems that *SC plays a major role in excluding nonradiative relaxation channels*

---

* Over the limited range where the oscillator strength has been measured, other functions such as power laws could also be fitted. The current exponential form should be considered as a guideline only.

in SiNCs. Understanding what possible mechanism could give rise to remarkably slower nonradiative relaxation rates is a major issue in the field and is expected not only to explain the origin of the efficient PL, but also to explore novel aspects of SC at the nanoscales. A possible mechanism that could prevent nonradiative relaxation processes at these length scales, and is related to SC, is presented and discussed in the following section (the vibron model). In principle, understanding this mechanism may open up new horizons concerning surface engineering of nonradiative processes at nanoscales.

25.4 The Vibron Model: The Relationship between Surface Polar Vibrations and Nonradiative Processes

The purpose of this section is to discuss basic properties of SiNCs surfaces, particularly the silicon–SiO$_2$ interface (Section 25.4.1) as most experimental data reported so far are related to SiNCs embedded in silicon dioxide matrices. The "vibron" model[19] (Section 25.4.2), which correlates surface vibrations at the silicon–SiO$_2$ interfaces to nonradiative processes, is proposed as a specific mechanism that passivates the nanocrystals against nonradiative relaxation channels. Such a mechanism can explain the role of SC and its influence on nonradiative relaxation processes in SiNCs. Finally, in Section 25.4.3, we describe recent experimental results that seem to support the vibron model.

25.4.1 The Silicon–Silicon Dioxide Interface

The silicon–SiO$_2$ interface is one of the most studied interfaces due to the vital role it plays in microelectronics, particularly in the MOS (metal-oxide-semiconductor) technology. In general, it is well known that thermally grown oxide on top of a crystalline silicon substrate has amorphous structure down to distances of about 1–3 nm away from the interface. The common picture of the interface is that of 1–2 monolayers of non-stoichiometric SiO$_x$ (with $1 < x < 2$), followed by 1–2 nm of strained SiO$_2$ and a remaining layer of stoichiometric, strain-free amorphous SiO$_2$.[1] The Si–SiO$_2$ interface plays a major role in the MOS technology as it can permanently trap charges (interface charges) acting as a source of voltage that shift the electrical characteristics of MOS devices. The amount of surface traps can be as high as 10^{15} cm^{-2} (meaning that essentially all surface atoms can trap charges); however, after hydrogen annealing that passivates most of surface dangling bonds, the amount of surface traps can be reduced to a level of ~10^{10} cm^{-2} or less. The electronic properties of bulk silicon crystals are almost insensitive to these imperfections due to the very small STV ratio of bulk materials (see, Figure 25.3). However, for small SiNCs, the circumstances are substantially different as the number of atoms (or molecules) belonging to the surface of the nanocrystal is similar to the number of silicon atoms of the "bulk" nanocrystal.

Several theoretical and experimental investigations have been conducted for exploring the nature of the interface transition region, of about 1–3 nm, where the structure changes from a perfect order of crystalline silicon to a disordered structure of amorphous SiO$_2$. Numerous reports[77–80] have shown that the interface may include few monolayers of compressed crystalline, epitaxial-oxide phase in the form of either cristoblite,[81] quartz, or even tridymite.[82] After few monolayers of strained crystalline oxide, the stress is released and amorphous phase of SiO$_2$ appears. A quite remarkable demonstration to the presence of crystalline SiO$_2$ phase at the Si–SiO$_2$ interface has been reported by Cho et al.[83] using thermally oxidized SOI (silicon-on-insulator) substrates to create a thin crystalline silicon layer embedded in between relatively thick amorphous SiO$_2$ (a 2D silicon quantum well*). The cross-section HRTEM images of these structures, which can be considered as 2D analogous[84] of 0D SiNCs, are shown in Figure 25.19a and b. Despite that the exact crystalline phase of SiO$_2$ could not be identified in this experiment, these images demonstrate the possibility of creating crystalline SiO$_2$ at the Si–SiO$_2$ transition region. Notice also that the model of surface passivation by Si=O double bonds (see Section 25.3.2) proposed by Wolkin et al.[17] has been shown to be consistent with the presence of cristoblite phase at the transition region of the interface.[85] The remarkable point about the TEM images shown in Figure 25.19a and b is the fairly thick crystalline-oxide layer, of ~2–3 nm, which may contain more than 10 monolayers of crystalline SiO$_2$. The situation is much more complicated when dealing with nonplanar surfaces such as the spherical surface of SiNCs. The EFTEM (energy filtered TEM) image of a single SiNC, shown in Figure 25.19c, was taken from Ref. [86] and is one of the best images of a single nanocrystal ever been reported. Daldosso et al.[86] provided clear evidences to the existence of a Si–SiO$_2$ spherical transition region, of about 1 nm in thickness, which is marked by the dashed lines in Figure 25.19c. While the signature of crystalline SiO$_2$ is absent in this image, it is worth noting that such a signature from few monolayers of a crystalline phase, thinner than 2 nm, has not been observed even for planar surfaces and definitely not for the much more challenging spherical surfaces discussed here. The presence of a spherical transition region made of crystalline SiO$_2$ is not a prerequisite for the vibron model to be discussed in Section 25.4.2. Yet, the existence of such a surface would provide a supplementary support to the model as coherent vibrations across the SiNCs surface, are expected to enhance the vibron effect (see Section 25.4.2) as compared to noncoherent vibrations. The presence of such a surface would also help to understand the analogy between the "classical" polaron problem and the vibron model as "phonons"

---

* Some readers may ask if such 2D silicon nanostructures could show similar quantum size effects (such as PL). However, as far as electronic properties are concerned, these 2D structures behave as 0D SiNCs. The reason is local fluctuations, on length scales of 1–2 monolayers, of the oxide layer that cannot be avoided during thermal oxidation. These fluctuations induce strong localization of the electronic wavefunctions that resembles the 0D characteristics of SiNCs (with lesser control of the additional confinement scales[84]).

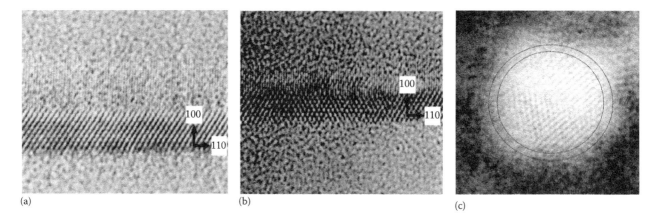

(a)   (b)   (c)

FIGURE 25.19   HRTEM images of crystalline SiO_2 on ultrathin silicon (100) layers with thicknesses of (a) 1.9–2.2 nm and (b) 1.4–1.6 nm. (c) High-resolution energy filtered TEM of a SiNC. The presence of silicon (111) reticular planes is clearly visible in the core region. The dotted lines mark the interface transition region. (Reprinted from Daldosso, N. et al., *Phys. Rev. B.*, 68, 085327, 2003, Cho, E.-C. et al., *J. Appl. Phys.*, 96, 3211, 2004. With permission.)

are natural vibrations of a crystalline SiO_2 phase. However, this is essentially a technical point and the vibron model can be justified even in the absence of a crystalline SiO_2 shell.

25.4.2 The Vibron Model

Following the conclusions of Section 25.3, particularly the conclusion that the high PL efficiency is due to the inhibition of nonradiative relaxation in SiNCs, we should look for a specific mechanism, associated with SC, which can "passivate" the nanocrystals against nonradiative relaxation processes. Such a mechanism has been proposed in Refs. [18,19,87,88] and is called the "vibron" model. Let us briefly explain this model that assigns the inhibition of nonradiative channels to a resonant coupling between electronic states of the nanocrystals and surface vibrations of the silicon–oxygen bonds. These vibrations can be either noncoherent, e.g., vibrations associated with the Si–O and/or the Si=O bonds on the surface of the SiNCs, or coherent vibrations of a spherical crystalline SiO_2 shell that wraps the inner crystalline silicon core of the nanocrystals.

In general, a major source of scattering and relaxation in any semiconductor is the electron scattering from lattice vibrations of its own crystal, known as phonons.[89] The phonon dispersion relation (e.g., the relation between the frequency and the wave-vector of the phonons) consists of two branches: an acoustic branch (where ω goes to zero as $\vec{k}$ approaches zero) and an optical branch of phonons that can interact with light (with non-zero value of ω at $\vec{k} \to 0$). In addition, each of these branches is divided into longitudinal and transversal phonon modes where, for longitudinal modes (LO and LA), the atoms vibrate parallel to the propagation direction while for transversal modes (TO and TA), the atoms vibrate perpendicular to the propagation direction of the phonons. In polar semiconductors (such as GaAs), where ions constructing a unit cell of the crystal carry positive and negatives (ionic) charges, longitudinal vibrations of the crystal generate a long-range polarization field. The polarization field is an electric field produced by the electric dipole moment of the

vibrating ions, which is responsible for the strong interaction between LO phonons and charged carriers (electrons and holes) in polar semiconductors, and is a major source of scattering and energy dissipation in these semiconductors.[90] In silicon, however, the situation is quite different as silicon is a covalent semiconductor having the same atoms in a unit cell of the crystal that do not carry a net charge. Therefore, neither longitudinal nor transversal phonon vibrations in silicon generate a polarization field, giving rise to significantly slower phonon scattering rates compared to polar semiconductors. The presence of charged impurities, vibrating bonds, or any other source of oscillating electric dipoles (or charges) on the surface of silicon could generate electric field that interacts with the charged carriers. However, these are small surface effects that have minor impact on the electronic properties of bulk silicon crystals.

The situation is substantially different when dealing with low-dimensional silicon nanostructures, particularly small SiNCs. In these nanocrystals, the surface can no longer be treated as a small perturbation since a significant portion of the atoms sit on the surface of the nanocrystals (see Figure 25.3). While vibrations of the core silicon crystal cannot produce a polarization field, polar vibrating bonds on the surface of the nanocrystals can generate a polarization field, in the vicinity of the nanocrystals, that interacts with the charged carriers. This is the basic mechanism that stands behind the vibron model[19]. Some readers may argue that this picture seems to contradict our goal, e.g., finding a mechanism that will allow to passivate the nanocrystals against nonradiative processes. Indeed, the smaller are the nanocrystals the bigger is the surface area giving rise to larger amount of nonradiative relaxation channels. This is also the case for SiNCs with a diameter larger than 10 nm. However, for smaller SiNCs with a diameter less than 10 nm, a new mechanism of resonant coupling between surface vibrations and electronic states appears and allows passivating the nanocrystals against nonradiative channels.

To follow the origin of this mechanism, let us consider the SiNC schematically illustrated in Figure 25.20, which consists

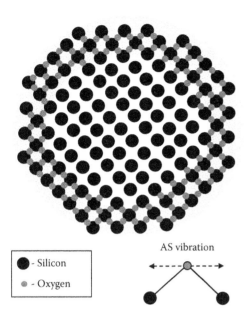

FIGURE 25.20 Schematic illustration of a spherical silicon nanocrystal terminated with silicon–oxygen bonds. The dashed area represents a surface region where a crystalline SiO₂ shell could be formed. The inset illustrates the Si–O–Si asymmetric stretching mode of vibration.

of (approximately) spherical nanocrystalline silicon core terminated with silicon–oxygen bonds on its surface. Traditionally, finding the electronic levels and the vibrational modes of a bulk semiconductor have been treated as two independent problems (the so-called adiabatic approximation) as the electronic energies are of the order of few eV while vibrational (or phonon) energies are an order of magnitude smaller. At present, we will follow this picture, but later on we will reexamine this assumption. The SiNCs spectrum of electronic levels has been calculated by several groups using various techniques such as tight-binding,[58,91] pseudo-potential,[59] and effective mass.[92,93] Most reports have focused on the lowest unoccupied conduction level (called the LUMO state) and the highest occupied valence level (the HOMO state) as these levels are involved in interband optical transitions contributing to the PL. However, the vibron model requires some knowledge about higher excited states of the system. In direct bandgap semiconductor nanocrystals, it is convenient to classify the electronic levels according to their orbital symmetry using the standard atomic orbital notations, e.g., S, P, D, etc.[2] Several factors make the classification for SiNCs more complex. First, the anisotropic effective mass of conduction electrons (with a transverse mass of $0.19m_0$ and longitudinal mass of $0.92m_0$) may gives rise to a splitting of the levels with lower energy levels associated with the heavier mass. Secondly, the six minima degeneracy of the conduction band in bulk silicon (near the X-point, along the family of (100) directions; see Figure 25.1), which should be preserved for an ideal spherical dot, will be lifted by any deviation from a perfect spherical symmetry. In addition, a unique splitting into two groups of levels that are symmetric/antisymmetric under inversion (e.g., $\vec{k} \rightarrow -\vec{k}$) are expected for small SiNCs due to

intervalley couplings.[91] Finally, mixing between different bands is expected in small SiNCs.

For the purpose of our model, we will ignore most of these fine splitting and mixing features, referring to the lowest two sublevels of each band as $1S_e$ and $1P_e$ for conduction electrons ($1S_h$ and $1P_h$ for holes), as schematically illustrated on the left side of Figure 25.21. E_g is the effective energy bandgap (the energy difference between the $1S_e$ and the $1S_h$ states) and ΔE_C and ΔE_V are the energy differences between lowest conduction and valence sublevels, respectively. On the right side of Figure 25.21, we schematically illustrate the configuration space diagram for the Si=O surface bonds (e.g., the variation of the potential energy versus the length of the Si=O bond) with a ground vibrational state and a first excited state having a vibrational energy of $\hbar\omega_{VB}$ (a common approximation here is taking the potential energy to be parabolic and the quantized vibrational levels and energies being the levels of a simple harmonic oscillator). The vibrational energies are measured upward for conduction electrons and downward for valence holes. Notice that vibrational energies are size independent (e.g., independent on the diameter of the nanocrystal), depending only on the type of bonding used to terminate the surface of the nanocrystal. On the other hand, the quantized electronic levels of the nanocrystal are size dependent and should increase with the decreasing diameter of the SiNCs. Hence, for (relatively) big SiNCs, one expects the conduction/valence energy differences to be fairly small compared to vibrational energies. With the decreasing size of the

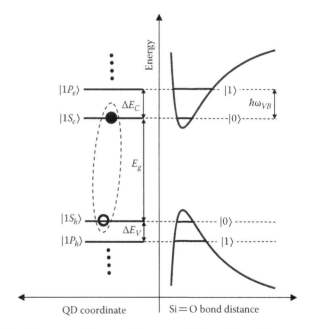

FIGURE 25.21 Schematic illustration of the condition for resonant coupling between electronic sublevels and surface vibrations. On the left, the lowest electronic sublevels of the conduction ($1S_e$ and $1P_e$) and the valence ($1S_h$ and $1P_h$) bands are shown. ΔE_C and ΔE_V are the energy differences between these sublevels. On the right, the ground and the first excited vibrational states are shown (with $\hbar\omega_{VB}$ being the vibrational energy). The solid lines represent the potential energy for surface vibrations (in the configuration space). Resonant coupling occurs once the vibrational energy is equal to ΔE_C or ΔE_V.

nanocrystals, ΔE_C (and ΔE_V) increases until, for a given diameter of the nanocrystals, the two energies (electronic and vibrational) coincide. This is the condition for *a resonant coupling between surface vibrations and the electronic states of the nanocrystals via a polarization field of the vibrations.*

The above model should slightly be modified when dealing with surface phonons of a crystalline SiO_2 shell, but the main conclusion concerning resonant coupling between electronic states and surface phonons, remains valid.* In this case, we will focus our attention on LO phonons as these phonons generate a considerably large polarization field. In general, the dispersion relation of the phonons should be taken into account (e.g., the dependence of the phonon energy on the phonon wavevector, $\vec{q}$), but, to a first order approximation we may take the LO phonon energy to be dispersionless and equal to $\hbar\omega_{LO}$. When the condition for a strong coupling is fulfilled, $\Delta E \approx \hbar\omega_{LO}$, the electrons and the phonons are not anymore in the weak coupling regime where scattering and energy dissipation occur. Instead, the strong coupling between electrons and phonons gives rise to a creation of virtually everlasting mixed electron–phonon modes, called polarons. In polar semiconductor quantum dots, these polarons are due to strong coupling between the electrons and the LO phonons of the same crystalline core while in the vibron case, a similar coupling occurs between surface phonons of the crystalline shell and the core electrons. These surface polarons "passivate" the surface against nonradiative relaxation processes. Let us emphasize that this process of creating surface polarons (vibrons) is expected to be particularly important in silicon where, as opposed to polar semiconductors, only surface phonons (or vibrations) can produce a polarization field explaining the crucial role of SC in SiNCs.

The polaron state of nanocrystals can be viewed as a "dressed" state of the electron–phonon system. A simple illustration of a polaron state is depicted in Figure 25.22 where the negatively charged electron produces a local disturbance to the polarization field of the lattice (which is created by longitudinal lattice vibrations). In the vibron case, the surface polaron is a "dressed" state of the surface, meaning that the electron and/or the hole are coupled to the surface of the SiNC. This provides a direct explanation and a specific mechanism to the model of Wolkin et al.[17] of surface-trapped electrons/holes/excitons discussed in Section 25.3.2. Here again, resonant coupling to surface vibrations can produce a localized (or trapped) electron-surface vibrations, hole-surface vibrations, or a localization of the entire exciton as suggested by Verzelen et al.[94] Another important characteristic of the polaron problem, which can be utilized for experimental verification of the model, is the polaron energy splitting due to confinement. To follow this phenomenon, one needs to solve the entire polaron Hamiltonian including the electron-polar phonon interaction, finding the new eigen-energies of the polaron that becomes a mixed state of the electrons (or the holes) and

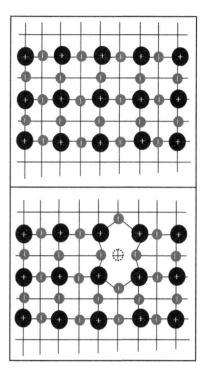

FIGURE 25.22 Schematic description of the polaron formation mechanism. On the left, the polarization field from the unperturbed lattice of charged ions is shown while on the right, the perturbed field (e.g., the polaron) created by a negatively charged electron (dotted circle) is presented.

the phonons. While the exact algebra is quite complicated and tedious, a simplified model has been proposed by Mahdouani et al.[95] by limiting the interaction into the subspace of the $|1S_{e(h)}, 1\vec{q}\rangle$ and $|1P_{e(h)}, 0\vec{q}\rangle$ uncoupled states, which are composed of the electronic ground state $(1S_{e(h)})$ + 1 phonon state $(1\vec{q})$, and the first excited electronic state $(1P_{e(h)})$ + 0 phonon state $(0\vec{q})$, respectively. In this subspace, the electron–phonon interaction Hamiltonian (known as the Fröhlich Hamiltonian[96]) takes the following form:

$$H_{POL} = \begin{pmatrix} E_S + \hbar\omega_{LO} & W \\ W^\star & E_P \end{pmatrix} \qquad (25.9)$$

where

E_S, E_P are the electronic energies of the $1S$ and the $1P$ sublevels, respectively

$\hbar\omega_{LO}$ is the LO phonon energy

The off-diagonal electron–phonon interaction term is given by

$$W = \langle 1S_{e(h)}, 1\vec{q} \,|\, H_{el-ph} \,|\, 1P_{e(h)}, 0\vec{q} \rangle$$

The solution to this problem (e.g., diagonalization of the Fröhlich Hamiltonian) yields the mixed electron (hole)–phonon states of the vibron and the energies of the vibron, which are schematically illustrated in Figure 25.23.

The lower energy branch (E_-) has a phonon-like behavior $(E_- \approx \hbar\omega_{LO})$ for small SiNCs and behaves as an electronic state $(1P_e)$ for large SiNCs (with $E_- \approx \Delta E_C \sim 1/d^\eta$; see Equation 25.3). The upper branch (E_+) is electronic-like for small SiNCs and phonon-like for

---

* For the sake of simplicity, we have ignored the effect of confinement on the spectrum of phonons and their dispersion relation. However, this topic could lead to additional novel phenomena that have not been studied so far.

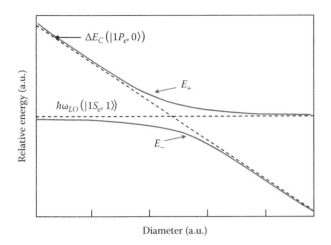

FIGURE 25.23 Schematic presentation of the vibron energies (solid lines) as a function of the nanocrystal diameter. The dashed lines represent the vibrational energy, which is not affected by the confinement (horizontal line), and the electronic energy (ΔE_C), which depends on the diameter of the nanocrystal.

large SiNCs. However, the most interesting behavior occurs for $\Delta E_C \approx \hbar\omega_{LO}$ (the resonant coupling condition), where the polaron states are mixed states of the form $\left|\psi_{POL}^\pm\right\rangle = a_\pm\left|1S_e, 1\vec{q}\right\rangle + b_\pm\left|1P_e, 0\vec{q}\right\rangle$ (with $|a_\pm|^2 + |b_\pm|^2 = 1$). Similar results can be obtained for the hole–phonon coupling[95] and for the exciton–phonon coupling.[94] In principle, the same formalism can be applied to noncoherent surface vibrations with $\hbar\omega_{VB}$ replacing $\hbar\omega_{LO}$. Yet, this problem requires further investigation as the application of the Fröhlich Hamiltonian for coupling to surface vibrations and/or amorphous phases has not been investigated so far. Finally, let us point out that measuring the vibron energies can provide a direct verification to the model. This is the subject of Section 25.4.3.

25.4.3 Experiments Supporting the Vibron Model

At first, let us mention several reports about polar optical phonons of the silicon–SiO₂ planar interfaces, which have been theoretically predicated[97] and experimentally observed.[98] In these planar Si–SiO₂ interfaces, polar optical phonons of the oxide layer give rise to enhanced scattering and energy dissipation, from charged carriers moving in the adjacent silicon channel of MOS devices and bipolar transistors having SiO₂ spacer layers.[98] In this case, the interaction with polar optical phonons causes a lower mobility of the carriers and a degradation of the current–voltage characteristics. As discussed in the previous section, this is the regime of weakly coupled electrons–phonons that is reflected by energy dissipation and enhanced nonradiative relaxation rates.

Among the first experiments providing direct evidence for vibrons in SiNCs is the one reported in Refs. [87,88]. In this experiment, the aim was to measure inter-sub-level (ISL) optical transitions between the quantized sublevels of SiNCs,[93] for example, the $|1S_e\rangle \rightarrow |1P_e\rangle$ and the $|1S_h\rangle \rightarrow |1P_h\rangle$ ISL transitions between the quantized states of the conduction and the valence bands, respectively (see Figure 25.21, left). Ignoring band mixing, let us indicate

that these are optical transitions between sublevels coming from the same band and therefore, having the same Bloch wavefunction (for example, the same conduction Bloch state for both the $|1S_e\rangle$ and the $|1P_e\rangle$ sublevels). Hence, dipole matrix elements for optical transitions between these sublevels are reduced to dipole matrix elements between envelope states of these sublevels.[99,100] Thus, the $|1S_e\rangle \rightarrow |1P_e\rangle$ transition is a dipole allowed optical transition (with the oscillator strength for this transition being close to 1[91]) that can be measured by infrared absorption spectroscopy. Let us emphasize that ISL optical transitions are a direct manifestation of QC as these sublevels do not exist in bulk semiconductors.[101] In addition, according to the vibron model, these sublevels are expected to interact with surface polar vibrations yielding modified energy spectra similar to those presented in Figure 25.23. Therefore, this experiment should be sensitive to both QC and SC in SiNCs.

The experimental setup for measuring ISL transitions[101] in SiNCs includes a visible pump laser exciting electron–hole pairs* (excitons) and a weak, tunable IR probe beam measuring the photoinduced absorption (PIA) spectra. Both the photoinduced transmittance, ΔT, and the linear IR transmittance, T, are measured (by modulating the pump laser beam in one experiment and the probe beam in a second experiment using a lock-in detection method[87]) to obtain PIA: PIA $= \Delta T/T$.† The same set of co-sputtered SiNC samples, with varying average diameter (and silicon volume content; see Figure 25.5), which were used for the PL experiments (see Figures 25.9 through 25.11), have also been exploited for PIA measurements to allow investigation of possible correlation between these experiments.

Figure 25.24 presents the linear IR absorption (dashed line) and the PIA (solid line) spectra for SiNCs having an average diameter

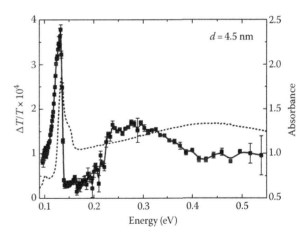

FIGURE 25.24 Room temperature PIA (symbols and a solid line) and linear IR absorption (dashed line) spectra for the sample with average diameter of 4.5 nm.

---

* Since our SiNCs are essentially undoped, the carriers must optically be induced. Once a technology to introduce dopants into SiNCs will emerge, it would be interesting to investigate direct ISL transitions from SiNCs having a single type of carriers.

† The linear and the photoinduced transmittance are given by $T = T_0 e^{-\alpha d}$ and $\Delta T = T_0[e^{-\alpha d} - e^{-(\alpha + \Delta\alpha)d}]$ where α is the linear absorption coefficient and $\Delta\alpha$ is the photoinduced absorption coefficient. Hence, we find PIA $= \Delta T/T = 1 - e^{-\Delta\alpha d} \cong \Delta\alpha \cdot d$ where we have assumed that $\Delta\alpha \cdot d \ll 1$.

of 4.5 nm ($x = 18\%$). The linear IR absorption spectrum includes a single absorption band, at about 130–150 meV (1050–1210 cm^{-1} in terms of wavenumbers), of the Si–O asymmetric stretching mode of vibration,[102,103] schematically illustrated in Figure 25.20. Similar linear IR absorption spectra have been obtained for all SiNCs embedded in SiO$_2$ matrices. The PIA spectrum reveals two photoinduced absorption bands; a relatively strong and narrow band at ~135 meV and a fairly broad, high-energy band at about 280 meV. The broadening of the PIA bands is expected here as absorption measurements average over all sizes of nanocrystals according to the size distribution of each sample (see Figure 25.5). In fact, the fairly narrow linewidth of the low-energy band (FWHM of ~20 meV) is quite surprising taking into account the size distribution. As the PIA technique is not sensitive to the polarity of the carriers, we may use the effective mass approximation for assigning the low-energy PIA band to valence-ISL (VISL) transitions and the high-energy band to conduction-ISL (CISL) optical transitions.[91–93] However, another explanation of assigning both bands to CISL transitions is possible, where the $1P_e$ state of the conduction splits into two sublevels, mainly due to differences between longitudinal and transversal conduction effective masses.[91] In the following, we follow the first scenario but the main conclusions of this experiment remain valid for the second scenario as well.

Next, in Figure 25.25, we show the PIA spectra for a series of SiNCs having variable average diameter, from 4 up to 6 nm. This is also the range of diameters where the two PIA bands could be resolved from the noise. Surprisingly, while the high-energy CISL

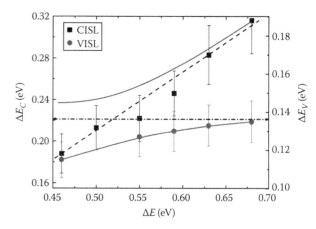

FIGURE 25.26 The measured ISL energies (left axis-CISL, right axis-VISL) versus the confinement energy. The horizontal dashed line indicates the energy of the Si–O asymmetric stretching mode of vibration.

absorption band shows a remarkable redshift with the increasing size of the nanocrystals, the low-energy VISL band presents a significantly weaker shift to the red, with the energy of the VISL transition (120–140 meV) being fairly close to the energy of the Si–O asymmetric stretching mode of vibration (~135 meV).

To correlate the results with the vibron model, we present in Figure 25.26 the PIA peak energies for both ISL bands (ΔE_C and ΔE_V) versus the confinement energy, ΔE (defined here as $\Delta E = E_{PL} - E_G$ with E_{PL} being the peak PL energy of the sample). The CISL energies display a linear dependence on ΔE as expected from the QC model (where both ΔE and $\Delta E_C \sim 1/d^8$ according to Equation 25.3). On the other hand, the VISL energies show a sublinear dependence on ΔE, approaching the vibrational energy of the Si–O stretching mode of vibration. The dashed line in Figure 25.26 represents the expected dependence of the VISL energies on ΔE according to the QC model (i.e., $\Delta E_V \sim \Delta E$). This line crosses the vibrational energy of the Si–O vibration, which is independent of the size of the nanocrystals, and is represented by the horizontal dashed-dotted line in Figure 25.26. Hence, the anti-crossing of the VISL energies and the vibration energies, which resembles the anti-crossing shown in Figure 25.23 (the vibron model), suggests that electronic states of the valence band are resonantly coupled to Si–O surface vibrations to create vibrons.

Notice also that the amplitude (or the strength) of the PIA bands increases with the decreasing size of the nanocrystals, up to a diameter of ~4.5 nm and then, decreases until the PIA spectra could not be resolved for SiNCs smaller then 2.5 nm. These results provide additional support to the vibron model. The correlation between the integrated PIA (e.g., the area below the VISL absorption bands shown in Figure 25.25) and the integrated PL, which is presented in Figure 25.27, suggests that the same mechanism, e.g., the formation of vibrons, is also responsible for the high PL efficiency. For SiNCs having a diameter in the range of 4–4.5 nm, the energy difference between the lowest valence sublevels becomes comparable to those of the Si–O surface vibrations. As a result, long-lived vibrons are created giving rise to slow nonradiative relaxation rates that, in turn, enhance

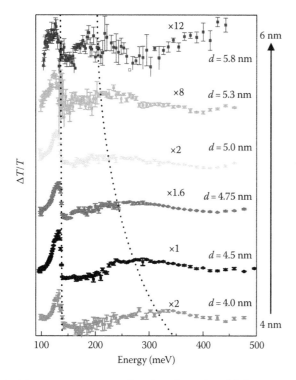

FIGURE 25.25 PIA spectra for selected samples of different average diameters from 4 up to 6 nm. The dashed lines indicate the remarkable redshift of the CISL band and the much weaker shift of the VISL band.

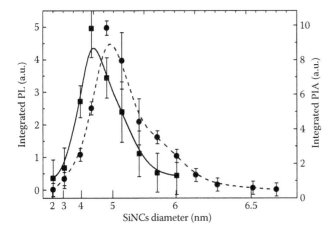

FIGURE 25.27 The correlation between the integrated PL (dashed line and circles) and the integrated PIA (solid line and squares) where both the PL and the PIA have a maximum in the range of 4.5–5 nm.

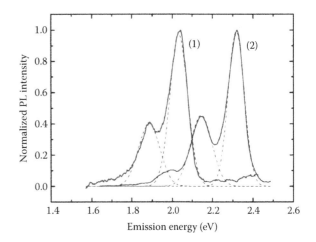

FIGURE 25.28 PL spectra of two, single SiNCs prepared by the laser pyrolysis techniques with vibronic features (low-energy shoulders). The dotted lines represent fitting to a double Gaussian lineshape. (Reprinted from Martin, J. et al., *Nano Lett.*, 8, 656, 2008. With permission.)

the PL efficiency. This mechanism, which takes into account the mutual role of QC and SC, also explains the correlation between the PIA and the PL shown in Figure 25.27. Finally, let us point out that the Si–O asymmetric vibration is related to a "longitudinal" mode of vibration where the oxygen atom vibrates back and forth along the line joining the axis of the two silicon atoms (see Figure 25.20). This vibrational mode is expected to generate a strong polarization field relative to other Si–O vibrations[103] such as the symmetric stretching mode (at about 810 cm[-1]), which can be seen in the linear IR absorption spectrum but without any correlation to the PIA spectra, and the Si–O rocking mode of vibration at ~450 cm[-1]. Notice that, it has been suggested that the asymmetric lineshape of the Si–O stretching mode at ~1100 cm[-1] (see Figure 25.24) is due to the two "quasi-phonon" modes of the amorphous SiO$_2$ medium,[104] with the AS$_1$ (first asymmetric stretching) mode at 1076 cm[-1] being related to "in-phase" motion of adjacent oxygen atoms while the AS$_2$ mode at ~1250 cm[-1] is related to "out-of-phase" motion of adjacent oxygen atoms.[103] Clearly, only the AS$_1$ mode is expected to generate a long-range polarization field, a result that can explain the strong asymmetry of the VISL band (see Figure 25.24), once approaching the high-energy shoulder of the linear Si–O stretching vibration.

Additional experimental results supporting the vibron model come from a single quantum dot spectroscopy of SiNCs fabricated by the laser pyrolysis technique (see Section 25.2). Martin et al.[105] have succeeded to dissolve SiNCs (surrounded by SiO$_2$), disperse them in polymer matrices and to spin-cast the nanocrystals on glass substrates for a single silicon nanoparticle spectroscopy. The PL spectra, shown in Figure 25.28 for two different sizes of nanocrystals, reveal vibronic features of low-energy PL satellites, at about 130–160 meV below the high-energy peaks. These satellite peaks have been interpreted as (surface) phonon-assisted optical transitions where the high-energy PL band is related to a zero-phonon transition (having homogeneous linewidth of ~100 meV) and a low-energy PL shoulder that is related to a phonon-assisted transition. Notice that phonon-assisted optical transitions (both interband[106] and ISL[58]) have been predicted for SiNCs; however,

in most cases the bulk phonons of silicon have been considered rather than polar surface vibrations (which could be either Si–O surface vibrations, or AS phonons of amorphous-SiO$_2$ or LO phonons of a crystalline SiO$_2$ shell). A similar phenomenon of vibronic satellites in the PL spectrum has also been reported for PS particles.[107] In another work, vibronic features due to surface-oxide absorption band using ellipsometric measurements[108] have been reported. Seraphin et al.[109] have shown that modifying the surface passivation of SiNCs does not alter the PL spectrum and only the PL intensity is changed, apparently due to modification of the surface bonds with a higher PL intensity associated with the more polar surface bonds.

Finally, let us discuss the recent magneto-PL experiment reported by Godefroo et al.[110] In this experiment, pulsed magnetic fields up to 50 T have been applied to SiNCs and the corresponding energy shift of the PL has been measured. The high magnetic field acts to further confine the exciton with $\ell_B = \sqrt{\hbar/eB}$ being the magnetic confinement length. Hence, free excitons of bulk nanocrystals are expected to show a blueshift in response to high enough B-fields, while localized or trapped excitons are not expected to show a similar shift. Indeed, no blueshift of the PL has been measured for SiNCs embedded in SiO$_2$. The authors have interpreted these observations as a PL originated from defect centers in the oxide layer that wraps the nanocrystals. However, according to the vibron model, surface-localized vibrons should not present a blueshift either. In another experiment, the authors annealed the samples in hydrogen, at temperatures low enough to avoid the creation of hydrogenated a:Si. In this experiment a blueshift of the PL has been observed, thus confirming the experimental results of Wolkin et al.[17] where H-terminated SiNCs have shown monotonic blueshift of the PL with the decreasing size of the nanocrystals (or with the increasing B-field in the case of the Godefroo experiment[110]). The topic of hydrogen-terminated SiNCs remains a puzzle that has not been addressed in this chapter for the very simple reason that no reliable methods to produce stable, H-terminated SiNCs have been reported so far. Yet, the Si–H bonds are expected to be considerably more polar than the Si–O bonds. Hydrogen atoms are highly mobile ions that can easily penetrate over few nanometers into the nanocrystalline silicon core. Even small amount of hydrogen in the silicon core is expected to produce considerable polarization field of the "core" nanocrystal. This topic calls for further investigation as the current knowledge about H-terminated SiNCs is insufficient for drawing any reliable conclusions.

25.5 Concluding Remarks

This chapter has covered one specific topic from the much broader, rapidly expanding field of silicon nanostructures and their applications, focusing on the question about the origin of the PL and the specific role of QC and SC in the eluding game played by the luminescence. Yet, this specific question is an excellent example to a situation where *small is really different*. Namely, not a property of matter that scales down in proportion to the size of the material, but rather a truly new phenomenon that appears at, and only at the nanometric length scales. The PL from silicon nanostructures appears to be such a property. The role of QC and SC is another fascinating example to the case where *physics meets chemistry* and both disciplines of science merge together into one, multidisciplinary field of nanoscience.

The questions regarding the origin of the PL from silicon nanostructures are far from being fully resolved. Many open questions still exist, some have been discussed in this chapter, some did not, and most likely, some new fundamental questions and puzzles will emerge soon. Yet, the basic direction in attempting to merge QC and SC, rather than choosing one as a sole winner, seems to be the proper approach. It is just a matter of time to see if this subjective opinion of the author of this chpater is indeed correct or not.

Acknowledgments

The author would like to thank many of his (former and current) graduate students at the Hebrew University of Jerusalem for invaluable contributions to this work, particularly, M. Dovrat, D. Krapf, Y. Shalibo, N. Arad, Y. Oppenheim-Goshen, and Y. Raichman. I wish to acknowledge the fruitful collaboration with my colleagues at the Hebrew University, including J. Jedrzejewski, J. Shappir, and especially Prof. Isaac Balberg for stimulating and helpful discussions and for critical reading of the manuscript. I wish to acknowledge the technical staff of the unit for nano-characterization (UNC) at the Hebrew University of Jerusalem, particularly Dr. Inna Popov for her assistance with the TEM measurements. I have enjoyed a fruitful and productive collaboration with X.-H. Zhang from the Chinese Academy of Sciences in Beijing and Prof. S.-T. Lee from the City University of Honk Kong during a joint project on silicon nanowires. This work has been supported by the Israel Science Foundation (ISF), the Israeli Ministry of Science, the binational Chinese–Israeli research grant provided by the ministries of science of both countries.

References

1. S. M. Sze, *Physics of Semiconductors*, 2nd edn., Wiley-Interscience, New York (1981).
2. Al. L. Efros and M. Rosen, The electronic structure of semiconductor nanocrystals, *Annu. Rev. Mater. Sci.* **30**, 475–521 (2000).
3. P. Moriarty, Nanostructured materials, *Rep. Prog. Phys.* **64**, 297–381 (2001).
4. C. J. Murphy and J. L. Coffer, Quantum dots: A primer, *Appl. Spectrosc.* **56**, 16A–27A (2002).
5. L. Pavesi, Will silicon be the photonic material of the third millennium? *J. Phys.: Condens. Matter* **15**, R1169–R1196 (2003).
6. A. G. Nassiopoulou, Silicon nanocrystals in SiO_2 thin layers, in *Encyclopedia of Nanoscience and Nanotechnology*, Vol. 9, H. S. Nalwa, Ed., pp. 793–813, American Scientific, Stevenson Ranch, CA (2004).
7. L.T. Canham, Silicon quantum wire array fabrication by electrochemical and chemical dissolution of wafers, *Appl. Phys. Lett.* **57**, 1046–1048 (1990).
8. A. G. Cullis and L. T. Canham, Visible light emission due to quantum size effects in highly porous crystalline silicon, *Nature* **353**, 335–338 (1991).
9. D. Kovalev, H. Heckler, G. Polisski, and F. Koch, Optical properties of Si nanocrystals, *Phys. Stat. Sol. (b)* **215**, 871–932 (1999).
10. G. Amato, C. Delerue, and H.-J. von Bardeleben, Eds., Structural and optical properties of porous silicon nanostructures, *Optoelectronic Properties of Semiconductor and Superlattices*, Vol. **5**, Gordon & Breach, Amsterdam, the Netherlands (1997) and references therein.
11. A. G. Cullis, L. T. Canham, and P. D. J. Calcott, The structural and luminescence properties of porous silicon, *J. Appl. Phys.* **82**, 909–965 (1997).
12. O. Bisi, S. Ossicini, and L. Pavesi, Porous silicon: A quantum sponge structure for silicon based optoelectronics, *Surf. Sci. Rep.* **38**, 1–126 (2000).
13. J. Gole and D. A. Dixon, Potential role of silanones in the photoluminescence-excitation, visible-photoluminescence-emission, and infrared spectra of porous silicon, *Phys. Rev. B.* **57**, 12002–12016 (1998).
14. J. J. Heitmann, F. Müller, M. Zacharias, and U. Gösele, Silicon nanocrystals: Size matters, *Adv. Mater.* **17**, 795–803 (2005).
15. M. Dovrat, Y. Oppenheim, J. Jedrzejewski, I. Balberg, and A. Sa'ar, Radiative versus nonradiative decay processes in silicon nanocrystals probed by time-resolved photoluminescence spectroscopy, *Phys. Rev. B.* **69**, 155311 (2004).
16. M. Dovrat, Y. Goshen, I. Popov, J. Jedrzejewski, I. Balberg, and A. Sa'ar, The role of radiative and nonradiative relaxation processes in the generation of light from silicon nanocrystals, *Phys. Stat. Sol. (c)* **2**, 3440–3444 (2005).
17. M. V. Wolkin, J. Jorne, P. M. Fauchet, G. Allan, and C. Delerue, Electronic states and luminescence in porous silicon quantum dots: The role of oxygen, *Phys. Rev. Lett.* **82**, 197–200 (1999).

18. A. Sa'ar, M. Dovrat, J. Jedrzejewski, and I. Balberg, Optical inter- and intra-band transitions in silicon nanocrystals: The role of surface vibrations, *Phys. E: Low-Dimensi. Syst. Nanostruct.* **38**, 122–127 (2007).

19. A. Sa'ar, Photoluminescence from silicon nanostructures: The mutual role of quantum confinement and surface chemistry, *J. Nanophoton* **3**, 032501 (2009).

20. A. Givant, J. Shappir, and A. Sa'ar, Photoluminescence anisotropy from laterally anodized porous silicon, *Appl. Phys. Lett.* **73**, 3150–3152 (1998).

21. B. Urbach, E. Axelrod, and A. Sa'ar, Correlation between transport, dielectric, and optical properties of oxidized and nonoxidized porous silicon, *Phys. Rev. B* **75**, 205330 (2007).

22. V. Lehmann, *Electrochemistry of Silicon*, Wiley-VCH, Weinheim, Germany (2002).

23. W. Theiβ, Optical properties of porous silicon, *Surf. Sci. Rep.* **29**, 91–192 (1997).

24. E. Sabatani, Y. Kalisky, A. Berman, Y. Golan, N. Gutman, B. Urbach, and A. Sa'ar, Photoluminescence of polydiacetylene membranes on porous silicon utilized for chemical sensors, *Opt. Mater.* **38**, 1766–1774 (2008).

25. B. Urbach, N. Korbakov, Y. Bar-David, S. Yitzchaik, and A. Sa'ar, Composite structures of polyaniline and mesoporous silicon: Electrochemistry, optical and transport properties, *J. Phys. Chem. C* **111**, 16586–16592 (2007).

26. S. Ben-Tabou de-Leon, R. Oren, M. E. Spira, S. Yitzchaik, and A. Sa'ar, Neurons culturing and biophotonic sensing using porous silicon, *Appl. Phys. Lett.* **84**, 4361–4363 (2004).

27. H. Lin, T. Gao, J. Fantini, and M. J. Sailor, A porous silicon–palladium composite film for optical interferometric sensing of Hydrogen, *Langmuir* **20**, 5104–5108 (2004).

28. S. Furukawa and T. Miyasato, Three-dimensional quantum well effects in ultrafine silicon particles, *Jpn. J. Appl. Phys.* **27**, L2207–L2209 (1988).

29. W. S. Cheong, N. M. Hwang, and D. Y. Yoon, Observation of nanometer silicon clusters in the hot-filament CVD process, *J. Cryst. Growth* **204**, 52–61 (1999).

30. S. Charvet, R. Madelon, R. Rizk, B. Garrido, O. Gonzàlez-Varona, M. López, A. Pérez-Rodríguez, and J. R. Morante, Substrate temperature dependence of the photoluminescence efficiency of co-sputtered Si/SiO₂ layers, *J. Lumin.* **80**, 241–245 (1988).

31. U. Kahler and H. Hofmaeister, Visible light emission from Si nanocrystalline composites via reactive evaporation of SiO, *Opt. Mater.* **17**, 83–86 (2001).

32. A. J. Kenyon, P. F. Trwoga, C. W. Pitt, and G. Rehm, Luminescence efficiency measurements of silicon nanoclusters, *Appl. Phys. Lett.* **73**, 523–525 (1998).

33. Z.-X. Ma, X.-B. Liao, J. He, W. C. Cheng, G. Z. Yue, Y. Q. Wang, and G. L. Kong, Annealing behaviors of photoluminescence from SiO$_x$:H, *J. Appl. Phys.* **83**, 7934–7939 (1998).

34. L. Ferraioli, M. Wang, G. Pucker, D. Navarro-Urrios, N. Dadosso, C. Kompocholis, and L. Pavesi, Photoluminescence of silicon nanocrystals in silicon oxide, *J. Nanomater.* **2007**, 43491 (2007).

35. T. Shimizu-Iwayama, S. Nakao, and K. Saitoh, Optical and structural characterization of implanted nanocrystalline semiconductors, *Nucl. Instrum. Methods B* **121**, 450–454 (1997).

36. K. S. Min, K. V. Scheglov, C. M. Yang, H. A. Atwater, M. L. Brongersma, and A. Polman, Defect-related versus excitonic visible light emission from ion beam synthesized Si nanocrystals in SiO₂, *Appl. Phys. Lett.* **69**, 2033–2035 (1996).

37. J. Linnros, N. Lalic, A. Galeckas, and V. Grivickas, Analysis of the stretched exponential photoluminescence decay from nanometer-sized silicon crystals in SiO₂, *J. Appl. Phys.* **86**, 6128–6134 (1999).

38. Y. Posada, I. Balberg, L. F. Fonseca, O. Resto, and S. Z. Wiesz, Diffusion length measurements of minority carriers in Si-SiO₂ using the photo-grating technique, in *Microcrystalline and Nanocrystalline Semiconductors—2000*, Boston, MA, Materials Research Society Symposia Proceedings, No. 638, F.14.44, pp. 1–6, M. Fauchet, J. M. Buriak, L.T. Canham, N. Koshida, and B. E. White, Jr. Eds., MRS, Warrendale, PA (2001).

39. I. Balberg, E. Savir, J. Jedrzeijewski, A. G. Nassiopoulou, and S. Gardelis, Fundamental transport processes in ensembles of silicon quantum dots, *Phys. Rev. B* **75**, 235329 (2007).

40. I. Balberg, E. Savir, and J. Jedrzeijewski, The mutual exclusion of luminescence and transport in nanocrystalline silicon networks, *J. Non-Cryst. Solids* **338–340**, 102–105 (2004).

41. G. Ledoux, J. Gong, F. Huisken, O. Guillois, and C. Reynaud, Photoluminescence of size-separated silicon nanocrystals: Confirmation of quantum confinement, *Appl. Phys. Lett.* **80**, 4834–4836 (2002).

42. D. J. Lockwood, Z. H. Lu, and J.-M. Baribeau, Quantum confined luminescence in Si/SiO₂ superlattices, *Phys. Rev. Lett.* **76**, 539–541 (1996).

43. V. Vinciguerra, G. Franzò, F. Priolo, F. Iacona, and C. Spinella, Quantum confinement and recombination dynamics in silicon nanocrystals embedded in Si/SiO₂ superlattices, *J. Appl. Phys.* **87**, 8165–8173 (2000).

44. Z. Ma, L. Wang, K. Chen, W. Li, L. Zhang, Y. Bao, X. Wang, J. Xu, X. Huang, and D. Feng, Blue light emission in nc-Si/SiO₂ multilayers fabricated using layer by layer plasma oxidation, *J. Non-Cryst. Solids* **299–302**, 648–652 (2002).

45. P. Photopooulos, A. G. Nassiopoulou, D. N. Kouvatsos, and A. Travlos, Photo- and electroluminescence from nanocrystalline silicon single and multilayer structures, *Mater. Sci. Eng. B* **69–70**, 345–349 (2000).

46. L. Tsybeskov, K. D. Hirschman, S. P. Duttagupta, M. Zacharias, P. M. Fauchet, J. P. McCaffrey, and D. J. Lockwood, Nanocrystalline-silicon superlattice produced by controlled recrystallization, *Appl. Phys. Lett.* **72**, 43–45 (1998).

47. M. Zacharias, J. Heitmann, R. Scholz, U. Kahler, M. Schmidt, and J. Bläsing, Size-controlled highly luminescent silicon nanocrystals: A SiO/SiO₂ superlattice approach, *Appl. Phys. Lett.* **80**, 661–663 (2002).

48. M. Dovrat, Y. Shalibo, N. Arad, S.-T. Lee, and A. Sa'ar, Fine structure and selection rules for excitonic transitions in silicon nanostructures, *Phys. Rev. B* **79**, 125306 (2009).

49. N. G. Shang, U. Vetter, I. Gerhards, H. Hofsäss, C. Ronning, and M. Seibt, Luminescence centres in silica nanowires, *Nanotechnology* **17**, 3215–3218 (2006).

50. K. Sato and K. Hirakuri, Influence of paramagnetic defects on multicolored luminescence from nanocrystalline silicon, *J. Appl. Phys.* **100**, 114303 (2006).

51. M. Dovrat, N. Arad, X.-H. Zhang, S.-T. Lee, and A. Sa'ar, Optical properties of silicon nanowires from cathodoluminescence imaging and time-resolved photoluminescence spectroscopy, *Phys. Rev. B* **75**, 205343 (2007).

52. M. Dovrat, N, Arad, S. T. Lee, and A. Sa'ar, Cathodoluminescence and photoluminescence of individual silicon nanowires, *Phys. Stat. Sol. (a)* **204**, 1512–1517 (2007).

53. C. Meier, A. Gondorf, S. Lüttjohann, A. Lorke, and H. Wiggers, Silicon nanoparticles: Absorption, emission, and the nature of the electronic bandgap, *J. Appl. Phys.* **101**, 103112 (2007).

54. J. A. Moreno, B. Garrido, P. Pellegrini, C. Garcia, J. Arbiol, J. R. Morante, P. Marie, F. Gourbilleau, and R. Rizk, Size dependence of refractive index of Si nanoclusters embedded in SiO_2, *J. Appl. Phys.* **98**, 013523 (2005).

55. M. Mansour, A. En Naciri, L. Johann, J. J. Grob, and M. Stchakovsky, Dielectric function and optical transitions of silicon nanocrystals between 0.6 eV and 6.5 eV, *Phys. Stat. Sol. (a)* **205**, 845–848 (2008).

56. A. En Naciri, M. Mansour, L. Johann, J. J. Grob, and H. Rinnert, Influence of the implantation profiles of Si^+ on the dielectric function and optical transitions in silicon nanocrystals, *J. Chem. Phys.* **129**, 184701 (2008).

57. C. Cimpean, V. Groenewegen, V. Kuntermann, A. Sommer, and C. Kryschi, Ultrafast exciton relaxation dynamics in silicon quantum dots, *Laser Photonics Rev.* **3**, 138–145 (2009).

58. C. Delerue, G. Allen, and M. Lannoo, Theoretical aspects of the luminescence of porous silicon, *Phys. Rev. B* **48**, 11024–11036 (1993).

59. A. Zunger and L. W. Wang, Theory of silicon nanostructures, *Appl. Surf. Sci.* **102**, 350–359 (1996).

60. J. Valenta, R. Juhasz, and J. Linnros, Photoluminescence spectroscopy of single silicon quantum dots, *Appl. Phys. Lett.* **80**, 1070–1072 (2002).

61. J. D. Holmes, K. J. Ziegler, C. Doty, L. E. Pell, K. P. Johnston, and B. A. Korgel, Highly luminescent silicon nanocrystals with discrete optical transitions, *J. Am. Chem. Soc.* **123**, 3743–3748 (2001).

62. R. M. Sankaran, D. Holunga, R. C. Flagan, and K. P. Giapis, Synthesis of Blue luminescent Si nanoparticles using atmospheric-pressure microdischarges, *Nano Lett.* **5**, 537–541 (2005).

63. R. J. Walters, J. Kalkman, A. Polman, H. A. Atwater, and M. J. A. de Dood, Photoluminescence quantum efficiency of dense silicon nanocrystal ensembles in SiO_2, *Phys. Rev. B* **73**, 132302 (2006).

64. T. Suemoto, K. Tanaka, and A. Nakajima, Interpretation of the temperature dependence of the luminescence intensity, lifetime, and decay profiles in porous Si, *Phys. Rev. B* **49**, 11005–11009 (1994).

65. D. Kovalev, H. Heckler, M. Ben-Chorin, G. Polisski, M. Schwartzkopff, and F. Koch, Breakdown of the *k*-conservation rule in Si nanocrystals, *Phys. Rev. Lett.* **81**, 2803–2806 (1998).

66. T. Takagahara and K. Takeda, Theory of the quantum confinement effect on excitons in quantum dots of indirect-gap materials, *Phys. Rev. B* **46**, 15578–15581 (1992).

67. H. Koyama and N. Koshida, Photo-assisted tuning of luminescence from porous silicon, *J. Appl. Phys.* **74**, 6365–6367 (1993).

68. F. Herman and R. V. Kasowski, Electronic structure of defects at Si/SiO_2 interfaces, *J. Vac. Sci. Technol.* **19**, 395–401 (1981).

69. M. L. Brongersma, P. G. Kik, A. Polman, K. S. Min, and H. A. Atwater, Size-dependent electron-hole exchange interaction in Si nanocrystals, *Appl. Phys. Lett.* **76**, 351–353 (2000).

70. X. Wen, L.V. Dao, P. Hannaford, E.-C. Cho, Y. H. Cho, and M. A. Green, Excitation dependence of photoluminescence in silicon quantum dots, *New J. Phys.* **9**, 337 (2007).

71. L. Pavesi, Influence of dispersive exciton motion on the recombination dynamics in porous silicon, *J. Appl. Phys.* **80**, 216–225 (1996).

72. J. Heitmann, F. Müller, L. Yi, M. Zacharias, D. Kovalev, and F. Eichhorn, Excitons in Si nanocrystals: Confinement and migration effects, *Phys. Rev. B* **69**, 195309 (2004).

73. I. Balberg, Electrical transport mechanisms in ensembles of silicon quantum dots, *Phys. Stat. Sol. (c)* **5**, 3771–3775 (2008).

74. M. Ben-Chorin, F. Müller, F. Koch, W. Schirmacher, and M. Eberhard, Hopping transport on a fractal: ac conductivity of porous silicon, *Phys. Rev. B* **51**, 2199–2213 (1995).

75. P. D. Calcott, K. J. Nash, L. T. Canham, M. J. Kane, and D. Brumhead, Identification of radiative transitions in highly porous silicon, *J. Phys. Condens. Matter* **5**, L91–L98 (1993).

76. J. M. Ziman, *Principles of the Theory of Solids*, 2nd edn., Cambridge University Press, Cambridge, U.K. (1972).

77. A. Ourmazd, D. W. Taylor, A. Rentschler, and J. Bevk, Si→SiO_2 transformation: Interfacial structure and mechanism, *Phys. Rev. Lett.* **59**, 213–216 (1987).

78. H. Kageshima, M. Uematsu, K. Akagi, S. Tsuneyuki, T. Akiyama, and K. Shiraishi, Theoretical study on atomic structures of thermally grown silicon oxide/silicon interfaces, *e-J. Surf. Sci. Nanotechnol.* **4**, 584–587 (2006).

79. V. V. Afanas'ev, A. Stesmans, and M. E. Twigg, Epitaxial growth of SiO_2 produced in silicon by oxygen ion implantation, *Phys. Rev. Lett.* **77**, 4206–4309 (1996).

80. C. Kaneta, T. Yamasaki, T. Uchiyama, T. Uda, and K. Kiyoyuki, Structure and electronic property of Si(100)/SiO_2 interface, *Microelectron. Eng.* **48**, 117–120 (1999).

81. T. Shimura, H. Misaki, M. Umeno, I. Takahashi, and J. Harada, X-ray diffraction evidence for the existence of epitaxial microcrystallites in thermally oxidized SiO_2 thin films on Si(111) surfaces, *J. Cryst. Growth* **166**, 786–791 (1995).

82. W. A. Dollase, The crystal structure at 220°C of ortho-rhombic high tridymite from the Steinbach meteorite, *Acta Crystallogr.* **23**, 617–623 (1967).

83. E.-C. Cho, M. A. Green, J. Xia, R. Corkish, and A. Nikulin, Atomistic structure of $SiO_2/Si/SiO_2$ quantum wells with an apparently crystalline silicon oxide, *J. Appl. Phys.* **96**, 3211–3216 (2004).

84. N. Pauc, V. Calvo, J. Eymery, F. Fournel, and N. Magnea, Electronic and optical properties of Si/SiO_2 nanostructures. II. Electron-hole recombination at the Si/SiO_2 quantum-well–quantum-dot transition, *Phys. Rev. B* **72**, 205325 (2005).

85. H. Kageshima and K. Shiraishi, Microscopic mechanism for SiO_2/Si interface passivation: Si=O double bond formation, *Surf. Sci.* **380**, 61–65 (1997).

86. N. Daldosso, M. Luppi, S. Ossicini, E. Degoli, R. Magri, G. Dalba, P. Fornasini, R. Grisenti, F. Rocca, L. Pavesi, S. Boninelli, F. Priolo, C. Spinella, and F. Iacona, Role of the inter-face region on the optoelectronic properties of silicon nano-crystals embedded in SiO_2, *Phys. Rev. B.* **68**, 085327 (2003).

87. A A. Saʼar, Y. Reichman, M. Dovrat, D. Krapf, J. Jedrzejewski, and I. Balberg, Resonant coupling between surface vibra-tions and electronic states in silicon nanocrystals at the strong confinement regime, *Nano Lett.* **5**, 2443–2447 (2005).

88. A. Saʼar, M. Dovrat, J. Jedrzejewski, I. Popov, and I. Balberg, The role of quantum confinement and surface chemistry in silicon nanocrystals at the strong confinement regime, *Phys. Stat. Sol. (a)* **204**, 1491–1496 (2007).

89. N. W. Ashcroft and N. D. Mermin, *Solid State Physics*, Saunders, Philadelphia, PA (1976).

90. J. Singh, *Physics of Semiconductors and Their Heterostructures*, McGraw-Hill, New York (1993).

91. G. Allen and C. Delerue, Efficient intraband optical transi-tions in Si nanocrystals, *Phys. Rev. B* **66**, 233303 (2002).

92. A. S. Moskalenko, J. Berakdar, A. A. Prokofiev, and I. N. Yassievich, Single-particle states in spherical Si/SiO_2 quan-tum dots, *Phys. Rev. B* **76**, 085427 (2007).

93. J. S. de Sousa, J.-P. Leburton, V. N. Freire, and E. F. da Silva, Intraband absorption in silicon nanocrystals: The combined effect of shape and crystal orientation, *Appl. Phys. Lett.* **87**, 031913 (2005).

94. O. O. Verzelen, R. Ferreira, and G. Bastard, Excitonic Polarons in semiconductor quantum dots, *Phys. Rev. Lett.* **88**, 146803 (2002).

95. M. Mahdouani, R. Bourguiga, and S. Jaziri, Polaronic states in Si nanocrystals embedded in SiO_2 matrix, *Physica E* **41**, 228–234 (2008).

96. G. D. Mahan, *Many-Particle Physics*, 3rd edn., Kluwer, New York (2000).

97. K. Hess and P. Vogl, Remote polar phonon scattering in silicon inversion layers, *Solid State Commun.* **30**, 797–799 (1979).

98. J.-Q. Lü and F. Koch, Polar phonon scattering at the $Si-SiO_2$ interface, *Microelectron. Eng.* **48**, 95–99 (1999).

99. For general reviews see: NATO ASI Series on *Quantum Well Intersubband Transitions: Physics and Devices*, E. Rosencher, B. Vinter, and B. Levine, Eds., *Series B: Physics*, Vol. **288**, Plenum Press, New York (1992); (b) *Series E: Applied Sciences*, Vol. **270**, Kluwer, Dordrecht, the Netherlands (1994); (c) *Intersubband Transitions in Quantum Wells: Physics and Devices*, S. S. Li and Yan-Kuin Su, Eds., Kluwer Academic, Boston, MA (1998).

100. A. Saʼar, S. Calderon, A. Givant, O. Ben-Shalom, E. Kapon, and C. Caneau, Energy subbands, envelope states, and intersubband optical transitions in one-dimensional quan-tum wires: The local-envelope-states approach, *Phys. Rev. B* **54**, 2675–2684 (1996).

101. D. Krapf, S. H. Kan, U. Banin, O. Millo, and A. Saʼar, Intersublevel optical transitions in InAs nanocrystals probed by photoinduced absorption spectroscopy: The role of thermal activation, *Phys. Rev. B* **69**, 073301 (2004).

102. H. J. Hrostowski and R. H. Kaiser, Infrared absorption of oxygen in silicon, *Phys. Rev.* **107**, 966–972 (1957).

103. C. T. Kirk, Quantitative analysis of the effect of disorder-induced mode coupling on infrared absorption in silica, *Phys. Rev. B* **38**, 1255–1273 (1988).

104. F. L. Galeener and G. Lucovsky, Longitudinal optical vibra-tions in glasses: GeO_2 and SiO_2, *Phys. Rev. Lett.* **37**, 1474–1478 (1976).

105. J. Martin, F. Cichos, F. Huisken, and C. von Borczyskowski, Electron–phonon coupling and localization of excitons in single silicon nanocrystals, *Nano Lett.* **8**, 656–660 (2008).

106. C. Delerue, G. Allan, and M. Lannoo, Electron-phonon coupling and optical transitions for indirect-gap semicon-ductor nanocrystals, *Phys. Rev. B* **64**, 193402 (2001).

107. M. D. Mason, G. M. Credo, K. D. Weston, and S. K. Buratto, Luminescence of individual porous Si chromophores, *Phys. Rev. Lett.* **80**, 5405–5408 (1998).

108. E. Lioudakis, A. Antoniou, A. Othonos, C. Christofides, A.G. Nassiopoulou, Ch.B. Lioutas, and N. Frangis, The role of surface vibrations and quantum confinement effect to the optical properties of very thin nanocrystalline silicon films, *J. Appl. Phys.* **102**, 083534 (2007).

109. A. A. Seraphin, S.-T. Ngiam, and K. D. Kolenbrander, Surface control of luminescence in silicon nanoparticles, *J. Appl. Phys.* **80**, 6429–6433 (1996).

110. S. Godefroo, M. Hayne, M. Jivanescu, A. Stesmans, M. Zacharias, O. I. Lebedev, G. Van Tendeloo, and V. V. Moshchalkov, Classification and control of the origin of photoluminescence from Si nanocrystals, *Nat. Nanotechnol.* **3**, 174–178 (2008).

Polarization-Sensitive Nanowire and Nanorod Optics

Harry E. Ruda
University of Toronto

Alexander Shik
University of Toronto

26.1 Introduction

Our goal is to present a large group of optical phenomena observed in anisotropic nanostructures such as nanowires (NWs) and nanorods (NRs) placed in an environment with a dielectric constant ε_0 different from that of the nanostructure ε. Actually, this situation occurs for most experimental work on NWs and NRs. For instance, in recent years, a large number of works have emerged devoted to the fabrication and the investigation of NWs grown in a free-standing fashion from catalyst particle nano-islands (Duan and Lieber 2000), or in a dielectric template with cylindrical pores (Mei et al. 2003), as well as for solution-grown NRs (Peng et al. 2000). Since the dielectric constant of the semiconductor ε is usually of the order of 10 or larger, this situation is realized for NWs and NRs free-standing in air, dissolved in liquids, or embedded in most dielectric matrices. Such effects of dielectric mismatch are observed not only for the semiconductor but also for metal nanostructures, where ε also differs noticeably from ε_0 and, besides, has a complex character and a strong frequency dispersion.

Due to the dielectric mismatch, the optical electric field in the vicinity of nanostructures is dramatically distorted by polarization effects (image forces), and the field intensity inside nanostructures, and hence all light-induced phenomena (optical absorption, luminescence, photoconductivity, etc.), strongly depend on the orientation of NWs and NRs related to the light polarization. Similarly, light emitted by such nanostructures is also strongly polarized.

We emphasize that the effects considered have a pure electrodynamic nature. They are not related to the presence or absence of size quantization and other possible size phenomena in nanostructures. The effects have a very general character and, as will be shown later, the only condition for their presence, besides the dielectric mismatch with the environment, is the smallness of the NW and the NR diameter in comparison with the light wavelength, which is well satisfied in most semiconductor and metal nanostructures.

26.2 Single Nanowires and Nanorods

26.2.1 Polarization of Absorption: General Concepts

26.2.1.1 Thin Nanowires and Nanorods

It is well known that at the interface of two media with different dielectric constants, the parallel component of electric field **E**, and the normal component of electric induction ε**E** must be continuous. If we have a cylindrical NW in the external field $\mathbf{E}_0$, this means that its parallel component, $E_{\parallel}$, remains the same inside an NW: $E_{\parallel} = E_{0\parallel}$. On the contrary, the field normal to the NW axis, $E_{\perp}$, is different inside and outside the NW. For a planar interface, we would have $E_{\perp} = \varepsilon_0 E_{0\perp}/\varepsilon$ but for a cylinder where $E_{\perp}$ contains components both parallel and perpendicular to the interface, the expression is more complicated (Landau and Lifshitz 1984):

$$E_{\perp} = \frac{2\varepsilon_0}{\varepsilon + \varepsilon_0} E_{0\perp}. \tag{26.1}$$

If, instead of a long (in theory—infinitely long) NW, we have an NR, which can be considered as a prolate ellipsoid of revolution

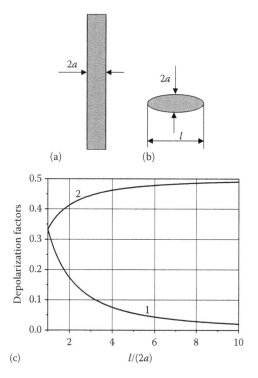

FIGURE 26.1 Schematic view of an NW (a) and an NR (b). Depolarization factors $n_\parallel$ (curve 1) and $n_\perp$ (curve 2) for an NR (c).

with semi-axes a and $l/2$ ($a < l/2$) (Figure 26.1b), the internal field is expressed in terms of the so-called depolarization factors $n_{\parallel,\perp}$:

$$E_{\parallel,\perp} = \frac{\varepsilon_0}{\varepsilon_0 + (\varepsilon - \varepsilon_0)n_{\parallel,\perp}} E_{0\parallel,\perp}. \qquad (26.2)$$

These factors depend only on the aspect ratio a/l (see Figure 26.1c) but not on the absolute dimensions and material of NRs and satisfy the conditions $n_\parallel + 2n_\perp = 1$ and $n_\parallel < n_\perp$, so that for $\varepsilon > \varepsilon_0$ the internal field, as in NWs, is maximal for the NR axis parallel to the electric field of light. For infinitely long NWs, $n_\parallel = 0$, $n_\perp = 1/2$, which gives us $E_\parallel = E_{0\parallel}$ and Equation 26.1. For simplicity, most of the results below are presented for the case of NWs but can be easily generalized to NRs by repeating the same calculations starting from Equation 26.2, rather than Equation 26.1.

For $\varepsilon > \varepsilon_0$, which is always the case for semiconductor nanostructures, Equations 26.1 and 26.2 have two important consequences:

1. The amplitude of the high-frequency electric field and, hence, the probability of optical transitions in a nanostructure depend dramatically on the light polarization, acquiring maximal values for light with a polarization parallel to its axis.
2. Illumination with unpolarized light creates a high-frequency electric field in the nanostructure with the predominant direction of the polarization parallel to its axis.

The practical consequence of the first effect is a strong dependence of the absorption coefficient α (both inter- and intraband) on the light polarization. The ratio of α in NWs for the two light polarizations is

$$\frac{\alpha_\parallel}{\alpha_\perp} = \left| \frac{\varepsilon + \varepsilon_0}{2\varepsilon_0} \right|^2. \qquad (26.3)$$

For most semiconductors, this factor exceeds 30 for free-standing NWs ($\varepsilon_0 = 1$) and 4.7 for NWs in the Al_2O_3 matrix ($\varepsilon_0 \cong 3$ in the optical spectral region). Since photoconductivity and photoluminescence are proportional to the number of optically generated carriers, their anisotropy is given by the same Equation 26.3.

Note that, we consider the classical electrodynamic effects caused exclusively by the difference of the refractive indices for nanostructures and the environment. In quantum nanostructures with a noticeable size quantization, the optical matrix element becomes anisotropic (Sercel and Vahala 1990, McIntyre and Sham 1992), which may quantitatively modify the described polarization effects so that Equation 26.3 acquires an additional factor equal to the square of the ratio of the matrix elements for the two perpendicular directions.

Though the given expressions are derived for static electric fields, they remain valid for high-frequency fields as well (with ε and ε_0 corresponding to the field frequency ω), as long as the NW radius a or both the NR dimensions a, l remain much less than the light wavelength λ. The situation when this inequality is violated will be referred to as the case of "thick" NWs and considered separately in Sections 26.2.1.2 and 26.2.3.2.

The direct measurement of optical absorption in a single NW is difficult but can be replaced by measurements of photoluminescence excitation or photoconductivity (for NWs provided with contacts). All the phenomena discussed should be polarization-sensitive acquiring their maximum for the exciting light polarization parallel to the NWs. The effect in its pure form was first observed and explained in (Wang et al. 2001). The authors observed photo-luminescence and photoconductivity in InP NWs increasing up to 49 times when the polarization of the exciting light changed from perpendicular to parallel (Figure 26.2). This number exactly corresponds to Equation 26.3 for InP in air. Similar effects with a slightly lesser degree of polarization (maybe, owing to the misorientation of the NWs, the role of that is discussed elsewhere (Ruda and Shik 2005a)) were also observed in the photoluminescence of Si NWs (Qi et al. 2003) and the photoconductivity of ZnO and GaN NWs (Fan et al. 2004, Han et al. 2004).

26.2.1.2 Thick Nanowires

The quasi-static approach of the previous section is adequate only for thin nanostructures with $a \ll \lambda$. At the same time, a number of optical experiments (Fan et al. 2004, Hsu et al. 2004, Philipose et al. 2005) are performed in semiconductor NWs with a diameter as large as 100 nm and more. This is comparable with

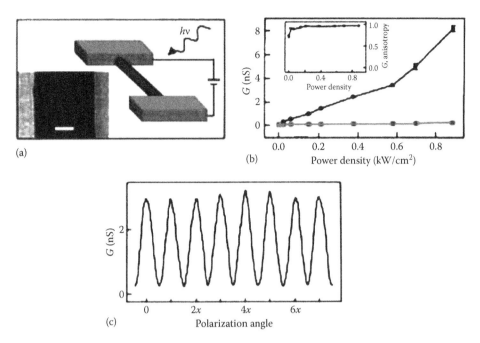

FIGURE 26.2 Photoconductivity versus light-polarization angle in an InP NW with $a = 10\,\text{nm}$. (a) Schematic depicting the use of an NW as a photodetector: inset, field-emission scanning electron microscopy image of an NW and contact electrodes, scale bar $2\,\mu\text{m}$. (b) Conductance, G, versus excitation power density for the illumination polarized parallel (upper curve) and perpendicular (lower curve) to the NW; inset, photoconductivity anisotropy versus excitation power, equal to 0.96 for the shown device. (c) Conductance versus polarization angle. The measurements were performed at room temperature for the excitation wavelength $514.5\,\text{nm}$. (From Wang, J. et al., *Science*, 293, 1455, 2001. With permission.)

the interband light wavelength, which makes the formulae of Section 26.2.1.1 inapplicable for this case.

Now we theoretically consider the anisotropy of light absorption and the emission caused by image forces in semiconducting NWs of an arbitrary radius a. For a comparable to the light wavelength, the high-frequency electric field inside NWs is no longer uniform, and the polarization characteristics acquire a noticeable, non-monotonic spectral dependence allowing one to observe a number of new effects in the optical properties of NWs with submicron thickness.

We consider a cylindrical NW placed in an external ac electric field $\mathbf{E}_0$ and find the real field distribution $\mathbf{E}(\mathbf{r})$ created by $\mathbf{E}_0$ together with the NW-induced image forces. Then we calculate the total absorbed power proportional to $|\mathbf{E}(\mathbf{r})|^2$ integrated over the NW volume and study the dependence of this characteristic on the angle between $\mathbf{E}_0$ and the NW axis, which gives us the polarization dependence of our interest.

The main distinction to Section 26.2.1.1 consists in the need to find $\mathbf{E}(\mathbf{r})$ not from the Laplace but from the Helmholtz wave equation. Using cylindrical coordinates (ρ, z, φ), it can be written inside the NW ($\rho < a$) as

$$\frac{1}{\rho}\frac{\partial}{\partial\rho}\left(\rho\frac{\partial E_i}{\partial\rho}\right)+\frac{1}{\rho^2}\frac{\partial^2 E_i}{\partial\varphi^2}+\frac{\partial^2 E_i}{\partial z^2}+\varepsilon\frac{\omega^2}{c^2}E_i=0 \qquad (26.4)$$

where c is the light velocity. Outside the NW ($\rho > a$) the equation differs from Equation 26.4 only by the replacement $\varepsilon \rightarrow \varepsilon_0$. The character of its solution depends on the mutual orientation of $\mathbf{E}_0$,

the light wave vector $\mathbf{k}$, and the NW axis $\mathbf{z}$. Below, we will give only the final results of calculations referring the reader to the original work (Ruda and Shik 2006).

For $\mathbf{k} \perp \mathbf{z}$ (say, $k = k_x$) and $\mathbf{E}_0 \parallel \mathbf{z}$, when $\mathbf{E}_0(\mathbf{r}) = E_0\,\mathbf{z}\,\exp[ik_0\rho\cos\varphi]$, the solution for $\mathbf{E}$ contains only a single component E_z given by the expression (Batygin and Toptygin 1978)

$$E_z(\rho,\varphi)=E_0\sum_{m=-\infty}^{\infty}i^m\frac{J_m'(ka)H_m^{(1)}(k_0a)-J_m(ka)H_m^{(1)'}(k_0a)}{\sqrt{\varepsilon}J_m'(ka)H_m^{(1)}(k_0a)-\sqrt{\varepsilon_0}J_m(ka)H_m^{(1)'}(k_0a)}$$

$$\times J_m(k\rho)\exp(im\varphi) \qquad (26.5)$$

where J_m and $H_m^{(1)}$ are the Bessel and the first-kind Hankel functions, $k = \sqrt{\varepsilon}\,\omega/c$, $k_0 = \sqrt{\varepsilon_0}\,\omega/c$.

For the same direction of $\mathbf{k}$ but perpendicular polarization, $\mathbf{E}_0 \parallel \mathbf{x}$, Equation 26.4 gives the following general solution inside the NW:

$$\mathbf{E}(\rho,\alpha)=\frac{2E_0 a\varepsilon_0\nabla\left[J_1\left(\dfrac{\sqrt{\varepsilon-\varepsilon_0}\,\omega\rho}{c}\right)\cos\varphi\right]}{\varepsilon_0 J_1\left(\dfrac{\sqrt{\varepsilon-\varepsilon_0}\,\omega a}{c}\right)+\varepsilon\dfrac{\sqrt{\varepsilon-\varepsilon_0}\,\omega a}{c}J_1'\left(\dfrac{\sqrt{\varepsilon-\varepsilon_0}\,\omega a}{c}\right)}. \qquad (26.6)$$

At small a Equation 26.6 transforms into the static expression (Equation 26.1).

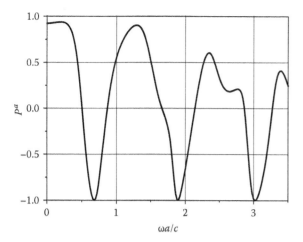

FIGURE 26.3 Spectral dependence of the polarization ratio $P^a = \left(I_\parallel^a - I_\perp^a\right)/\left(I_\parallel^a + I_\perp^a\right)$ in an NW for $\varepsilon = 9$, $\varepsilon_0 = 1$. (From Ruda, H.E. and Shik, A., *J. Appl. Phys.*, 100, 024314, 2006. With permission.)

Equations 26.5 and 26.6 show that, contrary to the case of a thin NW, the field distribution inside it is essentially nonuniform. However, real experimentally measured characteristics, such as the absorption, the photoconductivity, and the photoluminescence excitation spectra are determined not by details of this field distribution but by the total power absorbed, proportional to the integral of the light intensity over the NW cross section $I^a = \iint |E(\varphi,\alpha)|^2 \, d\varphi\rho d\rho$ (superscript *a* means absorption). Figure 26.3 shows the spectral dependence of the ratio $P^a = \left(I_\parallel^a - I_\perp^a/I_\parallel^a + I_\perp^a\right)$ characterizing the polarization dependence of the absorbed light intensity.

For $k_0 a \rightarrow 0$, this ratio tends to the value $(\varepsilon^2 + 2\varepsilon\varepsilon_0 - 3\varepsilon_0^2)/(\varepsilon^2 + 2\varepsilon\varepsilon_0 + 5\varepsilon_0^2)$ corresponding to Equation 26.3. However, at higher light frequencies (or for thicker NWs) we can no longer claim that light with a parallel polarization has a much higher absorption coefficient since $I_\parallel^a$ and $I_\perp^a$, as well as their ratio, demonstrate a strong frequency dependence. This dependence is especially dramatic for $I_\perp^a$, which tends to infinity in a series of critical points corresponding to roots of the denominator Equation 26.6 and $P^a = -1$ in Figure 26.3. In our example of $\varepsilon = 9$, $\varepsilon_0 = 1$, the first of these points are $k_0 a = 0.68, 1.89, 3.02, \ldots$ Considering a cylindrical NW as an optical fiber, we can express each of critical points as the cut-off of a fiber mode LP_{1m} ($m = 1, 2, 3\ldots$) with the same angular field distribution as Equation 26.6. At the cut-off, these modes become purely transverse and excited by an incident wave. The spectrum of $I_\parallel^a(\omega)$ has no singularities due to the absence of modes with a purely longitudinal polarization ($\mathbf{E} \parallel \mathbf{z}$).

Since an infinite value $I_\perp^a$ at some fixed ω is definitely nonphysical, we must discuss what initial theoretical assumptions are inadequate and what phenomena are responsible for the real height of the absorption peaks. In our calculations, we have considered one single NW in the electric field of an infinite plane wave. In this case, the field in an NW can, in theory, acquire arbitrary high values remaining equal to $\mathbf{E}_0$ at very large distances. For an array of NWs with an area density N, the electric

field will be redistributed between the NWs and their environment so that the total wave energy remains the same as that provided with external excitation in the absence of NWs and equal to $\varepsilon_0 E_0^2/4\pi$ per unit volume and the absorption coefficient is finite. There are also some other mechanisms of smoothening the mentioned singularities in $I_\perp^a$, e.g., non-rectilinearity and thickness fluctuations of NWs, which may become dominant at a low NW density.

The described frequency dependence of polarization-sensitive absorption has not yet been studied experimentally, except for the observation (Fan et al. 2004) that two different frequencies of excited light caused different degrees of photoconductivity anisotropy.

26.2.2 Metal Nanostructures: Plasmon Phenomena

26.2.2.1 Longitudinal and Transverse Plasmons

The conclusions of Section 26.2.1.1 and formulae presented there can be applied to both metal and semiconductor NWs and NRs but result in the different spectral dependences of optical anisotropy. The degree of polarization in thin NWs, determined exclusively by $\varepsilon/\varepsilon_0$, in semiconductor nanostructures does not explicitly depend on the light frequency ω (if we ignore the frequency dependence of ε, which in semiconductors is rather weak, even near the interband absorption edge). In metals, on the contrary, ε has a strong frequency dispersion caused by the electron plasma, which makes the polarization characteristics frequency-dependent even in the quasistatic limit of small nanostructures.

For an analytical description of these effects, we will use for ε the simplest Drude expression: $\varepsilon = 1 - \left(\omega_p^2/\omega(\omega - i\nu)\right)$ where ω_p is the plasma frequency in a bulk metal and ν is the scattering rate. Substituting this expression into Equation 26.2, we see that at a small ν (weak electron scattering), the optical electric field inside NRs increases dramatically at the frequency

$$\omega_{\parallel,\perp} = \frac{\omega_p \sqrt{n_{\parallel,\perp}}}{\varepsilon_0(1 - n_{\parallel,\perp}) + n_{\parallel,\perp}} \quad (26.7)$$

called the plasmon frequency,* which is smaller than ω_p and typically belongs to the visible spectral region. The optical absorption spectrum has a strong maximum near the plasmon frequency. In it seen that light polarized along the NR axis generates plasmons with the frequency $\omega_\parallel$ called longitudinal plasmons, while light with the perpendicular polarization generates transverse plasmons with the frequency $\omega_\perp$, larger than $\omega_\parallel$. In a random system of NRs with arbitrary orientations related to the light polarization (or for a nonpolarized optical excitation), the absorption spectrum has two absorption maxima: at $\omega_\parallel$ and $\omega_\perp$ (Kreibig and Vollmer 1995).

---

* $\omega_{\parallel,\perp}$ are often called the surface plasmon frequencies, to distinguish them from the plasmon frequency of bulk material ω_p.

We note that for long NWs $\omega_\parallel \to 0$, in other words, longitudinal plasmons cease to exist. Formally, it is due to the fact that in infinitely long cylinders, the parallel component of the electric field must be continuous at the interface and hence is uniform, without any structure characterizing plasma oscillations.

26.2.2.2 Transverse Plasmons in Nanowires

If we substitute the Drude expression for $\varepsilon(\omega)$ into Equation 26.3, we will see that the anisotropy of the optical absorption in metal NWs has an interesting spectral dependence. Far below the plasmon frequency, $|\varepsilon(\omega)|$ is very large, and the situation is similar to that in semiconductor NWs where the absorption is much higher for the light with parallel polarization. However, near the plasmon frequency, the absorption of the perpendicularly polarized light increases drastically, exceeding that for the parallel polarization and hence changing the sign of the anisotropy, as illustrated by Figure 26.4. At $\omega \to \infty$ the anisotropy tends to a constant value determined by ε_0 and equal to zero at $\varepsilon_0 = 1$.

Thus, the situation near the plasmon resonance is, to some extent, opposite to that of semiconductor NWs where, due to the large ε, the electric field of light is polarized mostly along the NW axis. In metal NWs, for light with the frequency in the transverse plasmon region polarized perpendicular to the NW axes, ε is close to zero. For this reason, the electric field in an NW has essentially normal polarization and very high amplitude—the effect often called the plasmon amplification (see also Section 26.2.5.3).

Though the described spectral dependences of absorption are very specific and interesting, their experimental observation is more difficult than for the effects in semiconductor NWs. In metals, we can measure neither the photoconductivity nor the luminescence excitation and have to restrict ourselves to direct measurements of optical absorption. The most straightforward method to employ is to investigate a large number of NWs suspended in some liquid optically inactive in the spectral region of

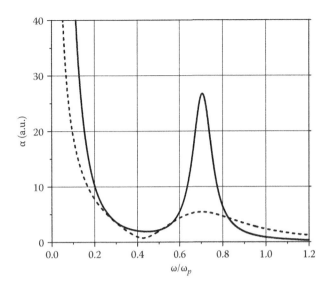

FIGURE 26.5 Absorption spectrum of randomly oriented metal NWs. Solid curve corresponds to $\nu = 0.1\omega_p$, dashed curves to $\nu = 0.5\omega_p$. (From Ruda, H.E. and Shik, A., *Phys. Rev. B*, 72, 115308, 2005a. With permission.)

interest. In this case, the orientation of NWs is random, and no polarization dependence should be observed. A strong anisotropy of absorbing objects will be revealed in a specific character of spectral dependence of the absorption coefficient (Figure 26.5). In this spectrum, strong low-frequency absorption is caused by NWs almost parallel to the light polarization while the plasmon maximum is formed by those with an orientation close to the perpendicular one.

26.2.3 Semiconductor Nanostructures: Polarization of Luminescence

26.2.3.1 Thin Nanowires and Nanorods

In the previous sections 26.2.1 through 26.2.2, we considered the physical phenomena related to the fact that external nonpolarized light becomes strongly polarized inside a nanostructure with a high refractive index. Now, we discuss the related problem of polarization of light emitted by an NW. As a first step, we solve the auxiliary problem of finding the electric field created by an electric dipole placed at the axis of a cylinder with a radius a and a dielectric constant ε, in an environment with ε_0. Details of these calculations can be found in Ruda and Shik (2005a), and here we present only the final results. For the dipole moment $\mathbf{d_0}$ parallel to the cylinder axis (z-axis), the electric field of the radiation far from the NW, has the same amplitude and configuration as would be created by a dipole with the moment $\mathbf{d_0}$ in free space. For the normal orientation of $\mathbf{d_0}$ (along x-axis), the corresponding effective moment is given by $\mathbf{d} = 2\varepsilon_0\mathbf{d_0}/(\varepsilon + \varepsilon_0)$.

As a result, even for NWs with isotropic interband matrix elements (which is the case for cubic semiconductors and a moderate a when size quantization is negligible), where luminescence would otherwise be nonpolarized, a strong polarization under the influence of image forces is acquired.

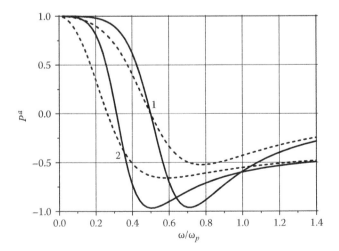

FIGURE 26.4 Absorption anisotropy $P^a = \left(k_\parallel - k_\perp\right)/\left(k_\parallel + k_\perp\right)$ in metal NWs for $\varepsilon_0 = 1$ (curves 1) and $\varepsilon_0 = 3$ (curves 2). Solid curves correspond to $\nu = 0.1\omega_p$, dashed curves to $\nu = 0.5\omega_p$. (From Ruda, H.E. and Shik, A., *Phys. Rev. B*, 72, 115308, 2005a. With permission.)

As a second step, we consider the net polarization of luminescence from a line x, $y = 0$ containing a mixture of different emitting dipoles. We take a point z along the wire axis containing three mutually perpendicular dipoles d_x, d_y, d_z with a frequency ω and calculate the resulting electric field far from the NW at some point $(x_0, 0, 0)$ using the standard formula for the electric field of an emitting dipole and adding contributions of all line segments by integrating over z. Since radiation from different segments is incoherent, we do not add electric fields, but Poynting vectors $\mathbf{S}$ containing no phase factors. The resulting $\mathbf{S}$ is directed along the x-axis and for the two different light polarizations is

$$(S_x)_\parallel = \frac{\eta\omega^4}{6\pi\varepsilon_0^2 c^3 x_0}\left(d_x^2 + 2d_z^2\right), \tag{26.8}$$

$$(S_x)_\perp = \frac{\eta\omega^4}{2\pi\varepsilon_0^2 c^3 x_0}d_y^2 \tag{26.9}$$

where η is the density of dipoles per unit line length.

Combining these two problems and substituting $d_{x,y} = 2\varepsilon_0 d_{0x,y}/(\varepsilon + \varepsilon_0)$; $d_z = d_{0z}$ into Equations 26.8 and 26.9, provides the resulting polarization of light emitted by an NW. The intensity ratio for different light polarizations in the case of isotropic matrix elements is

$$\frac{I_\parallel^e}{I_\perp^e} = \frac{d_x^2 + 2d_z^2}{3d_z^2} \tag{26.10}$$

where superscript e means emission. If the bare dipole moment components $d_{0x,y,z}$ do not coincide due to size quantization or/and crystal anisotropy, this modifies the polarization character introducing anisotropy even in the absence of image forces, similar to the case of absorption.

It is seen from Equation 26.10 that light emitted by an NW with a dielectric constant higher than that of its environment, will be strongly polarized in the wire direction, which was clearly demonstrated experimentally (Wang et al. 2001, Qi et al. 2003). The effect is qualitatively similar to the property of absorption considered in Section 26.2.1.1 but the magnitude of polarization is different. By comparing Equation 26.3 with Equation 26.10 one can see that the degree of polarization for luminescence is less than for absorption. It is caused by the fact that, according to Equation 26.8, light with parallel polarization is generated not only by d_z dipole moments but also by d_x that are weakened by image forces. For example, for InP NWs ($\varepsilon = 12.7$) where, as was already mentioned in Section 26.2.1.1, $\alpha_\parallel/\alpha_\perp \cong 49$, Equation 26.10 gives $I_\parallel^e/I_\perp^e \cong 30$, in good agreement with the experimentally observed values of $\alpha_\parallel/\alpha_\perp \cong 48$ and $I_\parallel^e/I_\perp^e \cong 24$ (Wang et al. 2001).

All the results obtained above for semiconductor NWs remain valid also for NRs, except for the numerical values of polarization coefficients, which for NRs depend not only on $\varepsilon/\varepsilon_0$ but also on the depolarization coefficients, or, in other words, on the NR aspect ratio.

26.2.3.2 Thick Nanowires

Now we consider the polarization properties of luminescence from thick NWs where, according to Section 26.2.1.2, the electric field cannot be assumed uniform. Though in fact luminescence occurs throughout the whole NW volume, we will model it by an effective emitting dipole at the axis ($\rho = 0$). (Precise results can be obtained, if necessary, by using general formulae for a field created in a system of coaxial cylindrical layers by a dipole with an arbitrary position $\mathbf{r}_0$ followed by integration over $\mathbf{r}_0$, to be performed only numerically.) The general approach to similar problems is described in (Chew 1990), which allows us to present only the final expression for the resulting electric field outside the NW.

For the dipole moment $\mathbf{d}_0$ oriented along the NW axis ($d_0 = d_{0z}$), the electric field E_z far from the NW (at $\rho \gg k_0^{-1}$) is given by

$$E_z \cong i\frac{d_{0z}\omega^2}{\varepsilon c^2 r}\left(\varepsilon - \varepsilon_0\cos^2\theta\right)\exp(ik_0 r - i\pi/4)$$

$$\times \frac{J_1(k_\rho^0 a)H_0^{(1)}(k_\rho^0 a) - J_0(k_\rho^0 a)H_1^{(1)}(k_\rho^0 a)}{J_1(k_\rho^0 a)H_0^{(1)}(k_{\rho 0}^0 a) - \frac{k_{\rho 0}^0}{k_\rho^0}J_0(k_\rho^0 a)H_1^{(1)}(k_{\rho 0}^0 a)} \tag{26.11}$$

where

$$k_\rho^0 = k_0\sqrt{\varepsilon/\varepsilon_0 - \cos^2\theta} = k\sqrt{1 - \varepsilon_0\cos^2\theta/\varepsilon}$$

$$k_{\rho 0}^0 = k_0\sin\theta$$

$$r = \sqrt{\rho^2 + z^2}$$

k_ρ^0 is the distance to the dipole

$\theta = \arccos(z/r)$ is the angle from the NW axis

If we consider the light emission normal to the dipole (along the y-axis) where $\theta = \pi/2$, the corresponding electric field will have the same coordinate dependence as that of a free dipole $\sim(d_{0z}\omega^2/c^2 r)\exp(ik_0 r)$ with an additional factor given by the fraction in Equation 26.11. Thus, the influence of image forces is equivalent to the replacement of $d_0 z$ with some effective dipole (Ruda and Shik 2006):

$$d_z = d_{0z}\sqrt{\varepsilon}\,\frac{J_1(ka)H_0^{(1)}(ka) - J_0(ka)H_1^{(1)}(ka)}{\sqrt{\varepsilon}J_1(ka)H_0^{(1)}(ka) - \sqrt{\varepsilon_0}J_0(ka)H_1^{(1)}(ka)} \equiv A_z d_{0z}. \tag{26.12}$$

For the perpendicular dipole orientation ($d_0 = d_{0x}$), similar calculations give

$$d_x = d_{0x}\sqrt{\varepsilon}\,\frac{J_1'(ka)H_0^{(1)}(ka) - J_0(ka)H_1^{(1)'}(ka)}{\sqrt{\varepsilon_0}J_1'(ka)H_0^{(1)}(ka) - \sqrt{\varepsilon}J_0(ka)H_1^{(1)'}(ka)} \equiv A_x d_{0x}. \tag{26.13}$$

Figure 26.6a shows the frequency dependence of the effective dipoles d_x and d_z. At $\omega \to 0$ they acquire the static values $d_z = d_{0z}$, $d_x = 2\varepsilon_0 d_{0x}/(\varepsilon + \varepsilon_0)$ given in Section 26.2.3.1, and then demonstrate strong oscillations. At a larger ω (or a larger a), the phases

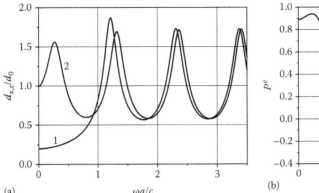
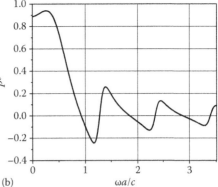

FIGURE 26.6 (a) Spectral dependence of the effective dipoles d_x (1) and d_z (2) for an NW with $\varepsilon = 9$, $\varepsilon_0 = 1$; (b) spectral dependence of the polarization ratio P^e for the same NW. (From Ruda, H.E. and Shik, A., *J. Appl. Phys.*, 100, 024314, 2006. With permission.)

of oscillations close in, d_z approaches d_x, so that the radiation becomes almost unpolarized.

So far, we have analyzed the emission characteristics of one single effective dipole in an NW. To obtain the radiation characteristics of the whole NW, we must perform an integration over its length assuming the dipoles distributed uniformly along z, exactly as it was done in Section 26.2.3.2. Substituting Equations 26.12 and 26.13 into the first equality of Equation 26.10 and assuming an isotropic internal emission ($d_{0z} = d_{0x}$), we obtain the spectral dependence of the resulting polarization ratio $P^e = \left(I_\parallel^e - I_\perp^e / I_\parallel^e + I_\perp^e \right)$ presented at Figure 26.6b. The latter is seen to have a strong oscillatory character changing sign at some critical frequencies differing from those describing the polarization dependence of absorption (Figure 26.3).

It is evident from Figure 26.6b that, if the luminescence spectrum of an NW material contains several lines with noticeably different frequencies, the polarizations of these lines can be essentially different and even have an opposite sign. It was confirmed by experiments (Hsu et al. 2004) in ZnO NWs. This material contains two strong luminescence lines with frequencies differed 1.3 times (Figure 26.7a). As is seen from Figure 26.7b, these lines have opposite angular dependences, indicating at opposite signs of luminescence anisotropy.

Note that the same polarization phenomena must be observed not only for photo- but for any other mechanisms of luminescence. In the case of photoluminescence, a change in the exciting light polarization (for a fixed ω_{ex}) does not change the described spectra $I_\parallel^e / I_\perp^e (\omega_{em})$, but modifies the amplitudes $I_\parallel^e$ and $I_\perp^e$ synchronously according to the polarization dependence of the absorbed light intensity described in Section 26.2.1.2.

The luminescence excitation spectrum in most physical systems (including thick NWs with microscopic electronic properties coinciding with those of bulk semiconductors) actually reproduces the optical absorption spectrum and hence is given by the formulae of Section 26.2.1.2. Thus, we may claim that the polarization dependence of the luminescence excitation spectrum for any emission frequency and polarization is given by Figure 26.3.

26.2.4 Nonlinear Phenomena in Metal and Semiconductor Nanowires

26.2.4.1 Theory

It is evident that the polarization dependence of the internal optical field in NWs and NRs must result in anisotropy not only of linear but also of nonlinear optical phenomena, such as, e.g., the second harmonic generation (SHG). Moreover, one may

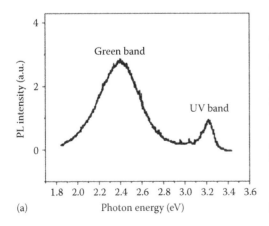
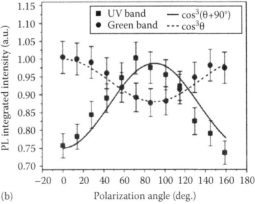

FIGURE 26.7 Luminescence spectrum of ZnO NW (a) with two emission lines and an angular dependence of the luminescence intensity (b) in these lines. (From Hsu, N.E. et al., *J. Appl. Phys.*, 96, 4671, 2004. With permission.)

expect the anisotropy of nonlinear effects to be larger than the linear ones, since the former are proportional to a higher degree of electric field.

The analysis of the SHG in NWs and NRs is less straightforward than that of linear effects where we took standard formulae for light absorption or emission in bulk materials and simply substituted the specific optical field distribution of nanostructures into them. On the contrary, in uniform bulk materials characterized by inversion symmetry, the SHG is completely absent. Hence, the theory of this effect in nanostructures should be developed from scratch.

We start from the classical expression connecting nonlinear polarization at the frequency 2ω with the spatial distribution of the electric field at the frequency ω (Blombergen et al. 1968):

$$\mathbf{P}^{2\omega} = \gamma \nabla (\mathbf{E}^{\omega})^2 + \beta \mathbf{E}^{\omega} (\nabla \mathbf{E}^{\omega}). \tag{26.14}$$

For nonconducting media, $\beta = -2\gamma$ and is determined by the square of linear susceptibility, while for metals $\beta = e/(8\pi m^{\star}\omega^2)$ and $\gamma = \beta[1 - \varepsilon(2\omega)]$ (for simplicity, $\varepsilon_0 = 1$ is assumed). Our task is to find $\mathbf{P}^{2\omega}$ for the $\mathbf{E}^{\omega}(\mathbf{r})$ determined for nanostructures in Section 26.2.1 and to use it for the calculation of the SHG intensity.

For the case of spherical nanocrystals, it was shown (Agarwal and Jha 1982) that the two terms on the right-hand side of Equation 26.14 correspond, respectively, to the volume component $\mathbf{P}_v^{2\omega}$ caused by the coordinate-dependent $\mathbf{E}^{\omega}$ inside the nanostructure and having a dipole symmetry, and the surface component $\mathbf{P}_s^{2\omega}$ caused exclusively by the nanostructure surface and with a quadrupole symmetry. Since the second term in Equation 26.14 for small nanocrystals essentially exceeds the first one, $\mathbf{P}_s^{2\omega}$ has a larger amplitude than $\mathbf{P}_v^{2\omega}$ but, at the same time, with a quadrupole, rather than a dipole symmetry, which emits less efficiently. For this reason, the relative role of these components is *a priori* not evident and depends on the direction and the polarization of light.

We restrict ourselves to the case of thin cylindrical NWs where $\omega a/c \ll 1$ can be used as a perturbation parameter. The particular expressions for $\mathbf{E}^{\omega}$ depend on the directions of the wave vector $\mathbf{k}$ and the electric field $\mathbf{E}_0^{\omega}$ of light related to the NW axis (z-axis) and can be obtained by expansion of the field distributions Equations 26.5 and 26.6 in terms of ω. There exist three possible optical geometries.

a. $\mathbf{k} \parallel \mathbf{z}$; $\mathbf{E}_0^{\omega} \perp \mathbf{z}$ (longitudinal propagation, transverse polarization)

In this case

$$E^{\omega} = E_x^{\omega} = \frac{2E_0^{\omega}}{[\varepsilon(\omega)+1]}(1+ikz) \tag{26.15}$$

($k = \omega/c$). Substitution of Equation 26.15 into Equation 26.14 shows that $\mathbf{P}_v^{2\omega}$ is determined only by the coordinate-dependent component of $\mathbf{E}^{\omega}(\mathbf{r})$, directed along the NW axis (z-axis) and has the value

$$P_v^{2\omega} = \frac{8i\gamma k[E_0^{\omega}]^2}{[\varepsilon(\omega)+1]^2} = const(\mathbf{r}). \tag{26.16}$$

Thus, each elementary volume of the NW dV can be considered as an elementary dipole $\mathbf{P}_v^{2\omega} dV$ parallel to the NW axis. For such dipole orientation, its radiation is not disturbed by the image forces and is the same as for free dipoles (Ruda and Shik 2005a), not depending on the NW dielectric constant at the emission frequency, $\varepsilon(2\omega)$. The resulting radiation electric field at the distance $r \gg a$ from NW can be found as a sum of elementary dipoles along the whole NW. Calculations, shown in more detail in Ruda and Shik (2007a), show that at a large distance r from an NW ($r \gg a$; $kr \gg 1$) the SHG intensity per unit NW length is

$$I_v^{2\omega} = \frac{\pi^3 \omega^3 a^4 [P_v^{2\omega}]^2}{c^2} = \frac{64\pi^3 \gamma^2 \omega^5 a^4 [E_0^{\omega}]^4}{c^4 [\varepsilon(\omega)+1]^4}. \tag{26.17}$$

The radiation is emitted isotropically (independent of a particular direction of light polarization).

The surface component of polarization $\mathbf{P}_s^{2\omega}$ is determined by an $\mathbf{E}^{\omega}$ discontinuity at the NW interface caused by a dielectric mismatch, and for its determination we can retain only the first, uniform term in Equation 26.15. After simple transformations, we have

$$P_s^{2\omega} = \frac{4\beta[\varepsilon(\omega)-1][E_0^{\omega}]^2}{[\varepsilon(\omega)+1]^2}\delta(\rho - a)\cos\alpha. \tag{26.18}$$

(ρ,α) are polar coordinates in the xy-plane. The direction of $\mathbf{P}_v^{2\omega}$ coincides with that of the $\mathbf{E}_0^{\omega}$ (x-axis). This distribution of polarization has a quadrupole symmetry. Due to the coordinate dependence of $P_s^{2\omega}$, the determination of the electric field of emitted light must contain integration not only over z, but over ρ and α as well. Besides, in this case, we deal with elementary dipoles directed not parallel (as for Equation 26.16) but perpendicular to the NW axis. Due to the image forces, the ac electric field of these dipoles is suppressed by the factor $2/[\varepsilon(2\omega) + 1]$ (see Section 26.2.3.1). Under the same conditions as before, we get the intensity of this surface-related SHG component:

$$I_s^{2\omega} = \frac{4\pi^3 \beta^2 \omega^5 a^4 [\varepsilon(\omega)-1]^2 [E_0^{\omega}]^4}{c^4 [\varepsilon(\omega)+1]^4 [\varepsilon(2\omega)+1]^2}\sin^2(2\varphi) \tag{26.19}$$

where φ is the angle between the exciting-light wave vector $\mathbf{k}$ and the direction to the observation point. As for any quadrupole emitter, the intensity $I_s^{2\omega}$ has four directional lobes.

b. $\mathbf{k} \perp \mathbf{z}$; $\mathbf{E}_0^\omega \parallel \mathbf{z}$ (transverse propagation, longitudinal polarization)

In this case

$$E^\omega = E_z^\omega = E_0^\omega \left(1 + ikx \frac{1 + \sqrt{\varepsilon(\omega)}}{2} \right) \tag{26.20}$$

and $\mathbf{P}_v^{2\omega}$, which is always colinear with the light wave vector, is directed along the x-axis normal to the NW and is equal to

$$P_v^{2\omega} = i\gamma k \left[E_0^\omega \right]^2 \left[1 + \sqrt{\varepsilon(\omega)} \right] = \text{const}(\mathbf{r}). \tag{26.21}$$

For such an orientation of elementary dipoles, the SHG is anisotropic with the radiation pattern directed perpendicularly to $\mathbf{k}$, and the ac electric field of these dipoles contains the factor $2/[\varepsilon(2\omega) + 1]$. Adding the fields of elementary dipoles in the same manner as before, we get

$$I_v^{2\omega} = \frac{4\pi^3 \omega^3 a^4 \left[P_v^{2\omega} \right]^2}{c^2 \left[\varepsilon(2\omega) + 1 \right]^2} \sin^2 \varphi$$

$$= \frac{4\pi^3 \gamma^2 \omega^5 a^4 \left[E_0^\omega \right]^4 \left| 1 + \sqrt{\varepsilon(\omega)} \right|^2}{c^4 \left[\varepsilon(2\omega) + 1 \right]^2} \sin^2 \varphi. \tag{26.22}$$

Since $\mathbf{P}_s^{2\omega}$ is caused only by the normal component of $\mathbf{E}_0^\omega$ suffering discontinuity at the NW interface, in the case considered it vanishes, and the SHG has only a dipole component given by Equation 26.22 and no quadrupole component.

c. $\mathbf{k} \perp \mathbf{z}$; $\mathbf{E}_0^\omega \perp \mathbf{z}$ (transverse propagation, transverse polarization)

$$E^\omega = E_x^\omega = \frac{2E_0^\omega}{\left[\varepsilon(\omega) + 1 \right]} (1 + iky). \tag{26.23}$$

This geometry combines characteristic features of the two previous cases. As in case (a), the SHG contains both dipole and quadrupole components. The dipole component of the polarization has the same value as Equation 26.16 but a perpendicular, rather than a longitudinal, orientation. For this reason, its radiation has an anisotropic pattern and is affected by the image force (as in case (b)) resulting in the additional dependence of the SHG on $\varepsilon(2\omega)$:

$$I_v^{2\omega} = \frac{4\pi^3 \omega^3 a^4 \left[P_v^{2\omega} \right]^2}{c^2 \left[\varepsilon(2\omega) + 1 \right]^2} \sin^2 \phi = \frac{256\pi^3 \gamma^2 \omega^5 a^4 \left[E_0^\omega \right]^4}{c^4 \left[\varepsilon(\omega) + 1 \right]^4 \left[\varepsilon(2\omega) + 1 \right]^2} \sin^2 \phi. \tag{26.24}$$

The surface polarization component caused by the first term in Equations 26.15 and 26.23 is exactly the same as in the case (a) (it results from the fact that this component is determined exclusively by light polarization being independent of the value and the direction of $\mathbf{k}$) and hence is described by Equation 26.19.

Let us summarize the results obtained thus far. The SHG was shown to consist of volume and surface components. They have, respectively, dipole and quadrupole symmetry and a different spatial distribution of emitted radiation at 2ω. It can be isotropic (Equation 26.17), bidirectional (Equations 26.22 and 26.24), or of a four-lobed shape (Equation 26.19). However, the integral radiation intensity, as in the case of spherical nanodots (Agarwal and Jha 1982), is given by qualitatively similar expressions differing by numerical coefficients and factors depending only on the NW dielectric constant. For this reason, SHG anisotropy is independent of the NW radius a (until the condition $a\omega/c \ll 1$ used in our calculations, remains valid). The explicit frequency dependence $I^{2\omega} \sim \omega^5$ is also the same in all formulae, and the possible spectral dependence of SHG anisotropy may occur only due to the frequency dispersion of the NW dielectric constant ε. The latter is of minor importance in semiconductors and dielectrics but may be very strong in metal NWs. That is why the characteristics of SHG anisotropy of these two classes of NW materials are very different and will be considered separately.

Semiconductor materials are characterized by high values of the dielectric constant (~10 or even more) with relatively weak frequency dependence. For this reason, in our estimates we do not distinguish between $\varepsilon(\omega)$ and $\varepsilon(2\omega)$ denoting both of them as ε. The ratio of the integral SHG for different NW orientations is independent not only of the NW material and the diameter but also of the light frequency ω, being a function of one single parameter ε. In the limit $\varepsilon \gg 1$, the ratio of the total SHG intensities $\overline{I_v^{2\omega}} + \overline{I_s^{2\omega}}$ (symbol $\bar{I}$ means averaging over ϕ) for all three cases considered is $36:\varepsilon^3:4$. It means that the SHG is maximal for the case (b) when the exciting light is polarized along the NW axis. The result is not unexpected since only in this geometry the optical electric field inside the NW is not weakened by NW polarization and is equal to the external field. It correlates with the linear optical phenomena (absorption, photoconductivity, photoluminescence) which are also maximal for this light polarization. Much less trivial is the dependence of the SHG in perpendicularly polarized light in the direction of the light wave vector. The SHG is essentially larger for light propagating along the NW since, for this geometry, the effective dipoles forming $\mathbf{P}_v^{2\omega}$ are oriented parallel to the NW and their emission is not weakened by NW polarization.

It is worth noting that in all three cases, the spatial distribution of the SHG has different characteristics. In case (a), the second harmonic is emitted in all directions perpendicular to the NW (the contribution of the quadrupole component is around 10%), while in case (b), the emission has a dipole and in case (c), a quadrupole symmetry. This fact should be taken into account while interpreting experimental results (see Section 26.2.4.2).

The main distinguishing feature of metal systems is the strong frequency dispersion of $\varepsilon(\omega)$, which, as in Section 26.2.2, will be approximated by the Drude formula. It is seen that, due to the factors $\varepsilon(\omega) + 1$ and $\varepsilon(2\omega) + 1$ in the denominators of Equations 26.17, 26.19, 26.22, and 26.24, the SHG is amplified dramatically near the frequencies $\omega_p/\sqrt{2}$ and $\omega_p/(2\sqrt{2})$ when the excitation frequency or its second harmonic coincides with the transverse plasmon frequency in a cylinder. In case (b), the resonance exists only for 2ω since the exciting light is polarized along the NW and does not interact with plasmons.

It is interesting to consider SHG anisotropy near these two resonance frequencies when the effect itself is anomalously large. As has been mentioned, at $\omega \cong \omega_p/\sqrt{2}$ the SHG increases dramatically for cases (a) and (c) remaining constant in case (b). In other words, for light with parallel polarization, the SHG is negligibly small compared to the case of perpendicular polarization (the situation opposite to semiconductor NWs). The surface quadrupole component (Equation 26.19) is in this case almost two orders of magnitude less than the volume dipole component. The formulae, Equations 26.17 and 26.24, describing the latter for two different geometries, differ by only 12%. This means that for perpendicular polarization, contrary to the semiconductor case, the total SHG intensity is practically independent of the direction of illumination, though in case (a), emission is isotropic and is polarized parallel to the NWs, while in case (c), it has a dipole symmetry and is polarized perpendicular to them.

At the plasmon resonance for emitted light, $\omega \cong \omega_p/(2\sqrt{2})$, all contributions to the SHG, except Equation 26.17, contain the diverging factor $[\varepsilon(2\omega) + 1]^2$, so that the SHG for all light directions and polarization increases at $\omega \to \omega_p/(2\sqrt{2})$ in a similar manner with the anisotropy remaining constant. Taking into account that $\varepsilon(\omega_p/(2\sqrt{2})) \cong -7$, it can be seen that in the vicinity of this low-frequency resonance, light with the parallel polarization (case (b)) causes a SHG three orders of magnitude larger than that with perpendicular polarization, qualitatively similar to the situation in semiconductor NWs.

26.2.4.2 Experiment

First experimental investigations of SHG in NWs (Barzda et al. 2008) were performed on ZnSe NWs with a length of 8–10 microns and a diameter of 80–100 nm, excited by a laser at a 1029 nm wavelength, that satisfies the quasi-static condition $\omega a/c \ll 1$. The angular dependence of the SHG efficiency $I_{SH}(\theta)$ measured by changing the angle θ between the sample axis and the excitation beam polarization is presented by experimental points at Figure 26.8. It is seen that this dependence has a very strong character with $I_{SH}(0)$ (parallel polarization) exceeding $I_{SH}(90°)$ (perpendicular polarization) by more than an order of magnitude.

An adequate quantitative interpretation must be based on the theoretical results of Section 26.2.4.1. In the experiment, the light wave vector $\mathbf{k}$ was always perpendicular to the NW, so that only conclusions for the cases (b) and (c) are relevant. It is claimed that the SHG caused by the optical field with a longitudinal polarization $E_\parallel$ can be written as the radiation

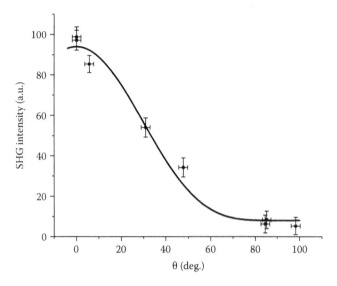

FIGURE 26.8 Polarization dependence of the SHG intensity (Barzda et al. 2008). The dots represent experimentally measured values, the solid line corresponds to the theoretical formulae (Equations 26.25 and 26.26). (From Barzda, V. et al., *Appl. Phys. Lett.*, 92, 113111, 2008. With permission.)

of an effective dipole parallel to the light wave vector with the intensity distribution Equation 26.22 proportional to $\sin^2\varphi$, while in the SHG caused by $E_\perp$, the dominating component has a quadrupole character (Equation 26.19) with the intensity proportional to $\sin^2(2\varphi)$. In the experiments, the detector collected radiation from a wide angle $-49° < \varphi < 49°$ (see Figure 26.9), so that Equations 26.19 and 26.22 should be integrated in these limits. This results in an additional form-factor equal to 0.36 for parallel and to 0.93 for perpendicular light polarization. By taking into account that $E_\parallel = E_0 \cos\theta$ and $E_\perp = 2E_0 \sin\theta/(\varepsilon + 1)$, we obtain the final result for the detected intensity of the SHG:

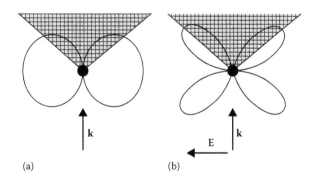

FIGURE 26.9 Schematic angular distribution of the SHG intensity for: (a) $\theta = 0$ (dipole mode) and (b) $\theta = 90°$ (quadrupole mode). $\mathbf{k}$ depicts the wave vector of the excitation light. The NW axis is perpendicular to the page. The orientation of the optical electric field $\mathbf{E}_0^\omega$ is perpendicular to the page in (a) and lies in the page plane in (b). The photodetector collects SHG radiation emitted in the forward direction within the shaded area. (From Barzda, V. et al., *Appl. Phys. Lett.*, 92, 113112, 2008. With permission.)

$$I_{SH} \propto E_0^4 \left[\cos^4 \theta + \frac{10.3(\varepsilon-1)^2 \sin^4 \theta}{(\varepsilon+1)^4 (1+\sqrt{\varepsilon})^2} \right]. \qquad (26.25)$$

For semiconductor NWs with $\varepsilon \gg 1$, Equation 26.25 predicts giant angular oscillations of the SHG with the maximum/minimum ratio of the order $\varepsilon^3/10$. For ZnSe, the value of ε reported by different authors varies from 5.9 to 7.2 (Grigoriev and Meilikhov 1997) but even for the smallest of these numbers, the theoretical ratio exceeds the corresponding experimental value from Figure 26.8, which is close to 20. The most evident explanation is attributed to deviations from an ideal cylindrical NW geometry used in deriving Equations 26.19 and 26.22. Free-standing NWs are inevitably bent, so that the local value of θ changes along the NW. After replacing θ by $\theta + \delta\theta$ and averaging over small $\delta\theta$ with Gaussian properties, Equation 26.25 acquires an additional term

$$\delta I_{SH} \propto 3E_0^4 \overline{(\delta\theta)^2} \sin^2 \theta \left[2\cos^2 \theta + \overline{(\delta\theta)^2} \sin^2 \theta \right]. \quad (26.26)$$

As a result, for $\overline{(\delta\theta)^2}\varepsilon^{3/2} > 2$ the maximum/minimum ratio becomes equal to $3\overline{(\delta\theta)^2}$, independent of the NW dielectric constant. The solid line in Figure 26.8 shows the theoretical angular dependence $I_{SH}(\theta)$ given by Equations 26.25 and 26.26 demonstrating a good agreement between theory and experiment. The best fitting of results was obtained for the value of the average NW corrugation $\overline{(\delta\theta)^2} \cong 0.2$.

26.2.5 Core–Shell Nanowires

26.2.5.1 Potential Distribution

So far, we have separately analyzed the optical properties of semiconductor and metal nanostructures and demonstrated their essential distinctions. In this connection, it is very interesting to investigate the properties of core–shell structures containing both types of layers, where luminescence in the semiconductor part can be amplified due to the plasmon effects in the metal part. Fabrication of such structures has been demonstrated experimentally (Lauhon et al. 2002, Choi et al. 2003).

As a starting point, we must find the formula for the intensity of the optical electric field inside such a structure, which would generalize Equation 26.1 to the case of an NW consisting of a central core with r_1 and a dielectric constant ε_1, and a shell with radius r_2 and a dielectric constant ε_2. Calculations show (Ruda and Shik 2005a) that an external electric field $\mathbf{E}_0$, perpendicular to the NW axis, creates in the core a uniform field

$$\mathbf{E}_1 = \frac{4\varepsilon_0 \varepsilon_2 \mathbf{E}_0}{2\varepsilon_2(\varepsilon_1 + \varepsilon_0) + p(\varepsilon_2 - \varepsilon_1)(\varepsilon_2 - \varepsilon_0)} \equiv A\mathbf{E}_0 \quad (26.27)$$

where $p = 1 - (r_1/r_2)^2$ is the relative shell volume. The field in the shell is a sum of a uniform field and that of a linear dipole:

$$\mathbf{E}_2 = \frac{2\varepsilon_0}{2\varepsilon_2(\varepsilon_1 + \varepsilon_0) + p(\varepsilon_2 - \varepsilon_1)(\varepsilon_2 - \varepsilon_0)}$$
$$\times \left\{ (\varepsilon_1 + \varepsilon_2)\mathbf{E}_0 + 2(\varepsilon_1 - \varepsilon_2)(1-p)r_1^2 \left[\frac{\mathbf{E}_0}{r^2} - \frac{2\mathbf{r}(\mathbf{r}\mathbf{E}_0)}{r^4} \right] \right\}.$$
$$(26.28)$$

For $\mathbf{E}_0$ parallel to the NW axis, $\mathbf{E}_1 = \mathbf{E}_2 = \mathbf{E}_0$. Formulae (Equations 26.27 and 26.28) can be used to determine the spectra and the polarization dependence of optical absorption in core–shell NWs. We restrict our analysis to the already mentioned case of a metal shell and a semiconductor core, promising luminescence amplification.

The specific features of the optical response of metal shells arise from the fact that the plasmon-related absorption maximum for the light with perpendicular polarization is split into two maxima with positions depending on the relationship between r_1 and r_2. The frequency regions near the plasmon resonances are characterized by an anomalously high electric field strength (light amplification) both in the core and in the shell. As it has been already pointed out for core–shell nanodots (Neeves and Birnboim 1989), this can be used for the enhancement of optical effects in a core (in NWs—only for light with perpendicular polarization). There is also a reciprocal effect: an oscillating dipole $\mathbf{d}_0$ in the core is created far from the wire radiation field that may be considerably larger than that in a uniform medium. This effect of light amplification will be considered later in Section 26.2.5.3. In the next section, we investigate the optical absorption near the plasmon resonances caused by the increased electric field $\mathbf{E}_2$ in the shell.

26.2.5.2 Spectra and Polarization Dependence of Absorption

To calculate the optical absorption in a core–shell NW caused by light with a perpendicular polarization, we must integrate the local absorption Im $\varepsilon_2 |\mathbf{E}_2(\mathbf{r})|^2$ over the whole shell. Using the Drude formula for ε_2 and Equation 26.28 for $\mathbf{E}_2(\mathbf{r})$, we obtain the absorption spectra shown by solid lines in Figure 26.10. These contain two maxima corresponding to plasmon resonances where the real part of the denominator in Equations 26.27 and 26.28 vanishes and the electric field in the NW becomes anomalously large. With an increasing p, the maxima approach each other and the low-frequency one diminishes so that at $p \to 1$ we get a single peak at $\omega = \omega_p/\sqrt{2}$ corresponding to the plasma resonance in a metal cylinder.

Similar phenomena in core–shell nanodots have been already observed experimentally (Zhou et al. 1994, Oldenburg et al. 1998) and have a good theoretical description (Neeves and Birnboim 1989, Ruda and Shik 2005b). It is important to emphasize that, contrary to nanodots, the described two-peak absorption spectra in NWs are observed only for light polarized perpendicularly to the NWs. For parallel polarization, when the electric field in the NWs is equal to the external one, we have simple monotonically decreasing spectra reflecting the low-frequency absorption given by Im ε_2.

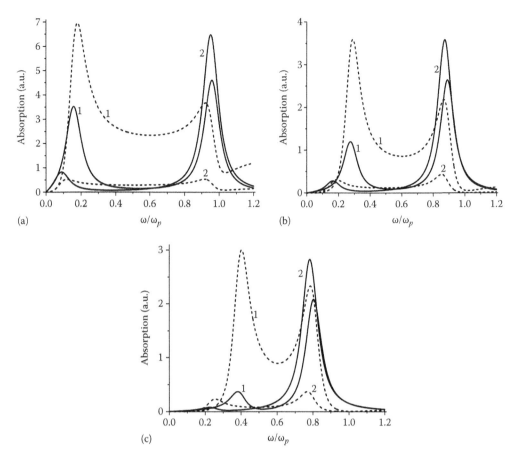

FIGURE 26.10 Absorption spectra of NW with a metal shell for $p = 0.2$ (a), 0.5 (b), 0.8 (c) and $v = 0.1\omega_p$ and a semiconductor core with $\varepsilon_1 = 3$ (curves 1) and $\varepsilon_1 = 10$ (curves 2) for light polarized perpendicularly to NWs. Solid curves correspond to the absorption in a metal shell, dashed curves—in a semiconductor core. (From Ruda, H.E. and Shik, A., *Phys. Rev. B*, 72, 115308, 2005a. With permission.)

If the core is a semiconductor with the bandgap $E_g < \hbar\omega$ then, besides the absorption described above, there is also interband absorption in the core proportional to $|\mathbf{E}_1(\mathbf{r})|^2$. The spectrum of this absorption can be easily calculated from Equation 26.27 and is shown by dashed lines in Figure 26.10. The spectrum also has a double maximum structure qualitatively similar to a metal shell one, but contrary to that, suffers a dramatic drop with an increasing ε_1 due to the extrusion of the electric field from the high-ε core. We emphasize that absorptions in a shell and a core depend on different parameters (i.e., the plasma frequency and the scattering rate in metals, and the bandgap width and the interband matrix element in semiconductors) and thus we can not compare the amplitudes of solid and dashed curves and say *a priori* what type of absorption is dominating. Moreover, it is also difficult to distinguish experimentally these two absorption components since the absorption spectra in a shell and a core have qualitatively similar characteristics. The problem can be solved by measuring the excitation spectrum for core luminescence, which is proportional only to the partial absorption in the core shown by the dashed lines in the figure.

As mentioned above, the dashed lines in Figure 26.10 actually show the spectral dependence of the electric field in a core $|\mathbf{E}_1(\mathbf{r})|^2$. As we have shown in Section 26.2.1, the semiconductor and the metal NWs have opposite characteristics of electrical anisotropy: the perpendicular component of the electric field is suppressed as compared with the parallel one, in a semiconductor NW, and enhanced in a metal NW in the vicinity of a plasmon frequency. In composite core–shell structures, both cases can be realized, depending on the shell thickness and the light frequency. The corresponding "phase diagram" using p-ω axes, is shown in Figure 26.11. The areas above the line(s) correspond to the "metal" case $E_\perp > E_{0\perp}$ while those below the line(s)—to the "semiconductor" case $E_\perp < E_{0\perp}$. In other words, these lines correspond to $|A| = 1$ where the amplification factor $A \equiv E_\perp/E_{0\perp}$ is given by Equation 26.27. At a small p, we naturally deal with the "semiconductor" regime. One might be surprised that at $p \to 1$, the regime is not "metal" but should be pointed so that the figure describes the field not in a metal shell, but in a semiconductor core, which in the limit of a pure metal nanowire $p \to 1$ simply disappears.

The condition $|A| > 1$ is satisfied in the vicinity of plasmon resonances, which means that for light-inducing interband absorption in the core, the intensity of the electron-hole pair generation, and hence, the intensity of the luminescence caused by these pairs is larger than in the absence of a metal shell. That is why, for A, we use the term "amplification factor."

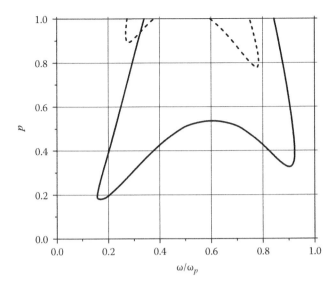

FIGURE 26.11 Values of p and ω, for which $E_\perp = E_{0\perp}$ at $\varepsilon_1 = 3$ (solid line) and $\varepsilon_1 = 10$ (dashed lines). The area above the line(s) corresponds to $E_\perp > E_{0\perp}$. (From Ruda, H.E. and Shik, A., *Phys. Rev. B*, 72, 115308, 2005a. With permission.)

26.2.5.3 Luminescence Amplification

Metal shells may cause amplification not only of the ac electric field of exciting light but also of the light emitted by a semiconductor core. The ac electric field emitted by a core, which can be considered as an effective dipole $\mathbf{d}_0$, is disturbed by a metal shell so that the field far from the nanostructure looks like one created by some other effective dipole $\mathbf{d}$. By analogy with the previous section, we may expect that in the vicinity of a plasmon resonance, the condition $|d/d_0| \gg 1$ can be realized, which means an effective amplification of the emitted radiation as well.

Calculations show that the luminescence of core–shell structures is characterized by exactly the same rules as absorption. The field of dipoles parallel to the NW axis is not disturbed by image forces, while the field of perpendicular dipoles acquires the same amplification factor A as given by Equation 26.27. As a result, for an isotropic distribution of effective dipoles, such as in a cubic semiconductor core in the absence of size quantization, the emission from an NW acquires a partial polarization parallel to its axis at $|A| < 1$ (as in Section 26.2.3) or perpendicular at $|A| > 1$. Since A has a strong frequency dependence, the components of the emitted light with a different polarization may have different spectra depending on the metal shell parameters.

From the applied point of view, the most important conclusion is as follows: by a proper choice of the metal-shell parameters, we can make one of the plasmon peaks coincide with the luminescence line in the core, which will result in an increase in the luminescence intensity (Ruda and Shik 2005b). It would be even more attractive to enhance the photoluminescence dramatically by a simultaneous amplification of both the exciting and the emitting light. Such a situation often occurs in surface-enhanced Raman scattering (Moskovits 1985) but in luminescence, its realization is more difficult since the difference

in frequencies of exciting ω_{ex} and emitted ω_{em} light is usually quite large and cannot be covered by a single plasmon peak. However, in core–shell structures, such double amplification can be reached by using both plasmon peaks as shown in Figure 26.10. In this case, the shell parameters should be chosen in such a way that the low-frequency peak corresponds to ω_{em} in a given NW core while the source of excitation is tuned for ω_{ex} to coincide with the second high-frequency peak (Ruda and Shik 2005a).

26.3 Arrays of Nanorods

26.3.1 Polarization Memory in Random Nanorod Arrays

26.3.1.1 Theory

The whole sections 26.2 was devoted to the optical properties of individual NWs or NRs and now we will discuss similar effects in NW arrays. All the polarization-dependent effects remain the same in such arrays as a system of parallel NWs (say, in porous dielectric matrices) with a relatively large distance between the NWs, but in the case of randomly oriented semiconductor NWs or NRs (say, in a polymer matrix or solution), the situation changes. Such a system has macroscopically isotropic optical properties but simultaneously possesses a very interesting property of polarization memory (Lavallard and Suris 1995, Ruda and Shik 2005a). If we excite photoluminescence in the system using polarized light, in accordance with Section 26.2.1, nonequilibrium carriers will be generated mostly in the NWs that are oriented close to and parallel to the light polarization. According to Section 26.2.3, light emitted by these NWs will have a preferable polarization parallel to their axes, or in other words, parallel to the polarization of the exciting radiation. In other words, an anisotropic random array of nanostructures "remembers" the polarization of the exciting radiation and emits luminescence in the same direction of polarization.

To calculate this effect in a random system of thin NWs, we assume that the NW array occupies the semi-space $x < 0$ and study the polarization of the net luminescence excited by z-polarized light and measured at some point $(x_0, 0, 0)$ outside the array (that is, $x_0 > 0$). The principle of calculation is as follows. We take some NWs with an orientation characterized by the spherical angles (θ, α) and consider it as a line of emitting dipoles with components d_0 along the NW axis and $2d_0\varepsilon_0/(\varepsilon + \varepsilon_0)$ in two other directions. Then, as in Section 26.2.3.1, we find the components $(S_x)_\parallel$ and $(S_x)_\perp$ of the Poynting vector created by these dipoles at the point $(x_0, 0, 0)$. Since, according to Equation 26.3, the intensity of the absorbed exciting light and hence the intensity of the luminescence depends on θ being proportional to $(\varepsilon + \varepsilon_0)^2\cos^2\theta + 4\varepsilon_0\sin^2\theta$, we integrate $(S_x)_\parallel$ and $(S_x)_\perp$ over θ and α with this weighting factor obtaining the total intensities $I_{\parallel,\perp}$ of light emission with polarizations parallel and perpendicular to the polarization of the exciting light.

Referring the reader to Ruda and Shik (2005a) for the details of calculations, we present only the final result:

$$\frac{I_{\parallel}}{I_{\perp}} =$$

$$\frac{9[(\varepsilon+\varepsilon_0)^2+2\varepsilon_0^2](\varepsilon+\varepsilon_0)^2+20[(\varepsilon+\varepsilon_0)^2+2\varepsilon_0^2]+21(\varepsilon+\varepsilon_0)^2+336\varepsilon_0^2}{3[(\varepsilon+\varepsilon_0)^2+2\varepsilon_0^2](\varepsilon+\varepsilon_0)^2+44[(\varepsilon+\varepsilon_0)^2+2\varepsilon_0^2]+63(\varepsilon+\varepsilon_0)^2+168\varepsilon_0^2}.$$

$$(26.29)$$

Equation 26.29 shows that maximal possible polarization, reached at $|\varepsilon/\varepsilon_0| \gg 1$, is equal to $I_{\parallel}/I_{\perp} = 3$. In terms of the polarization ratio $P = (I_{\parallel}-I_{\perp})/(I_{\parallel}+I_{\perp})$ it corresponds to $P = 0.5$. Figure 26.12 shows this polarization ratio for an arbitrary relationship between ε and ε_0. For InP NWs in air ($\varepsilon/\varepsilon_0 = 12.7$), we get $P = 0.44$. It is worth noting that at $\varepsilon < \varepsilon_0$ (e.g., in metal NWs near the plasmon resonance), when the electric field in the NWs has a preferably perpendicular orientation, the system also has a polarization memory. This effect is weaker than at a large ε, with the maximal $P = 6/68 = 0.088$.

As we have shown in Sections 26.2.1.2 and 26.2.3.2, for thick NWs, the polarization characteristics for absorption and emission vary dramatically with the light frequency. As a result, the polarization memory in these objects must demonstrate a strong dependence on the frequency of both the exciting (ω_{ex}) and the emitted (ω_{em}) light and even change its sign at some intervals of ω_{ex} and ω_{em}. The quantitative description of this effect (Ruda and Shik 2006) is based on the same approach as that for thin NWs resulting in Equation 26.29 but with two distinctions. First, the values of the longitudinal and the transverse components of the effective dipole moments in the NW are no longer constant but given by Equations 26.12 and 26.13 with $k = \omega_{em}/c$. Second, the intensity of the absorbed exciting light and hence the intensity of the luminescence (the weighting factor in the angular averaging) is now equal to $I_{\parallel}^a(\omega_{ex})\cos^2\theta + I_{\perp}^a(\omega_{ex})\sin^2\theta$ where the intensities of the absorbed light $I_{\parallel}^a$ and $I_{\perp}^a$ are determined from Equations 26.5 and 26.6 according to the procedure described in

Section 26.2.1.2. The final answer for the ratio of luminescence intensities with polarization parallel and perpendicular to that of the exciting light is

$$\frac{I_{\parallel}^e}{I_{\perp}^e} =$$

$$\frac{2I_{\perp}^a(\omega_{ex})[47d_x^2(\omega_{em})+10d_z^2(\omega_{em})]+3I_{\parallel}^a(\omega_{ex})[13d_x^2(\omega_{em})+12d_z^2(\omega_{em})]}{4I_{\perp}^a(\omega_{ex})[16d_x^2(\omega_{em})+11d_z^2(\omega_{em})]+3I_{\parallel}^a(\omega_{ex})[23d_x^2(\omega_{em})+4d_z^2(\omega_{em})]}.$$

$$(26.30)$$

To clarify, we emphasize the difference in notations between Equations 26.29, 26.30 and 26.10 where $I_{\parallel}^e$ and $I_{\perp}^e$ referred to the light polarized parallel and perpendicular to the NW axis. Figure 26.13 shows the dependences of the system polarization ratio $P^e = (I_{\parallel}^e - I_{\perp}^e)/(I_{\parallel}^e + I_{\perp}^e)$ on the frequencies of the exciting

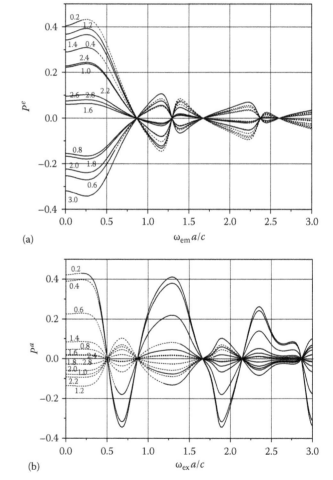

(a)

(b)

FIGURE 26.13 Polarization ratio of photoluminescence for a random NW system with $\varepsilon = 9$, $\varepsilon_0 = 1$. (a) Luminescence spectra $P^e(\omega_{em}a/c)$ for different ω_{ex}. Numbers at the curves indicate the values ω_{ex}/c. The curves corresponding to $\omega_{ex}a/c = 2.6$ and 2.8 practically coincide. (b) Excitation spectra $P^a(\omega_{ex}a/c)$ for different ω_{ems}. Numbers at the curves indicate the values of $\omega_{em}a/c$. The curves corresponding to $\omega_{em}a/c = 1.8$ and 2.8 practically coincide. For $\omega_{em}a/c = 2.6$ polarization is very close to zero and is not presented in the figure.

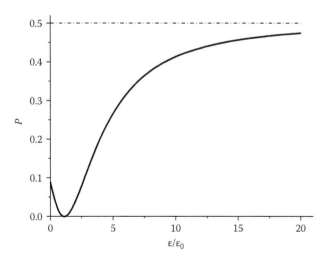

FIGURE 26.12 Luminescence polarization ratio in a random system of nanowires exited by polarized light. The dot-and-dash line corresponds to the limiting value $P = 0.5$. (From Ruda, H.E. and Shik, A., *Phys. Rev. B*, 72, 115308, 2005a. With permission.)

and the emitted light, given by Equation 26.30. Since in standard photoluminescence measurements ω_{ex} always exceeds ω_{em}, parts of the curves in Figure 26.13 where this condition is violated are given by dashed lines. At $\omega_{ex}, \omega_{em} \to 0$, polarization tends to its low-frequency limit shown in Figure 26.12 and for our choice $\varepsilon/\varepsilon_0 = 9$ equal to $P^e \cong 0.40$.

It is important to mention that if the NW material is characterized by several luminescence lines with essentially different frequencies, then these lines may have different degrees of polarization, in accordance with Figure 26.6. Since the curve at the figure crosses the level $I_\parallel^e = I_\perp^e$, there exists even the possibility when one of the luminescent lines may have. The latter case could be called "polarization anti-memory."

26.3.1.2 Experiment

The effect of polarization memory was first observed in porous Si (Kovalev et al. 1995) that has only a distant resemblance to a system of NWs or NRs. Hence, for demonstration of its main experimental features, we have chosen subsequently more detailed measurements in CdSe/ZnS NRs (Kravtsova et al. 2007) with the aspect ratio $l/2a \cong 2.4$ dissolved in a liquid and hence having a randomly distributed orientation. Figure 26.14 shows the luminescence spectrum consisting of two peaks corresponding to interband transitions between different size-quantized states. Polarization properties of the emitted light are presented in the form of luminescence spectra for different angular positions of the analyzer related to the direction of the exciting light polarization $\mathbf{E}_0$. The luminescence is seen to be

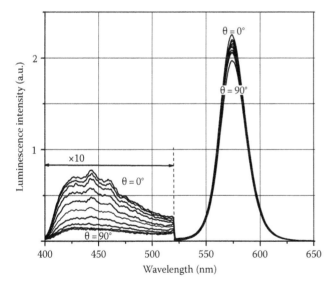

FIGURE 26.14 Photoluminescence spectrum of the NR solution for excitation by linearly polarized light with $\lambda = 366$ nm. Different curves correspond to the luminescence components with a polarization in the direction forming the angle θ with the polarization of excitation. θ increases from the upper to the lower curve with the step $10°$. The amplitude of the luminescence in the spectral region $\lambda < 520$ nm is shown with the magnification factor 10. (From Kravtsova, Y. et al., *Appl. Phys. Lett.*, 90, 083120, 2007. With permission.)

polarized mostly parallel to $\mathbf{E}_0$, in qualitative agreement with theoretical predictions.

The quantitative treatment of the results meets with some difficulties. According to Equation 26.29, the amplitude of polarization memory in small nanostructures has some universal value depending only on $\varepsilon/\varepsilon_0$. However, in the experiment, this amplitude differs noticeably for two different spectral lines in the same NRs. Besides, for the short wavelength peak, the amplitude of polarization memory exceeded the theoretically predicted one, while for the long wavelength peak, it was always less. These results could be explained if the matrix element responsible for the ground state 573 nm-long wavelength peak is larger for perpendicular light polarization (it is really the case for NRs with a small aspect ratio (Hu et al. 2001)) while that at 470 nm is either more isotropic or larger for parallel polarization. This is in agreement with the theoretical (Li and Wang 2003) and the experimental (Thomas et al. 2005) conclusions of essentially different polarization properties of different optical transitions in NRs.

26.3.2 Plasmon Spectra in Self-Assembled Nanorod Structures

26.3.2.1 Self-Assembling of Metal Nanorods

In the recent years, an impressive success has been achieved in fabricating various arrays of metal NRs by their self-assembling in solution. The most comprehensive results in this direction were obtained by using Au NRs with the lateral surface covered by a double layer of hydrophilic molecules and the ends terminated with polystyrene molecular chains (Nie et al. 2007, 2008). Such building blocks in solution can form different ordered arrays, such as chains, bundles, raft-like structures, rings, etc. (Figure 26.15), depending on the chemical composition of the solvent. In these systems, the distances between the NRs are very small, compared to their diameter, so that the electric fields of plasmons in neighboring NRs overlap and the plasmon spectrum of the system becomes different from that of individual NRs and depends on the NR array topology. Figure 26.16 shows the plasmon absorption spectra in some of these structures formed from identical NRs. They consist of two peaks (longitudinal and transverse plasmons) where the long wavelength one, corresponding to longitudinal plasmons (see Section 26.2.2.1), demonstrates a strong dependence on the array geometry.

Some general qualitative regularities were already formulated in literature and found their experimental confirmation in systems of NRs and nanodots. It was demonstrated (Rechenberger et al. 2003, Su et al. 2003) that by reducing the inter-nanoparticle separation in the direction of light polarization, a red shift in the plasmon resonance occurs, in contrast with a blue shift for the perpendicular direction of polarization. It has a simple qualitative explanation. The end-to-end alignment of NRs increases the effective length of an NR, which results in a decrease of the longitudinal depolarization factor $n_\parallel$ and, according to Equation 26.7, a decrease in the longitudinal plasmon frequency $\omega_\parallel$, while the side-to-side arrangement increases the effective $n_\parallel$ and $\omega_\parallel$.

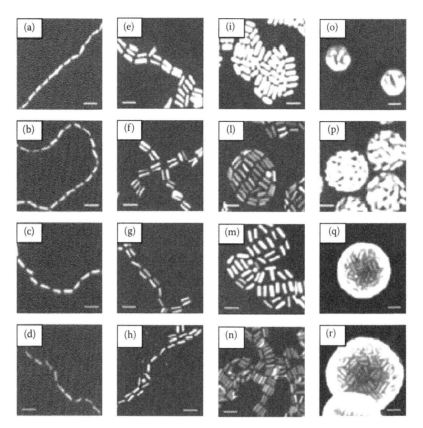

FIGURE 26.15 NR arrays obtained by self-assembling in different solvents. (From Fava, D. et al., *Adv. Mater.*, 20, 4318, 2008. With permission.)

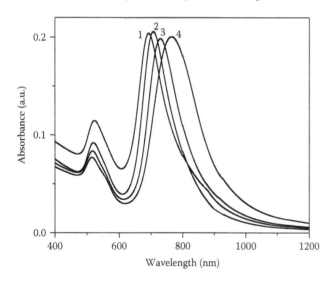

FIGURE 26.16 Optical absorption spectra for various NR arrays. (From Nie, Z.H. et al., *J. Am. Chem. Soc.*, 130, 3683, 2008. With permission.)

In the next sections, we present theoretical models allowing one to make some quantitative calculations of plasmon characteristics and their comparison with the experiment.

26.3.2.2 Longitudinal Plasmon Shift

Let us consider some arbitrary system of metal NWs or strongly anisotropic NRs (further we will mention them simply as NWs) and find the polarization of each NW in an external uniform electric field $\mathbf{E}_0$. We assume that the NW radius a is many times smaller than both their length l, so that for calculations of longitudinal plasmons, the radial distribution of the induced charges is inessential and we may characterize each NW by a linear charge density $q(z)$ depending on one single coordinate z measured along the NW axis. In the absence of a direct electrical contact between the NWs, their total charges remain zero: $\int_{-l/2}^{l/2} q(z)\,dz = 0$. For each metal NW, the charge distribution $q(z)$ is found from the condition of equipotentiality (Ruda and Shik 2007b). It means that the external field potential $-E_{0z}z$ is exactly compensated by the joint potential created by the charges of all NWs including the given one. To describe the potential created by $q(z)$ of the same NW in a medium with the dielectric constant ε, we use the approximate expression $2q(z)\ln(l/a)/\varepsilon$ valid for $l \gg a$ (Averkiev and Shik 1996). Potentials of all other NWs can be obtained by a simple Coulomb expression neglecting the finite thickness of NWs.

We demonstrate the described procedure for the case of two in-line NWs separated by the interval b and occupying the segments $(-l - b/2, -b/2)$ and $(b/2, l + b/2)$ of the z-axis. Noting the charge densities in the NWs as $q_-(z)$ and $q_+(z)$, respectively, we obtain the system of two equations:

$$\frac{2q_{\pm}(z)}{\varepsilon}\ln\left(\frac{l}{a}\right) - E_{0z}z + \int_{\mp b/2}^{\mp b/2 \mp l} \frac{q_{\mp}(z')}{\varepsilon(z'-z)}\,dz' = C_{\pm}. \quad (26.31)$$

Instead of $q_\pm$, we introduce two dimensionless functions $\eta_\pm(\xi)$ determined from

$$q_\pm(z) = \frac{\varepsilon E_{0z}}{2\ln(l/a)}\left[z + l\eta_\pm(\xi)\right] \qquad (26.32)$$

where ξ is the dimensionless coordinate measured from the center of the corresponding NWs: $\xi = [z \mp (b/2 + l/2)]/l$. It can be easily shown that $\eta_-(-\xi) = -\eta_+(\xi)$. Finding $C_\pm$ from the neutrality conditions: $\int_{-1/2}^{1/2}\eta_\pm(\xi)\mathrm{d}\xi = 0$, we reduce Equation 26.31 to one single equation for $\eta_+(\xi)$:

$$\begin{aligned}
2\ln\!\left(\frac{l}{a}\right)&\eta_+(\xi) - \int_{-1/2}^{1/2}\frac{\eta_+(\xi')\mathrm{d}\xi'}{\xi+\xi'+1+\beta} + \int_{-1/2}^{1/2}\eta_+(\xi')\ln\!\left(\frac{2\xi'+3+2\beta}{2\xi'+1+2\beta}\right)\mathrm{d}\xi' \\
&= \frac{1}{2} + \ln\!\left(\frac{2+\beta}{1+\beta}\right) - (\xi+1+\beta)\ln\!\left(\frac{2\xi+3+2\beta}{2\xi+1+2\beta}\right) \\
&+ \frac{\beta}{2}\ln\!\left[\frac{\beta(2+\beta)^3}{(1+\beta)^4}\right] + \frac{\beta^2}{2}\ln\!\left[\frac{\beta(2+\beta)}{(1+\beta)^2}\right].
\end{aligned} \qquad (26.33)$$

This equation was solved numerically for $l/a = 10$, which corresponds to the experimental results (Nie et al. 2007) to be discussed later. After $q(z)$ is found, the total dipole moment P of an NW is easily found: $P = \int zq(z)\mathrm{d}z$, which, in turn, determines the depolarization factor $n_\| = E_{0z}V/(4\pi P)$ (V is the NR volume) (Landau and Lifshitz 1984), connected with the longitudinal plasmon frequency $\omega_\|$ by Equation 26.7.

Using a similar approach, the frequency $\omega_\|$ was found (Ruda and Shik 2007b) for two other systems: the infinite periodic NR chain and a pair of two side-by-side NRs with the distance $d > 2a$ between the axes. The resulting dependences of the longitudinal plasmon frequency $\omega_\|$ on inter-NR distances b or d for all three cases are presented in Figure 26.17. It is seen that, in accordance with the above-mentioned qualitative considerations, $\omega_\|$ demonstrates a red shift for chains and a blue shift for bunches of NRs. Figure 26.17 also contains experimental points (Nie et al. 2007) partially taken from the curves of Figure 26.16. It is seen that the results for in-line structures (curves 1 and 2) demonstrate a good qualitative agreement but the experimental dependences are steeper than the theoretical. This effect can be caused by the fact that the polystyrene molecules connecting the NR tips are also polarized in the external electric field, so that the effective local dielectric constant of this gap exceeds that of the solvent. This distorts the electric field pattern and increases the intensity of inter-NR interaction and hence, the value of P, as compared to the ideal case considered above. For the side-by-side NR assemblies, practically all the lines of the electric field responsible for the inter-NR interaction are concentrated in the region filled by hydrophilic

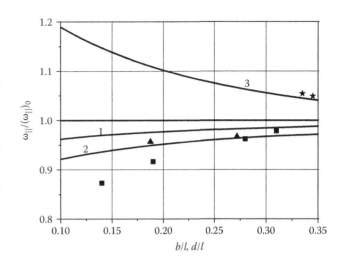

FIGURE 26.17 Dependence of the longitudinal plasmon frequency $\omega_\|$ (in units of $\omega_\|$ for individual NRs) on the distance between NRs with $l/a = 10$. 1, two in-line NRs (experimental points—triangles); 2, chain of NRs (experimental points—squares); 3, two parallel NRs (experimental points—stars).

molecules, so that the dielectric inhomogeneity of the system is irrelevant, and curve 3 demonstrates not only qualitative but also quantitative agreement.

26.4 Conclusion

We have described a large variety of polarization-sensitive optical phenomena in anisotropic nanostructures, such as NWs and NRs. The phenomena have a universal character being exclusively due to the difference in the dielectric constants between nanostructures and their environment. They include a strong dependence of linear (absorption, photoluminescence, photoconductivity, etc.) and nonlinear optical properties on polarization of the exciting light, as well as a high degree of polarization of luminescence emitted by these systems. The joint action of absorption and emission anisotropy results in a polarization memory in random arrays of NWs and NRs. In metal nanostructures, all the mentioned phenomena exhibit a dramatic dependence (including sometimes the change of sign) on the light frequency, related to plasmon effects. In composed metal-semiconductor nanostructures, these effects can be used for luminescence amplification. In ordered self-assembled arrays of metal NRs, the plasmon-related optical spectra are sensitive to the array geometry. All the described effects are provided with proper theoretical description and illustrated by experimental observations performed in recent years.

Acknowledgments

The authors would like to thank their collaborators, V. Barzda, R. Cisek, D. Fava, E. Kumacheva, Y. Kravtsova, U. Krull, L. Levina, S.F. Musikhin, Z.H. Nee, U. Philipose, and T.L. Spencer from the University of Toronto, whose experimental results are presented in this work.

References

Agarwal, G.S. and Jha, S.S. 1982. Theory of second harmonic generation at a metal surface with surface plasmon excitation. *Solid State Communications* 41: 499–501.

Averkiev, N.S. and Shik, A. 1996. Contact phenomena in quantum wires and porous silicon. *Semiconductors* 30: 112–116.

Barzda, V., Cisek, R., Spencer, T.L., Philipose, U., Ruda, H.E., and Shik, A. 2008. Giant anisotropy of second harmonic generation for a single ZnSe nanowire. *Applied Physics Letters* 92: 113111.

Batygin, V.V. and Toptygin, I.N. 1978. *Problems in Electrodynamics*. New York: Academic Press.

Blombergen, N., Chang, R.K., Jha, S.S., and Lee, C.H. 1968. Optical second-harmonic generation in reflection from media with inversion symmetry. *Physical Review* 174: 813–822.

Chew, W.C. 1990. *Waves and Fields in Inhomogeneous Media*. New York: Van Nostrand Reinhold.

Choi, H.-J., Johnson, J.C., He, R. et al. 2003. Self-organized GaN quantum wire UV lasers. *Journal of Physical Chemistry B* 107: 8721–8725.

Duan, X. and Lieber, C.M. 2000, General synthesis of compound semiconductor nanowires. *Advanced Materials* 12: 298–302.

Fan, Z., Chang, P., Lu, J.G. et al. 2004. Photoluminescence and polarized photodetection of single ZnO nanowires. *Applied Physics Letters* 85: 6128–6130.

Fava, D., Nie, Z.H., Winnik, M.A., and Kumacheva, E. 2008. Evolution of self-assembled structures of polymer-terminated gold nanorods in selective solvents. *Advanced Materials* 20: 4318–4322.

Grigoriev, I.S. and Meilikhov, E.Z. 1997. *Handbook of Physical Quantities*. Boca Raton, FL: CRC Press.

Han, S., Jin, W., Zhang, D. et al. 2004. Photoconduction studies on GaN nanowire transistors under UV and polarized UV illumination. *Chemical Physics Letters* 389: 176–180.

Hsu, N.E., Hung, W.K., and Chen, Y.F. 2004. Origin of defect emission identified by polarized luminescence from aligned ZnO nanorods. *Journal of Applied Physics* 96: 4671–4673.

Hu, J., Li, L., Yang, W., Manna, L., Wang, L., and Alivisatos, A.P. 2001. Linearly polarized emission from colloidal semiconductor quantum rods. *Science* 292: 2060–2063.

Kovalev, D., Ben Chorin, M., Diener, J. et al. 1995. Porous Si anisotropy from luminescence polarization. *Applied Physics Letters* 67: 1585–1587.

Kravtsova, Y., Krull, U., Musikhin, S.F., Levina, L., Ruda, H.E., and Shik, A. 2007. Polarization memory in a system of CdSe nanorods. *Applied Physics Letters* 90: 083120.

Kreibig, U. and Vollmer, M. 1995. *Optical Properties of Metal Clusters*. Berlin, Germany: Springer.

Landau, L.D. and Lifshitz, E.M. 1984. *Electrodynamics of Continuous Media*. New York: Pergamon Press.

Lauhon, L.J., Gudiksen, M.S., Wang, D., and Lieber, C.M. 2002. Epitaxial core-shell and core-multishell nanowire heterostructures. *Nature* 420: 57–61.

Lavallard, P. and Suris, R.A. 1995. Polarized photoluminescence of an assembly of non cubic microcrystals in a dielectric matrix. *Solid State Communications* 95: 267–269.

Li, J. and Wang, L. 2003. High energy excitations in CdSe quantum rods. *Nano Letters* 3: 101–105.

McIntyre, C.R. and Sham, L.J. 1992. Theory of luminescence polarization anisotropy in quantum wires. *Physical Review B* 45: 9443–9446.

Mei, X., Blumin, M., Sun, M. et al. 2003. Highly-ordered GaAs/AlGaAs quantum dot arrays on GaAs(100) substrate grown by molecular beam epitaxy using nanochannels alumina masks. *Applied Physics Letters* 82: 967–969.

Moskovits, M. 1985. Surface-enhanced spectroscopy. *Reviews of Modern Physics* 57: 783–826.

Neeves, A.E. and Birnboim, M.H. 1989. Composite structures for the enhancement of nonlinear optical susceptibility. *Journal of the Optical Society of America B* 6: 787–796.

Nie, Z.H., Fava, D., Kumacheva, E. et al. 2007. Self-assembly of metal-polymer analogues of amphiphilic triblock copolymers. *Nature Materials* 6: 609–614.

Nie, Z.H., Fava, D., Rubinstein, M., and Kumacheva, E. 2008. Supramolecular assembly of gold nanorods end-terminated with polymer "pom-poms". *Journal of the American Chemical Society* 130: 3683–3689.

Oldenburg, S.J., Averitt, R.D., Westcott, S.L., and Halas, N.J. 1998. Nanoengineering of optical resonances. *Chemical Physics Letters* 288: 243–247.

Peng, X., Manna, L., Yang, W. et al. 2000. Shape control of CdSe nanocrystals. *Nature* 404: 59–61.

Philipose, U., Ruda, H.E., Shik, A., de Souza, C.F., and Sun, P. 2005. Light emission from ZnSe nanowires. *Proceedings of SPIE* 5971: 597116–597128.

Qi, J., Belcher, A.M., and White, J.M. 2003. Spectroscopy of individual silicon nanowires. *Applied Physics Letters* 82: 2616–2618.

Rechenberger, W., Hohenau, A., Leitner, A. et al. 2003. Optical properties of two interacting gold nanoparticles. *Optics Communications* 220: 137–141.

Ruda, H.E. and Shik, A. 2005a. Polarization-sensitive optical phenomena in semiconducting and metallic nanowires. *Physical Review B* 72: 115308.

Ruda, H.E. and Shik, A. 2005b. Plasmon phenomena and luminescence amplification in nanocomposite structures. *Physical Review B* 71: 245328.

Ruda, H.E. and Shik, A. 2006. Polarization-sensitive optical phenomena in thick semi-conducting nanowires. *Journal of Applied Physics* 100: 024314.

Ruda, H.E. and Shik, A. 2007a. Nonlinear optical phenomena in nanowires. *Journal of Applied Physics* 101: 034312.

Ruda, H.E. and Shik, A. 2007b. Polarization and plasmon effects in nanowire arrays. *Applied Physics Letters* 90: 223106.

Sercel, P.C. and Vahala, K.J. 1990. Analytical technique for determining the polarization dependence of optical matrix elements in quantum wires with band-coupling effects. *Applied Physics Letters* 57: 545–547.

Su, K.-H., Wei, Q.-H., Zhang, X. et al. 2003. Interparticle coupling effects on plasmon resonances in nanogold particles. *Nano Letters* 3: 1087–1090.

Thomas, N.L., Herz, E., Schöps, O., Woggon, U., and Artemyev, M.V. 2005. Exciton fine structure in single CdSe nanorods. *Physical Review Letters* 94: 016803.

Wang, J., Gudiksen, M.S., Duan, X., Cui, Y., and Lieber, C.M. 2001. Highly polarized luminescence and photodetection from single InP nanowires. *Science* 293: 1455–1457.

Zhou, H.S., Honma, I., Komiyama, H., and Haus, J.W. 1994. Controlled synthesis and quantum-size effect in gold-coated nanoparticles. *Physical Review B* 50: 12052–12056.

the second-order nonlinear polarization induced in a crystal is expressed as

$$P^{(2)}(t) = \chi^{(2)} E(t)^2, \quad (27.3)$$

$$P^{(2)}(t) = \chi^{(2)}[E_1^2 e^{-2i\omega_1 t} + E_2^2 e^{-2i\omega_2 t} + 2E_1 E_2 e^{-i(\omega_1+\omega_2)t}$$

$$+ 2E_1 E_2^* e^{-i(\omega_1-\omega_2)t} + c.c] + 2\chi^{(2)}[E_1 E_1^* + E_2 E_2^*], \quad (27.4)$$

The first and second terms in Equation 27.4 represent the SHG, which is useful in the perspective of optoelectronic phenomenon such as tuning the frequency of the laser beam. The expression $2E_1 E_2 e^{-i(\omega_1+\omega_2)t}$ and $2E_1 E_2^* e^{-i(\omega_1-\omega_2)t}$ are associated with the sum-frequency generation (SFG) and the difference-frequency generation (DFG) respectively. The SFG is used to produce tunable radiation in the ultraviolet spectral region and the DFG produces tunable radiation in the infrared region. The process of the DFG is also used in the design of the optical parametric oscillator. The last term of Equation 27.4 does not show any frequency dependence and leads to a process known as optical rectification (OR).

The third order contribution to the nonlinear polarization is given by

$$P^{(3)}(t) = \chi^{(3)} E(t)^3, \quad (27.5)$$

In the case of the monochromatic applied electric field ($E(t) = \xi \cos \omega t$), the nonlinear polarization is expressed as

$$P^{(3)}(t) = \frac{1}{4}\chi^{(3)}\xi^3 \cos 3\omega t + \frac{3}{4}\chi^{(3)}\xi^3 \cos \omega t. \quad (27.6)$$

The first term of Equation 27.6 illustrates the third-harmonic generation (THG), which describes the response of the system at the frequency of 3ω. The second term in Equation 27.6 leads to a nonlinear contribution to the refractive index of the electromagnetic wave of frequency ω. This is recognized as the intensity-dependent refractive index (IDRI). In the presence of such nonlinearity, the refractive index becomes

$$n = n_0 + n_2 I, \quad (27.7)$$

where n_0 is the linear refractive index and

$$n_2 = \frac{12\pi^2}{n_0^2 c}\chi^{(3)}, \quad (27.8)$$

n_2 is an optical constant that characterizes the strength of the optical nonlinearity. I is the intensity of the incident wave expressed as $I = (n_0 c/8\pi)\xi^2$. The intensity-dependent refractive index (IDRI) is related to optical phenomenon, such as the self-focusing of the electromagnetic wave. Self-focusing occurs in a material in which n_2 is positive and the electromagnetic waves, with a nonuniform transverse intensity distribution, curve towards each other after passing through such material.

27.2 Quantum Formulation of Nonlinear Optics

A large number of quantum chemical techniques have been employed for the evaluation of the NLO properties of materials. Jensen et al. have developed a dipole interaction model. This model successfully describes the response properties of large aggregates of molecular clusters [17]. The density-matrix renormalization group (DMRG)-based formalism was proposed by Pati and coworkers and successfully applied in the description of the frequency-dependent NLO properties of π-conjugated systems [18]. Standard techniques like the time-density functional theory, the time-dependent Hartee Fock technique, as well as the finite-field methods [19] based on DFT are also very successful in describing the NLO responses of several classes of materials. The quantum chemical approach, like the density functional theory, had its limitations in the case of long-range dispersion interaction and has been improved by the introduction of the dispersion-corrected DFT DFD-D [20,21] technique. Higher-order wavefunction-based methods like the coupled cluster approach and the MP2 technique are also efficient but they are excessively expensive. In addition to these, the response properties are also evaluated through semi-empirical methods like ZINDO/MRDCI (multi reference doubles-configuration interaction) formalism with a correction vector. [22]. Another well-celebrated quantum chemical approach is the use of the two-state model proposed by Oudar and Chemla [23]. This model relates the nonlinear optical coefficients to the excited-state energy, the oscillator strength and the dipole moment of the molecule under consideration. The theoretical results of Zyss [24] and Andrews et al. [25] successfully established the two-state model of Oudar and Chemla. Datta and Pati [26] employed the two-state model to discuss the role of dipolar interaction and H-bonding in tuning the response properties in molecular aggregates. This model is simple and particularly useful in the case of charge-transfer systems [27]. In this two-state model, the expression of β is given by

$$\beta_{\text{two-level}} = \frac{3e^2}{2\hbar} \frac{\omega_{01} f \Delta\mu_{01}}{(\omega_{01}^2 - \omega^2)(\omega_{01}^2 - 4\omega^2)}$$

where

f is the oscillator strength

ω is the applied frequency

ω_{01} is the frequency of the optical transition between the ground state and the first dipole-allowed state

$\Delta\mu_{01}$ is the difference in moments between the ground state and the first dipole-allowed state

In the absence of an applied frequency, the above equation results in a first-order static hyperpolarizability tensor

$$\langle \beta(0;0,0) \rangle = 3\pi \frac{f \Delta\mu_{01}}{\omega_{01}^3}.$$

In two-state formalism, the parameters like f, ω_{01} and $\Delta\mu_{01}$ are estimated through standard excited approaches like the configuration interaction scheme (CIS) or the time-dependent density-functional theory (TDDFT).

Apart from all those techniques described in the above paragraph, another widely accepted quantum chemical technique of response calculation is the analytic approach. In this context, it will be useful to point out some of the basic differences between the finite-difference technique and the analytic approach. In the finite-difference technique, the energy is calculated at different values of the electric field; there after, the finite differentiation is performed to obtain the derivative of energies. The finite-difference method is suitable for any program capable of energy calculation in a perturbed system. However, the main limitation of this method is the prolonged computational time and it is not good enough when higher order derivatives of the energy are involved. On the contrary, in the analytical approach, physical properties are obtained by analytically evaluating the derivatives of the energy and are often preferred in response calculations. The main advantage of this technique is that it gives access to frequency-dependent properties and the data obtained are more accurate. This technique has been successfully applied by van Gisbergen et al. [28] in case of a set of small molecules. Sen et al. employed the same technique in case of CdSe clusters and Al_4M_4 clusters [11,13].

In the density-functional theoretical formulation of the analytic approach, the starting equation is a variation on the time-dependent Kohn–Sham (TDKS) equation. In order to obtain the starting equation of the analytic approach in the time-dependent density-functional theory (TDDFT), the action integral of the TDDFT is considered.

$$A = \int_{t_0}^{t_1} dt \langle \psi(t) | i \frac{\partial}{\partial t} - \hat{H}(t) | \psi(t) \rangle, \tag{27.9}$$

where ψ is the wave function of the system. In terms of time-dependent single-particle orbitals $\{\varphi_j(\vec{r}, t)\}$, the action integral becomes

$$A[\{\varphi_j\}] = \sum_n^j \int_{-\infty}^{t_1} dt \int d^3\vec{r} \varphi_j^\star(\vec{r}, t) \left(i \frac{\partial}{\partial t} + \frac{\nabla^2}{2} \right) \varphi_j(\vec{r}, t)$$

$$- \int_{-\infty}^{t_1} dt \int d^3\vec{r} \rho(\vec{r}, t) v_{ext}(\vec{r}, t)$$

$$- \frac{1}{2} \int_{-\infty}^{t_1} dt \int d^3\vec{r} \int d^3\vec{r}' \frac{\rho(\vec{r}, t)\rho(\vec{r}', t)}{|\vec{r} - \vec{r}'|} - A_{xc}[\{\varphi_j\}], \tag{27.10}$$

In Equation 27.10, the term $v_{ext}(\vec{r}, t)$ is the time-dependent exchange-correlation potential and A_{xc} is the exchange-correlation part of the action functional. On considering the orthonormal Kohn–Sham orbital, the action integral produces the time-dependent Kohn–Sham equation as

$$\sum_j \varepsilon_{ij}(t)\varphi_j(\vec{r}, t) + i \frac{\partial}{\partial t} \varphi_i(\vec{r}, t) = \left[-\frac{\nabla^2}{2} + v_s(\vec{r}, t) \right] \varphi_i(\vec{r}, t) = F_s\varphi_i(\vec{r}, t), \tag{27.11}$$

In Equation 27.11, $v_s(\vec{r}, t)$ is the time-dependent Kohn–Sham potential and $\varepsilon_{ij}(\vec{r}, t)$ is the Lagrangian multiplier. Different choices of Lagrangian multipliers are allowed. With $\varepsilon_{ij}(\vec{r}, t) = 0$ we attain the canonical form of the Kohn–Sham equation. In the case of orbitals varying rapidly with time, $\varepsilon_{ij}(\vec{r}, t)$ is suitably chosen to avoid an unphysical divergence in the Kohn–Sham equation. The time-dependent density $\rho(\vec{r}, t)$ is obtained from the relation

$$\rho(\vec{r}, t) = \sum_i^{occupied} |\varphi_i(\vec{r}, t)|^2, \tag{27.12}$$

When $\varphi_i(\vec{r}, t)$ is expanded in a fixed time-dependent basis set of atomic orbitals $\{\chi_\mu(\vec{r})\}$, the resulting expression becomes

$$\varphi_i(\vec{r}, t) = \sum_\mu \chi_\mu(\vec{r})C_\mu(t), \tag{27.13}$$

In Equation 27.13, the coefficient $C_\mu(t)$ manifests the time dependence of $\varphi_i(\vec{r}, t)$. Finally, Equations 27.11 through 27.13 produces the time-dependent Kohn–Sham equation in the matrix form given by

$$F_sC - i \frac{\partial}{\partial t} SC = SC\varepsilon, \tag{27.14}$$

where S is the overlap matrix of the atomic orbitals with

$$S_{\mu\nu} = \int d\vec{r} \chi_\mu^\star(\vec{r})\chi_\nu(\vec{r}), \tag{27.15}$$

The density matrix (D) is expressed in terms of the coefficient matrix C and the occupation number matrix n and is expressed as

$$D = CnC^\dagger. \tag{27.16}$$

The Kohn–Sham equation expressed in Equation 27.14 is the starting point of the perturbative expansion. Equation 27.14 is similar to the starting equation for the NLO calculation in the time-dependent Hartree–Fock technique (TDHF). In Equation 27.14, the term F_s represents the Kohn–Sham matrix or the Fock matrix. This is expressed as

$$F_s = h + DX\ 2J + v_{xc}. \tag{27.17}$$

In the above equation, h is the one-electron integral matrix comprising kinetic energy, the Coulomb field of the nuclei, and the applied external electric field. J is the four-index Coulomb supermatrices.

When all matrices are expanded in different orders of external perturbation and ε is chosen as the time-dependent zero-order matrix, the canonical Kohn–Sham equation for the ground-state DFT is obtained and given by the following expression:

$$F_s^{(0)}C^{(0)} = S^{(0)}C^{(0)}\varepsilon^{(0)}. \tag{27.18}$$

The Lagrangian multiplier matrix $[\epsilon^{(0)}]$ can be chosen at each order of perturbation. Now, if an external electric field of $E^a(\vec{r}, t)$ is applied, the mathematical form of $E^a(\vec{r}, t)$ being

$$E^a(\vec{r}, t) = E^a(\vec{r})X[1 + e^{i\omega_a t} + e^{-i\omega_a t}], \qquad (27.19)$$

then in the dipolar approximation, the external perturbation term (H) to the Kohn–Sham Hamiltonian becomes

$$H = \mu \cdot E(\vec{r}, t) \qquad (27.20)$$

where μ is the dipole-moment operator of the electrons.

The time dependence of the dipole moment produces various frequency-dependent hyperpolarizablity tensors. Frequency-dependent polarizability and hyperpolarizability tensors are obtained from the trace of the dipole moment matrix $[H^a]$ and the nth-order density matrix $[D^{(n)}]$ (where $n = 1$ for the linear polarizability α, $n = 2$ for the first-order hyperpolarizability tensor β, $n = 3$ for the second-order hyperpolarizability γ and so on). If the inducing electric field of frequency ω_a, ω_b, ω_c ... acts in the direction a, b, c... the tensors can be represented as

$$\alpha_{ab}(-\omega_\sigma; \omega_b) = -\mathrm{Tr}\left[H^a D^b(\omega_b)\right], \qquad (27.21)$$

$$\beta_{abc}(-\omega_\sigma; \omega_b, \omega_c) = -\mathrm{Tr}\left[H^a D^{bc}(\omega_b, \omega_c)\right], \qquad (27.22)$$

$$\gamma_{abcd}(-\omega_\sigma; \omega_b, \omega_c, \omega_d) = -\mathrm{Tr}\left[H^a D^{bcd}(\omega_b, \omega_c, \omega_d)\right], \qquad (27.23)$$

where $\omega_\sigma = \omega_b + \omega_c + \cdots$ These expressions are a set of generalized equations in terms of frequencies. The polarizability and the hyperpolarizability tensors defined above are determined through the iterative solution of the time-dependent Kohn–Sham (TDKS) equations given in Equation 27.14.

In ordinary TDDFT, the zeroth-order Kohn–Sham matrix is used, which is given by

$$F_s^{(0)} = h^{(0)} + D^{(0)}X(2J) + v_{xc}^{(0)}, \qquad (27.24)$$

where $h^{(0)}$ contains the external potential terms that are of zero order in the external field, kinetic energy and the nuclear Coulomb field. The Coulomb super matrix J is independent of the external electric field and results in a Coulomb term of the form $D^{ab...n}(\omega_a, \omega_b, ..., \omega_n)X(2J)$ in the nth-order Kohn–Sham matrix $F_s^{ab...n}(\omega_a, \omega_b, ..., \omega_n)$. The external perturbation appears only in the first-order Kohn–Sham matrices. Higher-order Kohn–Sham matrices consist of a Coulomb and an exchange-correlation part. Therefore, the general formula for the higher-order Kohn–Sham matrices becomes

$$F_s^{ab...n}(\omega_a, \omega_b, ..., \omega_n) = D^{ab...n}(\omega_a, \omega_b, ..., \omega_n)X(2J)$$
$$+ v_{xc}^{ab...n}(\omega_a, \omega_b, ..., \omega_n), \qquad (27.25)$$

The Taylor expansions of F_s, C, ϵ, and D are inserted in the time-dependent Kohn–Sham equation (Equation 27.14), the normalization condition, and the density matrix (Equation 27.16). This results in a first-order time-dependent Kohn–Sham equation as

$$F_s^a(\omega)C^{(0)} + F_s^0 C^a(\omega) + \omega S^0 C^a(\omega) = S^{(0)} C^a(\omega)\epsilon^{(0)} + S^{(0)} C^{(0)}\epsilon^a(\omega). \qquad (27.26)$$

Higher-order coupled equations are obtained by equating the left and the right hand sides of the TDKS equation, the normalization condition and the density matrix. The NLO properties are estimated through the iterative solution of the time-dependent Kohn–Sham equations up to a certain order n. Primarily, the static Kohn–Sham equations are solved that result in matrices $F_s^{(0)}$, $C^{(0)}$, $\epsilon^{(0)}$, $D^{(0)}$, and the converged SCF density $\rho^{(0)}$. These matrices are needed for the solution of the first-order Kohn–Sham equation, which produces the first-order density matrix. The first-order density matrix yields the frequency-dependent polarizability component $[\alpha_{ab}(-\omega_\sigma; \omega_b)]$ through Equation 27.21. The solution of the first-order equation provides matrices required for an iterative solution of the second-order equation. In either case, the technique adopted is called the $(2n + 1)$ theorem. The second-order equation is solved to get the second-order density matrix elements from which the frequency-dependent first-order hyperpolarizability tensor is obtained through Equation 27.23. If the external fields of the frequency zero and a common frequency ω are considered, a number of very important NLO properties become accessible. These are SHG $[\beta(-2\omega; \omega, \omega)]$, EOPE $[\beta(-\omega; \omega, 0)]$, OR $[\beta(0; \omega, -\omega)]$, and the static hyperpolarizability $[\beta(0; 0, 0)]$. Similar to the two previous cases, the $(2n + 1)$ theorem is also used to compute the higher-order hyperpolarizability tensors like γ, δ (third-order hyperpolarizability tensor), and the rest. When γ is calculated only at frequencies 0 and ω, THG $[\gamma(-3\omega; \omega, \omega, \omega)]$, EOKE $[\gamma(-\omega; \omega, 0, 0)]$, dc-SHG $[\gamma(-2\omega; \omega, \omega, 0)]$, and IDRI $[\gamma(-\omega; \omega, -\omega, \omega)]$ become accessible.

In practice, the average polarizability tensor is defined in terms of Cartesian components such as

$$\langle \alpha \rangle = \frac{\alpha_{xx} + \alpha_{yy} + \alpha_{zz}}{3}, \qquad (27.27)$$

where, α_{xx}, α_{yy}, and α_{zz} are the diagonal elements of the polarizability-tensor matrix. The first-order hyperpolarizability tensor is defined as the third derivative of the energy with respect to the electric field components, and hence, involves one additional field differentiation compared to polarizabilities. The average first-order hyperpolarizability is defined as

$$\langle \beta \rangle = \left(\sum_i \beta_i \beta_i^*\right)^{1/2}; \quad \beta_i = \frac{1}{3}\sum_j (\beta_{ijj} + \beta_{jij} + \beta_{jji}), \qquad (27.28)$$

where the sums are over the coordinates x, y, z ($i, j = x, y, z$), and $\beta_i{}^*$ refers to the conjugate of the vector β_i. The second-order hyperpolarizability tensor involves one additional field differentiation compared to the first-order analog. The average second-order hyperpolarizability is defined as

$$\langle\gamma\rangle = \frac{1}{15}\sum_{ij}(2\gamma_{iijj} + \gamma_{ijji}); \quad (i, j = x, y, z). \tag{27.29}$$

In some of the theoretical approaches second-order hyperpolarizability tensors are determined from a combination of analytical and finite-difference techniques [28]. All components of the γ tensor of interest (dc-SHG/EOKE/static second-order hyperpolarizability) are obtained from the analytical time-dependent calculation of the SHG/EOPE/static first-order hyperpolarizability in the presence of small electric fields. E.g., the relation used in the evaluation of EOKE is

$$\gamma_{abcd}(-\omega;\omega,0,0) = \lim_{E^d \to 0}\frac{\beta_{abc}(-\omega;\omega,0)\big|_{E=E^d}}{E^d}, \tag{27.30}$$

Frequency-dependent response properties can also be calculated from the time-averaged quasienergy [29]. In this formalism, the response properties are obtained as the derivative of the quasi energy and the same is estimated by using the variational criterion for the quasienergy and the time-averaged time-dependent Hellmann–Feynman theorem. Molecular properties are obtained by using the variational Lagrangian technique in accordance with the $2n + 1$ and the $2n + 2$ rules. Within this approach, the different frequency-dependent response properties are obtained from the simple extension of the variational perturbation theory to the Fourier component variational perturbation theory.

27.3 Nonlinear Optics with Materials

Much of the present research work in the field of nonlinear optics is motivated by certain nonlinear optical phenomena in suitable materials [14]. Some of these potentially useful phenomena include the ability to alter the frequency or color of light, to amplify one source of light with another, and to alter its transmission features through a medium. The nonlinear optical materials require unusual stability with respect to ambient conditions and high-intensity light sources. These materials can broadly be categorized into two major classes. The first is the class of inorganic materials, which includes semiconducting and metallic clusters, inorganic crystals, bulk materials, etc. The second belongs to the organic or, in general, molecular materials that include mainly organic crystals and polymers. For these systems, the optical nonlinearities are usually derived from their structures. In the present chapter, we will focus only on the first type, i.e., semiconducting and metallic clusters and their NLO behavior.

27.4 Why Are Clusters So Important?

Nonlinear optical processes in cluster materials provide the necessary information about the exact understanding of the quantum confinements and the surface effects in these systems [11]. In general, the properties at the nanoscale are usually nonmonotonic and oscillating in nature due to the quantum size effects, and as such, cluster materials can exhibit properties, which are quite different from those of the bulk [5,11,30]. More specifically, the energy band structure and the phonon distribution of cluster materials may differ from that of the bulk systems. Numerous theoretical as well as experimental investigations have been performed that show that the cluster-assembled materials have novel mechanical, electrical, and optical properties [5–16]. As mentioned earlier in this chapter, these NLO materials find huge applications in high-speed electro-optic devices for information processing and telecommunications [1–4]. In Section 27.5, we present a review on nonlinear optics with various cluster materials.

27.5 Review of Nonlinear Optics with Clusters

In this section, semiconducting clusters will be discussed first followed by metallic clusters. An example of a class of materials with a manifestation of unusual physicochemical and optical properties is cadmium selenide clusters [$(CdSe)_n$]. These $(CdSe)_n$ clusters are found to be the precursors of a wide range of low-dimensional materials such as quantum dots [7,10], tetrapods [31], and nanowires [32] that exhibit a variety of optical and electronic properties. The electronic properties of cadmium selenide clusters $(CdSe)_n$ and quantum dots are semiconducting in nature and these materials find tremendous application in the design and the development of novel nonlinear optical (NLO) materials. There have been a lot of experimental investigations [15,33] on the exploration of the NLO properties in CdSe clusters. The Hyper-Rayleigh scattering technique provides the experimental basis for NLO properties in CdSe clusters and nanoparticles [15]. Aktsipetrov et al. [33] have shown the size dependence of SHG from the surface of a composite material consisting of CdSe nanoparticles embedded in a glass matrix. The first-order hyperpolarizability in CdSe and CdS nanoparticles is reported to be very high [34].

Apart from the experimental works, theoretical *ab initio* investigations [8,11,12] have also been performed in order to calculate the NLO coefficients in CdSe clusters. Troparevsky and Chelikowsky [35], in their work, reported on the structural and the electronic properties (the energy gap between the highest occupied molecular orbital (HOMO) and the lowest unoccupied molecular orbital (LUMO), binding energies, and polarizabilities) of small $(CdS)_n$ and $(CdSe)_n$ clusters, where n ranges from 2 up to 8 using the pseudopotential method in real space. Karamanis et al. [36] have used DFT for the computation of static polarizability and anisotropy in the static polarizability of small CdSe clusters up to tetramer units (see Figure 27.1).

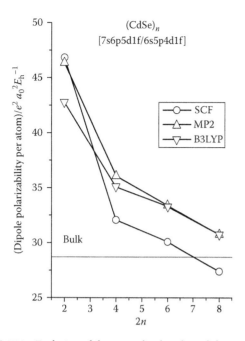

FIGURE 27.1 Evolution of the mean dipole polarizability with cluster size in a small CdSe cluster, $(CdSe)_n$. (From Karamanis, P. et al., *J. Chem. Phys.*, 124, 071101, 2006. With permission.)

Their investigations suggest that the optical properties of the bulk system can also be predicted from a plot of dipole polarizability per atom.

Later on, the same group of Karamanis extended their work on CdSe clusters, highlighting the importance of basis sets and electron correlation in these systems [12]. However, the calculations of the frequency-dependent optical polarizability, as well as that of hyperpolarizability tensors using theoretical methods remained an unresolved issue for CdSe clusters.

Recently, Sen and Chakrabarti [11], for the first time, investigated the frequency-dependent nonlinear optical properties (first- and second-order hyperpolarizability tensors) of $(CdSe)_n$ clusters up to the tetramer using the time-dependent density functional theory (TDDFT). Within the TDDFT, both the local-density approximation (LDA) and the generalized-gradient approximation (GGA) technique have been used. The more accurate LB94 (van Leeuwen and Baerends 94) [37] functional has been used under the GGA scheme. Response calculations have been implemented in the Amsterdam density functional package (ADF 2004.01) [38]. The analytical approach has been used for the evaluation of all the NLO coefficients. The calculated polarizability and hyperpolarizabilities depend on the choice of the basis set and the exchange-correlation potential. While the results obtained within the DFT are overestimated under the normal LDA and GGA functional, the coupled Hartree–Fock procedure exhibits exactly the opposite trend [39,40]. The static values of the average first-order hyperpolarizability ($\langle\beta\,(0;0,0)\rangle$) and the average second-order hyperpolarizability ($\langle\gamma\,(0;0,0,0)\rangle$) of CdSe, Cd_2Se_2, Cd_3Se_3, and Cd_4Se_4 are depicted in Table 27.1.

The frequency dependence of the SHG [$\beta\,(-2\omega;\omega,\omega)$], the EOPE [$\beta\,(-\omega;\omega,0)$], the EFISH [$\gamma\,(-2\omega;\omega,\omega,0)$], and the

TABLE 27.1   Average Static First and Second Order Static Hyperpolarizability in CdSe, Cd_2Se_2, Cd_3Se_3, and Cd_4Se_4

Sample	Average Static First-Order Hyperpolarizability, $\langle\beta\,(0;0,0)\rangle$ (a.u.)	Average Static Second-Order Hyperpolarizability, $\langle\gamma\,(0;0,0,0)\rangle$ (a.u.)
CdSe	−540.07 (LDA)	−556,680 (LDA)
Cd_2Se_2	Absent	7,306.5 (LDA)
		4,398.6 (LB94)
Cd_3Se_3	−0.12258 (LDA)	12,200 (LDA)
	−0.059125 (LB94)	8,929.3 (LB94)
Cd_4Se_4	Absent	89,565 (LDA)
		50,124 (LB94)

Source: Sen, S. and Chakrabarti, S., *Phys. Rev. B*, 74, 205435, 2006. With permission.

EOKE [$\gamma\,(-\omega;\omega,0,0)$] was paid premier attention in the investigation of Sen and Chakrabarti [11]. For a complete evaluation of the frequency dependence of NLO properties in different CdSe clusters, a wide range of frequencies (0 to 0.45 a.u.) was considered.

Figure 27.2 manifests the frequency behavior of $\langle\beta\,(-2\omega;\omega,\omega)\rangle$ and $\langle\beta\,(-\omega;\omega,0)\rangle$ in the case of CdSe and Cd_3Se_3 respectively. $\langle\beta\,(-2\omega;\omega,\omega)\rangle$, being a frequency-dependent property, is of paramount interest and has been investigated extensively in numerous previous investigations [15,41,42]. High degrees of fluctuations in $\langle\beta\,(-2\omega;\omega,\omega)\rangle$ are observed in all the cases. Experimentally, $\langle\beta\,(-2\omega;\omega,\omega)\rangle$ in CdSe nanocrystals and quantum dots has been observed by the HRS technique and its size dependence was also verified [15]. It is evident from Figure 27.2 that $\langle\beta\,(-2\omega;\omega,\omega)\rangle$ is negative over a wide range of frequencies. This makes CdSe and Cd_3Se_3 more significant in the perspective of quantum optics. The frequency variation in $\langle\beta\,(-2\omega;\omega,\omega)\rangle$ divulges one more significant information, i.e., the abrupt increase in $\langle\beta\,(-2\omega;\omega,\omega)\rangle$ at specific frequencies. At specific frequencies the magnitude of $\langle\beta\,(-2\omega;\omega,\omega)\rangle$ becomes very high. It is a common notion that larger values of $\langle\beta\,(-2\omega;\omega,\omega)\rangle$ are obtained at a near resonance of the input energy [5]. The presence of the near resonance indicates that linear absorption can occur at such frequencies.

Similar to the components of the first-order hyperpolarizability tensors, the components of the second-order hyperpolarizability tensors ($\langle\gamma\,(-2\omega;\omega,\omega,0)\rangle$ and $\langle\gamma\,(-\omega;\omega,0,0)\rangle$) were reported to be highly sensitive to frequency variation. Figure 27.3 demonstrates the frequency dependence of different γ tensors. It was concluded from both LDA and LB94 results that both $\langle\gamma\,(-2\omega;\omega,\omega,0)\rangle$ and $\langle\gamma\,(-\omega;\omega,0,0)\rangle$ are negative over a wide range of frequencies. The change in the sign of $\langle\gamma\,(-2\omega;\omega,\omega,0)\rangle$ and $\langle\gamma\,(-\omega;\omega,0,0)\rangle$, due to the change in frequency is a significant observation. This led to a frequency selection in assigning the optical activity of the particular cluster.

In a separate study, Chakrabarti and coworkers [43] investigated the evolution of electric polarizability and the anisotropy of the polarizability at static as well as dynamic (Nd:YAG laser) frequencies with an increase in the cluster size for small, as well

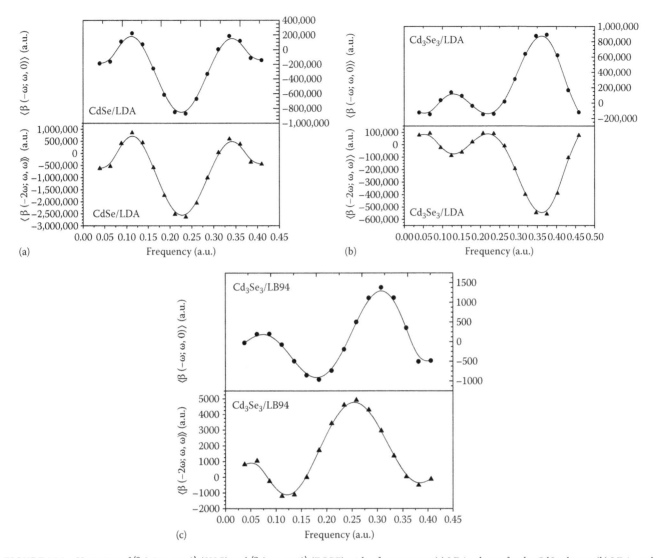

FIGURE 27.2 Variation of $\langle \beta\,(-2\omega;\,\omega,\,\omega)\rangle$ (SHG) and $\langle \beta\,(-\omega;\,\omega,\,0)\rangle$ (EOPE) with a frequency at (a) LDA scheme for the CdSe cluster, (b) LDA, and (c) GGA (LB94) scheme for the Cd_3Se_3 cluster. (From Sen, S. and Chakrabarti, S., *Phys. Rev. B*, 74, 205435, 2006. With permission.)

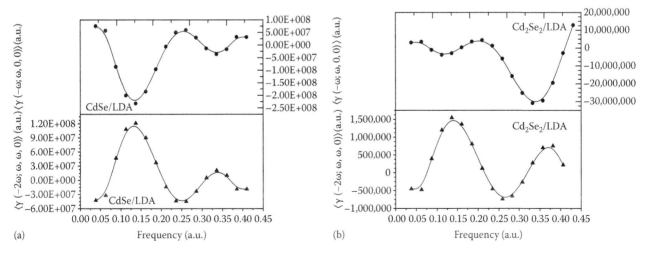

FIGURE 27.3 Variation of $\langle \gamma\,(-2\omega;\,\omega,\,\omega,\,0)\rangle$ and $\langle \gamma\,(-\omega;\,\omega,\,0,\,0)\rangle$ with a frequency at LDA. (a) Scheme for the CdSe cluster, (b) LDA and

(*continued*)

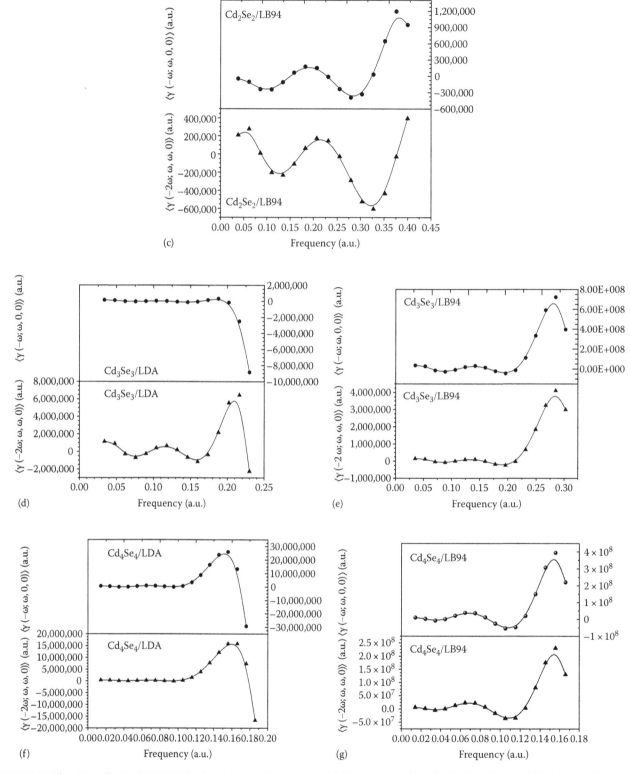

FIGURE 27.3 (continued)   (c) GGA (LB94) scheme for the Cd_2Se_2 cluster, (d) LDA and (e) GGA (LB94) scheme for the Cd_3Se_3 cluster, (f) LDA and (g) GGA (LB94) scheme for the Cd_4Se_4 cluster. (From Sen, S. and Chakrabarti, S., *Phys. Rev. B*, 74, 205435, 2006. With permission.)

as medium-sized $(CdSe)_n$ ($n = 1–16$) clusters within the density functional theory (DFT). The main motivation of that investigation was

1. None of the theoretical studies reported had gone beyond a cluster size of eight for the polarizability calculations for the CdSe clusters.
2. Except the study by Sen and Chakrabarti [11], no one was aware of any investigations where the dynamic hyperpolarizability for any system size was reported for the CdSe clusters.

Knowing the strong dependence of polarizability on the cluster diameter (the cluster size), they performed polarizability calculations beyond the cluster size of eight and calculated both static and dynamic polarizabilities.

Table 27.2 describes computed polarizability values at static and Nd:YAG laser frequencies. An even-odd oscillating behavior is observed in the anisotropy values between the dimer and the heptamer CdSe clusters.

Just like the CdSe clusters, GaAs is one such semiconducting cluster, which has gained a lot of importance over the past few years [5]. These cluster materials display long absorption tails in the low energy region [44] owing to the existence of free surfaces in it. Moreover, the static polarizabilities of these systems exceed the bulk value and follow a decreasing tendency with the increase in the cluster size [45]. The variable nature in the polarizabilities also indicates its "metallic-like" behavior. In the experimental investigation [46], it has been observed that the polarizabilities for small and medium-sized Ga_nAs_m clusters reside above and below the bulk limit. Albeit, there have been a lot of experimental and theoretical investigations [47–49] based on the NLO properties of GaAs bulk materials which suggest that GaAs should possess very large static second- and third-order susceptibilities, only a few of them deal with the theoretical evaluation of the hyperpolarizabilities of these cluster materials [50,51]. A recent work by Lan et al. [5] provides a systematic insight into the NLO properties of several small and medium Ga_nAs_m clusters, where $m + n$ runs up to 10. They employed the TDDFT method in combination with the sum-over-states (SOS) formalism to calculate the first- and second-order hyperpolarizability tensors. The results indicate the presence of large second- and third-order nonlinear susceptibilities in GaAs cluster materials similar to that of the bulk. These results also suggest that resonance absorption can be avoided in the frequency-dependent NLO experiments. Furthermore, the frequency dependence of second- and third-harmonic generation reveals that these clusters are good candidates for future NLO materials.

Besides these two clusters, investigations of the NLO properties have also been carried out on various other III–V atomic nanoclusters. Pineda and Karna [52] estimated the linear and the nonlinear polarizabilities of isolated GaN nanoclusters employing the *ab initio* time-dependent Hartree–Fock (TDHF) method. Their results suggest a strong dependence of the NLO properties on the size and the shape of the clusters. The linear and the nonlinear optical properties of $(CdS)_n$ clusters were analyzed and the merits of the DFT level calculations have been compared with different *ab initio* results, as well as the basis set dependence of the optical properties of these clusters has been evaluated by Maroulis and Pouchan [53].

Figure 27.4 depicts the variation in the mean polarizability and the mean second-order hyperpolarizability values in small $(CdS)_n$ clusters. The values suggest a reduction in the mean

TABLE 27.2 Average Values of Electric Polarizability per Atom ($\langle\alpha\rangle/2n$) and Anisotropy in Polarizability ($\Delta\alpha$) against the Cluster Size, $2n$ at Both Static (0.0 a.u.) and Nd:YAG Laser (0.04283 a.u.) Frequencies

			At 0.0 a.u. Frequency		At Nd:YAG Laser (0.04283 a.u.) Frequency	
Samples	n	$2n$	Polarizability per Atom ($\langle\alpha\rangle/2n$) (a.u.)	Anisotropy in Polarizability ($\Delta\alpha$) (a.u.)	Polarizability per Atom ($\langle\alpha\rangle/2n$) (a.u.)	Anisotropy in Polarizability ($\Delta\alpha$) (a.u.)
CdSe	1	2	34.0349	36.5416	31.3777	49.7528
Cd_2Se_2	2	4	30.9670	60.5853	32.5664	59.0703
Cd_3Se_3	3	6	29.8098	99.7527	30.5716	102.1434
Cd_4Se_4	4	8	29.5301	0.0000	30.8635	0.0000
Cd_5Se_5	5	10	32.4484	159.8706	33.3488	166.2891
Cd_6Se_6	6	12	29.1341	43.8215	30.0033	43.5338
Cd_7Se_7	7	14	29.0036	134.4511	29.9893	143.1402
Cd_8Se_8	8	16	29.5565	114.3660	30.4782	122.1772
Cd_9Se_9	9	18	29.4061	49.9355	30.2877	54.6105
$Cd_{10}Se_{10}$	10	20	29.8171	124.1952	30.7628	132.7159
$Cd_{11}Se_{11}$	11	22	29.8559	150.1088	30.7738	160.6296
$Cd_{12}Se_{12}$	12	24	28.4830	0.0165	29.3089	0.0169
$Cd_{13}Se_{13}$	13	26	30.9808	271.1190	32.0344	287.1187
$Cd_{14}Se_{14}$	14	28	30.3060	192.7016	31.2651	207.4891
$Cd_{15}Se_{15}$	15	30	30.2442	158.5999	31.1661	170.6002
$Cd_{16}Se_{16}$	16	32	33.5302	428.7594	34.9350	470.1254

Source: Jha, P.C. et al., *Comput. Mater. Sci.*, 44, 728, 2008. With permission.

Handbook of Nanophysics: Nanoelectronics and Nanophotonics

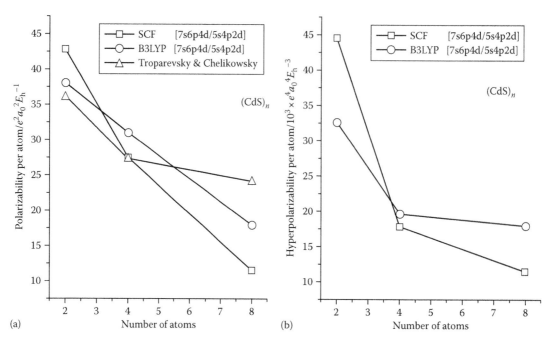

FIGURE 27.4 Mean (a) dipole polarizability and (b) second-order hyperpolarizability in $(CdS)_n$ clusters. (From Maroulis, G. and Pouchan, C., *J. Phys. Chem. B*, 107, 10683, 2003. With permission.)

dipole polarizability, as well as in the second-order hyperpolarizability values with an increase in the cluster size. Moreover, the results obtained from B3LYP and other conventional *ab initio* methods are well in accordance with each other and also with the earlier theoretical results [35]. However, in the case of the monomer geometry, the difference in the hyperpolarizability values between these methods is quite large. Champagne et al. estimated [54] average second-order hyperpolarizabilities of Si_n ($n = 3–38$) clusters using MP2, MP3, MP4, CCSD, and CCSD(T) levels of approximations and demonstrated the variation of polarizability and hyperpolarizability against the cluster size.

Apart from these semiconducting materials, cluster science has also made rapid progress in determining the electric properties of various metal clusters [55] including that of mixed metal. At first, a review of the some of the works based on the metal clusters containing only one type of atom is discussed, which is followed by some detailed analysis on mixed metal clusters. There are investigations on lithium [18,56,57], sodium [58], copper [59,60], nickel [61], niobium [62], and zinc [63] clusters. Maroulis and Xenides [57] reported highly accurate *ab initio* calculations on the polarizability and the hyperpolarizabilities of lithium tetramer, Li_4 with specially designed basis sets for Li. The molecule, Li_4 has got a very high anisotropic dipole polarizability and quite large second-order hyperpolarizability values. The main feature of this work is the extensive analysis of the basis set and the correlation effects in calculating the electric properties. The results indicate a discrepancy in dipole polarizabilities between theory and experiment, which might be resolved by measuring the anisotropy values experimentally. The same group also presented a thorough analysis of the performance of

several density functional theoretical (DFT) calculations of the (hyper) polarizability of Li_4. Their investigations elucidated that of the various DFT methods employed; only the mPW1PW91 and the O3LYP methods produce reliable results. Papadopoulos et al. [63] studied the (hyper) polarizabilities of small and medium-sized zinc clusters employing a hierarchy of basis sets and computational methods. The relativistic effect on the electric polarizabilities has also been investigated by employing the Douglas–Kroll approximation. It has been observed that the relativity contribution is significant in these cluster materials; however, the correlation effect plays a major part in influencing the electro-optic properties.

The NLO properties of Ag nanoparticles embedded in a Si_3N_4 matrix was investigated by Traverse et al. [64]. They performed a two-color sum-frequency generation (2c-SFG) spectroscopy experiment, which exhibited a surface plasmon resonance at a 421 nm wavelength in the visible region and absorption in the infrared region. The third-order optical nonlinearity of dielectrics in the presence of nanoparticles has been experimentally investigated by Flytzanis [65]. Chen and coworkers performed [66] high-level *ab-initio* calculations [HF, MP2, the fourth-order perturbation theory using single, double, quadrupole substitutions, CCSD, and QCISD] on $(HCN)_n...Li$ and $Li...$ $(HCN)_m$ clusters with electride characteristics. A high value of static first hyperpolarizability is reported in both samples. The static second-order hyperpolarizability of a series of tri-nuclear metal cluster $MS4(M'PPh_3)_2(M'PPh_3)$ (M = Mo, W; M′ = Cu, Ag, Au) have been estimated by [67] Chen and coworkers. They employed a finite-field approach using the hybrid density functional theory (B3LYP) and reported a very high value of second-order hyperpolarizability. The NLO properties of Au_n clusters

TABLE 27.3 Average Hirshfeld Charge on the Al_4 Ring, Ground State Dipole Moment, Magnitude of Average Static First- and Second-Order Hyperpolarizability in Al_4M_4 (M = Li, Na, and K)

Sample	Average Hirshfeld Charge on the Al_4 Ring	Ground-State Dipole Moment (Debye)	Average Static First-Order Hyperpolarizability, $\langle\beta(0; 0, 0)\rangle$ (a.u.)	Average Static Second-Order Hyperpolarizability, $\langle\gamma(0; 0, 0, 0)\rangle$ (a.u.)
Al_4Li_4 (2D)	−0.247	0.049	145.922	1.299×10^6
Al_4Li_4 (3D)	−0.202	0.018	33.079	2.406×10^6
Al_4Na_4 (2D)	−0.258	0.003	21.537	1.275×10^6
Al_4Na_4 (3D)	−0.229	0.054	187.843	1.961×10^6
Al_4K_4 (2D)	−0.303	0.080	316.781	3.380×10^6
Al_4K_4 (3D)	−0.297	6.513	8863.074	1.149×10^7

Source: Sen, S. et al., *Phys. Rev. B*, 76, 115414, 2007. With permission.

and their alloyed clusters, $Au_{n-m}M_m$ (M = Ag, Cu; m = 1, 2) have been theoretically investigated by Xu and coworkers [68]. They used the usual TD-DFT as well as the B3LYP hybrid density functional employing the LANL2DZ basis set and measured the first-order hyperpolarizabilites of such clusters. They suggested that the high degree of NLO properties of these clusters are due to the lack of a centro-symmetric feature in these clusters and an enhancement in the transition electric dipole moment in alloyed clusters.

The optical properties of several mixed metal clusters such as Li_nCu_m, Na_4Pb_4, Na_6Pb, $Al_{11}Fe$, and $Al_{18}Fe$, Al_4M_4 (M = alkali metals) have also been investigated extensively [69,70]. In this context, we particularly mention the NLO properties of aluminum metal clusters [Al_4M_4 (M = Li, Na, and K)]. Various investigations have been performed on these Al_4M_4 systems and their anions due to their unique characteristic features and their structural similarity with that of cyclobutadiene, C_4H_4 [6]. These cluster materials are also important since they are better polarized than their organic counterparts. Earlier, Datta and Pati [6] investigated the static linear and nonlinear polarizabilities of Al_4M_4 (M = Li, Na, and K) systems. They predicted that structural changes through the broken inversion symmetry of Al_4M_4 could lead to a unique polarization response and large optical coefficients. In their work, both the static linear and nonlinear optical properties of these Al_4M_4 were estimated through ZINDO-multireference doubles configuration interaction (MRDCI) calculations. They emphasized that the large NLO properties of Al_4M_4 are due to the charge transfer from alkali metals to the Al_4 ring. In spite of the fact that their work exhibits an exceptionally high magnitude of static linear and nonlinear optical coefficients, it lacks the description of its dynamic counterparts. It is to be noted that the importance of a material in nonlinear optics and its applications in device fabrication demands a complete description of the frequency dependence. Hence, it was highly instructive to explore the dynamic (frequency-dependent) NLO properties of these clusters for their proper assessment in optical device fabrication and to explain whether the large NLO properties are due to a charge transfer interaction or due to other excitations.

Recently Sen et al. [13] reported various components of first- (SHG, EOPE, static first-order hyperpolarizability) and second- (EFISH, EOKE, and static second-order

hyperpolarizability) order hyperpolarizability tensors at several near resonance frequencies. All these dynamic NLO properties of Al_4M_4 (M = Li, Na, and K) clusters have been explored using the TDDFT as a prime investigating tool. Within the TDDFT, the generalized gradient approximation (GGA) technique has been implemented with the use of a more accurate LB94 (van Leeuwen and Baerends 94) functional.

The magnitude of static $\langle\beta\rangle$ and $\langle\gamma\rangle$ along with an average Hirshfeld charge on the Al_4 ring and the ground-state dipole moments are shown in Table 27.3. It is quite clear from Table 27.3 that the magnitude of $\langle\beta(0; 0, 0)\rangle$ increases significantly from 2D to 3D in the case of Al_4Na_4 and Al_4K_4, on the other hand, a reverse trend is observed for Al_4Li_4.

Table 27.3 also suggests that the magnitudes of the average static $\langle\gamma(0; 0, 0)\rangle$ are very high (of the order of 10^6 and 10^7 a.u.) for all the clusters under investigation.

However, the central issue of the investigation of Sen and coworkers [13] was to explicate the frequency dependence of $\langle\beta(-2\omega; \omega, \omega)\rangle$, $\langle\beta(-\omega; \omega, 0)\rangle$, $\langle\gamma(-2\omega; \omega, \omega, 0)\rangle$, and $\langle\gamma(-\omega; \omega, 0, 0)\rangle$ in aluminum metal clusters. These components of the first- and second-order hyperpolarizability tensors were estimated in the vicinity of near-excitation frequencies with a substantial oscillator strength and presented in Table 27.4. A closer inspection of Table 27.4 reveals that $\langle\beta(-2\omega; \omega, \omega)\rangle$ and $\langle\beta(-\omega; \omega, 0)\rangle$ attain negative values at certain frequencies in all these cluster materials. Exceptionally high values [of the order of 10^6–10^8 a.u.] of $\langle\gamma(-2\omega; \omega, \omega, 0)\rangle$ and $\langle\gamma(-\omega; \omega, 0, 0)\rangle$ were also noticed. A significant observation was in the change in sign of $\langle\gamma(-2\omega; \omega, \omega, 0)\rangle$ and $\langle\gamma(-\omega; \omega, 0, 0)\rangle$ due to the change in frequency, which lead to a frequency selection in assigning the optical activity of a particular cluster.

In order to explain the observed high NLO coefficients in these aluminum metal clusters, an analysis of the magnitude of gross populations of molecular orbitals with the most significant SFO (Symmetrized Fragment Orbitals) was carried out. The SFO study divulged the role of charge transfer interactions. Molecular orbitals with the most significant SFO gross population indicated that except for Al_4K_4 (2D and 3D) and Al_4Li_4 (3D) there exists excitations other than a charge transfer interaction at near resonance frequencies.

Moreover, it is worth mentioning that the dynamic first-order hyperpolarizability tensors are functions of the excitation

TABLE 27.4 $\langle\beta(-2\omega;\omega,\omega)\rangle$, $\langle\beta(-\omega;\omega,0)\rangle$, $\langle\gamma(-2\omega;\omega,\omega,0)\rangle$, and $\langle\gamma(-\omega;\omega,0,0)\rangle$ Measured at Frequencies Nearer to Excitation Frequencies in Al_4M_4 (M = Li, Na and K)

Sample	Excitation Frequencies (a.u.)	Oscillator Strength	Frequency at which Data Are Taken (a.u.)	$\langle\beta(-2\omega;\omega,\omega)\rangle$ (a.u.)	$\langle\beta(-\omega;\omega,0)\rangle$ (a.u.)	$\langle\gamma(-2\omega;\omega,\omega,0)\rangle$ (a.u.)	$\langle\gamma(-\omega;\omega,0,0)\rangle$ (a.u.)
Al_4Li_4 (2D)	0.0205	0.0042	0.0200	−6081.100	12,741	8.988×10^6	-2.744×10^6
	0.0522	0.0034	0.0500	224.900	46.143	-5.239×10^5	1.155×10^6
	0.0558	0.0094	0.0540	−371.630	361.960	1.226×10^7	-7.558×10^5
Al_4Li_4 (3D)	0.0377	0.0441	0.0360	4057.900	−15,252	6.049×10^6	1.663×10^7
	0.0650	0.0863	0.0640	968.580	−902.360	2.729×10^7	-8.458×10^4
	0.0709	0.3736	0.0690	−638.610	236.140	-1.199×10^8	-1.866×10^7
Al_4Na_4 (2D)	0.0238	0.0068	0.0220	−429.590	2697.100	1.266×10^7	-7.639×10^7
	0.0403	0.0012	0.0390	51.085	−96.614	4.735×10^4	1.245×10^6
	0.0570	0.0852	0.0590	4641.500	−93.815	1.540×10^7	-6.290×10^6
Al_4Na_4 (3D)	0.0271	0.0301	0.0260	2193.700	−13333	1.675×10^7	-6.581×10^6
	0.0446	0.0060	0.0430	−452.340	−178.82	4.459×10^6	5.213×10^5
	0.0543	0.0099	0.0560	−7067.500	−168.550	1.242×10^7	1.401×10^6
Al_4K_4 (2D)	0.0244	0.0066	0.0230	305.930	864.090	1.286×10^6	6.640×10^6
	0.0427	0.0910	0.0410	−1465.600	3477.900	-3.240×10^7	3.698×10^7
	0.0431	0.0197	0.0440	−64908	−5213.200	-3.143×10^8	2.548×10^7
Al_4K_4 (3D)	0.0215	0.0010	0.0200	−231620	−16,582	1.842×10^7	-1.926×10^7
	0.0391	0.0105	0.0380	29403	150,940	9.758×10^6	2.791×10^7
	0.0416	0.0067	0.0400	−120770	380,220	-7.515×10^7	4.444×10^7

Source: Sen, S. et al., *Phys. Rev. B*, 76, 115414, 2007. With permission.

energy, the dipolar matrix, and the difference between the dipole moment of the ground and the excited states. On the other hand, components of the second-order hyperpolarizability tensor are associated with four dipolar matrices in the numerator and three excitation frequencies in the denominator. Hence, it is very difficult to find all the factors that play a dominant role in the high β and γ values in particular clusters.

27.6 Conclusions

Therefore, it is well established from both the theoretical as well as the experimental standpoint that cluster materials can be considered as promising candidates for future NLO devices due to their high nonlinear optical coefficient. However, the exploration of the fundamentals of the unique physicochemical behavior of these clusters is a challenge to modern physical and chemical research. A theoretical determination of a higher-order nonlinear optical coefficient is still a matter of extensive computation and there is need for further developments to provide a proper physical background for some of the higher-order optical coefficients. There is enough scope for future research work in the field of cluster science and the design of new cluster materials with interesting physical behavior.

References

1. M. Lee, E. H. Katz, C. Erben, D. M. Gill, P. Gopalan, J. D. Heber, and D. J. McGee, *Science* **298**, 1401 (2002).
2. Y. Shi, C. Zhang, H. Zhang, J. H. Bechtel, L. R. Dalton, B. H. Robinson, and W. H. Steier, *Science* **288**, 119 (2000).
3. J. Luo, M. Haller, H. Li, H.-Z. Tang, A. K.-Y. Jen, K. Jakka, C.-H. Chou, and C.-F. Shu, *Macromolecules* **37**, 248 (2004).
4. C. Zhang, L. R. Dalton, M. C. Oh, H. Zhang, and W. H. Steier, *Chem. Mater.* **13**, 3043 (2001).
5. Y.-Z. Lan, W.-D. Cheng, D.-S. Wu, J. Shen, S.-P. Huang, H. Zhang, Y.-J. Gong, and F.-F. Li, *J. Chem. Phys.* **124**, 094302 (2006).
6. A. Datta and S. K. Pati, *J. Phys. Chem. A* **108**, 9527 (2004).
7. A. Puzder, A. J. Williamson, F. Gygi, and G. Galli, *Phys. Rev. Lett.* **92**, 217401 (2004).
8. M. C. Troparevsky, L. Kronik, and J. R. Chelikowsky, *J. Chem. Phys.* **119**, 2284 (2003).
9. R. Jose, N. U. Zhanpeisov, H. Fukumura, Y. Baba, and M. Ishikawa, *J. Am. Chem. Soc.* **128**, 629 (2006).
10. J. Zhao, J. A. Bardecker, A. M. Munro, M. S. Liu, Y. Niu, I.-K. Ding, J. Luo, B. Chen, A. K.-Y. Jen, and D. S. Ginger, *Nano Lett.* **6**, 463 (2006).
11. S. Sen and S. Chakrabarti, *Phys. Rev. B* **74**, 205435 (2006).
12. P. Karamanis, G. Maroulis, and C. Pouchan, *Chem. Phys.* **331**, 19 (2006).
13. S. Sen, P. Seal, and S. Chakrabarti, *Phys. Rev. B* **76**, 115414 (2007).
14. P. N. Prasad and D. J. Williams in *Introduction to Nonlinear Optical Effects in Molecules and Polymers.* Wiley, New York (1991); S. P. Karna and A. T. Yeates in *Nonlinear Optical Materials*, Eds.; ACS Symposium Series 628, American Chemical Society, Washington, DC (1996).
15. M. Jacobsohn and U. Banin, *J. Phys. Chem. B* **104**, 1 (2000).
16. S.-i. Inoue and Y. Aoyagi, *Phys. Rev. Lett.* **94**, 103904 (2005).

17. L. Jensen, P.-O. Astrand, A. Osted, J. Kongsted, and K. V. Mikkelsen, *J. Chem. Phys.* **116**, 4001 (2002).

18. S. K. Pati, S. Ramashesha, Z. Shuai, and J. L. Bredas, *Phys. Rev. B: Condens. Matter* **59**, 14827 (1999).

19. K. B. Sophy and S. Pal, *J. Chem. Phys.* **118**, 10861 (2003); L. D. Freo, F. Terenziani, and A. Painelli, *J. Chem. Phys.* **116**, 755 (2002).

20. S. Grimme, *J. Comput. Chem.* **25**, 1463 (2004); Q. Wu and W. Yang, *J. Chem. Phys.* **116**, 515 (2002).

21. M. Elstner, P. Hobza, T. Frauenheim, S. Suhai, and E. Kaxiras, *J. Chem. Phys.* **114**, 5149 (2001).

22. S. K. Pati, T. J. Marks, and M. A. Ratner, *J. Am. Chem. Soc.* **123**, 7287 (2001); D. Beljonne, Z. Shuai, J. Cornil, D. dos Santos, and J. L. Bredas, *J. Chem. Phys.* **111**, 2829 (1999).

23. J. L. Oudar and D. S. Chemla, *J. Chem. Phys.* **66**, 2664 (1977); J. L. Oudar, *J. Chem. Phys.* **67**, 446 (1977).

24. J. J. Zyss, *Chem. Phys.* **71**, 909 (1979).

25. D. L. Andrews, L. C. Dávila Romero, and W. J. Meath, *J. Phys. B: At. Mol. Opt. Phys.* **32**, 1 (1999).

26. A. Datta and S. K. Pati, *Chem. Soc. Rev.* **35**, 1305 (2006).

27. P. Seal, P. C. Jha, and S. Chakrabarti, *J. Mol. Struct.: Theochem*, **855**, 64 (2008); D. R. Kanis, M. A. Ratner, and T. J. Marks, *Chem. Rev.* **94**, 195 (1994).

28. S. J. A. van Gisbergen, J. G. Snijders, and E. J. Baerends, *J. Chem. Phys.* **109**, 10644 (1998); S. J. A. van Gisbergen, J. G. Snijders, and E. J. Baerends, *J. Chem. Phys.* **109**, 10657 (1998).

29. O. Christiansen, P. JØrgensen, and C. HÄttig, *Int. J. Quantum Chem.* **88**, 1 (1998).

30. K. R. S. Chandrakumar, T. K. Ghanty, and S. K. Ghosh, *Int. J. Quantum Chem.* **105**, 166 (2005).

31. L. Manna, E. C. Scher, and A. P. Alivisatos, *J. Am. Chem. Soc.* **122**, 12700 (2000); J. Li and L. W. Wang, *Nano Lett.* **3**, 1357 (2003).

32. W. U. Huynh, J. J. Dittmer, and A. P. Alivisatos, *Science* **295**, 2425 (2002).

33. O. A. Aktsipetrov, P. V. Elyutin, A. A. Nikulin, and E. A. Ostrovskaya, *Phys. Rev. B* **51**, 17591 (1995).

34. B. S. Santos, G. A. L. Pereira, D. V. Petrov, and C. de Mello Donegá, *Opt. Commun.* **178**, 187 (2000).

35. M. C. Troparevsky and J. R. Chelikowsky, *J. Chem. Phys.* **114**, 943 (2001).

36. P. Karamanis, G. Maroulis, and C. Pouchan, *J. Chem. Phys.* **124**, 071101 (2006).

37. R. van Leeuwen and E. J. Baerends, *Phys. Rev. A* **49**, 2421 (1994).

38. http://www.scm.com

39. A. J. Cohen, N. C. Handy, and D. J. Tozer, *Chem. Phys. Lett.* **303**, 391 (1999).

40. R. M. Dickson and A. D. Becke, *J. Phys. Chem.* **100**, 16105 (1996).

41. D. V. Petrov, B. S. Santos, G. A. L. Pereira, and C. deMello-Donegá, *J. Phys. Chem. B* **106**, 5325 (2002).

42. N. Song, L. Men, J. P. Gao, Y. Bai, A. M. R. Beaudin, G. Yu, and Z. Y. Wang, *Chem. Mater.* **16**, 3708 (2004).

43. P. C. Jha et al., *Comput. Mater. Sci.* **44**, 728 (2008).

44. I. Vasiliev, S. Öğüt, and J. R. Chelikowsky, *Phys. Rev. B* **60**, R8477 (1999).

45. I. Vasiliev, S. Öğüt, and J. R. Chelikowsky, *Phys. Rev. Lett.* **78**, 4805 (1997).

46. R. Schäfer, S. Schlecht, J. Woenckhaus, and J. A. Becker, *Phys. Rev. Lett.* **76**, 471 (1996).

47. M. Zh. Huang and W. Y. Ching, *Phys. Rev. B* **47**, 9464 (1993).

48. Z. H. Levine and D. C. Allan, *Phys. Rev. Lett.* **66**, 41 (1991).

49. D. J. Moss, E. Ghahramani, J. E. Sipe, and H. M. van Driel, *Phys. Rev. B* **41**, 1542 (1990).

50. J. L. Wang, M. L. Yang, G. H. Wang, and J. J. Zhao, *Chem. Phys. Lett.* **367**, 448 (2003).

51. P. P. Korambath and S. P. Karna, *J. Phys. Chem. A* **104**, 4801 (2000).

52. A. C. Pineda and S. P. Karna, *Chem. Phys. Lett.* **429**, 169 (2006).

53. G. Maroulis and C. Pouchan, *J. Phys. Chem. B* **107**, 10683 (2003).

54. B. Champagne, M. Guillaume, D. Bégué, and C. Pouchan, *J. Comput. Methods Sci. Eng.* **7**, 297 (2007).

55. K. D. Bonin and V. V. Kresin, in *Electric Dipole Polarizabilities of Atoms, Molecules and Clusters*, World Scientific, Singapore (1997).

56. E. Benichou, R. Antoine, D. Rayane, B. Vezin, F. W. Dalby, Ph. Dugourd, M. Broyer et al., *Phys. Rev. A* **59**, R1 (1999).

57. G. Maroulis and D. Xenides, *J. Phys. Chem. A* **103**, 4590 (1999).

58. G. Tikhonov, V. Kasperovich, K. Wong, and V. V. Kresin, *Phys. Rev. A* **64**, 063202 (2001).

59. G. Maroulis, *J. Phys. Chem. A* **107**, 6495 (2003).

60. M. B. Knickelbein, *J. Chem. Phys.* **120**, 10450 (2004).

61. M. B. Knickelbein, *J. Chem. Phys.* **115**, 5957 (2001).

62. R. Moro, X. Xu, S. Yin, and W. A. de Heer, *Science* **300**, 1265 (2003).

63. M. G. Papadopoulos, H. Reis, A. Avramopoulos, S. Erkoç, and L. Amirouche, *J. Phys. Chem. B* **109**, 18822 (2005).

64. A. Traverse(a), C. Humbert, C. Six, A. Gayral, and B. Busson, E. *Phys. Lett.* **83**, 64004 (2008).

65. C. Flytzanis, *J. Phys. B* **38**, S661 (2005).

66. W. Chen, Z.-Ru Li, D. Wu, R. Li, and C. Sun, *J. Phys. Chem. B* **109**, 601 (2005).

67. X. Chen, K. Wu, J. G. Snijders, and C. Lin, *Chinese Chem. Lett.* **13**, 893 (2002).

68. Y. Xu, C. Xu, T. Zhou, and C. Cheng, *J. Mol. Struct. Theochem* **893**, 88 (2009).

69. P. Fuentealba, L. P.-Campos, and O. Reyes, *J. Comput. Methods Sci. Eng.* **4**, 589 (2004).

70. M. B. Torres and L. C. Balbás, *J. Comput. Methods Sci. Eng.* **4**, 517 (2004) and references therein.

28

Second-Harmonic Generation in Metal Nanostructures

Marco Finazzi
Politecnico di Milano

Giulio Cerullo
Politecnico di Milano

Lamberto Duò
Politecnico di Milano

28.1 Introduction

The explosive growth of nanoscience and nanotechnology during the last decade has led to great interest in the investigation of nanoscale optical fields and to the development of experimental tools for their study (Novotny and Hecht 2006). One of the most remarkable effects in light interaction with metal nanostructures is the strong and spatially localized field amplitude enhancement, due to lightning rod effects induced by the sharp curvatures (Novotny and Hecht 2006) and/or resonant excitation of collective electron oscillations (plasma oscillations) in single or coupled particles (Grand et al. 2003, Mühlschlegel et al. 2005). The resonance frequencies associated with such oscillations and thus the optical properties can be tuned in a broad spectral range according to the material and the shape of the nanostructure. Particularly, gold and silver may sustain plasma oscillations at optical frequencies.

Field enhancements are best observed by exploiting nonlinear optical effects such as second-harmonic generation (SHG) (Smolyaninov et al. 1997, 2000, Jakubczyk et al. 1999, Zayats et al. 2000, Shen et al. 2001, Biagioni et al. 2007, Breit et al. 2007)

or two-photon photo-luminescence (TPPL) (Jakubczyk et al. 1999, Bouhelier et al. 2003, 2005, Imura et al. 2005, 2006), which depend on the square of the light intensity, thus amplifying local field variations. To observe a significant nonlinear response, the samples are typically illuminated by high peak intensities, such as those associated with ultrashort laser pulses. Nonlinear optical microscopies, either in the far field or in the near field, are thus becoming powerful tools for the study of nanoscale field enhancements.

SHG was first experimentally demonstrated in 1961 (Franken et al. 1961) and has since then found many applications, especially in converting laser light to a different color. Another application more relevant to the content of this chapter is surface SHG. Although the sensitivity of the SHG process to boundaries was soon recognized (Bloembergen and Pershan 1962), it was only in the early 1980s that nonlinear optical spectroscopy of surfaces and interfaces started to develop as a well-established analysis tool in a wide area of fields (Shen 1989, Mc Gilp 1996, Bloembergen 1999, Lüpke 1999, Downer et al. 2001). The SHG ability to selectively probe surfaces and interfaces derives from a selection rule (see Section 28.2) that does not allow for dipole emission from the bulk of centrosymmetric materials. At the surface or interface, however, the symmetry of the bulk is broken and dipole generation is no longer forbidden. This peculiar property has determined the enormous spread of SHG-based techniques in the community of surface and interface physics. This technique has the advantages of being highly surface-sensitive, capable of remote sensing and in situ measurement, and applicable to any interface between two condensed media accessible by light, even if buried. Among the applications, surface SHG can be used to monitor the growth of monolayers on a surface or to probe the alignment of absorbed molecules.

This chapter presents theoretical and experimental results on SHG from metal nanostructures. It is organized as follows: Section 28.2 contains a brief description of the principles of nonlinear optics and SHG; Section 28.3 presents a survey of the recent literature on SHG from nanostructures; Section 28.4 introduces a theoretical analysis of SHG from nanostructures, allowing to derive selection rules for the process; Section 28.5 describes the light emission patterns and polarization, while Section 28.6 presents experimentally relevant illumination conditions. Section 28.7 reports experimental results on SHG from gold nanoparticles, while Sections 28.8 and 28.9 contain a discussion of possible future developments and conclusions, respectively.

28.2 Nonlinear Optics and Second Harmonic Generation

Nonlinear optics (Shen 1984, Boyd 2003) is the branch of physics that describes the phenomena that occur when the matter is illuminated by very intense light fields that have the capability of modifying its optical properties, thus generating an optical response that is a nonlinear function of the light intensity. Since electric fields strong enough to induce a nonlinear optical response can only be obtained by using the spatially and temporally coherent radiations provided by lasers, the birth of nonlinear optics closely coincides with the invention of the laser.

The physical origins of the nonlinear optical response of a material can be understood by recalling the Lorentz model, in which the atoms/molecules are modeled as harmonic oscillators that, driven by the external light field, become oscillating dipoles

emitting light at the same frequencies as the driving field. In a harmonic oscillator, the restoring force is a linear function of displacement; for large displacements, however, nonlinear terms in the restoring force start to become significant, thus leading to an anharmonic oscillation and emission of new frequency components not present in the driving field.

Let us consider an electromagnetic wave crossing a material: the time-varying electric field $E(t)$ induces a time-dependent polarization (dipole moment per unit volume) $P(t)$, which is responsible for light emission. At low light intensities, the polarization is a linear function of the driving field:

$$P(t) = \varepsilon_0 \chi^{(1)} E(t), \tag{28.1}$$

where
- ε_0 is the vacuum permittivity
- $\chi^{(1)}$ is known as the linear susceptibility

As the intensity increases, further nonlinear terms in which the polarization is proportional to higher powers of the electric field must be considered:

$$P(t) = \varepsilon_0 [\chi^{(1)} E(t) + \chi^{(2)} E^2(t) + \chi^{(3)} E^3(t) + \cdots], \tag{28.2}$$

where the quantities $\chi^{(2)}$ and $\chi^{(3)}$ are known as second- and third-order nonlinear susceptibilities, respectively. By considering the vectorial nature of E and P, $\chi^{(1)}$ becomes a second-rank tensor, $\chi^{(2)}$ a third-rank tensor, and so on.

The $\chi^{(2)}$ term is responsible for second-order nonlinear optical effects, such as sum frequency generation, difference frequency generation, and SHG. The following text briefly describes the SHG process. Let us consider the geometry shown in Figure 28.1a, in which a monochromatic wave at the fundamental frequency ω, $E(t) = E_0 \cos(\omega t)$, impinges on a crystal with nonzero second-order susceptibility. The second-order polarization can then be written as

$$P^{(2)}(t) = \varepsilon_0 \chi^{(2)} E^2(t) = \varepsilon_0 \chi^{(2)} E_0^2 \cos^2(\omega t) = \frac{1}{2} \varepsilon_0 \chi^{(2)} E_0^2 [1 + \cos(2\omega t)].$$

$$\tag{28.3}$$

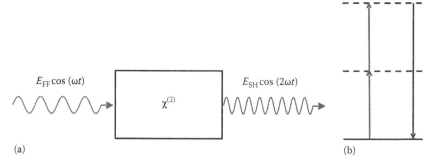

(a)   (b)

FIGURE 28.1 Scheme of SHG in a nanoparticle: two incoming FW photons are absorbed by the particle and their energy converted in a single photon with double frequency.

The nonlinear polarization thus contains, in addition to a constant term (optical rectification), a component at the second-harmonic (SH) frequency 2ω, which irradiates a field at the corresponding frequency. Note that, since the second-order nonlinear polarization depends on the square of the driving field amplitude E_0, the SHG efficiency is enhanced by high-peak power, pulsed illumination. The SHG process can also have a simple corpuscular interpretation, according to the energy-level scheme reported in Figure 28.1b: two fundamental wavelength (FW) photons at energy $\hbar\omega$ are absorbed to a virtual level of the material, and an SH photon with energy $2\hbar\omega$ is emitted, thus satisfying energy conservation.

In a centrosymmetric medium (i.e., a medium with a center of inversion), the $\chi^{(2)}$ term vanishes identically. In such a medium, in fact, by changing the sign of the driving field, also the sign of the nonlinear polarization must change. By substituting in Equation 28.3, one obtains

$$-P^2(t) = \varepsilon_0\chi^2[-E(t)]^2 = \varepsilon_0\chi^2 E^2(t), \qquad (28.4)$$

which can only be satisfied if $\chi^{(2)} \equiv 0$. Since gases, liquids, amorphous solids, and also several crystals possess an inversion symmetry, the second-order nonlinear susceptibility in such materials is zero, and thus, SHG is forbidden.

At the microscopic level, one can consider the single emitter (atom, molecule) as a nonlinear dipole radiating at the SH frequency 2ω. In order to generate a macroscopic SH signal, the fields emitted by the individual dipoles must interfere constructively, adding their contributions in phase. Since the phase of the nonlinear dipoles depends on the phase of the FW driving field, to achieve constructive interference between the nonlinear emissions at two different positions in the nonlinear medium, one must match the phase velocities of the FW field $v_{pFW} = \omega/k_{FW}$ and the SH field $v_{pSH} = 2\omega/k_{SH}$. From $v_{pFW} = v_{pSH}$, one obtains the condition $k_{SH} = 2k_{FW}$, also known as "phase-matching" condition, which implies that $n_{SH} = n_{FW}$, where n is the refractive index of the medium. If the phase-matching condition is satisfied, all the microscopic emissions add in phase and the SH field grows linearly (and the intensity quadratically) with the propagation distance. On the other hand, if $\Delta k = k_{SH} - 2k_{FW} \neq 0$, the SH field depends on the propagation coordinate x as $E_{SH} \propto \chi^{(2)} \sin(\Delta k\, x/2)$. It thus increases for propagation up to a "coherence length" L_c such that $\Delta k\, L_c = \pi$, and then starts to decrease due to destructive interference.

Achievement of the phase matching condition is made difficult by the fact that most materials display, in the optical frequency range, a refractive index increasing monotonously with frequency (normal dispersion), so that $n_{SH} > n_{FW}$. One can obtain phase-matching condition exploiting the birefringence of noncentrosymmetric crystals, that is, the dependence of the refractive index on the polarization and propagation direction. By polarizing the SH wave along the direction giving the lower refractive index, one can satisfy the condition $n_{SH} = n_{FW}$ and thus achieve phase matching. Under such conditions, the SHG process can become so efficient that nearly all the FW power is converted to the SH (Parameswaran et al. 2002).

28.3 Second Harmonic Generation in Nanosystems

As discussed in Section 28.2, SHG is forbidden, within the dipole approximation, from the bulk of a centrosymmetric material but is allowed at its surface, where the inversion symmetry is broken. For this reason, SHG has been extensively applied since its discovery to the investigation of the surface properties of centrosymmetric media. In more recent years, thanks to its surface selectivity, SHG has attracted a considerable interest also as a tool to investigate the properties of nonplanar surfaces that characterize nanoscale systems. The following text presents a survey of the recent literature concerning SHG in nanostructures.

28.3.1 Spherical Isolated Particles and Plane-Wave Illumination

Perfectly spherical small particles made of isotropic material under FW plane-wave illumination represent the simplest geometry to study SHG in nanostructured systems. Experiments have been conducted on spherical dielectric (Wang et al. 1996, Yang et al. 2001, Shan et al. 2006) or metallic (Vance et al. 1998, Nappa et al. 2005) particles dispersed in dilute suspensions, or on colloidal particles, ordered in a centrosymmetric crystalline lattice (Martorell et al. 1997). The influence of the particle diameter (Yang et al. 2001, Nappa et al. 2005, Shan et al. 2006), concentration (Wang et al. 1996, Vance et al. 1998), and the presence of adsorbates (Wang et al. 1996) on SHG have been investigated, by performing angle- and polarization-resolved measurements (Martorell et al. 1997, Yang et al. 2001, Nappa et al. 2005, Shan et al. 2006).

The theory is in good agreement with the experiments, confirming that, when the sphere is small compared with the wavelength of light (Rayleigh limit), the leading-order SH radiation is emitted by a locally excited electric quadrupole and by an electric dipole directed along the direction of the propagation of the incident FW light beam that requires a nonlocal excitation mechanism, in which the phase variation of the pump beam across the sphere needs to be considered (Dadap et al. 1999, 2004). The locally excited electric dipole term, analogous to the source for linear Rayleigh scattering, is absent for the nonlinear case because the FW field induces mutually canceling polarizations at opposite sides of a centrosymmetric spherical particle (Dadap et al. 1999, 2004). The theory predicts the absence of any SHG signal in the forward direction, with both dipole and quadrupole contributions producing SH radiation with an intensity of leading order $(ka)^6$, k being the wave vector and a the particle diameter (Dadap et al. 1999, 2004). Theoretical models have been applied to describe SHG in particles larger than the Rayleigh limit (Dewitz et al. 1996, Pavlyukh and Hübner 2004) and to explicitly account for the electromagnetic response of conducting nanoparticles (Hua and Gersten 1986, Dewitz et al. 1996, Panasyuk et al. 2008).

28.3.2 Low-Symmetry Particles

The fact that SHG is forbidden in the bulk of centrosymmetric materials makes SHG very sensitive to the particle symmetry (Sandrock et al. 1999). For instance, experimental results suggest that SHG can be used to characterize the symmetry and chirality of carbon nanotubes (Su et al. 2008). It has been theoretically (Bachelier et al. 2008) and experimentally demonstrated (Martorell et al. 1997, Nappa et al. 2006) that slight structural deviations from the spherical shape may lead to SH radiation and polarization properties of nanoparticles differing significantly from those of a sphere. For gold particles with a diameter smaller than 50 nm, SHG is dominated by a dipole contribution that is not due to defects in the crystalline structure of the particle but to the deviation of the particle shape from that of a perfect sphere (Nappa et al. 2006, Bachelier et al. 2008). For larger sizes, retardation effects in the interaction of the electromagnetic fields with the particles cannot be neglected any longer, and the response exhibits a strong quadrupolar contribution (Martorell et al. 1997, Nappa et al. 2006).

The importance of local defects in breaking the symmetry of the particles and, consequently, in influencing the SHG process has been verified also in noncentrosymmetric gold nanostructures lithographed on a glass substrate (Canfield et al. 2004, Canfield et al. 2006). SHG from such particles show a high degree of polarization sensitivity and reveals that responses forbidden to ideal, symmetric particles are not only present but are relatively large compared with the allowed response (Canfield et al. 2004). Indeed, with respect to spherical or ellipsoidal nanoparticles, very different polarization selection rules and SH emission directions are obtained when the symmetry of the particle is lowered (Neacsu et al. 2005, Finazzi et al. 2007).

Despite the nonlinear optical properties of a particle are very sensitive to local field enhancement, the presence of very strong fields alone may not be sufficient for efficient SHG from centrosymmetric nanoparticles, as a consequence of the high symmetry selectivity of SHG. An example is SHG from arrays of noncentrosymmetric T-shaped gold nanodimers with a nanogap (Canfield et al. 2007). In this case, SHG arises from asymmetry in the local fundamental field distribution and is not strictly related to the nanogap size, which determines the intensity of the electric field between the dimers. Calculations show that the local field contains polarization components that are not present in the exciting field, which yield the dominant SHG response. The strongest SHG responses occur through the local surface susceptibility of the particles for a fundamental field distributed asymmetrically at the particle perimeters. Weak responses result from more symmetric distributions despite high field enhancement in the nanogap. Nearly constant field enhancement persists for relatively large nanogap sizes (Canfield et al. 2007).

28.3.3 Nonuniform Illumination

The shape of particles participating in the SHG process is not the only geometrical parameter that might affect the SH emission modes and efficiency. Actually, also the spatial distribution of the exciting field can have a great importance to determine the nonlinear response of particles. For instance, for a nonuniform polarizing field, the cancellation of the SH fields generated at opposite sides of a centrosymmetric spherical particle (see Section 28.3.1) is no longer exact and lower order SH emission becomes possible (Brudny et al. 2000). This issue is particularly important since, in practical experimental geometries, the polarizing field is in general neither of a pure longitudinal character nor a simple plane wave, as in the case illumination is performed by focusing the light beam with a microscope objective.

In this context, it is worth mentioning an important application of SHG, namely, second-harmonic imaging microscopy (SHIM), which is a technique used for imaging living cells or tissue by detecting the SHG signal generated by noncentrosymmetric molecules (Campagnola and Loew 2003). By exploiting the quadratic intensity dependence of the SHG signal that confines it to the focal volume, SHIM has the capability of three-dimensional imaging deep within the tissue. SHIM offers several advantages: since it does not rely on absorption from a real electronic state, as in fluorescence microscopy, it does not give rise to photobleaching or phototoxicity. In addition, as it exploits the intrinsic hyperpolarizability of noncentrosymmetric molecules, it is a label-free technique that does not require the use of exogenous probes.

A general theory for the quadratic nonlinear response of a single small nonmagnetic centrosymmetric sphere illuminated by a nonhomogeneous electromagnetic field was developed by Mochán et al. (2003), while general symmetry-based selection rules for SHG with arbitrary illumination are given by Finazzi et al. (2007) and discussed in Section 28.4. The importance of the nature of FW illumination has also been demonstrated experimentally. In fact, it has been shown that the SH radiation generated in Si spherical nanocrystallites displays a peak along the forward direction with a very narrow angular aperture (Jiang et al. 2001, 2002). As explained in Section 28.3.1, this result contradicts the behavior expected for the SH radiation produced by a single sphere illuminated by a plane wave, which should identically vanish along the forward direction (Dadap et al. 1999, 2004). Nevertheless, this observation can be explained (Brudny et al. 2003) by considering the nonuniform illumination FW field distribution, consisting of a strongly focused Gaussian beam. Similarly, in Si nanocrystals embedded uniformly in an SiO_2 matrix, a configuration consisting of two noncollinear, orthogonally polarized FW beams is found to greatly enhance the SHG yield with respect to single-beam illumination (Figliozzi et al. 2005). This effect has been attributed to the strong inhomogeneities in the FW field under double-beam illumination (Figliozzi et al. 2005).

28.3.4 Second Harmonic Generation in Resonant Metal Particles

Although SHG is subject to the same geometry-based selection rules in both metal and dielectric nanoparticles (Finazzi et al. 2007), a relevant difference is represented by the fact that, in metal particles, the linear and nonlinear optical properties are governed by collective electronic excitations (plasma oscillations), which can be tuned in a broad spectral range by choosing the proper particle geometry. In nanosystems, the quasiparticles associated with the collective electron excitations are often addressed to as localized surface plasmons (LSPs), or as localized plasmon polaritons, to indicate the associated electric field. In particular, depending on the size and the shape, metal particles made of gold and silver may display LSP resonance in the visible region. Thanks to this tunability, the nonlinear optical properties of noble-metal nanoparticles have found applications for in vitro and in vivo imaging (Nagesha et al. 2007), high-resolution analysis of tumor tissues (Durr et al. 2007, Bickford et al. 2008, Park et al. 2008), and enhanced photothermal therapies (O'Neal et al. 2004, El-Sayed et al. 2006).

A relevant enhancement of the SHG efficiency is observed when either the FW (Hubert et al. 2007) or the SH (Antoine et al. 1997, 1998, Hao et al. 2002, Johnson et al. 2002, Abid et al. 2004, Russier-Antoine et al. 2004) frequency is set at the LSP resonance of the nanoparticles. When the SH wavelength is tuned in the vicinity of the surface plasmon resonance, the wavelength analysis indicates that the interband transitions do not play a significant role in the total SH response (Antoine et al. 1998). Rather, the nonlinear optical behavior of the clusters is dominated by the free-electron gas-like response of the conduction band, the surface plasmon resonance being much sharper than the interband transition contribution (Antoine et al. 1998, Russier-Antoine et al. 2004). Similarly, the SHG efficiency as a function of the FW frequency shows a curve that reproduces the LSP resonance (Hubert et al. 2007). Both electric dipole and electric quadrupole contributions to the SH-radiated power are observed (Hao et al. 2002).

28.3.5 Random Metal Nanostructures and Rough Metal Surfaces

The strong local field enhancements in random metal nanoparticles and rough metal surfaces can be exploited to improve the efficiency of the interaction of light with molecules approached to such nanostructures (Hartschuh et al. 2003), a phenomenon that can be exploited in surface-enhanced spectroscopies to obtain single-molecule sensitivity. Among these, surface-enhanced Raman scattering (SERS) makes use of the excitation provided by LSP and high local fields in such structures to enhance the Raman emission of adsorbed molecules by several orders of magnitudes (generally 10^6, up to 10^{12}). Surface enhanced spectroscopies such as SERS are, therefore, intimately correlated with the plasmon properties of the nanostructures, offering, at the same time, unique potential for ultrasensitive molecular identification (Nie and Emery 1997). LSP-based nanosensors of biomolecules, based on the shift of the plasmon resonance frequency of thin metallic films, have been developed and are already commercially available.

Since SHG is extremely sensitive to both local fields and LSP resonances, both theoretical (Agarwal and Jha 1982, Stockman et al.

2004, Beermann et al. 2006, Singh and Tripathi 2007) and experimental studies performed with either far-field (Bozhevolnyi et al. 2003, Beermann and Bozhevolnyi 2004) or near-field techniques (Smolyaninov et al. 1997, Zayats et al. 2000) have addressed the SHG process in such disordered systems characterized by LSP interacting with local defects. The latter act as plasmon-scattering centers and might also provide highly localized electric fields due to the lightning rod effect. LSP resonances emerge as a consequence of multiple interparticle light scattering (Stockman et al. 2004, Beermann et al. 2006) and exhibit very different strength, phase, polarization, and localization characteristics (Bozhevolnyi et al. 2003, Beermann and Bozhevolnyi 2004). SH emission is experimentally observed from small and very bright spots, whose locations depend on the light wavelength (Bozhevolnyi et al. 2003) and polarization (Zayats et al. 2000, Bozhevolnyi et al. 2003). According to simulations (Stockman et al. 2004), the spatial distributions of the fundamental frequency and SH local fields are very different, with highly enhanced SH hot spots corresponding to areas where FW and SH eigenmodes overlap (Beermann et al. 2006). Another feature of SHG in rough metal surfaces is that SH fields show a very rapid spatial decay and are strongly depolarized and incoherent (Stockman et al. 2004).

28.4 Leading Order Contributions to the Second Harmonic Generation Process in Nanoparticles

As illustrated in Section 28.3, the shape of the nanoparticle plays a significant role in determining the conditions for efficient SHG, which can also be significantly influenced by the illumination geometry. The latter represents a particularly important point that needs to be considered when microscopic techniques are employed to locally study field-enhancement processes. In this case, the FW illumination is not a plane wave anymore, and strong field gradients (and even a nonvanishing longitudinal field component when scanning near-field optical microscopy (SNOM) is used) can be present in the illumination area. Although nanoscale SHG has already been discussed in the literature for few particle geometries excited by either far (Zhu 2007, Dadap 2008) or near (Bozhevolnyi and Lozovski 2000, 2002) FW fields, a set of general validity selection rules for SHG might be helpful in addressing more complicated particle geometries and FW field distributions. The following text addresses the concepts that allow one to derive selection rules for SHG in isolated nanoparticles from angular momentum and parity conservation laws, as discussed by Finazzi et al. (2007).

28.4.1 Conservation of Parity and Angular Momentum

In the rest reference frame of the particle, the SHG process consists of two photons, belonging to the FW field interacting with the particle, that transfer their energy to a photon in the SH field oscillating at double frequency (see Figure 28.1b). Therefore, the energy of the electromagnetic field is conserved and the particle is left in the ground state, that is, the particle initial and final states in the SHG process coincide. This fact has relevant consequences if the particle is symmetric. In this case, the particle ground state is an eigenstate of parity and, possibly, angular momentum operators. In other words, the particle has a well-defined parity (or angular momentum). Since the particle initial and final states are the same, the electromagnetic field cannot transfer parity or angular momentum quanta to the particle, and the total parity or angular momentum of the field will be conserved. These concepts can be expressed more formally by considering the explicit expression of the SHG process cross section σ. By using perturbation theory, σ can be expressed as a third-order term of the form

$$
\sigma \propto \left| \sum_{u,w} \frac{\langle \varphi_0 | \hat{\mathbf{p}} \cdot \mathbf{A}_\omega | u \rangle \langle u | \hat{\mathbf{p}} \cdot \mathbf{A}_\omega | w \rangle \langle w | \hat{\mathbf{p}} \cdot \mathbf{A}_{2\omega} | \varphi_0 \rangle}{[(E_u - E_0) - \omega + i\Gamma_u][(E_w - E_u) - \omega + i\Gamma_w]} \right.
$$
$$
+ \frac{\langle \varphi_0 | \hat{\mathbf{p}} \cdot \mathbf{A}_\omega | u \rangle \langle u | \hat{\mathbf{p}} \cdot \mathbf{A}_{2\omega} | w \rangle \langle w | \hat{\mathbf{p}} \cdot \mathbf{A}_\omega | \varphi_0 \rangle}{[(E_u - E_0) - \omega + i\Gamma_u][(E_w - E_u) + 2\omega]}
$$
$$
\left. + \frac{\langle \varphi_0 | \hat{\mathbf{p}} \cdot \mathbf{A}_{2\omega} | u \rangle \langle u | \hat{\mathbf{p}} \cdot \mathbf{A}_\omega | w \rangle \langle w | \hat{\mathbf{p}} \cdot \mathbf{A}_\omega | \varphi_0 \rangle}{[(E_u - E_0) + 2\omega][(E_w - E_u) - \omega + i\Gamma_w]} \right|^2, \quad (28.5)
$$

where

  φ_0 is the particle ground state
  u and w are eigenfunctions corresponding to excited states of the unperturbed particle
  E_0, E_u, and E_w are the corresponding energies

Equation 28.5 consists of the sum of three terms, each corresponding to one of the Feynman diagrams shown in Figure 28.2. In the expression for σ, $\hat{\mathbf{p}}$ is the particle momentum operator, $\mathbf{A}_{2\omega}$ the SH vector potential, and $\mathbf{A}_\omega$ the vector potential of the *externally applied* FW field. Note that the field generated by the particle should not be added to $\mathbf{A}_\omega$ in Equation 28.5. In fact, the particle self-interactions, that is, the interactions between the particle and the electromagnetic field the particle produces are accounted for by u and w, which describe the electronic as well as the electromagnetic excitations of the particle. Retardation

effects that play an important role in determining SHG in nanoparticles (Dadap et al. 1999, 2004) are therefore implicitly accounted for in Equation 28.5.

As far as parity and angular momentum are concerned, it is convenient to expand the FW and SH fields appearing in Equation 28.5 in terms of electric (E) and magnetic (M) multipoles, which represent the photon angular momentum eigenstates. Let L'_ω and L''_ω be the L^2 quantum numbers of the two FW absorbed photons, M'_ω and M''_ω their L_z quantum numbers, z being an arbitrary quantization axis, and $L_{2\omega}$ and $M_{2\omega}$ the respective quantum numbers for the emitted SH photon (see Figure 28.3). The Feynman diagram in Figure 28.2a gives a contribution to Equation 28.5 that can be expressed as the sum of T terms defined as follows:

$$
T^{u,w}_{L'_\omega, M'_\omega, L''_\omega, M''_\omega, L_{2\omega}, M_{2\omega}}
$$
$$
\propto \langle \varphi_0 | \hat{\mathbf{p}} \cdot \mathbf{A}(L'_\omega, M'_\omega) | u \rangle \langle u | \hat{\mathbf{p}} \cdot \mathbf{A}(L''_\omega, M''_\omega) | w \rangle \langle w | \hat{\mathbf{p}} \cdot \mathbf{A}(L_{2\omega}, -M_{2\omega}) | \varphi_0 \rangle, \tag{28.6}
$$

where $\hat{\mathbf{p}} \cdot \mathbf{A}(L, M)$ represents the electric or magnetic multipole operator of order (L, M). $M_{2\omega}$ is taken with a negative sign since the emitted SH photon carries the corresponding L_z angular momentum component *away* from the particle. Similar terms can be given for the other Feynman diagrams in Figure 28.2. Each T term corresponds to an SHG interaction involving FW and SH photons characterized by a well-defined total angular momentum.

The selection rules that determine the conditions for which a given T term is identically zero depend on how the multipoles in Equation 28.6 transform under the particle symmetry group operations. When the particle has spherical symmetry, its ground as well as excited states are eigenstates of both L^2 and L_z. Therefore, the SH photon must carry away a total angular momentum equal to the sum of the angular momenta of the two FW incoming photons. In cylindrical symmetry, only the projection of the photon angular momentum along the symmetry axis (taken as the quantization axis z) needs to be conserved. The photon parity must be conserved in the case the particle displays inversion symmetry. If the particle is invariant under inversion symmetry with respect to a point (S_2 symmetry in Schoenflies notation, see Tinkham (1964)), then electric and magnetic multipoles of

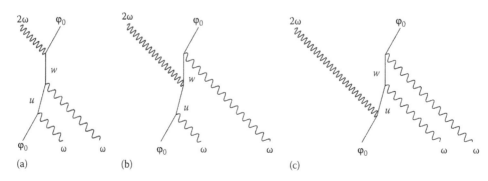

FIGURE 28.2 The three Feynman diagrams corresponding to SHG. (Reprinted from Finazzi, M. et al., *Phys. Rev. B*, 76, 125414, 2007. With permission.)

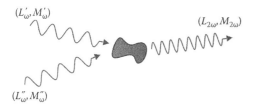

FIGURE 28.3 Scheme of SHG reporting the L and M quantum numbers that define the angular momentum and parity of the photons involved in the process.

order (L, M) have opposite contributions to the total parity of the electromagnetic field, equal to $(-1)^L$ and $(-1)^{L+1}$, respectively (Jackson 1975). In the case the particle displays inversion symmetry (C_2 symmetry) with respect to the quantization axis, electric and magnetic multipoles have the same parity, given by $(-1)^M$ (Jackson 1975). Finally, considering inversion with respect to a plane perpendicular to the quantization axis (C_{1h} symmetry), the parity of electric and magnetic multipoles of order (L, M) is given by $(-1)^{L+M}$ and $(-1)^{L+M+1}$, respectively (Jackson 1975).

By considering the electromagnetic field conservation laws imposed by the particle symmetry and contributions to angular momentum and parity associated with each multipole, one can obtain the selection rules listed in Table 28.1, which allow to

TABLE 28.1 SHG Selection Rules for the Different Particle Symmetries Illustrated in Figure 28.4

Symmetry and Point Group	Selection Rule
Spherical	$\begin{pmatrix} L'_\omega & L''_\omega & L_{2\omega} \\ M'_\omega & M''_\omega & -M_{2\omega} \end{pmatrix} \neq 0$ and $(-1)^{L'_\omega + L''_\omega + L_{2\omega} + m} = 1$
Cylindrical	$M'_\omega + M''_\omega - M_{2\omega} = 0$
Central, S_2	$(-1)^{L'_\omega + L''_\omega + L_{2\omega} + m} = 1$
Axial, C_2	$(-1)^{M'_\omega + M''_\omega - M_{2\omega}} = 1$
Reflection, C_{1h}	$(-1)^{L'_\omega + M'_\omega + L''_\omega + M''_\omega + L_{2\omega} - M_{2\omega} + m} = 1$

Source: Reprinted from Finazzi, M. et al., *Phys. Rev. B*, 76, 125414, 2007. With permission.

Note: Conservation of angular momentum for spherical symmetry corresponds to a condition involving a 3-*j* symbol (Tinkham 1964). The value *m* indicates the number of *magnetic* multipole transitions involved in the *T* term.

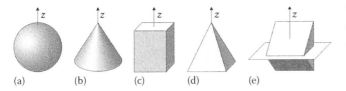

(a) (b) (c) (d) (e)

FIGURE 28.4 The particle symmetry groups discussed in the text and referred to the quantization axis *z*: (a) spherical, (b) cylindrical, (c) central, (d) axial, and (e) reflection symmetries across a mirror plane perpendicular to *z*. (Reprinted from Finazzi, M. et al., *Phys. Rev. B*, 76, 125414, 2007. With permission.)

determine the nonvanishing T terms that contribute to the SGH process for any combination of FW and SH multipoles (Finazzi et al. 2007). Thus, to apply the selection rules listed in Table 28.1, one has to determine the multipole expansion of the impinging FW electromagnetic field in order to understand which SH multipoles can contribute to the generated SH radiation.

If the symmetry of the particle results from a combination of symmetry groups, the corresponding conditions listed in Table 28.1 must be satisfied at the same time. For instance, SHG from particles with C_{2h} symmetry must obey both the selection rules for axial (C_2) and reflection (C_{1h}) symmetries with respect to the same quantization axis. For the same reason, when the particle exhibits two or three orthogonal symmetry axes, the corresponding selection rules must be simultaneously satisfied by the field multipole expansions expressed by using each symmetry axis as the quantization axis.

Note that, strictly speaking, the spherical and cylindrical symmetry groups would have to be excluded since they are not compatible with any of the symmetry properties of the various Bravais lattices. Moreover, crystallographic defects (such as grain boundaries, dislocations, and atomic vacancies) would disrupt the internal symmetry of the particles. Thus, it would seem that symmetry-based selection rules would not be respected by SHG in real nanostructures. However, such limitations are not expected to play a significant role in the metal structures that are usually investigated in nano-optics (Finazzi et al. 2007), which are typically realized in aluminum, silver, or gold. The electronic and electromagnetic properties of these materials are in fact well described by the free-electron model, so that the particle symmetry properties are independent from the particle crystallography but are just determined by the boundary conditions imposed by the particle surface.

28.4.2 Long-Wavelength Limit

In general, selection rules allow one to discriminate the transitions that, for symmetry reasons, cannot contribute to the cross section of a particular physical phenomenon, but they do not give any information about the magnitude of the nonvanishing contributions. However, by considering the multipole expansion of the electromagnetic FW and SH fields in the SHG process, it is possible not only to identify the emission channels that are ineffective to SHG but also to estimate a hierarchy among them, identifying the ones that mostly contribute to the SHG process. This is possible in the long-wavelength limit (Rayleigh limit) characterized by $ka \ll 1$, with k being the FW wave vector and a the particle lateral size. In fact, matrix elements of the form $\langle u \mid \hat{\mathbf{p}} \cdot \mathbf{A}(L, M) \mid w \rangle$ rapidly decrease when the multipole order L is increased (Jackson 1975). The ratio $R(L)$ between matrix elements for successive orders, L and $(L + 1)$, of either electric or magnetic multipoles of the same frequency is

$$R(L) \sim \frac{ka}{2L}, \tag{28.7}$$

while the magnetic multipoles of order L have cross sections of the same order in ka as electric multipoles of order $(L + 1)$ (Jackson 1975, Dadap et al. 2004). Since the expression of the SHG cross section σ reported in Equation 28.5 depends on the square value of the sum of products involving three matrix elements, we need to consider only the contributions associated with the lowest order multipoles.

By restricting the analysis to the lowest order terms, five distinct SHG emission modes can be indicated. Following the notation in Dadap et al. (2004), these are E1 + E1 → E1, E1 + E2 → E1, E1 + M1 → E1, E1 + E1 → E2, and E1 + E1 → M1. In this notation, the first two symbols refer to the nature of the interaction with the FW field, and the third symbol describes the SH emission. For example, the E1 + E2 → E1 interaction represents electric dipole (E1) SH emission that arises through combined FW electric dipole and electric quadrupole (E2) excitations. M1 indicates a magnetic dipole transition. The E1 + E1 → E1 transition corresponds to the leading order contribution to the SHG process, generating an SH field with a magnitude of a factor $(ka)^{-1}$ higher than the field generated by the other terms, resulting in a $(ka)^{-2}$ higher irradiated power. The E1 + E1 → E1 interaction violates parity conservation for any incident FW field distribution on centrosymmetric particles since $(-1)^{L'_\omega + L''_\omega + L_{2\omega}} = (-1)^{1+1+1} = -1$, thus violating the selection rule for S_2 symmetry particles reported in Table 28.1. This channel may, however, be allowed for lower symmetry particles.

28.5 Emission Patterns and Light Polarization

The emission modes that participate in the SHG process can be experimentally identified by considering SH emission pattern and light polarization. In fact, each multipole contributing to the SHG process is characterized by a well-defined angular dependence of the irradiated intensity and polarization.

28.5.1 Irradiated Intensity

Let $\mathbf{k}_{2\omega}$ be the SH wave vector and $\mathbf{r}$ the position with respect to the particle. According to Jackson (1975), in the far-field limit characterized by $k_{2\omega} \gg r^{-1}$, the SH electric fields $\mathbf{E}^{(E)}_{L_{2\omega}, M_{2\omega}}(\mathbf{r})$ and $\mathbf{E}^{(M)}_{L_{2\omega}, M_{2\omega}}(\mathbf{r})$, produced by either electric or magnetic multipoles of order $(L_{2\omega}, M_{2\omega})$, are given by

$$\mathbf{E}^{(E)}_{L_{2\omega}, M_{2\omega}}(\mathbf{r}) = -(-i)^{(L_{2\omega}+1)}\frac{e^{ik_{2\omega}r}}{k_{2\omega}r}(\hat{\mathbf{u}}_r \times \mathbf{r} \times \nabla Y_{L_{2\omega}, M_{2\omega}}), \quad (28.8a)$$

$$\mathbf{E}^{(M)}_{L_{2\omega}, M_{2\omega}}(\mathbf{r}) = -(-i)^{(L_{2\omega}+1)}\frac{e^{ik_{2\omega}r}}{k_{2\omega}r}(\mathbf{r} \times \nabla Y_{L_{2\omega}, M_{2\omega}}). \quad (28.8b)$$

In Equations 28.8 $\hat{\mathbf{u}}_r$ is a unit vector in the radial direction, and $Y_{L_{2\omega}, M_{2\omega}}$ is the spherical harmonic of order $(L_{2\omega}, M_{2\omega})$. Pure electric and magnetic SH multipoles have the same angular

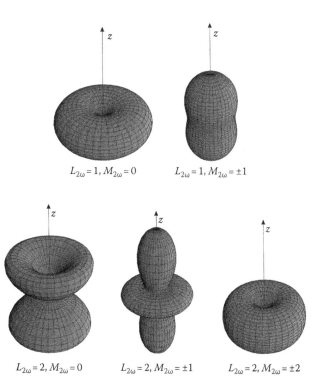

FIGURE 28.5 Graphical representation of emission patterns $dP/d\Omega(L_{2\omega}, M_{2\omega})$ for dipoles (upper row) and quadrupoles (lower row). The time-averaged radiated power per unit solid angle displays cylindrical symmetry around the quantization axis z. (Reprinted from Finazzi, M. et al., *Phys. Rev. B*, 76, 125414, 2007. With permission.)

distribution of the time-averaged radiated power per solid angle $dP/d\Omega(L_{2\omega}, M_{2\omega})$. A graphical representation of the radiated power is given in Figure 28.5 for dipoles ($L_{2\omega} = 1$) and quadrupoles ($L_{2\omega} = 2$) (Jackson 1975). If more than a single multiplet is responsible for the generation of SH radiation, interference among the SH radiation emitted by the contributing multipoles occurs and the emission pattern can be significantly different from those in Figure 28.5. In such cases, identifying the multipoles that are responsible for the SH emission might be quite complicated. Further information, however, can be extracted by analyzing the polarization dependence of the SH radiation as a function of the polarization of the impinging FW field.

28.5.2 Second Harmonic Polarization

From Equations 28.8, one can also obtain the polarization state of the emitted SH light. Table 28.2 reports the direction of the electric far field for pure electric and magnetic multipoles with $L_{2\omega} \leq 2$. The direction is indicated by a combination of the mutually perpendicular unit vectors $\hat{\mathbf{u}}_\theta$ and $\hat{\mathbf{u}}_\phi$ defined in a spherical coordinate system centered on the particle, in which θ and ϕ are the polar and azimuthal angles, respectively, the former being referred to the z quantization axis. An imaginary (real) ratio between the coefficients of $\hat{\mathbf{u}}_\theta$ and $\hat{\mathbf{u}}_\phi$ corresponds to a $\pi/2$ (zero) phase difference between the two field components and hence indicates elliptical (linear) polarization (Jackson 1975).

TABLE 28.2 SH Electric Field Polarization for Pure Electric or Magnetic Multipoles with $L_{2\omega} \leq 2$

	$M_{2\omega} = 0$	$M_{2\omega} = \pm 1$	$M_{2\omega} = \pm 2$
$L_{2\omega} = 1$ (E)	$\sin\theta\,\hat{\mathbf{u}}_\theta$	$e^{\pm i\phi}(\pm\cos\theta\,\hat{\mathbf{u}}_\theta \mp i\hat{\mathbf{u}}_\phi)$	
$L_{2\omega} = 1$ (M)	$\sin\theta\,\hat{\mathbf{u}}_\phi$	$-e^{\pm i\phi}(i\hat{\mathbf{u}}_\theta \mp \cos\theta\,\hat{\mathbf{u}}_\phi)$	
$L_{2\omega} = 2$ (E)	$-i\sin\theta\cos\theta\,\hat{\mathbf{u}}_\theta$	$e^{\pm i\phi}(\mp i\cos 2\theta\,\hat{\mathbf{u}}_\theta + \cos\theta\,\hat{\mathbf{u}}_\phi)$	$e^{\pm 2i\phi}\sin\theta(i\cos\theta\,\hat{\mathbf{u}}_\theta \pm \hat{\mathbf{u}}_\phi)$
$L_{2\omega} = 2$ (M)	$-i\sin\theta\cos\theta\,\hat{\mathbf{u}}_\phi$	$-e^{\pm i\phi}(\cos\theta\,\hat{\mathbf{u}}_\theta + i\cos 2\theta\,\hat{\mathbf{u}}_\phi)$	$e^{\pm 2i\phi}\sin\theta\,(\pm\hat{\mathbf{u}}_\theta + i\cos\theta\,\hat{\mathbf{u}}_\phi)$

Source: Reprinted from Finazzi, M. et al., *Phys. Rev. B*, 76, 125414, 2007. With permission.

Note: For each point in space, the unit vectors $\hat{\mathbf{u}}_\theta$ and $\hat{\mathbf{u}}_\phi$ are, respectively, parallel and perpendicular to the plane containing the z (quantization) axis, and are both perpendicular to the radial unit vector $\hat{\mathbf{u}}_r$.

28.6 Allowed and Forbidden Second Harmonic Emission Modes: Examples

This section discusses some examples of applications of the arguments discussed above. Two relevant cases of FW illumination are addressed, namely, illumination with a plane wave and illumination with a strongly laterally limited beam, for example, through a high numerical aperture or via a near-field tip.

28.6.1 Plane-Wave Illumination

In an arbitrarily polarized FW plane wave propagating along the z direction, the field vector $\mathbf{A}(z)$ can be expressed as $\mathbf{A}(0)e^{ikz}$, with $\mathbf{A}(0)$ being a vector perpendicular to the propagation direction, $\mathbf{A}(0) \cdot \hat{\mathbf{u}}_z = 0$. By taking the origin of the z axis coincident with the particle position, and truncating the multipole expansion around $z = 0$ of the matrix element $\langle u \mid \hat{\mathbf{p}} \cdot \mathbf{A} \mid w \rangle$ to second order in ka, one obtains

$$\langle u \mid \hat{\mathbf{p}} \cdot \mathbf{A} \mid w \rangle \frac{E_w - E_u}{\hbar\omega} \langle u \mid \mathbf{r} \cdot \mathbf{E}(0) \mid w \rangle + \langle u \mid \hat{\mathbf{L}} \cdot \mathbf{B}(0) \mid w \rangle$$

$$+ ik\langle u \mid \hat{p}_z[xA_x(0) + yA_y(0)] \mid w \rangle. \tag{28.9}$$

In this expression, $\mathbf{E}(0)$ and $\mathbf{B}(0)$ are complex vectors representing the amplitude and phase of the electric and magnetic fields,

respectively, at the particle position, while $A_x(0)$ and $A_y(0)$ are the complex components of $\mathbf{A}(0)$, and $\hat{\mathbf{L}} = \mathbf{r} \times \hat{\mathbf{p}}$. The three terms on the right-hand side of Equation 28.9 correspond to E1, M1, and E2 transitions, respectively.

Let us first discuss the case of circularly polarized light. In this case, it can be shown (Finazzi et al. 2007) that all the three FW absorption transitions described by the matrix elements in Equation 28.9 involve FW photons characterized by an L_z quantum number M_ω either equal to +1 or −1 according to the sign of the circular polarization (right or left) of the FW light impinging on the particle. By recalling that dipole terms are characterized by $L^2 = 1$ and quadrupole terms by $L^2 = 2$, the lowest allowed emission multipoles participating in the SHG process can be obtained from Table 28.1 in a straightforward manner. The allowed SH emission modes are listed in Table 28.3. As anticipated above, the lowest order transition E1 + E1 → E1 is forbidden under circularly polarized FW plane-wave illumination for particles displaying central symmetry, but it is allowed for noncentrosymmetric, noncylindrical particles. In this case, SH emission is given by an E1 term with $M_{2\omega} = 0$ or $M_{2\omega} = \pm 1$ for C_2 or C_{1h} symmetry, respectively. This emission mode corresponds to SHG from an electric dipole oriented parallel (for C_2 symmetry) or perpendicular (in the case of C_{1h} symmetry) to z. The lowest allowed SHG channel for spherical or cylindrical particles is E1 + E1 → E2 (Dadap et al. 1999, 2004), with $M_{2\omega} = \pm 2$. This SH emission mode corresponds to an irradiated SH power angular distribution displaying cylindrical symmetry around the propagation axis of the FW field, with a maximum in the plane perpendicular to z and a null in the forward and backward directions (Dadap et al. 1999, 2004), as displayed in Figure 28.5.

In the case of linearly polarized FW illumination, it is convenient to consider the selection rules that can be obtained by considering both the propagation direction of the FW plane wave (z axis) and the direction of the FW electric field vector (x axis) as the quantization axis. Therefore, one should calculate two multipole expansions of the FW wave, one for each choice of the quantization axis, and consider the possible values of both the L_z and L_x photon quantum numbers. Moreover, both axes should be taken into account to define the particle symmetry group. Let M_ω and $M_{2\omega}$ be the FW and SH photon L_z quantum numbers, and N_ω and $N_{2\omega}$ be the FW and SH photon L_x quantum numbers, respectively. It can be shown (Finazzi et al. 2007) that the multipole expansion

TABLE 28.3 SHG Selection Rules for the Different Particle Symmetries Illustrated in Figure 28.4, under Circularly Polarized Far-Field Plane-Wave Illumination

Symmetry and Point Group	E1 + E1 → E1	E1 + E1 → E2	E1 + E2 → E1	E1 + M1 → E1	E1 + E1 → M1
Spherical	Forbidden	Allowed for $M_{2\omega} = +2$ or -2	Forbidden	Forbidden	Forbidden
Cylindrical	Forbidden	Allowed for $M_{2\omega} = +2$ or -2	Forbidden	Forbidden	Forbidden
Central, S_2	Forbidden	Allowed	Allowed	Allowed	Allowed
Axial, C_2	Allowed for $M_{2\omega} = 0$	Allowed $M_{2\omega} = 0, \pm 2$	Allowed for $M_{2\omega} = 0$	Allowed for $M_{2\omega} = 0$	Allowed for $M_{2\omega} = 0$
Reflection, C_{1h}	Allowed for $M_{2\omega} = \pm 1$	Allowed for $M_{2\omega} = 0, \pm 2$	Allowed for $M_{2\omega} = 0$	Allowed for $M_{2\omega} = 0$	Allowed for $M_{2\omega} = 0$

Source: Reprinted from Finazzi, M. et al., *Phys. Rev. B*, 76, 125414, 2007. With permission.

Note: Every transition multipole is considered separately. The propagation direction (z axis) is chosen as the quantization axis.

TABLE 28.4 SHG Selection Rules for the Different Particle Symmetries Illustrated in Figure 28.4, under Linearly Polarized Far-Field Plane-Wave Illumination

Symmetry and Point Group	E1 + E1 → E1	E1 + E1 → E2	E1 + E2 → E1	E1 + M1 → E1	E1 + E1 → M1
Spherical	Forbidden	Allowed for $M_{2\omega} = 0, \pm 2$	Allowed for $M_{2\omega} = 0$	Allowed for $M_{2\omega} = 0$	Forbidden
Cylindrical	Allowed for $M_{2\omega} = 0$	Allowed for $M_{2\omega} = 0, \pm 2$	Allowed for $M_{2\omega} = 0$	Allowed for $M_{2\omega} = 0$	Forbidden
Central, S_2	Forbidden	Allowed	Allowed	Allowed	Allowed
Axial, C_2	Allowed for $M_{2\omega} = 0$	Allowed for $M_{2\omega} = 0, \pm 2$	Allowed for $M_{2\omega} = 0$	Allowed for $M_{2\omega} = 0$	Allowed for $M_{2\omega} = 0$
Reflection, C_{1h}	Allowed for $M_{2\omega} = \pm 1$	Allowed for $M_{2\omega} = 0, \pm 2$	Allowed for $M_{2\omega} = 0$	Allowed for $M_{2\omega} = 0$	Allowed for $M_{2\omega} = 0$

Source: Reprinted from Finazzi, M. et al., *Phys. Rev. B*, 76, 125414, 2007. With permission.

Note: The propagation direction (*z* axis) is chosen as the quantization axis.

TABLE 28.5 SHG Selection Rules for the Different Particle Symmetries Illustrated in Figure 28.4, under Linearly Polarized Far-Field Plane-Wave Illumination

Symmetry and Point Group	E1 + E1 → E1	E1 + E1 → E2	E1 + E2 → E1	E1 + M1 → E1	E1 + E1 → M1
Spherical	Forbidden	Allowed for $N_{2\omega} = 0$	Allowed for $N_{2\omega} = \pm 1$	Allowed for $N_{2\omega} = \pm 1$	Forbidden
Cylindrical	Allowed for $N_{2\omega} = 0$	Allowed for $N_{2\omega} = 0, \pm 2$	Allowed for $N_{2\omega} = \pm 1$	Allowed for $N_{2\omega} = \pm 1$	Allowed for $N_{2\omega} = 0$
Central, S_2	Forbidden	Allowed	Allowed	Allowed	Allowed
Axial, C_2	Allowed for $N_{2\omega} = 0$	Allowed for $N_{2\omega} = 0, \pm 2$	Allowed for $N_{2\omega} = \pm 1$	Allowed for $N_{2\omega} = \pm 1$	Allowed for $N_{2\omega} = 0$
Reflection, C_{1h}	Allowed for $N_{2\omega} = \pm 1$	Allowed for $N_{2\omega} = 0, \pm 2$	Allowed for $N_{2\omega} = \pm 1$	Allowed for $N_{2\omega} = \pm 1$	Allowed for $N_{2\omega} = 0$

Source: Reprinted from Finazzi, M. et al., *Phys. Rev. B*, 76, 125414, 2007.

Note: The electric field polarization direction (*x* axis) is chosen as the quantization axis.

of the FW plane wave corresponds to E1, M1, and E2 terms in Equation 28.9 that involve FW photons with an M_ω quantum number that can be either +1 or −1 but not 0. On the other hand, the expansion obtained by choosing the *x* axis as the quantization axis shows that the E1, M1, and E2 terms correspond to transitions involving FW photons with L^2 and L_x quantum numbers restricted to ($L_\omega = 1$, $N_\omega = 0$), ($L_\omega = 1$, $N_\omega = \pm 1$), and ($L_\omega = 2$, $N_\omega = \pm 1$), respectively. With these figures in mind and the general rules summarized in Table 28.1, one can find the selection rules listed in Tables 28.4 and 28.5. From these, one can see that the E1 + E1 → E1 SHG channel can be excited by linearly polarized FW plane-wave illumination in all noncentrosymmetric particles (Finazzi et al. 2007). In particular, if the particle exhibits C_2 symmetry either around the *z* or the *x* axis, SH emission will be given by an oscillating dipole parallel to the C_2 symmetry axis (Finazzi et al. 2007). In centrosymmetric particles illuminated by linearly polarized light, electric dipole SH emission is always given by a dipole parallel to *z*, while the quadrupole SH emission pattern is symmetric around *x* (see Figure 28.4) (Finazzi et al. 2007). M1 emission is forbidden (Dadap et al. 1999, 2004) because the selection rules would lead to $M_{2\omega} = N_{2\omega} = 0$, which correspond to a monopole term that cannot radiate (Finazzi et al. 2007).

28.6.2 Illumination with a Laterally Limited Light Beam

When the particle is illuminated by a laterally limited FW beam, which might be obtained by focusing with a high numerical-aperture objective or even with near-field illumination, and the beam waist is comparable with the particle size, the beam cannot be

considered as a plane wave any longer. In this case, other emission channels than those listed in Tables 28.3 through 28.5 become available. In these conditions, in fact, the particle experiences large longitudinal components of the FW field, which are absent in the case of illumination with a plane wave. The presence of such longitudinal components can be easily understood for a converging beam focused by an objective, where the FW light wave vector **k** has a broad angular distribution determined by the objective numerical aperture. Intense longitudinal components should also be expected in the proximity of near-field probes (Novotny and Hecht 2006). The consequence is that SHG can be induced by a combined FW photon absorption from perpendicular field components acting on the particle (Figliozzi et al. 2005), which is not possible for plane-wave illumination. For instance, illuminating with a laterally confined beam, absorption of FW photons with L_z quantum number $M_\omega = 0$ from a linearly polarized FW field becomes possible, which is excluded for plane-wave illumination (Finazzi et al. 2007). For the same reason, FW photons with $M_\omega = 0, \pm 1, \pm 2$ are available for E2 absorption (Finazzi et al. 2007).

Another important difference with respect to in-plane wave illumination is represented by the fact that the rapid spatial variations of the field intensity can increase the relative weight of the higher-order multipoles in the SHG process. This is readily understood in far-field illumination when the spot size, which is limited by diffraction to about $1/k$, is comparable with the particle size *a*, resulting in a violation of the long-wavelength approximation. Similarly, strong field gradients are expected at the apex of a near field probe (Novotny and Hecht 2006), again invalidating the long-wavelength approximation and implying that high-order multipoles cannot be a priori neglected.

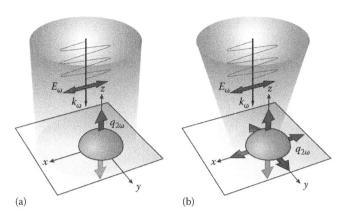

FIGURE 28.6 Allowed E1 second-harmonic emission channels for a particle with C_2 symmetry (axial particle over a substrate) excited by (a) a plane wave or (b) a laterally limited beam. The allowed directions of the particle electric dipole $q_{2\omega}$ generating the E1 second-harmonic radiation are indicated by the double-headed arrows on the particle. (Reprinted from Finazzi, M. et al., *Phys. Rev. B*, 76, 125414, 2007. With permission.)

As a relevant example, let us address SHG in lithographed particles on a substrate (see Section 28.7), such as those employed in surface-enhanced spectroscopy techniques. The presence of the substrate breaks the inversion symmetry with respect to the particle center. For ellipsoidal-shaped dots as those described by Grand et al. (2003) and Zavelani-Rossi et al. (2008), the particle symmetry group reduces to C_{2v} (see Figure 28.6). In this case, the selection rules governing SHG in C_2 symmetry around the z axis must apply, z being the substrate normal, which is taken parallel to the optical axis. The particles also belong to the C_{1h} symmetry group defined with respect to either the x or y axis, oriented parallel to the in-plane principal axes of the ellipsoids. However, this further symmetry does not provide more restrictive selection rules.

When such particles are excited with an FW plane wave with $\mathbf{k} \parallel z$ and considering only the lowest-order E1 + E1 → E1 transitions, M_ω can only assume the values ±1 both for linearly and circularly polarized light, so parity is conserved only for $M_{2\omega} = 0$, corresponding to SH emission from an electric dipole parallel to both $\mathbf{k}$ and z (see Figure 28.6) (Finazzi et al. 2007). When the particle is illuminated by a focused FW beam or by the near field of a tip, photons with $M_\omega = 0$ can be absorbed. This results in new SHG channels, namely the ones that correspond to SH radiation from electric dipoles lying in the plane of the substrate, becoming available (see Figure 28.6) (Finazzi et al. 2007). In this case, the details of the SH-radiated intensity and polarization will depend on the particle fine structure, which defines the relative strength of each channel.

28.7 Second Harmonic Generation in Single Gold Nanoparticles

As discussed in Section 28.3, the nonlinear optical response of nanostructured systems has been experimentally studied, so far, mainly by far-field techniques. These, however, are limited by diffraction and do not allow a direct mapping of the nanoscale field enhancement. These limitations can be overcome by SNOM, which can reach sub-100 nm lateral resolution. This technique consists of bringing the sample in interaction with the near field of a source (a tip or a backilluminated aperture in a metal-coated probe). In this case, the bandwidth of spatial frequencies associated with the evanescent waves from the probe is unlimited and the resolution can in principle be arbitrarily optimized. The response of the probe–sample interaction is recorded with standard far-field collection optics. State-of-the-art lateral resolution for aperture probes is around 50 nm. Near-field nonlinear optical microscopy is a powerful tool to characterize local field enhancements in metal nanostructures since it combines the great sensitivity of nonlinear optical response with the spatial superresolution of near-field microscopy. However, the technical challenges of combining scanning probe and ultrafast technologies and the typical low peak power available at the output of optical near field probes have so far limited the number of studies of SHG from single nanoparticles in closely packed arrangements (Biagioni et al. 2007, Breit et al. 2007, Celebrano et al. 2007, 2008a,b, Zavelani-Rossi et al. 2008).

This section describes SHG from single gold nanoparticles obtained with a SNOM setup. As illustrated in Section 28.6.2, the high lateral confinement typical of near-field illumination enables the observation and exploitation of unusual and peculiar SHG modes. Moreover, the acquisition of SH maps provides complementary information to FW images, which typically results from a complex interplay between scattering, absorption, and reflection, all contributing to light extinction (Novotny and Hecht 2006). The comparison between FW and SH maps therefore allows for a better interpretation of the optical response of metal nanoparticles and discriminating among different light extinction particle behaviors that would not be possible to address by just collecting the FW signals (Biagioni et al. 2007, Celebrano et al. 2007, 2008a,b, Zavelani-Rossi et al. 2008).

28.7.1 Near-Field Microscopy Setup

To observe SHG at the nanoscale, we combine high peak power and high spatial resolution by coupling femtosecond pulses to a hollow-pyramid aperture SNOM (see Figure 28.7). The pulses

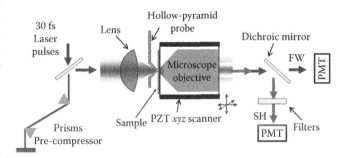

FIGURE 28.7 Schematics of our SNOM, with the ultrashort laser beam coupled to the measurement head: The collection optics, filters, and detector are also indicated. PMT indicates a photomultiplier tube.

are generated by a long-cavity mode-locked Ti:sapphire oscillator (26 MHz repetition rate), producing 20 nJ, 27 fs pulses at 800 nm. The SNOM probe consists of a silicon nitride cantilever with a hollow pyramid tip. The tip is aluminum-coated with a circular aperture at the apex, with diameter ranging between 100 and 200 nm. This probe offers several advantages compared with metal-coated tapered optical fibers, such as larger taper angle, lower absorption, preservation of light polarization (Biagioni et al. 2005), and pulse duration (Labardi et al. 2005) at the output. These improvements enable achieving peak powers more than two orders of magnitude higher in the near field, resulting in greatly enhanced nonlinear optical response of the sample. Typical tip throughputs at 800 nm range between ~10^{-4} for the 100 nm tips and 5×10^{-3} for the 200 nm ones. The tips with larger aperture were selected for the experiments presented in this section. Tip-sample distance is controlled by an optical lever that is sensitive enough to allow nondestructive contact mode stabilization on the samples.

The FW/SH light is collected in the far field by a long working distance microscope objective (numerical aperture = 0.75) and split by a dichroic filter. The FW and the SH are simultaneously detected by photomultiplier tubes. Bandpass filters are inserted in the SH beam to further reject the FW and TPPL from the sample. A 30 nm bandpass interference filter centered at 405 nm is inserted in the SH beam path to reject TPPL from the sample, which is negligible for wavelength shorter than 450 nm (Bouhelier et al. 2005, Imura et al. 2005, 2006). The quadratic dependence of SH intensity on the excitation beam power was verified. A mechanical chopper with a 1:6 duty cycle allows increasing the peak power for a given level of average power, typically 1 mW incident on the tip, and lock-in detection improves the signal-to-noise ratio, allowing for faster scans (integration time 30–100 ms per point).

Hereafter, we present two different types of gold nanoparticles. The first ones are triangles (height 15–25 nm) on a glass substrate obtained from a projection pattern ("Fischer pattern" (Fischer and Zingsheim 1981)), in a hexagonal array with 453 nm periodicity. Such a sample is chosen as an example of a SNOM study of SHG in a dense network of particles. The second type of particles are well-separated isolated ellipsoids (height about 60 nm) produced by electron beam lithography on a quartz substrate (Grand et al. 2003) in square arrays.

28.7.2 Fundamental Wavelength and Second Harmonic Maps

Figure 28.8a shows the topography of the Fischer pattern together with its SH optical image (Figure 28.8b). The topography shows the regular array of gold triangles: most of them are well separated, some are in contact with each other, and big defects cover portions of the sample. In the FW image (not shown), triangles appear dark, although the resolution is poor due to the quite large aperture diameter. Nevertheless, contrasted and well-resolved SHG maps from the gold triangles are detected. The background signal is attributed to SHG from both the glass substrate and the tip edges. The inset of Figure 28.8b shows a line profile of the SH image, demonstrating very good signal-to-noise ratio, high contrast (3:1), and good spatial resolution (better than 100 nm, see Figure 28.8c). Figure 28.8 highlights the unique capability of nonlinear SNOM to image SHG from closely packed metal nanostructures. It is interesting to note that not all the triangles observed in topography emit SH radiation with the same efficiency: some of them display an intense SH emission, while others are nearly dark. This supports the absence of topographical artifacts (Hecht et al. 1997) in the image, proved by the fact that in further measurements (not shown here), the topographic resolution was missing, yet a clear and well-resolved SHG image could still be collected. The high variability of the SHG signal is addressed further below when SHG in isolated nanoparticles is discussed.

The lithographed nanorods are about 60 nm high, with a short axis of about 70 nm and long axes equal to 100, 150, or 400 nm. The array period is 1 μm. According to the Mie theory, the plasmon resonances in elongated structures depend on their aspect ratio. Indeed, far-field extinction spectra (see Figure 28.9), with exciting white light beam polarization parallel to the major axis, display a peak around 690 nm for the 100 nm ellipsoids, around 800 nm (i.e., resonant with our FW) for the 150 nm ones and more than 1000 nm for the 400 nm ones (not shown). These three distinct particle geometries thus provide different resonating regimes that correspond to different linear and nonlinear optical behaviors and near-field properties. These are highlighted in Figure 28.10, showing the particle topography, FW near-field extinction, and SH emission obtained with the FW illumination linear polarization parallel to the nanorod long axis. The optical images at both

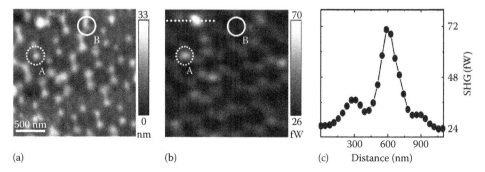

(a)   (b)   (c)   Distance (nm)

FIGURE 28.8   Projection pattern: (a) topography and (b) SH SNOM image (size: 5 × 2.9 μm²). The dashed circle (A) shows an SH emitting triangle and solid circle (B) shows a dark one. (c) Cross section from the raw data of the SH image, along the dotted line.

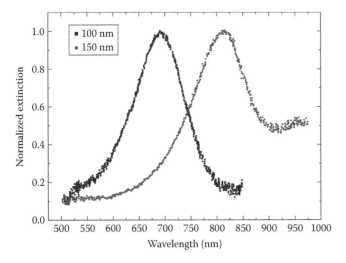

FIGURE 28.9 Extinction spectra of gold nanorods with a major axis length either equal to 100 (squares) or 150 nm (dots). The spectra have been normalized to each other at the maximum extinction. The light polarization is parallel to the particle long axis. The extinction spectra of the 400 nm-long nanorods also, discussed in the text, are not reported since the longitudinal plasmon resonance falls off the range of the analyzer. (Courtesy of J. Grand and P.-M. Adam.)

the FW and the SH wavelengths strongly depend on the particle size. In particular, in FW maps, all the 100 nm ellipsoids appear bright, while all the 150 and 400 nm ellipsoids appear dark. In SH images, (1) most of the 100 nm-long particles do not emit, (2) most of the resonant 150 nm-long particles emit uniformly with high contrast (see line profiles in Figure 28.10), and (3) the 400 nm-long particles appear darker than the substrate. A general picture for these results can be obtained by a combined analysis of the FW and SH maps of the metal nanostructures. As apparent from Figure 28.10, the SH maps provide complementary information with respect to the FW images. For instance, only by comparing the SH emission properties of the 150 and 400 nm-long particles, it is possible to recognize that these fall in different optical regimes, due to the FW excitation being on or off resonance, respectively, with respect to the particle LSP excitations.

The near-field FW properties of the particles can be understood by recalling that the particle-to-background contrast in the near-field extinction maps results from the interference between an FW wave propagating into the far-field and a nonpropagating near-field distribution (Mikhailowsky et al. 2003). As a general rule, a metal particle should appear brighter than the background when the frequency of the excitation is higher than that of the particle LSP resonance, while it should be darker when the frequency of the FW excitation is set below the resonance (Mikhailowsky et al. 2003). At the LSP resonance, the particle should show no contrast with respect to the background. The presence of these interference effects explains the trend observed in the FW maps displayed in Figure 28.10, where the particles have a well-controlled aspect ratio, but it may hinder the interpretation of FW near-field images in other contexts, especially when the frequency of the LSP resonances of the particles is not known. However, the resonating behavior of the 150 nm-long

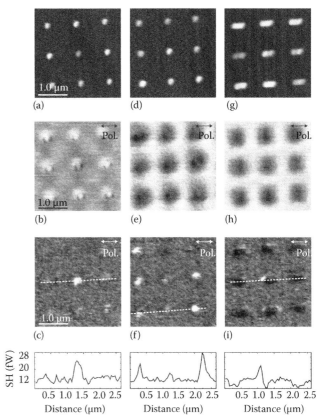

FIGURE 28.10 **(See color insert following page 21-4.)** Nanoparticles: (a), (d), and (g) topography; (b), (e), and (h) FW transmission; and (c), (f), and (i) SH emission SNOM images with corresponding cross sections along the dashed lines, from the raw data. Incident light is polarized parallel to the major axis. The particle major axis lengths are 100 nm (a)–(c), 150 nm (d)–(f), and 400 nm (g)–(i). Image size: 3 × 3 μm². (Reprinted from Zavelani-Rossi, M. et al., *Appl. Phys. Lett.*, 92, 093119, 2008. With permission.)

nanorods becomes apparent from the SH maps, since these particles efficiently emit SH radiation as a result of the presence of strongly enhanced and localized electric field associated with the particle LSP oscillations, which are effectively excited only for these particles. Instead, most of the 100 nm-long particles do not emit SH radiation, and the 400 nm-long particles, being larger than the tip aperture, strongly absorb/scatter the SH light generated at the tip and appear darker than the background.

The nonlinear optical behavior of off-resonance particles is further highlighted in Figure 28.11, where FW and SH maps are collected by exciting the nanorods with an FW field parallel to the rod minor axis. With this type of excitation, the LSP resonance frequencies are shifted toward the blue region of the spectrum and cannot be excited by the FW light. In this case, SHG is observed at correspondence with the high-curvature regions of the particle, as a consequence of highly enhanced and localized electric fields by the lightning rod effect, in agreement with the theoretical predictions of Beermann and Bozhevolnyi (2004) and Stockman et al. (2004). It is interesting to note that, in Figure 28.11, the bright spots in the SH maps seem to correspond to

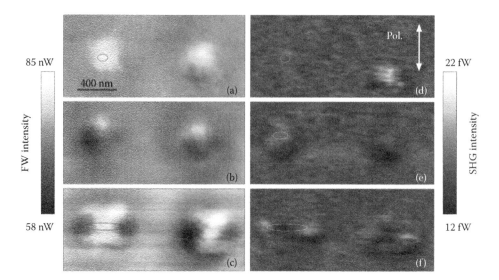

FIGURE 28.11   (a)–(c) FW transmission and (d)–(f) SH emission excited by incident FW light polarized perpendicular to the particle major axis. The particle major axis lengths are 100 nm (a) and (d), 150 nm (b) and (e), and 400 nm (c) and (f). (Reprinted from Zavelani-Rossi, M. et al., *Appl. Phys. Lett.*, 92, 093119, 2008. With permission.)

dark spots in the FW images, that is, the high-curvature regions where strong fields are localized also correspond to areas where larger FW scattering/absorption occurs, giving rise to a larger local linear extinction cross section. We believe that intense localized defects are also responsible for the high variability of the SHG efficiency displayed by nominally identical particles displaying quite similar FW (see Figures 28.8 and 28.10). In this case, local field enhancement would be associated with high-curvature defects in the particles.

To summarize, the SHG efficiency depends on the particle fine structure that determines (1) local FW field enhancements, induced either by LSP excitation or by lightning rod effects in areas with high-curvature or local imperfections and (2) the modes contributing to the emission process, which are described in the next section.

28.7.3 Second Harmonic Polarization Analysis

This section shows how the analysis of the polarization of the SH-emitted light might allow one to recognize which emission modes are contributing to the SHG process for a given particle geometry and illumination conditions. Such analysis will be applied to the same lithographed isolated gold particles, already described in Section 28.7.2.

The polarization state of the SH light emitted by the gold nanoellipsoids is analyzed as illustrated schematically in Figure 28.12. The FW light is polarized parallel to the major axes of the 150 nm-long particles to excite their strong longitudinal SPT resonance. The SH radiation polarization is analyzed by a polarizer on detection. Figure 28.13 shows two typical SH maps that are obtained with the analyzer parallel (Figure 28.13b) and perpendicular (Figure 28.13c) to the excitation polarization. From these maps, one can see that the SH light emitted by the nanoparticle is polarized parallel to its short axis

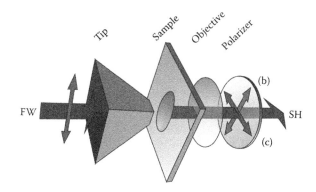

FIGURE 28.12   Experimental geometry of the polarization analysis setup in the SNOM microscope. The FW light is polarized parallel to the particle major axis. SH maps have been collected for the two analyzer directions (b) and (c), corresponding to SH linear polarizations parallel and perpendicular to the FW polarization, respectively. The corresponding SH maps are shown in Figure 28.13b and c, respectively.

(see Figure 28.13b,c,d). This particular polarization pattern of the emitted SH light is incompatible with the lowest allowed emission modes one should expect for such particles for an FW plane-wave illumination. In this case, in fact, SH radiation would be emitted by a dipole oscillating perpendicularly to the substrate sustaining the particle, as illustrated in Figure 28.6a. Such an emission mode would give the same SH maps independently from the direction of the analyzer axis. This is obviously in contradiction with the experimental results shown in Figure 28.13 and is a demonstration that a novel particle emission mode, which is forbidden for plane-wave illumination, becomes accessible. In fact, the presence of a symmetry-breaking substrate together with the strong longitudinal FW field component, typical of near-field illumination, allow for efficient excitation of an SH-emitting electric dipole perpendicular to both the incident FW light polarization and the detection axes,

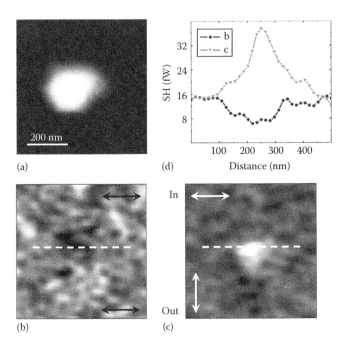

FIGURE 28.13 Nanoellipsoids with 150 nm major axis: topography (a), SH emission with a polarizer on detection parallel (b), and perpendicular (c) to the major axis (incident light is polarized parallel to the major axis). Image size: $0.6 \times 0.6 \, \mu m^2$. (d) Intensity profiles from the raw data of images (b) (circles) and (c) (triangles).

as illustrated in Figure 28.6b. From the analysis of the SH polarization pattern in Figure 28.13, one can conclude that only the SH-emitting dipole perpendicular to the particle minor axis is significantly excited by the FW field. For this particle geometry, the SHG selection rules discussed in Section 28.4 do not exclude the possibility that SHG could be due to a dipole oscillating in the direction of the long axis of the particles, so there is no simple symmetry-based justification to the fact that such emission is not observed. This behavior has to be ascribed to the particular shape and resonance conditions of the particles. Such characteristics can in effect influence in a complex way the particle SH emission properties. To give a tentative explanation of this behavior, we remark that a much larger portion of gold–air interface is available along the long sides of the ellipsoid, so that a major contribution of SH light oriented parallel to the short axis might actually be expected as a surface contribution to SHG. Finally, we would like to stress that the spatial uniformity of the polarized SHG pattern measured on several particles is a hint of the negligible influence of particle defects, which can also lower the symmetry of the particles and contribute to SHG with modes that would otherwise be forbidden by parity and angular momentum conservation in a particle with a perfectly controlled shape.

28.8 Perspectives

A very promising perspective for future nano-optics is the possibility to simultaneously control of the spatial and temporal properties of the optical near field in the vicinity of a nanostructure by illumination with broadband optimally polarization- and time-shaped femtosecond light pulses. Adaptive shaping of the phase and amplitude of femtosecond laser pulses has been developed into an efficient tool for the directed manipulation of interference phenomena, thus providing coherent control over various quantum-mechanical systems. The interest to extend coherent control methods to nanostructures consists in overcoming the spatial limitation due to diffraction. Coherent control of the spatial and temporal evolution of optical near-fields by amplitude and polarization pulse shaping in plasmonic nanostructures has recently been theoretically proposed (Stockman et al. 2002, Brixner et al. 2005, 2006, Sukharev and Seideman 2007) and experimentally demonstrated (Aeschlimann et al. 2007) by applying an adaptive method based on measuring the local fields with two-photon photoemission electron microscopy. Progress in the coherent control of electric fields in nanostructures thus might open the access to simultaneously ultrafast and nanoscale new physics.

Nonlinear near-field optical microscopy might thus play an extremely important role in the development of nanoscale coherent control, since it can be employed as a probe of local field enhancement. In this frame, the temporal dynamics of the nonlinear phenomenon is, however, extremely important. SHG represents an ideal candidate to this task since the conversion of two FW photons into an SH photon is itself a coherent process (Imura et al. 2005), at variance from, for example, TPPL, which is incoherent in noble metal nanostructures (Imura et al. 2005). Indeed, we have recently demonstrated (Biagioni et al. 2009) that the TPPL yield in gold nanowires becomes independent of the pulse duration for laser pulses shorter than about 1 ps, while an inverse proportionality with the pulse duration would be expected for a fixed pulse energy. The origin of this behavior consists of the fact that TPPL in gold nanostructures is obtained after a $3d$ hole is produced by two sequential absorption steps involving a single photon (Imura et al. 2005), and is governed by the relaxation time of the sp conduction band hole generated after the first photon absorption. This temporal limitation would advise against the utilization of TPPL in probing near-field enhancement for coherent control applications, leaving SHG as a better adapted tool.

28.9 Conclusions

Thanks to the combination of surface sensitivity and symmetry selectivity, SHG is a very powerful tool to investigate the nanoscale properties of matter. The application of SHG techniques for investigating the structure and various other properties of small particles is a subject of considerable current interest and has stimulated a conspicuous number of both theoretical and experimental works in a broad range of subjects. In particular, being SHG a nonlinear optical process whose yield depends on the square power of the FW field intensity, SHG has found many applications in the study of the electric field enhancement mechanisms in metal nanostructures, where the lightning rod effect associated with high-curvature particles or resonantly excited plasma oscillations can lead to extremely intense and highly localized electric fields. Since nanoscience is attracting an exponentially growing attention, SHG in nanostructured systems will be a very hot topic for many years to come.

Acknowledgments

We would like to particularly thank Prof. Klaus Sattler for his kind invitation to write this chapter. We are also indebted to many colleagues and friends, whom we had the opportunity to work and discuss with during their activity concerning this work. Among these, we would like to mention M. Allegrini, P.-M. Adam P. Biagioni, M. Celebrano, G. Grancini, J. Grand, M. Labardi, D. Polli, M. Savoini, and M. Zavelani-Rossi.

M. F. and L. D. are also affiliated to L-NESS (Laboratory for Nanometric Epitaxial Structures on Silicon and for Spintronics), which is an Interuniversity Centre between Politecnico di Milano and Università di Milano Bicocca. G. C. is also affiliated to ULTRAS (National Laboratory for Ultrafast and Ultraintense Optical Science) research center, which is founded by the National Institute of the Physics of Matter (INFM) of the National Research Council (CNR).

References

Abid, J.-P., Nappa, J., Girault, H. H., and Brevet, P.-F. 2004. Pure surface plasmon resonance enhancement of the first hyperpolarizability of gold core–silver shell nanoparticles. *J. Chem. Phys.* **121**: 12577.

Aeschlimann, M., Bauer, M., Bayer, D. et al. 2007. Adaptive subwavelength control of nano-optical fields, *Nature* **446**: 301–304.

Agarwal, G. S. and Jha, S. S. 1982. Theory of second harmonic generation at a metal surface with surface plasmon excitation. *Solid State Commun.* **41**: 499–501.

Antoine, R., Brevet, P. F., Girault, H. H., Bethell, D., and Schiffrin, D. J. 1997. Surface plasmon enhanced non-linear optical response of gold nanoparticles at the air/toluene interface. *Chem. Commun.* **1997**: 1901–1902.

Antoine, R., Pellarin, M., Palpant, B. et al. 1998. Surface plasmon enhanced second harmonic response from gold clusters embedded in an alumina matrix. *J. Appl. Phys.* **84**: 4532.

Bachelier, G., Russier-Antoine, I., Benichou, E., Jonin, C., and Brevet, P. 2008. Multipolar second-harmonic generation in noble metal nanoparticles. *J. Opt. Soc. Am. B* **25**: 955–960.

Beermann, J. and Bozhevolnyi, S. I. 2004. Microscopy of localized second-harmonic enhancement in random metal nanostructures. *Phys. Rev. B* **69**: 155429.

Beermann, J., Bozhevolnyi, S. I., and Coello, V. 2006. Modeling of nonlinear microscopy of localized field enhancements in random metal nanostructures. *Phys. Rev. B* **73**: 115408.

Biagioni, P., Polli, D., Labardi, M. et al. 2005. Unexpected polarization behavior at the aperture of hollow-pyramid near-field probes. *Appl. Phys. Lett.* **87**: 223112.

Biagioni, P., Celebrano, M., Polli, D. et al. 2007. Nonlinear optics and spectroscopy at the nanoscale with a hollow-pyramid aperture SNOM. *J. Phys. Conf. Ser.* **61**: 125–129.

Biagioni, P., Celebrano, M., Savoini, M., Grancini, G., Brida, D., Mátéfi-Tempfli, S., Mátéfi-Tempfli, M., Duò, L., Hecht, B., Cerullo, G., and Finazzi, M. 2009. Dependence of the two-photon photoluminescence yield of gold nanostructures on the laser pulse duration. *Phys. Rev. B* **80**: 045411.

Bickford, L., Sun, J., Fu, K. et al. 2008. Enhanced multi-spectral imaging of live breast cancer cells using immunotargeted gold nanoshells and two-photon excitation microscopy. *Nanotechnology* **19**: 315102.

Bloembergen, N. 1999. Surface nonlinear optics: A historical overview. *Appl. Phys. B: Lasers Opt.* **68**: 289–293.

Bloembergen, N. and Pershan, P. S. 1962. Light waves at the boundary of nonlinear media. *Phys. Rev.* **128**: 606.

Bouhelier, A., Beversluis, M. R., and Novotny, L. 2003. Characterization of nanoplasmonic structures by locally excited photoluminescence. *Appl. Phys. Lett.* **83**: 5041–5043.

Bouhelier, A., Bachelot, R., Lerondel, G., Kostcheev, S., Royer P., and Wiederrecht, G. P. 2005. Surface plasmon characteristics of tunable photoluminescence in single gold nanorods. *Phys. Rev. Lett.* **95**: 267405.

Boyd, R. W. 2003. *Nolinear Optics*. Academic Press, San Diego, CA.

Bozhevolnyi, S. I. and Lozovski, V. Z. 2000. Self-consistent model for second-harmonic near-field microscopy. *Phys. Rev. B* **61**: 11139.

Bozhevolnyi, S. I. and Lozovski, V. Z. 2002. Second-harmonic scanning optical microscopy of individual nanostructures. *Phys. Rev. B* **65**: 235420.

Bozhevolnyi, S. I., Beermann, J., and Coello, V. 2003. Direct observation of localized second-harmonic enhancement in random metal nanostructures. *Phys. Rev. Lett.* **90**: 197403.

Breit, M., Malkmus, S., Feldmann, J., and Danzebrink, H. U. 2007. Near-field second harmonic generation by using uncoated silicon tips. *Appl. Phys. Lett.* **90**: 093114.

Brixner, T., Garcia de Abajo, F. J., Schneider, J., and Pfeiffer, W. 2005. Nanoscopic ultrafast space-time-resolved spectroscopy. *Phys. Rev. Lett.* **95**: 093901.

Brixner, T., García de Abajo, F. J., Schneider, J., Spindler, C., and Pfeiffer, W. 2006. Ultrafast adaptive optical near-field control. *Phys. Rev. B* **73**: 125437.

Brudny, V. L., Mendoza, B. S., and Mochán, W. L. 2000. Second-harmonic generation from spherical particles. *Phys. Rev. B* **62**: 11152.

Brudny, V. L., Mochán, W. L., Maytorena, J. A., and Mendoza, B. S. 2003. Second harmonic generation from a collection of nanoparticles. *Phys. Stat. Sol. B* **240**: 518–526.

Campagnola, P. J. and Loew, L. M. 2003. Second-harmonic imaging microscopy for visualizing biomolecular arrays in cells, tissues and organisms. *Nat. Biotechnol.* **11**: 1356–1360.

Canfield, B. K., Kujala, S., Jefimovs, K., Turunen, J., and Kauranen, M. 2004. Linear and nonlinear optical responses influenced by broken symmetry in an array of gold nanoparticles. *Opt. Express* **12**: 5418–5423.

Canfield, B. K., Kujala, S., Laiho, K., Jefimovs, K., Turunen, J., and Kauranen, M. 2006. Chirality arising from small defects in gold nanoparticle arrays. *Opt. Express* **14**: 950–955.

Canfield, B. K., Husu, H., Laukkanen, J., Bai, B. et al. 2007. Local field asymmetry drives second-harmonic generation in noncentrosymmetric nanodimers. *Nano Lett.* **7**: 1251–1255.

Celebrano, M., Zavelani-Rossi, M., Biagioni, P. et al. 2007. Mapping local field distribution at metal nanostructures by near-field second-harmonic generation. *Proc. SPIE* **6641**: 66411E.

Celebrano, M., Zavelani-Rossi, M., Polli, D., et al. 2008a. Mapping local field enhancements at nanostructured metal surfaces by second-harmonic generation induced in the near field. *J. Microsc.* **229**: 233–239.

Celebrano, M., Biagioni, P., Finazzi, M., et al. 2008b. Near-field second-harmonic generation from gold nanoellipsoids. *Phys. Stat. Sol. C* **5**: 2657–2661.

Dadap, J. I. 2008. Optical second-harmonic scattering from cylindrical particles. *Phys. Rev. B* **78**: 205322.

Dadap, J. I., Shan, J., Eisenthal, K. B., and Heinz, T. F. 1999. Second-harmonic Rayleigh scattering from a sphere of centrosymmetric material. *Phys. Rev. Lett.* **83**: 4045.

Dadap, J. I., Shan, J., and Heinz, T. F. 2004. Theory of optical second-harmonic generation from a sphere of centrosymmetric material: Small-particle limit. *J. Opt. Soc. Am. B* **21**: 1328–1347.

Dewitz, J. P., Hübner, W., and Bennemann, K. H. 1996. Theory for nonlinear Mie-scattering from spherical metal clusters. *Z. Phys. D: At. Mol. Clusters* **37**: 75–84.

Downer, M. C., Mendoza, B. S., and Gavrilenko, V. I. 2001. Optical second harmonic spectroscopy of semiconductor surfaces: Advances in microscopic understanding. *Surf. Interface Anal.* **31**: 966–986.

Durr, N. J., Larson, T., Smith, D. K., Korgel, B. A., Sokolov, K., and Ben-Yakar, A. 2007. Two-photon luminescence imaging of cancer cells using molecularly targeted gold nanorods. *Nano Lett.* **7**: 941–945.

El-Sayed, I. H., Huang, X., and El-Sayed, M. A. 2006. Selective laser photo-thermal therapy of epithelial carcinoma using anti-EGFR antibody conjugated gold nanoparticles. *Cancer Lett.* **239**: 129–135.

Figliozzi, P., Sun, L., Jiang, Y. et al. 2005. Single-beam and enhanced two-beam second-harmonic generation from silicon nanocrystals by use of spatially inhomogeneous femtosecond pulses. *Phys. Rev. Lett.* **94**: 047401.

Finazzi, M., Biagioni, P., Celebrano, M., and Duò, L. 2007. Selection rules for second-harmonic generation in nanoparticles. *Phys. Rev. B* **76**: 125414.

Fischer, U. Ch. and Zingsheim, H. P. 1981. Submicroscopic pattern replication with visible light. *J. Vac. Sci. Technol.* **19**: 881–885.

Franken, P. A., Hill, A. E., Peters, C. W., and Weinreich, G. 1961. Generation of optical harmonics. *Phys. Rev. Lett.* **7**: 118.

Grand, J., Kostcheev, S., Bijeon, J.-L. et al. 2003. Optimization of SERS-active substrates for near-field Raman spectroscopy. *Synth. Met.* **139**: 621–624.

Hao, E. C., Schatz, G. C., Johnson, R. C., and Hupp, J. T. 2002. Hyper-Rayleigh scattering from silver nanoparticles. *J. Chem. Phys.* **117**: 5963.

Hartschuh, A., Pedrosa, H. N., Novotny, L., and Krauss, T. D. 2003. Simultaneous fluorescence and Raman scattering from single carbon nanotubes. *Science* **301**: 1354–1356.

Hecht, B., Bielefeldt, H., Inouye, Y., and Pohl, D. W. 1997. Facts and artifacts in near-field optical microscopy. *J. Appl. Phys.* **81**: 2492.

Hua, X. M. and Gersten, J. I. 1986. Theory of second-harmonic generation by small metal spheres. *Phys. Rev. B* **33**: 3756.

Hubert, C., Billot, L., Adam, P.-M. et al. 2007. Role of surface plasmon in second harmonic generation from gold nanorods. *Appl. Phys. Lett.* **90**: 181105.

Imura, K., Nagahara, T., and Okamoto, H. 2005. Near-field two-photon-induced photoluminescence from single gold nanorods and imaging of plasmon modes. *J. Phys. Chem. B* **109**: 13214–13220.

Imura, K., Nagahara, T., and Okamoto, H. 2006. Photoluminescence from gold nanoplates induced by near-field two-photon absorption. *Appl. Phys. Lett.* **88**: 023104.

Jackson, J. D. 1975. *Classical Electrodynamics*. Wiley, New York.

Jakubczyk, D., Shen, Y., Lal, M., Kim, K.S., Świątkewicz, J., and Prasad, P.N. 1999. Near-field probing of nanoscale nonlinear optical processes. *Opt. Lett.* **24**: 1151–1153.

Jiang, Y., Wilson, P. T., Downer, M. C., White, C. W., and Withrow, S. P. 2001. Second-harmonic generation from silicon nanocrystals embedded in SiO$_2$. *Appl. Phys. Lett.* **78**: 766.

Jiang, Y., Sun, L., and Downer, M. C. 2002. Second-harmonic spectroscopy of two-dimensional Si nanocrystal layers embedded in SiO$_2$ films. *Appl. Phys. Lett.* **81**: 3034.

Johnson, R. C., Li, J., Hupp, J. T., and Schatz, G. C. 2002. Hyper-Rayleigh scattering studies of silver, copper, and platinum nanoparticle suspensions. *Chem. Phys. Lett.* **356**: 534–540.

Labardi, M., Zavelani-Rossi, M., Polli, D. et al. 2005. Characterization of femtosecond light pulses coupled to hollow-pyramid near-field probes: Localization in space and time. *Appl. Phys. Lett.* **86**: 031105.

Lüpke, G. 1999. Characterization of semiconductor interfaces by second-harmonic generation. *Surf. Sci. Rep.* **35**: 75–161.

Martorell, J., Vilaseca, R., and Corbalán, R. 1997. Scattering of second-harmonic light from small spherical particles ordered in a crystalline lattice. *Phys. Rev. A* **55**: 4520.

Mc Gilp, J. F. 1996. A review of optical second-harmonic and sum-frequency generation at surfaces and interfaces. *J. Phys. D* **29**: 1812–1821.

Mikhailowsky, A. A., Petruska, M. A., Stockman, M. I., and Klimov, V. I. 2003. Broadband near-field interference spectroscopy of metal nanoparticles using a femtosecond white-light continuum. *Opt. Lett.* **28**: 1686–1688.

Mochán, W. L., Maytorena, J. A., Mendoza, B. S., and Brudny, V. L. 2003. Second-harmonic generation in arrays of spherical particles. *Phys. Rev. B* **68**: 085318.

Mühlschlegel, P., Eisler, H.-J., Martin, O. J. F., Hecht, B., and Pohl, D. W. 2005. Resonant optical antennas. *Science* **308**: 1607–1609.

Nagesha, D., Laevsky, G. S., Lampton, P. et al. 2007. In vitro imaging of embryonic stem cells using multiphoton luminescence of gold nanoparticles. *Int. J. Nanomed.* **2**: 813.

Nappa, J., Revillod, G., Russier-Antoine, I., Benichou, E., Jonin, C., and Brevet P. F. 2005. Electric dipole origin of the second harmonic generation of small metallic particles. *Phys. Rev. B* **71**: 165407.

Nappa, J., Russier-Antoine, I., Benichou, E., Jonin, Ch., and Brevet, P. F. 2006. Second harmonic generation from small gold metallic particles: From the dipolar to the quadrupolar response. *J. Chem. Phys.* **125**: 184712.

Neacsu, C. C., Reider, G. A., and Raschke, M. B. 2005. Second-harmonic generation from nanoscopic metal tips: Symmetry selection rules for single asymmetric nanostructures. *Phys. Rev. B* **71**: 201402(R).

Nie, S. M. and Emery, S. R. 1997. Probing single molecules and single nanoparticles by surface-enhanced Raman scattering. *Science* **275**: 1102–1106.

Novotny, L. and Hecht, B. 2006. *Principles of Nano-Optics*. Cambridge University Press, Cambridge, NY.

O'Neal, D. P., Hirsch, L. R., Halas, N. J., Payne, J. D., and West, J. L. 2004. *Cancer Lett.* **209**: 181.

Panasyuk, G. Y., Schotland, J. C., and Marke, V. A. 2008. Classical theory of optical nonlinearity in conducting nanoparticles. *Phys. Rev. Lett.* **100**: 047402.

Parameswaran, K. R., Kurz, J. R., Roussev, M. M., and Fejer, M. 2002. Observation of 99% pump depletion in single-pass second-harmonic generation in a periodically poled lithium niobate waveguide. *Opt. Lett.* **27**: 43–45.

Park, J., Estrada, A., Sharp, K. et al. 2008. Two-photon-induced photoluminescence imaging of tumors using near-infrared excited gold nanoshells. *Opt. Express* **16**: 1590–1599.

Pavlyukh, Y. and Hübner, W. 2004. Nonlinear Mie scattering from spherical particles. *Phys. Rev. B* **70**: 245434.

Russier-Antoine, I., Jonin, Ch., Nappa, J., Bénichou, E., and Brevet, P.-F. 2004. Wavelength dependence of the hyper Rayleigh scattering response from gold nanoparticles. *J. Chem. Phys.* **120**: 10748.

Sandrock, M. L., Pibel, C. D., Geiger, F. M., and Foss, Jr. C. A. 1999. Synthesis and second-harmonic generation studies of noncentrosymmetric gold nanostructures. *J. Phys. Chem. B* **103**: 2668–2673.

Shan, J., Dadap, J. I., Stiopkin, I., Reider, G. A., and Heinz, T. F. 2006. Experimental study of optical second-harmonic scattering from spherical nanoparticles. *Phys. Rev. A* **73**: 023819.

Shen, Y. R. 1984. *The Principles of Nonlinear Optics*. Wiley, New York.

Shen, Y. R. 1989. Optical second harmonic generation at interfaces. *Annu. Rev. Phys. Chem.* **40**: 327–350.

Shen, Y., Markowicz, P., Winiarz, J., Swiatkiewicz, J., and Prasad, P. N. 2001. Nanoscopic study of second-harmonic generation in organic crystals with collection-mode near-field scanning optical microscopy. *Opt. Lett.* **26**: 725–727.

Singh, D. B. and Tripathi, V. K. 2007. Surface plasmon excitation at second harmonic over a rippled surface. *J. Appl. Phys.* **102**: 083301.

Smolyaninov, I. I., Zayats, A. V., and Davis, C. C. 1997. Near-field second harmonic generation from a rough metal surface. *Phys. Rev. B* **56**: 9290.

Smolyaninov, I. I., Liang, H. Y., Lee, C. H., Davis, C. C., Aggarwal, S., and Ramesh, R. 2000. Near-field second-harmonic microscopy of thin ferroelectric films. *Opt. Lett.* **25**: 835–837.

Stockman, M. I., Faleev, S. V., and Bergman, D. J. 2002. Coherent control of femtosecond energy localization in nanosystems. *Phys. Rev. Lett.* **88**: 067402.

Stockman, M. I., Bergman, D. J., Anceau, C., Brasselet, S., and Zyss, J. 2004. Enhanced second-harmonic generation by metal surfaces with nanoscale roughness: Nanoscale dephasing, depolarization, and correlations. *Phys. Rev. Lett.* **92**: 057402.

Su, H. M., Ye, J. T., Tang, Z. K., and Wong, K. S. 2008. Resonant second-harmonic generation in monosized and aligned single-walled carbon nanotubes. *Phys. Rev. B* **77**: 125428.

Sukharev, M. and Seideman, T. 2007. Coherent control of light propagation via nanoparticle arrays. *J. Phys. B: At. Mol. Opt. Phys.* **40**: S283–S298.

Tinkham, M. 1964. *Group Theory and Quantum Mechanics*. McGraw-Hill, New York.

Vance, F. W., Lemon, B. I., and Hupp, J. T. 1998. Enormous hyper-Rayleigh scattering from nanocrystalline gold particle suspensions. *J. Phys. Chem. B* **102**: 10091–10093.

Wang, H., Yan, E. C. Y., Borguet, E., and Eisenthal, K. B. 1996. Second harmonic generation from the surface of centrosymmetric particles in bulk solution. *Chem. Phys. Lett.* **259**: 15–20.

Yang, N., Angerer, W. E., and Yodh, A. G. 2001. Angle-resolved second-harmonic light scattering from colloidal particles. *Phys. Rev. Lett.* **87**: 103902.

Zavelani-Rossi, M., Celebrano, M., Biagioni, P. et al. 2008. Near-field second-harmonic generation in single gold nanoparticles. *Appl. Phys. Lett.* **92**: 093119.

Zayats, A.V., Kalkbrenner, T., Sandoghdar, V., and Mlynek, J. 2000. Second-harmonic generation from individual surface defects under local excitation. *Phys. Rev. B* **61**: 4545.

Zhu, J. 2007. Theoretical study of the tunable second-harmonic generation (SHG) enhancement factor of gold nanotubes. *Nanotechnology* **18**: 225702.

Nonlinear Optics in Semiconductor Nanostructures

Mikhail Erementchouk
University of Central Florida

Michael N. Leuenberger
University of Central Florida

29.1 Introduction

The physics of semiconductor nonlinear response unites under the same cover a fascinating variety of different physical phenomena, theoretical ideas, and experimental approaches. It provides promising opportunities for technologies, for example, in the form of principally new light sources, and presents a convenient testing field for studying long-standing fundamental questions while raising new ones in fields ranging decoherence to low-energy quantum field theory.

It is not therefore, surprising, that the interest of researchers remains steadily persistent for already many decades. There are many books and reviews devoted to the details of different topics. However, as investigations move forward, new types of problems surface and the question of the basic principles of the theory of nonlinear response is raised again and again.

Perhaps the most clear demonstration of the recurring process of testing the background is the drastic change of the main theoretical methods, which took place in the 1990s. In the pre-1990s era, Green's functions method was the main technique used in the limit of zero, finite temperature, or nonequilibrium (the Keldysh technique). The main questions were formulated in terms of the spectral characteristics, and using the adiabatical approximation was the usual practice. With the development of new experimental methods and the improvement of the quality of the samples, the attention gradually moved toward the temporal behavior and the transient features of the excitations. As a result, in the post-1990s era, time dependence turned out to be the major question, and the typical theoretical approach became the method of the dynamical equations of motion.

In a sense the whole topic was reinvented, making the question of fundamentals of the theory especially important.

In the following sections, we present the derivation of the main dynamical equations for the electromagnetic field and the semiconductor excitations in the general context of quantum field theory. Such an approach is particularly useful because the resonant interaction of light with the semiconductor results in constantly changing the number of elementary excitations in the semiconductor bands. Therefore, the respective description must admit the creation and annihilation of particles on the fundamental level. Also, in order to cover at least partially the various physical situations relevant for the problem of nonlinear response, we avoid using specific approximations such as, for example, the spherical symmetry of the hole's bands or the translational symmetry of the structures. The latter circumstance compels to formulate the major part of considerations in the coordinate representation. Finally, the main equations are derived for operators that makes them applicable for studying quantum statistics and other complex questions going beyond the single-particle correlation functions (Green's functions).

The structure of the presentation is as follows. In Section 29.2 we derive the quantum-field Hamiltonian in the $\mathbf{k} \cdot \mathbf{p}$-approximation and briefly review the basic properties of the semiconductor bands. Section 29.3 is devoted to the derivation of the quantum equations of motion for the electromagnetic field and the semiconductor excitations. In Section 29.4, the perturbative approach of analyzing the quantum equations of motion is reviewed. Finally, in Section 29.5 we consider the semiconductor Bloch equation.

29.2 Semiconductor-Field Hamiltonian in k·p-Approximation

The quantum dynamics of the electrons moving in the periodic lattice field in the semiconductor under the action of the external field is described in SI units by the Hamiltonian

$$\mathcal{H}_{mat} = \sum_s \int d\mathbf{x}\, \psi_s^\dagger(\mathbf{x}) \left\{ \frac{1}{2m}[\mathbf{p} - e\mathbf{A}(\mathbf{x})]^2 + U_l(\mathbf{x}) + U_{ext}(\mathbf{x}) \right\} \psi_s(\mathbf{x})$$
$$+ \frac{1}{2} \sum_{s,s'} \int d\mathbf{x}_1\, d\mathbf{x}_2\, \psi_s^\dagger(\mathbf{x}_1)\psi_{s'}^\dagger(\mathbf{x}_2) V(\mathbf{x}_1 - \mathbf{x}_2)\psi_{s'}(\mathbf{x}_2)\psi_s(\mathbf{x}_1),$$
$$(29.1)$$

where

- $\psi_s(\mathbf{x})$ is the electron field operator
- s and s' are the electron spin indices
- $\mathbf{A}$ is the vector potential
- $U_l(\mathbf{x})$ is the periodic lattice potential
- $U_{ext}(\mathbf{x})$ is an external potential caused, for example, by the external static electric field
- $V(\mathbf{x}) = e^2/4\pi\epsilon_0\epsilon_b|\mathbf{x}|$, with ϵ_b being the background dielectric constant, is the potential of the Coulomb interaction between the electrons

In the following paragraphs we will be interested in establishing the relation with the standard k·p approach, therefore, we will omit both the external potential and the Coulomb interaction and will consider their effect later.

The main idea of the k·p approach is to employ the fact that we are interested in the weak and spatially smooth perturbations of the semiconductor near its ground state. Thus, we want to account exactly the effect of the periodic lattice potential forming the semiconductor ground state and to consider the deviation from it in the spirit of the perturbation theory.

In order to account for the effect of the periodic potential on the electron dynamics we need to develop an appropriate description of the respective electron single-particle states. The field operator $\psi^\dagger(\mathbf{x})$ can always be expanded in terms of the Bloch modes as

$$\psi_s^\dagger(\mathbf{x}) = \sum_n \int d\mathbf{k}\, \phi_{n,\mathbf{k}}^*(\mathbf{x}) a_{n,\mathbf{k}}^\dagger, \qquad (29.2)$$

where

- n enumerates the bands and includes the spin index
- the Bloch wave vector $\mathbf{k}$ lies in the first Brillouin zone
- the wave functions $\phi_{n,\mathbf{k}}(\mathbf{x})$ give the solutions of the respective Schrödinger equation in the periodic potential
- the operators $a_{n,\mathbf{k}}^\dagger$ are the operators, which create electrons in the respective Bloch states

According to the Bloch theorem the electron wave function $\phi_{n,\mathbf{k}}(\mathbf{x})$ can be presented as

$$\phi_{n,\mathbf{k}}(\mathbf{x}) = e^{i\mathbf{k}\cdot\mathbf{x}} u_{n,\mathbf{k}}(\mathbf{x}), \qquad (29.3)$$

where $u_{n,\mathbf{k}}(\mathbf{x})$ is a periodic function solving the equation

$$\frac{\hbar^2}{2m}(-i\nabla + \mathbf{k})^2 u_{n,\mathbf{k}}(\mathbf{x}) + U_l(\mathbf{x}) u_{n,\mathbf{k}}(\mathbf{x}) = \epsilon_n(\mathbf{k}) u_{n,\mathbf{k}}(\mathbf{x}). \qquad (29.4)$$

This equation is Hermitian and, hence, functions $u_{n,\mathbf{k}}(\mathbf{x})$ for a fixed $\mathbf{k} = \mathbf{k}_0$ constitute a complete set on the space of periodic functions. We can, therefore, represent functions corresponding to other $\mathbf{k} \neq \mathbf{k}_0$ as the respective Fourier series

$$u_{n,\mathbf{k}}(\mathbf{x}) = \sum_m \alpha_{n,m}(\mathbf{k};\mathbf{k}_0) u_{m,\mathbf{k}_0}(\mathbf{x}) \qquad (29.5)$$

with unitary $\alpha_{n,m}(\mathbf{k};\mathbf{k}_0)$. The latter follows from orthonormality of the amplitudes $u_{n,\mathbf{k}}(\mathbf{x})$

$$\frac{1}{\mathcal{V}} \int d\mathbf{x}\, u_{n',\mathbf{k}}^*(\mathbf{x}) u_{n',\mathbf{k}}(\mathbf{x}) = \frac{1}{(2\pi)^3}\delta_{nn'}, \qquad (29.6)$$

where $\mathcal{V}$ is the volume of the elementary cell. Indeed, substituting Equation 29.5 into Equation 29.6 we obtain

$$\sum_m \alpha_{n'm}^*(\mathbf{k};\mathbf{k}_0)\alpha_{nm}(\mathbf{k};\mathbf{k}_0) = \delta_{nn'}, \qquad (29.7)$$

which explicitly shows the unitarity of $\alpha_{n,m}(\mathbf{k};\mathbf{k}_0)$. This motivates the introduction of new (Luttinger–Kohn) function describing the electron states

$$\chi_{n,\mathbf{k}}(\mathbf{x}) = e^{i\mathbf{k}\cdot\mathbf{x}} u_{n,\mathbf{k}_0}(\mathbf{x}). \qquad (29.8)$$

These functions constitute a complete orthonormal set similarly to the electron wave functions $\phi_{n,\mathbf{k}}(\mathbf{x})$. The proof demonstrates the typical line of arguments and, therefore, it is useful to consider it in detail. The completeness of the Luttinger–Kohn functions

$$\sum_n \int d\mathbf{k}\, \chi_{n,\mathbf{k}}^*(\mathbf{x})\chi_{n,\mathbf{k}}(\mathbf{x}') = \delta(\mathbf{x} - \mathbf{x}'). \qquad (29.9)$$

follows from the respective relation for the electron wave functions

$$\sum_n \int d\mathbf{k}\, \phi_{n,\mathbf{k}}^*(\mathbf{x})\phi_{n,\mathbf{k}}(\mathbf{x}') = \delta(\mathbf{x} - \mathbf{x}') \qquad (29.10)$$

if we use Equations 29.5 and 29.7. For orthonormality we need to consider

$$(n,\mathbf{k}|n',\mathbf{k}') \equiv \int d\mathbf{x}\, \chi_{n,\mathbf{k}}^*(\mathbf{x})\chi_{n',\mathbf{k}'}(\mathbf{x}) = \int d\mathbf{x}\, e^{i(\mathbf{k}-\mathbf{k}')\cdot\mathbf{x}} u_{n,\mathbf{k}_0}^*(\mathbf{x}) u_{n',\mathbf{k}_0}(\mathbf{x}).$$
$$(29.11)$$

The product $u_{n,\mathbf{k}_0}^*(\mathbf{x})u_{n',\mathbf{k}_0}(\mathbf{x})$ is a periodic function and therefore can be expanded in the Fourier series

$$u_{n,\mathbf{k}_0}^*(\mathbf{x})u_{n',\mathbf{k}_0}(\mathbf{x}) = \sum_{\mathbf{Q}} B_{nn'}(\mathbf{Q})e^{i\mathbf{Q}\cdot\mathbf{x}}, \qquad (29.12)$$

where the sum runs over the vectors of the reciprocal lattice and

$$B_{nn'}(\mathbf{Q}) = \frac{1}{\mathcal{V}}\int d\mathbf{x}\, e^{-i\mathbf{Q}\cdot\mathbf{x}} u_{n,\mathbf{k}}^*(\mathbf{x})u_{n',\mathbf{k}}(\mathbf{x}). \qquad (29.13)$$

Using this representation in Equation 29.11 we obtain

$$(n,\mathbf{k}|n',\mathbf{k}') = (2\pi)^3 \sum_{\mathbf{Q}} B_{nn'}(\mathbf{Q})\delta(\mathbf{k}-\mathbf{k}'+\mathbf{Q}). \qquad (29.14)$$

The Bloch-wave-vectors $\mathbf{k}$ and $\mathbf{k}'$ are inside the first Brillouin zone and, therefore, the only possibility to reach the singularity point of the δ-function is when $\mathbf{Q} = 0$ leading to $(n,\mathbf{k}|n',\mathbf{k}') = (2\pi)^3 B_{nn'}(0)\delta(\mathbf{k}-\mathbf{k}')$. Taking into account Equation 29.6 we arrive at

$$(n,\mathbf{k}|n',\mathbf{k}') = \delta_{nn'}\delta(\mathbf{k}-\mathbf{k}'). \qquad (29.15)$$

Such introduced functions $\chi_{n',\mathbf{k}'}(\mathbf{x})$ provide a general background and are used in different contexts in solid-state physics (Luttinger and Kohn 1955, Johnson and Hui 1993). In what follows we limit ourselves to the case $\mathbf{k}_0 = 0$ and will omit the index corresponding to the Bloch-wave-vector, simplifying notations.

In terms of the functions $\chi_{n,\mathbf{k}}(\mathbf{x})$ the electron field operator is presented as

$$\psi_s^\dagger(\mathbf{x}) = \sum_n \int d\mathbf{k}\, \chi_{n,\mathbf{k}}^*(\mathbf{x})b_{n,\mathbf{k}}^\dagger. \qquad (29.16)$$

Substituting this representation into Equation 29.1, adopting the Coulomb gauge for the vector potential, $\nabla\cdot\mathbf{A}(\mathbf{x}) \equiv 0$, taking into account Equation 29.4 and using the same trick as for Equation 29.15 we obtain

$$\mathcal{H}_{mat} = \sum_{n,n'} \int d\mathbf{k}\left[\left(\epsilon_n \frac{\hbar^2 k^2}{2m}\right)\delta_{n,n'} + \frac{\hbar}{m}\mathbf{k}\cdot\mathbf{p}_{n,n'}\right]b_{n,\mathbf{k}}^\dagger b_{n',\mathbf{k}}$$

$$-\frac{e}{m}\sum_{n,n'}\int d\mathbf{k}\,d\mathbf{q}\left[\mathbf{A}(\mathbf{q})(\mathbf{p}_{n,n'}+\hbar\mathbf{k}\delta_{n,n'}) - \frac{e}{2}A_2(\mathbf{q})\delta_{n,n'}\right]$$

$$\times b_{n,\mathbf{k}+\mathbf{q}}^\dagger b_{n',\mathbf{k}}, \qquad (29.17)$$

where $A_2(\mathbf{q}) = (2\pi)^{-3/2}\int d\mathbf{x}\, A^2(\mathbf{x})e^{-i\mathbf{q}\cdot\mathbf{x}}$. The parameters of great importance introduced in Equation 29.17 are the matrix elements of the momentum operator between different bands

$$\mathbf{p}_{nn'} = -\frac{i\hbar}{\mathcal{V}}\int d\mathbf{x}\, u_n^*(\mathbf{x})\nabla u_{n'}(\mathbf{x}), \qquad (29.18)$$

which are responsible for the *interband* coupling and are the main parameters quantifying the strength of the light–matter interaction.

The first term in the r.h.s. of Equation 29.17 describes the dynamics of the semiconductor excitations, and the second term accounts for the interaction with the external field. The specific feature of the "eigen-part" of the Hamiltonian (that is the part which is independent of the external field) is that it is not diagonal with respect to bands. The reason for this is the special status given to the states corresponding to $\mathbf{k} = \mathbf{k}_0 = 0$. These states do not exhaust the eigenstates of the semiconductor and, therefore, the effective coupling between different bands should not be surprising. This coupling, however, is proportional to $\mathbf{k}$. Thus, if the relevant excitations are relatively weak and smooth, then the main contribution into the semiconductor dynamics would come from small $\mathbf{k}$ and we can treat the respective band coupling as a perturbation. As the first step we could just drop this term. This, however, gives obviously wrong results for the dynamics near the extreme points of the bands. Indeed, the resulting Hamiltonian would predict that the mass of the excitations in the conductance band is the mass of electron in empty space and that the energy of states in the valence band increases with $\mathbf{k}$. Both of these predictions contradict the experiment. Thus, it is necessary to develop a consistent procedure that allows taking the band coupling into account, similarly to the perturbation theory in quantum mechanics. Unfortunately, a straightforward implementation of this approach is significantly complicated by the complexity of the semiconductor band structure. It poses not principal, but technical difficulties, the resolution of which requires developing a special approach employing the symmetry properties of the semiconductor lattice. These questions are beyond the scope of the present consideration (see, e.g., Chapter III in Bir and Pikus (1974)), therefore we limit ourselves to a semiqualitative analysis, which clearly illustrates the main ideas and can be readily adopted for a particular situation, for example for studying the effect of the static magnetic field and so on.

In order to eliminate the off-diagonal elements of the eigen-part of the Hamiltonian we perform a unitary Schrieffer-Wolff transformation $T = e^S$

$$b_{n,\mathbf{k}} = \sum_m \int d\mathbf{q}(n,\mathbf{k}|T|m,\mathbf{q})c_{m,\mathbf{q}}, \quad b_{n,\mathbf{k}}^\dagger = \sum_m \int d\mathbf{q}(m,\mathbf{q}|T^\dagger|n,\mathbf{k})c_{m,\mathbf{q}}^\dagger \qquad (29.19)$$

such that the eigen-part of the transformed Hamiltonian $\tilde{\mathcal{H}}_{mat} = T^\dagger\mathcal{H}_{mat}T$ is *approximately* diagonal. In Bir and Pikus (1974) a general procedure is discussed, which allows developing the respective perturbation theory up to any order. Here, however, we restrict ourselves to the first relevant order, as it is usually done in the $\mathbf{k} \cdot \mathbf{p}$-approximation. In terms of the "generator" S the transformed Hamiltonian is written as

$$\tilde{\mathcal{H}}_{mat} = \mathcal{H}_{mat} + [\mathcal{H}_{mat}, S] + \frac{1}{2}[[\mathcal{H}_{mat}, S], S] + \cdots \quad (29.20)$$

where $[A, B] = AB - BA$ denotes the usual matrix commutator. We want to treat different contributions on the different basis and therefore we present the Hamiltonian as $\mathcal{H}_{mat} = \mathcal{H}_0 + \mathcal{H}_p + \mathcal{H}_A$, where $\mathcal{H}_0$ stands for the diagonal part of the first term in the r.h.s. of Equation 29.17, $\mathcal{H}_p$ is the $\mathbf{k} \cdot \mathbf{p}$ part and $\mathcal{H}_A$ is the term related to the external field. Using this presentation in Equation 29.20 we obtain

$$\tilde{\mathcal{H}}_{mat} = \mathcal{H}_0 + \mathcal{H}_p + \mathcal{H}_A + [\mathcal{H}_0, S] + [\mathcal{H}_p, S] + [\mathcal{H}_A, S]$$
$$+ \frac{1}{2}[[\mathcal{H}_0, S], S] + \frac{1}{2}[[\mathcal{H}_p, S], S] + \frac{1}{2}[[\mathcal{H}_A, S], S] + \cdots \quad (29.21)$$

Since the $\mathbf{k} \cdot \mathbf{p}$-coupling between the bands is small, the transformation diagonalizing the Hamiltonian is close to identical and, hence, the generator S is in some sense small $\sim \mathbf{k} \cdot \mathbf{p}$. Therefore, in the lowest order no contributions come from the terms not shown in Equation 29.21 and, moreover, we need to drop the term $[[\mathcal{H}_p, S], S]$.

We choose S in such a way that the off-diagonal part of $\mathcal{H}_p$ is canceled in the leading order of $\mathbf{k} \cdot \mathbf{p}$. Or, since the diagonal part of $\mathcal{H}_p$ is 0, we find such S that

$$\mathcal{H}_p + [\mathcal{H}_0, S] = 0. \quad (29.22)$$

Resolving this equation with respect to the matrix elements of S yields

$$(n, \mathbf{k}|S|n', \mathbf{k}') = -\frac{\hbar \mathbf{k} \cdot \mathbf{p}_{nn'}}{m \omega_{nn'}} \delta(\mathbf{k} - \mathbf{k}'), \quad (29.23)$$

where $\omega_{nn'} = \epsilon_n - \epsilon_{n'}$.

Using found S for the field related terms in Equation 29.21, we will have among the others the terms of the form $\mathbf{A}(\mathbf{q}) \otimes \mathbf{q}$, where $\otimes$ denotes the tensor product, so that in a Cartesian basis $(\mathbf{k} \otimes \mathbf{q})_{ij} = k_i q_j$. Generally the tensor $\mathbf{A}(\mathbf{q}) \otimes \mathbf{q}$ is not zero. Its trace vanishes because of the Coulomb gauge imposed on the vector potential, but there are no other general restrictions on its form. Such terms, however, can still be neglected because of the smoothness of the external field. Indeed, because of the slow spatial variation of $\mathbf{A}(\mathbf{x})$ the main contribution into $\mathbf{A}(\mathbf{q})$ arises from small $\mathbf{q}$, which leads to the essential reduction of the tensor $\mathbf{A}(\mathbf{q}) \otimes \mathbf{q}$ comparing to the field terms contained in $\mathcal{H}_A$ noncommuted with S in Equation 29.21. Thus, we obtain the matrix elements of the transformed Hamiltonian

$$\left(n, \mathbf{k} \mid \tilde{\mathcal{H}}_{mat} \mid n', \mathbf{k}'\right)$$

$$= \left(\epsilon_n + \frac{\hbar^2 k^2}{2m}\right)\delta_{nn'}\delta(\mathbf{k} - \mathbf{k}')$$

$$+ \frac{\hbar^2}{2m^2}\sum_{n''}\mathbf{k} \cdot \mathbf{p}_{nn''}\mathbf{k} \cdot \mathbf{p}_{n''n'}\left(\frac{1}{\omega_{nn''}} + \frac{1}{\omega_{n'n''}}\right)$$

$$- \frac{e}{m}\int d\mathbf{q}\left[\mathbf{A}(\mathbf{q}) \cdot (\mathbf{p}_{nn'} + \hbar \mathbf{k}'\delta_{nn'}) - \frac{e}{2}A_2(\mathbf{q})\delta_{nn'}\right]\delta(\mathbf{k} - \mathbf{k}' - \mathbf{q})$$

$$- \frac{\hbar e}{m^2}\sum_{n''}\int d\mathbf{q}\delta(\mathbf{k} - \mathbf{k}' - \mathbf{q})\left[\frac{\mathbf{k} \cdot \mathbf{p}_{n''n'}\mathbf{A}(\mathbf{q}) \cdot \mathbf{p}_{nn''}}{\omega_{n'n''}} + \frac{\mathbf{k} \cdot \mathbf{p}_{nn''}\mathbf{A}(\mathbf{q}) \cdot \mathbf{p}_{n''n'}}{\omega_{nn''}}\right]. \quad (29.24)$$

It should be noted that because of the smoothness property discussed above the field terms are symmetric with respect to the change $\mathbf{k} \leftrightarrow \mathbf{k}'$.

The Hamiltonian with the matrix elements given by Equation 29.24 looks quite complex and barely simpler than the initial Hamiltonian. In order to discuss the physical meaning of different contributions into Equation 29.24, we introduce the spatial field operators $c_n(\mathbf{x})$ and $c_n^\dagger(\mathbf{x})$ according to

$$c_{n,\mathbf{k}} = \frac{1}{(2\pi)^{3/2}}\int d\mathbf{x}\, c_n(\mathbf{x}) e^{-i\mathbf{k} \cdot \mathbf{x}} \quad (29.25)$$

and the tensors $\mathsf{M}_{nn'}$ of effective masses

$$\frac{1}{\mathsf{M}_{nn'}} = \frac{1}{m}\mathbf{1}\delta_{nn'} + \frac{1}{m^2}\sum_{n''}\left[\frac{\mathbf{p}_{n''n'} \otimes \mathbf{p}_{nn''}}{\omega_{nn''}} + \frac{\mathbf{p}_{nn''} \otimes \mathbf{p}_{n''n'}}{\omega_{n'n''}}\right]. \quad (29.26)$$

Additionally we need to analyze the different types of interband, $n \neq n'$, and intraband, $n = n'$, couplings induced by the external field. It can be seen that the main term of interest from the perspective of the semiconductor nonlinear response is $\propto \mathbf{A}(\mathbf{q}) \cdot \mathbf{p}_{nn'}$, while all others can be neglected. Indeed, the terms $\propto \mathbf{A} \otimes \mathbf{k}$ describing the interband coupling vanish in the limit $k \to 0$, and thus their contribution is negligible in the region of validity of the $\mathbf{k} \cdot \mathbf{p}$-approximation. In turn, the effect of the terms yielding the intraband coupling is small provided the gap Δ between the valence and the conduction bands is relatively wide. As will be shown below (see Equation 29.56), the strong interaction of light with the semiconductor occurs when the frequency of the electromagnetic field is tuned into the resonance with the gap, $\hbar\omega \approx \Delta$. Thus, the intraband coupling is $\propto \hbar(\Delta T)^{-1}$, where T is the duration of the external excitation field, and even at (reasonably) short time scales it can be neglected.

Taking into account these considerations the semiconductor Hamiltonian takes a compact form

$$\tilde{\mathcal{H}}_{mat} = \sum_{n,n'}\int d\mathbf{x}\, c_n^\dagger(\mathbf{x})\left[\epsilon_n\delta_{nn'} - \nabla \cdot \frac{\hbar^2}{2\mathsf{M}_{nn'}} \cdot \nabla - \mathbf{A}(\mathbf{x}) \cdot \mathbf{d}_{nn'}\right]c_{n'}(\mathbf{x}), \quad (29.27)$$

where $\mathbf{d}_{nn'} = \mathbf{p}_{nn'}e/m$. This representation clarifies the physical meaning of the operators $c_n^\dagger(\mathbf{x})$. These are the field operators of quasiparticles corresponding to different bands and characterized by specific tensors of the effective mass. The logic of perturbative treatment of the band coupling implies neglecting the off-diagonal elements $\mathsf{M}_{nn'}$ for $n \neq n'$ as the respective correction are of higher order in $\mathbf{k} \cdot \mathbf{p}$ than those kept for derivation of Equation 29.27. This naturally leads to the definition of the effective masses M_{nn} of the respective bands. The explicit expression for the tensors of the effective mass given by $n = n'$ elements of Equation 29.26 gives the correct prediction of significant reduction of the masses compared to the electron mass in the empty space and even the negative sign of the mass of the upper valence bands, it also describes correctly the dependence on the width of the gap. However, it lacks more subtle but nevertheless important features, such as the mixing of heavy and light holes.

In order to discuss these features, we inspect closely the semiconductor band structure (see, e.g., Chapter 2 in Yu and Cardona (2004)). If the electrons were spinless, the lowest energy bands were the valence and conduction bands. In semiconductors with diamond- or zinc-blende structures (e.g., Ge, GaAs, InP) the valence band would be threefold degenerate. It turns out that under rotations which map the lattice into itself the states belonging to this band transform similarly to the states with the orbital momentum $\ell = 1$. Because of this, the band is called* Γ_{4v}. The states in the conduction band transform trivially, and in group-theoretical terms the band is called Γ_{1c}. In addition to this one needs to take into account the twofold degeneracy of each state due to the electron spin. This makes the conduction band twofold and the valence band sixfold degenerate. Directly it does not imply divergence of the terms in Equations 29.23 and 29.26 corresponding to the degenerate bands because the respective matrix elements of the momentum operator are zero. This, however, does imply certain sensitivity of these bands to the external perturbations. The interaction, which takes advantage of the degeneracy, is the spin–orbit interaction originating from the dependence of the electron energy on the orientation of the spin in spatially inhomogeneous potential (see, e.g., Chapter XII in Schiff (1949)). The spin–orbit interaction leads to partial lifting of the degeneracy forming the semiconductor band structure schematically shown in Figure 29.1.

Thus, in order to be able to describe the complex picture of the band coupling in semiconductors, we need to consider the tensors of the effective masses as phenomenological parameters and, in particular, to keep the interband tensors $\mathsf{M}_{n,n'}$ with $n \neq n'$. The great simplification comes from the consideration of restrictions imposed on the form of these tensors due to the lattice symmetry. Since the semiconductor lattice turns into itself under the action of the point symmetry group so does Hamiltonian (29.27). For a general set of tensors of the effective masses $\mathsf{M}_{n,n'}$ it might not be the case, thus the symmetry implies that these

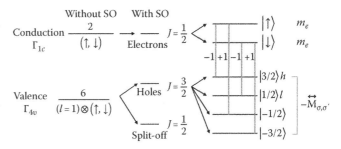

FIGURE 29.1 Effects of the spin–orbit interaction on the semiconductor band structure. The vertical lines indicate the subbands in the valence and conduction bands, between which the matrix elements of the momentum operator are not zero, together with the helicity of the respective transition. The convention for masses is shown.

tensors must have a special structure. This line of arguments was implemented and clearly explained in the series of papers by Luttinger and Kohn (1955, Luttinger 1956).

Studying these restrictions shows that in the conduction band the tensors of the effective masses are decoupled from all bands and are trivial $\mathsf{M}_{s,s'} = 1m_e\delta_{s,s'}$, where 1 is the unit tensor, and $\mathsf{M}_{s,\sigma} = \mathsf{M}_{\sigma,s} \equiv 0$. The tensors of the effective masses of the heavy and light holes are coupled $\mathsf{M}_{\sigma,\sigma'} \neq 0$ leading to the heavy–light hole hybridization. Additionally, the principal values of the tensors are not equal to each other leading to the anisotropy of the effective masses. It should be noted, however, that the anisotropy usually is not very strong and the mixing of the heavy and light holes is $\propto k^2$ and for small k the effect of hybridization is also small. This supports the popular approximation of spherical, decoupled heavy-hole and light-hole bands widely used for description of the semiconductor weak excitations. We also will use it for the qualitative discussions of some questions while keeping the main equations as general as possible.

So far we considered the case of a homogeneous semiconductor. In order to complete the derivation of the effective Hamiltonian we need to consider the spatial variation of the energies of the conduction and the valence bands due to the variation of composition and to include the terms responsible for the effect of the external static field and the Coulomb interaction between the electrons. To this end we take into consideration the last two terms in Hamiltonian (29.1) and substitute $\epsilon_n(\mathbf{x})$ for the variation of the bands edges. Generally, strong coordinate dependence of $\epsilon_n(\mathbf{x})$ would break the whole line of arguments, which allowed us to derive Equation 29.27 and to introduce the effective masses. It turns out, however, that as small as 10 elementary cells are sufficient for the approximation of effective mass to work. In particular, it can be safely applied for the description of the semiconductor quantum dots. Thus, the assumption of smoothness is fulfilled for composition-induced variations of the band edges (quantum heterostructures (Ivchenko 2005) or random interface fluctuations (Langbein et al. 2004)). It also usually holds for the external electric field (e.g., in the case of studying the quantum-confined Stark effect (Schmitt-Rink et al. 1989) on the optical semiconductor response). This assumption is employed in the standard manner. Taking into account that

---

* Strictly speaking, the correct formulation should sound like "the states transform according to irreducible representation Γ_4 of the point symmetry group of the lattice." Detailed and practical introduction into the group theory can be found in Yu and Cardona (2004) and Bir and Pikus (1974).

the main contribution into the dynamics comes from the small $\mathbf{k}$ part of Hamiltonian (29.24), one can show the validity of the important relation

$$\int d\mathbf{x}\, f(\mathbf{x})\chi_{n,\mathbf{k}}^{*}(\mathbf{x})\chi_{n',\mathbf{k}'}(\mathbf{x}) = \delta_{n,n'}\int \frac{d\mathbf{x}}{(2\pi)^3} f(\mathbf{x})e^{-i(\mathbf{k}-\mathbf{k}')\cdot\mathbf{x}} \qquad (29.28)$$

for sufficiently smooth $f(\mathbf{x})$. The key element is to observe that if the spatial Fourier spectrum $f(\mathbf{q})$ falls off fast enough, then $\delta(\mathbf{k}' - \mathbf{k} + \mathbf{q} = \mathbf{Q})$, with $\mathbf{Q}$ being the vector of the reciprocal lattice, reaches its singularity only when $\mathbf{Q} = 0$, or, in solid-state theory terms that the Umklapp processes can be neglected. Using relation (29.28) makes the transformation to the field operators $c_n(\mathbf{x})$, following Equations 29.16 and 29.19, trivial.

The Coulomb potential $V(\mathbf{x})$ is singular at the origin, therefore, generally its variation over an elementary cell is not small. However, the effect of the motion of electrons when they are extremely close to each other is not significant for the problem of the optical response. The strongest effect of the Coulomb interaction can be expected when it leads to the formation of bound states (excitons). The typical distance between the particles in such states significantly exceeds the lattice constants and the details of the behavior of the interaction potential at the origin are not important. Therefore, with good accuracy the Coulomb potential can be considered as slowly varying and one can apply relation (29.28).

Collecting these considerations, we can write down the full Hamiltonian in terms of the operators $c_n(\mathbf{x})$ and $c_n^\dagger(\mathbf{x})$

$$\mathcal{H}_{mat} = \sum_{n,m}\int d\mathbf{x}\, c_n^\dagger(\mathbf{x})\left[\epsilon_n(\mathbf{x})\delta_{nm} - \nabla\cdot\frac{\hbar^2}{2\mathsf{M}_{nm}}\cdot\nabla - \mathbf{A}(\mathbf{x})\cdot\mathbf{d}_{nm}\right]c_m(\mathbf{x})$$
$$+ \frac{1}{2}\sum_{n,m}\int d\mathbf{x}_1\, d\mathbf{x}_2\, c_n^\dagger(\mathbf{x}_1)c_m^\dagger(\mathbf{x}_2)V(\mathbf{x}_1-\mathbf{x}_2)c_m(\mathbf{x}_2)c_n(\mathbf{x}_1).$$
$$(29.29)$$

29.3 Quantum Equations of Motion

The main object of interest while studying the nonlinear response is the time dependence of observables such as the macroscopic electromagnetic field $\mathbf{A}(\mathbf{x}) = \langle\hat{\mathbf{A}}(\mathbf{x})\rangle$. These are determined by the state of the whole electromagnetic field–matter system. For example, if the system is in the coherent regime, that is, if the system can be described by the vector of state $|\Phi(t)\rangle$, then

$$\mathbf{A}(\mathbf{x}) = \langle\Phi(t)|\hat{\mathbf{A}}(\mathbf{x})|\Phi(t)\rangle = \langle\Phi(0)|U^\dagger(t;t_0)\hat{\mathbf{A}}(\mathbf{x})U(t;t_0)|\Phi(0)\rangle,$$
$$(29.30)$$

where $|\Phi(0)\rangle$ is the initial state of the system and the time dependence of the system state $|\Phi(t)\rangle = U(t;t_0)|\Phi(0)\rangle$ is written in terms of the evolution operator

$$U(t;t_0) = \mathcal{T}_+\exp\left[-\frac{i}{\hbar}\int_{t_0}^{t} dt'\mathcal{H}(t')\right] \qquad (29.31)$$

with $\mathcal{T}_+$ being the standard time-ordering operator. Here we have taken into account that the Hamiltonian may explicitly depend on time because of the presence of the external driving electromagnetic field.

The same ideology works also if the initial state is a mixture (e.g., thermal equilibrium) and therefore it should be described by the density matrix $\hat{\rho}(0)$. If there are no decohering processes, then one has

$$\mathbf{A}(\mathbf{x}) = \mathrm{Tr}\left[\hat{\rho}(t)\hat{\mathbf{A}}(\mathbf{x})\right] = \mathrm{Tr}\left[\hat{\rho}(0)U^\dagger(t;t_0)\hat{\mathbf{A}}(\mathbf{x})U(t;t_0)\right]\hat{\rho}(0).$$
$$(29.32)$$

Here we have explicitly spelled out the time dependence of the density matrix in the absence of the incoherent processes $\hat{\rho}(t) = U(t;t_0)\hat{\rho}(t)U^\dagger(t;t_0)$. The situation, however, becomes more complex when the decohering processes, for example, interaction with bath's phonons, is present. This problem can be treated rigorously using special methods (see, e.g., Chapter 3 in Mukamel (1995)), which are beyond the scope of the present consideration because they are somewhat excessive for the problem of finding the macroscopic electromagnetic field when a simple introduction of a phenomenological decay of the macroscopic polarization usually suffices.

The typical structure

$$\hat{O}(t) = U^\dagger(t;t_0)\hat{O}U(t;t_0) \qquad (29.33)$$

appearing in Equations 29.30 and 29.32 defines the *Heisenberg representation* of the field operator $\hat{O}$. Since this representation plays the same role whether the initial state is coherent or not, it is often more convenient to study its time dependence rather than directly the time dependence of the state of the system or its density matrix. It can be easily checked that the time evolution of $\hat{O}(t)$ is governed by the Heisenberg equations of motion.

$$i\hbar\frac{\partial\hat{O}}{\partial t} = \left[\hat{O},\mathcal{H}\right]. \qquad (29.34)$$

Since the explicit time dependence of the Hamiltonian does not change the derivation of the main equations of motion (all operators entering Equation 29.34 are taken at the same instant), we will omit the time argument in what follows.

29.3.1 Electromagnetic Field

We begin the general analysis of the semiconductor optical response from the derivation of the equations of motion of the electromagnetic field. The commutator in this equation is found using the commutation relation for the electromagnetic field in the Coulomb gauge (Cohen-Tannoudji et al. 1992)

$$\left[\hat{A}_j(\mathbf{x}_1),\ \epsilon_0\frac{\partial}{\partial t}\hat{A}_l(\mathbf{x}_2)\right] = \frac{i\hbar\delta_{jl}}{n^2(\mathbf{x}_1)}\hat{\delta}^\perp(\mathbf{x}_1-\mathbf{x}_2), \qquad (29.35)$$

where

n($\mathbf{x}$) is the refractive index, which is allowed to be spatially inhomogeneous

j and l denote the Cartesian coordinates

$\hat{\delta}^{\perp}(\mathbf{x})$ is the transverse δ-function (Cohen-Tannoudji et al. 1992)

As Helmholtz's theorem (Arfken and Weber 2005) states any vector field $\mathbf{V}(\mathbf{x})$ can be presented as the sum of irrotational and solenoidal parts, $\mathbf{U}(\mathbf{x}) = \mathbf{U}_l(\mathbf{x}) + \mathbf{U}_t(\mathbf{x})$ with $\nabla \times \mathbf{U}_l(\mathbf{x}) \equiv 0$ and $\nabla \cdot \mathbf{U}_t(\mathbf{x}) \equiv 0$. The convolution of the vector field with the transverse δ-function returns the value of the solenoidal component $\mathbf{V}_t(\mathbf{x})$ at the point of singularity

$$\int d\mathbf{x}' \hat{\delta}^{\perp}(\mathbf{x} - \mathbf{x}')\mathbf{U}(\mathbf{x}') = \mathbf{U}_t(\mathbf{x}). \tag{29.36}$$

The Hamiltonian of the electromagnetic field has the standard form

$$\mathcal{H}_F = \frac{\epsilon_0}{2} \int d\mathbf{x} \left[n^2(\mathbf{x})E^2(\mathbf{x},t) + c^2 B^2(\mathbf{x},t) \right], \tag{29.37}$$

where the transverse electric and magnetic fields are defined in terms of the vector potential as $\mathbf{E}(\mathbf{x}) = -\partial \mathbf{A}(\mathbf{x})/\partial t$ and $\mathbf{B}(\mathbf{x}) = \nabla \times \mathbf{A}(\mathbf{x})$. Substituting into Equation 29.34 $\mathcal{H} = \mathcal{H}_F + \mathcal{H}_{mat}$, where $\mathcal{H}_{mat}$ is given by Equation 29.29 one obtains the *quantum* equations of motion for the electromagnetic field

$$\frac{n^2(\mathbf{x})}{c^2}\ddot{\hat{\mathbf{A}}} = \nabla^2 \hat{\mathbf{A}} + \mu_0 \sum_{n,m} \int d^3\mathbf{x}' \hat{\delta}^{\perp}(\mathbf{x} - \mathbf{x}')\mathbf{d}_{nm}c_n^{\dagger}(\mathbf{x}')c_m(\mathbf{x}'). \tag{29.38}$$

This equation has the standard form of the inhomogeneous wave equation for the vector potential with the integral in the r.h.s. representing the current source.

Writing Equation 29.38 we have taken into account that $\nabla \cdot \mathbf{A}$ vanishes in the Coulomb gauge used for derivation of Equation 29.27. Since, generally, the vector field $\mathbf{U}(\mathbf{x}) = \mu_0 \sum_{n,m} \mathbf{d}_{nm}c_n^{\dagger}(\mathbf{x})c_m(\mathbf{x})$ is not necessarily solenoidal, the presence of the transverse δ-function in the r.h.s. of Equation 29.38 is important to ensure that at any instant $\hat{\mathbf{A}}(\mathbf{x})$ does not violate the Coulomb gauge. Clearly, if one replaces the transverse δ-function by the usual (local) δ-function it implies effective adding to the source the irrotational component of $\mathbf{U}(\mathbf{x})$. According to Equation 29.36, the convolution of $\mathbf{U}(\mathbf{x})$ with the transverse δ-function is $\int d\mathbf{x}' \tilde{\delta}^{\perp}(\mathbf{x} - \mathbf{x}')\mathbf{U}(\mathbf{x}') = \mathbf{U}(\mathbf{x}) - \mathbf{U}_l(\mathbf{x})$. Equation 29.38 is linear, therefore $\hat{\mathbf{A}}(\mathbf{x})$ can be presented as a superposition of two components, $\hat{\mathbf{A}}(\mathbf{x}) = \tilde{\mathbf{A}}(\mathbf{x}) + \hat{\mathbf{A}}_l(\mathbf{x})$, which satisfy the wave equation of the same form as Equation 29.38, with $\mathbf{U}(\mathbf{x})$ and $-\mathbf{U}_l(\mathbf{x})$ serving as sources. Since $\mathbf{U}_l(\mathbf{x})$ is irrotational it can be presented as a gradient of a scalar function $\mathbf{U}_l(\mathbf{x}) = -\nabla\psi(\mathbf{x})$. Furthermore, following the standard line of arguments as in electrostatics we obtain

$$\varphi(\mathbf{x}) = \frac{1}{4\pi} \int d\mathbf{x}' \frac{\nabla \cdot \mathbf{U}(\mathbf{x}')}{|\mathbf{x} - \mathbf{x}'|}. \tag{29.39}$$

Thus $\varphi(\mathbf{x})$ has the same form as the potential created by the charge spatially distributed with the density $\epsilon_0 \nabla \cdot \mathbf{U}(\mathbf{x})$. If $\hat{\tilde{\mathbf{A}}}(\mathbf{x})$ satisfies the same initial conditions as $\hat{\mathbf{A}}(\mathbf{x})$ the component $\hat{\mathbf{A}}_l(\mathbf{x})$ is early shown to be irrotational and, hence, it does not contribute to the outgoing radiation field but leads to a modification of the electron–electron interaction. This modification identically vanishes in many important physical situations, such as the excitation by the normally incident wave at a frequency tuned to the response with the heavy-hole excitons. While a rigorous study of the effect of $\hat{\mathbf{A}}_l(\mathbf{x})$ on the semiconductor dynamics is yet to be conducted, it can be conjectured that for the problems which will be studied below the respective correction of the electron–electron interaction is rather small and can be neglected; in other words one can approximate $\hat{\mathbf{A}}(\mathbf{x}) \approx \hat{\tilde{\mathbf{A}}}(\mathbf{x})$. This significantly simplifies Equation 29.38 since we can use the conventional δ-function instead of $\hat{\delta}^{\perp}(\mathbf{x} - \mathbf{x}')$ and perform the integration over $\mathbf{x}'$, obtaining

$$\frac{n^2(\mathbf{x})}{c^2}\ddot{\hat{\mathbf{A}}}(\mathbf{x}) = \nabla^2 \hat{\mathbf{A}}(\mathbf{x}) + \mu_0 \sum_{n,m} \mathbf{d}_{nm}c_n^{\dagger}(\mathbf{x})c_m(\mathbf{x}). \tag{29.40}$$

This equation completely describes the electromagnetic field driven by the semiconductor excitations. The macroscopic field, as has been discussed above, is found by averaging $\hat{\mathbf{A}}(\mathbf{x})$ over the state of the combined semiconductor-field system. It should be noted, however, that Equation 29.40 can also be used in a more general context to find higher-order moments of the field, for example, single- or two-photon density matrix and so on.

29.3.2 Semiconductor Dynamics

The current, which enters the equation of motion of the electromagnetic field as a source, is determined by the operator of the interband polarization $c_n^{\dagger}(\mathbf{x}_1)c_m(\mathbf{x}_2)$ taken at the diagonal $\mathbf{x}_1 = \mathbf{x}_2$. Its time evolution is governed by the Heisenberg equation of motion (29.34). The electron operators obey the canonical anti-commutation relation.

$$\left\{ c_n^{\dagger}(\mathbf{x}_1), c_m(\mathbf{x}_2) \right\} = \hbar\delta_{n,m}\delta(\mathbf{x}_1 - \mathbf{x}_2), \tag{29.41}$$

where $\{A, B\} = AB + BA$. The commutator $\left[c_n^{\dagger}(\mathbf{x}_1)c_m(\mathbf{x}_2), \mathcal{H} \right]$ is expressed in terms of the anticommutators using the relation, which holds for any four operators

$$[AB, CD] = A\{B,C\}D - \{A,C\}BD + CA\{B,D\} - C\{A,D\}B. \tag{29.42}$$

Substituting $\mathcal{H} = \mathcal{H}_F + \mathcal{H}_{mat}$ we obtain

$$
i\hbar \frac{\partial}{\partial t} c_n^\dagger(\mathbf{x}_1) c_m(\mathbf{x}_2) = \left[\epsilon_m(\mathbf{x}_2) - \epsilon_n(\mathbf{x}_1) \right] c_n^\dagger(\mathbf{x}_1) c_m(\mathbf{x}_2)
$$
$$
+ \sum_{j,k} \left[\delta_{km} \nabla_1 \cdot \frac{\hbar^2}{2M_{jn}} \cdot \nabla_1 - \delta_{jn} \nabla_2 \cdot \frac{\hbar^2}{2M_{mk}} \cdot \nabla_2 \right.
$$
$$
\left. + \delta_{mk} \hat{\mathbf{A}}(\mathbf{x}_1) \cdot \mathbf{d}_{jn} - \delta_{jn} \hat{\mathbf{A}}(\mathbf{x}_2) \cdot \mathbf{d}_{mk} \right] c_j^\dagger(\mathbf{x}_1) c_k(\mathbf{x}_2)
$$
$$
+ c_n^\dagger(\mathbf{x}_1) \sum_j \int d\mathbf{x}' \left[V(\mathbf{x}_2 - \mathbf{x}') - V(\mathbf{x}_1 - \mathbf{x}') \right]
$$
$$
\times c_j^\dagger(\mathbf{x}') c_j(\mathbf{x}') c_m(\mathbf{x}_2) \tag{29.43}
$$

where indices at ∇'s denote the number of the spatial variables they act on. The form of Equation 29.43 shows why it is not enough to consider directly the time dependence of the operator $c_n^\dagger(\mathbf{x}_1) c_m(\mathbf{x}_2)$ at the diagonal $\mathbf{x}_1 = \mathbf{x}_2$. Because of the presence of the spatial derivatives the rate of change of the operator at the diagonal is determined by the value the operator takes at the neighboring points requiring, thereby, studying the general case $\mathbf{x}_1 \neq \mathbf{x}_2$.

Equation 29.43 is presented in a general form unifying the conduction and the valence bands. This makes a constructive analysis difficult because the excitations belonging to different bands have distinctively different properties, for example, the mass of excitations in the conduction band (electrons) is positive while in the valence band (holes) it is negative. In order to account for these differences it is convenient to introduce special field operators for holes. It is done following the simple rule—creation of an electron in the valence band corresponds to annihilation of a hole and vice versa. As has been discussed above, different subbands in the valence and the conduction bands are enumerated by the projection of the angular momentum ($j = 1/2$ and $j = 3/2$ for the conduction and the valence bands, respectively). Thus, we introduce the hole operators according to $\upsilon_\sigma^\dagger(\mathbf{x}) \equiv c_{\upsilon,\sigma}(\mathbf{x})$ and $\upsilon_\sigma(\mathbf{x}) \equiv c_{\upsilon,\sigma}^\dagger(\mathbf{x})$. For the electrons in the conduction band we leave the same notation but will keep only the index corresponding to the spin of the electron so that $c_s^\dagger(\mathbf{x}) \equiv c_{c,s}^\dagger(\mathbf{x})$.

Rewriting Equation 29.43 in terms of the electron and hole operators is straightforward but is not free from the formal difficulties related to the fact that the holes exist on the background of all electrons filling the valence band. In a sense, in order to create a hole, one first needs to fill completely the valence band and then remove one electron. This naturally leads to divergences of the Coulomb energy. In order to demonstrate one kind of divergence let us consider the Coulomb term, which appears at the r.h.s. of Equation 29.43 for $n = \{c, s_1\}$, $m = \{c, s_s\}$ and $j = \{\upsilon, \sigma\}$

$$
c_{s_1}^\dagger(\mathbf{x}_1) \int d\mathbf{x}' \left[V(\mathbf{x}_2 - \mathbf{x}') - V(\mathbf{x}_1 - \mathbf{x}') \right] \upsilon_\sigma(\mathbf{x}') \upsilon_\sigma^\dagger(\mathbf{x}') c_{s_1}(\mathbf{x}_2). \tag{29.44}
$$

If one tries to rewrite this expression in the normal form, that is, with all the creation operators collected on the left, one gets an additional term $\propto \{\upsilon_\sigma(\mathbf{x}'), \upsilon_\sigma^\dagger(\mathbf{x}')\} = \hbar\delta(\mathbf{x}'-\mathbf{x}')$, which is infinite if not meaningless. There are different ways to work around this divergence. One of them is to consider Equation 29.44 as a continuous limit of the respective model formulated in a discrete space, where

the field operators taken at different cells of the space anticommute to Kronecker's δ so that $\{\upsilon_{\sigma,\mathbf{x}}, \upsilon_{\sigma,\mathbf{x}'}^\dagger\} = N\delta_{\mathbf{x},\mathbf{x}'}$ with some constant N. This gives for the term proportional to the anticommutator

$$
c_{s_1,\mathbf{x}_1}^\dagger c_{s_1,\mathbf{x}_2} N \sum_{\mathbf{x}'} \left[V_{\mathbf{x}_2 - \mathbf{x}'} - V_{\mathbf{x}_1 - \mathbf{x}'} \right]. \tag{29.45}
$$

Truncating the Coulomb potential $V_\mathbf{x}$ at infinity we see that this term vanishes regardless of the parameters of the truncation and the discretization. Thus, we can conclude that these divergencies do not contribute to the equations of motion.

A more different obstacle is the divergent band-gap renormalization, which enters the equations of motion in the form $\epsilon_\sigma \to \epsilon_\sigma + V(0)$. This divergence is of the same origin as the previous one but it cannot be discarded as easily. The reason is that we have introduced the band edges as the solutions of a single-electron problem, while the hole sees the band edges in the presence of all the electrons filling the valence band minus one. Therefore, the renormalization is a real physical effect. We incorporate it in the theory in a similar way as the elementary-charge renormalization is dealt with in quantum field theory. We will consider from now on that the hole energy offset $\epsilon_\sigma(\mathbf{x})$ in the equations of motion is the genuine one. This can be done because the actual band gap between the conduction and valence bands is directly observable (see, e.g., Equation 29.56).

Taking into account these considerations we obtain

$$
i\hbar \frac{\partial}{\partial t} \upsilon_\sigma^\dagger(\mathbf{x}_1) c_s^\dagger(\mathbf{x}_2)
$$
$$
= \upsilon_\sigma^\dagger(\mathbf{x}_1) c_s^\dagger(\mathbf{x}_2) \left[\epsilon_\sigma(\mathbf{x}_1) + \hat{U}_\Delta(\mathbf{x}_1) - \epsilon_s(\mathbf{x}_2) - \hat{U}_\Delta(\mathbf{x}_2) + V(\mathbf{x}_1 - \mathbf{x}_2) \right]
$$
$$
+ \frac{\hbar^2}{2} \left(\frac{1}{m_e} \nabla_2^2 + \sum_{\sigma'} \nabla_1 \cdot \frac{1}{M_{\sigma,\sigma'}} \cdot \nabla_1 \right) \times \upsilon_{\sigma'}^\dagger(\mathbf{x}_1) c_s^\dagger(\mathbf{x}_2)
$$
$$
- \hat{\mathbf{A}}(\mathbf{x}_1) \cdot \mathbf{d}_{\sigma,s} \delta(\mathbf{x}_1 - \mathbf{x}_2) + \sum_{s'} \hat{\mathbf{A}}(\mathbf{x}_1) \cdot \mathbf{d}_{\sigma,s'} c_s^\dagger(\mathbf{x}_2) c_{s'}(\mathbf{x}_1)
$$
$$
+ \sum_{\sigma'} \hat{\mathbf{A}}(\mathbf{x}_2) \cdot \mathbf{d}_{\sigma',s} \upsilon_{\sigma'}^\dagger(\mathbf{x}_1) \upsilon_\sigma(\mathbf{x}_2),
$$

$$
i\hbar \frac{\partial}{\partial t} c_{s_1}^\dagger(\mathbf{x}_1) c_{s_2}(\mathbf{x}_2)
$$
$$
= c_{s_1}^\dagger(\mathbf{x}_1) \left[\epsilon_{s_2}(\mathbf{x}_2) + \hat{U}_\Delta(\mathbf{x}_2) - \epsilon_{s_1}(\mathbf{x}_1) - \hat{U}_\Delta(\mathbf{x}_1) \right] c_{s_2}(\mathbf{x}_2)
$$
$$
+ \frac{\hbar^2}{2m_e} (\nabla_1^2 - \nabla_2^2) c_{s_1}^\dagger(\mathbf{x}_1) c_{s_2}(\mathbf{x}_2)
$$
$$
+ \sum_{\sigma'} \left[\hat{\mathbf{A}}(\mathbf{x}_2) \cdot \mathbf{d}_{s_2,\sigma'} \upsilon_{\sigma'}^\dagger(\mathbf{x}_2) c_{s_1}^\dagger(\mathbf{x}_1) - \hat{\mathbf{A}}(\mathbf{x}_1) \cdot \mathbf{d}_{\sigma',s_1} c_{s_2}(\mathbf{x}_2) \upsilon_{\sigma'}(\mathbf{x}_1) \right],
$$

$$
i\hbar \frac{\partial}{\partial t} \upsilon_{\sigma_2}^\dagger(\mathbf{x}_2) \upsilon_{\sigma_1}(\mathbf{x}_1)
$$
$$
= \upsilon_{\sigma_2}^\dagger(\mathbf{x}_2) \left[\epsilon_{\sigma_2}(\mathbf{x}_2) + \hat{U}_\Delta(\mathbf{x}_2) - \epsilon_{\sigma_1}(\mathbf{x}_1) - \hat{U}_\Delta(\mathbf{x}_1) \right] \upsilon_{\sigma_1}(\mathbf{x}_1)
$$
$$
+ \frac{\hbar^2}{2} \sum_{\sigma',\sigma''} \left(\nabla_1 \cdot \frac{\delta_{\sigma'',\sigma_2}}{M_{\sigma',\sigma_1}} \cdot \nabla_1 - \nabla_2 \cdot \frac{\delta_{\sigma',\sigma_1}}{M_{\sigma_2,\sigma''}} \cdot \nabla_2 \right) \upsilon_{\sigma''}^\dagger(\mathbf{x}_2) \upsilon_{\sigma'}(\mathbf{x}_1)
$$
$$
+ \sum_{s'} \left[\hat{\mathbf{A}}(\mathbf{x}_1) \cdot \mathbf{d}_{s',\sigma_1} \upsilon_{\sigma_2}^\dagger(\mathbf{x}_2) c_{s'}^\dagger(\mathbf{x}_1) - \hat{\mathbf{A}}(\mathbf{x}_2) \cdot \mathbf{d}_{\sigma_2,s'} c_{s'}(\mathbf{x}_2) \upsilon_{\sigma_1}(\mathbf{x}_1) \right],
$$

$$
\tag{29.46}
$$

where

$$\hat{U}_\Delta(\mathbf{x}) = \int d\mathbf{x}' V(\mathbf{x} - \mathbf{x}') \left[\sum_{s'} c_{s'}^\dagger(\mathbf{x}') c_{s'}(\mathbf{x}') - \sum_{\sigma'} \upsilon_{\sigma'}^\dagger(\mathbf{x}') \upsilon_{\sigma'}(\mathbf{x}') \right].$$

(29.47)

Equations 29.40 and 29.46 describe the coupled dynamics of the electromagnetic field and the semiconductor excitations in the endless variety of physical situations. They can be used for studying the modification of the radiative decay of the semiconductor excitations in photonic crystals (Sakoda 2005) or the speckle pattern caused by electron scattering on surface inhomogeneities of quantum wells and so on. In what follows we will be mostly interested in the effect of the many-body correlations on the nonlinear optical response and, therefore, we make the proper simplifications of the problem. We assume that the life time of a photon in the structure is not too long so that the polaritons, which are the coupled states of the photons and the semiconductor excitations, are not formed. In this case one can neglect the reabsorption of the photons emitted in the course of the radiative decay. This significantly simplifies the analysis of Equations 29.40 and 29.46 because now we can consider $\mathbf{A}(\mathbf{x})$ in Equation 29.46 as the external *classical* excitation field, solve Equation 29.46 for $c_n^\dagger(\mathbf{x}_1) c_m(\mathbf{x}_2)$, and then find the field produced by the semiconductor excitation using Equation 29.40.

29.3.3 Rotating Wave Approximation

The interaction of light with the semiconductor has a strong resonant character when the frequency of the excitation field, Ω, sweeps the vicinity of the gap, $\hbar\Omega \sim \Delta$. In order to see how the resonances show up in the equations of motion we consider Equations 29.46 in the linear approximation when we need only the equation with respect to the operator of the interband polarization

$$i\hbar \frac{\partial}{\partial t} \upsilon_\sigma^\dagger(\mathbf{x}_2) c_s^\dagger(\mathbf{x}_1) = -\hat{L}_{\sigma,s} \upsilon_\sigma^\dagger(\mathbf{x}_2) c_s^\dagger(\mathbf{x}_1) - \mathbf{A}(\mathbf{x}_1) \cdot \mathbf{d}_{\sigma,s} \delta(\mathbf{x}_1 - \mathbf{x}_2),$$

(29.48)

where the operator $\hat{L}_{\sigma,s}$ is defined as

$$\hat{L}_{\sigma,s} = \epsilon_s(\mathbf{x}_1) - \epsilon_\sigma(\mathbf{x}_2) - V(\mathbf{x}_1 - \mathbf{x}_2) - \frac{\hbar^2}{2} \left(\frac{1}{m_e} \nabla_1^2 + \nabla_2 \cdot \frac{1}{\mathsf{M}_\sigma} \cdot \nabla^2 \right).$$

(29.49)

We have also neglected the heavy-hole–light-hole mixing setting $\mathsf{M}_{\sigma',\sigma} = \delta_{\sigma',\sigma} \mathsf{M}_\sigma$. This does not make principle changes in the character of the following consideration, but greatly simplifies the notation.

Next we employ the spectral representation of the operator $\hat{L}_{\sigma,s}$

$$\hat{L}_{\sigma,s} f(\mathbf{x}_2, \mathbf{x}_1) = \sum_\mu E_\mu \phi_\mu^*(\mathbf{x}_2, \mathbf{x}_1) \int d\mathbf{x}_1' d\mathbf{x}_2' \, \phi_\mu(\mathbf{x}_2', \mathbf{x}_1') f(\mathbf{x}_2', \mathbf{x}_1'),$$

(29.50)

where the formal summation over μ implies summing over the discrete part and integrating over the continuous part of the spectrum of the operator $\hat{L}_{\sigma,s}$, and E_μ and $\phi_\mu(\mathbf{x}_2, \mathbf{x}_1)$ are, respectively, the eigenvalues and the eigenfunctions, $\hat{L}_{\sigma,s}\phi_\mu = E_\mu\phi_\mu$. The operator $\hat{L}_{\sigma,s}$ is (formally) self-adjoint and, hence, its eigenvalues are real and the eigenfunctions form a complete orthonormal set.

Using Equation 29.50 in Equation 29.48 and convoluting both sides of Equation 29.48 with ϕ_μ^* we obtain the ordinary differential equation

$$-i\hbar \frac{\partial}{\partial t} \hat{P}_\mu^\dagger(t) = E_\mu \hat{P}_\mu^\dagger(t) + \mathcal{A}_\mu(t),$$

(29.51)

where we have introduced

$$\hat{P}_\mu^\dagger(t) = \int d\mathbf{x}_1 \, d\mathbf{x}_2 \, \phi_\mu(\mathbf{x}_2, \mathbf{x}_1) \upsilon_\sigma^\dagger(\mathbf{x}_2) c_s^\dagger(\mathbf{x}_1)$$

(29.52)

and

$$\mathcal{A}_\mu(t) = \int d\mathbf{x} \, \phi_\mu(\mathbf{x}, \mathbf{x}) \mathbf{A}(\mathbf{x}, t) \cdot \mathbf{d}_{\sigma,s}.$$

(29.53)

The solution of Equation 29.51 is written as the sum of the general solution of the homogeneous equation (with $\mathcal{A}_\mu(t) \equiv 0$) and the partial solution of inhomogeneous

$$\hat{P}_\mu^\dagger(t) = \hat{P}_\mu^\dagger(0) e^{iE_\mu t/\hbar} + \frac{i}{\hbar} \int_0^t dt' \, e^{iE_\mu(t-t')/\hbar} \mathcal{A}_\mu(t').$$

(29.54)

The first term is determined by the initial value of the operators $\upsilon_\sigma^\dagger(\mathbf{x}_2) c_s^\dagger(\mathbf{x}_1)$ or, according to the definition of the Heisenberg representation (see Equation 29.33), by their bare values. If, for example, initially the semiconductor is in the vacuum state (empty conduction band and filled valence band) the average value of the first term vanishes (see Equation 29.30) and, hence, it does not contribute to the average interband polarization and, accordingly, to the macroscopic electromagnetic field, which, as follows from Equation 29.40, is driven by $\langle \upsilon_\sigma^\dagger(\mathbf{x}) c_s^\dagger(\mathbf{x}) \rangle$.

In order to discuss the effect of the external field, we first assume that the external field changes harmonically with time

$$\mathbf{A}(\mathbf{x}, t) = \mathbf{A}(\mathbf{x}, \Omega) e^{-i\Omega t} + \mathbf{A}^*(\mathbf{x}, \Omega) e^{i\Omega t},$$

(29.55)

which yields

$$\langle 0 | \hat{P}_{\sigma,s}^{(\mu)}(t) | 0 \rangle = \mathcal{A}_{\sigma,s}^{(\mu)}(\Omega) \frac{e^{iE_\mu t/\hbar} - e^{-i\Omega t}}{E_\mu + \hbar\Omega} + \bar{\mathcal{A}}_{\sigma,s}^{(\mu)}(\Omega) \frac{e^{iE_\mu t/\hbar} - e^{i\Omega t}}{E_\mu - \hbar\Omega},$$

(29.56)

where $\mathcal{A}_{\sigma,s}^{(\mu)}(\Omega)$ and $\bar{\mathcal{A}}_{\sigma,s}^{(\mu)}(\Omega)$ are found using in Equation 29.53 $\mathbf{A}(\mathbf{x}, \Omega)$ and $\mathbf{A}^*(\mathbf{x}, \Omega)$, respectively. If $\Omega \approx E_\mu/\hbar$, then at the time scale $(\Omega + E_\mu/\hbar)^{-1} \ll t \ll |\Omega - E_\mu/\hbar|^{-1}$ the second term in Equation 29.56 increases linearly with time reflecting the resonant semiconductor response and eventually prevails over the first term, so that one can neglect the nonresonant contribution. This constitutes the *rotating wave approximation* widely used in studying the semiconductor optical response. For estimates we note that the smallest value E_μ is determined by the width of the gap, $\Delta = \epsilon_s - \epsilon_\sigma$, minus binding energy of a possible bound state. The value of the binding energy cannot be expected to be significant compared to Δ and thus we can take $E_\mu \approx \Delta$. In GaAs $\Delta \approx 1.5\,\mathrm{eV}$ yielding the characteristic time scale

$$T_R = \hbar\pi/\Delta \approx 1\,\mathrm{fs}. \qquad (29.57)$$

Thus, if the excitation field is approximately tuned into the resonance with the semiconductor gap and the duration of the excitation pulse significantly exceeds 1 fs, one can use the resonant approximation and neglect the nonresonant contribution. The reverse, however, should also be mentioned. The experiments with pulses with duration of few femtoseconds are readily available now and one should be careful in applying the resonant approximation for the description of such experiments.

Next we consider a more general case wherein the external field changes harmonically with slowly varying amplitude, for example, when the excitation field is a pulse of finite duration much longer than T_R, so that the temporal Fourier transform is localized in a vicinity of Δ with both detuning $|\Omega - \Delta/\hbar|$ and the width of the distribution much smaller than Δ. Obviously, in this case we can again separate the resonant and nonresonant contributions. Indeed, this is true for each major harmonic and, hence, for the whole excitation pulse.

This motivates taking advantage of the resonance condition on the level of the equations of motion. This can be easily done. Let the external field be of the form

$$\mathbf{A}(\mathbf{x}, t) = \mathbf{A}_\Omega(\mathbf{x}, t) e^{-i\Omega t} + \mathrm{c.c.}, \qquad (29.58)$$

where $\mathbf{A}_\Omega(\mathbf{x}, t)$ is a function slowly changing with time. Following Equation 29.54, we use in Equation 29.46 the ansatz

$$\upsilon_\sigma(\mathbf{x}) \rightarrow e^{-i\Omega t/2} \upsilon_\sigma(\mathbf{x}), \quad c_s(\mathbf{x}) \rightarrow e^{-i\Omega t/2} c_s(\mathbf{x}). \qquad (29.59)$$

After canceling the factor $e^{i\Omega t}$ in the equation with respect to the operator of the interband polarization there will be two type of terms explicitly depending on time in the r.h.s. of Equation 29.46. The terms of the first type oscillate much faster than any typical time scale for the operators defined in Equation 29.59 and their contribution is small on the time scale $t > T_R$. Neglecting such terms corresponds to the rotating wave approximation and defining the operators according to Equation 29.59 is called the transition to the *rotating frame*.

We do not provide the form of the semiconductor equations of motion in the rotating frame but we would like to note that the width of the gap between the conduction and the valence bands in the equation for the operator of the interband polarization is renormalized by the frequency of the external field $\epsilon_s - \epsilon_\sigma \rightarrow \epsilon_s - \epsilon_\sigma - \hbar\Omega$. These equations can be derived as the Heisenberg equations of motion from the Hamiltonian $\mathcal{H} = \mathcal{H}_{e-h} + \mathcal{H}_{exc}$, where $\mathcal{H}_{e-h}$ is the electron–hole Hamiltonian written in the rotating frame

$$
\begin{aligned}
\mathcal{H}_{e-h} = & \sum_s \int d\mathbf{x}\, c_s^\dagger(\mathbf{x}) \left[\epsilon_s(\mathbf{x}) - \hbar\Omega/2 - \frac{\hbar^2}{2m_e}\nabla^2 \right] c_s(\mathbf{x}) \\
& - \sum_{\sigma,\sigma'} \int d\mathbf{x}\, \upsilon_\sigma^\dagger(\mathbf{x}) \left[\delta_{\sigma,\sigma'} \epsilon_\sigma(\mathbf{x}) - \delta_{\sigma,\sigma'}\hbar\Omega/2 - \nabla \cdot \frac{\hbar^2}{2M_{\sigma,\sigma'}} \cdot \nabla \right] \upsilon_{\sigma'}(\mathbf{x}) \\
& + \frac{1}{2}\sum_{s,s'} \int d\mathbf{x}_1 d\mathbf{x}_2\, c_s^\dagger(\mathbf{x}_1) c_{s'}^\dagger(\mathbf{x}_2) V(\mathbf{x}_1 - \mathbf{x}_2) c_{s'}(\mathbf{x}_2) c_s(\mathbf{x}_1) \\
& + \frac{1}{2}\sum_{\sigma,\sigma'} \int d\mathbf{x}_1 d\mathbf{x}_2\, \upsilon_\sigma^\dagger(\mathbf{x}_1) \upsilon_{\sigma'}^\dagger(\mathbf{x}_2) V(\mathbf{x}_1 - \mathbf{x}_2) \upsilon_{\sigma'}(\mathbf{x}_2) \upsilon_\sigma(\mathbf{x}_1) \\
& - \sum_{s,\sigma} \int d\mathbf{x}_1 d\mathbf{x}_2\, \upsilon_\sigma^\dagger(\mathbf{x}_1) c_s^\dagger(\mathbf{x}_2) V(\mathbf{x}_1 - \mathbf{x}_2) c_s(\mathbf{x}_2) \upsilon_\sigma(\mathbf{x}_1) \qquad (29.60)
\end{aligned}
$$

and $\mathcal{H}_{exc}$ is the Hamiltonian of interaction with the electromagnetic field in the rotating approximation

$$\mathcal{H}_{exc} = \sum_{s,\sigma} \int d\mathbf{x}[\mathbf{A}_\Omega(\mathbf{x}, t) \cdot \mathbf{d}_{s,\sigma}\upsilon_\sigma^\dagger(\mathbf{x}) c_s^\dagger(\mathbf{x}) + \mathbf{A}_\Omega^\star(\mathbf{x}, t) \cdot \mathbf{d}_{\sigma,s} c_s(\mathbf{x})\upsilon_\sigma(\mathbf{x})],$$

$$(29.61)$$

where $\mathbf{A}_\Omega(\mathbf{x}, t)$ is the amplitude slowly changing with time. It is worth noting that Hamiltonian (29.60) reflects that holes and electrons have opposite charges—the Coulomb interaction between electrons and between holes increases the energy and, hence, leads to repulsion while the Coulomb interaction between the electrons and holes decreases the energy and corresponds to attraction.

While we derived Equation 29.61, taking into account only external classical excitation field, it can be generalized to describe the interaction of the semiconductor with the quantized electromagnetic field in the rotating wave approximation. For this one should represent $\hat{\mathbf{A}}(\mathbf{x})$ in terms of the photon creation and annihilation operators and use in Equation 29.61 instead of $\mathbf{A}_\Omega(\mathbf{x}, t)$ the part containing only the annihilation operators and instead of $\mathbf{A}_\Omega^\star(\mathbf{x}, t)$ the part with the photon creation operators. This illustrates an evident but important feature of the light–semiconductor interaction. The process of absorption of a photon is accompanied with the creation of the electron–hole pair and the emission results in destruction of a pair, so that the *operator* of the difference between the total numbers of the electrons and holes is a constant of motion

$$\int d\mathbf{x} \left[\sum_s c_s^\dagger(\mathbf{x})c_s(\mathbf{x}) - \sum_s \upsilon_\sigma^\dagger(\mathbf{x})\upsilon_\sigma(\mathbf{x}), \mathcal{H} \right] = 0. \qquad (29.62)$$

29.4 Perturbation Theory: Four-Wave Mixing Response

The best understood way of analyzing the semiconductor non-linear response is the perturbation theory with respect to the external excitation field because it allows for a general model independent formulation. For example, in Ostreich et al. (1998) the powerful formalism of Hubbard operators is used, which, as has been demonstrated in Ostreich (2001), allows a relatively straightforward extension to higher orders. Here we choose a more traditional approach developed in the standard $T = 0$ quantum field theory. The derivation of the main equations, however, significantly benefits from the idea of describing the semiconductor excitations in terms of genuine eigenstates borrowed from Ostreich et al. (1998).

We limit our consideration to the case when there is no external static field, i.e., $U_{ext}(\mathbf{x}) \equiv 0$, and the spatial variations of the band edges correspond to a single quantum well grown along z direction, $\epsilon_n(\mathbf{x}) = \epsilon_n + \Delta\epsilon_n(z)$. As has been discussed above the states reached in the course of the resonance excitation of the semiconductor are characterized by the definite numbers of the electron–hole pairs. The simplest states contain one such pair and can be created by acting on vacuum with the operator

$$\hat{B}_\mu^\dagger = \sum_{\sigma_\mu, s_\mu} \int d\mathbf{x}_1 d\mathbf{x}_2 \phi_\mu(\mathbf{x}_1, \mathbf{x}_2) \upsilon_{\sigma_\mu}^\dagger(\mathbf{x}_1) c_{s_\mu}^\dagger(\mathbf{x}_2), \qquad (29.63)$$

where the index μ enumerates different states of the single electron–hole pair. The function $\phi_\mu(\mathbf{x}_1, \mathbf{x}_2)$ is chosen in such a way that the state $|\mu\rangle = \hat{B}_\mu^\dagger |0\rangle$ is the eigenstate of the semiconductor Hamiltonian without the excitation field

$$\mathcal{H}_{e-h} |\mu\rangle = E_\mu |\mu\rangle. \qquad (29.64)$$

Substituting Equations 29.60 and 29.63 into this equation, we find that $\phi_\mu(\mathbf{x}_1, \mathbf{x}_2)$ satisfies the differential equation, which has the familiar structure (compare with Equation 29.48)

$$E_\mu \phi_\mu(\mathbf{x}_1, \mathbf{x}_2) = \left[\epsilon_{s_\mu}(\mathbf{x}_2) - \epsilon_{\sigma_\mu}(\mathbf{x}_1) - \hbar\Omega \right.$$
$$\left. - \frac{\hbar}{2m_c}\nabla_2^2 - \nabla_1 \cdot \frac{\hbar^2}{2M_{\sigma',\sigma_\mu}} \cdot \nabla_1 - V(\mathbf{x}_2 - \mathbf{x}_1) \right] \phi_\mu(\mathbf{x}_1, \mathbf{x}_2). \qquad (29.65)$$

This equation has the meaning of the stationary Schrödinger equation for two particles with opposite charges (the hole at point $\mathbf{x}_1$ and the electron at the point $\mathbf{x}_2$) moving in potentials $\epsilon_{\sigma_\mu}(\mathbf{x}_1)$ and $\epsilon_{s_\mu}(\mathbf{x}_2)$. Thus, $\phi_\mu(\mathbf{x}_1, \mathbf{x}_2)$ has the physical meaning of the electron–hole wave function corresponding to the particular

stationary state. For a general form of potentials $\epsilon_{\sigma_\mu}(\mathbf{x}_1)$ and $\epsilon_{s_\mu}(\mathbf{x}_2)$. Equation 29.65 cannot be solved exactly and different approximate methods are developed for investigating the specific features of the solutions. We discuss the general properties of solutions of Equation 29.65 adopting the δ-functional approximation for the quantum well. In the framework of this approximation, the support of electron–hole wave function, that is, the region where the function is not identically zero along z direction is assumed to be limited by the position of the quantum well. Additionally, we neglect the off-diagonal interband elements and the anisotropy of the tensor of the holes effective mass, so we use $M_{\sigma',\sigma_\mu} = 1\delta_{\sigma',\sigma_\mu} M_{\sigma_\mu}$, where 1 is the unit tensor. This corresponds to neglecting the heavy-hole–light-hole mixing, which is valid if the hole momentum in plane of the quantum well is not too high (Zhu and Huang 1987). Under these approximations Equation 29.65 turns into the well-studied (and yet nontrivial (Parfitt and Portnoi 2002)) problem of the two-dimensional hydrogen atom.

The spectrum consists of the discrete and the continuous part. The discrete part corresponds to the bound electron–hole states, which are called *excitons*. Often this term is reserved for the bound states, however, often it is convenient to treat bound and unbound states on the equal ground and, therefore, the term "exciton" is also used as the short synonym of "electron–hole pair" regardless whether they are bound or not. From the perspective of Equation 29.64 operators $B_\mu^\dagger$ with the respective kernels $\phi_\mu(\mathbf{x}_1, \mathbf{x}_2)$ acting on vacuum create the exciton and thus $B_\mu^\dagger$ have the meaning of the exciton creation operators.

Because of the translational invariance of the r.h.s. of Equation 29.65 in the plane of the quantum well, the wave function admits the presentation

$$\phi_\mu(\mathbf{x}_1, \mathbf{x}_2) = e^{i\mathbf{K}_\mu \cdot \mathbf{R}} \tilde{\phi}_\mu(\mathbf{x}_1 - \mathbf{x}_2), \qquad (29.66)$$

where
  $\mathbf{K}_\mu$ is the momentum
  $\mathbf{R} = (M_{\sigma_\mu}\mathbf{x}_1 + m_e\mathbf{x}_2)/(M_{\sigma_\mu} + m_e)$ is the position of the center of mass of the pair

Similarly to the three-dimensional case, the eigenstates of the two-dimensional hydrogen atom are enumerated by the main quantum number n and the angular momentum ℓ. Thus, in this case, the index μ in Equation 29.63 combines all the quantum numbers characterizing the specific exciton state $\mu = \{\mathbf{K}_\mu, s_\mu, \sigma_\mu, n_\mu, \ell_\mu\}$. The important property of the eigenstates of the two-dimensional hydrogen atom is that the wave function corresponding to the states with $\ell \neq 0$ vanishes at $\mathbf{x}_1 = \mathbf{x}_2$. Thus, as follows from Equation 29.61, these states do not contribute to the coupling with the electromagnetic field. Among the states with $\ell = 0$ the most important one is the state with the lowest energy, whose wave function has the form

$$\tilde{\phi}_\mu(\mathbf{x}) = \sqrt{\frac{2}{\pi r_\mu^2}} e^{-|\mathbf{x}|/r_\mu}, \qquad (29.67)$$

where the Bohr radius of the two-dimensional exciton is given by $r_\mu = \epsilon_b/2m_{\sigma_\mu}e^2$ (notice that it is two times smaller than in three dimensions) with m_{σ_μ} being the electron–hole-reduced mass, $1/m_{\sigma_\mu} = 1/m_e + 1/M_{\sigma_\mu}$.

Using solutions of Equation 29.65 we can incorporate the single-pair states into the theory. The full set of solutions is complete

$$\sum_\mu \phi_\mu^*(\mathbf{x}_1, \mathbf{x}_2)\phi_\mu(\mathbf{x}_1', \mathbf{x}_2') = \delta(\mathbf{x}_1 - \mathbf{x}_1')\delta(\mathbf{x}_2 - \mathbf{x}_2'), \qquad (29.68)$$

where summation over μ implies summation over the discrete quantum numbers (e.g., ℓ and n for hydrogen-like bound excitons) and integration over continuous ones (such as the center of mass momentum $\mathbf{K}_\mu$). Relation (29.68) allows us to express any electron–hole operator in terms of the exciton operators, for example

$$\upsilon_\sigma^\dagger(\mathbf{x})c_s^\dagger(\mathbf{y}) = \sum_{\mu|s_\mu = s, \sigma_\mu = \sigma} \phi_\mu^*(\mathbf{x}, \mathbf{y})\hat{B}_\mu^\dagger \qquad (29.69)$$

and represent in terms of the exciton operators the Hamiltonian of the interaction of the semiconductor, Equation 29.61, with the electromagnetic field

$$\mathcal{H}_{exc} = \sum_\mu [\mathcal{A}_\mu^*(t)\hat{B}_\mu + \mathcal{A}_\mu(t)\hat{B}_\mu^\dagger], \qquad (29.70)$$

where

$$\mathcal{A}_\mu(t) = \int d\mathbf{x}\, \mathbf{A}_\Omega(\mathbf{x}, t) \cdot \mathbf{d}_{s_\mu, \sigma_\mu} \phi_\mu^*(\mathbf{x}, \mathbf{x}). \qquad (29.71)$$

The similarity between Equations 29.63 and 29.71 and Equations 29.52 and 29.53 is not accidental. If we act by the first equation in Equations 29.46 on vacuum $|0\rangle$ in the case of absent external field, we obtain the nonstationary Schrödinger equation for a single-pair state, so the operator $\hat{L}_{\sigma,s}$ defined in Equation 29.49 has the meaning of the single-pair Hamiltonian. The spectral representation of $\hat{L}_{\sigma,s}$ allows projecting on the eigenstates of the single-pair Hamiltonian, which are analogous to the projections on the genuine single-particle states defined by Equations 29.63 and 29.71. This demonstrates that the procedure shown above can be generalized to go beyond the adopted approximations.

The interaction Hamiltonian $\mathcal{H}_{exc}$ is considered as a perturbation and the theory is then built following the standard lines. First we account for the nonperturbed dynamics exactly introducing

$$|\tilde{\Phi}(t)\rangle = e^{i\mathcal{H}_{e-h}t/\hbar} |\Phi(t)\rangle, \qquad (29.72)$$

so that if there is no perturbation, then $|\tilde{\Phi}(t)\rangle$ does not change with time. Using the fact that the system state satisfies the

equation $i\hbar\partial |\Phi(t)\rangle/\partial t = \mathcal{H}|\Phi\rangle$, we find that the time evolution is governed by

$$i\frac{\partial}{\partial t}|\tilde{\Phi}(t)\rangle = \tilde{\mathcal{H}}_{exc}(t)|\tilde{\Phi}(t)\rangle, \qquad (29.73)$$

where

$$\tilde{\mathcal{H}}_{exc}(t) = e^{i\mathcal{H}_{e-h}t/\hbar}\mathcal{H}_{exc}(t)e^{-i\mathcal{H}_{e-h}t/\hbar}. \qquad (29.74)$$

The solution of Equation 29.73 is written in terms of the S-matrix

$$|\tilde{\Phi}(t)\rangle = S(t)|\tilde{\Phi}(0)\rangle, \qquad (29.75)$$

where

$$S(t) = 1 + \sum_{n=1}^\infty \left(-\frac{i}{\hbar}\right)^n \int_{0 \le t_1 \le \dots} \dots \int_{\le t_n \le t} dt_1 \dots dt_n \tilde{\mathcal{H}}_{exc}(t_1) \dots \tilde{\mathcal{H}}_{exc}(t_n). \qquad (29.76)$$

Often this expansion is presented in the compact form $S(t) = \mathcal{T}_+ \exp\left\{-i\hbar^{-1}\int_0^t dt' \tilde{\mathcal{H}}_{exc}(t')\right\}$ similarly to Equation 29.31; we, however, will need in what follows the explicit form of expansion (29.76).

Next, we express the observables

$$O(t) = \langle\hat{O}\rangle = \langle\Phi(t)|\hat{O}|\Phi(t)\rangle = \langle\tilde{\Phi}(t)| e^{i\mathcal{H}_{e-h}t/\hbar}\hat{O} e^{-i\mathcal{H}_{e-h}t/\hbar}|\tilde{\Phi}(t)\rangle \qquad (29.77)$$

and notice that in this equation and in Equation 29.74 appears the same structure

$$\hat{O}(t) = e^{i\mathcal{H}_{e-h}t/\hbar}\hat{O}e^{-i\mathcal{H}_{e-h}t/\hbar}, \qquad (29.78)$$

which is called the *interaction representation* of the operator $\hat{O}$. Taking into account Equation 29.75 we obtain

$$O(t) = \langle S^\dagger(t)\tilde{O}(t)S(t)\rangle_0, \qquad (29.79)$$

where $\langle\dots\rangle_0 \equiv \langle\Phi(0)|\dots|\Phi(0)\rangle$. Finally, the perturbation series is obtained substituting expansion (29.76) into Equation 29.79. The order of the perturbation is determined by the highest order of $\tilde{\mathcal{H}}(t)$ kept in the resulting expression.

We illustrate the perturbation theory in action finding the macroscopic exciton polarization, $P_\mu = \langle\hat{B}_\mu\rangle$ or

$$P_\mu = \langle S^\dagger(t)\tilde{B}_\mu(t)S(t)\rangle_0. \qquad (29.80)$$

Let the initial state of the system be vacuum $|\Phi(0)\rangle = |0\rangle$. Below we will consider only such initial state, therefore, in order to shorten notations we will omit the index 0 and write averages over vacuum simply $\langle...\rangle$. Since the semiconductor Hamiltonian $\mathcal{H}_{e-h}$ preserves both the number of electrons in the conduction band and the number of holes in the valence band, the exciton operator in the interaction representation acts in a similar way as in the Schrödinger or Heisenberg representations, namely, $\tilde{B}_\mu^\dagger(t)$ creates and $\tilde{B}_\mu(t)$ destroys an electron–hole pair, respectively. Hence, the terms in the expansion of the r.h.s. of Equation 29.80, which give nonzero contribution, must have balanced numbers of the exciton creation and annihilation operators. In particular, one can easily check that there are no contributions of even orders (zeroth, second, and so on).

There is only one term in the first order

$$
\begin{aligned}
P_\mu^{(1)}(t) &= -\frac{i}{\hbar}\sum_\nu \int_0^t dt_1 \left\langle \tilde{B}_\mu(t)\tilde{B}_\nu^\dagger(t_1)\right\rangle \mathcal{A}_\nu(t_1) \\
&= -\frac{i}{\hbar}\int_0^t dt_1 e^{-iE_\mu(t-t_1)/\hbar}\mathcal{A}_\mu(t_1),
\end{aligned}
\tag{29.81}
$$

where in the second equality we have taken into account Equation 29.64 and the fact that $\hat{B}_\mu\hat{B}_\nu^\dagger|0\rangle = \delta_{\mu,\nu}|0\rangle$, which follows from the commutation relation for the exciton operators

$$
\left[\hat{B}_\mu,\hat{B}_\nu^\dagger\right] = \hbar(\delta_{\mu,\nu} - \hat{C}_{\mu,\nu}),
\tag{29.82}
$$

where

$$
\begin{aligned}
\hat{C}_{\mu,\nu} &= \int d\mathbf{x}'\,d\mathbf{y}'\,d\mathbf{x}\,d\mathbf{y}\,\phi_\nu(\mathbf{x}',\mathbf{y}')\phi_\mu^*(\mathbf{x},\mathbf{y})[\upsilon_{\sigma_\nu}^\dagger(\mathbf{x}')\upsilon_{\sigma_\mu}(\mathbf{x})\delta(\mathbf{y}-\mathbf{y}')\delta_{s_\nu,s_\mu} \\
&\quad + c_{s_\nu}^\dagger(\mathbf{y}')c_{s_\mu}(\mathbf{y})\delta(\mathbf{x}-\mathbf{x}')\delta_{\sigma_\nu,\sigma_\mu}].
\end{aligned}
\tag{29.83}
$$

These operators describe the deviation of excitons from bosons. It should be noted, however, that the deviation depends on the number of the electrons and holes. Thus, in the limit of weak excitations, the excitons can be approximately considered as bosons. However, as the population of the valence and conduction bands increases, one has to take into account the full form of the commutation relation.

Differentiating Equation 29.81 with respect to time we find

$$
i\hbar\frac{\partial}{\partial t}P_\mu^{(1)}(t) = E_\mu P_\mu^{(1)}(t) + \mathcal{A}_\mu(t).
\tag{29.84}
$$

Comparing this with Equation 29.51 we can see that Equation 29.84 is nothing else but the exciton polarization of the linear response. This shows that the perturbation theory built with the help of the interaction representation correctly accounts for the nonperturbed dynamics of the semiconductor. With this regard it should be noted that this presents a convenient opportunity

to account for the incoherent processes, which take place in real systems. As has been noted before a rigorous introduction of the decoherence requires developing a complex technique. For many purposes, however, it suffices to introduce the imaginary part of the single-exciton energies, $E_\mu \to E_\mu - i\hbar\Gamma_\mu$.

Before we move further, we would like to discuss some basic features of the amplitudes $\mathcal{A}_\mu(t)$. Substituting Equation 29.66 within Equation 29.71, we obtain

$$
\mathcal{A}_\mu(t) = \tilde{\phi}_\mu^*(0)\int d\mathbf{x}\, e^{-i\mathbf{K}_\mu \cdot \mathbf{x}}\mathbf{A}_\Omega(\mathbf{x},t)\cdot\mathbf{d}_{s_\mu,\sigma_\mu},
\tag{29.85}
$$

where the integration is taken over the plane of the quantum well. Thus, if the excitation field has the form of a plain wave $\mathbf{A}_\Omega(\mathbf{x}) \propto \exp(i\mathbf{K}\cdot\mathbf{x})$, then $\mathcal{A}_\mu \propto \delta(\mathbf{K}_\parallel - \mathbf{K}_\mu)$, that is, the plane wave only excites the exciton states with the same center of mass momentum as the projection of $\mathbf{K}$ on the plane of the quantum well, $\mathbf{K}_\parallel$. Moreover, as follows from Equation 29.40 the same in-plane momentum will be passed to the outgoing electromagnetic field. Such conservation of the in-plane momentum is not a specific feature of the linear response. If the excitation field is a plane wave, the translations $\Delta\mathbf{x}$ in the plane of the quantum well lead to the amplitudes $\mathbf{A}_\Omega(\mathbf{x})$ acquiring a phase factor, $\mathbf{A}_\Omega(\mathbf{x}+\Delta\mathbf{x}) = \exp(i\mathbf{K}_\parallel\cdot\Delta\mathbf{x})\mathbf{A}_\Omega(\mathbf{x})$; in other words, the excitation field possesses the translational symmetry. If this is the case, then, as can be easily checked, the full system of the equations of motion, Equations 29.46, is also translationally invariant (more precisely it has the same translational symmetry as the external field) with the following transformation rules for the solutions

$$
\upsilon_\sigma^\dagger(\mathbf{x}_2+\Delta\mathbf{x})c_s^\dagger(\mathbf{x}_1+\Delta\mathbf{x}) = \upsilon_\sigma^\dagger(\mathbf{x}_2)c_s^\dagger(\mathbf{x}_1)e^{i\mathbf{k}_\parallel\cdot\Delta\mathbf{x}},
$$

$$
c_{s_1}^\dagger(\mathbf{x}_1+\Delta\mathbf{x})c_{s_2}(\mathbf{x}_2+\Delta\mathbf{x}) = c_{s_1}^\dagger(\mathbf{x}_1)c_{s_2}(\mathbf{x}_2),
\tag{29.86}
$$

$$
\upsilon_{\sigma_2}^\dagger(\mathbf{x}_2+\Delta\mathbf{x})\upsilon_{\sigma_1}(\mathbf{x}_1+\Delta\mathbf{x}) = \upsilon_{\sigma_2}^\dagger(\mathbf{x}_2)\upsilon_{\sigma_1}(\mathbf{x}_1).
$$

This also results in the transfer of the incoming in-plane momentum to the outgoing wave.

The picture becomes more complex if the external field is a superposition of two-plane waves $\mathbf{A}_\Omega(\mathbf{x}) = \mathbf{A}_1\exp(i\mathbf{K}^{(1)}\cdot\mathbf{x}) + \mathbf{A}_2\exp(i\mathbf{K}^{(2)}\cdot\mathbf{x})$ with noncollinear in-plane components of the momenta, $\mathbf{K}_\parallel^{(1)} \neq a\mathbf{K}_\parallel^{(2)}$ for any number a. In this case, the external field is invariant with respect to infinitesimal translations in the direction perpendicular to $\Delta\mathbf{K} = \mathbf{K}_\parallel^{(2)} - a\mathbf{K}_\parallel^{(1)}$, while in the direction along $\Delta\mathbf{K}$ it is invariant only with respect to *finite* shifts $\Delta R = 2\pi/\Delta K$. For the polarization of linear response it brings nothing new, as one could expect only the exciton polarizations with the center of mass momenta $\mathbf{K}_\parallel^{(1)}$ and $\mathbf{K}_\parallel^{(2)}$ are produced. However, this is not necessarily the case for the nonlinear system of equations. Generally, it is invariant only with respect to finite translations. As we know from the basic theory of crystals whenever one has such symmetry, the momentum conserves only up to the vectors of the reciprocal lattice, which in our case are $m\Delta\mathbf{K}$ with integer m. Thus, in addition to the initial momenta presented in the excitation field, one might expect to see additional momenta appearing in the nonlinear response. This is, indeed, the case and the exciton polarizations corresponding to the "new" momenta are called the polarizations of *multiwave mixing* response for reasons which will become evident later.

In order to see the multiwave mixing response one needs to consider next orders of the perturbation theory. In the third order we have four terms, which we show schematically as

$$P_\mu^{(3)} \sim \left\langle B_\mu B B^\dagger B^\dagger \right\rangle + \left\langle B_\mu B^\dagger B B^\dagger \right\rangle + \left\langle B B^\dagger B_\mu B^\dagger \right\rangle + \left\langle B B_\mu B^\dagger B^\dagger \right\rangle$$

(29.87)

The operators that appear to the right from B_μ originate from the expansion of $S(t)$ in Equation 29.80, and those to the left from B_μ come from $S^\dagger(t)$. After some algebra we obtain

$$\hbar \frac{\partial}{\partial t} P_\mu^{(3)}(t) = -(iE_\mu + \hbar\Gamma_\mu)P_\mu^{(3)}(t)$$

$$+ i\hbar \sum_{\nu_1\nu_2\nu_3} \left\langle \hat{B}_{\nu_2} \hat{C}_{\mu,\nu_3} \hat{B}_{\nu_1}^\dagger \right\rangle \mathcal{A}_{\nu_3}(t) P_{\nu_2}^{(1)*}(t) P_{\nu_1}^{(1)}(t)$$

$$+ \sum_{\nu_1,\nu_2,\nu_3} P_{\nu_2}^{(1)*}(t) \int_0^t dt' \left\langle \hat{D}_{\nu_2,\mu} e^{-i\mathcal{H}_{e-h}(t-t')/\hbar} \hat{B}_{\nu_3}^\dagger \hat{B}_{\nu_1}^\dagger \right\rangle \mathcal{A}_{\nu_3}(t') P_{\nu_1}^{(1)}(t'),$$

(29.88)

where we have introduced the operator

$$\hat{D}_{\nu,\mu} = \left[\hat{B}_\nu, \left[\hat{B}_\mu, \mathcal{H}_{e-h} \right] \right],$$

(29.89)

which appears in calculations as $\left\langle 0 \middle| \hat{B}_\nu^\dagger \hat{B}_\mu^\dagger (\mathcal{H}_{e-h} - E_\nu - E_\mu) = \left\langle 0 \middle| \hat{D}_{\nu,\mu} \right.$. In terms of the electron and hole operators $\hat{D}_{\nu,\mu}$ is written as

$$\hat{D}_{\nu,\mu} = \int d\mathbf{x}_1 d\mathbf{y}_1 d\mathbf{x}_2 d\mathbf{y}_2 \phi_\mu^*(\mathbf{x}_1,\mathbf{y}_1) \phi_\nu^*(\mathbf{x}_2,\mathbf{y}_2) V(\mathbf{x}_1,\mathbf{y}_1 : \mathbf{x}_2,\mathbf{y}_2)$$

$$\times c_{s_\mu}(\mathbf{y}_1) \upsilon_{\sigma_\mu}(\mathbf{x}_1) c_{s_\nu}(\mathbf{y}_2) \upsilon_{\sigma_\nu}(\mathbf{x}_2),$$

(29.90)

where $V(\mathbf{x}_1,\mathbf{y}_2;\mathbf{x}_2,\mathbf{y}_2) = V(\mathbf{x}_1 - \mathbf{x}_2) + V(\mathbf{y}_1 - \mathbf{y}_2) - V(\mathbf{x}_1 - \mathbf{y}_2) - V(\mathbf{x}_2 - \mathbf{y}_1)$ is the energy of the Coulomb interaction between two excitons.

We rewrite the last term in Equation 29.88 in order to express it solely in terms of the polarizations of the linear response. For this we notice that for a symmetric matrix $G_{\kappa,\lambda} = G_{\lambda,\kappa}$ one has

$$\frac{i\hbar}{2} \sum_{\kappa,\lambda} G_{\kappa,\lambda} e^{-i(E_\kappa + E_\lambda)t/\hbar} \frac{\partial}{\partial t} \left[P_\kappa^{(1)}(t) P_\lambda^{(1)}(t) e^{i(E_\kappa + E_\lambda)t/\hbar} \right]$$

$$= \sum_{\kappa,\lambda} G_{\kappa,\lambda} P_\kappa^{(1)}(t) \mathcal{A}_\lambda(t).$$

(29.91)

Using this relation for the terms under the time integral in Equation 29.88 (the symmetry is ensured by $[\hat{B}_\mu^\dagger, \hat{B}_\nu^\dagger] = 0$) and integrating by part we obtain

$$\frac{\partial}{\partial t} P_\mu^{(3)}(t) = -\left(i\frac{E_\mu}{\hbar} + \Gamma_\mu \right) P_\mu^{(3)}(t) + i\sum_{\kappa,\lambda,\nu} C_{\lambda,\mu}^{\nu,\kappa} \mathcal{A}_\nu(t) P_\lambda^{(1)*}(t) P_\kappa^{(1)}(t)$$

$$+ \frac{1}{2} \sum_{\kappa,\lambda,\nu} P_\lambda^{(1)*}(t) \Big[i\beta_{\lambda,\mu}^{\nu,\kappa} P_\kappa^{(1)}(t) P_\nu^{(1)}(t)$$

$$+ \frac{1}{\hbar} \int_0^t dt' e^{-(\Gamma_\kappa + \Gamma_\nu)(t-t')} F_{\lambda,\mu}^{\nu,\kappa}(t-t') P_\kappa^{(1)}(t') P_\nu^{(1)}(t') \Big],$$

(29.92)

where we have introduced the incoherent decay into the integral term assuming that the decay of the pair of excitons is not affected by the interaction between them. We also have introduced $C_{\lambda,\mu}^{\nu,\kappa} = \left\langle \hat{B}_\lambda \hat{C}_{\mu,\nu} \hat{B}_\kappa^\dagger \right\rangle$, $\beta_{\lambda,\mu}^{\nu,\kappa} = \left\langle \hat{D}_{\lambda,\mu} \hat{B}_\nu^\dagger \hat{B}_\kappa^\dagger \right\rangle$, and $F_{\lambda,\mu}^{\nu,\kappa}(\tau) = \left\langle \hat{D}_{\lambda,\mu} e^{-i\mathcal{H}_{e-h}\tau/\hbar} \hat{D}_{\nu,\kappa}^\dagger \right\rangle$.

The term $\propto C_{\lambda,\mu}^{\nu,\kappa}$ in Equation 29.92 accounts for the effect of Pauli blocking of electrons. The parameters $\beta_{\lambda,\mu}^{\nu,\kappa}$ and the memory function $F_{\lambda,\mu}^{\nu,\kappa}(\tau)$ describe the effect of the exciton–exciton correlations. It should be emphasized that since we have taken into account exactly the dynamics of the semiconductor nonperturbed by the external field these parameters exactly account for the Coulomb interaction between the excitons. More complex many-body states, for example, triples of excitons will contribute only to the higher orders of the perturbation theory.

The important feature of $C_{\lambda,\mu}^{\nu,\kappa}$, $\beta_{\lambda,\mu}^{\nu,\kappa}$, and $F_{\lambda,\mu}^{\nu,\kappa}$ is that all of them are proportional to $\delta(\mathbf{K}_\mu + \mathbf{K}_\lambda - \mathbf{K}_\nu - \mathbf{K}_\kappa)$. The proof consists of two steps. First, we notice that after we expand the definitions of the operators entering the respective mean values, we will have at hand the vacuum averages of a set of the electrons and holes creation and annihilation operators taken at different coordinates multiplied by some scalar functions. The vacuum averages, however, do not change if we uniformly shift in the plane of the quantum well all the coordinates of the creation and annihilation operators. Indeed, when we transform the operator under-averaging to the normal form, that is, with all creation operators collected on the left side we perform transpositions of neighboring operators, which are functions of, say, $\mathbf{x}_1$ and $\mathbf{x}_2$, taking into account their anticommutation relations. This results at most in the appearance of $\delta(\mathbf{x}_1 - \mathbf{x}_2)$; thus, the vacuum average can depend only on the differences between different coordinates and, therefore, it does not change after the uniform translation.

Next, we notice that the scalar functions, which are obtained after the expansion of all definitions, acquire the phase factor only due to the translational properties of the exciton wave functions (see Equation 29.66), all other functions including $\tilde{\Phi}$ and the Coulomb potential in the Hamiltonian $\mathcal{H}_{e-h}$ are translationally invariant, so eventually we end up with an integral of the form

$$I(\mathbf{K}_1,...,\mathbf{K}_n) = \int d\mathbf{x}_1...d\mathbf{x}_n \exp\left(i\sum_{j=1}^{n}\mathbf{K}_j\cdot\mathbf{x}_j\right)f(\mathbf{x}_1,...,\mathbf{x}_n),$$

(29.93)

where $f(\mathbf{x}_1 + \Delta\mathbf{x}, ..., \mathbf{x}_n + \Delta\mathbf{x}) = f(\mathbf{x}_1, ..., \mathbf{x}_n)$. We want to show that $I(\mathbf{K}_1, ..., \mathbf{K}_n) \propto \delta(\mathbf{K}_1 + \cdots + \mathbf{K}_n)$. To this end we present the integral in Equation 29.93 as

$$I(\mathbf{K}_1,...,\mathbf{K}_n) = \int d\mathbf{x}_1 e^{i\mathbf{K}_1\cdot\mathbf{x}_1}\int d\mathbf{x}_2...d\mathbf{x}_n \exp\left(i\sum_{j=2}^{n}\mathbf{K}_j\cdot\mathbf{x}_j\right)$$
$$\times f(0+\mathbf{x}_1,\mathbf{x}_2-\mathbf{x}_1+\mathbf{x}_1...,\mathbf{x}_n-\mathbf{x}_1+\mathbf{x}_1),$$

(29.94)

using the translational invariance of the function $f(\mathbf{x}_1, ..., \mathbf{x}_n)$ and change the variables of integration in the multiple integral. This yields

$$I(\mathbf{K}_1,...,\mathbf{K}_n) = \int d\mathbf{x}_1 \exp\left(i\sum_{j=1}^{n}\mathbf{K}_j\cdot\mathbf{x}_1\right)\int d\mathbf{x}_2...d\mathbf{x}_n \exp\left(i\sum_{j=2}^{n}\mathbf{K}_j\cdot\mathbf{x}_j\right)$$
$$\times f(0,\mathbf{x}_2...,\mathbf{x}_n).$$

(29.95)

The multiple integral does not depend on $\mathbf{x}_1$, so we can perform integration over $\mathbf{x}_1$ obtaining $I(\mathbf{K}_1, ..., \mathbf{K}_n) \propto \delta(\mathbf{K}_1 + \cdots + \mathbf{K}_n)$.

Thus, we prove that $P_\mu^{(3)} \propto \delta(\mathbf{K}_\mu + \mathbf{K}_\lambda - \mathbf{K}_\nu - \mathbf{K}_\kappa)$. Because four momenta have to add up to zero in order to have nonvanishing $P_\mu^{(3)}$, this type of response is called *four-wave mixing* response (see Figure 29.2). The possible values that $\mathbf{K}_\mu$ can take are obtained looking at different combinations of the momenta of the polarizations of the linear response, which are determined by the momenta of the incoming excitation, $\mathbf{K}_\parallel^{(1)}$ and $\mathbf{K}_\parallel^{(2)}$. We can see that in addition to $\mathbf{K}_\parallel^{(1)}$ and $\mathbf{K}_\parallel^{(2)}$ we also have $2\mathbf{K}_\parallel^{(2)} - \mathbf{K}_\parallel^{(1)} = \mathbf{K}_\parallel^{(2)} + \Delta\mathbf{K}$ and $\mathbf{K}_\parallel^{(1)} - \Delta\mathbf{K}$ in agreement with expected conservation of the momentum only up to multiples of $\Delta\mathbf{K}$.

It is important that the response in the four-wave mixing direction is created only owing to the many-body correlations in the semiconductor. As a result, the signal in the respective directions is not blurred by the nonabsorbed excitation field and the linear response. Therefore, four-wave mixing spectroscopy is a popular tool for experimentally studying the many-body properties of solids. Higher order of the perturbation theory will produce higher orders of wave mixing originating from the respective correlations, which recently started attracting significant attention.

We conclude the survey of the basics of the perturbation approach for developing a qualitative theory of the two-dimensional Fourier spectroscopy (Khalil et al. 2003, Jonas 2003), which recently became a powerful tool for studying the dynamics of the semiconductor excitations (Zhang et al. 2005, Li et al. 2006). The general scheme is the same as that of the standard four-wave mixing experiments with three pulses propagating

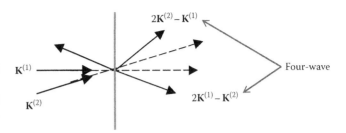

FIGURE 29.2 Linear and four-wave mixing responses. Bold solid arrows show the momenta of the external excitation field. Dashed lines show the linear. Notice that the horizontal component of momentum does not conserve and the outgoing wave exists in both half-spaces. The solid arrows show the outgoing waves produced by the exciton polarization of the four-wave mixing response.

along $\mathbf{K}^{(1)}$, $\mathbf{K}^{(2)}$, and $\mathbf{K}^{(3)}$ launched at $t = t_1$, t_2, and t_3, respectively. The difference from the standard technique is that measurements of the four-wave mixing signal are performed not at a fixed or at just a few values of the delay time $\tau = \min(t_2, t_3) - t_1$, but rather for a dense series of values lying in some interval. Subsequently, the Fourier transforms are done with respect to the delay time as well as with respect to the signal time. These two Fourier transforms constitute the two-dimensional Fourier spectrum.

Equation 29.92 can be directly applied for finding the exciton polarization produced by the excitation pulses, one only needs to take into account that the pulses with different directions of the wave vectors strike the quantum well at different times (Erementchouk et al. 2007). As the simplest approximation we neglect the contribution of the memory function $F(\tau)$ and assume that its effect can be accounted for by a renormalization of $\beta_{\lambda,\mu}^{\nu,\kappa}$. Let Ω and ω correspond to the Fourier transform with respect to the signal and delay times, respectively. As one would expect the two-dimensional spectrum $P_\mu^{(3)}(\Omega, \omega)$ has resonances at the frequencies corresponding to the light-hole and the heavy-hole excitons (see Figure 29.3). It turns out, however, that the form of the resonances depends on whether they are created due to the Pauli blocking $\propto C_{\lambda,\mu}^{\nu,\kappa}$ or due to the Coulomb interaction $\beta_{\lambda,\mu}^{\nu,\kappa}$. We denote the respective contributions to the spectrum by $P_P(\Omega, \omega)$ and $P_C(\Omega, \omega)$ respectively. One can show that near particular resonance

$$P_P(\Omega,\omega) \propto \frac{1}{(\Omega-\hbar^{-1}E_\kappa+i\Gamma_\kappa)(\omega+\hbar^{-1}E_\lambda+i\Gamma_\lambda)},$$

$$P_C(\Omega,\omega) \propto \frac{1}{(\Omega-\hbar^{-1}E_\kappa+i\Gamma_\kappa)(\Omega-\hbar^{-1}E_\kappa+i\Gamma_\kappa+2i\Gamma_\lambda)(\omega+\hbar^{-1}E_\lambda+i\Gamma_\lambda)},$$

(29.96)

where κ and λ stand for $1s$ state of either heavy- or light-hole excitons.

The contribution to the spectrum from the Pauli blocking has the Lorentz form along both the vertical and horizontal axes. The reason is that, as can be readily seen from Equation 29.92, the dependence on both the signal time $[P_\mu^{(3)}(t)]$ and delay

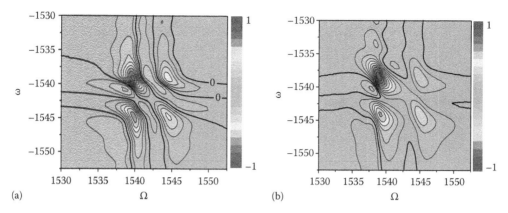

FIGURE 29.3 **(See color insert following page 21-4.)** The imaginary part of the two-dimensional Fourier spectrum, $P(\Omega, \omega)$. (a) The spectrum calculated using Equation 29.92. (b) The experimental results of Zhang et al. (2005).

time (through $P_\lambda^{(1)*}(t)$ in the r.h.s. of Equation 29.92) is essentially the free evolution of the polarization directly created by the short excitation pulse. This evolution has the form of oscillations, which produce a simple pole after the Fourier transform. At the same time the resonances created by the Coulomb interaction, $P_C(\Omega, \omega)$, fall off asymptotically as $\propto 1/\omega$ and $\propto 1/\Omega^2$, along the ω- and Ω-axes, respectively. This is a consequence of the fact that the polarization of the four-mixing response $P_\mu^{(3)}(t)$ is continuously excited by the polarizations of the linear response. The dependence on the signal time is found as a convolution of the respective Green function with the source, similarly to how it was done in Equation 29.54. After the Fourier transform with respect to signal time it yields the product of Fourier images of the Green function and the source. This results in the asymptotic form $\propto 1/\Omega^2$ because of the harmonic time dependence of these functions.

This simple consideration shows that the two-dimensional Fourier spectroscopy allows one to distinguish between different types of the many-body correlations, which is almost impossible to do with the standard approach. For example, in experiments reported in Li et al. (2006), the form of the two-dimensional spectra is clearly asymmetric along different axes (see Figure 29.3b) indicating that due to relatively low intensity of excitations the contribution of the Pauli blocking into the nonlinear response was small compared to the effect of the Coulomb interaction.

29.5 Nonperturbative Methods: Rabi Oscillations

The drawback of the perturbative treatment of interaction of the electromagnetic field with the semiconductor is overestimating the resonant response. As follows from Equation 29.81 if the excitation field is tuned exactly to the resonance with an exciton state, the amplitude of the respective exciton polarization will indefinitely increase linearly with time. This, however, is impossible since with increase of the population of the conduction and the valence bands it becomes more difficult to add additional

particles due to the Pauli blocking. Therefore, one would expect that the rate of change of the exciton polarization should change with time.

It turns out, however, that the effect of the environment is even stronger. The rate not only decreases, it reaches zero and then changes its sign, so the polarization starts to decrease until it completely vanishes. As a result the actual time dependence of the polarization has the form of oscillations, which are called *Rabi oscillations* by analogy with the effect of rotation of the magnetic moment in an external oscillating magnetic field.

In order to describe this effect we need to take into account the interaction with the electromagnetic field nonperturbatively. The problem of developing such nonperturbative description is that on the one hand there is no general way to study nonlinear operator equations other than using perturbation theory, while on the other hand there is no general way to map such equations onto the respective problem for regular functions. For example, averaging the first equation from Equations. 29.46 over vacuum (see e.g., Equation 29.30) one gets the dynamical equation for the interband polarization $p_{\sigma,s}^*(\mathbf{x}_1, \mathbf{x}_2) = \left\langle \upsilon_\sigma^\dagger(\mathbf{x}_1) c_s^\dagger(\mathbf{x}_2) \right\rangle$, which contains in the r.h.s. the correlation functions of higher order, for example, $\langle \upsilon^\dagger c^\dagger c c \rangle$, that is, the system will not be closed. An attempt to close the system by writing down the dynamical equation for new quantities will bring to attention correlation functions of even higher order and so on to infinity. Thus, one has to make the special approximation to truncate this hierarchy of the dynamical equations.

The simplest approximation of such kind is the Hartree–Fock approximation. It forces the closure at the level of the equation with respect to the interband polarization by applying the special decoupling scheme, which is illustrated by

$$
\begin{aligned}
\left\langle \upsilon_\sigma^\dagger(\mathbf{x}_1) c_{s_1}^\dagger(\mathbf{x}_2) c_{s_2}^\dagger(\mathbf{x}_3) c_{s_3}(\mathbf{x}_4) \right\rangle &\approx \left\langle \upsilon_\sigma^\dagger(\mathbf{x}_1) c_{s_1}^\dagger(\mathbf{x}_2) \right\rangle \left\langle c_{s_2}^\dagger(\mathbf{x}_3) c_{s_3}(\mathbf{x}_4) \right\rangle \\
&- \left\langle \upsilon_\sigma^\dagger(\mathbf{x}_1) c_{s_2}^\dagger(\mathbf{x}_3) \right\rangle \left\langle c_{s_1}^\dagger(\mathbf{x}_2) c_{s_3}(\mathbf{x}_4) \right\rangle
\end{aligned}
$$

$$(29.97)$$

Thus, we can express the four-point correlation functions in terms of the products of the two-point ones.

In the framework of the Hartree–Fock approximation we obtain from Equations 29.46 in the rotating wave approximation (here and below the summation over dashed spin indices is implied)

$$-i\hbar \frac{\partial}{\partial t} p_{\sigma,s}^*(\mathbf{x}_1, \mathbf{x}_2) = \hat{K}_{\sigma,s}[p] + \mathcal{A}_{\sigma,s}^*(\mathbf{x}_1)\delta(\mathbf{x}_1 - \mathbf{x}_2)$$

$$-\int d\mathbf{x}'[\tilde{\mathcal{A}}_{\sigma',s}^*(\mathbf{x}', \mathbf{x}_2)h_{\sigma,\sigma'}(\mathbf{x}_1, \mathbf{x}')$$

$$-\tilde{\mathcal{A}}_{\sigma,s'}^*(\mathbf{x}', \mathbf{x}_1)e_{s,s'}(\mathbf{x}_2, \mathbf{x}')],$$

$$-i\hbar \frac{\partial}{\partial t} e_{s_1,s_2}(\mathbf{x}_1, \mathbf{x}_2) = \hat{K}_{s_1,s_2}[e]$$

$$+\int d\mathbf{x}'[\tilde{\mathcal{A}}_{\sigma',s_1}^*(\mathbf{x}', \mathbf{x}_1)p_{\sigma',s_2}(\mathbf{x}', \mathbf{x}_2)$$

$$-\tilde{\mathcal{A}}_{\sigma',s_2}^*(\mathbf{x}', \mathbf{x}_2)p_{\sigma',s_1}^*(\mathbf{x}', \mathbf{x}_1)],$$

$$-i\hbar \frac{\partial}{\partial t} h_{\sigma_1,\sigma_2}(\mathbf{x}_1, \mathbf{x}_2) = \hat{K}_{\sigma_1,\sigma_2}[h]$$

$$+\int d\mathbf{x}'[\tilde{\mathcal{A}}_{\sigma_1,s'}^*(\mathbf{x}_1, \mathbf{x}')p_{\sigma_2,s'}(\mathbf{x}_2, \mathbf{x}')$$

$$-\tilde{\mathcal{A}}_{\sigma_2,s'}^*(\mathbf{x}_2, \mathbf{x}')p_{\sigma_1,s'}^*(\mathbf{x}_1, \mathbf{x}')], \quad (29.98)$$

where $\mathcal{A}_{\sigma,s}(\mathbf{x}) = \mathbf{A}_\Omega(\mathbf{x},t) \cdot d_{\sigma,s}$, the electron and hole correlation functions are $e_{s_1,s_2}(\mathbf{x}_1, \mathbf{x}_2) = \langle c_{s_1}^\dagger(\mathbf{x}_1) c_{s_2}(\mathbf{x}_2) \rangle$ and $h_{\sigma_1,\sigma_2}(\mathbf{x}_1, \mathbf{x}_2) = \langle \upsilon_{\sigma_1}^\dagger(\mathbf{x}_1) \upsilon_{\sigma_2}(\mathbf{x}_2) \rangle$, and we have introduced the modified coupling between the interband polarization and the charge densities

$$\tilde{\mathcal{A}}_{\sigma,s}(\mathbf{x},\mathbf{y}) = \mathcal{A}_{\sigma,s}(\mathbf{x})\delta(\mathbf{x} - \mathbf{y}) - V(\mathbf{x} - \mathbf{y})p_{\sigma,s}(\mathbf{x},\mathbf{y}). \quad (29.99)$$

The operators

$$\hat{K}_{\sigma,s}[p^*] = \hat{H}_{\sigma,\sigma'}p_{\sigma',s}^* + p_{\sigma,s'}^*\hat{H}_{s',s} - V(\mathbf{x}_1 - \mathbf{x}_2)p_{\sigma,s}^*(\mathbf{x}_1, \mathbf{x}_2)$$

$$-\hbar\Omega p_{\sigma,s}^*(\mathbf{x}_1, \mathbf{x}_2),$$

$$\hat{K}_{s_1,s_2}[e] = \hat{H}_{s_1,s'}e_{s',s_2} - e_{s_1,s'}\hat{H}_{s',s_2}, \quad (29.100)$$

$$\hat{K}_{\sigma_1,\sigma_2}[h] = \hat{H}_{\sigma_1,\sigma'}h_{\sigma',\sigma_2} - h_{\sigma_1,\sigma'}\hat{H}_{\sigma',\sigma_2}$$

are expressed in terms of the integro-differential operators $\hat{H}$. The action of these operators is defined by

$$\hat{H}_{s_1,s'}f_{\sigma'} = \left[-\frac{\hbar^2}{2m_e}\nabla_1^2 + \epsilon_{s_1}(\mathbf{x}_1) + U_\Delta(\mathbf{x}_1)\right]f_{s_1}(\mathbf{x}_1, \mathbf{x}_2)$$

$$-\int d\mathbf{x}'V(\mathbf{x}_1 - \mathbf{x}')e_{s_1,s'}(\mathbf{x}_1, \mathbf{x}')f_{s'}(\mathbf{x}', \mathbf{x}_2), \quad (29.101)$$

$$\hat{H}_{\sigma_1,\sigma'}f_{\sigma'} = -\nabla_1 \cdot \frac{\hbar^2}{2M_{\sigma_1,\sigma'}} \cdot \nabla_1 f_{\sigma'}(\mathbf{x}_1, \mathbf{x}_2) - [\epsilon_{\sigma_1}(\mathbf{x}_1) + U_\Delta(\mathbf{x}_1)]f_{\sigma_1}(\mathbf{x}_1, \mathbf{x}_2)$$

$$-\int d\mathbf{x}'V(\mathbf{x}_1 - \mathbf{x}')h_{\sigma_1,\sigma'}(\mathbf{x}_1, \mathbf{x}')f_{\sigma'}(\mathbf{x}', \mathbf{x}_2),$$

where $U_\Delta(\mathbf{x}) = \langle \hat{U}_\Delta(\mathbf{x}) \rangle$ (see Equation 29.47). It should be noted that for a semiconductor excited by a single-plane wave this potential vanishes in an overall neutral system due to the translational invariance. In the two-wave excitation setup the effect of U_Δ is small for bound exciton states if the order of the multiwave mixing response is not too high.

Equations 29.100 constitute the semiconductor Bloch equations, which are widely used for studying nonperturbative effects in the dynamics of optically excited semiconductors. It should be emphasized that the quantity of main interest is the exciton polarization P_μ since these are the excitons that constitute true single-particle semiconductor states. As follows from Equations 29.63 and 29.72, P_μ is related to the interband polarization $p_{\sigma,s}$ through

$$P_\mu = \int d\mathbf{x}\, d\mathbf{y}\, \phi_\mu^*(\mathbf{x},\mathbf{y})p_{\sigma_\mu,s_\mu}(\mathbf{x},\mathbf{y}). \quad (29.102)$$

For bound exciton states this naturally introduces a characteristic spatial scale, the exciton Bohr radius r_B (see Equation 29.67). At this point we assume that we deal with a deep and narrow quantum well, $\epsilon(\mathbf{x}) = \epsilon + \Delta\epsilon(z)$, and that the system is invariant with respect to translations along the quantum well, that is, there are no random fluctuations of the width of the well.

The existence of the characteristic spatial scale together with the assumption that the typical spatial variation of the external excitation in the plane of the quantum well is small compared to the Bohr radius $K^{(1,2)}r_B \ll 1$ allows one to simplify the semiconductor Bloch equation using the parametric approximation. For this we introduce new spatial variables $\mathbf{R} = (\mathbf{x} + \mathbf{y})/2$ and $\mathbf{r} = \mathbf{x} - \mathbf{y}$. Having in mind the consecutive convolution of the solutions with the exciton wave function we can neglect the terms $\propto \Delta K r_B$ and leave only the dependence on $\mathbf{R}$ running along the quantum well as a parameter. Thus, assuming for simplicity that $\mathbf{K}^{(1)} = 0$ we approximate Equations 29.98 by

$$i\hbar\dot{p}_{\sigma,s}(\mathbf{r},\mathbf{R}) = \tilde{K}_{\sigma,s}[p] + \mathcal{A}_{\sigma,s}(\mathbf{R})\delta(\mathbf{r})$$

$$-\int d\mathbf{r}'\Big[\tilde{\mathcal{A}}_{\sigma',s}(\mathbf{r}',\mathbf{R})h_{\sigma',\sigma}(\mathbf{r}' - \mathbf{r}, \mathbf{R}) - \tilde{\mathcal{A}}_{\sigma,s'}(\mathbf{r}',\mathbf{R})e_{s',s}(\mathbf{r}' - \mathbf{r}, \mathbf{R})\Big],$$

$$-i\hbar\dot{e}_{s_1,s_2}(\mathbf{r},\mathbf{R}) = \hat{K}_{s_1,s_2}[e]$$

$$+\int d\mathbf{r}'\Big[\tilde{\mathcal{A}}_{\sigma',s_1}^*(\mathbf{r}',\mathbf{R})p_{\sigma',s_2}(\mathbf{r}' + \mathbf{r}, \mathbf{R}) - \tilde{\mathcal{A}}_{\sigma',s_2}(\mathbf{r}',\mathbf{R})p_{\sigma',s_1}^*(\mathbf{r}' - \mathbf{r}, \mathbf{R})\Big],$$

$$-i\hbar\dot{h}_{\sigma_1,\sigma_2}(\mathbf{r},\mathbf{R}) = \hat{K}_{\sigma_1,\sigma_2}[h]$$

$$+\int d\mathbf{r}'\Big[\tilde{\mathcal{A}}_{\sigma_1,s'}^*(\mathbf{r}',\mathbf{R})p_{\sigma_2,s'}(-\mathbf{r} - \mathbf{r}', \mathbf{R}) - \tilde{\mathcal{A}}_{\sigma_2,s'}(\mathbf{r}',\mathbf{R})p_{\sigma_1,s'}^*(\mathbf{r} + \mathbf{r}', \mathbf{R})\Big].$$

$$(29.103)$$

Here $\tilde{\mathcal{A}}_{\sigma,s}(\mathbf{r},\mathbf{R}) = \mathcal{A}_{\sigma,s}(\mathbf{R})\delta(\mathbf{r}) - V(\mathbf{r})p_{\sigma,s}(\mathbf{r},\mathbf{R})$ and the integro-differential operators in Equations 29.100 are substituted by

$$\hat{H}_{s_1,s'}f_{s'} = \left[\frac{\hbar^2}{2m_e}\nabla_{\mathbf{r}}^2 + \epsilon_{s_1}(z)\right]f_{s_1}(\mathbf{r},\mathbf{R}) - \int d\mathbf{r}'V(\mathbf{r}')e_{s_1,s'}(\mathbf{r}',\mathbf{R})f_{s'}(\mathbf{r} - \mathbf{r}', \mathbf{R}),$$

$$\hat{H}_{\sigma_1,\sigma'}f_{\sigma'} = -\nabla_{\mathbf{r}} \cdot \frac{\hbar^2}{2M_{\sigma_1,\sigma'}} \cdot \nabla_{\mathbf{r}}f_{\sigma'}(\mathbf{r},\mathbf{R}) - \epsilon_{\sigma_1}(z)f_{\sigma_1}(\mathbf{r},\mathbf{R}) \quad (29.104)$$

$$-\int d\mathbf{r}'V(\mathbf{r}')h_{\sigma_1,\sigma'}(\mathbf{r}',\mathbf{R})f_{\sigma'}(\mathbf{r} - \mathbf{r}', \mathbf{R}).$$

Equations 29.103 are sufficiently general for studying various nonlinear effects. Before we turn to further approximations in order to extract the Rabi oscillations, we would like to sketch the procedure to find the polarizations of multiwave mixing response. As has been discussed earlier in the case of two-wave excitation the translational invariance of the semiconductor equations along the quantum well is broken. Instead one has the symmetry with respect to finite translations $\Delta R = 2\pi\Delta K/\Delta K^2$. The same holds for the semiconductor Bloch equations, which is especially evident in the parametric approximation. Indeed, the parametric dependence on R originates from $A(R)$, which is a periodic function of ΔR and, hence, so are the solutions of Equations 29.103. Thus, the interband polarization $p_{\sigma,s}(r = 0, R)$, which serves as the source of the outgoing re-emitted electromagnetic wave (see Equation 29.40), can be expanded in the Fourier series

$$p_{\sigma,s}(0,R) = \sum_{m=-\infty}^{\infty} p_{\sigma,s}^{(m)} e^{im\Delta K \cdot R}. \qquad (29.105)$$

The terms $m = 0$ and $m = 1$ correspond to the momenta brought by the excitation field, while the remaining terms are the polarizations of multiwave mixing response, for example, $m = -1, 2$ give the four-wave mixing polarizations, $m = -2, 3$ are the six-wave mixing components, and so on. In order to find the amplitudes of the multiwave mixing responses we merely use the formula of the inverse Fourier transform

$$p_{\sigma,s}^{(m)} = \int_{0}^{1} d\alpha \, p_{\sigma,s}(0, \alpha\Delta R) e^{-2\pi i \alpha m}. \qquad (29.106)$$

Thus, once the solution of Equation 29.103 is known, the polarizations of the multiwave mixing response can be found immediately.

In order to see the oscillatory behavior of the interband polarization driven by the external field we need to make drastic simplifications of Equation 29.103, which correspond to considering the case of very short excitation pulses of high intensity. Such simplification is possible owing to the existence of the characteristic spatial scale and the radius of exciton states responsible for the resonant response. We consider all the terms in the r.h.s. of Equation 29.103 besides the kinetic energy as sources. The response, therefore, is governed by the free-particle propagators for the correlation functions and the Coulomb propagator for the interband polarization. The fundamental property of these propagators is that in the short time limit $t \to 0$ they turn to spatial δ-functions owing to fast spatial oscillations of the kernel

$$K(r_1, r_2; t) = \sqrt{\frac{m}{2\pi i t}} \exp\left[im|r_1 - r_2|^2 / 2t\right]. \qquad (29.107)$$

For example, for the interband polarization in the short time limit, the Coulomb propagator can be approximated by the free-particle propagator (Blinder 1984, Kunikeev 2000) with $m \approx m_h = (m_e^{-1} + M_{hh})^{-1}$ and $m \approx m_l = (m_e^{-1} + M_{lh})^{-1}$, where m_σ are

the excitons' reduced masses. The effect of the dispersion of the electrons and holes on the *exciton* polarization is expressed as the distortion of the exciton wave function under the action of the free-particle propagators. The time scale, at which the distortion becomes essential, can be estimated as the time required for the initial δ-shape of the kernel to acquire the width of the order of the exciton Bohr radius. This leads to the estimate

$$T_c \sim \frac{m r_B^2}{4\pi}. \qquad (29.108)$$

The numerical value of the typical time scale for GaAs is determined by $m_{hh} = 0.45m_0$, $m_{lh} = 0.082m_0$, $m = m_\Delta$, and $r_B = r_{xh} = \epsilon_b/2m_{xh}e^2$, where m_0 is the electron mass in empty space and $\epsilon_b = 13$ is the background dielectric function. Using these values in Equation 29.108 we find $T_c \sim 20\,\text{fs}$. We would like to draw attention to the fact that $T_c \gg T_R \approx 1\,\text{fs}$ (see Equation 29.57), so that the rotating wave approximation is applicable at such time scales. Thus, for pulses with duration shorter than 20 fs and sufficiently high intensity, we can treat the semiconductor excitations as dispersionless. It can be shown that for the states in the discrete spectrum the condition is even less strict (Erementchouk and Leuenberger 2008).

In the dispersionless limit, Equations 29.103 take the simple form

$$i\hbar\dot{p}_{\sigma,s}(r,R) = \mathcal{A}_{\sigma,s}(R)\delta(r) - \mathcal{A}_{\sigma',s}(R)h_{\sigma',\sigma}(-r,R) - \mathcal{A}_{\sigma,s'}(R)e_{s',s}(r,R),$$

$$-i\hbar\dot{e}_{s_1,s_2}(r,R) = \mathcal{A}_{\sigma',s_1}^*(R)\rho_{\sigma',s_2}(r,R) - \mathcal{A}_{\sigma',s_2}(R)\rho_{\sigma',s_1}^*(-r,R), \qquad (29.109)$$

$$-i\hbar\dot{h}_{\sigma_1,\sigma_2}(r,R) = \mathcal{A}_{\sigma_1,s'}^*(R)\rho_{\sigma_2,s'}(-r,R) - \mathcal{A}_{\sigma_2,s'}(R)\rho_{\sigma_1,s'}^*(r,R),$$

where we assumed the strong resonance regime and neglected the small contribution $(\epsilon_s - \epsilon_\sigma - \hbar\Omega)p_{\sigma,s}$. The solutions of Equations 29.109 are singular, $\propto\delta(r)$, but by virtue of Equation 29.102 it does not pose any problem.

We limit ourselves to the consideration of the simplest case when the external field is circularly polarized with positive helicity (right polarized) and has the amplitude constant in the time, which corresponds to the continuous wave excitation tuned to resonance with the gap. The condition of circular polarization leaves only two nonvanishing interband polarizations corresponding to p_{h+} and p_{l+}, that is $\{\sigma = -3/2, s = -1/2\}$ and $\{\sigma = -1/2, s = 1/2\}$, respectively. Thus we find for the respective interband polarizations

$$p_{\sigma,s}(r,R;t) = -i\delta(r)\frac{1}{\hbar\Omega_{\sigma,s}(R)}\mathcal{A}_{\sigma,s}(R)\sin\left[\Omega_{\sigma,s}(R)t\right], \qquad (29.110)$$

where $\hbar\Omega_{\sigma,s}(R) = |\mathcal{A}_{\sigma,s}(R)|$. It should be noted that only when $\Omega_{\sigma,s}$ is uniform along the quantum well, that is, in the case of the single-wave excitation, it defines the frequency of the Rabi oscillations. When the semiconductor is excited by two incoming waves, the frequency of temporal oscillations of the polarizations of the multiwave mixing response at sufficiently long

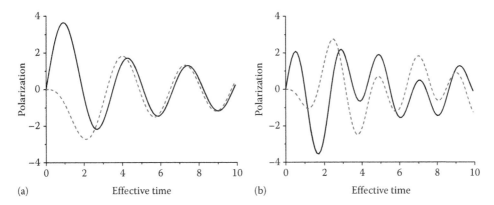

FIGURE 29.4 The exciton polarization in the case when the excitation consists of two plane waves with noncollinear wave vectors. The solid and dashed lines show the response in the forward ($m = 1$) and the four-wave mixing ($m = 2$) directions for (a) $\Omega_{\sigma,s}^{(1)}/\Omega_{\sigma,s}^{(2)} = 1$, (b) $\Omega_{\sigma,s}^{(1)}/\Omega_{\sigma,s}^{(2)} = 2$. The effective time is defined as $\tilde{t} = \Omega_{\sigma,s}^{(2)}t$.

time scales have two characteristic frequencies obtained as superpositions of the Rabi frequencies of each pulse $\left|\Omega_{\sigma,s}^{(1)} \pm \Omega_{\sigma,s}^{(2)}\right|$ (see Figure 29.4).

29.6 Conclusion

In order to keep the introduction into the theory of the semiconductor nonlinear response concise, we have concentrated on the problem of low energy excitation, when the dynamics of the semiconductor is determined by the bound states of the electron–hole pairs (excitons). In this regime the most natural description is provided by the effective quantum field theory, which accounts for the nonconstant number of particles in the semiconductor bands. The operator equations of motion, in turn, give access to all the information about the semiconductor and the electromagnetic fields. Using these equations one can either study the time evolution of the observables such as the macroscopic electromagnetic field $\langle \mathbf{A}(\mathbf{x},t) \rangle$ or the exciton polarization $\langle \hat{B}_\mu \rangle$; or one can turn, for example, to the single-particle exciton density matrix $\langle \hat{B}_\lambda^\dagger \hat{B}_\mu \rangle$ in order to incorporate into the picture the decoherence effect of the interaction with the phonon bath; or one can study the many-body properties of the semiconductor excitations looking at, say, the two-exciton density matrix $\langle \hat{B}_\kappa^\dagger \hat{B}_\lambda^\dagger \hat{B}_\mu \hat{B}_\nu \rangle$.

Unfortunately, the analysis of the equations of motion is tremendously hardened by their nonlinearity. The complexity is not of merely technical origin but reflects the concealed variety of physical phenomena. This should be taken into account when the equations of motion are simplified. As demonstrated in the main text, one can relatively easily develop the perturbational description of the semiconductor dynamics. However, one should be careful in applying the results of the perturbational approach to the situation, when the intensity of the excitation is significantly high, so that the properly defined pulse area approaches π. The perturbational series diverges in this case signifying the appearance of a new class of phenomena related to the reconstruction of the spectrum. The effect of the reconstruction on the semiconductor dynamics requires a non-perturbative treatment and has been illustrated above for the example of the Rabi oscillations.

We have covered the basics of the theory to the degree where it can be straightforwardly applied to various physical situations. However, we have left many details mentioned only on the formal level, such as the symmetries of the semiconductor bands, the specific form of the tensors of effective masses, and the matrix elements of the momentum operator between different bands. These details, while being easy to incorporate into the theory developed in the main text, may significantly simplify the analysis of the particular case. As a source of the fundamental insight into the physics of semiconductors we would like to refer the reader to the great book by Yu and Cardona (2004), the details of the symmetrical properties of the solids are provided in Bir and Pikus (1974), and the optical semiconductor response is considered in Haug and Koch (2004).

Acknowledgment

We would like to thank Prof. Xiaoqin Li for providing us with the data obtained in the experiments reported in Zhang et al. (2005). We also thank Sergio Tafur for useful comments. We acknowledge support from NSF-ECCS 0725514, the DARPA/MTO Young Faculty Award HR0011-08-1-0059, NSF-ECCS 0901784, and AFOSR FA9550-09-1-0450.

References

Arfken, G. B. and H. J. Weber. *Mathematical Methods for Physicists.* Academic Press, New York, 6th edn., 2005.

Bir, G. L. and G. E. Pikus. *Symmetry and Strain-Induced Effects in Semiconductors.* Wiley, New York, 1974.

Blinder, S. M. Semiclassical approximation for the nonrelativistic Coulomb propagator. *Phys. Rev. Lett.*, 52(20):1771–1773, 1984.

Cohen-Tannoudji, C., J. Dupont-Roc, and G. Grynberg. *Atom–Photon Interactions. Basic Processes and Applications.* Wiley, New York, 1992.

Erementchouk, M. V. and M. N. Leuenberger. Rabi oscillations in semiconductor multiwave mixing response. *Phys. Rev. B*, 78:075206, 2008.

Erementchouk, M. V., M. N. Leuenberger, and L. J. Sham. Many-body interaction in semiconductors probed with two-dimensional Fourier spectroscopy. *Phys. Rev. B*, 76:115307, 2007.

Haug, H. and S. W. Koch. *Quantum Theory of the Optical and Electronic Properties of Semiconductors*. World Scientific, Singapore, 2004.

Ivchenko, E. L. *Optical Spectroscopy of Semiconductor Nanostructures*. Alpha Science International, Oxford, U.K., 2005.

Johnson, N. F. and P. M. Hui. Theory of propagation of scalar waves in periodic and disordered composite structures. *Phys. Rev. B*, 48(14):10118–10123, 1993.

Jonas, D. M. Two-dimensional femtosecond spectroscopy. *Annu. Rev. Phys. Chem.*, 54:425–463, 2003.

Khalil, M., N. Demirdoven, and A. Tokmakoff. Coherent 2d IR spectroscopy: Molecular structure and dynamics in solution. *J. Phys. Chem. A*, 107:5258–5279, 2003.

Kunikeev, S. D. Coulomb propagator in the WKB approximation. *J. Phys. A Math. Gen.*, 33:5405–5428, 2000.

Langbein, W., G. Kocherscheidt, and R. Zimmermann. Probing localized excitons by speckle analysis of resonant light scattering. In H. Kalt and M. Hetterich (eds.), *Optics of Semiconductors and Their Nanostructures*, vol. 146 of *Springer Series in Solid-State Sciences*, pp. 47–72. Springer, Berlin, Germany, 2004.

Li, X., T. Zhang, C. N. Borca, and S. T. Cundiff. Many-body interactions in semiconductors probed by optical two-dimensional Fourier transform spectroscopy. *Phys. Rev. Lett.*, 96:057406, 2006.

Luttinger, J. M. Quantum theory of cyclotron resonance in semiconductors: General theory. *Phys. Rev.*, 102(4):1030–1041, 1956.

Luttinger, J. M. and W. Kohn. Motion of electrons and holes in perturbed periodic fields. *Phys. Rev.*, 97(4):869–883, 1955.

Mukamel, S. *Principles of Nonlinear Optical Spectroscopy*. Oxford University Press, New York, 1995.

Ostreich, T. Higher-order Coulomb correlation in the nonlinear optical response. *Phys. Rev. B*, 64:245203, 2001.

Ostreich, T., K. Schonhammer, and L. J. Sham. Theory of exciton–exciton correlation in nonlinear optical response. *Phys. Rev. B*, 58(19):12920–12936, 1998.

Parfitt, D. G. W. and M. E. Portnoi. The two-dimensional hydrogen atom revisited. *J. Math. Phys.*, 43(10):4681–4691, 2002.

Sakoda, K. *Optical Properties of Photonic Crystals*, vol. 80 of *Optical Sciences*. Springer, Berlin, Germany, 2nd edn., 2005.

Schiff, L. I. *Quantum Mechanics*. McGraw Hill, New York, 1949.

Schmitt-Rink, S., D. S. Chemla, and D. A. B. Miller. Linear and nonlinear optical properties of semiconductor quantum wells. *Adv. Phys.*, 38(2):89–188, 1989.

Yu, P. Y. and M. Cardona. *Fundamentals of Semiconductors: Physics and Materials Properties*. Springer, Berlin, Germany, 2004.

Zhang, T., C. N. Borca, X. Li, and S.T. Cundiff. Optical two-dimensional Fourier transform spectroscopy with active interferometric stabilization. *Opt. Express*, 13(19):7432–7431, 2005.

Zhu, B. and K. Huang. Effect of valence-band hybridization on the exciton spectra in GaAs–Ga$_{1-x}$Al$_x$As quantum wells. *Phys. Rev. B*, 36(15):8102–8108, 1987.

Light Scattering from Nanofibers

Vladimir G. Bordo
A.M. Prokhorov General
Physics Institute

30.1 Introduction

The progress in nanotechnology has led to a possibility of fabricating needlelike crystals that have transverse dimensions in the submicron range. They can be grown from different inorganic semiconductor materials such as InP (Duan et al., 2001; Wang et al., 2001), ZnO (Huang et al., 2001), GaN (Johnson et al., 2002), CdS (Duan et al., 2003), GaSb (Chin et al., 2006), as well as from p-phenylene oligomers (Balzer and Rubahn, 2001, 2005; Schiek et al., 2005). Alternatively, one can also obtain fibers with a diameter of a few tens of nanometers by means of drawing from silica (Tong et al., 2003). Such structures are called nanofibers or nanowires. The morphology of individual nanofibers and their mutual alignment can be controlled by the use of appropriate growth conditions and substrate surfaces (Balzer and Rubahn, 2005). The nanofibers can be detached from the substrate, thus allowing one to study their properties under a variety of different conditions, including free-floating nanoaggregates (Brewer et al., 2005). They act as optical waveguides (Balzer et al., 2003; Johnson et al., 2003; Tong et al., 2003; Barrelet et al., 2004) and demonstrate remarkable photoluminescence properties (Wang et al., 2001; Thilsing-Hansen et al., 2005). Under pumping conditions, they can operate as nanoscale lasers ("nanolasers") (Huang et al., 2001; Johnson et al., 2002, 2003; Duan et al., 2003; Agarwal et al., 2005; Quochi et al., 2005, 2006; Chin et al., 2006; Zimmler et al., 2008).

Light scattering by nanofibers plays an important role in optical processes at the nanoscale. Being grown on a substrate, nanofibers form an irregular array. The optical properties of such an ensemble are determined, on the one hand, by the optical response of individual aggregates and, on the other hand, by light scattering from nanofibers, which results in interaction between them. Light incident at a sample can launch the nanofiber optical modes or create excitons. Both processes manifest themselves in the intensity distribution of scattered light, thus providing valuable information on different nanofiber properties.

30.2 Background

In the simplest theoretical approach, a nanofiber is represented by an infinitely long dielectric circular cylinder. Light scattering by such an object was the first diffraction problem to be solved rigorously. Its solution dates back to Lord Rayleigh (Lord Rayleigh, 1881), who found the electromagnetic field diffracted from a cylinder of arbitrary radius and refractive index illuminated by a plane wave incident perpendicularly to its axis. Much later, that result was generalized to the case of oblique incidence (Wait, 1955). If α is the incidence angle relatively normal to the cylinder axis, then the scattered radiation is propagated along the surface of the cone with apical angle $(\pi - 2\alpha)$ (Kerker, 1969). The field amplitudes inside the cylinder and outside it are expressed in terms of a series of the Bessel functions of the first kind and Hankel functions, respectively. The result depends on the polarization of the incident wave. One can decompose it into two independent polarizations: (1) the transverse magnetic (TM) mode for which there is no magnetic vector component along the cylinder axis and (2) the transverse electric (TE) mode for which there is no electric field component along that axis. When the incident wave propagates perpendicularly to the cylinder axis, both the field inside the cylinder and the scattered field are of the same polarization as the incident wave. For oblique incidence, however, the incident TM waves are partially converted into scattered TE waves. Similarly, the incident TE waves are partially converted into scattered TM waves.

Light scattering in the far zone is characterized by the *scattering efficiency*, Q_{sca}, which is defined as the scattering cross section per unit length of the cylinder integrated over all the values of the scattering angle, θ, divided by the cylinder diameter, $2a$. If the polarization of the scattered radiation is not specified, light scattering is completely described by the two scattering efficiencies, Q_{sca}^{TM} and Q_{sca}^{TE}, where the superscript denotes the polarization of the incident wave. In the limit where the diameter of the cylinder is sufficiently small relative to the wavelength, all the results are considerably simplified. If the refractive index of the cylinder relative to that of the surrounding medium is n and the wave number of the incident wave is k, the scattering efficiencies are reduced to the following form (Kerker, 1969):

$$Q_{sca}^{TM} = \frac{\pi^2(ka)^3}{8}(n^2-1)^2; \quad Q_{sca}^{TE} = \frac{\pi^2(ka)^3}{4}\left(\frac{n^2-1}{n^2+1}\right)^2. \quad (30.1)$$

If the angle θ is not close to 0 or π,* the ratio of the intensities of the two components is

$$\rho = \frac{I^{TE}}{I^{TM}} = \frac{4\sin^2\theta}{(n^2+1)^2}. \quad (30.2)$$

This leads to the conclusion that the radiation scattered in both forward and backward directions will be partially polarized even if the incident radiation is completely unpolarized.

The abandonment of the assumption of an infinite cylinder length leads to the necessity to use numerical methods. For some ranges of the relevant parameters, simplified treatments can provide accurate results. If the cylinder radius a is much smaller than the wavelength λ, the integral equation for the electric field amplitude can be solved by the method of moments (Uzunoglu et al., 1978). In the case where the parameter $ka(n-1)$ is comparable with unity or smaller, an iterative method can be applied, which is quite effective for both long (Cohen et al., 1983) and short (Haracz et al., 1984) cylinders. For the general case of a finite cylinder length l, the approach based on solving the electric field integral equation by using the method of moments is not very efficient. A more appropriate way is to apply a Fourier transform, which leads to a Fredholm integral equation of the first kind. This method has been applied to the cylinders with aspect ratios $p = l/(2a)$ as large as 10 and the lengths parameters up to $kl \sim 7$ (Shepherd and Holt, 1983). An alternative approach exploits the extended boundary condition method (EBCM), also known as the T-matrix method, in which both the incident and scattered fields are expanded in spherical vector wavefunctions. The truncation of the expansions at N terms leads to the system of $2N$ linear equations for the scattered field coefficients. The EBCM calculations carried out for

the backscattering from a dielectric rod in the microwave region have demonstrated a satisfactory agreement with the existing experimental data (Ruppin, 1990). However, the EBCM method is tractable only when the size parameters ka and p are less than certain limits; e.g., for $n = 1.31$, $ka < 2.75$ and $p < 5.0$ (Kuik et al., 1994). Its convergence also becomes more difficult with increasing the dielectric constant of the cylinder.

A remarkable feature of light scattering from a dielectric cylinder at a fixed angle of incidence is the existence of resonant peaks in the scattering cross section plotted versus the size parameter (Van Bladel, 1977). Such peaks do not occur for a perfectly conducting cylinder and their sharpness increases with the dielectric constant of the cylinder. The resonant frequencies are determined by the poles of the expansion coefficients in the complex plane of the size parameter (Owen et al., 1981). The linewidths of the peaks are very sensitive to the circularity of the cylinder and to optical loss within or on the cylinder surface. These findings were confirmed experimentally in the scattering of either broadband radiation or light from a tunable laser by optical-communication fibers and used for the accurate determination of their diameters and shapes (Ashkin et al., 1981; Owen et al., 1981). The peaks of another kind are displayed in the dependence of the differential scattering efficiency on the angle between the incident light beam and the cylinder axis (Birkhoff et al., 1977). Fitting the angular positions of observed maxima in the scattered intensity provides a sensitive method for determining fiber radii, reproducible within ±3%–5%.

An essential feature of light scattering from nanofibers is reflection of the scattered radiation from the substrate surface. Therefore, it is desirable to take this effect into account when modeling nanofibers. If the cylinder axis is oriented parallelly to a flat reflecting surface, the diffraction problem is reduced to a two-dimensional one. The influence of the substrate can be calculated by introducing the Fresnel reflection coefficients. The corresponding solution can be found in the form of a series of the Bessel and Hankel functions. The procedure involves the truncation of the expansions for the field amplitudes at some index N, which is related to the size parameter by the rule $N \simeq 3ka$ (Borghi et al., 1997). Light scattering from nanofibers grown on a substrate involves the multiple-scattering effect. In some special cases, the nanofibers form an array of mutually parallel nanoaggregates oriented either parallelly or perpendicularly to the substrate surface. As it has been shown for an arbitrary configuration of parallel cylinders, the light scattering in such a system can be expressed as an infinite sum of the orders of scattering (Twersky, 1952). The first-order scattering originates from a diffraction of the incident wave on each cylinder. The second order results from a diffraction of the first-order-scattered radiation on each cylinder, and so on. A common approach to calculating the scattered field is to express the multiple-scattering linear equations in matrix form and then to find the corresponding scattering coefficients either by iterative method, or by direct matrix inversion, or by using a quadratic-programming algorithm. By this means, it has been found, in particular, that in an array of abutting cylinders, the resonant character of the cylinders with respect to the size parameter, ka, is damped (Bever and Allebach, 1992).

---

* We reckon the angle θ from the plane perpendicular to the plane of incidence (see Figure 30.1).

30.3 Basic Theory

In this section, we shall give the basic equations that describe light scattering from nanofibers (Bordo, 2006, 2007). We shall consider two models of a nanofiber: (1) an infinite circular dielectric cylinder (Figure 30.1) and (2) an infinite circular dielectric semicylinder placed on a perfectly reflecting surface (Figure 30.2). Although it might be possible that these models do not reproduce some details of light scattering, they nevertheless provide a fundamental understanding of this phenomenon. To match the conditions of many experiments with nanofibers, we assume that the incident radiation is represented by a pulsed Gaussian light beam.

30.3.1 Diffraction by a Cylindrical Nanofiber

Let us assume first that a nanofiber has a circular cross-section of radius a, both the nanofiber and surrounding medium are isotropic and are characterized by the dielectric functions ε_2 and ε_1, respectively. We choose the z axis along the nanofiber axis and the y axis in the plane containing both the z axis and the beam axis. Then the incident wave electric field has the form

$$\mathbf{E}_i(x,y,z,t) = \mathbf{E}_{i0}(x,y,z,t)\exp(-iq_{10}y + i\beta_0 z - i\omega_0 t), \quad (30.3)$$

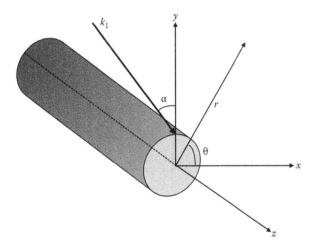

FIGURE 30.1 Model of a nanofiber: dielectric cylinder.

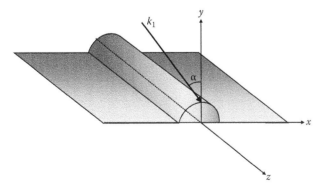

FIGURE 30.2 Model of a nanofiber: dielectric semicylinder on a perfectly reflecting surface.

where the field amplitude is given by

$$\mathbf{E}_{i0}(x,y,z,t) = \mathbf{E}_0 \exp\left[-\frac{x^2 + (y\sin\alpha + z\cos\alpha)^2}{2\sigma^2}\right]\exp\left(-\frac{t^2}{2\tau^2}\right),$$
$$(30.4)$$

where
- $-q_{10}$ and β_0 are the projections of the wave vector onto the axes y and z, respectively
- ω_0 is the frequency of the incident wave
- α is the angle of incidence with respect to the y axis
- σ and τ are the radius and pulse duration of the light beam, respectively

A similar expression is valid for the magnetic field of the incident beam, $\mathbf{H}_i(x, y, z, t)$.

In experiments, the beam radius is much larger than the nanofiber radius, i.e., $\sigma \gg a$. In such a case, in the vicinity of the nanofiber, Equation 30.4 is reduced to the following* form:

$$\mathbf{E}_{i0}(z,t) \approx \mathbf{E}_0 \exp\left(-\frac{z^2\cos^2\alpha}{2\sigma^2}\right)\exp\left(-\frac{t^2}{2\tau^2}\right). \quad (30.5)$$

On the other hand, to simplify the consideration, we shall assume that the beam radius is much less than the nanofiber length. Then one can neglect the edge effects and consider the nanofiber as infinitely long.

The total electric field has the form

$$\mathbf{E} = \begin{cases} \mathbf{E}_i + \mathbf{E}_1 & \text{if } r > a; \\ \mathbf{E}_2 & \text{if } r < a, \end{cases} \quad (30.6)$$

where $\mathbf{E}_1$ and $\mathbf{E}_2$ are the electric field vectors of the scattered field and the field inside the nanofiber, respectively. The total magnetic field vector, $\mathbf{H}$, is represented in a similar form. The fields $\mathbf{E}_1$, $\mathbf{E}_2$, $\mathbf{H}_1$, and $\mathbf{H}_2$ in turn can be found in terms of the Hertz vectors, Π_1 and Π_2, whose rectangular components satisfy the scalar wave equation. The Hertz vector for an infinite cylinder has the only nonzero component, $\Pi_z \equiv \psi$, which can be decomposed into two contributions, ψ^{TM} and ψ^{TE}, associated with TM and TE polarizations, respectively. Then, the electromagnetic field components in the cylindrical coordinates (r, θ, z) are found as (Stratton, 1941)

$$E_{jr} = \frac{\partial^2 \psi_j^{TM}}{\partial z \partial r} - \frac{1}{cr}\frac{\partial^2 \psi_j^{TE}}{\partial t \partial \theta}, \quad (30.7)$$

$$E_{j\theta} = \frac{1}{r}\frac{\partial^2 \psi_j^{TM}}{\partial z \partial \theta} + \frac{1}{c}\frac{\partial^2 \psi_j^{TE}}{\partial t \partial r}, \quad (30.8)$$

---

* This approximation is valid outside the nanofiber region $|z| \le a\tan\alpha$.

$$E_{jz} = \frac{\partial^2 \psi_j^{TM}}{\partial z^2} - \frac{\epsilon_j}{c^2} \frac{\partial^2 \psi_j^{TM}}{\partial t^2}, \tag{30.9}$$

$$H_{jr} = \frac{\epsilon_j}{r} \frac{\partial^2 \psi_j^{TM}}{\partial t \partial \theta} + \frac{\partial^2 \psi_j^{TE}}{\partial z \partial r}, \tag{30.10}$$

$$H_{j\theta} = -\epsilon_j \frac{\partial^2 \psi_j^{TM}}{\partial t \partial r} + \frac{1}{r} \frac{\partial^2 \psi_j^{TE}}{\partial z \partial \theta}, \tag{30.11}$$

$$H_{jz} = \frac{\partial^2 \psi_j^{TE}}{\partial z^2} - \frac{\epsilon_j}{c^2} \frac{\partial^2 \psi_j^{TE}}{\partial t^2}, \tag{30.12}$$

where

  c is the speed of light in vacuum

  the subscript j labels different media: $j = 1$ corresponds to the surrounding medium, whereas $j = 2$ denotes the nanofiber interior

We shall seek the functions ψ_j^μ (μ = TM or TE) in the form of the Fourier integral

$$\psi_j^\mu(x,y,z,t) = \frac{1}{(2\pi)^2} \int_{-\infty}^{\infty} \int_{-\infty}^{\infty} \tilde{\psi}_j^\mu(x,y;\beta,\omega) e^{i\beta z - i\omega t} d\beta d\omega. \tag{30.13}$$

The representation (30.13) can be considered as a superposition of an infinite number of the elementary waves $\tilde{\psi}_j^\mu e^{i\beta z - i\omega t}$, each of which should satisfy the wave equation. The Fourier-transformed quantities $\tilde{\psi}_j^\mu$, therefore, can be expanded in the series of the cylindrical functions as follows:

$$\tilde{\psi}_j^{TM}(r,\theta;\beta,\omega) = \frac{1}{q_j^2} \sum_{n=0}^{\infty} Z_n(q_j r) \left[a_{jn}(\beta,\omega) \sin(n\theta) + b_{jn}(\beta,\omega) \cos(n\theta) \right], \tag{30.14}$$

$$\tilde{\psi}_j^{TE}(r,\theta;\beta,\omega) = \frac{1}{q_j^2} \sum_{n=0}^{\infty} Z_n(q_j r) \left[c_{jn}(\beta,\omega) \sin(n\theta) + d_{jn}(\beta,\omega) \cos(n\theta) \right], \tag{30.15}$$

where the functions $Z_n(X)$ are determined as

$$Z_n(X) = \begin{cases} H_n^{(1)}(q_1 r) & \text{if } r > a; \\ J_n(q_2 r) & \text{if } r < a, \end{cases} \tag{30.16}$$

with

$$q_j = \sqrt{\frac{\omega^2}{c^2} \epsilon_j - \beta^2}, \tag{30.17}$$

J_n and $H_n^{(1)}$ are the Bessel functions of the first kind and Hankel functions, respectively. The choice of the Bessel functions for the cylinder interior ensures finiteness of the solution at its center, whereas the Hankel functions have proper behavior at infinite distance from the cylinder.

The continuity of the tangential components of the total fields, $E_{j\theta}$, E_{jz}, $H_{j\theta}$, and H_{jz}, across the boundary $r = a$ leads to the equations for the coefficients a_{jn}, b_{jn}, c_{jn}, and d_{jn}. Taking into account that $y = r \sin \theta$, one can use the expansion

$$e^{-iq_{10} r \sin \theta} = \sum_{n=-\infty}^{\infty} J_n(q_{10} r) e^{-in\theta} \tag{30.18}$$

in Equation 30.3. Then, combining the terms with the same θ-dependence in the boundary conditions and performing the inverse Fourier transform, one obtains the following equations:

$$\hat{M}_n \vec{A}_n = \vec{F}_n, \tag{30.19}$$

$$\hat{M}_n \vec{B}_n = \vec{G}_n, \tag{30.20}$$

where

$$\hat{M}_n = \begin{pmatrix} J_n(q_2 a) & 0 & -H_n^{(1)}(q_1 a) & 0 \\ \dfrac{i\beta n}{q_2^2 a} J_n(q_2 a) & -\dfrac{i\omega}{cq_2} J_n'(q_2 a) & -\dfrac{i\beta n}{q_1^2 a} H_n^{(1)}(q_1 a) & \dfrac{i\omega}{cq_1} H_n^{(1)\prime}(q_1 a) \\ \dfrac{i\omega\epsilon_2}{cq_2} J_n'(q_2 a) & -\dfrac{i\beta n}{q_2^2 a} J_n(q_2 a) & -\dfrac{i\omega\epsilon_1}{cq_1} H_n^{(1)\prime}(q_1 a) & \dfrac{i\beta n}{q_1^2 a} H_n^{(1)}(q_1 a) \\ J_n(q_2 a) & 0 & -H_n^{(1)}(q_1 a) \end{pmatrix} \tag{30.21}$$

and

$$\vec{A}_n = \begin{pmatrix} a_{2n} \\ d_{2n} \\ a_{1n} \\ d_{1n} \end{pmatrix}, \quad \vec{B}_n = \begin{pmatrix} -b_{2n} \\ c_{2n} \\ -b_{1n} \\ c_{1n} \end{pmatrix}. \tag{30.22}$$

Here, the prime above the Bessel and Hankel functions denotes differentiation with respect to their argument. The vector functions $\vec{F}_n$ and $\vec{G}_n$ depend on the polarization of the incident beam and are determined as

$$\vec{F}_n^{TM} = \begin{pmatrix} -2i\sigma_n \tilde{E}_{i0} \cos\alpha J_n(q_{10} a) \\ 2\sigma_n \tilde{E}_{i0} \sin\alpha \left[n J_n(q_{10} a)/q_{10} a \right] \\ 2\sigma_n \tilde{H}_{i0} J_n'(q_{10} a) \\ 0 \end{pmatrix}, \tag{30.23}$$

$$\vec{G}_n^{TM} = \begin{pmatrix} -2\tau_n(1-\sigma_n) \tilde{E}_{i0} \cos\alpha J_n(q_{10} a) \\ -2i\tau_n(1-\sigma_n) \tilde{E}_{i0} \sin\alpha \left[n J_n(q_{10} a)/q_{10} a \right] \\ -2i\tau_n(1-\sigma_n) \tilde{H}_{i0} J_n'(q_{10} a) \\ 0 \end{pmatrix}, \tag{30.24}$$

and

$$
\vec{F}_n^{\mathrm{TE}} = \begin{pmatrix} 0 \\ -2i\tau_n(1-\sigma_n)\tilde{E}_{i0}J_n'(q_{10}a) \\ -2i\tau_n(1-\sigma_n)\tilde{H}_{i0}\sin\alpha\left[nJ_n(q_{10}a)/q_{10}a\right] \\ 2\tau_n(1-\sigma_n)\tilde{H}_{i0}\cos\alpha J_n(q_{10}a) \end{pmatrix}, \quad (30.25)
$$

$$
\vec{G}_n^{\mathrm{TE}} = \begin{pmatrix} 0 \\ -2\sigma_n\tilde{E}_{i0}J_n'(q_{10}a) \\ -2\sigma_n\tilde{H}_{i0}\sin\alpha\left[nJ_n(q_{10}a)/q_{10}a\right] \\ -2i\sigma_n\tilde{H}_{i0}\cos\alpha J_n(q_{10}a) \end{pmatrix}, \quad (30.26)
$$

with

$$
\sigma_n = \begin{cases} 0 & \text{if } n \text{ is even;} \\ 1 & \text{if } n \text{ is odd,} \end{cases} \quad (30.27)
$$

and

$$
\tau_n = \begin{cases} 1/2 & \text{if } n = 0; \\ 1 & \text{if } n \neq 0. \end{cases} \quad (30.28)
$$

Here, we have introduced the Fourier-transformed amplitudes of the incident beam

$$
\tilde{E}_{i0}(\beta,\omega) = E_0 \frac{2\pi\sigma\tau}{\cos\alpha} \exp\left[-\frac{(\beta-\beta_0)^2\sigma^2}{2\cos^2\alpha}\right] \exp\left[-\frac{(\omega-\omega_0)^2\tau^2}{2}\right], \quad (30.29)
$$

and a similar expression for $\tilde{H}_{i0}(\beta,\omega)$.

Solving Equations 30.19 and 30.20, one obtains the Fourier-transformed electromagnetic field components both inside the nanofiber and outside it. We shall be interested, however, in the field scattered by the nanofiber, which is described by the coefficients a_{1n}, b_{1n}, c_{1n}, and d_{1n}. They are found as

$$
a_{1n}(\beta,\omega) = \frac{D_{1n}(\beta,\omega)}{D_n(\beta,\omega)}, \quad (30.30)
$$

$$
b_{1n}(\beta,\omega) = -\frac{D_{2n}(\beta,\omega)}{D_n(\beta,\omega)}, \quad (30.31)
$$

$$
c_{1n}(\beta,\omega) = \frac{D_{3n}(\beta,\omega)}{D_n(\beta,\omega)}, \quad (30.32)
$$

$$
d_{1n}(\beta,\omega) = \frac{D_{4n}(\beta,\omega)}{D_n(\beta,\omega)}, \quad (30.33)
$$

where

D_n is the determinant of the matrix $\hat{M}_n$

D_{1n} and D_{4n} are the determinants of the matrices obtained from $\hat{M}_n$ by replacing the third and fourth columns with the column given by $\vec{F}_n$, respectively

D_{2n} and D_{3n} are the determinants of the matrices obtained from $\hat{M}_n$ by replacing the third and fourth columns with the column given by $\vec{G}_n$, respectively

The corresponding Fourier-transformed electric and magnetic field amplitudes can be written in the form

$$
\tilde{E}_{1\mu}(r,\theta;\beta,\omega) = \tilde{E}_{i0}(\beta,\omega)
$$
$$
\times \sum_{n=0}^{\infty}\left[A_{\mu n}(r;\beta,\omega)\sin(n\theta) + B_{\mu n}(r;\beta,\omega)\cos(n\theta)\right], \quad (30.34)
$$

$$
\tilde{H}_{1\mu}(r,\theta;\beta,\omega) = \tilde{H}_{i0}(\beta,\omega)
$$
$$
\times \sum_{n=0}^{\infty}\left[C_{\mu n}(r;\beta,\omega)\sin(n\theta) + D_{\mu n}(r;\beta,\omega)\cos(n\theta)\right], \quad (30.35)
$$

where the subscript μ runs over the components r, θ, and z, and the functions $A_{\mu n}$, $B_{\mu n}$, $C_{\mu n}$, and $D_{\mu n}$ are given by the equations

$$
A_{rn}(r;\beta,\omega) = \frac{i\beta}{q_1}H_n^{(1)'}(q_1r)a_{1n}(\beta,\omega) - \frac{i\omega n}{cq_1^2 r}H_n^{(1)}(q_1r)d_{1n}(\beta,\omega), \quad (30.36)
$$

$$
B_{rn}(r;\beta,\omega) = \frac{i\beta}{q_1}H_n^{(1)'}(q_1r)b_{1n}(\beta,\omega) + \frac{i\omega n}{cq_1^2 r}H_n^{(1)}(q_1r)c_{1n}(\beta,\omega), \quad (30.37)
$$

$$
A_{\theta n}(r;\beta,\omega) = -\frac{i\beta n}{q_1^2 r}H_n^{(1)}(q_1r)b_{1n}(\beta,\omega) - \frac{i\omega}{cq_1}H_n^{(1)'}(q_1r)c_{1n}(\beta,\omega), \quad (30.38)
$$

$$
B_{\theta n}(r;\beta,\omega) = \frac{i\beta n}{q_1^2 r}H_n^{(1)}(q_1r)a_{1n}(\beta,\omega) - \frac{i\omega}{cq_1}H_n^{(1)'}(q_1r)d_{1n}(\beta,\omega), \quad (30.39)
$$

$$
A_{zn}(r;\beta,\omega) = H_n^{(1)}(q_1r)a_{1n}(\beta,\omega), \quad (30.40)
$$

$$
B_{zn}(r;\beta,\omega) = H_n^{(1)}(q_1r)b_{1n}(\beta,\omega). \quad (30.41)
$$

$$
C_{rn}(r;\beta,\omega) = \frac{i\omega\epsilon_1 n}{cq_1^2 r}H_n^{(1)}(q_1r)b_{1n}(\beta,\omega) + \frac{i\beta}{q_1}H_n^{(1)'}(q_1r)c_{1n}(\beta,\omega), \quad (30.42)
$$

$$D_{rn}(r;\beta,\omega) = -\frac{i\omega\epsilon_1 n}{cq_1^2 r}H_n^{(1)}(q_1 r)a_{1n}(\beta,\omega) + \frac{i\beta}{q_1}H_n^{(1)\prime}(q_1 r)d_{1n}(\beta,\omega),$$

(30.43)

$$C_{\theta n}(r;\beta,\omega) = \frac{i\omega\epsilon_1}{cq_1}H_n^{(1)\prime}(q_1 r)a_{1n}(\beta,\omega) - \frac{i\beta n}{q_1^2 r}H_n^{(1)}(q_1 r)d_{1n}(\beta,\omega),$$

(30.44)

$$D_{\theta n}(r;\beta,\omega) = \frac{i\omega\epsilon_1}{cq_1}H_n^{(1)\prime}(q_1 r)b_{1n}(\beta,\omega) + \frac{i\beta n}{q_1^2 r}H_n^{(1)}(q_1 r)c_{1n}(\beta,\omega),$$

(30.45)

$$C_{zn}(r;\beta,\omega) = H_n^{(1)}(q_1 r)c_{1n}(\beta,\omega),$$

(30.46)

$$D_{zn}(r;\beta,\omega) = H_n^{(1)}(q_1 r)d_{1n}(\beta,\omega).$$

(30.47)

Taking the Fourier transforms of Equations 30.34 and 30.35,

$$E_{1\mu}(r,\theta,z,t) = \frac{1}{(2\pi)^2}\int_{-\infty}^{\infty}\int_{-\infty}^{\infty}\tilde{E}_{1\mu}(r,\theta;\beta,\omega)e^{i\beta z - i\omega t}d\beta d\omega,$$

(30.48)

$$H_{1\mu}(r,\theta,z,t) = \frac{1}{(2\pi)^2}\int_{-\infty}^{\infty}\int_{-\infty}^{\infty}\tilde{H}_{1\mu}(r,\theta;\beta,\omega)e^{i\beta z - i\omega t}d\beta d\omega,$$

(30.49)

one obtains the solution of the problem under consideration.

The integrands in Equations 30.48 and 30.49 have the poles given by the zeros of the denominators $D_n(\beta,\omega)$ in Equations 30.30 through 30.33. On the other hand, these zeros determine the allowed values of pairs (β,ω) for the electromagnetic field in a free nanofiber, i.e., its normal modes (Stratton, 1941).

30.3.2 Nanofiber Normal Modes

Among the solutions of the equations

$$D_n(\beta,\omega) = 0,$$

(30.50)

there are those which are represented by pairs of real quantities $(\beta_{nm}, \omega_{nm})$. They give the propagation constants and frequencies of the waveguide (bound) modes. Such modes provide a complete description of the light propagation along the fiber in the steady-state regime far from the light source (Snyder and Love, 1983). Besides that, Equation 30.50 has solutions for which either β or ω, or both, have imaginary parts, i.e.,

$$\beta = \beta^r + i\beta^i,$$

(30.51)

$$\omega = \omega^r - i\omega^i.$$

(30.52)

The contribution of such poles to the integrals (30.48) and (30.49) leads to the transients along the fiber length or in time, or in both length and time, respectively. If β^i is small, the modes of the first type can propagate over long distances, and their

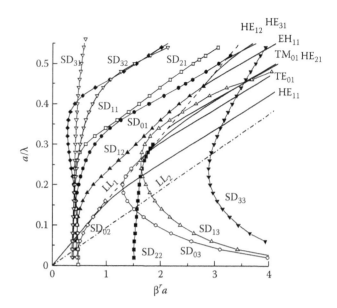

FIGURE 30.3 Dispersion curves of the SD modes. Real parts of the mode propagation constants. The modes with very large values of β^i are not shown in the figure. $\epsilon_1 = 1$, $\epsilon_2 = 2.89$. (Reprinted from Bordo, V., *J. Phys.: Condens. Matter*, 19, 236220-1, 2007. With permission.)

portions localized within or near the fiber core are known as leaky modes (Snyder and Love, 1983). In the following, to distinguish between different modes, we shall call the first-type modes as *space-decaying* (SD) modes, whereas the second-type ones as *time-decaying* (TD) modes. Figure 30.3 shows the dispersion curves $\beta^r(\omega)$ calculated numerically for the SD modes, SD_{nm}, with $n = 0$–3 and represented in dimensionless variables. It has been assumed that $\epsilon_1 = 1$ and $\epsilon_2 = 2.89$, which correspond to an isolated nanofiber obtained from an isotropic *para*-hexaphenyl material immersed in vacuum. The dispersion curves for the bound modes are also shown in Figure 30.3. They are disposed between the light lines $\omega = c\beta/\sqrt{\epsilon_1}$ (LL_1) and $\omega = c\beta/\sqrt{\epsilon_2}$ (LL_2), and are denoted as it is accepted in the optical waveguide theory (Snyder and Love, 1983). The dispersion curves for the quantities $\beta^i(\omega)$ are shown in Figure 30.4. Figure 30.5 represents the dispersion relations $\omega^r(\beta)$ for $n = 0$–3 calculated for the TD modes, TD_{nm}. The corresponding imaginary parts, $\omega^i(\beta)$, for the modes that are not strongly decaying are plotted in Figure 30.6. Here, the subscript m numerates the modes with a given n according to the order they approach the axis β^r for the SD-modes or the axis ω^r for the TD-modes. One can observe that some of these modes, both SD and TD, can be considered as a continuation of waveguide modes beyond cutoff, which coincides with the light line LL_1. The imaginary parts of the propagation constants or frequencies for such modes increase with the deviation of the corresponding real parts from cutoff.

30.3.3 Calculation of the Scattered Field

When carrying out the integration in Equations 30.48 and 30.49, one can take into account that the functions $\tilde{E}_{i0}(\beta,\omega)$ and $\tilde{H}_{i0}(\beta,\omega)$ in the integrands are essentially different from

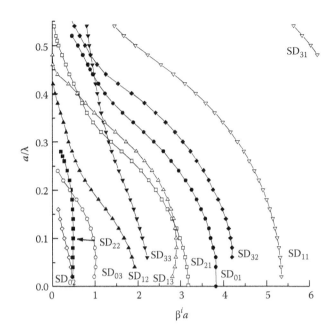

FIGURE 30.4 Same as Figure 30.3 but for the imaginary parts of the mode propagation constants. (Reprinted from Bordo, V., *J. Phys.: Condens. Matter*, 19, 236220-1, 2007. With permission.)

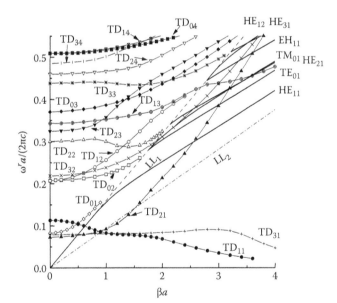

FIGURE 30.5 Dispersion curves of the TD modes. Real parts of the mode frequencies. $\varepsilon_1 = 1$, $\varepsilon_2 = 2.89$. (Reprinted from Bordo, V., *J. Phys.: Condens. Matter*, 19, 236220-1, 2007. With permission.)

zero around their maxima at $\beta = \beta_0$ and $\omega = \omega_0$. For typical experimental parameters, the widths of the corresponding peaks, $\Delta\beta \ll 1/a$ and $\Delta\omega \ll c/a$, and the slowly varying functions $A_{\mu n}(\beta, \omega)$, $B_{\mu n}(\beta, \omega)$, $C_{\mu n}(\beta, \omega)$, and $D_{\mu n}(\beta, \omega)$ can be taken off the integral at the point (β_0, ω_0).* The remaining integration leads to the result

---

* See Bordo (2007) for detail.

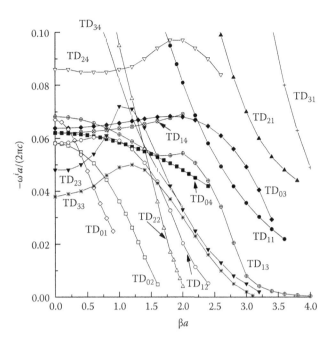

FIGURE 30.6 Same as Figure 30.5 but for the imaginary parts of the mode frequencies, which fall into the range $|\omega^i a/(2\pi c)| \le 0.1$. (Reprinted from Bordo, V., *J. Phys.: Condens. Matter*, 19, 236220-1, 2007. With permission.)

$$E_{1\mu}(r,\theta,z,t) = E_{i0}(z,t)\exp(i\beta_0 z - i\omega_0 t)$$
$$\times \sum_{n=0}^{\infty} \left[A_{\mu n}(r;\beta_0,\omega_0)\sin(n\theta) + B_{\mu n}(r;\beta_0,\omega_0)\cos(n\theta) \right],$$
$$(30.53)$$

$$H_{1\mu}(r,\theta,z,t) = H_{i0}(z,t)\exp(i\beta_0 z - i\omega_0 t)$$
$$\times \sum_{n=0}^{\infty} \left[C_{\mu n}(r;\beta_0,\omega_0)\sin(n\theta) + D_{\mu n}(r;\beta_0,\omega_0)\cos(n\theta) \right],$$
$$(30.54)$$

where $E_{i0}(z,t)$ and $H_{i0}(z,t)$ are the amplitudes of the incident wave electric and magnetic fields, respectively. In other words, under typical experimental conditions, the illumination by a pulsed light beam can be treated as the case of an incident plane monochromatic wave if instead of its amplitudes, E_0 and H_0, one substitutes the pulsed beam envelope functions, $E_{i0}(z,t)$ and $H_{i0}(z,t)$, respectively.

We shall be interested in the intensity of light scattered in the *xy* plane. This quantity is determined by the time-averaged radial component of the Poynting vector

$$S_r = \frac{c}{8\pi}\text{Re}(E_{1\theta}H_{1z}^* - E_{1z}H_{1\theta}^*), \qquad (30.55)$$

where the asterisk denotes complex conjugation. The substitution of the field components, as shown in Equations 30.53 and 30.54, into Equation 30.55 gives the intensity of the scattered electromagnetic field at any distance from the nanofiber, i.e., in both near-field and far-field regions. In the far zone where $r \gg \lambda$, the

Hankel functions and their derivatives can be replaced by their asymptotical expansions. Then in the leading order one obtains

$$S_r = S_r^{TM} + S_r^{TE}, \tag{30.56}$$

where the TM- and TE-polarized components are given by

$$S_r^{TM}(\theta) = \frac{\omega \epsilon_1}{4\pi^2 q_{10}^2 r} \left| \sum_{n=0}^{\infty} (-i)^n [a_{1n}(\beta_0, \omega_0) \sin(n\theta) + b_{1n}(\beta_0, \omega_0) \cos(n\theta)] \right|^2 \tag{30.57}$$

and

$$S_r^{TE}(\theta) = \frac{\omega}{4\pi^2 q_{10}^2 r} \left| \sum_{n=0}^{\infty} (-i)^n [c_{1n}(\beta_0, \omega_0) \sin(n\theta) + d_{1n}(\beta_0, \omega_0) \cos(n\theta)] \right|^2, \tag{30.58}$$

respectively. The total intensity scattered in all directions per unit length of the nanofiber is determined by the quantity

$$S_{tot} \equiv \int_0^{2\pi} S_r(\theta) d\theta = S_{tot}^{TM} + S_{tot}^{TE} \tag{30.59}$$

with

$$S_{tot}^{TM} = \frac{\omega \epsilon_1}{4\pi q_{10}^2 r} \left\{ \left| b_{10}(\beta_0, \omega_0) \right|^2 + \sum_{n=1}^{\infty} \left[\left| a_{1n}(\beta_0, \omega_0) \right|^2 + \left| b_{1n}(\beta_0, \omega_0) \right|^2 \right] \right\} \tag{30.60}$$

and

$$S_{tot}^{TE} = \frac{\omega}{4\pi q_{10}^2 r} \left\{ \left| d_{10}(\beta_0, \omega_0) \right|^2 + \sum_{n=1}^{\infty} \left[\left| c_{1n}(\beta_0, \omega_0) \right|^2 + \left| d_{1n}(\beta_0, \omega_0) \right|^2 \right] \right\}. \tag{30.61}$$

If in an experiment, one varies either the angle of incidence of the light beam at a fixed frequency, or tunes the light frequency at a fixed angle of incidence, the scattered intensity will also change. In the former case, the intensity (30.56) has maxima at the values $\beta_0 = \beta_{nm}^r$ corresponding to the SD modes. In the latter case, the maxima in intensity occur at $\omega_0 = \omega_{nm}^r$ and correspond to the TD modes. The widths of those peaks are determined by the quantities β_{nm}^i and ω_{nm}^i, respectively.*

---

* Besides the maxima relatively the light frequency discussed here, there may also be maxima originating from resonances in $\epsilon_2(\omega)$.

30.3.4 Semicylinder on an Ideally Reflecting Surface

The approach developed above can be equally applied to the nanofiber model represented by a circular semicylinder placed on an ideally reflecting plane. When the reflecting plane at $y = 0$ is present, the total electric field in the half-space $y > 0$ can be written as

$$\mathbf{E} = \mathbf{E}_i + \mathbf{E}_i^{ref} + \mathbf{E}_1 + \mathbf{E}_1^{ref}, \tag{30.62}$$

where $\mathbf{E}_i^{ref}$ and $\mathbf{E}_1^{ref}$ are the electric fields specularly reflected from the plane $y = 0$ originating from the incident wave and the field scattered by the nanofiber, respectively. A similar expression is valid for the magnetic field vector. The condition of an ideally reflecting surface implies that the tangential components of the total electric field at it are equal to zero. This requirement will be fulfilled if to introduce, instead of the reflecting plane, an image semicylinder illuminated by the image of the incident wave below the plane $y = 0$ so that $E_r = E_z = 0$ at both $\theta = 0$ and $\theta = \pi$ (Rao and Barakat, 1994). The same is true for the magnetic field components. The fields scattered by the image semicylinder, $\mathbf{E}_1^{im}$ and $\mathbf{H}_1^{im}$, are determined by Equations 30.34 through 30.47 with the substitutions

$$a_{1n}^{im} = -a_{1n}, \quad b_{1n}^{im} = -b_{1n}, \quad c_{1n}^{im} = -c_{1n}, \quad d_{1n}^{im} = -d_{1n}. \tag{30.63}$$

Then the components of the field scattered by the semicylinder in the presence of the reflecting plane are found as $\mathbf{E}_1(\theta) + \mathbf{E}_1^{im}(-\theta)$ and $\mathbf{H}_1(\theta) + \mathbf{H}_1^{im}(-\theta)$. Formally, the field components can be obtained from Equations 30.53 and 30.54 with the following substitutions:

$$A_{\mu n} \to 2A_{\mu n}, \quad B_{\mu n} \to 0, \quad C_{\mu n} \to 2C_{\mu n}, \quad D_{\mu n} \to 0. \tag{30.64}$$

Let us assume that the angle of incidence, α, differs from zero. Then the fields $\mathbf{E}_i^{ref}$ and $\mathbf{H}_i^{ref}$ do not contribute to the scattered intensity determined by S_r. In the far zone, its TM- and TE-polarized components are given by

$$S_r^{TM}(\theta) = \frac{\omega \epsilon_1}{\pi^2 q_{10}^2 r} \left| \sum_{n=1}^{\infty} (-i)^n a_{1n}(\beta_0, \omega_0) \sin(n\theta) \right|^2 \tag{30.65}$$

and

$$S_r^{TE}(\theta) = \frac{\omega}{\pi^2 q_{10}^2 r} \left| \sum_{n=1}^{\infty} (-i)^n c_{1n}(\beta_0, \omega_0) \sin(n\theta) \right|^2, \tag{30.66}$$

respectively. The corresponding total intensity scattered in all directions per unit length of the nanofiber is determined by the quantity

$$S_{tot} = \int_0^{\pi} S_r(\theta) d\theta = \frac{\omega}{2\pi q_{10}^2 r} \sum_{n=1}^{\infty} \left[\epsilon_1 \left| a_{1n}(\beta_0, \omega_0) \right|^2 + \left| c_{1n}(\beta_0, \omega_0) \right|^2 \right]. \tag{30.67}$$

Note that here the sum over the modes starts from $n = 1$, indicating that the modes with $n = 0$ are not excited in the course of light scattering. The suppression of such modes is conditioned by the presence of the reflecting plane.

Let us consider the special case of normal incidence with respect to the substrate surface ($\alpha = 0$). In such a case, $\beta_0 = 0$ and the equations for TM and TE polarizations are completely separated. This means that for TM incident wave polarization, one can set $c_{1n} = d_{1n} = 0$. Besides that, the vector $\vec{F}_n$, Equation 30.23, is nonzero only for odd indices n. Then the coefficients a_{1n} are nonzero for odd n and only such modes can be observed in scattering. For TE incident wave polarization, $a_{1n} = b_{1n} = 0$ and the vector $\vec{G}_n$, Equation 30.26, is nonzero for odd n. As a result, one obtains the same selection rule as for TM polarization.

30.4 Some Numerical Results

We shall illustrate the general theory considered above with some numerical examples. All the calculations have been carried out for $\varepsilon_2 = 2.89$ that corresponds to an isotropic *para*-hexaphenyl film. We shall compare the results obtained for two different models of a nanofiber.

30.4.1 Angular Distribution of Scattered Light

Let us consider first the dependence of the Poynting vector component S_r, Equation 30.55, calculated in the far zone on the detection angle, θ. This quantity is determined by the coefficients a_{1n}, b_{1n}, c_{1n}, and d_{1n}, as shown in Equations 30.30 through 30.33. Their variation with the nanofiber radius, a, is more rapid for larger values of $q_{10} = (\omega_0/c)\sqrt{\varepsilon_1}\cos\alpha$ and hence for smaller angles of incidence, α. Figures 30.7 and 30.8 show the angular distributions of light scattered by a cylindrical nanofiber for the TM and TE incident light polarizations, respectively, and for $\alpha = 45°$. One can see that in the case of the TM incident wave polarization, there are two lobes directed along the normal to the plane of incidence (at $\theta = 0°$ and $\theta = 180°$). For relatively large nanofiber radius, additional two lobes appear. In the case of the TE incident wave polarization, these lobes are more pronounced. Besides that, for small nanofiber radius, the angular distribution of scattered light is more isotropic as compared to the TM polarization.

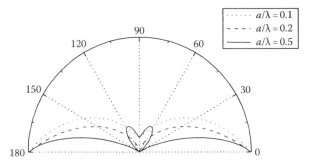

FIGURE 30.7 Angular distribution of light scattered from a cylindrical nanofiber of different radii. TM incident wave polarization. $\alpha = 45°$, $\varepsilon_1 = 1$, $\varepsilon_2 = 2.89$. All curves are normalized to their maximum values.

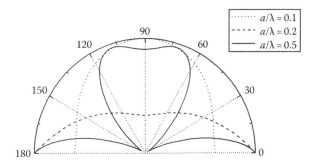

FIGURE 30.8 Same as Figure 30.7 but for TE incident wave polarization.

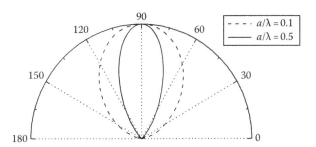

FIGURE 30.9 Same as Figure 30.7 but for a semicylindrical nanofiber on a perfectly reflecting surface.

The same quantity but calculated for a semicylindrical nanofiber placed on a perfectly reflecting substrate and for the TM incident wave polarization is depicted in Figure 30.9. The scattering diagram for the TE incident wave polarization has a similar form. In this case, the lobes directed along the substrate plane are suppressed and light is scattered mainly around the plane of incidence.

30.4.2 Total Scattered Intensity versus Incidence Angle

Let us consider the total scattered intensity, S_{tot}, as a function of the incidence angle, α, or, equivalently, the quantity $\beta_0 = (\omega_0/c)\sqrt{\varepsilon_1}\sin\alpha$. Figure 30.10 shows this dependence calculated for a cylinder and for the TM incident wave polarization with an account of the terms $n = 0$–10 in Equations 30.60 and 30.61. The individual contributions from the terms with $n = 0$, 1, and 2 are also shown in Figure 30.10, whereas the other contributions have much smaller amplitude. The intensity drop at $\beta_0 a \approx 2.2$ originates from the approach to the cutoff at which $q_{10} = 0$. The curve corresponding to $n = 2$ exhibits a shoulder around $\beta_0 a \approx 1.75$, which is also seen in the total intensity curve. Turning to Figure 30.3, one can identify it as being originating from the mode SD_{21}. The contributions of the other modes are not pronounced because of the large values of the propagation constant imaginary parts, β^i, corresponding to them. This plot can be compared with Figure 30.11, which represents the results of calculations for a semicylindrical nanofiber placed on an ideally reflecting surface. In this case, the zeroth term does not contribute to S_{tot} and the contribution of the term with $n = 2$ is equal

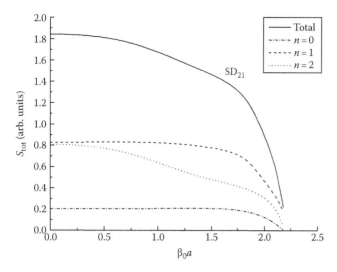

FIGURE 30.10 Light intensity scattered from a cylindrical nanofiber as a function of β_0 calculated for $\varepsilon_1 = 1$, $\varepsilon_2 = 2.89$, and $\omega_0 a/(2\pi c) = 0.35$. TM incident wave polarization.

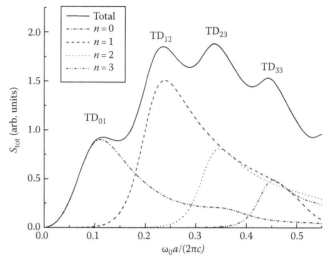

FIGURE 30.12 Light intensity scattered from a cylindrical nanofiber as a function of ω_0 calculated for $\varepsilon_1 = 1$, $\varepsilon_2 = 2.89$. Normal incidence. TM incident wave polarization.

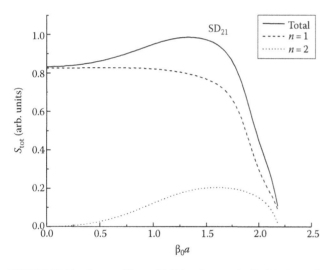

FIGURE 30.11 Same as Figure 30.10 but for a semicylindrical nanofiber on a perfectly reflecting surface.

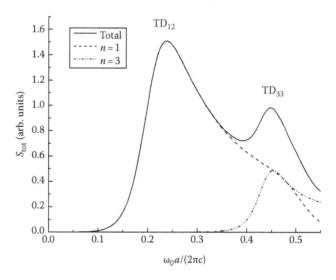

FIGURE 30.13 Same as Figure 30.12 but for a semicylindrical nanofiber on a perfectly reflecting surface. (Reprinted from Bordo, V., *J. Phys.: Condens. Matter*, 19, 236220-1, 2007. With permission.)

to zero at $\beta_0 = 0$ (normal incidence) in accordance with the selection rule discussed in Section 30.3.4. As a result, the shoulder in the total intensity originating from the mode SD_{21} becomes more distinct.

30.4.3 Total Scattered Intensity versus Frequency

Another situation when the light beam frequency is scanned at a fixed angle of incidence ($\alpha = 0$) is illustrated in Figure 30.12 for the case of scattering of a TM-polarized beam by a cylindrical nanofiber. This time the features in the total scattered intensity originate from the TD modes. They can be identified when comparing the positions of maxima, which occur for the curves with different n with the mode frequencies at $\beta_0 = 0$ in Figure 30.5.

In the case of a semicylinder placed on an ideally reflecting plane (Figure 30.13), only modes with odd n can be excited, as it is expected from the selection rule (see Section 30.3.4). It is worthwhile to note that the modes distinct from those for TM polarization can be observed in the scattering of a TE-polarized light beam (not shown).

30.4.4 Excitation near Exciton Resonance

So far we have implied that the frequency of the incident beam is far from any resonance in the dielectric function of the nanofiber and its frequency dispersion can be neglected. Let us assume

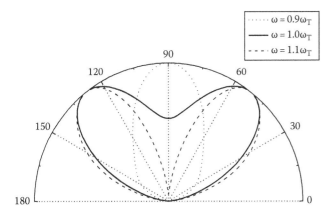

FIGURE 30.14 Angular distribution of scattered light in the vicinity of the exciton resonance. TM incident wave polarization. $\varepsilon_1 = 1$, $\varepsilon_\infty = 2.89$, $a = 4c/\omega_T$, $\omega_L = 1.2\omega_T$, $\Gamma = 10^{-3}\omega_T$, and $\alpha = 30°$.

now that the frequency ω_0 is close to an exciton resonance of the nanofiber. Then its dielectric function can be written as follows:

$$\varepsilon_2(\omega) = \frac{\omega_L^2 - \omega^2}{\omega_T^2 - \omega^2 + i\omega\Gamma} \varepsilon_\infty, \qquad (30.68)$$

where

  ω_T and ω_L are the frequencies of the transverse and longitudinal excitons, respectively
  Γ is the relaxation constant
  ε_∞ is the high-frequency dielectric constant, and the spatial dispersion has been neglected

Figure 30.14 shows the angular distribution of the scattered light in this case calculated for a semicylindrical nanofiber on an ideally reflecting surface. It is seen that the polar diagram of scattering changes dramatically in a narrow spectral interval when the frequency of the incident beam scans across the exciton resonance.

30.5 Experimental Implementation

Experimentally, one can observe light scattering from nanofibers in the following setup (Figure 30.15) (Fiutowski et al., 2008). A sample with deposited organic nanofibers is attached to a half-sphere made of fused silica ($n_s = 1.48$ at 325 nm) with a radius of 10 mm. The sample is oriented so that the nanofiber axes would be parallel to the plane of light incidence. A flat domain of parallelly oriented organic nanofibers is illuminated with a linearly polarized beam of a He-Cd laser ($\lambda = 325$ nm) from the half-sphere side, and the scattered light intensity is detected along the normal to the sample surface from the backside. The extinction ratio of the polarization system is 10^5. The light beam passes a pinhole of diameter 400 μm and is collimated before entering the half-sphere, which results in a beam divergence of less than 0.7° after leaving the half-sphere. The angles of incidence can be varied between 20° and 65° with respect to the normal axis of the

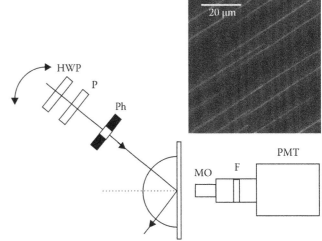

FIGURE 30.15 Experimental setup. HWP, half-wave plate; P, polarizer; Ph, pinhole; MO, microscope objective; F, filter; and PMT, photomultiplier. Inset: Fluorescence microscopy image of a sample domain. (Reprinted from Fiutowski, J. et al., *Appl. Phys. Lett.*, 92, 073302-1, 2008. With permission.)

plane of the half-sphere. A photomultiplier mounted on a goniometric table allows one to detect scattered light intensity distributions from a spot on the sample of a diameter of about 500 μm. The acceptance angle of the detection system is 44°. The absolute and relative accuracies of the incident and the detection angles are 1° and 0.5°, respectively. To observe only the wavelengths of interest, one can use band pass or interferometric filters in front of the photomultiplier.

The intensity of light scattered perpendicularly to the sample surface as a function of the incidence angle, α, is shown in Figure 30.16. Clear peaks at $\alpha = 40°$ in both TE and TM incident light polarizations are observed. Those peaks originate from the

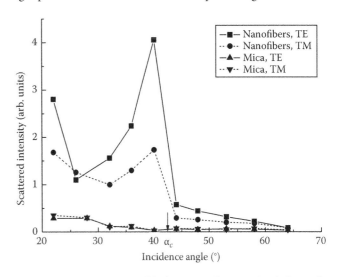

FIGURE 30.16 Intensity of light scattered perpendicularly to the sample surface as a function of the angle of incidence. The incident wave polarization is indicated in the inset. (Reprinted from Fiutowski, J. et al., *Appl. Phys. Lett.*, 92, 073302-1, 2008. With permission.)

nanofiber array exclusively, as seen from a comparison with the data obtained from a nanofiber-free mica surface. The scattered intensity decreases when the incidence angle exceeds the critical angle for total internal reflection at the quartz semisphere–air interface, $\alpha_c = 42.5°$. To describe the optical properties of a nanofiber grown on a substrate, we use the nanofiber model represented by a semicylinder placed on an ideally reflecting surface and considered above. Figure 30.17 shows the dispersion curves of the nanofiber normal modes plotted in the range of the size parameter, a/λ, corresponding to the experimentally observed typical nanofiber widths ($2a$) as well as the wavelength of the He-Cd laser used for excitation. The condition of the phase matching between the incident light and the nanofiber normal mode can be written as $\beta = (2\pi/\lambda)n_s \sin\alpha$. The corresponding dispersion line for the angle $\alpha = 40°$, at which the resonance in scattering is observed, is also shown in Figure 30.17.

The analysis of the polarization properties of the excited normal modes allows one to specify them. The sharp peak seen in the scattered light intensity at $\alpha = 40°$ has comparable amplitudes in both TE and TM incident wave polarizations. It can therefore be associated with the excitation of the hybrid SD_{12} mode. The increase in intensity at $\alpha < 26°$ originates probably from a resonance with the hybrid SD_{21} mode. The mode SD_{01} is suppressed in scattering due to the presence of the reflecting substrate (see Section 30.3.4).

The positions of the maxima in Figure 30.16 correspond to the most probable size parameter within the illuminated spot. The broadenings of the peaks observed in light scattering allow one to estimate the distribution of the nanofibers in the array over their widths. The angular broadening of about 7° corresponds to a scattering of 0.12λ in nanofiber diameters, which

gives approximately 40 nm for the relevant wavelength. This value agrees qualitatively with that measured via atomic force microscope (AFM).

30.6 Discussion

We have considered here the simplest theoretical models of nanofibers and applied them for the interpretation of the experimental results on light scattering from an array of organic nanoaggregates deposited on a mica substrate. Although the morphology of nanoaggregates is far from a cylindrical shape, light scattering from them can be similarly described in terms of excitation of their radiative normal modes. This conclusion is also supported by the measurements of photoluminescence excited in the course of light scattering (Fiutowski et al., 2008). In that case, the peaks in the photoluminescence intensity versus the angle of incidence originate from matching with the nanoaggregates' normal modes, both radiative and waveguiding. Such measurements being carried out at a varying frequency of incident light should allow one to determine the dispersion curves of the actual normal modes of nanoaggregates. Then, using an appropriate model, one could extract information on the morphology of nanofibers. However, it is necessary to keep in mind that the actual normal modes may differ from those of an isotropic cylindrical nanofiber. In particular, the optical anisotropy of nanoaggregates can cause distortion of the dispersion curves as well as their splitting.

Another point which has to be taken into account is the different morphology of nanoaggregates within the illuminated spot. As we have already mentioned, this leads to a broadening of the peaks in the intensity of scattered light. On the other hand, the data obtained from the scattering of a focused light beam reflect the local morphology of a sample and, thus, can be used for its local characterization.

30.7 Summary and Outlook

In this chapter, we have considered the theory which describes the scattering of a pulsed Gaussian light beam at an infinitely long dielectric cylinder. The results obtained for this model have been generalized to the model represented by a semicylinder placed on a perfectly reflecting plane. Although being rather simple, these approaches reproduce the main features of light scattering from nanofibers. In particular, this process can be understood in terms of the excitation of nanofiber radiative modes. As a result, the dependence of the scattered light intensity on the angle of incidence exhibits maxima corresponding to matching with the nanofiber SD modes.

These conclusions are in agreement with the measurements of light scattering from organic nanoaggregates grown on mica, which have demonstrated experimental evidence of mode launching in this system. Those modes show up as pronounced peaks in light scattering from the nanofibers. Using the model, one is able to identify specific modes that have been excited in the course of light scattering from the sample. The obtained

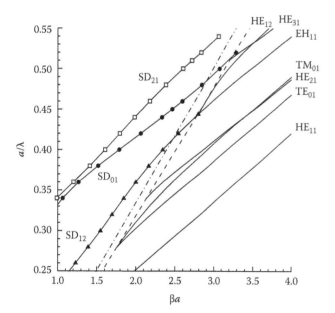

FIGURE 30.17 Dispersion curves of the SD modes plotted in the range of size parameters corresponding to measured nanofiber widths. The inclined dot-and-dash line indicates the dispersion for light incident at the angle $\alpha = 40°$.

results provide detailed information about possible electromagnetic mode propagation in nanosized, needle-shaped aggregates and also form the basis for a new way of optically characterizing the morphology of sub-wavelength-sized nanostructures via far-field scattering. The further development of this technique by using a tunable light source should allow one to determine the actual dispersion curves of nanofibers. On the other hand, one can use a broadband light source to illuminate the sample instead of a laser. In such a case, one should expect to observe peaks in the spectrum of scattered light, which originates from the excitation of TD nanofiber modes. The implementation of this technique with the use of a scanning near-field optical microscope for the registration of scattered field would provide an additional advantage. The data obtained in such a setup are not influenced by an averaging over the illuminated spot and, thus, are related to individual nanofibers. Another possible application of light scattering from nanofibers stems from correlation between the positions of peaks in scattered intensity and the dispersion of nanofiber normal modes. Any change in the optical properties of nanofibers caused by either heating, or mechanical tension, or adsorption of molecules from ambient medium would result in a shift of the peaks in light scattering. This opens up new opportunities for the creation of nanosensors of various kinds. Due to its generic nature, the method discussed in this chapter is not limited to organic nanofibers but can be equally well applied to other kind of light-emitting nanoaggregates.

Acknowledgments

The author is grateful to Prof. H.-G. Rubahn and Dr. L. Jozefowski for the discussions of the experimental aspects of light scattering from nanofibers.

References

Agarwal, R., Barrelet, C., and Lieber, C. (2005). Lasing in single cadmium sulfide nanowire optical cavity. *Nano Lett.*, 5:917–920.

Ashkin, A., Dziedzic, J., and Stolen, R. (1981). Outer diameter measurement of low birefringence optical fibers by a new resonant backscatter technique. *Appl. Opt.*, 20:2299–2303.

Balzer, F., Bordo, V., Simonsen, A., and Rubahn, H.-G. (2003). Optical waveguiding in individual nanometer-scale organic fibers. *Phys. Rev. B*, 67:115408-1–8.

Balzer, F. and Rubahn, H.-G. (2001). Dipole-assisted self-assembly of light-emitting *p*-nP needles on mica. *Appl. Phys. Lett.*, 79:3860–3862.

Balzer, F. and Rubahn, H.-G. (2005). Growth control and optics of organic nanoaggregates. *Adv. Funct. Mater.*, 15:17–24.

Barrelet, C., Greytak, A., and Lieber, C. (2004). Nanowire photonic circuit elements. *Nano Lett.*, 4:1981–1985.

Bever, S. and Allebach, J. (1992). Multiple scattering by a planar array of parallel dielectric cylinders. *Appl. Opt.*, 31:3524–3532.

Birkhoff, R., Ashley, J., Hubbel, H. Jr., and Emerson, L. (1977). Light scattering from micron-size fibers. *J. Opt. Soc. Am.*, 67:564–569.

Bordo, V. (2006). Light scattering from a nanofiber: Exact numerical solution of a model system. *Phys. Rev. B*, 73:205117-1–205117-7.

Bordo, V. (2007). Theory of nanofibre excitation by a pulsed light beam. *J. Phys. Condens. Matter*, 19:236220-1–236220-12.

Borghi, R., Santarsiero, M., Frezza, F., and Schettini, G. (1997). Plane-wave scattering by a dielectric circular cylinder parallel to a general reflecting flat surface. *J. Opt. Soc. Am. A*, 14:1500–1504.

Brewer, J., Maibohm, C., Jozefowski, L., Bagatolli, L., and Rubahn, H.-G. (2005). A 3D view on free-floating, space-fixed and surface-bound *para*-phenylene nanofibers. *Nanotechnology*, 16:2396–2401.

Chin, A., Vaddiraju, S., Maslov, A., Ning, C., Sunkara, M., and Meyyappan, M. (2006). Near-infrared semiconductor subwavelength-wire lasers. *Appl. Phys. Lett.*, 88:163115-1–163115-3.

Cohen, L., Haracz, R., Cohen, A., and Acquista, C. (1983). Scattering of light from arbitrary oriented finite cylinders. *Appl. Opt.*, 22:742–748.

Duan, X., Huang, Y., Agarwal, R., and Lieber, C. (2003). Single-nanowire electrically driven lasers. *Nature*, 421:241–245.

Duan, X., Huang, Y., Cui, Y., Wang, J., and Lieber, C. (2001). Indium phosphide nanowires as building blocks for nanoscale electronic and optoelectronic devices. *Nature*, 409:66–69.

Fiutowski, J., Bordo, V., Jozefowski, L., Madsen, M., and Rubahn, H.-G. (2008). Light scattering from an ordered array of needle-shaped organic nanoaggregates: Evidence for optical mode launching. *Appl. Phys. Lett.*, 92:073302-1–073302-3.

Haracz, R., Cohen, L., and Cohen, A. (1984). Perturbation-theory for scattering from dielectric spheroids and short cylinders. *Appl. Opt.*, 23:436–441.

Huang, M., Mao, S., Feick, H. et al. (2001). Room-temperature ultraviolet nanowire nanolasers. *Science*, 292:1897–1899.

Johnson, J., Choi, H.-J., Knutsen, K., Schaller, R., Yang, P., and Saykally, R. (2002). Single gallium nitride nanowire lasers. *Nat. Mater.*, 1:106–110.

Johnson, J., Yan, H., Yang, P., and Saykally, R. (2003). Optical cavity effects in ZnO nanowire lasers and waveguides. *J. Phys. Chem.*, 107:8816–8828.

Kerker, M. (1969). *The Scattering of Light and Other Electromagnetic Radiation*. Academic Press, New York.

Kuik, F., de Haan, J., and Hovenier, J. (1994). Single scattering of light by circular cylinders. *Appl. Opt.*, 33:4906–4918.

Lord Rayleigh (1881). On the electromagnetic theory of light. *Philos. Mag.*, 12:81–101.

Owen, J., Barber, P., Messinger, B., and Chang, R. (1981). Determination of optical-fiber diameter from resonances in the elastic scattering spectrum. *Opt. Lett.*, 6:272–274.

Quochi, F., Cordella, F., Mura, A., Bongiovanni, G., Balzer, F., and Rubahn, H.-G. (2005). One-dimensional random lasing in a single organic nanofiber. *J. Phys. Chem. B*, 109:21690–21693.

Quochi, F., Cordella, F., Mura, A., Bongiovanni, G., Balzer, F., and Rubahn, H.-G. (2006). Gain amplification and lasing properties of individual organic nanofibers. *Appl. Phys. Lett.*, 88:041106-1–041106-3.

Rao, T. and Barakat, R. (1994). Near-field scattering by a conducting cylinder partially buried in a conducting plane. *Opt. Commun.*, 111:18–25.

Ruppin, R. (1990). Electromagnetic scattering from finite dielectric cylinders. *J. Phys. D Appl. Phys.*, 23:757–763.

Schiek, M., Lützen, A., Koch, R. et al. (2005). Nanofibers from functionalized para-phenylene molecules. *Appl. Phys. Lett.*, 86:153107-1–153107-3.

Shepherd, J. and Holt, A. (1983). The scattering of electromagnetic radiation from finite dielectric circular cylinders. *J. Phys. A Math. Gen.*, 16:651–662.

Snyder, A. and Love, J. (1983). *Optical Waveguide Theory*. Chapman and Hall, London, U.K.

Stratton, J. (1941). *Electromagnetic Theory*. McGraw-Hill, New York.

Thilsing-Hansen, K., Neves-Petersen, M., Petersen, S., Neuendorf, R., Al-Shamery, K., and Rubahn, H.-G. (2005). Luminescence decay of oriented phenylene nanofibers. *Phys. Rev. B*, 72:115213-1–115213-7.

Tong, L., Gattass, R., Ashcom, J. et al. (2003). Subwavelength-diameter silica wires for low-loss optical wave guiding. *Nature*, 426:816–819.

Twersky, V. (1952). Multiple scattering of radiation by an arbitrary planar configuration of parallel cylinders. *J. Acoust. Soc. Am.*, 24:42–46.

Uzunoglu, N., Alexopoulos, N., and Fikioris, J. (1978). Scattering from thin and finite dielectric fibers. *J. Opt. Soc. Am.*, 68:194–197.

Van Bladel, J. (1977). Resonant scattering by dielectric cylinders. *IEEE J. Microwave Opt. Acoust.*, 1:41–50.

Wait, J. (1955). Scattering of a plane wave from a circular dielectric cylinder at oblique incidence. *Can. J. Phys.*, 33:189–195.

Wang, J., Gudiksen, M., Duan, X., Cui, Y., and Lieber, C. (2001). Highly polarized photoluminescence and photodetection from single indium phosphide nanowires. *Science*, 293:1455–1457.

Zimmler, M., Bao, J., Capasso, F., Müller, S., and Ronning, C. (2008). Laser action in nanowires: Observation of the transition from amplified spontaneous emission to laser oscillation. *Appl. Phys. Lett.*, 93:051101-1–051101-3.

Biomimetics: Photonic Nanostructures

Andrew R. Parker
The Natural History Museum

31.1 Introduction

Three centuries of research, beginning with Hooke and Newton, have revealed a diversity of optical devices at the nanoscale (or at least the submicron scale) in nature.[1] These include structures that cause random scattering, 2D diffraction gratings, 1D multilayer reflectors, and 3D liquid crystals (Figure 31.1a through d). In 2001, the first photonic crystal was identified as such in animals,[2] and since then the scientific effort in this subject has accelerated. Now we know of a variety of 2D- and 3D-photonic crystals in nature (e.g., Figure 31.1e and f), including some designs not encountered previously in physics.

Biomimetics is the extraction of good design from nature. Some optical biomimetic successes have resulted from the use of conventional (and constantly advancing) engineering methods to make direct analogues of the reflectors and antireflectors found in nature. However, recent collaborations between biologists, physicists, engineers, chemists, and material scientists have ventured beyond merely mimicking in the laboratory what happens in nature, leading to a thriving new area of research involving biomimetics via cell culture. Here, the nano-engineering efficiency of living cells is harnessed, and nanostructures such as diatom "shells" can be made for commercial applications via culturing the cells themselves.

31.2 Engineering of Antireflectors

Some insects benefit from antireflective surfaces, either on their eyes to see under low-light conditions, or on their wings to reduce surface reflections in transparent (camouflaged) areas. Antireflective surfaces, therefore, occur on the corneas of moth and butterfly eyes[3] and on the transparent wings of hawkmoths.[4]

These consist of nodules, with rounded tips, arranged in a hexagonal array with a periodicity of around 240 nm (Figure 31.2b). Effectively they introduce a gradual refractive index profile at an interface between chitin (a polysaccharide, often embedded in a proteinaceous matrix; r.i. 1.54) and air, and hence reduce reflectivity by a factor of 10.

This "moth-eye structure" was first reproduced at its correct scale by crossing three gratings at 120° using lithographic techniques, and employed as antireflective surfaces on glass windows in Scandinavia.[5] Here, plastic sheets bearing the antireflector were attached to each interior surface of triple-glazed windows using refractive-index-matching glue to provide a significant difference in reflectivity. Today the moth-eye structure can be made extremely accurately using e-beam etching,[6] and is employed commercially on solid plastic and other lenses.

A different form of antireflective device, in the form of a sinusoidal grating of 250 nm periodicity, was discovered on the cornea of a 45-million-year-old fly preserved in amber[7] (Figure 31.2a). This is particularly useful where light is incident at a range of angles (within a single plane, perpendicular to the grating grooves), as demonstrated by a model made in photoresist using lithographic methods.[7] Consequently it has been employed on the surfaces of solar panels, providing a 10% increase in energy capture through reducing the reflected portion of sunlight.[8] Again, this device is embossed onto plastic sheets using holographic techniques.

31.3 Engineering of Iridescent Devices

Many birds, insects (particularly butterflies and beetles), fishes, and lesser-known marine animals display iridescent (changing color with angle) and/or "metallic" colored effects resulting from

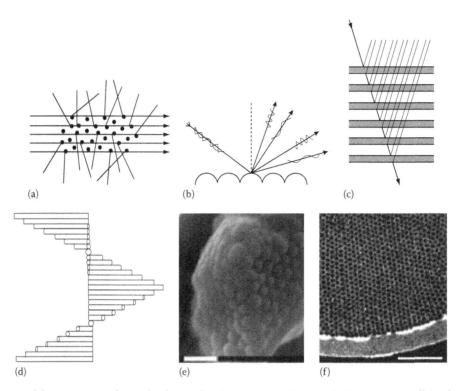

FIGURE 31.1 Summary of the main types of optical reflectors found in nature; a–d where a light ray is (generally) reflected only once within the system (i.e., they adhere to the single scattering, or first Born, approximation), and e and f where each light ray is (generally) reflected multiple times within the system. (a) An irregular array of elements that scatter incident light into random directions. The scattered (or reflected) rays do not superimpose. (b) A diffraction grating, a surface structure, from where light is diffracted into a spectrum or multiple spectra. Each corrugation is about 500 nm wide. Diffracted rays superimpose either constructively or destructively. (c) A multilayer reflector, composed of thin (ca. 100 nm thick) layers of alternating refractive index, where light rays reflected from each interface in the system superimpose either constructively or destructively. Some degree of refraction occurs. (d) A "liquid crystal" composed of nano-fibers arranged in layers, where the nano-fibers of one layer lie parallel to each other yet are orientated slightly differently to those of adjacent layers. Hence spiral patterns can be distinguished within the structure. The height of the section shown here—one "period" of the system—is around 200 nm. (e) Scanning electron micrograph of the "opal" structure—a close-packed array of submicron spheres (a "3D photonic crystal")—found within a single scale of the weevil *Metapocyrtus* sp.; scale bar = 1 μm. (f) Transmission electron micrograph of a section through a hair (neuroseta) of the sea mouse *Aphrodita* sp. (Polychaeta), showing a cross section through a stack of submicron tubes (a "2D photonic crystal"); scale bar = 5 μm.

photonic nanostructures. These appear comparatively brighter than the effects of pigments and often function in animals to attract the attention of a potential mate or to startle a predator. An obvious application for such visually attractive and optically sophisticated devices is within the anticounterfeiting industry. For secrecy reasons, work in this area cannot be described, although devices are sought at different levels of sophistication, from effects that are discernable by the eye to fine-scale optical characteristics (polarization and angular properties, for example) that can be read only by specialized detectors. However, new research aims to exploit these devices in the cosmetics, paint, printing/ink, and clothing industries. They are even being tested in art to provide a sophisticated color-change effect.

Original work on exploiting nature's reflectors involved copying the design but not the size, where reflectors were scaled up to target longer wavelengths. For example, rapid prototyping was employed to manufacture a microwave analogue of a *Morpho* butterfly scale that is suitable for reflection in the 10–30 GHz region. Here the layer thicknesses would be in the order of 1 mm

rather than 100 nm as in the butterfly, but the device could be employed as an antenna with broad radiation characteristics, or as an antireflection coating for radar. However, today techniques are available to manufacture nature's reflectors at their true size.

Nanostructures causing iridescence include photonic crystal fibers, opal and inverse opal, and unusually sculpted 3D architectures. Photonic crystals are ordered, often complex, sub-wavelength (nano) lattices that can control the propagation of light at the single-wave scale in the manner that atomic crystals control electrons.[9] Examples include opal (a hexagonal or square array of 250 nm spheres) and inverse opal (a hexagonal array of similar-sized holes in a solid matrix). Hummingbird feather barbs contain variation ultrathin layers with variations in porosity that cause their iridescent effects, due to the alternating nanoporous/fully dense ultrastructure.[10] Such layers have been mimicked using aqueous-based layering techniques.[10] The greatest diversity of 3D architectures can be found in butterfly scales, which can include micro-ribs with nano-ridges, concave multilayered pits, blazed gratings, and randomly punctate nano-layers.[11,12] The cuticle of many beetles contain

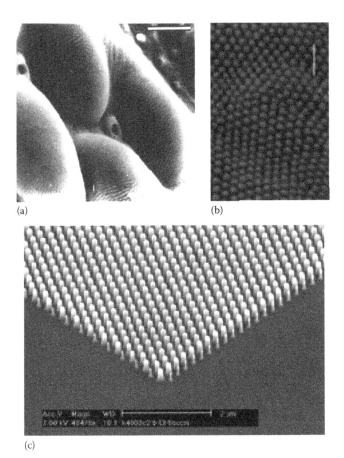

(a)

(b)

(c)

FIGURE 31.2 Scanning electron micrographs of antireflective surfaces. (a) Fly-eye antireflector (ridges on four facets) on a 45-million-year-old dolichopodid fly's eye. (Micrograph from Mierzejewski, P. With permission.) (b) Moth-eye antireflective surfaces. (c) Moth-eye mimic fabricated using ion-beam etching. Scale bars = 3 μm (a), 1 μm (b), 2 μm (c). (Micrograph by Boden, S.A. and Bagnall, D.M., Biomimetic subwavelength surfaces for near-zero reflection sunrise to sunset, *Proceedings 4th World Conference on Photovoltaic Energy, Conversion*, Waikoloa, HI, 2006. With permission.)

structurally chiral films that produce iridescent effects with circular or elliptical polarization properties.[13] These have been replicated in titania for specialized coatings,[13] where a mimetic sample can be compared with the model beetle and an accurate variation in spectra with angle is observed (Figure 31.3). The titania mimic can be nanoengineered for a wide range of resonant wavelengths; the lowest so far is a pitch of 60 nm for a circular Bragg resonance at 220 nm in a Sc_2O_3 film (Ian Hodgkinson, pers. com.).

Biomimetic work on the photonic crystal fibers of the *Aphrodita* sea mouse is underway. The sea mouse contains spines (tubes) with walls packed with smaller tubes of 500 nm, with varying internal diameters (50–400 nm). These provide a bandgap in the red region, and are to be manufactured via an extrusion technique. Larger glass tubes packed together in the proportion of the spine's nanotubes will be heated and pulled through a drawing tower until they reach the correct dimensions. The sea mouse fibre mimics will be tested for standard PCF applications (e.g., in telecommunications) but also for anticounterfeiting structures readable by a detector.

The analogues of the famous blue *Morpho* butterfly (Figure 31.4a) scales have been manufactured.[14,15] Originally, corners were cut. Where the *Morpho* wing contained two layers of scales—one to generate color (a quarter-wave stack) and another above it to scatter the light—the model copied only the principle.[14] The substrate was roughened at the nanoscale, and coated with 80 nm thick layers alternating in refractive index.[14] Therefore the device retained a quarter-wave stack centered in the blue region, but incorporated a degree of randomness to generate scattering. The engineered device closely matched the butterfly wing—the color observed changed only slightly with changing angle over 180°, an effect difficult to achieve and useful for a broad-angle optical filter without dyes.

A new approach to making the 2D "Christmas tree" structure (a vertical, elongated ridge with several layers of 70 nm thick side branches; Figure 31.4b) has been achieved using focused-ion-beam chemical-vapor deposition (FIB-CVD).[15] By combining the lateral growth mode with ion-beam scanning, the Christmas tree structures were made accurately (Figure 31.4c). However, this method is not ideal for low-cost mass production of 2D and 3D nanostructures, and therefore the ion-beam-etched Christmas trees are currently limited to high-cost items including nano- or micron-sized filters (such as "pixels" in a display screen or a filter). Recently further corners have been cut in manufacturing the complex nanostructures found in many butterfly scales, involving the replication of the scales in ZnO, using the scales themselves as templates[16] (Figure 31.4d and e).

31.4 Cell Culture

Sometimes nature's optical nanostructures have such an elaborate architecture at such a small (nano) scale that we simply cannot copy them using current engineering techniques. Additionally, sometimes they can be made as individual reflectors (as for the *Morpho* structure), but the effort is so great that commercial-scale manufacture would never be cost-effective.

An alternative approach to making nature's reflectors is to exploit an aspect other than design—that the animals or plants can make them efficiently. Therefore we can let nature manufacture the devices for us via cell-culture techniques. Animal cells are in the order of 10 μm in size and plant cells up to about 100 μm, and hence suitable for nanostructure production. The success of cell culture depends on the species and on type of cell from that species. Insect cells, for instance, can be cultured at room temperature, whereas an incubator is required for mammalian cells. Cell culture is not a straightforward method, however, since a culture medium must be established to which the cells adhere, before they can be induced to develop to the stage where they make their photonic devices.

The current work in this area centres on butterfly scales. The cells that make the scales are identified in chrysalises, dissected, and plated out. Then the individual cells are separated, kept alive in culture, and prompted to manufacture scales through the addition of growth hormones. Currently we have cultured blue *Morpho* butterfly scales in the lab that have identical optical and structural characteristics to natural scales. The cultured scales

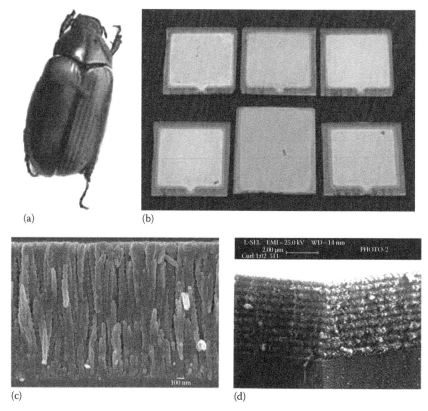

FIGURE 31.3 (a) A Manuka (scarab) beetle with (b) titania mimetic films of slightly different pitches. (c) Scanning electron micrograph of the chiral reflector in the beetle's cuticle. (d) Scanning electron micrograph of the titania mimetic film. (Reproduced from DeSilva, L. et al., *Electromagnetics*, 25, 391, 2005. With permission.)

FIGURE 31.4 (a) A *Morpho* butterfly with (b) a scanning electron micrograph of the structure causing the blue reflector in its scales. (c) A scanning electron micrograph of the FIB-CVD-fabricated mimic. A Ga$^+$ ion beam (beam diameter 7 nm at 0.4 pA; 30 kV), held perpendicular to the surface, was used to etch a precursor of phenanthrene ($C_{14}H_{10}$). Both give a wavelength peak at around 440 nm and at the same angle (30°). (From Watanabe, K. et al., *Jpn. J. Appl. Phys.*, 44, L48, 2005. With permission.) (d) Scanning electron micrograph of the base of a scale of the butterfly *Ideopsis similes*. (e) Scanning electron micrograph of a ZnO replica of the same part of the scale in (d). (Reproduced from Zhang, W. et al., *Bioinspir. Biomim.*, 1, 89, 2006. With permission.)

could be embedded in a polymer or mixed into a paint, where they may float to the surface and self-align. Further work, however, is required to increase the level of scale production and to harvest the scales from laboratory equipment in appropriate ways. A far simpler task emerges where the iridescent organism is single-celled.

31.5 Diatoms and Coccolithophores

Diatoms are unicellular photosynthetic microorganisms. The cell wall is called the frustule and is made of the polysaccharide pectin impregnated with silica. The frustule contains pores (Figure 31.5a through c) and slits that give the protoplasm access to the external environment. There are more than 100,000 different of species of diatoms, generally 20–200 μm in diameter or length, but some can be up to 2 mm long. Diatoms have been proposed to build photonic devices directly in 3D.[17] The biological function of the optical property (Figure 31.5d) is at present unknown, but may affect light collection by the diatom. This type of photonic device can be made in silicon using a deep photochemical-etching technique (initially developed by Lehmann[18]) (e.g., Figure 31.5e). However, there is a new potential here since diatoms carry the added advantage of exponential growth in numbers—each individual can give rise to 100 million descendents in a month.

Unlike most manufacturing processes, diatoms achieve a high degree of complexity and hierarchical structure under mild physiological conditions. Importantly, the size of the pores does not scale with the size of the cell, thus maintaining the pattern.

Fuhrmann et al.[17] showed that the presence of these pores in the silica cell wall of the diatom *Coscinodiscus granii* means that the frustule can be regarded as a photonic crystal slab waveguide. Furthermore, they present models to show that light may be coupled into the waveguide and give photonic resonances in the visible spectral range.

The silica surface of the diatom is amenable to simple chemical functionalization (e.g., Figure 31.6a through c). An interesting example of this uses a DNA-modified diatom template for the control of nanoparticle assembly.[19] Gold particles were coated with DNA complementary to that bound to the surface of the diatom. Subsequently, the gold particles were bound to the diatom surface *via* the sequence-specific DNA interaction. Using this method up to seven layers were added showing how a hierarchical structure could be built onto the template.

Porous silicon is known to luminesce in the visible region of the spectrum when irradiated with ultraviolet light.[20] This photoluminescence (PL) emission from the silica skeleton of diatoms was exploited by DeStafano[21] in the production of an optical gas sensor. It was shown that the PL of *Thalassiosira rotula* is strongly dependent on the surrounding environment. Both the optical intensity and peaks are affected by gases and organic vapors. Depending on the electronegativity and polarizing ability, some substances quench the luminescence, while others effectively enhance it. In the presence of the gaseous substances NO_2, acetone, and ethanol, the photoluminescence was quenched. This was because these substances attract electrons from the silica skeleton of the diatoms and hence quench the PL.

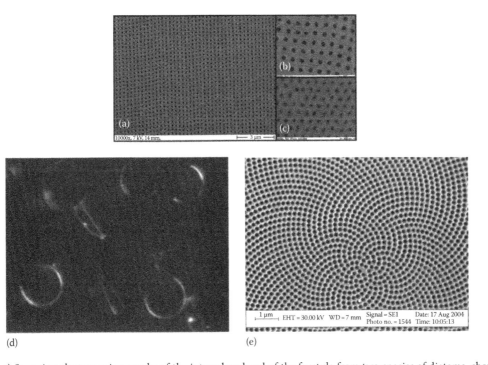

(d) (e)

FIGURE 31.5 (a–c) Scanning electron micrographs of the intercalary band of the frustule from two species of diatoms, showing the square array of pores from *C. granii* ((a) and (b)) and the hexagonal arrays of pores from *C. wailesii* (c). These periodic arrays are proposed to act as photonic crystal waveguides. (d) Iridescence of the *C. granii* girdle bands. (e) Southampton University mimic of a diatom frustule (patented for photonic crystal applications); scanning electron micrograph. (Micrograph by Parker, G. With permission.)

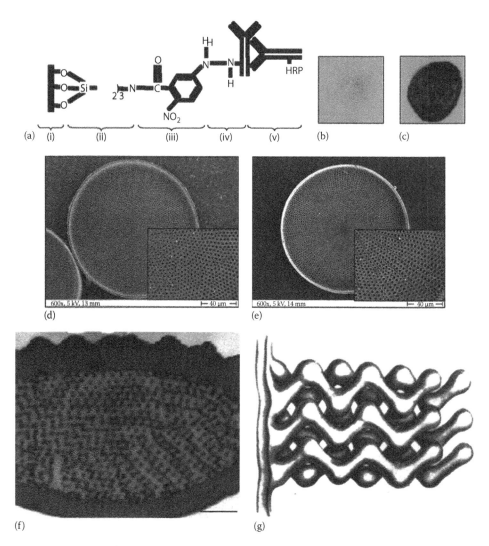

FIGURE 31.6 Modification of natural photonic devices. (a)–(c) Diatom surface modification. The surface of the diatom was silanized, then treated with a heterobifunctional cross-linker, followed by attachment of an antibody via a primary amine group. (a) (i) Diatom exterior surface (ii) APS (iii) ANB-NOS (iv) primary antibody (v) secondary antibody with HRP conjugate. Diatoms treated with primary and secondary antibody with (b) no surface modification (c) after surface modification. (d and e) Scanning electron micrographs showing the pore pattern of the diatom *C. wailesii* (d) and after growth in the presence of nickel sulphate (e). Note the enlargement of pores, and hence change in optical properties, in (e). (f) "Photonic crystal" of the weevil *Metapocyrtus* sp., section through a scale, transmission electron micrograph; scale bar: 1 μm (see Parker[33]). (g) A comparatively enlarged diagrammatic example of cell membrane architecture: tubular christae in mitochondria from the chloride cell of sardine larvae. Evidence suggests that preexisting internal cell structures play a role in the manufacture of natural nanostructures; if these can be altered then so will the nanostructure made by the cell. (From Threadgold, L.T., *The Ultrastructure of the Animal Cell*, Pregamon Press, Oxford, U.K., 1967.)

Nucleophiles, such as xylene and pyridine, which donate electrons, had the opposite effect, and increased PL intensity almost ten times. Both quenching and enhancements were reversible as soon as the atmosphere was replaced by air.

The silica inherent to diatoms does not provide the optimum chemistry/refractive index for many applications. Sandhage et al.[22] have devised an inorganic molecular conversion reaction that preserves the size, shape, and morphology of the diatom while changing its composition. They perfected a gas/silica displacement reaction to convert biologically derived silica structures such as frustules into new compositions. Magnesium was shown to convert SiO_2 diatoms by a vapor phase reaction at 900°C to MgO of identical shape and structure, with a liquid Mg_2Si byproduct. Similarly, when diatoms were exposed to titanium fluoride gas, the titanium displaced the silicon, yielding a diatom structure made up entirely of titanium dioxide; a material used in some commercial solar cells.

An alternative route to silica replacement hijacks that native route for silica deposition in vivo. Rorrer et al.[23] sought to incorporate elements such as germanium into the frustule—a semiconductor material that has interesting properties that could be of value in optoelectronics, photonics, thin film displays, solar cells, and a wide range of electronic devices. Using a two-stage cultivation process, the photosynthetic marine diatom *Nitzschia frustulum* was shown to assimilate soluble germanium and fabricate Si-Ge-oxide nanostructured composite materials.

Porous glasses impregnated with organic dye molecules are promising solid media for tunable lasers and nonlinear optical devices, luminescent solar concentrators, gas sensors, and active waveguides. Biogenic porous silica has an open sponge-like structure and its surface is naturally OH-terminated. Hildebrand and Palenik[24] have shown that rhodamine B and 6G are able to stain diatom silica in vivo, and determined that the dye treatment could survive the harsh acid treatment needed to remove the surface organic layer from the silica frustule.

Now attention is beginning to turn additionally to coccolithophores—single-celled marine algae, also abundant in marine environments. Here, the cell secretes calcitic photonic crystal frustules, which, like diatoms, can take a diversity of forms, including complex 3D architectures at the nano- and microscales.

31.6 Iridoviruses

Viruses are infectious particles made up of the viral genome packaged inside a protein capsid. The iridovirus family comprises a diverse array of large (120–300 nm in diameter) viruses with icosahedral symmetry. The viruses replicate in the cytoplasm of insect cells. Within the infected cell the virus particles produce a paracrystalline array that causes Bragg refraction of light. This property has largely been considered esthetic to date, but the research group of Vernon Ward (New Zealand), in collaboration with the Biomaterials laboratory at Wright–Patterson Air Force base, is using iridoviruses to create biophotonic crystals. These can be used for the control of light, with this laboratory undertaking large-scale virus production and purification as well as targeting the manipulation of the surface of iridoviruses for altered crystal properties. These can provide a structural platform for a broad range of optical technologies, ranging from sensors to waveguides.

Virus nanoparticles, specifically *Chilo* and *Wiseana* Invertebrate Iridovirus, have been used as building blocks for iridescent nanoparticle assemblies. Here, virus particles were assembled in vitro, yielding films and monoliths with optical iridescence arising from multiple Bragg scattering from closely packed crystalline structures of the iridovirus. Bulk viral assemblies were prepared by centrifugation followed by the addition of glutaraldehyde, a cross-linking agent. Long-range assemblies were prepared by employing a cell design that forced virus assembly within a confined geometry followed by cross-linking. In addition, virus particles were used as core substrates in the fabrication of metallodielectric nanostructures. These comprise a dielectric core surrounded by a metallic shell. More specifically, a gold shell was assembled around the viral core by attaching small gold nanoparticles to the virus surface using inherent chemical functionality of the protein capsid.[25] These gold nanoparticles then acted as nucleation sites for electroless deposition of gold ions from solution. Such nano-shells could be manufactured in large quantities, and provide cores with a

narrower size distribution and smaller diameters (below 80 nm) than currently used for silica. These investigations demonstrated that direct harvesting of biological structures, rather than biochemical modification of protein sequences, is a viable route to create unique, optically active materials.

31.7 The Mechanisms of Natural Engineering and Future Research

Where cell culture is concerned it is enough to know that cells *do* make optical nanostructures, which can be farmed appropriately. However, in the future an alternative may be to emulate the natural engineering processes ourselves, by reacting to the same concentrations of chemicals under the same environmental conditions, and possibly substituting analogous nano- or macro-machinery.

To date, the process best studied is the silica cell wall formation in diatoms. The valves are formed by the controlled precipitation of silica within a specialized membrane vesicle called the silica deposition vesicle (SDV). Once inside the SDV, silicic acid is converted into silica particles, each measuring approximately 50 nm in diameter. These then aggregate to form larger blocks of material. Silica deposition is molded into a pattern by the presence of organelles such as mitochondria spaced at regular intervals along the cytoplasmic side of the SDV.[26] These organelles are thought to physically restrict the targeting of silica from the cytoplasm, to ensure laying down of a correctly patterned structure. This process is very fast, presumably due to optimal reaction conditions for the synthesis of amorphous solid silica. Tight structural control results in the final species-specific, intricate exoskeleton morphology.

The mechanism whereby diatoms use intracellular components to dictate the final pattern of the frustule may provide a route for directed evolution. Alterations in the cytoplasmic morphology of *Skeletonema costatum* have been observed in cells grown in sublethal concentrations of Mercury and Zinc,[27] resulting in swollen organelles, dilated membranes, and vacuolated cytoplasm. Frustule abnormalities have also been reported in *Nitzschia liebethrutti* grown in the presence of mercury and tin.[28] Both metals resulted in a reduction in the length to width ratios of the diatoms, fusion of pores, and a reduction in the number of pores per frustule. These abnormalities were thought to arise from enzyme disruption either at the silica deposition site or at the nuclear level. We grew *Coscinodiscus wailesii* in sublethal concentrations of nickel and observed an increase in the size of the pores (Figure 31.6d and e), and a change in the phospholuminescent properties of the frustule. Here, the diatom can be "made to measure" for distinct applications such as stimuli-specific sensors.

Further, *trans*-Golgi-derived vesicles are known to manufacture the coccolithophore 3D "photonic crystals."[29] So the organelles within the cell appear to have exact control of (photonic) crystal growth ($CaCO_2$ in the coccolithophores) and packing (SiO_3 in the diatoms).[30,31] Indeed, Ghiradella[9] suggested that the employment of preexisting, intracellular structures lay behind the development of some butterfly scales, and Overton[32] reported the action of microtubules and microfibrils during butterfly-scale morphogenesis. Further evidence has been found to suggest that these mechanisms, involving the use of molds and nano-machinery (e.g., Figure 31.6f and g), reoccur with unrelated species, indicating that the basic "eukaryote" (containing a nucleus) cell can make complex photonic nanostructures with minimal genetic mutation.[33] The ultimate goal in the field of optical biomimetics, therefore, could be to replicate such machinery and provide conditions under which, if the correct ingredients are supplied, the optical nanostructures will self-assemble with precision.

For further information on the evolution of optical devices in nature, including those found in fossils, or when they first appeared on earth, see references.[34,35]

Acknowledgments

This work was funded by The Royal Society (University Research Fellowship), The Australian Research Council, European Union Framework 6 grant, and an RCUK Basic Technology grant.

References

1. Parker, A.R. 515 million years of structural colour. *J. Opt. A* **2**, R15–R28 (2000).

2. Parker, A.R., McPhedran, R.C., McKenzie, D.R., Botten, L.C., and Nicorovici, N.-A.P. Aphrodite's iridescence. *Nature* **409**, 36–37 (2001).

3. Miller, W.H., Moller, A.R., and Bernhard, C.G. The corneal nipple array. In C.G. Bernhard (ed.), *The Functional Organisation of the Compound Eye* (Pergamon Press, Oxford, U.K., 1966), pp. 21–33.

4. Yoshida, A., Motoyama, M., Kosaku, A., and Miyamoto, K. Antireflective nanoprotuberance array in the transparent wing of a hawkmoth *Cephanodes hylas. Zool. Sci.* **14**, 737–741 (1997).

5. Gale, M. Diffraction, beauty and commerce. *Phys. World* **2**, 24–28 (1989).

6. Boden, S.A. and Bagnall, D.M. Biomimetic subwavelength surfaces for near-zero reflection sunrise to sunset. *Proceedings of the 4th World Conference on Photovoltaic Energy, Conversion*, Waikoloa, HI (2006).

7. Parker, A.R., Hegedus, Z., and Watts, R.A. Solar-absorber type antireflector on the eye of an Eocene fly (45Ma). *Proc. R. Soc. Lond. B* **265**, 811–815 (1998).

8. Beale, B. Fly eye on the prize. *The Bulletin*, 46–48 (May 25, 1999).

9. Yablonovitch, E. Liquid versus photonic crystals. *Nature* **401**, 539–541 (1999).

10. Cohen, R.E., Zhai, L., Nolte, A., and Rubner, M.F. pH gated porosity transitions of polyelectrolyte multilayers in confined geometries and their applications as tunable Bragg reflectors. *Macromolecules* **37**, 6113 (2004).

11. Ghiradella, H. Structure and development of iridescent butterfly scales: Lattices and laminae. *J. Morph.* **202**, 69–88 (1989).

12. Berthier, S. *Les coulers des papillons ou l'imperative beauté. Proprietes optiques des ailes de papillons* (Springer, Paris, France, 2005), 142 pp.

13. DeSilva, L., Hodgkinson, I., Murray, P., Wu, Q., Arnold, M., Leader, J., and Mcnaughton, A. Natural and nanoengineered chiral reflectors: Structural colour of manuka beetles and titania coatings. *Electromagnetics* **25**, 391–408 (2005).

14. Kinoshita, S., Yoshioka, S., Fujii, Y., and Okamoto, N. Photophysics of structural color in the Morpho butterfly. *Forma* **17**, 103 (2002).

15. Watanabe, K., Hoshino, T., Kanda, K., Haruyama, Y., and Matsui, S. Brilliant blue observation from a *Morpho*-butterfly-scale quasi-structure. *Jpn. J. Appl. Phys.* **44**, L48–L50 (2005).

16. Zhang, W., Zhang, D., Fan, T., Ding, J., Gu, J., Guo, Q., and Ogawa, H. Bio-mimetic zinc oxide replica with structural color using butterfly (*Ideopsis similis*) wings as templates. *Bioinspir. Biomim.* 1, 89 (2006).

17. Fuhrmann, T., Lanwehr, S., El Rharbi-Kucki, M., and Sumper, M. Diatoms as living photonic crystals. *Appl. Phys. B* **78**, 257–260 (2004).

18. Lehmann, V. On the origin of electrochemical oscillations at silicon electrodes. *J. Electrochem. Soc.* **143**, 1313 (1993).

19. Rosi, N.L., Thaxton, C.S., and Mirkin, C.A. Control of nanoparticle assembly by using DNA-modified diatom templates. *Agnew Chem. Int. Ed.* **43**, 5500–5503 (2004).

20. Cullis, A.G., Canham, L.T., and Calcott, P.D.J. The structural and luminescence properties of porous silicon. *J. Appl. Phys.* **82**, 909–965 (1997).

21. De Stefano, L., Rendina, I., De Stefano, M., Bismuto, A., and Maddalena, P. Marine diatoms as optical chemical sensors. *Appl. Phys. Lett.* **87**, 233902 (2005).

22. Sandhage, K.H., Dickerson, M.B., Huseman, P.M., Caranna, M.A., Clifton, J.D., Bull, T.A., Heibel, T.J., Overton, W.R., and Schoenwaelder, M.E.A. Novel, bioclastic route to self-assembled, 3D, chemically tailored meso/nanostructures: Shape-preserving reactive conversion of biosilica (diatom) microshells. *Adv. Mater.* **14**, 429–433 (2002).

23. Rorer, G.L., Chang, C.H., Liu, S.H., Jeffryes, C., Jiao, J., and Hedberg, J.A. Biosynthesis of silicon-germanium oxide nanocomposites by the marine diatom Nitzschia frustulum. *J. Nanosci. Nanotechnol.* **5**, 41–49 (2004).

24. Hildebrand, M. and Palenik, B. Grant report *Investigation into the Optical Properties of Nanostructured Silica from Diatoms*, La Jolla, CA (2003).

25. Radloff, C., Vaia, R.A., Brunton, J., Bouwer, G.T., and Ward, V.K. Metal nanoshell assembly on a virus bioscaffold. *Nano Lett.* **5**, 1187–1191 (2005).

26. Schmid, A.M.M. Aspects of morphogenesis and function of diatom cell walls with implications for taxonomy. *Protoplasma* **181**, 43–60 (1994).

27. Smith, M.A. The effect of heavy metals on the cytoplasmic fine structure of Skeletonema costatum (Bacillariophyta). *Protoplasma* **116**, 14–23 (1983).

28. Saboski, E. Effects of mercury and tin on frustular ultrastructure of the marine diatom *Nitzschia liebethrutti*. *Water, Air Soil Pollut.* **8**, 461–466 (1977).

29. Corstjens, P.L.A.M. and Gonzales, E.L. Effects of nitrogen and phosphorus availability on the expression of the coccolith-vesicle v-ATPase (subunit C) of Pleurochrysis (Haptophyta). *J. Phycol.* **40**, 82–87 (2004).

30. Klaveness, D. and Paasche, E. Physiology of coccolithophorids. In: *Biochemistry and Physiology of Protozoa*, 2nd edn., vol. 1 (Academic Press, New York, 1979), pp. 191–213.

31. Klaveness, D. and Guillard, R.R.L. The requirement for silicon in *Synura petersenii* (Chrysophyceae). *J. Phycol.* **11**, 349–355 (1975).

32. Overton, J. Microtubules and microfibrils in morphogenesis of the scale cells of *Ephestia kuhniella*. *J. Cell Biol.* **29**, 293–305 (1966).

33. Parker, A.R. Conservative photonic crystals imply indirect transcription from genotype to phenotype. *Recent Res. Develop. Entomol.* **5**, 1–10 (2006).

34. Parker, A.R. *In the Blink of an Eye* (Simon & Schuster, London, U.K., 2003), 316 pp.

35. Parker, A.R. A geological history of reflecting optics. *J. R. Soc. Lond. Interface* **2**, 1–17 (2005).

V

Nanophotonic Devices

32

Photon Localization at the Nanoscale

Kiyoshi Kobayashi
University of Tokyo
Japan Science and Technology
University of Yamanashi

32.1 Introduction

You might think that the great success of quantum electrodynamics (QED) would settle the debate on the nature of light to provide a clear view of its behavior, where the photon is regarded as the unit of excitation associated with a quantized mode of the electromagnetic (radiation) field. However, Heisenberg's uncertainty principle tells us that a state of definite momentum, energy, and polarization associated with a plane wave used as a basis function of quantization must be completely indefinite in space and time. It suggests the difficulty of spatial localization of a photon as a particle. In fact, Newton and Wigner showed that a free photon, as a massless particle with spin 1, has no localized states on the basis of natural invariance requirements that localized states for which operators of the Lorentz group apply should be orthogonal to the undisplaced localized states, after a translation (Newton and Wigner 1949). According to them, one can obtain a general expression for a position operator for massive particles and for massless particles of spin 0 or 1/2, not for massless particles with finite spin, which indicates that there is no probability density for the position of the photon, and thus a position-representation wave function cannot be consistently introduced. It has also been shown that photons are not localizable, on the basis of imprimitive representations of the Euclidean group (Wightman 1962). It is now believed that photons are only weakly localizable, although single-photon states with arbitrarily fast asymptotic falloff of energy density exist, and that

a lack of strict localizability is directly related to the absence of a position operator for a photon in free space and a position-representation photon wave function (Hawton 1999).

On the other hand, several authors have claimed that a minimum modification of the naive route leads to a wave-function description of a photon, even though the probability density for the position of a photon and a position-representation wave function cannot be consistently introduced. For example, it has been shown to be possible to introduce a position-representation wave function $\vec{\psi}(\vec{r}, t)$ for a photon, which is the expectation value of the photon energy in a region $d\vec{r}$ about $\vec{r}$ (Sipe 1995, Scully and Zubairy 1997, Hawton 1999, Roychoudhuri and Roy 2003). Mandel et al., from another viewpoint, have found that a photon wave function as a probability amplitude is possible in a coarse-grained volume whose linear dimensions are larger than the photon wavelength (Mandel 1966, Mandel and Wolf 1995, Inagaki 1998). The localization of the photon energy density and photodetection rates having an exponential or arbitrarily fast asymptotic falloff have been discussed, as well as causality (Hegerfeldt 1974, Pike and Sarkar 1987, Hellwarth and Nouchi 1996, Adlard et al. 1997, Bialynicki-Birula 1998, Keller 2005).

When the interactions with matter were considered, different views opened up. Dressed states and operators, dressed and half-dressed sources in nonrelativistic QED, electromagnetic field correlations, and intermolecular interactions between molecules in either ground or excited states have been discussed (Compagno et al. 1988, 1995, Cohen-Tannoudji et al. 1989, 1992,

Power and Thirunamachandran 1993), focusing on the fact that a bare source interacting with a quantum field is surrounded by a cloud of virtual particles. It has been shown that the dressing of the source, or virtual cloud effects, can be detected by a test body (detector) located close to the source. Carniglia and Mandel (1971) proposed a complete basis for electromagnetic fields interacting with a material of refractive index n filled in a half space separated by vacuum in order to quantize the source fields, while Inoue and Hori (2001) discussed the detector modes and the behavior of a photon–atom interacting system near the material surface. Kobayashi et al. (2001) focused on the environmental effects on a nanomaterial interacting with photons and obtained a near-field optical interaction as an effective interaction between nanomaterials electronically disconnected, but closely located, in order to detect the cloud of virtual photons. They have also applied it to a discussion of nanophotonic devices (Sangu et al. 2004). Focusing on the photon degrees of freedom, on the other hand, photon hopping has been employed to discuss a photon–material interacting system (John and Quang 1995, Suzuura et al. 1996, Shojiguchi et al. 2003), and a photon dressing by material excitations has been recently discussed by using the photon-hopping model in real space and a quasiparticle model (Kobayashi et al. 2008).

Focusing on light–matter interactions at the nanoscale, we discuss a near-field optical interaction between nanomaterials surrounded by a macroscopic system, a dressing mechanism, and spatial localization of photons in this article, which is organized as follows. Section 32.2 is devoted to background issues for photon localization, in particular, the difficulty of the definition, effective methods, and free-field quantization. In Section 32.3 we discuss a dressing mechanism of photons and their localization in space, including virtual clouds of photons, electromagnetic field correlations and intermolecular interactions, effective near-field optical interactions, and phonons' effects on photon localization. Finally, a summary and a future outlook are presented.

32.2 Background

32.2.1 Difficulty in Defining a Wave Function of a Photon in Real Space

Since there is no probability density for the photon, and thus a position-representation wave function in free space cannot be consistently defined, one has to follow quantum electrodynamics (QED), that is, to redefine one- and a few-photon wave functions in a physically meaningful way in order to obtain fruitful insights into the photon–matter interacting system at the nanoscale. In the following subsections, we will give an overview of both approaches, after pointing out the difficulties involved with a position-representation wave function of a photon in free space.

We begin with Maxwell's equations in vacuum:

$$\nabla \times \vec{E}(\vec{r}, t) = -\frac{1}{c} \frac{\partial \vec{B}(\vec{r}, t)}{\partial t}, \tag{32.1a}$$

$$\nabla \times \vec{B}(\vec{r}, t) = \frac{1}{c} \frac{\partial \vec{E}(\vec{r}, t)}{\partial t}, \tag{32.1b}$$

where

c is the speed of light

$\vec{E}(\vec{r}, t)$ and $\vec{B}(\vec{r}, t)$ are electric and magnetic fields, respectively

Let us now define $\vec{\Phi}_\pm(\vec{r}, t)$ as $\vec{\Phi}_\pm(\vec{r}, t) = \vec{E}(\vec{r}, t) \pm i\vec{B}(\vec{r}, t)$; then, it follows from Maxwell's equations that $\vec{\Phi}_\pm(\vec{r}, t)$ should satisfy

$$i\hbar \frac{\partial \vec{\Phi}_\pm(\vec{r}, t)}{\partial t} = \pm c\hbar \nabla \times \vec{\Phi}_\pm(\vec{r}, t). \tag{32.2}$$

Using the Fourier transform of these functions, we introduce the vectors $\vec{\gamma}_\pm(\vec{p}, t)$ as

$$\vec{\Phi}_\pm(\vec{r}, t) = \int \frac{d^3 p}{(2\pi\hbar)^{3/2}} \vec{\gamma}_\pm(\vec{p}, t) e^{i\vec{p}\cdot\vec{r}/\hbar}, \tag{32.3}$$

which is associated with $\gamma_\pm(\vec{p}, t)$ as $\vec{\gamma}_\pm(\vec{p}, t) = \gamma_\pm(\vec{p}, t)\vec{e}_\pm(p)$ with

$$\vec{e}_\pm(\hat{p}) = \mp \frac{1}{\sqrt{2}} [\vec{e}_1(\hat{p}) \pm i\vec{e}_2(\hat{p})], \tag{32.4a}$$

where two unit vectors $\vec{e}_1(\hat{p})$ and $\vec{e}_2(\hat{p})$ are defined such that the unit vectors

$$[\vec{e}_1(\hat{p}), \vec{e}_2(\hat{p}), \hat{p} = \vec{p}/|\vec{p}|] \tag{32.4b}$$

form a right-handed triad. Then, it is easy to verify that $\gamma_\pm(\vec{p}, t)$ satisfy a Schrödinger-like equation in the momentum representation

$$i\hbar \frac{\partial \gamma_\pm(\vec{p}, t)}{\partial t} = cp\gamma_\pm(\vec{p}, t), \tag{32.5}$$

which indicates that $\gamma_\pm(\vec{p}, t)$ are probability amplitudes for photons of momentum $\vec{p}$, energy $E = cp$ with $p = |\vec{p}|$, and positive/negative helicity. Here note that

$$\vec{\gamma}_+^*(\vec{p}, t) \cdot \vec{\gamma}_+(\vec{p}, t) d\vec{p} = \vec{\gamma}_-^*(\vec{p} \cdot t) \cdot \vec{\gamma}_-(\vec{p}, t) d\vec{p} \tag{32.6a}$$

and

$$\vec{\gamma}_+^*(\vec{p}, t) \cdot \vec{\gamma}_-(\vec{p}, t) = 0 \tag{32.6b}$$

show the probability of detecting a photon of positive helicity and momentum $\vec{p}$ between $\vec{p}$ and $\vec{p} + d\vec{p}$, and likewise for a photon of negative helicity. Then the normalization condition is

$$\int \left[\vec{\gamma}_+^*(\vec{p}, t) \cdot \vec{\gamma}_+(\vec{p}, t) + \vec{\gamma}_-^*(\vec{p}, t) \cdot \vec{\gamma}_-(\vec{p}, t) \right] d^3 p = 1, \qquad (32.7)$$

which leads to

$$\int \left[\vec{\Phi}_+^*(\vec{r}, t) \cdot \vec{\Phi}_+(\vec{r}, t) + \vec{\Phi}_-^*(\vec{r}, t) \cdot \vec{\Phi}_-(\vec{r}, t) \right] d^3 r = 1, \quad (32.8)$$

and it follows from the dynamical equations that

$$\vec{\Phi}_\pm(\vec{r}, t) = \int \frac{d^3 p}{(2\pi\hbar)^{3/2}} \vec{\gamma}_\pm(\vec{p}, 0) e^{-icpt/\hbar} e^{i\vec{p}\cdot\vec{r}/\hbar}. \qquad (32.9)$$

Here we note that the sum of $\vec{\Phi}_\pm(\vec{r}, t)$ might be regarded as the position-representation wave function of a photon, but it cannot be regarded as such because Newton and Wigner and also Wightman have shown that the photon, being a massless particle, is not localizable in free space, and that there does not exist a probability amplitude and density for the position of the photon in the usual sense (see also (32.15a) and (32.15b)) (Newton and Wigner 1949, Wightman 1962).

32.2.2 Effective Spatial Wave Function of a Single Photon

In order to avoid the difficulties mentioned above, Mandel defined an operator representing the number of photons in a volume V as the integral over V of a so-called "detection operator," which led a simple formula for the probability that n photons are present in V, when the linear dimensions of V are larger than the wavelength of light used (Mandel 1966).

Sipe took another approach to this issue, seeking a probability amplitude $\vec{\Psi}(\vec{r}, t)$ for the photon energy to be detected about $d\vec{r}$ of $\vec{r}$ (Sipe 1995). Assuming that the integral of $\vec{\Psi}^*(\vec{r}, t) \cdot \vec{\Psi}(\vec{r}, t) d\vec{r}$ over all space is proportional to the photon energy, we normalize it as follows:

$$\int \vec{\Psi}^*(\vec{r}, t) \cdot \vec{\Psi}(\vec{r}, t) d^3 r = \int cp \left[\vec{\gamma}_+^*(\vec{p}, t) \cdot \vec{\gamma}_+^*(\vec{p}, t) \right.$$
$$\left. + \vec{\gamma}_-^*(\vec{p}, t) \cdot \vec{\gamma}_-^*(\vec{p}, t) \right] d^3 p. \qquad (32.10)$$

It is easily shown that if $\vec{\Psi}(\vec{r}, t)$ is set as

$$\vec{\Psi}(\vec{r}, t) = \vec{\Psi}_+(\vec{r}, t) + \vec{\Psi}_-(\vec{r}, t), \qquad (32.11a)$$

$$\vec{\Psi}_\pm(\vec{r}, t) = \int \frac{\sqrt{cp}}{(2\pi\hbar)^{3/2}} \vec{\gamma}_\pm(\vec{p}, t) e^{i\vec{p}\cdot\vec{r}/\hbar} d^3 p, \qquad (32.11b)$$

the normalization condition is satisfied. Here note that

$$\vec{\gamma}_+^*(\vec{p}, t) \cdot \vec{\gamma}_-(\vec{p}, t) = 0 = \vec{\gamma}_-^*(\vec{p}, t) \cdot \vec{\gamma}_+(\vec{p}, t) \qquad (32.12)$$

and

$$i\hbar \frac{\partial \vec{\Psi}_\pm(\vec{r}, t)}{\partial t} = \pm c\hbar \nabla \times \vec{\Psi}_\pm(\vec{r}, t). \qquad (32.13)$$

At the same time, $\vec{\Psi}_\pm(\vec{r}, t)$ should be chosen to satisfy an initial condition given by

$$\vec{\Psi}_\pm(\vec{r}, t) = \int \frac{\sqrt{cp}}{(2\pi\hbar)^{3/2}} \vec{\gamma}_\pm(\vec{p}, 0) e^{-icpt/\hbar} e^{i\vec{p}\cdot\vec{r}/\hbar} d^3 p. \qquad (32.14)$$

Since $\vec{\gamma}_\pm(\vec{p}, t)$ and $\vec{\Psi}_\pm(\vec{p}, \vec{r})$ are not the Fourier transform pairs, the arguments of the photon momentum $\vec{p}$ and the position $\vec{r}$ associated with the photon energy are not conjugate variables. When $\vec{\Phi}_\pm(\vec{r}, t)$ and $\vec{\Psi}_\pm(\vec{r}, t)$ are related by a local kernel, it can usually be regarded as a particle. However, $\vec{\Phi}_\pm(\vec{r}, t)$, the Fourier transform of $\vec{\gamma}_\pm(\vec{p}, t)$, is not a reasonable candidate for the position-representation wave function of the photon because of the following relation between $\vec{\Phi}_\pm(\vec{r}, t)$ and $\vec{\Psi}_\pm(\vec{r}, t)$

$$\vec{\Phi}_\pm(\vec{r}, t) = \int w(\vec{r} - \vec{r}') \vec{\Psi}_\pm(\vec{r}', t) d^3 r' \qquad (32.15a)$$

with the nonlocal kernel

$$w(\vec{r} - \vec{r}') = \int \frac{1}{(2\pi)^3} \frac{e^{i\vec{k}\cdot(\vec{r}-\vec{r}')}}{\sqrt{\hbar ck}} d^3 k. \qquad (32.15b)$$

Nevertheless, $\vec{\Psi}(\vec{r}, t)$ might be meaningful to describe the dynamics of a photon such as a spontaneous emission from an atom and the inverse process, or at least we can detect a photon, within the range of a detector's precision by placing a detector like an atom close to the source. In other words, it indicates that light–matter interactions near the source play an important role and should be treated consistently.

32.2.3 Canonical Field Quantization: Mode Functions, Field Operators, and Quantum States

It is natural to follow the canonical quantization of the electromagnetic field as a starting point for a discussion of light–matter interactions at the nanoscale. Since a lot of famous textbooks on

QED or quantum field theory have been published, we follow the essence of the theory and restrict ourselves to the free field that is free from charges and currents and whose scalar potential can be set to zero (Sakurai 1967, Roychoudhuri and Roy 2003). In the Coulomb gauge with $\nabla \cdot \vec{A}(\vec{r}, t) = 0$ for the vector potential $\vec{A}(\vec{r}, t)$, three basic equations we work with for the free-field case are

$$\vec{B}(\vec{r}, t) = \nabla \times \vec{A}(\vec{r}, t), \tag{32.16a}$$

$$\vec{E}(\vec{r}, t) = -\frac{1}{c}\frac{\partial \vec{A}(\vec{r}, t)}{\partial t}, \tag{32.16b}$$

$$\nabla^2 \vec{A}(\vec{r}, t) - \frac{1}{c^2}\frac{\partial^2 \vec{A}(\vec{r}, t)}{\partial t^2} = 0. \tag{32.16c}$$

We expand $\vec{A}(\vec{r}, t)$ into a complete set of mode functions $\vec{u}_{\vec{k},a}(\vec{r})$ defined by the Helmholtz equation

$$\left(\nabla^2 + \vec{k}^2\right)\vec{u}_{\vec{k},a}(\vec{r}) = 0, \tag{32.17}$$

and the boundary conditions set by the shape of a virtual cavity, for example, a box taken to be a cube of side $L = V^{1/3}$. It follows from the Coulomb gauge that the direction of $\vec{u}_{\vec{k},a}(\vec{r})$ has to be orthogonal to the wave vector $\vec{k}$, and thus there are two polarization degrees of freedom indicated by the index α. Note that the mode functions become more complicated, or even impossible for more sophisticated cavity shapes. The mode functions satisfy the orthonormality condition

$$\frac{1}{V}\int d^3r\, \vec{u}_\ell^*(\vec{r}) \cdot \vec{u}_{\ell'}(\vec{r}) = \delta_{\ell\ell'}, \tag{32.18}$$

where we have for simplicity combined the three components of the wave vector $\vec{k}$ and the polarization index α to one index ℓ. Using the normalization constant A_ℓ and the time-dependent amplitude $q_\ell(t)$, we have

$$\vec{A}(\vec{r}, t) = \sum_\ell \left[A_\ell q_\ell(t)\vec{u}_\ell(\vec{r}) + A_\ell^* q_\ell^*(t)\vec{u}_\ell^*(\vec{r}) \right], \tag{32.19}$$

where $q_\ell(t)$ follows from the differential equation of a harmonic oscillator of frequency $\Omega_\ell \equiv c|\vec{k}_\ell|$:

$$\frac{d^2 q_\ell(t)}{dt^2} + \Omega_\ell^2 q_\ell(t) = 0. \tag{32.20}$$

The Hamiltonian of the field is

$$H = \frac{1}{8\pi}\int d^3r(\vec{E}^2 + \vec{B}^2) = \frac{1}{8\pi}\int d^3r\left[\left(\frac{1}{c}\frac{\partial \vec{A}}{\partial t}\right)^2 + (\nabla \times \vec{A})^2\right]. \tag{32.21}$$

Noticing a typical term for the $\vec{E}^2$ integration

$$\int \left(\frac{1}{c}\frac{dq_\ell}{dt}\right)\left(\frac{1}{c}\frac{dq_{\ell'}^*}{dt}\right)\vec{u}_\ell(\vec{r}) \cdot \vec{u}_{\ell'}^*(\vec{r})d^3r = \left(\frac{\Omega_\ell}{c}\right)V|q_\ell|^2\,\delta_{\ell\ell'} \tag{32.22}$$

and for the $\vec{B}^2$ integration

$$\int (\nabla \times \vec{u}_\ell) \cdot (\nabla \times \vec{u}_{\ell'}^*)d^3r = \int \nabla \cdot (\vec{u}_\ell \times \nabla \times \vec{u}_{\ell'}^*)d^3r + \int \vec{u}_\ell \cdot \left[\nabla \times \nabla \times \vec{u}_{\ell'}^*\right]d^3r$$

$$= \int \vec{u}_\ell \cdot \left[\nabla(\nabla \cdot \vec{u}_{\ell'}^*) - \nabla^2\vec{u}_{\ell'}^*\right]d^3r = -\int \vec{u}_\ell \cdot \nabla^2\vec{u}_{\ell'}^*d^3r$$

$$= \left(\frac{\Omega_\ell}{c}\right)^2 V\delta_{\ell\ell'} \tag{32.23}$$

we obtain

$$H = \frac{V}{2\pi}\sum_\ell \left(\frac{\Omega_\ell}{c}\right)^2 (A_\ell q_\ell)^*(A_\ell q_\ell). \tag{32.24}$$

If we define

$$Q_\ell = \frac{1}{c}(q_\ell + q_\ell^*), \tag{32.25a}$$

$$P_\ell = -i\frac{\Omega_\ell}{c}(q_\ell - q_\ell^*), \quad A_\ell = \sqrt{\frac{4\pi}{V}}, \tag{32.25b}$$

the Hamiltonian can be expressed in terms of a collection of independent and uncoupled harmonic oscillators as

$$H = \sum_\ell \frac{1}{2}(P_\ell^2 + \Omega_\ell^2 Q_\ell^2). \tag{32.26}$$

Here, P_ℓ and Q_ℓ are seen to be canonical variables:

$$\frac{dP_\ell}{dt} = -\frac{\partial H}{\partial Q_\ell}, \tag{32.27a}$$

$$\frac{dQ_\ell}{dt} = \frac{\partial H}{\partial P_\ell}. \tag{32.27b}$$

The natural method to quantize the field is to replace the variable q_ℓ and its conjugate momentum $p_\ell \equiv dq_\ell/dt$ by operators $\hat{q}_\ell$ and $\hat{p}_\ell$ that satisfy the commutation relations $[\hat{q}_\ell, \hat{p}_\ell] = i\hbar\delta_{\ell\ell'}$, or $[Q_\ell, P_{\ell'}] = i\hbar\delta_{\ell\ell'}$. Then, the Hamiltonian operator for the quantized field can be written as follows:

$$\hat{H} = \sum_\ell \frac{1}{2}(\hat{P}_\ell^2 + \Omega_\ell^2 \hat{Q}_\ell^2). \qquad (32.28)$$

Next we consider linear combinations of $\hat{P}_\ell$ and $\hat{Q}_\ell$, given by

$$\hat{a}_\ell = \sqrt{\frac{\Omega_\ell}{2h}}\left(\hat{Q}_\ell + \frac{i}{\Omega_\ell}\hat{P}_\ell\right), \qquad (32.29a)$$

$$\hat{a}_\ell^\dagger = \sqrt{\frac{\Omega_\ell}{2h}}\left(\hat{Q}_\ell + \frac{i}{\Omega_\ell}\hat{P}_\ell\right), \qquad (32.29b)$$

where we insert a factor to make $\hat{a}_\ell$ and $\hat{a}_\ell^\dagger$ dimensionless and thus satisfy the following boson commutation relations:

$$\left[\hat{a}_\ell, \hat{a}_{\ell'}\right] = -\frac{i}{2h}[\hat{Q}_\ell, \hat{P}_{\ell'}] + \frac{i}{2h}[P_\ell, Q_{\ell'}] = \delta_{\ell\ell'}. \qquad (32.30)$$

Taking care of the order of $\hat{a}_\ell$ and $\hat{a}_{\ell'}^\dagger$, we obtain the Hamiltonian operator of the electromagnetic field as

$$\hat{H} = \sum_\ell \hbar\Omega_\ell \left(\hat{n}_\ell + \frac{1}{2}\right), \qquad (32.31)$$

where $\hat{n}_\ell \equiv \hat{a}_\ell^\dagger \hat{a}_\ell$ denotes the number operator. The contribution $1/2$ arises from the commutation relations and results in the familiar zero-point energy. Since there are infinitely many modes, the zero-point energy becomes infinite, but in general we drop this contribution by shifting the vacuum energy, which does not influence the dynamics. The quantization procedure is completed by writing the vector potential in terms of $\hat{a}_\ell$ and $\hat{a}_\ell^\dagger$ as a field operator

$$\hat{\vec{A}}(\vec{r},t) = \sum_\ell \sqrt{\frac{2\pi\hbar c^2}{V\Omega_\ell}}\left(\hat{a}_\ell \vec{u}_\ell + \hat{a}_\ell^\dagger \vec{u}_\ell^*\right). \qquad (32.32)$$

Thus, $\hat{a}_\ell^\dagger$ is called the creation operator for a photon specified in ℓ as corresponding to the quantum-mechanical excitations of the electromagnetic field, while $\hat{a}_\ell$ is interpreted as the annihilation operator for a photon in state ℓ. It is important to note that the $\vec{r}$ and t that appear in the quantized field $\hat{\vec{A}}(\vec{r},t)$ are not quantum-mechanical variables but just parameters on which the field operator depends, and, in particular, $\vec{r}$ and t should not be regarded as the space–time coordinates of the photon.

When we adopt a linearly polarized plane wave as the mode function

$$\vec{u}_\ell(\vec{r}) = \vec{\varepsilon}_{\vec{k}\alpha} e^{i\vec{k}\cdot\vec{r}} \qquad (32.33)$$

with the polarization vector $\vec{\varepsilon}_{\vec{k}\alpha}$, the energy and momentum of the photon are $\hbar\Omega_{\vec{k}} = \hbar|\vec{k}|c$ and $\hbar\vec{k}$, respectively. Therefore, the

mass of the photon is zero. In addition, since $\vec{\varepsilon}_{\vec{k}\alpha}$ transforms like a vector, the general theory of angular momentum encourages us to associate with it one unit of angular momentum, which means that the photon has one unit of spin angular momentum.

The field operators described above operate on quantum state vectors, and quantum states $|\Psi\rangle$ of the electromagnetic field are, in general, multimode states that involve quantum states $|\psi_\ell\rangle$ for each mode ℓ. One of the useful quantum states is photon number states denoted by $|n_\ell\rangle$, which are eigenstates of the number operator $\hat{n}_\ell$

$$\hat{n}_\ell |n_\ell\rangle = n_\ell |n_\ell\rangle \qquad (32.34)$$

with integer eigenvalues n_ℓ. At the same time, $|n_\ell\rangle$ are eigenstates of the Hamiltonian with eigenenergy $n_\ell \hbar\Omega_\ell$, that is, n_ℓ times the fundamental unit $\hbar\Omega_\ell$. It should be noted that n_ℓ quanta of energy $\hbar\Omega_\ell$ are in the mode, but the energy is distributed over the entire space, that is, not localized. Other useful quantum states used later are coherent states $|\alpha\rangle$, which are eigenstates of the annihilation operator $\hat{a}_\ell$ with eigenvalues α

$$\hat{a}_\ell |\alpha\rangle = \alpha |\alpha\rangle. \qquad (32.35)$$

The phase of the coherent states is completely determined, while the number of photons is completely undetermined.

In the subsequent sections, we will employ a quantum electromagnetic field discussed above in order to discuss the nature of light–matter interactions apparently exhibited at the nanoscale.

32.3 Dressing Mechanism and Spatial Localization of Photons

32.3.1 Virtual Photon Cloud Surrounding a Neutral Source (in Ground State or Excited State) in QED

A quantum source material system interacting with a quantum field is influenced by virtual processes such as emission and absorption of virtual quanta of the field, and the source can be described as a dressed source, that is, the "bare" source surrounded by a cloud of virtual particles (Compagno et al. 1988, 1995). It is true for a detector. The virtual-cloud effects are responsible for the modification of the values of fundamental constants. For example, in nonrelativistic quantum electrodynamics, the presence of a virtual cloud around a hydrogen atom in its ground state contributes to the Lamb shift. Dressed-source effects can also be seen in different physical systems, such as a nucleon coupled to the meson field, and an electron coupled to the optical phonon modes of a semiconductor (polaron). The virtual cloud around the source also modifies the energy density distribution of the electromagnetic field, and the detailed properties have direct physical significance. The energy density of the virtual photon cloud at a given point is in fact related to the van der Waals interaction experienced by a suitable test object as a detector at that point. The presence of a virtual cloud around a source can influence not only its energy levels but also its dynamics.

Let us roughly estimate the linear dimensions of the virtual cloud surrounding the source or the detector. Even when the source–field system is in the ground state, a fluctuation of the field leads to the possibility of absorption or the emission of photons, not necessarily to the conservation of energy. Such energy imbalance δE is constrained by the Heisenberg uncertainty principle $\delta E \sim \hbar/\tau$, where τ is the duration of the fluctuation. Since these fluctuations take place continuously, a steady-state cloud of virtual photons is continuously emitted and reabsorbed. The virtual photon can only attain a finite distance from the source or the detector given by

$$r \sim c\tau \sim \frac{\hbar c}{\delta E}, \tag{32.36}$$

where c is the speed of light. For a transition corresponding to one of typical visible light, we set $\delta E \sim 2\,\text{eV}$ and obtain a typical linear dimension of $100\,\text{nm}$. This indicates that dressing effects might be prominent at the nanoscale.

We have discussed the virtual clouds of the source or detector in its ground state. From now on, we discuss the virtual cloud of the source or the detector in excited states, which can decay by emission of real photons. The above discussion inclines us to use the perturbation theory, but it fails due to the near degeneracy of states that gives rise to vanishing energy denominators at all the orders of perturbation theory. Another difficulty is the description of decay processes in a consistent way. One of such attempts is based on an extension of the eigenvalue problem to the complex E-plane. The underlying theory, unfortunately, has not yet been established, but it seems to imply time symmetry breaking and irreversibility in the dynamics of the system Section 32.3.2 will be devoted to an approach focusing on field correlations and intermolecular interactions due to the virtual clouds.

32.3.2 Electromagnetic Field Correlations and Intermolecular Interactions between Molecules in Either Ground or Excited States

The London–van der Waals interaction between two molecules in their ground states located in free space is attractive with an R^{-6} power law, where R is the intermolecular separation (Power and Thirunamachandran 1993). When both molecules are excited, the potential energy gives a repulsive force. If one of the pair is excited, the sign of the potential depends on the relative magnitudes of the relevant transition energies of the two molecules. In both cases the power law is R^{-6} in the near zone. On the other hand, the power law in the far zone tends to R^{-7} for large intermolecular separations, which is called the Casimir–Polder potential, since the finite speed of propagation (retardation effect) plays an important role in the far zone.

In these kinds of studies, the multipolar quantum dynamics in Coulomb gauge is employed because all the interactions, except for the Coulomb binding within each molecule, are mediated by transverse photons, and at the same time the retarded effects are automatically satisfied (Power and Thirunamachandran 1993). For example, for upward transitions from the ground state $|0\rangle$ to an excited state $|m\rangle$, the electric–electric spatial correlation expectation value after spatial averaging of the molecular orientation is given by using the second-order perturbation method to include virtual-cloud effects:

$$\left\langle D_i(\vec{r}) D_j(\vec{r}) \right\rangle_{m \leftarrow 0} \sim \begin{cases} \dfrac{\left|\vec{\mu}^{m0}\right|^2}{6\pi k_0 r^7}(13\delta_{ij} + 7\hat{r}_i\hat{r}_j), & \text{for far zone } (k_0 r \gg 1), \\[2ex] \dfrac{\left|\vec{\mu}^{m0}\right|^2}{3r^6}(\delta_{ij} + 3\hat{r}_i\hat{r}_j), & \text{for near zone } (k_0 r \ll 1) \end{cases} \tag{32.37}$$

where $D_i(\vec{r})$ is the i-component of the transverse displacement vector field $\vec{D}(\vec{r})$, which satisfies $\nabla \cdot \vec{D}(\vec{r}) = 0$, $\vec{\mu}^{m0}$ is the electric dipole transition moment for the molecular states $|m\rangle$ and $|0\rangle$, and $k_0 \equiv k_{m0} \equiv (E_m - E_0)/(\hbar c)$ denotes the wave number associated with the $m \leftarrow 0$ transition of the molecule. The i-component of the unit vector $\hat{r}$ is designated by $\hat{r}_i$, while the absolute value of the position vector $\vec{r}$ is expressed by r. The Kronecker delta is denoted by δ_{ij}.

Similarly, for downward transitions from an excited state $|p\rangle$ to the ground state $|0\rangle$, the electric–electric correlation function is obtained:

$$\left\langle D_i(\vec{r}) D_j(\vec{r}) \right\rangle_{0 \leftarrow p} \sim \begin{cases} \dfrac{2p_0^4 \left|\vec{\mu}^{0p}\right|^2}{3r^2}(\delta_{ij} - \hat{r}_i\hat{r}_j), & \text{for far zone } (p_0 r \gg 1), \\[2ex] \dfrac{\left|\vec{\mu}^{0p}\right|^2}{3r^6}(\delta_{ij} + 3\hat{r}_i\hat{r}_j), & \text{for near zone } (p_0 r \ll 1), \end{cases} \tag{32.38}$$

where $p_0 \equiv p_{0p} \equiv (E_p - E_0)/(\hbar c)$ denotes the wave number associated with the $0 \leftarrow p$ transition of the molecule. It should be noted that the r^{-2} dependence arises from the real photon emission, while the r^{-6} dependence and the r^{-7} dependence are due to the virtual photon exchange. The far-zone behavior for the magnetic–magnetic correlation functions due to an electric-dipole source is also described by the same r^{-2} or r^{-7} dependence, while the near-zone result is different from its electric analog and the power law is r^{-5} instead of r^{-6}.

We have discussed the electric and magnetic correlation functions leading to the electric and magnetic energy densities associated with electric-dipole transitions in a source molecule, which can be detected by their effect on polarizable test bodies placed in the neighborhood of the source. This situation is analogous to an optical near-field system in which a nanometric material source connected to a macroscopic material and light source interacts with a nanometric detector connected to a macroscopic detector system. The difference is that the nanometric detector serving as a test body can disturb the field formed by the nanometric source in the case of the optical near-field system. We will move on to this topic in Section 32.3.3.

32.3.3 Effective Near-Field Optical Interaction between Nanomaterials Disconnected but Closely Separated

Several theoretical approaches to optical near-field problems, different from each other in viewpoints, have been proposed in the last two decades (Pohl and Courjon 1993, Ohtsu and Hori 1999). The optical near-field problems, including the application to nanophotonics, are ultimately regarded as how one should formulate a separated (more than two elements) composite system, each of which consists of a photon–electron–phonon interacting system on a nanometer scale as a source or a detector system and, which, at the same time, is connected with a macroscopic light–matter system. These questions must be clearly answered to achieve practical realization of nanophotonics. In order to provide a base for a variety of discussions in this research field, a new formulation has been developed within a quantum theoretical framework, putting matter excitations (electronic and vibrational) on an equal footing with photons (Kobayashi et al. 2001).

As discussed in Section 32.2, a "photon," whose concept has been established as a result of quantization of a free electromagnetic field, corresponds to a discrete excitation of electromagnetic modes in a virtual cavity. Unlike an electron, a photon is massless, and it is difficult to construct a wave function in the position representation that gives a picture of the photon as a spatially localized point particle like an electron. However, if there is a detector near the optical source, such as an atom, to absorb a photon in an area whose linear dimension is much smaller than the wavelength of light, it would be possible to detect a photon with the same precision as the detector size. In optical near-field problems, it is required to consider the interactions between light and nanomaterials surrounded by a macroscopic material system and detection of light by other nanomaterials on a nanometer scale. Then, a more serious question for the quantization of the field is how to define a virtual cavity, or which normal modes are to be used, since there exist more than two systems (nano-source and nano-detector) with arbitrary shape, size, and material in the nanometer region, which are still connected with a macroscopic material system, such as a source or a detector system. In this section, we describe a model and a theoretical approach to address the issue, which is essential to understand the operating principles nanophotonic devices, as well as nanofabrication using optical near fields (Ohtsu et al. 2008).

Let us consider a nanomaterial system surrounded by an incident light and a macroscopic material system, which electromagnetically interacts with one another in a complicated way, as schematically shown in Figure 32.1. Using the projection operator method, we can derive an effective interaction between the relevant nanomaterials in which we are interested (nano-source

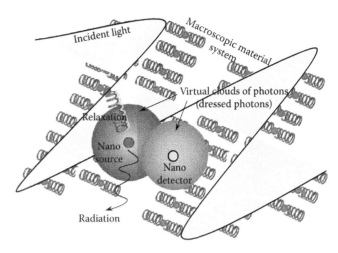

FIGURE 32.1 Nanometric source and detector system induced by incident light and a macroscopic material system.

and nano-detector—either one is in the excited state), as a result of renormalizing the other effects. It corresponds to an approach to describe "photons localized around nanomaterials," as if each nanomaterial would work as a detector and light source in a self-consistent way. The effective interaction related to optical near-fields is hereafter called a near-field optical interaction. As will be discussed in detail in this section, the near-field optical interaction between nanomaterials separated by R is as follows:

$$V_{\text{eff}} = \frac{\exp(-aR)}{R}, \tag{32.39}$$

where a^{-1} is the interaction range that represents the characteristic size of the nanomaterials and does not depend on the wavelength of the light used. It indicates that photons are localized around the nanomaterials (either of which is in the excited state) as a result of the interaction with matter fields, from which a photon, in turn, can acquire a finite mass. Therefore, we might consider that the near-field optical interaction is produced via localized photon hopping between nanomaterials. On the basis of the projection operator method, we will investigate formulation of an optical near-field system that was briefly mentioned above. Moreover, the explicit functional form of the near-field optical interaction will be given by using the effective interaction $m\hat{V}_{\text{eff}}$ in a perturbative way.

32.3.3.1 Relevant Nanometric Subsystem and Irrelevant Macroscopic Subsystem

As schematically illustrated in Figure 32.1, the optical near-field system consists of two subsystems: One is a macroscopic subsystem including the incident light, whose typical dimension is much larger than the wavelength of the incident light. The other is a nanometric subsystem (nano-source and nano-detector), whose constituents are, for example, a nanometric aperture or a protrusion at the apex of a near-field optical probe, and a nanometric sample. We call such an aperture or a protrusion a probe tip.

As a nanometric sample we mainly suppose a single atom/molecule, or quantum dot (QD). Two subsystems are interacting with each other, and it is very important to formulate the interaction consistently and systematically. Here let us call the nanometric subsystem a relevant subsystem n, and the macroscopic subsystem an irrelevant subsystem M. We are interested in the subsystem n; in particular, the interaction induced in the subsystem n. Therefore, it is the key to renormalize the effects originating from the subsystem M in a consistent and systematic way. Now we show a formulation based on the projection operator method, described below.

32.3.3.2 P Space and Q Space

It is preferable, for a variety of discussions, to express exact eigenstates $|\psi\rangle$ for the total system described by the total Hamiltonian $\hat{H}$ in terms of a small number of bases of as small a number of degrees of freedom as possible, which span P space. In the following, let us assume two states as the P-space components: $|\phi_1\rangle = |s^\star\rangle|p\rangle \otimes |0_{(M)}\rangle$ and $|\phi_2\rangle = |s\rangle|p^\star\rangle \otimes |0_{(M)}\rangle$, both of which are eigenstates of the unperturbed Hamiltonian $\hat{H}_0$. Here $|s\rangle$ and $|s^\star\rangle$ are eigenstates of the sample that is isolated from the others, while $|p\rangle$ and $|p^\star\rangle$ are eigenstates of the probe tip, which is also isolated. In addition, exciton polariton states, which are a mixture of photons and electron–hole pairs, are used as bases to describe the macroscopic subsystem M, and thus $|0_{(M)}\rangle$ represents the vacuum for exciton polaritons. Note that there exist photons and electronic matter excitations even in the vacuum state $|0_{(M)}\rangle$. The direct product is denoted by the symbol $\otimes$. The complementary space to the P space is called Q space, which is spanned by a huge number of bases of a large number of degrees of freedom not included in the P space, as schematically shown in Figure 32.2.

32.3.3.3 Effective Interaction Exerted in the Nanometric Subsystem

Noticing the relation between a bare interaction Hamiltonian $\hat{V} = \hat{H} - \hat{H}_0$ and an effective interaction Hamiltonian operator $\hat{V}_{\text{eff}}$,

namely, $\langle\psi|\hat{V}|\psi\rangle = \langle\psi|P\hat{V}_{\text{eff}}P|\psi\rangle$, we obtain the effective interaction Hamiltonian operator in the P space, given by

$$\hat{V}_{\text{eff}} = (P\hat{J}^\dagger \hat{J}P)^{-1/2}(P\hat{J}^\dagger \hat{V}\hat{J}P)(P\hat{J}^\dagger \hat{J}P)^{-1/2}, \tag{32.40}$$

and tracing out the other degrees of freedom gives an effective interaction Hamiltonian of the nanometric subsystem n after renormalizing the effects from the macroscopic subsystem M. Here, P is the projection operator, and Q is the complimentary operator defined by $Q = 1 - P$, both of which satisfy the following relations:

$$P = P^\dagger, \quad P^2 = P, \quad [P, \hat{H}_0] = 0, \tag{32.41a}$$

$$Q = Q^\dagger, \quad Q^2 = Q, \quad [Q, \hat{H}_0] = 0, \tag{32.41b}$$

$$PQ = QP = 0. \tag{32.41c}$$

The operator $\hat{J}$ is defined by

$$\hat{J} = [1 - (E - \hat{H}_0)^{-1}Q\hat{V}]^{-1}, \tag{32.42}$$

where E are the eigenvalues of the total Hamiltonian $\hat{H}$. Using the effective interaction Hamiltonian, one can forget the subsystem M as if the subsystem n were isolated and separated from the subsystem M.

To obtain an explicit expression of the effective interaction Hamiltonian, let us employ the bare interaction $\hat{V}$ between the two subsystems, which in the multipolar formalism (Craig and Thirunamachandran 1998) is given by

$$\hat{V} = -\left\{ \hat{\vec{\mu}}_s \cdot \hat{\vec{D}}^\perp(\vec{r}_s) + \hat{\vec{\mu}}_p \cdot \hat{\vec{D}}^\perp(\vec{r}_p) \right\}, \tag{32.43}$$

where the canonical momentum of the vector potential operator $\hat{\vec{A}}(\vec{r})$ is proportional to the transverse displacement vector field operator* $\hat{\vec{D}}^\perp(\vec{r})$, while the electric dipole operator is denoted as $\hat{\vec{\mu}}(\vec{r})$. It should be noted that there are no interactions, i.e., $\hat{V} = 0$, without incident photons in the macroscopic subsystem M. The subscripts s and p represent physical quantities related to the sample and the probe tip, respectively. Representative positions of the sample and the probe tip are chosen, for simplicity, by the vectors $\vec{r}_s$ and $\vec{r}_p$, respectively, but may be composed of several positions. In that case the quantities inside curly brackets in (32.43) should be read as a summation. The operator $\hat{\vec{\Pi}}(\vec{r})$ conjugate to $\hat{\vec{A}}(\vec{r})$ is expressed in terms of $\hat{\vec{D}}^\perp(\vec{r})$ as follows:

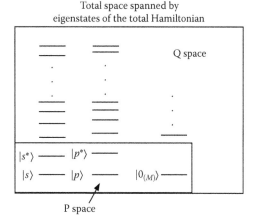

Total space spanned by
eigenstates of the total Hamiltonian

P space

FIGURE 32.2 Subdivision of the space spanned by eigenstates of the total Hamiltonian of the system.

---

* The transverse component is defined by $\nabla \cdot \vec{F}^\perp = 0$, while the longitudinal component is defined by $\nabla \times \vec{F}^\| = 0$, for an arbitrary vector field $\vec{F}(\vec{r})$.

$$\hat{\vec{\Pi}}(\vec{r}) = -\frac{1}{4\pi c}\hat{\vec{E}}^{\perp}(\vec{r}) - \frac{1}{c}\hat{\vec{P}}^{\perp}(\vec{r}) = -\frac{1}{4\pi c}\hat{\vec{D}}^{\perp}(\vec{r}),\qquad (32.44)$$

where $\hat{\vec{E}}^{\perp}(\vec{r})$ and $\hat{\vec{P}}^{\perp}(\vec{r})$ are the transverse electric field and the induced polarization field, respectively. With the help of the mode expansion of $\hat{\vec{A}}(\vec{r})$ and $\hat{\vec{\Pi}}(\vec{r})$, that is,

$$\hat{\vec{A}}(\vec{r}) = \sum_{\vec{k}}\sum_{\lambda=1}^{2}\left(\frac{2\pi\hbar c^2}{V\omega_{\vec{k}}}\right)^{1/2}\vec{e}_{\lambda}(\vec{k})\left\{\hat{a}_{\lambda}(\vec{k})e^{i\vec{k}\cdot\vec{r}} + \hat{a}_{\lambda}^{\dagger}(\vec{k})e^{-i\vec{k}\cdot\vec{r}}\right\},$$

$$(32.45)$$

and

$$\hat{\vec{\Pi}}(\vec{r}) = -\frac{i}{4\pi c}\sum_{\vec{k}}\sum_{\lambda=1}^{2}\left(\frac{2\pi\hbar\omega_{\vec{k}}}{V}\right)^{1/2}\vec{e}_{\lambda}(\vec{k})\left\{\hat{a}_{\lambda}(\vec{k})e^{i\vec{k}\cdot\vec{r}} - \hat{a}_{\lambda}^{\dagger}(\vec{k})e^{-i\vec{k}\cdot\vec{r}}\right\}$$

$$(32.46)$$

we can rewrite the transverse component of the electric displacement operator as

$$\hat{\vec{D}}^{\perp}(\vec{r}) = i\sum_{\vec{k}}\sum_{\lambda=1}^{2}\left(\frac{2\pi\hbar\omega_{\vec{k}}}{V}\right)^{1/2}\vec{e}_{\lambda}(\vec{k})\left\{\hat{a}_{\lambda}(\vec{k})e^{i\vec{k}\cdot\vec{r}} - \hat{a}_{\lambda}^{\dagger}(\vec{k})e^{-i\vec{k}\cdot\vec{r}}\right\},$$

$$(32.47)$$

where the plane waves are used for the mode functions, and the creation and annihilation operators of a photon with wave vector $\vec{k}$, angular frequency $\omega_{\vec{k}}$, and polarization component λ are designated by $\hat{a}_{\lambda}^{\dagger}(\vec{k})$ and $\hat{a}_{\lambda}(\vec{k})$, respectively. The quantization volume is V, and the unit vector related to the polarization direction is shown by $\vec{e}_{\lambda}(\vec{k})$. Note that the electric field outside the material corresponds to $\hat{\vec{D}}^{\perp}(\vec{r})$.

Since exciton polariton states are employed as bases to describe the macroscopic subsystem M, the creation and annihilation operators of a photon in (32.47) are rewritten using the exciton polariton operators $\hat{\xi}^{\dagger}(\vec{k})$ and $\hat{\xi}(\vec{k})$, and then they are substituted into (32.43). Using the electric dipole operator defined by

$$\hat{\vec{\mu}}_{\alpha} = \left(\hat{B}(\vec{r}_{\alpha}) + \hat{B}^{\dagger}(\vec{r}_{\alpha})\right)\vec{\mu}_{\alpha},\qquad (32.48)$$

with the creation and annihilation operators of excitation in subsystem n, $\hat{B}^{\dagger}(\vec{r}_{\alpha})$ and $\hat{B}(\vec{r}_{\alpha})$, and the transition dipole moments $\vec{\mu}_{\alpha}$ ($\alpha = s, p$), we obtain the bare interaction in the exciton polariton picture:

$$\hat{V} = -i\sum_{\alpha=s}^{p}\sum_{\vec{k}}\left(\frac{2\pi\hbar}{V}\right)^{1/2}\left(\hat{B}(\vec{r}_{\alpha}) + \hat{B}^{\dagger}(\vec{r}_{\alpha})\right)\left(K_{\alpha}(\vec{k})\hat{\xi}(\vec{k}) - K_{\alpha}^{*}(\vec{k})\hat{\xi}^{\dagger}(\vec{k})\right).$$

$$(32.49)$$

Here $K_{\alpha}(\vec{k})$ is the coupling coefficient between the exciton polariton and the nanometric subsystem n, given by

$$K_{\alpha}(\vec{k}) = \sum_{\lambda=1}^{2}\left(\vec{\mu}_{\varepsilon}\cdot\vec{e}_{\lambda}(\vec{k})\right)f(k)e^{i\vec{k}\cdot\vec{r}_{\alpha}}\qquad (32.50)$$

with

$$f(k) = \frac{ck}{\sqrt{\Omega(k)}}\sqrt{\frac{\Omega^2(k) - \Omega^2}{2\Omega^2(k) - \Omega^2 - (ck)^2}}.\qquad (32.51)$$

The complex conjugate of $K_{\alpha}(\vec{k})$ is denoted by $K_{\alpha}^{*}(\vec{k})$, while c, $\Omega(k)$, and Ω are light speed in vacuum and the eigenfrequencies of both exciton polariton and electronic polarization of the macroscopic subsystem M, respectively. The dispersion relation for a free photon, $\omega_{\vec{k}} = ck$, is used in (32.51). Note that the wave-number dependence of $f(k)$ characterizes a typical interaction range of exciton polaritons coupled to the nanometric subsystem n.

The next step is to evaluate the amplitude of the effective interaction exerted on the nanometric subsystem, for example, the effective sample–probe-tip interaction in the P space after tracing out the polariton degrees of freedom:

$$V_{\text{eff}}(2,1) \equiv \langle\phi_2|\hat{V}_{\text{eff}}|\phi_1\rangle.\qquad (32.52)$$

With the first-order approximation $\hat{J}^{(1)}$ of $\hat{J}$ in (32.42), we can explicitly write (32.52) in the following form:

$$V_{\text{eff}}(2,1) = \langle\phi_2|P\hat{V}Q\hat{V}(E_P^0 - E_Q^0)^{-1}P|\phi_1\rangle + \langle\phi_2|P(E_P^0 - E_Q^0)^{-1}\hat{V}Q\hat{V}P|\phi_1\rangle$$

$$= \sum_m \langle\phi_2|P\hat{V}Q|m\rangle\langle m|Q\hat{V}P|\phi_1\rangle\left(\frac{1}{E_{P1}^0 - E_{Qm}^0} + \frac{1}{E_{P2}^0 - E_{Qm}^0}\right).$$

$$(32.53)$$

The second line shows that a virtual transition from the initial state $|\phi_1\rangle$ in the P space to an intermediate state $|m\rangle$ in the Q space is followed by a subsequent virtual transition from the intermediate state $|m\rangle$ to the final state $|\phi_2\rangle$ in the P space. Here $E_{P1}^0(E_{P2}^0)$ and E_{Qm}^0 denote eigenenergies of $|\phi_1\rangle(|\phi_2\rangle)$ in the P space and that of $|\phi_m\rangle$ in the Q space, respectively. Now we can proceed to the next step by substituting the explicit bare interaction $\hat{V}$ in (32.49) with (32.50) and (32.51) into (32.53). First of all, note that the one-exciton polariton state among arbitrary intermediate states $|m\rangle$ can only contribute to nonzero matrix elements. Therefore, (32.53) can be transformed into

$$V_{\text{eff}}(2,1) = -\frac{\hbar}{(2\pi)^2}\int d^3k\left[\frac{K_p(\vec{k})K_s^{*}(\vec{k})}{\Omega(k) - \Omega_0(s)} + \frac{K_s(\vec{k})K_p^{*}(\vec{k})}{\Omega(k) + \Omega_0(s)}\right],$$

$$(32.54)$$

where the summation over $\vec{k}$ is replaced by $\vec{k}$-integration, i.e., $\frac{V}{(2\pi)^3}\int d^3k$, in the usual manner. Excitation energies of the sample (between $|s^*\rangle$ and $|s\rangle$) and the probe tip (between $|p^*\rangle$ and $|p\rangle$) are denoted as $E_s = \hbar\Omega_0(s)$ and $E_p = \hbar\Omega_0(p)$, respectively.

Exchanging the arguments 1 and 2, or the role of the sample and probe tip, we can similarly calculate $V_{eff}(1,2) \equiv \langle\phi_1|\hat{V}_{eff}|\phi_2\rangle$:

$$V_{eff}(1,2) = -\frac{\hbar}{(2\pi)^2}\int d^3k \left[\frac{K_s(\vec{k})K_p^*(\vec{k})}{\Omega(k)-\Omega_0(p)} + \frac{K_p(\vec{k})K_s^*(\vec{k})}{\Omega(k)+\Omega_0(p)}\right].$$

(32.55)

Therefore, the total amplitude of the effective sample–probe-tip interaction is given by the sum of (32.54) and (32.55), which includes the effects from the macroscopic subsystem M. We write this effective interaction for the nanometric subsystem n as $V_{eff}(\vec{r})$ in the following way:

$$V_{eff}(\vec{r}) = -\frac{1}{4\pi^2}\sum_{\lambda=1}^{2}\sum_{\alpha=p,s}\int d^3k\left[\left(\vec{\mu}_p\cdot\vec{e}_\lambda(\vec{k})\right)\left(\vec{\mu}_s\cdot\vec{e}_\lambda(\vec{k})\right)\right]$$
$$\times \hbar f^2(k)\left(\frac{e^{i\vec{k}\cdot\vec{r}}}{E(k)+E_\alpha} + \frac{e^{-i\vec{k}\cdot\vec{r}}}{E(k)-E_\alpha}\right)$$

(32.56)

where we have set $E(k) = \hbar\Omega(k)$, and $E_\alpha = \hbar\Omega_0(a^*) - \hbar\Omega_0(\alpha)$ for $\alpha = p$ and $\alpha = s$. The summation over polarization λ is performed as

$$\sum_{\lambda=1}^{2}\vec{e}_{\lambda i}(\vec{k})\vec{e}_{\lambda i}(\vec{k}) = \delta_{ij} - k_i k_j,$$

(32.57)

and thus the summation of $\left(\vec{\mu}_p\cdot\vec{e}_\lambda(\vec{k})\right)\left(\vec{\mu}_s\cdot\vec{e}_\lambda(\vec{k})\right)$ over λ can be reduced as follows:

$$\sum_{\lambda=1}^{2}\left(\vec{\mu}_p\cdot\vec{e}_\lambda(\vec{k})\right)\left(\vec{\mu}_s\cdot\vec{e}_\lambda(\vec{k})\right) = \sum_{\lambda=1}^{2}\sum_{i,j}\left(\vec{\mu}_{pi}\vec{e}_{\lambda i}(\vec{k})\right)\left(\vec{\mu}_{sj}\vec{e}_{\lambda j}(\vec{k})\right)$$
$$= \sum_{i,j}\mu_{pi}\mu_{sj}\left(\delta_{ij} - \hat{k}_i\hat{k}_j\right)$$

(32.58)

with the unit vector $\hat{k} \equiv \vec{k}/k$. Noticing that $d^3k = k^2 dk d\Omega = k^2 dk$ $\sin\theta d\theta d\varphi$ and

$$\delta_{ij}\int e^{\pm i\vec{k}\cdot\vec{r}}d\Omega = \delta_{ij}\int_0^{2\pi}\int_{-1}^{1}e^{\pm ikr\cos\theta}d(\cos\theta)d\varphi = \delta_{ij}\frac{2\pi}{ikr}(e^{ikr} - e^{-ikr}),$$

(32.59a)

$$-\int\hat{k}_i\hat{k}_j e^{\pm i\vec{k}\cdot\vec{r}}d\Omega = \frac{1}{k^2}\nabla_i\nabla_j\int e^{\pm i\vec{k}\cdot\vec{r}}d\Omega = \frac{2\pi}{ik^3}\nabla_i\nabla_j\left(\frac{e^{ikr}-e^{-ikr}}{r}\right),$$

(32.59b)

we find

$$\int(\delta_{ij} - \hat{k}_i\hat{k}_j)e^{\pm i\vec{k}\cdot\vec{r}}d\Omega = \delta_{ij}\frac{2\pi}{ik}\left(\frac{e^{ikr}-e^{-ikr}}{r}\right) + \frac{2\pi}{ik^3}\nabla_i\nabla_j\left(\frac{e^{ikr}-e^{-ikr}}{r}\right)$$
$$= 2\pi\left[\delta_{ij}\frac{(e^{ikr}-e^{-ikr})}{ikr} + (\delta_{ij} - 3\hat{r}_i\hat{r}_j)\left\{\frac{(e^{ikr}+e^{-ikr})}{k^2 r^2} - \frac{(e^{ikr}-e^{-ikr})}{ikr^3 r^3}\right\}\right.$$
$$\left. - \frac{(e^{ikr}-e^{-ikr})}{ikr}\hat{r}_i\hat{r}_j\right]$$

(32.60)

where $\hat{r}$ is the unit vector defined by $\hat{r} \equiv \vec{r}/r$, and the jth component is denoted by $\hat{r}_j$. Hence, the effective interaction can be rewritten as

$$V_{eff}(\vec{r}) = -\frac{1}{2\pi}\int_{-\infty}^{\infty}k^2 dk\hbar f^2(k)\sum_{a=s,p}\left(\frac{1}{E(k)+E_a} + \frac{1}{E(k)-E_a}\right)$$
$$\times\left\{(\vec{\mu}_s\cdot\vec{\mu}_p)e^{i\vec{k}\cdot\vec{r}}\left(\frac{1}{ikr} + \frac{1}{k^2 r^2} - \frac{1}{ik^3 r^3}\right) - (\vec{\mu}_s\cdot\hat{r})(\vec{\mu}_p\cdot\hat{r})e^{i\vec{k}\cdot\vec{r}}\right.$$
$$\left.\times\left(\frac{1}{ikr} + \frac{3}{k^2 r^2} - \frac{3}{ik^3 r^3}\right)\right\}$$

(32.61)

where the integration range is extended from $(0, \infty)$ to $(-\infty, \infty)$. When the dispersion relation of exciton polaritons, which have been chosen as a basis for describing the macroscopic subsystem M, is approximated as

$$E(k) = \hbar\Omega + \frac{(\hbar k)^2}{2m_{pol}} = E_m + \frac{(\hbar ck)^2}{2E_{pl}},$$

(32.62)

in terms of the effective mass of exciton polaritons, m_{pol}, or $E_{pl} = m_{pol}c^2$ and the electronic excitation energy of the macroscopic subsystem M, $E_m = \hbar\Omega$, (32.61) is further simplified:

$$V_{eff}(\vec{r}) = -\frac{1}{2\pi}\int_{-\infty}^{\infty}k^2 dk\hbar f^2(k)\sum_{\alpha=s,p}\frac{2E_{pl}}{(\hbar c)^2}$$
$$\times\left\{\frac{1}{(k+i\Delta_{\alpha+})(k-i\Delta_{\alpha+})} + \frac{1}{(k+i\Delta_{\alpha-})(k-i\Delta_{\alpha-})}\right\}$$
$$\times\left\{(\vec{\mu}_s\cdot\vec{\mu}_p)e^{i\vec{k}\cdot\vec{r}}\left(\frac{1}{ikr} + \frac{1}{k^2 r^2} - \frac{1}{ik^3 r^3}\right) - (\vec{\mu}_s\cdot\hat{r})(\vec{\mu}_p\cdot\hat{r})e^{i\vec{k}\cdot\vec{r}}\right.$$
$$\left.\times\left(\frac{1}{ikr} + \frac{3}{k^2 r^2} - \frac{3}{ik^3 r^3}\right)\right\}$$
$$\equiv \sum_{\alpha=s,p}\left[V_{eff,\alpha+}(\vec{r}) + V_{eff,\alpha-}(\vec{r})\right]$$

(32.63a)

with

$$\Delta_{\alpha\pm} \equiv \frac{1}{\hbar c}\sqrt{2E_{\mathrm{pl}}(E_m \pm E_\alpha)}, \quad (E_m > E_\alpha). \qquad (32.63b)$$

The k-integration can be performed with the residues evaluated at $k = i\Delta_{\alpha\pm}$, and we have

$$V_{\mathrm{eff},\alpha\pm}(\vec{r}) = \mp\frac{1}{2}\Bigg[(\vec{\mu}_s \cdot \vec{\mu}_p)\left\{\frac{(\Delta_{\alpha\pm})^2}{r} + \frac{\Delta_{\alpha\pm}}{r^2} + \frac{1}{r^3}\right\}W_{\alpha\pm}e^{-\Delta_{\alpha\pm}r}$$

$$- (\vec{\mu}_s \cdot \hat{r})(\vec{\mu}_p \cdot \hat{r})\left\{\frac{(\Delta_{\alpha\pm})^2}{r} + \frac{3\Delta_{\alpha\pm}}{r^2} + \frac{3}{r^3}\right\}W_{\alpha\pm}e^{-\Delta_{\alpha\pm}r}\Bigg], \qquad (32.64a)$$

where the constants $W_{\alpha\pm}$ are defined by

$$W_{\alpha\pm} \equiv \frac{E_{\mathrm{pl}}}{E_\alpha}\frac{E_m^2 - E_\alpha^2}{(E_m \pm E_\alpha)(E_m - E_{\mathrm{pl}} \mp E_\alpha) - E_m^2/2}. \qquad (32.64b)$$

If the angular average of $(\vec{\mu}_s \cdot \hat{r})(\vec{\mu}_p \cdot \hat{r})$ is taken, the expression

$$V_{\mathrm{eff}}(r) = -\frac{(\vec{\mu}_s \cdot \vec{\mu}_p)}{3}\sum_{\alpha=s,p}\left\{W_{\alpha+}(\Delta_{\alpha+})^2\frac{e^{-\Delta_{\alpha+}r}}{r} - W_{\alpha-}(\Delta_{\alpha-})^2\frac{e^{-\Delta_{\alpha-}r}}{r}\right\} \qquad (32.65)$$

is obtained for the effective interaction, or the near-field optical interaction $V_{\mathrm{eff}}(r)$, which consists of the sum of the Yukawa functions $Y(\Delta_{\alpha\pm}r) \equiv e^{-\Delta_{\alpha\pm}r}/r$ with a shorter interaction range $\Delta_{\alpha+}$ (heavier effective mass) and a longer interaction range $\Delta_{\alpha-}$ (lighter effective mass).

To sum up, we find that the major part of the effective interaction exerted in the nanometric subsystem n is the Yukawa potential after renormalizing the effects from the macroscopic subsystem. This interaction comes from the mediation of massive virtual photons corresponding to the counter-rotating term, or polaritons, where exciton polaritons have been employed in an explicit formulation, but in principle other types of polaritons would be applicable. In this section we have mainly focused on the effective interaction of the nanometric subsystem n, after tracing out the other degrees of freedom. It is certainly possible to have a formulation with a projection onto the P space that is spanned in terms of the degrees of freedom of the massive virtual photons, although this is left as the subject of future work. This kind of formulation emphasizes a "dressed photon" picture, in which photons are not massless but massive, due to light–matter interactions at the nanoscale.

32.3.4 Localization of a Photon Dressed by Matter Excitation in Nanomaterials at the Nanoscale

In this section, we consider a simple model system, for example, a pseudo one-dimensional near-field optical probe system, to discuss the mechanism of photon localization in space as well as the phonon's role. In order to focus on the photon–phonon interaction, the interacting part between photon and electronic excitation is first expressed in terms of a polariton, and is called a photon in the model. Then the model Hamiltonian that describes the photon and phonon interacting system is presented. Using the Davydov transformation, we rewrite the Hamiltonian in terms of quasiparticles. On the basis of the Hamiltonian, we present numerical results on the spatial distribution of photons and discuss the mechanism of photon localization due to phonons.

32.3.4.1 Model Hamiltonian

We consider a near-field optical probe, schematically shown in Figure 32.3, as an example system where light interacts with both phonons and electrons in the probe on a nanometer scale. Here, the interaction of a photon and an electronic excitation is assumed to be expressed in terms of a polariton basis, as discussed above, and is hereafter called a photon so that special attention is paid to the photon–phonon interaction. The system is simply modeled as a one-dimensional atomic or molecular chain coupled with photon and phonon fields. The chain consists of finite N molecules (representatively called), each of which is located at a discrete point (called a molecular site) whose separation represents a characteristic scale of the near-field system. Photons are expressed in the site representation and can hop to the nearest neighbor sites due to the short-range interaction nature of the optical near fields.

The Hamiltonian for the above model is given by

$$\hat{H} = \sum_{i=1}^{N}\hbar\omega\hat{a}_i^\dagger\hat{a}_i + \left\{\sum_{i=1}^{N}\frac{\hat{p}_i^2}{2m_i} + \sum_{i=1}^{N-1}\frac{k}{2}(\hat{x}_{i+1} - \hat{x}_i)^2 + \sum_{i=1,N}\frac{k}{2}\hat{x}_i^2\right\}$$

$$+ \sum_{i=1}^{N}\hbar\chi\hat{a}_i^\dagger\hat{a}_i\hat{x}_i + \sum_{i=1}^{N-1}\hbar J(\hat{a}_i^\dagger\hat{a}_{i+1} + \hat{a}_{i+1}^\dagger\hat{a}_i), \qquad (32.66)$$

where

$\hat{a}_i^\dagger$ and $\hat{a}_i$ correspondingly denote the creation and annihilation operators of a photon with energy $\hbar\omega$ at site i in the chain

$\hat{x}_i$ and $\hat{p}_i$ represent the displacement and conjugate momentum operators of the vibration, respectively

FIGURE 32.3 Simple one-dimensional model for a light–matter interacting system on a nanometer scale.

The mass of a molecule at site i is designated by m_i, and each molecule is assumed to be connected by springs with spring constant k. The third and fourth terms in (32.66) stand for the photon–vibration interaction with coupling constant χ and the photon hopping with hopping constant J, respectively. After the vibration field is quantized in terms of phonon operators of mode p and frequency Ω_p, $\hat{b}_p^\dagger$, and $\hat{b}_p$, the Hamiltonian (32.66) can be rewritten as

$$
\hat{H} = \sum_{i=1}^{N} \hbar\omega\hat{a}_i^\dagger\hat{a}_i + \sum_{p=1}^{N} \hbar\Omega_p\hat{b}_p^\dagger\hat{b}_p
$$
$$
+ \sum_{i=1}^{N}\sum_{p=1}^{N} \hbar\chi_{ip}\hat{a}_i^\dagger\hat{a}_i(\hat{b}_p^\dagger + \hat{b}_p) + \sum_{i=1}^{N-1} \hbar J(\hat{a}_i^\dagger\hat{a}_{i+1} + \hat{a}_{i+1}^\dagger\hat{a}_i),
$$

$$(32.67)$$

with the coupling constant χ_{ip} of a photon at site i and a phonon of mode p. It should be noted that the index p designates not the momentum but the mode number, because the translational invariance of the system is broken and the momentum is not a good quantum number. The site-dependent coupling constant χ_{ip} is related to the original coupling constant χ in terms of the transformation matrix as P_{ip} as

$$
\chi_{ip} = \chi P_{ip}\sqrt{\frac{\hbar}{2m_i\Omega_p}},
$$

$$(32.68)$$

and the creation and annihilation operators of a photon and a phonon satisfy the boson commutation relation as $\left[\hat{a}_i, \hat{a}_j^\dagger\right] = \delta_{ij}$ and $\left[\hat{b}_p, \hat{b}_q^\dagger\right] = \delta_{pq}$. The Hamiltonian (32.67), which describes the model system, is not easily handled because of the third order of the operators in the interaction term. To avoid this difficulty, this direct photon–phonon interaction term in (32.67) is eliminated by the Davydov transformation in the following subsection.

32.3.4.2 Davydov Transformation

Before going into the explicit expression, we discuss a unitary transformation $\hat{U}$ generated by an anti-Hermitian operator $\hat{S}$, defined as

$$
\hat{U} \equiv \exp(\hat{S}), \quad \text{with } \hat{S}^\dagger = -\hat{S}
$$

$$(32.69a)$$

and

$$
\hat{U}^\dagger = \hat{U}^{-1}.
$$

$$(32.69b)$$

Assume a Hamiltonian $\hat{H}$ that consists of a diagonalized part $\hat{H}_0$ and a non-diagonal interaction part $\hat{V}$:

$$
\hat{H} = \hat{H}_0 + \hat{V}.
$$

$$(32.70)$$

Transforming the Hamiltonian in (32.70) as

$$
\tilde{H} \equiv \hat{U}\hat{H}\hat{U}^\dagger = \hat{U}\hat{H}\hat{U}^{-1},
$$

$$(32.71)$$

we have

$$
\tilde{H} = \hat{H} + [\hat{S}, \hat{H}] + \frac{1}{2}\left[\hat{S}, [\hat{S}, \hat{H}]\right] + \cdots
$$
$$
= \hat{H}_0 + \hat{V} + [\hat{S}, \hat{H}_0] + [\hat{S}, \hat{V}] + \frac{1}{2}\left[\hat{S}, [\hat{S}, \hat{H}_0]\right] + \cdots. \quad (32.72)
$$

If the interaction $\hat{V}$ can be perturbative, and if the operator $\hat{S}$ is chosen so that the second and third terms in (32.72) are canceled out, then

$$
\hat{V} = -[\hat{S}, \hat{H}_0],
$$

$$(32.73)$$

the Hamiltonian (32.70) is rewritten as

$$
\tilde{H} = \hat{H}_0 - \frac{1}{2}\left[\hat{S}, [\hat{S}, \hat{H}_0]\right] + \cdots,
$$

$$(32.74)$$

and can be diagonalized within the first order of $\hat{V}$.

Now we apply the above discussion to the model Hamiltonian (32.67),

$$
\hat{H}_0 = \sum_{i=1}^{N} \hbar\omega\hat{a}_i^\dagger\hat{a}_i + \sum_{p=1}^{N} \hbar\Omega_p\hat{b}_p^\dagger\hat{b}_p,
$$

$$(32.75a)$$

$$
\hat{V} = \sum_{i=1}^{N}\sum_{p=1}^{N} \hbar\chi_{ip}\hat{a}_i^\dagger\hat{a}_i(\hat{b}_p^\dagger + \hat{b}_p),
$$

$$(32.75b)$$

tentatively neglecting the hopping term. Assuming the anti-Hermitian operator $\hat{S}$

$$
\hat{S} = \sum_i\sum_p f_{ip}\hat{a}_i^\dagger\hat{a}_i(\hat{b}_p^\dagger - \hat{b}_p),
$$

$$(32.76a)$$

we can determine f_{ip} from (32.73) as follows:

$$
f_{ip} = \frac{\chi_{ip}}{\Omega_p}.
$$

$$(32.76b)$$

This operator form of $\hat{S}$ leads us to not the perturbative but the exact transformation of the photon and phonon operators:

$$
\hat{\alpha}_i^\dagger \equiv \hat{U}^\dagger\hat{a}_i^\dagger\hat{U} = \hat{a}_i^\dagger \exp\left\{-\sum_{p=1}^{N} \frac{\chi_{ip}}{\Omega_p}\left(\hat{b}_p^\dagger - \hat{b}_p\right)\right\},
$$

$$(32.77a)$$

$$\hat{\alpha}_i \equiv \hat{U}^\dagger \hat{a}_i \hat{U} = \hat{a}_i \exp\left\{\sum_{p=1}^{N} \frac{\chi_{ip}}{\Omega_p}\left(\hat{b}_p^\dagger - \hat{b}_p\right)\right\}, \qquad (32.77b)$$

$$\hat{\beta}_p^\dagger \equiv \hat{U}^\dagger \hat{b}_p^\dagger \hat{U} = \hat{b}_p^\dagger + \sum_{i=1}^{N} \frac{\chi_{ip}}{\Omega_p}\hat{a}_i^\dagger \hat{a}_i, \qquad (32.77c)$$

$$\hat{\beta}_p \equiv \hat{U} \hat{b}_p \hat{U}^\dagger = \hat{b}_p + \sum_{i=1}^{N} \frac{\chi_{ip}}{\Omega_p}\hat{a}_i^\dagger \hat{a}_i. \qquad (32.77d)$$

These transformed operators can be regarded as the creation and annihilation operators of quasiparticles—dressed photons and phonons—that satisfy the same boson commutation relations as those of photons and phonons before the transformation, namely, $\left[\hat{\alpha}_i, \hat{\alpha}_j^\dagger\right] = \hat{U}^\dagger\left[\hat{a}_i, \hat{a}_j\right]\hat{U} = \delta_{ij}$ and $\left[\hat{\beta}_p, \hat{\beta}_q^\dagger\right] = \hat{U}^\dagger\left[\hat{b}_p, \hat{b}_q^\dagger\right]\hat{U} = \delta_{pq}$.

Using the quasiparticle operators, we can rewrite the Hamiltonian (32.67) as

$$\hat{H} = \sum_{i=1}^{N} \hbar\omega\hat{\alpha}_i^\dagger\hat{\alpha}_i + \sum_{p=1}^{N} \hbar\Omega_p\hat{\beta}_p^\dagger\hat{\beta}_p - \sum_{i=1}^{N}\sum_{j=1}^{N}\sum_{p=1}^{N} \frac{\hbar\chi_{ip}\chi_{jp}}{\Omega_p}\hat{\alpha}_i^\dagger\hat{\alpha}_i\hat{\alpha}_j^\dagger\hat{\alpha}_j$$

$$+ \sum_{i=1}^{N-1} \hbar\left(\hat{J}_i\hat{\alpha}_i^\dagger\hat{\alpha}_{i+1} + \hat{J}_i^\dagger\hat{\alpha}_{i+1}^\dagger\hat{\alpha}_i\right), \qquad (32.78a)$$

with

$$\hat{J}_i = J\exp\left\{\sum_{p=1}^{N} \frac{(\chi_{ip} - \chi_{i+1p})}{\Omega_p}\left(\hat{\beta}_p^\dagger - \hat{\beta}_p\right)\right\}, \qquad (32.78b)$$

where it is noted that the direct photon–phonon coupling term has been eliminated, while the quadratic form $\hat{N}_i\hat{N}_j$ with the number operator of $\hat{N}_i = \hat{\alpha}_i^\dagger\hat{\alpha}_i$ has emerged as well as the site-dependent hopping operator $\hat{J}_i$ in (32.78b). The number states of quasiparticles are thus eigenstates of each terms of the Hamiltonian (32.78a), except the last term that represents the higher-order effect of photon–phonon coupling through the dressed photon hopping. Therefore, it is a more appropriate form to discuss the phonon's effect on photon's behavior as localization.

32.3.4.3 Quasiparticle and Coherent State

In the previous section, we have transformed the original Hamiltonian with the Davydov transformation. In order to grasp the physical meanings of the quasiparticles introduced above, the creation operator $\hat{\alpha}_i^\dagger$ is applied to the vacuum state $|0\rangle$. Then, it follows from (32.77a) that

$$\hat{\alpha}_i^\dagger|0\rangle = \hat{a}_i^\dagger\exp\left\{-\sum_{p=1}^{N} \frac{\chi_{ip}}{\Omega_p}\left(\hat{b}_p^\dagger - \hat{b}_p\right)\right\}|0\rangle,$$

$$= \hat{a}_i^\dagger\exp\left\{-\sum_{p=1}^{N} \frac{1}{2}\left(\frac{\chi_{ip}}{\Omega_p}\right)^2\right\}\exp\left\{-\frac{\chi_{ip}}{\Omega_p}\hat{b}_p^\dagger\right\}|0\rangle, \qquad (32.79)$$

where a photon at site i is associated with phonons in a coherent state, i.e., a photon is dressed by an infinite number of phonons. This corresponds to the fact that an optical near field is generated from a result of interactions between the photon and matter fields.

When $\hat{\beta}_p^\dagger$ is applied to the vacuum state $|0\rangle$, we have

$$\hat{\beta}_p^\dagger|0\rangle = \hat{b}_p^\dagger|0\rangle, \qquad (32.80)$$

and it is expressed by only the bare phonon operator (before the transformation) in the same p mode. Therefore, we mainly focus on the quasiparticle expressed by $(\hat{\alpha}_i^\dagger, \hat{\alpha}_i)$ in the following subsection. Note that it is valid only if the bare photon number (the expectation value of $\hat{\alpha}_i^\dagger\hat{\alpha}_i$ is not so large that the fluctuation is more important than the bare photon number. In other words, the model we are considering is suitable for discussing the quantum nature of a few photons in an optically excited probe system.

In the coherent state of phonons, the number of phonons, as well as energy, is fluctuating. This fluctuation allows incident photons into the probe system to excite phonon fields. When all the phonons are in the vacuum state at time $t = 0$, the excitation probability $P(t)$ that a photon incident on site i in the model system excites the phonon mode p at time t is given by

$$P(t) = 1 - \exp\left\{2\left(\frac{\chi_{ip}}{\Omega_p}\right)^2\left[\cos(\Omega_p t) - 1\right]\right\}, \qquad (32.81)$$

where the photon-hopping term is neglected for simplicity. The excitation probability oscillates at a frequency of $2\pi/\Omega_p$ and has the maximum value at $t = \pi/\Omega_p$. The frequencies of the localized phonon modes are higher than those of the delocalized ones, and the localized modes at the earlier time are excited by the incident photons.

Figure 32.4 shows the temporal evolution of the excitation probability $P_{p_0}(t)$ calculated from

$$P_{p_0}(t) = \left[1 - \exp\left\{2\left(\frac{\chi_{ip_0}}{\Omega_{p_0}}\right)^2\left[\cos(\Omega_{p_0}t) - 1\right]\right\}\right]$$

$$\times \exp\left\{\sum_{p\neq p_0} 2\left(\frac{\chi_{ip}}{\Omega_p}\right)^2\left[\cos(\Omega_p t) - 1\right]\right\}, \qquad (32.82)$$

where a specific phonon mode p_0 is excited, while other modes are in the vacuum state. In Figure 32.4, the solid curve represents the probability that a localized phonon mode is excited as the p_0 mode, while the dashed curve illustrates how the lowest phonon mode is excited as the p_0 mode. It follows from the figure that the localized phonon mode is dominantly excited at an earlier time.

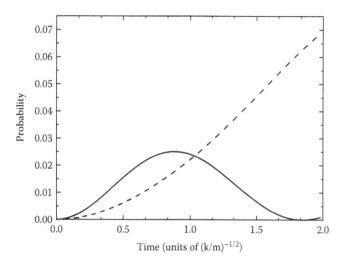

FIGURE 32.4 Temporal evolution of a specific phonon mode. The solid curve represents the probability for a localized phonon mode, while the dashed curve shows that of the lowest delocalized phonon mode.

32.3.4.4 Localization Mechanism of Dressed Photons

In this section, we discuss how phonons contribute to the spatial distribution of photons in the pseudo one-dimensional system under consideration. When there are no interactions between photons and phonons, the frequency and hopping constant are equal at all sites, and thus the spatial distribution of photons is symmetric. It means that no photon localization occurs at any specific site. However, if there are any photon–phonon interactions, spatial inhomogeneity or localization of phonons affects the spatial distribution of photons. On the basis of the Hamiltonian (32.78a), we analyze the contribution from the diagonal (the third term) and off-diagonal (the fourth term) parts of the Hamiltonian (32.78a) in order to investigate the localization mechanism of photons.

32.3.4.4.1 Contribution from the Diagonal Part

Let us rewrite the third term of the Hamiltonian (32.78a) with the mean field approximation as

$$-\sum_{i=1}^{N}\sum_{j=1}^{N}\sum_{p=1}^{N}\frac{\hbar\chi_{ip}\chi_{jp}}{\Omega_p}\hat{\alpha}_i^\dagger\hat{\alpha}_i\left\langle\hat{\alpha}_j^\dagger\hat{\alpha}_j\right\rangle \equiv -\sum_{i=1}^{N}\hbar\omega_i\hat{\alpha}_i^\dagger\hat{\alpha}_i, \quad (32.83a)$$

with

$$\omega_i = \sum_{j=1}^{N}\sum_{p=1}^{N}\frac{\chi_{ip}\chi_{jp}}{\Omega_p}\left\langle\hat{\alpha}_j^\dagger\hat{\alpha}_j\right\rangle. \quad (32.83b)$$

In addition, for the moment, we neglect the site dependence of the hopping operator $\hat{J}_i$, which is approximated as J. Then the Hamiltonian regarding the quasiparticles ($\hat{\alpha}$ and $\hat{\alpha}^\dagger$) can be expressed as

$$\hat{H} = \sum_{i=1}^{N}\hbar(\omega-\omega_i)\hat{\alpha}_i^\dagger\hat{\alpha}_i + \sum_{i=1}^{N-1}\hbar J\left(\hat{\alpha}_i^\dagger\hat{\alpha}_{i+1} + \hat{\alpha}_{i+1}^\dagger\hat{\alpha}_i\right), \quad (32.84)$$

or in matrix form as

$$\hat{H} = \hbar\hat{\vec{\alpha}}^\dagger\begin{pmatrix} \omega-\omega_1 & J & \cdots & 0 \\ J & \omega-\omega_2 & \ddots & \vdots \\ \vdots & \ddots & \ddots & J \\ 0 & \cdots & J & \omega-\omega_N \end{pmatrix}\hat{\vec{\alpha}} \quad (32.85a)$$

with

$$\hat{\vec{\alpha}} \equiv \left(\hat{\alpha}_1^\dagger, \hat{\alpha}_2^\dagger, \ldots, \hat{\alpha}_N^\dagger\right), \quad (32.85b)$$

where the effect from the phonon fields is involved in the diagonal elements ω_i. Denoting an orthonormal matrix to diagonalize the Hamiltonian (32.85a) as Q and the rth eigenvalue as E_r, we have

$$\hat{H} = \sum_{r=1}^{N}\hbar E_r\hat{A}_r^\dagger\hat{A}_r \quad (32.86a)$$

with

$$\hat{A}_r\sum_{i=1}^{N}(Q^{-1})_{ri}\hat{\alpha}_i = \sum_{i=1}^{N}Q_{ir}\hat{\alpha}_i, \quad (32.86b)$$

and

$$\left[\hat{A}_r, \hat{A}_s^\dagger\right] \equiv \hat{A}_r\hat{A}_s^\dagger - \hat{A}_s^\dagger\hat{A}_r = \delta_{rs}. \quad (32.86c)$$

Using the above relations (32.86a) through (32.86c), we can write down the time evolution of the photon number operator $\hat{N}_i$ at site i as follows:

$$\hat{N}_i(t) = \exp\left(i\frac{\hat{H}}{\hbar}t\right)\hat{N}_i\exp\left(-i\frac{\hat{H}}{\hbar}t\right)$$

$$= \sum_{r=1}^{N}\sum_{s=1}^{N}Q_{ir}Q_{is}\hat{A}_r^\dagger\hat{A}_s\exp\{i(E_r-E_s)t\}. \quad (32.87)$$

The expectation value of the photon number operator $\hat{N}_i(t)$ is then given by

$$\left\langle N_i(t)\right\rangle_j \equiv \left\langle\psi_j|\hat{N}_i(t)|\psi_j\right\rangle = \sum_{r=1}^{N}\sum_{s=1}^{N}Q_{ir}Q_{jr}Q_{is}Q_{js}\cos\{(E_r-E_s)t\},$$

$$(32.88)$$

in terms of one photon state at site j defined by

$$\left|\psi_j\right\rangle = \hat{\alpha}_j^\dagger \left|0\right\rangle = \sum_{r=1}^{N} Q_{jr} \hat{A}_r^\dagger \left|0\right\rangle. \quad (32.89)$$

Since the photon number operator $\hat{N}_i$ commutes with the Hamiltonian (32.84), the total photon number is conserved, which means that a polariton, called a photon in this section, conserves the total particle number within the lifetime. Moreover, $\langle N_i(t)\rangle_j$ can be regarded as the observation probability of a photon at an arbitrary site i and time t, initially populated at site j. This function is analytically expressed in terms of the Bessel functions $J_{j\pm i}$:

$$\left\langle N_i(t)\right\rangle_j = \left\{J_{j-i}(2Jt) - (-1)^i J_{j+i}(2Jt)\right\}^2, \quad (32.90)$$

when there are no photon–phonon interactions ($\omega_i = 0$) and the total site number N becomes infinite. Here the argument J is the photon hopping constant, and (32.90) shows that a photon initially populated at site j delocalizes to the whole system.

Focusing on the localized phonon modes, we take the summation in (32.83b) over the localized modes only, which means that an earlier stage is considered after the incident photon excites the phonon modes, or that the duration of the localized phonon modes that are dominant over the delocalized modes is focused on (see Figure 32.4). This kind of analysis provides us with an interesting insight into the photon–phonon coupling constant and the photon hopping constant, which is necessary for understanding the mechanism of photon localization.

The temporal evolution of the observation probability of a photon at each site is shown in Figure 32.5. Without the photon–phonon coupling ($\chi = 0$), a photon spreads over the whole system as a result of the photon hopping, as shown in Figure 32.5a.

Here the photon energy $\hbar\omega = 1.81$ eV and the hopping constant $\hbar J = 0.5$ eV are used in the calculation. Impurities are assumed to be doped at sites 3, 7, 11, 15, and 19, while the total number of sites N is 20, and the mass ratio of the host molecules to the impurities is 5. Figure 32.5b shows a result with $\chi = 1.4 \times 10^3$ fs^{-1} nm^{-1}; the other parameters used are the same as those in Figure 32.5a. It follows from the figure that a photon moves from one impurity site to another impurity site instead of delocalizing to the whole system. As the photon–phonon coupling constant becomes much larger than $\chi = 1.4 \times 10^3$ fs^{-1} nm^{-1}, a photon cannot move from the initial impurity site to others, but stays there.

The effect due to the photon–phonon coupling constant χ is expressed by the diagonal component in the Hamiltonian, while the off-diagonal component involves the photon hopping effect due to the hopping constant J. The above results indicate that a photon's spatial distribution depends on the competition between the diagonal and off-diagonal components in the Hamiltonian, i.e., χ and J, and that a photon can move among impurity sites and localize at those sites when both components are comparable under the condition

$$\chi \sim N\sqrt{\frac{kJ}{\hbar}}, \quad (32.91)$$

where the localization width seems very narrow.

32.3.4.4.2 Contribution from the Off-Diagonal Part

In the previous subsection, we have approximated J as a constant independent of the sites, in order to examine the photon's spatial distribution as well as the mechanism of the photon localization. Now let us treat the photon hopping operator $\hat{J}_i$ more rigorously, and investigate the site dependence of the off-diagonal contribution, which includes the inhomogeneity of the phonon fields. Noticing that a quasiparticle transformed from a photon

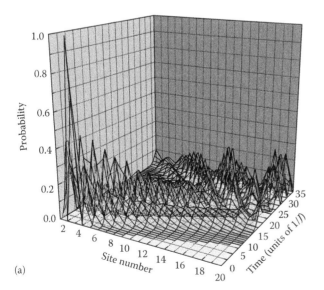

(a)

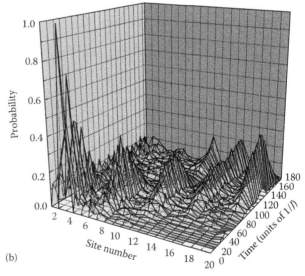

(b)

FIGURE 32.5 **(See color insert following page 21-4.)** Probability that a photon is found at each site as a function of time (a) without the photon–phonon coupling, and (b) with the photon–phonon coupling comparable to the photon hopping constant.

operator by the Davydov transformation is associated with phonons in the coherent state (see (32.77a) and (32.77b)), we take expectation values of $\hat{J}_i$ in terms of the coherent state of phonons $|\gamma\rangle$ as

$$J_i \equiv \langle\gamma|\hat{J}_i|\gamma\rangle. \qquad (32.92)$$

Here, the coherent state $|\gamma\rangle$ is an eigenstate of the annihilation operator $\hat{b}_p$ with eigenvalue γ_p and satisfies

$$\hat{b}_p|\gamma\rangle = \gamma_p|\gamma\rangle. \qquad (32.93)$$

Since the difference between the creation and annihilation operators of a phonon is invariant under the Davydov transformation, the following relation holds:

$$\hat{\beta}_p^\dagger - \hat{\beta}_p = \hat{b}_p^\dagger - \hat{b}_p. \qquad (32.94)$$

Using (32.78b), (32.93), and (32.94), we can rewrite the site-dependent hopping constant J_i in (32.92) as

$$J_i = J\left\langle\gamma\left|\exp\left\{\sum_{p=1}^N C_{ip}\left(b_p^\dagger - b_p\right)\right\}\right|\gamma\right\rangle$$

$$= J\exp\left(-\frac{1}{2}\sum_{p=1}^N C_{ip}^2\right)\left\langle\gamma\left|\exp\left(\sum_{p'=1}^N C_{ip'}\hat{b}_{p'}^\dagger\right)\exp\left(-\sum_{p''=1}^N C_{ip''}\hat{b}_{p''}\right)\right|\gamma\right\rangle$$

$$= J\exp\left(-\frac{1}{2}\sum_{p=1}^N C_{ip}^2\right), \qquad (32.95a)$$

where C_{ip} is denoted by

$$C_{ip} \equiv \frac{\chi_{ip} - \chi_{i+1p}}{\Omega_p}. \qquad (32.95b)$$

Figure 32.6 shows the site dependence of J_i in the case of $N = 20$. Impurities are doped at sites 4, 6, 13, and 19. The mass ratio of the host molecules to the impurities is 5, while $\hbar J = 0.5$ eV and $\chi = 14.0\,\text{fs}^{-1}\,\text{nm}^{-1}$ are used. It follows from the figure that the hopping constants are highly modified around the impurity sites and the edge sites. The result implies that photons are strongly affected by localized phonons and hop to the impurity sites to localize. Here we have not considered the temperature dependence of J_i, which is important for phenomena dominated by incoherent phonons. This is because coherent phonons weakly depend on the temperature of the system. However, there remains room to discuss a more fundamental issue, i.e., whether the probe system is in a thermal equilibrium state or not.

In Figure 32.7, we present a typical result that photons localize around the impurity sites in the system as the photon–phonon

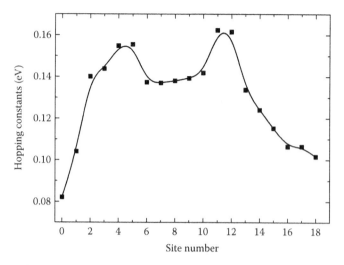

FIGURE 32.6 Site dependence of the hopping constant J_i in the case of $N = 20$. Impurities are doped at sites 4, 6, 13, and 19. The mass ratio of the host molecules to the impurities is 1–0.2, whereas $\hbar J = 0.5$ eV and $\chi = 40.0\,\text{fs}^{-1}\,\text{nm}^{-1}$ are used.

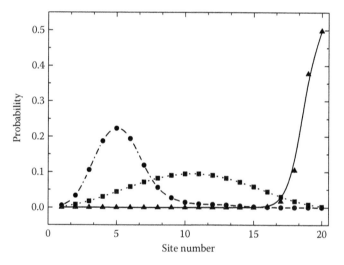

FIGURE 32.7 Probability of photons observed at each site. The filled squares, circles, and triangles represent the results for $\chi = 0$, 40.0, and 54.0 fs^{-1} nm^{-1}, respectively. Other parameters are the same as Figure 32.6.

coupling constants χ vary from 0 to $40.0\,\text{fs}^{-1}$ nm^{-1} or $54.0\,\text{fs}^{-1}$ nm^{-1}, while keeping $\hbar J = 0.5$ eV. As depicted by the filled squares in the figure, photons delocalize and spread over the system without the photon–phonon couplings. When the photon–phonon couplings are comparable to the hopping constants, $\chi = 40.0\,\text{fs}^{-1}$ nm^{-1}, photons can localize around the impurity site with a finite width and two sites at half width and half maximum (HWHM), as shown by the filled circles. This finite width of photon localization comes from the site-dependent hopping constants. As the photon–phonon couplings are larger than $\chi = 40.0\,\text{fs}^{-1}$ nm^{-1}, photons can localize at the edge sites with a finite width, as well as the impurity sites. In Figure 32.7, the photon localization at the edge site, shown by the filled triangles, originates from the finite size effect of the molecular chain. This kind of localization of photons dressed by the coherent state of phonons leads us to a simple understanding of phonon-assisted photodissociation using an optical near field: molecules in the electronic ground state approach the probe tip within the localization range of the dressed photons, and can be vibrationally excited by the dressed-photon transfer to the molecules, via a multi-phonon component of the dressed photons, which might be followed by electronic excitation. Thus, it leads to the dissociation of the molecules even if an incident photon energy less than the dissociation energy is used.

As a natural extension of the localized photon model, we have discussed the inclusion of the phonon's effects into the model. The study was initially motivated by experiments on the photodissociation of molecules by optical near fields, whose results show unique features different from the conventional one with far fields (Ohtsu et al. 2008). After clarifying whether the vibration modes in a pseudo one-dimensional system are delocalized or localized, we focused on the interaction between dressed photons and phonons by using the Davydov transformation. We have theoretically shown that photons are dressed by the coherent state of phonons, and found that the competition between the photon–phonon coupling constant and the photon hopping constant governs the photon localization or delocalization in space. The obtained results lead us to a simple understanding of an optical near field itself as an interacting system of photon, electronic excitation (induced polarization), and phonon fields in a nanometer space, which are surrounded by macroscopic environments, as well as phonon-assisted photodissociation using an optical near field.

32.4 Summary and Future Perspective

We have briefly outlined the difficulties in defining the position-representation wave function of a photon, followed by several trials to overcome these issues. On the basis of canonical quantization of the electromagnetic fields in free space, we have discussed a dressing mechanism and spatial localization of photons, which is a natural viewpoint from light–matter interactions, or from virtual photon clouds and field correlations. Finally, with the projection operator method, we have shown an effective interaction between nanomaterials electronically disconnected, but closely separated, which are also surrounded by a macroscopic system, in order to detect the virtual clouds. We have discussed the photon dressing by material excitation and pointed out the importance of the phonon's role for spatial localization of photons at the nanoscale.

The pace of development in photonics has accelerated, but most of the underlying science of the field is still Maxwell's classical electromagnetism, not field-quantized photons. In the near future, however, we are anticipating new breakthroughs in nano- and atom photonics, where the localized and quantized nature of photons, as well as an exact quantization formulation for an optical near-field system, including relaxation processes at the nanoscale, will play a critical role.

Acknowledgments

The author is grateful to M. Ohtsu (University of Tokyo) and H. Hori (University of Yamanashi) for stimulating discussions and valuable comments from the early stage of this study. He is also thankful to S. Sangu (Ricoh Co. Ltd.), A. Shojiguchi (NEC Co.), Y. Tanaka (Tokyo Institute of Technology), and A. Sato (Tokyo Institute of Technology) for discussions and collaborations. He greatly acknowledges the valuable guidance given by M. Tsukada (University of Tokyo, emeritus, Tohoku University), K. Kitahara (Tokyo Institute of Technology, emeritus, International Christian University), M. Kitano (Kyoto University), Y. Masumoto (University of Tsukuba), K. Cho (Osaka University, emeritus), and T. Yabuzaki (Kyoto University, emeritus). Finally but not the least, he expresses his gratitude to H. Ito (Tokyo Inst. Technology), T. Kawazoe (University of Tokyo), T. Yatsui (University of Tokyo), T. Saiki (Keio University), K. Matsuda (Kyoto University), H. Nejo (National Institute of Materials Science), M. Naruse (National Institute of Information and Communication Technology), M. Ikezawa (University of Tsukuba), I. Banno (University of Yamanashi), and H. Ishihara (Osaka Prefecture University).

References

Adlard, C., Pike, E. R., and Sarkar, S. 1997. Localization of one-photon states. *Phys. Rev. Lett.* 79: 1585–1587.

Bialynicki-Birula, I. 1998. Exponential localization of photons. *Phys. Rev. Lett.* 80: 5247–5250.

Carniglia, C. K. and Mandel, L. 1971. Quantization of evanescent electromagnetic waves. *Phys. Rev. D* 3: 280–296.

Cohen-Tannoudji, C., Dupont-Roc, I., and Grynberg, G. 1989. *Photons and Atoms*. New York: Wiley.

Cohen-Tannoudji, C., Dupont-Roc, I., and Grynberg, G. 1992. *Atom-Photon Interactions: Basic Processes and Applications*. New York: Wiley.

Compagno, G., Passante, R., and Persico, F. 1988. Dressed and half-dressed neutral sources in nonrelativistic QED. *Phys. Scr.* T21: 33–39.

Compagno, G., Passante, R., and Persico, F. 1995. *Atom-Field Interactions and Dressed Atoms*. Cambridge, NY: Cambridge University Press.

Craig, D. P. and Thirunamachandran, T. 1998. *Molecular Quantum Electrodynamics*. New York: Dover Pub.

Hawton, M. 1999. Photon wave functions in a localized coordinate space basis. *Phys. Rev. A* 59: 3223–3227.

Hegerfeldt, G. 1974. Remarks on causality and particle localization. *Phys. Rev. D* 10: 3320–3321.

Hellwarth, R. W. and Nouchi, P. 1996. Focused one-cycle electromagnetic pulses. *Phys. Rev. E* 54: 889–895.

Inagaki, T. 1998. Physical meaning of the photon wave function. *Phys. Rev. A* 57: 2204–2207.

Inoue, T. and Hori, H. 2001. Quantization of evanescent electromagnetic waves based on detector modes. *Phys. Rev. A* 63: 063805(1)–063805(16).

John, S. and Quang, T. 1995. Photon-hopping conduction and collectively induced transparency in a photonic band gap. *Phys. Rev. A* 52: 4083–4088.

Keller, O. 2005. On the theory of spatial localization of photons. *Phys. Rep.* 411: 1–232.

Kobayashi, K., Sangu, S., Ito, H. et al. 2001. Near-field optical potential for a neutral atom. *Phys. Rev. A* 63: 013806(1)–063806(9).

Kobayashi, K., Tanaka, Y., and Kawazoe, T. et al. 2008. Localized photon model including phonon's degrees of freedom. In *Progress in Nano-Electro-Optics*, vol. VI, ed. M. Ohtsu, pp. 41–66. Berlin, Germany: Springer.

Mandel, L. 1966. Configuration-space photon number operators in quantum optics. *Phys. Rev.* 144: 1071–1077.

Mandel, L. and Wolf, E. 1995. *Optical Coherence and Quantum Optics*. Cambridge, NY: Cambridge University Press.

Newton, T. D. and Wigner, E. P. 1949. Localized states for elementary systems. *Rev. Mod. Phys.* 21: 400–406.

Ohtsu, M. and Hori, H. 1999. *Near-Field Nano-Optics*. New York: Kluwer Academic/Plenum.

Ohtsu, M., Kobayashi, K., Kawazoe, T. et al. 2008. *Principles of Nanophotonics*. Boca Raton, FL: CRC Press.

Pike, E. R. and Sarkar, S. 1987. Spatial dependence of weakly localized single-photon wave packets. *Phys. Rev. A* 35: 926–928.

Pohl, D. W. and Courjon, D. 1993. *Near Field Optics*. Dordrecht, the Netherlands: Kluwer Academic.

Power, E. A. and Thirunamachandran, T. 1993. Quantum electrodynamics with nonrelativistic sources. V. Electromagnetic field correlations and intermolecular interactions between molecules in either ground or excited states. *Phys. Rev. A* 47: 2539–2551.

Roychoudhuri, C. and Roy, R. eds. 2003. The nature of light: What is a photon? *Opn. Trends* 3: S1–S35.

Sakurai, J. J. 1967. *Advanced Quantum Mechanics*. Reading, MA: Addison-Wesley.

Sangu, S., Kobayashi, K., Shojiguchi, A. et al. 2004. Logic and functional operations using a near-field optically coupled quantum-dot system. *Phys. Rev. B* 47: 115334(1)–115334(13).

Scully, M. O. and Zubairy, M. S. 1997. *Quantum Optics*. Cambridge, U.K.: Cambridge University Press.

Shojiguchi, A., Kobayashi, K., Sangu, S. et al. 2003. Superradiance and dipole ordering of an N two-level system interacting with optical near fields. *J. Phys. Soc. Jpn.* 72: 2984–3001.

Sipe, J. E. 1995. Photon wave functions. *Phys. Rev. A* 52: 1875–1883.

Suzuura, H., Tsujikawa, T., and Tokihiro, T. 1996. Quantum theory for exciton polaritons in a real-space representation. *Phys. Rev. B* 53: 1294–1301.

Wightman, A. S. 1962. On the localizability of quantum mechanical systems. *Rev. Mod. Phys.* 34: 845–872.

33
Operations in Nanophotonics

Suguru Sangu
Ricoh Company, Ltd.

Kiyoshi Kobayashi
University of Yamanashi
Japan Science and Technology

33.1 Introduction

Recently, a vast number of studies on nanophotonics have been published in the field of nanoscale optical measurement and bioimaging with super-resolution (Ohtsu 1998, Maheswari et al. 1999, Hosaka and Saiki 2001, Matsuda et al. 2003) and information processing technologies (Biolatti et al. 2000, Rinaldis et al. 2002, Troiani et al. 2002), owing to the unique characteristics of optical near field, which far exceed the technical limitations of conventional optics (Ohtsu et al. 2008). Especially, nanophotonics has shown promise and is expected to be the technology for next-generation nanoscale devices dealing with large amounts of information resources and low energy consumption (Ohtsu et al. 2002). In such devices, optical near field plays an important role in nanophotonic device operations, since the optical near field is a mixed state between photon and matter excitation, not only breaking the diffraction limit that is dependent on the wave nature of light, but also utilizing interesting characteristics inherent in nanophotonics, such as unidirectional energy transfer, optical forbidden transition (Kawazoe et al. 2002), and operations based on coherence of nanometric matters (Sangu et al. 2004), which have not been used in conventional optical devices.

Nanophotonic device is interpreted as a system that consists of several matter systems, localized photon field (optical near field) for driving carriers in the systems, and free photon (radiation) field for extracting some information, as illustrated in Figure 33.1. The important point of nanophotonic system is that the spatial distribution of photons is localized in a nanometric space rather than the matter itself being nanometer-sized. From this point, valuable device operations inconsiderable in conventional optical devices are expected. Relating to a theoretical viewpoint, some restrictions under the long

wavelength approximation is allowed in a nanophotonic device system (Cho 2004). Moreover, quantum nature appears in carrier dynamics.

In order to control the carriers, quantum systems with discrete energies, such as quantum dots and molecules, are adapted as matter systems because the processes with matter coherence and with energy dissipation are clearly distinguishable. And also, excitation energies and dissipation dynamics are readily determined by adjusting the sizes of quantum objects. In addition, optical allowed and forbidden transitions for external far-field light are usable to prepare initial excitations. In this chapter, a quantum-dot system is assumed as a matter system of nanophotonic device, and thus the signal carriers correspond to excitons, that is, electron–hole pairs, in the quantum dots. Although such systems resemble a quantum computation device, let us note that a nanophotonic device need not keep quantum coherence in the entire system, and is divided into several quantum coherent parts via dissipative process.

In this chapter, operation principles of typical nanophotonic operations are covered by describing exciton population dynamics theoretically and numerically on the basis of the density matrix formalism (Walls and Milburn 1994, Carmichael 1999, Breuer and Petruccione 2002). Basic formulations for energy transfer, dissipation, and exciton excitation are explained, and then, nanophotonic switch and carrier up-converter are numerically demonstrated in Section 33.2. In Section 33.3, nanophotonic device operations, such as logic gates and memory, are described, in which system coherence and spatial symmetries play important roles. As a summary, design concepts for nanophotonic device systems are discussed in Section 33.4.

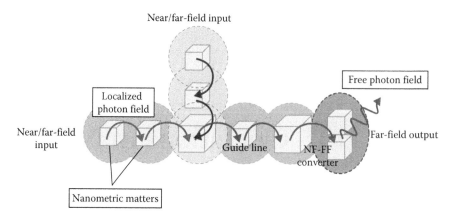

FIGURE 33.1 Schematic representation of a nanophotonic device system, in which carriers are transferred via localized photon field with interaction length in the order of a matter size.

33.2 Dissipation-Controlled Nanophotonic Devices

In this section, nanophotonic devices, where signal carrier in a single input terminal transfers to that in an output one, are formulated and evaluated numerically. Dynamics of signal transfer and control in a nanophotonic device can be described by density matrix formalism. Since quantum mechanical theory expresses an energy-conservative system, dissipation process should be introduced by some sort of approximation because a functional device operation needs to guarantee unidirectional signal transfer. After explanation of basic principles of exciton energy transfer and creation of excitation in a quantum-dot system with dissipation process, some concrete functional operations are explained in the later part of this section.

33.2.1 Basic Principles

33.2.1.1 Energy Transfer between Two Quantum Dots

As the simplest example of carrier dynamics, exciton energy transfer between two quantum dots with radiative and non-radiative relaxations is described by using density matrix formalism (Carmichael 1999, Sangu et al. 2003). Figure 33.2 denotes energy diagram for a two-quantum-dot system, where exciton ground state in QD-A and first exciton excited state in QD-B resonantly couple with each other. This resonant condition can be realized by adjusting the size of quantum dots. For example, in the case of a cubic quantum dot, such as CuCl quantum dot (Masumoto 2002, Kawazoe et al. 2003), the energy with the quantum numbers (n_x, n_y, n_z) is given as

$$E_{(n_x,n_y,n_z)} = \frac{\pi^2 \hbar^2}{2Ma^2}\left(n_x^2 + n_y^2 + n_z^2\right) + E_g, \quad (n_x, n_y, n_z = 1,2,3,\ldots),$$

$$(33.1)$$

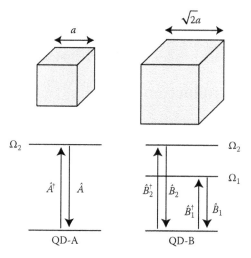

FIGURE 33.2 Energy diagram of a two-quantum-dot system, where the sizes of quantum dots are determined to satisfy the resonant condition, and are set as a (QD-A) and $\sqrt{2}a$ (QD-B). There are exciton sublevels in QD-B because of unidirectional energy transfer. Intra-sublevel relaxation originates from coupling between excitons and phonon reservoir.

where M and E_g denote the exciton mass and bulk exciton energy, respectively. Obviously, when the sizes of QD-A and QD-B are set as a and $\sqrt{2}a$, respectively, the $(1, 1, 1)$-level in QD-A and $(2, 1, 1)$, $(1, 2, 1)$, and $(1, 1, 2)$-levels in QD-B are the same energy and, thus, they couple resonantly. In the following, the energy levels for three quantum numbers are dealt with collectively and labeled as $\hbar\Omega_1$- and $\hbar\Omega_2$-levels, since degeneracy of several energy levels is not essential. Energy state of the system is described by the following non-perturbative and interaction Hamiltonians:

$$\hat{H}_0 = \hbar\Omega_2\left(\hat{A}^\dagger\hat{A} + \hat{B}_2^\dagger\hat{B}_2\right) + \hbar\Omega_1\hat{B}_1^\dagger\hat{B}_1,$$

$$\hat{H}_{int} = \hbar U\left(\hat{A}^\dagger\hat{B}_2 + \hat{A}\hat{B}_2^\dagger\right),$$

$$(33.2)$$

where

Ω₁ and Ω₂ are eigenfrequencies of QD-A and QD-B

U represents the optical near-field coupling strength, and creation and annihilation operators for excitons are depicted in Figure 33.2

The interaction Hamiltonian $\hat{H}_{\mathrm{int}}$ is well known as the Förster-type interaction (Förster 1965), which is often used by describing intermolecular and inter-quantum-dot interaction. Although the actual interaction between QD-A and QD-B occurs via intermediate virtual states of exciton-polaritons (Kobayashi et al. 2000), which are the coupled state of excitons and photons, such exciton-polariton degree of freedom is neglected in Equation 33.2 as is traced out and renormalized in the coupling strength U.

Exciton dynamics can be expressed by using density operator $\hat{\rho}(t)$, which is a projection operator expanding a certain energy state to appropriate bases depending on possible exciton states in a quantum-dot system. There are a three-exciton state, three two-exciton states, three one-exciton states, and a vacuum state in a two-quantum-dot system, as illustrated in Figure 33.3. Equation of motion of the density operator is given by the Liouville equation in the case without dissipation (Breuer and Petruccione 2002). Dynamics with dissipation is often expressed by so-called Lindblad type notation on the bases of the first-order Born approximation (Carmicheal 1999), in which free photon and phonon fields are considered as energy reservoirs as follows:

$$\frac{d\hat{\rho}(t)}{dt} = \frac{1}{i\hbar}\left[\hat{H}_0 + \hat{H}_{\mathrm{int}}, \hat{\rho}(t)\right] + \mathcal{D}_{\mathrm{photon}}\hat{\rho}(t) + \mathcal{D}_{\mathrm{phonon}}\hat{\rho}(t), \quad (33.3)$$

where $\mathcal{D}_{\mathrm{photon\,(phonon)}}$ is a superoperator defined as

$$\mathcal{D}_{\mathrm{photon}}\hat{\rho}(t) = \sum_{\alpha=A,B_1} \frac{\gamma_\alpha}{2}\left[2\hat{\alpha}\hat{\rho}(t)\hat{\alpha}^\dagger - \hat{\alpha}^\dagger\hat{\alpha}\hat{\rho}(t) - \hat{\rho}(t)\hat{\alpha}^\dagger\hat{\alpha}\right],$$

$$\mathcal{D}_{\mathrm{phonon}}\hat{\rho}(t) = \frac{\Gamma}{2}(n+1)$$
$$\times \left[2\hat{B}_2\hat{B}_1^\dagger\hat{\rho}(t)\hat{B}_2^\dagger\hat{B}_1 - \hat{B}_2^\dagger\hat{B}_1\hat{B}_2\hat{B}_1^\dagger\hat{\rho}(t) - \hat{\rho}(t)\hat{B}_2^\dagger\hat{B}_1\hat{B}_2\hat{B}_1^\dagger\right]$$
$$+ \frac{\Gamma}{2}n\left[2\hat{B}_2^\dagger\hat{B}_1\hat{\rho}(t)\hat{B}_2\hat{B}_1^\dagger - \hat{B}_2\hat{B}_1^\dagger\hat{B}_2^\dagger\hat{B}_1\hat{\rho}(t) - \hat{\rho}(t)\hat{B}_2\hat{B}_1^\dagger\hat{B}_2^\dagger\hat{B}_1\right],$$
$$(33.4)$$

n is the number of phonons in the reservoir, and the number of photons is assumed as zero because of external field being a vacuum state. As you readily understand from the bottom of Equation 33.4, energy transfer between the excited and ground states of an exciton occurs by mediating the intra-sublevel transition with non-radiative relaxation constant Γ, which guarantees unidirectional energy transfer in the two-quantum-dot system.

Since analytical solution of Equation 33.3 is complex and the physical meaning is unclear, the following is discussed on the basis of numerical calculations. For far-field light excitation, excitons in a two-quantum-dot system are created via mixed states of several bases, which are shown in Figure 33.3, because the two quantum dots cannot distinguish each other due to the diffraction limit of light. On the other hand, high spatial resolution of optical near field permits direct access for individual quantum dots. For example, an optical fiber probe possesses sub-100 nm spatial resolution (Hosaka and Saiki 2001,

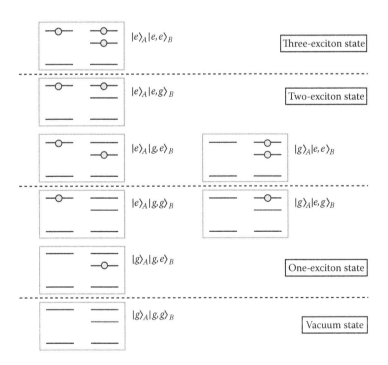

FIGURE 33.3 Appropriate base states in which excitation and ground states for an isolated quantum dot are used. There are eight bases, which include a three-exciton state, three two-exciton states, three one-exciton states, and a vacuum state.

Matsuda et al. 2003) and plasmonic excitation in metallic nano-structures are utilized as an interface in a nanophotonic device (Yatsui et al. 2001, Nomura et al. 2005). As an initial condition to demonstrate energy transfer between the two-quantum-dot systems, an exciton in QD-A is locally prepared that is expressed by density matrix elements of $\rho_{in}(t) = {_A}\langle e|_B \langle g, g|\hat{\rho}|e\rangle_A |g, g\rangle_B = 1$ and otherwise being zero. Output population is given by the density matrix element of $\rho_{out}(t) = {_A}\langle g|_B \langle g, e|\hat{\rho}|g\rangle_A |g, e\rangle_B$, which is the lower-level state excited in QD-B, and $(|g, e\rangle_B)^\dagger = {_B}\langle g, e|$. In this case, two- and three-exciton bases decouple from the output population dynamics, and it is enough to consider only four base states for calculating the dynamics.

In the following, numerical solutions in Equation 33.3 are explained, in which two types of typical dynamics appear depending on the optical near-field coupling strength. Figures 33.4 and 33.5 show the exciton population in the input (QD-A) and output (QD-B) energy levels, respectively, when the optical near-field

coupling strength are applied as $U^{-1} = 10\,\text{ps}$ (Figures 33.4a and 33.5a) and $50\,\text{ps}$ (Figures 33.4b and 33.5b). For Figure 33.4a, oscillating behavior still remains because the sublevel relaxation constant Γ is the same order of the coupling strength U. While in the case of $\Gamma > U$, coherence in the upper energy levels quickly reduces, and the exciton energy that is stored in the lowest energy level in the system is released as radiative relaxation. The exciton population given in Figure 33.5 is proportional to photoluminescence intensity from the coupled quantum-dot system.

Key point to design of a nanophotonic device is to determine energy transfer paths by using resonant coupling among plural energy states, and internal and external dissipation processes. Density matrix formalism discussed above is a useful tool to solve signal dynamics in an exciton–photon coupled or polariton-mediated system. In addition, multi-exciton process is important for functional operations, such as nanophotonic switch, which are given in Sections 33.2.2 and 33.2.3.

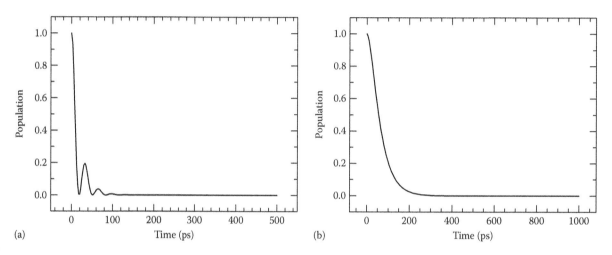

(a) Time (ps)

(b) Time (ps)

FIGURE 33.4 Temporal evolution of exciton population in the $\hbar\Omega_2$-level in QD-A (input). The optical near-field coupling strength are (a) $U^{-1} = 10\,\text{ps}$ and (b) $50\,\text{ps}$, and the other parameters are set as $\hbar\Omega_2 = 3.22\,\text{eV}$ ($M = 2.3m_e$, $E_g = 3.20\,\text{eV}$, $a = 5\,\text{nm}$ or $7.07\,\text{nm}$), $\hbar\Omega_1 = 3.21\,\text{eV}$ ($a = 7.07\,\text{nm}$), $\gamma_A^{-1} = 1\,\text{ns}$, $\gamma_B^{-1} = 0.5\,\text{ns}$, and $\Gamma^{-1} = 10\,\text{ps}$, where a CuCl quantum cube is assumed as a coupled two-quantum-dot system. For simplicity, zero temperature ($n = 0$) is assumed.

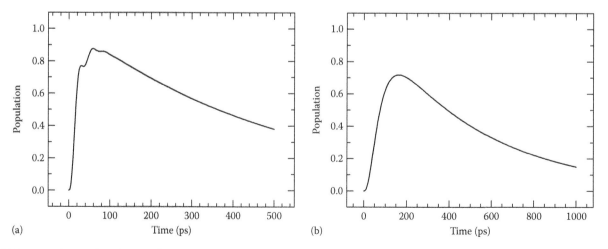

(a) Time (ps)

(b) Time (ps)

FIGURE 33.5 Temporal evolution of exciton population in the $\hbar\Omega_1$-level in QD-B (output). The optical near-field coupling strength are (a) $U^{-1} = 10\,\text{ps}$ and (b) $50\,\text{ps}$, and the other parameters are the same as in Figure 33.4.

33.2.1.2 Creation of Excitation by an External Field

Before explanation of multi-exciton process, a way to describe the creation of signal carriers in a quantum-dot system is discussed by means of the density matrix formalism. There are two types of carrier excitations: resonant and nonresonant. In the resonant excitation, energy of external photon field (laser pulse) corresponds to that of the exciton state (Carmicheal 1999), and the equation of motion, which is given in Equation 33.3, is modified with additional contribution of $\hat{H}_r$ as

$$\frac{d\hat{\rho}(t)}{dt} = \frac{1}{i\hbar}\left[\hat{H}_0 + \hat{H}_{int} + \hat{H}_r, \hat{\rho}(t)\right] + \mathcal{D}_{photon}\hat{\rho}(t) + \mathcal{D}_{phonon}\hat{\rho}(t),$$

$$\hat{H}_r = -\hbar g\left(\hat{B}_1^\dagger\langle\hat{a}\rangle + \langle\hat{a}^\dagger\rangle\hat{B}_1\right) \approx -\hbar V(t)\left(\hat{B}_1^\dagger + \hat{B}_1\right),$$

(33.5)

where $\hat{a}$ and $\hat{a}^\dagger$ represent annihilation and creation operators of external photon field, and on the basis of the semi-classical approximation, $\langle\hat{a}\rangle = \langle\hat{a}^\dagger\rangle \equiv V(t)/g$. It means that the annihilation and creation operators are renormalized in the parameter (Rabi frequency) $V(t)$. Equation 33.5 expresses the coherent process, which is well known as π-pulse excitation.

While in the case of the nonresonant excitation, coherence between an exciton and a photon is already lost because of interaction among a large number of energy states. This situation appears in such a case that an exciton creates mediated by continuous energy levels and surrounding barrier levels. The formulation of exciton excitation is equivalent to exciton creation from photon reservoir with finite temperature. According to Equation 33.4, the following term is added to Equation 33.3 for nonresonant excitation:

$$\mathcal{P}_{photon}\hat{\rho}(t) = \sum_{\alpha=A,B_1}\frac{V_\alpha(t)}{2}\left[2\hat{\alpha}^\dagger\hat{\rho}(t)\hat{\alpha} - \hat{\alpha}\hat{\alpha}^\dagger\hat{\rho}(t) - \hat{\rho}(t)\hat{\alpha}\hat{\alpha}^\dagger\right],$$

(33.6)

where $V_\alpha(t)$ is pumping rate in the optically allowed state in QD-α.

The above formulations are useful for making initial excited states in a quantum-dot system as well as for applying control signal to select energy transfer paths. In the following, exciton population dynamics for a two-quantum-dot system (Figure 33.2), which is driven by the external photon field, is explained by using numerical results of Equations 33.5 and 33.6. When the bases are restricted less than the one-exciton state, there are four states in this system (See Figure 33.3). Temporal evolution of exciton population in the $\hbar\Omega_2$-levels in QD-A and $\hbar\Omega_1$-level in QD-B are plotted in Figure 33.6a and b, respectively, which is in the case of resonant excitation. The results for nonresonant excitation are also given in Figure 33.7a and b. In both cases, the time width of pumping is set as 50 ps, which is denoted by the vertical dashed lines in Figures 33.6a and 33.7a. Temporal evolutions for resonant and nonresonant excitation are similar with each other except for those in the period when an external field is applied. Since the resonant oscillation keeps matter coherence, the population once increases and then decreases beyond π-pulse excitation. Figures 33.7b and 33.8b represent the population in QD-B, and the temporal profiles are similar in both cases, where the decay profiles are determined by the radiative relaxation constant in the lower energy level in QD-B, but total populations (the area of population curves) reflect the population for the early stage in QD-A.

Resonant and nonresonant exciton creations are important for projecting input signal to a quantum-dot system as well as for controlling signal carrier transiently in a nanophotonics device. Due to the above formulations for energy transfer, dissipation, and creation of excitation, some functional device operations can be demonstrated numerically, which are discussed in Sections 33.2.2 and 33.2.3.

33.2.2 Nanophotonic Switch

General switching device consists of three terminals, which correspond to input, output, and control. Similarly, a three-quantum-dot system can operate as a nanophotonics switch,

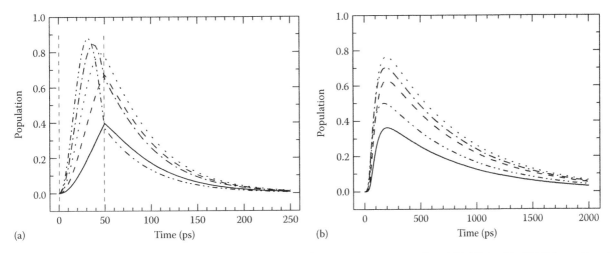

FIGURE 33.6 Temporal evolution of exciton population in (a) $\hbar\Omega_2$-level in QD-A (input terminal) and (b) $\hbar\Omega_1$-level in QD-B (output terminal), where $\hbar\Omega_2$-level in QD-A is resonantly excited. The optical near-field coupling strength is $U^{-1} = 50$ ps, and the other parameters are the same as in Figure 33.4. The solid, dashed, dotted, and dot-dash, dot-dot-dash curves represent the pulse area of 0.5π, 0.75π, 1.0π, 1.25π, and 1.5π, respectively.

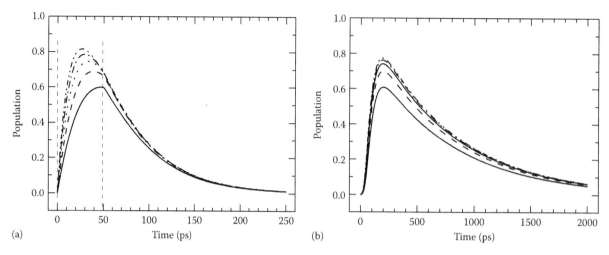

FIGURE 33.7 Temporal evolution of exciton population in (a) $\hbar\Omega_2$-level in QD-A (input port) and (b) $\hbar\Omega_1$-level in QD-B (output port), where $\hbar\Omega_2$-level in QD-A is nonresonantly excited. The parameters are the same as in Figure 33.6. The solid, dashed, dotted, dot-dash, and dot-dot-dash curves represent the pulse area of 0.5π, 0.75π, 1.0π, 1.25π, and 1.5π, respectively.

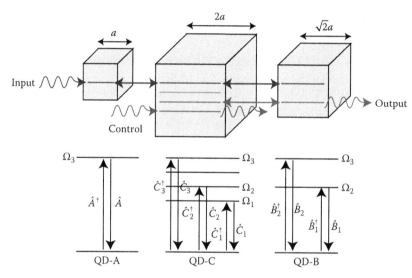

FIGURE 33.8 Energy diagram of a nanophotonic switch that consists of three quantum dots with the size ratio $A:B:C = 1:\sqrt{2}:2$. Input and control signals are injected in the $\hbar\Omega_3$-level in QD-A and the $\hbar\Omega_1$-level in QD-C, respectively, and output signal is detected from the $\hbar\Omega_2$-level in QD-B. By mediating intra-sublevel relaxations, unidirectional energy transfer is guaranteed in this system.

where three quantum dots correspond to input, output, and control terminals (Kawazoe et al. 2003). In Figure 33.8, energy diagram of the three-quantum-dot system is illustrated where the two resonant paths are labeled $\hbar\Omega_2$ and $\hbar\Omega_3$. The resonance conditions can be realized by choosing the quantum-dot sizes as $A:B:C = 1:\sqrt{2}:2$ (see Equation 33.1), where the interaction between QD-A and QD-B is not considered because interdot distance between them is assumed large enough for neglecting the interaction. The principle of the switching operation is as follows: when the control signal is injected in the $\hbar\Omega_1$-level in QD-C, the dissipation path toward the $\hbar\Omega_1$-level is blocked because of Fermionic feature of excitons in a system with discrete energies, an initial exciton in QD-A transfers to the second stable state of $\hbar\Omega_2$-levels in QD-B, and annihilates an exciton with a radiative photon. What is important for this type of functional operation,

in which dissipation paths are selectively determined, is to deal with a multi-exciton dynamics in a system.

Equation of motion of the nanophotonic switch can be written in accordance with the manner presented in Section 33.2.1, where the considered bases are extended in two-exciton states, and resonant exciton excitations are applied as an input signal in QD-A and a control signal in QD-C. Since output luminescent intensity is proportional to the $\hbar\Omega_2$-level population in QD-B, the population dynamics in QD-B is discussed in the following.

Figure 33.9 represents the numerical result of one and two-exciton population dynamics. The dashed curve shows the output population in QD-B, when rectangular optical pulse (π-pulse) with the duration of 10 ps is injected in the input terminal of QD-A. The exciton created in QD-A transfers to QD-C

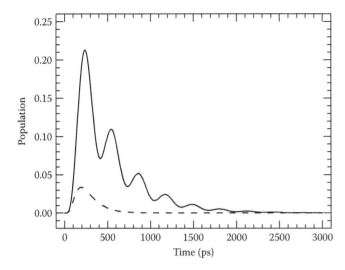

FIGURE 33.9 Temporal evolution of exciton population in the $\hbar\Omega_2$-level in QD-B (output terminal). The dashed and solid curves represent the cases where a single rectangular optical pulse is injected in QD-A (input terminal) and rectangular pulses are injected in the $\hbar\Omega_3$-level in QD-A (input terminal) and the $\hbar\Omega_1$-level in QD-C (control terminal). The parameters are set as follows: optical near-field coupling strength $U_{AC}^{-1} = U_{BC}^{-1} = 50$ ps, pulse duration $\Delta t = 10$ ps (π-pulse amplitude), radiative relaxation time $2^{-3}\gamma_A^{-1} = 2^{-3/2}\gamma_B^{-1} = \gamma_C^{-1} = 1000$ ps, intra-sublevel relaxation time $\Gamma_{B,32}^{-1} = 10$ ps, and $\Gamma_{B,21}^{-1} = \Gamma_{C,21}^{-1} = 20$ ps.

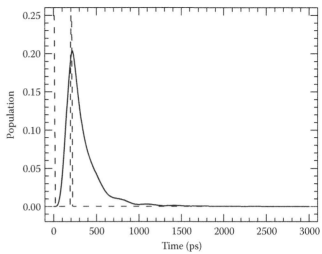

FIGURE 33.10 Temporal evolution of exciton population in the $\hbar\Omega_1$-level in QD-C (output port). The dashed curve represents the profile of the third rectangular optical pulse (π-pulse). The pumping pulse is injected with a delay of 200 ps and pulse duration of 10 ps. The other parameter values set are the same as in Figure 33.9.

via optical near-field coupling U_{AC} and then relaxes to the lowest energy level in QD-C because of fast intra-sublevel relaxation. Therefore, output population in QD-B is detected as very small value (OFF-state). While, the solid curve represents the case that two optical pulses with 10 ps-pulse duration are injected simultaneously in the lowest energy levels in QD-A and QD-C. In the early stage of the dynamics, a somewhat large signal with oscillation appears (ON-state). This is the result of occupation in the $\hbar\Omega_1$-level in QD-C, which is the optical nutation between lower resonant energy levels in QD-B and QD-C. The rise time or switching time is estimated less than 100 ps, which depends on the optical near-field coupling strength (Sangu et al. 2003).

The readers may see that the signal falling time from ON- to OFF-state is quite slower than the rising time from OFF- to ON-state. This is surely constrained by the radiative relaxation time in the control port of QD-C. However, the falling time is controllable by injection of the third control pulse, because π-pulse excitation can remove the exciton occupied in the $\hbar\Omega_1$-level in QD-C. Figure 33.10 represents the same result of exciton population dynamics shown in Figure 33.9, but additional rectangular optical pulse is injected with the delay of 200 ps in the control port of QD-C. The dashed curve represents timing of the control pulse. By injecting the third pulse, the optical nutation completely disappears and the dissipation path toward the lowest energy level in QD-C becomes effective.

Nanophotonic switches and nanophotonic devices, in which dissipation paths are controlled by the state-filling property of excitons, can realize quite low energy operations. Ideally, the energy loss is determined by intra-sublevel relaxation, which is

in the order of a few tens of meV. This is a big advantage not only for miniaturizing photonic devices, but also for heat dissipation and wiring in nanometric devices.

33.2.3 Carrier Manipulation (Up-Converter)

Unidirectional energy transfer and switching operation are guaranteed by considering the phonon reservoir system which is coupled with excitons. On the other hand, intense pulse excitation can create the finite phonons in the phonon reservoir system. This leads to blow back an exciton from lower to upper energy levels, instantaneously. Therefore, direction of energy transfer is also controllable by external optical pulse, which has been demonstrated experimentally in the recent publication (Yatsui et al. 2009). This up-conversion dynamics is formulated in a two-quantum-dot system, as shown in Figure 33.2. The intense optical pulse should be dealt with coherent interaction between excitons in the $\hbar\Omega_1$-level in QD-B and external photons. Therefore, the equation of motion described in Equation 33.5 is adopted. Since the atomic lattice in the reservoir system is transiently vibrated by the optical pulse, the phonon number of $n(\omega_0)$ in Equation 33.4 is given as the following function:

$$n(\omega_0) = \frac{1}{\exp\left[\hbar\omega_0 / k_B T(I_{in})\right] - 1}, \quad (33.7)$$

where ω_0, T, and k_B are the angular frequency of incident light, temperature of the system, and the Boltzmann constant, respectively. Equation 33.7 is the Bose–Einstein distribution function. For simplicity, the temperature is assumed to be proportional to the time integration of the optical pulse, I_{in}. The calculated results of the exciton population dynamics are

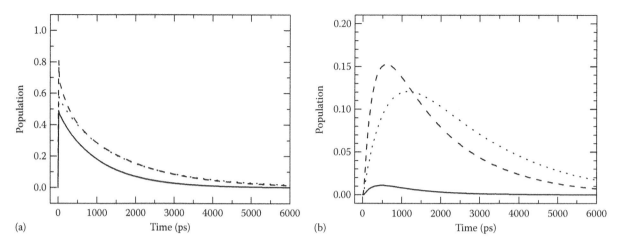

(a)   (b)

FIGURE 33.11 Temporal evolution of exciton population in (a) the $\hbar\Omega_1$-level in QD-B and (b) the $\hbar\Omega_2$-level in QD-A, when the optical pulse is injected at the time 10 ps. The solid, dashed, and dotted curves represent the pulse area of 0.5π, 1.0π, and 1.5π, respectively. The parameters are set as follows: optical near-field coupling strength $U^{-1} = 50$ ps, pulse duration $\Delta t = 10$ ps (π-pulse amplitude), radiative relaxation time $2^{-3/2}\gamma_A^{-1} = \gamma_B^{-1} = 1000$ ps, and intra-sublevel relaxation time $\Gamma = 10$ ps in QD-B. The phonon number in given by $[\exp(8.0/A^2) - 1]^{-1}$ where A denotes the pulse area.

plotted in Figure 33.11. Figure 33.11a and b represent the populations in the $\hbar\Omega_1$-level in QD-B and in the $\hbar\Omega_2$-level in QD-A which is the up-converted energy level. The solid, dashed, and dotted curves correspond to $\pi/2$-pulse, π-pulse, and $3\pi/2$-pulse, respectively. Owing to coherent excitation process by optical pulse injection, the population in the lower energy level in QD-B maximally increases for π-pulse excitation, and decreases for more intense excitation. Temporal evolution in this energy level obeys solely relaxation time of the lower energy level. The population dynamics in QD-A is more complex. The peak values depend on the population in the $\hbar\Omega_1$-level in QD-B for each optical pulse area while the number of phonons in the phonon reservoir monotonically increases depending on the power of the optical pulse. Therefore, population lifetime depicted in the dotted curve in Figure 33.11b becomes longer than the dashed one because of longer pulse injection period. Such an up-conversion signal cannot observe in an isolated quantum dot, since dipole inactive transition from QD-B to QD-A never occurs without optical near-field interaction.

Carrier dynamics driven by both exciton–photon and exciton–phonon interactions has large possibilities for some applications, such as novel light sources, optical near–far field interface devices, efficient photodetectors, and efficient photoelectric conversion devices (Yatsui et al. 2009).

33.3 Nanophotonic Devices Using Spatial Symmetries

In Section 33.2, nanophotonic devices, in which dissipation paths are selected by a state-filling condition in multi-exciton excitation states, were discussed. While another type of device using a coherently-coupled state shows interesting features. For such devices, the number of excitons in a system as well as the spatial symmetry of the system determine the device operations. In this section, some logical operations, memory operations, and

operations mediating quantum entangled states are explained with basic theoretical formulations.

33.3.1 Basic Principles

33.3.1.1 Symmetric and Antisymmetric States

When two identical quantum dots sharing an exciton are coupled with each other, as shown in Figure 33.12, appropriate bases for coupled states instead of those for individual quantum dots give a clear view for system dynamics. There are two coupled states via optical near-field interaction that are described by using mixed states between vacuum and exciton states for individual quantum dots, are as follows:

$$|S\rangle = \left(|e\rangle_A |g\rangle_B + |g\rangle_A |e\rangle_B\right)\big/\sqrt{2},$$
$$|A\rangle = \left(|e\rangle_A |g\rangle_B - |g\rangle_A |e\rangle_B\right)\big/\sqrt{2},$$

(33.8)

where the bases for the coupled states, $|S\rangle$ and $|A\rangle$ are called as symmetric and antisymmetric states as explained later in this section. At the beginning, energies in the symmetric and antisymmetric

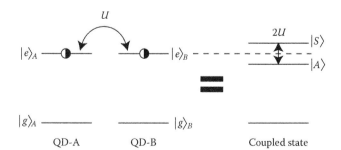

FIGURE 33.12 Energy diagram for two identical quantum dots (left). The system is expressed as a superposition of symmetric and antisymmetric states (right).

states are evaluated. Since the system Hamiltonian in a two-quantum-dot system with optical near-field coupling is given as

$$\hat{H} = \hat{H}_0 + \hat{H}_{\text{int}},$$

$$\hat{H}_0 = \hbar\Omega\hat{A}^\dagger\hat{A} + \hbar\Omega\hat{B}^\dagger\hat{B}, \tag{33.9}$$

$$\hat{H}_{\text{int}} = \hbar U\left(\hat{A}^\dagger\hat{B} + \hat{A}\hat{B}^\dagger\right),$$

where radiative relaxation terms are ignored for simplicity and assuming that the relaxation lifetime is longer than that of optical near-field interaction. The energies for these states are easily derived as

$$\langle S|\hat{H}|S\rangle = \hbar(\Omega + U),$$

$$\langle A|\hat{H}|A\rangle = \hbar(\Omega - U). \tag{33.10}$$

Equation 33.10 explains that the symmetric and antisymmetric states have positive and negative energy shift U from the energy in each quantum dot, respectively. In order to extract carrier excitation from the two identical quantum-dot systems, the third (output) quantum dot is set with resonance energy to the symmetric state or antisymmetric state. Detailed explanation of energy transfer properties in the three-quantum-dot system is given in Section 33.3.1.2.

In this stage, to examine total dipole moment in such a system is quite instructive. The total dipole moment is expressed by an expectation value of a summation of creation and annihilation operators, such as

$$\langle S|\mu\sum_{\alpha=A,B}\left(\alpha^\dagger + \alpha\right)|g\rangle_A|g\rangle_B = \sqrt{2}\mu,$$

$$\langle A|\mu\sum_{\alpha=A,B}\left(\alpha^\dagger + \alpha\right)|g\rangle_A|g\rangle_B = 0, \tag{33.11}$$

where μ denotes the dipole moment for QD-A and QD-B. Apparently, transition from vacuum state to the antisymmetric state is inactive and, thus, it cannot couple to the external far-field light; this is the reason this state is known as the antisymmetric state. In contrast, optical near field can excite the antisymmetric state because it has a spatial localization character beyond the diffraction limit of light, and can create an exciton in a one-side quantum dot. The state with an exciton is equal to a superposition between symmetric and antisymmetric states.

As described above, if the symmetric and antisymmetric states can be manipulated freely, coupling from near- to far-field light becomes controllable, which is useful for some functional operations, such as nanophotonic buffer memory. Furthermore, optical near-field coupling between the two identical quantum dots and the third output dot creates characteristic behaviors depending on spatial symmetry, where the symmetric and antisymmetric states play important roles, as explained in Section 33.3.1.2.

33.3.1.2 Selective Energy Transfer

For far-field light, excitation in each quantum dot is not distinguishable and radiation is proportional to total dipole moment, which is discussed in Section 33.3.1.1. However, optical near field can resolve the spatial distribution of exciton occupation in a nanometer scale. In order to access the excited states in a two-identical quantum-dot system by using optical near field, third quantum dot, QD-C, is considered as an output port. Figure 33.13 represents the energy transfer from two-identical quantum-dot system to the third quantum dot. The two identical quantum dots make symmetric and antisymmetric states via optical near-field interaction, and these states also couple to the third quantum dot with optical near-field interaction. Here, to evaluate transition moment between two identical quantum dots to the third quantum dot is useful for understanding the effect of spatial alignment of quantum dots. The interaction Hamiltonian is given as

$$\hat{H}_{\text{int},AC} = \hbar U_{AC}\left(\hat{A}^\dagger\hat{C} + \hat{C}^\dagger\hat{A}\right),$$

$$\hat{H}_{\text{int},BC} = \hbar U_{BC}\left(\hat{B}^\dagger\hat{C} + \hat{C}^\dagger\hat{B}\right), \tag{33.12}$$

where U_{AC} and U_{BC} represent optical near-field coupling strength between QD-A and QD-C, and QD-B and QD-C, respectively. Using these interaction Hamiltonians, the transition matrix element from symmetric state to excited state in QD-C reads

$$_A\langle g|_B\langle g|_C\langle e|\left(\hat{H}_{\text{int},AC} + \hat{H}_{\text{int},BC}\right)|S\rangle|g\rangle_C = \frac{\hbar(U_{AC} + U_{BC})}{\sqrt{2}}, \tag{33.13}$$

and that from antisymmetric state to excited state in QD-C is

$$_A\langle g|_B\langle g|_C\langle e|\left(\hat{H}_{\text{int},AC} + \hat{H}_{\text{int},BC}\right)|A\rangle|g\rangle_C = \frac{\hbar(U_{AC} - U_{BC})}{\sqrt{2}}. \tag{33.14}$$

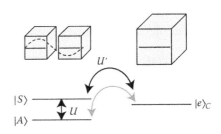

FIGURE 33.13 Schematic representation of energy transfer between two-identical quantum-dot system (input) and a larger quantum dot (output) than the other two. Antisymmetric state becomes dipole-allowed and dipole-forbidden states, depending on spatial symmetry of the output quantum dot.

From Equation 33.14, the transition matrix element can take a zero value when the initial state is an antisymmetric state and the coupling strength toward QD-C are set as $U_{AC} = U_{BC} \neq 0$. This condition can be realized by spatial symmetric alignment of quantum dots. Therefore, energy transfer mediating antisymmetric state is strongly dependent on spatial symmetry in a quantum-dot system.

The feature of selective energy transfer is useful for functional operations of a nanophotonic device, such as logic and memory operations. In the following sections, some functional operations using coherently coupled states, which are symmetric and antisymmetric states, are explained with numerical calculation results.

33.3.2 AND- and XOR-Logic Gates

Similar to Section 33.3.1.2, let us consider a three-quantum-dot system that consists of two identical input quantum dots (QD-A and QD-B) and the third output quantum dot (QD-C) with energy sublevels. QD-C is located in symmetrical position from QD-A and QD-B. Important points for this system are energy transfer reflecting spatial symmetry and difference of resonance conditions in one- and two-exciton states. By using these points, the principle of logic operations has been proposed (Sangu et al. 2004).

First, the case QD-C negatively detuned with energy shift $\hbar U$ is considered. Figure 33.14 represents energy diagram from initial state to final state in this system. The left (initial state) has no excitation in QD-C, and the right (final state) has an exciton in QD-C. Although negative energy shift creates resonance condition between the states of $|A\rangle|g\rangle_C$ and $|g\rangle_A|g\rangle_B|e\rangle_C$ for the one-exciton state, the energy transfer mediated by the antisymmetric state is not permitted in the quantum-dot system that is symmetrically aligned, as explained in Section 33.3.1.2. While, for the two-exciton state, the excited state $|e\rangle_A|e\rangle_B|g\rangle_C$ couples resonantly to the symmetric state $|S\rangle|e\rangle_C$, which is allowed transition. Therefore, this quantum-dot system permits the energy transfer

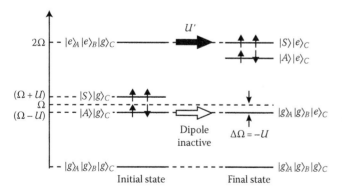

FIGURE 33.14 Energy diagram in three-quantum-dot system, which consists of two identical quantum dots (input port) and a third quantum dot (output port). The third quantum dot is negatively detuned against each input quantum dot. The energy transfer occurs only in the case of two-exciton state.

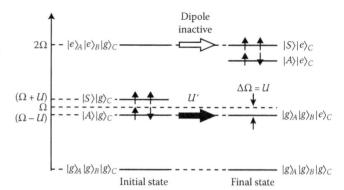

FIGURE 33.15 Energy diagram in three-quantum-dot system, which consists of two identical quantum dots (input port) and a third quantum dot (output port). The third quantum dot is positively detuned against each input quantum dot. The energy transfer occurs only in the case of one-exciton state.

only in the case of two-exciton state. This behavior represents just an AND-logic operation.

Second, the energy level in QD-C is set as positive energy shift $\hbar U$. The energy diagram is illustrated in Figure 33.15. In this case, two-exciton state is inactive because of antisymmetric state, while symmetric one-exciton state $|S\rangle|g\rangle_C$ can resonantly couple to the final state of $|g\rangle_A|g\rangle_B|e\rangle_C$. This means that the energy transfer occurs when either QD-A or QD-B is initially excited, where excitation in a quantum dot in two-identical quantum-dot system is equivalent to excite symmetric and antisymmetric states, simultaneously (see Equation 33.8). This corresponds to the XOR-logic operation.

Temporal evaluation of the above two logic operations can be derived analytically on the basis of the density matrix formalism. The system Hamiltonian is the same as Equation 33.9 except for additional third quantum dot terms, which include non-perturbed energy and non-radiative intra-sublevel relaxation, and radiative relaxation is ignored because optical near-field coupling and sublevel relaxation process are assumed to be fast enough. Final solution for one-exciton state, which is detected in lower energy level in QD-C, is written as follows (Sangu et al. 2004):

$$\rho_{gge,gge}(t) = \frac{1}{2} + \frac{4U'^2}{\omega_+^2 - \omega_-^2} e^{-\frac{\Gamma}{2}t}$$
$$\times \left[\cos\phi_+ \cos(\omega_+ t + \phi_+) - \cos\phi_- \cos(\omega_- t + \phi_-) \right],$$
$$(33.15)$$

where $\rho_{gge,gge}(t) \equiv {}_C\langle e|{}_B\langle g|{}_A\langle g| g|\rho(t)|g\rangle_A|g\rangle_B|e\rangle_C$, and the following abbreviations have been used:

$$\phi_\pm = \tan^{-1}\left(\frac{2\omega_\pm}{\Gamma}\right),$$

$$\omega_\pm^2 = \frac{1}{2}\left\{ (\Delta\Omega - U)^2 + W_+W_- \pm \sqrt{\left[(\Delta\Omega - U)^2 + W_+^2\right]\left[(\Delta\Omega - U)^2 + W_-^2\right]} \right\},$$

$$W_\pm = 2\sqrt{2}U' \pm \frac{\Gamma}{2},$$
$$(33.16)$$

where $\hbar\Delta\Omega$ is the energy shift in QD-C against that in QD-A or QD-B. In Equation 33.15, the second term denotes the energy transfer or nutation between two-identical quantum-dot system QD-A and QD-B, and the third of QD-C. Apparently, this term becomes the maximum value in the case of positive detuning $\Delta\Omega = U$, because the denominator $\omega_+^2 - \omega_-^2$ becomes minimum. This corresponds to dynamics in the XOR-logic operation, as is mentioned in Figure 33.15. The exciton population can reach maximum value of 0.5 not 1. This is because one-side quantum dot in QD-A or QD-B is excited as an initial excitation, which is same as coupled state between symmetric and antisymmetric states, and the antisymmetric state cannot couple to QD-C due to the symmetrically aligned system.

On the other hand, analytical solution for two-exciton state is given in a similar form as

$$\rho_{Se,Se}(t) + \rho_{gg2e,gg2e}(t)$$

$$= 2\left\{\frac{1}{2} + \frac{4U'^2}{\omega_+'^2 - \omega_-'^2} e^{-\frac{\Gamma}{2}t}\left[\cos\phi_+ \cos\left(\omega_+'t + \phi_+'\right) - \cos\phi_- \cos\left(\omega_-'t + \phi_-'\right)\right]\right\},$$

(33.17)

where $\rho_{Se,Se}(t) + \rho_{gg2e,gg2e}(t) \equiv {}_C\langle e|\langle S|\hat{\rho}(t)|S\rangle|e\rangle_C + {}_C\langle 2e|{}_B\langle g|{}_A\langle g|\rho(t) |g\rangle_A|g\rangle_B|2e\rangle_C$ is defined as the population in the lower level in QD-C and

$$\phi_\pm' = \tan^{-1}\left(\frac{2\omega_\pm'}{\Gamma}\right),$$

$$\omega_\pm'^2 = \frac{1}{2}\left\{\left(\Delta\Omega + U\right)^2 + W_+W_- \pm \sqrt{\left[\left(\Delta\Omega + U\right)^2 + W_-^2\right]\left[\left(\Delta\Omega + U\right)^2 + W_+^2\right]}\right\}.$$

(33.18)

From Equations 33.17 and 33.18, negative detuning $\Delta\Omega = -U$ makes resonance condition for two-exciton state,

while the negative detuning for one-exciton state is off-resonant to symmetric state (see Equations 33.15 and 33.16). Therefore, the energy transfer rate is quite small. This is the AND-logic operation.

Population dynamics in these devices are plotted in Figure 33.16, where exciton population in QD-C is analytically derived by using Equations 33.15 and 33.17. The solid and dashed curves represent one- and two-exciton states, respectively. Figure 33.16a is the result of the AND-logic gate with negative detuning $\Delta\Omega = -U$, where the strength of optical near-field interaction between QD-A and QD-B, and that between QD-A(B) and QD-C are set as $U^{-1} = 10\,\text{ps}$ and $U'^{-1} = 50\,\text{ps}$. In this operation, the two-exciton state in QD-A and QD-B is resonantly coupled with QD-C, while the one-exciton state is not, because of off-resonant condition. On the other hand, the XOR-logic operation is shown in Figure 33.16b. In this case, the one-exciton state in QD-A and QD-B is resonant to the energy level in QD-C, where initial population is prepared in QD-A, which is same as the coupled state between symmetric and antisymmetric states, and thus, the population reaches only the value of 0.5 because the antisymmetric state is a dark state in a spatially symmetric system. Although ON/OFF ratio is not so high, the XOR-logic gate operation is surely observed.

33.3.3 Nanophotonic Buffer Memory

As mentioned in Section 33.3.1.2, using energy transfer via a dipole inactive state, which can be controlled by energy detuning and quantum dot alignment or spatial symmetry, various functional operations inherent in nanophotonic devices are expected. In this section, one of interesting operations of a nanophotonic buffer memory is introduced. Quantum-dot alignment for realizing nanophotonic buffer memory is illustrated in Figure 33.17, in which two identical quantum dots (QD-A and QD-B) couple

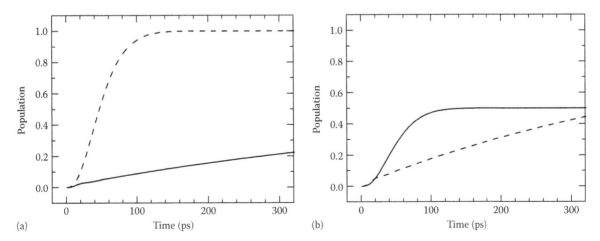

FIGURE 33.16 Temporal evolution of exciton population in three-quantum-dot systems, which consist of input-side two identical quantum dots (QD-A and QD-B) and an output-side quantum dot (QD-C) with intra-sublevel relaxation. (a) and (b) represent the cases where upper energy level is negatively (AND-logic gate) and positively (XOR-logic gate) detuned. The solid and dashed curves correspond to the one-exciton and two-exciton states, respectively. The calculation parameters in Equations 33.15 through 33.18 are set as $U^{-1} = 10\,\text{ps}$, $U'^{-1} = 50\,\text{ps}$, and $\Gamma^{-1} = 10\,\text{ps}$.

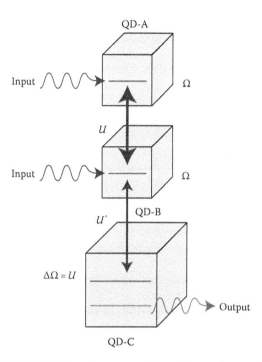

FIGURE 33.17 Schematic representation of nanophotonic buffer memory, which consists of two identical quantum dots, and the larger quantum dot with positive detuning $\Delta\Omega = U$. The three quantum dots are aligned maximally asymmetric and, thus, the coupling strength between QD-A and QD-C can be neglected.

to the third quantum dot (QD-C) with maximal asymmetric position. The third quantum dot is detuned positively with the energy $\hbar(\Omega + U)$, that is, $\Delta\Omega = U$, which has the same energy diagram as in Figure 33.15, but energy transfer via antisymmetric state is allowed. When both QD-A and QD-B are excited as an initial condition, the two-exciton state in the input side resonantly couples to antisymmetric state with excitation in QD-C and then relaxes to the one-exciton antisymmetric state. This antisymmetric state decouples with outer far-field light because of dipole inactive (dark) state. This is just buffer memory operation.

Dynamics of buffer memory operation is numerically evaluated by using density matrix formalism, as explained above. The calculation results are plotted in Figure 33.18, where the vertical axis represents population relating to the two-exciton coupled states, and the coupling strengths via optical near field are set as $U^{-1} = 10$ ps for QD-A and QD-B, and $U'^{-1} = 50$ ps for QD-B and QD-C (QD-A and QD-C are decoupled). The solid and dashed curves represent antisymmetric and symmetric states, respectively. In this calculation, radiative relaxation is ignored because the radiative relaxation time is longer than exciton population dynamics via optical near field and, thus, two-exciton antisymmetric state becomes the final state. Actually, the two-exciton state decays into the one-exciton antisymmetric state with a photon radiation from QD-C, and an exciton is retained in the system.

In Section 33.3.2, the dark state is used for suppressing energy transfer, leading to logical operations, while it is used to create

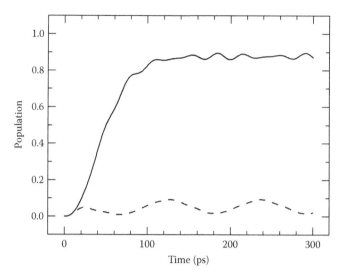

FIGURE 33.18 Temporal evolution of exciton population in three-quantum-dot systems, which have maximal asymmetric configuration (see Figure 33.17). The solid and dashed curves represent the populations for the antisymmetric two-exciton state and the symmetric two-exciton state, respectively. The antisymmetric state has a dipole inactive nature and, thus, the excitation is maintained in the three-quantum-dot system. The calculation parameters are set as $U^{-1} = 10$ ps, $U'^{-1} = 50$ ps, and $\Gamma^{-1} = 10$ ps.

excitation in a dark state in this section, which enables us to control far- and near-field conversion. It is one of the characteristic features of nanophotonic devices.

33.3.4 Manipulation of Quantum-Entangled States

As mentioned in Sections 33.3.1 through 33.3.3, a nanophotonic device system consists of coherent energy transfer process and dissipation process. Since the coherent energy transfer process maintains quantum coherence, this system can be used for quantum information processing as well as interface between quantum and classical computations. As an instructive example, a typical device that identifies quantum mixed states is numerically demonstrated in this section. Figure 33.19 schematically shows spatial alignment of quantum dots, where two identical quantum dots couple resonantly, and the excitations are led to two output quantum dots with positive (QD-C) and negative (QD-D) detuning. Both QD-C and QD-D are set at symmetric and asymmetric positions, respectively. Apparently, QD-C resonantly couples to the symmetric state in the input quantum dots, and QD-D couples to the antisymmetric state. Therefore, this device operates that input signal with arbitral quantum state divides two quantum bases of symmetric and antisymmetric states into two different output signals with different optical frequencies as well as different positions in a nanometric space. Temporal evolution of an input quantum mixed state is numerically calculated in Figure 33.20. When an input state $c_1|S\rangle + c_2|A\rangle$ is injected, where one-exciton state is assumed, the output exciton population in QD-C and QD-D

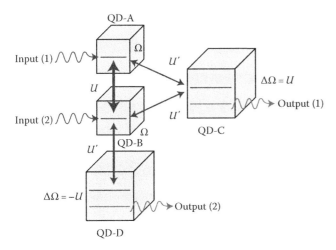

FIGURE 33.19 Spatial arrangement of a four-quantum-dot system, which has symmetric and asymmetric configurations simultaneously. QD-C resonantly couples with symmetric state in QD-A and QD-B, and QD-D couples with antisymmetric state.

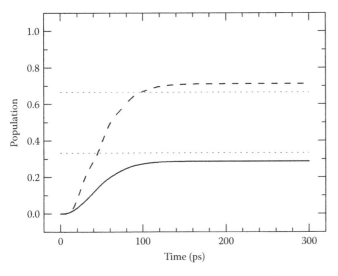

FIGURE 33.20 Population dynamics in four-quantum-dot systems that have two identical quantum dots (QD-A and QD-B) for input ports of an arbitral quantum state, and output quantum dots (QD-C and QD-D) for identifying the input quantum states. The solid and dashed curves represent the output population of QD-C and QD-D, respectively, and the dotted lines correspond to the coefficients of an initial quantum mixed state. The calculation parameters in Equations 33.15 through 33.18 are set as $U^{-1} = 10\,ps$, $U'^{-1} = 50\,ps$, and $\Gamma^{-1} = 10\,ps$.

reflects the coefficients of c_1 and c_2 of an input state. In Figure 33.20, the coefficients of initial quantum mixed state are set as $c_1 = 1/3$ and $c_2 = 2/3$, and thus, the output populations approach these values asymptotically.

Although general quantum information processing devices are built upon quantum coherent states in whole device systems, nanophotonic devices are characterized by the control of dissipation processes, but keeping the quantum coherence locally. Such a mixed operation of classical and quantum information processing may open up future device technologies.

33.4 Summary

In this chapter, operation of a nanophotonic device, in which characteristics of the light localized among nanometric objects, that is, optical near fields, are positively used, has been demonstrated theoretically and numerically. Resonant energy transfer between quantum systems mediated by optical near-field interaction and fast energy relaxation in intra-energy levels via exciton–phonon interaction are fundamental for a switching device and excitation carrier manipulations, as discussed in Section 33.2. By using coupled states, which are made from resonant optical near-field interaction between quantum systems, nanophotonic logic-gate, memory, and quantum-classical interface devices can be realized with the help of selective energy transfer depending on energies of the coupled state and a selection rule for optical near and far fields. These nanophotonic devices are taken as extensions of classical devices, in which simple functions are tandemly arranged, to novel-type devices locally utilized the quantum nature. Here, the optical near fields play important roles for controlling quantum coherence and/or decoherence.

As a summary, key fundamental design concepts for nanophotonic inherent operations are enumerated in the following:

- Externally switchable dissipation paths
- Dependence of the number of excitation carriers
- Selection rule for optical near field based on spatial arrangement of quantum systems
- Selection rule for optical far field based on total dipole moment

The above may not explain all concepts for designing nanophotonic devices—additional unique natures may be hidden. Accomplished device systems might be realized, which have possible advantages such as low energy consumption, low heat liberation, large parallelism, miniaturization, and environmental tolerance, by comprehending physical phenomena inherent in nanophotonics correctly and designing nanophotonic devices from a standpoint throughout from nano- to macro-scales.

Acknowledgments

The authors are grateful to M. Ohtsu, T. Kawazoe, and T. Yatsui from the University of Tokyo for fruitful discussions. This work was mainly carried out at the project of ERATO, Japan Science and Technology Agency, from 1998 to 2003. The authors would like to thank the persons concerned.

References

Biolatti, E., Iotti, R. C., Zanardi, P., and Rossi, F. 2000. Quantum information processing with semiconductor macroatoms. *Phys. Rev. Lett.* 85: 5647–5650.

Breuer, H.-P. and Petruccione, F. 2002. *The Theory of Open Quantum Systems*. New York: Oxford University Press.

Carmichael, H. J. 1999. *Statistical Methods in Quantum Optics 1*. Berlin/Heidelberg, Germany: Springer-Verlag.

Cho, K. 2004. *Optical Response of Nanostructures: Microscopic Nonlocal Theory (Springer Series in Solid-State Sciences)*. Tokyo, Japan: Springer-Verlag.

Förster, T. 1965. Delocalized excitation and excitation transfer. In *Modern Quantum Chemistry*. O. Sinanoglu (Ed.), pp. 93–137. London, U.K.: Academic Press.

Hosaka, N. and Saiki, T. 2001. Near-field fluorescence imaging of single molecules with a resolution in the range of 10 nm. *J. Microsc.* 202: 362–364.

Kawazoe, T., Kobayashi, K., Lim, J., Narita, Y., and Ohtsu, M. 2002. Direct observation of optically forbidden energy transfer between CuCl quantum cubes via near-field optical spectroscopy. *Phys. Rev. Lett.* 88: 067404-1–067404-4.

Kawazoe, T., Kobayashi, K., Sangu, S., and Ohtsu, M. 2003. Demonstration of a nanophotonic switching operation by optical near-field energy transfer. *Appl. Phys. Lett.* 82: 2957–2959.

Kobayashi, K., Sangu, S., Ito H., and Ohtsu, M. 2000. Near-field optical potential for a neutral atom. *Phys. Rev. A* 63: 013806-1–013806-9.

Maheswari, R. U., Mononobe, S., Yoshida, K., Yoshimoto, M., and Ohtsu, M. 1999. Nanometer level resolving near field optical microscope under optical feedback in the observation of a single-string deoxyribo nucleic acid. *Jpn. J. Appl. Phys.* 38: 6713–6720.

Masumoto, Y. 2002. Persistent spectral hole burning in semiconductor quantum dots. In *Semiconductor Quantum Dots: Physics, Spectroscopy and Applications*. Y. Masumoto and T. Takagahara (Eds.), pp. 209–244. Berlin/Heidelberg, Germany: Springer-Verlag.

Matsuda, K., Saiki, T., Nomura, S., Mihara, M., Aoyagi, Y., Nair, S., and Takagahara, T. 2003. Near-field optical mapping of exciton wave functions in a GaAs quantum dot. *Phys. Rev. Lett.* 91: 177401–177404.

Nomura, W., Ohtsu, M., and Yatsui, T. 2005. Nanodot coupler with a surface plasmon polariton condenser for optical far/near-field conversion. *Appl. Phys. Lett.* 86: 181108-1–181108-3.

Ohtsu, M. 1998. *Near-Field Nano/Atom Optics and Technology*. Tokyo, Japan: Springer-Verlag.

Ohtsu, M., Kobayashi, K., Kawazoe, T., Sangu, S., and Yatsui, T. 2002. Nanophotonics: Design, fabrication, and operation of nanometric devices using optical near fields. *IEEE J. Sel. Top. Quantum Electron.* 8: 839–862.

Ohtsu, M., Kobayashi, K., Kawazoe, T., Yatsui, T., and Naruse, M. 2008. *Principles of Nanophotonics*. London, U.K.: Taylor & Francis.

Rinaldis, S. D., D'Amico, I., and Rossi, F. 2002. Exciton–exciton interaction engineering in coupled GaN quantum dots. *Appl. Phys. Lett.* 81: 4236–4238.

Sangu, S., Kobayashi, K., Kawazoe, T., Shojiguchi, A., and Ohtsu, M. 2003. Excitation energy transfer and population dynamics in a quantum dot system induced by optical near-field interaction. *J. Appl. Phys.* 93: 2937–2945.

Sangu, S., Kobayashi, K., Shojiguchi, A., and Ohtsu. M., 2004. Logic and functional operations using a near-field optically coupled quantum-dot system. *Phys. Rev. B* 69, 115334-1–115334-13.

Troiani, F., Hohenester, U., and Molinari, E. 2002. Electron–hole localization in coupled quantum dots. *Phys. Rev. B* 65: 161301-1–161301-4.

Walls, D. F. and Milburn, G. J. 1994. *Quantum Optics*. Berlin/Heidelberg, Germany: Springer-Verlag.

Yatsui, T., Kourogi, M., and Ohtsu, M. 2001. Plasmon waveguide for optical far/near-field conversion. *Appl. Phys. Lett.* 79: 4583–4585.

Yatsui, T., Sangu, S., Kobayashi, K., Kawazoe, T. et al. 2009. Nanophotonic energy up-conversion using ZnO nanorod double-quantum-well structures. *Appl. Phys. Lett.* 94: 083113-1–083113-3.

34

System Architectures for Nanophotonics

Makoto Naruse
*National Institute of Information
and Communications Technology
The University of Tokyo*

34.1 Introduction

To accommodate the continuously growing amount of digital data and qualitatively new requirements demanded by industry and people in society, optics is expected to be highly integrated and to play a wider role in enhancing system performance. However, many technological difficulties remain to be in overcome in adopting optical technologies in critical information and communication systems; one problem is the poor integrability of optical hardware due to the diffraction limit of light (Pohl and Courjon 1993, Ohtsu and Hori 1999).

Nanophotonics, on the other hand, which is based on local interactions between nanometer-scale matters via optical near fields, offers ultrahigh-density integration since it is not constrained by the diffraction limit. Fundamental nanophotonic processes, such as optical excitation transfer via optical near fields between nanometer-scale matters, have been studied in detail (Ohtsu et al. 2002, 2008, Maier et al. 2003). Moreover, this higher integration density is not the only advantage that optical near fields have over conventional optics and electronics. From a system architectural perspective, nanophotonics drastically changes the fundamental design rules of functional optical systems, and suitable architectures may be built to exploit this. As a result, it also gives qualitatively strong impacts on information and communication systems.

This chapter discusses system architecture for nanophotonics considering the unique physical principles of optical near-field interactions as well as their experimental verification based on technological vehicles, such as quantum dots and engineered metal nanostructures. In particular, two unique physical processes in light–matter interactions in the nanometer scale are exploited. One is optical excitation transfer via optical near-field interactions and the other is the hierarchical property in optical near-field interactions, which are explained in Sections 34.2 and 34.3, respectively.

The overall structure of this chapter is outlined in Figure 34.1. Through these architectural and physical insights, nanophotonic information and communication systems are demonstrated, which overcome the integration-density limit imposed by the diffraction of light with ultralow-power operation, as well as provide unique functionalities that are only achievable using optical near-field interactions.

34.2 System Architectures Based on Optical Excitation Transfer

34.2.1 Optical Excitation Transfer via Optical Near-Field Interactions and Its Functional Features

In this section, optical excitation transfer processes involving optical near-field interactions are reviewed from a system perspective. Here, their fundamental principles are first briefly reviewed and their functional features are introduced for later discussion.

The interaction Hamiltonian between an electron and an electric field is given by

$$\hat{H}_{\text{int}} = -\int \hat{\psi}^{\dagger}(\vec{r})\vec{\mu}\hat{\psi}(\vec{r}) \cdot \hat{\vec{D}}(\vec{r})d\vec{r}, \tag{34.1}$$

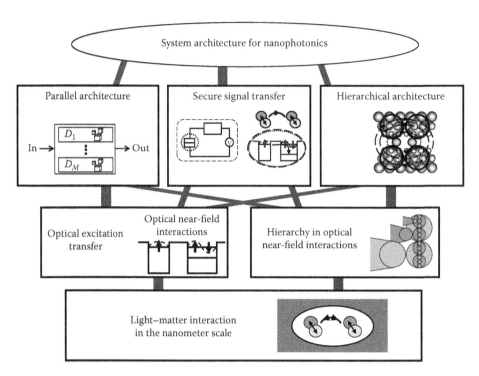

FIGURE 34.1   Overview of this section: from light–matter interactions on the nanometer scale to system architectures for nanophotonics.

where

  $\vec{\mu}$ is a dipole moment

  $\hat{\psi}^{\dagger}(\vec{r})$ and $\hat{\psi}(\vec{r})$ are, respectively, creation and annihilation operators of an electron at $\vec{r}$

  $\hat{\vec{D}}(\vec{r})$ is the operator of electric flux density

In usual light–matter interactions, the operator $\hat{\vec{D}}(\vec{r})$ is a constant since the electric field of propagating light is considered to be constant on the nanometer scale. Therefore, as is well known, one can derive optical selection rules by calculating a transfer matrix of an electrical dipole. As a consequence, in the case of cubic quantum dots, for instance, transitions to states described by quantum numbers containing an even number are prohibited. In the case of optical near-field interactions, on the other hand, due to the steep electric field of optical near fields in the vicinity of nanoscale matter, an optical transition that violates conventional optical selection rules is allowed.

Optical excitations in nanostructures, such as quantum dots, can be transferred to neighboring ones via optical near-field interactions (Ohtsu et al. 2002, 2008). For instance, assume two cubic quantum dots, QD$_A$ and QD$_B$, whose side lengths L are a and $\sqrt{2}a$, respectively (see Figure 34.2a). Suppose that the energy eigenvalues for the quantized exciton energy level specified by quantum numbers (n_x, n_y, n_z) in a QD with side length L are given by

$$E_{(n_x,n_y,n_z)} = E_B + \frac{\hbar^2\pi^2}{2ML^2}(n_x^2 + n_y^2 + n_z^2), \qquad (34.2)$$

where

  E_B is the energy of the bulk exciton

  M is the effective mass of the exciton

According to Equation 34.2, there exists a resonance between the level of quantum number (1, 1, 1) for QD$_A$ and that of quantum number (2, 1, 1) for QD$_B$. There is an optical near-field interaction, which is denoted by U, due to the steep electric field in the vicinity of QD$_A$. Therefore, excitons in QD$_A$ can move to the (2, 1, 1)-level in QD$_B$. Note that such a transfer is prohibited for propagating light since the (2, 1, 1)-level in QD$_B$ contains an even number (Tang et al. 1993). In QD$_B$, the exciton sees a sublevel energy relaxation, denoted by Γ, which is faster than the near-field interaction, and so the exciton goes to the (1, 1, 1)-level of QD$_B$. It should be emphasized that the sublevel relaxation determines the unidirectional exciton transfer from QD$_A$ to QD$_B$.

Now, several unique functional aspects should be noted in the above excitation transfer processes. First, as already mentioned, the transition from the (1, 1, 1)-level in QD$_A$ to the (2, 1, 1)-level in QD$_B$ is usually a dipole-forbidden transfer. In contrast, the optical near field allows such processes. Second, in the resonant energy levels of those quantum dots, optical excitation can go back and forth between QD$_A$ and QD$_B$, which is called optical nutation. The direction of excitations is determined by the energy dissipation processes. Therefore, based on the above mechanisms, the flow of optical excitations can be controlled in quantum-dot systems via optical near-field interactions.

From an architectural standpoint, such a flow of excitations directly leads to digital processing systems and computational

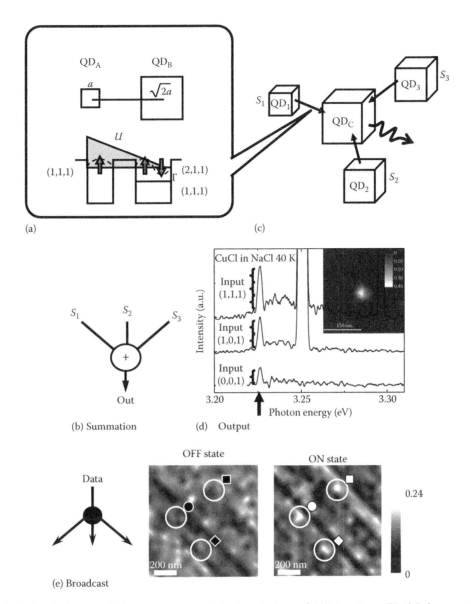

FIGURE 34.2 (a) Optical excitation transfer between quantum dots via optical near-field interactions. (b) Global summation: a basic function for memory-based architectures. (c) Quantum-dot arrangement for summation via an optical near field. (d) Intensity for three different input combinations and the spatial intensity distribution of the output photon energy. (e) Broadcast interconnects for parallel processing. Spatial intensity distribution of the output of 3-dot AND gates.

architectures. First of all, two different physical states appear by controlling the dissipation processes in the larger dot; this is the principle of the nanophotonic switch (Kawazoe et al. 2003). Also, such a flow control itself allows an architecture known as a binary decision diagram, where an arbitrary combinatorial logic operation is determined by the destination of a signal flowing from a root (Akers 1978).

Such optical excitation transfer processes also lead to unique system architectures. In this regard, Section 34.2.2 discusses a massively parallel architecture and its nanophotonic implementations. Also, Section 34.2.3 demonstrates that optical excitation transfer provides higher tamper resistance against attacks than conventional electrically wired devices, by focusing on environmental factors for signal transfer.

34.2.2 Parallel Architecture Using Optical Excitation Transfer

34.2.2.1 Memory-Based Architecture

This section discusses a memory-based architecture where computations are regarded as a table lookup or database search problem, which is also called content addressable memory (CAM) (Liu 2002). The inherent parallelism of this architecture is well matched with the physics of optical excitation transfer, and provides performance benefits in high-density, low-power operations.

In this architecture, input signal (content) serves as a query to a lookup table, and the output is the address of the data matching the input. This architecture plays a critical role in various systems for example in a data router where the output port for an incoming packet is determined based on the lookup tables.

All optical means for implementing such functions have been proposed, for instance, by using planar lightwave circuits (Grunnet-Jepsen et al. 1999). However, since separate optical hardware for each table entry is needed if based on today's known methods, if the number of entries in the routing table is on the order of 10,000 or more, the overall physical size of the system becomes impractically large. On the other hand, by using diffraction-limit-free nanophotonic principles, huge lookup tables can be configured compactly.

Then, it is important to note that the table lookup problem is equivalent to an inner product operation. Assume an N-bit input signal $\mathbf{S} = (s_1, \ldots, s_N)$ and reference data $\mathbf{D} = (d_1, \ldots, d_N)$. Here, the inner product $\mathbf{S} \cdot \mathbf{D} = \sum_{i=1}^{N} s_i \cdot d_i$ will provide a maximum value when the input perfectly matches the reference data with an appropriate modulation format (Naruse et al. 2004). Then, the function of a CAM is to derive j, which maximizes the $\mathbf{S} \cdot \mathbf{D}_j$.

34.2.2.2 Global Summation Using Near-Field Interactions

As discussed in Section 34.2.2.1, the inner product operations are the key functionalities of the memory-based architecture. The multiplication of two bits, namely, $x_i = s_i \cdot d_i$, has already been demonstrated by a combination of three quantum dots (Kawazoe et al. 2003). Therefore, one of the key operations remaining is the summation, or data gathering scheme, denoted by $\sum x_i$, where all data bits should be taken into account, as schematically shown in Figure 34.2b.

In known optical methods, wave propagation in free-space or in waveguides, using focusing lenses or fiber couplers, for example, well matches such a data gathering scheme because the physical nature of propagating light is inherently suitable for the collection or distribution of information, such as global summation. However, the level of the integration of these methods is restricted due to the diffraction limit of light. In nanophotonics, on the other hand, the near-field interaction is inherently physically local, although functional global behavior is required.

The global data gathering mechanism, or summation, is realized based on the unidirectional energy flow via an optical near field, as schematically shown in Figure 34.2c, where surrounding excitations are transferred toward a quantum dot QD_C located at the center (Kawazoe et al. 2005, Naruse et al. 2005b). This is based on the excitation transfer processes presented in Section 34.2.1 and in Figure 34.2a, where an optical excitation is transferred from a smaller dot (QD_A) to a larger one (QD_B) through a resonant energy sublevel and a sublevel relaxation process occurring at a larger dot. In the system shown in Figure 34.2c, similar energy transfers may take place among the resonant energy levels in the dots surrounding QD_C so that excitation transfer can occur. The lowest energy level in each quantum dot is coupled to a free photon bath to sweep out the excitation radiatively. The output signal is proportional to the $(1, 1, 1)$-level population in QD_B.

A proof-of-principle experiment was performed to verify the nanoscale summation using CuCl quantum dots in a NaCl matrix, which has also been employed for demonstrating nanophotonic switches (Kawazoe et al. 2003) and optical nano-fountains (Kawazoe et al. 2005). A quantum-dot arrangement where small QDs (QD_1–QD_3) surrounded a large QD at the center (QD_C) was chosen. Here, at most three light beams with different wavelengths, 325, 376, and 381.3 nm, are irradiated, which excite the quantum dots QD_1–QD_3 having sizes of 1, 3.1, and 4.1 nm, respectively. The excited excitons are transferred to QD_C, and their radiation is observed by a near-field fiber probe tip. Notice the output signal intensity at a photon energy level of 3.225 eV in Figure 34.2d, which corresponds to a wavelength of 384 nm, or a QD_C size of 5.9 nm. The intensity varies approximately as 1:2:3 depending on the number of excited QDs in the vicinity, as observed in Figure 34.2d. The spatial intensity distribution was measured by scanning the fiber probe, as shown in the inset of Figure 34.2d, where the energy is converged at the center. Hence, this architecture works as a summation mechanism, counting the number of input channels, based on exciton energy transfer via optical near-field interactions.

Such a quantum-dot-based data-gathering mechanism is also extremely energy efficient compared to other optical methods such as focusing lenses or optical couplers. For example, the transmittance between two materials with refractive indexes n_1 and n_2 is given by $4n_1n_2/(n_1 + n_2)^2$; this gives a 4% loss if n_1 and n_2 are 1 and 1.5, respectively. The transmittance of an N-channel-guided wave coupler is $1/N$ from the input to the output if the coupling loss at each coupler is 3 dB. In nanophotonic summation, the loss is attributed to the dissipation between energy sublevels, which is significantly smaller. Incidentally, it is energy- and space-efficient compared to electrical CAM VLSI chips (Lin and Kuo 2001, Arsovski et al. 2003, Naruse et al. 2005a).

34.2.2.3 Broadcast Interconnects

For the parallel architecture shown in Section 34.2.2.2, it should also be noted that the input data should be commonly applied to all lookup table entries. In other words, broadcast interconnect is another important requirement for parallel architectures. Broadcast is also important in applications such as matrix-vector products (Goodman et al. 1978, Guilfoyle and McCallum 1996) and switching operations, for example, broadcast-and-select architectures (Li et al. 2001). Optics is in fact well suited to such broadcast operations in the form of simple imaging optics (Goodman et al. 1978, Guilfoyle and McCallum 1996) or in optical waveguide couplers; thanks to the nature of wave propagation. However, the integration density of this approach is physically limited by the diffraction limit, which leads to bulky system configurations.

The overall physical operation principle of broadcast using optical near fields is as follows. Suppose that the arrays of nanophotonic circuit blocks are distributed within an area whose size is comparable to the wavelength. For broadcasting, multiple input QDs simultaneously accept identical input data carried by diffraction-limited far-field light by tuning their optical frequency so that the light is coupled to dipole-allowed energy sublevels.

The far- and near-field coupling mentioned above is explained based on a model assuming cubic quantum dots, which was introduced in Section 34.2.1. According to Equation 34.2, there exists a resonance between the quantized exciton energy sublevel of quantum number $(1, 1, 1)$ for the QD with effective side length a and that of quantum number $(2, 1, 1)$ for the QD with effective side length $\sqrt{2}a$. Energy transfer from the smaller QD to the larger one occurs via optical near fields, which is forbidden for far-field light (Kawazoe et al. 2003).

The input energy level for the QDs, that is, the $(1, 1, 1)$-level, can also couple to the far-field excitation. This fact can be utilized for data broadcasting. One of the design restrictions is that energy sublevels for input channels do not overlap with those for output channels. Also, if there are QDs internally used for near-field coupling, dipole-allowed energy sublevels for those QDs cannot be used for input channels since the inputs are provided by far-field light, which may lead to the misbehavior of internal near-field interactions if resonant levels exist. Therefore, frequency partitioning among the input, internal, and output channels is important. The frequencies used for broadcasting, denoted by $\Omega_i = \{\omega_{i,1}, \omega_{i,2},..., \omega_{i,A}\}$, should be distinct values and should not overlap with the output channel frequencies $\Omega_o = \{\omega_{o,1}, \omega_{o,2},..., \omega_{o,B}\}$. A and B indicate the number of frequencies used for input and output channels, respectively. Also, there will be frequencies needed for internal device operations, which are not used for either input or output, denoted by $\Omega_n = \{\omega_{n,1}, \omega_{n,2},..., \omega_{n,C}\}$, where C is the number of those frequencies. Therefore, the design criteria for global data broadcasting is to exclusively assign input, output, and internal frequencies, Ω_i, Ω_o, and Ω_n, respectively.

In a frequency multiplexing sense, this interconnection method is similar to multi-wavelength chip-scale interconnection (De Souza et al. 1995). Known methods, however, require a physical space comparable to the number of diffraction-limited input channels due to wavelength demultiplexing, whereas in the nanophotonic scheme, the device arrays are integrated on the sub-wavelength scale, and multiple frequencies are multiplexed in the far-field light supplied to the device.

To verify the broadcasting method, the following experiments were performed using CuCl QDs inhomogeneously distributed in a NaCl matrix at a temperature of 22 K (Naruse et al. 2006). To operate a 3-dot nanophotonic switch (2-input AND gate) in the device, at most two input light beams (IN1 and IN2) were irradiated. When both inputs exist, an output signal is obtained from the positions where the switches exist, as described above. In the experiment, IN1 and IN2 were assigned to 325 and 384.7 nm, respectively. They were irradiated over the entire sample (global irradiation) via far-field light. The spatial intensity distribution of the output, at 382.6 nm, was measured by scanning a near-field fiber probe within an area of approximately $1\,\mu m \times 1\,\mu m$. When only IN1 was applied to the sample the output of the AND gate was ZERO (OFF state). When both inputs were irradiated the output was ONE (ON state). Note the regions marked by ■, ●, and ◆ in Figure 34.2e. In those regions, the output signal

levels were, respectively, low and high, which indicate that multiple AND gates were integrated at densities beyond the scale of the globally irradiated input beam area. That is to say, broadcast interconnects to nanophotonic switch arrays are accomplished by diffraction-limited far-field light.

Combining this broadcasting mechanism with the summation mechanism discussed in Section 34.2.2.2 will allow the development of the nanoscale integration of massively parallel architectures, which have conventionally resulted in bulky configurations.

34.2.3 Secure Signal Transfer in Nanophotonics

In addition to breaking through the diffraction limit of light, such local interactions of optical near fields also have important functional aspects, such as in security applications, which particularly tamper resistance against attacks (Naruse et al. 2007a). One of the most critical security issues in present electronic devices is so-called side-channel attacks, by which information is tampered either invasively or noninvasively. This may be achieved, for instance, merely by monitoring their power consumption (Kocher et al. 1998).

In this section, it is shown that devices based on optical excitation transfer via near-field interactions are physically more tamper-resistant than their conventional electronic counterparts. The key is that the flow of information in nanoscale devices cannot be completed unless they are appropriately coupled with their environment (Hori 2001), which could possibly be the weakest link in terms of their tamper resistance. A theoretical approach is presented to investigate the tamper resistance of optical excitation transfer, including a comparison with electrical devices.

Here, the tampering of information is defined as involving simple signal transfer processes, since the primary focus is on their fundamental physical properties.

In order to compare the tamper resistance, an electronic system based on single-charge tunneling is introduced here, in which a tunnel junction with capacitance C and tunneling resistance R_T is coupled to a voltage source V via an external impedance $Z(\omega)$, as shown in Figure 34.3a. In order to achieve

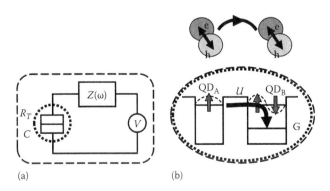

(a) (b)

FIGURE 34.3 Model of tamper resistance in devices based on (a) single charge tunneling and (b) optical excitation transfer. Dotted curves show the scale of a key device and dashed curves show the scale of the environment required for the system to work.

single charge tunneling, besides the condition that the electrostatic energy $E_C = e^2/2C$ of a single excess electron be greater than the thermal energy $k_B T$, the environment must have appropriate conditions, as discussed in detail in Ingold and Nazarov (1992). For instance, with an inductance L in the external impedance, the fluctuation of the charge is given by

$$\langle \delta Q^2 \rangle = \frac{e^2}{4\rho} \coth\left(\frac{\beta \hbar \omega_S}{2}\right), \tag{34.3}$$

where

$\rho = E_C/\hbar\omega_S$
$\omega_S = (LC)^{-1/2}$
$\beta = 1/k_B T$

Therefore, charge fluctuations cannot be small even at zero temperature unless $\rho \gg 1$. This means that a high-impedance environment is necessary, which makes tampering technically easy, for instance by adding another impedance circuit.

Here, let us define two scales to illustrate tamper resistance: (1) the scale associated with the key device size and (2) the scale associated with the environment required for operating the system, which are, respectively, indicated by dotted and dashed curves in Figure 34.3. In the case of Figure 34.3a, scale I is the scale of a tunneling device, whereas scale II covers all of the components. It turns out that the low tamper resistance of such wired devices is due to the fact that scale II is typically the macroscale, even though scale I is the nanometer scale.

In contrast, in the case of the optical excitation transfer shown in Figure 34.3b, the two quantum dots and their surrounding environment are governed by scale I. It is also important to note that scale II is the same as scale I. More specifically, the transfer of an exciton from QD_A to QD_B is completed due to the non-radiative relaxation process occurring at QD_B, which is usually difficult to tamper with. Theoretically, the sublevel relaxation constant is given by

$$\Gamma = 2\pi \, |g(\omega)|^2 \, D(\omega), \tag{34.4}$$

where

$\hbar g(\omega)$ is the exciton–phonon coupling energy at frequency ω
$\hbar$ is Planck's constant divided by 2π
$D(\omega)$ is the phonon density of states (Carmichael 1999)

Therefore, tampering with the relaxation process requires somehow "stealing" the exciton–phonon coupling, which would be extremely difficult technically.

It should also be noted that the energy dissipation occurring in the optical excitation transfer, derived theoretically as $E_{(2,1,1)} - E_{(1,1,1)}$ in QD_B based on Equation 34.2, should be larger than the exciton-phonon coupling energy of $\hbar\Gamma$, otherwise, the two levels in QD_B cannot be resolved. This is similar to the fact that the condition $\rho \gg 1$ is necessary in the electron-tunneling example, which means that the mode energy $\hbar\omega_S$ is smaller than the required

charging energy E_C. By regarding $\hbar\Gamma$ as a kind of mode energy in the optical excitation transfer, the difference between the optical excitation transfer and a conventional wired device is the physical scale at which this mode energy is realized: nanoscale for the optical excitation transfer and macroscale for electric circuits.

34.3 Hierarchical Architectures in Nanophotonics

34.3.1 Physical Hierarchy in Nanophotonics and Functional Hierarchy

In this section, another feature of nanophotonics, the inherent hierarchy in optical near-field interactions, is exploited. As schematically shown in Figure 34.4a, there are multiple layers associated with the physical scale between the macroscale world and the atomic-scale world, which are primarily governed by propagating light and electron interactions, respectively. Between these two extremes, typically in scales ranging from a few nanometers to wavelength size, optical near-field interactions play a crucial role. In this section, such hierarchical properties in this mesoscopic or sub-wavelength regime are exploited.

Such physical hierarchy in optical near-field interactions will be analyzed by a simple dipole–dipole interaction model and an angular spectrum representation of optical near fields, as shown in Section 34.3.2. Before going into the details of the physical processes, functionalities required for system applications are briefly reviewed in terms of hierarchy.

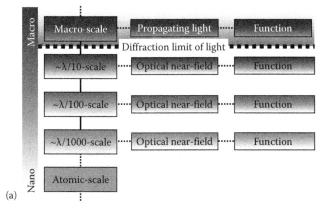

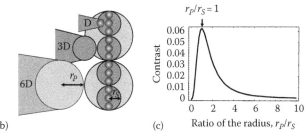

FIGURE 34.4 (a) Hierarchy in optical near-field interactions. (b) Dipole–dipole interaction. (c) Signal contrast as a function of the ratio of the radius of the sample and the probe.

One of the problems for ultrahigh-density nanophotonic systems is interconnection bottlenecks, which have been addressed previously in Section 34.2.3 regarding broadcast interconnects. In fact, a hierarchical structure can be found in these broadcast interconnects by relating far-field effects at a coarser scale and near-field effects at a finer scale.

In this regard, it should also be mentioned that such physical differences in optical near-field and far-field effects can be used for a wide range of applications. The behavior of usual optical elements, such as diffractive optical elements, holograms, or glass components, is associated with their optical responses in optical far fields. In other words, nanostructures can exist in such optical elements as long as they do not affect the optical responses in far fields. Designing nanostructures accessible only via optical near fields provides additional, or hidden, information recorded in those optical elements, while maintaining the original optical responses in far fields. In fact, a *"hierarchical hologram"* or *"hierarchical diffraction grating"* has been experimentally demonstrated, as schematically shown in Figure 34.5 (Tate et al. 2008).

Since there is more hierarchy in the optical near-field regime, further applications should be possible, for example, it should be possible for nanometer-scale high-density systems to be gradually hierarchically connected to coarse layer systems.

Hierarchical functionalities are also important for several aspects of memory systems. One is related to recent high-density, huge-capacity memory systems, in which data retrieval or searching from entire memory archives is made even more difficult. Hierarchy is one approach for solving such a problem by making systems hierarchical, that is, by recording abstract data, metadata, or tag data in addition to the original raw data.

Hierarchy in nanophotonics provides a physical solution to achieve such functional hierarchy. As will be introduced below, low-density, rough information is read out at a coarser scale, whereas high-density, detailed information is read out at a finer scale. Sections 34.3.2 and 34.3.3 will show physical mechanisms for such hierarchical information retrieval.

Another issue in hierarchical functionalities will be security. High-security information is recorded at a finer scale, whereas less-critical security information is associated with a coarse layer. Also, in addition to associating different types of information with different physical scales, another kind of information can also be related to one or more layers of the physical hierarchy, for instance, traceability, history, or aging of information. Section 34.3.3 will demonstrate a *traceable* memory as an example.

34.3.2 Hierarchical Memory Retrieval

This section describes a physical model of optical near-field interactions based on dipole–dipole interactions (Ohtsu and Kobayashi 2004). Suppose that a probe, which is modeled by a sphere of radius r_P, is placed close to a sample to be observed, which is modeled as a sphere of radius r_S. Figure 34.4b shows three different sizes for the probe and the sample. When they are illuminated by incident light, whose electric field is $\mathbf{E}_0$, electric dipole moments are induced in both the probe and the sample; these moments are denoted by $\mathbf{p}_P = \alpha_P \mathbf{E}_0$ and $\mathbf{p}_S = \alpha_S \mathbf{E}_0$, respectively. The electric dipole moment induced in the sample $\mathbf{p}_S$, then generates an electric field, which changes the electric dipole moment in the probe by an amount $\Delta \mathbf{p}_P = \Delta \alpha_P \mathbf{E}_0$. Similarly, $\mathbf{p}_P$ changes the electric dipole moment in the sample by $\Delta \mathbf{p}_S = \Delta \alpha_S \mathbf{E}_0$. These electromagnetic interactions are called

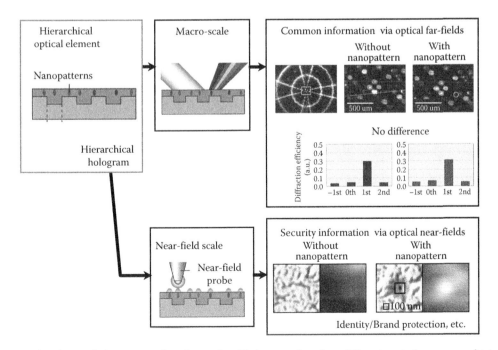

FIGURE 34.5 Hierarchical optical elements, such as *hierarchical holograms*, based on different optical responses obtained in optical far and near fields.

dipole–dipole interactions. The scattering intensity induced by these electric dipole moments is given by

$$I = \left| \mathbf{p}_P + \Delta\mathbf{p}_P + \mathbf{p}_S + \Delta\mathbf{p}_S \right|^2$$
$$\approx (\alpha_P + \alpha_S)^2 |E_0|^2 + 4\Delta\alpha(\alpha_P + \alpha_S)|E_0|^2 \quad (34.5)$$

where $\Delta\alpha = \Delta\alpha_S = \Delta\alpha_P$ (Ohtsu and Kobayashi 2004). The second term in Equation 34.5 shows the intensity of the scattered light generated by the dipole–dipole interactions, containing the information of interest, which is the relative difference between the probe and the sample. The first term in Equation 34.5 is the background signal for the measurement. Therefore, the ratio of the second term to the first term of Equation 34.5 corresponds to a signal contrast, which will be maximized when the sizes of the probe and the sample are the same ($r_P = r_S$), as shown in Figure 34.4c. (A detailed derivation is found in Ohtsu and Kobabashi (2004).) Thus, one can see a scale-dependent physical hierarchy in this framework, where a small probe, say $r_P = D/2$, can nicely resolve objects with a comparable resolution, whereas a large probe, say $r_P = 3D/2$, cannot resolve a detailed structure but it can resolve a structure with a resolution comparable to the probe size. Therefore, although a large diameter probe cannot detect smaller-scale structure, it could detect certain features associated with its scale.

Based on the above simple hierarchical mechanism, a hierarchical memory system is constructed. Consider, for example, a maximum of N nanoparticles distributed in a region of a subwavelength scale. These nanoparticles can be nicely resolved by a scanning near-field microscope if the size of its fiber probe tip is comparable to the size of individual nanoparticles; in this way, the *first-layer* information associated with each distribution of nanoparticles is retrievable, corresponding to 2^N-different codes. By using a larger-diameter fiber probe tip instead, the distribution of the particles cannot be resolved, but a mean-field feature with a resolution comparable to the size of the probe can be extracted, namely, the number of particles within an area comparable to the size of the fiber probe tip. Thus, the *second-layer* information associated with the number of particles, corresponding to ($N + 1$)-different level of signals, is retrievable. Therefore, one can access different set of signals, 2^N or $N + 1$, depending on the scale of observation. This leads to hierarchical memory retrieval by associating this information hierarchy with the distribution and the number of nanoparticles using an appropriate coding strategy, as schematically shown in Figure 34.6a.

For example, in encoding N-bit information, ($N − 1$)-bit signals can be encoded by the distributions of nanoparticles while associating the remaining 1-bit with the number of nanoparticles. The details of encoding/decoding strategies will be found in Naruse et al. (2005c).

Simulations were performed assuming ideal isotropic metal particles to see how the second-layer signal varies depending on the number of particles using a finite-difference time-domain simulator (*Poynting for Optics*, a product of Fujitsu, Japan).

Here, 80 nm-diameter particles are distributed over a 200 nm-radius circular grid at constant intervals. The solid circles in Figure 34.6d show calculated scattering cross sections as a function of the number of particles. A linear correspondence to the number of particles was observed. This result supports the simple physical model described above.

In order to experimentally demonstrate such principles, an array of Au particles, each with a diameter of around 80 nm, was distributed over a SiO$_2$ substrate in a 200 nm-radius circle. These particles were fabricated by a liftoff technique using electron-beam (EB) lithography with a Cr buffer layer. Each group of Au particles was spaced by 2 µm. A scanning electron microscope (SEM) image is shown in Figure 34.6b in which the values indicate the number of particles within each group. In order to illuminate all Au particles in each group and collect the scattered light from them, a near-field optical microscope (NOM) with a large-diameter-aperture (500 nm) metallized fiber probe was used in an illumination collection setup. The light source used was a laser diode with an operating wavelength of 680 nm. The distance between the substrate and the probe was maintained at 750 nm. Figure 34.6c shows an intensity profile captured by the probe, from which the second-layer information is retrieved. The solid squares in Figure 34.6d indicate the peak intensity of each section, which increased linearly. These results show the validity of hierarchical memory retrieval from nanostructures.

34.3.3 Design of *Unscalable* Hierarchical Response

The coarse graining process is usually an averaging process, meaning that the signal in the coarser layer is obtained in terms of a mean-field approximation of the fine-grained, lower-layer signals, which is also the case shown in Section 34.3.2. However, it should be noted that the optical near-field amplitude can be distributed independently at different scales of observation. In other words, the coarse graining of the optical near fields in fact provides an optical property independent of the lower-layer feature; such an *unscalable* hierarchical property of optical near fields will be analyzed below.

Here, the angular spectrum representation of the electromagnetic field is used for discussing the hierarchy in optical near fields (Wolf and Nieto-Vesperinas 1985, Inoue and Hori 2005, Naruse et al. 2007b). This allows an analytical treatment and gives an intuitive picture of the localization of optical near fields and represents relevance/irrelevance in optical near-field interactions at different scales of observation since it describes electromagnetic fields as a superposition of evanescent waves with different decay length and corresponding spatial frequency.

Suppose, for example, that there is an oscillating electric dipole, $\mathbf{d}^{(k)} = d^{(k)}(\cos\phi^{(k)}, 0)$, on the xz plane, which is oriented parallel to the x axis. Now, consider the electric field of radiation observed at a position displaced from the dipole by $\mathbf{R} = (r_{\parallel}^{(k)}\cos\varphi^{(k)}, z^{(k)})$. The angular spectrum representation of the z-component of the optical near field is given by

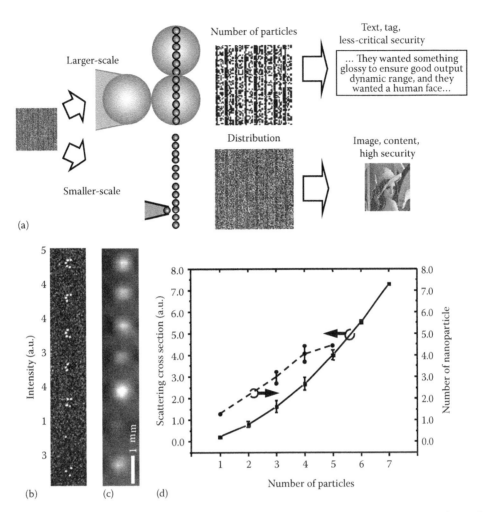

FIGURE 34.6 (a) Hierarchical memory retrieval from nanostructures. (b) SEM picture of an array of Au nanoparticles. Each section consists of up to seven nanoparticles. (c) Intensity pattern captured by a fiber probe tip whose diameter is comparable to the size of each section of nanoparticles. (d) Square marks: calculated scattering cross sections in each section, circular marks: peak intensity of each section in the intensity profile shown in (c).

$$E_z(\mathbf{R}) = \left(\frac{iK^3}{4\pi\varepsilon_0} \right) \int\limits_1^\infty ds_\| \frac{s_\|}{s_z} f_z(s_\|, \mathbf{d}^{(1)}, \ldots, \mathbf{d}^{(N)}) \qquad (34.6)$$

where

$$f_z(s_\|, \mathbf{d}^{(1)}, \ldots, \mathbf{d}^{(N)}) = \sum_{k=1}^N d^{(k)} s_\| \sqrt{s_\|^2 - 1} \, \cos(\phi^{(k)} - \varphi^{(k)}) J_1\left(K r_\|^{(k)} s_\| \right)$$

$$\times \exp\left(-K z^{(k)} \sqrt{s_\|^2 - 1} \right). \qquad (34.7)$$

Here, $s_\|$ is the spatial frequency of an evanescent wave propagating parallel to the x axis, and $J_n(x)$ represents Bessel functions of the first kind. Here, the term $f_z(s_\|, \mathbf{d}^{(1)}, \ldots, \mathbf{d}^{(N)})$ is called the angular spectrum of the electric field.

In the following, a two-layer system is introduced where (1) by observing very close to the dipoles, two items of first layer information are retrieved, and (2) by observing relatively far from the dipoles, one item of second-layer information is retrieved.

Suppose that there are two closely spaced dipole pairs (so there are four dipoles in total). The dipoles $\mathbf{d}^{(1)}$ and $\mathbf{d}^{(2)}$ are oriented in the same direction, namely, $\varphi^{(1)} = \varphi^{(2)} = 0$, and another dipole pair, $\mathbf{d}^{(3)}$ and $\mathbf{d}^{(4)}$, are both oriented in the opposite direction to $\mathbf{d}^{(1)}$ and $\mathbf{d}^{(2)}$, namely, $\varphi^{(3)} = \varphi^{(4)} = \pi$. These four dipoles are located at positions shown in Figure 34.7a. Here, at a position close to the x axis equidistant from $\mathbf{d}^{(1)}$ and $\mathbf{d}^{(2)}$, such as at the position A_1 in Figure 34.7a, the electric field is weak (logical ZERO) since (1) the angular spectrum originating from $\mathbf{d}^{(1)}$ and $\mathbf{d}^{(2)}$ vanishes, and (2) the electric field originating from $\mathbf{d}^{(3)}$ and $\mathbf{d}^{(4)}$ is small because they are far from the position A_1. In fact, as shown by the dashed curve in Figure 34.7b, since the angular spectrum at position A_1 oscillates, the integral of the angular spectrum, which is correlated to the field intensity at that point, will be low.

For the second layer retrieval, consider the observation at an intermediate position between the dipole pairs, such as the position B in Figure 34.7a. From this position, the four dipoles effectively appear to be two dipoles that are oriented in opposite directions to each other. As shown by the solid curve in Figure 34.7b, the angular spectrum involving relatively low

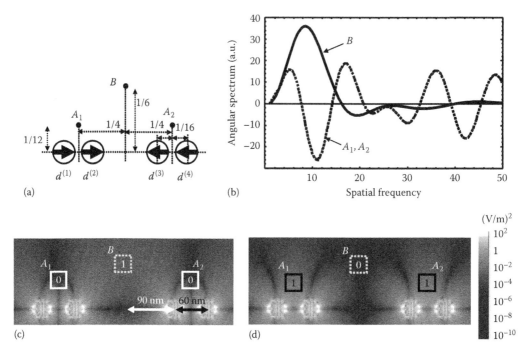

FIGURE 34.7 Examples of *unscalable* hierarchy in optical near fields. (a) The positions and orientations of four electric dipole systems. (b) And the corresponding angular spectrum observed at positions A_1, A_2, and B, indicating that the first-layer signals (obtained at A_1 and A_2) are logical ZEROs, while the second-layer signal (obtained at B) is logical ONE. (c) Calculated electric field intensity distribution agreed with theoretical predictions shown in (b). (d) Another electric field distribution where logical ONEs are retrieved at the first layer (A_1 and A_2) and logical ZERO is retrieved at the second layer (B).

spatial-frequency components shows a single peak, indicating that the electric field in the xy-plane is localized to the degree determined by its spectral width so that a logical ONE is retrievable at position B. Meanwhile, the angular spectrum observed at position A_2, shown in Figure 34.7a, is identical to that observed at position A_1, meaning that the electric field at A_2 is also at a low level.

To summarize the above mechanisms, a logical level of ONE in the second layer can be retrieved even though the two items of information retrieved in the first layer are both ZEROs; therefore, an unscalable hierarchy is achieved.

As described above, one of the two first-layer signals, the electric field at A_1, primarily depends on the dipole pair $\mathbf{d}^{(1)}$ and $\mathbf{d}^{(2)}$, and the other, the electric field at A_2, depends primarily on the dipole pair $\mathbf{d}^{(3)}$ and $\mathbf{d}^{(4)}$. The second-layer signal is determined by all of those dipoles. Concerning such a hierarchical mechanism, it was shown that a total of eight different signal combinations were achieved by appropriately orienting the four dipoles (Naruse et al. 2007b).

Numerical simulations were performed based on finite-difference time-domain methods to see how they agree with the theoretical analysis based on the angular spectrum. Four silver nanoparticles (of radius 15 nm) containing a virtual oscillating light source were assumed in order to simulate dipole arrays. Their positions are shown in Figure 34.7c. The first and the second layers were located 40 and 80 nm away from the dipole plane, respectively. The operating wavelength was 488 nm. The electric fields obtained at A_1, A_2, and B agree with the combinations of the

first- and second-layer signals to be retrieved, as shown in Figure 34.7c. As another unscalable hierarchy example, Figure 34.7d represents a situation where logical ONEs are obtained at the first layer, while logical ZERO is obtained at the second layer. These agree with the theoretical analysis based on the angular spectrum.

34.3.4 Versatile Functionalities Based on Hierarchy in Optical Near-Fields

The hierarchical nature has been further exploited by combining other physical mechanisms in nanophotonics. For example, we can associate one of the hierarchical layers with energy dissipation processes. Specifically, a two-layer system is demonstrated where (1) at smaller scale, called *Scale* 1, the system should exhibit a unique response, and (2) at a larger scale, called *Scale* 2, the system should output two different signals. Such a hierarchical response can be applied to functions like the *traceability* of optical memory in combination with a localized energy dissipation process at Scale 1 (Naruse et al. 2008c). Optical access to this memory will be automatically recorded due to energy dissipation occurring locally in Scale 1, while at the same time, information will be read out based on Scale 2 behavior. Therefore, such hierarchy enables the traceability of optical memory, which will be important for the security (confidentiality is ensured) and management of digital content.

Shape-engineered metal nanostructures can achieve the hierarchy required for traceable memory (Naruse et al. 2008e). Here, two types of shapes are assumed. The first one (*Shape* I) has two triangular metal plates aligned in the same direction, and the other one (*Shape* II) has them facing each other, as shown in Figure 34.8a. The metal is gold, the gap between the two apexes is 50 nm, the horizontal length of one triangular plate is 173 nm, the angle at the apex is 30°, and the thickness is 30 nm. An incident uniform plane wave with a wavelength of 680 nm is assumed for input light. The polarization is parallel to the *x* axis in Figure 34.8b. Now, Scale 1 is associated with the scale around the gap of the triangles, and Scale 2 is associated with the scale covering both of the triangles, as shown in Figure 34.8a.

Regarding the optical response at Scale 1, as shown in Figure 34.8b, the electric field near the surface (1 nm away from the metal surface) shows an intensity nearly five orders of magnitude higher than the surrounding area. It should also be noted that nearly comparable electric-field enhancements are observed near the apexes of Shapes I and II, which are, respectively, denoted by the squares and circles in Figure 34.8b.

On the other hand, Shapes I and II exhibit different responses at Scale 2. As shown in Figure 34.8c, Shape I exhibits larger scattering cross section compared to Shape II. This indicates that a digital output is retrievable by observing the scattering from the entire structure (Scale 2), where, for example, digital 1 and 0 are associated with Shape I and Shape II, respectively.

In order to experimentally demonstrate the principle, Shapes I and II were fabricated in gold metal plates on a glass substrate by a liftoff technique using EB lithography. A NOM in an illumination collection setup was used with an apertured fiber probe having a diameter of 500 nm, as shown in Figure 34.8d. The light

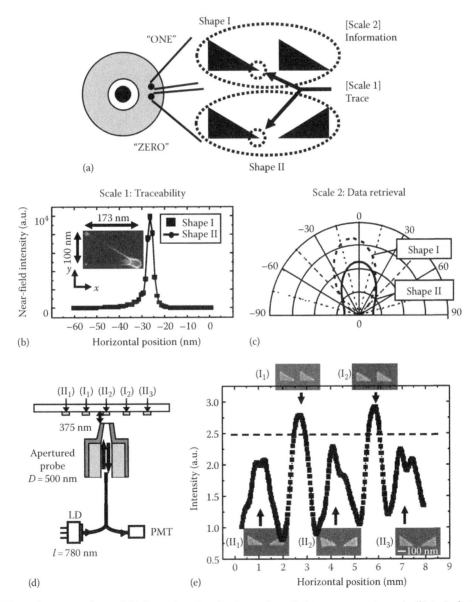

FIGURE 34.8 (a) Hierarchy in optical near fields by engineering the shape of metal plates at nanometer scale. (b) In Scale 1, both shapes exhibit comparable electric-field enhancement. (c) In Scale 2, they exhibit different system responses. (d) Experimental setup for Scale 2 signal retrieval. (e) Electric field intensity for Scale 2 signals for both Shapes I and II.

source used was a laser diode with an operating wavelength of 780 nm. The distance between the substrate and the probe was maintained at 375 nm. Figure 34.8e shows the electric field intensity depending on the shape of the metal plates, where the Shape I series exhibited larger values compared to the Shape II series, as expected.

Hierarchical nature in optical near-field interactions provides other unique functions. For example, the near-field photoluminescence of semiconductor quantum dots exhibits a hierarchical nature and its spectra diversity is maximized at an optimal scale of the optical near fields. This leads to a novel *non-pixelated memory architecture*, which can simultaneously retrieve a sequence of bits, as opposed to conventional bit-sequential pixelated architecture. The principles have been experimentally verified by InAs quantum dots (Naruse et al. 2008a). Also, a nanophotonic *lock-and-key* system has been demonstrated for the authentication of applications based on shape-engineered nanostructures and their associated optical near fields (Naruse et al. 2008b,d).

34.4 Summary

In this chapter, fundamental nanophotonic system architectures were presented along with two principal physical features of nanophotonics: one is optical excitation transfer and the other is hierarchy in optical near-field interactions. Both of these physical features originate from light–matter interactions on the nanometer scale. It should be emphasized that those basic features provide versatile applications and functionalities besides the example demonstrations shown in the above sections. Also, there are many other degrees-of-freedom in the nanometer scale that need to be deeply understood for systems. Further exploration and attempts to exploit nanophotonics for future devices and systems will certainly be exciting.

References

Akers, S. B. 1978. Binary decision diagrams. *IEEE Trans. Comput.* C-27:509–516.

Arsovski, I., Chandler, T., and Sheikholeslami, A. 2003. A ternary content-addressable memory (TCAM) based on 4T static storage and including a current-race sensing scheme. *IEEE J. Solid-State Circuits* 38:155–158.

Carmichael, H. J. 1999. *Statistical Methods in Quantum Optics I*. Berlin, Germany: Springer-Verlag.

De Souza, E. A., Nuss, M. C., Knox, W. H. et al. 1995. Wavelength-division multiplexing with femtosecond pulses. *Opt. Lett.* 20:1166–1168.

Goodman, J. W., Dias, A. R., and Woody, L. M. 1978. Fully parallel, high-speed incoherent optical method for performing discrete Fourier transforms. *Opt. Lett.* 2:1–3.

Grunnet-Jepsen, A., Johnson, A. E., Maniloff, E. S. et al. 1999. Fibre Bragg grating based spectral encoder/decoder for lightwave CDMA. *Electron. Lett.* 35:1096–1097.

Guilfoyle, P. S. and McCallum, D. S. 1996. High-speed low-energy digital optical processors. *Opt. Eng.* 35:436–442.

Hori, H. 2001. Electronic and electromagnetic properties in nanometer scales. In *Optical and Electronic Process of Nano-Matters*, ed. M. Ohtsu, pp. 1–55. Tokyo: KTK Scientific/Dordrecht, the Netherlands: Kluwer Academic.

Ingold, G.-L. and Nazarov, Y. V. 1992. Charge tunneling rates in ultrasmall junctions. In *Single Charge Tunneling*, eds. H. Grabert and M. H. Devoret, pp. 21–107. New York: Plenum Press.

Inoue, T. and Hori, H., 2005. Quantum theory of radiation in optical near field based on quantization of evanescent electromagnetic waves using detector Mode. In *Progress in Nano-Electro-Optics IV*, ed. M. Ohtsu, pp. 127–199. Berlin, Germany: Springer-Verlag.

Kawazoe, T., Kobayashi, K., Sangu, S. et al. 2003. Demonstration of a nanophotonic switching operation by optical near-field energy transfer. *Appl. Phys. Lett.* 82:2957–2959.

Kawazoe, T., Kobayashi, K., and Ohtsu, M. 2005. Optical nanofountain: A biomimetic device that concentrates optical energy in a nanometric region. *Appl. Phys. Lett.* 86:103102 1–3.

Kocher, P., Jaffe, J., and Jun, B. 1998. Introduction to differential power analysis and related attacks. Cryptography Research. http://www.cryptography.com/resources/whitepapers/DPATechInfo.pdf

Li, B., Qin, Y., Cao, X. et al. 2001. Photonic packet switching: Architecture and performance. *Opt. Netw. Mag.* 2:27–39.

Lin, P.-F. and Kuo, J. B. 2001. A 1-V 128-kb four-way set-associative CMOS cache memory using wordline-oriented tag-compare (WLOTC) structure with the content-addressable-memory (CAM) 10-transistor tag cell. *IEEE J. Solid-State Circuits* 36:666–675.

Liu, H. 2002. Routing table compaction in ternary CAM, *IEEE Micro* 22:58–64.

Maier, S. A., Kik, P. G., Atwater, H. A. et al. 2003. Local detection of electromagnetic energy transport below the diffraction limit in metal nanoparticle plasmon waveguides, *Nat. Mater.* 2:229–232.

Naruse, M., Mitsu, H., Furuki, M. et al. 2004. Terabit all-optical logic based on ultrafast two-dimensional transmission gating. *Opt. Lett.* 29:608–610.

Naruse, M., Miyazaki, T., Kawazoe, T. et al. 2005a. Nanophotonic computing based on optical near-field interactions between quantum dots. *IEICE Trans. Electron.* E88-C:1817–1823.

Naruse, M., Miyazaki, T., Kubota, F. et al. 2005b. Nanometric summation architecture using optical near-field interaction between quantum dots. *Opt. Lett.* 30:201–203.

Naruse, M., Yatsui, T., Nomura, W. et al. 2005c. Hierarchy in optical near-fields and its application to memory retrieval *Opt. Express* 13:9265–9271.

Naruse, M., Kawazoe, T., Sangu, S. et al. 2006. Optical interconnects based on optical far- and near-field interactions for high-density data broadcasting. *Opt. Express* 14:306–313.

Naruse, M., Hori, H., Kobayashi, K. et al. 2007a. Tamper resistance in optical excitation transfer based on optical near-field interactions. *Opt. Lett.* 32:1761–1763.

Naruse, M., Inoue, T., Hori, H. 2007b. Analysis and synthesis of hierarchy in optical near-field interactions at the nanoscale based on Angular Spectrum. *Jpn. J. Appl. Phys.* 46:6095–6103.

Naruse, M., Nishibayashi, K., Kawazoe, T. et al. 2008a. Scale-dependent optical near-fields in InAs quantum dots and their application to non-pixelated memory retrieval. *Appl. Phys. Express* 1:072101 1–3.

Naruse, M., Yatsui, T., Hori, H. et al. 2008b. Polarization in optical near- and far-field and its relation to shape and layout of nanostructures. *J. App. Phys.* 103:113525 1–8.

Naruse, M., Yatsui, T., Kawazoe, T. et al. 2008c. Design and simulation of a nanophotonic traceable memory using localized energy dissipation and hierarchy of optical near-field interactions *IEEE Trans. Nanotechnol.* 7:14–19.

Naruse, M., Yatsui, T., Kawazoe, T., Tate, N. et al. 2008d. Nanophotonic matching by optical near-fields between shape-engineered nanostructures. *Appl. Phys. Express* 1:112101 1–3.

Naruse, M., Yatsui, T., Kim, J. H. et al. 2008e. Hierarchy in optical near-fields by nano-scale shape engineering and its application to traceable memory *Appl. Phys. Express* 1:062004 1–3.

Ohtsu, M. and Hori, H. 1999. *Near-Field Nano-Optics*. New York: Kluwer Academic/Plenum Publishers.

Ohtsu, M. and Kobayashi, K. 2004. *Optical Near Fields*. Berlin, Germany: Springer-Verlag.

Ohtsu, M., Kobayashi, K., Kawazoe, T. et al. 2002. Nanophotonics: Design, fabrication, and operation of nanometric devices using optical near fields. *IEEE J. Sel. Top. Quantum Electron.* 8:839–862.

Ohtsu, M., Kobayashi, K., Kawazoe, T. et al. 2008. *Principles of Nanophotonics*. Boca Raton, FL: Taylor & Francis.

Pohl, D. W. and Courjon, D. eds. 1993. *Near Field Optics*. Dordrecht, the Netherlands: Kluwer Academic.

Tang, Z. K., Yanase, A., Yasui, T. et al. 1993. Optical selection rule and oscillator strength of confined exciton system in CuCl thin films. *Phys. Rev. Lett.* 71:1431–1434.

Tate, N., Nomura, W., Yatsui, T. et al. 2008. Hierarchical hologram based on optical near- and far-field responses. *Opt. Express* 16:607–612.

Wolf, E. and Nieto-Vesperinas, M. 1985. Analyticity of the angular spectrum amplitude of scattered fields and some of its consequences. *J. Opt. Soc. Am. A.* 2:886–890.

Nanophotonics for Device Operation and Fabrication

Tadashi Kawazoe
The University of Tokyo

Motoichi Ohtsu
The University of Tokyo

35.1 Introduction

The optical near field is an electromagnetic field that mediates the interaction between nanometric particles located in close proximity to each other. Nanophotonics utilizes this field to realize novel devices, fabrications, and systems. That is, a photonic device with a novel function can be operated by transferring the optical near-field energy between nanometric particles and subsequent dissipation. In such a device, the optical near field transfers a signal and carries the information. Novel photonic systems become possible by using these novel photonic devices. Furthermore, if the magnitude of the transferred optical near-field energy is sufficiently large, structures or conformations of nanometric particles can be modified, which suggests the feasibility of novel photonic fabrications.

Note that the true nature of nanophotonics is to realize "qualitative innovation" in photonic devices, fabrications, and systems by utilizing novel functions and phenomena caused by optical near-field interactions, which are impossible as long as conventional propagating light is used. On reading this note, one may understand that the advantage of going beyond the diffraction limit, that is, "quantitative innovation," is no longer essential, but only a secondary nature of nanophotonics. In this sense, one should also note that optical near-field microscopy, that is, the methodology used for image acquisition and interpretation in a nondestructive manner, is not an appropriate application of nanophotonics because the magnitude of the optical near-field energy transferred between the probe and sample must be extrapolated to zero to avoid destroying the sample.

Quantitative innovation has already been realized by breaking the diffraction limit. Examples include the optical–magnetic hybrid disk storage systems, nanophotonic devices and systems, and photochemical vapor deposition and photolithography. However, it is important to note that these examples also realize qualitative innovation. Details of these examples are described in this chapter.

35.2 Excitation Energy Transfer in Nanophotonic Devices

Kagan et al. observed the energy transfer among CdSe quantum dots (QDs) coupled via a dipole–dipole inter-dot interaction [1]. Crooker et al. also studied the dynamics of the exciton energy transfer in close-packed assemblies of monodisperse and mixed-size CdSe nanocrystal QDs and reported the energy-dependent transfer rate of excitons from smaller to larger dots [2]. These examples are based on the optical near-field interaction. The physical model for the unidirectional resonant energy transfer between QDs via the optical near-field interaction has been presented, and the optically forbidden energy transfer among randomly dispersed CuCl QDs has been demonstrated experimentally using optical near-field spectroscopy [3]. The theoretical analysis and temporal evolution of the energy transfer via the optical near-field interaction were discussed in Ref. [4]. This section reviews the principles of nanophotonic devices and experimental works involving the direct observation of energy transfer from the exciton state in a CuCl QD to the optically forbidden exciton state in another CuCl QD using optical near-field spectroscopy.

Cubic CuCl QDs embedded in an NaCl matrix have the potential to be an optical near-field coupling system that exhibits the optically forbidden energy transfer. This is made possible because for this system, other forms of energy transfer, such as carrier tunneling and Coulomb coupling, can be neglected as the carrier wave function is localized in the QDs; this occurs because the potential depth exceeds 4 eV and the Coulomb interaction is weak due to the small exciton Bohr radius of 0.68 nm in CuCl [3]. The energy transfer via a propagating light is also negligible, since the optically forbidden transition in nearly perfect cubic CuCl QDs is used; that is, the transition to the confined exciton energy levels has an even principal quantum number [5].

So far, this type of energy transfer has not been observed directly because such a nanometric system is usually extremely complex. However, CuCl QDs in a NaCl matrix is a very simple system. The translational motion of the exciton center of mass is quantized due to the small exciton Bohr radius for CuCl QDs, and CuCl QDs become cubic in a NaCl matrix [6–8]. The potential barrier of CuCl QDs in a NaCl crystal can be regarded as infinitely high, and the energy eigenvalues E_{n_x,n_y,n_z} for the quantized Z_3 exciton energy level (n_x, n_y, n_z) in a CuCl QD with side length L depends on the values of quantum numbers n_x, n_y, n_z, and the effective side length $d = (L - a_B)$ found after considering the dead layer correction [6], where a_B is an exciton Bohr radius. The exciton energy levels with even quantum numbers are dipole-forbidden states, which are optically forbidden [5]. However, the optical near-field interaction is finite for such coupling to the forbidden energy state [9].

Figure 35.1 shows schematic drawings of different-sized cubic CuCl QDs (A and B) and the confined exciton Z_3 energy levels. Here, d and $\sqrt{2}d$ are the effective side lengths of cubic QDs A and B, respectively. The quantized exciton energy levels of (1,1,1) in QD A and (2,1,1) in QD B resonate with each other. Under this

resonant condition, the coupling energy of the optical near-field interaction is given by the following Yukawa function [9,10]:

$$V(r) = \frac{A}{r} \exp\left(-\frac{\pi\sqrt{3m_p}}{a\sqrt{m_e}} r \right),$$

where

 A is a proportional constant
 r is the separation between the two QDs
 a is the size of the QD
 m_p is the effective mass of an exciton–polariton
 m_e is the effective mass of an electron

Assuming that the two CuCl QDs in the NaCl matrix have side lengths 5 and 7 nm (a size ratio of $1 : \sqrt{2}$) and the inter-dot distance is 6.1 nm, then the coupling energy $V(r)$ is 5.05 μeV. This corresponds to an energy transfer time of 130 ps due to the optical near-field coupling, which is much shorter than the exciton lifetime of a few nanoseconds. In addition, the inter-sublevel transition τ_{sub}, from higher exciton energy levels to the lowest, as shown in Figure 35.1, is generally less than a few picoseconds [11] and is much shorter than the transfer time τ_{et}. Therefore, most of the energy of the excitation in a cubic CuCl QD with a side length of d is transferred to the lowest exciton energy level in a neighboring QD with a side length of $\sqrt{2}d$ and recombines radiatively in the lowest level.

The CuCl QDs embedded in NaCl matrix used experimentally were fabricated using the Bridgman method and successive annealing, and the average size of the QDs was found to be 4.3 nm. The sample was cleaved just before the near-field optical spectroscopy experiment to keep the sample surface clean. The cleaved surface of a 100 μm-thick sample was sufficiently flat for the experiment; that is, its roughness was less than 50 nm, at least within a few microns squared. A 325 nm He–Cd laser was used as the light source. A double-tapered fiber probe was fabricated using chemical etching and a 150 nm gold coating was applied [12]. A 50 nm aperture was fabricated using the pounding method [13].

The spatial distributions of the luminescence intensity, that is, near-field optical microscope images, clearly show anti-correlation features in their intensity distributions. This anti-correlation feature can be clarified by noting that these spatial distributions in luminescence intensity represent not only the spatial distributions of the QDs, but also some kind of resonant interaction between the QDs. This interaction induces energy transfer from QDs A ($L = 4.6$ nm) to QDs B ($L = 6.3$ nm) because most of the 4.6 nm QDs located close to 6.3 nm QDs cannot emit light, but instead transfer the energy to the 6.3 nm QDs. As a result, in the region containing embedded 6.3 nm QDs, the luminescence intensity from 4.6 nm QDs is low, while the intensity from the 6.3 nm QDs is high at the corresponding position. This anti-correlation feature originates from the near-field energy transfer, which appears for every pair of QDs with

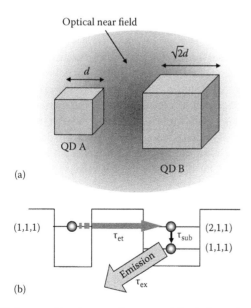

FIGURE 35.1 (a) Schematic drawings of closely located cubic CuCl QDs A and B with effective side lengths ($L - a_B$) of d and $\sqrt{2}d$, respectively, where L and a_B are the side lengths of the cubic quantum dots and the exciton Bohr radius, respectively. (b) Their exciton energy levels. n_x, n_y, and n_z represent quantum numbers of an exciton. E_B is the exciton energy level in a bulk crystal.

different sizes to satisfy the resonant conditions of the confinement exciton energy levels. This is the first spatially resolved observation of energy transfer between QDs via an optical near field. This evidence of the near-field energy transfer between QDs can give rise to a variety of applications, as shown in the following sections.

35.3 Device Operation

Optical fiber transmission systems require increased integration of photonic devices for higher data transmission rates. Since conventional photonic devices, for example, diode lasers and optical waveguides, have to confine light waves within their cavities and core layers, respectively, their minimum sizes are limited by the diffraction of light [14]. Therefore, they cannot meet the size requirement, which is beyond this diffraction limit. An optical near field is free from the diffraction of light and enables the operation and integration of nanometric optical devices. That is, by using a localized optical near field as the carrier, which is transmitted from one nanometric element to another, the above requirements can be met. Based on this idea, nanometer-sized photonic devices have been proposed, which are called *nanophotonic devices* [15]. A nanometric all-optical AND gate (i.e., a nanophotonic switch) is one of the most important devices for realizing nanophotonic integrated circuits, and the operation of a nanophotonic AND gate has been already demonstrated using a coupled QD system.

A logic gate, for example, an AND gate and a NOT gate, is a block in a digital system. Logic gates have some inputs and some outputs, and every terminal is under one of two binary conditions, low (0) or high (1), given by different optical intensities for the optical device. The logic state of the input terminal is controlled by the optical input signal, and the logic state of the output terminal changes depending on the logic state of the input terminals. An intensity of approximately zero and a much higher intensity are preferable in the low and high logic states, respectively, such that the ratio of high to low intensity exceeds 30 db. For nanophotonic devices, the high state (1), which is called "true," is defined simply as the higher intensity state and the low state (0), which is called "false," is defined as the lower intensity state.

Section 35.3.1 presents the operation of nanophotonic AND gates using three CuCl QDs [16,17], while a nanophotonic NOT gate using CuCl QDs is outlined in Section 35.3.2. The optically forbidden energy transfer between neighboring nanostructures via the optical near-field interaction, which was reviewed in Section 35.2, is a key phenomenon for these operations.

35.3.1 Nanophotonic AND Gate

Operation of a nanophotonic AND gate using cubic CuCl QDs embedded in a NaCl matrix has been demonstrated [16,17]. When closely spaced QDs with quantized energy levels resonate with each other, near-field energy is transferred between them, even if the transfer is optically forbidden, as noted in

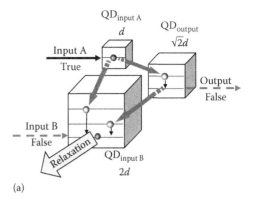

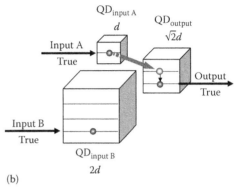

FIGURE 35.2 Principle of AND-gate operation. (a) and (b) The "false" and "true" states of the nanophotonic AND gate, respectively.

Section 35.2. The output is "true" when both inputs are "true"; otherwise, the output is "false." Figure 35.2a and b explains the "false" and "true" states of the proposed nanophotonic AND gate. Three cubic QDs, $QD_{input\ A}$, $QD_{input\ B}$, and QD_{output}, are used as the two inputs and output ports of the AND gate, respectively. Assuming an effective size ratio of $1 : \sqrt{2} : 2$, the quantized energy levels (1,1,1) in $QD_{input\ A}$, (2,1,1) in QD_{output}, and (2,2,2) in $QD_{input\ B}$ resonate with each other. Furthermore, energy levels (1,1,1) in QD_{output} and (2,1,1) in $QD_{input\ B}$ also resonate. In the "false" state operation (Figure 35.2a), for example, input A is "true" and input B is "false," almost all of the exciton energy in $QD_{input\ A}$ is transferred to the (1,1,1) level in the neighboring QD_{output}, and then to the (1,1,1) level in $QD_{input\ B}$. Therefore, the input energy escapes to $QD_{input\ B}$, and consequently no optical output signals are generated from QD_{output}. This means that the output is "false." In the "true" state (Figure 35.2a) when inputs A and B are both "true," the escape route to $QD_{input\ B}$ is blocked by the excitation of QD_{inputB} due to state filling in $QD_{input\ B}$ on applying the input B signal. Therefore, the input energy is transferred to QD_{output} and an optical output signal is generated. This means that the output is "true." These operating principles are realized with the condition $\tau_{ex} > \tau_{et} > \tau_{sub}$, where τ_{ex}, τ_{et}, and τ_{sub} are the exciton lifetime, energy transfer time between QDs, and inter-sublevel transition time, respectively. Since τ_{ex}, τ_{et}, and τ_{sub} are a few nanoseconds, 100 ps, and a few picoseconds, respectively, for the CuCl QDs used in a NaCl matrix, the condition of operation described in Section 35.2 is satisfied.

In an experiment using CuCl QDs embedded in a NaCl matrix, a double-tapered UV fiber probe was fabricated using chemical etching and coated with 150 nm-thick aluminum (Al) film. An aperture less than 50 nm in diameter was formed by the pounding method [13]. To confirm the AND-gate operation, the fiber probe was used to search for a trio of QDs that had an effective size ratio of $1 : \sqrt{2} : 2$. Since the homogeneous linewidth of a CuCl QD increases with the sample temperature [18,19], the allowance in the resonatable size ratio is 10% at 15 K. The separation of the QDs must be less than 30 nm for operation of the proposed AND gate because the energy transfer time increases with the separation; moreover, it must be shorter than the exciton lifetime. It is estimated that at least one trio of QDs that satisfies these conditions exists in a $2 \times 2\,\mu\text{m}$ scan area. To demonstrate AND-gate operation, a QD trio had to be found in the sample, as shown in Figure 35.2.

Near-field photoluminescence (PL) pump-probe spectroscopy was used to find the QD trio. Figure 35.3 shows the PL spectrum obtained at the position where the QD trio exists. In this figure, peaks appear at the positions of the (1,1,1) levels in the 4.6 and 3.5 nm QDs. The appearance of the satellite peaks means that the AND-gate system proposed in Figure 35.2 was present in the area under the probe. In other words, a trio of cubic QDs with sizes of 3.5, 4.6, and 6.3 nm existed. Since their effective respective sizes $L - a_B$ were 2.8, 3.9, and 5.6 nm (a_B: an exciton Bohr radius of 0.7 nm in CuCl), the size ratio was close to $1 : \sqrt{2} : 2$ and they could be used as QD_{inputA}, QD_{output}, and QD_{inputB}, respectively. The pumping to the 6.3 nm QD blocks the energy transfer from the 3.5 and 4.6 nm QDs to the 6.3 nm QD due to state filling of the 6.3 nm QD, and the 3.5 and 4.6 nm QDs emit light that results in the peaks in Figure 35.3. Therefore, a QD trio for a nanophotonic AND gate was found. The PL peak from the 4.6 nm QD corresponds to the output signal in Figure 35.2b.

The PL intensity from the 4.6 nm QD was 0.05–0.02 times the PL intensity from the 6.3 nm QD, which was obtained with the probe laser only. This value is quite reasonable considering the pumping pulse energy density of $10\,\mu\text{W/cm}^2$ because the probability density of excitons in a 6.3 nm QD is 0.1–0.05 [18], which is close to the PL intensity from the 4.6 nm QD. This result indicates that the internal quantum efficiency of the AND-gate system is close to unity.

In the experiment examining the AND-gate operation, the second harmonics of Ti:sapphire lasers (wavelengths 379.5 and 385 nm), which were tuned to the (1,1,1) exciton energy levels of QD_{inputA} and QD_{inputB}, respectively, were used as the signal light sources for inputs A and B. The output signal was collected by the fiber probe, and its intensity was measured using a cooled microchannel plate after passing through three interference filters of 1 nm bandwidth tuned to the (1,1,1) exciton energy level in QD_{output} at 383 nm. Figure 35.4a and b shows the spatial distribution of the output signal intensity in the "false" state (i.e., with one input signal only) and in the "true" state (i.e., with both input A and input B signals) using near-field spectroscopy at 15 K. The insets in this figure are schematic drawings of the existing QD trio used for the AND gate, which was confirmed by the near-field PL spectra. Here, separation of the QDs by less than 20 nm was estimated theoretically from time-resolved PL measurements, as explained in the next paragraph (see the illustrations in Figure 35.4). In the "false" state, no output signal was observed because the energy of the input signal was transferred to QD_{inputB} and swept out as PL at 385 nm. To quench the output signal in the "false" state, which was generated by accumulating excitons in QD_{inputB}, the input signal density to QD_{inputA} was regulated to less than 0.1 excitons in QD_{inputB}. In the "true" state, a clear output signal was obtained in the dashed circle of Figure 35.4b. The output signal was proportional to the intensity of the control signal, which had a density of 0.01–0.1 excitons in QD_{inputB}.

Next, the dynamic properties of the nanophotonic AND gate were evaluated using the time-correlation single-photon counting method. As a pulse-input B signal source, the 385 nm second harmonic of a mode-locked Ti:sapphire laser was used. The repetition rate of the laser was 80 MHz. To avoid cross talk originating from spectral broadening of the pulse duration between input signals

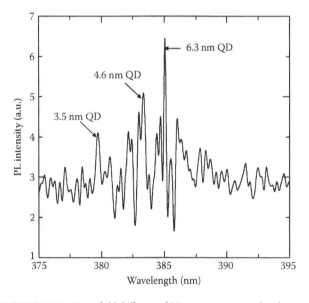

FIGURE 35.3 Near-field differential PL spectra measured at the position of a QD trio acting as a nanophotonic AND gate.

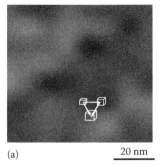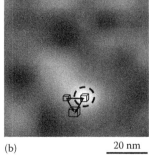

FIGURE 35.4 Spatial distribution of the output signal from the nanophotonic AND gate in the "false" (a) and "true" (b) states measured using near-field microscopy.

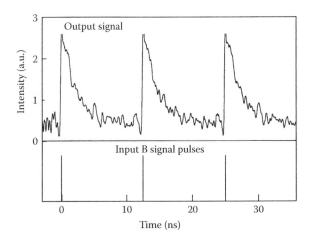

FIGURE 35.5　Temporal evolution of the output signal from the nanophotonic AND gate located in the dashed circle in Figure 35.4b. The duration and repetition rate of the control pulse were 10 ps and 80 MHz, respectively.

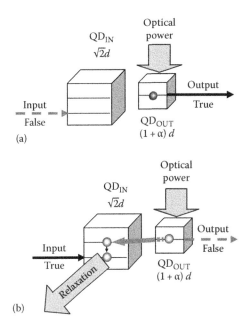

FIGURE 35.6　A nanophotonic NOT gate. (a) and (b) Schematic explanation of the "true" and "false" states using cubic QDs.

A and B, the pulse duration of the mode-locked laser was set to 10 ps. The time resolution of the experiment was 15 ps. Figure 35.5 shows the temporal evolution of the input B pulse signal (lower part) applicable to QD_{inputB} and the output signal (upper part) from QD_{output}. The output signal increases synchronously with the input B pulse within less than 100 ps, which agrees with the theoretically expected result based on the Yukawa model [9]. As this signal rise time is determined by the energy transfer time between the QDs, the separation between the QDs can be estimated from the rise time as being less than 20 nm; the rise time can be shortened to a few picoseconds by decreasing the separation of the QDs. Since the decay time of the output signal is limited by the exciton lifetime, this nanophotonic AND gate can be operated at a few hundred megahertz, and the operating frequency can be increased to several gigahertz by exciton quenching using plasmon coupling [20]. The output signal ratio between "true" and "false" was about 10, which is sufficient for use as an all-optical AND gate, and can be increased using a saturable absorber and electric field enhancement of the surface plasmon [21].

The advantages of this nanophotonic device are its small size and high-density integration capability based on the locality of the optical near field. The figure of merit (FOM) of an optical AND gate should be more important than the switching speed. Here, the FOM is defined as $F = C/Vt_{sw}P_{sw}$, where C, V, t_{sw}, and P_{sw} are the "true"–"false" (ON–OFF) ratio, volume of the device, switching time, and switching energy, respectively. The FOM of the nanophotonic AND gate is 10–100 times higher than that of conventional photonic gates because its volume and switching energy are 10^{-5} times and 10^{-3} times those of conventional photonic gates, respectively.

35.3.2 Nanophotonic NOT Gate

A nanophotonic NOT gate is a key device for realizing a functionally complete set of logic gates for nanophotonic systems, and its operation is demonstrated in this section using CuCl QDs [22]. Figure 35.6 shows a schematic explanation of the nanophotonic NOT gate. QD_{IN} and QD_{OUT} correspond to the input and output terminals of the NOT gate, respectively. Assuming a pair of QDs with a size ratio of $1+\alpha : \sqrt{2}(\alpha \ll 1)$, the quantized energy levels with the set of quantum numbers (2,1,1) in QD_{IN} and (1,1,1) in QD_{OUT} are slightly nonresonant with each other. The energy from the optical power supply generates an exciton in QD_{OUT}. Without the input signal (i.e., the input is "false"), the exciton in QD_{OUT} disappears and emits a photon, which is observed as an output signal, as shown in Figure 35.6a. That is, input = "false" and output = "true." Conversely, by applying the input signal (i.e., the input is "true"), the energy level (2,1,1) in QD_{IN} becomes resonant to (1,1,1) in QD_{OUT} due to broadening of its line width. This broadening was confirmed experimentally for CuCl QDs; anti-holes appear on both sides of the spectral hole in the absorption spectrum of CuCl QDs observed in a far-field hole-burning experiment at 5 K. This experimental result shows that the excitation broadens the homogeneous line width of QDs, which might arise from shortening of the phase relaxation time of the excitons in QDs due to carrier–carrier scattering. Consequently, the exciton energy in QD_{OUT} is transferred to QD_{IN} via an optical near-field interaction [3], which suppresses output signal generation (Figure 35.6b, i.e., input = "true" and output = "false"). As a result, the temporal behavior of the output signal is the inverse of that of the input signal. By selecting a suitable threshold to distinguish "true" and "false," these behaviors are used for a NOT gate.

CuCl QDs embedded in a NaCl matrix were used to verify operation of the NOT gate, as CuCl QDs offer discrete energy levels similar to the exciton described in Figure 35.6 [6]. The mean size of the QDs was 4.1 nm and the mean distance between the QDs was 25 nm. In the experiment, the second harmonics of a continuous-wave (CW) Ti:sapphire laser ($\hbar\omega = 3.2704\,eV$) and a mode-locked Ti:sapphire laser ($\hbar\omega = 3.2195\,eV$) were used as the optical power supply and input signal pulse, respectively. The respective power densities were 1 and $2\,W/cm^2$ at the sample surface. Under the excitation condition, fewer than 0.1 excitons occurred in a QD. These lasers excited the sample from its back and the output signal was observed using a near-field spectrometer in collection mode. The sample temperature was controlled at 15 K. To find the QD pair acting as a nanophotonic NOT gate, the QD positions were mapped by measuring the luminescence distribution on the sample, which was obtained using a near-field spectrometer using He–Cd laser excitation ($\hbar\omega = 3.81\,eV$). The signal was observed after several QD pairs were selected as candidates for nanophotonic NOT-gate operation. Finally, a nanophotonic NOT gate was found, with the probability of about one device per $1\,\mu m^2$ of scanning area. Figure 35.7a shows the spatial distribution of the optical near-field output-signal intensity when an input signal is "false" (i.e., with the optical power supply only). Figure 35.7b shows the distribution when the input signal is "true." The insets in this figure are schematic drawings of an existing QD pair that function as a NOT gate, which was confirmed from the PL spectra. The sizes of the two QDs estimated from the wavelengths of their luminescence were 5.0 and 6.3 nm, which satisfy the NOT-gate operation condition, as shown in Figure 35.6. Note that the photon energy of the optical power supply ($\hbar\omega = 3.2704\,eV$) is maintained nonresonant to the (1,1,1) exciton level in the 5.0 nm QD ($\hbar\omega = 3.2304\,eV$) to decrease the artifact originating from the laser by using narrowband optical filters and observing a clear output signal. Sufficiently low optical power was supplied to the 5.0 nm QD from neighboring QDs [3,23]. The NOT-gated signal appears at the center of the dashed circle in Figure 35.7a, from which the size of the device was estimated to be 20 nm. Comparison of Figure 35.7a and b clearly demonstrates the operation of a NOT gate.

The dynamic behavior of the NOT gate was observed using the time-correlation single-photon counting method. Figure 35.8 shows the temporal evolution of the output signal. The horizontal dashed line indicates the output signal level without the

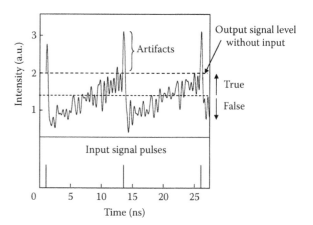

FIGURE 35.8 The temporal evolution of the output signal from the nanophotonic NOT gate circled in Figure 35.7.

input signal pulses (i.e., the input is "false"). Without the input pulses, the signal level is constant, since a CW laser was used as the optical power supply. This signal level is defined as "true." With the input pulses, the output signal increases within a time period shorter than the time resolution of 20 ps due to the artifact of the input pulses, and it decreases to a level lower than the initial level. Here, the signal level lower than the dotted line is defined as "false." Then, the output signal level becomes "false" 50 ps after the "true" input pulse. The fall time of the output signal to the minimum level is about 100 ps, which corresponds to the energy transfer time between QDs. The "false" output signal level recovers to the "true" level within 10 ns. The recovery time is longer than the exciton lifetime in 6.3 nm CuCl QDs ($\tau_{ex} \sim 1\,ns$) because the energy transfer from the optical power supply affects the recovery. That is, the recovery time depends on the competition between exciton annihilation via recombination and exciton creation via energy transfer.

35.3.3 Interconnection with Photonic Devices

An interconnection device needs to be developed to collect the incident propagating light and drive the nanophotonic device for efficient operation of the system [15,24,25]. Conventional far-field optical devices, such as convex lenses and concave mirrors, cannot be used for this interconnection because of their diffraction-limited operation. This section demonstrates a novel optical device, the *optical nanofountain*, which concentrates optical energy in a nanometric region using optical near-field energy transfer among QDs. This nanometric optical device enables highly efficient interconnection to nanophotonic devices. The optical nanofountain is operated using the energy transfer between QDs via an optical near-field interaction, as shown in Figure 35.1 [9–11]. The unidirectional energy transfer from smaller to larger QDs concentrates the optical energy in a nanometric region. When different sized QDs with resonant energy sublevels are distributed, as shown in Figure 35.9a, energy transfer occurs via the optical near field, as illustrated by the arrows. Light incident to the QD array is ultimately concentrated in

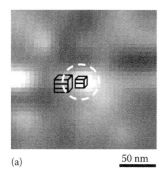
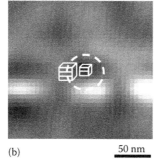

(a)　　　50 nm　　(b)　　　50 nm

FIGURE 35.7 The spatial distribution of the output signal from a nanophotonic NOT gate measured using near-field microscopy at Input = "false" (a) and Input = "true" (b).

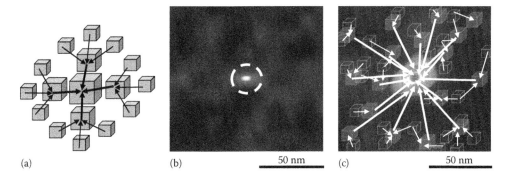

FIGURE 35.9 Optical nanofountain. (a) Schematic explanation of the optical nanofountain and unidirectional energy transfer. (b) Spatial distribution of the PL intensity in an optical nanofountain. The bright spot surrounded by a dashed circle is the focal spot. (c) Spatial distribution of the PL intensity of CuCl QDs for 2.5 nm ≤ L ≤ 10 nm (3.215 eV ≤ E_p ≤ 3.350 eV) for the same area as in (b). The cubes represent the positions estimated from the PL intensity distribution.

the largest QD. The area of optical energy concentration corresponds to the size of the QD.

The name optical nanofountain was proposed because light spurts from the largest QD after it is concentrated by stepwise energy transfer from smaller neighboring QDs, so that this device looks like a fountain in a basin. From the experimental tests of nanophotonic AND-gate operation in Section 35.3.1, the concentration efficiency of this device should be close to unity because no other possible relaxation paths exist in the system. The operation of an optical nanofountain was demonstrated using CuCl cubic QDs embedded in a NaCl matrix. The average QD size was 4.2 nm and the average separation was less than 20 nm. Although the QDs have an inhomogeneous size distribution and are arranged in the matrix randomly, the operation can be confirmed if an appropriate QD group is found using a nanometric-resolution near-field spectrometer. For the operation, the optimum sample temperature T was 40 K. At $T < 40$ K, the resonant condition becomes tight due to narrowing of the homogeneous line width of the quantized energy sublevels, while at $T > 40$ K, the unidirectional energy transfer is obstructed by the thermal activation of excitons in the QDs. A 325 nm He–Cd laser was used as the excitation light source. A double-tapered UV fiber probe with an aperture 20 nm in diameter was fabricated using chemical etching and coated with a 150 nm-thick Al film to ensure sufficiently high detection sensitivity [12].

Figure 35.9b shows the typical spatial distribution of the PL from QDs operating as an optical nanofountain. The bright spot inside the dashed circle corresponds to a spurt from an optical nanofountain, i.e., the focal spot of the nanometric optical condensing device. The diameter of the focal spot was less than 20 nm, which was limited by the spatial resolution of the near-field spectrometer. From the Rayleigh criterion (i.e., resolution = 0.61 · λ/NA) [26], its numerical aperture (NA) was estimated to be 12 for λ = 385 nm. To demonstrate the detailed operating mechanism of the optical nanofountain, we show the size-selective PL intensity distribution; that is, the photon energy is shown in Figure 35.9c. Here, the collected PL photon energy, E_p, was 3.215 eV ≤ E_p ≤ 3.350 eV, which corresponds to the PL from QDs of size 2.5 nm ≤ L ≤ 10 nm. The area scanned by the probe are

equivalent to those in Figure 35.9b. The PL intensity distribution is shown using a gray scale.

This device can also be used as a frequency selector based on the resonant frequency of the QDs, which can be applied, for example, to frequency domain measurements, multiple optical memories, multiple optical signal processing, and frequency division multiplexing.

Practical nanophotonic devices for room-temperature operation are under development using III–V compound semiconductor QDs [27] and ZnO nanorods [28].

35.4 Nanophotonics Fabrication

This section presents the nonadiabatic processes involved in optical chemical vapor deposition (CVD) and photolithography. These methods have realized qualitative innovation in nanofabrications by utilizing the spatially localized nature of optical near fields.

35.4.1 Nonadiabatic Near-Field Optical CVD

Conventional optical CVD utilizes a two-step process: photodissociation and adsorption. For photodissociation, a propagating light must resonate the reacting molecular gases to excite molecules from the ground state to an excited electronic state. The Franck–Condon principle holds that this resonance is essential for excitation. The excited molecules then relax to the dissociation channel, and the dissociated atoms adsorb to the substrate surface. However, a nonadiabatic photodissociation process is observed in near-field optical chemical vapor deposition (NFO-CVD) under the nonresonant condition of the electronic transition, which violates the Franck–Condon principle. This section discusses the nonadiabatic NFO-CVD of nanometric Zn dots and presents experimental results based on the exciton–phonon–polariton (EPP) model.

Figure 35.10 shows the cross-sectional profiles of the shear-force topographical images after NFO-CVD for photon energies of 5.08 eV (λ = 244 nm; broken curve) and 2.54 eV (λ = 488 nm; solid curve) of Zn dots deposited on a sapphire substrate in

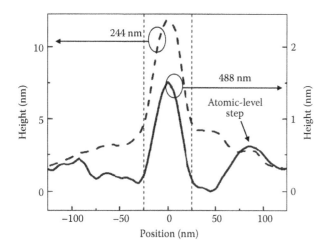

FIGURE 35.10 Cross-sectional profiles of the shear-force topographical images of the deposited Zn patterns.

atomic-level steps [29]. In the experiment, diethylzinc (DEZn) was used as the CVD gas source. For the broken curve, the laser power was 1.6 μW and the irradiation time was 60 s. Before carrying out NFO-CVD, 0.4 nm-high atomic-level step structures were clearly observed on the sapphire substrate. After NFO-CVD, they disappeared and a deposited Zn dot less than 50 nm in diameter was seen at the center of the image. This occurred because the optical near field deposited the Zn dot directly under the apex of the fiber probe. Furthermore, since high-intensity propagating light leaks from a bare fiber probe, that is, one without a metallic coating, and is absorbed by the DEZn, a Zn layer was deposited on top of the atomic-step structures.

For the solid curve, the laser power was 150 μW and the irradiation time was 75 s. The photon energy (2.54 eV) was higher than the dissociation energy of DEZn, but it was still lower than the absorption edge of DEZn [30]. Therefore, it was not absorbed by DEZn. However, a Zn dot less than 50 nm in diameter appears at the center. While using conventional CVD with propagating light, a Zn film cannot be grown using a light source with a photon energy lower than the absorption edge (4.13 eV: λ = 300 nm) [31]. Deposited Zn dots were observed on the substrate just below the apex of the fiber probe using NFO-CVD. The atomic-level steps in this figure are still observed, despite the leakage of the propagating light from the bare fiber probe. These curves confirm that Zn dots with a full-width at half-maximum (FWHM) of 30 nm were deposited in the region where the optical near field is dominant.

The dashed curve has 4 nm-high tails on both sides of the dot, which represent the deposition caused by the leaked propagating light. This deposition process is based on the conventional adiabatic photochemical process. In contrast, the solid curve has no tails; therefore, it is clear that the leaked 488 nm propagating light did not deposit a Zn layer. Note that a Zn dot 30 nm in diameter without tails was deposited under nonabsorbed conditions (λ = 488 nm).

Figure 35.11 shows shear-force topographical images of the sapphire substrate after NFO-CVD using an optical near

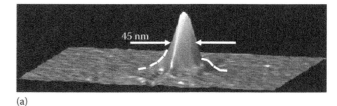

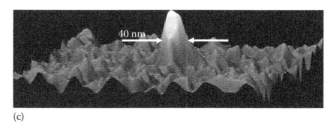

FIGURE 35.11 Shear-force topographical images after NFO-CVD at wavelengths of λ = 325 nm (a), 488 nm (b), and 684 nm (c). The laser output power and irradiation time for deposition were 2.3 μW and 60 s (a), 360 μW and 180 s (b), and 1 mW and 180 s (c), respectively.

field with photon energies of 3.81 eV (λ = 325 nm) (a), 2.54 eV (λ = 488 nm) (b), and 1.81 eV (λ = 684 nm) (c). The respective laser power and irradiation time were (a) 2.3 μW and 60 s, (b) 360 μW and 180 s, and (c) 1 mW and 180 s. The high quality of the deposited Zn was confirmed by x-ray photoelectron spectroscopy. Furthermore, PL was observed from ZnO dots, which were fabricated by oxidizing the Zn dots deposited by NFO-CVD [8]. In Figure 35.11a, the photon energy ($\hbar\omega$) exceeds the dissociation energy (E_d) of DEZn, and is close to the absorption band edge (E_{abs}) of DEZn, i.e., $\hbar\omega > E_d$ and $\hbar\omega \cong E_{abs}$ [30]. The diameter (FWHM) and height of the topographical image were 45 and 26 nm, respectively. The small tail (shown by the dotted curve) represents a Zn layer less than 2 nm thick, and was deposited by the propagating light leaking from the bare fiber probe. This deposition is possible because the DEZn absorbs some of the propagating light at $\hbar\omega = 3.18$ eV. The very high peak suggests that the optical near field enhances the photodissociation rate at this photon energy because its intensity increases rapidly at the apex of the fiber probe. In Figure 35.11b, the photon energy still exceeds the dissociation energy of DEZn, but is lower than the absorption band edge of DEZn, i.e., $E_{abs} > \hbar\omega > E_d$ [30]. The diameter and height of the image were 50 and 24 nm, respectively. This image has no tail because Zn was not deposited by the high-intensity propagating light leaking from the bare fiber probe. This confirmed that the photodissociation of DEZn and Zn deposition occurred only with an optical near field of $\hbar\omega = 2.54$ eV. Figure 35.11c represents the cases $\hbar\omega < E_d$ and $\hbar\omega < E_{abs}$. Zn dots were deposited successfully at these low photon energies.

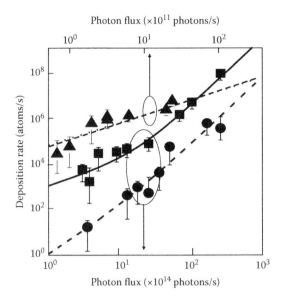

Photon flux ($\times 10^{11}$ photons/s)

Photon flux ($\times 10^{14}$ photons/s)

FIGURE 35.12 The relationship between the photon flux and the rate of Zn deposition. The dotted, solid, and dashed curves represent the calculated values fitted to the experimental results.

The topographical image showed dots with a diameter of 40 nm and a height of 2.5 nm. The experimental results in Figure 35.11 demonstrate dissociation based on a nonadiabatic photochemical process that violates the Franck–Condon principle.

To discuss this novel dissociation process quantitatively, Figure 35.12 shows the relationship between the photon flux (I) and the deposition rate of Zn (R). For $\hbar\omega = 3.81$ eV (▲), R is proportional to I. For $\hbar\omega = 2.54$ eV (■) and 1.81 eV (●), higher order dependencies appear and R is fitted by the third-order function $R = a_{\hbar\omega}I + b_{\hbar\omega}I^2 + c_{\hbar\omega}I^3$. The respective values of $a_{\hbar\omega}$, $b_{\hbar\omega}$, and $c_{\hbar\omega}$ are $a_{3.81} = 5.0 \times 10^{-6}$, $b_{3.81} = 0$, and $c_{3.81} = 0$ for $\hbar\omega = 3.81$ eV, $a_{2.54} = 4.1 \times 10^{-12}$, $b_{2.54} = 2.1 \times 10^{-27}$, and $c_{2.54} = 1.5 \times 10^{-42}$ for $\hbar\omega = 2.54$ eV, and $a_{1.81} = 0$, $b_{1.81} = 4.2 \times 10^{-29}$, and $c_{1.81} = 3.0 \times 10^{-44}$ for $\hbar\omega = 1.81$ eV. The results of fitting are shown with the solid, dashed, and dotted curves in Figure 35.12. Since no conventional photochemical processes, for example, the Raman process and two-photon absorption, can explain these experimental results, the discussion below uses a unique theoretical model based on the discussion in Ref. [32].

Figure 35.13 shows the potential curves of an electron in a DEZn molecular orbital drawn as a function of the internuclear distance of the C–Zn bond, which is involved in photodissociation [30]. The relevant energy levels of the molecular vibration mode are indicated by the horizontal broken lines in each potential curve. When a propagating light is used, photo-absorption (indicated by the white arrow) triggers the dissociation of DEZn [33]. With an optical near field nonresonant to the electronic state, there are three possible origins of photodissociation [34]: (1) the multiple-photon absorption process; (2) a multiple-step transition process via the intermediate energy level induced by the fiber probe; and (3) multiple-step transition via an excited state of the molecular vibration mode. Case (1) is negligible because the optical power density was less than 10 kW/cm².

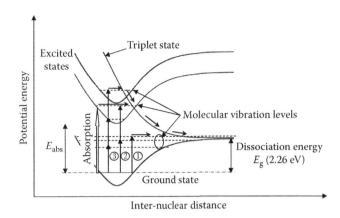

FIGURE 35.13 Potential curves of an electron in DEZn molecular orbitals. The relevant energy levels of the molecular vibration modes are indicated by the horizontal broken lines.

Case (2) is also negligible because the DEZn was dissociated by ultraviolet–near-infrared light, although DEZn does not have relevant energy levels over such a broad wavelength region. As a result, the experimental results strongly support Case (3). That is, the photodissociation is caused by a transition to an excited state via a molecular vibration mode, which involves three multiple-step excitation processes, as shown in Figure 35.13. Since the system is strongly coupled with the vibration state, it must be considered a nonadiabatic system.

For this consideration, an exciton–phonon polariton (EPP) model was presented in Ref. [32]. The EPP model holds that the optical near fields excite the molecular vibration mode due to the steep spatial gradient. Since the optical near-field energy distribution is spatially inhomogeneous in a molecule due to its gradient, the electrons respond inhomogeneously. As a result, the molecular vibration modes are excited because the molecular orbital changes and the molecule is polarized as a result of the inhomogeneous response of the electrons. The EPP model describes this excitation process quantitatively. The EPP is a quasiparticle, which is an exciton–polariton carrying the phonon (lattice vibration) generated by the steep spatial gradient of the optical field energy distribution. In contrast, since the propagating light energy distribution is homogeneous in a molecule, only the electrons in the molecule respond to the electric field of the propagating light. Therefore, the propagating light cannot excite the molecular vibration.

Zinc-bis(acetylacetonate) (Zn(acac)₂) has never been used for conventional optical CVD due to its low optical activity. With NFO-CVD, however, the optical near field can activate the molecule nonadiabatically and the dissociated Zn atom is adsorbed under the fiber probe. Figure 35.14a shows a shear-force topographical image of Zn deposited on a sapphire substrate. The laser power and irradiation time were 1 mW and 15 s, respectively. The Zn dot was 70 nm in diameter and 24 nm high [35,36]. The chemical stability of Zn(acac)₂ keeps the substrate surface clean and helps to fabricate an isolated nanostructure. Figure 35.14b shows the shear-force topographical image of a deposited Zn dot that is among the smallest ever fabricated using

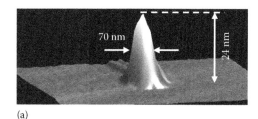

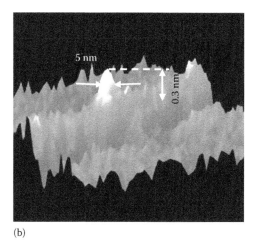

FIGURE 35.14 Shear-force topographical images after NFO-CVD using Zn(acac)$_2$ with a 457 nm-wavelength light source. (a) A deposited Zn dot with a diameter of 70 nm and height of 24 nm. (b) A deposited Zn with a diameter of 5 nm and height of 0.3 nm.

NFO-CVD (5 nm in diameter and 0.3 nm high). The deposition conditions consisted of Zn(acac)$_2$ at a pressure of 70 mTorr in the CVD chamber and a laser wavelength, power, and irradiation time of 457 nm, 65 μW, and 30 s, respectively.

35.4.2 Nonadiabatic Near-Field Photolithography

Section 35.4.1 reviewed a unique nonadiabatic photochemical reaction, which was explained using the EPP model. According to this model, the nonadiabatic photochemical reaction can be considered a universal phenomenon and is applicable to several photochemical processes. This section reviews the application of the nonadiabatic photochemical reaction to photolithography, which can be called *nonadiabatic photolithography* [36,37]. For the mass production of photonic and electronic devices, nonadiabatic photolithography can be used because conventional photolithographic components can be applied to this system.

The wave properties of propagating light cause problems for high-resolution photolithography due to diffraction and the dependence on the coherency and polarization of the light source. To fabricate high-density corrugations, the optical coherent length is too long compared to the separation between adjacent corrugation elements, even when a Hg lamp is used. In addition, the absorption by the photoresist is insufficient to suppress interference of scattered light. Furthermore, since the intensity of the propagating light transmitted through a

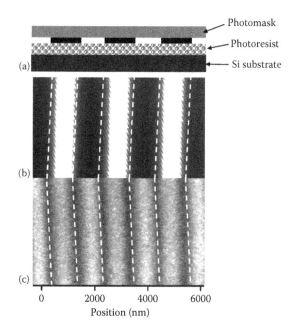

FIGURE 35.15 Experimental results of nonadiabatic photolithography. (a) A schematic of the photomask and Si substrate on which the photoresist (OFPR-800) was spin-coated. (b) Atomic force microscopy images of photoresist OFPR-800 exposed to the g-line of a Hg lamp. (c) AFM images of photoresist OFPR-800 developed after a 4 h exposure with a 672 nm laser.

photomask strongly depends on its polarization, the photomask must be designed while considering these dependences. In contrast, the outstanding advantage of nonadiabatic photolithography is that it is free from these problems.

Figure 35.15a shows a schematic configuration of the photomask and the Si substrate on which the photoresist (OFPR-800: Tokyo-Ohka Kogyo) was spin-coated. They were used in contact mode. Figure 35.15b and c shows atomic force microscopy (AFM) images of the photoresist surface after development. Figure 35.15b shows the result obtained using conventional photolithography. The g-line (436 nm) from a Hg lamp was used as the light source. The fabricated pattern of corrugation was an exact replica of the photomask. Conversely, with nonadiabatic photolithography using a 672 nm-wavelength light source, the grooves on the photoresist appeared along the edges of the Cr mask pattern, as shown in Figure 35.15c. The corrugated pattern was 30 nm deep. The line width was 150 nm, which was narrower than the wavelength of the light source. On the photomask, a steep spatial gradient of optical energy distribution is expected due to optical near fields, while direct irradiation with 672 nm light cannot expose the photoresist. This demonstrated that the photoresist was patterned using a nonadiabatic process.

Figure 35.16 shows AFM images of another photoresist surface (TDMR-AR87 for the 365 nm-wavelength i-line from a Hg lamp: Tokyo-Ohka Kogyo) after development. Figure 35.16a shows the corrugated pattern fabricated using linearly polarized g-line light. Two-dimensional arrays of circles and T-shapes have also been fabricated successfully on this photoresist

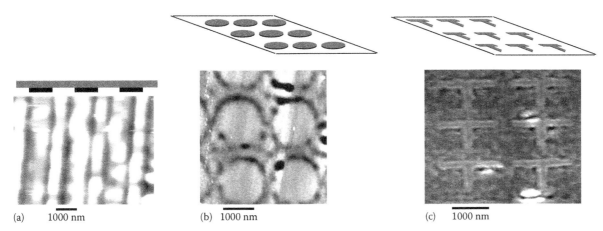

FIGURE 35.16 Experimental results of nonadiabatic photolithography. (a) AFM images of photoresist TDMR-AR87 exposed to the linearly polarized g-line of a Hg lamp for 3 s. (b) AFM images of photoresist TDMR-AR87 exposed to the linearly polarized g-line of a Hg lamp for 10 s using a circle-shaped array photomask. (c) AFM images of photoresist OTDMR-AR87 developed after a 40 s exposure to the g-line of a Hg lamp using a T-shaped array photomask.

(see Figure 35.16b and c). This would be impossible using adiabatic photolithography due to its polarization-dependent nature and interference effects.

An optically inactive electron beam (EB) resist film (ZEP-520: ZEON) can also be patterned nonadiabatically. Figure 35.17 shows an AFM image of the developed EB resist surfaces. The light source was the third harmonic of a Q-switched Nd:YAG laser and the exposure time was 5 min. A two-dimensional array of 1 μm-diameter disks was fabricated successfully, even on the EB resist, which would be impossible using propagating light. The developed pattern had a depth of 70 nm, which is sufficient for the subsequent etching of the substrate. Since the EB resist film has an extremely smooth surface, the homogeneity in the contact with the photomask was improved. This suggests that a smooth organic or inorganic thin film can be used as a photoresist irrespective of its optical inactivity.

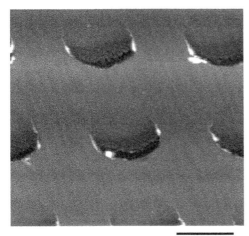

FIGURE 35.17 AFM image of the surface of an electron beam resist exposed for 5 min using a Q-switched laser (355 nm) and a circle-shaped (1 μm diameter) array photomask.

35.5 Summary

After nanophotonics was proposed by M. Ohtsu in 1993 [38], it now exists as a novel field of optical technology in nanometric space. However, the name "nanophotonics" is occasionally used for photonic crystals [39], plasmonics [40], metamaterials [41,42], silicon photonics [43], and QD lasers [44] using conventional propagating lights. For example, plasmonics utilizes the resonant enhancement of the light in a metal by exciting free electrons. The letters "on" in the word "plasmon" represent the quanta, or the quantum mechanical picture of the plasma oscillation of free electrons in a metal. However, plasmonics utilizes the classical wave optical picture using conventional terminology, such as the refractive index, wave number, and guided mode. Even when a metal is irradiated with light that obeys the laws of quantum mechanics, the quantum mechanical property is lost because the light is converted into the plasma oscillation of electrons, which has a short phase relaxation time. Furthermore, the energy transferred via this interaction must be dissipated in the nanometric particles or adjacent macroscopic materials to fix the position and magnitude of the transferred energy. Since plasmonics does not deal with this local dissipation of energy, it is irrelevant for quantitative innovation by breaking the diffraction limit, or for qualitative innovation. Local energy transfer and its subsequent dissipation have become possible only in nanophotonics by using optical near fields [45,46].

Here, we should consider the stern warning by C. Shannon on the casual use of the term "information theory," which was a trend in the study of information theory during the 1950s [47]. The term "nanophotonics" has been used in a similar way, although some work in "nanophotonics" is not based on optical near-field interactions. For the true development of nanophotonics, one needs deep physical insights into the virtual exciton–polariton and the nanometric subsystem composed of electrons and photons.

References

1. Kagan, C. R., Murray, C. B., Nirmal, M., and Bawendi, M. G., 1996. Electronic energy transfer in CdSe quantum dot solids. *Phys. Rev. Lett.* **76**: 1517–1520.

2. Crooker, S. A., Hollingsworth, J. A., Tretiak, S., and Klimov, V. I., 2002. Spectrally resolved dynamics of energy transfer in quantum-dot assemblies: Towards engineered energy flows in artificial materials. *Phys. Rev. Lett.* **89**: 186802-1–186802-4.

3. Kawazoe, T., Kobayashi, K., Lim, J., Narita, Y., and Ohtsu, M., 2002. Direct observation of optically-forbidden energy transfer between CuCl quantum cubes via optical near-field. *Phys. Rev. Lett.* **88**: 067404-1–067404-4.

4. Sangu, S., Kobayashi, K., Shojiguchi, A., Kawazoe, T., and Ohtsu, M., 2006. Theory and principles of operation of nanophotonic functional devices. In M. Ohtsu (ed.) *Progress in Nano-Electro-Optics V*, Springer-Verlag, New York, pp. 1–62.

5. Tang, Z. K., Yanase, A., Yasui, T., Segawa, Y., and Cho, K., 1993. Optical selection rule and oscillator strength of confined exciton system in CuCl thin films. *Phys. Rev. Lett.* **71**: 1431–1434.

6. Sakakura, N. and Masumoto, Y., 1997. Persistent spectral-hole-burning spectroscopy of CuCl quantum cubes. *Phys. Rev. B* **56**: 4051–4055.

7. Ekimov, A. I., Eflos, AI. L., and Onushchenko, A. A., 1985. Quantum size effect in semiconductor microcrystals. *Solid State Commun.* **56**: 921–924.

8. Itoh, T., Yano, S., Katagiri, N., Iwabuchi, Y., Gourdon, C., and Ekimov, A. I., 1994. Interface effect on the properties of confined excitons in CuCl microcrystals. *J. Lumin.* **60&61**: 396–399.

9. Kobayashi, K., Sangu, S., Ito, H., and Ohtsu, M., 2001. Near-field optical potential for a neutral atom. *Phys. Rev. A* **63**: 013806-1–013806-9.

10. Ohtsu, M., 1998. *Near-Field Nano/Atom Optics and Technology*. Springer, Tokyo, Japan.

11. Suzuki, T., Mitsuyu, T., Nishi, K., Ohyama, H., Tomimasu, T., Noda, S., Asano, T., and Sasaki A., 1996. Observation of ultrafast all-optical modulation based on intersubband transition in n-doped quantum wells by using free electron laser. *Appl. Phys. Lett.* **69**: 4136–4138.

12. Saiki, T., Mononobe, S., Ohtsu, M., Saito, N., and Kusano, J., 1996. Tailoring a high-transmission fiber probe for photon scanning tunneling microscope. *Appl. Phys. Lett.* **68**: 2612–2614.

13. Saiki, T. and Matsuda, K., 1999. Near-field optical fiber probe optimized for illumination-collection hybrid mode operation. *Appl. Phys. Lett.* **74**: 2773–1775.

14. Yariv, A., 1971. *Introduction to Optical Electronics*. Holt, Rinehart & Winston, New York.

15. Ohtsu, M., Kobayashi, K., Kawazoe, T., Sangu, S., and Yatsui, T., 2002. Nano-photonics: Design, fabrication, and operation of nanometric devices using optical near fields. *IEEE J. Select. Top. Quantum Electron.* **8**: 839–862.

16. Kawazoe, T., Kobayashi, K., Sangu, S., and Ohtsu, M., 2003. Demonstration of a nanophotonic switching operation by optical near-field energy transfer. *Appl. Phys. Lett.* **82**: 2957–2959.

17. Kawazoe, T., Kobayashi, K., Sangu, S., and Ohtsu, M., 2003. Demonstrating nanophotonic switching using near-field pump-probe photoluminescence spectroscopy of CuCl quantum cubes. *J. Microsc.* **209**: 261–266.

18. Madelung, O., Schulg, M., and Weiss H., (eds), 1982. *Landolt-Bornstein, Physics of II-VI and I-VII Compounds, Semimagnetic Semiconductors*. vol. 17b. Springer-Verlag, Berlin, Germany.

19. Masumoto, Y., Kawazoe, T., and Matsuura, N., 1998. Exciton-confined-phonon interaction in quantum dots. *J. Lumin.* **76&77**: 189–192.

20. Neogi, A., Lee, C. W., Everitt, H. O., Kuroda, T., Tackeuchi, A., and Yablonovitch, E., 2002. Enhancement of spontaneous recombination rate in a quantum well by resonant surface plasmon coupling. *Phys. Rev. B* **66**: 153305-1–153305-4.

21. Raether, H., 1988. *Surface Plasmons, Vol. III of Springer Tracts in Modern Physics*. Springer-Verag, Berlin, Germany.

22. Kawazoe, T., Kobayashi, K., Akahane, K., Naruse, M., Yamamoto, N., and Ohtsu, M., 2006. Demonstration of nanophotonic NOT gate using near-field optically coupled quantum dots. *Appl. Phys. B* **84**: 243–246.

23. Kawazoe, T., Kobayashi, K., and Ohtsu, M., 2005. Optical nanofountain: A biomimetic device that concentrates optical energy in a nanometric region. *Appl. Phys. Lett.* **86**: 103102-1–103102-3.

24. Nomura, W., Ohtsu, M., and Yatsui, T., 2005. Nanodot coupler with a surface plasmon polariton condenser for optical far/near-field conversion. *Appl. Phys. Lett.* **86**: 181108-1–181108-3.

25. Nomura, W., Yatsui, T., and Ohtsu, M., 2006. Efficient optical near-field energy transfer along an Au nanodot coupler with size-dependent resonance. *Appl. Phys. B* **84**: 257–259.

26. Born, M. and Wolf, E., 1983. *Principles of Optics*, 6th edn. Pergamon Press, Oxford, U.K.

27. Nishibayashi, K., Kawazoe, T., Akahane, K., Yamamoto, N., and Ohtsu, M., 2008. Observation of interdot energy transfer between InAs quantum dots. *Appl. Phys. Lett.* **93**: 042101-1–042101-3.

28. Yatsui, T., Sangu, S., Kawazoe, T., Ohtsu, M., An, S.-J., Yoo, J., and Yi, G.-C., 2007. Nanophotonic switch using ZnO nanorod double-quantum-well structures. *Appl. Phys. Lett.* **90**: 223110-1–223110-3.

29. Yoshimoto, M., Maeda, T., Ohnishi, T., Koinuma, H., Ishiyama, O., Shinohara, M., Kubo, M., Miura, R., and Miyamoto, A., 1995. Atomic-scale formation of ultrasmooth surfaces on sapphire substrates for high-quality thin film fabrication. *Appl. Phys. Lett.* **67**: 2615–2617.

30. Calvert, J. G. and Patts, J. N. Jr., 1996. *Photochemistry*. Wiley, New York.

31. Shimizu, M., Kamei, H., Tanizawa, M., Shiosaki, T., and Kawabata, A., 1988. Low temperature growth of ZnO film by photo-MOCVD. *J. Cryst. Growth* **89**: 365–370.

32. Kobayashi, K., Kawazoe, T., and Ohtsu, M., 2008. Localized photon model including phonons' degrees of freedom. In M. Ohtsu (ed.) *Progress in Nano-Electro-Optics VI*, Springer-Verlag, Berlin, Germany, pp. 41–66.

33. Kawazoe, T., Kobayashi, K., Takubo, S., and Ohtsu, M., 2005. Nonadiabatic photodissociation process using an optical near field. *J. Chem. Phys.* **122**: 024715-1–024715-5.

34. Kawazoe, T., Yamamoto, Y., and Ohtsu, M., 2001. Fabrication of a nanometric Zn dot by nonresonant near-field optical chemical vapor deposition. *Appl. Phys. Lett.* **79**: 1184–1186.

35. Kawazoe, T., Kobayashi, K., and Ohtsu, M., 2006. Near-field optical chemical vapor deposition using $Zn(acac)_2$ with a non-adiabatic photochemical process. *Appl. Phys. B* **84**: 247–251.

36. Kawazoe, T. and Ohtsu, M., 2004. Adiabatic and nonadiabatic nanofabrication by localized optical near fields. *Proc. SPIE* **5339**: 619–630.

37. Yonemitsu, H., Kawazoe, T., Kobayashi, K., and Ohtsu, M., 2007. Nonadiabatic photochemical reaction and application to photolithography. *J. Lumin.* **122–123**: 230–233.

38. Ohtsu, M., (ed). 2006. *Progress in Nano-Electro-Optics V*, Spinger-Verlag, Preface to Volume V: Based on "nano-photonics" proposed by Ohtsu in 1993, OITDA (Optical Industry Technology Development Association, Japan) organized the nanophotonics technical group in 1994, and discussions on the future direction of nanophotonics were started in collaboration with academia and industry.

39. Ho, K. M., Chan, C. T., and Soukoulis, C. M., 1990. Existence of a photonic gap in periodic dielectric structures. *Phys. Rev. Lett.* **65**: 3152–3155.

40. Podolskiy, V. A., Sarychev, A. K., and Shalaev, V. M., 2003. Plasmon modes and negative refraction in metal nanowire composites. *Opt. Express* **11**: 735–745.

41. Shelby, R. A., Smith, D. R., and Shultz, S., 2001. Experimental verification of a negative index of refraction. *Science* **292**: 77–79.

42. Pendry, J. B., 2000. Negative refraction makes a perfect lens. *Phys. Rev. Lett.* **85**: 3966–3969.

43. Rong, H., Liu, A., Nicolaescu, R., and Paniccia, M., 2004. Raman gain and nonlinear optical absorption measurements in a low-loss silicon waveguide. *Appl. Phys. Lett.* **85**: 2196–2198.

44. Arakawa, Y. and Sakaki, H., 1982. Multidimensional quantum well laser and temperature-dependence of its threshold current. *Appl. Phys. Lett.* **40**: 939–941.

45. Ohtsu, M. and Kobayashi, K., 2003. *Optical Near Fields*, vol. 109–150. Springer-Verlag, Berlin, Germany.

46. Ohtsu, M., 2007. Nanooptics. In F. Traeger (ed.) *Handbook of Lasers and Optics*, Springer-Verlag, Berlin, Germany, pp. 1079–1090.

47. Shannon, C., 1956. The bandwagon. *IEEE Trans. Inform. Theory* **2**: 3.

36

Nanophotonic Device Materials

Takashi Yatsui
University of Tokyo

Wataru Nomura
University of Tokyo

36.1 Introduction

Systems of optically coupled quantum structures should be applicable to quantum information processing (Bayer et al. 2001, Stinaff et al. 2006). Additional functional devices, i.e., nanophotonic devices (Ohtsu et al. 2002, Kawazoe et al. 2003, 2005, 2006), can be realized by controlling the exciton excitation in quantum dots (QDs) and quantum-well structures (QWs). This chapter reviews the recent achievements with nanophotonic devices based on colloidal QDs (Section 36.2) and nanorod QWs (Section 36.3).

36.2 Nanophotonic Devices Based on Quantum Dots

36.2.1 Discrete Energy Levels in Spherical Quantum Dots

The translational motion of the exciton center of mass is quantized in nanoscale semiconductors when the size is decreased so that it is as small as an exciton Bohr radius. If the QDs are assumed to be spheres with radius R, with the following potential

$$V(x) = \begin{cases} 0 & \text{for } |x| \leq R \\ \infty & \text{for } |x| > R \end{cases} \quad (36.1)$$

the quantized energy levels are given by a spherical Bessel function as

$$R_{nl}(r) = A_{nl} j_l\left(\rho_{n,l} \frac{r}{R}\right) \quad (36.2)$$

Figure 36.1 shows the lth order of the spherical Bessel function. Note that an odd quantum number of l has an odd function and it is a dipole-forbidden energy state. To satisfy the boundary conditions as

$$R_{nl}(R) = A_{nl} j_l(\rho_{n,l}) = 0 \quad (36.3)$$

the quantized energy levels are calculated using

$$E(n,l) = E_B + \frac{\hbar^2 \pi^2}{2mR^2} \xi_{n,l}^2 \quad (36.4)$$

where $\pi \xi_{n,l} = \rho_{n,l}$ is the nth root of the spherical Bessel function of the lth order. The principal quantum number n and the angular momentum quantum number l take values $n = 1,2,3,\ldots$, and $l = 0,1,2,\ldots$, respectively. $\xi_{n,l}$ takes values $\xi_{1,0} = 1$, $\xi_{1,1} = 1.43$, $\xi_{1,2} = 1.83$, $\xi_{2,0} = 2$, and so on (see Table 36.1) (Sakakura and Masumoto 1997).

Figure 36.2 show schematic drawings of different-sized spherical QDs (X and Y) and confined exciton energy levels. Here, R and $1.43R$ are the radii of spherical QDs X and Y, respectively. According to Equation 36.4, the quantized exciton energy levels of $E(1,0)$ in QD X and $E(1,1)$ in QD Y resonate with each other. Although the energy state $E(1,1)$ is a dipole-forbidden state, the optical near-field interaction is finite for such coupling to the forbidden energy state (Kobayashi et al. 2000). In addition, the intersublevel transition, τ_{sub}, from higher exciton energy levels to the lowest one, is generally less than a few picoseconds and is much shorter than the transition time due to optical near-field coupling (τ_{ET}) (Suzuki et al. 1996). Therefore, most of the energy of the exciton in a QD X with radius R transfers to the lowest exciton energy level in the neighboring QD Y with a radius of $1.43R$ and recombines radiatively at the lowest level. In this manner, unidirectional energy flow is achieved.

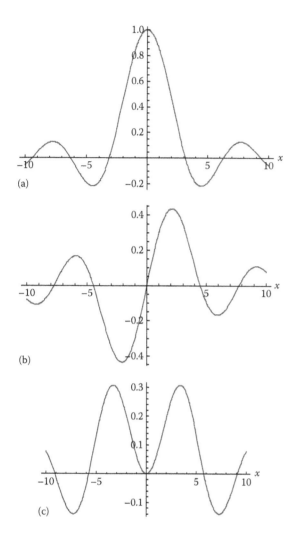

(a)

(b)

(c)

FIGURE 36.1 *l*th order of spherical Bessel function. (a) $l = 0$ ($j_0(x) = (\sin x)/x$), (b) $l = 1$ ($j_1(x) = (1/x)(((\sin x)/x) - \cos x)$), and (c) $l = 2$ ($j_2(x) = (1/x)(((3 - x^2)/x^2) \sin x) - ((3/x) \cos x)$).

TABLE 36.1 Calculated $\xi_{n,l}$ to Satisfy the Condition of $j_l(\pi\xi_{n,l}) = 0$

	$l = 0$	$l = 1$	$l = 2$	$l = 3$	$l = 4$	.
$n = 1$	1	1.43	1.83	2.22	2.61	.
$n = 2$	2	2.46	2.90	3.31	·	.
$n = 3$	3	·	·	·	·	.
·	·	·	·	·	·	.

36.2.2 The Observation of Dissipated Optical Energy Transfer between CdSe QDs

To evaluate the energy transfer and the energy dissipation, we used CdSe/ZnS core–shell QDs from *Evident Technologies*. As described in Section 36.2.1, assuming that the respective diameters, D, of the QD_{a1} and QD_{a2} were 2.8 and 4.1 nm, the ground energy level in the QD_{a1} and the excited energy level in the QD_{a2} resonate (Trallero-Giner et al. 1998). A solution of QDs_{a1} ($D = 2.8$ nm) and QDs_{a2} ($D = 4.1$ nm) in 1-feniloctane at a density

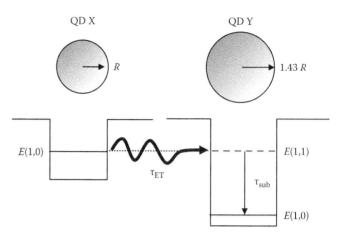

FIGURE 36.2 Schematic drawings of different-sized spherical QDs (X and Y) and the confined exciton energy levels.

of 1.0 mg/mL was dropped on mica substrate (see Figure 36.3a), such that areas A and C consisted of QDs_{a1} and QDs_{a2}, respectively, while there were both QDs_{a1} and QDs_{a2} in area B. Using transmission electron microscopy (TEM), we confirmed the mean center-to-center distance of each QD was maintained at about less than 10 nm in all areas, due to the 2 nm-thick ZnS shell and surrounding ligands (2 nm-length long chain amine) of the QDs (Figure 36.3b).

In the following experiments, the light source used was the third harmonic of a mode-locked Ti:sapphire laser (wavelength 306 nm, frequency 80 MHz, and pulse duration 2 ps). The incident power of the laser was 0.6 mW and the spot size was 1×10^{-3} cm². The density of QDs was less than 3.5×10^{12} cm^{-2} and the quantum yield of CdSe/ZnS QD was 0.5. Under these conditions, the probability of exciton generation by one laser pulse in each QD was calculated to be 1.6×10^{-2}. Therefore, we assumed single-exciton dynamics in the following experiments.

The energy transfer was confirmed using micro-photoluminescence (PL) spectroscopy. Temperature-dependent micro-PL spectra were obtained. In the spectral profile of the PL emitted from area A, we found a single peak which originated from the ground state of QD_{a1} at a wavelength of $\lambda = 540$ nm, from room temperature to 30 K (broken line in Figure 36.4). From area C, the single peak, which originated from the ground state of QD_{a2}, was found at $\lambda = 600$ nm (dotted line in Figure 36.4). By contrast, area B had two peaks at room temperature, as shown by solid line in Figure 36.4. This figure also shows that the PL intensity peak at $\lambda = 540$ nm decreased relative to that of at $\lambda = 600$ nm on decreasing the temperature. This relative decrease in the PL intensity originated from the energy transfer from the ground state in the QD_{a1} to the excited state in the QD_{a2} and the subsequent rapid dissipation to the ground state in the QD_{a2}. This is because the coupling between the resonant energy levels becomes stronger due to the increase in the exciton decay time on decreasing the temperature (Itoh et al. 1990). Furthermore, although nanophotonic device operation using CuCl quantum cubes (Kawazoe et al. 2003, 2005, 2006) and ZnO quantum wells (Yatsui et al. 2007) has been reported at 15 K, we observed the

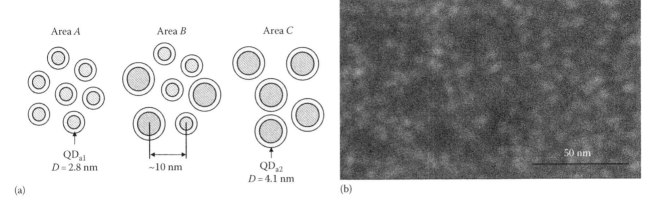

FIGURE 36.3 (a) Schematic images of CdSe QDs dispersed substrate. Areas A, B, and C are covered by QDs$_{a1}$, both QDs$_{a1}$ and QDs$_{a2}$, and QDs$_{a2}$, respectively. (b) TEM image of dispersed CdSe/ZnS core–shell QDs in area B.

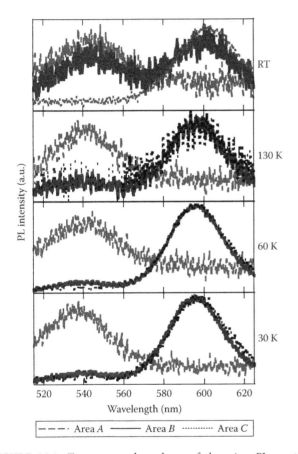

FIGURE 36.4 Temperature dependence of the micro-PL spectra. Broken, solid, and dotted lines show the spectra from areas A, B, and C, respectively.

decrease in the PL intensity at $\lambda = 540$ nm at temperatures as high as 130 K, which is advantageous for the high-temperature operation of nanophotonic devices.

To confirm this energy transfer from QD$_{a1}$ to QD$_{a2}$ at the temperature under 130 K, we evaluated the dynamic property of the energy transfer using time-resolved spectroscopy with the time correlation single photon counting method. Circles A_{a1}, squares

B_{a1}, and triangles B_{a2} in Figure 36.5a represent the respective time-resolved micro-PL intensities (60 K) from ground energy level in QD$_{a1}$ ($D = 2.8$ nm) in area A, QD$_{a1}$ ($D = 2.8$ nm) in area B, and QD$_{a2}$ ($D = 4.1$ nm) in area B. The peak intensities at $t = 0$ were normalized to unity. Note that B_{a2} decreases faster than A_{a1}, although these signals were generated from QDs of the same size. In addition, although the exciton lifetime decreases on increasing the QD size, owing to the increased oscillator strength, B_{a2} decreased more slowly than A_{a1} over the range $t < 0.2$ ns (see the inset of Figure 36.5a). Furthermore, as we did not see any peak in the power spectra of A_{a1}, B_{a1}, and B_{a2}, we believe that the temporal signal changes originated from the optical near-field energy transfer and subsequent dissipation. Since the QDs$_{a1}$ in area B were near QDs$_{a2}$ whose excited energy level resonates with the ground energy level of the QD$_{a1}$ (see Figure 36.5b), near-field coupling between the resonant levels resulted in the energy transfer from the QD$_{a1}$ to the QD$_{a2}$ and the consequent faster decrease in the excitons of the QD$_{a1}$ in area B compared with area A. Furthermore, as a result of inflow of the carriers from the QD$_{a1}$ to the QD$_{a2}$, the PL intensity from the QD$_{a2}$ near the QD$_{a1}$ decayed more slowly than that of the QD$_{a1}$.

For comparison, we also obtained time-resolved PL profiles of different pairs of CdSe/ZnS QDs. Their diameters were $D = 2.8$ nm (QD$_{a1}$) and 3.2 nm (QD$_{b1}$), which means that their energy levels did not resonant with each other. Figure 36.6a shows a schematic image of a sample named area D, where QDs$_{a1}$ and QDs$_{b1}$ are mixed with a mean center-to-center distance of less than 10 nm. Circles A_{a1} and diamonds D_{a1} in Figure 36.6b show the time-resolved PL intensity (30 K) from the ground energy level in QDs$_{a1}$ from areas A and D, respectively. There is no difference in the decay profile. This indicates that the excited carriers in QDs$_{a1}$ did not couple with QDs$_{b1}$ due to their off-resonance and, consequently, no energy was transferred (Figure 36.6c). This supports the postulate that Figures 36.4 and 36.5 demonstrate energy transfer and subsequent dissipation due to near-field coupling between the resonant energy levels of the QDs$_{a1}$ and QDs$_{a2}$.

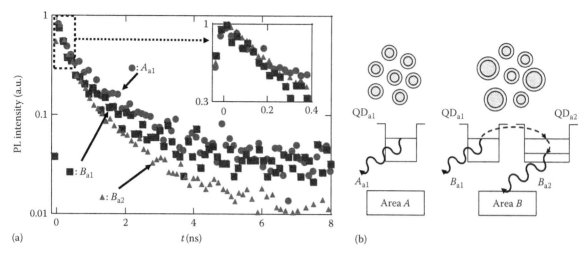

(a)

FIGURE 36.5 (a) Time-resolved PL intensity profiles from QDs$_{a1}$ in area A (circles A_{a1}), QDs$_{a1}$ in area B (squares B_{a1}), and QDs$_{a2}$ in area B (triangles B_{a2}). The peak intensities were normalized at $t = 0$. (b) Schematic of the respective system configurations in areas A and B.

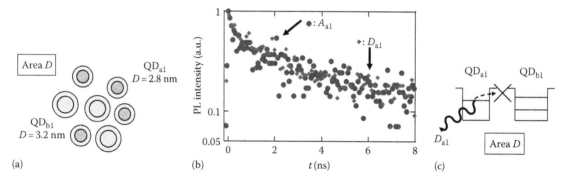

FIGURE 36.6 (a) Schematic image of CdSe QDs dispersed substrate area D. (b) Time-resolved PL intensity profiles from QDs$_{a1}$ in area A (circles A_{a1}) and QDs$_{a1}$ in area D (diamonds D_{a1}). The peak intensities were normalized at $t = 0$. (c) Schematic of the system configuration in area D.

To discuss the exciton energy transfer from QD$_{a1}$ to QD$_{a2}$ quantitatively, we investigated the exciton dynamics by fitting multiple exponential decay curve functions to curves A_{a1}, B_{a1}, and B_{a2} (Bawendi et al. 1992, Crooker et al. 2003):

$$A_{a1} = R_{S1} \exp\left(\frac{-t}{\tau_{s1}}\right) + R_{S2} \exp\left(\frac{-t}{\tau_{s2}}\right), \qquad (36.5)$$

$$B_{a1} = R_S \cdot A_{a1} + R_{rS} \exp\left(\frac{-t}{\tau_t}\right), \qquad (36.6)$$

and

$$B_{a2} = R_{L1} \exp\left(\frac{-t}{\tau_{L1}}\right) + R_{L2} \exp\left(\frac{-t}{\tau_{L2}}\right). \qquad (36.7)$$

We used double exponential decay for A_{a1} and B_{a2} (Equations 36.5 and 36.7), which corresponds to the non-radiative lifetime (fast decay: τ_{S1} and τ_{L1}) and radiative lifetime of free-carrier recombinations (slow decay: τ_{S2} and τ_{L2}). Given the imperfect homogeneous

distribution of the QDs$_{a1}$ in area B, some QDs$_{a1}$ lack energy transfer routes to QD$_{a2}$. However, we introduced the mean energy transfer time τ_t from QDs$_{a1}$ and QDs$_{a2}$ in Equation 36.6. In these equations, we neglected the energy dissipation time of about 1 ps (Guyot-Sionnest et al. 1999) because that is much smaller than exciton lifetimes and energy transfer time. Figure 36.7 shows the best-fitted numerical results and experimental data. Here, we used exciton lifetimes of $\tau_{S2} = 2.10$ ns and $\tau_{L2} = 1.79$ ns. The mean energy transfer time was $\tau_t = 135$ ps, which is comparable with the observed energy transfer time (130 ps) in CuCl quantum cubes (Kawazoe et al. 2003) and ZnO QWs (Yatsui et al. 2007). Furthermore, the relation $\tau_t < \tau_{S2}$ agrees with the assumption that most of the excited excitons in QDs$_{a1}$ transfer to excited exciton energy level in a QD$_{a2}$ before being emitted from QD$_{a1}$.

36.2.3 Controlling the Energy Transfer between Near-Field Optically Coupled ZnO QDs

ZnO is a promising material for room-temperature operation of a nanophotonic device because of its large exciton binding energy (Ohtomo et al. 2000, Huang et al. 2001, Sun et al. 2002). Here, we

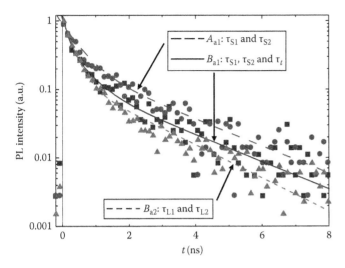

FIGURE 36.7 Experimental results (circles, squares and triangles) and fitting curves (broken, solid, and short broken lines) using Equations 36.5 through 36.7 for the PL intensity profiles. The fitting parameters are $R_{S1} = 0.560$, $\tau_{S1} = 2.95 \times 10^{-10}$, $R_{S2} = 0.329$, $\tau_{S2} = 2.10 \times 10^{-9}$, $R_S = 0.740$, $R_{tS} = 0.330$, $\tau t = 1.35 \times 10^{-10}$, $R_{L1} = 0.785$, $\tau_{L1} = 2.94 \times 10^{-10}$, $R_{L2} = 0.201$, $\tau_{L2} = 1.79 \times 10^{-9}$.

used chemically synthesized ZnO QD to realize a highly integrated nanophotonic device. We observed the energy transfer from smaller ZnO QDs to larger QDs with mutually resonant energy levels. The energy transfer time and energy transfer ratio between the two QDs were also calculated from the experimental results (Yatsui et al. 2008).

ZnO QDs were prepared using the sol–gel method (Spanhel and Anderson 1991, Meulenkamp 1998).

1. A sample of 1.10 g (5 mmol) of $Zn(Ac)_2 \cdot 2H_2O$ was dissolved in 50 mL of boiling ethanol at atmospheric pressure, and the solution was then immediately cooled to 0°C. A sample of 0.29 g (7 mmol) of $LiOH \cdot H_2O$ was dissolved in 50 mL of ethanol at room temperature in an ultrasonic bath and cooled to 0°C. The hydroxide-containing solution was then added dropwise to the $Zn(Ac)_2$ suspension with vigorous stirring at 0°C. The reaction mixture became transparent after approximately 0.1 g of LiOH had been added. The ZnO sol was stored at 0°C to prevent particle growth.
2. A mixed solution of hexane and heptane, with a volume ratio of 3:2, was used to remove the reaction products (LiAc and H_2O) from the ZnO sol.
3. To initiate the particle growth, the ZnO solution was warmed to room temperature. The mean diameter of ZnO QD was determined from the growth time, T_g.

Figure 36.8a shows a TEM image of synthesized ZnO dots after the second step. Dark areas correspond to the ZnO QD. This image suggested that monodispersed single crystalline particles were obtained.

To check the optical properties and diameters of our ZnO QD, we measured the photoluminescence (PL) spectra using He–Cd laser ($\lambda = 325$ nm) excitation at 5 K. We compared the PL spectra of ZnO QD with $T_g = 0$ and 42 h (solid and dashed curves in Figure 36.8b, respectively). A redshifted PL spectrum was obtained, indicating an increase in the QD diameter. Figure 36.8c shows the growth time dependence of the QD diameter. This was determined from the effective mass model, with peak energy in the PL spectra, $E_g^{bulk} = 3.35$ eV, $m_e = 0.28$, $m_h = 1.8$, and $\varepsilon = 3.7$ (Brus 1984). This result indicated that the diameter growth rate at room temperature was 1.1 nm/day.

Assuming that the diameters, D, of the QDS and the QDL were 3.0 and 4.5 nm, respectively, E_{S1} in the QDS and E_{L2} in the QDL resonated (Figure 36.9a). An ethanol solution of QDS and QDL was dropped onto a sapphire substrate. The mean surface-to-surface separation of the QD was approximately 3 nm.

The spectra S and L in Figure 36.9b correspond to the QDS and the QDL, with spectral peaks of 3.60 and 3.44 eV, respectively. The curve A in Figure 36.9b shows the spectrum from the QDS and QDL mixture with $R = 1$, where R is the ratio (number of QDS)/(number of QDL). The spectral peak of 3.60 eV, which corresponded to the PL from the QDS, was absent from this curve. This peak was thought to have disappeared due to energy transfer from the QDS to the QDL, because the first excited state of the QDL resonated with the ground state of the QDS. Our hypothesis was supported by the observation that when R was increase by eightfold, the spectral peak from the QDS reappeared (see spectrum C in Figure 36.9b).

To confirm this energy transfer from the QDS to the QDL at 5 K, we evaluated dynamic effects using time-resolved spectroscopy with the time-correlated single photon counting method. The light source used was the third harmonic of a mode-locked Ti:sapphire laser (photon energy 4.05 eV, frequency 80 MHz, and pulse duration 2 ps). We compared the signals from mixed samples with ratios $R = 2$, 1, and 0.5. The curves T_A ($R = 2$), T_B ($R = 1$), and T_C ($R = 0.5$) in Figure 36.10 show the respective time-resolved PL intensities from the ground state of the QDS (E_{S1}) at 3.60 eV. We investigated the exciton dynamics quantitatively by fitting multiple exponential decay curve functions (Bawendi et al. 1992, Crooker et al. 2003):

$$TRPL = A_1 \exp\left(\frac{-t}{\tau_1}\right) + A_2 \exp\left(\frac{-t}{\tau_2}\right) \qquad (36.8)$$

We obtained average τ_1 and τ_2 values of 144 ps and 443 ps, respectively (see Table 36.2). Given the disappearance of the spectral peak at 3.60 eV in the PL spectra, it is likely that these values corresponded to the energy transfer time from the QDS to the QDL and the radiative decay time from the QDS, respectively. This hypothesis was supported by the observation that the average value of τ_1 (144 ps) was comparable with the observed energy transfer time in CuCl quantum cubes (130 ps) (Kawazoe et al. 2003).

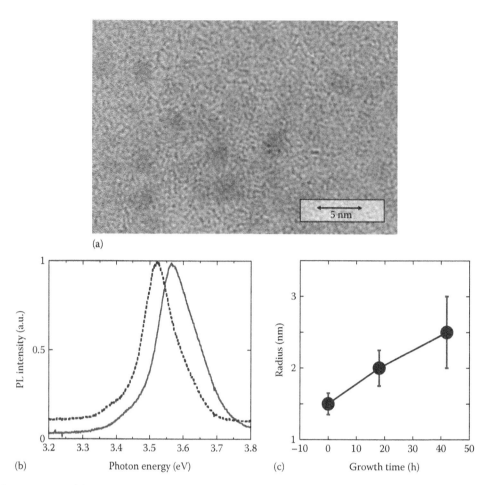

FIGURE 36.8 (a) TEM image of the ZnO QD. The dark areas inside the white dashed circles correspond to the ZnO QD. (b) The PL spectra observed at 5 K. The solid and dashed curves indicate growth time $T_g = 0$ and 42 h, respectively. (c) The growth time dependence of the mean ZnO QD diameter.

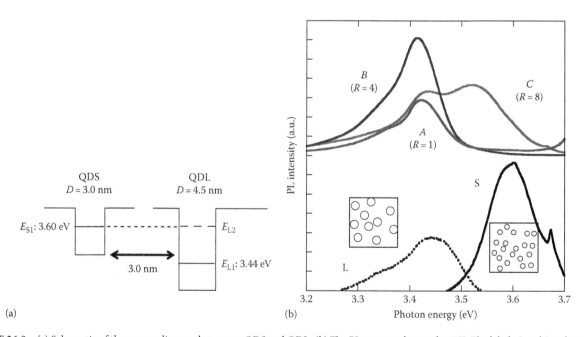

FIGURE 36.9 (a) Schematic of the energy diagram between a QDS and QDL. (b) The PL spectra observed at 5 K. The labels S and L indicate QDS and QDL, respectively. The labels A, B, and C indicate mixes with R ratios of 1, 4, and 8, respectively.

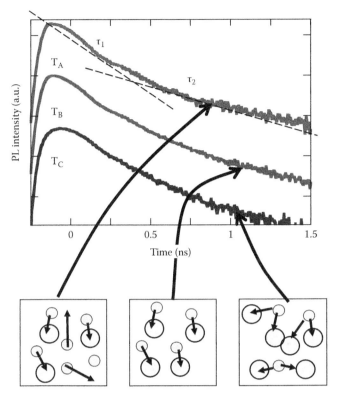

FIGURE 36.10 Time-resolved PL spectra observed at 5 K. The values of R were 2, 1, and 0.5 for curves T_A, T_B, and T_C, respectively.

TABLE 36.2 Dependence of the Time Constants (τ_1 and τ_2) on R as Derived from the Two Exponential Fits of the Time-Resolved PL Signals and the Coefficient Ratio A_1/A_2

R = QDS/QDL	τ_1 (ps)	τ_2 (ps)	A_1/A_2
2	133	490	12.4
1	140	430	13.7
0.5	160	410	14.4
Average	144	443	

We also investigated the value of coefficient ratio A_1/A_2 (see Table 36.2); this ratio was inversely proportional to R, hence proportional to the number of QDL. This result indicated that an excess QDL caused energy transfer from QDS to QDL, instead of direct emission from the QDS.

We observed the dynamic properties of exciton energy transfer and dissipation between ZnO QD *via* an optical near-field interaction, using time-resolved PL spectroscopy. Furthermore, we successfully increased the energy transfer ratio between the resonant energy state, instead of the radiative decay from the QD. Chemically synthesized nanocrystals, both semiconductor QD and metallic nanocrystals (Brust and Kiely 2002), are promising nanophotonic device candidates, because they have uniform sizes, controlled shapes, defined chemical compositions, and tunable surface chemical functionalities.

36.3 Nanophotonic AND-Gate Device Using ZnO Nanorod Double-Quantum-Well Structures

ZnO/ZnMgO nanorod heterostructures have been fabricated and the quantum confinement effect has been observed from single QW structures (Park et al. 2003, 2004). In this section, we review the time-resolved near-field spectroscopy to demonstrate the switching dynamics that result from controlling the optical near-field energy transfer in ZnO nanorod double-quantum-well structures (DQWs). We observed nutation of the population between the resonantly coupled exciton states of DQWs, where the coupling strength of the near-field interaction decreased exponentially as the separation increased (Yatsui et al. 2007).

To evaluate the energy transfer, three samples were prepared (Figure 36.11a): (1) Single-quantum-well structures (SQWs) with a well-layer thickness of L_w = 2.0 nm (SQWs), (2) DQWs with L_w = 3.5 nm with 6 nm separation (1-DQWs), and (3) three pairs of DQWs with L_w = 2.0 nm with different separations (3, 6, and 10 nm), where each DQW was separated by 30 nm (3-DQWs). ZnO/ZnMgO quantum-well structures (QWs) were fabricated on the ends of ZnO nanorods with a mean diameter of 80 nm using catalyst-free metalorganic vapor phase epitaxy (Park et al. 2002). The average concentration of Mg in the ZnMgO layers used in this study was determined to be 20 atm %.

The far-field PL spectra were obtained using a He–Cd laser (λ = 325 nm) before detection using near-field spectroscopy. The emission signal was collected with an achromatic lens (f = 50 mm). The near-field photoluminescence (NFPL) spectra were obtained using a He–Cd laser (λ = 325 nm), collected with a fiber probe with an aperture diameter of 30 nm, and detected using a cooled charge-coupled device through a monochromator. Blueshifted PL peaks were observed at 3.499 (I_S), 3.429 (I_{1D}), and 3.467 (I_{3D}) eV in the far- and near-field PL spectra (Figure 36.12a). These peaks originated from the respective ZnO QWs because their energies are comparable to the predicted ZnO well-layer thicknesses of 1.7 (I_S), 3.4 (I_{1D}), and 2.2 (I_{3D}) nm, respectively, calculated using the finite square-well potential of the quantum confinement effect in ZnO SQWs (see Figure 36.12b) (Park et al. 2004).

To confirm the near-field energy transfer between QWs, we compared the time-resolved near-field PL (TR$_{NFPL}$) signals at the I_S, I_{1D}, and I_{3D} peaks. For the time-resolved near-field spectroscopy, the signal was collected using a micro-channel plate through a band-pass filter with 1 nm spectral width. Figure 36.13 shows the typical TR$_{NFPL}$ of SQWs (TR$_S$), 1-DQWs (TR$_{1D}$), and 3-DQWs (TR$_{3D}$).

We calculate the exciton dynamics using quantum mechanical density-matrix formalism (Coffey and Friedberg 1978, Kobayashi et al. 2005), where the Lindblad-type dissipation is assumed for the relaxation due to exciton–photon and exciton–phonon couplings:

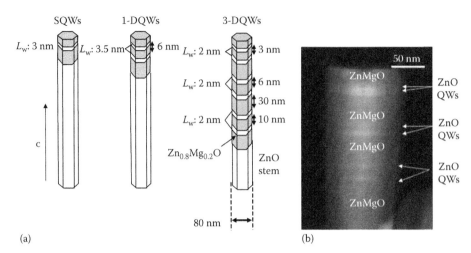

FIGURE 36.11   ZnO/ZnMgO nanorod quantum-well-structures. c: *c*-axis of the ZnO stem. (a) Schematic of ZnO/ZnMgO SQWs, DQWs (1-DQWs), and triple pairs of DQWs (3-DQWs). (b) Z-contrast TEM image of 3-DQWs clearly shows the compositional variation, with the bright layers representing the ZnO well layers. Scale bar: 50 nm.

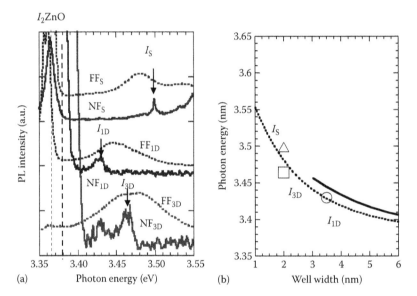

FIGURE 36.12   Near-field time-resolved spectroscopy of ZnO nanorod DQWs at 15 K. (a) NF_S, NF_{1D}, NF_{3D}: near-field PL spectra. FF_S, FF_{1D}, FF_{3D}: far-field PL spectrum of ZnO SQWs ($L_w = 2.0$ nm), 1-DQWs ($L_w = 3.5$ nm, 6 nm separation), and 3-DQWs ($L_w = 2.0$ nm; 3, 6, and 10 nm separation). (b) Well width dependent on the exciton ground state and the first excited state. SQWs: open triangle, 1-DQWs: open circle, 3-DQWs: open square.

$$\dot{\rho} = -\frac{i}{\hbar}[H,\rho] + \sum_n \frac{\gamma_n}{2}\left(2A_n\rho A_n^{\dagger} - A_n^{\dagger}A_n\rho - \rho A_n^{\dagger}A_n\right) \quad (36.9)$$

where

- ρ is the density operator
- H is the Hamiltonian in the considered system
- $A_n^{\dagger}$ and A_n are creation and annihilation operators for an exciton energy level labeled n
- γ_n is the photon or phonon relaxation constant for the energy level

The exciton population is calculated using matrix elements for all exciton states in the system considered. First, we apply the calculation to a three-level system of SQWs (Figure 36.14a), where the continuum state $\hbar\Omega_C$ is initially excited using a 10 ps laser pulse. Then, the initial exciton population in ZnO QWs is created in $\hbar\Omega_{1S}$, where the energy transfer from $\hbar\Omega_C$ to $\hbar\Omega_{1S}$ is expressed phenomenologically as a Gaussian input signal with a temporal width of $2\sigma_{1S}$ (an incoherent excitation term is added in Equation 36.9), because non-radiative relaxation paths via exciton–phonon coupling make a dephased input signal, statistically. Finally, an exciton carrier relaxes due to the electron–hole recombination with relaxation constant γ_{1S}. Figure 36.14b shows a numerical result and experimental data. Here, we used $2\sigma_{1S} = 100$ ps, and γ_{1S} was evaluated as 460 ps.

A similar calculation was applied for DQWs. We used two three-level systems, coupled via an optical near-field with a

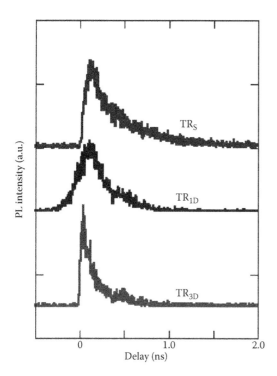

FIGURE 36.13 TR$_S$, TR$_{1D}$, and TR$_{3D}$ show TR$_{NFPL}$ signal obtained at I$_S$, I$_{1D}$, and I$_{3D}$, respectively, using the 4.025 eV (λ = 308 nm) light with a pulse of 10 ps duration to excite the barrier layers of ZnO QWs.

coupling strength of U_{12} (Figure 36.14c). Figure 36.14d shows the numerical results for the exciton population in QWs$_1$ and the experimental data. Here $2\sigma_{1D}$ and $2\sigma_{2D}$ were set at 200 ps, which is twice the value for SQWs, because the relaxation paths extend the barrier energy state in the two quantum wells (QWs). γ_{1D} and γ_{2D} are evaluated as 200 ps. We believe that the faster relaxation for DQWs compared with SQWs reflects the lifetime of the coupled states mediated by the optical near-field. Furthermore, the characteristic behavior that results from near-field coupling appears as the oscillatory decay in Figure 36.14d. This indicates that the timescale of the near-field coupling is shorter than the decoherence time, and that coherent coupled states, such as symmetric and antisymmetric states (Sangu et al. 2004), determine the system dynamics. Furthermore, nutation never appears unless unbalanced initial exciton populations are prepared for $\hbar\Omega_{1D}$ and $\hbar\Omega_{2D}$. In the far-field excitation, only the symmetric state is excited because the antisymmetric state is dipole-inactive. Then, the exciton populations of the two quantum wells are equal and they have the same decay rate. By contrast, in the near-field excitation, both the symmetric and antisymmetric states are excited due to the presence of a near-field probe. Since the symmetric and antisymmetric states have different eigenenergies, the interference of these states generates a detectable beat signal. The unbalanced excitation rate is given by $A_1/A_2 = 10$ here. From

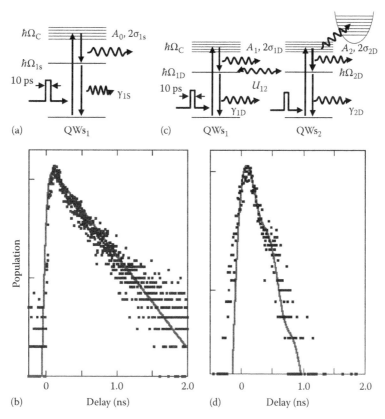

FIGURE 36.14 The schematic depict the (a) SQWs and (c) 1-DQWs system configurations. $\hbar\Omega_C$: barrier energy state with a central energy. Theoretical results on the transient exciton population dynamics (solid curves) and experimental PL data (filled squares) of (b) SQWs (same as curve TR$_S$ in Figure 36.13) and (d) 1-DQWs (same as curve TR$_{1D}$ in Figure 36.13).

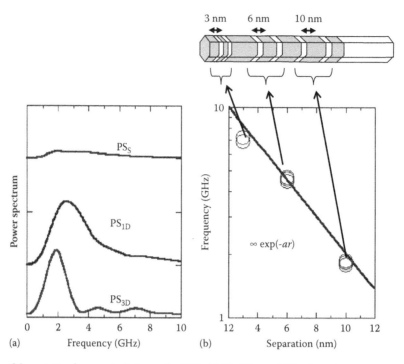

FIGURE 36.15 Evaluation of the nutation frequencies between the QWs. (a) PS_S, PS_{1D}, and PS_{3D} show the power spectrum of TR_S, TR_{1D}, and TR_{3D}, respectively. (b) Separation D dependence of frequency of the nutation.

the period of nutation, the strength of the near-field coupling is estimated to be $U_{12} = 7.7\,ns^{-1}$ ($= 4.9\,\mu eV$).

We evaluated nutation frequencies using Fourier analysis. In Figure 36.15a, the power spectral density of SQWs (PS_S) does not exhibit any peaks, indicating a monotonic decrease. By contrast, the power spectral density of 1-DQWs (PS_{1D}) had a strong peak at a frequency of $2.6\,ns^{-1}$. Furthermore, that of 3-DQWs (PS_{3D}) had three peaks at 1.9, 4.7, and $7.1\,ns^{-1}$. Since, the degree of the coupling strength, which is proportional to the frequency of the nutation, increases as the separation decreases, the three peaks correspond to the signals from DQWs with separations of 10, 6, and 3 nm, respectively. Since the coupling strength $\hbar U$ [eV] is given by $\hbar\pi f$ (f: nutation frequency), $\hbar U$ is estimated as 4.0, 9.9, and $14.2\,\mu eV$ for DQWs with respective separations of 10, 6, and 3 nm. These values are comparable to that estimated above ($U_{12} = 4.9\,\mu eV$). Furthermore, the peak intensity for the DQWs with 3 nm separation is much lower than for those with 10 nm separation,

which might be caused by decoherence of the exciton state due to penetration of the electronic carrier. Considering the carrier penetration depth, the strong peak of DQWs with 10 nm separation originates from the near-field coupling alone. The solid line in Figure 36.15b shows the separation dependence of the peak frequency. The exponentially decaying dependence represented by this line supports the origin of the peaks in the power spectra from the localized near-field interaction between the QWs.

Next, we performed the switching operation. Figure 36.16a and b explains the "OFF" and "ON" states of the proposed nanophotonic switch, consisting of two coupled QWs. QW_1 and QW_2 are used as the input/output and control ports of the switch, respectively. Assuming $L_w = 3.2$ and 3.8 nm, the ground exciton state in QW_1 and the first excited state in QW_2 resonate. In the "OFF" operation (Figure 36.16a), all the exciton energy in QW_1 is transferred to the excited state in the neighboring QW_2 and relaxes rapidly to the ground state. Consequently, no output

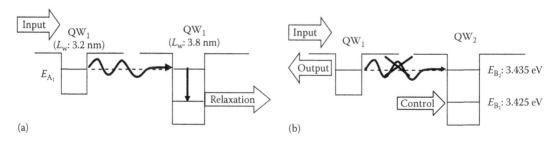

FIGURE 36.16 The switching operation by controlling the exciton excitation. Schematic of the nanophotonic switch of (a) "OFF" state and (b) "ON" state.

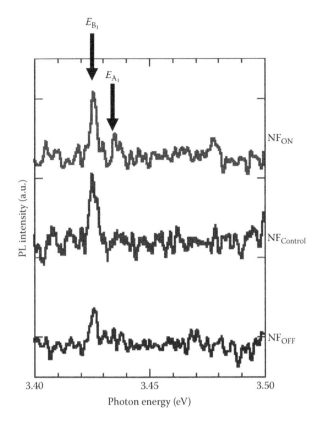

FIGURE 36.17 (a) NF_{ON}, $NF_{Control}$, and NF_{OFF} show NFPL signal obtained with the illumination of input laser alone, control laser alone, and input and control laser, respectively.

signals are generated from QW_1. In the "ON" operation (Figure 36.16b), the escape route to QW_2 is blocked by the excitation of QW_2 owing to state filling in QW_2 on applying the control signal; therefore, an output signal is generated from QW_1.

Figure 36.17 shows the NFPL for the three pairs of DQWs with $L_w = 3.2$ and $3.8\,nm$ with different separations (3, 6, and 10 nm). Curve NF_{OFF} was obtained with continuous input light illumination from a He–Cd laser (3.814 eV). No emission was observed from the exciton ground state of QW_1 (E_{A1}) or the excited state of QW_2 (E_{B2}) at a photon energy of 3.435 eV, indicating that the excited energy in QW_1 was transferred to the excited state of QW_2. Furthermore, the excited state of QW_2 is a dipole-forbidden level. Curve $NF_{Control}$ shows the NFPL signal obtained with control light excitation of 3.425 eV with a 10 ps pulse. Emission from the ground state of QW_2 at a photon energy of 3.425 eV was observed. Both input and control light excitation resulted in an output signal with an emission peak at 3.435 eV, in addition to the emission peak at 3.425 eV (curve NF_{ON}), which corresponds to the ground state of QW_2. Since the excited state of QW_2 is a dipole-forbidden level, the observed 3.435 eV emission indicates that the energy transfer from the ground state of QW_1 to the excited state of QW_2 was blocked by the excitation of the ground state of QW_2.

Finally, the dynamic properties of the nanophotonic switching were evaluated. We observed TR_{NFPL} signals at 3.435 eV with both input and control laser excitation (see Figure 36.18). The

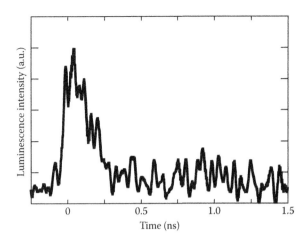

FIGURE 36.18 Near-field time-resolved PL signal with "ON" state.

decay time constant was found to be 483 ps. The output signal increased synchronously, within 100 ps, with the control pulse. Since the rise time is considered equal to one-quarter of the nutation period τ (Sangu et al. 2003), the value agrees with those obtained for DQWs with the same well width in the range from τ/4 = 36 ps (3 nm separation) to τ/4 = 125 ps (10 nm separation).

We observed the nutation between DQWs and demonstrated the switching dynamics by controlling the exciton excitation in the QWs. Examination of the electronic coupling between QWs is now in progress to analyze the detailed switching dynamics. For room-temperature operation, since the spectral width reaches thermal energy (26 meV), a higher Mg concentration in the barrier layers and narrower L_w are required so that the spectral peaks of the first excited state (E_2) and ground state (E_1) do not overlap. This can be achieved by using two QWs with $L_w = 1.5\,nm$ (QW_1) and $2\,nm$ (QW_2) with a Mg concentration of 50%, where the energy difference between E_2 and E_1 in QW_2 is 50 meV (Park 2001).

References

Bawendi, M. G., Carroll, P. J., William, L. W., and Brus, L. E. 1992. Luminescence properties of CdSe quantum crystallites: Resonance between interior and surface localized states. *J. Chem. Phys.* 96: 946–954.

Bayer, M., Hawrylak, P., Hinzer, K. et al. 2001. Coupling and entangling of quantum states in quantum dot molecules. *Science* 291: 451–453.

Brus, L. E. 1984. Electron–electron and electron–hole interactions in small semiconductor crystallites: The size dependence of the lowest excited electronic state. *J. Chem. Phys.* 80: 4403–4409.

Brust, M. and Kiely, C. J. 2002. Some recent advances in nanostructure preparation from gold and silver particles: A short topical review. *Colloids Surf. A* 202: 175–186.

Coffey, B. and Friedberg, R. 1978. Effect of short-range Coulomb interaction on cooperative spontaneous emission. *Phys. Rev. A* 17: 1033–1048.

Crooker, S. A., Barrick, T., Hollingsworth, J. A., and Klimov, V. I. 2003. Multiple temperature regimes of radiative decay in CdSe nanocrystal quantum dots: Intrinsic limits to the dark-exciton lifetime. *Appl. Phys. Lett.* 82: 2793–2795.

Guyot-Sionnest, P., Shim, M., Matranga, C., and Hines, M. 1999. Intraband relaxation in CdSe quantum dots. *Phys. Rev. B* 60: R2181–R2184.

Huang, M. H., Mao, S., and Feick, H. 2001. Room-temperature ultraviolet nanowire nanolasers. *Science* 292: 1897–1899.

Itoh, T., Furumiya, M., Ikehara, T., and Gourdon, C. 1990. Size-dependent radiative decay time of confined excitons in CuCl microcrystals. *Solid State Commun.* 73: 271–274.

Kawazoe, T., Kobayashi, K., Akahane, K., Naruse, M., Yamamoto, N., and Ohtsu, M. 2006. Demonstration of nanophotonic NOT gate using near-field optically coupled quantum dots. *Appl. Phys. B* 84: 243–246.

Kawazoe, T., Kobayashi, K., and Ohtsu, M. 2005. Optical nano-fountain: A biomimetic device that concentrates optical energy in a nanometric region. *Appl. Phys. Lett.* 86: 103102.

Kawazoe, T., Kobayashi, K., Sangu, S., and Ohtsu, M. 2003. Demonstration of a nanophotonic switching operation by optical near-field energy transfer. *Appl. Phys. Lett.* 82: 2957–2959.

Kobayashi, K., Sangu, S., Itoh, H., and Ohtsu, M. 2000. Near-field optical potential for a neutral atom. *Phys. Rev. A* 63: 013806.

Kobayashi, K., Sangu, S., Kawazoe, T., and Ohtsu, M. 2005. Exciton dynamics and logic operations in a near-field optically coupled quantum-dot system. *J. Lumin.* 112: 117–121.

Meulenkamp, E. A. 1998. Synthesis and growth of ZnO nanoparticles. *J. Phys. Chem. B* 102: 5566–5572.

Ohtomo, A., Tamura, K., Kawasaki, M. et al. 2000. Room-temperature stimulated emission of excitons in ZnO/(Mg, Zn)O superlattices. *Appl. Phys. Lett.* 77: 2204–2206.

Ohtsu, M., Kobayashi, K., Kawazoe, T., Sangu, S., and Yatsui, T. 2002. Nanophotonics: Design, fabrication, and operation of nanometric devices using optical near fields. *IEEE J. Sel. Top. Quantum Electron.* 8: 839–862.

Park, W. I., Yi, G.-C., and Jang, M. 2001. Metalorganic vapor-phase epitaxial growth and photoluminescent properties of $Zn_{1-x}Mg_xO$ (0 < x < 0.49) thin films. *Appl. Phys. Lett.* 79: 2022–2024.

Park, W. I., An, S. J., Long, Y.-J. et al. 2004. Photoluminescent properties of $ZnO/Zn_{0.8}Mg_{0.2}O$ nanorod single-quantum-well structures. *J. Phys. Chem. B* 108: 15457–15460.

Park, W. I., Kim, D. H., Jung, S.-W., and Yi, G.-C. 2002. Metalorganic vapor-phase epitaxial growth of vertically well-aligned ZnO nanorods. *Appl. Phys. Lett.* 80: 4232–4234.

Park, W. I., Yi, G.-C., Kim, M. Y., and Pennycook, S. J. 2003. Quantum confinement observed in ZnO/ZnMgO nanorod heterostructure. *Adv. Mater.* 15: 526–529.

Sakakura, N. and Masumoto, Y. 1997. Persistent spectral-hole-burning spectroscopy of CuCl quantum cubes. *Phys. Rev. B* 56: 4051–4055.

Sangu, S., Kobayashi, K., Kawazoe, T., Shojiguchi, A., and Ohtsu, M. 2003. Quantum-coherence effect in a quantum dot system coupled by optical near fields. *Trans. Mater. Res. Soc. Jpn.* 28: 1035–1038.

Sangu, S., Kobayashi, K., Shojiguchi, A., and Ohtsu, M. 2004. Logic and functional operations using a near-field optically coupled quantum-dot system. *Phys. Rev. B* 69: 115334.

Spanhel, L. and Anderson, M. A. 1991. Semiconductor clusters in the sol-gel process; quantized aggregation, gelation, and crystal growth in concentrated ZnO colloids. *J. Am. Chem. Soc.* 113: 2826–2833.

Stinaff, E. A., Scheibner, M., Bracker, A. S. et al. 2006. Optical signatures of coupled quantum dots. *Science* 311: 636–639.

Sun, H. D., Makino, T., Segawa, Y. et al. 2002. Enhancement of exciton binding energies in ZnO/ZnMgO multiquantum wells. *J. Appl. Phys.* 91: 1993–1997.

Suzuki, T., Mitsuyu, T., Nishi, K. et al. 1996. Observation of ultra-fast all-optical modulation based on intersubband transition in n-doped quantum wells by using free electron laser. *Appl. Phys. Lett.* 69: 4136–4138.

Trallero-Giner, C., Debernardi, A., Cardona, M., Menendez-Proupin, M., and Ekimov, A. I. 1998. Optical vibrons in CdSe dots and dispersion relation of the bulk material. *Phys. Rev. B* 57: 4664–4669.

Yatsui, T., Jeong, H., and Ohtsu, M. 2008. Controlling the energy transfer between near-field optically coupled ZnO quantum dots. *Appl. Phys. B* 93: 199–202.

Yatsui, T., Sangu, S., Kawazoe, T., Ohtsu, M., An, S. J., Yoo, J., and Yi, G.-C. 2007. Nanophotonic switch using ZnO nano-rod double-quantum-well structures. *Appl. Phys. Lett.* 90: 223110.

Waveguides for Nanophotonics

Jan Valenta
Charles University

Tomáš Ostatnický
Charles University

Ivan Pelant
*Academy of Sciences
of the Czech Republic*

37.1 Introduction

The term *nanophotonics* may be understood as abbreviation from "photonics of nanostructures." This rapidly evolving research area deals with light interaction in nanostructured materials and their applications in photonic devices like light sources, modulators, detectors, etc.

37.1.1 Basic Types of Waveguides for Nanophotonics

Optical waveguides are structures that are able to confine and guide optical electromagnetic field. *Classical waveguides* (optical fibers, Figure 37.1a) use refractive index difference between the core and the cladding layer to guide light by total internal reflection. When the size of waveguide approaches the wavelength of light, the confinement decreases and losses increase. Therefore, the size of practical classical waveguides is limited to several hundreds of nanometers. Somewhat better confinement can be achieved in more complex waveguides with guiding properties determined by the formation of photonic bands due to their regular spatial structure (Figure 37.1b and c): (1) *photonic crystal waveguides* using "defects" in periodic structures of photonic crystals to confine light and (2) *plasmonic waveguides* based on metallic nanostructures with plasmon resonance (see, e.g., Pavesi and Guillot 2006). Due to limited space, we are going to describe only a special type of the "classical" waveguides formed by luminescing silicon nanocrystals (Si-nc).

37.1.2 Silicon Nanophotonics

The recent years can be characterized by the association of microelectronics with optoelectronics or photonics. While the microelectronics is based almost exclusively on silicon (indirect band-gap semiconductor), photonic light sources are currently made out of direct band-gap III–V compounds (family of GaAs materials). In order to reduce the material diversity, an effort to develop an efficient, electrically pumped silicon-compatible light-emitting device is becoming very strong (Pavesi and Lockwood 2004). This tendency to "siliconize photonics" is driven also by the need to reduce the overheating of silicon integrated circuits due to the ohmic resistance of excessively long multilevel metal interconnects (Pavesi and Guillot 2006). Supplementary advantages like the reduction in charging (RC) time constant (speeding up the circuit performance) and prevention from crosstalks can be obtained as a bonus.

Since bulk silicon, as an indirect band-gap semiconductor, is a very bad light emitter, nanometer-sized silicon nanocrystals, brightly luminescent at room temperature, represent one of the possible solutions (Pavesi and Lockwood 2004). It is then expected that the light signal originating in Si-nc and carrying the required information propagates in a low-loss medium, which assures, at the same time, a good directionality of the radiation. It has been discovered recently that slabs of fused quartz SiO_2 "doped" with Si-nc are able to accomplish both the role of a light emitter and that of a waveguide (Khriachtchev et al. 2001, Valenta et al. 2003a, Khriachtchev et al. 2004). We shall call this type of nanophotonic waveguides (which generate the luminous signal themselves and therefore there is no need to

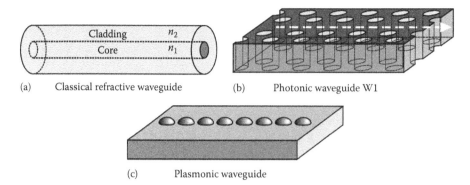

FIGURE 37.1 Nanophotonic waveguides: (a) "Classical" waveguide based on total internal reflection using the core with higher refractive index than the cladding. (b) Photonic crystal waveguide formed by a "defect" in the regular structure—here is the W1 waveguide formed by a row of missing holes. (c) Plasmonic waveguide formed by a row of metallic nanocrystals.

couple the light to them from an external source) "active waveguides." In this chapter, we shall describe the preparation methods of active nanocrystalline waveguides, their experimentally observed properties, and relevant theoretical description. We shall also briefly mention their various application possibilities.

37.2 Fabrication of Planar and Rib Waveguides

There have been many various methods of how to grow thin sheets of luminescent Si-nc embedded in an optically transparent matrix: (1) *Implantation of fused quartz* (SiO_2) slabs with Si^+ ions (Figure 37.2) (Cheylan et al. 2000), (2) *reactive Si deposition onto fused quartz* (Khriachtchev et al. 2002), (3) *plasma-enhanced chemical vapor deposition* (PECVD) of substoichiometric silicon oxide SiO_x thin films on a Si substrate (Iacona et al. 2000), (4) *co-sputtering of a Si wafer and a piece of glass* using fused quartz plates or Si wafers as a substrate (Imakita et al. 2005),

(5) *crystallization of Si/SiO$_2$ superlattices* with nm-thick amorphous Si or SiO layers (Riboli et al. 2004), and (6) *embedding of porous silicon grains into sol-gel derived SiO$_2$ layers* (Luterová et al. 2004), to list at least the most frequently used. Most of these techniques comprise high-temperature (1100°C–1200°C) annealing of deposited SiO_x films in order to achieve the phase separation between Si-nc and the SiO_2 matrix. The thickness of the resulting sheets containing Si-nc can vary from hundreds of nanometers to tens of micrometers. Due to the difference in refractive index between the matrix ($n_{silica} \approx 1.45$) and silicon ($n_{Si} \approx 3$), such sheets act as planar or rib waveguides. Interestingly, the attractive waveguiding features that we are going to discuss, namely, the wavelength selective guiding of light (which can be also called "spectral filtering") and microcavity-like behavior, are critically dependent upon the preparation method: Till now, they have been discovered only in samples fabricated using the first two methods. We describe them in detail.

Implantation of accelerated Si^+ ions can be applied either to fused quartz (silica) slabs or to a thin SiO_2 layer thermally

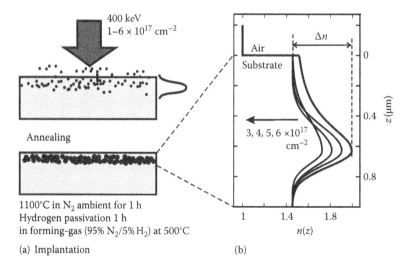

FIGURE 37.2 (a) The schematics of a Si-nc planar waveguide preparation. (b) Refractive index profiles (n as a function of the depth z beneath the surface) of silicon nanocrystalline waveguides (implanted with 400 keV Si^+ ions and different implant fluences), extracted from optical transmission measurements. (After Valenta, J. et al., *J. Appl. Phys.*, 96, 5222, 2004.)

grown on a silicon wafer. Implantation energy varies usually between 30 and 600 keV, and implant fluences (ion doses) are of the order of 10^{16}–10^{17} cm^{-2}. The attractive waveguide properties of fabricated planar waveguides, i.e., pronounced emission line narrowing and high output beam directionality in the visible region, have been found in waveguides prepared from 1 mm thick silica (Infrasil) slabs with optically polished surface and edges, using the implantation energy of 400 keV and the implant fluences 1×10^{17} cm^{-2} to 6×10^{17} cm^{-2}. Because the Si$^+$ ions are almost monoenergetic, their stopping distance beneath the silica slab surface has only a small dispersion, which results in narrow and slightly skewed implant distribution, as reflected in the resulting refractive index profiles shown in Figure 37.2 (Valenta et al. 2004). Peak excess Si concentration was up to 26 atomic%. Spherical silicon nanocrystals with diameter distribution roughly from 4 to 6 nm were formed by annealing the implanted samples at 1100°C in a nitrogen ambient for 1 h. To enhance several times the luminescence intensity of Si-nc, an additional anneal at 500°C in forming gas ($N_2 + H_2$) for 1 h was applied (Cheylan and Elliman 2001). The last operation promotes enhanced hydrogen passivation of non-radiative defects.

The reactive silicon deposition has yielded substoichiometric SiO_x films on circular 1 mm-thick substrate silica plates (Khriachtchev et al. 2001, 2004). Electron beam evaporation and radio frequency cells were used as silicon and oxygen sources. The SiO_x layer thickness was ~2 μm, and the value of the x parameter varied around $x = 1.70$. The as-grown material contained amorphous Si inclusions. Annealing of these "suboxide" films at 1100°C in nitrogen atmosphere resulted in the formation of well-defined Si-nc with diameters of 3–4 nm, as evidenced by Raman spectroscopy. Extensive investigations

of this type of silica waveguides containing Si-nc discovered a small effective optical birefringence inside the layers, however, without radically influencing the properties of the waveguides (Khriachtchev et al. 2007). Refractive index profile across the thickness of these waveguides is flat, unlike the preceding case of the implanted layers.

The reason why Si-nc waveguides fabricated using other techniques do not show the spectral filtering effect are not fully clear at present. Of course, the condition *sine non qua* is asymmetry of the waveguide, i.e., the waveguiding layer must not be sandwiched between two materials (cladding layers) having the same effective refractive index, as we shall see in the next section. One can speculate about several further critical parameters: suitable composition and optical quality of the matrix; suitable density of Si-nc (in the order of 10^{18} cm^{-3}); favorable layer thickness (around 1 μm); excellent flatness and parallelism of both interfaces; and, last but not least, certain optimum value of propagation losses (several dB/cm). The dominant losses in the waveguides under discussion are probably due to self-absorption and Mie light scattering on Si-nc clusters, not due to surface roughness (Pelant et al. 2006).

The above discussion has dealt implicitly with planar waveguides. Rib-loaded waveguides containing nanocrystals seem, in principle, even more desirable for nanophotonic applications. They can be also prepared using the ion implantation technique in either thin SiO_2 films thermally grown on Si or in polished fused quartz slabs, as described above. The rib structure can be formed (using standard photolithography and etching) by two possible means: either before or after the implantation procedure. Figure 37.3a represents schematics of such a rib waveguide structure.

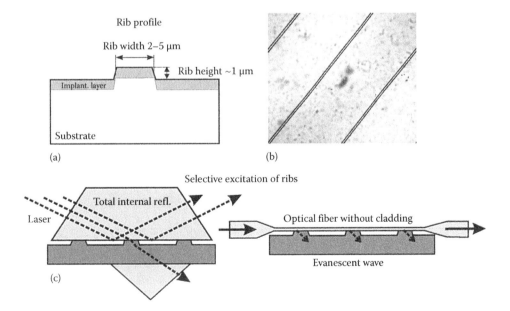

(a) (b) (c)

FIGURE 37.3 (a) Sketch of a rib waveguide structure with silicon nanocrystals, prepared on a polished silica slab. The ribs were fabricated by optical lithography before Si$^+$ implantation. (b) Microscopic (reflection) image of a part of the structure. The spacing between the ribs is 100 μm. (c) Methods to selectively excite the rib waveguides through the TIR prism or the evanescent field of a fiber. (After Skopalová, E., Mode structure in the light emission from planar waveguides with silicon nanocrystals, Diploma thesis, Charles University, Prague, Czech Republic, 2007.)

Lateral confinement of the light in a narrow rib is expected, in comparison with the planar waveguides, to have some advantages: saving considerable space in photonics circuits and offering much wider variability in their design. On the other hand, the transition from the two-dimensional to the one-dimensional case can degrade the optical quality of the waveguide and introduce additional losses, such as those due to sidewall roughness or even due to considerably deformed rib cross section. Indeed, measurements of propagation losses in similar waveguides have given values above 10 dB/cm, and this loss coefficient even increases for rib widths below 4 μm (Pellegrino et al. 2005). Therefore, in what follows, we discuss predominantly the two-dimensional nanocrystalline waveguides, and limited space only will be given to the rib structures.

37.3 Experimental Techniques

37.3.1 Techniques to Study Internal Photoluminescence Propagation in Waveguides

One of the most important advantages of active waveguides is that light sources are embedded in the waveguide. In our material, Si nanocrystals forming the waveguide core can efficiently emit luminescence in the orange-red-infrared spectral range when excited with the UV-blue light (usually focused laser beam). Such internally produced light is automatically coupled to all possible modes of the waveguide: radiation, substrate, and

guided modes (see Section 37.5). The emission modes may be distinguished by measuring their propagation direction, spectra, and polarization, which imposes requirements to experimental setup.

Two types of setups can be conveniently applied to study luminescence from active waveguides: (1) The micro-imaging-spectroscopy setup is based on a microscope connected to an imaging spectrograph with a CCD camera (Figure 37.4a). For good angular resolution, low numerical aperture (NA) objective lenses should be used (in our experiments, we used mostly the lens with 2.5 × magnification and NA = 0.075, i.e., an angular resolution of about 8.6°). The sample is fixed to an $x-y-z$ table with a rotating holder and excited by a laser beam (325 nm from the cw He–Cd laser), approximately perpendicular to the objective axis. (2) In order to achieve better angular resolution, an experimental arrangement based on a goniometer is employed (Figure 37.4b). Here, the sample is fixed to the center of the goniometer, and the photoluminescence emission is collected by a silica optical fiber (core diameter 1 mm) rotated around the sample at a distance of 50 mm, giving an angular resolution slightly less than 1° (NA ~ 0.01). The output of the fiber can be coupled to the same detection system (a spectrograph with a CCD camera), as described above.

Typical images observed with the microscopic setup (Figure 37.4a) are illustrated in Figure 37.4c and d. Here, the diameter of the excitation spot, located about 1 mm from the sample edge, is roughly 1 mm. One can easily recognize PL emission from the excited spot as a bright ellipsoid. However, there is also a second

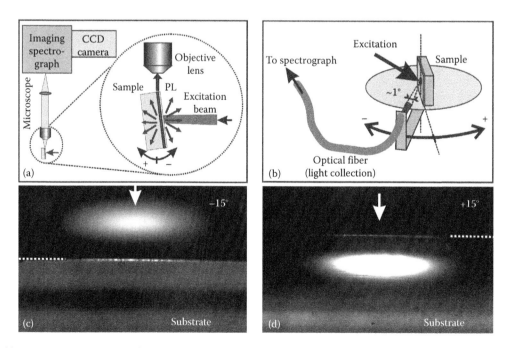

FIGURE 37.4 (a) Micro-spectroscopy setup for studying spatially resolved emission from waveguides, (b) collection of signals from a waveguide using optical fiber mounted on a goniometer. (c,d) Luminescence microimages of the active planar waveguide (obtained with the setup illustrated in panel a), where the elliptical spot is the spot excited by the laser beam and the light line is emission emanating from the sample facet. The inclination of the sample with regard to optical axis is −15° and +15°, respectively. (Adapted from Valenta, J. et al., *J. Appl. Phys.*, 96, 5222, 2004.)

contribution emanating from the facet of the sample. This light is obviously guided in the implanted layer or close to it. The images in Figure 37.4c and d were collected for sample inclination angles of −15° and +15°, respectively, i.e., in a geometry for which the excited spot was observed either directly (Figure 37.4c) or through the substrate (Figure 37.4d). The experimental arrangement shown in Figure 37.4a enables the detection of the PL either from the excited spot or from the edge of sample by positioning the entrance slit of the spectrograph to different locations of the PL image. All experiments described in this chapter were performed at room temperature.

37.3.2 Techniques to Study Propagation of Light from External Sources

The coupling of external light to narrow submicrometer waveguides is a difficult task. The most used approaches are as follows: (1) *Prism coupling of light* from the surface of the sample (Figure 37.5a). Light from Xe lamp, halogen lamp, or LED is collimated into a prism in contact with waveguide. For better optical contact, an immerse liquid should be dropped between the prism and the sample (the best optical contact—i.e., minimizing reflections on interface—is achieved when refractive index of the immersion liquid is between the values for the materials to be connected). A second prism may be placed on the opposite surface of waveguide to couple out the passing beam to avoid total internal reflection. (2) *Direct coupling into the facet* (Figure 37.5b, sometimes called *end-fire coupling*). The edge of the sample can be polished at some angle (here 70°) in order to separate light refracted to the higher index waveguide from light

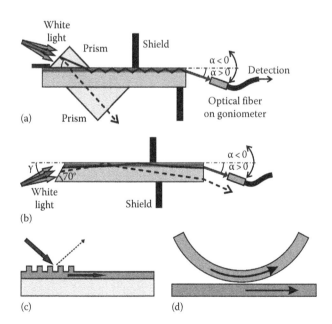

(a)

(b)

(c)   (d)

FIGURE 37.5 External light coupling into a narrow waveguide: (a) prism coupling, (b) end-fire coupling, (c) grating coupling, and (d) evanescent-wave coupling. (Adapted from Janda, P. et al., *J. Lumin.*, 121, 267, 2006; Pavesi, L. and Guillot, G. (Eds.), *Optical Interconnects. The Silicon Approach*, Springer-Verlag, Berlin, Germany, 2006.)

entering lower index substrate. For both the external light coupling setups, the signal can be collected by an optical fiber and guided to the entrance slit of the imaging spectrometer with a CCD detector. The other coupling methods described in literature are *grating coupling* and *evanescent wave coupling* (sometimes called waveguide-to-waveguide coupling) (see Figure 37.5c and d).

37.3.3 Techniques to Study Losses and Optical Amplification in Waveguides

The losses in waveguide are divided into *insertion (coupling) losses* and *propagation losses*, which contain *scattering, radiation*, and *absorption* (Figure 37.6a). If the waveguide material can be pumped to reach population inversion, the guided light may be amplified by stimulated emission. In this case, the absorption coefficient α becomes negative and it is called the gain coefficient g ($g = -\alpha$).

The most used methods to measure losses in waveguides are (Figure 37.6b through e) (1) *"cutback" method* detecting light output from waveguide with different length; (2) *scattering detection* from different points of waveguide surface; (3) *Fabry–Perot resonance* method, which can be applied to waveguides with good facets; and (4) *shifting excitation spot* (SES) method; which can be applied to active waveguides.

Optical gain is measured by the *variable stripe-length* (VSL) method or by the *pump-and-probe* (PP) method. The principle of the VSL method consists in excitation of a narrow stripe-like region at the edge of a sample (Figure 37.7a) (Shaklee and Leheny 1971, Valenta et al. 2003b, Dal Negro et al. 2004). If the pumping is able to induce population inversion in studied material, photons passing through the excited region can be amplified by stimulated emission. It corresponds to a single-pass optical amplifier. If the emission from the sample edge $I^{VSL}(l,\lambda)$ is detected as a function of the stripe length l, the net optical gain $G(\lambda)$ is calculated from a simple equation

$$I^{VSL}(l,\lambda) = \frac{I_{sp}(\lambda)}{G(\lambda)}\Big[\exp(G(\lambda)\cdot l) - 1\Big], \qquad (37.1)$$

where

I_{sp} is the intensity of spontaneous emission
$G(\lambda) = [g(\lambda) - K]$ is net optical gain (K stands for losses and $g(\lambda)$ is material gain coefficient, i.e., negative absorption coefficient) (Valenta et al. 2002)

Equation 37.1 is valid only when $I_{sp}(\lambda)$ and $g(\lambda)$ are constant over the whole excited stripe. It means that the exciting power density and properties of sample must be uniform. Also, the coupling of the output emission to a detector must be constant, independent of x, for any part of the stripe. In reality, the geometry and emission properties of the sample are never perfectly constant, and the coupling of light is influenced by directionality of emitted light, limited collection angle (NA), the confocal effect of imaging optics, etc. (Valenta et al. 2003b). Therefore, the SES method has been proposed to correct the VSL method (Valenta et al.

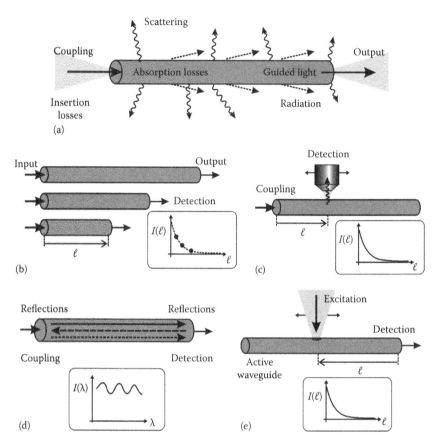

FIGURE 37.6 (a) Schematic illustration of losses in a waveguide. Methods to measure losses: (b) cutback, (c) scattering light detection, (d) Fabry–Perot resonance, and (e) SES methods.

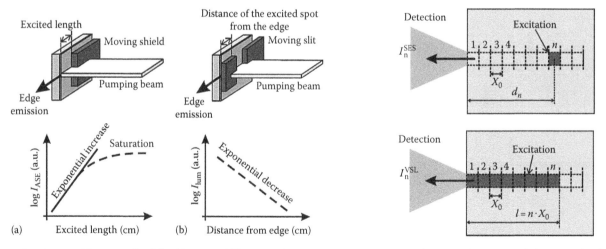

FIGURE 37.7 The principle of the (a) VSL and (b) SES methods and typical dependence of the detected signal as function of the excited length and the distance from edge, respectively (lower panel). (Adapted from Park, J.H. and Steckl, A.J., *Appl. Phys. Lett.*, 85, 4588, 2004.)

FIGURE 37.8 Schematic illustration of the principle of combining the SES and VSL experiments.

2002). The principle of the SES method takes advantage of the fact that instead of the whole stripe, only small segments are excited and measured separately (Figure 37.7b).

Figure 37.8 illustrates the procedure to compare the VSL and SES results. The most transparent approach consists of SES measurement with the spot width equal to the shift step. Let us assume that the VSL measurement is done using the same elemental steps x_0 as the SES experiment (obviously, also the stripe width and excitation density must be the same for both SES and VSL). Then we can compare the VSL signal I_n^{VSL} for a stripe consisting of n elemental steps ($l = n \cdot x_0$) with integrated SES signal I_n^{iSES}, i.e., sum of SES signals from n steps

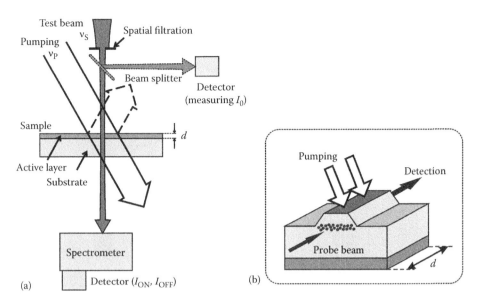

FIGURE 37.9 The basic arrangement of the PP experiment for a (a) planar and (b) rib waveguide.

$$I_n^{\text{iSES}} = \sum_{k=1}^{n} I_k^{\text{SES}}.$$

Integrated SES signal should contain the same amount of spontaneous emission as the VSL measurement (including losses and effects of imperfect experimental conditions), the only difference being that in VSL experiment some photons go through the excited area and might be amplified by stimulated emission or affected by induced absorption. Which effect takes place is clear from plotting both I_n^{VSL} and I_n^{iSES} in one figure.

Similarly, we can calculate the differential VSL intensity I_n^{dVSL}, i.e., the difference of signals obtained for stripe lengths $l = n \cdot x_0$ and $l = (n-1) \cdot x_0$

$$I_n^{\text{dVSL}} = I_n^{\text{VSL}} - I_{n-1}^{\text{VSL}}$$

and compare it with I_n^{SES}. Again, the detected signal comes from spontaneous emission in the nth excited spot that is passing through the excited or unexcited area for the dVSL and SES, respectively. In the ideal case, we can use equations $I_n^{\text{dVSL}} = \text{const} \exp[(g-K)d_n]$ and $I_n^{\text{SES}} = \text{const} \exp(-Kd_n)$ to derive the net optical gain $g(\lambda)$

$$g(\lambda) = \frac{\ln\left(I_n^{\text{dVSL}}(\lambda)\,/\,I_n^{\text{SES}}(\lambda)\right)}{d_n},$$

where d_n is the distance of the center of the nth spot from the edge.

The PP technique is based on the application of two beams: the first one is pumping the active sample and the second one is testing the state created by the pumping beam. In case of studying stimulated emission, the test beam may be amplified by stimulated emission. In general, the test beam spot must be situated inside the pumping spot in order to test area pumped as homogeneously as possible. In case of pulsed beams, the temporal coincidence of both pumping and testing pulses must be controlled (the test beam may be delayed after pumping pulses to study decay of the effect). If the thickness of the active layer is small, the induced effect on the test beam may be difficult to detect. Therefore, for waveguide samples, the test beam is often coupled to the waveguide, which is excited from the surface (Figure 37.9b). Then, changes of the outcoupled light with and without pumping are detected.

37.4 Main Experimental Observations in Active Si-nc Waveguides

The most important observations for active planar waveguides are (1) spectral filtering; (2) TE (transverse electric) and TM (transverse magnetic) mode splitting; and (3) high directionality of TE, TM mode emission. These effects are illustrated in Figure 37.10; the panel (a) shows that the PL leaving the edge of waveguide layer is substantially different from the PL perpendicular to the layer. The perpendicular PL is formed by a single band that corresponds to the well-known emission of Si nanocrystals, but the edge PL contains two narrow PL bands. Both narrow lines are linearly polarized, one in the direction of the waveguide layer (TE mode) and the second one perpendicular to the layer (TM mode). The spectral position of the TE and TM modes depends on the exact profile of the refractive index, it means on the implantation fluence (see also Figure 37.16, which shows spectra of another set of samples). The panels (b) and (c) illustrate that the TE and TM modes are emitted only in an extremely narrow angle (a few degrees), basically in the direction of the waveguide plane (only slightly inclined to the substrate, i.e., under positive angles in our notation).

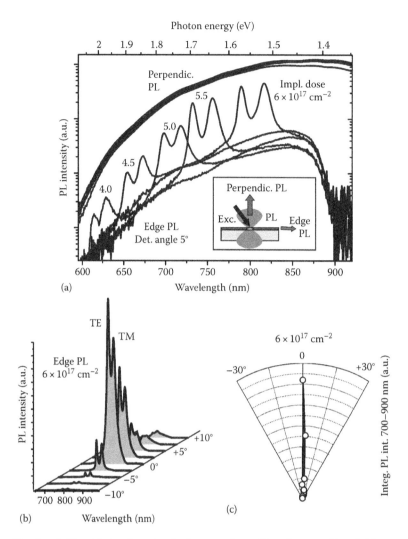

FIGURE 37.10 PL spectra of five fused silica slabs implanted to the fluences of 4×10^{17} cm^{-2} to 6×10^{17} cm^{-2}. (a) Upper curves (a single wide band) correspond to PL emitted in a direction perpendicular to the waveguide, while lower spectra with doublet peaks are facet-PL detected in a direction $\alpha = 5°$ (a sketch of the experimental arrangement is shown in the inset). (b) Angle-resolved facet PL spectra of the sample 6×10^{17} cm^{-2}. (c) Polar representation of integrated PL intensity of the angle-resolved facet spectra from the panel b. Most of the PL intensity is emitted in a direction close to 0°. (Adapted from Janda, P. et al., *J. Lumin.*, 121, 267, 2006.)

The coupling and propagation of *external* light in active Si-nc waveguides were studied by the direct (end-fire) and the prism coupling (Figure 37.5a and b). The results for one sample are illustrated in Figure 37.11. For prism coupling, two broad transmission bands (in the blue and red spectral region) are observed in the measured spectral range. The positions of both bands coincide with those of the PL modes (Figure 37.10a). Our calculations show that the red and blue bands corresponds to the second and third order of substrate modes (the first order being in the infrared region), see Section 37.5 and Figure 37.14. Broadening of the mode structure may be a consequence of the very low number of reflections undertaken by coupled light before escaping to the substrate. Coupling of external light through a truncated facet (Figure 37.5b) gives the best result for a coupling angle $\gamma \sim 20°$, as expected (Figure 37.11, upper curves). In this configuration, the narrow and polarization-split peaks at an output angle $\alpha \sim 2°$ are detected. The peaks

are, however, not transmission but absorption peaks. This can be understood if it is assumed that the detected light comes not from substrate modes (which represent a small portion of transmitted light) but from filtered transmitted light propagating almost parallel to the Si-nc waveguide from which a part of power escaped to the substrate modes that are, however, gradually absorbed in the waveguide core. The blue third order modes are much stronger compared to second order because of higher absorption in the blue spectral region.

Experimental observation of rib waveguides indicates (Figure 37.12) that normal guiding of light is improved—see increased intensity of the long-wave edge of PL spectra in Figure 37.12a and b. This is due to the introduction of confinement in the second lateral direction of waveguide (in this case, the confining refractive index profile is symmetrical and the contrast higher (compared to the implanted layer profile) as the surrounding medium is air). On the other hand, roughness of

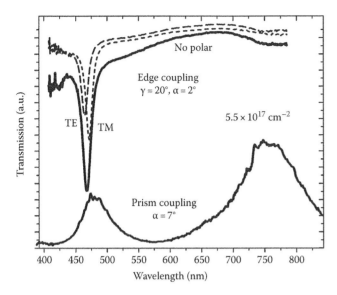

FIGURE 37.11 Comparison of transmission spectra of a sample 5.5×10^{17} cm^{-2} obtained by the direct facet-coupling (upper curves, solid line—no polarization, dashed and dotted lines correspond to TE and TM polarization, respectively) and by the prism coupling (lower spectrum). (Adapted from Janda, P. et al., *J. Lumin.*, 121, 267, 2006.)

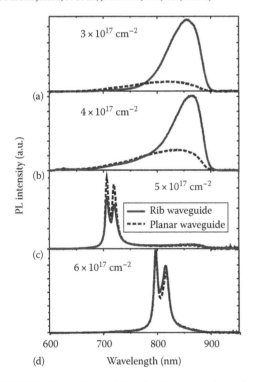

FIGURE 37.12 Comparison of the edge PL spectra from planar and rib waveguides (dashed and continuous line, respectively) in samples implanted to a fluence of (a) 3×10^{17} cm^{-2}, (b) 4×10^{17} cm^{-2}, (c) 5×10^{17} cm^{-2}, and (d) 6×10^{17} cm^{-2}.

the rib edges can introduce significant losses due to scattering. The influence of rib structure on the substrate (leaky) modes is less significant (see Figure 37.12c and d). This is because these modes propagate mostly in substrate, where the effect of side walls is negligible.

37.5 Theoretical Description of Active Lossy Waveguides

Light coupling to a waveguide and its further propagation may be understood in terms of coupling between the waveguide modes and nanocrystals emitting photons. Light field inside the waveguide may be expressed as a coherent superposition of excited waveguide modes, which arise as a solution of a homogeneous wave equation with proper boundary conditions [the waveguide modes are an orthonormal basis of the waveguide optical field (Snyder and Love 1983)]. It is therefore necessary to distinguish between various types of the waveguide modes in order to explain different types of energy transport in the waveguides.

Let us consider a common model structure with three transparent layers (see Figure 37.13): the top *cladding* layer with the refractive index n_1, the bottom *substrate* with the refractive index n_3, and the *core* with the refractive index $n_2 > n_3 > n_1$ and thickness d. We may imagine the modes as rays emerging from inside the waveguide core and propagating toward one of the core boundaries. At the boundary, the ray may be either totally reflected back to the core or it may be partially reflected and partially refracted. The ray then travels toward the other boundary where it totally or partially reflects again. Only those modes that are totally reflected twice during one round-trip may propagate in the core without losses: these are the common *guided modes*. There are, in addition, lossy modes that are called the *substrate radiation modes* (they are refracted only into the substrate) and the *radiation modes* refracted both to the substrate and the cladding.

In many applications, considering traveling of light at large distances compared to its wavelength (and thus working in the far-field limit), we may consider only the guided modes as efficient carriers of energy between a source and a detector. In silicon-based nanophotonics devices, however, the situation may be even more complicated. The typical distances between components on a chip may be comparable to the light wavelength and the far-field approach fails. It is therefore important to analyze an influence of the substrate radiation modes on the response. We may rule out the radiation modes at this stage: compared to the substrate modes, they play only a negligible role in the overall system behavior.

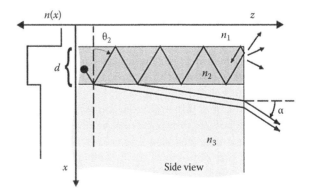

FIGURE 37.13 Schematic representation of propagation and decoupling of substrate modes.

Once light is emitted by a nanocrystal inside the waveguide mode, it undergoes internal reflections at the core/cladding and the core/substrate boundaries. The overall intensity of electric field of an excited mode inside the waveguide core may be written as superposition of the partial waves:

$$E_{mode}(\lambda, \theta_2) = E_0(\lambda)\left(1 + r_{21}r_{23}\exp\left[\frac{4\pi i n_2 d\cos(\theta_2)}{\lambda}\right] + \cdots\right)$$

$$= \frac{E_0(\lambda)}{1 - r_{21}r_{23}\exp[4\pi i n_2 d\cos(\theta_2)/\lambda]}. \quad (37.2)$$

Here, r_{21} and r_{23} are the reflection coefficients at the respective boundaries, which depend upon the propagation angle θ_2. The wavelength is λ and $E_0(\lambda)$ is an effective emission amplitude of a nanocrystal to the waveguide mode. Considering the guided modes and thus $|r_{21}| = |r_{23}| = 1$, the electric field intensity reveals sharp resonances implying there are well-defined discrete guided modes. The energy carried by each mode is finite, indeed. Unlike the guided modes, the electric field intensity is finite for the substrate modes and there are no sharp resonances in the spectra. We may, however, resolve some weak resonances, depending on the phase factor of the term $r_{21}r_{23}\exp[4\pi i n_2 d\cos(\theta_2)/\lambda]$: the electric field is maximum if this term is positive and real.

It is clear from the ray optics that there are substrate modes that refract to the substrate at the angles near $\pi/2$, which means that they propagate nearly parallel to the core/substrate boundary; these modes may be understood as a crossover between the guided modes and the substrate radiation modes which decouple rapidly from the core, possessing mixed characteristics of the both types of modes. As the modes may be assigned to a distinct guided mode of the nth order (energy of the substrate mode is slightly below the cutoff energy of the nth guided mode), we denote each series of the substrate modes to be of the nth order (see Figure 37.14). The Fresnel formulae give us $|r_{21}| = 1$ and $|r_{23}| \approx 1$ in such a case (Figure 37.15), and it is then clear from Equation 37.2 that these substrate modes reveal sharp resonances in the spectra (for a fixed angle $\theta_2 < \arcsin(n_2/n_3)$). Energy carried by the substrate modes may be therefore comparable to the energy carried by guided modes. The most important point here is that this type of substrate modes cannot be experimentally distinguished from the guided modes at short distances from the source as they propagate near the core/substrate boundary. The substrate modes therefore significantly contribute to the system response and they may play an important role in many experiments as well as in real on-a-chip devices.

Although the guided and the substrate modes propagate in a similar direction, they may behave in very different ways because they propagate in two different environments. In our particular case, the guided modes are mostly localized inside the waveguide core doped by Si-nc, providing a possibility of reabsorption or optical amplification. The substrate modes, on the contrary, propagate in a transparent substrate and they are therefore neither amplified nor attenuated. Rigorous derivation of PL spectra in real waveguides accounting for losses or

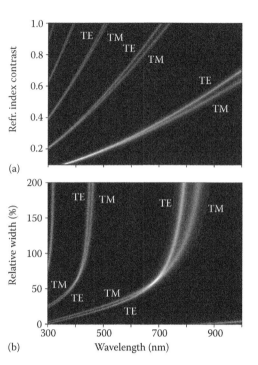

(a)

(b)

FIGURE 37.14 (See color insert following page 21-4.) Calculated spectral positions of the substrate modes as a function of (a) the refractive index contrast $\Delta n = n_2 - n_3$ and (b) the relative thickness of the waveguide core compared to the sample 5×10^{17} Si cm^{-2}. Gray scale indicates intensity from black up to white for the highest intensity. Several orders of modes are seen starting from the first one in infrared region.

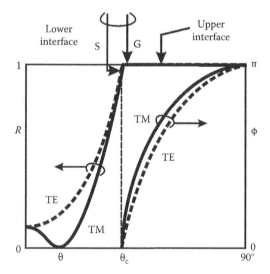

FIGURE 37.15 Reflectance $R = |r_{ij}|^2$ and phase shifts ϕ on the planar boundary between two dielectric media plotted for TE and TM modes versus incident angle θ. S and G stand for substrate and guided modes, respectively. (Adapted from Pelant, I. et al., *Appl. Phys. B*, 83, 87, 2006.)

amplification, decoupling of the light from the waveguide and real detection geometry is presented elsewhere (Ostatnický et al. 2008); here, we summarize only the main results.

Considering lossy waveguide core, the substrate modes become dominant in the optical response of a waveguide when detected in the direction parallel to the core/substrate boundary using a

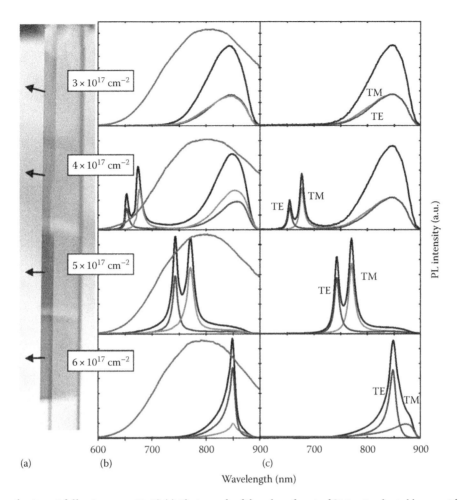

FIGURE 37.16 (See color insert following page 21-4.) (a) Photograph of the edge of a set of Si+ ion implanted layers with direction of PL indicated by arrows, the edge is on the left. (b) Measured PL from samples implanted to different Si ion fluences in standard (the broadest curves) and waveguiding geometry (black lines, the slightly lighter gray lines stand for TE and TM resolved polarizations). (c) Theoretically calculated PL spectra. We note that these results were obtained on different set of samples than in Figure 37.10. The mode positions are not exactly the same for samples with identical implantation dose because the annealing conditions were slightly different. Therefore, the refractive index profiles are not identical. (Adapted from Pelant, I. et al., *Appl. Phys. B*, 83, 87, 2006.)

detector with a small numerical aperture. This situation is clearly illustrated through the comparison of our experimental data with the theoretical estimates in Figure 37.16 (we considered a continuous refractive index profile in our calculations). The broad part of the emission spectra originates from the (attenuated) guided modes and the narrow peaks are due to the substrate modes. The double-peak structure is further resolved as two single peaks with the respective TE and TM polarizations. In order to explain this point, we should remark that the coefficient r_{23} is real for the substrate modes while the coefficient r_{21} is in general complex. Considering a fixed angle of incidence θ_2, r_{21} gives different values of the phase shift for the respective TE and TM polarizations (see Figures 37.14 and 37.15). Therefore, the resonance condition is met at different wavelengths for different polarizations. From these considerations, the necessity of waveguide asymmetry for the appearance of the TE–TM polarization–resolved modes follows.

The TE–TM splitting was also reported in Khriachtchev et al. (2004), but the respective peaks have an opposite order in the spectra compared to our experiments. Our numerical calculations have shown that the TE–TM splitting depends strongly on the refractive index profile of the waveguide core. For the three-layer structures, the TE mode is positioned always at the high-energy side of the PL spectra. It is, nevertheless, possible to fabricate a structure with multiple layers or with a graded refractive index profile, which provides an arbitrary TE–TM splitting (including both negative and positive values and also their degeneracy); this feature is well illustrated in Figure 37.14b, where the respective TE and TM modes interchange depending on the thickness of the guiding layer.

The appearance of the substrate modes is well controllable by the parameters of the particular layers of the waveguide. Clearly, the optical thickness of the core d determines the resonance condition through the exponent in the denominator in Equation 37.2 and the ratio n_2/n_1 has influence on the phase of the r_{21} term. The latter may be used in detectors integrated on a chip—the cladding can be made of a material that changes its refractive index due to the changes of the surrounding environment conditions, and the position of the emission peak may then

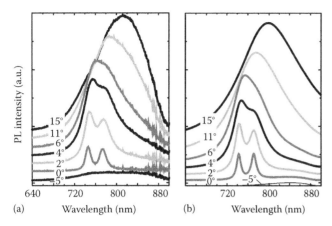

(a) Wavelength (nm)   (b) Wavelength (nm)

FIGURE 37.17 (a) Measured PL spectra at different detection angles α relative to the waveguide axis. (b) Numerical simulation of the measurements from the panel a. The sample with an implantation fluence of 5×10^{17} cm^{-2}. (Adapted from Ostatnický, T. et al., Guiding and amplification of light due to silicon nanocrystals embedded in waveguides, in Khriachtchev, L. (ed.), *Silicon Nanophotonics*, World Scientific, Singapore, 2008, pp. 267–296.)

be correlated with some external variable (temperature, pH, etc., see Figure 37.20) (Luterová et al. 2006).

As the sharp resonance in the spectra of the substrate modes is restricted only to the region where $r_{23} \approx 1$, we may expect disappearance of the substrate modes at large detection angles, i.e., when the substrate modes do not propagate parallel to the core/substrate boundary. This feature is illustrated in Figure 37.17 in comparison to the experimental and the theoretical data. We see a very good agreement justifying the correctness of our model.

An important aspect of nanodevices is the magnitude of the decoupling length for the substrate modes, i.e., the distance at which the substrate modes completely leak out from the waveguide core. Our calculations (Ostatnický et al. 2008) show that this distance may be few micrometers but also it may be as large as several hundreds of micrometers. As a consequence, the appearance of the substrate modes and their decoupling from the waveguide core may be of a big importance in nanodevices on the contrary to the fiber optics where the optical response to any excitation is provided solely by the guided modes at the typical distances of the order from meters to kilometers.

37.6 Application of Active Nanocrystalline Waveguides

37.6.1 Amplification of Light

The significant narrowing of substrate leaky mode emission suggests immediately that optical amplification could be responsible for this observation. The VSL method (described in Section 37.3.3) has been applied to the planar asymmetric Si-nc waveguides in several laboratories (Khriachtchev et al. 2001, Ivanda et al. 2003, Luterová et al. 2005). It is tempting to interpret the frequently observed initial weak exponential growth of the VSL curve (Figure 37.18a) as a manifestation of

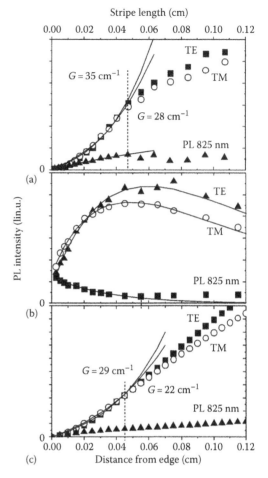

(a)

(b)

(c) Distance from edge (cm)

FIGURE 37.18 VSL and SES measurements on a sample implanted to the dose of 4×10^{17} cm^{-2} under continuous wave excitation 325 nm, 0.26 W/cm^2: (a) VSL measurement at the peak of TE and TM modes and for non-guided PL around 825 nm. The fits (lines) give values of $G = 35$ and 28 cm^{-1} for TE and TM modes, respectively, and losses of 11 cm^{-1} for non-guided PL. (b) Results of the SES measurement performed under identical conditions as the VSL. (c) Integration of data from panel (b). The gain fits (lines) give values of $G = 29$ and 22 cm^{-1} for TE and TM modes, respectively. (Adapted from Valenta, J. et al., *Appl. Phys. Lett.*, 81, 1396, 2002.)

the occurrence of optical gain. High output directionality of the substrate mode emission apparently supports such interpretation. However, it has turned out that, as shown above, it is very difficult if not impossible to evaluate correctly the optical gain magnitude of the substrate modes because the nonlinear growth can originate in the mode leaking itself. Figure 37.18 demonstrates that the evaluation of the VSL experiments can yield a false optical gain if proper comparison between the VSL and SES results is not taken into account. On the other hand, VSL measurements employing a high pulsed laser excitation, performed on a similar sample, revealed a characteristic switch from the light attenuation ($G < 0$) to amplification ($G > 0$) with increasing excitation energy density (Figure 37.19). It may be also of interest here to call the reader's attention to recent articles reporting firmly positive optical gain on leaky

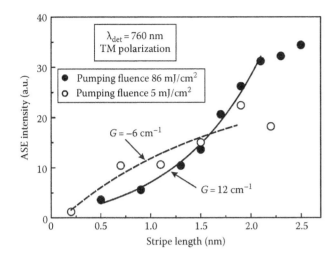

FIGURE 37.19 Time-resolved VSL measurement of optical amplification at position of the TM mode (760 nm) under pumping by 6 ns pulses, 355 nm. The fit with Equation 37.1 gives the net gain coefficient of (-6 ± 6) and (12 ± 2) cm^{-1} for pumping fluence of 5 and 86 mJ/cm^2, respectively. The threshold for positive gain is about 50 mJ/cm^2. Sample was implanted to a dose of 4×10^{17} cm^{-2}. (After Luterova, K. et al., *Phys. Status Solidi (c)*, 2, 3429, 2005.)

substrate modes in asymmetric thin-film organic waveguides (Nakanotani et al. 2007, Yokoyama et al. 2008).

At present, checking the reliability of the VSL method in the case of active asymmetric thin waveguides represents an issue requiring further investigation.

37.6.2 Other Applications

Although the benefit of the waveguide spectral filtering for nanocrystalline thin-film laser design is still controversial, the substrate modes may be attractive for other nanophotonic applications. Firstly, they provide a way how to *generate easily the spectrally narrow, polarization resolved and directional emission* in the wavelength range 650–950 nm without the necessity of building optical (micro)cavities. Moreover, this emission can be spectrally tuned simply by engineering the silicon excess content, as shown in Figures 37.10 and 37.16. Potential feasibility of these properties for *demultiplexing optical signal in photonic circuits* is evident, but improvement in quality of the rib waveguides is anticipated.

Secondly, spectral sensitivity of the radiative substrate modes to surrounding (organic) compounds can be utilized in *photonic sensing* (Figure 37.20): Magnitude of the refractive index

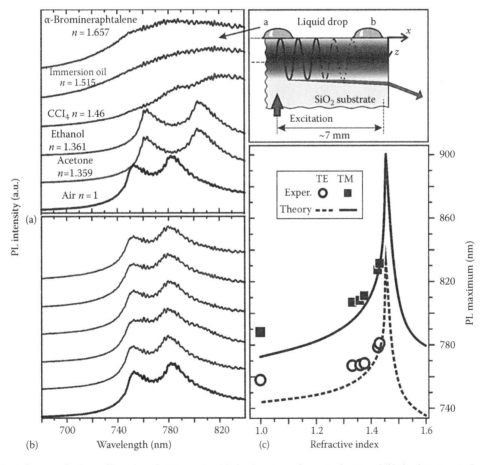

FIGURE 37.20 The influence of a drop of liquid on the PL spectra: (a) the drop is on the excited spot and (b) the drop is out of excited spot (between the spot and the sample edge). (c) Comparison of experimental (points) and calculated (line) positions of PL maxima for different refractive indices of liquid drops. Sample is implanted to a dose of 5×10^{17} cm^{-2}. (Adapted from Pelant, I. et al., *Appl. Phys. B*, 83, 87, 2006; Luterová, K. et al., *J. Appl. Phys.*, 100, 074307, 2006.)

n_1 of the compound (playing role of the cladding layer) affects markedly the spectral position of the TE–TM doublet. When n_1 exceeds the magnitude of the core refractive index ($n_2 \approx 1.45$), the condition of total reflection $|r_{21}| = 1$ is canceled and the TE–TM doublet completely disappears. A more systematic research in this direction is, however, missing.

Finally, the *active waveguides* give the chance to circumvent the *coupling problem* connected with discrete photonic elements, i.e., the problem of how to inject efficiently light from an external source in a waveguide (Orobtchouk 2006). The *active waveguides eliminate the need of any optical couplers* and enable it to design integrated photonic circuits. Theoretical analysis of active Si-nc devices integrating optical emission and waveguiding, compatible with silicon VLSI processing technology, have been submitted quite recently (Milgram et al. 2007, Redding et al. 2008).

37.7 Conclusions

Active waveguides formed by densely packed silicon nanocrystals represent a promising type of nanophotonic waveguides. In contrast to other types of nanowaveguides (based on photonic waveguides, plasmonic structures, etc.), it can be prepared by various technological approaches available and well mastered in many research and industrial laboratories. Its potential applications range from optical amplifiers and filters to optical sensors. The descriptions of characterization techniques and theory of waveguide modes in this chapter are applicable to many other waveguide structures.

Acknowledgments

This work was supported by the Czech Ministry of Education, Youth and Sports through the research center LC510 and research plans MSM0021620835 and 60840770022, the project 202/07/0818 of the Grant Agency of the Czech Republic, and the projects IAA101120804 and KAN401220801 of the Grant Agency of the Academy of Sciences. Research carried out in the Institute of Physics was supported by the Institutional Research Plan AV0Z10100521. The authors thank Prof. R.G. Elliman (Australian National University, Canberra) and Ing. V. Jurka for the fabrication of waveguides.

References

Cheylan, S. and Elliman, R. G. 2001. Effect of hydrogen on the photoluminescence of Si nanocrystals embedded in a SiO$_2$ matrix. *Applied Physics Letters* 78: 1225–1227.

Cheylan, S., Langford, N., and Elliman, R. G. 2000. The effect of ion-irradiation and annealing on the luminescence of Si nanocrystals in SiO$_2$. *Nuclear Instruments and Methods in Physical Research: Section B* 166–167: 851–856.

Dal Negro, L., Bettotti, P., Cazzanelli, M., Pacifici, D., and Pavesi, L. 2004. Applicability conditions and experimental analysis of the variable stripe length method for gain measurements. *Optics Communications* 229: 337–348.

Iacona, F., Franzò, G., and Spinella, C. 2000. Correlation between luminescence and structural properties of Si nanocrystals. *Journal of Applied Physics* 87: 1295–1303.

Imakita, K., Fujii, M., Yamaguchi, Y., and Hayashi, S. 2005. Interaction between Er ions and shallow impurities in Si nanocrystals. *Physical Review B* 71: 115440-1–115440-7.

Ivanda, M., Desnica, U. V., White, C. W., and Kiefer, W. 2003. Experimental observation of optical amplification in silicon nanocrystals. In L. Pavesi, S. Gaponenko, and L. Dal Negro (Eds.), *Towards the First Silicon Laser*, NATO Science Series, II: Mathematics, Physics and Chemistry, Vol. 93, pp. 191–96. Dordrecht, the Netherlands: Kluwer Academic Publishers.

Janda, P., Valenta, J., Ostatnický, T., Skopalová, E., Pelant, I., Elliman, R. G., and Tomasiunas, R. 2006. Nanocrystalline silicon waveguides for nanophotonics. *Journal of Luminescence* 121: 267–273.

Khriachtchev, L., Räsänen, M., Novikov, S., and Sinkkonen, J. 2001. Optical gain in Si/SiO$_2$ lattice: Experimental evidence with nanosecond pulses. *Applied Physics Letters* 79: 1249–1251.

Khriachtchev, L., Novikov, S., and Lahtinen, J. 2002. Thermal annealing of Si/SiO$_2$ materials: Modification of structural and photoluminescence emission properties. *Journal of Applied Physics* 92: 5856–5862.

Khriachtchev, L., Räsänen, M., Novikov, S., and Lahtinen, J. 2004. Tunable wavelength-selective waveguiding of photoluminescence in Si-rich silica wedges. *Journal of Applied Physics* 95: 7592–7601.

Khriachtchev, L., Navarro-Urios, D., Pavesi, L., Oton, C. J., Capuj, N. E., and Novikov, S. 2007. Spectroscopy of silica layers containing Si nanocrystals: Experimental evidence of optical birefringence. *Journal of Applied Physics* 101: 044310-1–044310-6.

Luterová, K., Dohnalová, K., Švrček, V., Pelant, I., Likforman, J.-P., Crégut, O., Gilliot, P., and Hönerlage, B. 2004. Optical gain in porous silicon grains embedded in sol-gel derived SiO$_2$ matrix under femtosecond excitation. *Applied Physics Letter* 84: 3280–3282.

Luterová, K., Navarro, D., Cazzanelli, M., Ostatnicky, T., Valenta, J., Cheylan, S., Pelant, I., and Pavesi, L. 2005. Stimulated emission in the active planar optical waveguide made of silicon nanocrystals. *Physica Status Solidi (c)* 2: 3429–3434.

Luterová, K., Skopalová, E., Pelant, I., Rejman, M., Ostatnický, T., and Valenta, J. 2006. Active planar optical waveguides with silicon nanocrystals: Leaky modes under different ambient conditions. *Journal of Applied Physics* 100: 074307.

Milgram, J. N., Wojcik, P., Mascher, P., and Knights, A. P. 2007. Optically pumped Si nanocrystal emitter integrated with low loss silicon nitride waveguides. *Optics Express* 15: 14679–14788.

Nakanotani, H., Adachi, C., Watanabe, S., and Katoh, R. 2007. Spectrally narrow emission from organic films under continuous-wave excitation. *Applied Physics Letters* 90: 231109-1–231109-3.

Orobtchouk, R. 2006. On chip optical waveguide interconnect: The problem of the in/out coupling. In L. Pavesi and G. Guillot (Eds.), *Optical Interconnects. The Silicon Approach*, pp. 263–290. Berlin, Germany: Springer-Verlag.

Ostatnický, T., Rejman, M., Valenta, J., Herynková, K., and Pelant, I. 2008. Guiding and amplification of light due to silicon nanocrystals embedded in waveguides. In L. Khriachtchev (Ed.), *Silicon Nanophotonics*, pp. 267–296. Singapore: World Scientific.

Park, J. H. and Steckl, A. J. 2004. Laser action in Eu-doped GaN thin-film cavity at room temperature. *Applied Physics Letters* 85: 4588–4590.

Pavesi, L. and Guillot, G. (Eds.). 2006. *Optical Interconnects. The Silicon Approach*. Berlin, Germany: Springer-Verlag.

Pavesi, L. and Lockwood, D. J. (Eds.). 2004. *Silicon Photonics*. Berlin, Germany: Springer-Verlag.

Pelant, I., Ostatnický, T., Valenta, J., Luterova, K., Skopalova, E., Mates, T., and Elliman, R. G. 2006. Waveguides cores containing silicon nanocrystals as active spectral filters for silicon-based photonics. *Applied Physics B* 83: 87–91.

Pellegrino, P., Garrido, B., Garcia, C., Arbiol, J., Morante, J. R., Melchiorri, M., Daldosso, N., Pavesi, L., Schedi, E., and Sarrabayrouse, G. 2005. Low-loss rib waveguides containing Si nanocrystals embedded in SiO_2. *Journal of Applied Physics* 97: 074312-1–074312-8.

Redding, B., Shi, S., Creazzo, T., and Prather, D. W. 2008. Electromagnetic modeling of active silicon nanocrystal waveguides. *Optics Express* 16: 8792–8799.

Riboli, F., Navarro-Urios, D., Chiasera A., Daldosso, N., Pavesi, L., Oton, C. J., Heitmann, J., Yi, L. X., Scholz, R., and Zacharias, M. 2004. Birefringence in optical waveguides made by silicon nanocrystal superlattices. *Applied Physics Letter* 85: 1268–1270.

Shaklee, K. L. and Leheny, R. F. 1971. Direct determination of optical gain in semiconductor crystals. *Applied Physics Letters* 18: 475–477.

Skopalová, E. 2007. Mode structure in the light emission from planar waveguides with silicon nanocrystals. Diploma thesis. Prague, Czech Republic: Charles University.

Snyder, A. W. and Love, J. D. 1983. *Optical Waveguide Theory*. London, U.K.: Chapman & Hall.

Valenta, J., Pelant, I., and Linnros, J. 2002. Waveguiding effects in the measurement of optical gain in a layer of Si nanocrystals. *Applied Physics Letters* 81: 1396–1398.

Valenta, J., Pelant, I., Luterová, K., Tomasiunas, R., Cheylan, S., Elliman, R., Linnros, J., and Honerlage, B. 2003a. Active planar optical waveguide made from luminescent silicon nanocrystals. *Applied Physics Letters* 82: 955–957.

Valenta, J., Luterová, K., Tomašiunas, R., Dohnalová, K., Hőnerlage, B., and Pelant, I. 2003b. Optical gain measurements with variable stripe length technique. In L. Pavesi, S. Gaponenko, and L. Dal Negro (Eds.), *Towards the First Silicon Laser*, NATO Science Series, II: Mathematics, Physics and Chemistry, Vol. 93, pp. 223–242. Dordrecht, the Netherlands: Kluwer Academic Publishers.

Valenta, J., Ostatnický, T., Pelant, I., Elliman, R. G., Linnros, J., and Hőnerlage, B. 2004. Microcavity-like leaky mode emission from a planar optical waveguide made of luminescent silicon nanocrystals. *Journal of Applied Physics* 96: 5222–5225.

Yokoyama, D., Moriwake, M., and Adachi, C. 2008. Spectrally narrow emissions from edges of optically and electrically pumped anisotropic organic films. *Journal of Applied Physics* 103: 123104-1–123104-13.

Biomolecular Neuronet Devices

Grigory E. Adamov
*Central Scientific Research
Institute of Technology
"Technomash"*

Evgeny P. Grebennikov
*Central Scientific Research
Institute of Technology
"Technomash"*

38.1 Introduction

Due to the constant development of nanotechnologies and functional nanomaterials used in information systems, researchers need to look for new ideas and radically new constructive solutions to ensure efficient device operation, record-breaking high speed, and integration of elements. On the other hand, functional elements being formed in nanometer range require landmark approaches. It refers especially to the formation of information-logical devices at the molecular level. That is, the formation of nanodevices by direct impact on separate molecules or atoms is not efficient in the long run. The total time taken for operations with separate molecules leads to tremendous amount of time taken to form the device and huge costs.

Another techno-systematic problem is the necessity to use "macro–nano" and "nano–macro" interfaces to put in the original information and deduce final results. The advantages of the small-size molecular elements cannot help to ensure access to them. Probably the only effective way to arrange the bonding with molecular system is to use optical interaction.

Under the circumstances, it seems quite reasonable to refer to the most competent specialist, whose experience in creating molecular systems cannot raise any doubts, that is, to the nature itself and try to apply bionic principles to engineering and design. The solution seems to be connected with hierarchic structural and functional self-organization of molecular systems that use assemblies of molecules as a basis for molecular information-logical appliances.

Among all the information systems, bionic approach seems to be the most effective, harmonic, and holistic one; it is important to point out molecular neural network appliances. Both adaptive self-organization principles of data processing and self-organization principles of functional nanostructures can be applied to those systems.

Below you will find a detailed description of techno-systematic approaches aimed at implementation of biomolecular neural network appliances. Mentioned below, biological material—protein "bacteriorhodopsin"—has unique technological potential and its optical properties allow the use of optical input–output devices and create molecular appliances based on self-organization principles.

38.1.1 Bacteriorhodopsin Is an Advanced Material for Molecular Nanophotonics

Bacteriorhodopsin (BR)—light-sensitive protein—is similar to the optic rhodopsin of the human eye. BR is obtained from halobacteria containing BR in cellular membranes (so-called purple membranes). During the separation from the bacterial cells, purple membranes save the entire structure (Vsevolodov, 1988; Oesterhelt et al., 1991). The typical size of purple membranes is 500–1000 nm. It is the unique biocrystalline structure capable of saving its permanent properties for a number of years, which consists of dry and polymer films with the thickness from 5 nm (monolayer) up to a few tens of micrometers.

The fundamental BR property is the photochemical cycle availability: after the light quantum absorption, BR molecule passes through the sequence of states and spontaneously returns to the primary form (Figure 38.1). At that point, in compliance with the cycling of BR molecule state, the light-induced cycling

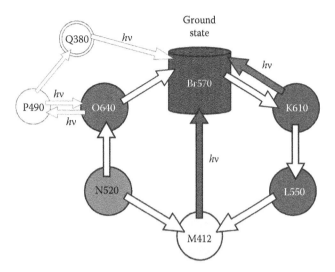

FIGURE 38.1 BR photocycle.

of optical characteristics (refraction and absorption indices) occurs (Haronian and Lewis, 1991; Zeisel and Hampp, 1992). Each of the interstitial states is identified as the intermediate according to its absorption spectrum. The existence of branched photocycles is typical for some BR types (Birge et al., 1999).

The main BR function in purple membranes is light-dependent proton transfer (H⁺) over the purple membrane that results in electrochemical hydrogen potential formation on halobacteria membrane. The potential energy is utilized by cell. Ejection of H⁺ occurs outside the cell membrane, and the H⁺ is captured inside the cell (from cytoplasm). Supposedly, it happens during the formation and the disappearance of intermediate M412 (Siebert et al., 1982; Haronian and Lewis, 1991).

BR spectral sensitivity lies in the optical band (Figure 38.2). Absorption maximum in primary state of BR570 corresponds to wave length 570 nm. M412 in the main intermediate state reaches its absorption maximum at wave length 412 nm (Birge et al., 1999). The absorption of optical emission by BR-containing medium is characterized by certain peculiarities. That happens due to the change of adsorption sites concentration (molecules in form BR570), which is the result of light quantum absorption at wave

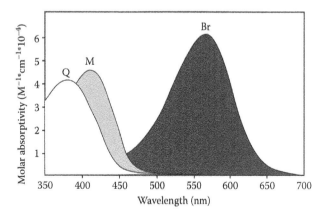

FIGURE 38.2 Absorption spectra of BR and of the main photocycle intermediates.

length 570 nm by these sites and transition to form M412 with a low absorption at wave length 570 nm. As a result, absorption in the yellow range reduces; medium becomes more transparent—bleached. The intensity of BR-containing medium bleach effect depends particularly on the time of intermediate M412 molecules relaxation to form BR570. Relaxation time is characterized by half-value period of intermediate M412 molecules. The light effect at wave length 412 nm results in fast coercive transition of molecules to the primary state. The values of photocycle time parameters lie in the range from microseconds up to tens of seconds.

Thus, BR behaves as a photochromic medium with a quick time of information storage. Optical and dynamic BR characteristics change in wide range by production conditions and matrix composition (medium).

38.2 Some Techniques of Neuro-Molecular and Molecular Information Processing Using Bacteriorhodopsin (Basic Processes, Constructions, Technology)

38.2.1 Basic Process of Classic Optical Neural Network in Bacteriorhodopsin Medium

BR properties listed above allow us to illustrate one of the available basic processes of data transformation in BR-containing media. It is based on reversible light-sensitive changes of absorption index and appears in optical effects considered below.

38.2.1.1 Nonlinear Absorption of Optical Radiation: Medium Bleaching

The absorption of optical radiation in substance is described by some known classic equations. Generalized Bouguer–Beer law associates intensities of the incident light and the light transmitted through the substance layer with the thickness of the layer and molecular concentration of absorption agent:

$$I = I_0 \cdot e^{-D} = I_0 \cdot e^{-\alpha d} = I_0 \cdot e^{-\varepsilon c d}, \tag{38.1}$$

where

  I is the transmitted light intensity
  I_0 is the incident light intensity
  D is the optical density of the substance
  α is the absorption index of the substance
  d is the thickness of the substance layer
  c is the molar concentration of absorbing substance molecules
  ε is the extinction coefficient, characteristic feature of absorbing substance molecule

The values of ε, α, and D depend on the wave length of incident light. When the values of coefficients in Equation 38.1 are invariable, the transmitted light intensity is in direct proportion to the incident light intensity.

Nonlinear absorption of optical radiation by BR-containing media is connected to the change of absorption centers concentration (molecules in form BR570) as a result of light quanta absorption by these centers at wave length 570 nm and transformation to M412 with low absorption at wave length 570 nm. Finally, as was mentioned above, absorption in the yellow range ($\lambda = 570$ nm) decreases, medium becomes more transparent—bleached.

38.2.1.2 Indirect Interaction of Optical Radiation Fluxes in Bacteriorhodopsin-Containing Media

Mediated interaction of optical radiation fluxes appears during the sequential or combined transmission through the same part of BR-containing medium and is at its clearest for monochromatic radiation with wave length 412 and 570 nm.

As a result of interaction between BR and the radiation with wave length 570 nm during the transmission through the medium, the energy of the light flux is absorbed, and in BR-containing medium, the photo-induced allocation of the absorption index forms. The light-induced allocation of the absorption index variation corresponds to energy distribution along the surface of the transmitted light wave front (Figure 38.3A).

Non-modulated along the front surface, the light pulse with the wave length of 570 or 412 nm (as actuating or inhibiting signal) absorbs spatially and nonuniformly in compliance with the changed value of the absorption index (Figure 38.3B and C).

Thus, the energy distribution along the preceding pulse front surface, indirectly, over the BR-containing medium, modulates the energy distribution along the following pulse front surface.

As a result, BR-containing medium is capable of accumulating (summing up) effects signed "+" and "–," correspondingly, at wave lengths 570 and 412 nm. Predicating upon the indirect interaction of optical radiation in BR-containing media, the method of formal neuron creation in such media is available to offer. At that point, all the main functions can be realized by optical technique.

38.2.1.3 One of the Possible Methods of Formal Neuron Main Functions Realization in Bacteriorhodopsin Medium

Realization of the main operations of neuronet algorithms—weighing of input signal vector according to the matrix of weighing coefficients of synaptic bonds; composition of weighed values of input signals; realization of activation (threshold) function by optical method without optoelectronic buffering—permits to simplify the construction and the technological realization of multilayered optical neuronet, to increase the integration of neuro-like elements in device, and to solve the problem of areal density limitation inherent in microelectronic elements and electric connections.

38.2.1.4 Formation of the Neuro-Like Element

The neuro-like element is formed under the exposure of optical emission with the spectrum, corresponding to the absorption spectrum of BR molecules' initial state and BR-containing material medium. This element is a part of the BR-containing medium with photo-induced absorption index. The threshold properties of such neurons are defined by the concentration ratio of molecules in primary and photo-induced forms, and the interaction of neuro-like elements is provided by optical emission.

The example of the similar neuronet realization, based on information conversion of basic process in BR-containing media (Figure 38.4), is considered below.

Construction for optical neuronet formation includes the following:

1. The source of the plane light front (transparent for normal incident light flux) providing the signal formation and transfer to neurons.
2. Photo-detecting layer based on BR-containing material for imaging of the input optical information by photo-induced variation (according to light energy distribution along the surface of input light front) of absorption/transmission in BR-containing medium.

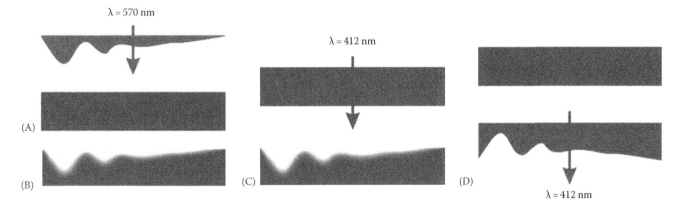

FIGURE 38.3 The indirect interaction of optical radiation fluxes in BR-containing media: (A) green light pulse (modulated along the front surface) acts on a layer of BR-containing medium (nontransparent for green light); (B) modulated allocation of molecules in form M412 (transparent for green light) and BR570 (nontransparent for green light), (C) the action of unmodulated blue light pulse, (D) modulated blue pulse as a result of transmission via the modulated medium (BR-containing medium layer recovered the primary state).

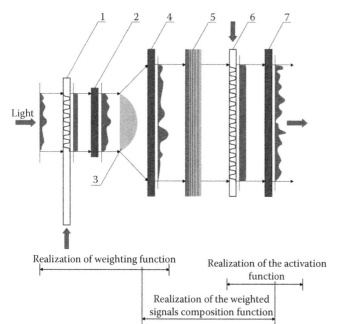

FIGURE 38.4 Neuronet organization based on the basic process: 1, 6—plane waveguides, including gitters for emission input/output; 2, 4, 7—BR-containing layers (photo-detecting layer, layer of weighting coefficients and neuron layer); 3, 5—cylindrical lenses.

3. Light flux expanding lens.
4. Layer of synaptic bond synthesis (matrix of weighing coefficients) made of BR-containing material.
5. The lens, focusing the light flux on layer 7 and forming the combination of input optical signals based on neuro-like elements in that manner.
6. The source of flat light front (transparent for normal incident light flux) for parallel comparison with thresholds, formation, and transfer to other layers of output neuron signals in layer 7 as light front, modulated by intensity along front surface according to the values of absorption/transmission area of BR-containing medium (corresponding to neuro-like elements).
7. The layer of neuro-like optical elements (being obtained on the basis of BR-containing material) composes input optical signals, traversing matrix 4 and realizing the activation function during the light front traverse from the source 6, and forms thereby output signals.

The input optical information in the form of light flux (input vector) effects the photo-detecting layer 2 that results in the absorption of light flux energy in BR-containing medium of photo-detecting layer and the distribution of the photo-induced absorption index forms.

Distribution of the photo-induced BR-containing medium absorption index along the surface and in depth corresponds to the power distribution along the surface of the effective light front.

Plain light front from the source 1, modulated by intensity according to the contour of absorption index of photo-detecting layer 2 by lens 3, allocates on the surface of weighing coefficients layer 4.

38.2.1.5 Realization of Weighing Function

The input vector weighing function comes around during the transfer process of input light signal—the components of input vector—over the matrix of weighing coefficients 4. By the component of input vector, we mean the quantity of the light energy (the intensity multiplied by exposure—exposition) affecting the matrix section.

Weighing coefficient (by which the input vector component is multiplied) is a transmission coefficient of the corresponding section of BR-containing matrix:

$$I_{w\,ij} \cdot t = \omega_{ij} \cdot I_{in\,i} \cdot t, \tag{38.2}$$

where
  $I_{w\,ij}$ is the light intensity transmitted over the ij matrix section or weighed component of the input vector
  $I_{in\,i}$ is the light intensity of the light front section or a component of the input vector
  t is the exposition time of the corresponding component of the input vector
  ω_{ij} is the weighing coefficient or transmission of the corresponding ij-section of the BR-containing matrix

Transmission of the ij matrix section is defined according to Bouguer–Beer law

$$\omega_{ij} = e^{-\varepsilon d c_{i/f\,ij}}, \tag{38.3}$$

where
  ε is the BR absorption factor
  d is the thickness of BR-containing layer
  $c_{i/f\,ij}$ is the concentration of the BR molecules in the initial state in the ij matrix section

The properties of the input vector weighed components are formed by lens 5 in a light flux that gets to the inputs of the corresponding neuro-like elements of the layer 7.

38.2.1.6 Realization of the Weighed Signals Composition Function

Composition function of the input signals is realized in the BR-containing medium in layer 7 by the converging cylindrical lens 5 as a result of the combined effect at the same area of BR-containing medium of the light energy exposition by corresponding components of the input vector.

Every component of the input vector contributes to the formation of the molecules ensemble in photo-induced state in proportion to the intensity and exposure time:

$$\Delta c_{j\,p/i} = k \cdot \sum I_{w\,ij} \cdot t, \tag{38.4}$$

where

  $\Delta c_{jp/i}$ is the concentration of BR molecules in photo-induced spectral state

  k is the coefficient of proportionality depending on the concentration of BR molecules in the initial state, cross-section interaction, photo response of BR-molecules transition from the initial to the photo-induced form, and the effecting light wave length

  I_{wij} is the intensity of the input vector weighed ij-component

  t is the exposure time of the input vector i-component

Thus, $\Delta c_{jp/i}$ contains information about the value of the weighed input interactions sum on the j-neuron.

38.2.1.7 Realization of the Activation Function

Magnitude $\sum_i I_{wij} \cdot t$ (the total dose of light energy effecting on the j-area of BR-containing medium) assigns the point at the graph (dependence of the transmission value on the sum of weighed input effects) and defines the transmission of the light signal over the j-neuro-like element.

The changed transmission $\omega_{n/e}$ value of the BR-containing medium j-section according to the j-neuro-like element assigns the value of the activation function and the output signal of the neuro-like element in layer 7, according to Figure 38.5. This magnitude depends on the number of molecules possessing the changed spectral properties of the medium section corresponding to the neuro-like element of the layer 7 according to

$$\omega_{n/ej} = e^{-\varepsilon d(c_{il f} - \Delta c_{j\,p/i})}. \tag{38.5}$$

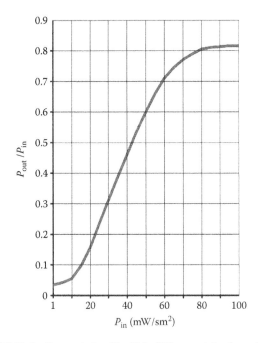

FIGURE 38.5 Transmission (P_{out}/P_{in}) of BR-containing layer depending on the effective emission energy (P_{in}).

The graphic chart on the dependence of the BR-containing layer relative transmission on the emission energy consists of the area with the initial transmission value (unequal to zero), the area of almost linear transmission change, and the saturation area (Figure 38.5). In general, the curve corresponds to the activation function proposed for the neuronet realization by Grossberg (Wasserman, 1989). The similar compressive function automatically provides the output signal range from 0 to 1 and corresponds to the necessary requirements for realization of reconversion algorithm during the neuronet learning, for example, according to scheme (Wasserman, 1989).

The output signal formation of neuro-like element (threshold comparison and realization of the activation function) is carried out by the light front of the specified intensity and duration induced by the source-former 6.

The output signal of the neuro-like j-element is formed as an energy portion of the active light signal being transferred over to the corresponding section of BR-containing medium in the layer of neuro-like elements 7 (according to the transmission of the section considered) in conformity with the formula

$$I_{out\,n/e} \cdot t_{active} = W_{n/e} \cdot I_{active} \cdot t_{active}. \tag{38.6}$$

The minimum value of the output signal is fixed by the transmission of non-firing neuron (the input signals sum value is close to zero) and corresponds to the initial transmission of photochromic medium, and the maximum value is close to the value of active front energy and corresponds to the saturation area of the curve (Figure 38.5) and to the maximum excited state (transmission) of firing neuron. Output signals of neuro-like elements form the continued light front being modulated according to the activation function at every point of BR-containing layer 7. System learning (formation of matrix weighing coefficients) corresponds to the formation of adequate values of transmission coefficients of matrix sections based on BR-containing medium that can be achieved by the inverse transformation method.

The optical version of the inverse transformation method can be simply and effectively realized (failing optoelectronic transformations) by combined presentation of learning pair: the input image in its usual direction and the ordered output as the light front in the counter direction. Due to the reversibility of the light passing, both of light fronts will affect the matrix made of BR-containing material and will change the matrix transmission corresponding to the intensity distribution.

38.2.2 Neuronets Based on Multilayered Optical Structures Including Polymeric Bacteriorhodopsin-Containing Layers

The considered approach of neuro-computer element base formation takes into account the cyclicity of processes in living systems appearing, for example, in spontaneous activity of pacemaking (assigning the rhythm of functioning) neurons. As was conclusively shown in Prigogine (1980), the cyclicity proceeds as a result of the processes' self-organization in open nonlinear nonequilibrium systems and the origination of stable dissipative structures due to which the coordination of trophic processes (providing cell nutrition) is possible in living systems.

The similarity of cyclic processes in the living cell membranes (including neurons) and the processes under optical emission exposure in isolated purple membranes is obvious, particularly, taking into consideration that in that case one of the halobacteria trophic cycles is reproduced.

If the neuron is represented as a structure population (purple membranes) allocated in media and the interactions between the neurons are carried out by light fluxes, the "neurolike medium" in question can be considered as the alternative to the net of neuro-like elements (Grebennikov, 1997).

In the interpretation being stated, the problem of neuronet formation is in simulation of nonequilibrium nonlinear systems in BR-containing media with allocated parameters by optical emission of two different wave lengths corresponding to the absorption maxima of two "long-living" intermediates. It can be solved in the following consequence:

1. Using the spectral sensitivity of BR in optical range to initiate the cyclic light-dependent processes by the light flux and to control the processes by the application of emission with various wave lengths.
2. Stationary and dynamic structuring of BR-containing films by modulated light fluxes to create the conditions of formation and development of neuro-like relations in neuro-like medium and to correlate the multitude of cyclic processes.

Data input and processing in that case is "deformation" of the correlated cyclic processes in BR-containing medium, and the commands of system behavior control are the transient processes originating in the organized neuro-like medium.

On the other hand, analyzing the development tendencies of neuronet technology, the best perspective can be emphasized, and their combination in the same device can provide essential expansion of neuronet data processing:

1. *Optical mechanism of data transmission and processing*, which will allow to construct three-dimensional nets that function simultaneously at high speed. The problem of wide application is in large dimensions, peculiar to optical systems, and the inevitable efficiency losses due to multiple intermediate optoelectronic conversions and the finite size of the optoelectronic elements.

The application of continuous photochromic media with high resolution close to molecular level of data representation and processing can become a solution to the problem. BR is the most available and well-investigated (at present) photochromic material with sufficiently high cyclicity (>10^6) and is suitable for the considered purposes. Using BR enables to carry out data processing in optical mode without intermediate optoelectronic conversions and to increase the areal density of neuro-like elements value comparable to the native neuron systems.

2. *Neuro-like structure construction on the principles of self-organization of the processes in the open nonlinear allocated dissipative systems like biological objects*. Although in model version, systems require substance transfer proceeding for nonequilibrium conditions maintenance.

BR application can solve the problem of substance transfer that complicates the construction and limits the continuous working life of neuro-like media processing. Light-dependent properties of that material allow the open nonlinear allocated dissipative systems to simulate effectively. Optical exposure in the range of 520–650 nm is like the input flux and dissipative properties, and in that case can be provided by the component of trophic cycle that remains invariable in composition of artificial medium (for example, in polymeric matrix) and also by emission exposure in blue light range ($\lambda = 400$–420 nm).

3. *Element base adaptability* to losses of some elements during the preparation and exploitation compensated by self-organizing and self-modification of neuro-like structures. Application of continuous media based on BR in conjunction with optical methods permits to form neuro-like elements, and the bonds between it suit requirements according to the light energy allocation.

Classic methods of the light fluxes formation by lens systems result in the loss of advantages expected from significant density of neuro-like elements and dimension restrictions of elements and systems, peculiar to optical computers. Moreover, when the volume of BR-containing medium is greater, the probable process integration is higher. However, the emission access to all the molecule groups is more unfavorable as the medium absorption increases.

In the field of optoelectronic technology, efforts are made to deflate the construction dimension of optical neuro-computers by the application of multilayered structures. Multilayered structures forming neuronets and containing the layer of spatial light modulators as the liquid crystal matrix, the layer of photoconducting material (Engel, 1990) or material with the photovoltaic effect (Akiyama et al., 1995), and the layer of electric transducer-amplifiers (forming and transducing the commands to liquid crystal panels) are proposed.

Essential disadvantage of the proposed constructions is the facility of intermediate conversion of light exposure to electric current or voltage used for the following changes of optical transmission modulator. Furthermore, neuronet function realization requires the application for these purposes of optoelectronic and microelectronic elements. Consequently, the application of metallic conductors in inter-cell links for realization of neuro-systems with a great number of neurons results in delays in communication lines and neuronet processing deceleration due to the capacity influence of inter-cell communication lines. The existing areal density restrictions of optoelectronic elements and electrical bonds inevitably limit the spatial resolution of constructions.

38.2.2.1 Multilayered Constructions, Including Bacteriorhodopsin-Containing Films

We proposed multilayered structures including layers based on the BR for the realization of neuronet medium in nonequilibrium nonlinear dynamically allocated dissipative systems.

Supposedly, the multilayered structures would allow the data processing to continue at the level of BR molecules groups by forming neuro-like elements using optical methods in BR-containing media.

The basic processes in BR medium are defined by light-dependent changes in absorption index allocation profile along the surface of BR-containing polymeric films. In multilayered structures, many light fluxes circulate without interaction. This property is usually proved as an advantage of optical methods enabling data processing and transmission in the three-dimensional space. At the same time, the information arrays in the form modulate indirectly by intensity light fronts over the reciprocal fluctuation of absorption index local value of BR-containing media sections, and the local intensity value of the light front sections realize concurrent information interactions in the three-dimensional space of the multilayered structure.

We would like to consider a possibility of neuro-like element net organization in BR medium by optical method using multilayered structures (Figure 38.6) including BR-based layers, wave guide layers, and reflecting layers. To allocate light fluxes, the system of waveguides, transparent in the optical range, is formed in BR-containing medium. It is possible to input the controlled emission in the form of light front in BR-containing medium, activating at that point the groups of neuro-like elements, by producing the sections with the disturbed conditions of total internal reflection in waveguides.

It is expected that the multilayered structure will provide not only functioning and interaction of neurons and their ensembles but also the generation of new neurons and links (emission output from one layer and penetrating to the other layers) between single neurons and neuronets, and will permit, according to the information (image) at system output, to connect and to correlate the cyclical processes originating in BR medium. At that

point, the process of self-organization of data processing system will continue.

The adaptability principles realization of data processing elements and system self-organization will permit to essentially reduce the requirements of the elements and facility as a whole to provide reliable functioning in case of the single-element failure.

Reduction of technological requirements is achieved by that neuro-like elements and links that are formed in continuous (uniform, i.e., not divided into constructive matrix elements) transparent (without optical dispersion) layer of photochromic material according to the light energy allocation.

38.2.2.2 Expected Parameters of the Element Base

According to the traditional criteria, it is acceptable to evaluate the number of the neuro-like elements in medium containing BR at area $10\,mm^2$ in quantity not less than 10^6. At the area in question, not less than 10^{11} bonds per second are realized (circuit time 0.1 ms, coefficient of bond formation 10).

38.2.2.3 The Basic Elements of the Multilayered Structures

The multilayered structures (Figure 38.6) for the realization of the basic neuronet data processing include: the system of flat waveguides, the elements of optical emission input as gitter, and the devices of the surface light front formation (Figure 38.7).

BR-containing polymeric films are meant for neuro-like elements formation by the change of the absorption/transmission surface geometry by modulated light flux effect.

The elements of optical emission input in flat waveguides and the elements of output are made as diffraction lattices (gitters) (Zlenko et al., 1975; Unger, 1980). The angles of radiation input and output depend on different layer indexes and glitter spacing; therefore, those angles can be different for various multilayered waveguides.

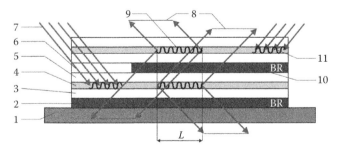

FIGURE 38.6 Fragment of multilayer structure: 1—substrate (glass K-8); 2, 10—layers, containing BR; 3, 5—boundary layers of flat waveguide; 4—guide layer of flat waveguide; 6—emission input area; 7—input emission; 8—output emission; 9—output emission gitter; 11—input emission gitter; L is the length of the output emission gitter.

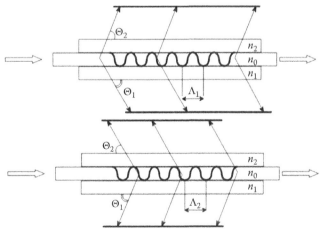

FIGURE 38.7 Device for surface light front formation: Θ_1, Θ_2—angles of emission input; Λ_1, Λ_2—diffraction lattice spacing; n_0, n_1, n_2—refraction indices.

The device for surface light front formation (Figure 38.7) generates directed light fluxes for effective allocation of the light energy in BR-containing polymeric films for neuro-like elements concurrent formation, their concurrent interaction, and output of the data processing results in neuro-like element medium as optical signals.

38.2.2.4 Requirements of BR-Containing Polymeric Films

The following are the requirements of optical and geometrical properties of BR-containing polymeric films.

For photochromic effects (induced changes of refraction and absorption indexes) to significantly appear during functioning process, high optic density and consequently substantial BR concentration in polymeric films are needed. Optical density of such films should be 0.8–1.3. In those conditions, basic characteristics of BR-containing media are utilized under optimum light flux density so that the media could be used for technical purposes. Exposure to light of the fluxes with radiation density equal to 10–100 mW/cm² induces the films to experience changes in their absorption and transmission characteristics as much as 10%–50% from the original indexes in as long as 0.1–10 s.

Functioning of films under induced changes in absorption level requires quite homogenous distribution of BR concentration, as optical heterogeneity infringes the conversion of optical information.

Besides, we must ensure repeatability of the main structural parameters: film thickness, surface finish, homogenous BR distribution throughout film surface (1–10 cm²), etc.

Requirements to physical and chemical properties of polymeric matrix. Selection of matrix material.

To minimize the influence of diffraction divergence on information conversion processes, the thickness of BR-containing films must be 6–14 μm. To reach the specified optical density with that thickness, the volume of BR content in polymeric films should be 40%–50%. Far from all, the polymers transparent in optical range can meet the requirements above. Moreover, only water-soluble polymers can be used to form BR-containing polymeric films.

The comparative studies held to form polyvinyl alcohol- and gelatine-based BR-containing polymeric films have proved that gelatin-based polymeric matrixes have obvious advantages.

Gelatin-based films could apparently have the highest possible BR concentration (up to 50 vol. %) without aggregation of PM fragments due to thermodynamic peculiarities of gelatin polymerization process. Gelatin properties allow us to avoid destruction of BR protein structure while making polymeric mixture and further polymerization.

PM fragments embedded into a gelatin matrix are long-lasting and resistant to many technological factors. Polymerized gelatin creates optimum conditions for BR to function while retaining enough water needed for photochromic cycle. For the same reason, gelatin matrices make it possible to place environment-modifying water-soluble components and to change the photo cycle time frame.

38.2.2.5 Obtaining of Bacteriorhodopsin-Containing Polymeric Films for Multilayered Structures

It is important for processing and conversion of optical information to take into consideration the dispersion of optical emission in BR-containing media conditioned by purple membranes size (500–1000 nm)—comparable to the wave length of the optical range emission. Therefore, the primary suspensions of PM and BR-containing films based on them are optically nonhomogeneous, which results in the functional properties loss.

The dispersion demagnification is reached consequently: at the step of preparation of purple membranes suspension—by PM fragments separation; at the steps of polymeric mixture preparation and polymerization—by elimination of the aggregation process of the purple membranes fragments.

38.2.2.5.1 Preparation of Bacteriorhodopsin Suspension

For the preparation of PM suspension, the triple centrifugal purification (3000 rpm, 5 min) was carried out; pH value and BR concentration in suspension were measured. The pH value is significant for the following polymeric solutions and film obtaining, since the investigations showed that at pH less than 4.1 PM aggregated, the optical transparency of the suspension was not achieved. It was determined that the ultrasound exposure results in the decrease of pH value in suspension at 0.2–0.4. The control of pH value was managed by the addition of 0.01 M borax buffer solution $Na_2B_4O_7 \cdot 10H_2O$, pH = 9.18. During the ultrasound treatment, the suspension temperature must not exceed 36°C.

As the result of technological investigations of ultrasound treatment, optically transparent homogeneous BR suspensions were obtained without detergent addition. The side effect can be the partial melting of protein. Optically transparent PM suspensions were obtained with BR concentration up to 15 mg/mL.

Size evaluation (8.7 ± 0.5 μm) of PM fragments in treated suspensions was carried out by the intensity of Rayleigh scattering and showed that the applied technological mode of suspension treatment permits to separate PM into naturally minimum fragments—trimers without BR protein destruction and the principle possibility of BR-containing medium optical resolution at the level of a few thousand lines per millimeter can be considered.

38.2.2.5.2 Preparation of Bacteriorhodopsin-Containing Mixture

During the preparation of polymeric mixture based on BR, the last exhibited the property of aggregation on the polymer molecules that lead to optical heterogeneity of films. As the result of technological experiments combining thermal parameters and operating pH of components, the conditions selected under that aggregation were not observed and transparent optically homogeneous BR-containing polymeric mixtures were obtained. At the step of polymeric mixture preparation, pH control of gelatine solution was carried out because pH value in gelatine solution depends either on the obtaining method or on gelatine concentration. The component stirring in the mixture was also carried out by the ultrasound exposure. All the modifying components were put into the polymeric mixture at the last step under condition: the final pH value of the polymeric mixture had to be >4.1.

38.2.2.6 The Properties of BR-Containing Polymeric Films for Multilayered Structures

According to the elaborated technology, there are transparent and optically homogeneous BR-containing polymeric films (thickness 6–14 μm with optical density 0.8–1.3 D at λ = 570 nm) on substrates of glass K-8 and fused quartz (area up to 60×48 mm^2) and also on Si plates (76 mm in diameter).

38.2.2.6.1 Evaluation of PM Size Embedded to Film

Since during the film polymerization from the polymeric mixture fragment aggregation is possible, the evaluation of PM fragments size is being embedded into film. The placement of films with the embedded PM fragments in detection system between crossed polarizers resulted in no changes in the initial zero signal of photodetector. It means either no rotation of polarization in the light-pass direction, or no significant dispersion that confirms the PM fragments size to be much less than 0.63 μm.

Optical heterogeneity specified by surface geometry and allocation of BR concentration to the film surface is given below.

The distribution of optical absorption heterogeneity coefficient of BR-containing films is specified by the product of two values: the allocation heterogeneity of BR bulk concentration to the film surface and quality of the film surface as the local dilatation from the average thickness value. In the aggregate it results in the local dilatation of so-called surface concentration and, correspondingly, the optical density.

The optical homogeneity of the film being specified only by the surface quality is provided comparatively easily both for the films being realized by glazing method and for the films being obtained by centrifugation method. It was determined that the typical thickness deviation of BR-containing films being obtained equals less than 50 nm at length 10 mm, that, for example, at film thickness >5 μm equals <1%.

The support of the optical homogeneity being concerned with the allocation of BR bulk concentration along the film surface is hampered by the molecules migration in polymeric solutions during the polymerization process to the area of increased surface tension. Nevertheless, the attained deviation from the average value for allocation of BR bulk concentration does not exceed 3% since the changes of surface concentration and optical density lie in the same range.

38.2.2.6.2 The Structure of Bacteriorhodopsin-Containing Films

It is ascertained that the structure of BR-containing polymeric films surface depends on the obtaining conditions. Depending on pH, the surface is either smooth (Figure 38.8A) or it has punctual (100–300 nm) and rectilinear (400–1000 nm lengthwise, 100–200 nm wide) protuberances (Figure 38.8B and C). The cutting of the surface geometry elements is observed (Figure 38.8B). These investigations confirmed the necessity of pH value maintenance higher than 4.1.

Surface roughness (step height R_z of surface geometry) of the film with particle size of BR phase <10 nm amounts to $R_z \approx 0.1$ μm, with particle size of BR phase ~100–1000 nm $R_z \approx 0.6$–3.8 μm.

The analysis of the films with thickness 20 μm (containing BR-phase cutting) by Roentgen diffractometry led to the missing crystalline phase. Only amorphous gallo is observed on roentgenograms of BR-containing polymeric films, being formed under various conditions in the range of Bragg angles $2\theta = 5°$–8° ($k\alpha$—copper emission).

Clearly defined diffraction maximum appears on roentgenograms of some samples for angles $2\theta = 7.5°$–7.8° (Figure 38.9) corresponding to interplanar spacing $d/n = 1.179$–1.133 nm. For a number of samples, diffraction maximum appears on roentgenograms for angles $2\theta = 18°$–19° ($d/n = 0.467$–0.493 nm). That indicates the ordered placement of protein complexes in polymeric (gelatinous) matrix. Apparently, the axial texturing of lamellar protein complexes exists by the axis normal to the surface of BR plates.

38.2.2.6.3 Refraction Index of BR and BR-Containing Polymeric Films

Information about refraction index in literature (Hampp, 2000) is inconsistent and not exact enough to solve the problem of the multilayered structure construction. Therefore, we made our own measurements.

(A)

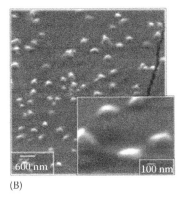
(B)

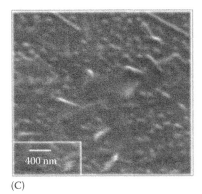
(C)

FIGURE 38.8 The structure of BR-containing polymeric film surface being formed from the polymeric mixture: (A) pH = 4.2–4.5; (B) pH = 3.8–4.0 (the enlarged surface fragment is presented at the insertion); (C) pH = 3.5–3.7.

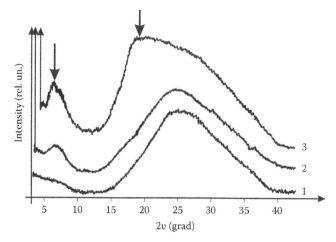

FIGURE 38.9 Roentgenograms, obtained on the samples: 1—glass K-8; 2—glass K-8 with the deposited gelatinous film; 3—glass K-8 with the deposited BR-containing polymeric film.

The refraction index measurement was carried out using refractometer IRF-454B capable of measuring the refraction index in the range from 1.2 to 1.7 in reflected and transmitted light. BR-containing film was formed on the surface of measuring prism by the method of suspension glazing (concentration 15–20 mg/mL) and the following drying at 22°C temperature, relative 60% humidity. The film thickness reached 20 ± 5 μm.

The value 1.534 ± 0.002 of refraction index was obtained in the white light. This result is well reproducible on every one of the samples being prepared from the three different halobacteria strains and conforms to the results of the following refraction index measurements in polymeric BR-containing films. The value of the refraction index in that case is defined by BR concentration in the film.

The refraction index measurement of gelatin was held in order to predict the parameters of the polymeric BR-containing films under construction depending on BR concentration in the film. The refraction index value of polymeric film based on gelatin without BR (formation by the glazing method on the surface of the refractometer measuring prism) got equal to 1.543 ± 0.001. Refraction index was measured at 22°C temperature and relative 60% humidity.

The refraction index value of BR-containing gelatinous film with thickness 50 ± 10 μm lies in the range of 1.539–1.542 depending on BR concentration in gelatin.

38.3 Nanostructuring of Bacteriorhodopsin-Containing Molecular Media

Functional parameters of BR-based multilayered structures on the base of BR (the increase of photochromic sensitivity, management of photocycle duration) can reach their maximum capacity in use by additional nanostructuring of molecular media. Enhancement of functional properties of multilayered structures is caused by introduction of new functional elements—hybrid nanostructures—making it possible to further realize a new class of optical information systems based on self-organizing oscillatory and auto wave hierarchical processes in BR-containing nanocomposites.

Both nanotechnological and nanotechno-systematic approaches are described below.

While developing BR-based media used to process information with neural network methods, two ways of material structuring are applied. The first way means introduction of modifying organic and nonorganic additives into the system to improve functional characteristics of BR. Modification is done on molecular level that is why such material is a molecular one. The second way means building of hybrid functional structures with every component performing its own function. In this way, structuring goes at nanometric range instead of molecular level. Hybrid nanostructures are systems made of three components—colloidal nanoparticles (metal or semiconductor structures based on transition element [TM] chalcogenide), bridge molecules (spacers) with various linear dimensions, and BR molecules.

38.3.1 Procedure of Complex Estimation of Functional Characteristics of Bacteriorhodopsin-Containing Materials

To characterize and compare functional characteristics of BR-based films made by various methods and difference in thickness, optical absorption, structure, and content (including concentration of BR molecules), it is convenient to apply the designated factor for quantity estimation of photochromic sensitivity. For optimization of experimental investigations, we developed the simulator of photo-dependent processes and designated factor $k_{570}(t)$ of the photo-induced transitions of BR molecules from the primary state BR570:

$$k_{570}(t) = \frac{N_2(t)}{N_0} = \frac{N_0 - N_1(t)}{N_0}, \qquad (38.7)$$

where
  N_1 is concentration of BR570
  N_2 is the concentration of BR412
  $N_1 + N_2 = N_0$ is the total concentration of BR molecules

That factor $k_{570}(t)$ named by us as the coefficient of photo-induced transition of BR molecules from basic state BR570 was determined from experimental data of changing optical absorption at wavelength 570 nm under the influence of active inducing radiation. This complex parameter takes into consideration the temporary characteristics of excitation and relaxation processes, quantum efficiency, and cross-section of BR molecules contained in the nanocomposite nanostructured films (media).

While studying the influence of technological formation processes on properties of BR-containing materials, it is very labor-consuming to measure each of the above specified characteristics. Therefore, we use a model to describe photo-dependent processes that pass in BR-based materials under radiation.

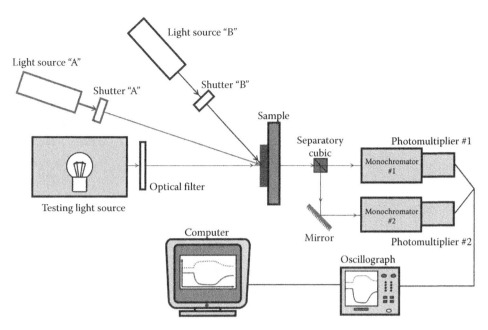

FIGURE 38.10 Schematic diagram of a soft- and hardware complex designed to study functional characteristics of BR-based materials.

We have also designed a method to assess the functional characteristics of those media after one measurement. To carry out all necessary measurements, we can use a soft- and hardware complex (Figure 38.10), making it possible to combine light exposure with two wave lengths in the band of main intermediate absorption (at 570 and 412 nm, respectively).

The light radiated by the source of continuous testing radiation with up to 2 mW/cm² intensity passes through the sample and falls onto the monochromator where a narrow spectrum line (at 570 nm) is allocated. The signal goes to the photo-electronic multiplier being adjusted to convert the signal and transfer it to the oscillograph. The sample is exposed to light pulsed by the source of exciting radiation with intensity at least 10 times more than the intensity of testing radiation. The signal registered with the oscillograph goes to the computer for mathematic curve processing as per the below physical and mathematic ratios.

Based on the Buger–Lambert–Beer law, taking into consideration photo-induced changing concentrations N_1 and N_2, the expression has been obtained that allows to calculate the value of $k_{570}(t)$ from the experimental data recording testing light $\lambda = 570$ nm transmission changing under illumination of exciting radiation:

$$k_{570}(t) = \frac{\lg\left(I(t)/I_1\right)}{\lg\left(I_1/I_0\right)}, \qquad (38.8)$$

where

I_0 is the intensity of the incident test light
I_1 is the intensity of the transmitted test light in the absence of exciting radiation
$I(t)$ is the intensity of the transmitted test light under exciting radiation at the time moment t

Also the proposed method of the complex estimation is based on the kinetic equation evaluating concentration distribution of BR molecules between forms of BR570 (N_1) and M412 (N_2):

$$\frac{dN_1}{dt} = -\sigma_1 \cdot A_1 \cdot \frac{P}{h \cdot \nu} \cdot N_1 + \frac{1}{\tau} \cdot N_2, \qquad (38.9)$$

where

σ_1 is the absorption cross-section of BR570 form (on wavelength of acting light)
A_1 is the quantum yield of photoreaction
P is the power density of acting light
τ is the lifetime of M412 form
h is the Plank constant
ν is the acting light frequency

From the same experimental data (Figure 38.11) based on the solution of equation evaluating concentration distribution of BR molecules between forms BR570 (N_1) and M412 (N_2), the estimation of quantum yield of photoreaction A_1 from the value of the derivative at the point $t = 0$ (the beginning of exciting radiation action) and the estimation of the lifetime of M412 form τ from value of derivative at the point $t = t^*$ (the end of exciting radiation action) may be obtained. As time goes by, film transmission value being influenced by the relaxation of excited molecules approaches to the original transmission level.

38.3.2 Functional Characteristics of Nanostructured Bacteriorhodopsin Films

To form the effective functional structures out of biomaterials, we need to apply various technological methods. In particular, we can use BR-containing materials to prove the necessity to use

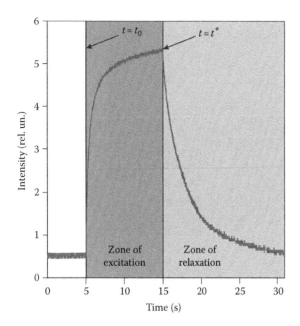

FIGURE 38.11 Typical curve of photo-induced changes of optical transmission of BR-containing media.

the additional nanostructuring of functional biomolecular films subject to storage and neural network processing with optical methods.

As a result of non-modified BR films' investigations, we established that immediately after obtaining the values, $k_{570}(t)$ are in interval 0.5–0.7 and then decrease to value 0.2 for 3–5 h and demonstrate this value during a few years. In the primary formed films, BR molecules constitute the structure that in time is destroyed in consequence of the heat oscillations, resulting in disordered molecule orientation relative to each other and destruction of hydrogen bonds. To keep the high photochromic sensitivity for the whole period, we can create additional linkages (covalent or hydrogenous) between protein molecules by using chemical reagents.

For example, our use of diamine (1.4-diaminobenzen [DAB]) in the ratio of BR:DAB = 1:3, 1:6 и 1:9 (molecular ratio) has resulted in increased values of factor $k_{570}(t)$ (0.35–0.5) in comparison with non-modified BR films (0.2–0.25) during 12 days. The stabilization of the primary formed BR structure occurs due to the fact that 1.4-diaminobenzen is an aromatic amine and capable to interact with the carboxyl groups of glutamic and aspartic acids. In our opinion, the high values $k_{570}(t)$ for BR:DAB films are the result of raising the lifetime of M412 form owing to the fact that 1,4-diaminobenzen screens extract proton groups.

At the same time the use of glutaric dialdehyde (GA) with sodium tetraborate for the alkaline catalysis of the reaction of bonding between amine groups of BR lysine residues and aldehydic groups of GA (the Schiff base) for the samples BR:GA = 1:5 had more high values of factor $k_{570}(t) = 0.55$ (Figure 38.12), whereas BR:GA = 1:10 and 1:30 had $k_{570}(t) = 0.4$–0.5, which is explained by the high content of GA oxidation products by atmospheric O_2 (glutaric acid) that partially destroy BR molecules. The dynamics change of decay factor $k_{570}(t)$

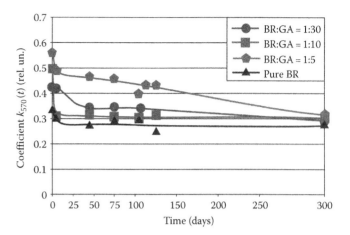

FIGURE 38.12 Changing values $k_{570}(t)$ in the operation process for the films. BR:GA = 1:30, 1:10, and 1:5.

value was studied during 1 year for BR:GA films (Figure 38.12). The samples BR:GA = 1:5 kept the high value of factor $k_{570}(t)$ ($\geq$0.45) for 120 days. After 330 days for all the samples of BR:GA $k_{570}(t) = 0.3$.

Introduction of amino acids in the content of BR-containing films leads to even better results. The films of BR: glycine, BR: isoleucine, and BR: lysine in molecular ratio from 1:1 to 1:25 had been made. The BR: glycine and BR: isoleucine films were first inhomogeneous optically and characterized by the high light scattering, and consequently had low values of factor $k_{570}(t)$ within the limits of 0.05–0.1. The BR: lysine films were optically transparent and homogeneous. The dynamics of changing value $k_{570}(t)$ for the films with ratio BR: lysine = 1:10 and 1:25 was shown in Figure 38.13. The samples with the ratio BR: lysine = 1:25 kept stable higher values during not less than

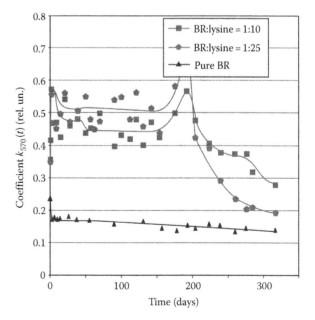

FIGURE 38.13 Changing values $k_{570}(t)$ in the operation process for the films. BR: lysine = 1:10 and 1:25.

80 days—$k_{570}(t)$ (0.4–0.55)—in comparison with the check sample (0.15–0.2). At that optical density and transparency, films did not change throughout that time. The samples with ratio BR: lysine = 1:10 demonstrated higher values $k_{570}(t)$ (0.35–0.5) in comparison with the check sample (0.15–0.2) for 330 days.

The above results prove that there are considerable opportunities to improve functional characteristics of BR-containing materials by nanostructuring (by forming additional bonds between BR molecules).

38.3.3 Metal Nanoparticles and Bacteriorhodopsin-Based Hybrid Nanostructures

The other way to expand the functional opportunities of molecular nanophotonics materials is to form hybrid nanostructures. We suggest the system use of some physical phenomena common for nanoscale structures as the basic processes for photonic information systems. In particular, we can use the effect that nano-objects have on the lifetime of excited atoms and molecules located closer than radiation wavelength. This effect is rapidly enhanced near nanoparticles when strong electromagnetic fields locate at plasma resonance frequency. Strong local fields influence the speed of electron transitions (i.e., the processes of light absorption and spontaneous radiation). Such fields can also change spectral characteristics of those processes and lead to substantial enhancement of various nonlinear optical effects. In particular due to strong local fields we can manage tumescence enhancement and effective suppression (depending on the distance), have radiationless energy transition, and manage the quantum efficiency of BR photoreaction and the lifetime of its spectral intermediates. The specified mechanisms could be useful to design photonic devices, which perform basic functions of information systems on the basis of nonlinear dissipative media with distributed feedbacks. It is to be taken into account that the effect nanoparticles have on the excited state of the functional molecule becomes different with the change of the distance between them. This circumstance creates wide opportunities to crease hybrid nanostructure-based composition materials with bridge molecules of various lengths.

In general, hybrid nanostructures are the systems made of three components—colloidal nanoparticles, bridge molecules (spacers) with various linear dimensions, and functional (photochromic, photoluminescence, electroluminescence, etc.) molecules. Below is the description and diagram (Figure 38.14) to show different aspects of hybrid nanostructures' formation with metal nanoparticles.

Formation of the hybrid nanostructures with adjusted distance between a nanoparticle and a functional molecule includes several key stages:

1. Formation and stabilization of nanoparticles in the polar or nonpolar solvent.
2. Modification of nanoparticle surface in order to functionalize and to ensure selective and self-organizing hybrid nanostructures.
3. Binding of functional molecules with nanoparticle surface.

Formation of nanoparticles compatible with BR photochromic protein has its own peculiarities due to the biological nature of the material as water is used as the main solvent.

The number of spacers used to form BR-based hybrid structures is quite limited due to the chemical properties of both components. First, the spacers must be hydrophilic substances, well soluble in water. Second, functional groups are also very limited. As BR is a protein, its structure is based on polypeptide amino acid consequence, including among others asparagic and glutamic acid residues. Those residues have free carboxylate groups taking no part in the formation of peptide bonds. pH area, where BR demonstrates its functional characteristics, is limited to pH > 5. Under those conditions the molecule is negatively charged. Therefore, to bind with BR functional spacer, the groups must be positively charged. The most efficient spacers look like molecules, including free aliphatic positively charged amino acids. Those substances include, for example, symmetrical diamines (1.6-diaminohexane, 1.7-diaminoheptane) or lysine (2.5-diaminopentane acid). Besides, in low-molecular compounds, one can use polymers, including amino acids in their structure. Polylysine is one of such polymers.

Gold nanoparticle-based nanohybrid structures are relatively easy to form as gold nanoparticles are chemically inert in aqueous suspensions. At the first technological stage we apply the borohydride method, which makes it possible to get smaller particles with more homogenous size distribution than nanoparticles obtained by using the citrate method. Besides to reach that we do not need to use any additional technological tricks. To apply the borohydride method, add 1 mL 1% $HAuCl_4$, 0,5 mL 0.1 M K_2CO_3 to 100 mL distilled water. While stirring the mixture vigorously with a magnetic stirrer, add 140 μL 0.1 M $NaBH_4$ solution (in some 10 μL portions during 10 min).

The technological methods of forming hybrid structures that we are developing are based on mechanisms of special bonding and self-organization activated by interaction of spacers with the appropriate functional groups. To build hybrid nanostructures based on formed gold nanoparticles, we have to modify nanoparticle surface with amino-containing spacers, which are able to actively interact with both BR molecules and the surface of nanoparticle. We have selected symmetrical aliphatic diamines, in particular, 1.6-diaminohexane. The surface of nanoparticles is modified by introducing 1.6-diaminohexane water solution with up to 5000 spacer molecules to 1 nanoparticle. The intensity of spacer molecules bonding with the surface

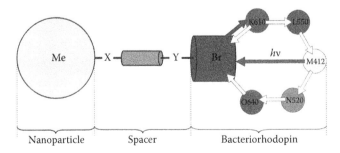

FIGURE 38.14 Schematic composition of hybrid nanostructures.

of nanoparticles can be indicated by the fact that after introducing 1.6-diaminohexane, excess gold nanoparticles are quick to aggregate (sometimes with residue). The results of that process can be seen while extracting films out of suspensions by methods of atom microscopic investigation (Figure 38.15).

Comprehensive studies have proved that we need to not only modify the surface of gold nanoparticles but also remove the suspension reacting system components that have failed to react during nanoparticle synthesis as those components destruct BR molecules and obstruct synthesis of hybrid structures. To solve this problem, we have suggested the use of dialyze. Place nanoparticle-containing suspension in the so-called dialysis bag, which is a hollow plastic container with some nanometer-size holes. Sink the dialysis bag into the solution of the substance to be introduced into the system (concentration of the solution should be enough to modify the surface of nanoparticles effectively). The volume of that solution is usually at least 10 times more than the volume of the original suspension.

During the process, small organic and nonorganic molecules migrate through the hollow walls of the dialysis bag, depending on the concentration gradient of the hole. Nanoparticles that are bigger than bag holes stay inside the dialysis bag. Some time (around 2 days) later thermodynamic balance is reached. That means that concentration of organic and nonorganic substances get homogenous throughout the volume. Therefore, the solution inside the dialysis bag experiences dilution due to contaminating impurities to an extent and solution volume relates to the volume of the original suspension. Concentration of the substance to modify nanoparticle surface is the same as that of the original solution. Using this method, we have obtained gold nanoparticles with 1.6-diaminohexane-modified surface. Those suspensions are stable. The studies of spectral characteristics of such suspensions have proved that dialysis has no effect on the spectrum of gold nanoparticles form (Figure 38.16).

We have synthesized hybrid structures with nanoparticles to BR molecules ratio 1:3600. BR suspension (51 mg/mL) is used as an original one. Hybrid structures are formed by vigorously stirring the suspension with a magnetic stirrer for 2 days at the temperature of the reaction mixture equal to 20°C–25°C. Figure 38.17 demonstrates spectral characteristics of the hybrid nanostructures suspension based on gold nanoparticles, 1.6-diaminohexane, and BR. We have to stress that characteristic spectral maximums of certain components do not decompose and therefore make it more complicated to study functional characteristics of those structures. By using extraction, we have obtained films with the surface characterized by atom microscopic investigation (Figure 38.18).

The use of silver nanoparticles providing optimum combination of spectral characteristics in the nanoparticle–BR system is of more interest. At the same time, silver nanoparticles demonstrate substantial chemical and photochemical activity in water suspension and require special stabilization measures.

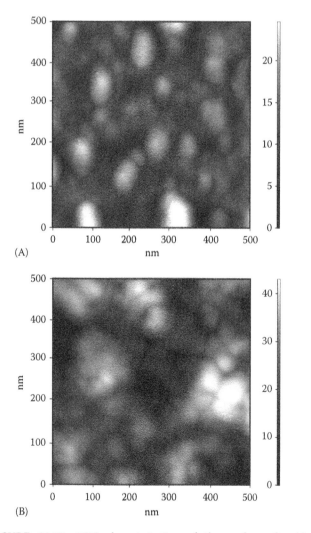

(A)

(B)

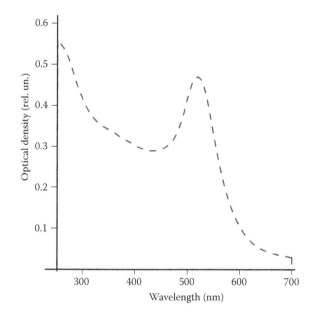

FIGURE 38.15 AFM characterization of the surface of gold nanoparticle-based films extracted from (A) a suspension with optimum 1.6-diaminohexane level; (B) from a suspension with excessive 1.6-diaminohexane level.

FIGURE 38.16 Spectrum of gold nanoparticles after dialysis with 1.6-diaminohexane solution.

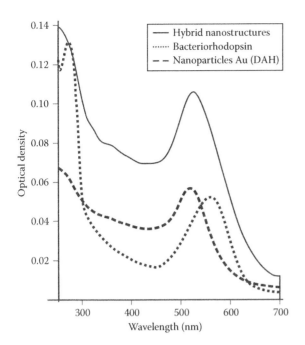

FIGURE 38.17 Absorption spectrum for suspension of hybrid nanostructures based on gold nanoparticles, 1.6-diaminohexane, and BR.

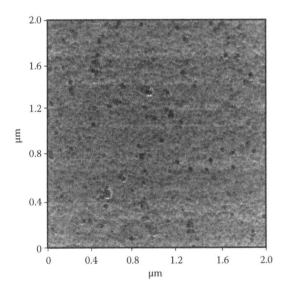

FIGURE 38.18 AFM characterization of the surface of hybrid nanostructures films based on gold nanoparticles, 1.6-diaminohexane, and BR.

To solve this problem, we introduce bridge molecules with various functional groups into the reaction system used to synthesize nanoparticles. This single technological operation includes two stages: formation and stabilization of nanoparticles and modification of their surface.

By using this method, we have obtained water suspensions of hybrid nanostructures based on BR and silver nanoparticles while using cysteine, arginine, lysine, and polylysine as spacers.

To form water colloidal suspension of silver nanoparticles we have used the following original water solutions:

1. Cysteine, arginine, and lysine with concentrations $c = 0.09\%$, 0.118%, and 0.122% (weight) correspondingly
2. Polylysine with concentration $c = 0.125\%$ (weight)
3. Silver nitrate with concentration $c = 0.024\%$ (weight)
4. Water solution of sodium boron hydride with concentration $c = 0.4\%$ (weight) (to stabilize particles we have added 2.5% (vol.) 1 M solution of sodium hydroxide)

To form nanoparticles, first we add the specified volume of spacer solution (up to 0.2% total volume) into silver nitrate solution. Then we introduce 120 μL boron hydride solution (divided into several 30 μL portions to be introduced every 5 min) while stirring the mixture vigorously with a magnetic stirrer. Characteristic absorption maximums in the spectrums of silver nanoparticles suspensions stabilized with lysine, arginine, cysteine, and polylysine are at wavelength 390–400 nm (Figure 38.19). Comparing the bands of absorption of silver nanoparticles stabilized with monomer amino acids and stabilized with polylysine, we can conclude that nanoparticles stabilized with amino acids are smaller in size than nanoparticles stabilized with polylysine. That conclusion is proved by the fact that the spectral maximum of the polylysine-stabilized nanoparticles is located in a longer wavelength area, at $\lambda = 398$ nm, whereas absorption maximums for amino-acid-stabilized nanoparticles are located at $\lambda = 390$–392 nm. Nanoparticles of smaller sizes are likely to occur due to the higher mobility of monomer molecules as synthesis is a quick

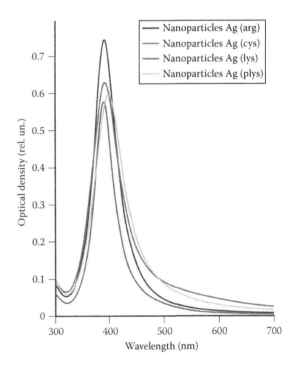

FIGURE 38.19 Absorption spectra for suspensions of silver nanoparticles stabilized with various amino acids (arg—arginine; cys—cysteine; lys—lysine; plys—polylysine).

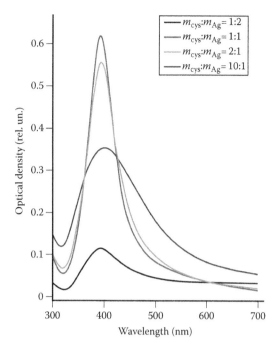

FIGURE 38.20 Absorption spectra for suspensions of cysteine-stabilized silver nanoparticles depending on various spacer concentrations.

reaction, whereas polymer molecules of polylysine are too big to take optimal confrontation that fast. It could also be explained by the fact that silver nanoparticles quite easily form nonspherical structures. This can lead to absorption maximums moving to the long wavelength area (Krutyakov et al., 2008).

The spectral characteristics (location of absorption maximum and absorption intensity) of the silver nanoparticle suspensions also depend on spacer concentration. Nanoparticles of the most homogenous sizes have been formed in suspensions with silver into spacer ratio close to 1:1. The fact is proved by the smaller semi-width of characteristic peaks in the nanoparticle spectrum (Figure 38.20).

Figure 38.19 demonstrates that characteristic maximums in the spectrum of absorption of silver nanoparticles and BR are divided into 160–170 nm to make it easier for the analysis and application of hybrid nanostructure optical properties to form functional nanocomposites to happen.

As mentioned above, the technological methods of forming hybrid nanostructures are based on the mechanisms of special bonding and self-organization activated by selecting appropriate bridge molecules for various functional groups. For example, during formation and stabilization of silver nanoparticle molecules of cysteine, arginine, lysine, and polylysine bond with the surface with the help of thio- (cysteine), guanydo- (arginine), or amino groups (lysine and polylysine). At the same time those molecules are spacers as they include additional amino groups which are able to actively interact with carboxyl groups of glutamic and asparagic acid residues, included in the amino acid consequence of BR. So, by using silver nanoparticles and BR molecules we can obtain hybrid nanostructures.

Such hybrid structures are synthesized with nanoparticles into BR molecules ratio equal to 1:3600. We use 5 mg/mL BR suspension as the original one. The hybrid nanostructures are formed by vigorously stirring the suspension with a magnetic stirrer for 1 day at the temperature of the reaction mixture equal to 22°C–25°C.

Figures 38.21 and 38.22 demonstrate spectral characteristics of hybrid nanostructures using lysine and polylysine as spacers.

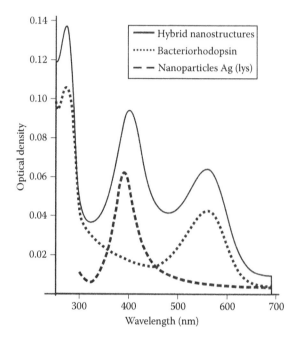

FIGURE 38.21 Absorption spectrum for suspension of hybrid nanostructures based on silver nanoparticles, lysine, and BR.

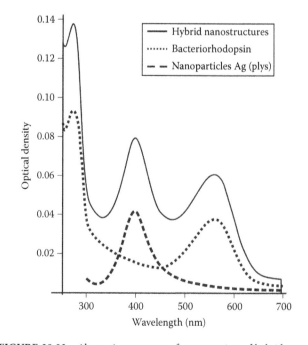

FIGURE 38.22 Absorption spectrum for suspension of hybrid nanostructures based on silver nanoparticles, polylysine, and BR.

While analyzing the technological results of obtaining hybrid nanostructures, we have to pay attention to the following:

The spectral characteristics of hybrid nanostructure suspensions demonstrate two characteristic spectral maximums, corresponding to certain components. It proves that BR and nanoparticle bonding causes no destruction of materials and BR molecules keep their functioning.

1. The characteristic spectral maximums of certain components of hybrid nanostructures may be offset. For example, the suspension of hybrid nanostructures based on silver nanoparticles, lysine, and BR demonstrates absorption maximum offset for silver nanoparticles from 391 to 408 nm (Figure 38.21), whereas the location of absorption maximum for BR stays the same—568 nm. The suspension of hybrid nanostructures based on the silver nanoparticles, polylysine, and BR demonstrates absorption maximum offset for BR from 568 to 560 nm, whereas the location of absorption maximum for the silver nanoparticles stays almost the same 400 nm (Figure 38.22). It could possibly happen because of the different distances between nanostructure components due to various spacer molecules (≈0.75 nm of lysine and ≈1.5 nm of polylysine), the type of bonding with nanoparticle surface, and BR.

2. After measuring the spectral characteristics of hybrid nanostructures, we have found out considerably increased absorption by every component of hybrid nanostructure against the adaptive spectral characteristic calculated by the simple adding of the spectrum of separate components in the same concentration they have in the suspension of hybrid nanostructures. The absorption increase reaches 13% at $\lambda = 394$ nm (cysteine-stabilized silver nanoparticles) and 26% at $\lambda = 568$ nm (BR) (Figure 38.23).

Therefore, synthesized hybrid nanostructures demonstrate functionality, mutual influence of certain components on spectral characteristics of each other, as well as an influence on BR photocycle.

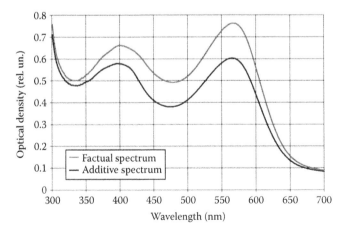

FIGURE 38.23 Factual spectrum of hybrid nanostructure suspension based on silver nanoparticles, cysteine, bacteriorhodopsin, and additive spectrum from spectra of separated components.

38.4 Summary

The use of molecules with conformation-adjusted relative position (distance and angle) of molecular groups including bridge functional groups as spacers can contribute to the development of nanotechnological and nanosystem approaches based on self-organization. Using the above method, we can regulate dipole–dipole interaction of a nanoparticle and a functional molecule. The effect is more obvious for nonspherical particles. Taking into the account that BR molecule performs photo-dependent transit of charge (H⁺), the above method of regulating photochromic photoreaction and the lifetime of intermediates must be quite effective. The suggested approach can also be applied for hybrid nanostructures with different composition.

Therefore, we have demonstrated the system and technological aspects of self-organization of functional nanostructures of molecular photonics realizing bionic principles of neural network appliances operation.

The systems designed are expected to have some functional characteristics, making them to sense and analyze like live intellectual neural systems. For example, apart from having distributed and content-addressed memory and the ability to learn and identify steady images, those systems will be able to identify dynamic images, i.e., images of evolving objects. Such abilities are based on new functions built upon self-organization of structures and processes, including: adaptive sensing with feedbacks, regulating activity of various parts of artificial retina; adaptive change of sampling rate and of retina coverage; modeling of tunable retinal receptive fields with spatial filtering; and formation of structures sensitive to the direction and speed of object movements.

References

Akiyama, K., Takimoto, A., and Ogawa, H. 1995. Spatial light modulator and neural network. U.S. Patent 5428711.

Birge, R.R., Gillespie, N.B., Izaguirre, E.W., Kusnetzow, A., Lawrence, A.F., Singh, D., Song, Q.W., Schmidt, E., Stuart, J.A., Seetharaman, S., and Wise, K.J. 1999. Biomolecular electronic: Protein-based associative processors and volumetric memories. *Journal of Physical Chemistry B* 103: 10746–10766.

Engel, S.J. 1990. Neural network processing system. European Patent EP0382230A2.

Grebennikov, E.P. 1997. Light-radiation-induced structure formation of bacteriorhodopsin films for the development of self-organizing information processing systems. *Proceedings of SPIE* 3402: 460–465.

Hampp, N. 2000. Bacteriorhodopsin as a photochromic retinal protein for optical memories. *Chemical Review* 100: 1755–1776.

Haronian, D. and Lewis, A. 1991. Elements of a unique bacteriorhodopsin neural network architecture. *Applied of Optics* 30(5): 597–608.

Krutyakov, Yu.A., Kudrinskiy, A.A., Olenin, A.Yu., and Lisichkin, G.V. 2008. Synthesis and properties of silver nanoparticles: Advances and prospects. *Russian Chemical Review* 77(3): 233–257.

Oesterhelt, D., Brauchle, C., and Hampp, N. 1991. Bacteriorhodopsin: A biological material for information processing. *Quarterly Reviews of Biophysics* 24(4): 425–478.

Prigogine, I. 1980. *From Being to Becoming: Time and Complexity in the Physical Sciences.* San Francisco, CA, W. H. Freeman & Co.

Siebert, F., Mantele, W., and Kreutz, W. 1982. Evidence for the protonation of two internal carboxylic groups during the photocycle of bacteriorhodopsin. *FEBS Letters* 41: 82–87.

Unger, H.G. 1980. *Planar Optical Waveguides and Fibers.* Oxford, U.K., Oxford University Press.

Vsevolodov, N.N. 1988. *Biopigment-Photographic Recorders. Photomaterial Based on Bacteriorhodopsin.* Moscow, Science Publ. (in Russian).

Wasserman, F. 1989. *Neural Computing: Theory and Practice.* New York, Van Nostrand Reinhold Co.

Zeisel, D. and Hampp, N. 1992. Spectral relationship of light-induced refractive-index and absorption changes in bacteriorhodopsin films containing BR-WT and the variant BR-D96N. *Journal of Physical Chemistry* 96(19): 7787–7792.

Zlenko, A.A., Kiselev, V.A., Prokhorov, A.M., Spikhal'skii, A.A., and Sychugov, V.A. 1975. Emission and reflection of light by a corrugated section of a waveguide. *Soviet Journal of Quantum Electronics* 5: 1325–1328.

VI

Nanoscale Lasers

39

Nanolasers

Marek S. Wartak
Wilfrid Laurier University

39.1 Introduction: Properties of Laser Radiation

A laser is a device that emits light (or more generally electromagnetic (EM) radiation) through a process called stimulated emission. The term "laser" is an acronym for light amplification by stimulated emission of radiation [1]. The first working laser was demonstrated on May 16, 1960, by Theodore Maiman at Hughes Research Laboratories. It was a pulsed ruby laser. The main characteristics of laser beam are monochromaticity, coherence, divergence, and brightness.

Monochromaticity is associated with the fact that in a laser, only an EM radiation at a particular frequency (or wavelength) is amplified. In practice, it has a certain range, called linewidth, which is determined by homogeneous and inhomogeneous broadening factors. The linewidth is very small compared with typical light sources. Additionally, due to the laser cavity, which forms a resonant system, the oscillation can occur only at the resonance frequencies of this cavity. This leads to the further narrowing of laser linewidth, often by several orders of magnitude.

Coherence is associated with the phase difference. For any EM wave, there are two kinds of coherence, namely, spatial and temporal coherence. Typical laser source emission is very coherent. Most other light sources emit incoherent light, which has a phase that varies randomly with time and position.

Let us illustrate first the spatial coherence. For that, consider two points of EM wave, which, at time $t = 0$, have phase difference ϕ_0. If for any time $t > 0$ the phase difference of EM wave at those points remains ϕ_0, we say the EM wave has perfect coherence between the two points. If this is true for any two points of the wave front, we say the wave has perfect spatial coherence.

Now, consider a fixed point on the EM wave front. If at any time the phase difference between time t and time $t + \Delta t$ remains the same, where Δt is some time delay, we say that the EM wave has temporal coherence over a time Δt. If Δt can be any value, we say the EM wave has a perfect temporal coherence. If this happens only in a range $0 < \Delta t < t_0$, we say it has partial temporal coherence, with a coherence time equal to t_0. Laser light is highly coherent, and this property has been widely used in measurement, holography, etc.

Laser beam is also highly *directional*, which implies that laser light has very small divergence. This is a direct consequence of the fact that laser beam comes from the resonant cavity, and only waves propagating along the optical axis can be sustained in the cavity. The directionality is described by the light beam divergence angle.

The brightness of a light source is defined as the power emitted per unit surface area per unit solid angle. Brightness is inversely proportional to the square of laser divergence; therefore laser light is much brighter than normal light source.

39.2 Overview of Lasers

In gas lasers, the active medium can be regarded effectively as an ensemble of absorption or amplification centers (like, e.g., atoms or molecules) with only electronic energy levels, which couple to the resonant optical field. Other electronic states are used to excite or pump the system.

The generic laser structure is shown in Figure 39.1. It consists of a resonator (cavity) formed by two mirrors and gain medium where the amplification of electromagnetic radiation (light) takes place. A laser is an oscillator analogously like an oscillator in electronics and requires a resonator, which provides feedback. Feedback is provided by two mirrors. Mirrors confine light and provide optical feedback. One of the mirrors is partially transmitting, which allows the light to escape from the device. There must be an external energy provided into the gain medium (process known as pumping). The most popular (practical) pumping mechanisms are by optical or electrical means.

The gain medium can be created in several ways. Conceptually, the simplest one is the collection of gas molecules. The pumping process excites these molecules into a higher energy level. The popular visualization of such collection of molecules is as two-level systems (TLS) (see Figure 39.2). Only two energies (out of many in the case of a molecule) are selected and the transitions are considered within these two energies. As illustrated, three basic processes are possible: absorption, spontaneous emission, and stimulated emission.

Such TLS are found very often in nature. Generally, for an atomic system, for the case under consideration, we can always separate just two energy levels: upper level and ground state, thereby forming TLS.

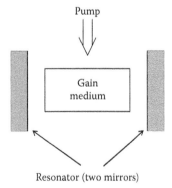

FIGURE 39.1 Generic laser structure: two mirrors with a gain medium in between. The two mirrors form a cavity, which confines the light and provides the optical feedback. One of the mirrors is partially transmitted and thus allows light to escape. The resulting laser light is directional, with a small spectral bandwidth.

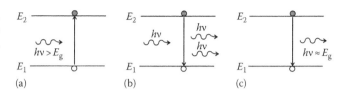

FIGURE 39.2 Possible electronic transitions in two-level system. (a) Absorption, (b) Stimulated emission, and (c) Spontaneous emission.

Electron can be excited into upper level due to external interactions (for lasers through a process known as pumping). Electrons can lose their energies radiatively (emitting photons) or nonradiatively, say by collisions with phonons.

For laser action to occur, the pumping process must produce population inversion meaning that there are more molecules in the excited state (here upper level with energy E_2) than in the ground state. If population inversion is present in the cavity, the incoming light can be amplified by the system (see Figure 39.2b) where one incoming photon generates two photons as the output.

The way how TLS is practically utilized results in various types of lasers, like gaseous, solid state, or semiconductor. Also, different types of resonators are possible as will be discussed in subsequent sections.

39.3 Semiconductor Lasers

A significant percentage of today's lasers are fabricated using the semiconductor technology. Those devices are known as semiconductor lasers. Over the last 15 years or so, several excellent books describing different aspects and different types of semiconductor lasers have been published [2,3].

The operation of semiconductor lasers as sources of electromagnetic radiation is based on the interaction between EM radiation and the semiconductor. Typical semiconductor laser structures are shown in Figure 39.3. These lasers can be classified as in-plane laser where light propagates in a parallel direction [4] and vertical cavity surface emitting laser (VCSEL) [5].

The largest dimension of in-plane structures is typically in the range of 250 μm (longitudinal direction) and as such cannot be considered as a nanolaser. The structure of interest to us is the one where light propagates perpendicularly to wafer's surface and it is known as VCSEL (see Figure 39.3b). The typical diameter of VCSEL cylinder is about 10 μm.

The basic semiconductor laser is just a p–n junction (see Figure 39.4) in which the cross-section along the lateral–transversal directions is shown. Current flows (holes on p-side and electrons on n-side) along the vertical direction, whereas the light travels horizontally and leaves the device on both sides.

Light propagation with amplification is illustrated in Figure 39.5. Mathematically it is described by assuming that there is no phase change on reflection at either end (left and right). The left end is defined as $z = 0$ and right end as $z = L$. At the right facet, the forward optical wave has a fraction r_R reflected (amplitude

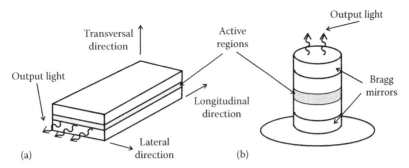

FIGURE 39.3 In-plane laser (a) and VCSEL (b).

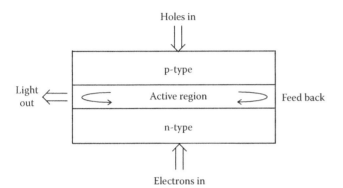

FIGURE 39.4 The basic p–n junction laser.

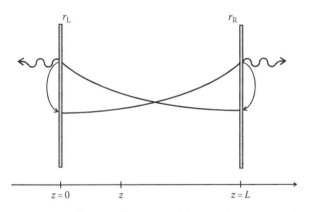

FIGURE 39.5 Schematic illustration of the amplification in a Fabry–Perot (FP) semiconductor laser with homogeneously distributed gain.

reflection) and after reflection the fraction travels back (from right to left).

In order to form a stable resonance, the amplitude and phase of the single round trip must match the amplitude and phase of the starting wave. At arbitrary point z inside the cavity (see Figure 39.5) the forward wave is

$$E_0 e^{gz} e^{-j\beta z}$$

where we have dropped $e^{j\omega t}$ common term and $g = g_m - \alpha_m$. Here r_R and r_L are, respectively, right and left reflectivities, g is gain (and loss), L length of the cavity, and β the propagation constant.

The wave traveling one full round will be

$$\{E_0 e^{gz} e^{-j\beta z}\}\ \{e^{g(L-z)}e^{-j\beta(L-z)}\}\{r_R e^{gL}e^{-j\beta L}\}\times\{r_L e^{gz}e^{-j\beta z}\} \quad (39.1)$$

The above terms are interpreted as follows. In the first bracket, there is an original forward propagating wave at z, in the second bracket, there is wave traveling from z to L, third bracket describes wave propagating from $z = L$ to $z = 0$, and the last one contains wave traveling from $z = 0$ to the starting point z. At that point, the wave must match original wave and thus one obtains condition for stable oscillations:

$$r_R r_L e^{2gL}e^{-2j\beta L} = 1 \quad (39.2)$$

That condition can be split into amplitude condition

$$r_R r_L e^{2(g_m - \alpha_m)L} = 1 \quad (39.3)$$

and phase condition

$$e^{-2j\beta L} = 1. \quad (39.4)$$

From the amplitude condition, the following relation is obtained:

$$g_m = \alpha_m + \frac{1}{2L}\ln\frac{1}{r_R r_L} \quad (39.5)$$

From the phase condition, it follows that

$$2\beta L = 2\pi n \quad (39.6)$$

where n is an integer. The last equation determines wavelengths of oscillations since

$$\beta = \frac{2\pi}{\lambda_m} \quad (39.7)$$

with λ_m being the wavelength.

In VCSEL, the cavity is formed by the so-called Bragg mirrors and an active region typically consists of several quantum well layers separated by barrier layers (see Figure 39.3). Bragg mirrors consist of several layers of different semiconductors, which have different values of refractive index. Due to the Bragg reflection,

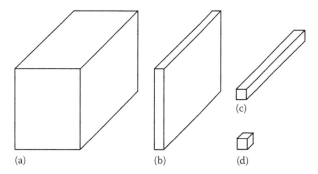

FIGURE 39.6 Illustration of various active regions with different dimensionality. (a) three-dimensional (3D) or bulk structure, (b) two-dimensional (2D) structure known as quantum well, (c) one-dimensional (1D) structure known as quantum wire, and (d) zero-dimensional (0D) structure known as quantum dot. (Adapted from Arakawa, Y. and Sakaki, H., *Appl. Phys. Lett.*, 40, 939, 1982.)

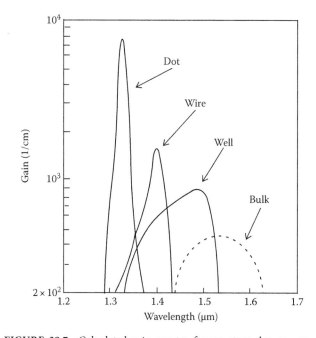

FIGURE 39.7 Calculated gain spectra for quantum dot, quantum wire and quantum well. For comparison, results for bulk crystal are also shown as dotted line. (Adapted from Asada, M. et al., *IEEE J. Quantum Electron.*, 22, 1915, 1986.)

such a structure shows a very large reflectivity (around 99.9%). Such large values are needed because a very short distance of propagation of light does not allow to build enough amplification when propagating between mirrors.

From the electromagnetic analysis of the optical cavity, the quality factor Q of the cavity [6] can be determined as

$$Q = \frac{\Delta\omega}{\omega_0} \tag{39.8}$$

where
$\Delta\omega$ is the cavity linewidth
ω_0 the resonant frequency of the cavity

Gain spectra (also other properties) strongly depend on size of the active region. Different gain media are possible, consisting of 0D, 1D, 2D, 3D (bulk) structures (see Figure 39.6). The typical thickness of a quantum well is about 10 nm. The dimensionality of these structures have profound consequences on laser properties. An illustration of gain spectra obtained with gain media for structures of different dimensionalities is shown in Figure 39.7. As can be seen, the shape of spectrum becomes sharper with increasing quantization dimension. This is due to the variation of the density of states.

39.4 Rate Equation Approach

In a typical laser, there exists two types of subsystems: photons and carriers. A quantitative description of a laser system is given in terms of rate equations. One introduces the number of photons S inside the cavity and number of carriers N (can be also a number of excited molecules for some systems). The rate equations describe the time evolution of S and N, as follows:

$$\frac{dN}{dt} = \eta_i \frac{I}{qV} - \frac{N}{\tau} - v_g g(N)S \tag{39.9}$$

$$\frac{dS}{dt} = \Gamma v_g g(N)S - \frac{S}{\tau_p} + \Gamma\beta R_{sp} \tag{39.10}$$

We have explicitly indicated that gain g depends on the carrier's concentration.

In Equation 39.9 the first term is responsible for pumping (in this case electrical), with I being the current, the second term accounts phenomenologically for losses, and the last one describes coupling to the photon system.

The last term in Equation 39.10 accounts for spontaneous emission (with the coefficient β). The β coefficient describes the amount of spontaneous emission that contributes to the lasing mode. The β factor is inversely proportional to the number of available modes into which the gain medium can spontaneously emit photons. In typical in-plane lasers, it is a small number (like 10^{-4}) [3]. The value of β is between 0 and 1.

The typical power–current characteristic of semiconductor laser is shown in Figure 39.8 for two extreme cases of β. When $\beta = 1$, all the spontaneously emitted photons end up in the lasing mode. For small values of β, the laser has a well-defined threshold. At threshold, the optical gain compensates for the losses. Above the threshold, the laser operates in a stimulated emission mode and below the threshold, spontaneous emission dominates.

A large β is the key factor in single-photon laser sources [7,8]. The (still hypothetical) case of $\beta = 1$ is often referred to as a thresholdless laser [9]. In the thresholdless laser, all photons participate in the stimulated emission. Such a device would require a small amount of energy to operate. Its dimensions should be very small, say ~λ.

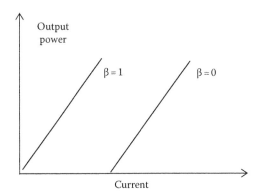

FIGURE 39.8 Current–power characteristics for two values of spontaneous emission factor.

Several issues must be addressed before one could fabricate thresholdless laser [10,11]. Those include: (a) optical modes that induce undesired spontaneous emission should be suppressed where possible, (b) creation of a single-cavity mode with a sufficiently high Q factor and a small modal volume is essential, and (c) excited carriers should be concentrated to emit light coupled to the single-mode cavity.

39.5 Definition of Nanolaser

The need to reduce the size of the semiconductor laser is one of the most active and challenging areas of modern optoelectronics. On the theoretical side, size reduction is important for understanding the basic laser concepts and fundamental light–matter interactions. On the practical side, smaller lasers will find various applications as light sources in integrated optical systems.

If photonics should be compatible with VLSI as for as lasers are concerned, photonic devices must shrink in size to 100 nm (or less) length scales. As indicated in the previous section, the smallest lasers available commercially today are VCSELs. However, in the last few years, the new type of even smaller devices is emerging, namely, nanowire lasers. Their cylindrical dimensions range from few tens up to hundreds of nanometers, whereas their lengths are typically within a few hundreds of microns.

The typical volume of conventional miniature lasers such as VCSELs is determined by the volume of the cavity mode. The effective wavelength in the dielectric should be of the order of the characteristic length of the device. This leads to the existence of the effective modal volume $V > (\lambda/2n)^3$, a condition known as the diffraction limit. Reduction of the volume of the active region is the potential factor for lowering the threshold current.

Recent advances in technology allow the fabrication of optical nanoscale devices where the wave nature of photons becomes one of the most critical variables. It provides the challenges in the realization of a tiny coherent photon source. The localization of the wave is difficult when wavelengths of photons become much larger than the spatial variation of the confinement structure.

As the dimensions of optical nanodevices scale down, devices can be fabricated with effectively only one optical emission mode. These structures could be termed *nanolasers* (see [8]). Alternatively, nanolasers can be defined as structures having dimensions smaller than the wavelength of light in all three dimensions [12]. Recent summaries of research on nanolasers can be found in Refs. [12,13].

In the following sections, we will discuss examples of recently the developed nanolasers (and related) structures:

- Nanowire lasers
- Plasmonic lasers
- Photonic crystal lasers
- Scattering lasers
- Organic lasers

We conclude this chapter with a brief discussion of applications of nanolasers in sensing and medicine.

39.6 Nanowire Lasers

Various types of nanocavities have been fabricated and a coherent laser emission from such structures has been observed. Among others, lasing has been demonstrated in droplets [14,15], silica [16], and polystyrene spheres [17], semiconductor microdisks [18–20], micropillars [21], photonic crystal cavities [22], nanoribbons [23], ZnO arrays [24], GaN nanowires [25], and single-crystal ZnO nanowires [26].

Several methods of the synthesis of semiconductor nanowire heterostructures have been developed, including chemical vapor deposition and the vapor–liquid–solid growth of crystalline semiconductor nanowires. Recent progress in the development of semiconductor nanowires was reviewed by Lauhon et al. [27].

The schematics of a typical cylindrical nanowire laser is shown in Figure 39.9 and the possible schemes of current injection are schematically shown in Figure 39.10. These structures provide photon confinement in volumes of a few cubic wavelengths. A typical lateral dimension of a nanowire is between 20 and 400 nm, with a length in the range of 2–40 μm. As was shown in the journal articles cited earlier, it is possible to grow nanowire arrays with tight control over size (diameter <20 nm) and uniformity (<±10%).

A significant reduction in size of nanometer laser was possible due to large spatial overlap between the active medium (the wire itself) and the guided mode, which propagates along the axial direction. A strong lateral optical confinement is also created. It originates due to the size of the structure itself and also due to large dielectric contrast (difference) between the nanowire material itself and the surrounding material. The end facets of the nanowire form axial Fabry–Perot-type cavity necessary for laser action.

Early experimental evidence for guided modes in nanowires surrounded by air has been reported by Johnson et al. [25] and Duan et al. [28]. Lasing was observed in structures that are rather thick (radius in the 80–200 nm range). In these structures, quantization effects are absent. Such structures can function as the active medium and waveguide at the same time. Structures

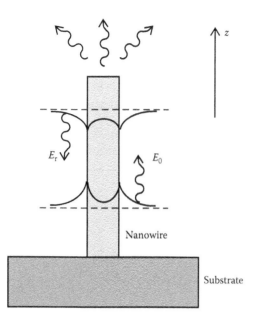

FIGURE 39.9 Side view of a nanowire laser. (Adapted from Maslov, A.V. and Ning, C.Z., *Appl. Phys. Lett.*, 83, 1237, 2003.)

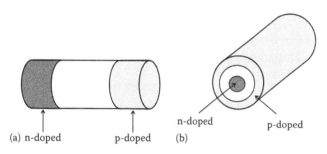

(a) n-doped   p-doped   (b)

FIGURE 39.10 Possible schemes of current injection in nanowires: (a) through the ends of the nanowire and (b) from the core and shell regions. (Adapted from Maslov, A.V., *J. Appl. Phys.*, 99, 024314, 2006.)

with much smaller dimensions (radius about 1 nm) have also been fabricated [29]. In such samples, quantization effects play a dominant role. As an example, resonant nanowire cavity fabricated by Huang et al. [24] achieved gain and lasing through an excitonic lasing action in ZnO with a threshold of $40\,kW/cm^2$ under optical excitation. (In this case, because of optical pumping, the density of optical power is provided instead of value of electrical current.) A summary of the properties of basic materials used to fabricate nanowires is shown in Table 39.1.

TABLE 39.1 Basic Properties of Materials Used to Fabricate Nanowires

Materials	Emission Wavelength (nm)	Threshold (nJ/cm²)	Gain (cm⁻¹)	Reference
GaN	370–390	500	400–1000	[85]
ZnO	370–400	70	1000–3000	[85]
CdS	510	—	—	[28]

39.6.1 Ring Resonator Nanolasers

Microrings are common building blocks in photonics [30]. Formation of nanoring resonator is also possible by converting the nanowire into a ring [31].

This change of shape results in a significant modification of the optical spectra of the cavity. For example, the photoluminescence displays Fabry–Perot (FP) resonances that match those calculated for ring resonance. Also, the dielectric imperfection at the overlapping junction causes each FP mode to split into a doublet, thus breaking the degeneracy between clockwise and anticlockwise photon propagation (see Figure 39.11b).

Microring nanolasers were fabricated and analyzed by Berkeley group led by Yang [31]. The ring structure was fabricated from a linear GaN nanowire (see Figure 39.11). We show here a GaN nanowire before manipulation into a ring configuration (a) and after creation of ring geometry (b). It is not a subwavelength device as can be seen from that figure. In the ring after coupling, a dielectric discontinuity is introduced (since each ends of the original wire are now positioned side by side). This side-by-side overlap is responsible for evanescent coupling between cavity arms (see Figure 39.12).

The circulating optical modes within a resonator cavity containing a defect are theoretically equivalent to a photonic molecule [32] (compare Figure 39.11c). Dielectric discontinuity after coupling leads to perturbation within the cavity and breaks the resonance degeneracy into clockwise and counterclockwise mode propagations. It is known that such perturbed cavity is theoretically equivalent to two perfect cavities that are coupled.

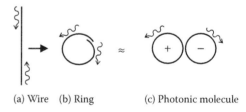

(a) Wire   (b) Ring   (c) Photonic molecule

FIGURE 39.11 Transition from wire linear cavity to a ring resonator and the theoretically equivalent photonic molecule. (Adapted from Pauzauskie, P.J. et al., *Phys. Rev. Lett.*, 96,143903, 2006.)

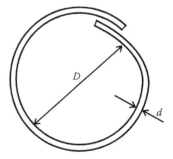

FIGURE 39.12 Cross-section view of the ring structure showing the side-by-side overlap that enables evanescent coupling between cavity modes.

The lasing behavior between nanowires and their resonator counterpart is also markedly different. The emission maximum of ring laser is red shifted up to 10 nm relative to the wire (see [31]).

39.6.2 Theoretical Analysis of Nanowires

For a mathematically oriented reader, we provide here the main elements of theory of nanolasers. This section can be skipped in the first reading.

Theoretical and numerical analysis of nanowire lasers was reported in a series of papers by Maslov and Ning [33–37] and also by Chen and Towe [38]. From experiments, we know that lasing occurs through the end facets of the nanowire (see Figure 39.9 (from Ref. [33].)

The nanowire can support transverse electric (TE_{0m}) and transverse magnetic (TM_{0m}) modes, which have only three field components and no dependence on ϕ and hybrid modes (HE_{nm} and EH_{nm}), which have all six field components [33]. In the axial approximation, the field components are of the type [34]

$$\tilde{E}_z^n(r) = \begin{cases} A_n J_n(\kappa_2 r), & r < r' \\ B_n J_n(\kappa_2 r) + C_n Y_n(\kappa_2 r), & r' < r < R \\ D_n H_n(\kappa_1 r), & r > R \end{cases} \quad (39.11)$$

$$\tilde{H}_z^n(r) = \begin{cases} F_n J_n(\kappa_2 r), & r < r' \\ G_n J_n(\kappa_2 r) + N_n Y_n(\kappa_2 r), & r' < r < R \\ M_n H_n(\kappa_1 r), & r > R \end{cases} \quad (39.12)$$

where J_n and Y_n are the Bessel functions of the first and second kind, respectively. $H_n = H_n^{(1)} = J_n + iY_n$ is the Bessel function of the third kind (Hankel's function). Also, the following definitions of transverse wave numbers were introduced: $\kappa_{1,2}^2 = \varepsilon_{1,2}\omega^2/c^2 - h^2$. The eight unknown coefficients A, B, C, D, F, G, N, and M should be determined from the boundary conditions at the interfaces $r = r'$ and $r = R$.

The numerical approach is based on solving Maxwell's equations using finite difference time-domain (FDTD) method in cylindrical coordinates [35]. Typical computational window is shown in Figure 39.13.

The modes supported in such nanowire structures are similar to those of optical fibers (for a discussion of modes in fiber, see [39]) but are more localized due to high refractive index contrast between the nanowires and the surrounding air.

The Maslov–Ning analysis suggests that the natural facets of the nanowires provide very low quality factors (of the order of hundreds) for nanowires of about 10 μm in length. These factors are very sensitive to the mode type and nanowire radius.

Far-field patterns of the emitted radiation were also discussed [34,35]. It was determined how the radiation pattern depends on the lasing mode. The radiation is emitted in a very broad range of angles with respect to the nanowire axis. Also, the directionality weakens with an increase of the nanowire radius.

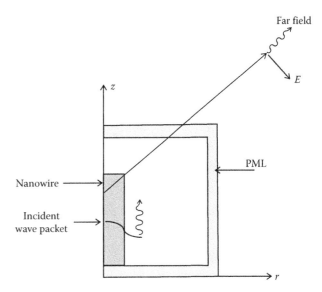

FIGURE 39.13 Schematic of the FDTD computational domain. (Adapted from Maslov, A.V. and Ning C.Z., *Opt. Lett.*, 29, 572, 2004.)

Using their methods, Maslov and Ning recently reported on the numerical analysis of semiconductor nanowire covered with a metal as a possible laser waveguide [40]. Their analysis opens the possibilities of fabricating even smaller nanowire-based lasers. Maslov and Ning analyzed the possible advantages of using a semiconductor nanowire encased in a metal as a laser waveguide. They showed that despite large Joule loss, such structure can be a good candidate for subwavelength laser operating in TM_{01} mode.

Coupled drift–diffusion simulations of nanowire lasers have been recently reported by Chen and Towe [38]. They extended the FDTD approach by including carrier effects. Their method is based on numerical solution of the steady-state 2D drift–diffusion carrier transport equations, which are coupled with the photon generation rate equations. The basic system of equations is as follows:

$$\nabla \cdot \left[-\varepsilon \nabla \psi(x,y) \right] = q \left[p(x,y) - n(x,y) + N_D^+ - N_A^- \right] \quad (39.13)$$

$$\nabla \cdot \mathbf{J}_n = -q \left[G(n,p) - R_{sp}(n,p) - R_{st}(n,p) - R_{Auger}(n,p) - R_{SRH}(n,p) \right] \quad (39.14)$$

$$\nabla \cdot \mathbf{J}_p = q \left[G(n,p) - R_{sp}(n,p) - R_{st}(n,p) - R_{Auger}(n,p) - R_{SRH}(n,p) \right] \quad (39.15)$$

$$\mathbf{J}_n = q D_n \nabla n(x,y) - q\mu_n n(x,y) \nabla \psi(x,y) \quad (39.16)$$

$$\mathbf{J}_p = -q D_p \nabla p(x,y) - q\mu_p p(x,y) \nabla \psi(x,y) \quad (39.17)$$

$$G_m S_m - \frac{S_m}{\tau_{p,m}} + R_{sp,m} = 0 \quad (39.18)$$

where $G(n, p)$ is the carrier generation rate, $R_{sp}(n, p)$ is the local spontaneous recombination rate, $R_{st}(n, p)$ is the local stimulated recombination rate, $R_{\mathrm{SRH}}(n,p)$ represents Shockley–Read–Hall (SRH) dark recombination rate, G_m, S_m, $\tau_{p,m}$, $R_{sp,m}$ are, respectively, the modal gain, the photon number in the cavity, photon lifetime and for the mth lasing mode, and the spontaneous emission rate that couples to the mth lasing mode.

39.6.3 Recent Work on Nanowire Lasers

We conclude the section on nanowires with a description of recent works that go beyond single cylindrical nanowire. First, we describe nanowires with metal coating and then the combination of nanowires and quantum wells.

The dimensions of semiconductor nanolasers can be further shrunk by using metal-coated nanowires. Such a structure was first fabricated in 2007 by Hill et al. [41] and this structure was shown to be the smallest electrically pumped nanolaser. They coated a thin semiconductor heterostructure post made of In/InGaAs/InP with a layer of gold as shown in Figure 39.14. The diameter was 210 nm. Electrons are injected through the top of the pillar and holes are injected through a large-area lateral contact.

The device lased in the near-infrared at a wavelength of 1408 nm. The mode was an HE_{11}-like mode oscillating near the InGaAs region cutoff frequency.

All previously described studies have concentrated on fabrication of nanolasers from homogeneous semiconductors such as gallium nitride (GaN). This means that the laser wavelength is determined by the band gap of the used material. In such a design, there is no possibility to tune the properties of the laser.

However, with proper technological tools, structures in which gain and cavity functionalities are decoupled can be designed and fabricated. The purpose of decoupling the gain medium and the cavity is not to separate but to combine the best properties of both subsystems. Thus quantum wells provide the optical gain medium whereas the nanowire acts as the optical cavity.

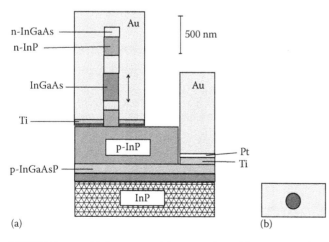

FIGURE 39.14 (a) Structure of the cavity formed by a semiconductor pillar encapsulated in gold. (b) Cross-section of the pillar. (Adapted from Hill, M.T. et al., *Nat. Photonics*, 1, 589, 2007.)

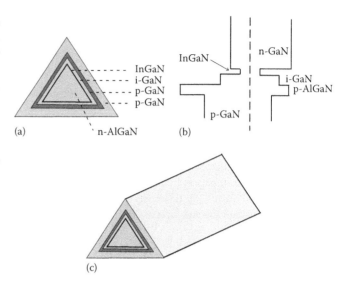

FIGURE 39.15 Cross-sectional view of a (a) GaN-based core/multi-shell nanowire structure and (b) the corresponding energy band diagram. (c) InGaN multiple quantum well (MQW) nanowire structure. (Adapted from Qian, F. et al., *Nat. Mater.*, 7, 701, 2008; Qian, F. et al., *Nano Lett.*, 5, 2287, 2005.)

This new type of nanolasers was described by Qian et al. [42]. The structure is shown in Figure 39.15. It consists of a GaN nanowire core that acts as the optical cavity surrounded by InGaN/GaN multiple quantum well (MQW) shells that serve as a composition-tunable gain medium. By varying the indium content, the emission wavelength can be tuned between 365 and 494 nm, with all devices operating at room temperature. Their nanowire heterostructure contains 3 to 26 quantum wells. Typical nanowires were 200–400 nm in "diameter" and 20–60 μm in length. Although the nanolasers were optically pumped, the authors believe that electrical injection is possible.

39.7 Plasmonic Lasers

Surface plasmon mode confinement has been used to achieve subwavelength modal dimensions at the expense of optical loss [43–45]. Several structures that involve plasmons were proposed. A comprehensive introduction to the main physical aspects involved in plasmonic devices were recently summarized by Dragoman and Dragoman [46].

A summary of the analysis of the plasmonic nanoresonators has recently been provided by Maier [47]. Plasmon cavities can confine electromagnetic energy into both physical and effective mode volumes far below the diffraction limit.

Manolatou et al. [48,49] proposed and analyzed a family of nanoscale cavities for electrically pumped surface-emitting semiconductor lasers that use surface plasmons to provide optical confinement. The analyzed cavities have radii between 100 and 300 nm and the heights of the dielectric part of the cavity between 100 and 250 nm. The metal layers were assumed to be thick enough to prohibit light transmission. The typical circular nanopatch laser analyzed by Manolatou and Rana [49] is shown in Figure 39.16.

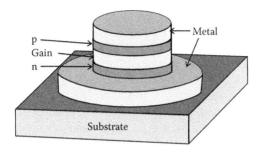

FIGURE 39.16 Circular nanopatch laser.

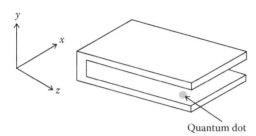

FIGURE 39.17 Horseshoe optical nanoantenna. (Adapted from Sarychev, A.K. and Tartakovsky, G., *Phys. Rev. B*, 75, 085436, 2007.)

Sarychev and Tartakovsky [50,51] proposed plasmonic nanolaser where the metal nanoantenna operates similarly to a resonator. The structure is shown in Figure 39.17. In this type of laser, metallic horseshoe-shaped nanoantenna interacts with a two-level amplifying system (TLS). TLS can represent quantum dot and can be pumped optically or electrically. The size of the proposed plasmonic nanolaser is much smaller than the wavelength.

39.7.1 Bowtie Structures

Bowtie structure consists of two opposing tip-to-tip metallic nanotriangles separated by a gap with an active element in the form of quantum dot in-between or quantum wells below it. The interaction of a single quantum dot with a bowtie antenna was demonstrated by Farahani et al. [52] for visible light. The enhancement of the electromagnetic field in such a structure was described by Sundaramurthy et al. [53]. They also conducted FDTD analysis. The theory of the electrically pumped plasmonic nanolaser based on bowtie structure was recently reported by Chang et al. [54]. They considered both quantum dot and quantum well.

The geometry of a bowtie nanolaser is shown in Figure 39.18. It consists of a metallic bowtie separated (typically) by about 20 nm. Multiple quantum wells are located below the metallic bowtie, which at optical frequencies have negative dielectric constant. The bowtie tips reduce the effective volume of the cavity mode and lead to the field enhancement around the bowtie tips. This results in an increase of the stimulated and spontaneous emission rates and significant decrease of the threshold current.

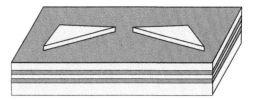

FIGURE 39.18 The geometry of a bowtie plasmonic nanolaser. Typical separation between metallic nanotriangles is 20 nm. (Adapted from Chang, S.-W. et al., *Opt. Express*, 16, 10580, 2008.)

39.7.2 Dipole Nanolaser

Diple nanolaser (DNL) has been proposed by Protsenko et al. [55,56]. It consists of a metallic nanoparticle of size r_0 and a two-level system (TLS) of size r_2 formed by a quantum dot, separated by a small distance r (see Figure 39.19). The device does not need an optical cavity and may have a volume much smaller than λ, the lasing wavelength. An incoherent pump provides population inversion in the TLS. The system can be pumped optically or electrically, in which case pumping is provided by injection of carriers from the bands of a semiconductor material surrounding an embedded quantum dot.

The transition frequency of TLS ω_2 is close to the plasmon resonance frequency ω_p of the metallic nanoparticle. There is a strong interaction (a dipole interaction) between those two systems through a near field. It is known that such coupled interaction significantly modifies optical emission.

To understand the operation of DNL, let us remind ourselves that the usual lasing mechanism involves stimulated emission into the mode of a cavity (for the single mode operation) from a medium in which there is a population inversion. Above the threshold condition, when stimulated emission (gain) exceeds absorption plus internal losses, the energy from incoherent pump is transferred into coherent laser radiation with a narrow lasing spectrum.

In DNL, instead of electromagnetic field, one deals with linear oscillations of polarization of a medium. Such a medium can be excited in such a way that the total energy flux into polarization exceeds losses. This leads to polarization oscillations with a narrow spectrum. The analysis of dipole nanolaser conducted by Protensko et al. [55] is based on equations of motion identical to Maxwell–Bloch equations for a TLS in the electromagnetic field of a cavity. Lasing conditions are created when the population

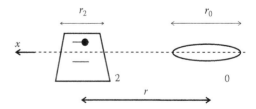

FIGURE 39.19 Dipole nanolaser. Nanoparticle (here on the right) is labeled with "0," TLS (here on the left) has label "2." (Adapted from Protsenko, I.E. et al., *Phys. Rev. A*, 71, 063812, 2005.)

inversion D of energy states of TLS is maintained by optical pumping or injection current to be [56]

$$D > D_{th} \qquad (39.19)$$

where $D_{th} = (r/r_{cr})^6$, $r_{cr} = (4|\alpha_0||\alpha_2|)^{1/6}$ and α_0 and α_2 are polarizabilities of the nanoparticle and the TLS. When D exceeds the threshold value D_{th}, the macroscopic dipole moment of two particles appears spontaneously and they emit coherent dipole radiation at the frequency $\omega \approx \omega_{plr}$, the nanoparticle plasmon resonance frequency. Condition (39.19) is fulfilled if the distance between particles is small, which means a strong interaction between them through the near field.

39.8 Photonic Crystal Lasers

Photonics crystal (PC) structure consists of a drilled repeating pattern of holes through the laser material. This pattern is called a photonic crystal. One can deliberately introduce an irregularity, or defect, into the crystal pattern, for example, by slightly shifting the positions of two holes. The photonic crystal structure and the defect prevent light of most frequencies from existing in the structure, with the exception of a small band of frequencies that can exist in the region near the defect.

Photonic crystal defect microcavities can provide extremely small mode volume [22,57]. The schematic of the laser proposed by Painter et al. [22] is shown in Figure 39.20. The interhole separation is 515 nm. It was fabricated from InGaAsP grown by MOCVD on an InP substrate. The active region consists of four (here only two are shown) 9 nm 0.85% compressively strained InGaAsP quantum wells. Two-dimensional photonic crystal hexagonal lattice was formed by etching, resulting in air holes that penetrate through the active region and into an underlying sacrificial InP layer.

Over the last few years, various devices have been fabricated and characterized [58–61]. Such structures are associated with single-photon sources, which are required for quantum computing and quantum communications. Single-photon sources are recent applications of the Purcell effect in quantum-dot microcavities.

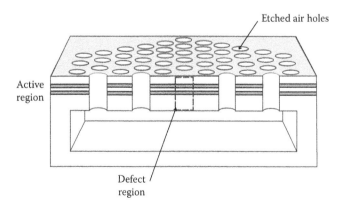

FIGURE 39.20 Cross-section through the middle of the photonic crystal microcavity. A defect is formed (shown in the middle) by removing a single hole. (Adapted from Painter, O. et al., *Science*, 284, 1819, 1999.)

Recently, scientists from the Yokohama National University in Japan [61] have reported interesting results concerning PC nanolasers. They demonstrated high-performing room-temperature nanolaser in the form of PC slab. The laser is made of a GaInAsP. This ultrasmall laser has a modal volume close to the diffraction limit. When operating in a high-Q mode (about 20,000), it will be useful for optical devices in optical integrated circuits. In a moderate-Q (1500) configuration, the nanolaser needs only an extremely small amount of external power to bring the device to the threshold of producing laser light. In this near-thresholdless operation, it might permit the emission of very low light levels, even single photons.

39.8.1 Edge-Emitting Photonic Crystal Nanolaser

Tiny resonators and waveguides based on photonic crystals provide a promising approach to fabricating high-density photonic integrated circuits. Recently, Yang et al. [62] reported the first demonstration of an edge-emitting photonic crystal nanocavity laser integrated with a photonic crystal waveguide. The structure is based on a double-heterostructure photonic crystal nanocavity with a InAs quantum dot active region. The device consists of four photonic crystal sections, each with a slightly different lattice constant. Five layers of InAs quantum dots, each with a quantum dot density of about $2 \times 10^{10}/cm^2$, were embedded in the 220 nm-thick GaAs membrane. An output waveguide is butt coupled to the mirror with the fewer number of periods. The device was optically pumped with a semiconductor laser diode at 850 nm wavelength. The threshold peak pump power absorbed by the cavity was estimated to be 12 μW.

39.8.2 Silicon Nanocrystals

Compatibility with CMOS materials have stimulated research on nanoscale silicon laser by Jaiswal and Norris [63]. They have theoretically analyzed and numerically simulated various designs based on photonic crystal concept and utilizing Si nanocrystals embedded in SiO_2. Their studies were motivated by observations of optical gain in silicon nanocrystals [64].

Pavesi et al. [64] observed light amplification in silicon itself. Silicon nanocrystals in the form of quantum dots were dispersed in a silicon dioxide matrix. Net optical gain was seen in both waveguide and transmission configurations, with the material gain being of the same order as that of direct band gap quantum dots. Their findings open the possibility for the fabrication of a silicon laser.

39.9 Scattering Lasers

In scattering lasers, the feedback is provided through scattering of light instead of a cavity [65]. These lasers are of two basic types: Mie lasers consisting of only one sphere and random lasers that are formed by many scattering particles. In Mie lasers, the surface of the sphere serves as multiple scatterer, while in random lasers, scattering is provided by randomly distributed particles.

In both types of scattering lasers, the average particle size used in typical experiments was about 5–10 µm [66,67]. They are, therefore, not of the size of true nanolasers.

39.9.1 Mie Nanolasers

In a Mie nanolaser, the mirrors of a conventional laser are replaced by the boundary of a microsphere. Light is multiply scattered at the boundary, and along the boundary, whispering gallery modes (WGMs) at a certain wavelength exist for specific sizes of the sphere.

WGMs occur at particular resonant wavelengths of light for a given droplet size (see Figure 39.21). At these wavelengths, the light undergoes total internal reflection at the particle surface and, after one roundtrip, interferes constructively. It becomes trapped within the particle for timescales of the order of nanoseconds.

The nomenclature of these modes derive from the observation of Lord Rayleigh in the dome in St. Paul's Cathedral in London. He observed sound ("whispers") propagating along the walls and circling around the dome several times.

Theoretical and experimental study of spherical resonators, which form the basis of Mie lasers, is still an open field of research. This interest stems from the analysis of fundamental processes such as scattering, energy propagation through disordered media, and cavity quantum electrodynamics, and from the large number of applications in photonics, chemistry, meteorology, astronomy, and sensing.

39.9.2 Random Lasers

In random lasers, the conventional optical cavity is replaced by light scattering from many particles. They do not have mirrors or optical elements.

In random lasers, the feedback is provided by multiple scattering of light at many scattering points. Random laser does not have a clear feedback mechanism like a conventional laser. It is nonresonant, even disordered and works in qualitatively different way than feedback by a resonant cavity (see [68] for a recent review). A comparison of conventional and random laser is shown in Figure 39.22. The wavelengths of random lasers span from the ultraviolet to the mid-infrared region. The materials used in random lasers include inorganic dielectrics, semiconductors, polymers, and liquids. The size of random lasers can vary from a cubic micrometer to hundreds of cubic millimeters.

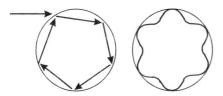

FIGURE 39.21   Whispering gallery mode inside a microsphere.

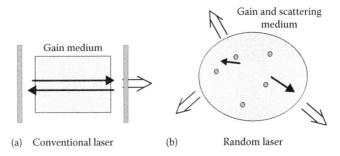

(a)   Conventional laser   (b)   Random laser

FIGURE 39.22   Comparison of conventional and random laser. (Adapted from Wiersma, D., *Nature*, 406, 132, 2000.)

A short history of the development of random lasers is as follows:

- In 1968, Letokhov [69] described an idea of scattering with "negative absorption."
- In 1994, Lawandy et al. [70] conducted experiments on scatterers in laser dye and observed threshold and line narrowing.
- In 1995, Wiersma et al. [67] conducted multiple scattering experiments in Ti:Al$_2$O$_3$ random lasers.

Recently a definite reference on solid state random lasers was published by Noginov [71] who tried to answer the question of what is a random laser by reviewing many types of random lasers. On the theoretical side, we mention two approaches to describe scattering lasers [72,73].

Fratalocchi et al. [72] derived a 3D vector set of Maxwell–Bloch equations, which were solved numerically by employing FDTD method. They performed a series of numerical experiments by investigating the process of laser emission from a single nanosphere covered by a layer of active material for different pumping rates.

Novel theoretical approach to diffusive random lasers has recently been developed by Tuereci et al. [73]. They developed a model by considering all the possible ways in which light can reflect back and forth in the medium. In their approach, they considered coexistence of gain, nonlinear interactions, and overlapping resonances.

39.10 Organic Nanolasers

Ease of fabrication make organic materials attractive for various optoelectronic devices, including lasers. O'Carroll et al. [74] used the flexibility of structuring organic materials to fabricate subwavelength optical nanowire lasers.

Using a melt-assisted template wetting method, they synthesized arrays of semicrystalline nanowires with diameters in a range of 150–400 nm, and with typical length of ~6 µm. Their structures are cylindrical wires with optically flat end facets, which form optical cavities. Lasers were optically pumped by an external pump laser. The observed that lasing wavelength (about λ = 460 nm) was determined by a standard standing wave condition.

Organic lasers can potentially have various applications [75] as inexpensive and lightweight sources of coherent radiation.

39.11 Applications

To conclude this chapter, we briefly discuss a few recent applications of nanolasers focused on nanowire and photonic crystal-based devices.

39.11.1 Nanolaser Device Detects Cancer in Single Cells

Investigators at Sandia National Laboratories in New Mexico reported a method to rapidly detect cancer in a single cell using nanolaser techniques [76,77]. Their technique, which allows the investigator to distinguish between malignant and normal cells, has the potential of detecting cancer at a very early stage, a development that could change profoundly the way cancer is diagnosed and treated.

The method of determination of malignant cells is based on a technique that rapidly assesses the properties of cells flown through a nanolaser. They observed biophotonic differences in normal and cancer mouse liver cells by using intracellular mitochondria as biomarkers for disease. This difference arises from the fact that mitochondria, the internal organelles that produce energy in a cell, are scattered in a chaotic, unorganized manner in malignant cells, while they form organized networks in healthy cells. This difference produces a marked change in the way that malignant cells scatter laser light.

39.11.2 Biological and Chemical Detection

Optofluidic integration of a new type of photonic crystal nanolaser incorporated into a microfluidic chip was described by Kim et al. [78]. The proposed nanolasers are an ideal platform for high-fidelity biological and chemical detection tools in micro-total-analytical or lab-on-a-chip systems. Its operation is based on the wavelength tunability induced by the refractive index variation of the fluid.

A record high sensitivity utilizing photonic crystal nanolaser was recently reported by Kita et al. [79]. The index resolution limit of their sensor could be <10^{-6}.

39.11.3 Ultrahigh Data Storage

Ikkawi et al. [80] reported the development of a near-field optical system capable of focusing light into a spot with diameter of less than 30 nm. To fabricate this type of nanolaser, they deposited a 100 nm thick aluminum film on the emitting edge of a laser diode and then used a focused ion beam to etch various apertures into the film. Their nanolaser technique is capable of storing data beyond 10 Tbit/in^2.

39.12 Conclusions

In this chapter, we provided an introduction to the concept of nanolasers, which are considered to be applicable as main elements, among others, in nanoscale optoelectronics. We discussed numerous examples of recently developed nanolasers.

The field of nanolasers shows rapid progress both on the research side as well as in applications. Examples include nanowire lasers with diameters as small as 100 nm. They find applications in medicine and in biological and chemical sensing. There is no doubt that field of nanolasers is set for interesting research in future and that nanolaser devices will find new and yet unforeseen applications.

Acknowledgment

I would like to acknowledge the financial support provided by the Natural Science and Engineering Research Council of Canada (NSERC).

References

1. A. Yariv. *Quantum Electronics*, 3rd edn. Wiley, New York, 1989.
2. B. Mroziewicz, M. Bugajski, and W. Nakwaski. *Physics of Semiconductor Lasers*. Polish Scientific Publishers/North-Holland, Warszawa, Amsterdam, the Netherlands, 1991.
3. L.A. Coldren and S.W. Corzine. *Diode Lasers and Photonic Integrated Circuits*. Wiley, New York, 1995.
4. G.P. Agrawal and N.K. Dutta. *Semiconductor Lasers*, 2nd edn. Kluwer Academic Publishers, Boston, MA/Dordrecht, the Netherlands/London, U.K., 2000.
5. C.W. Wilmsen, H. Temkin, and L.A. Coldren, (eds.). *Vertical-Cavity Surface-Emitting Lasers*. Cambridge University Press, Cambridge, U.K., 1999.
6. R. Coccioli, M. Boroditsky, K.W. Kim, Y. Rahmat-Samii, and E. Yablonovitch. Smallest possible electromagnetic mode volume in a dielectric cavity. *IEE Proc. Optoelectron.*, 145:391–397, 1998.
7. C. Santori, D. Fattal, J. Vuckovic, G.S. Solomon, and Y. Yamamoto. Indistinguishable photons from a single-photon device. *Nature*, 419:594–597, 2002.
8. S. Strauf, K. Hennessy, M.T. Rakher, Y.-S. Choi, A. Badolato, L.C. Andreani, E.L. Hu, P.M. Petroff, and D. Bouwmeester. Self-tuned quantum dot gain in photonic crystal lasers. *Phys. Rev. Lett.*, 96:127404, 2006.
9. H. Yokoyama. Physics and device applications of optical microcavities. *Science*, 256:66–70, 1992.
10. E. Yablonovitch. Inhibited spontaneous emission in solid-state physics and electronics. *Phys. Rev. Lett.*, 58:2059–2062, 1987.
11. S. Noda. Seeking the ultimate nanolaser. *Science*, 314:260–261, 2006.
12. P.L. Gourley. Nanolasers. *Sci. Am.*, 278:56–61, 1998.
13. S.S. Mao. Nanolasers: Lasing from nanoscale quantum wires. *Int. J. Nanotechnol.*, 1:42–85, 2004.
14. S.-X. Qian, J. B. Snow, H.-M. Tzeng, and R.K. Chang. Lasing droplets: Highlighting the liquid–air interface by laser emission. *Science*, 231:486–488, 1986.

15. H.B. Lin, J.D. Eversole, and A.J. Campillo. Spectral properties of lasing microdroplets. *J. Opt. Soc. Am. B*, 9:43–50, 1992.

16. M. Cai, O. Painter, K.J. Vahala, and P.C. Sercel. Fiber-coupled microsphere laser. *Opt. Lett.*, 25:1430–1432, 2000.

17. M. Kuwata-Gonokami, K. Takeda, H. Yasuda, and K. Ema. Laser emission from dye-doped polystyrene microsphere. *Jpn. J. Appl. Phys. Lett.*, 2, 32:L99–L101, 1992.

18. B. Gayral, J.M. Gerard, A. Lemaitre, C. Dupuis, L. Manin, and J.L. Pelouardb. High-q wet-etched GaAs microdisks containing InAs quantum boxes. *Appl. Phys. Lett.*, 75:1908–1910, 1999.

19. S.L. McCall, A.F.J. Levi, R.E. Slusher, S.J. Pearton, and R.A. Logan. Whispering-gallery mode microdisk lasers. *Appl. Phys. Lett.*, 60:289–291, 1992.

20. N.B. Rex, R.K. Chang, and L.J. Guido. Threshold lowering in GaN micropillar lasers by means of spatially selective optical pumping. *IEEE Photon. Technol. Lett.*, 13:1–3, 2001.

21. J.L. Jewell, S.L. McCall, Y.H. Lee, A. Scherer, A.C. Gosasard, and J.H. English. Lasing characteristics of GaAs microresonators. *Appl. Phys. Lett.*, 54:1400–1402, 1989.

22. O. Painter, R.K. Lee, A. Scherer, A. Yariv, J.D. O'Brien, P.D. Dapkus, and I. Kim. Two-dimensional photonic band-gap defect mode laser. *Science*, 284:1819–1821, 1999.

23. H. Yan, J. Johnson, M. Law, R. He, K. Knutsen, J.R. McKinney, J. Pham, R. Saykally, and P. Yang. ZnO nanoribbon microcavity lasers. *Adv. Mater.*, 15:1907–1911, 2003.

24. M.H. Huang, S. Mao, H. Feick, H. Yan, Y. Wu, H. Kind, E. Weber, R. Russo, and P. Yang. Room-temperature ultraviolet nanowire nanolasers. *Science*, 292:1897–1899, 2001.

25. J.C. Johnson, H.-J. Choi, K.P. Knutsen, R.D. Schaller, P. Yang, and R.J. Saykally. Single gallium nitride nanowire lasers. *Nat. Mater.*, 1:106–110, 2002.

26. M.A. Zimmler, J. Bao, F. Capasso, S. Mller, and C. Ronning. Laser action in nanowires: Observation of the transition from amplified spontaneous emission to laser oscillation. *Appl. Phys. Lett.*, 93:051101, 2008.

27. L.J. Lauhon, M.S. Gudiksen, and C.M. Lieber. Semiconductor nanowire heterostructures. *Phil. Trans. R. Soc. Lond. A*, 362:1247–1260, 2004.

28. X. Duan, Y. Huang, R. Agarval, and C.M. Lieber. Single-nanowire electrically driven lasers. *Nature*, 421:241, 2003.

29. Y. Gu, I.L. Kuskovsky, M. Yin, S. O'Brien, and G.F. Neumark. Quantum confinement in ZnO nanorods. *Appl. Phys. Lett.*, 85:3833, 2004.

30. C.R. Pollock and M. Lipson. *Integrated Photonics*. Kluwer Academic Publishers, Boston, MA, 2003.

31. P.J. Pauzauskie, D.J. Sirbuly, and P. Yang. Semiconductor nanowire ring resonator laser. *Phys. Rev. Lett.*, 96:143903, 2006.

32. M. Bayer, T. Gutbrod, J.P. Reithmaier, and A. Forchel. Optical modes in photonic molecules. *Phys. Rev. Lett.*, 81:2582–2585, 1998.

33. A.V. Maslov and C.Z. Ning. Reflection of guided modes in a semiconductor nanowire laser. *Appl. Phys. Lett.*, 83:1237–1239, 2003.

34. A.V. Maslov, M.I. Bakunov, and C.Z. Ning. Distribution of optical emission between guided modes and free space in a semiconductor nanowire. *J. Appl. Phys.*, 99:024314, 2006.

35. A.V. Maslov and C.Z. Ning. Far-field emission of a semiconductor nanowire laser. *Opt. Lett.*, 29:572–574, 2004.

36. A.V. Maslov and C.Z. Ning. Modal gain in a semiconductor nanowire laser with anisotropic bandstructure. *IEEE J. Quantum Electron.*, 40:1389–1397, 2004.

37. A.V. Maslov and C.Z. Ning. Band structure and optical absorption of GaN nanowires grown along c axis. *Phys. Rev. B*, 72:125319, 2005.

38. L. Chen and E. Towe. Coupled optoelectronic modeling and simulation of nanowire lasers. *J. Appl. Phys.*, 100:044305, 2006.

39. J.W. Buck. *Fundamentals of Optical Fibers*, 2nd edn. Wiley-Interscience, Hoboken, NJ, 2004.

40. A.V. Maslov and C.Z. Ning. Metal-encased semiconductor nanowires as waveguides for ultra-small lasers. In *Conference on Lasers and Electro-Optics, CLEO*, 6–11 May 2007, Baltimore, MD, pp. 1–2, 2007.

41. M.T. Hill, Y.-S. Oei, B. Smalbrugge, Y. Zhu, T. De Vries, P.J. Van Veldhoven, F.W.M. Van Otten, T.J. Eijkemans, J.P. Turkiewicz, H. De Waardt, E.J. Geluk, S.-H. Kwon, Y.-H. Lee, R. Noetzel, and M. K. Smit. Lasing in metallic-coated nanocavities. *Nat. Photon.*, 1:589–594, 2007.

42. F. Qian, Y. Li, S. Gradecak, H.-G. Park, Y. Ding, Z.L. Wang, and C.M. Lieber. Multi-quantum-well nanowire heterostructures for wavelength-controlled lasers. *Nat. Mater.*, 7:701–706, 2008.

43. W.L. Barnes, A. Dereux, and T.W. Ebbesen. Surface plasmon subwavelength optics. *Nature*, 424:824–830, 2003.

44. M.L. Brongersma, J.W. Hartman, and H.A. Atwater. Electromagnetic energy transfer and switching in nanoparticle chain arrays below the diffraction limit. *Phys. Rev. B*, 62:R16358, 2000.

45. J.A. Dionne, L.A. Sweatlock, H.A. Atwater, and A. Polman. Plasmon slot waveguides: Towards chip-scale propagation with subwavelength-scale localization. *Phys. Rev. B*, 73:035407, 2006.

46. M. Dragoman and D. Dragoman. Plasmonics: Applications to nanoscale terahertz and optical devices. *Prog. Quantum Electron.*, 32:1–41, 2008.

47. S.A. Maier. Effective mode volume of nanoscale plasmon cavities. *Opt. Quantum Electron.*, 38:257–267, 2006.

48. F. Rana, C. Manolatou, and S.G. Johnson. Nanoscale semiconductor plasmon lasers. In *Conference on Lasers and Electro-Optics, CLEO*, 6–11 May 2007, Baltimore, MA, pp. 1–2, 2007.

49. C. Manolatou and F. Rana. Subwavelength nanopatch cavities for semiconductor plasmon lasers. *IEEE J. Quantum Electron.*, 44:435–447, 2008.

50. A.K. Sarychev and G. Tartakovsky. Magnetic plasmonic metamaterials in actively pumped host medium and plasmonic nanolaser. *Phys. Rev. B*, 75:085436, 2007.

51. A.K. Sarychev, A.A. Pukhov, and G. Tartakovsky. Metamaterial comprising plasmonic nanolaser. *PIERS Online*, 3:1264–1267, 2007.

52. J.N. Farahani, D.W. Pohl, H.-J. Eisler, and B. Hecht. Single quantum dot coupled to a scanning optical antenna: A tunable superemitter. *Phys. Rev. Lett.*, 95:017402, 2005.

53. A. Sundaramurthy, K.B. Crozier, G.S. Kino, D.P. Fromm, P.J. Schuck, and W.E. Moerner. Field enhancement and gap-dependent resonance in a system of two opposing tip-to-tip Au nanotriangles. *Phys. Rev. B*, 72:165409, 2005.

54. S.-W. Chang, C.-Y.A. Ni, and S.L. Chuang. Theory of bowtie plasmonic nanolaser. *Opt. Express*, 16:10580–10595, 2008.

55. I.E. Protsenko, A.V. Uskov, O.A. Zaimidoroga, V.N. Samoilov, and E.P. O'Reilly. Dipole nanolaser. *Phys. Rev. A*, 71:063812, 2005.

56. I.E. Protsenko, N.F. Starodubtsev, V.M. Rudoy, O.V. Dementieva, N.N. Naumov, O.A. Zaimidoroga, and V.N. Samoilov. Plasmon resonance, laser generation and photo-effect in thin heterogeneous layers and nanostructures. In V.V. Samartsev (ed.), *Photon Echo and Coherent Spectroscopy*, vol. 6181 of Proc. SPIE, Bellingham, WA, 2005, p. 618116.

57. K.J. Vahala. Optical microcavities. *Nature*, 424:839–846, 2003.

58. H.-G. Park, C.J. Barrelet, Y. Wu, B. Tian, F. Qian, and C.M. Lieber. A wavelength-selective photonic-crystal waveguide coupled to a nanowire light source. *Nat. Photon.*, 2008.

59. K. Nozaki, H. Watanabe, and T. Baba. Photonic crystal nanolaser monolithically integrated with passive waveguide for effective light extraction. *Appl. Phys. Lett.*, 92:021108, 2008.

60. T. Yoshie, M. Loncar, K. Okamoto, Y. Qiu, O.B. Shchekin, H. Chen, D.D. Deppe and A. Scherer. Photonic crystal nanocavities with quantum well or quantum dot active material. In A. Abidi, A. Scherer, and S.-Y. Lin (eds.), *Photonic Crystal Materials and Devices II*, vol. 5360 of Proc. SPIE, Bellingham, WA, 2004, pp. 16–23.

61. K. Nozaki, S. Kita, and T. Baba. Room temperature continuous wave operation and controlled spontaneous emission in ultrasmall photonic crystal nanolaser. *Opt. Express* 15:7506–7514, 2007.

62. T. Yang, A. Mock, J.D. OBrien, S. Lipson, and D.G. Deppe. Edge-emitting photonic crystal double-heterostructure nanocavity lasers with InAs quantum dot active material. *Opt. Lett.*, 32:1153–1155, 2007.

63. S.L. Jaiswal and P.M. Norris. Variations in a design for a nanoscale silicon laser. In P.V. Farrell, F.-P. Chiang, C.R. Mercer, and G. Shen (eds.), *Optical Diagnostics for Fluids Solids, and Combustion II*, vol. 5191 of Proc. SPIE, Bellingham, WA, 2003, pp. 226–233.

64. L. Pavesi, L. Dal Negro, C. Mazzolenim, G. Franzo, and F. Priolo. Optical gain in silicon nanocrystals. *Nature*, 408:440–444, 2000.

65. K.L. van der Molen. Experiments on scattering lasers from Mie to random. PhD thesis, University of Twente, 2007.

66. K.L. van der Molen, P. Zijlstra, A. Lagendijk, and A.P. Mosk. Laser threshold of Mie resonances. *Opt. Lett.*, 31:1432–1434, 2006.

67. D.S. Wiersma, M.P. van Albada, and A. Lagendijk. Coherent backscattering of light from amplifying random media. *Phys. Rev. Lett.*, 75:1739–1743, 1995.

68. D. Wiersma. The physics and applications of random lasers. *Nat. Phys.*, 4:359–367, 2008.

69. V.S. Letokhov. Generation of light by a scattering medium with negative resonance absorption. *Sov. Phys. JETP*, 26:835 840, 1968.

70. N.M. Lawandy, R.M. Balachandran, A.S.L. Gomes, and E. Sauvain. Laser action in strongly scattering media. *Nature*, 368:436–438, 1994.

71. M. Noginov. *Solid-State Random Lasers*. Springer-Verlag, New York, 2005.

72. A. Fratalocchi, C. Conti, and G. Ruocco. Mode competitions and dynamical frequency pulling in Mie nanolasers: 3D ab-initio Maxwell-Bloch computations. *Opt. Express* 16:8342–8349, 2008.

73. H.E. Tuereci, L. Ge, S. Rotter, and A.D. Stone. Strong interactions in multimode random lasers. *Science*, 320:643–646, 2008.

74. D. O'Carroll, I. Lieberwirth, and G. Redmond. Microcavity effects and optically pumped lasing in single conjugated polymer nanowires. *Nat. Nanotechnol.*, 2:180–184, 2007.

75. R.J. Holmes. Nanowire lasers go organic. *Nat. Nanotechnol.*, 2:141–142, 2007.

76. P.L. Gourley, J.K. Hendricks, A.E. McDonald, R.G. Copeland, K.E. Barrett, C.R. Gourley, and R.K. Naviaux. Ultrafast nanolaser flow device for detecting cancer in single cells. *Biomed Microdevices*, 7:331–9, 2005.

77. P.L. Gourley and R.K. Naviaux. Optical phenotyping of human mitochondria in a biocavity laser. *IEEE J. Sel. Top. Quantum Electron.*, 11:818–826, 2005.

78. S.-H. Kim, J.-H. Choi, S.-K. Lee, S.H. Kim, S.-M. Yang, Y.-H. Lee, C. Seassal, P. Regrency and P. Viktorovitch. Optofluidic integration of a photonic crystal nanolaser. *Opt. Express*, 16:6515–6527, 2008.

79. S. Kita, K. Nozaki, and T. Baba. Refractive index sensing utilizing cw photonic crystal nanolaser and its arrayed configuration. In *OSA/CLEO/QELS*, San Jose, CA, page paper CM06, 2008.

80. R. Ikkawi, N. Amos, A. Krichevsky, R. Chomko, D. Litvinov, and S. Khizroev. Nanolasers to enable data storage beyond 10 tbit/in². *Appl. Phys. Lett.*, 91:153115, 2007.

81. Y. Arakawa and H. Sakaki. Multidimensional quantum well laser and temperature dependence of its threshold current. *Appl. Phys. Lett.*, 40:939–941, 1982.

82. M. Asada, Y. Miyamoto, and Y. Suematsu. Gain and the threshold of three-dimensional quantum-box lasers. *IEEE J. Quantum Electron.*, 22:1915–1921, 1986.

83. F. Qian, S. Gradecak, Y. Li, C.-Y. Wen, and C.M. Lieber. Core/multishell nanowire heterostructures as multicolor, high-efficiency light-emitting diodes. *Nano Lett.*, 5:2287–2291, 2005.

84. D. Wiersma. The smallest random laser. *Nature*, 406:132–133, 2000.

85. P.J. Pauzauskie and P.D. Yang. Nanowire photonics. *Mater. Today*, 9:36–45, 2006.

40

Quantum Dot Laser

Frank Jahnke

University of Bremen

40.1 Introduction

The success of optoelectronics is largely based on the availability of efficient and reliable semiconductor lasers. Most prominent among the many applications are optical data storage solutions on CDs, DVDs, and BDs or the fiber-based optical communication. The initial success started with conventional laser diodes (Casey and Panish, 1978a,b; Agrawal and Dutta, 1993; Chuang, 1995). More recent microcavity lasers (Li and Iga, 2003) provide a new level of miniaturization and novel emission properties.

In general, the laser emission is based on population inversion of the electronic states, which contribute to the radiative recombination. For semiconductor lasers, it has been realized early (Dingle and Henry, 1976) that the carrier confinement in quantum wells leads to improved laser properties, like a laser threshold reduction. The successful demonstration of this effect was based on advances in the material growth (Tsang, 1982). The use of quantum dots (QDs) as an active material promises even greater benefits. First of all, one gains more freedom to engineer the emission wavelength (Bimberg et al., 1999). Another advantage is the enhancement of the single-particle density of states. For states experiencing a carrier confinement in all three dimensions, discrete electronic energies similar to the spectrum of atomic systems are obtained. This has led to predictions of further increased efficiency for laser applications, which are based on less temperature-dependent operational parameters (Arakawa and Sakaki, 1982), further reduced threshold currents

and higher differential gain (Asada et al., 1986), higher modulation bandwidth (Kim et al., 2004) and reduced anti-guiding effects (Newell et al., 1999), as well as a reduced sensitivity to material defects. Also, the miniaturization of the active material in connection with new microcavity laser resonators opens the door for novel applications in quantum information technologies.

The goal of this chapter is to provide an introduction to the physics underlying QD lasers. We will discuss the fundamental properties of the QDs, which are relevant for laser applications, as well as the important interaction processes that determine the laser properties.

40.1.1 Material Systems

Various techniques can be employed for the fabrication of QDs. Each approach has advantages for particular applications. Self-assembled QDs, which are typically obtained in the Stranski–Krastanov growth mode, are the most widely used for optoelectronic devices. Colloidal semiconductor nanostructures, which can be functionalized as biological markers, are less practical for the device integration. Lateral QDs, which are fabricated by etching of quantum wells following electron beam lithography, have the advantage of better controllable size and position. They are favorable for quantum-transport studies. The slow preparation process, the low dot density, and defect formation prevent the broad use of this method in optoelectronics.

For emission in the near infrared and red, including the frequencies for optical fiber telecommunication networks at 1300 and 1500 nm, the GaAs material system (GaAs/AlGaAs, InGaAs/GaAs) is routinely used. The self-organized growth of QDs is well established for this material system. The fabrication starts with the strained growth of a two-dimensional wetting layer (WL) on a substrate with a higher band gap and lower lattice constant. After a critical thickness is reached, partial strain relaxation changes the morphology to a thinner two-dimensional layer and small three-dimensional islands. These islands constitute the self-assembled QDs after their overgrowth with a higher band-gap material.

II–VI compounds with a wider band gap like CdSe, CdTe, or ZnSe provide emission in the green and yellow spectral range (Klude et al., 2004; Sebald et al., 2009). However, issues of the device lifetime due to degradation effects have so far prevented the broad application in light emitting devices. Also the growth mechanism often differs from that in III–V compounds. For example, CdTe/ZnTe has a similar lattice mismatch to InAs/GaAs, but the growth of CdTe QDs starts with two-dimensional platelets on which small dots are formed. Isolated dots are growing, when the material thickness exceeds a critical value.

The group-III nitride material system, like GaN/AlN or InGaN/GaN, received much attention for applications in light emitters due to their extended range of emission frequencies from visible light to UV emission (Nakamura and Chichibu, 2000). While the nonradiative loss of carriers due to trapping at threading dislocations lowers the efficiency of group-III quantum-well light emitters, this effect is reduced by the three-dimensional carrier confinement in QDs. Nitride-based heterostructures with a wurtzite crystal structure are known to have built-in electrostatic fields due to the spontaneous polarization and piezoelectric effects, which lower the spatial overlap of the electron and hole wave functions thereby reducing the interband recombination efficiency.

40.1.2 Edge Emitter Laser Diodes

Two different types of semiconductor lasers need to be distinguished in general. They vary in design, properties, and applications. In conventional edge-emitting laser diodes, the feedback for the laser mode is provided by an optical cavity, which is formed by the cleaved end faces of the device (Agrawal and Dutta, 1993; Chuang, 1995). Correspondingly, the emission occurs from the edge of the semiconductor chip. The low mirror reflectivity due to the jump of the refractive index from the semiconductor material to air results in a short cavity lifetime (large cavity damping). To compensate for the cavity losses, a large propagation length of the laser mode through the active material is necessary. The large cavity length in the range of a few hundred micrometers in turn leads to a small frequency spacing between the cavity modes, so that many modes appear within the gain region of the active material. As a result, edge emitters typically operate multimode. Also the spontaneous emission of the active material is distributed over a large number

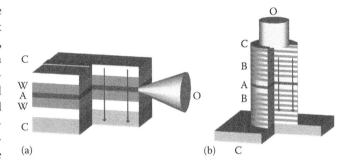

FIGURE 40.1 Edge-emitting laser (left) and vertical-cavity surface-emitting laser (right) with active region (A), waveguide layers (W), upper and lower Bragg mirrors (B), contacts (C), and light output (O). The arrows indicate the direction of the current flow.

of nonlasing modes, which leads to a pronounced laser threshold as discussed in Section 40.5.

The optical gain is achieved by current-injection pumping of the active material. For an edge-emitting laser diode, it is outlined in Figure 40.1 that the electrical injection current flows perpendicular to the optical cavity modes and the emission direction, which is characteristic for this type of devices. To provide an efficient confinement of the pumped charge carriers to the active region, a heterostructure design is used, where the active layer is positioned between materials with wider band gaps. A common heterostructure has a GaAs active layer between AlGaAs barriers. The carrier confinement is increased by designing the laser structure to be very thin in the direction of current flow, which is also the epitaxial growth direction. Additional cladding layers can be used to confine the optical mode in the transverse direction by means of a waveguide geometry and to increase the spatial overlap of the optical mode with the active region. Such a design is called double heterostructure and was independently proposed by Kroemer as well as Alferov and Kazarinov (Casey and Panish, 1978a). The conventional laser configuration works very well, as is evident from its wide ranging applications in many electro-optical systems.

This design allows for a natural integration of self-organized QDs as active material. Since the dots are grown in layers, these can serve as the planes in which the optical mode propagates. To increase the modal gain, often several layers are staggered. Accompanied with the large number of QDs and unavoidable fluctuations of the individual QD properties (mostly geometry and composition) is a large inhomogeneous broadening of the emission spectrum of the active material.

On the other hand, the combination of edge emission and active layer shape give rise to several disadvantages (independent on whether quantum wells or QDs are used as active material). For edge emission, the laser transverse and lateral modes depend on the cross section of the gain region, which is very thin in the transverse dimension for carrier confinement and wide in the lateral dimension for output power. The resulting near field of the laser emission is highly elongated, which does not match well to the circular cross section of optical fibers for possible out-coupling. Since the transverse beam dimension is very small, typically of the order of 1 μm or less, the transverse beam divergence is rather high (typically ≈50° full angle) because of diffraction. Therefore, the far field of the laser emission is also highly elongated. The beam astigmatism together with the high divergence makes design and fabrication of coupling optics rather challenging. Approaches to single-mode operation involve embedded gratings or buried heterostructures, which increase the unit costs and limit the output power.

40.1.2.1 Microcavity Lasers

Microcavity lasers represent a new type of devices. In analogy to the carrier confinement, one utilizes a strong confinement of the electromagnetic field. Cavity dimensions of the order of the light wavelength are characteristic of microcavity lasers. Due to the strongly reduced gain length, this requires very high mirror reflectivities larger 99%. To characterize microresonators, one uses the cavity quality factor $Q = \omega_c/\Delta\omega_c$ with the cavity frequency ω_c and the spectral linewidth $\Delta\omega_c$. The combination of small mode volume and high Q-factor leads to modifications in the density of states of the electromagnetic field. According to Fermi's golden rule, this allows to enhance or inhibit the spontaneous recombination probability, which is known as Purcell effect (Vahala, 2003; Lodahl et al., 2004). Particularly important for lasers is that the small cavity dimensions strongly increase the frequency spacing between the cavity modes. Microcavity lasers can be designed in a way to find only a single high-Q cavity mode within the gain spectrum of the active material, thus allowing genuine single-mode laser operation. The Purcell effect can also be used to suppress the spontaneous emission into other low-Q modes (e.g., modes propagating in other directions). This leads to a strongly reduced laser threshold, and the possibility of thresholdless lasers has been discussed (Yokoyama and Brorson, 1989; Rice and Carmichael, 1994) as a consequence of the novel physics.

The foundation of microcavity lasers lies in the improved engineering of optical resonators in combination with the new physics in the field of cavity-quantum electrodynamics. This includes mastering Maxwell's equations in even more and more complicated materials and geometries and still being able to predict the electromagnetic properties and the response of devices. The photon confinement is based either on Bragg or total internal reflection. A central role plays the concept of photonic bandgap materials. Here, one uses the analogy to the electronic band structure, which emerges for the electron motion in a lattice periodic potential. Photonic structures with periodic modulations of the refractive index can be designed in a way that a photonic band gap for the light propagation, the so-called stop band appears.

For practical device applications, vertical-cavity surface-emitting lasers (VCSELs) are frequently used, since many of the above mentioned disadvantages of the edge-emitting design are avoided. The VCSEL constitutes of a Fabry-Perot resonator, in which the very high reflectivity is achieved with distributed Bragg reflectors (DBRs), consisting of epitaxially grown periodic layers of different refractive indices (Jewell et al., 1991). The typical DBR has alternating layers that are each λ/4 thick so as to provide constructive interference of the reflected waves from each interface. The spacer layer between the DBRs acts as a defect giving rise to the high-Q cavity mode, which provides the photon confinement.

The basic VCSEL structure is that of an effectively one-dimensional resonator with a length of a few μm. The active region between the mirrors consists either of one or a few quantum wells or layers with QDs, which are embedded between cladding layers. The short cavity length of this structure allows to design a single high-Q (longitudinal) cavity mode within the spectral gain region of the active material. For laser structures, current confinement to a small area of the active region has been introduced by means of oxide apertures.

Since the optical cavity axis is in the vertical (epitaxial growth) direction (see Figure 40.1), the laser emission for a VCSEL occurs out of the surface of the wafer. This surface emitting configuration has several desirable features. Being independent of the gain layer cross section, there is more flexibility in the shape and size of the transverse optical mode. In particular, a circular beam can be realized. With an approximately 6 μm diameter circular output aperture, lowest order (Gaussian-like) transverse mode operation with less than 10° beam divergence is possible (Chang-Hasnain et al., 1991). Other advantages of a VCSEL are even more practical in nature. Unlike an edge-emitter, the VCSEL mirrors are fabricated during the epitaxial growth of the entire wafer, so that the mirrors of hundreds of lasers can be manufactured simultaneously. Edge emitters can only be combined to one-dimensional laser arrays, whereas with surface emitters truly two-dimensional arrays are routinely possible. To test an edge emitter, one needs to expose the edges, which means cleaving the wafer and applying the necessary facet preparations. These are time consuming and therefore costly steps that have to be repeated for each laser diode. Surface emitters can be fabricated, tested, and operated at the wafer level.

The VCSEL design has its own challenges. This gain-length reduction has to be compensated by the high cavity Q, putting great demands on the quality of the DBR mirrors. Having only one high-Q cavity resonance in the spectral gain region, the laser threshold current is very sensitive to the position of this resonance in relation to the peak of the material gain spectrum. From this, a related complication arises: the cavity-gain alignment can be maintained typically only within a small temperature range, because gain spectrum and cavity resonances have different temperature dependences. The gain spectrum shifts in frequency because of the temperature dependence of the bandgap energy and the carrier distributions. The cavity resonances shift with temperature due to thermally-induced changes in the refractive indices of the mirror material and the material within the optical cavity. The result is a temperature dependence of the cavity-gain alignment, which plays an important role in the sensitivity of a VCSEL output to temperature variations.

An additional transverse mode confinement in VCSEL structures can be introduced with the fabrication of micropillars, e.g., due to a combination of electron-beam lithography and various etching methods. The resulting three-dimensional confinement of the optical mode to a very small volume of the order of the cubic light wavelength strongly increases the efficiency of the light–matter interaction. With the three-dimensional photon confinement, the Purcell effect can be used for a strong suppression of the spontaneous emission into non-lasing modes. Even higher cavity-Q factors for three-dimensional mode confinement have been achieved in microdisks or microspheres. They lack, however, the directionality of the light emission. The ultimate control of the photon confinement and record values for the cavity-Q has been possible in photonic crystal structures.

Both in conventional edge-emitters and in microcavity lasers, the QD physics that determines the properties of the gain material is the same and the effects discussed in the subsequent chapters equally apply. Nonetheless, the miniaturization of microcavity lasers can be pushed to the extreme case where only a single QD resonantly interacts with a single high-Q optical mode. This system is the analogous to the single-atom micromaser or the one-atom laser of quantum optics and opens the door for novel applications in quantum-information technologies like quantum cryptography (Fattal et al., 2004) and quantum computers (Scholz et al., 2006). Essential physical effects or mechanisms are the generation of single-photons on demand (Michler et al., 2000) or correlated photons (Moreau et al., 2001), as well as the demonstration of vacuum field Rabi oscillations (strong coupling) (Reithmaier et al., 2004; Yoshie et al., 2004).

40.2 Electronic States of Quantum Dots

The single-particle states will be taken as the starting point for the description of the electronic and optical properties of semiconductor QDs as active material in lasers. In terms of quantum mechanics, single-particle states constitute the basis for the analysis of interaction processes. Interaction matrix elements can be formulated with the corresponding wave functions, and transition processes can be described in terms of the single-particle states.

40.2.1 Localized and Delocalized States in Self-Assembled Structures

Due to the three-dimensional carrier confinement, QDs possess localized states in contrast to the band structure of the bulk material. Nevertheless, the single-particle states are closely related to the properties of the bulk material. Especially the optical selection rules and symmetry properties of the interaction matrix elements reflect properties of the underlying crystal lattice. Also band mixing effects and built-in crystal fields are results of the lattice properties.

Depending on material composition and growth conditions like temperature and material flux, a variety of different QD geometries can be fabricated (Jacobi, 2003). In the following, we consider self-assembled QD systems in a characteristic geometry, which consist of lens-shaped caps positioned on a residual two-dimensional WL, as shown in the left part of Figure 40.2. For the frequently used InGaAs/GaAs material system, one finds typically a diameter of 10–30 nm and a height-to-diameter aspect ratio of 0.3–0.4. The random distribution of QDs on the WL plane has a density of 10^{10}–10^{11} cm^{-2}.

From the geometry of the QD-islands on top of the two-dimensional WL, the single-particle energy spectrum can be understood qualitatively. The WL provides a confinement of carriers in the growth direction and—in the absence of QDs—a free carrier motion in the perpendicular WL plane, as in a quantum well. At the position of the islands, the height of the lower band-gap material increases, thus reducing the ground-state energy and therefore giving rise to an additional in-plane confinement potential. The resulting effective three-dimensional carrier confinement in the QDs leads to a localized state with discrete energies for electrons and holes. These discrete states are located energetically below a quasi-continuum of delocalized states, corresponding to the two-dimensional motion of carriers in the WL. At even higher energies, one finds the barrier states, which are delocalized in all three dimensions. The energy levels of the combined QD-WL system are sketched in the right part of Figure 40.2. The actual number of discrete QD states is limited by the height of the confinement potential as

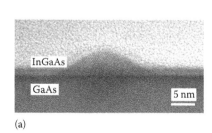

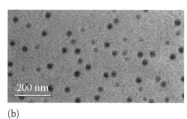

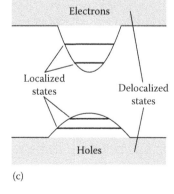

(a)   (b)   (c)

FIGURE 40.2 Cross section (a) and top view (b) TEM image of self-assembled InGaAs/GaAs QDs. The corresponding energy spectrum (c) contains localized states due to the three-dimensional carrier confinement, which are positioned energetically below a quasi-continuum of delocalized states corresponding to the two-dimensional motion of carriers in the WL. (From Anders, S. et al., *Phys. Rev. B*, 66, 125309, 2002.)

well as the geometry of the dots. As we will discuss in Section 40.3, the delocalized states play an important role for the scattering processes in light-emitters based on QDs and, hence, need to be included in the analysis of the single-particle states.

40.2.2 Envelope-Function Approximation and Effective-Mass Approach

The simplest approach to the carrier confinement in semiconductors is based on the envelope-function approximation. The starting point is the Bloch theorem, which represents the wave functions in a three-dimensional crystal in the form

$$\Phi_k(r) = e^{ik\cdot r}u_\lambda(r), \qquad (40.1)$$

describing plane waves modulated with the lattice-periodic Bloch function $u_\lambda(r)$, where λ is the band index and k is the carrier wave vector. In a confinement situation, one assumes that the three-dimensional plane wave can be replaced with envelope wave functions that are localized in one, two, or three dimensions.

In effective-mass approximation, one assumes that the actual QD and the two-dimensional WL consist of the same semiconductor material, in which the carrier motion experiences a constant potential energy and a kinetic energy of the carriers that can be expressed in terms of an effective mass m^*. The structure is embedded in a buffer material with a larger band gap. This is described by a step-wise increase of the confinement potential $V(r)$ to the outside region. Typically, the effective mass is also different in the buffer material, which is expressed with a step-wise constant $m^*(r)$. The Schrödinger equation

$$\left[\frac{\hbar^2}{2m^*(r)}\Delta + V(r)\right]\phi_n(r) = \varepsilon_n\phi_n(r) \qquad (40.2)$$

defines the confinement wave functions $\phi_n(r)$, which replace the plane waves in Equation 40.1, and provides the corresponding energies ε_n for the single-particle states.

For the geometry displayed in the left part of Figure 40.2, several simplifications of the solution of Equation 40.2 are possible. The rotational symmetry allows for a classification of the single-particle states according to their angular momentum and to effectively reduce the three-dimensional problem to a two-dimensional one. Details for the solution of this problem are discussed in Wojs et al. (1996). Additionally, when the QD height-to-diameter aspect ratio is small, one can distinguish a weak (strong) confinement situation in the WL plane (growth direction), and in good approximation the confinement wave function can be separated into the corresponding components

$$\Phi_n(r) = \varphi_{l,\lambda}(\varrho)\xi_{\sigma,\lambda}(z)u_\lambda(r), \qquad (40.3)$$

with $r = (\varrho, z)$. The in-plane envelope function $\varphi_{l,\lambda}(\varrho)$ represents both localized and delocalized states for the QD and WL contributions, respectively. In a simple but often useful approximation, the in-plane confinement of flat QDs with cylindrical symmetry

is described by a two-dimensional harmonic confinement potential with a characteristic equidistant spacing of the corresponding energies (Wojs et al., 1996). In the WL, the in-plane momentum k is a good quantum number if the effect of the localized states on the continuum is neglected. Thus, the quasi-continuum of the WL states (in the absence of QDs) is modeled by using plane waves for the corresponding in-plane envelope functions $\varphi_{l,\lambda}(\varrho)$. However, the localized QD states and the delocalized WL states are solutions of a single-particle problem for one common confinement potential and must, therefore, form an orthogonal basis. By performing a separate ansatz for the QD and WL contributions, this orthogonality is not ensured. It can be enforced, for example, by an orthogonalization procedure of the continuum states. The obtained so-called orthogonal plane waves (OPWs) represent a more correct description of the WL states. For details, we refer the reader to Nielsen et al. (2004, 2005).

40.2.3 Continuum Approaches and Atomistic Models

The effective-mass approach has several shortcomings. Band-mixing effects usually modify the electronic states, and a more detailed description of the valence-band structure is often desired. This can be included in semiempirical continuum approaches like k·p-models or atomistic tight-binding approaches. The latter account also for symmetry properties of the atoms in the crystal unit cell, which can influence optical transition rules, and allows to study systematically the combined influence of structure and chemical morphology including alloy fluctuations and defects.

These effects are particularly important in nitride-based heterostructures, and for these systems extensive studies of the electronic states have been performed. The valence band of a wurtzite nitride semiconductor is strongly nonparabolic due to the small spin–orbit and crystal-field splitting (Chuang and Chang, 1996; Vurgaftman and Meyer, 2003). The hexagonal symmetry of the crystal leads to the subbands having the same effective mass along the c-axis. Therefore, the confinement of the holes in this direction due to the WL is not able to separate these subbands, and strong subband mixing, and nonparabolicity effects remain present. Also modifications of the wave functions due to built-in fields and their screening due to excited carriers are important.

First-principle calculations are available for the computation of the electronic structure of small QD systems (200–400 atoms). Due to the large number of atoms together with the absence of translational symmetry, which greatly simplifies bulk calculations, density functional theory-based approaches are computationally too demanding to date.

40.3 Scattering Processes of Carriers in Quantum Dots

To utilize the specific optical properties of QDs, we are interested in recombination processes between the conduction and valence band states associated with the three-dimensional carrier confinement. Efficient light emitters like LEDs rely on a large

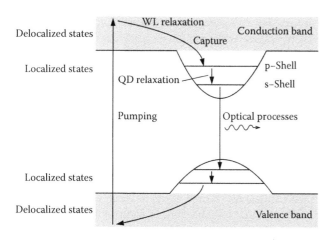

FIGURE 40.3 Fundamental carrier scattering processes in QD-based light emitters.

population of the QD states with electrons and holes. For laser applications, the realization of optical gain even requires population inversion. Electrical current injection leads to the generation of carriers in the delocalized states. These carriers need to be scattered into the localized QD states (carrier capture). Also efficient carrier scattering between the QD states (carrier relaxation) is necessary to obtain a large ground-state population. The corresponding transition processes are schematically shown in Figure 40.3. Likewise, optically pumped QD lasers depend on fast carrier-scattering processes. The two important intrinsic scattering mechanisms contributing to the capture and relaxation are the carrier–carrier Coulomb interaction, which can be assisted by carriers in bound (QD) or extended (WL) states, and the carrier–phonon interaction. Both mechanisms are discussed separately in Sections 40.3.1 and 40.3.2. These scattering processes are also an important source for optical dephasing, which strongly influences the optical properties. Furthermore, this underscores the integral part the WL states play for the physics and application of self-assembled QDs.

Experimental insight into carrier scattering processes can be obtained by ultrafast optical pump–probe spectroscopy, ranging from absorption measurements to advanced two-color pump–probe and four-wave mixing experiments (Shah, 1999). Both at room temperature and at lower temperatures, efficient carrier scattering processes have been demonstrated.

40.3.1 Carrier–Carrier Interaction

The most simple and at the same time most efficient form of carrier–carrier collisions involves the energy transfer between two particles. To analyze the corresponding re-distribution of the carrier population within a quantum-statistical treatment, one can investigate the dynamics of the carrier population $f_\nu(t)$ for the state ν as a function of time t. The corresponding equation of motion for $f_\nu(t)$ can be formulated on different levels of refinement. Frequently used are kinetic equations with Boltzmann scattering rates, which describe the temporal change of the

carrier population $f_\nu(t)$ due to various scattering processes into and out of the states ν according to

$$\frac{\partial}{\partial t} f_\nu = (1 - f_\nu)S_\nu^{\text{in}} - f_\nu S_\nu^{\text{out}}. \tag{40.4}$$

To obtain this form, a Markov approximation has been used, where one assumes that the population changes at a given time t depend explicitly only on the population functions $f_\nu(t)$ at the same time. Generally, the scattering rates are proportional to the population f of the initial states and to the nonoccupation $1 - f$ of the final states. Correspondingly, on the RHS of Equation 40.4, the in-scattering (out-scattering) rate, which increases (decreases) the population, is proportional to the non-occupation probability $1 - f_\nu$ (occupation probability f_ν).

For the Coulomb interaction, where two carriers are scattered from the states ν_1 and ν_3 into ν and ν_2 and vice versa, the scattering integral S_ν^{in} is given by

$$S_\nu^{\text{in}} = \frac{2\pi}{\hbar} \sum_{\nu_1,\nu_2,\nu_3} W_{\nu\nu_2\nu_3\nu_1}[W^*_{\nu_2\nu_3\nu_1} \quad - W^*_{\nu\nu_2\nu_1\nu_3}]$$
$$\times f_{\nu_1}(1 - f_{\nu_2})f_{\nu_3}\delta(\tilde{\varepsilon}_\nu - \tilde{\varepsilon}_{\nu_1} + \tilde{\varepsilon}_{\nu_2} - \tilde{\varepsilon}_{\nu_3}). \tag{40.5}$$

A similar expression for the out-scattering contribution S_ν^{out} is obtained by replacing $f \to 1 - f$. The population factors account for the availability of scattering partners, as discussed above. The sum involves all available initial states for in-scattering processes as well as all possible scattering partners that can be provided in QD or WL states.

In Markov approximation, the delta function accounts for strict energy conservation of the scattering processes. This approximation is valid in the long time limit, where scattering events are considered for times that are much larger than the inverse scattering rates. Even in this case, it is necessary to account for renormalizations of the single-particle energies $\tilde{\varepsilon}_\nu$ by the Coulomb interaction instead of using free-carrier energies ε_ν as done in the perturbation theory.

The scattering cross section is also determined by the matrix elements of the screened Coulomb interaction $W_{\nu\nu_2\nu_3\nu_1}$. The first and second terms in the square brackets correspond to direct and exchange Coulomb scattering, respectively.

A treatment beyond the Markov approximation is possible within a quantum-kinetic description (Haug and Jauho, 1996). Then the delta-function is replaced by the spectral functions of the involved carriers, and time-integrals over the past of the system account for memory effects. The most simple form of a quantum-kinetic equation requires two further approximations: (1) Within the generalized Kadanoff–Baym ansatz (GKBA), closed equations for single-time population functions are obtained (see Haug and Jauho, 1996 for detailed discussions). (2) The quasi-particle properties of the involved carriers are described by renormalized energies and damping (finite lifetime of the quasiparticle state), which can be summarized in complex energies $\hat{\varepsilon}_\nu$. The latter approximation is related to the Fermi liquid theory

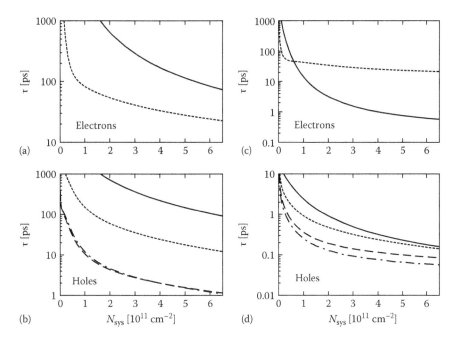

FIGURE 40.4 Capture times for electrons (a) and holes (b) and relaxation times for electrons (c) and holes (d) as a function of the total carrier density in the system N_{sys} for the following quantum-dot states of an InGaN/GaN material system and room temperature: s-shell (solid lines), p-shell (dotted lines), d_0-shell (dashed-dotted lines), and $d_{\pm}$-shell (dashed lines). (From Nielsen, T.R. et al., *Phys. Rev. B*, 72, 235311, 2005. With permission.)

and valid for large carrier densities, where excitonic effects can be neglected. The quantum-kinetic equation with the discussed approximations has the form

$$\frac{\partial}{\partial t} f_v(t) = \int_{-\infty}^{t} dt' \left\{ [1 - f_v(t')] S_v^{in}(t,t') - f_v(t') S_v^{out}(t,t') \right\}, \quad (40.6)$$

where the in-scattering rate, which now depends on two time arguments, is given by

$$S_v^{in}(t,t') = \frac{1}{\hbar^2} \text{Re} \sum_{v_1, v_2, v_3} W_{vv_2v_3v_1}(t) \left[W_{vv_2v_3v_1}^*(t') - W_{vv_2v_1v_3}^*(t') \right]$$

$$\times f_{v_1}(t') [1 - f_{v_2}(t')] f_{v_3}(t')$$

$$\times \exp\left(-\frac{i}{\hbar} (\hat{\varepsilon}_v - \hat{\varepsilon}_{v_1} + \hat{\varepsilon}_{v_2} - \hat{\varepsilon}_{v_3})(t-t') \right). \quad (40.7)$$

The out-scattering rate is again obtained from $f \to 1 - f$. It turns out, that especially for the calculation of dephasing and its influence on optical absorption and gain spectra, the quantum-kinetic description leads to more realistic results. As discussed in Lorke et al. (2006a), for the description of dephasing in optical spectra, equations similar to (40.6) and (40.7) are used for the polarization dynamics, which will be discussed in Section 40.4.3.

The results for the capture of carriers from the WL into the QD states and for the relaxation between the QD levels are shown in Figure 40.4. Capture and relaxations times have been defined as those times τ_v, in which the system returns to thermal equilibrium due to the respective processes after a small perturbation. In relaxation-time approximation, one finds $1/\tau_v = S_v^{in} + S_v^{out}$, where the scattering rates are evaluated for quasi-equilibrium population functions. With increasing carrier density in the system, the scattering times are decreasing due to the larger number of available scattering partners. For the energetically lower s-shell, direct carrier capture from the WL is less efficient than capture into the p-shell and subsequent relaxation into the s-shell. For further details, we refer the reader to Nielsen et al. (2005).

In the following, we provide additional information on the analysis of carrier–carrier scattering processes. The determination of the Coulomb interaction matrix elements, which enter in Equations 40.5 and 40.7, involves two steps. The bare (unscreened) interaction matrix elements can be calculated from the single-particle wavefunctions $\Phi_v(\mathbf{r})$, see Section 40.2, of the involved electronic states according to

$$V_{vv_2v_3v_1} = \int d^3r d^3r' \Phi_v^*(\mathbf{r}) \Phi_{v_2}^*(\mathbf{r}') v(\mathbf{r} - \mathbf{r}') \Phi_{v_3}(\mathbf{r}') \Phi_{v_1}(\mathbf{r}), \quad (40.8)$$

where $v(\mathbf{r} - \mathbf{r}') = e^2/(4\pi\varepsilon_0 \epsilon |\mathbf{r} - \mathbf{r}'|)$ is the Coulomb potential with the background dielectric function ϵ. The electron charge and the vacuum dielectric constant are given by e and ε_0, respectively.

A second step involves the inclusion of screening effects due to excited carriers. From the nature of screening as well as from the excitation conditions at elevated temperatures (considered for optoelectronic applications), it appears natural to account only for screening due to the carriers in the delocalized WL states. A consistent theory of screening for the QD-WL system is developed in Nielsen et al. (2005). On these grounds, a simplified treatment can be justified, which consists in the replacement of $v(\mathbf{r} - \mathbf{r}')$ in Equation 40.8 by a two-dimensionally screened Coulomb potential. The time-dependence of the screened Coulomb matrix elements $W_{v v_2 v_3 v_1}(t)$ is introduced via the carrier population functions $f_v(t)$, which determine the screening for the present excitation conditions via a longitudinal dielectric function.

The renormalized single-particle energies, that enter in the scattering rates (40.5) and (40.7), can be obtained from the free-carrier energies by adding Hartree, exchange, and correlation self-energy contributions. The Hartree self-energy involves the electrostatic interaction with all other excited electrons and holes. Only in a system with local charge neutrality, the Hartree terms cancel. In QDs, local charging appears as a result of different electron and hole populations, different envelope wave functions for electrons and holes, as well as due to the charge separation of electrons and holes in a built-in electrostatic fields (Nielsen et al., 2005).

40.3.2 Carrier–Phonon Interaction

For low carrier densities, when carrier–carrier Coulomb scattering is less important, the interaction of carriers with phonons provides the only remaining efficient scattering mechanism. At higher carrier densities, the carrier–phonon interaction remains important, since the Coulomb scattering is unable to dissipate kinetic energy from the carrier system. Note that the carrier–carrier scattering leads to an evolution of the carrier system towards a quasi-equilibrium state for electrons and holes. However, it conserves the average kinetic energy of the carriers. Pumping of carriers with above-average kinetic energy and the preferred recombination of carriers in the highly populated states with below-average kinetic energy effectively heats up the carrier system. This excess kinetic energy can only be dissipated to the crystal lattice via the emission of phonons. A thermalization of the carriers with respect to the crystal lattice is essential for efficient laser operation, since the increasing carrier temperature diminishes the optical gain via a reduction of the population inversion.

40.3.2.1 Perturbation Theory versus Polaron Picture

The early work on carrier–phonon interaction in semiconductor QDs has been based on the application of time-dependent perturbation theory. This method has previously been applied to bulk semiconductors or quantum wells. In these cases, its success is based on the application to carriers in continuum states. For the discussion of carrier–phonon interaction involving localized QD states with discrete energies, one has to consider the fact that the transition matrix elements provide only efficient coupling to phonons with small momenta (Bockelmann and Bastard, 1990; Inoshita and Sakaki, 1997). The strongest contribution to the interaction of carriers with lattice vibrations

is typically provided by LO-phonons due to the Coulomb-like coupling of carriers to lattice displacements in polar crystals. The nearly constant LO-phonon energy, at least for the considered small phonon momenta, and the energy-conservation requirement from Fermi's golden rule led to the prediction of inefficient carrier transitions between the QD states, except for the unlikely coincidence of transition energies between the QD states with the LO-phonon energy. The other, usually less important interaction process is provided via deformation potential coupling to LA-phonons. For small phonon momenta, their energy is too small to facilitate efficient carrier scattering processes between the QD states. On this ground, particularly due to the application of perturbation theory, inefficient carrier–phonon scattering processes in QD systems have been predicted (Benisty et al., 1991). To prevent this so-called *phonon bottleneck*, only much less efficient higher-order processes, like those involving combinations of LA- and LO-phonons (Inoshita and Sakaki, 1992; Jiang and Singh, 1998) or the contributions of carrier–carrier scattering processes at elevated carrier densities have been proposed.

For a more advanced description of the interaction with LO-phonons, one has to take into account that freely moving charge carriers in ionic crystals are surrounded by a cloud of lattice distortions. The renormalization of electrons and holes introduced by the corresponding interaction leads to the emergence of a quasiparticle called polaron. In the past, polarons have mostly been studied for semiconductor structures with a quasi-continuum of electronic states, like bulk materials or quantum wells. Then, electronic transitions are already possible within a free-carrier picture. For the usually considered materials with weak polar coupling, this explains why the application of perturbation theory already provides the leading contributions of polaron effects. They consist in a lowering of the carrier energy (polaron shift) in a renormalization of the carrier mass and the dielectric constant for the Coulomb interaction of carriers, and in a small broadening of the electronic states (Schäfer and Wegener, 2002).

On the contrary, it has been pointed out in Inoshita and Sakaki (1997) and Kral and Khas (1998), and Verzelen et al. (2002) that even in materials with weak polar coupling, the interaction of carriers in localized QD states with LO-phonons leads to strong modifications of the electronic properties. In this case, the phonon satellites of one state hybridize with other discrete states, even if the level spacing does not match the LO-phonon energy. The resulting new quasiparticle of the strongly interacting system are called QD-polarons. They cannot be treated within perturbation theory. Therefore, a simple discussion of carrier–phonon interaction based on Fermi's golden rule is not applicable to QDs. Related to this is an ongoing controversy in the QD literature regarding experimental evidence for (Minnaert et al., 2001; Urayama et al., 2001; Xu et al., 2002) and against (Tsitsishvili et al., 2002; Peronne et al., 2003; Quochi et al., 2003) the phonon bottleneck.

Early investigations of QD-polarons have been based on the "random phase approximation" (RPA) (Inoshita and Sakaki, 1997; Kral and Khas, 1998) or on the direct diagonalization of a restricted state space (Verzelen et al., 2002), where only localized states are considered. For a single discrete electronic state, an analytic treatment of the interaction with LO-phonons is possible within the "independent boson model" (Mahan, 1990). The numerical extension to several discrete electronic states is discussed in Stauber et al., (2000). However, in this work, the influence of delocalized WL states with a quasi-continuum of energies, which plays a key role for carrier generation in optoelectronic devices, has been neglected. When investigations are restricted to localized states with discrete energies, RPA results deviate from a more accurate direct diagonalization of a restricted system. On the other hand, the inclusion of WL states provides a natural source of broadening of the electronic states, which justifies the application of the RPA.

The next step in the analysis of carrier–phonon scattering in QD systems was the development of a theory that accounts for the scattering of renormalized quasiparticles (QD-polarons). This became possible within a quantum-kinetic theory that has been applied to the problem of carrier capture and relaxation (Seebeck et al., 2005). Such a quantum-kinetic description naturally incorporates both non-Markovian effects on ultrafast timescales and a nonperturbative treatment of the carrier–phonon interaction in the polaron picture. Polaronic renormalization effects have been found to be the main origin for fast carrier capture and relaxation processes, as well as efficient carrier thermalization on a picosecond timescale. The results strongly deviate from a perturbative treatment using Fermi's golden rule.

40.3.2.2 Polaron States and Kinetics

In the following, the retarded Green's function $G_\alpha^r(\tau)$ with the time τ and the state index α is used to describe the quasiparticle renormalizations of QD electrons and holes in the single-particle states α due to the nonperturbative interaction with LO-phonons. In the absence of interaction, this function oscillates with the free-carrier energy ε_α according to

$$G_\alpha^r(\tau) = \frac{i}{\hbar}\Theta(\tau)e^{-\frac{i}{\hbar}\varepsilon_\alpha\tau} \qquad (40.9)$$

and the Fourier transform

$$G_\alpha^r(\omega) = \frac{1}{\hbar\omega - \varepsilon_\alpha + i\eta}, \qquad (40.10)$$

with the frequency ω has a pole at the free-carrier energy with $\eta > 0$, $\eta \to 0$. Quasiparticle renormalizations are obtained from the solution of the Dyson equation:

$$\left[i\hbar\frac{\partial}{\partial\tau} - \varepsilon_\alpha\right]G_\alpha^r(\tau) = \delta(\tau) + \int_0^\tau d\tau' G_\alpha^r(\tau')\sum_\beta G_\beta^r(\tau-\tau')D_{\alpha\beta}^>(\tau-\tau'), \qquad (40.11)$$

where the sum involves all available states β. The Dyson equation contains a self-energy for the many-body description of

interaction processes. In Equation 40.11, the so-called random-phase approximation has been used to formulate the interaction term with the phonon propagator

$$D_{\alpha\beta}^{\gtrless}(\tau) = \sum_{\vec{q}} |M_{\alpha\beta}(\vec{q})|^2 \times [n_{LO}e^{\pm i\omega_{LO}\tau} + (1+n_{LO})e^{\mp i\omega_{LO}\tau}], \qquad (40.12)$$

where

  n_{LO} is a Bose–Einstein function for the population of the phonon modes (assumed to act as a bath in thermal equilibrium)

  ω_{LO} is the LO-phonon frequency

The sum over the three-dimensional phonon wave vector $\vec{q}$ involves all phonon modes. For an introduction into the many-body theory in general and Green's functions in particular, we refer the reader to Schäfer and Wegener (2002) and Mahan (1990).

The Dyson equation (Equation 40.11) takes into account all possible virtual transitions due to carrier–phonon interaction for a carrier in the state α. The single-particle energies for both localized and delocalized states are considered, and the corresponding wave functions enter via the interaction matrix elements $M_{\alpha\beta}$.

The Fourier transform of the retarded Green's function directly provides the spectral function $\hat{G}_\alpha(\omega) = -2\,\mathrm{Im}\,G_\alpha^r(\omega)$ that reflects the density of states (DOS) for a carrier in the state α. Only for noninteracting carriers, the spectral function contains a delta-function at the free-particle energies, expressing that each state is associated with a single energy. This picture changes due to the quasiparticle renormalizations.

The dynamics of the carrier population due to interaction with LO-phonons can be described with a quantum-kinetic equation (Schafer and Wegener, 2002; Seebeck et al., 2005)

$$\frac{\partial}{\partial t}f_\alpha(t) = 2\,\mathrm{Re}\sum_\beta\int_{-\infty}^t dt' G_\beta^r(t-t')[G_\alpha^r(t-t')]^*$$

$$\times\left\{f_\alpha(t')\left[1-f_\beta(t')\right]D_{\alpha\beta}^<(t'-t) - [1-f_\alpha(t')]f_\beta(t')D_{\alpha\beta}^>(t'-t)\right\}. \qquad (40.13)$$

In comparison to Boltzmann scattering rates, the delta-functions are replaced by convolutions of the phonon propagator with retarded Green's functions. The latter account for the polaronic renormalizations of the initial and final carrier states. Furthermore, memory effects are included by the time dependence of the population factors on the past evolution of the system.

Boltzmann scattering integrals follow as a limiting case, when the Markov approximation is applied (population functions $f_\alpha(t')$ are taken at the external time t) and when quasiparticle renormalizations are neglected with the use of free-carrier retarded Green's functions (40.9).

Examples for ultrafast carrier scattering processes in the polaron picture are displayed in Figure 40.5. Even if the level

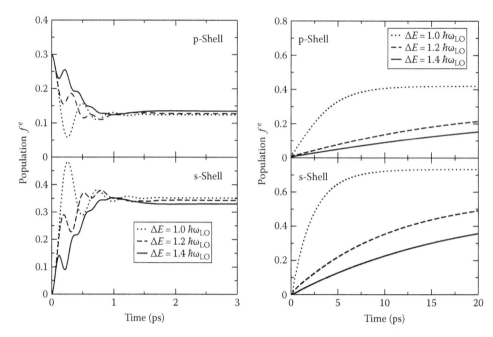

FIGURE 40.5 Temporal evolution of the QD electron population due to scattering of carriers with LO-phonons. Calculations have been performed in the polaron picture for the InGaAs/GaAs material system at room temperature. Left: carrier relaxation from the initially populated p-shell (top) into the initially empty s-shell (bottom). Right: carrier capture from the WL into the initially empty p-shell (top) and s-shell (bottom). Different energy spacings ΔE between the s- and p-shell in units of the LO-phonon energy are compared. (From Seebeck, J. et al., *Phys. Rev. B*, 71, 125327, 2005. With permission.)

spacing between s and p-shell and between p-shell and WL band-edge exceeds the LO-phonon energy by 40%, efficient carrier relaxation and carrier capture are observed, while Boltzmann scattering integrals would predict vanishing carrier transitions for the interaction with LO-phonons.

40.4 Optical Gain of the Active Material

Of central importance for the design process and for various emission properties of lasers are the optical gain of the active material and the corresponding refractive index changes. Both quantities are linked via the optical susceptibility, which is introduced in this section.

In the following, we give some examples for the role of the optical gain. The threshold current of a laser is determined by the transparency carrier density at which the active material switches from absorption to optical gain. The carrier-density dependence of the gain influences the modulation bandwidth, and the temperature dependence of the optical gain is the main factor for the temperature stability of the laser operation. It is also important to study the physical mechanisms of gain saturation, which limits the achievable optical gain.

For the edge-emitting laser structures, the optical mode propagates in the QD plane and refractive index changes, which are induced by excited carriers, directly influence the mode properties like frequency chirp or filamentation, as well as the laser linewidth. In the following, we establish a microscopic picture to determine these gain and refractive index properties in QD systems.

40.4.1 Optical Susceptibility

From Maxwell's equations, a wave equation for the optical field $E(r, t)$ can be derived, which describes how the emitted field from the sample can be traced back to the macroscopic polarization $P(r, t)$ inside the sample:

$$\left(-\nabla \times \nabla \times -\frac{1}{c^2}\frac{\partial^2}{\partial t^2} \right) E(r, t) = \mu_0 \frac{\partial^2}{\partial t^2} P(r, t). \quad (40.14)$$

The polarization represents the macroscopic dipole density in the medium, which is induced by the field itself, as expressed by the optical susceptibility

$$P(r, t) = \int d^3 r' \int dt' \chi(r, r', t, t') E(r', t'). \quad (40.15)$$

In order to fulfill Equations 40.14 and 40.15 simultaneously, a self-consistent solution is necessary: the field determined from the wave equation should be the same function entering the calculation of the polarization. For simplicity, we neglect the tensorial character of χ and consider given polarization directions for P and E.

From Equation 40.15 can be inferred that the susceptibility is a response function describing the answer of the medium to the optical field. Mathematically this can be expressed with the functional derivative

$$\chi(r, r', t, t') = \frac{\delta P(r, t)}{\delta E(r', t')}. \quad (40.16)$$

The optical absorption spectrum is commonly defined via the *linear response* of the medium. In this case, χ depends only on the difference of the time arguments (as can be shown from the explicit quantum mechanical calculation of the macroscopic polarization). Then one obtains in Equation 40.15 a convolution in time, which translates into a product in Fourier space and to give

$$\chi_{QD}(\omega) = \frac{P_{QD}(\omega)}{E_{QD}(\omega)}. \tag{40.17}$$

For notational simplicity, we have no longer written the space dependence of the functions. In the considered case of an active material of QDs, the optical susceptibility is given as the macroscopic QD polarization P_{QD} divided by the linear optical test field at the QD position E_{QD}.

The absorption spectrum $\alpha(\omega)$ as well as the refractive index changes of the medium due to resonant excitation $\delta n(\omega)$ of the QD system is given by

$$-K\delta n(\omega) + i\alpha(\omega) = \frac{\omega}{\varepsilon_0 n_B c L}\chi_{QD}(\omega), \tag{40.18}$$

where

- c and ε_0 are the speed of light and permittivity in vacuum
- n_B is the background refractive index
- L is the thickness of the QD sheet
- K is the wave number of the optical field

40.4.2 Interband Transitions and Macroscopic Polarization

The macroscopic optical QD polarization can be determined from the quantum mechanical expectation value of the dipole operator $\langle d \rangle = \mathrm{Tr}\{d\rho\} = \sum_{v_1,v_2} d_{v_1,v_2}\rho_{v_2,v_1}$, which is calculated with the statistical operator ρ. Since d is a single-particle operator, a partial trace over the statistical operator can be performed in a way that only the single-particle statistical operator with the matrix elements ρ_{v_1,v_2} contributes. The sum involves all single-particle states of the chosen basis, as discussed for QD systems in Section 40.2.

In the following, we further specify the single-particle density-matrix elements. The diagonal elements represent the population of the state v and have already been abbreviated by f_v in Section 40.3. The off-diagonal elements Ψ_{v_1,v_2} are transition amplitudes between the corresponding states. Then the single-particle density matrix has the form

$$\rho_{v_1,v_2} = \begin{pmatrix} f_{v_1} & \Psi_{v_1,v_2} \\ \Psi_{v_1,v_2}^* & f_{v_2} \end{pmatrix}. \tag{40.19}$$

For the calculation of the optical polarization, we use that the diagonal dipole-matrix elements vanish exactly. Also, we consider only interband transition contributions to the dipole coupling. For the final calculation of the macroscopic QD

polarization, we assume that the optical field averages over sufficiently many QDs to obtain

$$P_{QD}(t) = n_{QD}\langle d \rangle = n_{QD}\sum_{v_1,v_2} d_{v_1,v_2}\Psi_{v_1,v_2}(t), \tag{40.20}$$

with the (sheet) density of QDs n_{QD}.

40.4.3 Optical Gain Calculations

A consistent microscopic theory for the calculation of coherent optical properties (related to the interaction of the active material with a classical optical field) can be formulated on the basis of equations of motion for the single-particle density matrix elements. These equations can be derived from the Hamiltonian of the QD system, which contains the contributions of free-carriers, their dipole interaction with the optical fields, as well as further relevant interaction processes like the Coulomb interaction of carriers and the carrier–phonon interaction.

When only the free-carrier Hamiltonian and the dipole interaction with the optical field are considered, the density matrix elements obey the optical Bloch equations (Meystre and Sargent III, 1991). We obtain coupled equations for the interband transition amplitude $\Psi_{v_1,v_2}(t)$ and for the occupation probabilities $f_{v_{1,2}}(t)$.

$$i\hbar\frac{\partial}{\partial t}\Psi_{v_1,v_2} = \left(\epsilon_{v_1} - \epsilon_{v_2} - \frac{i\hbar}{T_2}\right)\Psi_{v_1,v_2} + (f_{v_1} - f_{v_2})d_{v_1,v_2}E_{QD},$$

$$\tag{40.21}$$

$$\frac{\partial}{\partial t}f_{v_1} = \frac{i}{\hbar}\sum_{v_2}[d_{v_1,v_2}E_{QD}\Psi_{v_1,v_2}^* - c.c.] - \frac{f_{v_1} - F_{v_1}}{T_1}, \tag{40.22}$$

$$\frac{\partial}{\partial t}f_{v_2} = -\frac{i}{\hbar}\sum_{v_1}[d_{v_1,v_2}E_{QD}\Psi_{v_1,v_2}^* - c.c.] - \frac{f_{v_2} - F_{v_2}}{T_1}. \tag{40.23}$$

A phenomenological way to include the influence of carrier scattering and dephasing is via time constants $T_{1,2}$, respectively. For the last term in Equations 40.22 and 40.23, a relaxation-time approximation has been used, which describes the evolution of the occupation probabilities $f_{v_{1,2}}$ toward quasi-equilibrium functions $F_{v_{1,2}}$ on a time-scale T_1. Damping of the interband transition amplitude Ψ_{v_1,v_2}, which determines the linewidth of the interband transitions, is accounted for with the dephasing time T_2 in Equation 40.21.

The stationary solution of Equation 40.21 together with Equations 40.17 and 40.20 allows to calculate the optical susceptibility in the form

$$\chi_{QD}(\omega) = n_{QD}\sum_{v_1,v_2}|d_{v_1,v_2}|^2(f_{v_1} - f_{v_2})$$

$$\times \left[\frac{1}{\hbar\omega - \varepsilon_{v_{12}} + \frac{i}{T_2}} - \frac{1}{\hbar\omega - \varepsilon_{v_{12}} + \frac{i}{T_2}}\right]. \tag{40.24}$$

The imaginary part of $\chi_{QD}(\omega)$ describes the absorption spectrum of the system and reveals discrete lines for interband transitions between the QD states $v_{1,2}$ with a linewidth $\propto 1/T_2$ at the interband energies $\varepsilon_{v_{1,2}} = \varepsilon_{v_1} - \varepsilon_v$. When the population in the upper state exceeds that in the lower state, the population factor $f_{v_1} - f_{v_2}$ changes its sign and the system switches from absorption to optical gain.

It should be noted that the optical Bloch equations (Equations 40.21 through 40.23) and the resulting susceptibility (40.24) represent only a free-carrier picture, which is usually a poor approximation for a quantitative analysis. Important is the additional inclusion of Coulomb interaction effects in the optical Bloch equations for semiconductor systems (Haug and Koch, 2004), which is responsible for different types of effects. First of all, the interband Coulomb exchange interaction between electrons and holes accounts for excitonic effects at low excitation densities and interband Coulomb enhancement of the optical transitions. Secondly, energy renormalization shows up as excitation-dependent shifts of the transition energies, of which the band-gap renormalization is the most prominent signature. A closer inspection reveals, however, nonrigid shifts of all states. As a third modification, the microscopic replacement for scattering terms, expressed by the relaxation approximation in Equations 40.22 and 40.23, has been provided in Section 40.3. For consistency, similar terms need to be used in Equation 40.21 to describe the excitation-induced dephasing, which replaces the constant dephasing time T_2. Furthermore, screening of the Coulomb interaction contributes to a nonlinear dependence of the optical properties on the excitation conditions.

40.4.4 Gain Saturation, Refractive Index, and α-Factor

A microscopic theory for gain calculations in QD systems, which includes excitonic effects as well as excitation-induced energy renormalization and dephasing due to the screened Coulomb interaction and carrier–phonon interaction, has been developed in Lorke et al. (2006a). An example for the transition of a QD system from absorption to gain with increasing excitation density is shown in Figure 40.6. The calculations are performed at room temperature and assume a quasi-equilibrium distribution of the carriers over the QD and WL states due to efficient carrier-scattering processes. For a low excitation density (dotted line), QD s-shell and p-shell transitions as well as the WL exciton resonance at $E - E_G \approx -138$, -80, and -17 meV, respectively, are broadened due to carrier–phonon and carrier–carrier scattering processes. With increasing carrier density, the transition lines are bleached (due to phase-space filling) and further broadened (due to increased scattering efficiency). Note that bleaching reduces the oscillator strength, which needs to be distinguished from the increasing the linewidth (broadening). With population inversion of the QD states, optical gain is realized. Just above the transparency carrier density, only the s-shell transition exhibits optical gain. At higher carrier densities, the optical gain of the p-shell and of the WL are taking over. The WL gain is due

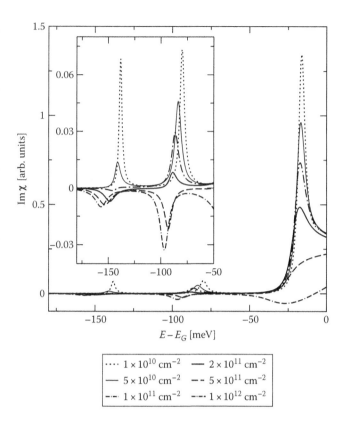

FIGURE 40.6 Imaginary part of the optical susceptibility, representing the absorption spectrum, versus energy relative to the band edge of the WL for a InGaAs/GaAs QD system. The spectrum contains two interband transitions due to confined states, which are magnified in the inset. (From Lorke, M. et al., *Phys. Rev. B*, 73, 085324, 2006. With permission.)

to the increased population of the corresponding quantum-well-like states. The combination of band-gap shrinkage and band filling positions the WL gain near the low-density WL exciton resonance. The related small excitation-dependent shifts of the quantum well gain are discussed in Chow and Koch (1999).

When the QD states are completely populated with electrons and holes, a further increase of the total carrier density in the system does not lead to a larger optical gain at the corresponding transitions. This is clearly visible at the s-shell transition for the largest carrier densities in Figure 40.6. A higher carrier density in the assumed quasi-equilibrium situation only increases the population of the higher QD states and the WL states. The complete filling of the QD states is the origin for the observed gain saturation. Adding even more carriers to the system finally starts to reduce the QD ground-state gain due to further increasing dephasing (Lorke et al., 2006b).

At low carrier densities, the QD interband transition lines can be associated with QD excitons. The terminology might be viewed with reservations, since the electron–hole pairs are restricted to the QDs by the confinement potential and not—like quantum-well or bulk-semiconductor excitons—bound by the Coulomb interaction. The interplay of direct and exchange Coulomb interaction even allows for "anti-bound" states; a biexciton transition with an energy larger than that of two exciton transitions is the most prominent example. With the term "QD exciton," one emphasizes the role of the Coulomb interaction for characterizing the states. For quantum well and bulk semiconductor excitons, it is known that the delicate balance of energy renormalization (mainly band-gap shrinkage) and reduction of the exciton binding energy due to phase-space filling and screening leads to a nearly constant energetic position of the exciton resonance, which is bleached and broadened for increasing carrier density. The same behavior is observed for the WL exciton in Figure 40.6. However, the QD excitons show a pronounced red shift, which persists as a shift of the gain peaks at elevated carrier densities. When the energy renormalization is determined within a many-body theory formulated in a single-particle basis, it is the result of a partial compensation of state-diagonal and state-nondiagonal self-energies. It can be shown that this partial compensation is strongly reduced in QD systems (Lorke et al., 2006a).

The physics of QD gain spectra is expected to play an important role for QD-microcavity lasers with a small number of dots. In edge-emitting devices, inhomogeneous broadening can mask the discussed effects until the sample quality improves or the laser behavior (saturation effects, gain dynamics) is examined in greater detail.

The excitation-induced refractive index changes are directly linked to the corresponding absorption spectra via the Kramers–Kronig relation (Haug and Koch, 2004), since both can be derived from the complex optical susceptibility according to Equation 40.18. Refractive index changes directly influence the mode properties of edge-emitting lasers, including filamentation, as well as frequency chirp and emission spectra. In the past, the α-factor (linewidth-enhancement factor, antiguiding parameter) has been used to characterize the importance of the excitation-density-induced refractive index changes. In the free-carrier picture outlined above, the discrete nature of the QD states leads to symmetric absorption lines and vanishing refractive index changes at the absorption peak. The resulting value of zero for the α-factor has led to the prediction of many beneficial properties of QD lasers. Calculations based on microscopic semiconductor models show deviations from this simple picture, but also smaller α-factors than for quantum-well lasers (Lorke et al., 2007).

40.5 Laser Emission Properties

The optical gain characterizes the active material itself, while the steady-state and dynamical emission properties of lasers depend more generally on the interplay of the photon and carrier systems. The optical gain determines the rate of stimulated emission of photons into the laser mode. Together with the spontaneous emission rate, which is related to the photoluminescence spectrum, and the cavity losses, these processes govern the laser field inside the laser resonator and the light output. The carrier dynamics is determined by the interplay of the pump process, the transfer of carriers into the laser levels, as well as the optical processes.

The most intuitive approach to the steady-state and dynamical properties of lasers is provided by rate equations, which

will be introduced in Section 40.5.1. Such a theory is based on restricting the information about the laser output to the mean photon number in the laser mode, as well as on a convenient parametrization of the rates for various processes. A generalization, which allows the systematic inclusion of semiconductor effects, is outlined in Section 40.5.2. The statistical properties of the light emission as well as the control of light–matter interaction in microcavity lasers are addressed in Section 40.5.3.

40.5.1 Rate Equations for Quantum-Dot Systems

In a rate-equation description (Yokoyama and Brorson, 1989; Rice and Carmichael, 1994), one uses two coupled dynamical equations for the number of photons in the laser mode n and the number of excited emitters N, which are QDs for the present purpose:

$$\frac{d}{dt}N = P - \frac{nN}{\tau_l} - \frac{N}{\tau_{sp}}, \qquad (40.25)$$

$$\frac{d}{dt}n = -2\kappa n + \frac{(n+1)N}{\tau_l}. \qquad (40.26)$$

The pump rate P increases the emitter number, while the cavity loss rate 2κ reduces the photon number. The loss rate is directly connected to the Q-factor of the laser mode, $Q = \hbar\omega_l/2\kappa$, with the laser frequency ω_l. The total rate of spontaneous emission N/τ_{sp}, which includes emission into all available lasing and nonlasing modes, reduces the number of excited emitters, while only the spontaneous emission directed into the laser mode N/τ_l contributes to the increase of the respective photon number. The simulated emission is additionally proportional to the mean photon number n, and the corresponding rate nN/τ_l appears in both rate equations with opposite signs.

An important parameter to characterize laser resonators is the β-factor, which is defined as the ratio of the spontaneous emission rate into the laser mode to the total spontaneous emission according to

$$\beta = \frac{1/\tau_l}{1/\tau_{sp}}. \qquad (40.27)$$

In edge-emitting laser diodes, the large number of nonlasing modes leads to small values of $\beta = 10^{-6} \dots 10^{-5}$, while in microcavity lasers the spontaneous emission into nonlasing modes can be strongly suppressed and $\beta \approx 1$ is approached.

A general evaluation of the rate equations can be readily performed by means of a direct numerical solution in time for a given pump rate P. If the initial condition is the unexcited system and the pump rates are switched on to a constant value, the solution either shows a smooth evolution to the steady-state or damped relaxation oscillations (Milonni and Eberly, 1991), depending on the chosen parameters.

The results for the steady-state solution and various β-factors are shown in Figure 40.7. A set of parameters referring to

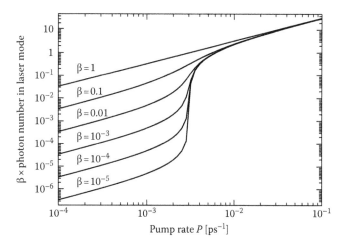

FIGURE 40.7 Calculated input–output curves from the laser rate equations for $\beta = 1$ to 10^{-5} from top to bottom. The photon number and the pump rate are scaled with β in order to have the thresholds appear at equal pump intensities for better comparison. (From Gies, C. et al., *Phys. Rev. A*, 75, 013803, 2007. With permission.)

QD-microcavity lasers as been used: $\tau_{sp} = 50\,\text{ps}$ and $\kappa = 20\,\mu\text{eV}$. The corresponding cavity lifetime is about 17 ps, yielding a Q-factor of $\approx 30{,}000$. The curves show a typical intensity jump $\propto \beta^{-1}$ from below to above threshold. In the limit $\beta = 1$, the kink in the input–output curve disappears.

The above discussed form of the rate equations is based on various assumptions. Only one optical mode is subject to stimulated emission; otherwise separate rate equations for the photon numbers in different laser modes need to be used and their modal gain determining τ_l needs to be distinguished. The derivation of the rate equations from a microscopic theory is based on the adiabatic elimination of transition amplitudes as well as on the factorization of carrier and photon correlations (Rice and Carmichael, 1994). Furthermore, in the rate equations, one assumes that the spontaneous and stimulated recombination rates are proportional to number of excited emitters N. Our discussion of the optical gain in Section 40.4.3 has already revealed that this is only a rough estimate neglecting optical nonlinearities and saturation effects. Furthermore, each QD is assumed to possess only two possible configurations, the excited and de-excited state. In reality, QDs often contain more than two localized states that can be occupied by several charge carriers. This can lead to various modifications of the simple "atomic approach."

The form of the spontaneous recombination term in the laser rate equations predicts an exponential decay of the photoluminescence after pulsed excitation and in the absence of stimulated emission. Recent experiments with self-assembled InGaAs/GaAs QDs embedded in GaAs-based micropillars (Schwab et al., 2006) showed, however, a nonexponential decay of the time-resolved photoluminescence. Furthermore, this decay was shown to be accompanied by a strong dependence on the excitation intensity. These two effects make it impossible to assign a single spontaneous emission lifetime to the QDs and question the simple picture used in the rate equations. This further

underscores the importance of semiconductor models for a more detailed microscopic description of photoluminescence and gain in QD systems.

40.5.2 Microscopic Generalization: Semiconductor Effects

A semiconductor theory for QD lasers can be formulated on the basis of equations of motions for the carrier occupation probabilities of the QD levels and the photon number in the laser mode. For a microscopic derivation, one starts from the Hamiltonian describing the quantized electromagnetic field, the carrier system in second quantization, and the carrier–photon interaction in dipole approximation. Further interaction processes, like the Coulomb interaction of carriers and the carrier–phonon interaction, can be included systematically. Heisenberg equations of motion for the carrier and photon operators can be used to derive dynamical equations for the mean photon number ($n = b^\dagger b$) and the carrier occupation functions $f_v^e = \langle c_v^\dagger c_v \rangle, f_v^h = 1 - \langle v_v^\dagger v_v \rangle$. In these expectation values, $b^\dagger$ and b are photon creation and annihilation operators, $c_v^\dagger$ and c_v are creation and annihilation operators for conduction-band carriers in the states v and $v_v^\dagger v_v$ are the corresponding valence-band operators, respectively. The contribution of the interaction of carriers with the laser mode then leads to

$$\left(\frac{d}{dt} + 2\kappa\right) n = \frac{2}{\hbar} |g|^2 \sum_{v'} \text{Re} \langle b^\dagger v_{v'}^\dagger c_{v'} \rangle, \qquad (40.28)$$

$$\frac{d}{dt} f_v^{e,h} \bigg|_{\text{opt}} = -\frac{2}{\hbar} |g|^2 \, \text{Re} \langle b^\dagger v_v^\dagger c_v \rangle, \qquad (40.29)$$

with the light–matter coupling strength g. It can be seen that the corresponding dynamics of the photon number and carrier populations are determined by the photon-assisted polarization $\langle b^\dagger v_v^\dagger c_v \rangle$, that describes the expectation value for a correlated event, where a photon is created in connection with an interband transition of an electron from the conduction to the valence band. The sum over v involves all possible interband transitions from various QDs.

The time evolution of the photon-assisted polarization follows from its equation of motion,

$$\left(\frac{d}{dt} + \kappa + \Gamma + i(\tilde{\varepsilon}_v^e + \tilde{\varepsilon}_v^h - \hbar\omega_l)\right) \langle b^\dagger v_v^\dagger c_v \rangle$$

$$= f_v^e f_v^h - (1 - f_v^e - f_v^h) n$$

$$+ i(1 - f_v^e - f_v^h) \sum_{\alpha} V_{v\alpha v\alpha} \langle b^\dagger v_\alpha^\dagger c_\alpha \rangle$$

$$+ \sum_{\alpha} C_{\alpha v v \alpha}^x + \delta \langle b^\dagger b c_v^\dagger c_v \rangle - \delta \langle b^\dagger b v_v^\dagger v_v \rangle. \qquad (40.30)$$

Here, the free evolution of $\langle b^\dagger v_v^\dagger c_v \rangle$ is determined by the detuning of the QD transitions at the renormalized energies $\tilde{\varepsilon}_v^{e,h}$ from the optical mode ω_l. Cavity losses κ and dephasing processes represented by Γ lead to a damping of the time evolution and to a broadening of the spectral components of the optical processes, which are spontaneous and stimulated emission. When deriving the equation of motion for $\langle b^\dagger v_v^\dagger c_v \rangle$, one finds that the source term of spontaneous emission is described by an expectation value of four carrier operators $\langle c_\alpha^\dagger v_\alpha v_v^\dagger c_v \rangle$. For uncorrelated carriers, the Hartree–Fock factorization of this source term leads to $f_v^e f_v^h$, which appears as the first term on the right hand side of Equation 40.30. It describes a spontaneous recombination probability proportional to the occupation probabilities of electrons and holes, as expected in a free-carrier picture, but opposed to the situation in an atomic system, where the recombination depends only on the electron population. However, electrons and holes will not just contribute as independent carriers. Correlation contributions to the spontaneous emission are included in $C_{\alpha v v \alpha}^x = \delta \langle c_\alpha^\dagger v_v^\dagger c_v v_\alpha \rangle$. These correlations describe the joint probability of a two-particle process, where two interband carrier transitions between the states α and v take place. As mentioned above, the indices include not only the electronic states but also the QD position. Consequently, one must distinguish between correlated transitions within one QD, which can be due to an excitonic population, and correlations of transitions within two separate QDs. The latter case is connected to the phenomenon of superfluorescence or superradiant coupling of different emitters. The Coulomb interaction contributes to the excitonic correlations of electrons and holes, which, in turn, are weakened by screening, phase-space filling, and dephasing effects of the excited carriers. The detailed microscopic description of these correlations is an intricate problem of many-body theory, and for further details we refer the interested reader to Baer et al. (2006).

In a similar fashion, stimulated emission or re-absorption of photons is represented by $(1 - f_v^e - f_v^h)n$ in the absence of carrier–photon correlations. The latter are represented by $\delta \langle b^\dagger b c_v^\dagger c_v \rangle$ and $\delta \langle b^\dagger b v_v^\dagger v_v \rangle$, which require their own equations of motion. Furthermore, the interband Coulomb–exchange interaction with the matrix elements $V_{v\alpha v\alpha}$ is responsible for excitonic effects similar to its appearance in the semiconductor Bloch equations (Haug and Koch, 2004).

While the semiconductor theory offers the potential to account for various interaction effects in a much more detailed way, this also complicates a numerical analysis significantly. Results from the semiconductor theory for the input–output curves are shown in the lower part of Figure 40.8. Deviations from the rate-equation results involve a different height in the intensity jump at the laser threshold and saturation effects. The semiconductor theory also opens the possibility to calculate parameters entering rate equation models based on single-particle properties and interaction effects.

40.5.3 Quantum-Dot Microcavity Lasers: Modifications of the Photon Statistics

Latest advances in the growth and design of semiconductor-QD microcavity lasers have now attained the regime of β-values

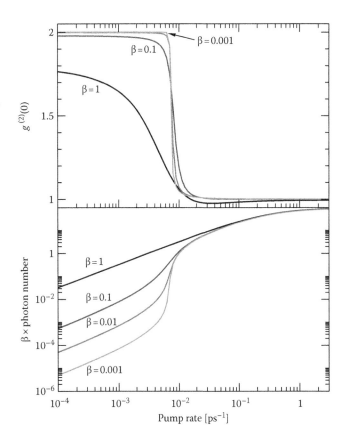

FIGURE 40.8 Input–output curves and auto-correlation functions $g^{(2)}$ ($\tau = 0$) for various β-factors calculated with the semiconductor model. (From Gies, C. et al., *Phys. Rev. A*, 75, 013800, 2007. With permission.)

close to unity (Strauf et al., 2006; Ulrich et al., 2007). The strongly increased cavity-Q (corresponding to a long lifetime of photons in the cavity) allows to fabricate QD-based lasers with a small number of dots in the active region. The strongly enhanced light-matter coupling and the operation with a small number of photons in the laser mode leads to novel emission properties that are directly related to the quantum-mechanical nature of light. In particular, the photon statistics of the emitted radiation does no longer exhibit the transition from thermal to coherent radiation for increasing pumping, and "nonclassical" properties on the level of few photons can be realized.

To characterize these systems, one needs to study the coherence properties of the emitted light and the statistical properties of the photons. Following Glauber, the quantum states of light can be characterized in terms of photon correlation functions. Coherence properties of the electromagnetic field itself are reflected by the (normalized) correlation function of first order,

$$g^{(1)}(\tau) = \frac{\langle b^\dagger(t) b(t+\tau) \rangle}{\langle b^\dagger(t) b(t) \rangle}. \tag{40.31}$$

Its decay in τ is determined by the coherence time of the emitted light $\tau_c = \int |g^{(1)}(\tau)|^2 \, d\tau$. Here, $b^\dagger$ and b are again the creation and annihilation operators for photons in the laser mode. Information about the statistical properties of the emitted light can be deduced from the correlation function of second order at zero delay time

$$g^{(2)}(\tau = 0) = \frac{\langle n^2 \rangle - \langle n \rangle}{\langle n \rangle^2} = \frac{\langle b^\dagger b b^\dagger b \rangle - \langle b^\dagger b \rangle}{\langle b^\dagger b \rangle^2}. \qquad (40.32)$$

The function $g^{(2)}(\tau = 0)$ reflects the possibility of the correlated emission of two photons at the same time. Often discussed are the limiting cases of light emission from a thermal, a coherent, and a single-emitter light source. Thermal light is characterized by an enhanced probability that two photons are emitted at the same time (bunching), reflected in a value of $g^{(2)}(0) = 2$. For coherent light emission with Poisson statistics, one finds $g^{(2)}(0) = 1$. An ideal single-photon emitter exhibits antibunching with $g^{(2)}(0) = 0$. This, the second-order correlation function, can be used to characterize the emission and to analyze the transition from thermal (or even sub-Poissonian) to coherent light emission.

For atomic systems, various methods have been established to analyze photon correlation functions. A master equation can be used to describe an ensemble of emitters interacting with a single high-Q laser cavity mode and a bath of non-lasing modes (Rice and Carmichael, 1994). For a single-atom laser, a direct analysis of the von Neumann equation for the statistical operator including the coupling to dissipative systems is possible (Mu and Savage, 1992). Semiconductor effects can be included in generalized equations of motion for higher-order carrier and photon correlation functions (Gies et al., 2007). Results of a semiconductor theory for the second-order photon correlation function are displayed in the upper part of Figure 40.8. For smaller β-factors referring to conventional lasers, the output-intensity jump at the laser threshold is accompanied with a sudden change of the correlation function between values representing thermal and coherent light. For increasing values of β, the gradually disappearing jump in the photon number is connected with a smoother transition of $g^{(2)}(0)$ and the presence of correlations for small pump rates.

Acknowledgments

The author thanks P. Gartner and J. Wiersig for the intense and productive collaboration on the covered subjects. The contributions from our PhD students T. Nielsen, N. Baer, J. Seebeck, M. Lorke, and C. Gies are also gratefully acknowledged. This work was financially supported by the Deutsche Forschungsgemeinschaft. We also thank P. Gartner and C. Gies for the critical reading of the manuscript and A. Beuthner for the preparation of Figure 40.1.

References

Agrawal, G. P. and N. K. Dutta. 1993. *Semiconductor Laser.* Norwell, MA: Kluwer.

Anders, S., C. S. Kim, B. Klein, M. W. Keller, R. P. Mirin, and A. G. Norman. 2002. Bimodal size distribution of self-assembled In_xGa_{1-x} As quantum dots. *Phys. Rev. B* **66**:125309.

Arakawa, Y. and H. Sakaki. 1982. Multidimensional quantum well laser and temperature dependence of its threshold current. *Appl. Phys. Lett.* **40**:939.

Asada, M., Y. Miyamoto, and Y. Suematsu. 1986. Gain and the threshold of three-dimensional quantum-box lasers. *IEEE J. Quantum Electron.* **QE-22**:1915.

Baer, N., C. Gies, J. Wiersig, and F. Jahnke. 2006. Luminescence of a semiconductor quantum dot system. *Eur. Phys. J. B* **50**:411.

Benisty, H., C. M. Sotomayor-Torres, and C. Weisbuch. 1991. Intrinsic mechanism for the poor luminescence properties of quantum-box systems. *Phys. Rev. B* **44**:10945.

Bimberg, D., M. Grundmann, and N. N. Ledentsov. 1999. *Quantum Dot Heterostructures.* Chichester, U.K.: John Wiley & Sons.

Bockelmann, U. and G. Bastard. 1990. Phonon scattering and energy relaxation in two-, one-, and zero-dimensional electron gases. *Phys. Rev. B* **42**:8947.

Casey, H. C. and M. B. Panish. 1978a. *Heterostructure Lasers—Part A: Fundamental Principles.* New York: Academic Press.

Casey, H. C. and M. B. Panish. 1978b. *Heterostructure Lasers—Part B: Materials and Operating Characteristics.* New York: Academic Press.

Chang-Hasnain, C. J., J. P. Harbison, G. Hasnain, A. C. von Lehmen, L. T. Florenz, and N. G. Stoffel. 1991. Dynamic, polarization, and transverse mode characteristics of vertical cavity surface emitting lasers. *IEEE J. Quantum Electron.* **27**:1402.

Chow, W. W. and S. W. Koch. 1999. *Semiconductor-Laser Fundamentals.* Berlin, Germany: Springer.

Chuang, S. L. 1995. *Physics of Optoelectronic Devices.* New York: Wiley.

Chuang, S. L. and C. S. Chang. 1996. k·p method for strained wurtzite semiconductors. *Phys. Rev. B* **54**:2491.

Dingle, R. and C. H. Henry. 1976. Quantum effects in heterostructure lasers. U.S. Patent 3982207.

Fattal, D., E. Diamanti, K. Inoue, and Y. Yamamoto. 2004. Quantum teleportation with a quantum dot single photon source. *Phys. Rev. Lett.* **92**:037904.

Gies, C., J. Wiersig, M. Lorke, and F. Jahnke. 2007. Semiconductor model for quantum-dot-based microcavity lasers. *Phys. Rev. A* **75**:013803.

Haug, H. and A.-P. Jauho. 1996. *Quantum Kinetics in Transport & Optics of Semiconductors*, 1st edn. Berlin, Germany: Springer.

Haug, H. and S. W. Koch. 2004. *Quantum Theory of the Optical and Electronic Properties of Semiconductors*, 4th edn. Singapore: World Scientific Publishing Company.

Inoshita, T. and H. Sakaki. 1992. Electron relaxation in a quantum dot: Significance of multiphonon processes. *Phys. Rev. B* **46**:7260.

Inoshita, T. and H. Sakaki. 1997. Density of states and phonon-induced relaxation of electrons in semiconductor quantum dots. *Phys. Rev. B* **56**:4355.

Jacobi, K. 2003. Atomic structure of InAs quantum dots on GaAs. *Prog. Surf. Sci.* **71**:185.

Jewell, J. L., J. P. Harbison, A. Scherer, Y. H. Lee, and L. T. Florez. 1991. Vertical-cavity surface-emitting lasers: Design, growth, fabrication, characterization. *IEEE J. Quantum Electron.* **27**:1332.

Jiang, H. and J. Singh. 1998. Self-assembled semiconductor structures: Electronic and optoelectronic properties. *IEEE J. Quantum Electron.* **34**:1188.

Kim, S. M., Y. Wang, M. Keever, and J. S. Harris. 2004. High-frequency modulation characteristics of 1.3 μm InGaAs quantum dot lasers. *IEEE Photon. Technol. Lett.* **16**:377.

Klude, M., T. Passow, R. Kröger, and D. Hommel. 2004. Electrically pumped lasing from CdSe quantum dots. *Electron. Lett.* **37**:1119.

Kral, K. and Z. Khas. 1998. Electron self-energy in quantum dots. *Phys. Rev. B* **57**:2061.

Li, H. E. and K. Iga, eds. 2003. *Vertical-Cavity Surface-Emitting Laser Devices*, 1st edn. Berlin, Germany: Springer.

Lodahl, P., A. F. van Driel, I. S. Nikolaev, A. Irman, K. Overgaag, D. Vanmaekel-bergh, and W. L. Vos. 2004. Controlling the dynamics of spontaneous emission from quantum dots by photonic crystals. *Nature* **430**:654.

Lorke, M., T. R. Nielsen, J. Seebeck, P. Gartner, and F. Jahnke. 2006a. Influence of carrier-carrier and electron-phonon correlations on optical absorption and gain in quantum-dot systems. *Phys. Rev. B* **73**:085324.

Lorke, M., W. W. Chow, T. R. Nielsen, J. Seebeck, P. Gartner, and F. Jahnke. 2006b. Anomaly in the excitation dependence of optical gain in semiconductor quantum dots. *Phys. Rev. B* **74**:035334.

Lorke, M., F. Jahnke, and W. Chow. 2007. Excitation dependence of gain and carrier induced refractive index changes in quantum dots. *Appl. Phys. Lett.* **90**:051112.

Mahan, G. D. 1990. *Many-Particle Physics*, 2nd edn. New York: Plenum.

Meystre, P. and M. Sargent III. 1991. *Elements of Quantum Optics*, 2nd edn. New York: Springer.

Michler, P., A. Kiraz, C. Becher, W. V. Schoenfeld, P. M. Petroff, L. Zhang, E. Hu, and A. Imamoglu. 2000. A quantum dot single-photon turnstile device. *Science* **290**:2282.

Milonni, P. and J. H. Eberly. 1991. *Lasers*, 1st edn. New York: John Wiley & Sons.

Minnaert, A. W. E., A. Yu. Silov, W. van der Vleuten, J. E. M. Haverkort, and J. H. Wolter. 2001. Frhlich interaction in InAs/GaAs self-assembled quantum dots. *Phys. Rev. B* **63**:075303.

Moreau, E., I. Robert, L. Manin, V. Thierry-Mieg, J. M. Gerad, and I. Abram. 2001. Quantum cascade of photons in semiconductor quantum dots. *Phys. Rev. Lett.* **87**:183601.

Mu, Y. and C. M. Savage. 1992. One-atom lasers. *Phys. Rev. A* **46**:5944.

Nakamura, S. and S. F. Chichibu, eds. 2000. *Introduction to Nitride Semiconductor Blue Lasers and Light Emitting Diodes*, 1st edn. Boca Raton, FL: CRC Press.

Newell, T. C., D. J. Bossert, A. Stintz, B. Fuchs, K. J. Malloy, and L. F. Lester. 1999. Gain and linewidth enhancement factor in InAs quantum-dot laserdiodes. *IEEE Photon. Technol. Lett.* **11**:1527.

Nielsen, T. R., P. Gartner, and F. Jahnke. 2004. Many-body theory of carrier capture and relaxation in semiconductor quantum-dot lasers. *Phys. Rev. B* **69**:235314.

Nielsen, T. R., P. Gartner, M. Lorke, J. Seebeck, and F. Jahnke. 2005. Coulomb scattering in nitride-based self-assembled quantum dot systems. *Phys. Rev. B* **72**:235311.

Peronne, E., F. Fossard, F. H. Julien, J. Brault, M. Gendry, B. Salem, G. Bremond, and A. Alexandrou. 2003. Dynamic saturation of an intersublevel transition in self-organized InAs/InAlAs quantum dots. *Phys. Rev. B* **67**:205329.

Quochi, F., M. Dinu, L. N. Pfeiffer, K. W. West, C. Kerbage, R. S. Windeler, and B. J. Eggleton. 2003. Coulomb and carrier-activation dynamics of resonantly excited InAs/GaAs quantum dots in two-color pump-probe experiments. *Phys. Rev. B* **67**:235323.

Reithmaier, J. P., G. Sek, A. Löffler, C. Hofmann, S. Kuhn, S. Reitzenstein, L. V. Keldysh, V. D. Kulakovskii, T. L. Reinecke, and A. Forchel. 2004. Strong coupling in a single quantum dot-semiconductor microcavity system. *Nature* **432**:197.

Rice, P. R. and H. J. Carmichael. 1994. Photon statistics of cavity-QED lasers. *Phys. Rev. A* **50**:4318.

Schäfer, W. and M. Wegener. 2002. *Semiconductor Optics and Transport Phenomena*, 1st edn. Berlin, Germany: Springer-Verlag.

Scholz, M., T. Aichele, S. Ramelow, and O. Benson. 2006. Deutsch-Jozsa algorithm using triggered single photons from a single quantum dot. *Phys. Rev. Lett.* **96**:180501.

Schwab, M., H. Kurtze, T. Auer, T. Berstermann, M. Bayer, J. Wiersig, N. Baer et al. 2006. Radiative emission dynamics of quantum dots in a single cavity micropillar. *Phys. Rev. B* **74**:045323.

Sebald, K., C. Kruse, and J. Wiersig. 2009. Properties and prospects of blue-green emitting II-VI-based monolithic micro-cavities. *Phys. Stat. Sol. (b)* **246**:255.

Seebeck, J., T. R. Nielsen, P. Gartner, and F. Jahnke. 2005. Polarons in semiconductor quantum-dots and their role in the quantum kinetics of carrier relaxation. *Phys. Rev. B* **71**:125327.

Shah, J. 1999. *Ultrafast Spectroscopy of Semiconductors and Semiconductor Nanostructures*. Berlin, Germany: Springer-Verlag.

Stauber, T., R. Zimmermann, and H. Castella. 2000. Electron-phonon interaction in quantum dots: A solvable model. *Phys. Rev. B* **62**:7336.

Strauf, S., K. Hennessy, M. T. Rakher, Y.-S. Choi, A. Badolato, L. C. Andreani, E. L. Hu, P. M. Petroff, and D. Brouwmeester. 2006. Self-tuned quantum dot gain in photonic crystal lasers. *Phys. Rev. Lett.* **96**:127404.

Tsang, W. T. 1982. Extremely low threshold (AlGa)As graded-index waveguide separate-confinement heterostructure lasers grown by molecular beam epitaxy. *Appl. Phys. Lett.* **40**:217.

Tsitsishvili, E., R. V. Baltz, and H. Kalt. 2002. Temperature dependence of polarization relaxation in semiconductor quantum dots. *Phys. Rev. B* **66**:161405.

Ulrich, S. M., C. Gies, J. Wiersig, S. Reitzenstein, C. Hofmann, A. Löffler, A. Forchel, F. Jahnke, and P. Michler. 2007. Photon statistics of semiconductor microcavity lasers. *Phys. Rev. Lett.* **98**:043906.

Urayama, J., T. B. Norris, J. Singh, and P. Bhattacharya. 2001. Observation of phonon bottleneck in quantum dot electronic relaxation. *Phys. Rev. Lett.* **86**:4930.

Vahala, K. 2003. Optical microcavities. *Nature* **424**:839.

Verzelen, O., R. Ferreira, G. Bastard, T. Inoshita, and H. Sakaki. 2002. Polaron effects in quantum dots. *Phys. Stat. Sol. (a)* **190**:213.

Vurgaftman, I. and J. R. Meyer. 2003. Band parameters for nitrogen-containing semiconductors. *J. Appl. Phys.* **94**:3675.

Wojs, A., P. Hawrylak, S. Fafrad, and L. Jacak. 1996. Electronic structure and magneto-optics of self-assembled quantum dots. *Phys. Rev. B* **54**:5604.

Xu, S., A. A. Mikhailovsky, J. A. Hollingsworth, and V. I. Klimov. 2002. Hole intra-band relaxation in strongly confined quantum dots: Revisiting the phonon bottleneck problem. *Phys. Rev. B* **65**:045319.

Yokoyama, H. and S. D. Brorson. 1989. Rate equation analysis of microcavity lasers. *J. Appl. Phys.* **66**:4801.

Yoshie, T., A. Scherer, J. Hendrickson, G. Khitrova, H. M. Gibbs, G. Rupper, C. Ell, O. B. Shchekin, and D. G. Deppe. 2004. Vacuum Rabi splitting with a single quantum dot in a photonic crystal nanocavity. *Nature* **432**:200.

Mode-Locked Quantum-Dot Lasers

Maria A. Cataluna
University of Dundee

Edik U. Rafailov
University of Dundee

41.1 Introduction

Over the last three decades, laser physics has advanced dramatically. Starting from lasers operated in a continuous wave (cw) regime, scientists have developed techniques for generating periodic sequences of optical pulses with ultrashort durations—between a few picoseconds (1×10^{-12} s) and a few femtoseconds (1×10^{-15} s). To put this into perspective, 1 fs compared to 1 s is the same as 1 s compared to 32 million years! Such ultrafast lasers have important applications in medicine, micromachining, optical communications, spectroscopy, and anything else that requires studying physics at extremely high powers or extremely short timescales. For instance, these lasers have been successfully adapted in eye surgery, because ultrashort pulses can make extremely precise cuts with minimum thermal damage.

However, despite the wide range of important areas that can benefit from ultrafast lasers, the use of these lasers is constrained due to several limitations. The ultrafast lasers currently available are often bulky, expensive, and difficult to operate. The ideal ultrafast laser would be a low-cost, handheld, and turnkey laser—features which could be offered by semiconductor lasers. Semiconductor lasers cannot yet directly generate the sub-100 fs pulses routinely available from crystal-based lasers, but they represent the most compact and efficient sources of picosecond

and sub-picosecond pulses. Furthermore, the bias can be easily adjusted to determine the pulse duration and the optical power, thus offering, to some extent, electrical control of the characteristics of the output pulses. These lasers also offer the best option for the generation of high-repetition rate trains of pulses, owing to their small cavity size. Ultrafast diode lasers have thus been favored over other laser sources for high-frequency applications such as optical data/telecommunications. Being much cheaper to fabricate and operate, ultrafast semiconductor lasers also offer the potential for dramatic cost savings in a number of applications that traditionally use solid-state lasers. The deployment of high-performance ultrafast diode lasers would therefore have a significant economic impact, by enabling ultrafast applications to become more profitable, and even facilitate the emergence of new applications.

Novel nanomaterials such as quantum dots (QDs) have enhanced the characteristics of semiconductor lasers, greatly improving their performance. QDs are tiny clusters of semiconductor material with dimensions of only a few nanometers. At these small sizes, materials behave very differently, giving QDs distinctive physical properties of quantum nature—for instance, the emission wavelength or "color" depends on the size of the dot! These nanomaterials afford major advantages in ultrafast science and technology, and they can form the basis for very compact and efficient lasers delivering short pulses of the order of hundreds of femtoseconds (Rafailov et al., 2007).

In this chapter, we show how QDs have enabled the generation of ultrashort pulses from compact optical sources based on semiconductor laser diodes. In Section 41.2, the necessary background information is presented on ultra-short-pulse generation from diode lasers. The concept of mode locking is introduced, and an overview of the mode-locking techniques available for semiconductor lasers is provided. The unique properties of QD materials and their suitability for ultra-short-pulse diode lasers are explained in Section 41.3. Finally, a summary of the state of the art in the field of QD mode-locked laser diodes is provided in Section 41.4. The chapter is finalized by a summary and an outlook on the future perspectives of this fascinating field.

41.2 Ultrafast Laser Diodes

41.2.1 Basics of Mode Locking

Mode locking is a technique that involves the locking of the phases of the longitudinal modes in a laser. This results in the generation of a sequence of pulses with a repetition rate corresponding to the cavity round-trip time. This well-established technique enables the production of the shortest pulse durations and the highest repetition rates available from ultrafast lasers, whether they are semiconductor or crystal-based laser systems. In a standing-wave resonator, the pulse repetition rate f_R is given by

$$f_R = \frac{c}{2nL}$$

where

c is the speed of light in vacuum
n is the refractive index
L is the length of the laser cavity

In terms of Fourier analysis, there is an inverse proportionality between the duration of a mode-locked pulse and the corresponding bandwidth of its optical spectrum. The product of both the pulse duration $\Delta\tau$ and the optical frequency bandwidth $\Delta\nu$ is called the time-bandwidth product (TBWP). For a given frequency bandwidth, there is a minimum corresponding pulse duration—if this is the case and the optical spectrum is symmetrical, then the pulse is said to be transform-limited, and the TBWP equals a constant K, whose value depends on the shape of the pulse, whether it is Gaussian, hyperbolic, secant, squared, or Lorentzian. By measuring the full-width at half maximum from an optical spectrum $\Delta\lambda$, it is easy to calculate the TBWP of a given pulse:

$$\Delta\nu \cdot \Delta\tau = K \quad \Rightarrow \quad \frac{c}{\lambda^2}\Delta\lambda \cdot \Delta\tau = K$$

Another important property of mode-locked lasers is that the energy that was dispersed in several modes while in cw

operation, is now concentrated in short pulses of light. This implies that although the output average power P_{av} may be low, the pulse peak power P_{peak} can be significantly higher:

$$P_{peak} = \frac{E_p}{\Delta\tau} \quad \Rightarrow \quad P_{peak} = \frac{1}{\Delta\tau} \cdot \frac{P_{av}}{f_R}$$

where E_p is the pulse energy.

41.2.2 Mode-Locking Techniques in Semiconductor Lasers

In recent years, mode-locked laser diodes have been at the center of a quest for ultrafast, transform-limited, and high-repetition-rate lasers. To achieve these goals, a variety of mode-locking techniques and semiconductor device structures have been demonstrated and optimized (Vasil'ev, 1995). The three main forms of mode locking can be described as active, passive, and hybrid techniques, as outlined below.

Active mode locking relies on the direct modulation of the gain with a frequency equal to the repetition frequency of the cavity, or to a sub-harmonic of this frequency. The main advantages of this approach are the resultant low jitter and the ability to synchronize the laser output with the modulating electrical signal. These features are especially relevant for optical transmission and signal-processing applications. However, high repetition frequencies are not readily obtained through directly driven modulation of lasers because fast RF (radiofrequency) modulation of the drive current becomes progressively more difficult with increase in frequency.

The frequency limitation imposed by electronic drive circuits can be overcome by employing passive mode-locking techniques. This scheme typically utilizes a saturable absorbing region in the laser diode. In a saturable absorber, the loss decreases as the optical intensity increases. This feature acts as a discriminator between cw and pulsed operation and can facilitate a self-starting mechanism for mode locking. Most importantly, saturable absorption plays a crucial role in shortening the duration of the circulating pulses, as will be explained, thus providing the shortest pulses achievable by all three techniques and the absence of a RF source simplifies the fabrication and operation considerably. Passive mode locking also allows for higher pulse repetition rates that are determined solely by the cavity length.

Inspired by active and passive mode locking, the technique of hybrid mode locking meets the best of both worlds because the pulse generation is initiated by an RF current imposed in the gain or absorber section, while further shaping and shortening is assisted by saturable absorption. The next section explores in more detail the physical mechanisms behind passive mode locking.*

---

* From this point, mode locking will implicitly mean passive mode locking, unless otherwise stated.

41.2.3 Passive Mode Locking: Physics and Devices

So far, a simple frequency-domain picture for mode locking has been provided, where the relative phases are of primary relevance. A physical model for passive mode locking can alternatively and equivalently be described in terms of the temporal broadening and narrowing mechanisms.

Upon startup of laser emission, the laser modes initially oscillate with relative phases that are random such that the radiation pattern consists of noise bursts. If one of these bursts is energetic enough to provide a fluence that matches the saturation fluence of the absorber, it will bleach the absorption. This means that around the peak of the burst where the intensity is higher, the loss will be smaller, while the low-intensity wings become more attenuated. The pulse generation process is thus initiated by this family of intensity spikes that experience lower losses within the absorber carrier lifetime.

The dynamics of absorption and gain play a crucial role in pulse shaping. In steady state, the unsaturated losses are higher than the gain. When the leading edge of the pulse reaches the absorber, the loss saturates more quickly than the gain, which results in a net gain window, as depicted in Figure 41.1. The absorber then recovers from this state of saturation to the initial state of high loss, thus attenuating the trailing edge of the pulse. It is thus easy to understand why the saturation fluence and the recovery time of the absorber are of primary importance in the formation of mode-locked pulses.

This temporal scenario can be connected to the previously described frequency domain description of mode locking. The burst of noise is the result of an instantaneous phase locking

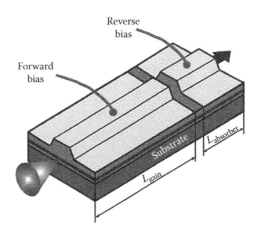

FIGURE 41.2 A schematic of a two-section semiconductor laser diode.

occurring among a number of modes. The self-saturation at the saturable absorber then helps to sustain and strengthen this favorable combination, by discriminating against the lower power cw noise.

In practical terms, a saturable absorber can be integrated monolithically into a semiconductor laser, by electrically isolating one section of the device (Figure 41.2). By applying a reverse bias to this section, the carriers that are photogenerated by the pulses can be more efficiently swept out of the absorber, thus enabling the saturable absorber to recover more quickly to its initial state of high loss. An increase in the reverse bias serves to decrease the absorber recovery time, and this will have the effect of further shortening the pulses.

41.2.4 Requirements for Successful Passive Mode Locking

Ultrafast carrier dynamics are fundamental for successful mode locking in semiconductor lasers, particularly in the saturable absorber, because the absorption should saturate faster and recover faster. Indeed, the absorption recovery time is one of the determining factors for obtaining ultrashort pulses. In particular, for high-repetition-rate lasers, the absorber recovery time should be much shorter than the cavity period so that the absorber can return to a state of total attenuation prior to the incidence of each incoming pulse. The fast absorption recovery also prevents the appearance of satellite pulses within the window of the net gain. On the other hand, the gain recovery time should be shorter than the cavity round-trip time. Thus, for lasers operating at pulse repetition rates of 20 GHz or more, this means that the recovery times of both gain and absorption should be much shorter than 40 ps.

The saturation dynamics represents another crucial aspect for successful mode locking, as shown schematically in Figure 41.3. Such dynamics can be translated in terms of saturation fluence F_{sat} or saturation energy $E_{sat} = A \cdot F_{sat}$, where A is the optical mode cross-sectional area. The saturation energy is an indication of how much energy is necessary to saturate the absorption or the gain. Indeed, to achieve robust mode locking, the saturation

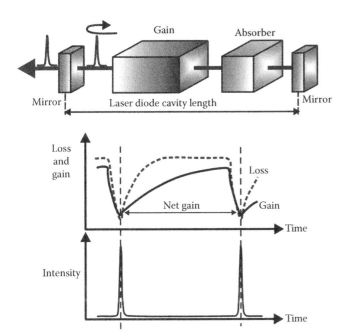

FIGURE 41.1 A schematic diagram of the main components that forms a two-section laser diode (top). Loss and gain dynamics that lead to pulse generation (bottom).

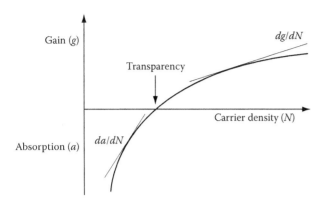

FIGURE 41.3 Dependence of absorption/gain with carrier concentration in a semiconductor laser.

energy of the absorber E_{sat}^a should be as small as possible and smaller than the saturation energy of the gain E_{sat}^g:

$$E_{sat}^a = \frac{h\nu A}{\dfrac{\partial a}{\partial N}} < E_{sat}^g = \frac{h\nu A}{\dfrac{\partial g}{\partial N}}$$

where

    h is Planck's constant

    ν is the optical frequency

    $\partial a/\partial N$ and $\partial g/\partial N$ are the differential loss and gain, respectively

The special dependence of the loss/gain with carrier density in a semiconductor laser allows $\partial a/\partial N > \partial g/\partial N$, as shown in Figure 41.3.

This condition implies that the absorber will saturate faster than the gain for a given pulse fluence, thus enabling the creation of the net gain window as already mentioned. The ratio between saturation energies should also be as large as possible to ensure that the losses saturate more strongly than the gain.

41.2.5 Self-Phase Modulation and Dispersion

In a semiconductor material, both the refractive index and gain (or loss) depend on the carrier density and are thus strongly coupled. As the pulse propagates in the gain section,* the carrier density and thus the gain is depleted across the pulse, as the carriers recombine through stimulated emission. This leads to a dynamic increase of the refractive index, which then introduces a phase modulation on the pulse, changing the instantaneous frequency across the pulse. This phenomenon is called self-phase modulation (SPM) and is one of the main nonlinear effects associated with pulse propagation in semiconductor media.

To understand the mechanism of SPM, consider the simple and illustrative example of a plane wave $E(t, x)$:

---

* In the following discussion, reference is made mostly to gain to simplify the description. However, all this reasoning can be applied equally well to the absorber.

$$E(t,x) = E_0 \exp i\Phi(t) = E_0 \exp i(\omega_0 t - kx), \quad k = \frac{\omega_0}{c} n(t)$$

where

    $\Phi(t)$ is the time-varying phase

    k is the wave vector

    ω_0 is the optical carrier frequency

    c is the speed of light

    $n(t)$ is the time-varying refractive index

The instantaneous frequency is the time derivative of the phase and thus can be written as

$$\omega(t) = \frac{\partial}{\partial t}\Phi(t) = \omega_0 - \frac{\omega_0}{c}\frac{\partial n(t)}{\partial t}x$$

From this expression, it is clear that if the refractive index varies with time, then the instantaneous frequency of the plane wave will vary relative to ω_0 and in a manner proportional to the temporal derivative of the index. The time dependence of this instantaneous frequency is called the frequency chirp. An up-chirp[†] (down-chirp) means that the frequency increases (decreases) with time. An example of a frequency up-chirped pulse is illustrated in Figure 41.4.

SPM is not dispersive in itself, but the pulse will not remain transform-limited when it propagates in a dispersive material such as the laser medium. The effect of dispersion manifests itself in the variation of refractive index for different wavelengths which means that different spectral components will travel at different speeds. For an up-chirped pulse, the frequency is higher in the trailing edge than in the leading edge. When the pulse propagates through a material exhibiting positive (normal) dispersion, the trailing edge of the pulse propagates more slowly than the leading edge of the pulse and so this results in a temporal broadening of the pulse.

In a monolithic two-section mode-locked semiconductor laser, where a saturable absorber and a gain section coexist, the resulting chirp is a balance between the effects caused by the absorber and the gain. In the gain section, a frequency up-chirp results, while the saturable absorber helps to further

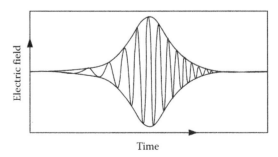

FIGURE 41.4 Illustration of an electric field of a strongly up-chirped pulse, where the instantaneous frequency increases with time.

[†] Up-chirp is also known as blue-chirp or positive chirp.

shape the pulse by contributing with a negative chirp. With a suitable balance between both sections, the chirp can be close to zero thereby leading to transform-limited pulses. Unfortunately, this is the exception rather than the rule, because this usually only occurs for a limited set of bias conditions and/or for given ratios of absorber/gain lengths. Therefore, up-chirp prevails for passively mode-locked lasers, leading to significant pulse broadening as the pulse propagates. The combined effect of SPM and dispersion impose the strongest limitation in the achievable shortest duration of pulses from mode-locked semiconductor diode lasers.

The mechanism of SPM implies that in addition to the original frequency ω_0, there are now more frequencies inside the pulse envelope. This richer spectral content is not necessarily unhelpful because it can provide bandwidth support for shorter pulses, if the chirp of a pulse can be removed by provision of a suitable dispersion-induced chirp of the opposite sign. For up-chirped pulses, a dispersion compensation setup can be configured such that a negative (anomalous) group velocity dispersion is able to slow down the leading edge of the pulse and speed the blueshifted trailing edge to such an extent that at a certain point both edges propagate simultaneously and the pulse is shorter.

To routinely generate pulses that are nearly transform-limited, an alternative could be found in the choice of a material that exhibits lower coupling between refractive index and gain, as described by linewidth enhancement factor (LEF), or α-factor:

$$\alpha = -\frac{4\pi}{\lambda}\frac{dn/dN}{dg/dN}$$

A higher α-factor implies a more significant coupling between gain and refractive index changes with carrier concentration and thus the possibility for higher levels of SPM and frequency chirp.

41.3 Quantum Dots: Distinctive Advantages for Ultrafast Diode Lasers

41.3.1 The Role of Dimensionality in Semiconductor Lasers

The history of semiconductor laser materials has been punctuated by dramatic revolutions. Everything started with the proposal of p-n junction semiconductor lasers in 1961, followed by experimental realization on different semiconductor materials (Basov et al., 1961; Basov, 1964). However, the lasers fabricated at that time exhibited an extremely low efficiency due to high optical and electrical losses. In fact, until the mid-1960s, only bulk materials were used in semiconductor devices, which were functionalized by introducing a doping profile. At the time, pioneers like Alferov and Herbert Kroemer independently considered the hypothesis of building heterostructures, consisting

of layers of different semiconductor materials (Alferov, 2001). The classic heterostructure example consists of a lower bandgap layer surrounded by a higher bandgap semiconductor material. Such design results in electronic and optical confinement, because a higher bandgap semiconductor also exhibits a higher refractive index. The enhanced confinement improved notably the operational characteristics of laser diodes, in particular the threshold current density, which decreased by two orders of magnitude.

But another revolution was about to come when it was realized that the confinement of electrons in lower dimensional semiconductor structures translated into completely new optoelectronic properties, when compared to bulk semiconductors. And how small should this confinement be? In order to answer this question, let us recall the concept of the de Broglie wavelength of thermalized electrons, λ_B:

$$\lambda_B = \frac{h}{p} = \frac{h}{\sqrt{2m^\star E}}$$

where

h is the Planck's constant
p is the electron momentum
$m^\star$ is the electron effective mass
E is the energy

In the case of III–V compound semiconductors, λ_B is typically of the order of tens of nanometers (Saleh and Teich, 1991). If one of the dimensions of a semiconductor is comparable or less than λ_B, the electrons will be strongly confined in one dimension, while moving freely in the remaining two dimensions—this is the case of a quantum well (QW). A quantum wire is a one-dimensional confined structure, while a QD is confined in all the three dimensions. QDs are thus tiny clusters of semiconductor material with dimensions of only a few nanometers, surrounded by a semiconductor matrix that has a higher bandgap.

The spatial confinement of the carriers in lower dimensional semiconductors leads to dramatically different energy–momentum relations in the directions of confinement, which results in completely new density of states, when compared to the bulk case, as depicted in Figure 41.5. As dimensionality decreases, the density of states is no longer continuous or quasi-continuous but becomes quantized. In the case of QDs, the charge carriers occupy only a restricted set of energy levels rather like the electrons in an atom, and for this reason, QDs are sometimes referred to as "artificial atoms."

For a given energy range, the number of carriers necessary to fill out these states reduces substantially as the dimensionality decreases, which implies that it becomes easier to achieve transparency and inversion of population—with the resulting reduction of threshold current density. In fact, this reduction has been quite spectacular over the years, with sudden jumps whenever the dimensionality is decreased (Alferov, 2001).

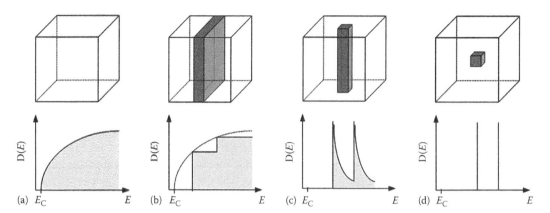

FIGURE 41.5 Schematic structures of bulk and low-dimensional semiconductors and corresponding density of states. The density of states in different confinement configurations: (a) bulk; (b) quantum well; (c) quantum wire; and (d) quantum dot.

41.3.2 Quantum Dots: Materials and Growth

The group of QD materials that has shown particular promise is based on III–V QDs epitaxially grown on a semiconductor substrate. For instance, InGaAs/InAs QDs on a GaAs substrate emit in the 1–1.3 μm wavelength range, which could be extended to 1.55 μm. Alternatively, InGaAs/InAs QDs can be grown on an InP substrate that covers emission in the 1.4–1.9 μm wavelength range (Ustinov et al., 2003).

The remarkable achievements in QD epitaxial growth have enabled the fabrication of QD lasers, amplifiers, and saturable absorbers offering excellent performance characteristics. To date, the most promising results have been achieved using the spontaneous formation of three-dimensional islands during strained layer epitaxial growth in a process known as the Stranski–Krastanow mechanism (Goldstein et al., 1985; Ustinov et al., 2003). In this process, when a film A is epitaxially grown over a substrate B, the initial growth occurs layer by layer, but beyond a certain critical thickness, three-dimensional islands start to form—the quantum dots. A continuous film lies underneath the dots, and is called the wetting layer. The most important condition in this technique is that the lattice constant of the deposited material is larger than the one of the substrate. This is the case of an InAs film (lattice constant of 6.06 Å) on a GaAs substrate (lattice constant of 5.64 Å), for example.

In spite of being an extremely complex process, the Stranski–Krastanow mode is now widely used in the self-assembly of QDs. An advantage of this technique is that films can be grown using the well-known techniques of molecular beam epitaxy (MBE) and metal organic chemical vapor deposition (MOCVD), and therefore the science of QDs growth has benefited immensely from all the previous knowledge gained with this technology. These are also good news for commercialization, because manufacturers do not have to invest in new epitaxy equipment to fabricate these structures.

Due to the statistical fluctuations occurring during growth, there is a distribution in dot size, height, and composition but, at the moment, epitaxy techniques have evolved to such an extent that the amount of fluctuations can be reasonably controlled, and can be as small as a few percent.

If the dots are grown on a plane surface, their lateral positions will be random. An example of such structure is shown in Figure 41.6. In the self-assembly process, there is no standard way of arranging the dots in a planar ordered way, unless they are encouraged to grow at particular positions in a pre-patterned substrate.

At present, the densities of QDs lie typically between 10^9 cm^{-2} and 10^{11} cm^{-2}. The sparse distribution of QDs results in a low value of optical gain. Thus, the levels of gain and optical confinement provided by a single layer of QDs may not be enough for the optimal performance of a laser. In order to circumvent this problem, QDs can also be grown in stacks, which allows an increase in the modal gain without increasing the internal optical mode loss (Smowton et al., 2001), where the various layers are usually separated by GaAs barriers. The GaAs separators are responsible for transmitting the tensile strain from layer to layer, inducing the formation of ordered arrays of QDs aligned on top of each other. Further optical confinement is enabled through the cladding of such arrays within layers of higher refractive index and bandgap energy, therefore forming a heterostructure.

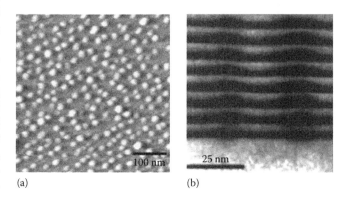

FIGURE 41.6 Photographs of an InGaAs quantum dots grown on GaAs substrate: (a) A TEM image of a single sheet of quantum dots. (b) A TEM image of a cross section of an 8-layer thick stack of quantum dots in GaAs layers.

41.3.3 Broad Gain Bandwidth

A QD laser was proposed in 1976 (Dingle and Henry, 1976) and the first theoretical treatment was published in 1982 (Arakawa and Sakaki, 1982). The main motivation was to conceive a design for a low-threshold, single-frequency, and temperature-insensitive laser, owing to the discrete nature of the density of states. In fact, practical devices exhibit the predicted outstandingly low thresholds (Kovsh et al., 2004; Liu et al., 2005), but the spectral bandwidths of such lasers are significantly broader than those of conventional QW lasers (Rafailov et al., 2007). This results from the self-organized growth of QDs, leading to a Gaussian distribution of dot sizes, with a corresponding Gaussian distribution of emission frequencies. Additionally, lattice strain may vary across the wafer, thus further affecting the energy levels in the quantum dots. These effects lead to the inhomogeneous broadening of the gain—a useful phenomenon in the context of ultrafast applications, because a very wide bandwidth is available for the generation, propagation, and amplification of ultrashort pulses. The effects of inhomogeneous broadening on the density of states are schematically illustrated in Figure 41.7. However, it is important to stress that a highly inhomogeneously broadened gain also encompasses a number of disadvantages, because it partially defeats the purpose of a reduced dimensionality, by broadening the density of states. Indeed, the fluctuation in the size of the QDs has the effect of increasing the transparency current and reducing the modal and differential gain (Qasaimeh, 2003; Dery and Eisenstein, 2005). Therefore, much effort has been put into improving the dots uniformization by engineering the growth and post-growth processes (Ustinov et al., 2003).

The extremely broad bandwidth available in QD mode-locked lasers offers potential for generating sub-100 fs pulses provided all of the bandwidth can be engaged coherently and dispersion effects suitably minimized.

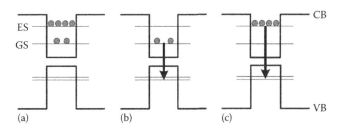

FIGURE 41.8 Schematic of the energy levels in a QD material (a), and radiative transitions via GS—ground state (b) and ES—excited state (c). CB—Conduction band; VB—valence band.

Indeed, it has been shown that there is usually some gain in narrowing/filtering effects in mode-locked QW lasers (Delfyett et al., 1998). With the inhomogeneously broadened gain bandwidth exhibited by QDs, there is support for more bandwidth and this can oppose the effect of pulse broadening that may arise from spectral narrowing. Additionally, due to the particular nature of QD lasers, many possibilities open up in respect of the exploitation of ground-state (GS) and excited-state (ES) bands,[*] as schematically represented in Figure 41.8. Such versatility has been successfully exploited in a multiple-wavelength-band switchable mode locking (Cataluna et al., 2006c). On the other hand, the interplay between GS and ES can be deployed in novel mode-locking regimes (Cataluna et al., 2006a). Using an external cavity, it is possible to set up tunable mode-locked sources that can operate in the wavelength range that extends from the GS to the ES transition bands (Kim et al., 2006a).

41.3.4 Ultrafast Carrier Dynamics

In the initial studies of QD materials, it was thought that their carrier dynamics would be significantly slower than those in QW materials due to a phonon bottleneck effect (Mukai et al., 1996). Interestingly, experiments have demonstrated quite the opposite. As a consequence of access to a number of recombination paths for the carriers, QD structures exhibit ultrafast recovery both under absorption and gain conditions (Borri et al., 2006). In two evaluations, the absorber dynamics of surface and waveguided QD structures were investigated by using a pump-probe technique (Borri et al., 2000; Rafailov et al., 2004b). This showed the existence of at least two distinct time constants for the recovery of the absorption. A fast recovery of around 1 ps is followed by a slower recovery process that extends over 100 ps (Rafailov et al., 2004b).

More recently, sub-picosecond carrier recovery was measured directly in a QD absorption modulator when a reverse bias was applied (Malins et al., 2006). Absorption recovery times ranged from 62 ps down to 700 fs and showed a decrease by nearly two orders of magnitude when the reverse bias applied to the structure was changed from 0 V to −10 V. This important observation provides significant promise for ultrafast modulators that can operate

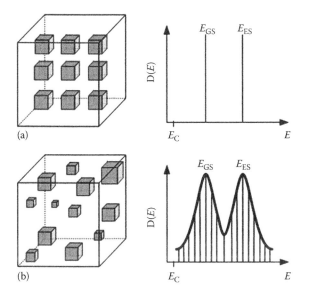

FIGURE 41.7 Schematic morphology and density of states for charge carriers in (a) an ideal quantum-dot system and (b) a real quantum-dot system, where inhomogeneous broadening is illustrated.

---

[*] Ground and excited states are also available in quantum wells. However, the δ-like density of states associated with quantum dots enables an easier access to the ES, owing to the faster saturation of the GS in quantum dots.

above 1 THz and for the optimization of saturable absorbers used for the passive mode locking of semiconductor lasers at high repetition rates, where the absorption recovery should occur within the round-trip time of the cavity. Crucially, the shaping mechanism of the fast absorption recovery also enhances the shortening of the mode-locked pulses, and thus QD lasers have the potential for generating shorter pulses than their QW counterparts.

41.3.5 Low Absorption Saturation Fluence

QD-based saturable absorbers exhibit lower saturation fluence than QW-based materials due to their delta-like density of states. For example, in a QD one electron is enough to achieve transparency and two to achieve inversion. This characteristic facilitates the self-starting of mode locking at modest pulse energies. This feature is particularly important in high-repetition-rate lasers where the optical energy available in each pulse is small. Indeed, it has also been observed that the saturation power is at least 2–5 times smaller for a QD saturable absorber than for a QW-based counterpart when integrated in a monolithic mode-locked laser (Thompson et al., 2004a). In this paper, the authors pointed out that saturation would further depend on the density of dots, reverse bias, and inhomogeneous broadening.

41.3.6 Low Threshold Current and Low Temperature Sensitivity

As devices, QD diode lasers have the advantage of requiring a very low threshold current to initiate lasing (Ustinov et al., 2003). This attribute applies also to operation in the mode-locking regime, because most QD lasers exhibit mode-locked operation right from the threshold of laser emission. (Bistability between the non-lasing state and the onset of lasing/mode locking might be present, as has been shown experimentally (Huang et al., 2001; Thompson et al., 2006b) and numerically (Viktorov et al., 2006).) A low threshold current is clearly advantageous because this can represent a device that is compatible as an efficient and compact source of ultrashort pulses where the demand for electrical power can be very low. Furthermore, having a low threshold avoids the need for higher carrier densities for pumping the laser and this implies less amplified spontaneous emission and reduced optical noise in the generated pulse sequences.

Due to the discrete nature of their density of states, QD lasers also exhibit low-temperature sensitivity (Mikhrin et al., 2005), making them excellent candidates for applications where resilience to temperature effects is important.

41.3.7 Low Linewidth Enhancement Factor

One of the main motivations for the enthusiastic investigation of QD materials in the last few years has been the theoretically predicted potential for very low values of LEF, owing to the symmetry of the gain associated with QD structures. The possibility of a low LEF is very attractive for a number of performance aspects, such as lower frequency chirp in directly modulated lasers,

lower sensitivity to optical feedback effects, and suppressed beam filamentation. The potential of a lower effect of SPM in QD lasers also held a promise for the generation of transform-limited pulses. However, disparate reports have been published in the last 3 years, with some reports of LEF values of nearly zero (Newell et al., 1999), and others with values of LEF similar (Ukhanov et al., 2004) or significantly higher than in QW structures (Dagens et al., 2005). Ultimately, the LEF is a characteristic that is highly dependent on the operation conditions of the laser, and as such, its meaning always needs to be contextualized for a set of particular conditions. This is a topic that is currently under intense investigation.

41.4 Mode-Locked Quantum-Dot Lasers: State of the Art

41.4.1 Pulse Duration

The first demonstration of a QD mode-locked laser was reported in 2001, with pulse durations of ~17 ps at 1.3 μm and repetition rate of 7.4 GHz, using passive mode locking (Huang et al., 2001). Hybrid mode locking at the same wavelength was demonstrated in 2003 by the Cambridge University group (Thompson et al., 2003); they reported an upper limit estimation of 14.2 ps for the shortest pulses measured at a repetition rate of 10 GHz. Later in 2004, the same group demonstrated Fourier-transform-limited 10 ps pulses at 18 GHz repetition rate, using passive mode locking (Thompson et al., 2004b).

In 2004, we demonstrated the generation of sub-picosecond pulses directly from a QD laser where the shortest pulse durations were measured to be 390 fs, without any form of supplementary pulse compression (Rafailov et al., 2004a, 2005). These pulses were generated by a two-section passively mode-locked QD laser and this was the first time that sub-picosecond pulses were generated directly from such a monolithic laser.

The generation of sub-picosecond pulses was reported later by several groups (Laemmlin et al., 2006; Thompson et al., 2006b). In one of these reports (in 2006), Thompson and coworkers demonstrated the generation of pulses as short as 790 fs, by using a flared waveguide configuration in a two-section QD laser (Thompson et al., 2006b). Because the beam mode size in the saturable absorber section was much smaller than that in the gain section, the ratio of saturation fluences in the absorber and gain sections was increased. This enhanced the pulse formation mechanisms and allowed for better pulse shaping and shortening.

There has also been much effort in designing QD-based mode-locked sources that could be deployed in the 1.55 μm band (Lelarge et al., 2007). Ultra-short-pulse generation has been achieved from single-section lasers based either on InAs QDs (Renaudier et al., 2005) or quantum dashes (Gosset et al., 2006) grown on an InP substrate. These authors have suggested that there are no fundamental differences between the QDs and dashes in the context of mode-locked laser sources.

41.4.2 Toward Higher Pulse Repetition Rates

To achieve higher repetition rates in mode-locked lasers, it is necessary to decrease the cavity length. This poses a significant challenge to QD lasers because of their lower gain and the operation in short cavities may shift the emission to the ES band (Markus et al., 2003). To avoid this problem, a higher number of QD layers should be deployed in the active region. Using this simple approach, the highest repetition rate directly generated from a passively mode-locked QD two-section laser was 80 GHz (Laemmlin et al., 2006), when a 15-layer structure was used. Another method to boost the repetition rate of mode-locked lasers is to use colliding pulse mode locking. This technique is similar to passive mode locking, but the saturable absorber region is placed at the precise center of the gain section. Two counter-propagating pulses from each outer gain section therefore meet in the saturable absorber region, bleaching it much more efficiently than if just one pulse was present. This process can also result in shorter and more stable pulses. Owing to the device geometry, mode locking is achieved at the second harmonic of the fundamental (round-trip) frequency, and the pulse repetition rate is doubled. A variation of colliding pulse mode locking is harmonic mode locking, where more than two pulses circulate in the cavity, the number being equal to the harmonic. Colliding-pulse mode-locking was first demonstrated for QD lasers in 2005 (Thompson et al., 2005), resulting in a modest repetition rate of 20 GHz. Harmonic mode locking has also been demonstrated with repetition rates of approximately 40, 80, 120, and 240 GHz (Rae et al., 2006).

41.4.3 Temperature Resilience of Mode-Locked QD Lasers

Due to their delta-function-like density of states, QDs offer great potential for designing temperature-resilient devices. If their high-speed performance is also proven to be resilient to temperature, QD lasers can become the next generation of sources for ultrafast optical telecoms and datacoms, because the constraint of using thermoelectric coolers can be avoided, thus decreasing cost and complexity. In this context, we have demonstrated stable passive mode-locked operation of a two-section QD laser over an extended temperature range (from 20°C to 80°C) at relatively high output average powers (Cataluna et al., 2006b).

Additionally, to meet the requirements for high-speed communications, it is important to investigate the temperature dependence of the pulse duration. For instance, in communication systems with transmission rates of 40 Gb/s or more, the temporal interval between pulses is less than 25 ps and so the duration of the optical pulses should be well below this value at any operating temperature. We have shown that the pulse duration and the spectral width decrease significantly as the temperature is increased up to 70°C (Cataluna et al., 2007). The combination of all these effects resulted in a sevenfold decrease of the time-bandwidth product (the pulses were still highly chirped due to the strong SPM and dispersion effects in the semiconductor material).

To account for the decrease in pulse duration with temperature, a model for mode locking in QD lasers was used. It was found that the pulse durations are determined principally by the escape rate of the carriers in the absorber section, which lead to a decrease of absorber recovery time with increasing temperature, thus inducing a decrease in the pulse durations. This has been verified recently using ultrafast spectroscopy to probe the absorber recovery time as a function of temperature (Malins et al., 2007).

41.4.4 Mode Locking Involving Excited-State Transitions

It has been observed that laser emission in QD lasers can access the transitions in GS, ES or both (Markus et al., 2003), as represented in Figure 41.8. Furthermore, sub-picosecond gain recovery has been demonstrated for both GS and ES transitions in electrically pumped QD amplifiers (Schneider et al., 2005). In this reported work, the LEF was shown to decrease significantly for wavelengths below the GS transition, even becoming negative at ES thereby implying a potential for chirp-free operation for the range of wavelengths involved. Laser emission in the ES is also characterized by a higher differential gain than GS, with associated benefits for ultrafast QD lasers. We have demonstrated an optical gain-switched QD laser, where pulses were generated from both GS and ES, and where the ES pulses were shorter than those generated by GS alone (Rafailov et al., 2006). The potential for shorter and chirp-free pulses from ES transitions motivated us to investigate the mode-locked operation of QD lasers in this band. We demonstrated, for the first time, passive mode locking via GS (1260 nm) or ES (1190 nm) in a QD laser, at repetition frequencies of 21 and 20.5 GHz, respectively (Cataluna et al., 2006c). The switch between these two states in the mode-locking regime was easily achieved by changing the electrical biasing conditions, thus providing full control of the operating spectral band. It is important to stress that the average power in both operating modes was relatively high and exceeded 25 mW. In the range of bias conditions explored in this study, the shortest pulse duration measured for ES transitions was ~7 ps, where the spectral bandwidth was 5.5 nm, at an output power of 23 mW. These pulse durations are similar to those generated by GS mode locking at the same power level.

Although pulse durations from ES spectral band have been below expectations so far, it is our opinion that exploitation of the ES transitions—a unique feature of QD lasers—can lead to a new generation of high-speed sources, where mode locking involves electrically switchable GS or ES transitions that are spectrally distinct. This could enable a range of applications extending from time-domain spectroscopy, through to optical interconnects, wavelength-division multiplexing, and ultrafast optical processing.

41.5 Summary and Outlook

41.5.1 Critical Discussion

In this chapter, we have presented the physics and reported on the progress of ultrafast laser diodes based on QD materials.

The results presented in the literature show that monolithic passively mode-locked QD lasers can currently surpass the performance of similar QW lasers in terms of pulse duration (Rafailov et al., 2005, Thompson et al., 2006b). There are other particular features where QD lasers have already been shown to have a superior performance, notably in the case of pulse timing jitter where record low values have been reported (Choi et al., 2006, Thompson et al., 2006a).

We strongly believe that the appeal of QD lasers also resides in the novel functionalities that are distinctive of QDs. These are the exploitation of an ES level as a means to achieve novel mode-locking regimes; the temperature resilience offered by the quantized density of states; lower threshold and higher output power levels; and access to the enlarged spectral bandwidths associated with the inhomogeneously broadened gain features. These characteristics are not only useful from an operational point of view, but also provide some insights into a more comprehensive understanding of the underlying physical mechanisms of mode locking in QD lasers.

41.5.2 Future Perspectives

Although there have been many advances in the control of the growth of QD laser having ultra-low threshold current and temperature resilience, it is not yet understood what is the most advantageous QD structure layout to be used in the regime of mode locking. In particular, it is not clear what is the optimum level of inhomogeneous broadening that results in shorter and higher peak power pulses. Therefore, it is relevant to investigate if and how the inhomogeneously broadened spectral modes are engaged coherently in the generation of ultrashort pulses and how that effect could be used to improve the performance of the lasers toward sub-picosecond pulse durations. Exploiting novel QD materials based on p-doped and tunnel injection structures could also bring advantages in minimizing the effect of any deleterious SPM effects in mode-locked lasers.

Comparison between theory and experiment of undoped and p-doped lasers has shown how this technique can improve the LEF (Kim and Chuang, 2006b). By tuning the level of doping, lasers can exhibit zero and even negative LEF at low current densities (Alexander et al., 2007). Tunnel injection QD structures can also be of great interest for use in mode-locked lasers, as the injection of cold carriers may bring many benefits to the operation of mode-locked lasers (Delfyett, 2006). The LEF has been recently calculated and has been demonstrated to be much less than that reported for other lasers (Mi and Bhattacharya, 2007). Selective excitation of population in these lasers has been demonstrated, which could lead to a mitigation of the inhomogeneous broadening effects and contribute to a narrower spectrum and the production of transform-limited pulses (Bret and Gires, 1964).

The ES spectral band can also be exploited in tunable lasers using QD materials where the inhomogeneous broadening is controlled so as to maximize the overlap between GSs and ESs. While in cw operation, QD lasers have been demonstrated with tunability ranges up to 200 nm (Varangis, 2000), by exploiting the gain available from the GSs and ESs. Combining this tunability with the possibility of generating ultrashort pulses, it will be possible to achieve a new generation of versatile lasers emitting pulses across a wide range of wavelengths—as if we had compressed many different lasers into a single laser! Such disruptive characteristics will offer endless possibilities and enable applications never seen before in science and technology.

Acknowledgments

We wish to thank Dr. D. Livshits, Dr. I. Krestnikov, and Dr. A. Kovsh from Innolume GmbH (Dortmund) for helping to prepare samples and for stimulating discussions.

This work was supported in part by the European Community's Seventh Framework Programme FAST-DOT under grant agreement 224338.

References

Alexander, R. R., Childs, D., Agarwal, H. et al. 2007. Zero and controllable linewidth enhancement factor in p-doped 1.3 μm quantum dot lasers. *Japanese Journal of Applied Physics* 46: 2421–2423.

Alferov, Z. I. 2001. Nobel Lecture: The double heterostructure concept and its applications in physics, electronics, and technology. *Reviews of Modern Physics* 73: 767.

Arakawa, Y. and Sakaki, H. 1982. Multidimensional quantum well laser and temperature dependence of its threshold current. *Applied Physics Letters* 40: 939–941.

Basov, N. G. 1964. Nobel lecture: Semiconductor lasers. http://www.nobel.se/physics/laureates/1964/basov-lecture.html.

Basov, N. G., Krokhin, O. N., and Popov, Y. M. 1961. The possibility of use of indirect transitions to obtain negative temperature in semiconductors. *Soviet Physics JETP* 12: 1033.

Borri, P., Langbein, W., Hvam, J. M., Heinrichsdorff, F., Mao, M. H., and Bimberg, D. 2000. Spectral hole-burning and carrier-heating dynamics in InGaAs quantum-dot amplifiers. *IEEE Journal of Selected Topics in Quantum Electronics* 6: 544–551.

Borri, P., Schneider, S., Langbein, W., and Bimberg, D. 2006. Ultrafast carrier dynamics in InGaAs quantum dot materials and devices. *Journal of Optics A-Pure and Applied Optics* 8: S33–S46.

Cataluna, M. A., Mcrobbie, A. D., Sibbett, W., Livshits, D. A., Kovsh, A. R., and Rafailov, E. U. 2006a. New mode locking regime in a quantum-dot laser: Enhancement by simultaneous CW excited-state emission. *Conference on Lasers and Electro-Optics/Quantum Electronics and Laser Science Conference paper CThH3*: CThH3, Long Beach, CA.

Cataluna, M. A., Rafailov, E. U., Mcrobbie, A. D., Sibbett, W., Livshits, D. A., and Kovsh, A. R. 2006b. Stable mode-locked operation up to 80°C from an InGaAs quantum-dot laser. *IEEE Photonics Technology Letters* 18: 1500–1502.

Cataluna, M. A., Sibbett, W., Livshits, D. A., Weimert, J., Kovsh, A. R., and Rafailov, E. U. 2006c. Stable mode locking via ground- or excited-state transitions in a two-section quantum-dot laser. *Applied Physics Letters* 89: 81124–3.

Cataluna, M.-T., Viktorov, E. A., Mandel, P. et al. 2007. Temperature dependence of pulse duration in a mode-locked quantum-dot laser. *Applied Physics Letters* 90: 101102–3.

Choi, M.-T., Kim, J.-M., Lee, W., and Delfyett, P. J. 2006. Ultralow noise optical pulse generation in an actively mode-locked quantum-dot semiconductor laser. *Applied Physics Letters* 88: 131106–3.

Dagens, B., Markus, A., Chen, J. X. et al. 2005. Giant linewidth enhancement factor and purely frequency modulated emission from quantum dot laser. *Electronics Letters* 41: 323–324.

Delfyett, P. J. 2006. Personal communication.

Delfyett, P. J., Shi, H., Gee, S., Barty, C. P. J., Alphonse, G., and Connolly, J. 1998. Intracavity spectral shaping in external cavity mode-locked semiconductor diode lasers. *IEEE Journal of Selected Topics in Quantum Electronics* 4: 216–223.

Dery, H. and Eisenstein, G. 2005. The impact of energy band diagram and inhomogeneous broadening on the optical differential gain in nanostructure lasers. *IEEE Journal of Quantum Electronics* 41: 26–35.

Dingle, R. and Henry, C. H. 1976. Quantum effects in heterostructure lasers. U.S. Patent.

Goldstein, L., Glas, F., Marzin, J. Y., Charasse, M. N., and Roux, G. L. 1985. Growth by molecular beam epitaxy and characterization of InAs/GaAs strained-layer superlattices. *Applied Physics Letters* 47: 1099–1101.

Gosset, C., Merghem, K., Martinez, A. et al. 2006. Subpicosecond pulse generation at 134 GHz using a quantum-dash-based Fabry-Perot laser emitting at 1.56 μm. *Applied Physics Letters* 88: 241105–3.

Huang, X. D., Stintz, A., Li, H., Lester, L. F., Cheng, J., and Malloy, K. J. 2001. Passive mode-locking in 1.3 μm two-section InAs quantum dot lasers. *Applied Physics Letters* 78: 2825–2827.

Kim, J., Choi, M. T., Lee, W., and Delfyett, P. J. 2006a. Wavelength tunable mode-locked quantum-dot laser. *Enabling Photonics Technologies for Defense, Security, and Aerospace Applications II*: 6243: M2430.

Kim, J. and Chuang, S. L. 2006b. Theoretical and experimental study of optical gain, refractive index change, and linewidth enhancement factor of p-doped quantum-dot lasers. *IEEE Journal of Quantum Electronics* 42: 942–952.

Kovsh, A. R., Ledentsov, N. N., Mikhrin, S. S. et al. 2004. Long-wavelength (1.3–1.5 μm) quantum dot lasers based on GaAs. *Physics and Simulation of Optoelectronic Devices XII*: 5349: 31–45.

Laemmlin, M., Fiol, G., Meuer, C. et al. 2006. Distortion-free optical amplification of 20–80 GHz modelocked laser pulses at 1.3 μm using quantum dots. *Electronics Letters* 42: 697–699.

Lelarge, F., Dagens, B., Renaudier, J. et al. 2007. Recent advances on InAs/InP quantum dash based, semiconductor lasers and optical amplifiers operating at 1.55 mu m. *IEEE Journal of Selected Topics in Quantum Electronics* 13: 111–124.

Liu, H. Y., Childs, D. T., Badcock, T. J. et al. 2005. High-performance three-layer 1.3-μm InAs-GaAs quantum-dot lasers with very low continuous-wave room-temperature threshold currents. *IEEE Photonics Technology Letters* 17: 1139–1141.

Malins, D. B., Gomez-Iglesias, A., White, S. J., Sibbett, W., Miller, A., and Rafailov, E. U. 2006. Ultrafast electroabsorption dynamics in an InAs quantum dot saturable absorber at 1.3 μm. *Applied Physics Letters* 89: 171111–3.

Malins, D. B., Gomez-Iglesias, A., Cataluna, M. A., Rafailov, E. U., Sibbett, W., and Miller, A. 2007. Temperature dependence of electroabsorption dynamics in an InAs quantum dot saturable absorber at 1.3 μm. *CLEO-Europe'07*: CF-6, Munich, Germany.

Markus, A., Chen, J. X., Paranthoen, C., Fiore, A., Platz, C., and Gauthier-Lafaye, O. 2003. Simultaneous two-state lasing in quantum-dot lasers. *Applied Physics Letters* 82: 1818–1820.

Mi, Z. and Bhattacharya, P. 2007. Analysis of the linewidth-enhancement factor of long-wavelength tunnel-injection quantum-dot lasers. *IEEE Journal of Quantum Electronics* 43: 363–369.

Mikhrin, S. S., Kovsh, A. R., Krestnikov, I. L. et al. 2005. High power temperature-insensitive 1.3 μm InAs/InGaAs/GaAs quantum dot lasers. *Semiconductor Science and Technology* 20: 340–342.

Mukai, K., Ohtsuka, N., Shoji, H., and Sugawara, M. 1996. Phonon bottleneck in self-formed $In_xGa_{1-x}As$/GaAs quantum dots by electroluminescence and time-resolved photoluminescence. *Physical Review B* 54: R5243–R5246.

Newell, T. C., Bossert, D. J., Stintz, A., Fuchs, B., Malloy, K. J., and Lester, L. F. 1999. Gain and linewidth enhancement factor in InAs quantum-dot laser diodes. *IEEE Photonics Technology Letters* 11: 1527–1529.

Qasaimeh, O. 2003. Effect of inhomogeneous line broadening on gain and differential gain of quantum dot lasers. *IEEE Transactions on Electron Devices* 50: 1575–1581.

Rae, A. R., Thompson, M. G., Penty, R. V. et al. 2006. Harmonic mode-locking of a quantum-dot laser diode. *LEOS 2006*: ThR5, Montreal, QC, Canada.

Rafailov, E. U., Cataluna, M. A., Sibbett, W. et al. 2004a. High-power ultrashort pulses output from a mode-locked two-section quantum-dot laser. *Conference on Lasers and Electro-Optics/International Quantum Electronics Conference*: CPDB5, post-deadline, San Francisco, CA.

Rafailov, E. U., White, S. J., Lagatsky, A. A. et al. 2004b. Fast quantum-dot saturable absorber for passive mode-locking of solid-state lasers. *IEEE Photonics Technology Letters* 16: 2439–2441.

Rafailov, E. U., Cataluna, M. A., Sibbett, W. et al. 2005. High-power picosecond and femtosecond pulse generation from a two-section mode-locked quantum-dot laser. *Applied Physics Letters* 87: 81107–3.

Rafailov, E. U., Mcrobbie, A. D., Cataluna, M. A., O'faolain, L., Sibbett, W., and Livshits, D. A. 2006. Investigation of transition dynamics in a quantum-dot laser optically pumped by femtosecond pulses. *Applied Physics Letters* 88: 41101–3.

Rafailov, E. U., Cataluna, M. A., and Sibbett, W. 2007. Mode-locked quantum-dot lasers. *Nature Photonics* 1: 395–401.

Renaudier, J., Brenot, R., Dagens, B. et al. 2005. 45 GHz self-pulsation with narrow linewidth in quantum dot Fabry-Perot semiconductor lasers at 1.5 μm. *Electronics Letters* 41: 1007–1008.

Saleh, B. E. A. and Teich, M. C. 1991. *Fundamentals of Photonics*. New York: Wiley.

Schneider, S., Borri, P., Langbein, W. et al. 2005. Excited-state gain dynamics in InGaAs quantum-dot amplifiers. *IEEE Photonics Technology Letters* 17: 2014–2016.

Smowton, P. M., Herrmann, E., Ning, Y., Summers, H. D., Blood, P., and Hopkinson, M. 2001. Optical mode loss and gain of multiple-layer quantum-dot lasers. *Applied Physics Letters* 78: 2629–2631.

Thompson, M. G., Marinelli, C., Tan, K. T. et al. 2003. 10 GHz hybrid modelocking of monolithic InGaAs quantum dot lasers. *Electronics Letters* 39: 1121–1122.

Thompson, M. G., Marinelli, C., Chu, Y. et al. 2004a. Properties of InGaAs quantum dot saturable absorbers in monolithic mode-locked lasers. *IEEE* 19th *International Semiconductor Laser Conference, Conference Digest*, Matsue-shi, Japan.

Thompson, M. G., Tan, K. T., Marinelli, C. et al. 2004b. Transform-limited optical pulses from 18 GHz monolithic modelocked quantum dot lasers operating at 1.3 μm. *Electronics Letters* 40: 346–347.

Thompson, M. G., Marinelli, C., Zhao, X. et al. 2005. Colliding-pulse modelocked quantum dot lasers. *Electronics Letters* 41: 248–250.

Thompson, M. G., Larson, D., Rae, A. et al. 2006a. Monolithic hybrid and passive mode-locked 40 GHz quantum dot laser diodes. 32nd *European Conference on Optical Communication*: We4.6.3, Cannes, France.

Thompson, M. G., Rae, A., Sellin, R. L. et al. 2006b. Subpicosecond high-power mode locking using flared waveguide monolithic quantum-dot lasers. *Applied Physics Letters* 88: 133119–3.

Ukhanov, A. A., Stintz, A., Eliseev, P. G., and Malloy, K. J. 2004. Comparison of the carrier induced refractive index, gain, and linewidth enhancement factor in quantum dot and quantum well lasers. *Applied Physics Letters* 84: 1058–1060.

Ustinov, V. M., Zhukov, A. E., Egorov, A. Y., and Maleev, N. A. 2003. *Quantum Dot Lasers*. New York: Oxford University Press.

Varangis, P. M., Li, H., Liu, G. T., Newell, T. C., Stintz, A., Fuchs, B., Malloy, K. J., and Lester, L. F. 2000. Low-threshold quantum dot lasers with 201 nm tuning range. *Electronics Letters* 36: 1544–1545.

Vasil'ev, P. 1995. *Ultrafast Diode Lasers: Fundamentals and Applications*. Boston, MA: Artech House.

Viktorov, E. A., Mandel, P., Vladimirov, A. G., and Bandelow, U. 2006. Model for mode locking in quantum dot lasers. *Applied Physics Letters* 88: 201102–3.

Zhang, L., Cheng, L., Gray, A. L. et al. 2005. Low timing jitter, 5 GHz optical pulses from monolithic two-section passively mode-locked 1250/1310 nm quantum dot lasers for high-speed optical interconnects. *Optical Fiber Communication Conference, OFC/NFOEC* OWM4, Anaheim, CA.

Index

Printed and bound by CPI Group (UK) Ltd, Croydon, CR0 4YY

22/10/2024

01777526-0003